Dennis II



# Selected Papers in
# Digital Signal Processing, II

OTHER IEEE PRESS BOOKS

A Guide for Better Technical Presentations
*Edited by R. M. Woelfle*
Career Management: A Guide to Combating Obsolescence
*Edited by H. G. Kaufman*
Energy and Man: Technical and Social Aspects of Energy
*Edited by G. Morgan*
Magnetic Bubble Technology: Integrated-Circuit Magnetics for Digital Storage and Processing
*Edited by H. Chang*
Frequency Synthesis: Techniques and Applications
*Edited by J. Gorski-Popiel*
Literature in Digital Signal Processing: Author and Permuted Title Index (Revised and Expanded Edition)
*Edited by H. D. Helms, J. F. Kaiser, and L. R. Rabiner*
Data Communications via Fading Channels
*Edited by K. Brayer*
Nonlinear Networks: Theory and Analysis
*Edited by A. N. Willson, Jr.*
Computer Communications
*Edited by P. E. Green, Jr. and R. W. Lucky*
Stability of Large Electric Power Systems
*Edited by R. T. Byerly and E. W. Kimbark*
Automatic Test Equipment: Hardware, Software, and Management
*Edited by F. Liguori*
Key Papers in the Development of Coding Theory
*Edited by E. R. Berlekamp*
Technology and Social Institutions
*Edited by K. Chen*
Key papers in the Development of Information Theory
*Edited by D. Slepian*
Computer-Aided Filter Design
*Edited by G. Szentirmai*
Laser Devices and Applications
*Edited by I. P. Kaminow and A. E. Siegman*
Integrated Optics
*Edited by D. Marcuse*
Laser Theory
*Edited by F. S. Barnes*
Digital Signal Processing
*Edited by L. R. Rabiner and C. M. Rader*
Minicomputers: Hardware, Software, and Applications
*Edited by J. D. Schoeffler and R. H. Temple*
Semiconductor Memories
*Edited by D. A. Hodges*
Power Semiconductor Applications, Volume II: Equipment and Systems
*Edited by J. D. Harnden, Jr., and F. B. Golden*
Power Semiconductor Applications, Volume I: General Considerations
*Edited by J. D. Harnden, Jr., and F. B. Golden*
A Practical Guide to Minicomputer Applications
*Edited by F. F. Coury*
Active Inductorless Filters
*Edited by S. K. Mitra*
Clearing the Air: The Impact of the Clean Air Act on Technology
*Edited by J. C. Redmond, J. C. Cook, and A. A. J. Hoffman*

# Selected Papers in
# Digital Signal Processing, II

## Edited by the
## Digital Signal Processing Committee
### IEEE Acoustics, Speech, and Signal Processing Society


Committee members:

A.V. Oppenheim (Chairman), Massachusetts Institute of Technology
G.D. Bergland, Bell Laboratories
J.W. Cooley, IBM Research Center
R.E. Crochiere, Bell Laboratories
J.R. Fisher, Signal Processing Systems, Inc.
Bernard Gold, M.I.T. Lincoln Laboratory
H.D. Helms, Bell Laboratories
L.B. Jackson, University of Rhode Island
J.F. Kaiser, Bell Laboratories
S.A. Lerman, Itek Corporation
R.M. Mersereau, Massachusetts Institute of Technology
L.R. Rabiner, Bell Laboratories
C.M. Rader, M.I.T. Lincoln Laboratory
R.W. Schafer, Georgia Institute of Technology
H.F. Silverman, IBM Research Center
Kenneth Steiglitz, Princeton University
Clifford Weinstein, M.I.T. Lincoln Laboratory



A volume in the IEEE PRESS Selected Reprint Series, prepared under the sponsorship of the IEEE Acoustics, Speech, and Signal Processing Society

IEEE PRESS

The Institute of Electrical and Electronics Engineers, Inc. New York

International Standard Book Numbers: Clothbound: 0-87942-059-6
Paperbound: 0-87942-060-X

Library of Congress Catalog Card Number 75-22925

PRINTED IN THE UNITED STATES OF AMERICA

# Contents

# Preface

During the past decade, the field of digital signal processing has been an extremely dynamic and active field. Advances in integrated circuit technology have had a major impact on the technical areas to which digital signal processing techniques and hardware are being applied. Research in fundamental signal processing techniques and algorithms has led to dramatic improvements in efficiency of signal processing systems.

Corresponding to the expanding activity in the field has been an expansion of the technical literature. In 1971, in an effort to make a number of the important papers in the field easily accessible, the Digital Signal Processing Committee of the IEEE Acoustics, Speech, and Signal Processing Group (then the Audio and Electroacoustics Group) undertook the preparation of the IEEE Press reprint book, *Digital Signal Processing*. This book was published in 1972 under the editorship of the Committee Chairmen at that time, Lawrence Rabiner and Charles Rader. Since the publication of that first reprint book, there have been many important developments in the field with regard to theory, techniques, and hardware. Thus it appeared timely to prepare this second volume of selected reprints on digital signal processing to accompany the first.

In selecting the papers to include in this reprint book, we necessarily had to limit the scope and number of the papers, and in order to include as many new papers as possible, we decided to make this a companion to the first volume, rather than a revision. Thus none of the papers included in the first volume has been reprinted here. Furthermore, in deciding among a number of papers covering a similar topic, we tended toward including those which were the best combination of length, representation, and comprehensiveness. In a number of cases, when a good review paper was available, it was included in place of the original source papers. In limiting the scope of the papers included, we have decided to focus on basic digital signal processing theory, techniques, and hardware. There are a wide variety of important applications of digital signal processing techniques, including speech and seismic data processing, radar and sonar processing, image enhancement and processing, and many others. We have specifically avoided considering the inclusion of any of the many excellent applications-oriented papers.

As is evidenced in the table of contents for both the first volume and this second volume, the literature in digital signal processing appears in a variety of technical journals published both within and outside the IEEE. In selecting the papers to include, we have tried to be as thoughtful as possible, and have solicited and received suggestions and opinions from many of our colleagues, both domestic and foreign, representing a number of IEEE Groups and Societies. As was stressed in the first volume, papers included are not necessarily more meritorious than papers omitted.

The papers are organized into seven parts; preceding each part is a brief discussion of the papers in that part. Part 1 deals with some general signal processing techniques. Part 2 deals with one-dimensional digital filtering, including both finite impulse response and infinite impulse response filter design. Part 3 concerns two-dimensional digital signal processing, including a presentation of some general theory and issues of design and stability for two-dimensional digital filters. Parts 4 and 5 concern the related topics of finite register length effects and digital network structures. Part 6 presents a number of papers relating to hardware considerations for digital signal processing. Part 7 consists of one paper on digital signal processing terminology, written by this Committee in 1972. For the past several years this paper has served a useful purpose, both as a tutorial presentation of some of the basic digital signal processing concepts and as a summary of some of the commonly used terminology. While terminology has changed somewhat, and many new techniques, results, and attitudes have emerged since 1972, we felt that it would be useful to include this paper as a tutorial, covering many of the basic concepts of digital signal processing.

### Acknowledgment

We gratefully acknowledge the suggestions and comments received from a number of our colleagues, including J. Aggarwal, University of Texas, Austin; S. Burrus, Rice University, Houston Tex.; D. Dudgeon, Bolt, Beranek and Newman, Cambridge, Mass.; A. Fettweis, Ruhr-Universitat Bochum, Bochum, Germany; O. Herrmann, University of Erlangen, Erlangen, Germany; S. Mitra, University of California, Davis; T. Parks, Rice University, Houston, Tex.; W. Schuessler, University of Erlangen, Erlangen, Germany; J. Shanks, Amoco Production Company, Tulsa, Okla.; and T. Stockham, University of Utah, Salt Lake City.

# Part 1
# General

The predecessor of this volume of collected papers on digital signal processing might be thought of as an account of the discovery and invention of new theory and algorithms for the digital processing of signals. When electronic computers and peripheral data acquisition and display equipment made the time ripe, the rediscovery and subsequent development of fast Fourier transform (FFT) algorithms gave an enormous impetus to the use of digital rather than analog equipment in signal processing. These were significant events and are covered by the collection of papers in *Digital Signal Processing*. The subsequent development of digital techniques has not only brought about the simulation and replacement of analog techniques, but has also led to entirely new theories and methods which exploit the discrete nature of the data and have no counterpart in older theories. Due to the wide applicability of this work, interest goes far beyond the subscribers of the journals in which these papers originally appeared. This probably accounts for the success of the first volume.

The present collection may be considered to be the research and development stages of the work initiated by the papers of the first volume.

Part 1 deals with number-theoretic transforms. This subject has some real novelty and is attracting the attention of both engineers and theoretical mathematicians. It presents the answer to a paradox introduced by the FFT algorithm which seemed to show that it is computationally most efficient to calculate convolutions of sequences in terms of irrational numbers, i.e., sines and cosines, even when the sequences consist of integers. It is shown here that there is a more efficient way to compute convolutions in integer arithmetic without sines and cosines and that it can yield exact results.

The first paper, by Rader, describes a particular class of number-theoretic transforms and tells how they can be implemented on a computer and what advantages they have to offer. The second paper, by Agarwal and Burrus, is more general in that it considers a broad class of number-theoretic transforms and derives important theorems concerning them. One of these theorems is extremely useful in choosing the parameters determining the word length, data sequence length, and the modulus of the residue class arithmetic that is to be used. Another important result demonstrates that the word length must be proportional to the data sequence length. This limitation was the incentive for a suggestion by Rader that a convolution be computed by expressing it as a two-dimensional convolution and that number-theoretic transforms be used for computing the convolutions in each dimension. The third paper, by Agarwal and Burrus, investigates and develops this idea thoroughly. It also shows how special fast formulas for short convolutions can be used in one of the dimensions.

To process a signal on digital equipment, one must necessarily represent it as a finite sequence of discrete numbers, each given as a finite sequence of digits. There are many ways other than pointwise sampling to do this. The fourth paper, by Cooley *et al.*, shows how to relate Fourier series and integral transform pairs to discrete Fourier transform (DFT) pairs in terms of "aliased" versions of the former. This gives useful relations between sampling rates and aliasing error without appealing to distribution theory or mathematically fictitious impulse functions. The fifth paper, by Oppenheim and Johnson, considers other representations of signals which yield preferred distributions of sampling points. The discretization of a signal permits one to use numerical algorithms (such as the FFT) which exploit the discrete nature of the problem.

To take advantage of the speed advantage in the FFT algorithm by doing processing with the DFT, one must cope with the inconveniences brought about by the fact that the DFT is periodic. The sixth paper, by Helms, deals with the fundamental problems and suggests what have now become standard numerical procedures for computing convolutions and doing frequency-domain filtering by using DFT's and the FFT algorithm. These methods are fairly basic and have received wide usage, as described except where one must perform a deconvolution. In the time domain, this is, in effect, a problem of solving a system of linear equations where the matrix is a cyclic Toeplitz matrix, i.e., a matrix whose rows are all shifted copies of the same periodic sequence. In general, such a calculation requires a number of operations proportional to $N$ cubed, while FFT methods reduce this to something proportional to $N \log N$. Unfortunately, the difficulties caused by the cyclic nature of the DFT make it impossible to do a direct deconvolution of an indefinitely long sequence. An iterative algorithm for doing this is described and analyzed in the seventh paper, by Silverman and Pearson.

It was felt that the literature on linear predicive coding in speech signal processing contains material of interest in other areas, such as control theory and statistics. The last paper, by Makhoul, was chosen from among many other good papers on the subject because it presents the general algorithms and ideas independently of applications to speech processing.

# Discrete Convolutions via Mersenne Transforms

CHARLES M. RADER

*Abstract*—A transform analogous to the discrete Fourier transform is defined in the ring of integers with a multiplication and addition modulo *a* Mersenne number. The arithmetic necessary to perform the transform requires only additions and circular shifts of the bits in a word. The inverse transform is similar. It is shown that the product of the transforms of two sequences is congruent to the transform of their circular convolution. Therefore, a method of computing circular convolutions without quantization error and with only very few multiplications is revealed.

*Index Terms*—Convolution, fast Fourier transforms, Fermat numbers, Mersenne numbers, number theoretic transform, transforms.

## Introduction

DISCRETE Fourier transforms have been defined in finite (Galois) fields [1] and, much more familiarly, in the complex number field [2]. In both cases, they are of interest as a means of computing convolutions. We introduce here an anlog of a discrete Fourier transform in certain associative, commutative rings[1] (such a ring is like a field except that there is not necessarily a multiplicative inverse for every element). Specifically, the rings we shall consider are the integers with addition and multiplication modulo Mersenne numbers, hence we propose the name Mersenne transform. We shall see that a circular convolution of two $p$-point sequences can be performed with only $2p$ real multiplications using an arithmetic modulo *a* Mersenne number. However, the process requires a number of additions, and cyclic shifts of bits within a word, which is proportional to $p^2$. An interesting advantage of the process is that the convolution is obtained with perfect accuracy.[2]

## Definitions

Let $p$ be a prime. Let $M_p=2^p-1$. $M_p$ is called a Mersenne number. Mersenne numbers are usually of interest to mathematicians only if they are prime. We shall be interested in Mersenne composites also.

We are used to algebra in a field. Multiplication and addition mod $M_p$ work about as we expect them to, and commutativity, associativity, and distributivity also hold, but *division is undefined*, and therefore some numbers besides zero have no multiplicative inverse modulo $M_p$ (unless $M_p$ is indeed prime). The name for an algebraic structure like a field with division undefined is a *ring*. The integers under multiplication and addition mod $M_p$ form a ring.

To simplify notation, we adopt the following.[3] Any quantity enclosed by superfluous double parentheses $((x))$ is to be replaced by its value modulo $M_p$, so that

$$((x)) = x + rM_p$$

where $r$ is the integer such that

$$0 \leq ((x)) < M_p. \tag{1}$$

Any quantity enclosed by $\langle\ \rangle$ is to be replaced by its value modulo $p$, so that

$$\langle x \rangle = x + rp$$

where $r$ is the integer such that

$$0 \leq \langle x \rangle < p. \tag{2}$$

A congruence modulo $M_p$ is indicated by $\equiv$.

We now define the Mersenne transform of a set of $p$ integers $\{a_n\}$ as a set of $p$ integers $\{A_k\}$, each given by

$$A_k = \left(\left(\sum_{n=0}^{p-1} a_n 2^{nk}\right)\right), \qquad k = 0, 1, \cdots, p-1. \tag{3}$$

Preceding any discussion of an inverse Mersenne transform we need a quantity $q$ to play the role played by $1/N$ in the equation for an inverse discrete Fourier transform.

Letting $q=M_p-(M_p-1)/p$, we have a solution to the equation

$$((pq)) = 1 \tag{4}$$

if we can show that $(M_p-1)/p$ is an integer. But $M_p-1=2^p-2$. It is asserted that $p$ divides $2^p-2$. In fact this is a special case of Fermat's theorem.

*Fermat's Theorem:* For every prime $p$ and every integer $a$, $p$ divides $a^p-a$. This proves that $q$ is an integer. Equation (4) follows since $pq=(p-1)M_p+1\equiv 1$.

Having settled the question of the existence of $q$, we define an inverse Mersenne transform by

$$r_m = \left(\left(q\sum_{k=0}^{p-1} A_k 2^{\langle -mk \rangle}\right)\right), \qquad m = 0, 1, \cdots, p-1 \tag{5}$$

Manuscript received February 15, 1972; revised July 27, 1972. This work was sponsored by the Department of the Air Force.

The author is with the Lincoln Laboratory, Massachusetts Institute of Technology, Lexington, Mass. 02173.

[1] The fourth section of [1] discusses the transform in a ring, quite generally but quite tersely.

[2] The reader must be cautioned here. Perfect accuracy implies long word lengths and in practice this can be burdensome rather than advantageous [see (19)–(21)].

[3] The normal notation for algebra in a ring is actually very simple. However, we are interested in dealing with integers, rather than with equivalence classes, and we are therefore forced to write equations whose quantities are free of ambiguity.

Reprinted from *IEEE Trans. Comput.*, vol. C-21, pp. 1269–1273, Dec. 1972.

although we have yet to establish that this inverse satisfies $r_m = a_m$. We must first discuss the term $2^{\langle -mk \rangle}$. We note that

$$2^p \equiv 1. \tag{6}$$

From this it follows, for $x$ positive, that

$$((2^x)) = ((2^{\langle x \rangle})). \tag{7}$$

By definition we will extend (7) to negative $x$. This makes $2^{-mk}$ meaningful even though there is not necessarily a meaning to negative powers of other numbers. Therefore, with no further ado we rewrite (5) as

$$r_m = \left(\left( q \sum_{k=0}^{p-1} A_k 2^{-mk} \right)\right). \tag{8}$$

We could prove that this is indeed an inverse but the proof plays no important role in what follows, and can be seen as a special case of the more important result, namely the following theorem.

*Theorem:* If $\{A_k\}$ is the Mersenne transform of $\{a_n\}$, and $\{B_k\}$ in the Mersenne transform of $\{b_n\}$, then, with $\{C_k\}$ given by $C_k = ((A_k B_k))$, the inverse Mersenne transform $\{c_n\}$ of $\{C_k\}$ is

$$c_n \equiv \left(\left( \sum_{m=0}^{p-1} a_m b_{\langle n-m \rangle} \right)\right). \tag{9}$$

To prove the theorem we write the complete expression for $c_n$

$$c_n = \left(\left( q \sum_{k=0}^{p-1} \left( \sum_{m=0}^{p-1} a_m 2^{mk} \right) \left( \sum_{l=0}^{p-1} b_l 2^{lk} \right) 2^{-nk} \right)\right). \tag{10}$$

Since the commutative, associative, and distributive laws all hold in the ring, we are able to manipulate sums and products, etc., giving

$$c_n \equiv \left(\left( \sum_{m=0}^{p-1} \sum_{l=0}^{p-1} a_m b_l \left( q \sum_{k=0}^{p-1} 2^{(m+l-n)k} \right) \right)\right). \tag{11}$$

Clearly for combinations of $m$, $l$, $n$ such that $\langle m+l-n \rangle = 0$ the sum over $k$ is a sum of $p$ ones and is therefore equal to $p$. For $\langle m+l-n \rangle = r \neq 0$, however, we have

$$\sum_{k=0}^{p-1} 2^{\langle rk \rangle} \tag{12}$$

and since the set of exponents $(rk)$ mod $p$ is merely a permutation of the integers $0, 1, \cdots, p-1$, the sum is

$$\sum_{k=0}^{p-1} 2^{\langle rk \rangle} \equiv \sum_{k=0}^{p-1} 2^k = M_p \tag{13}$$

and it is therefore congruent to zero. Thus

$$\left(\left( q \sum_{k=0}^{p-1} 2^{(m+l-n)k} \right)\right) = \begin{cases} ((pq)), & \langle m+l-n \rangle = 0 \\ 0, & \langle m+l-n \rangle \neq 0. \end{cases} \tag{14}$$

$((pq))$, of course, is unity, $q$ having been chosen earlier to meet that requirement. We are now able to eliminate either the sum on $l$, using

$$l = \langle n - m \rangle \tag{15}$$

or the sum on $m$, using

$$m = \langle n - l \rangle. \tag{16}$$

The former case gives

$$c_n = \left(\left( \sum_{m=0}^{p-1} a_m b_{\langle n-m \rangle} \right)\right) \tag{17}$$

and the latter gives

$$c_n = \left(\left( \sum_{l=0}^{p-1} b_l a_{\langle n-l \rangle} \right)\right). \tag{18}$$

The two are, of course, equivalent. The proof that the claimed inverse transform is indeed an inverse transform now follows if $b_n$ is chosen to be

$$b_n = \begin{cases} 1, & n = 0 \\ 0, & n \neq 0 \end{cases}$$

in which case $B_k \equiv 1$, $C_k = A_k B_k = A_k$, and $c_n$, the inverse Mersenne transform of $A_k$, is seen to be equal to $a_n$, as we would expect from an inverse transform.

## Discussion of Mersenne Transform

We have so far shown that there is a transform which we can perform in arithmetic modulo $M_p$, that it is invertible, and that it maps circular convolution modulo $M_p$ into multiplication modulo $M_p$. The author was led to investigate such a transform, along with similar transforms defined in the arithmetic of finite fields, because of the observation that truncation and rounding effects, which necessarily lead to quantization errors when performing convolution by transform methods in the complex number field, may be avoided, and indeed have no natural meaning, when performing computations in a finite field or ring. Thus if two sequences are to be convolved exactly, there is an advantage to performing the convolution in a finite field or ring. On the other hand, the choice of the field or ring will depend on how easy it is to do operations like addition or multiplication in such a field or ring. In the general case, arithmetic modulo some arbitrary modulus can be performed as ordinary binary arithmetic, followed by a division of the result by the modulus—the remainder is the answer. The methods of [1] therefore, permit exact convolution of integer sequences using FFT methods in a finite field. In practice the need to divide is a major complexity, and one is tempted to consider only those moduli for which there are easy arithmetic methods. The easiest arithmetic is modulo $2^k$ but no transforms seem to be available in this arithmetic. The second easiest case is arithmetic modulo $2^k - 1$. Letting 2 play the role of $e^{-j2\pi/N}$, transforms exist for this arithmetic when $k$ is a prime. Having chosen this arithmetic, addition modulo $M_p$ is simply one's complement addition in a $p$-bit word.

Multiplication modulo $M_p$ may be accomplished by forming the $2p$-bit product of two words and adding the $p$ least significant bits to the $p$ most significant bits (one's complement addition, of course). However, multiplication by a power of two modulo $M_p$ is particularly simple, amounting to nothing more than a rotation of the bits within the $p$-bit word. Thus the computation of a Mersenne transform requires no multiplications, only additions and bit rotations. The same statement can be made for inverse Mersenne transforms, except that it is necessary to multiply the $p$ results by the constant $q$ (either before or after the transform computation).

In summary, a Mersenne transform requires $p(p-1)$ adds and $(p-1)^2$ shifts, and a circular convolution requires 3 transforms and $2p$ multiplications.[4]

If the output of the ordinary (rather than mod $M_p$) convolution $\tilde{c}_n$ can be guaranteed to obey

$$R \leq \sum_{m=0}^{p-1} a_m b_{n-m} < R + M_p \tag{19}$$

then $\tilde{c}_n$ is known exactly, given $c_n$ computed in the ring of integers mod $M_p$. In practice with $R = -M_p/2$, suppose

$$\begin{aligned} |a_n| &\leq \alpha \\ |b_n| &\leq \beta. \end{aligned} \tag{20}$$

Then if

$$\alpha\beta < 2^{p-1}/p \tag{21}$$

the ordinary circular convolution can be computed without error using the Mersenne transform approach. If aperiodic convolution is desired, both sequences must be padded out with zeros as in [3], [4]. This relaxes the bound of (21) even further.

## Properties of a Mersenne Transform

The Mersenne transform has properties similar to the DFT. Among these properties are the following.

*Even Sequences:* If

$$a_n = a_{\langle -n\rangle}, \quad A_k = A_{\langle -k\rangle}. \tag{22}$$

*Odd Sequences:* If

$$a_n = -\, a_{\langle -n\rangle}, \quad A_k = -\, A_{\langle -k\rangle}. \tag{23}$$

*Parseval Theorem:*

$$\begin{aligned} \sum_{n=0}^{p-1} a_n^2 &\equiv q \sum_{k=0}^{p-1} A_k^2 \\ \sum_{k=0}^{p-1} A_k^2 &\equiv p \sum_{n=0}^{p-1} a_n^2. \end{aligned} \tag{24}$$

[4] If $a_n$ is to be convolved with a known sequence $b_n$, $B_k$ is computed once and for all and stored. If one stores, instead, $qB_k$, the multiplications by $q$ need not be part of the inverse Mersenne transform computation and the total number of multiplications becomes $p$. If multiplication by $q$ cannot be so avoided, it is still possible to make use of the fact that it is a known constant.

## Permutations

If $A_k$ is the Mersenne transform of $\{a_n\}$, then $A_{\langle rk\rangle}$ is the Mersenne transform of the permuted sequence $\{a\langle_{sn}\rangle\}$, where $\langle rs\rangle = 1$. This means that the same network or algorithm which computes $A_1$ can compute all the other transform points except $A_0$, assuming only that the inputs can be applied to the network in a permuted order.

## Practical Considerations

If one were to be computing Mersenne transforms by digital computer, it is frustrating that there are few, if any, computers for which a circular shift of $p$ bits is an elementary operation. However, if hardware were to be built, this would be the case. In addition, although $p^2$ additions are required for the transform, only $p$ of these need be full additions, the others being amenable to carry-save addition techniques. Fig. 1 shows a Mersenne transform configuration for $p=5$ making use of carry-save adders.

Another approach is a sort of recursive filter, as in Fig. 2, using

$$y_n = ((2^{\langle -k\rangle} y_{n-1} + a_n)) \tag{25}$$

and therefore,

$$y_p \equiv ((2^{\langle -k\rangle} a_{p-1} + 2^{\langle -2k\rangle} a_{p-2} + \cdots + 2^{\langle k\rangle} a_1 + a_0)) = A_k. \tag{26}$$

This method is easier to program since the amount necessary to shift need not be updated in the main loop. A variation similar to the Goertzel algorithm [5] will work for the Mersenne transform, but introduces the need for multiplication by a less trivial quantity than $2^k$.

An objection to the use of Mersenne transforms may well be that only very short convolutions make practical sense. However, a long convolution may be computed with the aid of an appropriate two-dimensional convolution, and a multidimensional Mersenne transform is a practical affair. Another application could be the multiplication of polynomials.

## Double Length Convolutions

If we define a slightly different transform

$$\hat{A}_k \equiv \sum_{n=0}^{2p-1} a_n((-2))^{nk} \tag{27}$$

we can prove an analog to (9).[5] The inverse used is

$$r_m \equiv s \sum_{k=0}^{2p-1} \hat{A}_k((-2))^{(-mk \bmod 2p)} = a_m \tag{28}$$

where

$$s = M_p - \frac{M_p - 1}{2p}. \tag{29}$$

[5] The proof is omitted.

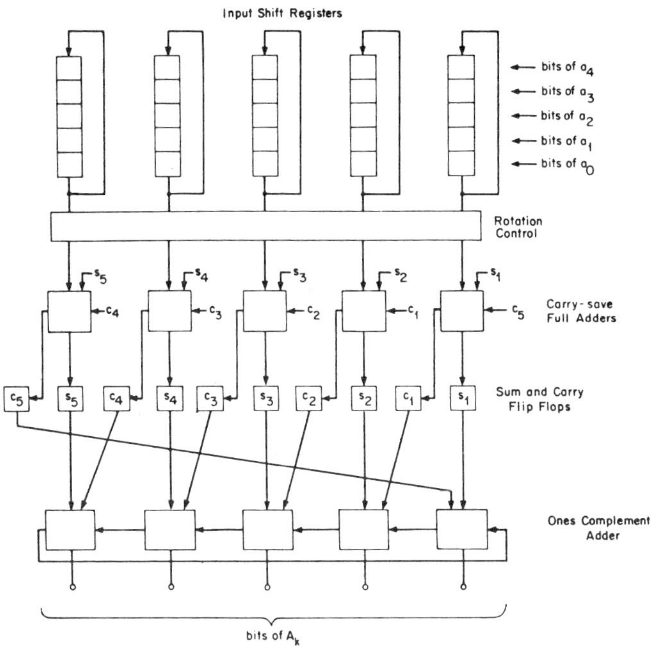


Fig. 1. Possible hardware for the Mersenne transform.

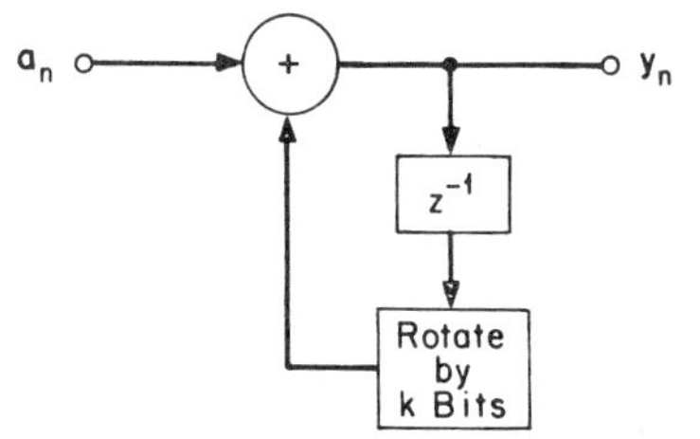


Fig. 2. "Recursive" computational Mersenne transform.

With $\hat{B}_k$ defined from $b_n$ as in (27) and $\hat{C}_k = \hat{A}_k \hat{B}_k$ the inverse transform (28) of $\hat{C}_k$ is

$$c_n \equiv \sum_{m=0}^{2p-1} a_m b_{n-m \bmod 2p}. \tag{30}$$

Implementing (27) or (28) requires multiplications by $((-2))^{nk}$. This is only slightly more complicated than before, involving rotation of the $p$-bit word by $\langle nk \rangle$ positions, as before, and complementing all the bits if $nk$ is odd.

Since $2p$ is composite (barely) a minor savings is possible using an FFT-type decomposition.

## Convolutions in the Ring of Integers Modulo a Fermat Number

The Mersenne transform, as defined, precludes the use of an FFT-type algorithm in its calculation. Had $p$ been replaced by $\mu$, a power of two,

$$\mu = 2^{m+1} \tag{31}$$

with

$$M_\mu = 2^\mu - 1, \tag{32}$$

the calculation of the revised (3) and (5) would be susceptible to the simplications of an FFT-type algorithm (although (4) would no longer be a meaningful definition of $q$). However, (3) and (5) as so modified are not inverses of one another.

A useful result can still be obtained if we introduce the ring of integers modulo a Fermat number $F_m$ defined as

$$F_m = 1 + 2^{2^m}. \tag{33}$$

The first five Fermat numbers are prime, and for these cases the results given below follow trivially from [1]. No other prime $F_m$ are known. We assert that $\mu$-point circular convolutions modulo $F_m$ may be computed using multiplicationless transforms like (3) and (5) and that an FFT-type calculation is available as well. The author has rigorously proved these assertions but the proofs will be omitted here.

We have defined $\mu$, $M_\mu$, and $F_m$ by (31), (32), and (33). Since

$$M_\mu = F_m(F_m - 2) \tag{34}$$

it is possible to carry out computations in the ring of integers modulo $M_\mu$ and obtain results of arithmetic carried out modulo $F_m$. Specifically,

$$\chi \text{ modulo } F_m = (\chi \text{ modulo } M_\mu) \text{ modulo } F_m. \tag{35}$$

The conversion (35) is simplified enormously by the following relation. If

$$\chi \text{ modulo } M_\mu = a + 2^m b$$

then

$$\chi \equiv a - b \text{ modulo } F_m. \tag{36}$$

Therefore, in what follows we reassign the symbols

$$((\chi)) = \chi \text{ modulo } F_m$$
$$\langle \chi \rangle = \chi \text{ modulo } \mu. \tag{37}$$

We visualize computations modulo $F_m$ as being carried out modulo $M_\mu$ with the conversion (35) whenever convenient.

Note that 2 is of order $\mu$ modulo $F_m$ (and also modulo $M_\mu$). We now define the transform[6] of a sequence $\{a_n\}$ as

$$A_k = \left(\left(\sum_{n=0}^{\mu-1} a_n 2^{nk}\right)\right), \qquad k = 0, 1, \cdots, \mu - 1 \tag{38}$$

which we can compute by FFT techniques. Only $\mu \log_2 \mu = (m+1)\mu$ additions and circular bit shifts are required. We now define

$$q = 2^\mu/\mu = 2^{\mu-m-1} \tag{39}$$

and an inverse transform

$$a_n = \left(\left(q \sum_{k=0}^{\mu-1} A_k 2^{\langle -nk \rangle}\right)\right) \tag{40}$$

[6] We resist naming still another transform.

which also can be computed by FFT techniques. The proof of (40) requires the (omitted) proof that

$$\left(\left(q \sum_{k=0}^{\mu-1} 2^{\langle nk \rangle}\right)\right) = \begin{cases} 1, & \langle n \rangle = 0 \\ 0, & \langle n \rangle \neq 0 \end{cases} \tag{41}$$

and (41) also supplies the essential step in showing that the product of two transforms is the transform of a $\mu$-point circular convolution.

## Summary

Convolutions may be computed by transform techniques in a finite field, and in certain finite rings. The Mersenne transform, defined herein, involves computations in the ring of integers with arithmetic modulo $2^p-1$, with $p$ a prime. It may be computed with about $p^2$ additions and circular shifts of the bits of a $p$-bit word. It is a simply invertible transform, and the inverse transform is computed similarly. The product of two Mersenne transforms is the Mersenne transform of the $p$-point circular convolution of the two sequences. Such a convolution could be computed by Mersenne transform techniques using $3p^2$ additions and $2p$ multiplications. A modified transform, susceptible to FFT-type computation, is defined in the ring of integers with arithmetic modulo $2^{(2^m)}+1$. This transform may be carried out with $(m+1)2^{m+1}$ additions and circular bit shifts for a sequence of $2^{m+1}$ points. A $\mu$-point circular convolution using the modified transform requires about $3\mu \log_2 \mu$ additions and circular bit shifts, and $\mu$ multiplications.

Both of these transforms have the difficulty that the word length required is proportional to the length of the sequences being convolved. It is conceivable that they may be of value when sequences must be convolved *exactly*, when multiplication is a very costly operation, or when special-purpose hardware for modest length convolutions is being considered.

## References

[1] J. M. Pollard, "The fast Fourier transform in a finite field," *Math. Comput.*, vol. 25, pp. 365–374, Apr. 1971.
[2] J. W. Cooley and J. W. Tukey, "An algorithm for the machine computation of complex Fourier series," *Math. Comput.*, vol. 19, pp. 297–301, Apr. 1965.
[3] T. G. Stockham, Jr., "High speed convolution and correlation," in *1966 Spring Joint Comput. Conf., AFIPS Conf. Proc.*, vol. 28. Washington, D. C.: Spartan, 1966, pp. 229–233.
[4] H. D. Helms, "Fast Fourier transform method of computing difference equations and simulating filters," *IEEE Trans. Audio Electroacoust.*, vol. AU-15, pp. 85–90, June 1967.
[5] B. Gold and C. M. Rader, *Digital Processing of Signals.* New York: McGraw-Hill, 1969, pp. 171–172.

# Number Theoretic Transforms to Implement Fast Digital Convolution

RAMESH C. AGARWAL AND C. SIDNEY BURRUS, MEMBER, IEEE

*Invited Paper*

*Abstract*—Transforms using number theoretic concepts are developed as a method for fast and error-free calculation of finite digital convolution. The transforms are defined on finite fields and rings of integers with arithmetic carried out modulo an integer and it is shown that under certain conditions this gives the same results as conventional digital convolution. Because of these characteristics they are ideally suited to digital computation by taking into account quantization of amplitude as well as time in their definition. When the modulus is chosen as a Fermat number a transform results that requires only on the order of $N \log N$ additions and word shifts but no multiplications. In addition to being efficient, they have no roundoff error and do not require storage of basis functions. There is a restriction on sequence length imposed by word length and a problem with overflow but methods for overcoming these are presented. Results of an implementation on an IBM 370/155 are presented and compared with the fast Fourier transform showing a substantial improvement in efficiency and accuracy. Variations on the basic number theoretic transforms are also presented.

## I. Introduction

FINITE DIGITAL convolution is a numerical procedure defined by

$$y(n) = \sum_{m=0}^{N-1} h(n-m)\,x(m), \qquad n = 0, 1, 2, \ldots \tag{1}$$

and symbolically denoted

$$y(n) = h(n) * x(n)$$

where $x(n)$, $h(n)$, and $y(n)$ are digital number sequences. This operation has many very powerful applications. It is used to implement nonrecursive or finite-impulse-response digital filters either directly or with sectioning or block techniques [1], [2] and recursive or infinite-impulse-response digital filter by block methods [3]. It also is used to carry out auto and cross correlation as well as for computations such as polynomial multiplication and multiplication of very large integers [4]–[6].

There are several methods to implement finite convolution that differ in the amount of computation required, the effects of arithmetic roundoff, and the amount of storage required. It is somewhat difficult to compare various algorithms because of the tradeoffs between these various factors that depend on the hardware or software that is available. However, because of the complexity of performing multiplication, the number of multiplications necessary to implement convolution is often an important factor to be minimized.

Manuscript received September 5, 1974; revised October 15, 1974. This work was supported in part by the National Science Foundation under Grant GK-23697.

R. C. Agarwal was with the Department of Electrical Engineering, Rice University, Houston, Tex. He is now with the IBM Thomas J. Watson Research Center, Yorktown Heights, N.Y. 10598.

C. S. Burrus is with the Department of Electrical Engineering, Rice University, Houston, Tex. 77001.

The use of transform methods has proven to be useful when an application allows sequences to be processed in blocks. The most versatile transform is the discrete Fourier transform (DFT) defined by

$$\mathrm{DFT}[x] \triangleq X(k) = \sum_{n=0}^{N-1} x(n) \exp(-j2\pi nk/N), \qquad k = 0, 1, \cdots, N-1 \tag{2}$$

and the inverse transform

$$\mathrm{IDFT}[X] \triangleq x(n) = N^{-1} \sum_{k=0}^{N-1} X(k) \exp(j2\pi nk/N), \qquad n = 0, 1, \cdots, N-1. \tag{3}$$

The property of this transform that is important here is the cyclic convolution property (CCP) which states that

$$\mathrm{DFT}[h * x] = \mathrm{DFT}[h] \cdot \mathrm{DFT}[x].$$

This implies that a convolution can be calculated by

$$y(n) = \mathrm{IDFT}\{\mathrm{DFT}[h] \cdot \mathrm{DFT}[x]\} \tag{4}$$

using two transforms, $N$ multiplications, and one inverse transform. The convolution implemented by (4) is called cyclic convolution since it evaluates (1) as if $h(n)$ and $x(n)$ were periodically extended outside of the range from 0 to $(N-1)$ or, equivalently, the indices are evaluated mod $N$. Normal finite convolution can be calculated by cyclic convolution if zeros are appended to $x(n)$ and $h(n)$ to prevent folding or aliasing [1], [2].

This transform approach became useful only when Cooley and Tukey [7] introduced a very efficient algorithm known as the fast Fourier transform (FFT) for calculating the DFT and its inverse in (2) and (3). The number of multiplications necessary to calculate the FFT of a number sequence of length $N$ is on the order of $N \log_2 N$. Implementation of convolution using the FFT results in a considerable savings in multiplication when lengths are approximately above $N = 32$. The disadvantage of this approach is in the form of significant amounts of roundoff error [8], storage or generation of the complex basis functions that have to be rounded and still a considerable amount of multiplying.

If one looks for the properties that a general transform with the DFT structure

$$X(k) = \sum_{n=0}^{N-1} x(n)\,\alpha^{nk} \tag{5}$$

must have to have the CCP, it is found [9], [10] that $\alpha$ is a

Reprinted from *Proc. IEEE*, vol. 63, pp. 550-560, Apr. 1975.

root of unity of order $N$, i.e., $N$ is the least positive integer such that

$$\alpha^N = 1. \tag{6}$$

This analysis shows that in the complex number field, the conventional DFT with $\alpha = \exp(-j2\pi/N)$ is the only transform given by (5) with the CCP. If, however, other fields and arithmetic systems are used, new transforms become possible with very interesting properties. This is pursued by considering mathematical systems that are fundamentally compatible with digital computing capability.

In any practical situation, or when working with digital machines, the data are available only with some finite precision, and therefore, without loss of generality, the data can be considered to be integers with some upper bound. To compute convolution in this digital domain, operations in the complex number field of the continuous domain can be imitated in a finite field or, more generally, in a finite ring of integers under additions and multiplications modulo some integer $M$. An integer $\alpha$ of order $N$ replaces $\exp(-j2\pi/N)$ used in a DFT. In this ring, when two integer sequences $x(n)$ and $h(n)$ are convolved, the output integer sequence $y(n)$ is congruent to the conventional convolution of $x(n)$ and $h(n)$ mod $M$. In the ring of integers mod $M$, conventional integers can be unambiguously represented if their absolute value is less than $M/2$. If the input integer sequences $x(n)$ and $h(n)$ are so scaled that $|y(n)|$ never exceeds $M/2$, we would get the same results by implementing convolution in the ring of integers mod $M$ as that obtained with normal arithmetic. This is similar to the overflow constraint in fixed-point digital machines. In most digital filtering applications, $h(n)$ represents the impulse response and is known *a priori*; also the maximum magnitude of the input signal is usually known.

By working in a finite field or ring of integers with arithmetic carried out modulo an integer $M$, a large class of transforms exist that have the CCP. By special choices of the length $N$, the mod $M$, and the value $\alpha$, it is possible to have transforms that need only word shifts and additions but no multiplications, that have an FFT type fast algorithm, that do not require storage of complex values for $\alpha$, and that have no roundoff errors. These transforms are called number theoretic transforms (NTT) and they look very promising in the evaluation of finite convolutions. Their main disadvantage seems to be a relation of the sequence length $N$ to the required word length that can require long word lengths for long sequence lengths.

These number theoretic transforms are truly digital transforms, taking into account the quantization in amplitude and the finite precision of digital signals. They bear the same relation to digital signals as the DFT does to discrete-time or sampled data signals and the Fourier or Laplace transforms do to continuous-time signals. In the same manner that the relation of discrete-time signals to continuous-time signals through sampling involves a possible folding or aliasing in the frequency domain, the relation of calculations with the DFT to calculations with the number theoretic transforms involves a possible folding of the amplitude that must be taken into account.

The literature on transforms of these types is fairly recent. Knuth [4] has proposed the use of transforms in finite fields. Pollard [5] discussed transforms having the CCP in a finite field and also gives conditions for having transforms with the CCP in a finite ring of integers. Good [11] also mentioned the use of transforms in a finite ring of integers. Schonhage and Strassen [6] defined transforms having the CCP modulo a Fermat number and discussed their application to fast multiplication of very large integers. Knuth [12] elaborated on the work of Schonhage and Strassen. Nicholson [10] presented an algebraic theory of FFT's in any ring and established fast FFT-type algorithms to compute these transforms. Rader [13], [14] proposed number theoretic transforms in rings of integers modulo both Mersenne and Fermat numbers. He first proposed the application to digital signal processing, showed that the transforms could be calculated using only additions and bit shifting, showed the word length constraint, and suggested two-dimensional transforms as a possible relaxation of that constraint. Agarwal and Burrus [9], [15] discussed number theoretic transforms in detail, defined Fermat number transforms and also proposed their application for fast digital convolution. They also suggested possible hardware and software implementations. Their implementation on the IBM 370/155 showed a factor of 3 to 5 speed improvement over efficient FFT implementations of cyclic convolution for lengths up to 256. An earlier article by Takahasi and Ishibashi [30] was recently brought to our attention by Dr. J. W. Cooley of IBM.

## II. Modular Arithmetic

In this section, some of the basic concepts of modular arithmetic from number theory relevant to NTT will be discussed. This can be found in most basic books on number theory [16], [17].

Two integers $a$ and $b$ are said to be congruent mod $M$ if

$$a = b + kM \tag{7}$$

where $k$ is some integer and $M$ is the modulus. This is written as

$$a \equiv b \pmod{M}. \tag{8}$$

All integers are congruent mod $M$ to some integer in the finite set $\{0, 1, 2, \cdots, M-1\}$ which is called the set of integers mod $M$ and denoted by $Z_M$. $Z_M$ is also known as the ring of integers mod $M$. If in a ring of integers multiplicative inverses exist for all nonzero integers, this ring becomes a field and it can be shown that $Z_M$ is a field if and only if $M$ is a prime. We will use the symbol "$Z_M$" and the expression "the ring of integers mod $M$" for rings as well as fields since a field is also a ring. The following basic arithmetic operations are permissible with modular arithmetic.

a) Addition: Example, $7 + 12 = 19 = 2 \pmod{17}$.
b) Negation: Example, $-7 = -7 + 17 = 10 \pmod{17}$.
c) Subtraction: Example, $7 - 12 = 7 + (-12) = 7 + 5 = 12 \pmod{17}$.
d) Multiplication: Example, $7 \times 12 = 84 = 16 \pmod{17}$.
e) Multiplicative Inverse: Multiplicative inverse of an integer $b$ in $Z_M$ exists if and only if $b$ and $M$ are relatively prime. In that case $b^{-1}$ is an integer such that $b \times b^{-1} = 1 \pmod{M}$. Example, $7^{-1} = 5 \pmod{17}$; $7 \times 5 = 35 = 1 \pmod{17}$.
f) Division: $a/b$ exists if and only if $b$ has an inverse. In that case $a/b = a \times b^{-1}$. Example, $12/7 = 12 \times 5 = 9 \pmod{17}$; $7 \times 9 = 12 \pmod{17}$.

This may seem like a rather peculiar way to do arithmetic but it is used quite often by everyone. In discussing the day of the week, one uses an arithmetic mod 7 or in stating the time, one is calculating mod 12 or perhaps 24. Indeed the mantissa of a number in scientific notation is evaluated mod 10.

Because of the nature of modular arithmetic, numbers do not have sizes or magnitude. We can not say that a particular number is larger than another or that two numbers are close. Tuesday may not be close to Wednesday or come before Wednesday if they occur in different weeks.

As was mentioned in the introduction, for the existence of transforms with the DFT structure given in (5) and having the CCP, it is necessary that an integer exist that is the $N$th root of unity. We will now consider this problem using modular arithmetic. First Euler's $\varphi$ function is defined as $\varphi(M)$, the number of integers in $Z_M$ that are relative primes to $M$. For $M$ a prime, $\varphi(M) = M - 1$. If $M$ is composite and its prime factored form is denoted by $M = p_1^{r_1} p_2^{r_2} \cdots p_l^{r_l}$ then the general expression for $\varphi$ is

$$\varphi(M) = M(1 - 1/p_1)(1 - 1/p_2) \cdots (1 - 1/p_l).$$

An important theorem known as Euler's theorem states that for every $\alpha$ relatively prime to $M$

$$\alpha^{\varphi(M)} = 1 \qquad (\text{mod } M). \tag{9}$$

For $M$ prime this reduces to Fermat's theorem

$$\alpha^{M-1} = 1 \qquad (\text{mod } M) \tag{10}$$

which holds for all nonzero elements of $Z_M$ since they are all relatively prime to $M$ if $M$ is prime.

There are certain roots of unity that are of particular interest. If $N$ is the least positive integer such that

$$\alpha^N = 1 \qquad (\text{mod } M) \tag{11}$$

then $\alpha$ is said to be a root of unity of order $N$, or simply of order $N$. In some of the literature $\alpha$ is said to belong to the exponent $N$ or $N$ is the exponent to which $\alpha$ belongs. Another terminology says $\alpha$ is a primitive $N$th root of unity.

If the order of $\alpha$ (the exponent to which $\alpha$ belongs) is equal to $\varphi(M)$, then $\alpha$ is called a primitive root (do not confuse with a primitive $N$th root of unity). If $M$ is prime and $\alpha$ is a primitive root, the set of integers $\{\alpha^k \text{ (Mod } M), k = 0, 1, 2, \cdots, M - 2\}$ is the total set of nonzero elements in $Z_M$. Thus all nonzero integers in $Z_M$ can be generated by powers of a primitive root. This characterizes the entire field.

Euler's theorem implies that if $\alpha$ is of order $N$ then $N$ must divide $\varphi(M)$, denoted by $N | \varphi(M)$. If $M$ is prime it can be shown that roots of order $N$ exist if and only if $N | (M - 1)$ and the roots are given by

$$\alpha = \alpha_\varphi^{(M-1)/N} \tag{12}$$

where $\alpha_\varphi$ denotes a primitive root. More generally, if $\alpha$ is a root of order $N$ then

$$\alpha^k \text{ is of order } N/k \text{ if } k | N$$

$$\alpha^k \text{ is of order } N \text{ if } N \text{ and } k \text{ are relatively prime.} \tag{13}$$

This implies the number of roots of order $N$ is given by $\varphi(N)$ and, therefore, the number of primitive roots is $\varphi(\varphi(M))$. These relations will allow one to calculate all of the roots of all possible orders from one primitive root. Tables will often list primes and the smallest primitive root for each.

These ideas will become clearer by looking at an example. Consider the field $Z_7$ with arithmetic mod 7. First we will give the first few evaluations of Euler's function.

$$\varphi(1) = 1 \;\; \varphi(2) = 1 \;\; \varphi(3) = 2 \;\; \varphi(4) = 2$$
$$\varphi(5) = 4 \;\; \varphi(6) = 2 \;\; \varphi(7) = 6. \tag{14}$$

Consider raising each element of $Z_7$ to powers from 1 to 6 (mod 7).

| $N = 0$ | 1 | 2 | 3 | 4 | 5 | 6 |
|---|---|---|---|---|---|---|
| $1^N = 1$ | 1 | 1 | 1 | 1 | 1 | 1 |
| $2^N = 1$ | 2 | 4 | 1 | 2 | 4 | 1 |
| $3^N = 1$ | 3 | 2 | 6 | 4 | 5 | 1 |
| $4^N = 1$ | 4 | 2 | 1 | 4 | 2 | 1 |
| $5^N = 1$ | 5 | 4 | 6 | 2 | 3 | 1 |
| $6^N = 1$ | 6 | 1 | 6 | 1 | 6 | 1 |

This illustrates several very interesting features. Consider the various roots of order $N$.

| $N$ | Roots of order $N$ |
|---|---|
| 1 | 1 |
| 2 | 6 |
| 3 | 2, 4 |
| 6 | 3, 5 |

Only those $N$ that divide $\varphi(M) = \varphi(7) = 6$ have roots that belong to them. The number of roots is given by $\varphi(N)$ and the number of primitive roots is $\varphi(\varphi(M)) = 2$ and they are 3 and 5. Note that both of the primitive roots generate all the nonzero elements of the field while the other roots generate cyclic subsets with $N$ distinct members. Also note that Euler's theorem (Fermat's theorem in this case) is indeed satisfied in that all elements raised to the 6th power are congruent to unity and (13) does generate all the roots of order $N$ from the primitive roots. Also note that every nonzero integer $\alpha$ has an inverse $x^{M-2}$. For a nonprime $M$, $\alpha$ has an inverse given by $\alpha^{\phi(M)-1}$ if $\alpha$ and $M$ are relatively prime.

By considering a similar example with $M$ a composite rather than a prime, one observes several differences. First $Z_M$ is not a field since all elements will not have inverses. There is no primitive root that will generate the entire ring, only subsets with $\varphi(M)$ elements.

When considering a nonprime mod $M$, $Z_M$ is a ring and inverses exist only for integers relatively prime to $M$. Let $M$ have the following unique prime power factorization.

$$M = p_1^{r_1} \cdot p_2^{r_2} \cdots p_l^{r_l}. \tag{15}$$

When the arithmetic is done mod $M$, it is in effect done modulo each prime power $p_i^{r_i}$ simultaneously [4], [18]. A set of arithmetic operations can be done either modulo each $p_i^{r_i}$ separately and the final result mod $M$ obtained using the Chinese remainder theorem [4], [16], [18], or alternatively all the operations may be done mod $M$, but, they must be valid operations mod for each $p_i^{r_i}$. An integer $\alpha$ is said to be of order $N$ in $Z_M$ if and only if it is of order $N$ in each $Z_{p_i^{r_i}}$. Here we present some basic results.

$$a = b \qquad (\text{mod } M) \tag{16}$$

is true if and only if

$$a = b \qquad (\text{mod } p_i^{r_i}), \qquad i = 1, 2, \cdots, l. \tag{17}$$

If we know the residues of an integer $a$ modulo each $p_i^{r_i}$, we can uniquely reconstruct the integer $a$ (mod $M$) using the Chinese remainder theorem given in the following.

Let

$$a = a_i \; (\text{mod } p_i^{r_i})$$
$$d_i \triangleq M/(p_i^{r_i})$$

and

$$d_i^{-1} \triangleq (d_i \bmod p_i^{r_i})^{-1} \pmod{p_i^{r_i}}$$

then

$$a = \left(\sum_{i=1}^{l} d_i d_i^{-1} a_i\right) \pmod{M}. \tag{18}$$

## III. Number Theoretic Transforms

In this section, the definition and basic conditions for the existence of the NTT will be presented and in particular the allowed relations between the modulus $M$ and the transform length $N$ and basis function $\alpha$ are spelled out.

If we have a length $N$ sequence of numbers, then a transform pair of the form given by

$$X(k) = \sum_{n=0}^{N-1} x(n)\, \alpha^{nk} \tag{19}$$

$$x(n) = N^{-1} \sum_{k=0}^{N-1} X(k)\, \alpha^{-nk} \tag{20}$$

is said to have a DFT structure. By requiring that application of the transform method in (4) results in cyclic convolution, the following theorem can be proven.

*Theorem 1:* A length $N$ transform having the DFT structure will implement cyclic convolution if and only if there exists an inverse of $N$ and an element $\alpha$, a root of unity of order $N$, i.e., $N$ is the least positive integer such that

$$\alpha^N = 1.$$

This is a very general result applying to both rings and fields that are finite or infinite and it has been developed from a variety of points of view [5], [9], [10]. In addition to the CCP, transforms of this type also allow fast computation algorithms of the FFT type when $N$ is highly composite [10].

Theorem 1 is somewhat difficult to use when investigating various possible moduli with modular arithmetic, so an alternate set of conditions will be developed. Let $Z_M$ represent the ring of integers $\{0, 1, 2, \cdots, M-1\}$ with arithmetic carried out mod $M$.

Let $M$ have the following unique prime power factorization

$$M = p_1^{r_1} p_2^{r_2} \cdots p_l^{r_l} \tag{21}$$

where the $p_i$'s are distinct primes. As pointed out in Section II, when we carry our arithmetic mod $M$, we are in effect doing it modulo each $p_i^{r_i}$ simultaneously.

Therefore, the length $N$ number theoretic transform having the CCP in $Z_M$ must also have the CCP in $Z_{p_i^{r_i}}$ for $i = 1, 2, \cdots, l$. This requires that $\alpha \pmod{p_i^{r_i}}$ an integer of order $N$ must exist in $Z_{p_i^{r_i}}$, i.e., $N$ is the least positive integer such that

$$\alpha^N = 1 \pmod{p_i^{r_i}}, \qquad i = 1, 2, \cdots, l. \tag{22}$$

Furthermore, since the inverse transform requires $N^{-1}$; the inverse of $N$ should exist in $Z_{p_i^{r_i}}$, or, $N$ should be relatively prime to $M$. Now we investigate the existence of an $\alpha$ of order $N$, in each $Z_{p_i^{r_i}}$. By Euler's theorem (9) and (22), we have

$$N \mid \varphi(p_i^{r_i}), \qquad i = 1, 2, \cdots, l \tag{23}$$

or $N \mid p_i^{r_i-1}(p_i - 1)$ because $\varphi(p_i^{r_i}) = p_i^{r_i-1}(p_i - 1)$. Since $N$ is relatively prime to $M$ (or $p_i$'s)

$$N \mid (p_i - 1), \qquad i = 1, 2, \cdots, l$$

$$N \mid \gcd\{p_1 - 1, p_2 - 1, \cdots, p_l - 1\}.$$

We define $0(M)$ as the greatest common divisor (gcd) of the $(p_i - 1)$

$$0(M) \triangleq \gcd\{p_1 - 1, p_2 - 1, \cdots, p_l - 1\}. \tag{24}$$

Therefore,

$$N \mid 0(M). \tag{25}$$

Equation (25) gives the necessary condition for the existence of a transform of length $N$ in the arithmetic mod $M$. Now consider the converse of it. If $N \mid 0(M)$ or $N \mid \varphi(p_i^{r_i})$, then there exist integers $\alpha_i \pmod{p_i^{r_i}}$ of order $N$ in $Z_{p_i^{r_i}}$. Using these $\alpha_i$ we can construct transforms $(\bmod\ p_i^{r_i})$ which have the DFT structure of (19) and are invertible. Combining these transforms by the Chinese remainder theorem (18) one can obtain a transform $(\bmod\ M)$ having the CCP in $Z_M$. Alternatively, one can combine the $\alpha_i$'s by the Chinese remainder theorem to obtain an $\alpha \pmod{M}$ of order $N$ in $Z_M$ and construct the final transform using this $\alpha$. The results will be identical. Therefore, (25) is the necessary and sufficient condition for the existence of an invertible transform of length $N$ which has the CCP mod $M$. This is stated in the form of a theorem [9], [15].

*Theorem 2:* A length $N$ transform having the DFT structure will implement cyclic convolution mod $M$ if and only if

$$N \mid 0(M). \tag{26}$$

This also establishes the maximum transform length in $Z_M$ as

$$N_{\max} = 0(M).$$

This is a very important theorem that states exactly what the possible transform lengths for a given modulus are.

Although both theorems as stated here assume the DFT structure of (19), they hold for any general transform having the CCP [10]. It is possible in a ring with a composite modulus to have a transform with the CCP but not the DFT structure.[1] Transforms of this sort do not allow an FFT-type fast algorithm and, therefore, do not seem promising.

For number theoretic transforms to be attractive in comparison to other implementations of convolution, they should be computationally efficient. There are three requirements that will be considered. First, $N$ should be highly composite (preferably a power of 2) for a fast FFT-type algorithm to exist and should be large enough for practical sequence lengths. Second, since complex multiplications take most of the computational effort in calculating the FFT, it is important that the multiplication by powers of $\alpha$ be a simple operation. This is possible if the powers of $\alpha$ have binary representations with very few bits; preferably also be a power of two, where multiplication by a power of $\alpha$ reduces to a word shift. Third, in order to facilitate arithmetic mod $M$, $M$ should also have a binary representation with a very few bits and should be large enough to prevent overflow.

Although the class of all possible number theoretic transforms seems very large at first consideration, closer examination shows that very few seem to satisfy the aforementioned criteria. The parameters that must be chosen are $M$, $N$, and $\alpha$. Unfortunately the conditions given by Theorems 1 and 2 do

[1] An example of a transform in $Z_{15}$ having the CCP but not the DFT structure was produced by G. Kopec at M.I.T. and led to this observation.

not give a systematic way of determining the "best" choices. As a result one must use intuition, insight, and a bit of searching. Usually an $M$ is selected and the resulting possible $N$ and $\alpha$ are then examined.

First we see that if $M$ is even, it has a factor of 2 and, therefore, $0(M)$ and $N_{max}$ are 1 which implies $M$ should be odd. If $M$ is a prime then $0(M) = M - 1$ which is as large as one could hope for in a field of $M$ integers. For $M = 2^k - 1$, let $k$ be a composite $PQ$, where $P$ is prime. Then $2^P - 1$ divides $2^{PQ} - 1$ and the maximum possible length of the transform will be governed by the length possible for $2^P - 1$. Therefore, only the prime $k$ need to be considered interesting. Numbers of this form are known as Mersenne numbers and Radar [14] has discussed convolution using Mersenne numbers in detail. For Mersenne number transforms, it can be shown that transforms of length at least $2P$ exist and the corresponding $\alpha$ is $-2$. Mersenne number transforms are not of as much interest because $2P$ is not highly composite and, therefore, we do not have fast FFT-type computational algorithms.

| $N =$ | 0 | 1 | 2 | 3 | 4 | 5 | 6 | 7 | 8 | 9 | 10 | 11 | 12 | 13 | 14 | 15 | 16 |
|---|---|---|---|---|---|---|---|---|---|---|---|---|---|---|---|---|---|
| $2^N =$ | 1 | 2 | 4 | 8 | 16 | 15 | 13 | 9 | 1 | 2 | 4 | 8 | 16 | 15 | 13 | 9 | 1 |
| $3^N =$ | 1 | 3 | 9 | 10 | 13 | 5 | 15 | 11 | 16 | 14 | 8 | 7 | 4 | 12 | 2 | 6 | 1 |
| $4^N =$ | 1 | 4 | 16 | 13 | 1 | 4 | 16 | 13 | 1 | 4 | 16 | 13 | 1 | 4 | 16 | 13 | 1 |
| $6^N =$ | 1 | 6 | 2 | 12 | 4 | 7 | 8 | 14 | 16 | 11 | 15 | 5 | 13 | 10 | 9 | 3 | 1 |

For $M = 2^k + 1$ and $k$ odd, 3 divides $2^k + 1$ and the largest possible transform length is 2, thus we consider only $k$ even. Let $k$ be $s2^t$, where $s$ is an odd integer. Then $2^{2^t} + 1$ divides $2^{s2^t} + 1$ and the length of the possible transform will be governed by the length possible for $2^{2^t} + 1$. Therefore, integers of the form $M = 2^{2^t} + 1$ are of interest. These numbers are known as Fermat numbers and will be discussed in detail in this paper. Fermat numbers seem to be optimum in the sense of having transforms whose length is interesting while the word size is moderate. Numbers of the form $2^{s2^t} + 1$ are also of limited interest and are discussed in Section IX. A systematic investigation of those $M$ which require more than two bit representation is difficult. Our preliminary investigation in that direction has not been very encouraging.

## IV. Fermat Number Transforms

In this section, we consider one of the most promising number theoretic transforms where the modulus is chosen to be a Fermat number

$$M = F_t = 2^{2^t} + 1$$
$$= 2^b + 1, \qquad b = 2^t \tag{27}$$

and $F_t$ is called the $t$th Fermat number. Originally, Fermat conjectured [16] that these numbers were all prime but unfortunately not only was the conjecture wrong, it seems that only $F_0$ through $F_4$ are prime and all the others are composite. The first few values are:

$$\left.\begin{array}{l} F_0 = 3 \\ F_1 = 5 \\ F_2 = 17 \\ F_3 = 257 \\ F_4 = 65\,537 \end{array}\right\} \text{prime}$$
$$F_5 = 4\,294\,967\,297 = 641 \times 6\,700\,417$$
$$F_6 \simeq 1.84 \times 10^{19} = 274\,177 \times 67\,280\,421\,310\,721.$$

Number theoretic transforms with a Fermat number as a modulus are called Fermat number transforms (FNT).

As discussed in the last section, for the FNT of length $N$ to exist, $N$ must divide $0(F_t) = N_{max}$. Now, we consider transform lengths possible in arithmetic modulo various Fermat numbers and also give the corresponding values of $\alpha$.

Since Fermat numbers up to $F_4$ are prime, $0(F_t) = 2^b$, and we can have an FNT for any length $N = 2^m$, $m \leqslant b$. For these Fermat primes the integer 3 is an $\alpha$ of order $N = 2^b$, allowing the largest possible transform length. There are $2^{b-1} - 1$ other integers also which are of order $2^b$ and can be obtained from (13). The integer 2 is of order $N = 2b = 2^{t+1}$. If $\alpha$ is taken as 2 or a power of 2, all the powers of $\alpha$ would be some powers of 2, and for these cases, as discussed in the last section and in [14], the FNT can be computed very efficiently and is called the Rader transform (RT).

To better see the character of these prime moduli consider an example for $F_2$ similar in manner to that in Section II. If the modulus is $M = F_2 = 17$ then

Here we see that 3 and 6 are primitive roots that will generate the entire field $Z_{17}$. The value 2 is of order 8 and 4 is of order 4. Also note that $6 = \sqrt{2}$ in the sense that $6^2 = 2 \pmod{17}$.

For digital filtering applications, the composites $F_5 (b = 32)$ and $F_6 (b = 64)$ also seem to be practical. Lucas [19] has proven that every prime factor of a composite $F_t$ is of the form $K2^{t+2} + 1$. Therefore, $2^{t+2}$ divides $0(F_t)$, for $t > 4$. In particular it can be verified that for $F_5$ and $F_6$, $0(F_t) = 2^{t+2}$. Therefore, for these choices of Fermat numbers, the maximum possible transform length is $N = 2^{t+2} = 4b$. Also, we assert that $\alpha_{4b}$ given by (28) is of order $4b$ in $Z_{F_t}$, $t \geqslant 2$.

$$\sqrt{2} \triangleq \alpha_{4b} = 2^{b/4}(2^{b/2} - 1). \tag{28}$$

We denote this $\alpha_{4b}$ as $\sqrt{2}$ because

$$\alpha_{4b}^2 = 2 \qquad (\text{mod } F_t).$$

The proof that $\alpha_{4b}$ given by (28) is of order $4b$ with respect to any factor of $F_t$ is given in [9]. Any odd power of $\sqrt{2}$ will also be of order $2^{t+2}$. By raising $\sqrt{2}$ to $(2^{t+2-m})$th power, we obtain an integer $\alpha$ of order $2^m$, $m \leqslant t + 2$.

Table I below gives values of $N$ for the two most important values of $\alpha$ and also gives the maximum possible $N$ for the most practical values of $b$.

For FNT's with a prime or composite modulus we see $\alpha = 2$ or a power of 2 is possible for sequence lengths up to $N = 2b = 2^{t+1}$. This is a very desirable situation since $N$ is highly composite allowing an FFT type algorithm and all multiplications by powers of $\alpha$ are simple word shifts. If $\alpha = \sqrt{2}$ is used then sequences of length $N = 4b = 2^{t+2}$ are possible but one stage in the FFT algorithm will require two shifts [9]. This $\alpha = \sqrt{2}$ and the resulting $N = 4b$ give the maximum length possible for $F_5$ and $F_6$, however, for prime $F_t$ further increases in $N$ are possible up to $N = 2^b$ if more stages of the FFT algorithm are allowed to have multiplication rather than simple word shifts. From this example it is seen that $\alpha = \sqrt{2} = 6$ for $M = F_2$ gives the maximum possible $N$.

TABLE I
PARAMETERS FOR SEVERAL POSSIBLE IMPLEMENTATIONS FOR FNT'S

| $t$ | $b$ | $F_t$ | $N^a$ $\alpha = 2$ | $N$ $\alpha = \sqrt{2}$ | $N_{max}$ | $\alpha$ for $N_{max}$ |
|---|---|---|---|---|---|---|
| 3 | 8 | $2^8 + 1$ | 16 | 32 | 256 | 3 |
| 4 | 16 | $2^{16} + 1$ | 32 | 64 | 65536 | 3 |
| 5 | 32 | $2^{32} + 1$ | 64 | 128 | 128 | $\sqrt{2}$ |
| 6 | 64 | $2^{64} + 1$ | 128 | 256 | 256 | $\sqrt{2}$ |

[a]This case corresponds to the Rader Transform.

Because of the nature of modular arithmetic discussed in Section II, the FNT coefficients do not seem to have any physical meaning. Although the signal for which the FNT is being taken may be very small, its FNT coefficients may lie anywhere between 0 and $F_t - 1$. This is because the concept of magnitude does not exist. This also means that the concept of "closeness" of two numbers does not exist in the modular arithmetic. Therefore, approximations or roundings are not allowed in the modular arithmetic. A seemingly small approximation in the transform domain may introduce serious error in the final result. But, because of the nature of the modular arithmetic, there is no need for approximation. During various stages of the computation each accumulation of signal "overflows" many times. But still the end result of the convolution will be exact if the input signals are properly bounded. Some of the properties of the FNT's are given in [9, appendix A].

*Example*

To make the ideas of this section more clear, we now present an example. This example will illustrate several points: treatment of negative values in the data, the structure of the transform and the inverse transform matrix, negative powers of $\alpha$, frequent "overflow" during computation, meaninglessness of the transform values, and exactness of the final answer. This example will not demonstrate the efficient implementation of the FNT using the binary arithmetic.

Consider two sequences $x = (2, -2, 1, 0)$ and $h = (1, 2, 0, 0)$, whose convolution is desired. From the overflow consideration, it is sufficient if we work modulo $F_2 = 17$. We want $N = 4$, for $F_2$ the integer 2 is of order 8, therefore $2^2 = 4$ is an $\alpha$ of order 4.

The transformation matrix $T$ is given by

$$T = \begin{bmatrix} 1 & 1 & 1 & 1 \\ 1 & 4 & 4^2 & 4^3 \\ 1 & 4^2 & 4^4 & 4^6 \\ 1 & 4^3 & 4^6 & 4^9 \end{bmatrix}$$

$$\equiv \begin{bmatrix} 1 & 1 & 1 & 1 \\ 1 & 4 & -1 & -4 \\ 1 & -1 & 1 & -1 \\ 1 & -4 & -1 & 4 \end{bmatrix}$$

$$= \begin{bmatrix} 1 & 1 & 1 & 1 \\ 1 & 4 & 16 & 13 \\ 1 & 16 & 1 & 16 \\ 1 & 13 & 16 & 4 \end{bmatrix} \pmod{17}.$$

Since $4^{-1} = -4 \pmod{17}$, the inverse transformation matrix $T^{-1}$ is given by

$$T^{-1} = 4^{-1} \begin{bmatrix} 1 & 1 & 1 & 1 \\ 1 & 4^{-1} & 4^{-2} & 4^{-3} \\ 1 & 4^{-2} & 4^{-4} & 4^{-6} \\ 1 & 4^{-3} & 4^{-6} & 4^{-9} \end{bmatrix}$$

$$= -4 \begin{bmatrix} 1 & 1 & 1 & 1 \\ 1 & -4 & -1 & 4 \\ 1 & -1 & 1 & -1 \\ 1 & 4 & -1 & -4 \end{bmatrix}$$

$$= 13 \begin{bmatrix} 1 & 1 & 1 & 1 \\ 1 & 13 & 16 & 4 \\ 1 & 16 & 1 & 16 \\ 1 & 4 & 16 & 13 \end{bmatrix} \pmod{17}.$$

The transforms of $x$ and $h$ are given by

$$X = Tx = \begin{bmatrix} 1 & 1 & 1 & 1 \\ 1 & 4 & 16 & 13 \\ 1 & 16 & 1 & 16 \\ 1 & 13 & 16 & 4 \end{bmatrix} \begin{bmatrix} 2 \\ 15 \\ 1 \\ 0 \end{bmatrix}$$

$$= \begin{bmatrix} 18 \\ 78 \\ 243 \\ 213 \end{bmatrix} = \begin{bmatrix} 1 \\ 10 \\ 5 \\ 9 \end{bmatrix} \pmod{17}.$$

Note that in $x$, $-2$ was represented by $-2 + 17 = 15$. Similarly,

$$H = (3, 9, 16, 10) \quad \text{and} \quad Y = X \cdot H = (3, 90, 80, 90)$$
$$= (3, 5, 12, 5) \pmod{17}.$$

Taking the inverse transform of $Y$,

$$y = (2, 2, 14, 2) \pmod{17}.$$

According to our assumption, integers are supposed to lie between $-8$ and 8. Therefore, 14 must be represented as $14 - 17 = -3$. This gives $y = (2, 2, -3, 2)$, which is the correct answer. Also, note that $y$ is a symmetric sequence, therefore, $Y$ is also a symmetric sequence. Other than this, the transform values seem to have no interpretation.

For many applications a direct application of the FNT to implement convolution will result in a significant improvement over any alternative methods. There are many other situations where the constraints of the transform are too severe.

If data magnitude or machine constraints dictate a certain word length and hence a certain $F_t$, the allowed sequences length $N$ may be too short. If input data magnitude and filter length indicate a possible output magnitude that would exceed $F_t/2$, then overflow becomes a problem. If $b$ bit words are used with a modulus of $F_t = 2^b + 1$, the machine can represent $2^b$ integers but the transform needs $2^b + 1$. We now consider several partial solutions to these problems.

## V. METHODS FOR CONVOLVING LONG SEQUENCES AND FOR AVOIDING OVERFLOW

Arithmetic mod $F_t$ can be implemented using $b = 2^t$ bit representation of integers with some provision for representing $2^b$. We have seen the maximum length of sequences which can

TABLE II
MAXIMUM ONE-DIMENSIONAL CYCLIC CONVOLUTION LENGTHS USING TWO-DIMENSIONAL FNT OR RT

| Word Length $b$ | $N$ for $\alpha = 2$ | $N$ for $\alpha = \sqrt{2}$ |
|---|---|---|
| 16 | 512 | 2048 |
| 32 | 2048 | 8192 |
| 64 | 8192 | 32768 |

be cyclically convolved using the FNT with $\alpha = 2$ is $N = 2b$ and therefore the length of sequences which can be convolved is proportional to the word length in bits. Thus, for long sequences, word length requirement may be excessive. Rader [14] suggested using a two-dimensional convolution scheme to convolve long one-dimensional sequences and Agarwal and Burrus [15], [23] presented such a two-dimensional convolution scheme. Using this scheme, cyclic convolution of length $N = LP$ is implemented as a two-dimensional cyclic convolution of length $2L$ by $P$. This two-dimensional cyclic convolution can be implemented using a two-dimensional FNT [15], [23] defined similar to the one-dimensional transform. Using this two-dimensional scheme, the word length required is proportional to the square root of the length of the sequences to be convolved which would give for a maximum sequence length $8b^2$, rather than $4b$. If $P$ is taken as the maximum possible length $4b$, and $2L$ is a small integer, than either direct convolution or another high-speed algorithm could be employed [23], to compute convolution along the short dimension and the one-dimensional FNT could be used along the long dimension. Computationally this combination can be very efficient as will be shown in the implementation in Section VIII. Table II lists the maximum lengths for two-dimensional transforms. This approach requires approximately a factor of two increase in computation and storage requirements over a direct one-dimensional implementation.

Another approach to achieving longer transforms than allowed with $\alpha = 2$ is to use $\alpha = \sqrt{2}$ in (28). It can be shown [9, 15] that $\alpha = \sqrt{2}$ is of order $N = 4b = 2^{t+2}$ and that multiplication times $\sqrt{2}$ requires two word shifts.

Examination of the FFT algorithm [1] shows that if $\alpha = \sqrt{2}$ is used, only one stage will require multiplication by odd powers of $\sqrt{2}$ and from [9] it is shown this can be done with two word shifts. The other stages will multiply by even powers of $\sqrt{2}$ and, therefore, use a single shift as for the case with $\alpha = 2$. This modification is relatively simple and allows a doubling of the allowed $N$. Note in the example for $F_2$ that $\sqrt{2} = 6$ which also gives $N = N_{\max}$.

Each additional square root of $\alpha$ results in a doubling of the allowed $N$ (up to $N = N_{\max}$) and adds an additional stage of calculation to the FFT algorithm. Unfortunately, beyond $\sqrt{2}$ each stage will require a general multiplication. For the case where $F_t$ is a prime, if a few stages of multiplication can be allowed, then $N$ can be increased [25]. For $F_5$ and $F_6$ $\alpha = \sqrt{2}$ gives the maximum $N$.

This use of $\alpha \neq 2$ can be combined with the two-dimensional methods to give various desired $N$.

Another possible problem arises because of the modular arithmetic. In the ring of integers mod $M$, conventional integers can be unambiguously represented only if their absolute value is less than $M/2$. If the input integer sequences $x(n)$ and $h(n)$ are so scaled that $|y(n)|$ never exceeds $M/2$, we would get the same results by implementing convolution in the ring of integers modulo $M$ as that obtained with normal arithmetic. This is similar to the overflow constraint in fixed-point digital machines. In most digital filtering applications, $h(n)$ represents the impulse response and is known *a priori*; also the maximum magnitude of the input signal is usually known. In this situation, we can bound the peak output magnitude by

$$|y_n| \leqslant |x_n|_{\max} \sum_{k=0}^{N-1} |h_k| . \tag{29}$$

This may well require a longer word length than is possible or practical.

One possible solution to this overflow problem involves segmenting the words into shorter blocks and convolving them separately [9]:

$$\begin{aligned} x(n) &= x_2(n) + x_1(n)2^k, \qquad |x_2(n)| < 2^k \\ h(n) &= h_2(n) + h_1(n)2^k, \qquad |h_2(n)| < 2^k \end{aligned} \tag{30}$$

$$\begin{aligned} y = x * h &= (x_1 * h_1)2^{2k} \\ &\quad + (x_1 * h_2 + x_2 * h_1)2^k + x_2 * h_2. \end{aligned} \tag{31}$$

Now, since $x_1, h_1, x_2$, and $h_2$ have roughly half the number of bits, it should be possible to convolve them using approximately half the number of bits. If necessary, a more precise analysis of the above situation could be easily performed. In (31), the last term, in comparison to the first term, is very small and can be neglected. We need to take two transforms for $x$ and two transforms for $h$, the summation shown within the parentheses can be performed in the transform domain. Finally, we need to take two inverse transforms, one for $x_1 * h_1$ and the other for $(x_1 * h_2 + x_2 * h_1)$.

There is another alternative to this problem suggested by Rader [24] and Parks [25]. This is based on the Chinese remainder theorem. The convolution is done modulo two different integers $M_1$ and $M_2$ where $M_1$ and $M_2$ are such that the cyclic convolution in $Z_{M_1}$ and $Z_{M_2}$ is easily implemented on the same machine. The final result mod $M_1 \cdot M_2$ is obtained by the Chinese remainder theorem. $M_1$ is usually a Fermat number and the cyclic convolution in $Z_{M_1}$ is computed using the FNT. Rader [24] suggested using $M_2$ a power of 2, in that case the convolution in $Z_{M_2}$ is computed by taking the sequences mod $M_2$ and then convolving them in $Z_{M_1}$ using the FNT and then reducing them back to $Z_{M_2}$. In this case $M_2$ should be small enough so that no error is introduced by implementing convolution mod $M_2$, in $Z_{M_1}$. This requires

$$NM_2^2 < M_1. \tag{32}$$

Parks [25] suggested that $M_2$ could be a Fermat number, just smaller than $M_1$.

In this case the convolution in $Z_{M_2}$ can also be done using an FNT. Furthermore, the same machine can be utilized to compute the FNT in $Z_{M_2}$. Because $M_2 | (M_1 - 2)$, therefore, arithmetic in $Z_{M_2}$ can be carried out in $Z_{(M_1 - 2)}$. Example: $M_1 = 2^{16} + 1, M_1 - 2 = 2^{16} - 1$, and $M_2 = 2^8 + 1$. These ideas can be further extended.

Still another approach to solving the sequence length $N$ and word length constraints would be to use block processing [3]. By breaking the sequence of length $N$ into smaller blocks and scaling and processing them separately with the FNT one can combine the results to get the desired output. This can be viewed as a type of two-dimensional processing.

## VI. OVERFLOW AND QUANTIZATION CONSIDERATIONS

As mentioned in Section V, we could perform cyclic convolution modulo integer $M$ and obtain the correct result if the absolute value of the output never exceeds $M/2$. If this condition is violated, the resulting error is rather serious. Because of the nature of the modular arithmetic we obtain folding or aliasing of the signal amplitude. This situation could be avoided if the signals are properly quantized (or normalized).

Let $x(n)$ and $h(n)$ represent the original signals. They may have fractional parts (bits to the right of the binary point). To make use of the number theoretic transforms these sequences must be integer sequences. This is easily accomplished by merely shifting the binary position all the way to the right. This introduces scale factors in the sequences. The integer sequences $\tilde{x}(n)$ and $\tilde{h}(n)$ are given by

$$\tilde{x}(n) = x(n)\, 2^{b_1} \tag{33}$$

$$\tilde{h}(n) = h(n)\, 2^{b_2} \tag{34}$$

where $b_1$ and $b_2$, respectively, represent the number of bits to the right of the binary point in $x(n)$ and $h(n)$ sequences.

$$\begin{aligned}\tilde{y}(n) &= \tilde{x}(n) * \tilde{h}(n)\\ &= 2^{b_1+b_2}\, x(n) * h(n)\\ &= 2^{b_1+b_2}\, y(n).\end{aligned} \tag{35}$$

Now we give some upper bounds on the output $\tilde{y}(n)$. These are due to Jackson [21]. The $L_p$ norm of a signal is defined by

$$\|x\|_p = \left(\frac{1}{N}\sum_{n=0}^{N-1} |x(n)|^p\right)^{1/p}, \quad p \geqslant 1. \tag{36}$$

The output of the cyclic convolution is bounded by

$$|\tilde{y}(n)| \leqslant N\|\tilde{x}\|_p \|\tilde{h}\|_q, \quad \frac{1}{p} + \frac{1}{q} = 1, \quad p, q \geqslant 1. \tag{37}$$

In particular

$$|\tilde{y}(n)| \leqslant N\|\tilde{x}\|_2\, \|\tilde{h}\|_2 \tag{38}$$

$$|\tilde{y}(n)| \leqslant N\|\tilde{x}\|_1\, \|\tilde{h}\|_\infty \tag{39}$$

$$|\tilde{y}(n)| \leqslant N\|\tilde{x}\|_\infty\, \|\tilde{h}\|_1. \tag{40}$$

All these are valid upper bounds, but, they differ in the amount of computation required to use them. The computation of $L_2$ norm requires $N$ multiplications, the computation of $L_1$ norm requires $N$ additions, and the computation of $L_\infty$ norm requires $N$ comparisons because $\|x\|_\infty = |x(n)|_{\max}$. Depending on the particular situation any of these bounds can be used. To avoid aliasing error $|\tilde{y}(n)|_{\max}$ should be less than $M/2$. If the bound thus computed exceeds maximum allowable value, i.e., $M/2$, either or both of the signals are scaled down by truncating or rounding the low-order bits. We choose new integers $\tilde{b}_1$ and $\tilde{b}_2$ which are less than or equal to $b_1$ and $b_2$, respectively, and obtain integer sequences $\tilde{x}(n)$ and $\tilde{y}(n)$ as follows:

$$\tilde{x}(n) = [x(n)\, 2^{\tilde{b}_1}] \tag{41}$$

$$\tilde{h}(n) = [h(n)\, 2^{\tilde{b}_2}] \tag{42}$$

where $[\cdot]$ represents rounding to the nearest integer. This introduces roundoff noise sources $\epsilon_x(n)$ and $\epsilon_h(n)$.

$$\tilde{x}(n) = x(n)\, 2^{\tilde{b}_1} + \epsilon_x(n) \tag{43}$$

$$\tilde{h}(n) = h(n)\, 2^{\tilde{b}_2} + \epsilon_h(n). \tag{44}$$

The output

$$|\tilde{y}(n)| \simeq |y(n)\, 2^{\tilde{b}_1+\tilde{b}_2}| \leqslant M/2. \tag{45}$$

Eq. (45) suggests that we have some freedom in choosing $\tilde{b}_1$ and $\tilde{b}_2$ with the constraint that $\tilde{b}_1 + \tilde{b}_2$ is constant. This freedom can be utilized to minimize the effect of the quantization of the sequences on the output noise to signal ratio. If the bound (38) is used to minimize the output noise to signal ratio, $\tilde{b}_1$ and $\tilde{b}_2$ should be so selected [22] that

$$\|\tilde{x}\|_2 \simeq \|\tilde{h}\|_2. \tag{46}$$

The quantization of the sequences can be done on a block to block basis.

The output $\tilde{y}(n)$ thus obtained would roughly have twice the number of bits. Not all these bits are meaningful. Roughly speaking lower half of the bits represent noise. Using quantization noise models, an analysis can be made for the roundoff noise in $\tilde{y}(n)$ (due to the quantization of $x(n)$ and $h(n)$) and an appropriate number of low order bits of $\tilde{y}(n)$ can be discarded for further processing.

## VII. Arithmetic Considerations

Since the structure of the FNT is similar to that of the DFT, and $N$ is a power of 2, a fast implementation of the FNT similar to the FFT with radix 2 exists. As a matter of fact, all we have to do is replace $W = \exp(-j2\pi/N)$ by $\alpha$ in any FFT algorithm [1], [7].

In computing the FNT, arithmetic is done mod $2^b + 1$. In this arithmetic the only allowed integers are $0, 1, \cdots, 2^b$ and all integers whose absolute values do not exceed $2^{b-1}$ can be represented unambiguously. Negative integers are represented by adding $2^b + 1$ to them; this is similar to twos complement representation of negative integers. Using a $b$-bit register, all integers from 0 to $2^b - 1$ can be represented. A very unfortunate problem arises in that $2^b$ (which is also $-1$) cannot be represented.

If $-1$ is encountered in the data it is rounded either to 0 or to $-2$. This is equivalent to introducing some initial quantization noise and will not have any serious effect on the result. But, if during the course of the computation $2^b$ appears as a result of some arithmetic operation (mod $F_t$) it can not be represented correctly and will give rise to serious error in the output for that block. If the data are uncorrelated, the probability that this number will appear after an arithmetic operation is approximately $2^{-b}$. For digital filtering applications, $b$ would typically be 32 or 64; in these cases the probability of occurrence of $2^b$ is extremely small. If $b = 32$ and $N = 64$, the probability that a particular block is erroneous is roughly $10^{-6} \sim 10^{-7}$. For some filtering operations, this may be permissible. If, however, this possible error cannot be allowed, an extra bit will be required to represent $2^b$ at the expense of a slightly more complicated hardware.

The arithmetic unit has to be wired to do arithmetic modulo $F_t$. Therefore, whenever a carry bit ($2^b = -1 \bmod F_t$) is encountered it should be subtracted from the data. The arithmetic will be of carry subtract type (similar to carry-add arithmetic required for 1's complement hardware). This is the only major difference as compared to the conventional hardware units. If this is implemented, the hardware will automatically do arithmetic modulo Fermat numbers.

In [9], it is explained how to do various basic arithmetic operations modulo Fermat numbers. If $\alpha$ is taken as 2 or a power of 2 the only multiplications involved in computing the fast FNT are those by some powers of 2. These multiplications are particularly simple to implement in arithmetic modulo $F_t$. They are implemented as a double register shift followed by the subtraction of the higher order bits [9]. This is much simpler than complex multiplication which is the main reason why the FNT is much faster than the FFT. For implementation of the

TABLE III
CYCLIC CONVOLUTION TIMINGS FOR LENGTH $N$ REAL SEQUENCES

| $N$ | FFT (ms) | FNT or RT (ms) |
|---|---|---|
| 32 | 16 | 3.3 |
| 64 | 31 | 7.4 |
| 128 | 60 | 16.6[a] |
| 256 | 123 | 40.0[b] |
| 256 | 123 | 80.0[c] |
| 512 | 245 | 166.0[c] |
| 1024 | 530 | 340.0[c] |
| 2048 | 1260 | 720.0[c] |

[a]Using $\alpha = \sqrt{2}$.
[b]Using 2 by 128 convolution.
[c]Using two-dimensional RT.

fast FNT, unlike the FFT, we do not need to store the powers of $\alpha$ (if $\alpha$ is taken as 2 or a power of 2).

## VIII. COMPARISON WITH THE FFT

As noted in the previous section, computing the FNT is a very simple operation on a binary machine. Now let us compare the complexity of various basic operations involved in computing the FNT vis-à-vis the FFT. If the two sequences $x(n)$ and $h(n)$ have $b_1$ and $b_2$ bit representations, respectively, and are of length $N$, then the output $y(n)$ would need no more than a $(b_1 + b_2 + \log_2 N)$ bit representation. To obtain the correct result $b \geqslant b_1 + b_2 + \log_2 N$. In Section V, we have given a better bound on the output. In Section VI, we have given other bounds. Roughly speaking, we need twice the number of bits to carry out the convolution using the FNT as compared to the fixed-point FFT implementation of the convolution. But in the DFT, every data point is treated as a complex number and therefore requires two words, one for the real part and one for the imaginary part. Thus, in effect, the hardware requirement for two transforms is about the same. Although for real data it is possible to make use of the symmetry properties of the DFT's they require extra computation and for the purpose of comparison it will be ignored, even though we have taken this into account for our IBM 370/155 implementation to be discussed later. Therefore, we shall assume that in the FFT implementation, each data point is represented by a $b/2$ bit real part and a $b/2$ bit imaginary part.

One $b/2$ bit complex addition is equivalent to two $b/2$ bit real additions, which are comparable to a $b$-bit addition mod $F_t$. Thus, the complexity of addition/subtraction is the same in both the transforms. Similarly, it can be shown that a $b/2$ bit complex multiplication is comparable to a $b$-bit multiplication mod $F_t$. Computation of the RT requires multiplications by powers of 2, which implemented as bit shifts and subtractions become much simpler operations compared to complex multiplications required in the FFT implementation.

To compute a length $N$ fast RT, $N \log_2 N$ additions/subtractions, and $(N/2) \log_2 N/2$ multiplications by some powers of 2 are required which are implemented as bit shifts and subtractions. To compute the convolution using the FFT, most of the time is taken in computing the complex multiplications required to compute the transforms. A comparison with RT reveals that these complex multiplications are replaced by bit shifts and subtractions which are much faster operations. This results in considerable computational savings in the implementation of convolution. The computation required to multiply the two transforms is about the same for both the implementations. To convolve long sequences using the two-dimensional RT the computational effort and required storage increases by, at the most, a factor of 2. Still, the FNT implementation of convolution is much faster as compared to the FFT implementation.

These transforms were implemented in assembler language on an IBM 370/155 which has a 32-bit word length [9]. The results were compared with an efficient FFT program for computing convolution which makes use of the symmetry of the DFT for real data (see Table III) [20].

## IX. GENERALIZATIONS, VARIATIONS, AND OTHER RESULTS

### A. Other Choices for M

In this paper, we have primarily discussed number theoretic transforms in the rings of integers modulo Fermat numbers. These numbers seem to be the best choice for implementation on binary computers. Nevertheless, any odd integer $M$ can be used as was discussed in Section III. Rader [14] proposed the use of Mersenne numbers $M_p = 2^p - 1$, where $p$ is a prime. Mersenne number transforms for $\alpha = -2$ have $N = 2p$, and, therefore, do not have an FFT-type fast computational algorithm.

FNT's require the computing word length to be a power of 2. Many computers do not have word length a power of 2. Transforms similar to FNT's exist for many of these situations. For example, on a 24-bit machine, we may perform convolution modulo $M = 2^{24} + 1$. For this $M$, $\alpha = 2^3$ gives $N = 16$, and $\alpha = 2^{3/2} = 2^7(2^{12} - 1)$ gives $N = 32$ as the maximum length.

In general one could take $M = 2^{s2^t} + 1$, $s$ is an odd integer. In that case, $\alpha = 2^s$ would give $N = 2^{t+1}$, and $\alpha = (2^s)^{1/2} = 2^{[(s-1)/2 + s2^{t-2}]}(2^{s2^{t-1}} - 1)$ would give $N = 2^{t+2}$. In many situations, it may be possible to have transforms of greater length than $2^{t+2}$. For example, taking $M = 2^{40} + 1$, the maximum possible transform length is 256, and taking $M = 2^{80} + 1$, the maximum possible transform length is 1024, but the corresponding $\alpha$'s may not be simple.

For many computers whose word length is not a power of 2, if the above formulation is used to compute transforms analogous to the FNT, the maximum transform length is very small. But, if we are willing to sacrifice the "effective word length" to some extent, we can increase the maximum transform length significantly. Let $b$, a multiple of 4 be the word length of the machine. Let

$$M = 2^b + 1 = M_1 M_2. \tag{47}$$

$M_1$ and $M_2$ may be nonprimes. It can be easily proved that

$$0(M) = \gcd \{0(M_1), 0(M_2)\}. \tag{48}$$

It may so happen that $M_1$ is a small integer and $0(M_2) >> 0(M)$, therefore, because of the presence of $M_1$ the maximum transform length is being considerably reduced while at the same time $M_1$ may not be increasing the maximum allowable output ($y_{max}$) or the "effective word length" significantly. In this situation we can compute transforms in $Z_{M_2}$ with maximum transform length $0(M_2)$ and maximum allowable output magnitude as $M_2/2$. At the expense of reduced output range, we have increased the transform length. Furthermore, arithmetic mod $M_2$ can be conveniently carried out as arithmetic mod $M$, because $M_2$ is a factor of $M$. At the end of the computation, we have to reduce the result mod $M_2$.

Table IV shows this factorization of $M$ for several values of $b$. $\log_2 M_2$ shows the "effective word length" of the machine.

TABLE IV
FACTORIZATION OF $M = 2^b + 1$ AS $M_1 \cdot M_2$ AND THE MAXIMUM TRANSFORM LENGTH CORRESPONDING TO $M_2$

| Machine Word Length | | | | Effective Word Length | | In $Z_{M_2}$ | |
|---|---|---|---|---|---|---|---|
| $b$ | $0(M)$ | $M_1$ | $M_2$ | $\log_2 M_2$ approx. | $N_{max} = 0(M_2)$ | $N$ for $\alpha = 2$ $2b$ | $N$ for $\alpha = \sqrt{2}$ $4b$ |
| 12 | 16 | 17 | 241 | 8 | 240 | 24 | 48 |
| 20 | 16 | 17 | 61681 | 16 | 61680 | 40 | 80 |
| 24 | 32 | 257 | 97 X 673 | 16 | 96 | 48 | 96 |
| 28 | 16 | 17 | 15790321 | 24 | 15790320 | 56 | 112 |
| 36 | 16 | 17 X 241 | 433 X 38737 | 24 | 144 | 72 | 144 |
| 40 | 256 | 257 | 4278255361 | 32 | 4278255360 | 80 | 160 |
| 48 | 64 | 65537 | 193 X 22253377 | 32 | 192 | 96 | 192 |
| 56 | 32 | 257 | 5153 X 54410972897 | 48 | 224 | 112 | 224 |
| 60 | 16 | 17 X 241 X 61681 | 4562284561 | 32 | 4562284560 | 120 | 240 |
| 72 | 32 | 97 X 257 X 673 | 577 X 487824887233 | 48 | 576 | 144 | 288 |
| 80 | 1024 | 65537 | 44479210368001 X 414721 | 64 | 15360 | 160 | 320 |

TABLE V
SOME PARAMETERS FOR NUMBER THEORETIC TRANSFORMS IN DECIMAL ARITHMETIC

| Machine Digits | | | | Effective Digits | | |
|---|---|---|---|---|---|---|
| $b$ | $0(M)$ | $M_1$ | $M_2$ | $\log_{10} M_2$ approx. | $N_{max} = 0(M_2)$ | $N$ for $\alpha = 10$ |
| 4 | 8 | 1 | 73 X 137 | 4 | 8 | 8 |
| 6 | 100 | 101 | 9901 | 6 | 9900 | 12 |
| 8 | 16 | 17 | 5882353 | 7 | 5882352 | 16 |
| 10 | 20 | 101 | 3541 X 27961 | 8 | 60 | 20 |
| 12 | 8 | 73 X 137 | 99990001 | 8 | 99990000 | 24 |
| 16 | 32 | 1 | 353 X 449 X 641 X 1409 X 69857 | 16 | 32 | 32 |

Note that for the choices of $M_2$'s shown in Table IV, 2 is always of order $2b$ and also there exists an integer $\sqrt{2} = 2^{b/4}(2^{b/2} - 1)$ which is of the order $4b$ in $Z_{M_2}$. This leads to an efficient implementation of these transforms, because powers of $\alpha$ are simple and arithmetic mod $M$ is also simple. For these transforms the word length is not a power of 2, but, still the transform length $4b$ is highly composite. The case $b = 60$ needs special attention, for this if $M_1$ is taken as 17 for remaining $M_2$, $0(M_2) = 240$ but $\sqrt{2}$ is not of order 240. This is because $\sqrt{2}$ is of order 48 in $Z_{241}$ and of order 80 in $Z_{61681}$. For this case although one can find an integer $\alpha$ which is of order 240 in $Z_{M_2}$ it is not likely to be simple.

Thus far, our discussion was based on the assumption that the computer is a binary computer. Many computers have decimal representation of integers and for these computers, it will be efficient if the arithmetic is done mod $M = 10^b + 1$ and $\alpha$ and powers of $\alpha$ are powers of 10. We have compiled Table V to be used for decimal computers. Similar tables can be compiled for other radices also.

*B. Complex Number Theoretic Transforms*

The integer field $Z_M$ (assuming $M$ to be prime) can be extended to complex integer field denoted by $Z_M^c$, if the following equation does not have a solution in $Z_M$:

$$x^2 + 1 = 0. \tag{49}$$

This means $(-1)$ does not have a square root in $Z_M$ or equivalently a root of order 4 does not exist in $Z_M$. This implies

$$4 \nmid 0(M) = M - 1. \tag{50}$$

Equation (50) is the only condition required for $Z_M^c$ to exist. In $Z_M^c$ every integer is represented as $a + jb$; $a, b \in Z_M$. All the arithmetical operations are done as in the normal complex arithmetic with $j^2 = -1$. Both real and complex parts are evaluated mod $M$, separately. The concept of magnitude and phase does not exist in $Z_M^c$. Complex number theoretic transforms (CNT) similar to NTT exist in $Z_M^c$, and can be used to compute the cyclic convolution of two complex integer sequences. To avoid error due to aliasing both real and imaginary parts of the output should be separately bounded to $M/2$. The idea of the CNT has been considered by Reed [26].

*Theorem:* A transform having the cyclic convolution property in $Z_M^c$ exists if and only if

$$N | (M^2 - 1). \tag{51}$$

We will not give a formal proof of this theorem, but we will outline a procedure to find a complex integer $\alpha$ of order $N$ in $Z_M^c$, if $N$ divides $(M^2 - 1)$.

*Theorem:*

$$(a + jb)^{M+1} = a^2 + b^2 \text{ in } Z_M^c; a, b \in Z_M. \tag{52}$$

This theorem can be easily proved. This theorem implies that every complex integer is at most an $(M + 1)$th root of a real integer, in $Z_M^c$. Let $a + jb$ be an $N$th root of a real integer, i.e., $N$ is the least positive integer such that

$$(a + jb)^N = \text{a real integer} \tag{53}$$

then by (52)

$$N | (M + 1). \tag{54}$$

*How to find an $\alpha$ of order $M^2 - 1$ in $Z_M^c$:*
Consider complex integers of the form $(1 + jb)$ and search over $b \in Z_M$ such that $(1 + jb)$ is an $(M + 1)$th root of a real integer (proof for the existence of such a $b$ can be given).

Then,

$$(1 + jb)^{M+1} = 1 + b^2. \tag{55}$$

Let $\alpha_{M-1}$ be a root of order $(M - 1)$ in $Z_M$. Then $1 + b^2$ can be written as some power of $\alpha_{M-1}$:

$$(1 + b^2) = \alpha_{M-1}^x. \tag{56}$$

It can be shown that $x$ is odd. Then, $\alpha_{M^2-1}$ given by (57) is of order $(M^2 - 1)$ in $Z_M^c$.

$$\alpha_{M^2-1} = (1 + jb)\, \alpha_{M-1}^{-((x-1)/2)}. \tag{57}$$

This can be easily proved. By raising $\alpha_{M^2-1}$ by $((M^2 - 1)/N)$th power, we can find a complex integer $\alpha_N$ of order $N$ in $Z_M^c$ if $N | (M^2 - 1)$. It can be easily proved that

$$\gcd (M - 1, M + 1) = 2. \tag{58}$$

Let

$$N_1 = \gcd (M - 1, N)$$
$$N = N_1 \times N_2$$

then $\alpha_N$ will be complex, but $\alpha_{N_1} = \alpha_N^{N_2}$ will be real.

In the fast FFT algorithm to compute the CNT the part corresponding to $N_2$ will require complex arithmetical operations but the part corresponding to $N_1$ will require only real arithmetical operations. CNT are good in theory as they offer more choice in transform lengths. But, so far no CNT have been found for which powers of $\alpha$ are simple. CNT do not exist for Fermat numbers, but they exist for Mersenne numbers.

The only extension we have investigated is when

$$M = p_1 p_2 \cdots p_l \tag{59}$$

where $p_i$'s are distinct primes. We have not investigated the case when $M$ contains prime powers. $Z_M$ can be extended to $Z_M^c$ as before if

$$4 \nmid (p_i - 1), \qquad i = 1, 2, \cdots, l. \tag{60}$$

Also the Chinese remainder theorem can be used in $Z_M^c$. It is applied separately to the real and imaginary parts. For CNT to exist in $Z_M^c$, they must exist in $Z_{p_i}^c$ also. This gives the following theorem.

*Theorem:* CNT of length $N$ in $Z_M^c$ exists, if and only if

$$N | \gcd \{p_1^2 - 1, p_2^2 - 1, \cdots, p_l^2 - 1\}. \tag{61}$$

We find $\alpha_i$ of order $N$ in $Z_{p_i}^c$ and then combine them by the Chinese remainder theorem to obtain an $\alpha$ of order $N$ in $Z_M^c$.

### *C. Application to Two-Dimensional Filtering*

Rader [27] has recently discussed the application of the FNT to two-dimensional filtering. In 2-*D* applications, the length of the impulse response along each dimension is not too large, therefore the FNT's are ideally suited for this application, because, for these applications the length constraint of the FNT's is not important. Other choices of $M$ discussed in this section can also be used for this application.

## REFERENCES


[1] B. Gold and C. M. Rader, *Digital Processing of Signals.* New York: McGraw-Hill, 1969.

[2] T. G. Stockham, "High speed convolution and correlation," in *AFIPS Conf. Proc., 1966 Joint Computer Conf.*, vol. 28, pp. 229-233 (also in [28]).

[3] C. S. Burrus, "Block realization of digital filters," *IEEE Trans. Audio Electroacoust.*, vol. AU-20, pp. 230-235, Oct. 1972.

[4] D. E. Knuth, *The Art of Computer Programming*, vol. 2, *Seminumerical Algorithms.* Reading, Mass.: Addison-Wesley, 1969.

[5] J. M. Pollard, "The fast Fourier transform in a finite field," *Math. Comput.*, vol. 25, pp. 365-374, Apr. 1971.

[6] A. Schonhage and V. Strassen, "Fast multiplication of large numbers," *Comput.* (in German), vol. 7, pp. 281-292, 1971.

[7] J. W. Cooley and J. W. Tukey, "An algorithm for machine calculation of complex Fourier series," *Math. Comput.*, vol. 19, pp. 297-301, 1966 (also in [28]).

[8] A. V. Oppenheim and C. Weinstein, "Effects of finite register length in digital filtering and the fast Fourier transform," *Proc. IEEE*, vol. 60, pp. 957-976, Aug. 1972.

[9] R. C. Agarwal and C. S. Burrus, "Fast convolution using Fermat number transforms with applications to digital filtering," *IEEE Trans. Acoustics, Speech, and Signal Processing*, vol. ASSP-22, pp. 87-97, Apr. 1974.

[10] P. J. Nicholson, "Algebraic theory of finite Fourier transforms," *J. Comput. Syst. Sci.*, vol. 5, pp. 524-547, 1971.

[11] I. J. Good, "The relation between two fast Fourier transforms," *IEEE Trans. Comput.*, vol. C-20, pp. 310-317, Mar. 1971.

[12] D. E. Knuth, "The art of computing programming—errata et addenda," Comput. Sci. Dep., Stanford Univ., Stanford, Calif., Rep. STAN-CS-71-194, pp. 21-26, Jan. 1971.

[13] C. M. Rader, "The number theoretic DFT and exact discrete convolution," presented at IEEE Arden House Workshop on Digital Signal Processing, Harriman, N.Y., Jan. 11, 1972.

[14] —, "Discrete convolution via Mersenne transforms," *IEEE Trans. Comput.*, vol. C-21, pp. 1269-1273, Dec. 1972.

[15] R. C. Agarwal and C. S. Burrus, "Fast digital convolution using Fermat transforms," in *Southwest IEEE Conf. Rec.*, pp. 538-543, Apr. 1973.

[16] O. Ore, *Number Theory and Its History.* New York: McGraw-Hill, 1948.

[17] G. H. Hardy and E. M. Wright, *The Theory of Numbers.* Oxford, England: Oxford Univ. Press, 1960.

[18] N. S. Szabo and R. I. Tanaka, *Residue Arithmetic and Its Applications to Computer Technology.* New York: McGraw-Hill, 1967.

[19] L. E. Dickson, *History of the Theory of Numbers*, vol. I. Washington, D.C.: Carnegie Institute, 1919, p. 376.

[20] R. C. Singleton, "An algorithm for computing the mixed radix fast Fourier transform," *IEEE Trans. Audio Electroacoust.*, vol. AU-17, pp. 93-103, June 1969 (also in [28]).

[21] L. B. Jackson, "On the interaction of round-off noise and dynamic range in digital filters," *Bell Syst. Tech. J.*, vol. 49, pp. 159-184, Feb. 1970 (also in [28]).

[22] R. C. Agarwal, "On realization of digital filters," Ph.D. dissertation, Dep. Elec. Eng., Rice Univ., Houston, Tex., Dec. 1973.

[23] R. C. Agarwal and C. S. Burrus, "Fast one-dimensional digital convolution by multi-dimensional techniques," *IEEE Trans. Acoust., Speech, and Signal Processing*, vol. ASSP-22, pp. 1-10, Feb. 1974.

[24] C. M. Rader, Private Commun.

[25] T. W. Parks, Private Commun.

[26] I. S. Reed, Private Commun.

[27] C. M. Rader, "On the application of the number theoretic transforms of high speed convolution to two-dimensional filtering," *IEEE Trans. Circuit Theory*, to be published.

[28] L. R. Rabiner and C. M. Rader, Eds., *Digital Signal Processing.* New York: IEEE Press, 1972.

[29] C. M. Rader, "A note on exact discrete Fourier transforms," *IEEE Trans. Audio Electroacoust.* (Corresp.), vol. AU-21, pp. 558-559, Dec. 1973.

[30] H. Takahasi and Y. Ishibashi, "A new method for 'exact calculation' by a digital computer," *Inform. Process. Jap.*, vol. I, pp. 28-42, 1962.

# Fast One-Dimensional Digital Convolution by Multidimensional Techniques

RAMESH C. AGARWAL and CHARLES S. BURRUS, Member, IEEE

*Abstract*—This paper presents two formulations of multidimensional digital signals from one-dimensional digital signals so that multidimensional convolution will implement one-dimensional convolution of the original signals. This has reduced an important word length restriction when used with the Fermat number transform. The formulation is very general and includes block processing and sectioning as special cases and, when used with various fast algorithms for short length convolutions, results in improved multiplication efficiency.

## I. Introduction

There are several advantages to formulating a one-dimensional digital convolution as a two- or higher dimensional problem. The first involves the Mersenne and Fermat number transforms which have recently been defined [1], [2] and which seem to have some advantages over the discrete Fourier transform (DFT) for implementing convolution on a digital computer. They can be computationally faster than the fast Fourier transform (FFT) implementation of the DFT and result in no roundoff error. These transforms have one limitation: the number of bits required for each word in an implementation is proportional to the length of the sequences to be convolved [1], [2]. It is the purpose of this paper to present a scheme whereby long sequences can be convolved by a two-dimensional convolution as mentioned by Rader [1]. This two-dimensional convolution can be implemented by a two-dimensional transform to allow a high-speed error-free convolution with the word length proportional to the square root of the length of the sequences.

This formulation also allows use of special short convolution algorithms similar to those proposed by Pitassi [7], Rayner [8], Davis [9], and Allwright [10]. These can be extended and combined with others to provide a very general and versatile format for doing fast digital convolution. The case of sequences of different lengths and the case where only part of the output sequence is desired are covered as special cases of the general problem.

Manuscript received March 15, 1973; revised July 10, 1973 and September 20, 1973. This work was supported by the National Science Foundation under Grant GK-23697.

The authors are with the Department of Electrical Engineering, Rice University, Houston, Tex. 77001.

## II. Two-Dimensional Convolution Based on Overlap-Save

Consider the cyclic convolution of two sequences, $x(n)$ and $h(n)$, giving an output sequence $y(n)$, all of length $N$.

$$h(n) * x(n) = y(n). \tag{1}$$

This is defined by

$$y(n) = \sum_{q=0}^{N-1} h(n-q)\,x(q)$$

$$n = 0, 1, \cdots N-1 \tag{2}$$

where $h(n)$ and $x(n)$ are periodically extended outside their original domains of definition (or their indices evaluated modulo $N$).

In order to convert this one-dimensional problem to two dimensions a change of variables is made. First, it must be possible to factor $N$ into integer factors

$$N = LM. \tag{3}$$

If a change of variables is made such that

$$\left.\begin{array}{l} n = l + mL \\ q = k + pL \end{array}\right\} \quad \left\{\begin{array}{l} k, l = 0, 1, \cdots L-1 \\ p, m = 0, 1, \cdots M-1 \end{array}\right.$$

then (2) becomes

$$y(l+mL) = \sum_{k=0}^{L-1} \sum_{p=0}^{M-1} h(l + mL - k - pL)\,x(k + pL). \tag{4}$$

Define now a two-dimensional $L \times M$ array $\hat{X}$ from the original length $N = LM$ signal, $x(n)$, by

$$\hat{X}(l, m) = x(l + mL) \tag{5}$$

where columns of $\hat{X}$ are the sections or blocks of $x$ and the rows are samples of $x$ taken every $L$ values of $n$. In a similar way $\hat{H}$ and $\hat{Y}$ are defined by

$$\hat{H}(l, m) = h(l + mL)$$

$$\hat{Y}(l, m) = y(l + mL). \tag{6}$$

In terms of the two-dimensional signals, (4) becomes

$$\hat{Y}(l, m) = \sum_{p=0}^{M-1} \sum_{k=0}^{L-1} \hat{H}(l-k, m-p)\,\hat{X}(k, p) \tag{7}$$

which is a two-dimensional convolution. Note that values of $\hat{H}$ outside the $L \times M$ array are required. These values of $\hat{H}$ that are needed can be seen from (7) where $1 - L \leqslant l - k \leqslant L - 1$ and $1 - M \leqslant m - p \leqslant M - 1$. Values of $\hat{H}(l, m)$ outside the $L \times M$ array are defined by (6). We therefore define suitably extended

Reprinted from *IEEE Trans. Acoust., Speech, Signal Processing*, vol. ASSP-22, pp. 1-10, Feb. 1974.

arrays $\hat{\hat{H}}$ and $\hat{\hat{X}}$ so that two-dimensional convolution will give the desired answer. The extension of $\hat{H}$ is analogous to the overlap extensions used in the overlap-save algorithm for doing sectioning or block processing [3], [4]. Note that along the $m$ dimension the values of $\hat{H}$ are periodic with period $M$, i.e., the desired convolution in (7) is cyclic in the $m$ dimension (this is because the original desired convolution in (2) was defined as cyclic). Considering the values of $\hat{H}$ along $l$ shows the convolution in that dimension is not cyclic.

An important factor when considering multidimensional convolution is the number of multiplications required in an implementation. If in (7) $m$ and $p$ are held constant then (7) becomes a scalar convolution along the $l$ dimension of a length-$L$ sequence of $x$ with a length-$(2L - 1)$ sequence of $h$ giving a length-$L$ sequence of the output. There will be one of these length-$L$ scalar convolutions for each value of $m$ and $p$ in (7) so that along the $m$ dimension each "operation" will be a length-$L$ convolution rather than a simple scalar multiplication. This is a sort of convolution of convolutions [3]. To count the total number of multiplications to compute (7) with $l$ and $k$ constant are found. But, with $l$ and $k$ not constant, this will be the number of length-$L$ convolutions necessary and the total number of multiplications will be the number of length-$L$ convolutions times the number of multiplications for a length-$L$ convolution. In this case along the $m$ dimension the number of multiplications is $M^2$ and along the $l$ dimension it is $L^2$, so the total is $M^2L^2$ or $N^2$, which is the same as a direct calculation of (2) would require. We will later use transforms and other schemes to reduce this number. Note the convolution can be carried out in either order.

If (7) is to be carried out by transforms and therefore cyclic convolution, $\hat{X}$ must be augmented with zeros so that aliasing of the noncyclic convolution along $l$ does not occur and so that all arrays are the same size.

Consider the $2L - 1 \times M$ array $\hat{\hat{X}}$ formed by appending $(L - 1)$ rows of zeros to the bottom of $\hat{X}$.

$$\hat{\hat{X}} = \begin{bmatrix} x(0) & x(L) & \cdots & x(N-L) \\ x(1) & x(L+1) & & \\ \cdot & \cdot & & \\ \cdot & \cdot & & \\ \cdot & \cdot & & \\ x(L-1) & & & x(N-1) \\ 0 & 0 & & 0 \\ 0 & 0 & & 0 \\ \cdot & \cdot & & \cdot \\ \cdot & \cdot & & \cdot \\ \cdot & \cdot & & \cdot \end{bmatrix}. \tag{8}$$

$\hat{\hat{H}}$ is formed so that the columns of $\hat{\hat{H}}$ contain the periodic extension of the original $h(n)$ with period $N$.

$$\hat{\hat{H}} = \begin{bmatrix} h(N-L+1) & & h(N-2L+1) \\ \cdot & \cdot & \\ \cdot & \cdot & \\ \cdot & \cdot & \\ h(N-1) & h(L-1) & \\ h(0) & h(L) \cdots & h(N-L) \\ h(1) & h(L+1) & \\ \cdot & \cdot & \\ \cdot & \cdot & \\ \cdot & \cdot & \\ h(L-1) & & h(N-1) \end{bmatrix}. \tag{9}$$

If two-dimensional cyclic convolution is carried out we have

$$\hat{\hat{Y}} = \hat{\hat{H}} * \hat{\hat{X}} \tag{10}$$

where the lower $L \times M$ partition of $\hat{\hat{Y}}$ is $\hat{Y}$ and the columns of $\hat{Y}$ are the desired blocks of $y(n)$ in (2). Because of ease in implementation with transforms, the arrays would usually be extended one additional row to be $2L \times M$ rather than the minimum $(2L - 1) \times M$.

### III. Two-Dimensional Transform Implementation

The two-dimensional transform of $\hat{\hat{X}}$ is defined as

$$T\{\hat{\hat{X}}\} = F(j,k) = \sum_{m=0}^{M-1} \sum_{l=0}^{2L-1} \hat{\hat{X}}(l,m)\, \alpha_{2L}^{lj}\, \alpha_M^{mk} \tag{11}$$

and the inverse transform

$$T^{-1}\{F\} = \hat{\hat{X}}(l,m) = (2N)^{-1} \sum_{k=0}^{M-1} \sum_{j=0}^{2L-1} \cdot F(j,k)\, \alpha_{2L}^{-lj}\, \alpha_M^{-mk} \tag{12}$$

where $\alpha_M$ is of an order $M$ (i.e., $M$ is the least positive integer such that $(\alpha_M)^M = 1$ [2]). Applying the transform to (10) it can be shown that

$$T\{\hat{\hat{Y}}\} = T\{\hat{\hat{H}}\}\, T\{\hat{\hat{X}}\} \tag{13}$$

so that, similar to the one-dimensional case, (10) can be carried out by

$$\hat{\hat{Y}} = T^{-1}[T\{\hat{\hat{H}}\}\, T\{\hat{\hat{X}}\}]. \tag{14}$$

If the transform is the DFT, then

$$\alpha_M = e^{-j2\pi/M}.$$

To compare multiplication efficiencies we assume the DFT of $\hat{\hat{H}}$ is already known and the number of multiplications for one $2L \times M$ transform, one $2L \times M$ complex multiplication, and one $2L \times M$ inverse transform are calculated. The number of complex multiplications is approximately $(2N \log 2N + 2N)$ as compared to $(N \log N + N)$ for a one-dimensional implementation and therefore one would not use the two-dimensional approach with the DFT for improved multiplication efficiency.

The computational advantage appears when used with the Fermat number transform [2] where word length requirements are a possible restriction. The Fermat number transform is defined in [2] and, although not named, is defined in a restricted form in the latter part of [1, eq. (38)]. It is a transform defined in a finite ring of integers with arithmetic performed modulo Fermat numbers ($2^b + 1$, $b = 2^t$), with $\alpha_{2b} = 2$ and $\alpha_{4b} = \sqrt{2}$, and having the property that multiplication of Fermat number transforms corresponds to conventional cyclic convolution (2) modulo $2^b + 1$. To perform convolution with the transform requires $N$ real multiplications and a number of additions and word shifts proportional to $N \log N$. Unfortunately the transform requires word lengths proportional to the length of the sequences to be convolved [1], [2]. Since the lengths of the two dimensions are $2L$ and $M$ rather than $N = LM$ for the one-dimensional signal, the word-length requirement using the two-dimensional transform is proportional to the square root of $N$ rather than to $N$ as for the one-dimensional problem. It is this reduction in the necessary word-length that makes two-dimensional formulation attractive with the Fermat number transform.

The consequences of this reduction in word length is of considerable practical importance. For example, using a word length of 16 b and the Fermat number transform [2] with $\alpha = 2$ to compute the complete noncyclic convolution of two sequences of equal length, the one-dimensional implementation restricts the sequence length to a maximum of 16 where the two-dimensional implementation increases the maximum to 256. A summary of the restrictions on sequence lengths is shown in Table I for the most practical word lengths and for two values of $\alpha$. Note that one-dimensional length restrictions would be too severe for many applications but the two-dimensional restrictions would include most practical filters. If cyclic convolution is desired, then all the length restrictions are doubled since the addition of zeros to prevent aliasing is unnecessary.

The two-dimensional transform—or inverse transform—can be taken in either order. There is, however, a computational advantage in taking the transform first along the $m$ direction (length $M$) and then along the $l$ direction (length $2L$); half the $\hat{\hat{X}}$ sequences along the $m$ direction are zero and half the $\hat{\hat{H}}$ sequences along the $m$ direction are cyclically shifted by one position of the other half sequences. Also, while taking the inverse transform, there is an advantage to first taking the inverse transform along $l$, then along $m$, because we need only half the $\hat{\hat{Y}}$ sequences and therefore only half the sequences need be inverted along $m$.

TABLE I
**Sequence Length Restrictions for One- and Two-Dimensional Implementation of Noncyclic Convolution by the Fermat Number Transform**

| Word Length (Bits) | Transform Basis $\alpha$ | Maximum Sequence 1-*D* | Length—$N/2$ 2-*D* |
|---|---|---|---|
| 16 | 2 | 16 | 256 |
| 16 | $\sqrt{2}$ | 32 | 1024 |
| 32 | 2 | 32 | 1024 |
| 32 | $\sqrt{2}$ | 64 | 4096 |
| 64 | 2 | 64 | 4096 |
| 64 | $\sqrt{2}$ | 128 | 16 384 |

### IV. Generalizations and the Inverse Problem

A generalization of the approach in this paper to higher orders is fairly obvious. For example, $N$ could be factored into three integer factors $N = LMP$, as was done with two factors in (3). The signals $x(n)$ and $h(n)$ would then be redefined as three-dimensional $L \times M \times P$ arrays and (2) converted to a three-dimensional convolution by a change of variables as was done in two dimensions in (4)–(7). For $x$ this would be

$$\tilde{X}(l, m, p) = x(l + mL + pML) \tag{15}$$

and with similar definitions for $\tilde{H}$ and $\tilde{Y}$ and after augmentation to prevent aliasing, (2) would become a three-dimensional convolution.

$$\tilde{\tilde{\tilde{Y}}} = \tilde{\tilde{\tilde{H}}} * \tilde{\tilde{\tilde{X}}}. \tag{16}$$

We would then have one-dimensional cyclic convolution of $N$-length sequences being carried out by three-dimensional cyclic convolution with dimensions of lengths $2L$, $2M$, and $P$. Use of orders higher than two does not seem needed with the Fermat number transform at this time, but will be exploited with other schemes in the next sections.

Still another variation would apply to the case where the filter is periodically time-varying. This, when converted to a two-dimensional problem with $L$ equal the period of the filter, becomes time-invariant along one dimension, $m$, and time-varying along the other, $l$ [5]. The DFT or Fermat transform could be applied to the two-dimensional signal along the time-invariant dimension and either direct calculation or another type transform applied to the other dimension.

The inverse of the above problem can be considered where one wishes to implement two-dimensional convolution by one-dimensional methods. This involves reversing the process presented in Section II by constructing one-dimensional sequences from the given arrays. MacAdams [6] has presented a scheme for doing this that can be seen to be the inverse of the problem addressed in this paper and can be applied to cyclic or noncyclic two-dimensional convolution.

### V. *M*-Dimensional Convolution

If the length of the signals to be cyclically convolved in (2) can be written $N = 2^M$, then the methods used in (5) and (15) can be extended to define an $M$-dimensional signal with each dimension of length two. As before, this is done by a change of variables.

$$\tilde{X}(l, m, p, \cdots) = x(l + 2m + 4p + \cdots), \qquad l, m, p, \cdots = 0, 1. \quad (17)$$

After a similar definition for $\tilde{H}$ and $\tilde{Y}$ cyclic convolution in (2) becomes

$$\tilde{Y}(l, m, p, \cdots) = \sum_{k=0}^{1} \sum_{j=0}^{1} \sum_{q=0}^{1} \cdots \cdot \tilde{H}(l-k, m-j, p-q, \cdots)\tilde{X}(k, j, q, \cdots) \quad (18)$$

for $l, m, p, \cdots = 0, 1$.

Both $\tilde{X}$ and $\tilde{Y}$ are $M$-dimensional with each dimension of length two and $\tilde{H}$ is also $M$-dimensional but must be defined with dimensions of length three in order to carry out (18). This is the same as was required in two dimensions where $\hat{H}$ was defined in (9). Since the original convolution in (2) was defined as cyclic, the convolution along the last dimension in (18) is also cyclic so that $\tilde{H}$ need not be extended along that dimension but can be evaluated for this index modulo 2.

This is an extremely general and versatile formulation for the original problem that can be used to improve computational efficiency. The convolution along each dimension of length two can be written in terms of scalar variables as

$$\begin{bmatrix} h_0 & h_{-1} \\ h_1 & h_0 \end{bmatrix} \begin{bmatrix} x_0 \\ x_1 \end{bmatrix} = \begin{bmatrix} y_0 \\ y_1 \end{bmatrix}. \quad (19)$$

The convolution of (18) can be viewed as $M$ nested length 2 convolutions, each separately of the form shown in (19) requiring four multiplications. Using the same reasoning that was explained for the two-dimensional formulation, the total number of multiplications is $F = 4^M$. Since $N = 2^M$, this becomes $F = N^2$, which is again the same as would be required by directly calculating (2).

### VI. A High Speed Algorithm

With this formulation of scalar convolution in terms of multidimensional short convolutions, various tricks for efficient short convolution can be used. Consider one of several possible algorithms suggested by Pitassi [7] where, rather than directly calculating (19), three intermediate numbers are found:

$$\begin{aligned} g_0 &= (h_0 + h_{-1})x_1 \\ g_1 &= h_0(x_0 - x_1) \\ g_2 &= (h_1 + h_0)x_0. \end{aligned} \quad (20)$$

The desired outputs are then calculated by

$$\begin{aligned} y_0 &= g_0 + g_1 \\ y_1 &= g_2 - g_1. \end{aligned} \quad (21)$$

This approach uses three multiplications in comparison with four multiplications for a direct calculation.

Using this result, the total number of multiplications to calculate (18) becomes

$$F = 3^M. \quad (22)$$

If the fact pointed out earlier, that the convolution along the last dimension is cyclic, is used, then a further reduction is possible.

Length-two cyclic convolution is given by

$$\begin{bmatrix} h_0 & h_1 \\ h_1 & h_0 \end{bmatrix} \begin{bmatrix} x_0 \\ x_1 \end{bmatrix} = \begin{bmatrix} y_0 \\ y_1 \end{bmatrix}. \quad (23)$$

This can be calculated from two intermediate values by

$$\begin{aligned} f_0 &= (h_0/2 + h_1/2)(x_0 + x_1) \\ f_1 &= (h_0/2 - h_1/2)(x_0 - x_1) \end{aligned}$$

to give

$$\begin{aligned} y_0 &= f_0 + f_1 \\ y_1 &= f_0 - f_1 \end{aligned} \quad (24)$$

requiring two multiplications (assuming the factors of $\frac{1}{2}$ are either precalculated or obtained by shifting). Using this result on the last dimension reduces (22) to

$$F = 2 \cdot 3^{M-1} \quad (25)$$

which is the same number obtained by Pitassi [7] and Rayner [8].

Pitassi developed his algorithm by relating the cyclic convolution of two sequences to the convolution of subsequences in the same manner that the FFT can be developed based on decimation in time. This was extended by Davis [9] to an approach similar to decimation in frequency where the subsequences are halves of the original sequences. Both of these are special cases of the multidimensional formulation since the values along any dimension are samples or blocks of the original sequence.

Another form of convolution that is sometimes desired is of the same form as (2) but with $h(n)$ not being periodically extended, rather having independent values for negative indices. The transmission matrix formulation for $N = 3$ is

$$\begin{bmatrix} h_0 & h_{-1} & h_{-2} \\ h_1 & h_0 & h_{-1} \\ h_2 & h_1 & h_0 \end{bmatrix} \begin{bmatrix} x_0 \\ x_1 \\ x_2 \end{bmatrix} = \begin{bmatrix} y_0 \\ y_1 \\ y_2 \end{bmatrix} \tag{26}$$

and because of its structure the operator is called a constant diagonal convolution matrix. This requires $2N - 1$ values of $h(n)$ from $n = -N + 1$ to $n = N - 1$ and $N$ values of $x(n)$ to give $N$ value of $y(n)$.

The same approach that was used for cyclic convolution is applicable here except the reduction described in (23) and (24) does not work since the $M$-dimensional formulation of (18) is no longer cyclic along any dimension. Therefore the number of multiplications necessary to implement a length-$N$ constant diagonal convolution is

$$F = 3^M \tag{27}$$

which is the same as obtained by Allwright [10], using a matrix factorization approach.

Note that cyclic convolution in (2) can be viewed as a special case of constant diagonal convolution. Causal noncyclic convolution where $h(n)$ is zero for $n < 0$ is also a special case.

$$\begin{bmatrix} h_0 & 0 & 0 \\ h_1 & h_0 & 0 \\ h_2 & h_1 & h_0 \end{bmatrix} \begin{bmatrix} x_0 \\ x_1 \\ x_2 \end{bmatrix} = \begin{bmatrix} y_0 \\ y_1 \\ y_2 \end{bmatrix}. \tag{28}$$

The most common form of convolution desired is causal noncyclic convolution where all of the output sequence is obtained rather than only the first $N$ points as in (28). For this case two length-$N$ sequences are convolved to give a length-$(2N - 1)$ output. The transmission matrix formulation for $N = 3$ is

$$\begin{bmatrix} h_0 & 0 & 0 \\ h_1 & h_0 & 0 \\ h_2 & h_1 & h_0 \\ 0 & h_2 & h_1 \\ 0 & 0 & h_2 \end{bmatrix} \begin{bmatrix} x_0 \\ x_1 \\ x_2 \end{bmatrix} = \begin{bmatrix} y_0 \\ y_1 \\ y_2 \\ y_3 \\ y_4 \end{bmatrix}. \tag{29}$$

To apply the results from cyclic convolution, all sequences are extended with zeros to length $2N$. Cyclic convolution then gives the desired output of (29) and uses

$$F = 2 \cdot 3^M \tag{30}$$

multiplications.

A further reduction is possible by recognizing that along the last dimension only one multiplication is necessary rather than two for the cyclic case that gave (25) or three for the constant diagonal noncyclic case that gave (27). Consider the $(M + 1)$-dimensional convolution of (18) from the extended sequences with all indices except the last one held constant. The resulting length-two scalar convolution is

$$\hat{y}(p) = \sum_{j=0}^{1} \hat{h}(p - j)\hat{x}(j), \qquad p = 0, 1 \tag{31}$$

which becomes

$$\begin{aligned} \hat{y}_0 &= \hat{h}_0 \hat{x}_0 + \hat{h}_{-1} \hat{x}_1 \\ \hat{y}_1 &= \hat{h}_1 \hat{x}_0 + \hat{h}_0 \hat{x}_1 . \end{aligned} \tag{32}$$

The $\hat{x}_1$ terms are always zero since the last half of the length-$2N$ sequence $x(n)$ has been added as zeros. Close examination shows that because $h(n)$ has also been extended with zeros, either $\hat{h}_0$ or $\hat{h}_1$ will always be zero, depending on values of the constant indices. Therefore the length-two convolution of (31) will require only one multiplication and the resulting total number of multiplications necessary to noncyclically convolve two length-$N$ sequences giving a length-$2N$ output is

$$F = 3^M \tag{33}$$

which is the same as for the length-$N$ constant diagonal convolution with a length-$N$ output.

## VII. Multidimensional Convolution Based on Overlap-Add

Another two-dimensional formulation can be developed that is a generalization of the overlap-add algorithm [3], [4]. In (7) the $\hat{H}$ function was augmented so that the desired output $y(n)$ is in blocks that are the columns of $\hat{Y}$. For this formulation $\hat{H}$ will not be augmented but $\hat{Y}$ will be and the desired $y(n)$ will be obtained by adding the overlapped columns of $\hat{Y}$. This formulation is based on noncyclic convolution given by

$$y(n) = \sum_{q=0}^{N-1} h(n - q)x(q), \qquad n = 0, 1, \cdots 2N - 1 \tag{34}$$

where $h(n)$ and $x(n)$ are sequences of length $N$ and $y(n)$ of length $2N - 1$ and all are defined to be zero outside these lengths. The transmission matrix formulation for $N = 3$ is given by (29).

Using a similar factoring of $N$ and change of variables as was done in (3) and (5) we have (34) becoming

$$\begin{aligned} \hat{Y}(l, m) &= \sum_{p=0}^{M-1} \sum_{k=0}^{L-1} \hat{H}(l - k, m - p)\hat{X}(k, p) \\ l &= 0, 1, \cdots 2L - 2 \\ m &= 0, 1, \cdots 2M - 2. \end{aligned} \tag{35}$$

In this case both $\hat{X}$ and $\hat{H}$ are $L \times M$ arrays and $\hat{Y}$ is $(2L - 1) \times (2M - 1)$ with $\hat{X}$ and $\hat{H}$ having their columns the blocks of $x(n)$ and $h(n)$ but the columns of $\hat{Y}$ and the blocks of $y(n)$ have a more complicated relation than in (6). Both $\hat{X}$ and $\hat{H}$ are defined to be zero outside the domain of definition. Here we can show

$$y(l+mL) = \hat{Y}(l,m) + \hat{Y}(l+L, m-1) \quad (36)$$

for

$$l = 0, 1, \cdots L-1$$
$$m = 0, 1, \cdots 2M-1$$

and $\hat{Y}(l,m) = 0$ outside its domain of definition of

$$l = 0, 1, \cdots 2L-2$$
$$m = 0, 1, \cdots 2M-2. \quad (37)$$

This formulation gives the implementation of one-dimensional convolution by the sum of overlapped columns of an array obtained by noncyclic two-dimensional convolution. This can be viewed as a generalization of the overlap-add algorithm [3], [4] used for sectioning or block processing.

If $N = 2^M$, the extension to $M$-dimensional convolution is similar to that done for the overlap-save technique in (17) and (18). Both $\tilde{X}$ and $\tilde{H}$ are $M$-dimensional with each dimension of length two and $\tilde{Y}$ is $M$-dimensional but with each of length three. For

$$\tilde{X}(l,m,p,\cdots) = x(l+2m+4p+\cdots)$$
$$\tilde{H}(l,m,p,\cdots) = h(l+2m+4p+\cdots) \quad (38)$$

(34) becomes

$$\tilde{Y}(l,m,p,\cdots) = \sum_{k=0}^{l} \sum_{j=0}^{m} \sum_{q=0}^{p} \cdots$$
$$\cdot \tilde{H}(l-k, m-j, p-q, \cdots)\tilde{X}(k,j,q,\cdots) \quad (39)$$

for $l, m, p, \cdots = 0, 1, 2.$

The calculation of $y(n)$ from $\tilde{Y}$ is a bit complicated but is a generalization of (36) and involves only additions. For example if $N = 2^3 = 8$ the $\tilde{Y}$ function would have three dimensions with a total of 27 elements. Along the $l$ dimension there would be nine length-three blocks that, when overlapped and added along the $m$ dimension, would give three length-seven blocks. These, when overlapped and added, would give the single length-fifteen sequence that would be $y(n)$.

An example for $N = 4$ would have a two-dimensional three $\times$ three $\tilde{Y}$

$$\tilde{Y} = \begin{bmatrix} \tilde{y}_{00} & \tilde{y}_{01} & \tilde{y}_{02} \\ \tilde{y}_{10} & \tilde{y}_{11} & \tilde{y}_{12} \\ \tilde{y}_{20} & \tilde{y}_{21} & \tilde{y}_{22} \end{bmatrix} \quad (40)$$

and $y(n)$ would be found by

$$\begin{aligned} y_0 &= \tilde{y}_{00} \\ y_1 &= \tilde{y}_{10} \\ y_2 &= \tilde{y}_{20} + \tilde{y}_{01} \\ y_3 &= \tilde{y}_{11} \\ y_4 &= \tilde{y}_{21} + \tilde{y}_{02} \\ y_5 &= \tilde{y}_{12} \\ y_6 &= \tilde{y}_{22}. \end{aligned} \quad (41)$$

Along each dimension of (39) the matrix formulation of the length-two convolution is

$$\begin{bmatrix} h_0 & 0 \\ h_1 & h_0 \\ 0 & h_1 \end{bmatrix} \begin{bmatrix} x_0 \\ x_1 \end{bmatrix} = \begin{bmatrix} y_0 \\ y_1 \\ y_2 \end{bmatrix} \quad (42)$$

which, if done directly, requires four multiplications and one addition. If an intermediate step is added for $y_1$, then

$$\begin{aligned} y_0 &= h_0 x_0 \\ y_1 &= (h_0 + h_1)(x_0 + x_1) - y_0 - y_1 \\ y_2 &= h_1 x_1 \end{aligned}$$

gives the three outputs with three multiplications and four additions/subtractions. Using this algorithm on each dimension of (39) and counting the number of required multiplications by the method used to find (33) we find that (39) can be computed with

$$F = 3^M \quad (44)$$

multiplications which are the same as that obtained by the overlap-save method in (33).

## VII. Sequences of Different Lengths and Partial Outputs

Two modifications of the usual formulation of convolution are often desired. The first occurs when the two sequences are significantly different in lengths and the second when one desires only a portion of the output rather than all of it. Both cases can be formulated in terms of multidimensional convolutions and computational savings can be realized.

If the $h(n)$ is assumed to be the shorter sequence and if its length can be expressed as $L = 2^S$ the length of the longer sequence $x(n)$ will be expressed as $R = 2^S \cdot M$. It may be necessary to add zeros to both $x$ and $h$. The multidimensional signals are formed to give $S$ length-two dimensions and the last dimension of length $M$. If $m$ is the index for the length-$M$ dimension, then $\tilde{H}$ is formulated to be zero for all $m$ other than $m = 0$. Therefore, if the fast algorithm of (20) or (43) is used along the $S$ length-two dimensions, there will be only $M$ multiplications required along the $M$ dimension. The total number of multiplications is then

$$F = 3^S \cdot M. \quad (45)$$

This could also be seen by considering the problem as requiring $M$ length-$2^S$ noncyclic convolutions with the outputs overlapped and added.

If only a portion of the total output from the convolution is desired, then a similar saving can be obtained. Assume that both $h(n)$ and $x(n)$ are of length $N = LM$, with a noncyclic output $y(n)$ of length $2N - 1$. Out of this only a block of $y(n)$ of length $L$ is desired. First, $N$ extra zeros are appended to both the sequences and cyclic convolution of length $2N$ is formulated. This one-dimensional cyclic convolution is reformulated as a two-dimensional convolution as in (5)–(7). The first dimension is of length $L$ and the convolution in this dimension is noncyclic and can be carried out as cyclic convolution of length $2L - 1$. In the second dimension the convolution is cyclic and of length $2M$. The two-dimensional arrays are formulated so that the desired length $L$ block of $y(n)$ appears as a column of $\hat{Y}$ in (6). Thus, $\hat{Y}$ along the second dimension has to be computed only for one index. This replaces cyclic convolution of length $2M$ with a summation of $2M$ terms along the second dimension. The desired output block can be written as a summation of $2M$ convolutions of length $L$, where each convolution represents convolution of two sequences of lengths $2L - 1$ and $L$, respectively, giving a sequence of length $L$. Because zeros were appended to both $x(n)$ and $h(n)$, out of these $2M$ convolutions, between one and $M$ convolutions would be nonzero, depending on the second index of $\hat{Y}$ for which the output is desired.

The convolutions along the first dimension of length $L$ can be carried out either by transform techniques or multidimensional techniques discussed in this paper. If the transform techniques are used, the summation of convolutions can be carried out in the transform domain, thus requiring only one inverse transform to obtain the desired output. Rader [11] has discussed a similar technique for the particular case of estimation of autocorrelation function for the first few lag values. If $L = 2^S$ and the multidimensional methods are used, the number of multiplications are at most the same as given in (45).

The formulation just discussed can be extended for the situation where sampled output is desired, sampled at every $L$th value. In this case for the two-dimensional formulation, the output appears as a row of $\hat{Y}$, and as before, the two-dimensional convolution again reduces to a summation of convolutions. If the output $y(n)$ is a narrow-band signal as compared to the sampling frequency, to reconstruct the analog signal, the samples of $y(n)$ at a lower sampling rate are sufficient. In this situation the formulation discussed here can result in computational savings.

If a multidimensional formulation is considered, we can obtain partial outputs as combinations of blocks and samples.

## IX. Relations to Logical Convolution and Walsh Transforms

Consider logical convolution [12] of two sequences $x(n)$ and $h(n)$ of length $N = 2^M$, each giving an output sequence $y(n)$ also of length $N$. Logical convolution is defined similar to the cyclic convolution of (2), but the addition and subtraction of indices is done differently. All indices are represented in the binary form as an $M$-bit index. When indices have to be added or subtracted, they are added or subtracted bit by bit, modulo 2. Note that in logical convolution, addition and subtraction of indices are equivalent. We can convert this one-dimensional logical convolution problem to an $M$-dimensional convolution as in (17) and (18). Along any dimension, convolution appears as in (19), but since logical convolution is desired, $h(-1) = h(1)$. Therefore, if (18) is implemented as length-two cyclic convolutions along all the dimensions, $y(n)$ thus obtained is logical convolution of $x(n)$ and $h(n)$. Alternatively, if (18) is implemented as a noncyclic convolution along all the dimensions, we obtain noncyclic convolution of $x(n)$ and $h(n)$ as in (26).

Length-two cyclic convolution can be implemented using just two multiplications as in (24), which is a length-two DFT implementation. Thus (18) can be implemented as an $M$-dimensional cyclic convolution using $M$-dimensional DFT's of $\tilde{X}$ and $\tilde{H}$, where each dimension is of length two. Alternatively length-$N$ logical convolution can also be implemented using length-$N$ Walsh transforms of $x(n)$ and $h(n)$ [12]. The preceding development shows that length-$(N = 2^M)$ Walsh transform is equivalent to the $M$-dimensional DFT. Thus the $M$-dimensional approach establishes the logical convolution theorem for the Walsh transforms and it also establishes the fast Walsh transform algorithm as an $M$-dimensional DFT. These facts have been noted before in the literature.

For a particular formulation of (20), Pittasi [7] and Davis [9] have shown that some of the intermediate products correspond to multiplication of the Walsh transforms of the two sequences.

## X. Generalizations and Applications

There are several modifications and generalizations that are possible with the formulation used here. The first will illustrate that fast algorithms exist for sequences of length other than two.

Consider the noncyclic convolution of two length-three sequences.

$$\begin{bmatrix} h_0 & 0 & 0 \\ h_1 & h_0 & 0 \\ h_2 & h_1 & h_0 \\ 0 & h_2 & h_1 \\ 0 & 0 & h_2 \end{bmatrix} \begin{bmatrix} x_0 \\ x_1 \\ x_2 \end{bmatrix} = \begin{bmatrix} y_0 \\ y_1 \\ y_2 \\ y_3 \\ y_4 \end{bmatrix}. \tag{46}$$

This would normally require nine multiplications and four additions. If six intermediate variables are calculated by

$$g_0 = h_0 x_0 \quad g_3 = (h_0 + h_1)(x_0 + x_1)$$
$$g_1 = h_1 x_1 \quad g_4 = (h_1 + h_2)(x_1 + x_2)$$
$$g_2 = h_2 x_2 \quad g_5 = (h_2 + h_0)(x_2 + x_0) \tag{47}$$

then the desired output can be obtained by

$$y_0 = g_0$$
$$y_1 = g_3 - g_0 - g_1$$
$$y_2 = g_5 + g_1 - g_0 - g_2$$
$$y_3 = g_4 - g_1 - g_2$$
$$y_4 = g_2 \tag{48}$$

requiring six multiplications.

If two sequences of length $N = 3^M$ were to be convolved, then using an implementation with $M$ dimensions of length three would require for total multiplications

$$F = 6^M. \tag{49}$$

For short sequences of this length, this approach is faster than adding zeros and using the next larger power of two-length with (39).

Similar results are possible with other lengths and a very general scheme can be developed for sequences of length

$$N = 2^M \cdot 3^S \cdots \tag{50}$$

using different algorithms along different dimensions in a manner similar to using a multiple radix FFT.

Because of the generality of the multidimensional formulation there are mixtures of high speed techniques that can be used.

In some situations, it may be advantageous to combine transform techniques with the fast algorithms for short convolutions discussed in this paper. One such situation uses the Fermat number transforms [2]. As discussed in Section III, using a $b$-bit implementation of Fermat number transforms, the maximum cyclic convolution length is $4b$. To cyclically convolve sequences longer than this, we need to formulate a two-dimensional convolution as in Section II. Assume $x(n)$, $h(n)$, and $y(n)$ are sequences of length-($N = 8b$) each. Consider the formulation of (7), with $M = 4b$ and $L = 2$. Equation (7) can be rewritten as

$$\hat{Y}(0, m) = \sum_{p=0}^{M-1} \hat{H}(0, m-p)\hat{X}(0,p) + \sum_{j=0}^{M-1} \hat{H}(-1, m-j)\hat{X}(1, j)$$
$$\hat{Y}(1, m) = \sum_{p=0}^{M-1} \hat{H}(1, m-p)\hat{X}(0,p) + \sum_{j=0}^{M-1} \cdot \hat{H}(0, m-j)\hat{X}(1,j), \qquad m = 0, 1, \cdots, M-1. \tag{51}$$

In these equations, each summation represents cyclic convolution of length $4b$, which can be carried out by Fermat transforms. Taking length $4b$ Fermat transforms of all the sequences in (51), we obtain

$$T\{\hat{Y}(0, l)\} = T\{\hat{H}(0, l)\}\, T\{\hat{X}(0, l)\} + T\{\hat{H}(-1, l)\}T\{\hat{X}(1, l)\}$$
$$T\{\hat{Y}(1, l)\} = T\{\hat{H}(1, l)\}\, T\{\hat{X}(0, l)\} + T\{\hat{H}(0, l)\}T\{\hat{X}(1, l)\}. \tag{52}$$

These equations are similar to (19) and we could employ the tricks of (20) and (21) along the first dimension, giving

$$T\{\hat{Y}(0, l)\} = [T\{\hat{H}(0, l)\} + T\{\hat{H}(-1, l)\}]\, T\{\hat{X}(1, l)\} + [T\{\hat{X}(0, l)\} - T\{\hat{X}(1, l)\}]\, T\{\hat{H}(0, l)\}$$
$$T\{\hat{Y}(1, l)\} = [T\{\hat{H}(0, l)\} + T\{\hat{H}(1, l)\}]\, T\{\hat{X}(0, l)\} - [T\{\hat{X}(0, l)\} - T\{\hat{X}(1, l)\}]\, T\{\hat{H}(0, l)\}. \tag{53}$$

Note that $\hat{H}(-1, m) = \hat{H}(1, m-1)$, thus $\hat{H}(-1, l)$ is the cyclic shift by one position of $\hat{H}(1, l)$. We could employ the cyclic shift theorem to compute $T\{\hat{H}(-1, l)\}$ from $T\{\hat{H}(1, l)\}$. Assuming the $\hat{H}$ transforms are precalculated, this method requires computation of 2 length-($4b$) $\hat{X}$ transforms, $12b$ multiplications and $12b$ additions/subtractions to compute the $\hat{Y}$ transforms using (53), and two length-($4b$) $\hat{Y}$ inverse transforms. This is efficient because the only extra computation is $4b$ extra multiplications and is better than the two-dimensional Fermat number transform approach, which requires roughly twice the amount of computation. This use of $2 \times M$ convolution has a small computational advantage even when used with the FFT where, in effect, the last stage of the FFT algorithm is replaced by one of the fast length-two algorithms.

There are many other possibilities of combinations of Fermat and Fourier transforms of various lengths and dimensions with short convolution algorithms or with the use of special hardware. The arbitrary order of the various operations can also be used to advantage. As pointed out by Rader [11], and illustrated in our discussion of partial outputs and in (53), it is often more efficient to take transforms along one dimension than convolve along another before taking the inverse transform.

To illustrate the efficiencies of some of the techniques of this paper, a comparison of a particular case will be made with direct and FFT implementations. Consider the problem of noncyclic convolution of two length-($N = 2^M$) sequences to give an output of length $2N - 1$ as described in (34). The number of multiplications per output point will be calculated for three implementations. First, consider a direct implementation which requires $N^2$ multiplication for

$2N-1$ output points (we will approximate this by $2N$ for simplification). This gives for multiplication per output point

$$F_0 = \tfrac{1}{2}N.$$

If the FFT is used in an efficient way, taking advantage of the fact that the data are real and using a mixed radix algorithm (2, 4, and 8), the number of multiplications necessary per output point can be calculated as described by Singleton [13] and is denoted $F_1$.

Using the $M$-dimensional formulation with the fast algorithm of (33) or (44) gives for the multiplications per output

$$F_2 = \frac{1}{2N} 3^M.$$

Table II compares these functions for various lengths up to 1024. Note the multidimensional implementation is more efficient than a direct implementation for all lengths and requires fewer multiplications than the FFT for lengths up to 128. If a less efficient FFT implementation requiring $4 \log N + 4$ multiplications per output point is used, the crossover length is above 2048.

The multidimension approach can also be used in the same way the FFT is used to implement on going processing or sectioning [4]. In contrast, to use with the FFT, this approach is most efficient when the length of the section or block is the same as the length of the convolution operator.

Initial observations indicate that block implementation of recursive filters [3] becomes more attractive when used with the techniques described in this paper. To illustrate this, we will again consider the multiplication efficiencies for three realizations. First consider a recursive filter with an equal order numerator and denominator of $N = 2^M$. The multiplications per output points for a direct implementation

$$F_0 = 2N.$$

Using efficient FFT algorithms, the results given by the methods in [3] are denoted $F_1$. Using three convolutions by (33) or (44) to implement the block recursive filter with the block length equal the order, gives as the multiplications per output point

$$F_2 = \frac{1}{N} \cdot 3^{M+1}.$$

Table III compares these multiplication efficiencies for orders up to 128. Note the multidimensional approach is more efficient than the direct for orders above three and more efficient than the FFT for orders up to about 256. This is yet to be explored in detail.

TABLE II
**A Comparison of Multiplication Efficiencies for Three Implementations of Length-$N$ Convolution**

| $N$ | $F_0$ | $F_1$ | $F_2$ |
|---|---|---|---|
| 2 | 1 | 1.5 | 0.75 |
| 4 | 2 | 2.25 | 1.12 |
| 8 | 4 | 4.12 | 1.70 |
| 16 | 8 | 5.06 | 2.53 |
| 32 | 16 | 6.03 | 3.80 |
| 64 | 32 | 8.01 | 5.70 |
| 128 | 64 | 9.00 | 8.54 |
| 256 | 128 | 10.00 | 12.81 |
| 512 | 256 | 12.00 | 19.22 |
| 1024 | 512 | 13.00 | 28.83 |

TABLE III
**A Comparison of Multiplication Efficiencies for Three Implementations of Block Recursive Filters of Order $N$**

| $N$ | $F_0$ | $F_1$ | $F_2$ |
|---|---|---|---|
| 2 | 4 | 14 | 4.5 |
| 4 | 8 | 21 | 6.7 |
| 8 | 16 | 31 | 10.1 |
| 16 | 32 | 38 | 15.2 |
| 32 | 64 | 45 | 22.8 |
| 64 | 128 | 53 | 34.2 |
| 128 | 256 | 58 | 51.2 |

## XI. Conclusions

This paper has presented two formulations of convolution in terms of multidimensional convolution—one based on a generalization of the overlap-save algorithm for sectioning and the other on the overlap-add algorithm. The first proved to be well suited for cyclic and constant-diagonal convolution and the second for noncyclic convolution. Fast algorithms were developed based on length-two and -three convolution that lead to an improvement in multiplication efficiency. The reduction in required word lengths proved to be a valuable feature when used with the Fermat number transform. The formulation proved to be well suited for the special cases where unequal-length sequences were convolved or where only a portion of the output was desired. It was further shown that various mixtures of algorithms could be used along the different dimensions to achieve certain advantages or to fit particular requirements. Finally, examples were presented to compare the multiplication efficiencies of a few implementations.

The formulation is so general that a complete and systematic investigation of all possible applications is difficult. The main ideas and relations to other works that we know of have been presented here. The investigation of word length and storage requirements and a more complete consideration of recursive implementations is still to be made.

## Acknowledgment

The authors would like to thank R. A. Meyer for valuable discussions.

## References


[1] C. M. Rader, "Discrete convolutions via Mersenne transforms," *IEEE Trans. Comput.*, vol. C-21, pp. 1269-1273, Dec. 1972.

[2] R. C. Agarwal and C. S. Burrus, "Fast digital convolutions using Fermat transforms," in *Southwestern IEEE Conf. Rec.*, Houston, Tex., Apr. 1973, pp. 538-543.

[3] C. S. Burrus, "Block realization of digital filters," *IEEE Trans. Audio Electroacoust.*, vol. AU-20, pp. 230-235, Oct. 1972.

[4] B. Gold and C. M. Rader, *Digital Processing of Signals.* New York: McGraw-Hill, 1969, pp. 208-211.

[5] R. A. Meyer and C. S. Burrus, "Certain properties of periodically time-varying digital filters," in *Southwestern IEEE Conf. Rec.*, Houston, Tex., Apr. 1973, pp. 529-535.

[6] D. P. MacAdam, "Image restoration by constrained deconvolution," *J. Opt. Soc. Amer.*, vol. 60, pp. 1617-1627, Dec. 1970.

[7] D. A. Pitassi, "Fast convolution using the Walsh transforms," in *Proc. Conf. Applications of Walsh Functions*, Washington, D.C., Apr. 1971, pp. 130-133.

[8] P. J. W. Rayner, "A fast cyclic convolution algorithm," presented at Symp. Digital Filtering, Imperial College, London, England, Aug. 1971.

[9] W. F. Davis, "A class of efficient convolution algorithms," in *Proc. Symp. Applications of Walsh Functions*, Washington, D.C., Mar. 1972, pp. 318-329.

[10] J. C. Allwright, "Real factorization of noncyclic convolution operators with applications to fast convolution," *Electron. Lett.*, vol. 7, pp. 718-719, Dec. 1971.

[11] C. M. Rader, "An improved algorithm for high speed autocorrelation with applications to spectral estimation," *IEEE Trans. Audio Electroacoust.*, vol. AU-18, pp. 439-441, Dec. 1970.

[12] G. S. Robinson, "Logical convolution and discrete Walsh and Fourier power spectra," *IEEE Trans. Audio Electroacoust.*, vol. AU-20, pp. 271-280, Oct. 1972.

[13] R. C. Singleton, "An algorithm for computing the mixed radix fast Fourier transform," *IEEE Trans. Audio Electroacoust.*, vol. AU-17, pp. 93-103, June 1969.

# Application of the Fast Fourier Transform to Computation of Fourier Integrals, Fourier Series, and Convolution Integrals

JAMES W. COOLEY, PETER A. W. LEWIS, AND PETER D. WELCH

***Abstract*—The fast Fourier transform is a computational procedure for calculating the finite Fourier transform of a time series. In this paper, the properties of the finite Fourier transform are related to commonly used integral transforms including the Fourier transform and convolution integrals. The relationship between the finite Fourier transform and Fourier series is also discussed.**

## Introduction

MANY PROBLEMS of current interest require the use of a variety of numerical methods for their solution. A technique that has found wide applicability is the integral transform method. However, numerical problems are generally solved with the aid of a digital computer which is not designed to handle the continuous waveforms that occur when integral transform methods are used. For this reason, it is necessary to convert continuous time series or other functions to a series of discrete data samples, and to perform numerical operations such as the finite Fourier transform on these samples. The fast Fourier transform (FFT) algorithm has reduced the time required to compute finite Fourier transforms by the fraction $\log_2 N/N$ where $N$ is the number of discrete data samples. For large values of $N$, this reduction is important. It is therefore necessary to relate the properties of continuous integral transforms to the properties of the finite Fourier transform in order to take advantage of digital computers and the fast Fourier transform algorithm. In this paper, the correspondence between the finite Fourier transform and the Fourier integral is described, and a method that can be used to compute Fourier integrals is discussed in detail. Other problems covered are the computation of convolution integrals and integral equations of the convolution type. Finally, the relationship between the finite Fourier transform and the Fourier series is discussed in order to apply the FFT to problems that involve harmonic analysis and synthesis.

Manuscript received January 26, 1967; revised March 6, 1967.
The authors are with the IBM Research Center, Yorktown Heights, N. Y.

## Relationship between the Finite Fourier Transform and the Fourier Transform: Use of the Algorithm to Calculate Fourier Integrals

Suppose we have a function $x(t)$ which has a Fourier transform

Reprinted from *IEEE Trans. Audio Electroacoust.*, vol. AU-15, pp. 79–84, June 1967.

$$a(f) = \int_{-\infty}^{\infty} x(t)e^{-2\pi i f t}dt. \tag{1}$$

Then, the inverse relationship is

$$x(t) = \int_{-\infty}^{\infty} a(f)e^{2\pi i f t}df. \tag{2}$$

If $x(t)$ is sampled at intervals of length $\Delta t$, then (2) expressed at the sampling points $j\cdot\Delta t, j=0, \pm 1, \pm 2, \cdots$, can be written

$$x(j\cdot\Delta t) = \int_{-\infty}^{\infty} a(f)e^{2\pi i j f/F}df$$
$$= \sum_{k=-\infty}^{\infty} \int_{kF}^{(k+1)F} a(f)e^{2\pi i j f/F}df \tag{3}$$

where $F=1/\Delta t$. $F$ is twice the Nyquist frequency, $1/(2\Delta t)$. Now the exponential function $e^{2\pi i j f/F}$ is a periodic function of $f$ with period $F$ so that by changing the variable of integration one can put (3) in the form

$$x(j\cdot\Delta t) = \int_0^F a_p(f)e^{2\pi i j f/F}df \tag{4}$$

where

$$a_p(f) = \sum_{k=-\infty}^{\infty} a(f+kF). \tag{5}$$

Here, and in what follows, the subscript $p$ on a function will denote the periodic function formed by the superposition of the nonperiodic function shifted by all multiples of a fundamental period. The function $a_p(f)$ is said to be an "aliased" version of $a(f)$, with the aliasing occurring relative to the Nyquist frequency $F/2$.

Since $a_p(f)$ is a periodic function of $f$ it has a Fourier series expansion. Further, we see from (4) that the coefficients of this expansion are given by $1/F$ times the sequence $x(j\cdot\Delta t)$. Hence, (4) has the reciprocal equation

$$a_p(f) = \frac{1}{F}\sum_{j=-\infty}^{\infty} x(j\cdot\Delta t)e^{-2\pi i f j/F}. \tag{6}$$

In this relationship between $a_p(f)$ and $x(j\cdot\Delta t)$ the usual roles of time and frequency are interchanged; i.e., a periodic, continuous function of frequency corresponds to a sequence of time coefficients. Now, (6) is a discrete Fourier transform but it is not finite. However, if we consider the values of $a_p(f)$ at $N$ equally spaced points between 0 and $F$, i.e., sample $a_p(f)$ at intervals $\Delta f = F/N = 1/(N\Delta t)$, we obtain

$$a_p(n\Delta f) = \frac{1}{F}\sum_{j=-\infty}^{\infty} x(j\Delta t)e^{-2\pi i jn/N}$$
$$= \frac{1}{F}\sum_{j=0}^{N-1}\left\{\sum_{l=-\infty}^{\infty} x(j\Delta t + lN\Delta t)\right\}e^{-2\pi i jn/N}. \tag{7}$$

The latter result follows from the fact that $e^{-2\pi i jn/N}$ is a periodic sequence of $j$ with period $N$. Hence, we finally have

$$a_p(n\Delta f) = \frac{1}{F}\sum_{j=0}^{N-1} x_p(j\Delta t)e^{-2\pi i jn/N} \tag{8}$$

where

$$x_p(t) = \sum_{l=-\infty}^{\infty} x(t+lT) \tag{9}$$

is periodic of period $T=N\Delta t=1/\Delta F$. It is only a matter of a multiplicative constant to put (8) in the form of the finite Fourier transform and prove the following important theorem.

*Theorem 1*

If $x(t)$, $-\infty<t<\infty$, and $a(f)$, $-\infty<f<\infty$, are a Fourier integral transform pair,

$$x(t) \leftrightarrow a(f),$$

then $Tx_p(j\Delta t)$, $j=0, 1, \cdots, N-1$, and $a_p(n\Delta f)$, $n=0, 1, 2, \cdots, N-1$, are a finite Fourier transform pair,

$$Tx_p(j\Delta t) = T\sum_{l=-\infty}^{\infty} x(j\Delta t + lT) \leftrightarrow \sum_{k=-\infty}^{\infty} a(n\Delta f + kF)$$
$$= a_p(n\Delta f)$$

where $\Delta f = 1/(N\Delta t) = 1/T$.

It is to be remembered that the finite Fourier transform pair

$$X(j) \leftrightarrow A(n)$$

is defined by

$$X(j) = \sum_{n=0}^{N-1} A(n)W_N{}^{jn}, \qquad W_N = e^{2\pi i/N}.$$

Thus, if two functions are Fourier transforms of one another, then the sequences obtained from them by aliasing and sampling in this fashion are finite Fourier transforms of one another.

An interesting application of this theorem is to the problem of determining the vector $X(j)$ which corresponds to the normalized Gaussian function $x(t)=e^{-\pi t^2}$, the function having the unique property that it is its own Fourier transform, i.e.,

$$x(t) = e^{-\pi t^2} \leftrightarrow a(f) = e^{-\pi f^2}. \tag{11}$$

We must have $\Delta t = \Delta f$, which implies that $T=F=\sqrt{N}$. Theorem 1 now gives us Corollary 1.

*Corollary 1:*

$$X(j) = Tx_p(j\Delta t) = \sqrt{N}\sum_{k=-\infty}^{\infty} e^{-\pi(j+kN)^2/N} \tag{12}$$

and

$$A(n) = a_p(n\Delta f) = \frac{1}{\sqrt{N}}X(n), \quad j, n = 0, 1, \cdots, N-1.$$

form a finite Fourier transform pair.

The factor $\sqrt{N}$ arises because, for computational efficiency, we did not define the discrete Fourier transform so as to preserve the Euclidean norm in the transformation. Otherwise, if we had defined $X(j)$ and $A(n)$ with a factor $1/\sqrt{N}$ before the summation sign, we could have obtained a transform pair of identical functions which, for $-T/2 < j\Delta t < T/2$, approximate the normalized Gaussian function (11) for large $N$.

A further word is in order at this point concerning the relationship between $a(f)$ and $a_p(f)$. Suppose $\Delta t$ is chosen so that the aliasing involved in constructing $a_p(f)$ is negligible, i.e., $a(f) \approx 0$ for $|f| > \frac{1}{2}F$. As we said before, $a_p(f)$ is periodic of period $F$ and if we view it from $-\frac{1}{2}F$ to $\frac{1}{2}F$ then from (5)

$$a(f) \approx a_p(f) \qquad \text{for } |f| \leq \tfrac{1}{2}F.$$

However, if we view it from 0 to $F$ (as we do in the finite Fourier transform approach) then the negative $f$ half of $a(f)$ is produced to the right of the positive $f$ half of $a(f)$:

$$a(f) \approx a_p(f) \qquad 0 \leq f \leq \tfrac{1}{2}F$$

$$a(f-F) \approx a_p(f) \qquad \tfrac{1}{2}F \leq f \leq F.$$

In other words, $a_p(f)$ does not approximate $a(f)$ for $0 \leq f \leq F$ but rather for $|f| \leq \frac{1}{2}F$. The relationship between $a_p(f)$ and $a(f)$ is illustrated in Fig. 1.

We will now discuss the application of Theorem 1 to the problem of using the fast Fourier transform algorithm for the evaluation of Fourier integrals. We will discuss the problem of evaluating $a(f)$ given $x(t)$, but the procedure for going in the other direction is a parallel one. Consider again the continuous function $x(t)$. In our formulation [see (7) and (8)] we have replaced the problem of calculating the frequency function $a(f)$ by the problem of calculating, from the sequence $x(j\Delta t)$, the periodic function, $a_p(f)$, constructed by adding the set of functions $a(f+kF)$ for $k=0, \pm 1, \pm 2, \cdots$. Hence, the error in approximating $a(f)$ by $a_p(f)$, in the range $-\frac{1}{2}F < f < \frac{1}{2}F$, is the sum of the $a(f+kF)$'s for $k \neq 0$. Thus, by choosing $\Delta t$ small enough, $F=1/\Delta t$ can be made sufficiently large so that this error is negligible in the frequency range of interest. This proper choice of $\Delta t$ is the first choice to make in applying the algorithm. If one is interested in the entire frequency range then $\Delta t$ should be chosen so that the contributions of $a(f)$ for all $|f| > \frac{1}{2}F$ are negligible.

It is worthy of note here that (8) is precisely the formula one would have obtained in approximating the integral in (1) by the trapezoidal rule. The advantage of the present discussion is that the "integration error" is seen to be the difference between $a(f)$ and $a_p(f)$. Thus, if one has an estimate of the frequency function, one can predict the magnitude of the error introduced by the numerical integration.

With $\Delta t$ chosen, one now has the freedom to choose the interval $\Delta f = 1/(N\Delta t)$ at which one wants to sample

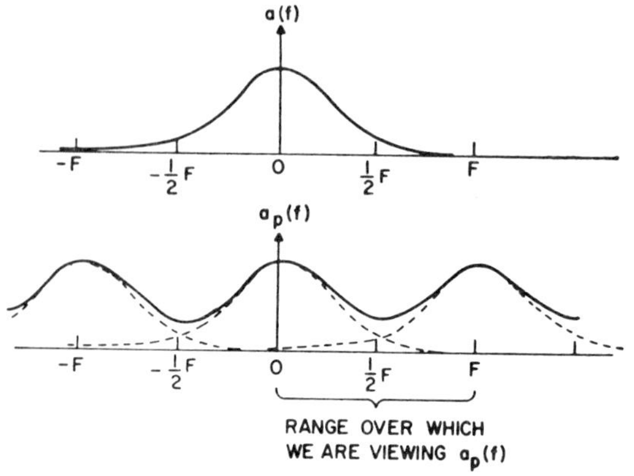


Fig. 1. Relationship between the frequency function $a(f)$ and the periodic aliased function $a_p(f)$.

$a_p(f)$. One does this by selecting $N$, and, for the sake of the fast Fourier transform algorithm, $N$ should be highly composite (preferably a power of 2). In addition, $N$ must be large enough so that the resulting $\Delta f$ gives sufficiently fine spacing of the frequency estimates. With the choice of $N$ we then form the aliased sequence $x_p(j\Delta t) = \sum_{l=-\infty}^{\infty} x(j\Delta t + lT)$ and apply the algorithm. In the application of the algorithm the sequence $X(j) = Tx_p(j\Delta t)$ would be formed. If we let $A(n)$ be the finite Fourier transform of $X(j)$ then $A(n) = a_p(n\Delta f)$.

With the formation of $x_p(j\Delta t)$, we have done in the time domain what was originally done in the frequency domain: we have formed a periodic sequence $x_p(j\Delta t)$ by displacing $x(j\Delta t)$ by $0, \pm N, \pm 2N, \cdots$, points and adding the resulting sequences. The difference between $x_p(j\Delta t)$ and $a_p(n\Delta f)$ is that while $\Delta t$ is chosen so as to make the frequency period $F=1/\Delta t$ just large enough to encompass the region where $a(f)$ is significantly different from zero, $N$ need not be chosen so that $N\Delta t$ just encompasses the nonzero extent of $x(j\Delta t)$. Rather, $N$ might be chosen to be less than or greater than the nonzero extent of $x(j\Delta t)$, depending on the frequency resolution desired. A choice of $N$ for which $N\Delta t$ is less than the nonzero range of $x(j\Delta t)$ is illustrated in Fig. 2.

The method outlined above gives independent control over the upper frequency ($\frac{1}{2}F$) and frequency resolution ($\Delta f = 1/T$) parameters. If you do not follow this method and do not do the aliasing of $x(j\Delta t)$ leading to $x_p(j\Delta t)$ you will, in some cases, obtain greater frequency resolution than you want and, hence, do more computing than necessary. Also note that the positioning of $x(t)$ in the interval $[0, N\Delta t = T]$ is automatically taken care of in the aliasing operation. Equation (9) places $x(t)$ in the interval $[0, T]$ without disturbing the phases of its components. Finally, stressing again a point made earlier, the finite Fourier transform will produce a sampling of the periodic function $a_p(f)$ from 0 to $F$, in which the negative $f$ half is reproduced to the right of the positive $f$ half. This might be confusing since experimenters are accustomed to viewing $a_p(f)$ from $-\frac{1}{2}F$ to $\frac{1}{2}F$.

In the preceding discussion we have assumed that

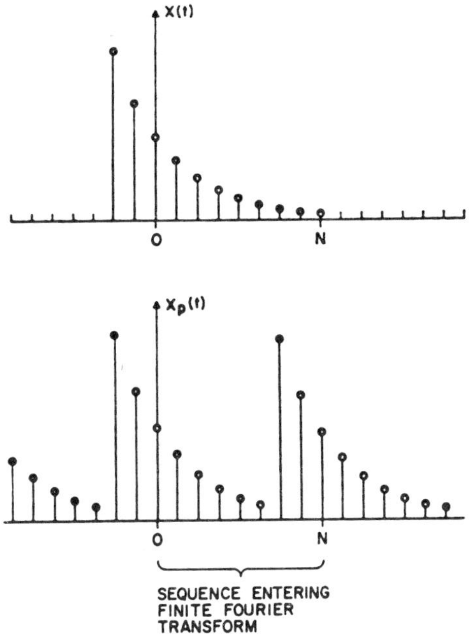


Fig. 2. Relationship between sampled values of the time function $x(t)$ and the periodic aliased function $x_p(t)$.

$x(t)$ is such that the infinite summation involved in obtaining $x_p(j\Delta t)$ could be performed [e.g., that $x(t)$ was nonzero only over a finite interval]. If the summation involved in $x_p(j\Delta t)$ must be truncated then the estimated frequency function will be convolved with a $\sin \pi f/\pi f$ shaped spectral window. This spectral window is not a desirable one and one might choose to apply the smoothing techniques used in spectral analysis for similarly truncated covariance functions.

Mention was also made above of the equivalence between trapezoidal rule integration and (7) for the discrete Fourier transform. Actually, trapezoidal rule integration yields better accuracy than one would anticipate for the type of functions considered here. To demonstrate this, consider the Euler–MacLaurin sum formula for integrating an arbitrary function $f(t)$:

$$\int_{t_0}^{t_N} f(t)dt = \Delta t(\tfrac{1}{2}f_0 + f_1 + \cdots + f_{N-1} + \tfrac{1}{2}f_N) - \frac{(\Delta t)^2}{12}(f_N' - f_0') + \frac{(\Delta t)^4}{720}(f_N''' - f_0''') - \frac{(\Delta t)^6}{30240}(f_N^{v} - f_0^{v}) + \cdots, \tag{13}$$

where primes denote derivatives. Here it can be seen that if the integrand and odd derivatives at the end points $t_0$, $t_N$ are equal and/or negligible, then the discrete Fourier transform (7), terminated at indices corresponding to $t_0$ and $t_N$, becomes identical with the Euler–MacLaurin sum formula (13) applied to the integrand

$$f(t) = x(t)e^{-2\pi itf}$$

with an integration interval $\Delta t$.

## Relationship between Fourier Series and the Finite Fourier Transform: Application of the Algorithm to Harmonic Analysis and Synthesis

Suppose, we have a function $x(t)$ which is periodic of period $T$. Then $x(t)$ has a Fourier series expansion

$$x(t) = \sum_{n=-\infty}^{\infty} c(n)e^{2\pi i(nt/T)} \tag{14}$$

where the $c(n)$ are given by

$$c(n) = \frac{1}{T}\int_0^T x(t)e^{-2\pi i(nt/N)}dt. \tag{15}$$

Now, if we sample $x(t)$ at $N$ equally spaced points between 0 and $T$, we generate the sequence $x(j\Delta t)$ where $\Delta t = T/N$. This sequence is periodic of period $N$; substituting in (14), we obtain

$$\begin{aligned} x(j\Delta t) = x(jT/N) &= \sum_{n=-\infty}^{\infty} c(n)e^{2\pi i(nj/N)} \\ &= \sum_{n=0}^{N-1}\left\{\sum_{l=-\infty}^{\infty} c(n+Nl)\right\} e^{2\pi i(nj/N)} \\ &= \sum_{n=1}^{N-1} c_p(n)e^{2\pi i(nj/N)}. \end{aligned} \tag{16}$$

Thus, we see that $x(j\Delta t)$ is the finite Fourier transform of

$$c_p(n) = \sum_{l=-\infty}^{\infty} c(n + Nl).$$

This is summarized by Theorem 2.

*Theorem 2*

If the periodic function $x(t)$ with period $T$ has the Fourier series expansion $c(n)$,

$$x(t) \leftrightarrow c(n)$$

then the periodic sequence $x(j\Delta t)$ of period $N$, where $\Delta t = T/N$, has the finite Fourier transform $c_p(n)$:

$$x(j\Delta t) \leftrightarrow c_p(n) = \sum_{l=-\infty}^{\infty} c(n + lN).$$

From this we see that in using the algorithm for harmonic analysis we should pick an $N$ such that the error due to aliasing in the approximation of $c(n)$ by $\sum_{l=-\infty}^{\infty} c(n+lN)$ is acceptable. Then let $\Delta t = T/N$ form $x(j\Delta t)$, and take its finite Fourier transform. Again, as with the Fourier transform, if we let

$$c_p(n) = \sum_{l=-\infty}^{\infty} c(n + lN)$$

then $c_p(n) \approx c(n)$ for $n = 0, \cdots, N/2$,

and $c_p(N-n) \approx c(-n)$ for $n = -1, \cdots, -N/2$.

## The Calculation of Convolution Integrals

Consider the problem of computing the convolution integral

$$z(t) = \int_{-\infty}^{\infty} x(t-\tau)y(\tau)d\tau. \tag{17}$$

The convolution theorem of Fourier theory states that if

$$x(t) \leftrightarrow a(f)$$
$$y(t) \leftrightarrow b(f)$$
$$z(t) \leftrightarrow c(f)$$

then

$$c(f) = a(f)b(f). \tag{18}$$

The integral (17) may be thought of as describing a process which has an input function, or signal $x(t)$, which is averaged over points $t-\tau$ around each point $t$, with weights given by $y(\tau)$, the "impulse response function." The transform of $y(t)$, $b(f)$, is known as the transfer function and the output of the process is the convolution $z(t)$. For stable systems, the impulse response function $y(t)$ must decrease to zero or to negligible values for large $|t|$. Therefore, let us assume that $y(t)=0$ for $|t|>T_y$. In this case, (17) can be written

$$z(t) = \int_{-T_y}^{T_y} x(t-\tau)y(\tau)d\tau. \tag{19}$$

In a practical situation, one obtains signals such as $x(t)$ in segments of a finite length, say $0\leq t\leq T$. For such a segment, the output can be determined by (17) or (19) for $T_y\leq t\leq T-T_y$.

One can, of course, approximate (19) by numerical integration. The simplest and most convenient formula would be the trapezoidal rule which, applied to (19), gives

$$z(j\Delta t) \approx \Delta t \sum_{k=-K}^{k=K} x[(j-k)\Delta t]y(k\Delta t) \tag{20}$$

where $\Delta t=T/N$, $K=T_y/\Delta t$, and $j=0, 1, 2, \cdots, N-t$. From the convolution property of finite Fourier transforms we know that if $A(n)$ and $B(n)$ are the $N$-point Fourier transforms of $x(j\Delta t)$ and $y(j\Delta t)$, with indexes interpreted modulo $N$, then, the $N$-point Fourier transform of $z(j\Delta t)$ will be given, except for a wrap-around error, by

$$\Delta t N A(n)B(n). \tag{21}$$

This wrap-around error affects values of $j$ in (20) where the index $j-k$ of $x[(j-k)\Delta t]$ goes out of the range $(0, N-1)$. The convolution, so computed, will have been computed as though $x(j\Delta t)$ repeated itself outside the $(0, N-1)$ interval; i.e., for $j<0$, the identity $x(j\Delta t)=x[(N+j)\Delta t]$ is assumed. Thus, $z(j\Delta t)$ is incorrect for $0\leq j<K$ and $N-K\leq j<N$. Valid results are obtained, however, for $K\leq j<N-K$. Therefore, if the next segment of signal is taken to start at $j=N-2K$, and the procedure is repeated, one will have valid outputs for a continuous set of $2(N-2K)$ points. This can be continued on such overlapping segments.

It is desirable to estimate the integration error in the use of the trapezoidal rule as was done earlier for Fourier integrals. To do so, we can assume that $x_p(t)$ and $y_p(t)$ are just periodic repetitions, with period $T$, of $x(t)$ and $y(t)$, respectively, in $[0, T]$ and that no "time-aliasing" is present. Next, we note that the values of $z_p(t)$, the "aliased" function $z(t)$ at sampling points $t_j=j\Delta t$, $j=0, 1, 2, \cdots, N-1$, are given by Theorem 1 as the Fourier synthesis of the "aliased" finite Fourier transform of $z(t)$,

$$Tz_p(j\Delta t) \leftrightarrow c_p(n\Delta f) \text{ (finite Fourier transform).} \tag{22}$$

The procedure considered here to obtain $z(t)$ would be to compute the finite Fourier transforms, $a_p(n\Delta f)$ and $b_p(n\Delta f)$,

$$Tx_p(j\Delta t) \leftrightarrow a_p(n\Delta f)$$
$$Ty_p(j\Delta t) \leftrightarrow b_p(n\Delta f), \tag{23}$$

and multiply them to get an approximation

$$\hat{c}_p(n\Delta f) = a_p(n\Delta f)b_p(n\Delta f) \tag{24}$$

to $c_P(f)$ at the points $f=n\Delta f$, $n=0, 1, \cdots, N-1$.

We define the aliasing error

$$e_a(f) = a_p(f) - a(f),$$

and similarly for $b(f)$ and $c(f)$. In most practical applications, one can assume that

$$e_a(f) \approx a(f+F) + a(f-F). \tag{25}$$

The error in the transform of the sequence $zp(j\Delta t)$ is

$$\hat{c}_p(f) - c_p(f) = [a(f)+e_a(f)][b(f)+e_b(f)] - c_p(f)$$
$$\approx a(f)e_b(f) + b(f)e_a(f), \tag{26}$$

where we neglect the product of the aliasing errors. This error is small if both $e_a(f)$ and $e_b(f)$ are small. Applying Parsevals' theorem this yields a small mean-square error in the convolution.

The second source of error, in going from $c(f)$ to $z(t)$, was just shown to affect only the end regions of $z(t)$. It is eliminated by proper overlap.

The procedure just discussed may be summarized as follows:

1) Determine the sampling rate $\Delta t=1/F$ such that the frequency domain aliasing, given by (26), is small.

2) Segment $x(t)$ into intervals of length $T$ which yield efficient application of the algorithm.

3) Let $X(j)=x(j\Delta t)$ be the input to the FFT subroutine. The output will be

$$A(n) = \frac{1}{T} a_p(n\Delta f).$$

4) Let $Y(j)=y(j\Delta t)$ be the input to the FFT subroutine. The output will be

$$B(n) = \frac{1}{T} b_p(n\Delta f).$$

Perhaps some confusion can be avoided by pointing out again that indexes are to be interpreted modulo $N$ so that $y(j\Delta t)$, for negative $j$, will appear in the input vector in the position corresponding to $Y(N+j)$, i.e., in the top half of the vector.

5) Compute

$$C(n) = TA(n)B(n) = \frac{1}{T} \hat{c}_p(n\Delta f).$$

6) With $C(n)$ as input to the FFT subroutine, do Fourier synthesis to get

$$Z(j) = z_p(j\Delta t), \qquad j = 0, 1, \cdots, N-1.$$

7) Let the next segment of signal start at $T-2T_y$ and repeat the whole procedure. Note that the transform of the weight function $y(t)$ need not be recomputed.

This discussion applies to a system where the impulse response function $y(t)$ is nonzero for both positive and negative $t$. This is appropriate in cases where the input is a disturbance of a physical system or a spatial distribution of illumination, as in image processing. In the latter case, $t$ is a space variable. When $t$ is a time variable and the convolution describes a causal relationship between signal and response, $y(t)$ must be zero for $t<0$; i.e., cause must always precede effect. The only change to be made in the previous discussion is that $y(t)=0$ for $t<0$ so that the lower limit of the integral in (17) is zero. Another effect of this is that there is no wrap-around error for $N-K \leq j < N$, and the output values of $Z(j)$ for these indices are correct. This means that the overlap of segments of signal need only be by an amount $T_y$ rather than $2T_y$, or $K$ instead of $2K$ sampling points.

Finally, one may indeed have a signal which has started at some point in time; i.e., $x(t)=0$ for $t<0$. In this case, the upper limit of integration in (17) is $t$

$$Z(t) = \int_0^t x(t-\tau)y(\tau)d\tau. \tag{27}$$

This is the situation usually treated by Laplace transform methods. Here, it is a special case of the above procedure. Whether treating the first block of a segmented signal or taking the whole signal as a single segment, one would merely set

$$x(t) = 0 \qquad \text{for } -T_y \leq t < 0$$

or, equivalently,

$$x(j\Delta t) = 0 \qquad \text{for } N-K \leq j < N.$$

## The Solution of Integral Equations of the Convolution Type

When dealing with processes described by the convolution integral (17) one often has one or both of two other associated problems: 1) to determine the impulse response $y(t)$ of a system when given an input $x(t)$ and an output $z(t)$; and 2) to determine the input $x(t)$ when given the output $z(t)$ and the impulse response.

These two problems are, in principle, the same, since $x(t)$ and $y(t)$ are interchangeable in the expression of the convolution. However, in practice, there is a difference in the time-domain aliasing which comes from the size of the regions over which $x(t)$ and $y(t)$ are significantly different from zero. Let us, for now, however, suppose that we are solving for $y(t)$ when given $x(t)$ and $z(t)$.

Again, there is a computational advantage is using Fourier transform methods. One computes the Fourier transform $a(f)$ and $c(f)$ and divides to get the transfer function

$$b(f) = c(f)/a(f). \tag{28}$$

If the impulse response function $y(t)$ is wanted, one need only perform Fourier synthesis to compute it from $b(f)$. However, it is often $b(f)$ that is really wanted for the subsequent signal processing.

Now, if $z(t)$ and $x(t)$ are given then we assume $z_p(t)=z(t)$ and $x_p(t)=x(t)$ in the interval $(0, T)$ and, by discrete Fourier transformations, we obtain $c_p(n\Delta f)$ and $a_p(n\Delta f)$, $n=0, 1, \cdots, N-1$. The error in the frequency domain, between the desired transfer function $b_p(f)$ and the computed transfer function $c_p(f)/a_p(f)$, is

$$b_p(f) - \frac{c_p(f)}{a_p(f)} \approx \frac{b(f)e_a(f) + a(f)e_b(f)}{a_p(f)}. \tag{29}$$

This result shows that the "integration error" amounts to an error in the frequency domain which is a linear function of the "overlap" of the frequency spectrum $a(f)$ of the input signal with the "aliasing" error $e_b(f)$ of the transfer function, the "overlap" of the transfer function $b(f)$ and the aliasing error $e_a(f)$ of the input signal, and the aliasing in $c_p(f)$. The error is also inversely proportional to the "aliased" input frequency function $a_p(f)$. If $a(f)$ is very small or zero for some $f$, then there is little or no input at the frequency $f$ and the output signal can convey little or no information about the response of the system to such a frequency.

The result of this analysis is to evaluate the error incurred in converting the problem to a discrete one. It is seen that the numerator reflects the integration error—the aliasing error resulting from the size of $\Delta t$. The second cause of error is the smallness of the input frequency functions. The first of these yields criteria for selecting the sampling rate; the second yields criteria for evaluating or perhaps selecting an input function $x(t)$ for the purpose of accurately determining the transfer function of a system.

# Fast Fourier Transform Method of Computing Difference Equations and Simulating Filters

HOWARD D. HELMS, MEMBER, IEEE

*Abstract*—Two methods for using the fast Fourier transform to reduce the number of arithmetic operations and, therefore the time required for computing discrete, preformulated, and finite convolutions are listed and justified. Under the idealistic assumption that the impulse response of a preformulated difference equation terminates, a theorem is proved that these two methods can be modified to compute such difference equations. This theorem makes plausible the application of these methods when the impulse response does not terminate, provided that the impulse response decays to a small value. In such cases, the fast Fourier transform can be used to compute approximations to the solutions, although usually this use of the fast Fourier transform offers no reduction in the amount of time required for computing the definition of the difference equation. However, if a filtering operation is specified as a frequency response, the fast Fourier transform can be used to compute the filtering operation directly without need of formulating a difference equation, although this simplification is achieved at the cost of a moderate increase (e.g., twice) in the amount of computation time.

## Glossary

$a_k$—convolution sequence appearing in the right side of (1)

$b_k$—sequence derived from $a_k$ as shown in (5)

$B_j$—discrete Fourier transform of $b_k$

$c_n$—recursion sequence appearing in the left side of (1)

$d_n$—sequence derived from $c_n$ as shown before (6)

$D_j$—discrete Fourier transform of $d_k$

DFT—discrete Fourier transform, defined just before (2)

FFT—fast Fourier transform

$g_k$—sequence representing impulse response of (1)

$\tilde{h}_k$—periodic sequence derived from $g_k$ as shown before (7)

IDFT—inverse discrete Fourier transform, defined in (2)

$L$—length of convolution sequence $a_k$

$L'$—length of recursion sequence $c_n$

$L''$—length of impulse response sequence $g_k$

$M$—length of sequence of values of $x_l$

$N$—number of terms and frequencies appearing in the discrete Fourier transform and inverse discrete Fourier transform defined in (2)

$W$—exp $(2\pi\sqrt{-1}/N)$

$X_j$—DFT of $\tilde{x}_l$, as shown just before (2)

$x_l$—independent variable in (1)

$\tilde{x}_l$—sequence derived from $x_l$ and repeating with period $N$; two different definitions relating $\tilde{x}_l$ to $x_l$ are given in the "select-saving" and "overlap-adding" sections

$y_m$—dependent variable in (1)

$z_m$—IDFT of the product of two DFT's, as shown in (3).

Manuscript received January 16, 1967; revised February 22, 1967.

The author is with Bell Telephone Laboratories, Inc., Whippany, N. J.

## Introduction

The following class of preformulated difference equations involving large numbers of terms can often be computed much more rapidly in the frequency domain than in the time domain:

$$\sum_{n=0}^{L'-1} c_n y_{m-n} = \sum_{k=0}^{L-1} a_k x_{m-k}, \tag{1}$$

where $c_n$ and $a_k$ are previously determined finite constants (possibly complex numbers), with $c_0 \neq 0$, the $x$'s (also possibly complex) are the input, with $x_l$ being arbitrary finite numbers for $l=0, 1, \cdots, M-1$, and $x_l=0$ for $l=M, \cdots, \infty$, and $y_m$ the output, with $m=0, \cdots, \infty$. Difference equations of this class are commonly used in digital simulations of filters.

Sande [1] and Stockham [2] have discovered the time savings resulting from using the fast Fourier transform [3] for computing the finite and discrete convolution defined by the right side of the above difference equation. In particular, Stockham discussed this procedure on the assumption that the number $M$ of values of $x_m$ is much greater than $L$. Under the same assumption, the purpose of this paper is to restate one of Stockham's methods (the "overlap-adding" method) and to publish another of his methods (the "select-saving" method discovered independently by the author). Also, this paper discusses how these methods might be applied to computing *both* sides of (1). Frequency response methods for using the fast Fourier transform to simulate filters without even formulating (1) are described.

"Fast Fourier transform" (FFT) denotes a rapid technique [3] for computing the "discrete Fourier transform" (DFT) $X_j$ of $\tilde{x}_l$, as follows:

$$X_j \doteq \sum_{l=0}^{N-1} \tilde{x}_l W^{-jl} \qquad W \doteq \exp\frac{2\pi\sqrt{-1}}{N},$$

$$j = 0, \cdots, N-1.$$

The $\tilde{x}_l$ often are equally spaced samples of a function $\tilde{x}(t)$, so that $l$ represents time, and $j$ represents fre-

Reprinted from *IEEE Trans. Audio Electroacoust.*, vol. AU-15, pp. 85–90, June 1967.

quency. There exists an inverse DFT (IDFT), which is

$$\bar{x}_l = \frac{1}{N}\sum_{j=0}^{N-1} X_j W^{jl}, \qquad l = 0, \cdots, N-1. \tag{2}$$

If $l$ in (2) is replaced by a continuous variable, the resulting equation interpolates an $N$th-degree trigonometric polynomial through the $N$ original values of $\bar{x}$. The FFT can also be used to compute the IDFT, because the IDFT of $X_j$ can be rearranged into a DFT by replacing the subscript $j$ by the subscript $k=N-j$, giving

$$\frac{1}{N}\sum_{j=0}^{N-1} X_j W^{jl} = \frac{1}{N}\sum_{k=0}^{N-1} X_{N-k} W^{(N-k)} = \frac{1}{N}\sum_{k=0}^{N-1} X_{N-k} W^{-kl}.$$

Thus, the IDFT of a sequence is $1/N$ times the DFT of the reverse of the sequence.

Instead of requiring of the order of $N^2$ operations, the FFT requires only of the order of $N \log_2 N$ operations, resulting in proportionate savings in computation time. The integer $N$ appearing in the DFT must be a highly composite number (i.e., the number must have many factors). In the most commonly used programs for the FFT, $N$ is an integral power of 2. Roundoff errors arising in the computation of the FFT are usually negligible whenever the FFT is computed using floating point arithmetic [2].

## THE "FAST CONVOLUTION"

The following theorem permits using the FFT to compute the right side of (1) more rapidly than computing the right side of (1) straightforwardly. This use of the FFT is sometimes called the "fast convolution."

*Theorem:* The inverse DFT (IDFT) of the product of two DFT's produces a convolution in which the terms appear cyclically with period $N$.

*Proof:* Let $X_j$ be defined as above and $B_j$ be the DFT of some sequence $b_l$, where $l=0, \cdots, N-1$. Then, letting $z$ represent the IDFT,

$$z_m \doteq \frac{1}{N}\sum_{j=0}^{N-1} B_j X_j W^{jm}, \tag{3}$$

where $m=0, 1, \cdots, N-1$.

Substituting for $B_j$ and $X_j$ and reversing the summations gives

$$z_m = \sum_{k=0}^{N-1}\sum_{l=0}^{N-1} b_k \bar{x}_l \frac{1}{N}\sum_{j=0}^{N-1} W^{j(m-k-l)}.$$

But

$$\sum_{j=0}^{N-1} W^{j(m-k-l)}$$

is a finite geometric series, for which

$$\sum_{j=0}^{N-1} W^{j(m-k-l)} = \frac{1 - W^{N(m-k-l)}}{1 - W^{(m-k-l)}}$$

$$= \begin{cases} N & \text{if } m-k-l = pN, \quad p = \text{an integer}, \\ 0 & \text{otherwise.} \end{cases}$$

Since $m$, $k$, and $l$ vary from 0 to $N-1$, the only permissible values of $p$ become 0 and $-1$. Substituting this result into the previous equation for $z_m$ and summing over all values of $l$ gives

$$z_m = \sum_{k=0}^{m} b_k \bar{x}_{m-k} + \sum_{k=m+1}^{N-1} b_k \bar{x}_{N+m-k}. \tag{4}$$

The last term on the right side of this equation contains the "cyclical" part of the convolution. As derived here, $x$ is the cyclical variable passing from $\bar{x}_0$ to $\bar{x}_{N-1}$ as $k$ passes from $m$ to $m+1$. This convolution can be made cyclical in $b$ instead of $\bar{x}$ by interchanging $\bar{x}$ and $b$ in all the previous steps. Q.E.D.

The "select-saving" and "overlap-adding" methods of avoiding the cyclical nature of this convolution and of minimizing the required number of computer operations are presented next. Both of these methods construct the right side of (1) by decomposing it into many smaller convolutions. In these methods, the juxtaposition or summation of these smaller convolutions produces the desired right side of (1).

## "SELECT-SAVING" METHOD

The term "select-saving" denotes the following method for using the fast Fourier transform to compute the right side of (1):

1) A number $N$ is chosen, where $N \geq L$. (The choice of $N$ is discussed in a later section.)

2) $N-L$ zeros are added to the sequence $a_k$, producing a sequence $b_k$ defined as

$$b_k = \begin{cases} a_k & \text{for } k = 0, 1, \cdots, L-1 \\ 0 & \text{for } k = L, L+1, \cdots, N-1. \end{cases} \tag{5}$$

3) The DFT's of $b_k$ and a sequence $\bar{x}_l$ defined as the first $N$ values of $x_l$ are calculated using the FFT and are multiplied together at corresponding frequencies.

4) The IDFT of the products in (3) are computed by the FFT.

5) The values of $z_m$ are regarded as valid data and are retained only for the following $N-L+1$ values of $m$:

$$m = L-1, L, L+1, \cdots, N-1.$$

The results are regarded as invalid and are discarded for the following values of $m$:

$$m = 0, 1, \cdots, L-2.$$

The term "select-saving" arises from this operation.

6) The sequence $\bar{x}_l$ is redefined by discarding the first $N-L+1$ values of $\bar{x}_l$ and the corresponding values of $x_l$, then by reducing the subscripts of these $L-1$ remaining members of $\bar{x}_l$ by $N-L+1$, and finally by ap-

pending the next $N-L+1$ values of $x_l$ to the remaining original $L-1$ members of the sequence $\tilde{x}_l$.

7) Steps 3) through 6) are repeated until at least $M+L-1$ values of $x_l$ have been used.

8) All of the values of $z_m$ retained in the repetitions of step 5) are juxtaposed to form a single sequence.

*Theorem:* The sequence obtained in step 8) is the sequence determined by the right side of (1).

*Remark:* All values of the right side of (1) are obtained except for $L-1$ values at the beginning. If desired, these values can be obtained by prefacing the $M$ values of $x_l$ with $L-1$ zeros, thereby defining a sequence of $M+L-1$ values of $x_l$ which is used in place of the original sequence of values of $x_l$.

*Proof:* The second term on the right side of (4) vanishes if $m$ is restricted to the interval $L-1 \leq m \leq N-1$, because $k>m$ in that term, and $b_k=0$ for $k \geq L$. The first term in (4) provides a subsequence of $N-L+1$ values of the right side of (1). The juxtaposition of these subsequences provides the whole sequence of values of the right side of (1). Q.E.D.

## "Overlap-Adding" Method

The term "overlap-adding" denotes the following method of computing the right side of (1):

Steps 1) and 2) are the same as in the "select-saving" method.

3) The DFT's of $b_k$ and the sequence $\tilde{x}_l$ created by appending $L-1$ zeros to the first $N-L+1$ values of $x_l$ are calculated using the FFT and are multiplied together at corresponding frequencies.

4) The IDFT (3) is computed by the FFT.

5) The whole sequence of $N$ values of $z_m$ is shifted $N-L+1$ units forward and added to the previous accumulation of $z_m$'s. (The accumulation is assumed to be initially zero.) The shifting and accumulation in this step are the origin of the term "overlap-adding."

6) The sequence $\tilde{x}_l$ is redefined by replacing the previously used $N-L+1$ values of $x_l$ by the next $N-L+1$ values of $x_l$.

7) Steps 3) through 6) are continued until at least $M+L-1$ values of $x_l$ are exhausted.

*Theorem:* The final accumulation of values of $z_m$ is the sequence determined by the right side of (1).

*Proof:* This is in two parts. First, each of the subsequences of $N$ values of $z_m$ obtained from step 4) represent all of the nontrivial values of $z_m$ which arise in response to the $N-L+1$ values of $x_m$ used. This can be observed from (4) whose second term is zero because $b_k$ and $\tilde{x}_l$ are defined to have nonzero terms only at the beginning of their sequences. Specifically, as $k$ increases $b_k$ begins to be zero no later than when $\tilde{x}_{N+m-k}$ ceases being nonzero. This can be verified from the fact that when $k=L-1$, $b_k$ is at its last nonzero value, and the index $N+m-k$ of $\tilde{x}$ is at least as great as $N-L+1$, and $\tilde{x}_{N-L+1}$ is the beginning of the portion of $\tilde{x}_l$ which is defined to be zero. Second, the final sequence of accumulated values of $z_m$ represents the whole right side of (1), because (1) is linear in $x_m$.

## Choice of $N$ to Minimize the Number of Computations

The number of real (as opposed to complex) multiplications required for computing the FFT for $N$ complex numbers is asymptotically $2N \log_2 N$, if $N$ is chosen to be an integral power of 2. The number of additions is $3N \log_2 N$. The running time is then approximately $(2\alpha+3\beta)N \log_2 N$, where $\alpha$ and $\beta$ are the times required for a multiplication and an addition operation, respectively. The FFT must be repeated approximately $2M/(N-L+1)$ times. $N$ can be chosen as an integral power of 2 in a way which minimizes the ratio $N \log_2 N/(N-L+1)$. This was done numerically; the results are presented in Table I. This table indicates that the optimum values of $N$ is approximately $L \log_2 L$. The optimum value of $2MN \log_2 N/(N-L+1)$ is approximately $3M \log_2 L$, corresponding to a running time of approximately $(6\alpha+9\beta)M \log_2 L$. This can be compared with the running time of $4(\alpha+\beta)ML$ required to compute the right side of (1), if the sequences $x_l$ and $a_k$ are composed of complex numbers.

Departures from the optimum value of $N$ can be made without greatly increasing the running time [7]. For example, if $N$ is chosen equal to $2L$, the running time is increased by only a factor of 4/3. The use of such departures from the optimum may be forced by limitations on the amount of computer memory available.

Sande [1] showed that if the sequences $x_l$ and $a_k$ are composed of purely real numbers, a factor of 2 reduction in the number of operations can be achieved by combining pairs of successive subsequences of values of $x_l$ into complex sequences by multiplying one subsequence by $\sqrt{-1}$ and adding it to the other subsequence. The real and imaginary parts of $z_m$ then represent two separate subsequences of values of $z_m$, each of which is treated as a sequence of values of $z_m$ occurring in the "select-saving" or "overlap-adding" method.

TABLE I

Table of Optimum Values of $N$ and the Resulting Optimum Numbers of Operations, on the Assumption that $N$ is a Power of 2

| $L_{min}$ through $L_{max}$ | $n_{opt}$ | $N_{opt}$ | $\dfrac{2 \log_2 N_{opt}}{1 - \dfrac{L-1}{N_{opt}}}$ |
|---|---|---|---|
| <11 | 5 | 32 | <13.9 |
| 11– 17 | 6 | 64 | 14.22–16 |
| 18– 29 | 7 | 128 | 16–18 |
| 30– 52 | 8 | 256 | 18–20 |
| 53– 94 | 9 | 512 | 20–22 |
| 95– 171 | 10 | 1024 | 22–24 |
| 172– 310 | 11 | 2048 | 24–26 |
| 310– 575 | 12 | 4096 | 26–28 |
| 575–1050 | 13 | 8192 | 28–30 |
| 1050–2000 | 14 | 16 384 | 30–32 |
| 2000–3800 | 15 | 32 768 | 32–34 |
| 3800–7400 | 16 | 65 536 | 34–36 |
| >7400 | 17 | 131 072 | ≥36 |

Example: Let the number $L$ of terms equal 100. Then, by looking at the entries corresponding to the 95–171 entry (under $L_{min}$ through $L_{max}$), it is seen that the optimum value of $2N \log_2 N/(N-L+1)$ is approximately 22.

### Application of "Overlap-Adding" and "Select-Saving" to Computing Both Sides of (1)

There are two cases to be considered: 1) The impulse response of $y_m$ terminates in a finite number of samples; and 2) the impulse response of $y_m$ decays geometrically to below a value arbitrarily small but never completely terminates. The following theorem refers to the first case. In the section following this section, this theorem provides guidance and plausibility for the treatment of the second case.

*Theorem:* If the values of $y_m$ in response to $x_0=1$ and $x_l=0$, with $l\neq 0$, vanish for all $m\geq L''$, thereby making $L=L'+L''-1$, and if $N$ is chosen so that $N\geq L=L'+L''-1$, then $y_m$ can be computed by either the "overlap-adding" or "select-saving" method, which are both modified so that $y_m$ replaces $z_m$ and is equal to the IDFT of $B_jX_j/D_j$, where $D_j$ is the DFT of a sequence $d_n$ defined as

$$d_n = \begin{cases} c_n & \text{for } n = 0, 1, \cdots, L'-1 \\ 0 & \text{for } n = L', L'+1, \cdots, N-1. \end{cases}$$

*Proof:* In the following, the sequence $g_k$ represents the (impulse) response of $y_m$ to $x_0=1$ and $x_l=0$, $l\neq 0$. Therefore, solutions $y_m$ of (1) can be written

$$y_m = \sum_{k=0}^{L''-1} g_k x_{m-k}. \qquad (6)$$

Defining

$$\tilde{h}_k = \begin{cases} g_k, & k = 0, \cdots, L''-1 \\ 0, & k = L'', \cdots, N-1, \end{cases}$$

where the tilde indicates that the sequence is assumed to repeat with period $N$, values of $y_m$ given by the above equation can be obtained from

$$\tilde{y}_m = \sum_{k=0}^{N-1} \tilde{h}_k \tilde{x}_{m-k}, \qquad (7)$$

provided that 1) for "select-saving," only those values of $y_m$ for $m=L''-1, \cdots, N-1$ are retained, or 2) for "select-adding," $\tilde{x}_m\neq 0$ only for $m=0, \cdots, N-L''$.

*Lemma:* All values of $\tilde{y}_m$ (including the invalid values discarded in the "select-saving" method) determined by (7) can be obtained as solutions to the equation

$$\sum_{k=0}^{N-1} d_k \tilde{y}_{m-k} = \sum_{q=0}^{N-1} b_q \tilde{x}_{m-q}, \quad m = 0, 1, \cdots, N-1, \qquad (8)$$

provided that $N\geq L$.

*Proof:* If (6) is substituted into (1) and if coefficients of the same values of $x_m$ are equated, the result is

$$\sum_{k=\max(0,m-L''+1)}^{\min(m,L'-1)} c_k g_{m-k} = a_m, \quad m = 0, \cdots, L-1. \qquad (9)$$

In view of the definition of $b_m$, $d_m$, and $\tilde{h}_m$, this equation can be written as

$$\sum_{k=0}^{N-1} d_k \tilde{h}_{m-k} = b_m, \qquad (10)$$

where now $m=0, \cdots, N-1$, provided that $N\geq L$. Equation (10) is equivalent to (9) because

$$\sum_{k=0}^{N-1} d_k \tilde{h}_{m-k} = \sum_{k=0}^{m} d_k \tilde{h}_{m-k} + \sum_{k=m+1}^{N-1} d_k \tilde{h}_{N+m-k} = \sum_{k=\max(0,m-L''+1)}^{\min(m,L'-1)} c_k g_{m-k},$$

since

$$\sum_{k=m+1}^{N-1} d_k \tilde{h}_{N+m-k}$$

is always zero (for the same reason that the last term in (4) is zero in the proof of the "overlap-adding" method). The only difference that could exist between (9) and (10) occurs when $N>L$, but no difference exists since both sides of (10) reduce to zero for $m=L, L+1, \cdots, N-1$.

Replacing $m$ by $q$, multiplying both sides by $\tilde{x}_{m-q}$, and summing from $q=0$ to $q=N-1$ transforms (10) into

$$\sum_{k=0}^{N-1} \sum_{q=0}^{N-1} d_k \tilde{h}_{q-k} \tilde{x}_{m-q} = \sum_{q=0}^{N-1} b_q \tilde{x}_{m-q}.$$

It is possible to define $l=q-k$ and, because of the periodicity of $\tilde{x}$, to perform the summation over $l$ instead of $q$, producing

$$\sum_{k=0}^{N-1} d_k \sum_{l=0}^{N-1} \tilde{h}_l \tilde{x}_{m-l-k} = \sum_{q=0}^{N-1} b_q \tilde{x}_{m-q}. \qquad (11)$$

Substituting (7) into (11) produces (8), proving the lemma.

To conclude the proof of the theorem, the IDFT of $B_jX_j/D_j$ gives the sequence $\tilde{y}_m$ for $m=0, \cdots, N-1$. This can be seen by substituting the DFT of $\tilde{y}_m$ and $\tilde{x}_m$ into (8), producing after summing over $k$ and $q$

$$D_jY_j = B_jX_j, \quad j = 0, \cdots, N-1. \qquad (12)$$

Concerning whether it is permissible to obtain $Y_j$ by dividing both sides of (12) by $D_j$, (12) was obtained by invertible substitutions into (8), and the lemma implies that all solutions of $\tilde{y}$ obtained from (8) are identical to the $N$ values of $\tilde{y}$ obtained from (7). Therefore, because each of the $N$ values of $Y$ is determined as the DFT of $\tilde{y}$, (12) must possess solutions for all values of $j$ and, thus, $D_j$ cannot be equal to zero. Q.E.D

In summary, the theorem just proved permits computing a preformulated difference equation by the "select-saving" or "overlap-adding" method, as modified in the three following ways: 1) The output $z_m$ (actually $y_m$) is obtained from the DFT of $B_jX_j/D_j$ instead of $B_jX_j$; 2) $L''$ is used in place of $L$; and 3) there is a new but usually inconsequential requirement on $N$, viz., $N\geq L'+L''-1$.

### Computing Both Sides of (1) If Its Impulse Response Does Not Terminate

In the usual case in which the response of (1) to an impulse does not terminate, the method contained in the statement of the previous theorem would seem to be capable of being applied as before, except that $L''$ would be the maximum number of contiguous values of $g_m$ which are not "very small." An appropriate value of $L''$ might be found by inspecting $g_m$, with $g_m$ possibly being obtained by computing the IDFT's of $B_j/D_j$ for a sufficiently large value of $N$. If $L''$ is chosen large enough, the error contributed by the last term in (4), due to the nonzero values of $g_m$ for $m \geq L''$, should be negligible.

The errors contributed by the last term in (4) can of course be avoided by setting $g_m$ (obtained from the IDFT of $B_j/D_j$) equal to zero for the $N-L''$ contiguous values of $m$ for which $g_m$ is "very small." Application of the "overlap-adding" or "select-saving" method to this truncated sequence of values of $g_m$ results in no contribution from the last term in (4). A suitable value of $L''$ can be found by increasing $L''$ until further increases no longer produce significant changes in the output $y_m$.

To summarize, the errors (or sins) of commission in the method of the first paragraph are replaced by the errors (or sins) of omission in the method of the second paragraph.

### Methods of Simulating Filters

If a filtering operation is specified as an impulse response which decays to a "small enough" value, the "select-saving" or "overlap-adding" method can be applied as in the previous section. If a filtering operation is specified as a frequency response (i.e., an amplitude response or a phase response or both), it is often possible to obtain from the frequency response an impulse response which can be treated as in the previous section. For example, Ormsby [4] describes a method of choosing the number of terms $L$ and the sequence $g_m$ so that lowpass or bandpass filtering is achieved with given tolerances on the error in the passband and stopband.

Another alternative for the situation in which the filtering operation is specified as a frequency response is to use the FFT to compute the impulse response and then to apply the methods of the previous section. The following section summarizes this use of the FFT combined with using the truncation method described in the second paragraph of the previous section.

### "Four-T's" Method

The "Four-T's" (i.e., transform-truncate-transform-test) method for using the FFT to simulate filters is:

1) If only the amplitude response or the phase response of the filtering operation is preformulated, the missing response is formulated arbitrarily, and the frequency response is computed by multiplying the amplitude response by $\exp(\sqrt{-1}\theta)$, where $\theta$ is the phase response.

2) If the frequency response is not already available as a preformulated sequence of samples, a suitably large number $N$ of samples is chosen and the frequency response is sampled at $N$ equally spaced values of frequency.

3) Subscripts from 0 through $N-1$ are assigned to the samples. For most applications, the first $N/2$ samples will be assigned subscripts from $N/2$ through $N-1$ (e.g., Fig. 1, with its abscissa multiplied by $N/2$), so that the spectrum will have complex conjugate symmetry around zero frequency.

4) The IDFT of the frequency response is computed. This produces an impulse response $a_k$, $k=0, \cdots, N-1$.

5) A number $L$ having been chosen arbitrarily, $N-L$ contiguous values of the impulse response $a_k$ are set equal to zero. (The values $a_0$ and $a_{N-1}$ are considered contiguous.) The subscripts of the remaining $L$ values of the impulse response $a_k$ are cyclically increased or decreased by the same amount (modulus $N$) so that these $L$ values occupy the first $L$ places of the sequence, thereby producing a sequence $b_k$ as defined in (5).

6) The DFT of the sequence $b_k$ obtained in the previous step is computed, producing a frequency response $B_j$.

7) The amplitude response and/or phase response or both is obtained as $B_j$ and/or arctan Im $B_j$/Re $B_j$, respectively, and is compared with the predetermined amplitude and/or phase response.

8) If the agreement between the amplitude and/or phase response obtained in the previous step is not close enough, steps 5) through 8) are repeated with a larger value of $L$ (e.g., twice as large). This process terminates when the test comprising the previous step is passed, so that there is satisfactory agreement between the two frequency responses.

9) The filtering operation is performed using the "select-saving" or "overlap-adding" method, with the value of $L$ used in the last repetition of step 5) and with the values of $B_j$ produced by the last repetition of step 6).

An example of the "Four-T's" approach is contained in Fig. 1, whose $L=\infty$ curve depicts a seventh-order, $\frac{1}{2}$-dB ripple, and Chebyshev approximation [5] to an ideal lowpass characteristic (no attenuation below the cutoff frequency, infinite attenuation above the cutoff frequency, and no phase shift). Figure 1 demonstrates an unsophisticated but apparently satisfactory criterion for the comparison in step 7) of the "Four-T's" method. This criterion is that one's eyes see no difference between graphs of the two frequency responses. Thus, the $L=\infty$ curve appears to be duplicated by the $L=133$ curve but does not appear to be duplicated by the $L=73$ curve. (These three curves are displaced 5dB apart for ease in viewing.)

The FFT is efficient enough so that a filtering operation with a value of $L$ as large or larger than 133 could

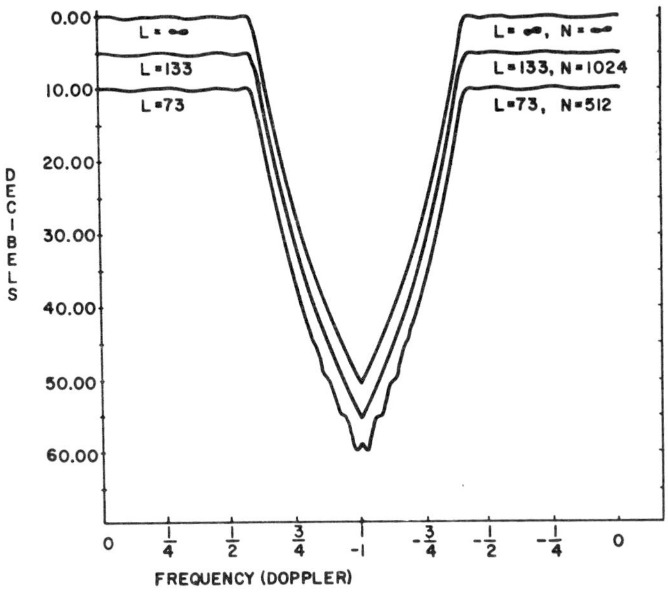


Fig. 1. Example of the "Four-T's" method.

be computed in only moderately more time than that required for computing a difference equation designed to perform a similar operation. For instance, the $L=133$ value in the above example would require approximately $\frac{1}{2}(6\alpha+9\beta)\log_2 L \doteq 21\alpha+32\beta$ units of time. If $\alpha=2\beta$, as it is in digital computers such as the IBM 7094, this is nearly twice the $7(2\alpha+2\beta)=14\alpha+14\beta$ units of time required for computing the response of the parallel representation of a recursive difference equation producing the Chebyshev characteristic nearly identical to the $L=\infty$ curve in Fig. 1 [7]. The comparison of these two methods assumes that the input sequence $x_l$ consists of real numbers. The factor of $\frac{1}{2}$ multiplying the parenthesis comes from taking advantage of this assumption and also of the fact that the filtering sequence was con structed to consist of real numbers. Consequently, by constructing $\bar{x}$ as $N-L+1$ values of $x_l$ plus $\sqrt{-1}$ times the next $N-L+1$ values of $x_l$, and using the real and imaginary parts of $z_m$ as separate output sequences to be combined by the "overlap-adding" method, the $M$ values of $x_l$ are exhausted twice as rapidly.

The advantage in using the "Four-T's" method is its simplicity. Thus, no tables of $z$ transforms need to be consulted. Also, factoring of the denominator of the filtering characteristic is unnecessary as a means of avoiding the roundoff problems arising in using the $z$ transform to formulate a difference equation [8]. Moreover, if a numerical specification of the filtering characteristic is specified, it is unnecessary to determine the poles and zeros of a transfer function that approximates this characteristic.

## Conclusion

If the desired frequency response of a filtering operation is specified, the "Four-T's" method is capable of formulating a filtering operation so that a simulation of it can be computed by the "select-saving" or "overlap-adding" version of the fast Fourier transform. At moderate increase in computing time, this procedure avoids the complexities in algebra and theory associated with determining the coefficients of a convolution or difference equation.

## Acknowledgment

Many of the ideas contained herein were stimulated by conversations with Dr. P. D. Welch of IBM and Dr. T. G. Stockham of Lincoln Labs. Many suggestions by Dr. J. F. Kaiser of Bell Telephone Labs. were included in this paper.

## References

[1] G. Sande, "On an alternative method of calculating covariance functions," unpublished paper, July 1965.
[2] T. G. Stockham, "High-speed convolution and correlation," *Proc. of 1966 Spring Joint Computer Conf., AFIPS Proc.*, vol. 28. Washington, D. C.: Spartan, 1966, pp. 229–233.
[3] J. W. Cooley and J. W. Tukey, "An algorithm for the machine calculation of Fourier series," *Math. of Comput.*, vol. 19 pp. 297–301, April 1965.
[4] J. F. A. Ormsby, "Design of numerical filters with applications to missile data processing," *J. ACM*, vol. 8, pp. 440–466, July 1961.
[5] L. Weinberg, *Network Analysis and Synthesis*. New York: McGraw-Hill, 1962, p. 488.
[6] J. W. Cooley, "Applications of the fast Fourier transform method," presented at the IBM Scientific Computing Symp. on Digital Simulation of Continuous Systems, June 20–22, 1966.
[7] J. F. Kaiser, private communication, January 8, 1967.
[8] J. F. Kaiser, *System Analysis by Digital Computer*. New York: Wiley, 1966, sect. 7.6.1. and 7.6.2.

# Discrete Representation of Signals

ALAN V. OPPENHEIM, SENIOR MEMBER, IEEE, AND DONALD H. JOHNSON

*Abstract*—**In processing continuous-time signals by digital means, it is necessary to represent the signal by a digital sequence. There are many ways other than periodic sampling for obtaining such a sequence. The requirements for such representations and some examples are discussed within the framework of simulating linear time-invariant systems. The representation of digital sequences by other digital sequences is also discussed, with particular emphasis on the use of such representations to implement a nonlinear warping of the digital frequency axis. Some applications and hardware implementation of this digital-frequency warping are described.**

## I. Introduction

With the increasing speed and decreasing cost of digital hardware, digital methods are playing an important role in signal processing. In some applications, the input and output are continuous-time signals but the processing is digital. In such cases, the input must first be represented by a sequence; after this sequence is processed, the output sequence must then be reconverted to a continuous-time signal. In other instances (such as spectral analysis), the objective of the signal processing is a set of measurements on a continuous-time signal; the input must be converted to a digital sequence but it is unnecessary to convert back to a continuous-time signal. There is, of course, a variety of applications in which the signals to be processed are inherently discrete signals. If conceptually useful, they can be viewed as representing continuous-time signals; however, this often introduces additional complications that may not be warranted. An example of such a signal is arithmetic roundoff noise or limit cycles generated in a digital filter.

When the signal processing involves a continuous-time input or output, the continuous-time signal must be represented by a discrete-time signal, i.e., a sequence. The most common procedure when the continuous-time signal is band-limited is to choose the sequence values to be samples of the continuous-time function equally spaced in time. This representation is commonly referred to as periodic sampling. There are, however, many ways other than periodic sampling in which a continuous-time function can be represented by a sequence.

In this paper a class of representations of continuous-time functions by sequences is discussed. In addition, a parallel notion is developed—the representation of a sequence by other sequences. Many of the results presented appear elsewhere by the present authors and by others. Thus to some extent this paper can be considered as a review of these results. A major objective is to provide this review within a common framework and, toward this end, results of other authors are combined with our own results and point of view.

Manuscript received November 12, 1971; revised March 13, 1972. This work was supported in part by the Joint Services Electronics Program under Contract DAAB07-71-C-0300, in part by the Advanced Research Projects Agency under Lincoln Laboratory Purchase Order CC-570, and in part by the National Science Foundation under Grant GK-31353.

The authors are with the Department of Electrical Engineering and Research Laboratory of Electronics, Massachusetts Institute of Technology, Cambridge, Mass. 02139.

In Section II, a discussion of the discrete representation of continuous-time signals for the digital simulation of linear time-invariant continuous-time filters is presented, and several examples of such representations are discussed. In Section III, discrete representation of discrete-time signals is discussed, corresponding to the represention of one digital sequence by another digital sequence. The development of the representation of discrete-time signals parallels closely the development of the representation of continuous-time signals. For the representation of continuous-time signals, it is shown that requiring the representation to map linear time-invariant continuous-time systems to linear shift-invariant discrete systems is equivalent to requiring that the representation map the Laplace transform of the continuous-time signals to the $z$-transform of the discrete-time signal by a substitution of variables. Consequently, in addition to its application to simulation, such a representation is useful in instances such as spectral analysis where it is desirable for the analog frequency axis to map to the digital-frequency axis.

In the representation of discrete-time signals by other discrete-time signals, the notion of mapping the $z$-transform by a substitution of variables assumes the major emphasis. In particular, it is possible to maintain the *form* of the spectrum for a discrete-time signal but transform the frequency axis in a nonlinear manner. The ability to do this has potential application in a number of contexts, such as unequal resolution and vernier spectral analysis, and the correction of the frequency distortion inherent in divers' speech due to the effect of pressure and the content of their breathing mixture.

## II. Discrete Representation of Continuous-Time Signals

In this section, the discussion is directed toward the simulation of signal processing techniques which can be carried out with a linear time-invariant filter. While there are many examples of signal processing which do not fall into this category, linear filtering does represent a wide class of signal-processing problems. Just as linear time-invariant filtering plays an important role in continuous-time signal processing, linear shift-invariant digital filtering plays an important role in digital signal processing. The reason for this lies partly in the fact that linear time-invariant systems, either in the continuous-time or in the discrete-time cases, are analytically manageable and consequently much insight into processing signals with such systems has been gained. For this reason it is desirable to choose a discrete representation for continuous-time signals which will permit linear time-invariant systems to be implemented digitally with linear shift-invariant discrete systems. This is similar to the condition imposed by Steiglitz in demonstrating the isomorphism between digital and analog signal processing [1].

A linear time-invariant filter is conveniently characterized in terms of the convolution integral so that if $x(t)$, $y(t)$, and $h(t)$ are the system input, output, and impulse response, respectively, then

Reprinted from *Proc. IEEE*, vol. 60, pp. 681–691, June 1972.

$$g(t) = \int_{-\infty}^{+\infty} f(\tau)h(t-\tau)d\tau \triangleq f(t) \otimes h(t) \tag{1}$$

where $\otimes$ is used to denote continuous-time convolution. Let us denote the discrete representation of $f(t)$, $g(t)$, and $h(t)$ by $f_n$, $g_n$, and $h_n$, respectively. We want $f_n$, $g_n$, and $h_n$ to correspond to the input, output, and unit sample response of a discrete linear shift-invariant system, so that $g_n$ must be the discrete convolution of $f_n$ and $h_n$:

$$g_n = \sum_{k=-\infty}^{+\infty} f_k h_{n-k} \triangleq f_n * h_n \tag{2}$$

where $*$ is used to denote discrete convolution.

The validity of (2) for any $f_n$ and $h_n$ implies that the representation of a continuous-time function by a sequence must be linear, i.e., if $x_n$ and $y_n$ are the representations of $x(t)$ and $y(t)$, then $x_n+cy_n$ is the representation of $x(t)+cy(t)$ where $c$ is an arbitrary constant. This "linearity property" can be derived as follows. Let $f_n$ denote the discrete representation of $f(t)=x(t)+cy(t)$. Then from (1) and (2),

$$g(t) = f(t) \otimes h(t)$$
$$g_n = f_n * h_n$$

since

$$f(t) = x(t) + cx(t),$$
$$f(t) = x(t) \otimes h(t) + cy(t) \otimes h(t)$$

so that

$$\begin{aligned} f_n &= x_n * h_n + cy_n * h_n \\ &= [x_n + cy_n] * h_n. \end{aligned} \tag{3}$$

Since we require (3) to hold for any $h_n$, it follows that

$$f_n = x_n + cy_n.$$

Now let us express an arbitrary sequence $f_n$ as a linear combination of weighted delayed unit samples:

$$f_n = \sum_{k=-\infty}^{+\infty} f_k \delta_{n-k}$$

where $\delta_{n-k}$ represents a sequence whose values are zero except for $n=k$ so that $\delta_0=1$. Let $\phi_k(t)$ denote the continuous-time function represented by the sequence $\delta_{n-k}$. Then by virtue of the linearity property previously derived,

$$f(t) = \sum_{k=-\infty}^{+\infty} f_k \phi_k(t)$$

or, for convenience, changing the summation index,

$$f(t) = \sum_{n=-\infty}^{+\infty} f_n \phi_n(t). \tag{4}$$

Alternatively, with $F_L(s)$ and $\Phi_n(s)$ designating the Laplace transforms of $f(t)$ and $\phi_n(t)$, then

$$F_L(s) = \sum_{n=-\infty}^{+\infty} f_n \Phi_n(s). \tag{5}$$

The relationship between $f(t)$ and $f_n$ in (4) can be viewed as an expansion of $f(t)$ in terms of a set of functions $\{\phi_n(t)\}$; for convenience we will require that given the set of functions $\{\phi_n(t)\}$, the sequence $f_n$ representing $f(t)$ is unique. This then requires that the set of functions $\{\phi(t)\}$ be linearly independent, i.e., that any one of the functions is not expressible as a linear combination of the others. In this discussion, we have not assumed that the set of functions $\{\phi_n(t)\}$ is complete, but we have assumed that $f(t)$ is in the space spanned by $\{\phi_n(t)\}$.

Equations (1) and (2) can now be used to derive necessary and sufficient conditions on the set of functions $\{\Phi_n(s)\}$ in (5). Toward this end, let $G_L(s)$, $F_L(s)$, and $H_L(s)$ denote the Laplace transforms of $g(t)$, $f(t)$, and $h(t)$, respectively. Then (1) can be rewritten as

$$G_L(s) = F_L(s)H_L(s). \tag{6}$$

From (5) and its counterparts for $G_L(s)$ and $H_L(s)$, it follows that

$$\sum_{n=-\infty}^{+\infty} g_n\Phi_n(s) = \sum_{r=-\infty}^{+\infty} \sum_{k=-\infty}^{+\infty} f_k h_r \Phi_r(s)\Phi_k(s). \tag{7}$$

If we carry out the substitution of variables $n=r+k$ on the right-hand side of (7), the equation can be rewritten as

$$\sum_{n=-\infty}^{+\infty} g_n\Phi_n(s) = \sum_{n=-\infty}^{+\infty} \sum_{k=-\infty}^{+\infty} f_k h_{n-k} \Phi_{n-k}(s)\Phi_k(s). \tag{8}$$

But using (2) to express the left-hand side of (8) in terms of $f_n$ and $h_n$, we obtain

$$\sum_{n=-\infty}^{+\infty} \sum_{k=-\infty}^{+\infty} f_k h_{n-k}\Phi_n(s) = \sum_{n=-\infty}^{+\infty} \sum_{k=-\infty}^{+\infty} f_n h_{n-k} \Phi_{n-k}(s)\Phi_k(s). \tag{9}$$

In order that (9) may hold for any sequences $f_k$ and $h_{n-k}$, we require that

$$\Phi_n(s) = \Phi_{n-k}(s)\Phi_k(s) \tag{10}$$

or equivalently

$$\Phi_n(s)\Phi_m(s) = \Phi_{n+m}(s) \tag{11}$$

from which a general form for the functions $\{\Phi_n(s)\}$ can be derived. In particular,

$$\Phi_{n+1}(s) - \Phi_1(s)\Phi_n(s) = 0$$

which is a difference equation in $\Phi_n(s)$ with a general solution of the form

$$\Phi_n(s) = c[\Phi_1(s)]^n. \tag{12}$$

Furthermore, from (11), $\Phi_0^2(s)=\Phi_0(s)$, which is satisfied only for $c=0$ and $c=1$. Thus the nontrivial solution to (12) is

$$\Phi_n(s) = [\Phi_1(s)]^n. \tag{13}$$

Consequently, (13) must be satisfied if the expansion in (5) is to result in a discrete representation for which continuous convolution is mapped to discrete convolution. The condition on $\Phi_1(s)$ under which the set of functions $\{\phi_n(t)\}$, corresponding to $\{\Phi_n(s)\}$ as given in (13), is complete has been discussed by Masry, Steiglitz, and Liu [2].

Since a continuous-time function can be represented by its Laplace transform and a sequence can be represented by its z-transform, the discrete representation as expressed in (4) and (5) can be thought of as a mapping from the Laplace transform of the continuous-time function to the $z$-transform of the sequence. However, a mapping from the Laplace transform to the $z$-transform cannot always be expressed as a mapping from the $s$ plane to the $z$ plane ($z$ cannot be written as a function of $s$). If the mapping also preserves convolution, i.e., if (13) is satisfied, then the discrete representation of (4) or (5) will correspond to expressing $z$ as a function of $s$. To show this result, let the $z$-transform of the sequence $f_n$ be denoted by $F_D(z)$ so that

$$F_D(z) = \sum_{n=-\infty}^{+\infty} f_n z^{-n}. \qquad (14)$$

Since $\Phi_n(s) = [\Phi_1(s)]^n$, then (5) can be rewritten as

$$F_L(s) = \sum_{n=-\infty}^{+\infty} f_n \{[\Phi_1(s)]^{-1}\}^{-n}. \qquad (15)$$

Comparing (14) and (15), it is clear that the variables $z$ and $s$ are related by

$$z = [\Phi_1(s)]^{-1} \triangleq M(s). \qquad (16)$$

A common example of a discrete representation that preserves convolution for band-limited continuous-time signals is periodic sampling. In this case, the continuous-time signal is represented by a sequence $f_n$ consisting of samples of the signal equally spaced in time so that

$$f_n = Tf(nT) \qquad (17)$$

where $T$ designates the sampling period. It is well known that periodic sampling can be viewed as an expansion of the continuous-time function $f(t)$ in the form of (4) with the functions $\{\phi_n(t)\}$ given by

$$\phi_n(t) = \sin \frac{\pi}{T}(t - nT)/\pi(t - nT). \qquad (18)$$

The Laplace transform of the functions in (18) converge only on the $j\omega$ axis. On the $j\omega$ axis their transforms are

$$\Phi_n(j\omega) = \begin{cases} e^{-j\omega nT} = [e^{j\omega T}]^{-n}, & |\omega T| \leq \pi \\ 0, & |\omega T| > \pi. \end{cases} \qquad (19)$$

Thus we see that the transform of these functions has the desired form as given in (13). The advantage to a discrete representation based on periodic sampling is that the coefficients in the expansion are easily obtained. Furthermore, as opposed to many other representations, the function $f(t)$ is not required for all $t$ in order to obtain each of the sequence values of $f_n$. The major disadvantage to this representation is that it requires that the function $f(t)$ be band-limited. When we are considering simulating a continuous-time system digitally, the frequency response of the continuous-time linear system is usually not truly band-limited. In addition, periodic sampling corresponds to a linear relationship between analog frequencies and digital frequencies. As we will see in a later discussion, it is sometimes advantageous to have a nonlinear relationship between analog and digital frequencies.

Another example of a discrete representation that preserves convolution is the Poisson transform. For a continuous-time function $f(t)$ that is zero for $t<0$, its Poisson transform is defined as

$$f_n = \int_0^\infty \frac{t^n}{n!} e^{-t} f(t) dt, \qquad n = 0, 1, 2, \cdots. \qquad (20)$$

The sequence $f_n$ then provides a discrete representation of the continuous-time function $f(t)$. Piovoso and Bolgiano [3] have shown that the Poisson transform has the desired property of mapping continuous convolution to discrete convolution, i.e., the Poisson transform of the convolution of two time functions is the discrete convolution of the Poisson transforms of the components of the convolution.

In the case of periodic sampling, the function $M(s)$ of (16) which specifies the mapping from the $s$ plane to the $z$ plane is $M(j\omega) = e^{j\omega T}$. When $\omega$ is real, corresponding to the imaginary axis in the $s$ plane, the magnitude of $M(j\omega)$ is unity corresponding to the unit circle in the $z$ plane. Thus the representation resulting from periodic sampling maps the spectrum of the analog signal as viewed on the imaginary axis of the $s$ plane to the spectrum of the digital signal as observed on the unit circle. In contrast, the function $M(s)$ resulting from the use of the Poisson transform is $M(s) = 1/(1-s)$. It is straightforward to verify that the $j\omega$ axis in the $s$ plane maps to a circle in the $z$ plane with radius one half and center at $z = 1/2$ so that it passes through the points $z=0$ and $z=1$. Consequently, it is tangent to the unit circle at $z=1$ but otherwise does not coincide with the unit circle. Thus when an analog function is mapped to the $z$ plane by use of the Poisson transform, its Fourier transform does not map onto the unit circle, but rather onto the circle to which the $j\omega$ axis in the $s$ plane is mapped. This does not, of course, affect the validity of the Poisson transform for simulation since both the system input and the system impulse response are mapped in the same way.

As discussed by Steiglitz [1], another example of a discrete representation with the desired property is based on the use of the bilinear transformation in relating the $s$ and $z$ planes. The bilinear transformation is commonly used for designing frequency selective digital filters [4]. This transformation has the property that no aliasing results when it is used to map a continuous-time filter to a discrete-time filter. On the other hand, the use of the bilinear transformation introduces a distortion in the frequency axis; the imaginary axis in the $s$ plane is mapped nonlinearly onto the unit circle in the $z$ plane. A common design procedure in discrete-time filtering is to derive the filter from a continuous-time design by means of the bilinear transformation while the discrete-time signal is obtained from the continuous-time signal by means of periodic sampling. Because of the distortion in the frequency axis resulting from the bilinear transformation, this procedure is generally restricted to situations in which the distortion in the frequency axis is not important. This is true, for example, in the design of digital filters with piecewise constant frequency characteristics.

If we wish to use the bilinear transformation for simulation when the filter characteristics are not piecewise constant, the same nonlinear distortion must be applied to the frequency characteristics of the input signal and the output signal; *both* the signals and the systems must be mapped to discrete signals and systems using the bilinear transformation. Since the bilinear transformation corresponds to a mapping from the $s$ plane to the $z$ plane with

$$z = M(s) = \left[\frac{a + s}{a - s}\right] \qquad (21)$$

it can also be viewed as an expansion of a time function $f(t)$ in terms of a series with the form of (4). $\Phi_n(s)$, the Laplace transform of $\phi_n(t)$, is from (13) and (16)

$$\Phi_n(s) = \left[\frac{a+s}{a-s}\right]^{-n}. \tag{22}$$

The functions $\{\phi_n(t)\}$ corresponding to (22) are

$$\phi_n(t) = \begin{cases} 2a(-1)^{n-1}e^{-at}L_{n-1}{}^{(1)}(2at)u_{-1}(t) \\ \qquad + (-1)^n u_0(t), & n > 0 \\ u_0(t), & n = 0 \\ 2a(-1)^{n-1}e^{at}L_{-n-1}{}^{(1)}(-2at)u_{-1}(-t) \\ \qquad + (-1)^n u_0(t), & n < 0 \end{cases} \tag{23}$$

where

$$L_{n-1}{}^{(1)}(x) = -\frac{d}{dx}L_n(x)$$

$$L_n(x) = \frac{e^x}{n!}\frac{d^n}{dx^n}(x^n e^{-x}). \tag{24}$$

The coefficients $f_n$ in the expansion of $f(t)$ can be obtained by utilizing the fact that, for the set of functions given in (23),

$$\int_{-\infty}^{+\infty} t\phi_n(t)\phi_m(t)dt = \begin{cases} n, & n = m \\ 0, & n \neq m \end{cases} \tag{25}$$

so that

$$f_n = \frac{1}{n}\int_{-\infty}^{+\infty} tf(t)\phi_n(t)dt, \qquad n \neq 0. \tag{26}$$

We can obtain $f_0$ from the $f_n$ by noting that if $f(t)$ is bounded, so that it contains no impulses at $t=0$, then

$$\sum_{n=-\infty}^{+\infty} (-1)^n f_n = 0. \tag{27}$$

If we restrict $f(t)$ to be zero for $t<0$, then from (26) and (27) it will follow that the coefficients $f_n$ can be obtained by exciting a linear time-invariant network with $f(-t)$. To derive this result, we note that from (23) and (26), if $f(t)$ is zero for $t<0$, $f_n$ will be zero for $n<0$. From (26), it now follows that for $n>0$,

$$f_n = \frac{1}{n}\int_0^{+\infty} f(t)[t\phi_n(t)]dt \tag{28a}$$

$$= \frac{1}{n}\int F_L(-s)[-\Phi_n'(s)]\frac{ds}{2\pi} \tag{28b}$$

where the prime denotes differentiation with respect to $s$. Since $\Phi_n(s) = [(a-s)/(a+s)]^n$,

$$-\Phi_n'(s) = n\frac{2a}{(s+a)^2}\left[\frac{a-s}{a+s}\right]^{n-1}$$

so that

$$f_n = \int F_L(-s)\frac{2a}{(s+a)^2}\left[\frac{a-s}{a+s}\right]^{n-1}\frac{ds}{2\pi}. \tag{29}$$

The integral of (28a) therefore corresponds to the response at $t=0$ of a linear time-invariant filter with impulse response $h_n(t) = (1/n)t\phi_n(t)$ or system function

$$H_n(s) = \frac{2a}{(s+a)^2}\left[\frac{a-s}{a+s}\right]^{n-1} \tag{30}$$

to an input $f(-t)$. To obtain $f_0$ we note from (27) that

$$f_0 = -\sum_{n=1}^{\infty}(-1)^n f_n$$

$$= -\sum_{n=1}^{\infty}(-1)^n \int F_L(-s)\frac{2a}{(s+a)^2}\left[\frac{a-s}{a+s}\right]^{n-1}\frac{ds}{2\pi}.$$

Interchanging the order of summation and integration, we find that

$$f_0 = \int \frac{ds}{2\pi}F_L(-s)\frac{2a}{(s+a)^2}\sum_{n=1}^{\infty}(-1)^{n-1}\left[\frac{a-s}{a+s}\right]^{n-1}$$

$$= \int \frac{ds}{2\pi}F_L(-s)\frac{2a}{(s+a)^2}\sum_{n=0}^{\infty}\left[\frac{s-a}{s+a}\right]^n$$

$$= \int \frac{ds}{2\pi}F_L(-s)\frac{2a}{(s+a)^2}\frac{(s+a)}{2a}$$

$$f_0 = \int \frac{ds}{2\pi}F_L(-s)\frac{1}{s+a}.$$

Consequently, $f_0$ is obtained as the output at $t=0$ of a linear filter with input $f(-t)$ and system function $1/(s+a)$. In summary, the sequence $f_n$ can be obtained by means of the network in Fig. 1 where $\tilde{f}_n(t)$ denote the outputs at the indicated taps and $f_n = \tilde{f}_n(0)$. The network of Fig. 1, which is similar, but not identical, to a Laguerre network [5], is used to obtain the coefficients $f_n$ from the continuous-time function $f(t)$. To "desample," i.e., to obtain $f(t)$ from the $f_n$, a network similar to that in Fig. 1 can be used. In particular, we recall that

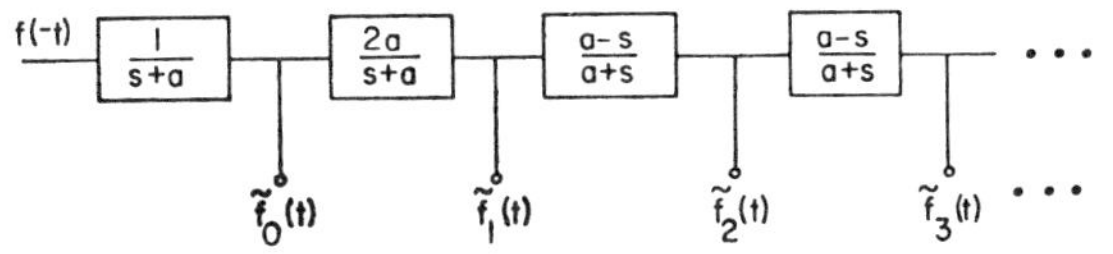


Fig. 1. All-pass network for conversion of a continuous-time signal $f(t)$ to its bilinear representation $f_n$. The representation $f_n$ is related to the output of each stage by $f_n = \tilde{f}_n(0)$.

$$f(t) = \sum_{n=0}^{\infty} f_n\phi_n(t)$$

$$F_L(s) = \sum_{n=0}^{\infty} f_n\Phi_n(s)$$

where $\Phi_n(s) = [(a-s)/(a+s)]^n$. Consequently, the function $\phi_n(t)$ is obtained as the impulse response of a network with system function $[(a-s)/(a+s)]^n$, and thus $f(t)$ can be recovered from the sequence $f_n$ by the means shown in Fig. 2.

In general, the discrete sequence representing a continuous-time signal will be of infinite length. For practical reasons it is usually necessary to truncate the representation in some way so that only a finite number of nonzero values is used. To consider the effect, in the case of the bilinear transformation, of modifying the representation to contain only a finite number of nonzero terms, let us denote the approximation to

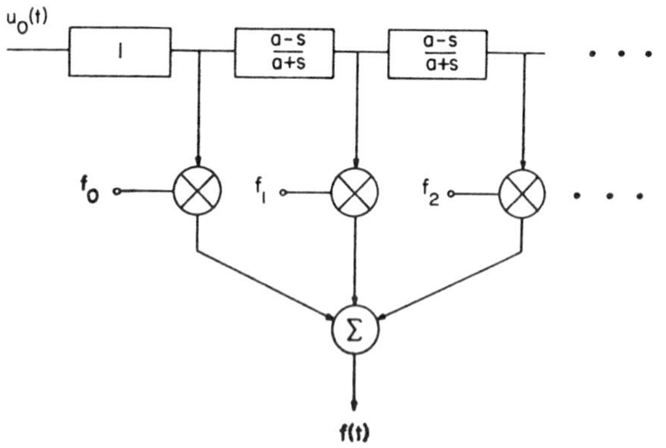


Fig. 2. All-pass network for reconstruction of a continuous-time signal $f(t)$ from its bilinear representation $f_n$.

$f(t)$ using only a finite number of nonzero terms by

$$f^A(t) = \sum_{n=0}^{N-1} f_n{}^A \phi_n(t)$$

with the exact representation of $f(t)$ given as

$$f(t) = \sum_{n=0}^{\infty} f_n \phi_n(t).$$

If the coefficients $f_n{}^A$ are chosen to equal the coefficients $f_n$ for $n=0, 1 \cdots, N-1$, and since

$$\int_{-\infty}^{+\infty} t\phi_n(t)\phi_m(t)dt = \begin{cases} n, & n = m \\ 0, & n \neq m \end{cases} \tag{31}$$

then $f^A(t)$ will correspond to an approximation to $f(t)$ that minimizes the weighted integral square error given by

$$E = \text{error} = \int_0^{+\infty} t[f(t) - f^A(t)]^2 dt. \tag{32}$$

Hence, generally speaking, truncation weights errors for large $t$ more than for small $t$ so that we would expect $f^A(t)$ to approximate $f(t)$ more closely as $t$ increases.

More generally, rather than obtaining the finite duration sequence $f_n{}^A$ from $f_n$ by simple truncation, it can be obtained by "windowing":

$$f_n{}^A = w_n f_n \tag{33}$$

where $w_n = 0$ for $n < 0$ and $n \geq N$. For the case of simple truncation $w_n = 1$ for $0 \leq n < N$. To describe the effect of applying the window $w_n$ to the sequence, let $F^A(\Omega)$, $W(\Omega)$, and $F(\Omega)$ represent the $z$-transforms on the unit circle of the sequences $f_n{}^A$, $w_n$, and $f_n$, respectively. Then

$$F^A(\Omega) = \frac{1}{2\pi} \int_{-\pi}^{\pi} W(\Omega - \alpha) F(\alpha) d\alpha \tag{34}$$

i.e., multiplication of the squence $f_n$ by the window $w_n$ corresponds to "smearing" the transform of the sequence $f_n$ with the transform of $w_n$. Alternatively, we can view this as a "smearing" of the spectrum of the original time function in such a way that spectral resolution decreases as the frequency increases. To see that this is so, let $f^A(t)$ denote the continuous-time function represented by the sequence $f_n w_n$ and let $G^A(\omega)$ and $G(\omega)$ denote the Fourier transforms of $f^A(t)$ and $f(t)$, respectively. From (21) with $s = j\omega$ and $z = e^{j\Omega}$, the relationship between $\Omega$ and $\omega$ is given by

$$\Omega = 2 \arctan \frac{\omega}{a} \cdot \tag{35}$$

Therefore,

$$G^A(\omega) = F^A\left(2 \arctan \frac{\omega}{a}\right)$$

$$G(\omega) = F\left(2 \arctan \frac{\omega}{a}\right)$$

so that (34) can be rewritten as

$$G^A(\omega) = \frac{1}{2\pi} \int_{-\infty}^{+\infty} W\left[2 \arctan \frac{\omega}{a} - 2 \arctan \frac{\beta}{a}\right] \cdot G(\beta) \frac{2a}{a^2 + \beta^2} d\beta. \tag{36}$$

As an approximation, let us assume that the effective width of the spectral window $W(\Omega)$ is sufficiently small so that over the width of the window, the bilinear frequency transformation of (35) can be approximated by a linear characteristic. Then (36) can be approximated by

$$G^A(\omega) = \frac{1}{2\pi} \int_{-\infty}^{+\infty} W\left[(\omega - \beta) \frac{2a}{a^2 + \omega^2}\right] \cdot G(\beta) \frac{2a}{a^2 + \omega^2} d\beta. \tag{37}$$

The argument $(\omega - \beta)[2a/(a^2+\omega^2)]$ can be interpreted as a linear scaling of the spectral window, while maintaining the shape so that the window becomes wider as $\omega$ increases. Thus the distortion due to truncating the representation by the application of a window can be viewed in terms of "smearing" the analog spectrum in such a way that spectral resolution decreases as the analog frequency increases.

One of the advantages of a discrete representation based on the bilinear transformation, as developed above in contrast to periodic sampling, is that the analog signal need not be band-limited. On the other hand, periodic sampling is easier to implement and the periodic samples can be obtained without buffering, in the sense that periodic sampling is a memoryless transformation. Bilinear sampling requires the entire waveform to be available before any of the values in the discrete sequence can be obtained, corresponding to infinite buffering of the input.

For the simulation of an analog filter it is possible to avoid the need for infinite buffering of the input by sectioning. To illustrate the procedure, let us consider an analog filter with impulse response $h(t)$, system function $H(s)$, and input $x(t)$. The discrete filter used for the simulation will have a unit sample response $h_n$ where, assuming the system is causal,

$$h(t) = \sum_{n=0}^{\infty} h_n \phi_n(t).$$

Now the entire input $x(t)$ can be expanded in terms of the $\{\phi_n(t)\}$, yielding a sequence of coefficients $x_n$, this sequence filtered, and the resulting output $y_n$ used to construct the continuous-time output $y(t)$ as

$$y(t) = \sum_{n=-\infty}^{+\infty} y_n \phi_n(t).$$

However, as has already been discussed, this requires infinite buffering since no output values can be obtained until all of the input is available. As an alternative, let us divide the input $x(t)$ into sections of equal duration $T$ so that, assuming $x(t)=0$, $t<0$,

$$x(t) = \sum_{k=0}^{+\infty} x_k(t - kT) \tag{38}$$

where $x_k(t)=0$ for $t<0$ and for $t>T$. Now each of the sections $x_k(t)$ can be expanded as

$$x_k(t) = \sum_{n=0}^{\infty} x_{n,k}\phi_n(t) \tag{39}$$

where we note that the sequence $x_{n,k}$ is zero for $n<0$ since each of the sections is zero for $t<0$. Since the system is linear, the response of the continuous-time filter can be expressed as the sum of the responses to each of the sections, i.e., if $y_k(t-kT)$ is the response to $x_k(t-kT)$, then

$$y(t) = \sum_{k=0}^{+\infty} y_k(t - kT).$$

To implement the simulation we obtain the discrete representation $\{x_{n,k}\}$ for each section, process this sequence with a discrete linear system having a unit sample response $h_n$, and use the resulting sequence to construct $y_k(t-kT)$. In general, although each section of the input is of finite duration $T$, the sequence representing it will be of infinite duration as will the output for each section. In a practical implementation, it would be necessary to truncate both the representation $\{x_{n,k}\}$ for each section and the output for each section $y_k(t-kT)$. The effect of this truncation must be taken into account in evaluating this procedure.

## III. Discrete Representation of Sequences

The foregoing examples provide several means for obtaining a discrete representation of continuous-time functions consistent with the requirement that the representation preserve convolution. It also is possible to convert from one discrete representation to another by means of a suitable transformation of the sequences. In particular, let us consider $f(t)$ represented by two different sequences $f_n$ and $g_k$ so that

$$f(t) = \sum_{n=-\infty}^{+\infty} f_n\phi_n(t) \tag{40a}$$

$$f(t) = \sum_{k=-\infty}^{+\infty} g_k\lambda_k(t). \tag{40b}$$

If we assume that the set of functions $\{\phi_n(t)\}$ is complete so that each $\lambda_k(t)$ can be expanded in terms of the $\{\phi_n(t)\}$, then

$$\lambda_k(t) = \sum_{n=-\infty}^{+\infty} \psi_{k,n}\phi_n(t). \tag{41}$$

The sequences $f_n$ and $g_k$ are then related by

$$f_n = \sum_{k=-\infty}^{+\infty} g_k\psi_{k,n}. \tag{42}$$

Equation (42) can be viewed as an expansion of the sequence $f_n$ in terms of a set of sequences $\{\psi_{k,n}\}$, with the sequence $g_k$ representing the coefficients in the expansion. Thus the sequences $\{\psi_{k,n}\}$ play the same role in (42) that the functions $\{\phi_n(t)\}$ play in (40a).

Since the mapping from $f(t)$ to $f_n$ and also from $f(t)$ to $g_k$ preserves convolution, it follows that an expansion of (42) must also preserve convolution, i.e.. if $f_n = f_{1,n} * f_{2,n}$ where $*$ denotes discrete convolution, and if $g_{1,k}$ and $g_{2,k}$ denote the coefficients in the expansion of $f_{1,n}$ and $f_{2,n}$, respectively, then according to (42) the sequence $g_k$ will equal the convolution of $g_{1,k}$ and $g_{2,k}$. The properties required of the set of sequences $\{\psi_{k,n}\}$ in order that convolution be preserved can be determined in a manner analogous to (5) through (14) with the result that if $\Psi_k(z)$ denotes the $z$-transform of $\psi_{k,n}$, then the $\{\Psi_k(z)\}$ must satisfy the relation

$$\Psi_r(z)\Psi_{k-r}(z) = \Psi_k(z) \tag{43}$$

so that $\Psi_k(z)$ must be given by

$$\Psi_k(z) = [\Psi_1(z)]^k. \tag{44}$$

Thus (44) plays the same role in mapping sequences to preserve discrete convolution that (13) plays in mapping continuous convolution to discrete convolution. When the $\{\Psi_k(z)\}$ satisfy (44), the $z$-transforms of the sequences $f_n$ and $g_k$ can be related by a substitution of variables. Specifically, let $F(z)$ denote the $z$-transform of $f_n$ and $G(\hat{z})$ denote the $\hat{z}$-transform of $g_k$. Then with $\hat{z} = [\Psi_1(z)]^{-1} \triangleq m(z)$, it follows from (42), in a manner similar to the derivation of (16), that

$$F(z) = G[m(z)]. \tag{45}$$

Thus changing from the representation $f_n$ to the representation $g_k$ is equivalent to mapping the complex frequency variable $z$ for $f_n$ to the complex frequency variable $\hat{z}$ for $g_k$ by means of a substitution of variables $\hat{z} = m(z)$.

Thus far we have been concerned with the representation of a continuous-time function by a sequence so that linear time-invariant systems could be simulated by linear shift-invariant systems. We found that we can achieve this result if we map the $s$ plane to the $z$ plane in an invertible fashion. Similarly, we can convert from one representation to another, each of which preserves convolution, by mapping one $z$ plane to another. There are some applications, however, in which we are concerned not so much with simulation as with the mapping of one complex plane to another. We would now like to turn our attention to the discussion of such problems.

### A. Digital Warping of Spectra [6]

In a variety of applications, it is useful to transform a sequence to a new sequence whose Fourier transform is equal to the Fourier transform of the original sequence on a distorted (warped) frequency scale. In this section, the basis for carrying out such a transformation using the ideas presented in the foregoing will be discussed, followed by several applications of frequency warping.

We found in the previous section that the transformation between the sequences $f_n$ and $g_k$ as expressed by (42) is equivalent to the mapping of one complex frequency plane to another if (44) is satisfied. We did not, however, focus upon any particular choice for the mapping $\hat{z} = m(z)$. Let us require that the function $\hat{z} = m(z)$ map the unit circle in the $z$ plane to the unit circle in the $\hat{z}$ plane. Letting $\Omega$ be angular frequency in the

$z$ plane and $\hat{\Omega}$ be angular frequency in the $\hat{z}$ plane, we want $\hat{z}=m(z)$ to satisfy

$$e^{j\hat{\Omega}} = m[e^{j\Omega}] \tag{46}$$

or

$$\hat{\Omega} = \theta(\Omega) \tag{47}$$

where $m[e^{j\Omega}]=e^{j\theta(\Omega)}$.

If the function $\hat{z}=m(z)$ satisfies (46), the Fourier transform of $g_k$ will be the Fourier transform of $f_n$ on a warped frequency axis.

As the $z$-transform of the sequence $\psi_{k,n}$ must satisfy (44), (46), and (47), the $z$-transform of $\psi_{k,n}$ evaluated on the unit circle is found by

$$\Psi_k(e^{j\Omega}) = e^{-jk\theta(\Omega)}. \tag{48}$$

Therefore, if the spectrum of $g_k$ is to be a warped version of the spectrum of $f_n$, the Fourier transform of the sequence $\psi_{k,n}$ must have an all-pass characteristic, i.e., the magnitude of the spectrum of $\psi_{k,n}$ must be unity independent of frequency. The *negative* of the phase of $\Psi(e^{j\Omega})$ will then correspond to the mapping between the frequency axes. In particular, if we want the function $\hat{z}=m(z)$ to be rational and also map the circle $z=e^{j\Omega}$ to the circle $\hat{z}=e^{j\hat{\Omega}}$, then $m(z)$ must be of the form

$$\hat{z} = m(z) = \prod_k \frac{1 - a_k z^{-1}}{z^{-1} - a_k^*}$$

where $a_k^*$ denotes the complex conjugate of $a_k$.

In order that $m(z)$ may be invertible and in the form already given, the coefficients $a_k$ must be such that the interval $-\pi<\Omega\leq\pi$ maps one-to-one to the interval $-\pi<\hat{\Omega}\leq\pi$. A necessary (but not sufficient) condition for this to be true is that the number of zeros minus the number of poles of $m(z)$ which lie inside the unit circle be plus or minus unity. A useful choice for $m(z)$, to which we will restrict the remaining discussion, is a first-order all-pass of the form

$$\hat{z} = m(z) = \frac{1 - az^{-1}}{z^{-1} - a^*} \tag{49}$$

where $0<|a|<1$. Then the $z$-transforms of the sequences $\{\psi_{k,n}\}$ used in the expansion of $f_n$ must be

$$\Psi_k(z) = \left(\frac{z^{-1} - a^*}{1 - az^{-1}}\right)^k. \tag{50}$$

We see that the $z$-transform of the sequence $\psi_{k,n}$ corresponds to a $k$th-order all-pass having unity gain with $k$ poles at $z=a$ and $k$ zeros at $z=1/a^*$. For real values of the parameter $a$, the mapping between the frequency variables $\hat{\Omega}$ and $\Omega$ is given by

$$\hat{\Omega} = \theta(\Omega) = \arctan\left[\frac{(1 - a^2)\sin\Omega}{(1 + a^2)\cos\Omega - 2a}\right]. \tag{51a}$$

An alternate form of (51a) is

$$\hat{\Omega} = \theta(\Omega) = \Omega + 2\arctan\left[\frac{a\sin\Omega}{1 - a\cos\Omega}\right]. \tag{51b}$$

To determine the inverse of the relationships in (51), we note

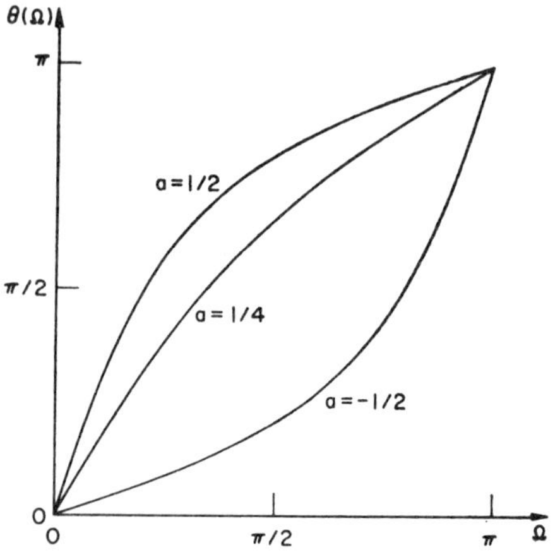


Fig. 3. Frequency-warping function of (51) for several values of parameter $a$.

that the inverse of (49) is

$$z = \frac{1 + a\hat{z}^{-1}}{\hat{z}^{-1} + a^*}$$

which corresponds to replacing $a$ by $-a$ and interchanging $z$ and $\hat{z}$ in (49). Consequently, the inverse of (51a) is obtained by replacing $a$ by $-a$ in that equation:

$$\Omega = \theta^{-1}(\hat{\Omega}) = \arctan\left[\frac{(1 - a^2)\sin\hat{\Omega}}{(1 + a^2)\cos\hat{\Omega} + 2a}\right].$$

A plot of the frequency-warping function of (51a) for several values of the parameter $a$ is given in Fig. 3. We see that the frequency mapping corresponding to this choice of sequences $\{\psi_{k,n}\}$ is inherently nonlinear. Therefore, the Fourier transform of the sequence $g_k$, the sequence derived from $f_n$, will be the Fourier transform of $f_n$ evaluated on a nonlinear scale:

$$F[e^{j\Omega}] = G[e^{j\theta(\Omega)}].$$

In particular, computing $G[e^{j\hat{\Omega}}]$ at a set of uniformly spaced samples in $\hat{\Omega}$ (using the fast Fourier transform algorithm, for instance) corresponds to the evaluation of $F(e^{j\Omega})$ at a set of nonuniformly spaced samples. If the parameter $a$ is picked to be real and between 0 and 1, the effect on the spectrum of $f_n$ will be to sample with higher resolution at low frequencies and with lower resolution at higher frequencies. If instead, $a$ is negative between 0 and $-1$, the effect is reversed, i.e., the spectrum of $f_n$ is evaluated with greater resolution at high frequencies than at low frequencies. By letting the parameter $a$ assume complex values, the point of maximal resolution can be placed at any desired frequency.

The sequence $g_k$ can be determined from the sequence $f_n$ by using the fact that the set of sequence $\{\psi_{k,n}\}$ is orthogonal with a weighting sequence of $n$:

$$\sum_{n=-\infty}^{\infty} n\psi_{r,n}\psi_{k,n} = \begin{cases} k, & k = r \\ 0, & k \neq r. \end{cases} \tag{52}$$

This relationship can be derived by applying Parseval's theorem to the left side of (52). Consequently,

$$g_k = \frac{1}{k}\sum_{n=-\infty}^{\infty} n\psi_{k,n}f_n, \qquad k \neq 0. \tag{53}$$

In the remainder of the discussion, we will assume, for convenience, that the original sequence $f_n$ is zero for $n<0$. The results are easily extended to the more general case. With this restriction, (53) becomes

$$g_k = \begin{cases} \frac{1}{k} \sum_{n=0}^{\infty} n\psi_{k,n} f_n, & k > 0 \\ 0, & k < 0. \end{cases} \tag{54}$$

This relationship is similar to (28a) which was used to determine the coefficients in the expansion of a continuous-time function based on the bilinear transformation. It was shown from (28a) that the coefficients can be obtained by processing the continuous-time function with the cascade chain of networks of Fig. 1. In a similar manner, the coefficients $g_k$ can be obtained by processing the sequence $f_n$ with a cascade chain of digital networks. Specifically, the sum in (54) corresponds to the response at $n=0$ of a linear discrete system to an input $f_{-n}$, with the unit sample response of the system being $h_{k,n}=(1/k)n\psi_{k,n}$ or system function

$$H_k(z) = \frac{-z}{k} \frac{d\Psi_k(z)}{dz}$$

$$H_k(z) = \frac{(1-a^2)z^{-1}}{(1-az^{-1})^2}\left[\frac{z^{-1}-a}{1-az^{-1}}\right]^{k-1}, \qquad k = 1, 2, \cdots. \tag{55}$$

Equation (54) cannot be used to determine $g_0$. However, from (44),

$$g_0 = f_0 - \sum_{k=1}^{\infty} g_k \psi_{k,0}$$

or since $\psi_{k,0}=(-a)^k$,

$$g_0 = f_0 - \sum_{k=1}^{\infty} g_k(-a)^k.$$

Thus, using (54),

$$g_0 = f_0 - \sum_{n=0}^{\infty} f_n n \sum_{k=1}^{\infty} \frac{1}{k}(-a)^k \psi_{k,n}.$$

Let

$$\sum_{k=1}^{\infty} \frac{1}{k}(-a)^k \psi_{k,n} = v_n \tag{56}$$

so that

$$g_0 = f_0 - \sum_{n=0}^{\infty} f_n n v_n$$

$$= \sum_{n=0}^{\infty} f_n(\delta_n - nv_n) \tag{57}$$

where $\delta_n$ is a unit sample, i.e., $\delta_n=1$ for $n=0$, and $\delta_n=0$ otherwise. Thus, from (57), $g_0$ can be obtained as the response at $n=0$ to an input $f_{-n}$ of a linear discrete system having a unit sample response $h_{0,n}=(\delta_n - nv_n)$. From (55) and (56), the $z$-transform of $nv_n$ is given by

$$\sum_{n=0}^{\infty} nv_n z^{-n} = -\frac{az^{-1}}{1-az^{-1}}.$$

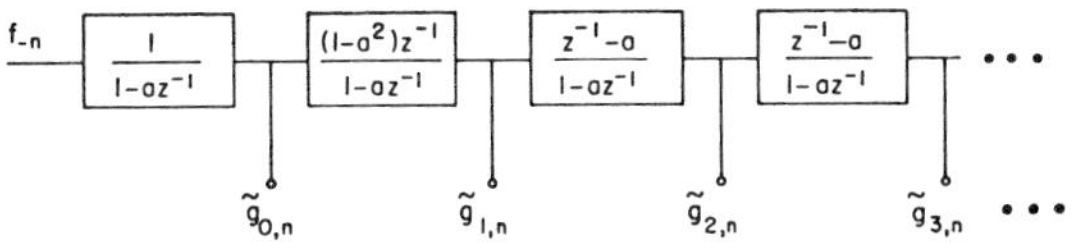


Fig. 4. All-pass network used to implement spectral warping of discrete-time signals. The sequence $g_k$, whose spectrum is the warped version of the input sequence $f_n$, is related to the output of each stage by $g_k=\tilde{g}_{k,0}$.

Therefore, the system function corresponding to the unit sample response $h_{0,n}$ is

$$H_0(z) = \frac{1}{1-az^{-1}}.$$

In summary, we can find the sequence $g_k$ by passing the sequence $f_{-n}$ through a linear shift-invariant system $H_k(z)$ given by

$$H_k(z) = \begin{cases} \frac{(1-a^2)z^{-1}}{(1-az^{-1})^2}\left(\frac{z^{-1}-a}{1-az^{-1}}\right)^{k-1}, & k > 0 \\ \frac{1}{1-az^{-1}}, & k = 0. \end{cases} \tag{58}$$

The output of the system $H_k(z)$ at $n=0$ is then the desired coefficient $g_k$. We note from (58) that for $k>2$, the $k$th system corresponds to cascading the $(k-1)$th with an all-pass network and that $H_1(z)$ corresponds to $H_0(z)$ in cascade with a first-order network. Thus the set of systems $\{H_k(z)\}$ can be implemented as a cascade chain as illustrated in Fig. 4. With the output of each section denoted by $\tilde{g}_{k,n}$, the desired coefficients $g_k$ are given by $g_k=\tilde{g}_{k,0}$. We note that replacing $a$ by $-a$ in (58) is equivalent to multiplying $H_k(z)$ by $(-1)^k$ and replacing $z$ by $-z$, which in turn corresponds to multiplying $h_{k,n}$ by $(-1)^n(-1)^k$. Since

$$g_k = \sum_{n=0}^{\infty} f_n h_{k,n}$$

multiplying $h_{k,n}$ by $(-1)^{n+k}$ is equivalent to multiplying $f_n$ by $(-1)^n$ and $g^k$ by $(-1)^k$. Thus if the input to the network of Fig. 4 is multiplied by $(-1)^n$ and the resulting sequence $g_k$ is multiplied by $(-1)^k$, then we have equivalently implemented the network of Fig. 4 with $a$ replaced by $-a$. In a hardware realization of this network, for example, only positive values of $a$ need to be implemented.

In the use of the bilinear transformation for the discrete representation of continuous-time signals, we noted that, in general, the representation would be of infinite length and would, in a practical implementation, need to be truncated. Truncation of the sequence corresponded to a "smearing" of the spectrum of the original time function with a spectral window that increases in width as frequency increases.

Truncation of the representation based on the sequences expressed in (44) causes a similar type of effect. In particular, we will assume that the truncation is carried out by multiplying $g_k$ by a finite duration window so that the truncated sequence is represented by $g_k{}^A=w_k g_k$. With $f_n{}^A$ denoting the sequence represented by $g_k{}^A$:

$$f_n{}^A = \sum_{k=0}^{\infty} g_k{}^A \psi_{n,k}$$

and with $F^A(\Omega)$ and $F(\Omega)$ denoting the Fourier transforms of

$f_n^A$ and $f_n$, we have, paralleling the derivation of (36),

$$F^A(\Omega) = \frac{1}{2\pi}\int_{-\pi}^{\pi} F(\beta)W[\theta(\Omega) - \theta(\beta)] \frac{d\theta(\beta)}{d\beta} d\beta$$

or with $\theta(\Omega)$ given by (51a)

$$F^A(\Omega) = \frac{1}{2\pi}\int_{-\pi}^{\pi} F(\beta)W[\theta(\Omega) - \theta(\beta)] \frac{1 - a^2}{1 + a^2 - 2a\cos\beta} d\beta.$$

As with (37), we will assume that the window is sufficiently narrow so that over the width of the window the characteristic $\theta(\Omega)$ is approximately linear. Then

$$W[\theta(\Omega) - \theta(\beta)] = W\left[(\Omega - \beta)\frac{1 - a^2}{1 + a^2 - 2a\cos\Omega}\right]$$

so that

$$F^A(\Omega) = \frac{1}{2\pi}\int_{-\pi}^{\pi} F(\beta)W\left[(\Omega - \beta)\frac{1 - a^2}{1 + a^2 - 2a\cos\Omega}\right] \cdot \frac{1 - a^2}{1 + a^2 - 2a\cos\Omega} d\beta.$$

As with (37), the factor of $[(\Omega - \beta)(1 - a^2)]/(1 + a^2 - 2a\cos\Omega)$ can be interpreted as a linear scaling of the shape of the spectral window. For positive $a$, this corresponds to "smearing" the spectrum of the original sequence in such a way that spectral resolution decreases as frequency increases. For negative $a$, spectral resolution decreases as frequency decreases.

### *B. Applications of Digital Frequency Warping*

*1) Evaluation of Spectra at Nonuniformly Spaced Samples* [6]: The ability to be able to distort the frequency variable in the Fourier transform of a sequence has proved to be useful in several contexts. The first to be discussed is directed toward warping the spectrum to achieve spectral analysis with unequal resolution.

In many applications, we are concerned with the problem of computing samples of the $z$-transform of a sequence on the unit circle. To obtain samples equally spaced around the unit circle, the most efficient procedure is to compute the discrete Fourier transform (DFT) using the fast Fourier transform (FFT) algorithm. Often we would like to obtain samples that are unequally spaced, corresponding, for example, to a constant-$Q$ spectral analysis of the original sequence. An algorithm for accomplishing this nonlinearly spaced spectral analysis with an efficiency similar to that achievable with the FFT algorithm is not known. One procedure sometimes used is to evaluate explicitly the samples at the desired frequencies, using, for example, the Goertzel algorithm [7]. Another approximate procedure is to sum equally spaced frequency samples in bands. A related procedure corresponds to performing the spectral analysis of a sequence with a recursive or nonrecursive filter bank.

As an alternative to these methods, unequally spaced samples of the spectrum can be obtained by implementing an equal resolution spectral analysis on a distorted frequency scale. To carry out such an analysis , the sequence $f_n$ whose spectrum is desired is processed by the network of Fig. 4 and the DFT of the output sequence $g_k$ is computed using the FFT algorithm. To illustrate the effect of the warping, a

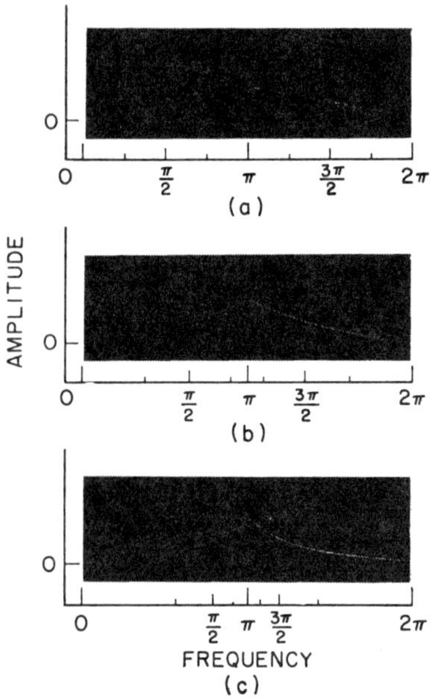


Fig. 5. Example of digital warping of spectra. (a) Original spectral magnitude. (b) Spectral magnitude of warped version of the original with $a = 1/4$. (c) Spectral magnitude of warped version of the original with $a = 1/2$.

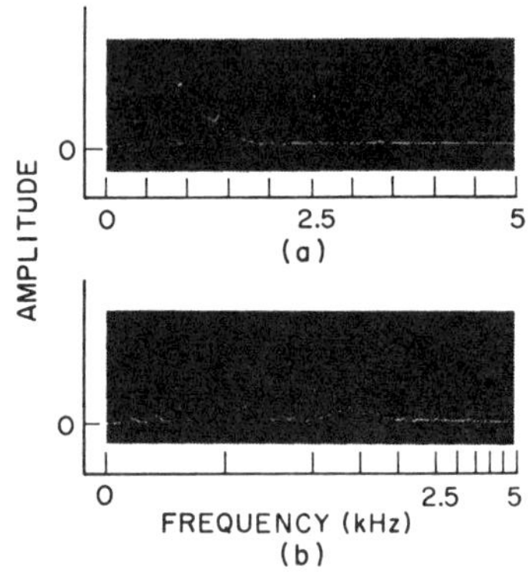


Fig. 6. Example of digital frequency warping on a sample of speech. (a) Original spectral magnitude. (b) Warped spectral magnitude with $a = 1/2$.

sequence whose spectrum is linear was processed in the manner just described. Specifically, the spectrum of the original sequence is indicated in Fig. 5(a). The DFT of the sequence obtained from the network of Fig. 4 is shown in Fig. 5(b) and (c) for two different values of parameter $a$. Another example is shown in Fig. 6. Fig. 6(a) represents the DFT of 51.2 ms of speech weighted with a Hanning window; Fig. 6(b) shows the DFT of the sequence obtained by processing the same speech sample with the network of Fig. 4.

The fact that the original sequence must first be processed by the network of Fig. 4 means that if the original sequence is of length $N$ and the transformed sequence is of length $M$, then the computation time is proportional to $NM$ excluding the time required for computation of the DFT. The use of the Goertzel algorithm to evaluate spectra at unequally spaced samples would likewise require computation proportional to $NM$. However, the network of Fig. 4 involves a single parameter. Consequently, a hardware implementation of frequency warping can utilize a fixed coefficient multiplier, which offers a considerable advantage over the use of algorithms requiring a coefficient memory and a more general multiplier.

*2) Vernier Spectrum Analysis:* In some applications it is desirable to carry out a vernier spectrum analysis corresponding to a high-resolution sampling of the spectrum over a small frequency range. One algorithm that has been proposed for such analysis is the chirp $z$-transform [8]. Another common procedure consists of modulating the signal to center the frequency band on which the vernier analysis is to be carried out around $\Omega=0$, low-pass filtering and sampling the modulated signal, and transforming the result. As an alternative, vernier analysis can be carried out with the use of frequency warping. Referring to Fig. 3, we note that for $a$ positive, the slope of the warping function $\theta(\Omega)$ near the origin is greater than unity. Thus the transform of a sequence processed with the all-pass network of Fig. 4 with $a$ positive has highest resolution near $\Omega=0$. As $a$ increases toward unity, the effect is to provide a vernier analysis around $\Omega=0$. In general, of course, we would like to obtain a vernier analysis around an arbitrary frequency $\Omega$. This is accomplished by multiplying the original sequence by the complex exponential sequence $e^{j\Omega_0 n}$ prior to processing with the network of Fig. 4, to center the desired frequency range at $\Omega=0$.

*3) Helium Speech Translation:* In the previous discussion, spectral warping was used to compute spectra on a nonlinear frequency scale. Another class of applications is directed toward the use of spectral warping to correct for a nonlinear distortion in frequency. An example of such an application is the correction of the speech of divers [9]. It is well known that when underwater divers breathe helium-rich gas mixtures at ambient pressures higher than atmospheric pressure, their speech is highly unintelligible. The effect of the change in breathing mixture and pressure results in a nonlinear shift in the resonant frequencies of the vocal tract. However, the fundamental frequency of the speech remains close to normal; consequently, when correcting for the frequency distortion, the fundamental frequency of the diver's speech should not be changed. This can be accomplished by extracting individual pitch periods of the speech, processing each period of the speech to compensate for the nonlinear frequency distortion, and then resynthesizing the speech with the original periodicity. Zue [10] has investigated the application of frequency warping within this context with encouraging results.

### C. *Hardware Implementation of Digital Frequency Warping*

One of the advantages of the use of the network of Fig. 4 is that it can be implemented with digital hardware in a relatively straightforward and inexpensive manner. Thus, for example, it can be used peripherally to a digital computer in conjunction with an FFT program to implement a spectral analysis with unequal resolution.

To indicate one possible form of the hardware configuration for the network of Fig. 4, the network is redrawn in Fig. 7 indicating a realization with multipliers, adders, and delays. The set of equations by means of which the network of Fig. 7 is iterated is

$$\tilde{g}_{0,n} = a[\tilde{g}_{0,n-1} - 0] + f_{-n} \tag{59a}$$

$$\tilde{g}_{1,n} = a[\tilde{g}_{1,n-1} - 0] + \tilde{g}_{0,n-1} \tag{59b}$$

$$\tilde{g}_{k,n} = a[\tilde{g}_{k,n-1} - \tilde{g}_{k-1,n}] + \tilde{g}_{k-1,n-1}, \quad k = 2, 3, \cdots. \tag{59c}$$

With the difference equations written in the form of (59), the arithmetic operations involved are identical for each iteration

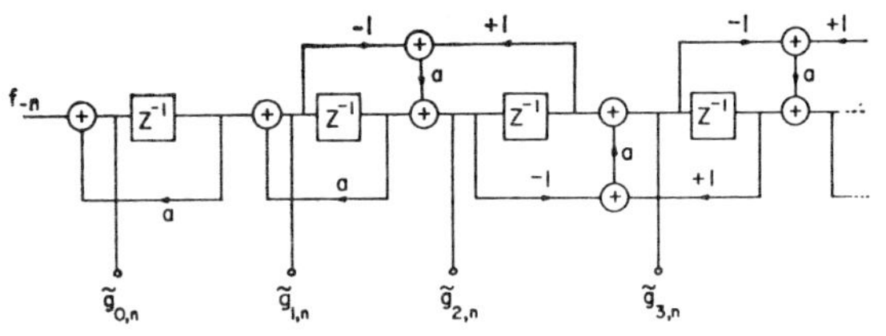


Fig. 7. Detailed realization of network in Fig. 4.

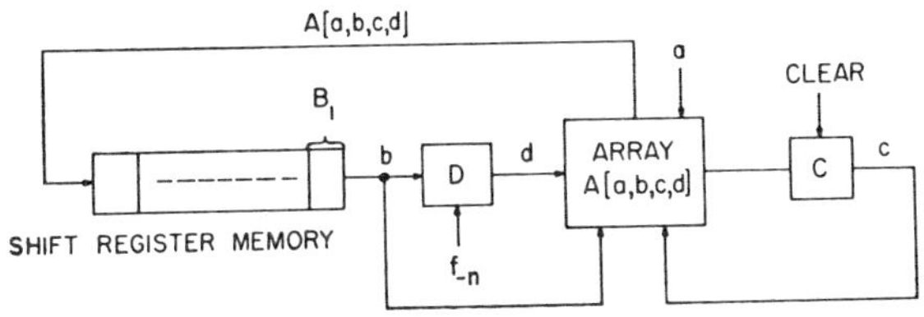


Fig. 8. Block diagram of a hardware realization of Fig. 7.

in both $k$ and $n$. In particular, our implementation of the networks utilizes an arithmetic array with inputs $a$, $b$, $c$, and $d$. With $A[a, b, c, d]$ denoting the array output, the array implements the arithmetic function

$$A[a, b, c, d] = a(b - c) + d. \tag{60}$$

Thus the input $a$ to the array represents the coefficient that controls the amount of spectral warping obtained. For the computation of $\tilde{g}_{0,n}$, the inputs $b$, $c$, and $d$ are $\tilde{g}_{0,n-1}$, 0, and $f_{-n}$, respectively, so that

$$\tilde{g}_{0,n} = A[a, \tilde{g}_{0,n-1}, 0, f_{-n}].$$

Similarly,

$$\tilde{g}_{1,n} = A[a, \tilde{g}_{1,n-1}, 0, \tilde{g}_{0,n-1}]$$

$$\tilde{g}_{k,n} = A[a, \tilde{g}_{k,n-1}, \tilde{g}_{k-1,n}, \tilde{g}_{k-1,n-1}], \qquad k = 2, 3, \cdots.$$

The network of Fig. 7 must be iterated in both $k$ and $n$. The iteration in the index $k$ corresponds to computing the outputs $\tilde{g}_{k,n}$ along the network chain for a fixed $n$. The iteration in $n$ then corresponds to updating each of these outputs for a new input sample. With the network iterated in $k$ and then $n$, so that for each value of $n$ all of the outputs $\tilde{g}_{k,n}$ are updated, the network states, corresponding to the delay outputs in the network of Fig. 7, can be stored in shift register memory.

A general structure that can be used for the implementation using a shift register memory is indicated in Fig. 8. Values in the shift register memory correspond to the outputs for each of the delay registers in the network of Fig. 7. For $k=0$, register $D$ contains $f_{-n}$, otherwise it is connected to the output of the shift register memory. For $k=0$ and $k=1$, the register $C$ is cleared, otherwise it is connected to the output of the arithmetic array. The length of the shift register memory is equal to the length of the sequence $g_k$, which will be denoted by $M$. For each value of $n$, the network of Fig. 8 is iterated $M$ times. In Table I the contents of the registers as a function of $k$ are indicated.

The hardware configuration indicated in Fig. 8 has been implemented using TTL logic and dynamic shift register memory. With an 18-bit register length and $M=512$, the hardware implementation contains 100 integrated-circuit packages. The configuration can operate at more than a 2-MHz clock rate. Consequently, to transform 512 data

TABLE I

CONTENTS OF THE SPECIFIED REGISTER IN THE HARDWARE REALIZATION OF THE DIGITAL WARPING NETWORK (FIG. 8) AS A FUNCTION OF $k$

| $k$ | $B_1$ | $D$ | $C$ |
|---|---|---|---|
| 0 | $\tilde{g}_{0,n-1}$ | $f_{-n}$ | 0 |
| 1 | $\tilde{g}_{1,n-1}$ | $\tilde{g}_{0,n-1}$ | 0 |
| 2 | $\tilde{g}_{2,n-1}$ | $\tilde{g}_{1,n-1}$ | $\tilde{g}_{1,n}$ |
| 3 | $\tilde{g}_{3,n-1}$ | $\tilde{g}_{2,n-1}$ | $\tilde{g}_{2,n}$ |
| ⋮ | ⋮ | ⋮ | ⋮ |

*Note:* $k$ = the number of the stage in the all-pass network.

points to obtain a new 512-point sequence requires approximately 0.13 s. Computation time for a 512-point DFT using the FFT algorithm, implemented on a fixed-point 18-bit computer such as a DEC PDP-9, requires between 0.5 and 1 s; thus the inclusion of frequency warping to obtain unequal resolution using this piece of peripheral hardware increased computation time by from 10 to 20 percent over that required for an FFT alone. As the application warrants it, faster circuitry can, of course, be used.

## IV. CONCLUSIONS

This paper has been directed toward a discussion and review of the discrete representation of signals within two contexts. The first related to the representation of continuous-time signals and systems by sequences for digital simulation. The second was the representation of a sequence by another to implement a frequency warping. Some of the ideas and results in the paper have found practical application while others are more speculative. It is hoped that as these ideas are further developed, additional applications will arise.

## ACKNOWLEDGMENT

Much of the material in Section III-A was developed in collaboration with Prof. K. Steiglitz and other parts have benefited from discussions with him. The hardware implementation in Section III-C was carried out with the assistance of J. Tierney. Dr. R. Schafer, Dr. L. Rabiner, and Prof. H. Grebe offered many helpful comments on early drafts of the paper. The work has also benefited from the interest, suggestions, and encouragement of Dr. B. Gold.

## REFERENCES

[1] K. Steiglitz, "The equivalence of digital and analog signal processing," *Inform. Contr.*, vol. 8, pp. 455–467, Oct. 1965.

[2] E. Masry, K. Steiglitz, and B. Liu, "Bases in Hilbert space related to the representation of stationary operators," *SIAM J. Appl. Math.*, vol. 16, pp. 552–562, May 1968.

[3] M. Piovoso and L. Bolgiano, "Digital simulation using Poisson transform sequences," in *Proc. Symp. Computer Processing in Communication.* Brooklyn, N. Y.: Polytechnic Press, 1969, pp. 505–514.

[4] B. Gold and C. Rader, *Digital Processing of Signals.* New York: McGraw-Hill, 1969.

[5] Y. W. Lee, "Synthesis of optimum linear systems by means of orthonormal functions," in *Statistical Theory of Communication.* New York: Wiley, 1960, ch. 19.

[6] A. Oppenheim, D. Johnson, and K. Steiglitz, "Computation of spectra with unequal resolution using the fast Fourier transform," *Proc. IEEE* (Lett.), vol. 59, pp. 299–301, Feb. 1971.

[7] C. R. Arnold, "Spectral estimation for transient waveforms," *IEEE Trans. Audio Electroacoust.*, vol. AU-18, pp. 248–257, Sept. 1970.

[8] L. R. Rabiner, R. W. Schafer, and C. M. Rader, "The chirp z-transform algorithm," *IEEE Trans. Audio Electroacoust.*, vol. AU-17, pp. 86–92, June 1969.

[9] G. Fant and J. Lindqvist, "Pressure and gas mixture effects on divers' speech," Speech Transmission Lab., Royal Inst. Tech. (KTH), Stockholm, Sweden, Quart. Progr. and Status Rep., Apr. 1968, pp. 7–17.

[10] V. Zue, "Translation of divers' speech using digital frequency warping," M.I.T. Res. Lab. Electron., Cambridge, Mass., Quart. Progr. Rep. 101, Apr. 15, 1971, pp. 175–182.

# On Deconvolution Using the Discrete Fourier Transform

HARVEY F. SILVERMAN and ALLAN E. PEARSON

*Abstract*—A solution is given to the problem of deconvolving two time sequences using discrete Fourier transform (DFT) techniques when one of the sequences is of infinite duration. Both input- and impulse-response deconvolution problems are considered. Results are summarized of a computer study of the algorithm that performs the deconvolution iteratively, using the fast Fourier transform (FFT) algorithm at each stage.

Manuscript received October 30, 1972. This research was supported in part by the U. S. Air Force Office of Scientific Research under Contract AF-AFOSR-71-2078, and in part by the National Science Foundation under Grant GK-31073X.

H. F. Silverman is with the IBM T. J. Watson Research Center, Yorktown Heights, N. Y. 10598.

A. E. Pearson is with the Division of Engineering and Center for Dynamical Systems, Brown University, Providence, R. I.

## I. Introduction

The advantage of applying the discrete Fourier transform (DFT) to calculate the output sequence of the discrete convolution

$$y(n) = \sum_{i=0}^{M-1} h(i)u(n-i) = h(n) \otimes u(n) \tag{1}$$

where $h(i)$ is nonzero over $[0, M-1]$, has been explored and discussed by several authors [1]–[3]. Computation is saved by applying a fast Fourier transform (FFT) algorithm in cases when $M$ is larger than an implementation dependent parameter $M_0$, which is typically in the range 32–64. However, the DFT convolution is circular in nature; that is, it describes ordinary convolution only in cases where the sequence $u(n)$ can be considered to be periodic and nonzero only over an initial portion of each cycle. This is shown in Fig. 1(a) under the ideal condition where $u(n)$ is periodic, with period at least $L+M-1$ and nonzero only over $[0, L-1]$.

For this case it is easy to show that the discrete convolution theorem holds; namely, that for time domain sequences $h(n)$, $u(n)$, and $y(n)=h(n)\otimes u(n)$, defined over $[0, L+M+2]$, with $h(n)$ and $u(n)$ nonzero only over $[0, M-1]$ and $[0, L-1]$, respectively, as implied in Fig. 1(a),

Reprinted from *IEEE Trans. Audio Electroacoust.*, vol. AU-21, pp. 112–118, Apr. 1973.

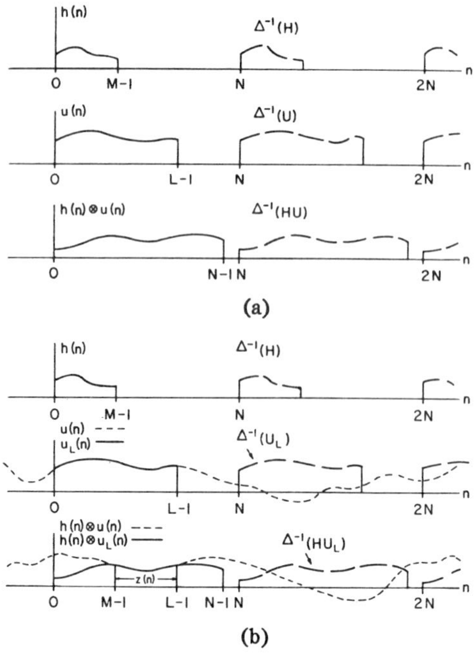


Fig. 1. (a) Circular convolution under ideal conditions. (b) Circular convolution with truncated signal $U_L$.

$$
\begin{aligned}
y(n) &= \Delta^{-1}[\Delta(h)\Delta(u)] \\
&= \Delta^{-1}[HU] \\
n &= 0, 1, \cdots, L+M-2
\end{aligned}
\tag{2}
$$

where $\Delta$ and $\Delta^{-1}$ denote the DFT operator and its inverse. The DFT operators are defined by

$$
\begin{aligned}
H(r) &= \Delta(h) = \sum_{h=0}^{N-1} h(n) W^{-rn} \\
h(n) &= \Delta^{-1}(H) = \frac{1}{N}\sum_{r=0}^{N-1} H(r) W^{rn}
\end{aligned}
\tag{3}
$$

where $W = e^{j2\pi/N}$, and $N$ is an integer greater than or equal to $L+M-1$. In the sequel all figures are drawn for $N=L+M-1$, although any integer satisfying $N \geq L+M-1$ will suffice provided the appropriate number of zero values are appended to the given data. Equation (2) will be referred to as the idealized discrete convolution theorem [4], [5].

In the nonideal situation one of the sequences, say $u(n)$, is of infinite or very long duration, thus necessitating its artificial truncation to a finite interval $[0, L-1]$ as shown in Fig. 1(b). In this case the inverse function $\Delta^{-1}(HU_L)$, where $U_L = \Delta(u_L)$ and $u_L(n)$ denotes the truncation of $u(n)$ to $[0, L-1]$, yields correct values for the convolution only in an internal interval (the interval $M-1 \leq n \leq L-1$ in Fig. 1(b). Specifically, if $w(n)$ denotes the rectangular window function[1]

$$
w(n) = \begin{cases} 1, & \text{for } M-1 \leq n \leq L-1 \\ 0, & \text{otherwise} \end{cases}
\tag{4}
$$

and $z(n) = w(n) \cdot y(n)$ denotes the truncation of the original convolved signal to the interval, i.e.,

$$
z(n) = \begin{cases} y(n) = h(n) \otimes u(n), & \text{for } M-1 \leq n \leq L-1 \\ 0, & \text{otherwise.} \end{cases}
\tag{5}
$$

Then, the circular convolution involving the artificially truncated signal $u_L(n)$ yields the function $z(n)$ when properly restricted:

$$
z(n) = w(n)\Delta^{-1}(HU_L).
\tag{6}
$$

Additional segments of the original convolved signal $y(n) = h(n) \otimes u(n)$ then can be obtained by shifting and truncating $u(n)$ appropriately, using the circular convolution in a step-by-step manner.

The inverse problem of deconvolving the equation $y(n) = h(n) \otimes u(n)$, in order to recover either $h(n)$ or $u(n)$, will have a direct solution by DFT techniques only in the ideal situation of Fig. 1(a), in which the three signals $(h(n), u(n), y(n))$ are identically zero outside the intervals $\{[0, M-1], [0, L-1], [0, L+M-2]\}$. Then, $h(n) = \Delta^{-1}(Y/U)$ and $u(n) = D^{-1}(Y/H)$ yield the correct functions for $0 \leq n \leq N-1$, assuming that the transforms $U$ and $H$ contain all pertinent frequencies. However, there exist many applications in which deconvolution is desired, but the given signals are of very long (perhaps infinite) duration. The discrete linear system identification problem is one such application in which the excitation (input) sequence $u(n)$ may be of infinite duration, and it is desired to compute the impulse-response (kernel) sequence $h(n)$, based upon a knowledge of the input–output $(u(n), y(n))$ observed over a finite time interval.[2] Mathematically, this problem is one of solving the equation

$$
z(n) = w(n)[h(n) \otimes u_L(n)]
\tag{7}
$$

for $h(n)$, given the truncated signals $u_L(n)$ and $z(n)$.

The dual of the system identification problem is the input deconvolution problem, in which a filtered signal $y(n)$ is observed as the output of a smoothing filter with very long memory, and it is desired to compute the input $u(n)$ using only a finite segment $h_L(n)$ of the filter impulse-response function $h(n)$ and a truncation $z(n)$ of the response data $y(n)$; hence, solve

$$
\tilde{z}(n) = \tilde{w}(n)[h_L(n) \otimes u(n)]
\tag{8}
$$

for $u(n)$, given $h_L(n)$ and $\tilde{z}(n)$.

Equations (7) and (8) can be solved directly by matrix inversion of a suitably defined Toeplitz matrix. However, this approach can be inefficient if the number

[1] Other window functions have been investigated. However, straightforward results are obtained only with the rectangular window function because of its nonoverlapping decomposition property. See, e.g., Fig. 2.

[2] The problem of continuous-time linear system identification is effectively treated if it can be assumed that all signals are band limited to the same bandwidth.

of samples needed to represent the various signals is large, e.g., 100 or 1000. As an alternative to matrix inversion, the formulations of the next section leads to operator equations in the frequency domain that are equivalent to (7) and (8) and suggest iterations based on the contraction mapping principle. These iterations employ the DFT circular convolution at each step, and converge to the desired function under conditions involving the given data that can be tested beforehand. Computational results using the FFT algorithm to compute the DFT are discussed in Section III. Computational efficiency is discussed in Section IV.

## II. Formulations for Deconvolution via DFT

Consider first the impulse-response deconvolution problem represented by (7). Assume that the upper bound on the possible memory lengths for the unknown kernel sequence $h(n)$ is given by $M$, i.e., $h(n)=0$ for all $n \geq M$.[3] Also, assume that the input data length $L$ is at least as great as $2M-1$, i.e., $L \geq 2M-1$. (If $L<2M-1$ there will be fewer nonzero data points in $z(n)$ than in $h(n)$, and a unique solution for arbitrary $u_L(n)$ will not exist.) Referring to Fig. 2, where $v_1(n)$ and $v_2(n)$ are decompositions of the truncated input $u_L(n)$, i.e., $v_1(n)+v_2(n)=u_L(n)$ for all $n$, and $p(n)$ and $q(n)$ are the corresponding responses, the following identity holds.

$$z(n) = \Phi(p(n)) + q(n), \qquad 0 \leq n \leq L+M-2 \quad (9)$$

where $\Phi$ is the truncation operator defined by $\Phi(p(n))=w(n)\cdot p(n)$, and $w(n)$ is given by (4). Notice that $p(n)=h(n)\otimes v_1(n)$ and $q(n)=h(n)\otimes v_2(n)$ satisfy the ideal conditions for the DFT circular convolution, in that $\Delta^{-1}(HV_1)$ and $\Delta^{-1}(HV_2)$ are defined to yield correct values for $p(n)$ and $q(n)$ on the interval $[0, L+M-2]$. Then, applying the idealized discrete convolution theorem to $p(n)$ and $q(n)$ and taking the DFT of (9), it follows that $Q$ satisfies the operator equation

$$Q = Z - \Delta\Phi\Delta^{-1}\frac{V_1}{V_2}Q. \quad (10)$$

Equivalently, since $Q=HV_2$, the DFT of the unknown kernel function $h(n)$ satisfies the equation

$$H = \hat{H} - \frac{1}{V_2}[\Delta\, \Phi\Delta^{-1}V_1 H] \quad (11)$$

where

$$\hat{H} = \frac{Z}{V_2}. \quad (12)$$

The function $\hat{h}(n)=\Delta^{-1}(\hat{H})$ can be interpreted as an estimate of the impulse-response that would be obtained if the error due to truncation of the input data were ignored and the DFT were applied directly to deconvolve

[3] The effect of assuming too small a value for $M$ will be discussed in Section III.

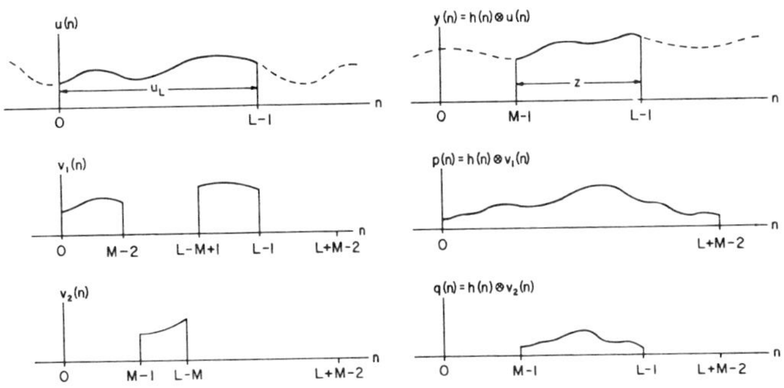


**Fig. 2.** Formulation for kernel function deconvolution.

the input–output relation. It may be noted that one necessary condition for deconvolution is that no element of $V_2$ have magnitude zero.

Equations (10) and (11) are operator equations of the form $x=\Gamma(x)$ that suggest the iterations $x_{n+1}=\Gamma(x_n)$, $n=0, 1, 2, \cdots$, in obtaining a solution. If the norm of the operator $\Gamma$ is a contraction [6], [7], i.e., if there exists a scalar $0<\lambda<1$ such that $\|\Gamma(x_1)-\Gamma(x_2)\| \leq \lambda\|x_1-x_2\|$ for all $x_1$ and $x_2$, then the iterations $x_{k+1}=\Gamma(x_k)$ leads to a convergent sequence that converges to the unique solution $x^*$ at a rate indicated by $\|x_k-x^*\| \leq (\lambda^k/1-\lambda)\|x_1-x_0\|$. The contraction mapping theorem is proved in Appendix A.

The operator $\Delta\,\Phi\Delta^{-1}(V_1/V_2)$ in (10) can be viewed as a matrix operator $\theta$

$$\theta = \Delta\Phi\Delta^{-1}\left(\frac{V_1}{V_2}\right) \quad (13)$$

acting in the discrete frequency domain of $N$ dimensions. Here $\Delta$ and $\Delta^{-1}$ are the $N \times N$ DFT matrices [8], $\Phi$ is a diagonal matrix of 0's and 1's with unity in the entries corresponding to the samples of the interval $M-1 \leq n \leq L-1$, and $V_1/V_2$ is a diagonal matrix with the diagonal entries $V_1(r)/V_2(r)$, $r=0, 1, 2, \cdots, N-1$. Since it is customary to define the DFT with integers ranging over $[0, N-1]$, as in (3), the diagonal entries for $\Phi$ are given by

$$\Phi_{ii} = \begin{cases} 1, & \text{for } M \leq i \leq L \\ 0, & \text{for } 1 \leq i \leq M-1 \text{ and } L+1 \leq i \leq N. \end{cases}$$

Two succeeding steps of the iterated version of (9) yield

$$Q_{k+1} - Q_k = \theta(Q_k) - \theta(Q_{k-1}) = \theta(Q_k - Q_{k-1}). \quad (14)$$

By the definition of the norm of a linear operator, a contraction is guaranteed when

$$0 < \|\theta\| < 1. \quad (15)$$

Several different norms for matrices may be applied [6]. Two of these are (where $\theta^*$ is the conjugate transpose of $\theta$) as follows.

*1) Spectral Norm—Square Root of the Largest Eigenvalue of $\theta^*\theta$:*

$$\|\theta\|_S \leq \|\theta\|_{SB} = \max_{0 \leq r \leq N-1} \left| \frac{V_1(r)}{V_2(r)} \right|.^4 \qquad (16)$$

*2) Trace Norm—Square Root of the Trace of $\theta\theta^*$:*

$$\|\theta\|_{TR} = \left[ \frac{2(L-M+1)}{N} \sum_{r=0}^{N-1} \left| \frac{V_1(r)}{V_2(r)} \right|^2 \right]^{\frac{1}{2}}. \qquad (17)$$

It can be shown [9] that the spectral norm bound will yield a smaller value than the trace norm for all $L$ and $M$ of reasonable size. Therefore, when

$$\max_{0 \leq r \leq N-1} \left| \frac{V_1(r)}{V_2(r)} \right| < 1 \qquad (18)$$

the operator $\theta$ will be a contraction, and $Q_{k+1} = Z - \theta Q_k$ for $k = 0, 1, \cdots$ will converge to the unique solution. Also, since the matrix operator

$$\theta' = (V_2)^{-1} \Delta \Phi \Delta^{-1} V_1 \qquad (19)$$

in (11) is related to the matrix $\theta$ by a similarity transformation, i.e., $V_2 \theta' (V_2)^{-1} = \theta$, the eigenvalues of $\theta$ and $\theta'$ are the same. Hence, the spectral norm bound of (16) will suffice as a predictor of convergence for either (10) or (11). As a practical matter, convergence can be obtained with a larger value of $\|\theta\|_{SB}$, as will be discussed in Section III.

Consider next the input deconvolution problem of solving (8) for $u(n)$, using the artificially truncated filter kernel function $h_L(n)$. It will be assumed that the unknown input $u(n)$ is identically zero outside a given interval $[0, M-1]$, which implies that the system is initially at rest. The dual of the kernel function deconvolution problem is formulated with reference to Fig. 3. The given filter kernel function $h(n)$ is decomposed into three parallel filters with kernels $h_1(n)$, $h_2(n)$, and $h_3(n)$, i.e., $h(n) = h_1(n) + h_2(n) + h_3(n)$, in such a way that the observation interval $[0, L-1]$ is at least as great as the settling time of the first filter when subject to the input $u(n)$, i.e., it is assumed that $L \geq M + \delta - 1$, as shown in Fig. 3. Under these circumstances the following identity holds, involving the responses of the first two filters.

$$\tilde{z}(n) = \tilde{q}(n) + \tilde{\Phi}(\tilde{p}(n)), \qquad 0 \leq n \leq L + M - 2 \qquad (20)$$

where $\tilde{\Phi}$ is the truncation operator defined by $\tilde{\Phi}(\tilde{p}(n)) = \tilde{w}(n) \cdot \tilde{p}(n)$, with the window function $\tilde{w}(n)$ given by

$$\tilde{w}(n) = \begin{cases} 1, & \text{for } 0 \leq n \leq L-1 \\ 0, & \text{otherwise.} \end{cases} \qquad (21)$$

The ideal conditions for the application of the circular convolution of order $N \geq L + M$ are seen to be upheld with respect to the convolved signals $\tilde{q}(n) = h_1(n) \otimes u(n)$ and $\tilde{p}(n) = h_2(n) \otimes u(n)$. The analogs of (10) and (11) are the operator equations

[4] This bound follows from the fact that the maximum eigenvalues of $\Delta\Phi\Delta^{-1}$ are precisely those of $\Phi$, which means that $\|\Delta\Phi\Delta^{-1}\|_{SB} = 1$.

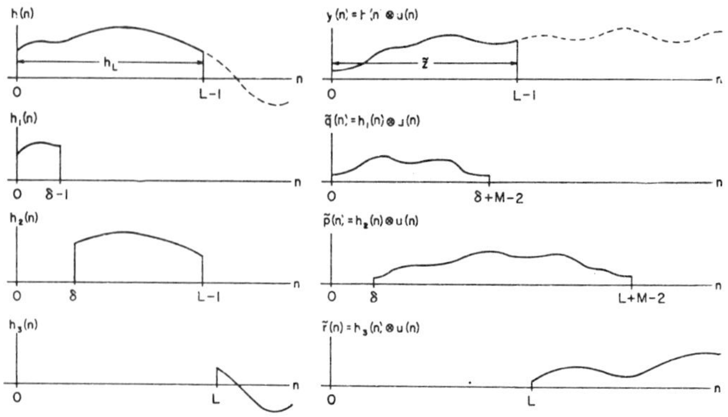


Fig. 3. Formulation for input function deconvolution.

$$\tilde{Q} = \tilde{Z} - \Delta\tilde{\Phi}\Delta^{-1} \left( \frac{H_2}{H_1} \right) \tilde{Q} \qquad (22)$$

$$U = \hat{U} - \frac{1}{H_1} [\Delta\tilde{\Phi}\Delta^{-1} H_2 U] \qquad (23)$$

where $U = \tilde{Q}/H_1$ and $\hat{U} = \tilde{Z}/H_1$. With the appropriate matrix interpretation to the operator $\tilde{\theta} = \Delta\tilde{\Phi}\Delta^{-1}(H_2/H_1)$, the spectral norm bound analogous to (16) is

$$\|\tilde{\theta}\|_{SB} = \max_{0 \leq r \leq L+N-1} \left| \frac{H_2(r)}{H_2(r)} \right|. \qquad (24)$$

Convergence of the iterations $\tilde{Q}_{k+1} = \tilde{Z} - \tilde{\theta}\tilde{Q}_k$, $k = 0, 1, \cdots$, applied to (22), or the similar iterative process applied to (23), is guaranteed to converge to the unique solution if the decomposition of $h_L(n)$ is chosen such that the norm in (24) is less than unity.

## III. Computer Results

An extensive computer investigation of the contraction map iterations applied to the operator equations of the previous section has been carried out in order to verify the utility of the algorithms. A standard FFT algorithm was used to compute the two DFT's at each stage of the iterations. Most of the following discussion refers to the operator equation (10), the conclusions of which apply with equal validity to (11), (22), and (23), in view of the relationships involved.

*1) A Particular Run:* An example of kernel function deconvolution to solve (10), for a particular kernel $h(n)$, is shown in Fig. 4. A 512-point DFT was used in this particular run ($N = 512$) with a value of $M = 80$. The truncation interval was $L = 4M = 320$. The first iterate $H_1(r)$ is the element $\hat{H}$ in (12), which is the kernel obtained by ignoring the error due to truncation. The effect of this error is graphically displayed in the log magnitude and kernel function plots of Fig. 4. Convergence of the algorithm was obtained in essentially ten iterations, with the error reduction indicated by the Euclidean norm (rms error) values: $\|H_1 - H\| = 50.28$, $\|H_{10} - H\| = 0.10$, $\|H_{15} - H\| = 0.01$. The spectral norm bound (16) was 2.04 for this run.

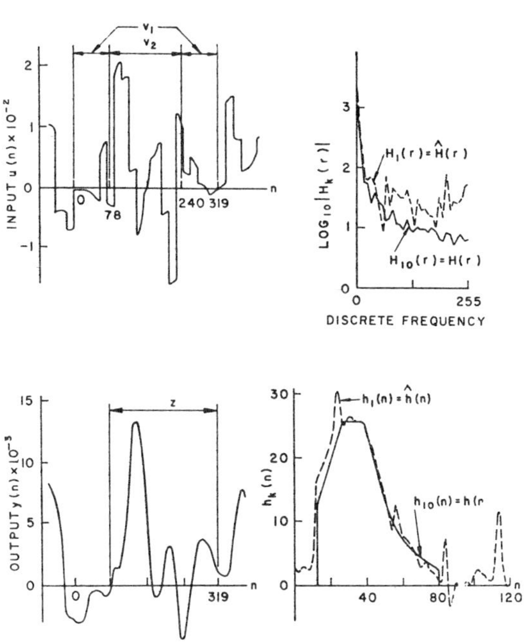


Fig. 4. Example of kernel function deconvolution.

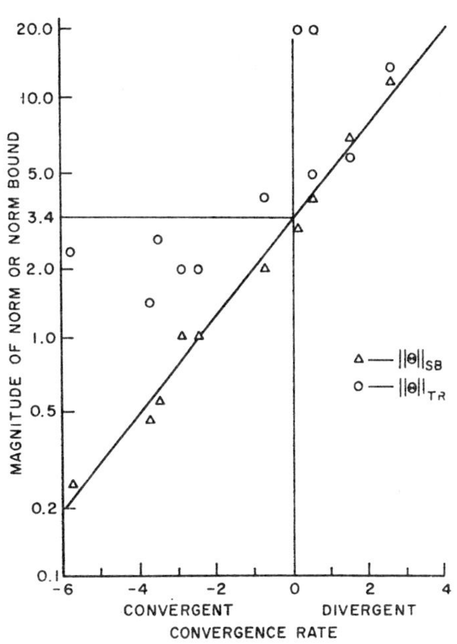


Fig. 6. Empirical correlation of norms to convergence.

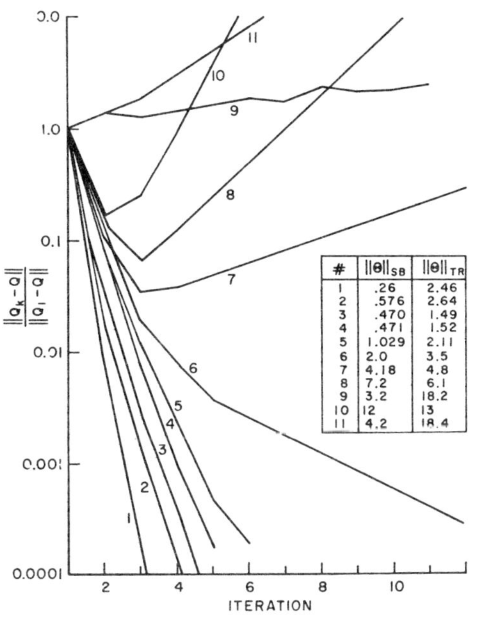


| # | $\|\Theta\|_{SB}$ | $\|\Theta\|_{TR}$ |
|---|---|---|
| 1 | .26 | 2.46 |
| 2 | .576 | 2.64 |
| 3 | .470 | 1.49 |
| 4 | .471 | 1.52 |
| 5 | 1.029 | 2.11 |
| 6 | 2.0 | 3.5 |
| 7 | 4.18 | 4.8 |
| 8 | 7.2 | 6.1 |
| 9 | 3.2 | 18.2 |
| 10 | 12 | 13 |
| 11 | 4.2 | 18.4 |

Fig. 5. Convergence of contraction map iterations.

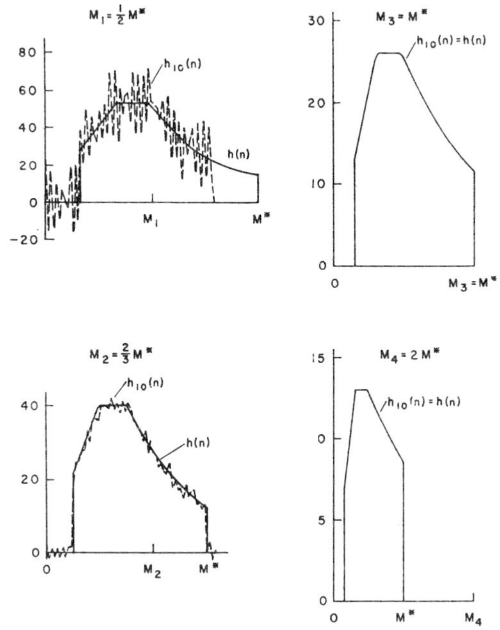


Fig. 7. Effect of assumed memory time.

*2) Summarized Data:* Using a variety of inputs (both deterministic and random) and a variety of systems (lumped parameter and otherwise), the results for approximately two hundred runs are summarized in Fig. 5. Values of the spectral norm bound and the trace norm were computed for each input and are shown in the table of Fig. 5 for the selected runs. In Fig. 6 the values of the spectral norm bound and the trace norm are plotted as a function of the convergence rate—the slope of the linear asymptotic portion of the curves shown in Fig. 5. These data reveal a very close correlation between the value of the spectral norm bound and the actual convergence of the iterations. While there is some correlation between the trace norm and convergence, it is not nearly as consistent as in the case of the spectral norm bound.

A variety of initial iterates $Q_o$ and kernel functions $h(n)$ (see Fig. 8) were used in the study in order to verify the contraction map property that convergence should be independent of these quantities.

*3) Effect of Incorrect Value for M:* The results summarized in Figs. 5 and 6 were obtained with the assumed memory length $M$ greater than or equal to the actual memory length $M^*$. The effect of an incorrect value of $M$ is demonstrated in Fig. 7 for kernel function deconvolution in which an input satisfying the empirical conditions for convergence was used in each case, i.e., $\|\theta\|_{SB} < 3.4$. The solution obtained by iteration possesses a noise-like character if $M < M^*$, while there is no deleterious effect if $M \geq M^*$. The error due to assuming $M < M^*$ is essentially an aliasing error in the time domain incurred by the necessity of working with time-limited signals. The magnitude of this error is a function of the relative energy contained in the ignored tail portion of $h(n)$, for $n \rightarrow M^*$. As seen from the curves of Fig. 7, a gross error in assuming too small a value for $M$ can be easily detected, if the input satisfies the conditions of convergence and the input-output data is essentially noise free.

## IV. Computational Complexity

The deconvolution problem may also be solved directly in the time domain by matrix inversion. Consider (1) in matrix notation as

$$y = Uh \tag{25}$$

where $y$ and $h$ are $M$-dimensional column vectors and $U$ is an $M \times M$ symmetric Toeplitz matrix

$$U = \begin{bmatrix} u(0) & u(-1) & u(-2) & \cdots & u(-(M-1)) \\ u(1) & u(0) & u(-1) & \cdots & u(-(M-2)) \\ & & \cdots & & \\ u(M-1) & u(M-2) & \cdots & u(0) & \end{bmatrix}. \tag{26}$$

If $U$ is nonsingular, (25) may be rearranged to yield

$$h = U^{-1}y. \tag{27}$$

Normal matrix inversion (e.g., Gaussian elimination) requires on the order of $M^3$ additions and multiplications. A method by Trench [10] for inverting Toeplitz matrices requires on the order of $M^2$ additions and multiplications, but may develop problems due to the requirement that successful inversion depends upon each of the minors being nonsingular.[5]

The complexity of DFT deconvolution is largely contained in calculating the FFT algorithm. It is necessary to calculate three DFT's to get the iteration started, $V_1$, $V_2$, and $Z$, and from there, two DFT's per iteration. If it is assumed, to simplify notation, that $L+M-1 = 2^m$, then for $I$ iterations the method requires $(2I+3) \cdot (L+M-1)m$ operations.[6] At the nominal value of ten iterations and for $L = kM+1$, one need compare $23m(k+1)M$ operations to $M^3$ or $23m(k+1)$ to $M^2$. Some examples are shown in Table I. It appears that for $M$ of nominal size, i.e., $>50$, that, when measured by complexity, the DFT technique is considerably better than Gaussian elimination, and somewhat worse than Trench's method.

[5] One may test for this condition being satisfied, but on the order of $M^3$ operations are required for the test! Only the iterative DFT method provides an inexpensive test.

[6] The DFT method operations are complex multiply-adds, and thus, for many computers, a factor of about four might multiply the given result. If an efficient FFT algorithm is used, however, much of this factor can be removed by applying symmetry conditions, etc.

TABLE I
Comparison of Computational Complexity

| M | L+M | $M^2$ | 23m(k+1) |
|---|---|---|---|
| 100 | 256 | $10\times10^3$ | 478 |
| | 512 | $10\times10^3$ | 1000 |
| 200 | 512 | $40\times10^3$ | 1000 |
| | 1024 | $40\times10^3$ | 2300 |
| 500 | 1024 | $250\times10^3$ | 2300 |
| | 2048 | $250\times10^3$ | 5000 |

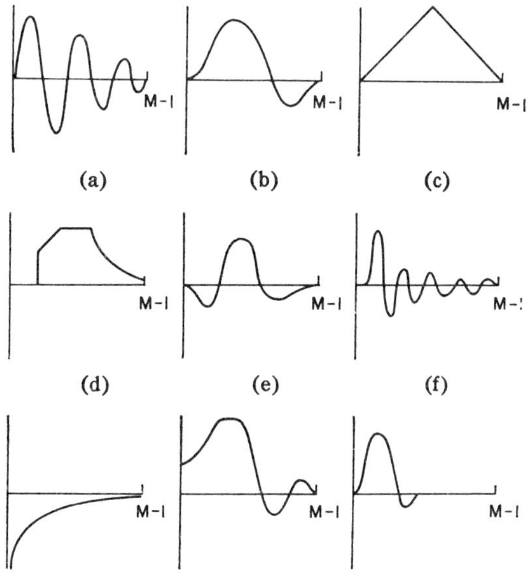


Fig. 8. Sample of kernel functions used for tests.

## V. Concluding Remarks

The method of this paper allows for the deconvolution of two time signals via discrete Fourier transform techniques when one of the signals is of infinite duration. The spectral norm bound, (16) or (24), has been shown to provide a practical indicator of convergence for the contraction map iterations. On the other hand, other iterative schemes for solving operator equations could be considered although only the contraction map iterations were investigated in this study. It is interesting, and in fact important, to point out that none of the numerical difficulties mentioned in [11] were encountered in this study, even though very smooth kernels, $h(n)$ in the neighborhood of $n=0$, were among those used in the study [see Fig. 8(b), (e), (f) and (i)]. In addition, small amounts of noise superimposed on the output data did not lead to instabilities in obtaining the solution. Ad hoc methods for handling large amounts of noise were also investigated [9], but will not be reported here. Other considerations of related interest can be found in [9]. These include a generalization to the multiinput multioutput case, and a more complete description of the computational complexity of the algorithm in relation to direct matrix inversion.

## Appendix A

### The Contraction Mapping Theorem

*Given:* $X$, a complete normed linear space (Banach space) with a norm $\|x\|$. $\Gamma$ is a contraction operator, if for $\Gamma: X \rightarrow X$ there is a number $\lambda$, with $0<\lambda<1$, such that for $x, y \in X$

$$\|\Gamma(x) - \Gamma(y)\| \leq \lambda \|x - y\|.$$

*Contraction Mapping Theorem:* If $X$ is as above and $\Gamma$ is a contraction in $X$, then $\Gamma$ has a fixed point and is unique, i.e., there is a unique $x^* \in X$ such that $\Gamma(x^*) = x^*$. Furthermore, the iterations $x_{k+1} = \Gamma(x_k)$ converge to $x^*$ at a rate indicated by $\|x_k - x^*\| \leq \lambda^k / 1 - \lambda \|x_1 - x_0\|$.

*Proof:*

*Uniqueness:* Suppose $\Gamma(x^*) = x$ *and* $\Gamma(y) = y$. Then if $x \neq y$, $\|x-y\| = \|\Gamma(x) - \Gamma(y)\|$, which is impossible by the contraction condition. Hence $x = y$.

*Existence:* Consider $x$, $\Gamma(x)$, $\Gamma^2(x), \cdots, \Gamma^n(x) = \Gamma(\Gamma^{n-1}(x))$ where $\Gamma^2 = \Gamma(\Gamma(x))$, etc. (We show $\Gamma^n(x)$ is Cauchy.) For $n>1$ and $p>0$

$$\begin{aligned} \|\Gamma^{n+p}(x) - \Gamma^n(x)\| &\leq \lambda \|\Gamma^{n+p-1}(x) - \Gamma^{n-1}(x)\| \leq \cdots \\ &\leq \lambda^n \|\Gamma^p(x) - x\| \leq \lambda^n \|\Gamma^p(x) - \Gamma^{p-1}(x) \\ &\quad + \Gamma^{p-1}(x) - \Gamma^{p-2}(x) \cdots - x\|. \end{aligned}$$

By the triangle inequality

$$\begin{aligned} &\leq \lambda^n \{\|\Gamma^p(x) - \Gamma^{p-1}(x)\| + \|\Gamma^{p-1}(x) \\ &\quad - \Gamma^{p-2}(x)\| + \cdots + \|\Gamma(x) - x\|\} \\ &\leq \lambda^n \|\Gamma(x) - x\| \{\lambda^{p-1} + \lambda^{p-2} + \cdots + \lambda + 1\} \end{aligned}$$

$$\|\Gamma^{k+p}(x) - \Gamma^k(x)\| < \frac{\lambda^n}{1-\lambda} \|\Gamma(x) - x\|.$$

Let $x^* = \lim_{k\to\infty} \Gamma^k(x_0)$. Since $\Gamma$ is a contraction, the interchange of operations in $\Gamma(x^*)^{k\leftarrow\infty} = \Gamma \lim_{k\to\infty} \Gamma^k(x^*) = \lim_{k\to\infty} \Gamma^{k+1}(x^*) = x^*$ is easily justified that $x^*$ is a fixed point. Also, as for the iterations $\Gamma^k(x) = x_k$, the rate of convergence is bounded by

$$\|x^* - x_k\| \leq \frac{\lambda^k}{1-\lambda} \|x_1 - x_0\|.$$

Q.E.D.

## References


[1] J. W. Cooley, P. A. W. Lewis, and P. D. Welch, "Application of the fast Fourier transform to computation of Fourier integrals, Fourier series, and convolution integrals," *IEEE Trans. Audio Electroacoust.*, vol. AU-15, pp. 79–84, June 1967.

[2] H. D. Helms, "Fast Fourier transform method of computing difference equations and simulating filters," *IEEE Trans. Audio Electroacoust.*, vol. AU-15, pp. 85–90, June 1967.

[3] T. G. Stockham, "High-speed convolution and correlation with applications to digital filtering," in *Digital Processing of Signals*, B. Gold and C. M. Rader, Ed. New York: McGraw-Hill, 1969, ch. 7.

[4] H. F. Silverman, "Fast Fourier transform methods," IBM Res. Rep. RC-3747, Feb. 1972.

[5] B. R. Hunt, "A matrix theory proof of the discrete convolution theorem," *IEEE Trans. Audio Electroacoust.*, vol. AU-19, pp. 285–288, Dec. 1971.

[6] L. Collatz, *Functional Analysis and Numerical Mathematics.* New York: Academic, 1966.

[7] C. Goffman and G. Pedrich, *First Course in Functional Analysis.* Englewood Cliffs, N. J.: Prentice-Hall, 1965.

[8] E. O. Brigham and R. E. Morrow, "The fast Fourier transform," *IEEE Spectrum*, vol. 4, pp. 63–70, Dec. 1967.

[9] H. F. Silverman, "Identification of linear systems using fast Fourier transform techniques," Ph. D. dissertation, Brown Univ., Providence, R. I., June 1971.

[10] W. F. Trench, "An algorithm for the inversion of finite Toeplitz matrices," *J. Soc. Ind. Appl. Math.*, vol. 12, pp. 515–522, 1964.

[11] B. R. Hunt, "A theorem on the difficulty of numerical deconvolution," *IEEE Trans. Audio Electroacoust.* (Corresp.), vol. AU-20, pp. 94–95, Mar. 1972.

# Linear Prediction: A Tutorial Review

JOHN MAKHOUL, MEMBER, IEEE

*Invited Paper*

*Abstract*—This paper gives an exposition of linear prediction in the analysis of discrete signals. The signal is modeled as a linear combination of its past values and present and past values of a hypothetical input to a system whose output is the given signal. In the frequency domain, this is equivalent to modeling the signal spectrum by a pole-zero spectrum. The major part of the paper is devoted to all-pole models. The model parameters are obtained by a least squares analysis in the time domain. Two methods result, depending on whether the signal is assumed to be stationary or nonstationary. The same results are then derived in the frequency domain. The resulting spectral matching formulation allows for the modeling of selected portions of a spectrum, for arbitrary spectral shaping in the frequency domain, and for the modeling of continuous as well as discrete spectra. This also leads to a discussion of the advantages and disadvantages of the least squares error criterion. A spectral interpretation is given to the normalized minimum prediction error. Applications of the normalized error are given, including the determination of an "optimal" number of poles. The use of linear prediction in data compression is reviewed. For purposes of transmission, particular attention is given to the quantization and encoding of the reflection (or partial correlation) coefficients. Finally, a brief introduction to pole-zero modeling is given.

## I. INTRODUCTION

### A. Overview

THE MATHEMATICAL analysis of the behavior of general dynamic systems (be they engineering, social, or economic) has been an area of concern since the beginning of this century. The problem has been pursued with accelerated vigor since the advent of electronic digital computers over two decades ago. The analysis of the outputs of dynamic systems was for the most part the concern of "time series analysis," which was developed mainly within the fields of statistics, econometrics, and communications. Most of the work on time series analysis was actually done by statisticians. More recently, advances in the analysis of dynamic systems have been made in the field of control theory based on state-space concepts and time domain analysis.

This paper is a tutorial review of one aspect of time series analysis: linear prediction (defined here). The exposition is based on an intuitive approach, with emphasis on the clarity of ideas rather than mathematical rigor. Although the large body of related literature available on this topic often requires advanced knowledge of statistics and/or control theory concepts, this paper employs no control theory concepts *per se* and only the basic notions of statistics and random processes. For example, the very important statistical concepts of *consistency* and *efficiency* [74], [75] in the estimation of parameters will not be dealt with. It is hoped this paper will serve as a simple introduction to some of the tools used in time series analysis, as well as be a detailed analysis of those aspects of linear prediction of interest to the specialist.

Manuscript received July 21, 1974; revised November, 1974. This work was supported by the Information Processing Techniques Branch of the Advanced Research Projects Agency under Contract DAHC15-17-C-0088.

The author is with Bolt Beranek and Newman, Inc., Cambridge, Mass. 02138.

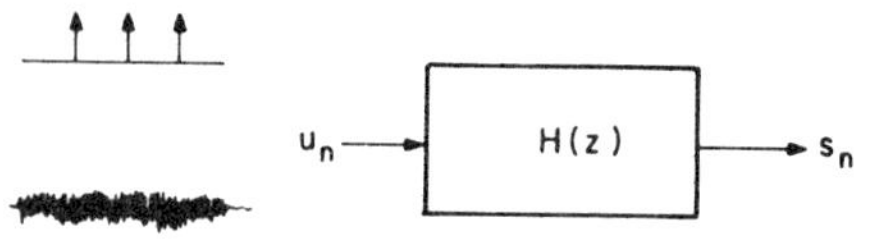


Fig. 1. Discrete speech production model.

### B. Current Applications

Before we delve into signal analysis, we shall give three examples of the types of problems that are of current interest.[1] These examples will then serve to illustrate some of the concepts that are developed.

*Neurophysics [15], [32], [36], [102]:* The spontaneous electrical brain activity is normally measured by means of electrodes placed on the patient's scalp. The recordings, known as electroencephalograms (or EEG signals), show certain periodicities (sharp resonances) accompanied by some randomness. These signals are believed to carry information about the medical status of the brain and are used by physicians as a means of diagnosis. It is of interest to detect the presence, position, and strength of the different resonances, known as rhythms. The three most common rhythms are known as the alpha, beta, and delta rhythms. Therefore, the basic interest here is to describe the spectrum in a simple mathematical manner that would yield the characteristics of the different rhythms.

*Geophysics [84]-[87], [114]:* In one of the successful methods of oil exploration, a charge of dynamite is exploded in the earth, and the resulting vibrations at various points on the surface of the ground are recorded by a seismograph as seismic traces. The job of the geophysicist is to use these traces in the determination of the structure of the sedimentary rock layers. Such information is then used to decide on the presence of oil in that area. Of interest here are the direct arrival times and strengths of the deep reflections of the explosion, which are then used to determine the layered structure. If somehow one is able to remove (deconvolve) the impulse response of the structure from the seismic trace, the desired arrival times should appear as impulses of different phases and amplitudes.

*Speech Communication [10], [33], [47], [50], [51], [62], [68], [89]:* In EEG analysis, the spectrum of the recorded signal was of interest. In seismic analysis, the spectral properties of the seismic trace were of interest only to facilitate the deconvolution process in order to obtain the desired impulses. In the analysis of speech, both types of information are of interest.

Fig. 1 shows a rather successful model of speech production. The model consists of a filter that is excited by either a quasi-periodic train of impulses or a random noise source. The periodic source produces voiced sounds such as vowels and

[1] For applications to economic and industrial time series, see for example [17].

Reprinted from *Proc. IEEE*, vol. 63, pp. 561–580, Apr. 1975.

nasals, and the noise source produces unvoiced or fricated sounds such as the fricatives (*f*, *th*, *s*, *sh*). The parameters of the filter determine the identity (spectral characteristics) of the particular sound for each of the two types of excitation.

Given a particular speech signal, it is of interest to determine the general type of sound it is, voiced or fricated, and if voiced what the pitch period is (i.e., distance between pitch pulses). In addition, one is interested in the identity of the sound which can be obtained from the spectrum. Such derived information can then be used in an automatic speech recognition system or a speech compression system.

### C. Linear Prediction

In applying time series analysis to the aforementioned applications, each continuous-time signal $s(t)$ is sampled to obtain a discrete-time[2] signal $s(nT)$, also known as a time series, where $n$ is an integer variable and $T$ is the sampling interval. The sampling frequency is then $f_s = 1/T$. (Henceforth, we shall abbreviate $s(nT)$ by $s_n$ with no loss in generality.)

A major concern of time series analysis [6], [11], [12], [14], [17], [43], [45], [46], [54], [105], [112] has been the estimation of power spectra, cross-spectra, coherence functions, autocorrelation and cross-correlation functions. A more active concern at this time is that of system modeling. It is clear that if one is successful in developing a parametric model for the behavior of some signal, then that model can be used for different applications, such as prediction or forecasting, control, and data compression.

One of the most powerful models currently in use is that where a signal $s_n$ is considered to be the output of some system with some unknown input $u_n$ such that the following relation holds:

$$s_n = -\sum_{k=1}^{p} a_k s_{n-k} + G \sum_{l=0}^{q} b_l u_{n-l}, \qquad b_0 = 1 \tag{1}$$

where $a_k$, $1 \leqslant k \leqslant p$, $b_l$, $1 \leqslant l \leqslant q$, and the gain $G$ are the parameters of the hypothesized system. Equation (1) says that the "output" $s_n$ is a linear function of past outputs and present and past inputs. That is, the signal $s_n$ is *predictable* from *linear* combinations of past outputs and inputs. Hence the name *linear prediction*.

Equation (1) can also be specified in the frequency domain by taking the $z$ transform on both sides of (1). If $H(z)$ is the transfer function of the system, as in Fig. 1, then we have from (1):

$$H(z) = \frac{S(z)}{U(z)} = G \frac{1 + \sum_{l=1}^{q} b_l z^{-l}}{1 + \sum_{k=1}^{p} a_k z^{-k}} \tag{2}$$

where

$$S(z) = \sum_{n=-\infty}^{\infty} s_n z^{-n} \tag{3}$$

is the $z$ transform of $s_n$, and $U(z)$ is the $z$ transform of $u_n$. $H(z)$ in (2) is the general *pole-zero model*. The roots of the numerator and denominator polynomials are the zeros and poles of the model, respectively.

There are two special cases of the model that are of interest:

1) all-zero model: $a_k = 0$, $1 \leqslant k \leqslant p$

2) all-pole model: $b_l = 0$, $1 \leqslant l \leqslant q$.

The all-zero model is known in the statistical literature as the *moving average* (MA) model, and the all-pole model is known as the *autoregressive* (AR) model [17]. The pole-zero model is then known as the *autoregressive moving average* (ARMA) model. In this paper we shall use the pole-zero terminology since it is more familiar to engineers.

The major part of this paper will be devoted to the all-pole model. This has been, by far, the most widely used model. Historically, the first use of an all-pole model in the analysis of time series is attributed to Yule [115] in a paper on sunspot analysis. Work on this subject, as well as on time series analysis in general, proceeded vigorously after 1933 when Kolmogorov laid a rigorous foundation for the theory of probability. Later developments by statisticians, such as Cramér and Wold, culminated in the parallel and independent work of Kolmogorov [58] and Norbert Wiener [107] on the prediction and filtering of stationary time series. For a bibliography on time series through the year 1959, see the encyclopedic work edited by Wold [113]. For a discussion of all-pole (autoregressive) models see, for example, [17], [45], [105], [112].

Much of the recent work on system modeling has been done in the area of control theory under the subjects of estimation and system identification. Recent survey papers with extensive references are those of Åström and Eykhoff [8] and Nieman *et al.* [73]. The December 1974 issue of the IEEE TRANSACTIONS ON AUTOMATIC CONTROL is devoted to the subject of system identification. Another relevant survey paper is that of Kailath [55] on linear filtering theory. Related books are those of Lee [60], Sage and Melsa [88], and Eykhoff [30].

### D. Paper Outline

Sections II–V deal exclusively with the all-pole model. In Section II, the estimation of model parameters is derived in the time domain by the method of least squares. The resulting normal equations are obtained for deterministic as well as random signals[3] (both stationary and nonstationary). Direct and iterative techniques are presented for the computation of the predictor coefficients, and the stability of the all-pole filter $H(z)$ is discussed. The response of the all-pole filter is then analyzed for two important types of input excitation: a deterministic impulse and statistical white noise.

In Section III, the all-pole modeling of a signal is derived completely in the frequency domain. The method of least squares translates into a spectral matching method where the signal spectrum is to be matched or fitted by a model spectrum. This formulation allows one to perform arbitrary spectral shaping before modeling. This viewpoint has special relevance today with the availability of hardware spectrum analyzers and fast Fourier transform techniques [21]. (We point out that all-pole modeling by linear prediction is identical to the method of maximum entropy spectral estimation [18], [96].)

Section IV gives a detailed discussion of the advantages and disadvantages of the least squares error criterion. The properties of the normalized error are reviewed. Its use is discussed

[2] See [80] for an exposition of the terminology in digital signal processing.

[3] See [12] for a description of deterministic and random signals.

in measuring the ill-conditioning of the normal equations, and in determining an optimal number of poles.

Section V discusses the use of linear prediction in data compression. Alternate representations of the linear predictor are presented and their properties under quantization are discussed. Particular emphasis is given to the quantization and encoding of the reflection (or partial correlation) coefficients.

Finally, in Section VI, a brief discussion of pole-zero modeling is given, with emphasis on methods presented earlier for the all-pole case.

## II. Parameter Estimation

### A. *All-Pole Model*

In the all-pole model, we assume that the signal $s_n$ is given as a linear combination of past values and some input $u_n$:

$$s_n = -\sum_{k=1}^{p} a_k s_{n-k} + G u_n \tag{4}$$

where $G$ is a gain factor. This model is shown in Fig. 2 in the time and frequency domains. The transfer function $H(z)$ in (2) now reduces to an all-pole transfer function

$$H(z) = \frac{G}{1 + \sum_{k=1}^{p} a_k z^{-k}}. \tag{5}$$

Given a particular signal $s_n$, the problem is to determine the predictor coefficients $a_k$ and the gain $G$ in some manner.

The derivations will be given using an intuitive least squares approach, assuming first that $s_n$ is a deterministic signal and then that $s_n$ is a sample from a random process. The results are identical to those obtained by the method of maximum likelihood [6], [74], [75] with the assumption that the signal is Gaussian [60], [73]. The reader is reminded of the existence of more general least squares methods such as weighted and *a priori* least squares [16], [90].

### B. *Method of Least Squares*

Here we assume that the input $u_n$ is totally unknown, which is the case in many applications, such as EEG analysis. Therefore, the signal $s_n$ can be predicted only approximately from a linearly weighted summation of past samples. Let this approximation of $s_n$ be $\tilde{s}_n$, where

$$\tilde{s}_n = -\sum_{k=1}^{p} a_k s_{n-k}. \tag{6}$$

Then the error between the actual value $s_n$ and the predicted value $\tilde{s}_n$ is given by

$$e_n = s_n - \tilde{s}_n = s_n + \sum_{k=1}^{p} a_k s_{n-k}. \tag{7}$$

$e_n$ is also known as the *residual*. In the method of least squares the parameters $a_k$ are obtained as a result of the minimization of the mean or total squared error with respect to each of the parameters. (Note that this problem is identical to the problem of designing the optimal one-step prediction digital Wiener filter [85].)

The analysis will be developed along two lines. First, we assume that $s_n$ is a deterministic signal, and then we give analogous derivations assuming that $s_n$ is a sample from a random process.

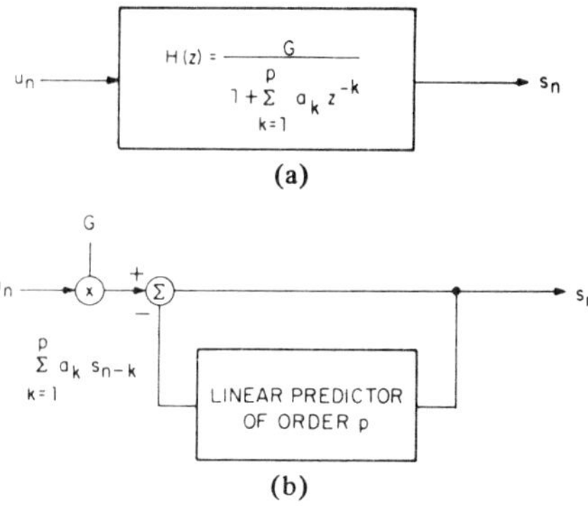


Fig. 2. (a) Discrete all-pole model in the frequency domain. (b) Discrete all-pole model in the time domain.

*1) Deterministic Signal:* Denote the total squared error by $E$, where

$$E = \sum_n e_n^2 = \sum_n \left(s_n + \sum_{k=1}^{p} a_k s_{n-k}\right)^2. \tag{8}$$

The range of the summation in (8) and the definition of $s_n$ in that range is of importance. However, let us first minimize $E$ without specifying the range of the summation. $E$ is minimized by setting

$$\frac{\partial E}{\partial a_i} = 0, \qquad 1 \leqslant i \leqslant p. \tag{9}$$

From (8) and (9) we obtain the set of equations:

$$\sum_{k=1}^{p} a_k \sum_n s_{n-k} s_{n-i} = -\sum_n s_n s_{n-i}, \qquad 1 \leqslant i \leqslant p. \tag{10}$$

Equations (10) are known in least squares terminology as the *normal equations*. For any definition of the signal $s_n$, (10) forms a set of $p$ equations in $p$ unknowns which can be solved for the predictor coefficients $\{a_k, 1 \leqslant k \leqslant p\}$ which minimize $E$ in (8).

The minimum total squared error, denoted by $E_p$, is obtained by expanding (8) and substituting (10). The result can be shown to be

$$E_p = \sum_n s_n^2 + \sum_{k=1}^{p} a_k \sum_n s_n s_{n-k}. \tag{11}$$

We shall now specify the range of summation over $n$ in (8), (10), and (11). There are two cases of interest, which will lead to two distinct methods for the estimation of the parameters.

*a) Autocorrelation method:* Here we assume that the error in (8) is minimized over the infinite duration $-\infty < n < \infty$. Equations (10) and (11) then reduce to

$$\sum_{k=1}^{p} a_k R(i-k) = -R(i), \qquad 1 \leqslant i \leqslant p \tag{12}$$

$$E_p = R(0) + \sum_{k=1}^{p} a_k R(k) \tag{13}$$

where

$$R(i) = \sum_{n=-\infty}^{\infty} s_n s_{n+i} \tag{14}$$

is the autocorrelation function of the signal $s_n$. Note that $R(i)$ is an even function of $i$, i.e.,

$$R(-i) = R(i). \tag{15}$$

Since the coefficients $R(i - k)$ form what often is known as an autocorrelation matrix, we shall call this method the *autocorrelation method*. An autocorrelation matrix is a symmetric Toeplitz matrix. (A Toeplitz matrix is one where all the elements along each diagonal are equal [42].)

In practice, the signal $s_n$ is known over only a finite interval, or we are interested in the signal over only a finite interval. One popular method is to multiply the signal $s_n$ by a *window* function $w_n$ to obtain another signal $s'_n$ that is zero outside some interval $0 \leqslant n \leqslant N - 1$:

$$s'_n = \begin{cases} s_n w_n, & 0 \leqslant n \leqslant N-1 \\ 0, & \text{otherwise.} \end{cases} \tag{16}$$

The autocorrelation function is then given by

$$R(i) = \sum_{n=0}^{N-1-i} s'_n s'_{n+i}, \qquad i \geqslant 0. \tag{17}$$

The shape of the window function $w_n$ can be of importance. The subject is discussed further in Section III.

*b) Covariance method:* In contrast with the autocorrelation method, here we assume that the error $E$ in (8) is minimized over a finite interval, say, $0 \leqslant n \leqslant N - 1$. Equations (10) and (11) then reduce to

$$\sum_{k=1}^{p} a_k \varphi_{ki} = -\varphi_{0i}, \qquad 1 \leqslant i \leqslant p \tag{18}$$

$$E_p = \varphi_{00} + \sum_{k=1}^{p} a_k \varphi_{0k} \tag{19}$$

where

$$\varphi_{ik} = \sum_{n=0}^{N-1} s_{n-i} s_{n-k} \tag{20}$$

is the covariance of the signal $s_n$ in the given interval. The coefficients $\varphi_{ki}$ in (18) form a covariance matrix, and, therefore, we shall call this method the *covariance method*. From (20) it can be easily shown that the covariance matrix $\varphi_{ik}$ is symmetric, i.e., $\varphi_{ik} = \varphi_{ki}$. However, unlike the autocorrelation matrix, the terms along each diagonal are not equal. This can be seen by writing from (20)

$$\varphi_{i+1,k+1} = \varphi_{ik} + s_{-i-1} s_{-k-1} - s_{N-1-i} s_{N-1-k}. \tag{21}$$

Note from (21) also that values of the signal $s_n$ for $-p \leqslant n \leqslant N - 1$ must be known: a total of $p + N$ samples. The covariance method reduces to the autocorrelation method as the interval over which $n$ varies goes to infinity.

We point out here that the covariance method is similar to the method of Prony [49], [71] where a signal is approximated by the summation of a set of damped exponentials.

*2) Random Signal:* If the signal $s_n$ is assumed to be a sample of a random process, then the error $e_n$ in (7) is also a sample of a random process. In the least squares method, we minimize the expected value of the square of the error. Thus

$$E = \mathcal{E}(e_n^2) = \mathcal{E}\left(s_n + \sum_{k=1}^{p} a_k s_{n-k}\right)^2. \tag{22}$$

Applying (9) to (22), we obtain the normal equations:

$$\sum_{k=1}^{p} a_k \mathcal{E}(s_{n-k} s_{n-i}) = -\mathcal{E}(s_n s_{n-i}), \qquad 1 \leqslant i \leqslant p. \tag{23}$$

The minimum average error is then given by

$$E_p = \mathcal{E}(s_n^2) + \sum_{k=1}^{p} a_k \mathcal{E}(s_n s_{n-k}). \tag{24}$$

Taking the expectations in (23) and (24) depends on whether the process $s_n$ is stationary or nonstationary.

*a) Stationary case:* For a stationary process $s_n$, we have

$$\mathcal{E}(s_{n-k} s_{n-i}) = R(i - k) \tag{25}$$

where $R(i)$ is the autocorrelation of the process. Equations (23) and (24) now reduce to equations identical to (12) and (13), respectively. The only difference is that here the autocorrelation is that of a stationary process instead of a deterministic signal. For a stationary (and ergodic) process the autocorrelation can be computed as a time average [12]. Different approximations have been suggested in the literature [54] for estimating $R(i)$ from a finite known signal $s_n$. One such approximation is given by (17).[4] Using this estimate in the stationary case gives the same solution for the coefficients $a_k$ as the autocorrelation method in the deterministic case.

*b) Nonstationary case:* For a nonstationary process $s_n$, we have

$$\mathcal{E}(s_{n-k} s_{n-i}) = R(n - k, n - i) \tag{26}$$

where $R(t, t')$ is the nonstationary autocorrelation between times $t$ and $t'$. $R(n - k, n - i)$ is a function of the time index $n$. Without loss of generality, we shall assume that we are interested in estimating the parameters $a_k$ at time $n = 0$. Then, (23) and (24) reduce to

$$\sum_{k=1}^{p} a_k R(-k, -i) = -R(0, -i) \tag{27}$$

$$E'_p = R(0, 0) + \sum_{k=1}^{p} a_k R(0, k). \tag{28}$$

In estimating the nonstationary autocorrelation coefficients from the signal $s_n$, we note that nonstationary processes are not ergodic, and, therefore, one cannot substitute the ensemble average by a time average. However, for a certain class of nonstationary processes known as *locally stationary processes* [12], [92], it is reasonable to estimate the autocorrelation function with respect to a point in time as a short-time average. Examples of nonstationary processes that can be considered to be locally stationary are speech and EEG signals.

[4] Usually the estimate given by (17) is divided by $N$, but that does not affect the solution for the predictor coefficients.

In a manner analogous to the stationary case, we estimate $R(-k, -i)$ by $\varphi_{ik}$ in (20). Using this approximation for the nonstationary autocorrelation leads to a solution for the parameters $a_k$ in (27) that is identical to that given by (18) in the covariance method in the deterministic case.

Note that for a stationary signal: $R(t, t') = R(t - t')$, and therefore, the normal equations (27) and (28) reduce to (12) and (13).

*3) Gain Computation:* Since in the least squares method we assumed that the input was unknown, it does not make much sense to determine a value for the gain $G$. However, there are certain interesting observations that can be made.

Equation (7) can be rewritten as

$$s_n = -\sum_{k=1}^{p} a_k s_{n-k} + e_n. \tag{29}$$

Comparing (4) and (29) we see that the only input signal $u_n$ that will result in the signal $s_n$ as output is that where $Gu_n = e_n$. That is, the input signal is proportional to the error signal. For any other input $u_n$, the output from the filter $H(z)$ in Fig. 2 will be different from $s_n$. However, if we insist that whatever the input $u_n$, the energy in the output signal must equal that of the original signal $s_n$, then we can at least specify the total energy in the input signal. Since the filter $H(z)$ is fixed, it is clear from the above that the total energy in the input signal $Gu_n$ must equal the total energy in the error signal, which is given by $E_p$ in (13) or (19), depending on the method used.

Two types of input that are of special interest are: the deterministic impulse and stationary white noise. By examining the response of the filter $H(z)$ to each of these two inputs we shall gain further insight into the time domain properties of the all-pole model. The input gain is then determined as a byproduct of an autocorrelation analysis.

*a) Impulse input:* Let the input to the all-pole filter $H(z)$ be an impulse or unit sample at $n = 0$, i.e. $u_n = \delta_{n0}$, where $\delta_{nm}$ is the Kronecker delta. The output of the filter $H(z)$ is then its impulse response $h_n$, where

$$h_n = -\sum_{k=1}^{p} a_k h_{n-k} + G\delta_{n0}. \tag{30}$$

The autocorrelation $\hat{R}(i)$ of the impulse response $h_n$ has an interesting relationship to the autocorrelation $R(i)$ of the signal $s_n$. Multiply (30) by $h_{n-i}$ and sum over all $n$. The result can be shown to be [10], [62]:

$$\hat{R}(i) = -\sum_{k=1}^{p} a_k \hat{R}(i-k), \qquad 1 \leqslant |i| \leqslant \infty \tag{31}$$

$$\hat{R}(0) = -\sum_{k=1}^{p} a_k \hat{R}(k) + G^2. \tag{32}$$

Given our condition that the total energy in $h_n$ must equal that in $s_n$, we must have

$$\hat{R}(0) = R(0) \tag{33}$$

since the zeroth autocorrelation coefficient is equal to the total energy in the signal. From (33) and the similarity between (12) and (31) we conclude that [62]

$$\hat{R}(i) = R(i), \qquad 0 \leqslant i \leqslant p. \tag{34}$$

This says that the first $p + 1$ coefficients of the autocorrelation of the impulse response of $H(z)$ are identical to the corresponding autocorrelation coefficients of the signal. The problem of linear prediction using the autocorrelation method can be stated in a new way as follows. Find a filter of the form $H(z)$ in (5) such that the first $p + 1$ values of the autocorrelation of its impulse response are equal to the first $p + 1$ values of the signal autocorrelation, and such that (31) applies.

From (32), (34), and (13), the gain is equal to

$$G^2 = E_p = R(0) + \sum_{k=1}^{p} a_k R(k) \tag{35}$$

where $G^2$ is the total energy in the input $G\delta_{n0}$.

*b) White noise input:* Here the input $u_n$ is assumed to be a sequence of uncorrelated samples (white noise) with zero mean and unit variance, i.e., $\mathcal{E}(u_n) = 0$, all $n$, and $\mathcal{E}(u_n u_m) = \delta_{nm}$. Denote the output of the filter by $\hat{s}_n$. For a fixed filter $H(z)$, the output $\hat{s}_n$ forms a stationary random process:

$$\hat{s}_n = -\sum_{k=1}^{p} a_k \hat{s}_{n-k} + Gu_n. \tag{36}$$

Multiply (36) by $\hat{s}_{n-i}$ and take expected values. By noting that $u_n$ and $\hat{s}_{n-i}$ are uncorrelated for $i > 0$, the result can be shown [17] to be identical to (31) and (32), where $\hat{R}(i) = \mathcal{E}(\hat{s}_n \hat{s}_{n-i})$ is the autocorrelation of the output $\hat{s}_n$. Therefore, (31) and (32) completely specify an all-pole random process as well. Equations (31) are known in the statistical literature as the *Yule-Walker equations* [17], [98], [115].

For the random case we require that the average energy (or variance) of the output $\hat{s}_n$ be equal to the variance of the original signal $s_n$, or $\hat{R}(0) = R(0)$, since the zeroth autocorrelation of a zero-mean random process is the variance. By a reasoning similar to that given in the previous section, we conclude that (34) and (35) also apply for the random case.

From the preceding, we see that the relations linking the autocorrelation coefficients of the output of an all-pole filter are the same whether the input is a single impulse or white noise. This is to be expected since both types of input have identical autocorrelations and, of course, identical flat spectra. This dualism between the deterministic impulse and statistical white noise is an intriguing one. Its usefulness surfaces very elegantly in modeling the speech process, as in Fig. 1, where both unit impulses as well as white noise are actually used to synthesize speech.

### C. Computation of Predictor Parameters

*1) Direct Methods:* In each of the two formulations of linear prediction presented in the previous section, the predictor coefficients $a_k$, $1 \leqslant k \leqslant p$, can be computed by solving a set of $p$ equations with $p$ unknowns. These equations are (12) for the autocorrelation (stationary) method and (18) for the covariance (nonstationary) method. There exist several standard methods for performing the necessary computations, e.g., the Gauss reduction or elimination method and the Crout reduction method [49]. These general methods require $p^3/3 + O(p^2)$ operations (multiplications or divisions) and $p^2$ storage locations. However, we note from (12) and (18) that the matrix of coefficients in each case is a covariance matrix. Covariance matrices are symmetric and in general positive semidefinite, although in practice they are usually positive definite. Therefore, (12) and (18) can be solved more efficiently by the square-root or Cholesky decomposition method [31], [39], [59], [110]. This method requires about half the computation $p^3/6 + O(p^2)$ and about half the storage $p^2/2$ of the general methods. The numerical stability properties of this method are well understood [109], [111]; the method is considered to be quite stable.

Further reduction in storage and computation time is possible in solving the autocorrelation normal equations (12) because of their special form. Equation (12) can be expanded in matrix form as

$$\begin{bmatrix} R_0 & R_1 & R_2 & \cdots & R_{p-1} \\ R_1 & R_0 & R_1 & \cdots & R_{p-2} \\ R_2 & R_1 & R_0 & \cdots & R_{p-3} \\ \vdots & \vdots & \vdots & & \vdots \\ R_{p-1} & R_{p-2} & R_{p-3} & \cdots & R_0 \end{bmatrix} \begin{bmatrix} a_1 \\ a_2 \\ a_3 \\ \vdots \\ a_p \end{bmatrix} = - \begin{bmatrix} R_1 \\ R_2 \\ R_3 \\ \vdots \\ R_p \end{bmatrix} \tag{37}$$

Note that the $p \times p$ autocorrelation matrix is symmetric and the elements along any diagonal are identical (i.e., a Toeplitz matrix). Levinson [61] derived an elegant recursive procedure for solving this type of equation. The procedure was later reformulated by Robinson [85]. Levinson's method assumes the column vector on the right hand side of (37) to be a general column vector. By making use of the fact that this column vector comprises the same elements found in the autocorrelation matrix, another method attributed to Durbin [25] emerges which is twice as fast as Levinson's. The method requires only $2p$ storage locations and $p^2 + O(p)$ operations: a big saving from the more general methods. Durbin's recursive procedure can be specified as follows:

$$E_0 = R(0) \tag{38a}$$

$$k_i = -\left[R(i) + \sum_{j=1}^{i-1} a_j^{(i-1)} R(i-j)\right] \Big/ E_{i-1} \tag{38b}$$

$$a_i^{(i)} = k_i$$

$$a_j^{(i)} = a_j^{(i-1)} + k_i a_{i-j}^{(i-1)}, \qquad 1 \leqslant j \leqslant i-1 \tag{38c}$$

$$E_i = (1 - k_i^2) E_{i-1}. \tag{38d}$$

Equations (38b)–(38d) are solved recursively for $i = 1, 2, \cdots, p$. The final solution is given by

$$a_j = a_j^{(p)}, \qquad 1 \leqslant j \leqslant p. \tag{38e}$$

Note that in obtaining the solution for a predictor of order $p$, one actually computes the solutions for all predictors of order less than $p$. It has been reported [78] that this solution is numerically relatively unstable. However, most researchers have not found this to be a problem in practice.

It should be emphasized that, for many applications, the solution of the normal equations (12) or (18) does not form the major computational load. The computation of the autocorrelation or covariance coefficients require $pN$ operations, which can dominate the computation time if $N >> p$, as is often the case.

The solution to (37) is unaffected if all the autocorrelation coefficients are scaled by a constant. In particular, if all $R(i)$ are normalized by dividing by $R(0)$, we have what are known as the *normalized autocorrelation coefficients* $r(i)$:

$$r(i) = \frac{R(i)}{R(0)} \tag{39}$$

which have the property that $|r(i)| \leqslant 1$. This can be useful in the proper application of scaling to a fixed point solution to (37).

A byproduct of the solution in (38) is the computation of the minimum total error $E_i$ at every step. It can easily be shown that the minimum error $E_i$ decreases (or remains the same) as the order of the predictor increases [61]. $E_i$ is never negative, of course, since it is a squared error. Therefore, we must have

$$0 \leqslant E_i \leqslant E_{i-1}, \qquad E_0 = R(0). \tag{40}$$

If the autocorrelation coefficients are normalized as in (39), then the minimum error $E_i$ is also divided by $R(0)$. We shall call the resulting quantity the *normalized error* $V_i$:

$$V_i = \frac{E_i}{R(0)} = 1 + \sum_{k=1}^{i} a_k r(k). \tag{41}$$

From (40) it is clear that

$$0 \leqslant V_i \leqslant 1, \qquad i \geqslant 0. \tag{42}$$

Also, from (38d) and (41), the final normalized error $V_p$ is

$$V_p = \prod_{i=1}^{p} (1 - k_i^2). \tag{43}$$

The intermediate quantities $k_i$, $1 \leqslant i \leqslant p$, are known as the *reflection coefficients*. In the statistical literature, they are known as *partial correlation coefficients* [6], [17]. $k_i$ can be interpreted as the (negative) partial correlation between $s_n$ and $s_{n+i}$ holding $s_{n+1}, \cdots, s_{n+i-1}$ fixed. The use of the term "reflection coefficient" comes from transmission line theory, where $k_i$ can be considered as the reflection coefficient at the boundary between two sections with impedances $Z_i$ and $Z_{i+1}$. $k_i$ is then given by

$$k_i = \frac{Z_{i+1} - Z_i}{Z_{i+1} + Z_i}. \tag{44}$$

The transfer function $H(z)$ can then be considered as that of a sequence of these sections with impedance ratios given from (44) by

$$\frac{Z_{i+1}}{Z_i} = \frac{1 + k_i}{1 - k_i}, \qquad 1 \leqslant i \leqslant p. \tag{45}$$

The same explanation can be given for any type of situation where there is plane wave transmission with normal incidence in a medium consisting of a sequence of sections or slabs with different impedances. In the case of an acoustic tube with $p$ sections of equal thickness, the impedance ratios reduce to the inverse ratio of the consecutive cross-sectional areas. This fact has been used recently in speech analysis [10], [52], [97]. Because of the more familiar "engineering interpretation" for $k_i$, we shall refer to them in this paper as reflection coefficients.

*2) Iterative Methods:* Beside the direct methods for the solution of simultaneous linear equations, there exist a number of iterative methods. In these methods, one begins by an initial guess for the solution. The solution is then updated by adding a correction term that is usually based on the gradient of some error criterion. In general, iterative methods require more computation to achieve a desired degree of convergence than the direct methods. However, in some applications [100] one often has a good initial guess, which might lead to the solution in only a few iterations. This can be a big saving over direct methods if the number of equations is large. Some of the iterative methods are the gradient method, the steepest descent method, Newton's method, conjugate gradient method and the stochastic approximation method [81], [108].

Up till now we have assumed that the whole signal is given all at once. For certain real time applications it is useful to be able to perform the computations as the signal is coming in. Adaptive schemes exist which update the solution based on every new observation of the signal [106]. The update is usually proportional to the difference between the new observation and the predicted value given the present solution. Another application for adaptive procedures is in the processing of very long data records, where the solution might converge long before all the data is analyzed. It is worth noting that Kalman filtering notions [56] are very useful in obtaining adaptive solutions [60].

*3) Filter Stability:* After the predictor parameters are computed, the question of the stability of the resulting filter $H(z)$ arises. Filter stability is important for many applications. A causal all-pole filter is stable if all its poles lie inside the unit circle (in which case it is also a filter with minimum phase). The poles of $H(z)$ are simply the roots of the denominator polynomial $A(z)$, where

$$A(z) = 1 + \sum_{k=1}^{p} a_k z^{-k} \tag{46}$$

and

$$H(z) = \frac{G}{A(z)}. \tag{47}$$

$A(z)$ is also known as the *inverse filter*.

If the coefficients $R(i)$ in (12) are positive definite [79] (which is assured if $R(i)$ is computed from a nonzero signal using (17) or from a positive definite spectrum[5]), the solution of the autocorrelation equation (12) gives predictor parameters which guarantee that all the roots of $A(z)$ lie inside the unit circle, i.e., a stable $H(z)$ [42], [85], [104]. This result can also be obtained from orthogonal polynomial theory. In fact, if one denotes the inverse filter at step $i$ in iteration (38) by $A_i(z)$, then it can be shown that the polynomials $z^i A_i(z)$ for $i = 0, 1, 2, \cdots$, form an orthogonal set over the unit circle [35], [42], [93]:

$$\frac{1}{2\pi}\int_{-\pi}^{\pi} P(\omega)\, A_n(e^{j\omega}) A_m(e^{-j\omega})\, e^{j(n-m)\omega}\, d\omega = E_n \delta_{nm}, \qquad n, m = 0, 1, 2, \cdots \tag{48}$$

where $E_n$ is the minimum error for an $n$th order predictor, and $P(\omega)$ is any positive definite spectrum whose Fourier transform results in the autocorrelation coefficients $R(i)$ that are used in (12). The recurrence relation for these polynomials is as follows:

$$A_i(z) = A_{i-1}(z) + k_i z^{-i} A_{i-1}(z^{-1}) \tag{49}$$

which is the same as the recursion in (38c).

The positive definiteness of $R(i)$ can often be lost if one uses a small word length to represent $R(i)$ in a computer. Also, roundoff errors can cause the autocorrelation matrix to become ill-conditioned. Therefore, it is often necessary to check for the stability of $H(z)$. Checking if the roots of $A(z)$ are inside the unit circle is a costly procedure that is best avoided One method is to check if all the successive errors are positive. In fact, the condition $E_i > 0$, $1 \leqslant i \leqslant p$, is a necessary and sufficient condition for the stability of $H(z)$. From (38d) and (40) it is clear that an equivalent condition for the stability of $H(z)$ is that

$$|k_i| < 1, \qquad 1 \leqslant i \leqslant p. \tag{50}$$

Therefore, the recursive procedure (38) also facilitates the check for the stability of the filter $H(z)$.

The predictor parameters resulting from a solution to the covariance matrix equation (18) cannot in general be guaranteed to form a stable filter. The computed filter tends to be more stable as the number of signal samples $N$ is increased, i.e., as the covariance matrix approaches an autocorrelation matrix. Given the computed predictor parameters, it is useful to be able to test for the stability of the filter $H(z)$. One method is to compute the reflection coefficients $k_i$ from the predictor parameters by a backward recursion, and then check for stability using (50). The recursion is as follows:

$$k_i = a_i^{(i)}$$

$$a_j^{(i-1)} = \frac{a_j^{(i)} - a_i^{(i)} a_{i-j}^{(i)}}{1 - k_i^2}, \qquad 1 \leqslant j \leqslant i - 1 \tag{51}$$

where the index $i$ takes values $p, p-1, \cdots, 1$ in that order. Initially $a_j^{(p)} = a_j$, $1 \leqslant j \leqslant p$. It is interesting to note that this method for checking the stability of $H(z)$ is essentially the same as the Lehmer–Schur method [81] for testing whether or not the zeros of a polynomial lie inside the unit circle. An unstable filter can be made stable by reflecting the poles outside the unit circle inside [10], such that the magnitude of the system frequency response remains the same. Filter instability can often be avoided by adding a very small number to the diagonal elements in the covariance matrix.

A question always arises as to whether to use the autocorrelation method or covariance method in estimating the predictor parameters. The covariance method is quite general and can be used with no restrictions. The only problem is that of the stability of the resulting filter, which is not a severe problem generally. In the autocorrelation method, on the other hand, the filter is guaranteed to be stable, but problems of parameter accuracy can arise because of the necessity of windowing (truncating) the time signal. This is usually a problem if the signal is a portion of an impulse response. For example, if the impulse response of an all-pole filter is analyzed by the covariance method, the filter parameters can be computed accurately from only a finite number of samples of the signal. Using the autocorrelation method, one cannot obtain the exact parameter values unless the whole infinite impulse response is used in the analysis. However, in practice, very good approximations can be obtained by truncating the impulse response at a point where most of the decay of the response has already occurred.

[5] A spectrum that can be zero at most at a countable set of frequencies.

## III. Spectral Estimation

In Section II, the stationary and nonstationary methods of linear prediction were derived from a time domain formulation. In this section we show that the same normal equations can be derived from a frequency domain formulation. It will become clear that linear prediction is basically a correlation type of analysis which can be approached either from the time or frequency domain. The insights gained from the frequency domain analysis will lead to new applications for linear predictive analysis. This section and the following are based mainly on references [62]-[64].

### A. Frequency Domain Formulations

*1) Stationary Case:* The error $e_n$ between the actual signal and the predicted signal is given by (7). Applying the $z$ transform to (7), we obtain

$$E(z) = \left[1 + \sum_{k=1}^{p} a_k z^{-k}\right] S(z) = A(z)\, S(z) \tag{52}$$

where $A(z)$ is the inverse filter defined in (46), and $E(z)$ and $S(z)$ are the $z$ transforms of $e_n$ and $s_n$, respectively. Therefore, $e_n$ can be viewed as the result of passing $s_n$ through the inverse filter $A(z)$. Assuming a deterministic signal[6] $s_n$, and applying Parseval's theorem, the total error to be minimized is given by

$$E = \sum_{n=-\infty}^{\infty} e_n^2 = \frac{1}{2\pi} \int_{-\pi}^{\pi} |E(e^{j\omega})|^2 \, d\omega \tag{53}$$

where $E(e^{j\omega})$ is obtained by evaluating $E(z)$ on the unit circle $z = e^{j\omega}$. Denoting the power spectrum of the signal $s_n$ by $P(\omega)$, where

$$P(\omega) = |S(e^{j\omega})|^2 \tag{54}$$

we have from (52)-(54)

$$E = \frac{1}{2\pi} \int_{-\pi}^{\pi} P(\omega) A(e^{j\omega}) A(e^{-j\omega}) \, d\omega. \tag{55}$$

Following the same procedure as in Section II, $E$ is minimized by applying (9) to (55). The result can be shown [64] to be identical to the autocorrelation normal equations (12), but with the autocorrelation $R(i)$ obtained from the signal spectrum $P(\omega)$ by an inverse Fourier transform

$$R(i) = \frac{1}{2\pi} \int_{-\pi}^{\pi} P(\omega) \cos(i\omega) \, d\omega. \tag{56}$$

Note that in (56) the cosine transform is adequate since $P(\omega)$ is real and even. The minimum squared error $E_p$ can be obtained by substituting (12) and (56) in (55), which results in the same equation as in (13).

*2) Nonstationary Case:* Here the signal $s_n$ and the error $e_n$ are assumed to be nonstationary. If $R(t, t')$ is the nonstationary autocorrelation of $s_n$, then we define the nonstationary two-dimensional (2-$D$) spectrum $Q(\omega, \omega')$ of $s_n$ by [12], [64], [67], [79]

$$Q(\omega, \omega') = \sum_{t'=-\infty}^{\infty} \sum_{t=-\infty}^{\infty} R(t, t') \exp[-j(\omega t - \omega' t')]. \tag{57}$$

[6] A similar development assuming a random signal gives the same results.

$R(t, t')$ can then be recovered from $Q(\omega, \omega')$ by an inverse 2-$D$ Fourier transform

$$R(t, t') = \left(\frac{1}{2\pi}\right)^2 \int_{-\pi}^{\pi} \int_{-\pi}^{\pi} Q(\omega, \omega') \exp[j(\omega t - \omega' t')] \, d\omega \, d\omega'. \tag{58}$$

As in the time domain formulation, we are interested in minimizing the error variance for time $n = 0$, which is now given by [64]

$$E = \left(\frac{1}{2\pi}\right)^2 \int_{-\pi}^{\pi} \int_{-\pi}^{\pi} Q(\omega, \omega') A(e^{j\omega}) A(e^{-j\omega'}) \, d\omega \, d\omega'. \tag{59}$$

Applying (9) to (59) results in equations identical to the nonstationary normal equations (27), where $R(t, t')$ is now defined by (58). The minimum error is then obtained by substituting (27) and (58) in (59). The answer is identical to (28).

### B. Linear Predictive Spectral Matching

In this section we shall examine in what manner the signal spectrum $P(\omega)$ is approximated by the all-pole model spectrum, which we shall denote by $\hat{P}(\omega)$. From (5) and (47):

$$\hat{P}(\omega) = |H(e^{j\omega})|^2 = \frac{G^2}{|A(e^{j\omega})|^2} = \frac{G^2}{\left|1 + \sum_{k=1}^{p} a_k e^{-jk\omega}\right|^2}. \tag{60}$$

From (52) and (54) we have

$$P(\omega) = \frac{|E(e^{j\omega})|^2}{|A(e^{j\omega})|^2}. \tag{61}$$

By comparing (60) and (61) we see that if $P(\omega)$ is being modeled by $\hat{P}(\omega)$, then the error power spectrum $|E(e^{j\omega})|^2$ is being modeled by a flat spectrum equal to $G^2$. This means that the actual error signal $e_n$ is being approximated by another signal that has a flat spectrum, such as a unit impulse, white noise, or any other signal with a flat spectrum. The filter $A(z)$ is sometimes known as a "whitening filter" since it attempts to produce an output signal $e_n$ that is white, i.e., has a flat spectrum.

From (53), (60), and (61), the total error can be written as

$$E = \frac{G^2}{2\pi} \int_{-\pi}^{\pi} \frac{P(\omega)}{\hat{P}(\omega)} \, d\omega. \tag{62}$$

Therefore, minimizing the total error $E$ is equivalent to the minimization of the integrated ratio of the signal spectrum $P(\omega)$ to its approximation $\hat{P}(\omega)$. (This interpretation of the least squares error was proposed in a classic paper by Whittle [103]. An equivalent formulation using maximum likelihood estimation has been given by Itakura [50], [51].) Now, we can back up and restate the problem of linear prediction as follows. Given some spectrum $P(\omega)$, we wish to model it by another spectrum $\hat{P}(\omega)$ such that the integrated ratio between the two spectra as in (62) is minimized. The parameters of the model spectrum are computed from the normal equations (12), where the needed autocorrelation coefficients $R(i)$ are easily computed from $P(\omega)$ by a simple Fourier transform. The gain factor $G$ is obtained by equating the total energy in the two

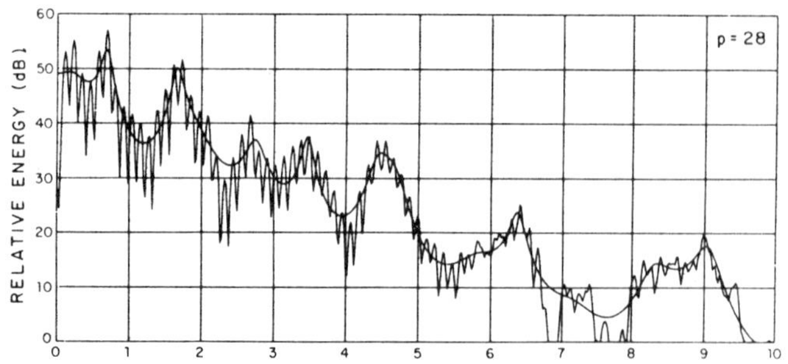


Fig. 3. A 28-pole fit to an FFT-computed signal spectrum. The signal was sampled at 20 kHz.

spectra, i.e., $\hat{R}(0) = R(0)$, where

$$\hat{R}(i) = \frac{1}{2\pi} \int_{-\pi}^{\pi} \hat{P}(\omega) \cos(i\omega)\, d\omega. \tag{63}$$

Note that $\hat{R}(i)$ is the autocorrelation of the impulse response of $H(z)$, which is given by (31) and (32). As then, the gain is computed from (35).

The manner in which the model spectrum $\hat{P}(\omega)$ approximates $P(\omega)$ is largely reflected in the relation between the corresponding autocorrelation functions. From (34), we have $\hat{R}(i) = R(i)$, $0 \leqslant i \leqslant p$. Since $P(\omega)$ and $\hat{P}(\omega)$ are Fourier transforms of $R(i)$ and $\hat{R}(i)$, respectively, it follows that increasing the value of the order of the model $p$ increases the range over which $R(i)$ and $\hat{R}(i)$ are equal, resulting in a better fit of $\hat{P}(\omega)$ to $P(\omega)$. In the limit, as $p \to \infty$, $\hat{R}(i)$ becomes identical to $R(i)$ for all $i$, and hence the two spectra become identical:

$$\hat{P}(\omega) = P(\omega), \qquad \text{as } p \to \infty. \tag{64}$$

This statement says that we can approximate any spectrum arbitrarily closely by an all-pole model.

Another important conclusion is that since linear predictive analysis can be viewed as a process of spectrum or autocorrelation matching, one must be careful how to estimate the spectrum $P(\omega)$ or the corresponding autocorrelation that is to be modeled. Since the signal is often weighted or windowed[7] before either the autocorrelation or the spectrum is computed, it can be quite important to properly choose the type and width of the data window to be used. The choice of window depends very much on the type of signal to be analyzed. If the signal can be considered to be stationary for a long period of time (relative to the effective length of the system impulse response), then a rectangular window suffices. However, for signals that result from systems that are varying relatively quickly, the time of analysis must necessarily be limited. For example, in many transient speech sounds, the signal can be considered stationary for a duration of only one or two pitch periods. In that case a window such as Hamming or Hanning [14] is more appropriate. See [13], [14], [26], [54], [64], [99], [101] for more on the issue of windowing and spectral estimation in general.

An example of linear predictive (LP) spectral estimation is shown in Fig. 3, where the original spectrum $P(\omega)$ was obtained by computing the fast Fourier transform (FFT) of a 20-ms, Hamming windowed, 20-kHz sampled speech signal. The speech sound was the vowel [æ] as in the word "bat." The harmonics due to the periodicity of the sound are evident in the FFT spectrum. Fig. 3 also shows a 28-pole fit ($p = 28$) to the signal spectrum. In this case the autocorrelation coefficients needed to solve the normal equations (12) were computed directly from the time signal. The all-pole spectrum $\hat{P}(\omega)$ was computed from (60) by dividing $G^2$ by the magnitude squared of the FFT of the sequence: $1, a_1, a_2, \cdots, a_p$. Arbitrary frequency resolution in computing $\hat{P}(\omega)$ can be obtained by simply appending an appropriate number of zeros to this sequence before taking the FFT. An alternate method of computing $\hat{P}(\omega)$ is obtained by rewriting (60) as

$$\hat{P}(\omega) = \frac{G^2}{\rho(0) + 2\sum_{i=1}^{p} \rho(i) \cos(i\omega)} \tag{65}$$

where

$$\rho(i) = \sum_{k=0}^{p-i} a_k a_{k+i}, \qquad a_0 = 1, \quad 0 \leqslant i \leqslant p \tag{66}$$

is the autocorrelation of the impulse response of the inverse filter $A(z)$. From (65), $\hat{P}(\omega)$ can be computed by dividing $G^2$ by the real part of the FFT of the sequence: $\rho(0)$, $2\rho(1)$, $2\rho(2), \cdots, 2\rho(p)$. Note that the slope of $\hat{P}(\omega)$ is always zero at $\omega = 0$ and $\omega = \pi$.

Another property of $\hat{P}(\omega)$ is obtained by noting that the minimum error $E_p = G^2$, and, therefore, from (62) we have

$$\frac{1}{2\pi} \int_{-\pi}^{\pi} \frac{P(\omega)}{\hat{P}(\omega)}\, d\omega = 1. \tag{67}$$

(This relation is a special case of a more general result (48) relating the fact that the polynomials $A_0(z), A_1(z), \cdots, A_p(z), \cdots$, form a complete set of orthogonal polynomials with weight $P(\omega)$.) Equation (67) is true for all values of $p$. In particular, it is true as $p \to \infty$, in which case from (64) we see that (67) becomes an identity. Another important case where (67) becomes an identity is when $P(\omega)$ is an all-pole spectrum with $p_0$ poles, then $\hat{P}(\omega)$ will be identical to $P(\omega)$ for all $p \geqslant p_0$. Relation (67) will be useful in discussing the properties of the error measure in Section IV.

The transfer functions $S(z)$ and $H(z)$ corresponding to $P(\omega)$ and $\hat{P}(\omega)$ are also related. It can be shown [62] that as $p \to \infty$, $H(z)$ is given by

$$H_\infty(z) = \frac{G}{1 + \sum_{k=1}^{\infty} a_k z^{-k}} = \sum_{n=0}^{N-1} h_\infty(n) z^{-n}, \qquad p \to \infty \tag{68}$$

where $h_\infty(n)$, $0 \leqslant n \leqslant N-1$, is the minimum phase sequence corresponding to $s_n$, $0 \leqslant n \leqslant N-1$. Note that the minimum phase sequence is of the same length as the original signal. Fig. 4 shows a signal ($N = 256$) and its approximate minimum phase counterpart, obtained by first performing a LP analysis for $p = 250$, and then computing the sequence $h(n)$ by long division.

### *C. Selective Linear Prediction*

The major point of the previous section was that LP analysis can be regarded as a method of spectral modeling. We had tacitly assumed that the model spectrum spans the same frequency range as the signal spectrum. We now generalize the

[7] Note that here we are discussing *data* windows which are applied directly to the signal, as opposed to *lag* windows, which statisticians have traditionally applied to the autocorrelation.

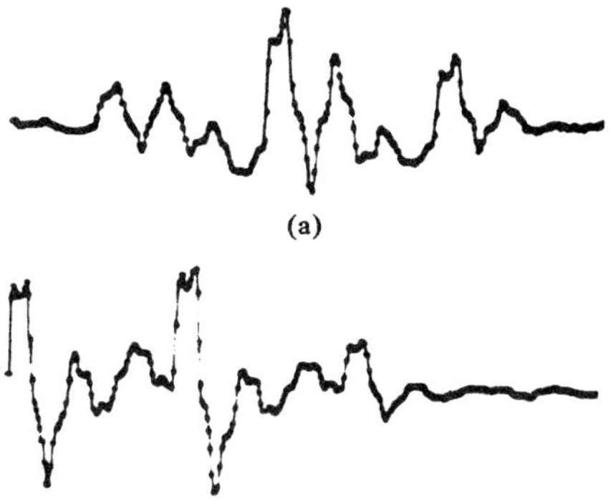


Fig. 4. (a) A 256-sample windowed speech signal. (b) The corresponding approximate minimum phase sequence obtained using a linear predictor of order $p = 250$.

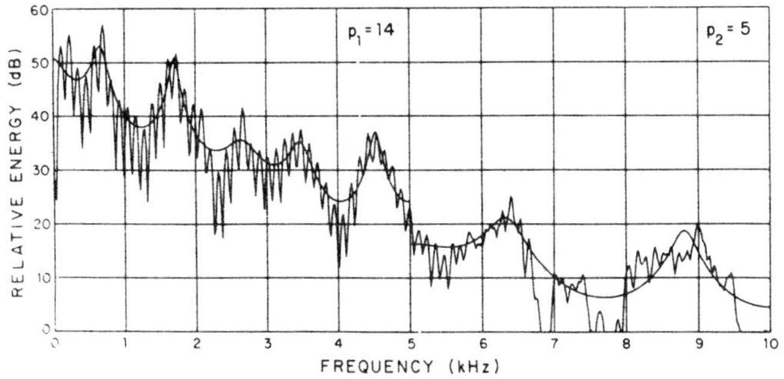


Fig. 5. Application of selective linear prediction to the same signal spectrum as in Fig. 3, with a 14-pole fit to the 0–5 kHz region and a 5-pole fit to the 5–10 kHz region.

LP spectral modeling method to the case where we wish to fit only a selected portion of a given spectrum.

Suppose we wish to model the spectrum $P(\omega)$ only in the region[8] $\omega_\alpha \leqslant \omega \leqslant \omega_\beta$ by an all-pole spectrum given by (60). Call the signal spectrum in that region $P'(\omega)$. In order to compute the parameters of the model spectrum $\hat{P}(\omega)$, we first perform a linear mapping of the given region onto the upper half of the unit circle in the $z$ plane. This can be accomplished by the mapping $\omega' = \pi(\omega - \omega_\alpha)/(\omega_\beta - \omega_\alpha)$, so that the given region is mapped onto $0 \leqslant \omega' \leqslant \pi$. In addition, let $P'(-\omega') = P'(\omega')$ define the spectrum over the lower half of the unit circle. The model parameters are then computed from the normal equations (12), where the autocorrelation coefficients are obtained by a Fourier transform with $P'(\omega')$ replacing $P(\omega)$ and $\omega'$ replacing $\omega$ in (56).

Selective linear prediction has had applications in speech recognition and speech compression [63]. An example of its usage is shown in Fig. 5. For speech recognition applications, the 0–5 kHz region is more important than the 5–10 kHz. Even when the 5–10 kHz region is important, only a rough idea of the shape of the spectrum is sufficient. In Fig. 5, the signal spectrum is the same as in Fig. 3. The 0–5 kHz region is modeled by a 14-pole spectrum, while the 5–10 kHz region is modeled independently by only a 5-pole model.

An important point, which should be clear by now, is that since we assume the availability of the signal spectrum $P(\omega)$, any desired frequency shaping or scaling can be performed directly on the signal spectrum before linear predictive modeling is applied.

[8] The remainder of the spectrum is simply neglected.

### D. Modeling Discrete Spectra

Thus far we have assumed that the spectrum $P(\omega)$ is a continuous function of frequency. More often, however, the spectrum is known at only a finite number of frequencies. For example, FFT-derived spectra and those obtained from many commercially available spectrum analyzers have values at equally spaced frequency points. On the other hand, filter bank spectra, and, for example, third-octave band spectrum analyzers have values at frequencies that are not necessarily equally spaced. In order to be able to model these discrete spectra, only one change in our analysis need be made. The error measure $E$ in (62) is defined as a summation instead of an integral. The rest of the analysis remains the same except that the autocorrelation coefficients $R(i)$ are now computed from

$$R(i) = \frac{1}{M} \sum_{m=0}^{M-1} P(\omega_m) \cos(i\omega_m) \tag{69}$$

where $M$ is the total number of spectral points on the unit circle. The frequencies $\omega_m$ are those for which a spectral value exists, and they need not be equally spaced. Below we demonstrate the application of LP modeling for filter bank and harmonic spectra.

Fig. 6(a) shows a typical 14-pole fit to a spectrum of the fricative [s] that was FFT computed from the time signal. Fig. 6(b) shows a similar fit to a line spectrum that is typical of filter bank spectra. What we have actually done here is to simulate a filter bank where the filters are linearly spaced up to 1.6 kHz and logarithmically spaced thereafter. Note that the all-pole spectrum for the simulated filter bank is remarkably similar to the one in the top figure, even though the number of spectral points is much smaller.

The dashed curve in Fig. 7(a) is a 14-pole spectrum. If one applied LP analysis to this spectrum, the all-pole model for $p = 14$ would be identical to the dashed spectrum. The situation is not so favorable for discrete spectra. Let us assume that the dashed spectrum corresponds to the transfer function of a 14-pole filter. If this filter is excited by a periodic train of impulses with fundamental frequency $F_0$, the spectrum of the output signal will be a discrete line spectrum with spectral values only at the harmonics (multiples of $F_0$). The line spectrum for $F_0 = 312$ Hz is shown in Fig. 7(a). Note that the dashed spectrum is an envelope of the harmonic spectrum. The result of applying a 14-pole LP analysis to the harmonic spectrum is shown as the solid curve in Fig. 7(a). The discrepancy between the two all-pole spectra is obvious. In general, the types of discrepancies that can occur between the model and original spectra include merging or splitting of pole peaks, and increasing or decreasing of pole frequencies and bandwidths. Pole movements are generally in the direction of the nearest harmonic. As the fundamental frequency decreases, these discrepancies decrease, as shown in Fig. 7(b) for $F_0 = 156$ Hz.

It is important to note in Fig. 7 that the dashed curve is the *only* possible 14-pole spectrum that coincides with the line spectrum at the harmonics.[9] It is significant that the all-pole spectrum resulting from LP modeling does not yield the spectrum we desire. The immediate reason for this is that the solution for the model parameters from (12) depends on the values of the signal autocorrelation, which for the periodic signal are

[9] In general this is true only if the period between input impulses is greater than twice the number of poles in the filter.

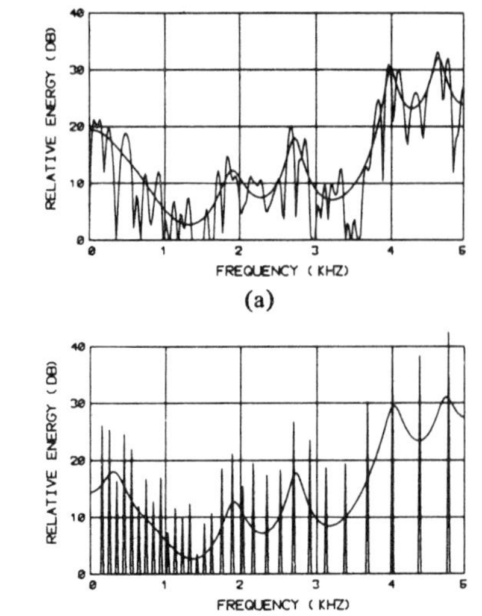


Fig. 6. Application of LP modeling to a filter bank spectrum. (a) A 14-pole fit to the original spectrum. (b) A 14-pole fit to the simulated filter bank spectrum.

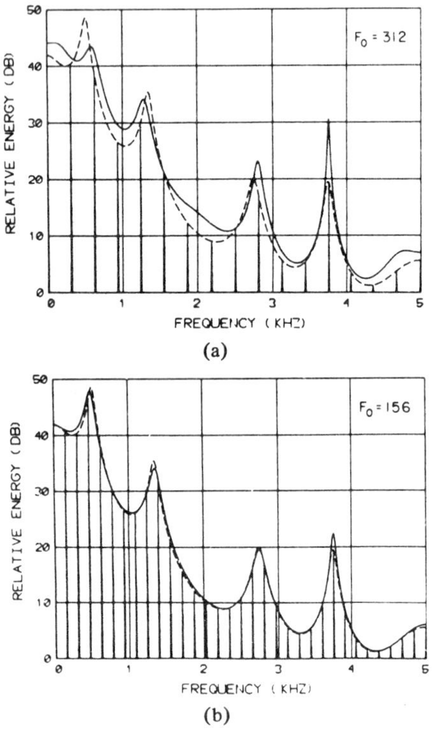


Fig. 7. Application of LP modeling to harmonic spectra. Dashed curve: 14-pole filter spectrum. Vertical lines: Corresponding harmonic spectrum for (a) and (b). (a) $F_0$ = 312 Hz. (b) $F_0$ = 156 Hz. Solid curve: 14-pole fit to the discrete harmonic spectrum. (For display purposes, the energy in the model spectrum (solid curve) was set equal to the energy in the filter spectrum (dashed curve).)

different from that for the single impulse response. However, the major underlying reason lies in the properties of the error measure used. This is the topic of the next section.

## IV. Error Analysis

An important aspect of any fitting or matching procedure is the properties of the error measure that is employed, and whether those properties are commensurate with certain objectives. In this section we shall examine the properties of the error measure used in LP analysis and we shall discuss its strengths and weaknesses in order to be able to fully utilize its capabilities. The analysis will be restricted to the stationary (autocorrelation) case, although the conclusions can be extrapolated to the nonstationary (covariance) case.

The error measure used in Section II-B to determine the predictor parameters is the least squares error measure due to Gauss, who first reported on it in the early 1800's. This error measure has been used extensively since then, and is quite well understood. Its major asset is its mathematical tractability. Its main characteristic is that it puts great emphasis on large errors and little emphasis on small errors. Purely from the time domain, it is often difficult to say whether such an error measure is a desirable one or not for the problem at hand. Many would probably agree that it does not really matter which error measure one uses as long as it is a reasonable function of the magnitude of the error at each point. For the linear prediction problem, we are fortunate that the error measure can also be written in the frequency domain and can be interpreted as a goodness of fit between a given signal spectrum and a model spectrum that approximates it. The insights gained in the frequency domain should enhance our understanding of the least squares error criterion.

### *A. The Minimum Error*

For each value of $p$, minimization of the error measure $E$ in (62) leads to the minimum error $E_p$ in (13), which is given in terms of the predictor and autocorrelation coefficients. Here we derive an expression for $E_p$ in the frequency domain, which will help us determine some of its properties. Other properties will be discussed when we discuss the normalized minimum error.

Let

$$\hat{c}_0 = \frac{1}{2\pi}\int_{-\pi}^{\pi} \log \hat{P}(\omega)\, d\omega \tag{70}$$

be the zeroth coefficient (quefrency) of the cepstrum (inverse Fourier transform of log spectrum) [38], [77] corresponding to $\hat{P}(\omega)$. From (60), (70) reduces to

$$\hat{c}_0 = \log G^2 - \frac{1}{2\pi}\int_{-\pi}^{\pi} \log |A(e^{j\omega})|^2\, d\omega. \tag{71}$$

$A(z)$ has all its zeros inside the unit circle. Therefore, the integral in (71) is equal to zero [64], [69], [103]. Since $G^2 = E_p$, we conclude from (71) that

$$E_p = e^{\hat{c}_0}. \tag{72}$$

From (72) and (70), $E_p$ can be interpreted as the geometric mean of the model spectrum $\hat{P}(\omega)$. From (40) we know that $E_p$ decreases as $p$ increases. The minimum occurs as $p \to \infty$, and is equal to

$$E_{\min} = E_\infty = e^{c_0} \tag{73}$$

where $c_0$ is obtained by substituting $P(\omega)$ for $\hat{P}(\omega)$ in (70).[10] Therefore, the absolute minimum error is a function of $P(\omega)$ only, and is equal to its geometric mean, which is always positive for positive definite spectra.[11] This is a curious result, because it says that the minimum error can be nonzero even

[10] If $P(\omega)$ is a $p_0$-pole spectrum then $E_p = E_{\min}$ for all $p \geqslant p_0$.

[11] $E_{\min}$ is equal to zero only if $P(\omega)$ is zero over a noncountable set of frequencies (i.e., over a line segment). In that case, the signal is perfectly predictable and the prediction error is zero [107].

when the matching spectrum $\hat{P}(\omega)$ is identical to the matched spectrum $P(\omega)$. Therefore, although $E_p$ is a measure of fit of the model spectrum to the signal spectrum, it is not an absolute one. The measure is always relative to $E_{min}$. The nonzero aspect of $E_{min}$ can be understood by realizing that, for any $p$, $E_p$ is equal to that portion of the signal energy that is not predictable by a $p$th order predictor. For example, the impulse response of an all-pole filter is perfectly predictable *except* for the initial nonzero value. It is the energy in this initial value that shows up in $E_p$. (Note that in the covariance method one can choose the region of analysis to exclude the initial value, in which case the prediction error would be zero for this example.)

### B. Spectral Matching Properties

The LP error measure $E$ in (62) has two major properties:[12] a global property and a local property.

*1) Global Property:* Because the contributions to the total error are determined by the *ratio* of the two spectra, the matching process should perform uniformly over the whole frequency range, irrespective of the general shaping of the spectrum.

This is an important property for spectral estimation because it makes sure that the spectral match at frequencies with little energy is just as good, on the average, as the match at frequencies with high energy (see Fig. 3). If the error measure had been of the form $\int |P(\omega) - \hat{P}(\omega)|\, d\omega$, the spectral matches would have been best at high energy frequency points.

*2) Local Property:* This property deals with how the match is done in each small region of the spectrum.

Let the ratio of $P(\omega)$ to $\hat{P}(\omega)$ be given by

$$E(\omega) = \frac{P(\omega)}{\hat{P}(\omega)}. \qquad (74)$$

Then from (67) we have

$$\frac{1}{2\pi}\int_{-\pi}^{\pi} E(\omega)\, d\omega = 1, \qquad \text{for all } p. \qquad (75)$$

$E(\omega)$ can be interpreted as the "instantaneous error" between $P(\omega)$ and $\hat{P}(\omega)$ at frequency $\omega$. Equation (75) says that the arithmetic mean of $E(\omega)$ is equal to 1, which means that there are values of $E(\omega)$ greater and less than 1 such that the average is equal to 1.[13] In terms of the two spectra, this means that $P(\omega)$ will be greater than $\hat{P}(\omega)$ in some regions and less in others such that (75) applies. However, the contribution to the total error is more significant when $P(\omega)$ is greater than $\hat{P}(\omega)$ than when $P(\omega)$ is smaller, e.g., a ratio of $E(\omega) = 2$ contributes more to the total error than a ratio of $1/2$. We conclude that:

> after the minimization of error, we expect a better fit of $\hat{P}(\omega)$ to $P(\omega)$ where $P(\omega)$ is greater than $\hat{P}(\omega)$, than where $P(\omega)$ is smaller (on the average).

For example, if $P(\omega)$ is the power spectrum of a quasi-periodic signal (such as in Fig. 3), then most of the energy in $P(\omega)$ will exist in the harmonics, and very little energy will reside between harmonics. The error measure in (62) insures that the approximation of $\hat{P}(\omega)$ to $P(\omega)$ is far superior at the harmonics than between the harmonics. If the signal had been generated by exciting a filter with a periodic sequence of impulses, then the system response of the filter must pass through all the harmonic peaks. Therefore, with a proper choice of the model order $p$, minimization of the LP error measure results in a model spectrum that is a good approximation to that system response. This leads to one characteristic of the local property:

> minimization of the error measure $E$ results in a model spectrum $\hat{P}(\omega)$ that is a good estimate of the *spectral envelope* of the signal spectrum $P(\omega)$.

Fig. 6 shows that this statement also applies in a qualitative way when the excitation is random noise. It should be clear from the above that the importance of the local property is not as crucial when the variations of the signal spectrum from the spectral envelope are much less pronounced.

In the modeling of harmonic spectra, we showed an example in Fig. 7(a) where, although the all-pole spectrum resulting from LP modeling was a reasonably good estimate of the harmonic spectral envelope, it did not yield the unique all-pole transfer function that coincides with the line spectrum at the harmonics. This is a significant *disadvantage* of LP modeling, and is an indirect reflection of another characteristic of the local property: *the cancellation of errors*. This is evident from (75) where the instantaneous errors $E(\omega)$ are greater and less than 1 such that the average is 1. To help elucidate this point, let us define a new error measure $E'$ that is the logarithm of $E$ in (62):

$$E' = \log\left[\frac{1}{2\pi}\int_{-\pi}^{\pi}\frac{P(\omega)}{\hat{P}(\omega)}\, d\omega\right] \qquad (76)$$

where the gain factor has been omitted since it is not relevant to this discussion. It is simple to show that the minimization of $E'$ is equivalent to the minimization of $E$. For cases where $P(\omega)$ is smooth relative to $\hat{P}(\omega)$ and the values of $P(\omega)$ are not expected to deviate very much from $\hat{P}(\omega)$, the logarithm of the average of spectral ratios can be approximated by the average of the logarithms, i.e.,

$$E' \cong \frac{1}{2\pi}\int_{-\pi}^{\pi}\log\frac{P(\omega)}{\hat{P}(\omega)}\, d\omega. \qquad (77)$$

From (77) it is clear that the contributions to the error when $P(\omega) > \hat{P}(\omega)$ *cancel* those when $P(\omega) < \hat{P}(\omega)$.

The above discussion suggests the use of an error measure that takes the magnitude of the integrand in (77). One such error measure is

$$E'' = \frac{1}{2\pi}\int_{-\pi}^{\pi}\left[\log\frac{P(\omega)}{\hat{P}(\omega)}\right]^2 d\omega$$
$$= \frac{1}{2\pi}\int_{-\pi}^{\pi}[\log P(\omega) - \log\hat{P}(\omega)]^2\, d\omega. \qquad (78)$$

$E''$ is just the mean squared error between the two log spectra. It has the important property that the minimum error of zero occurs if and only if $\hat{P}(\omega)$ is identical to $P(\omega)$. Therefore, if we use (the discrete form of) $E''$ in modeling the harmonic spectrum in Fig. 7(a), the resulting model spectrum (for $p = 14$) will be identical to the dashed spectrum, since the minimum error of zero is achievable by that spectrum. However, while the error measure $E''$ solves one problem, it introduces an-

[12] Itakura [50], [51] discusses a maximum likelihood error criterion having the same properties.

[13] Except for the special case when $P(\omega)$ is all-pole, the condition $E(\omega) = 1$ for all $\omega$ is true only as $p \to \infty$.

other. Note that the contributions to the total error in (78) are equally important whether $P(\omega) > \hat{P}(\omega)$ or vice versa. This means that if the variations of $P(\omega)$ are large relative to $\hat{P}(\omega)$ (such as in Fig. 3), the resulting model spectrum will *not* be a good estimate of the spectral envelope. In addition, the minimization of $E''$ in (78) results in a set of nonlinear equations that must be solved iteratively, thus increasing the computational load tremendously.

Our conclusion is that the LP error measure in (62) is to be preferred in general, except for certain special cases (as in Fig. 7(a)) where an error measure such as $E''$ in (78) can be used, provided one is willing to carry the extra computational burden.

The global and local properties described here are properties of the error measure in (62) and do not depend on the details of $P(\omega)$ and $\hat{P}(\omega)$. These properties apply *on the average* over the whole frequency range. Depending on the detailed shapes of $P(\omega)$ and $\hat{P}(\omega)$, the resulting match can be better in one spectral region than in another. For example, if $\hat{P}(\omega)$ is an all-pole model spectrum and if the signal spectrum $P(\omega)$ contains zeros as well as poles, then one would not expect as good a match at the zeros as at the poles. This is especially true if the zeros have bandwidths of the same order as the poles or less. (Wide bandwidth zeros are usually well approximated by poles.) On the other hand, if $\hat{P}(\omega)$ is an all-zero spectrum then the preceding statement would have to be reversed.

### C. *The Normalized Error*

The normalized error has been a very useful parameter for the determination of the optimal number of parameters to be used in the model spectrum. This subject will be discussed in the following section. Here we shall present some of the properties of the normalized error, especially as they relate to the signal and model spectra.

*1) Relation to the Spectral Dynamic Range:* The normalized error was defined in Section II as the ratio of the minimum error $E_p$ to the energy in the signal $R(0)$. Keeping in mind that $R(0) = \hat{R}(0)$, and substituting for $E_p$ from (72), we obtain

$$V_p = \frac{E_p}{R(0)} = \frac{e^{\hat{c}_0}}{\hat{R}(0)}. \tag{79}$$

Also, from (73), we have in the limit as $p \to \infty$:

$$V_{\min} = V_\infty = \frac{e^{c_0}}{R(0)}. \tag{80}$$

Therefore, the normalized error is always equal to the normalized zero quefrency of the model spectrum. From (40) and (79) it is clear that $V_p$ is a monotonically decreasing function of $p$, with $V_0 = 1$ and $V_\infty = V_{\min}$ in (80). Fig. 8 shows plots of $V_p$ as a function of $p$ for two speech sounds (sampled at 10 kHz) whose spectra are similar to those in Figs. 3 and 6.

It is instructive to write $V_p$ as a function of $\hat{P}(\omega)$. From (63) and (70), (79) can be rewritten as

$$V_p = \frac{\exp\left[\frac{1}{2\pi}\int_{-\pi}^{\pi} \log \hat{P}(\omega)\, d\omega\right]}{\frac{1}{2\pi}\int_{-\pi}^{\pi} \hat{P}(\omega)\, d\omega}. \tag{81}$$

It is clear from (81) that $V_p$ depends completely on the shape of the model spectrum, and from (80), $V_{\min}$ is determined solely by the shape of the signal spectrum. An interesting way

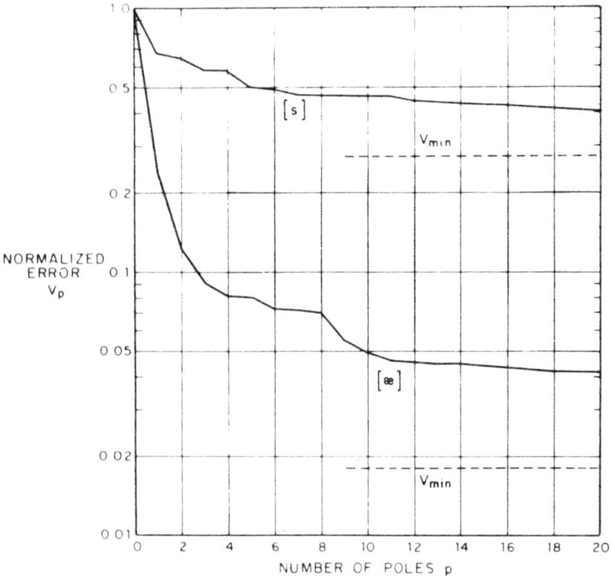


**Fig. 8.** Normalized error curves for the sounds [s] in the word "list" and [æ] in the word "potassium."

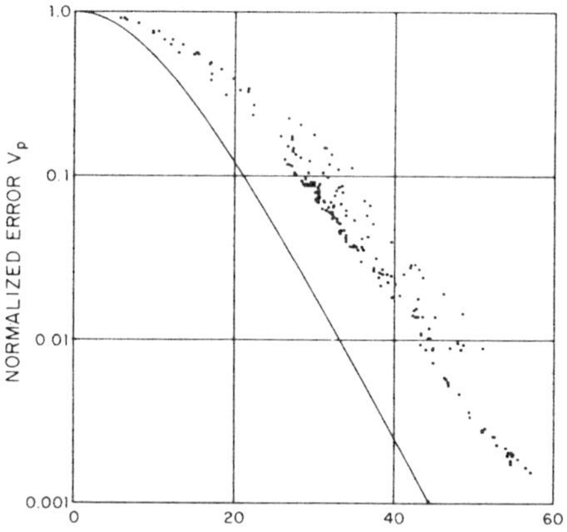


**Fig. 9.** Two-pole normalized error versus spectral dynamic range for 200 different two-pole models. The solid curve is $V_a$, the absolute lower bound on the normalized error.

to view (81) is that $V_p$ is equal to the ratio of the geometric mean of the model spectrum to its arithmetic mean. This ratio has been used in the past as a measure of the spread of the data [22], [48]. When the spread of the data is small, the ratio is close to 1. Indeed, from (81) it is easy to see that if $\hat{P}(\omega)$ is flat, $V_p = 1$. On the other hand, if the data spread is large, then $V_p$ becomes close to zero. Again, from (81) we see that if $\hat{P}(\omega)$ is zero for a portion of the spectrum (hence a large spread), then $V_p = 0$. (Another way of looking at $V_p$ is in terms of the flatness of the spectrum [40].)

Another measure of data spread is the dynamic range. We define the spectral dynamic range $d$ as the ratio of the highest to the lowest amplitude points on the spectrum:

$$d = H/L$$

where

$$H = \max_\omega \hat{P}(\omega) \qquad L = \min_\omega \hat{P}(\omega). \tag{82}$$

The relation between the normalized error and the spectral dynamic range is illustrated in Fig. 9. The dark dots in the figure are plots of the normalized error versus the spectral dynamic range (in decibels) for 2-pole models of 200 different speech spectra. The solid curve in Fig. 9 is an absolute lower bound on the geometric-to-arithmetic mean ratio for *any* spectrum with a given dynamic range. The curve is a plot of the following relation [22], [64]

$$V_a = \gamma e^{(1-\gamma)} \tag{83}$$

where

$$\gamma = \frac{\log d}{d-1} \tag{84}$$

and $V_a$ stands for the absolute lower bound on $V_p$ for a given $d$. The overall impression from Fig. 9 is that the normalized error generally decreases as the dynamic range of the spectrum increases. This is apparent in Fig. 8 where $V_p$ for the vowel [æ] is less than that for the fricative [*s*], and [æ] has a much higher spectral dynamic range than [*s*].

*2) A Measure of Ill-Conditioning:* In solving the autocorrelation normal equations (12), the condition of the autocorrelation matrix is an important consideration in deciding the accuracy of the computation needed. An ill-conditioned matrix can cause numerical problems in the solution. An accepted measure of ill-conditioning in a matrix is given by the ratio

$$d' = \lambda_{\max}/\lambda_{\min} \tag{85}$$

where $\lambda_{\max}$ and $\lambda_{\min}$ are the maximum and minimum eigenvalues of the matrix [27], [81]. Grenander and Szegö [41], [42] have shown that all the eigenvalues of an autocorrelation matrix lie in the range $\lambda_i \in [H, L]$, $1 \leqslant i \leqslant p$, where $H$ and $L$ are defined in (82). In addition, as the order of the matrix $p$ increases, the eigenvalues become approximately equal to $\hat{P}(\omega)$ evaluated at equally spaced points with separation $2\pi/(p+1)$. Therefore, the ratio $d'$ given in (85) can be well approximated by the dynamic range of $\hat{P}(\omega)$:

$$d' \cong d. \tag{86}$$

Therefore, the spectral dynamic range is a good measure of the ill-conditioning of the autocorrelation matrix. The larger the dynamic range, the greater is the chance that the matrix is ill-conditioned.

But in the previous section we noted that an increase in $d$ usually results in a decrease in the normalized error $V_p$. Therefore, $V_p$ can also be used as a measure of ill-conditioning: the ill-conditioning is greater with decreased $V_p$. The problem becomes more and more serious as $V_p \to 0$, i.e., as the signal becomes highly predictable.

If ill-conditioning occurs sporadically, then one way of patching the problem is to increase the values along the principal diagonal of the matrix by a small fraction of a percent. However, if the problem is a regular one, then it is a good idea if one can reduce the dynamic range of the signal spectrum. For example, if the spectrum has a general slope, then a single-zero filter of the form $1 + az^{-1}$ applied to the signal can be very effective. The new signal is given by

$$s_n' = s_n + a s_{n-1}. \tag{87}$$

An optimal value for $a$ is obtained by solving for the filter $A(z)$ that "whitens" (flattens) $s_n'$. This is, of course, given by the first order predictor, where

$$a = -\frac{R(1)}{R(0)}. \tag{88}$$

$R(1)$ and $R(0)$ are autocorrelation coefficients of the signal $s_n$. The filtered signal $s_n'$ is then guaranteed to have a smaller spectral dynamic range. The above process is usually referred to as *preemphasis.*

One conclusion from the above concerns the design of the low-pass filter that one uses before sampling the signal to reduce aliasing. In order to ensure against aliasing, it is usually recommended that the cutoff frequency of the filter be lower than half the sampling frequency. However, if the cutoff frequency is appreciably lower than half the sampling frequency, then the spectral dynamic range of the signal spectrum increases, especially if the filter has a sharp cutoff and the stop band is very low relative to the pass band. This increases problems of ill-conditioning. Therefore, if one uses a lowpass filter with a sharp cutoff, the cutoff frequency should be set as close to half the sampling frequency as possible.

### *D. Optimal Number of Poles*

One of the important decisions that usually has to be made in fitting of all-pole models is the determination of an "optimal" number of poles. It is a nontrivial exercise to define the word "optimal" here, for as we have seen, the fit of the model "improves" as the number of poles $p$ increases. The problem is where to stop. Clearly we would like the minimum value of $p$ that is adequate for the problem at hand, both to reduce our computation and to minimize the possibility of ill-conditioning (which increases with $p$ since $V_p$ decreases).

If the signal spectrum is an all-pole spectrum with $p_0$ poles, then we know that $V_p = V_{p_0}$, $p \geqslant p_0$, and $k_p = 0$, $p > p_0$, i.e., the error curve remains flat for $p > p_0$. Therefore, if we expect the signal spectrum to be an all-pole spectrum, a simple test to obtain the optimal $p$ is to check when the error curve becomes flat. But, if the signal is the output of a $p_0$-pole filter with white noise excitation, then the suggested test will not work, because the estimates of the poles are based on a finite number of data points and the error curve will not be flat for $p > p_0$. In practice, however, the error curve will be almost flat for $p > p_0$. This suggests the use of the following threshold test

$$1 - \frac{V_{p+1}}{V_p} < \delta. \tag{89}$$

This test must succeed for several consecutive values before one is sure that the error curve has actually flattened out.

The use of the ratio $V_{p+1}/V_p$ has been an accepted method in the statistical literature [6], [17], [112] for the determination of the optimal $p$. The test is based on hypothesis testing procedures using maximum likelihood ratios. A critical review of hypothesis testing procedures has been given recently by Akaike [5]. Akaike's main point is that model fitting is a problem where multiple decision procedures are required rather than hypothesis testing. The fitting problem should be stated as an estimation problem with an associated measure of fit. Akaike suggests the use of an information theoretic criterion that is an estimate of the mean log-likelihood [3], [5].[14] This is given by

$$I(p) = -2 \log (\text{maximum likelihood}) + 2p. \tag{90}$$

The value of $p$ for which $I(p)$ is minimum is taken to be the optimal value. In our problem of all-pole modeling, if we assume that the signal has a Gaussian probability distribution,

[14] An earlier criterion used by Akaike is what he called the "final prediction error" [1], [2], [37].

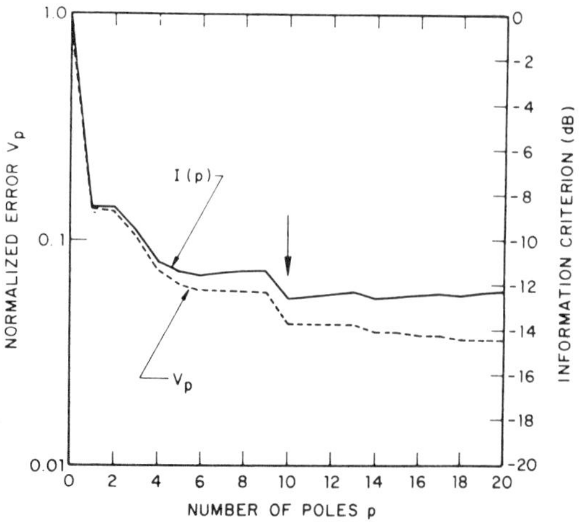


Fig. 10. A plot of Akaike's information criterion versus the order of the predictor $p$. Here, $I(p) = 10 \log_{10} V_p + (8.686p/0.4N)$, ($N$ = 200, Hamming windowed). The "optimal" value of $p$ occurs at the minimum of $I(p)$, shown by the arrow at $p = 10$.

then (90) reduces to (neglecting additive constants and dividing by $N$) [6], [17], [51]

$$I(p) = \log V_p + \frac{2p}{N_e} \tag{91}$$

where $N_e$ is the "effective" number of data points in the signal. The word "effective" is used to indicate that one must compensate for possible windowing. The effective width of a window can be taken as the ratio of the energy under the window relative to that of a rectangular window. For example, for a Hamming window, $N_e = 0.4N$.

Note that the first term in (91) decreases as a function of $p$, and the second term increases. Therefore, a minimum can occur. In practice, there are usually several local minima, then the value of $p$ corresponding to the absolute minimum of $I(p)$ is taken as the optimal value. Usually $I(p)$ is computed up to the maximum value of interest,[15] and the minimum of $I(p)$ is found in that region.

Fig. 10 shows an example of the application of Akaike's criterion. The dotted curve is the usual error curve and the solid curve is a plot of $I(p)$ in (91) multiplied by $10 \log_{10} e$ to obtain the results in decibels. In Fig. 10, the optimal predictor order is $p = 10$. Note that $I(p)$ for $p > 10$ slopes upward, but very gently. This indicates that the actual absolute minimum is quite sensitive to the linear term in (91). In practice, the criterion in (91) should not be regarded as an absolute, because it is based on several assumptions which might not apply for the signal of interest. For example, the assumptions of uncorrelated noise excitation and Gaussian distributions might not hold. Therefore, the experimenter should feel free to adjust the criterion to suit one's application. One simple way of "tuning" the criterion is to multiply $N_e$ by an appropriate factor.

## V. Data Compression by Linear Prediction

The methods outlined in Section II for the modeling of the behavior of a signal can be very useful in data compression. The process of signal or system modeling is essentially one of redundancy removal, which is the essence of data compression.

[15] Akaike informed me that he usually recommends $p_{max} < 3N^{1/2}$ as the maximum value of $p$ that should be used if one is interested in a reliable estimate.

The idea of attempting to predict the value of a signal from previous sample values has been labeled in communications as "predictive coding" [28]. Adaptive linear prediction has been used extensively in speech and video transmission [7], [9], [23], [44], [50], [57], [83]. For the purposes of transmission one must quantize and transmit the predictor parameters or some transformation thereof. It has been known for some time that the quantization of the predictor parameters themselves is quite inefficient since a large number of bits is required to retain the desired fidelity in the reconstructed signal at the receiver [72]. Below, several equivalent representations of the predictor are presented and their quantization properties are discussed. We shall continue to assume that $H(z)$ is always stable, and hence minimum phase. $A(z)$ is, of course, then also minimum phase.

### A. *Alternate Representations of Linear Predictor*

The following is a list of possible sets of parameters that characterize uniquely the all-pole filter $H(z)$ or its inverse $A(z)$.

1) (a) Impulse response of the inverse filter $A(z)$, i.e., predictor parameters $a_k$, $1 \leqslant k \leqslant p$.

(b) Impulse response of the all-pole model $h_n$, $0 \leqslant n \leqslant p$, which is defined in (30). Note that the first $p + 1$ coefficients uniquely specify the filter.

2) (a) Autocorrelation coefficients of $a_k$, $\rho(i)$, $0 \leqslant i \leqslant p$, as defined in (66).

(b) Autocorrelation coefficients of $h_n$, $\hat{R}(i)$, $0 \leqslant i \leqslant p$, as defined in (31) and (32).

3) Spectral coefficients of $A(z)$, $\Gamma_i$, $0 \leqslant i \leqslant p$ (or equivalently spectral coefficients of $H(z)$, $G^2/\Gamma_i$):

$$\Gamma_i = \rho(0) + 2 \sum_{j=1}^{p} \rho(j) \cos \frac{2\pi i j}{2p+1}, \qquad 0 \leqslant i \leqslant p \tag{92}$$

where $\rho(i)$ are as defined in (66). In other words, $\{\Gamma_i\}$ is obtained from $\{\rho(i)\}$ by a discrete Fourier transform.

4) Cepstral coefficients of $A(z)$, $c_n$, $1 \leqslant n \leqslant p$ (or equivalently cepstral coefficients of $H(z)$, $-c_n$):

$$c_n = \frac{1}{2\pi} \int_{-\pi}^{\pi} \log A(e^{j\omega})\, e^{jn\omega}\, d\omega. \tag{93}$$

Since $A(z)$ is minimum phase, (93) reduces to [38], [77]

$$c_n = a_n - \sum_{m=1}^{n-1} \frac{m}{n} c_m a_{n-m}, \qquad 1 \leqslant n \leqslant p. \tag{94}$$

Equation (94) is an iterative method for the computation of the cepstral coefficients directly from the predictor coefficients.

5) Poles of $H(z)$ or zeros of $A(z)$, $z_k$, $1 \leqslant k \leqslant p$, where $\{z_k\}$ are either real or form complex conjugate pairs. Conversion of the roots to the $s$ plane can be achieved by setting each root $z_k = e^{s_k T}$, where $s_k = \sigma_k + j\omega_k$ is the corresponding pole in the $s$ plane, and $T$ is the sampling period. If the root $z_k = z_{kr} + jz_{ki}$, then

$$\omega_k = \frac{1}{T} \arctan \frac{z_{ki}}{z_{kr}}$$

$$\sigma_k = \frac{1}{2T} \log (z_{kr}^2 + z_{ki}^2) \tag{95}$$

where $z_{kr}$ and $z_{ki}$ are the real and imaginary parts of $z_k$, respectively.

6) Reflection coefficients $k_i$, $1 \leq i \leq p$, which are obtained as a byproduct of the solution of the autocorrelation normal equations, as in (38), or from the backward recursion (51).

Some of the preceding sets of parameters have $p + 1$ coefficients while others have only $p$ coefficients. However, for the latter sets the gain $G$ needs to be specified as well, thus keeping the total number of parameters as $p + 1$ for all the cases.

For purposes of data transmission, one is usually interested in recovering the predictor coefficients from the parameters that are chosen for transmission. The required transformations are clear for most of the above parameters, except perhaps for the parameters $\rho(i)$ and $\Gamma_i$. Through an inverse DFT, the spectral coefficients $\Gamma_i$ can be converted to autocorrelation coefficients $\rho(i)$. One method of recovering the predictor parameters from $\{\rho(i)\}$ is as follows. Apply a DFT to the sequence $\{\rho(i)\}$ after appending it with an appropriate number of zeros to achieve sufficient resolution in the resulting spectrum of $A(z)$. Divide $G^2$ by this spectrum to obtain the spectrum of the filter $H(z)$. Inverse Fourier transformation of the spectrum of $H(z)$ yields the autocorrelation coefficients $\hat{R}(i)$. The first $p + 1$ coefficients $\hat{R}(i)$, $0 \leq i \leq p$, are then used to compute the predictor coefficients via the normal equations (12) with $R(i) = \hat{R}(i)$.

### B. Quantization Properties

Although the sets of parameters given above provide equivalent information about the linear predictor, their properties under quantization are different. For the purpose of quantization, two desirable properties for a parameter set to have are: 1) filter stability upon quantization and 2) a natural ordering of the parameters. Property 1) means that the poles of $H(z)$ continue to be inside the unit circle even after parameter quantization. By 2), we mean that the parameters exhibit an inherent ordering, e.g., the predictor coefficients are ordered as $a_1, a_2, \cdots, a_p$. If $a_1$ and $a_2$ are interchanged then $H(z)$ is no longer the same in general, thus illustrating the existence of an ordering. The poles of $H(z)$, on the other hand, are not naturally ordered since interchanging the values of any two poles does not change the filter. When an ordering is present, a statistical study on the distribution of individual parameters can be used to develop better encoding schemes. Only the poles and the reflection coefficients insure stability upon quantization, while all the sets of parameters except the poles possess a natural ordering. Thus only the reflection coefficients possess both of these properties.

In an experimental study [63] of the quantization properties of the different parameters, it was found that the impulse responses $\{a_k\}$ and $\{h_n\}$ and the autocorrelations $\{\rho(i)\}$ and $\{\hat{R}(i)\}$ are highly susceptible to causing instability of the filter upon quantization. Therefore, these sets of parameters can be used only under minimal quantization, in which case the transmission rate would be excessive.

In the experimental investigation of the spectral and cepstral parameters, it was found that the quantization properties of these parameters are generally superior to those of the impulse responses and autocorrelation coefficients. The spectral parameters often yield results comparable to those obtained by quantizing the reflection coefficients. However, for the cases when the spectrum consists of one or more very sharp peaks (narrow bandwidths), the effects of quantizing the spectral coefficients often cause certain regions in the reconstructed spectrum (as described in the previous section) to become negative, which leads to instability of the computed filter. Quantization of the cepstral coefficients can also lead to instabilities. It should be noted here that the quantization properties of these parameters give better results if the spectral dynamic range of the signal is limited by some form of preprocessing.

Filter stability is preserved under quantization of the poles. But poles are expensive to compute, and they do not possess a natural ordering.

The conclusion is that, of the sets of parameters given in the preceding, the reflection coefficients are the best set to use as transmission parameters. In addition to ease of computation, stability under quantization, and natural ordering, the values of the reflection coefficients $k_i$, $i < p$, do not change as $p$ is increased, unlike any of the other parameters. In the following, we discuss the optimal quantization of the reflection coefficients.

### C. Optimal Quantization [53], [65]

Optimal quantization of the reflection coefficients depends on the fidelity criterion chosen. For many applications, it is important that the log spectrum of the all-pole model be preserved. In this case, it is reasonable to study the sensitivity of the log spectrum with respect to changes in the reflection coefficients. In a recent study [65], a spectral sensitivity curve was plotted versus each of the reflection coefficients $k_i$ for many different all-pole models obtained by analyzing a large number of speech samples. The results of the study show that each sensitivity curve versus $k_i$ has the same general shape, irrespective of the index $i$. Each sensitivity curve is U-shaped; it is even-symmetric about $k_i = 0$, with large values when $|k_i| \to 1$, and small values when $|k_i|$ is close to zero. These properties indicate that linear quantization of the reflection coefficients is not desirable, especially if some of them take values very close to 1, which happens when the spectrum contains sharp resonances. Nonlinear quantization of $k_i$ is equivalent to a linear quantization of another parameter, say $g_i$, which is related to $k_i$ by a nonlinear transformation. The requirement that the spectral sensitivity of the new parameters be flat resulted in the following optimal transformation [65]:

$$g_i = \log \frac{1 + k_i}{1 - k_i}, \quad \text{all } i. \tag{96}$$

It is interesting to note from (45) that $g_i$ is simply the logarithm of the hypothetical impedance ratios corresponding to $k_i$.

The optimality of the preceding transformation was based on a specific spectral fidelity criterion. Other transformations would result if other quantization fidelity criteria were adopted.[16]

The transmission rate can be reduced further without affecting the fidelity by proper encoding of each parameter. Variable word length encoding [34] (such as Huffman) can be used for this purpose if the statistical distributions of each of the parameters is known. These distributions can be obtained very simply from a representative sample of signals.

## VI. Pole-Zero Modeling

Given the spectrum of some arbitrary signal, it is generally not possible to determine for certain the identity of the system that generated the signal in terms of a set of poles and

[16] Using a log likelihood criterion, Itakura informed me that a transformation proportional to arcsin $(k_i)$ is optimal.

zeros. The problem is inherently nondeterministic, for a zero can be approximated arbitrarily closely by a large number of poles and vice versa. Indeed, we have seen in this paper that the all-pole model spectrum can approximate the signal spectrum arbitrarily closely by simply increasing the number of poles. However, if there are a number of influential zeros in the signal spectrum, the number of model poles can become very large. For data compression applications, this is an undesirable situation. Also, there are applications where the identification of the zeros is important. Therefore, it is useful to be able to model a spectrum in terms of poles and zeros.

Much effort is currently being expended on the problem of pole-zero modeling [8], [17], [19], [20], [29], [30], [88], [91]. Most of these methods are purely in the time domain. However, there seems to be a growing interest in frequency domain methods [4], [46], [94], partly due to the speed offered by the FFT. Add to this the market availability of spectrum analyzers and special hardware FFT processors. Of course, the time and frequency domain approaches should give similar results since LP analysis is actually performed in the autocorrelation domain.

The beauty of all-pole modeling is that it is relatively simple, straightforward, well understood, inexpensive, and "always" works. Unfortunately, none of these properties apply to pole-zero modeling. The main difficulty is that the pole-zero problem is nonlinear. We show this below for the stationary case. Then, we sketch out representative schemes for iterative and noniterative estimation of the pole and zero parameters. No exhaustive analysis is attempted; the reader is referred to the aforementioned references.

### A. Normal Equations

The transfer function of the pole-zero model is given by $H(z)$ in (2). The corresponding model spectrum is given by

$$\hat{P}(\omega) = |H(e^{j\omega})|^2 = G^2 \frac{|B(e^{j\omega})|^2}{|A(e^{j\omega})|^2} = G^2 \frac{N(\omega)}{D(\omega)} \tag{97}$$

where $B(z)$ and $A(z)$ are the numerator and denominator polynomials in $H(z)$, and the all-zero spectra $N(\omega)$ and $D(\omega)$ form the numerator and denominator of $\hat{P}(\omega)$ and are given by

$$N(\omega) = \left| 1 + \sum_{l=1}^{q} b_l e^{-jl\omega} \right|^2 \tag{98}$$

and

$$D(\omega) = \left| 1 + \sum_{k=1}^{p} a_k e^{-jk\omega} \right|^2. \tag{99}$$

The matching error between the signal spectrum $P(\omega)$ and $\hat{P}(\omega)$ is given by (62), and from (97) is equal to

$$E = \frac{1}{2\pi} \int_{-\pi}^{\pi} P(\omega) \frac{D(\omega)}{N(\omega)} d\omega. \tag{100}$$

$E$ can be interpreted as the residual energy obtained by passing the signal through the filter $A(z)/B(z)$. The problem is to determine $\{a_k\}$ and $\{b_l\}$ such that $E$ in (100) is minimized.

In the sequel we shall make use of the following two relations:

$$\frac{\partial N(\omega)}{\partial b_i} = 2 \sum_{l=0}^{q} b_l \cos (i-l)\,\omega, \qquad b_0 = 1 \tag{101}$$

$$\frac{\partial D(\omega)}{\partial a_i} = 2 \sum_{k=0}^{p} a_k \cos (i-k)\,\omega, \quad a_0 = 1. \tag{102}$$

In addition, we shall use the notation $R_{\alpha\beta}(i)$ to represent the autocorrelation defined by

$$R_{\alpha\beta}(i) = \frac{1}{2\pi} \int_{-\pi}^{\pi} P(\omega) \frac{[D(\omega)]^{\beta}}{[N(\omega)]^{\alpha}} \cos (i\omega)\, d\omega. \tag{103}$$

Thus, $R_{00}(i)$ is simply the Fourier transform of $P(\omega)$.[17] Taking $\partial E/\partial a_i$ in (100) one obtains

$$\frac{\partial E}{\partial a_i} = 2 \sum_{k=0}^{p} a_k R_{10}(i-k), \qquad 1 \leqslant i \leqslant p. \tag{104}$$

Similarly, one can show that

$$\frac{\partial E}{\partial b_i} = -2 \sum_{l=0}^{q} b_l R_{21}(i-l), \qquad 1 \leqslant i \leqslant q. \tag{105}$$

In order to minimize $E$, we set $\partial E/\partial a_i = 0$, $1 \leqslant i \leqslant p$, and $\partial E/\partial b_i = 0$, $1 \leqslant i \leqslant q$, simultaneously. These, then, comprise the normal equations.

From (103), it is clear that $R_{10}(i-k)$ is not a function of $a_k$. Therefore, setting (104) to zero results in a set of linear equations, identical in form to the autocorrelation normal equations (12). However, $R_{21}(i-l)$ in (105) is a function of $b_l$, as can be deduced from (103) with $\alpha = 2$ and $\beta = 1$. Therefore, setting (105) to zero results in a set of nonlinear equations in $b_l$. If one wishes to solve for $\{a_k\}$ and $\{b_l\}$ simultaneously, then one solves a set of $p+q$ *nonlinear* equations.

Note that if the signal is assumed to be nonstationary, the above analysis can be modified accordingly in a manner similar to that in Section III-A for the all-pole case. The resulting equations will be very similar in form to the preceding equations, with nonstationary autocorrelations replacing the stationary autocorrelations.

### B. Iterative Solutions

Since the minimization of $E$ in (100) leads to a set of nonlinear equations, the problem of minimizing $E$ must then be solved iteratively. There are many methods in the literature for finding the extrema of a function [30], [108], many of which are applicable in our case. In particular, gradient methods are appropriate here since it is possible to evaluate the error gradient, as in (104) and (105). One such method was used by Tretter and Steiglitz [94] in pole-zero modeling. Other schemes, such as the Newton–Raphson method, require the evaluation of the Hessian (i.e., second derivative). This can be very cumbersome in many problems, but is straightforward in our case. This is illustrated below by giving a Newton–Raphson solution.

Let $x' = [a_1 a_2 \cdots a_p b_1 b_2 \cdots b_q]$ be the transpose of a column vector $\boldsymbol{x}$ whose elements are the coefficients $a_k$ and $b_l$. If $\boldsymbol{x}(m)$ is the solution at iteration $m$, then $\boldsymbol{x}(m+1)$ is given by

$$\boldsymbol{x}(m+1) = \boldsymbol{x}(m) - J^{-1} \left.\frac{\partial E}{\partial \boldsymbol{x}}\right|_{\boldsymbol{x}=\boldsymbol{x}(m)} \tag{106}$$

where $J$ is the $(p+q) \times (p+q)$ symmetric Hessian matrix

[17] For a discrete spectrum $P(\omega_n)$ the integrals in (100) and (103) are replaced by summations.

given by $J = \partial^2 E/\partial x \partial x'$. Setting $a' = [a_1 a_2 \cdots a_p]$ and $b' = [b_1 b_2 \cdots b_q]$, (106) can be partitioned, with $x' = [a'b']$, as

$$\begin{bmatrix} a(m+1) \\ b(m+1) \end{bmatrix} = \begin{bmatrix} a(m) \\ b(m) \end{bmatrix} - \begin{bmatrix} \dfrac{\partial^2 E}{\partial a \partial a'} & \dfrac{\partial^2 E}{\partial a \partial b'} \\ \dfrac{\partial^2 E}{\partial b \partial a'} & \dfrac{\partial^2 E}{\partial b \partial b'} \end{bmatrix}^{-1}_{\substack{a=a(m)\\ b=b(m)}} \begin{bmatrix} \dfrac{\partial E}{\partial a} \\ \dfrac{\partial E}{\partial b} \end{bmatrix}_{\substack{a=a(m)\\ b=b(m)}} \tag{107}$$

The elements of the first-order partial derivatives in (107) are given by (104) and (105). The elements of the second partial derivatives can be shown to be equal to

$$\frac{\partial^2 E}{\partial a_i \partial a_j} = 2R_{10}(i-j) \tag{108}$$

$$\frac{\partial^2 E}{\partial a_i \partial b_j} = -2 \sum_{k=0}^{p} \sum_{l=0}^{q} a_k b_l \cdot [R_{20}(j+i-l-k) + R_{20}(j-i-l+k)] \tag{109}$$

$$\frac{\partial^2 E}{\partial b_i \partial b_j} = -2R_{21}(i-j) + 4 \sum_{k=0}^{p} \sum_{l=0}^{q} b_k b_l \cdot [R_{31}(j+i-l-k) + R_{31}(j-i-l+k)]. \tag{110}$$

Given the estimates $a(m)$ and $b(m)$, one can compute $N(\omega)$ and $D(\omega)$ from (98) and (99) using FFT's, and then use (103) to compute the autocorrelations $R_{10}$, $R_{20}$, $R_{21}$, and $R_{31}$, which can then be used in (107)–(110) to compute the new estimates $a(m+1)$ and $b(m+1)$. The iterations are halted when the error gradient goes below some prespecified threshold. The minimum error is then computed from (100).

The Newton–Raphson method works very well if the initial estimate is close to the optimum. In that case, the Hessian $J$ is positive definite and the convergence is quadratic [30]. In the next section we discuss noniterative methods which can be used to give these good initial estimates.

### C. Noniterative Solutions

One property that is common to noniterative methods is that a good estimate of the number of poles and zeros seems to be necessary for a reasonable solution. Indeed, in that case, there is not much need to go to expensive iterative methods. However, in general, such information is unavailable and one is interested in obtaining the best estimate for a given $p$ and $q$. Then, noniterative methods can be used profitably to give good initial estimates that are necessary in iterative methods.

*1) Pole Estimation:* Assume that the signal $s_n$ had been generated by exciting the pole-zero filter $H(z)$ in (2) by either an impulse or white noise. Then it is simple to show that the signal autocorrelation obeys the autocorrelation equation (31) for $i > q$. Therefore, the coefficients $a_k$ can be estimated by solving (31) with $q + 1 \leqslant i \leqslant q + p$.

The effect of the poles can now be removed by applying the inverse filter $A(z)$ to the signal. In the spectral domain this can be done by computing a new spectrum $P_1(\omega) = P(\omega) D(\omega)$. The problem now reduces to the estimation of the zeros in $P_1(\omega)$.

*2) Zero Estimation:* A promising noniterative method for pole-zero estimation is that of *cepstral prediction* [76], [95]. The basic idea is that the poles of $nc_n$, where $c_n$ is the complex cepstrum, comprise the poles and zeros of the signal [77]. Therefore, for zero estimation, the problem reduces to finding the poles of $nc_n$ which can be computed by the method just described above, where $c_n$ here is the cepstrum corresponding to $P_1(\omega)$.

Another method for zero estimation is that of *inverse LP modeling* [63]. The idea is quite simple: Invert the spectrum $P_1(\omega)$ and apply a $q$-pole LP analysis. The resulting predictor coefficients are then good estimates of $b_l$. This method gives good results if $P_1(\omega)$ is smooth relative to the model spectrum. Problems arise if the variations of the signal spectrum about the model spectrum are large. The reason is that LP modeling attempts to make a good fit to the spectral envelope, and the envelope of the inverted spectrum is usually different from the inverse of the desired spectral envelope. One solution is to smooth the spectrum $P_1(\omega)$ before inversion. Spectral smoothing is usually performed by applying a low-pass filter to the spectrum (autocorrelation smoothing) or to the log spectrum (cepstral smoothing). Another method is all-pole smoothing. Indeed, all-pole modeling can be thought of as just another method of smoothing the spectrum, where the degree of smoothing is controlled by the order of the predictor $p$, which is usually chosen to be much larger than the number of zeros in the model $q$. We point out that zero estimation by inverse LP modeling with all-pole smoothing is similar to the method of Durbin [24], [25] in the time domain.

## VII. Conclusion

Linear prediction is an autocorrelation-domain analysis. Therefore, it can be approached from either the time or frequency domain. The least squares error criterion in the time domain translates into a spectral matching criterion in the frequency domain. This viewpoint was helpful in exploring the advantages and disadvantages of the least squares error criterion.

The major portion of this paper was devoted to all-pole modeling. This type of modeling is simple, inexpensive and effective; hence its wide applicability and acceptance. In contrast, pole-zero modeling is not simple, generally expensive, and is not yet well understood. Future research should be directed at acquiring a better understanding of the problems in pole-zero modeling and developing appropriate methodologies to deal with these problems.

## Acknowledgment

The author wishes to thank the following friends and colleagues who have read and commented on earlier versions of this paper: H. Akaike, R. Barakat, C. Cook, T. Fortmann, F. Itakura, G. Kopec, A. Oppenheim, L. Rabiner, R. Viswanathan, and V. Zue. He is especially grateful to R. Viswanathan for the fruitful discussions they have had throughout the writing of the paper and for his help in supplying many of the references. He also wishes to thank B. Aighes, C. Williams, and R. Schwartz for their assistance in the preparation of the manuscript.

## References

[1] H. Akaike, "Power spectrum estimation through autoregressive model fitting," *Ann. Inst. Statist. Math.*, vol. 21, pp. 407–419, 1969.

[2] ——, "Statistical predictor identification," *Ann. Inst. Statist. Math.*, vol. 22, pp. 203–217, 1970.

[3] ——, "Information theory and an extension of the maximum likelihood principle," in *Proc. 2nd Int. Symp. Information Theory* (Supplement to Problems of Control and Information Theory), 1972.

[4] —, "Maximum likelihood identification of Gaussian autoregressive moving average models," *Biometrika*, vol. 60, no. 2, pp. 255-265, 1973.

[5] —, "A new look at statistical model identification," *IEEE Trans. Automat. Contr.*, vol. AC-19, pp. 716-723, Dec. 1974.

[6] T. W. Anderson, *The Statistical Analysis of Time Series.* New York: Wiley, 1971.

[7] C. A. Andrews, J. M. Davies, and G. R. Schwartz, "Adaptive data compression," *Proc. IEEE*, vol. 55, pp. 267-277, Mar. 1967.

[8] K. J. Åström and P. Eykhoff, "System identification–A survey," *Automatica*, vol. 7, pp. 123-162, 1971.

[9] B. S. Atal and M. R. Schroeder, "Adaptive predictive coding of speech signals," *Bell Syst. Tech. J.*, vol. 49, no. 6, pp. 1973-1986, Oct. 1970.

[10] B. S. Atal and S. L. Hanauer, "Speech analysis and synthesis by linear prediction of the speech wave," *J. Acoust. Soc. Amer.*, vol. 50, no. 2, pp. 637-655, 1971.

[11] M. S. Bartlett, *An Introduction to Stochastic Processes with Special Reference to Methods and Applications.* Cambridge, England: Cambridge Univ. Press, 1956.

[12] J. S. Bendat and A. G. Piersol, *Random Data: Analysis and Measurement Procedures.* New York: Wiley, 1971.

[13] C. Bingham, M. D. Godfrey, and J. W. Tukey, "Modern techniques of power spectrum estimation," *IEEE Trans. Audio Electroacoust.*, vol. AU-15, pp. 56-66, June 1967.

[14] R. B. Blackman and J. W. Tukey, *The Measurement of Power Spectra.* New York: Dover, 1958.

[15] T. Bohlin, "Comparison of two methods of modeling stationary EEG signals," *IBM J. Res. Dev.*, pp. 194-205, May 1973.

[16] S. F. Boll, "*A priori* digital speech analysis," Computer Science Div., Univ. Utah, Salt Lake City, UTEC-CSC-73-123, 1973.

[17] G. E. Box and G. M. Jenkins, *Time Series Analysis Forecasting and Control.* San Francisco, Calif.: Holden-Day, 1970.

[18] J. P. Burg, "The relationship between maximum entropy spectra and maximum likelihood spectra," *Geophysics*, vol. 37, no. 2, pp. 375-376, Apr. 1972.

[19] C. S. Burrus and T. W. Parks, "Time domain design of recursive digital filters," *IEEE Trans. Audio Electroacoust.*, vol. AU-18, pp. 137-141, June 1970.

[20] J. C. Chow, "On estimating the orders of an autoregressive moving-average process with uncertain observations," *IEEE Trans. Automat. Contr.*, vol. AC-17, pp. 707-709, Oct. 1972.

[21] W. T. Cochran *et al.*, "What is the fast Fourier transform?," *IEEE Trans. Audio Electroacoust.*, vol. AU-15, pp. 45-55, June 1967.

[22] H. Cox, "Linear versus logarithmic averaging," *J. Acoust. Soc. Amer.*, vol. 39, no. 4, pp. 688-690, 1966.

[23] L. D. Davisson, "The theoretical analysis of data compression systems," *Proc. IEEE*, vol. 56, pp. 176-186, Feb. 1968.

[24] J. Durbin, "Efficient estimation of parameters in moving-average models," *Biometrika*, vol. 46, Parts 1 and 2, pp. 306-316, 1959.

[25] —, "The fitting of time-series models," *Rev. Inst. Int. Statist.*, vol. 28, no. 3, pp. 233-243, 1960.

[26] A. Eberhard, "An optimal discrete window for the calculation of power spectra," *IEEE Trans. Audio Electroacoust.*, vol. AU-21, pp. 37-43, Feb. 1973.

[27] M. P. Ekstrom, "A spectral characterization of the ill-conditioning in numerical deconvolution," *IEEE Trans. Audio Electroacoust.*, vol. AU-21, pp. 344-348, Aug. 1973.

[28] P. Elias, "Predictive coding, Parts I and II," *IRE Trans. Inform. Theory*, vol. IT-1, p. 16, Mar. 1955.

[29] A. G. Evans and R. Fischl, "Optimal least squares time-domain synthesis of recursive digital filters," *IEEE Trans. Audio Electroacoust.*, vol. AU-21, pp. 61-65, Feb. 1973.

[30] P. Eykhoff, *System Identification: Parameter and State Estimation.* New York: Wiley, 1974.

[31] D. K. Faddeev and V. N. Faddeeva, *Computational Methods of Linear Algebra.* San Francisco, Calif.: Freeman, 1963.

[32] P. B. C. Fenwick, P. Michie, J. Dollimore, and G. W. Fenton, "Mathematical simulation of the electroencephalogram using an autoregressive series," *Bio-Med. Comput.*, vol. 2, pp. 281-307, 1971.

[33] J. L. Flanagan, *Speech Analysis Synthesis and Perception*, 2nd Edition. New York: Springer-Verlag, 1972.

[34] R. G. Gallager, *Information Theory and Reliable Communication.* New York: Wiley, 1968.

[35] L. Y. Geronimus, *Orthogonal Polynomials.* New York: Consultants Bureau, 1961.

[36] W. Gersch, "Spectral analysis of EEG's by autoregressive decomposition of time series," *Math. Biosci.*, vol. 7, pp. 205-222, 1970.

[37] W. Gersch and D. R. Sharpe, "Estimation of power spectra with finite-order autoregressive models," *IEEE Trans. Automat. Contr.*, vol. AC-18, pp. 367-369, Aug. 1973.

[38] B. Gold and C. M. Rader, *Digital Processing of Signals.* New York: McGraw-Hill, 1969.

[39] G. H. Golub and C. Reinsch, "Singular value decomposition and least squares solutions," *Numer. Math.*, vol. 14, pp. 403-420, 1970.

[40] A. H. Gray and J. D. Markel, "A spectral-flatness measure for studying the autocorrelation method of linear prediction of speech analysis," *IEEE Trans. Acoust., Speech, Signal Processing*, vol. ASSP-22, pp. 207-216, June 1974.

[41] R. M. Gray, "On the asymptotic eigenvalue distribution of Toeplitz matrices," *IEEE Trans. Inform. Theory*, vol. IT-18, pp. 725-730, Nov. 1972.

[42] U. Grenander and G. Szegö, *Toeplitz Forms and Their Applications.* Berkeley, Calif.: Univ. California Press, 1958.

[43] U. Grenander and M. Rosenblatt, *Statistical Analysis of Stationary Time Series.* New York: Wiley, 1957.

[44] A. Habibi and G. S. Robinson, "A survey of digital picture coding," *Computer*, pp. 22-34, May 1974.

[45] E. J. Hannan, *Time Series Analysis.* London, England: Methuen, 1960.

[46] —, *Multiple Time Series.* New York: Wiley, 1970.

[47] J. R. Haskew, J. M. Kelly, R. M. Kelly, Jr., and T. H. McKinney, "Results of a study of the linear prediction vocoder," *IEEE Trans. Commun.*, vol. COM-21, pp. 1008-1014, Sept. 1973.

[48] R. L. Hershey, "Analysis of the difference between log mean and mean log averaging," *J. Acoust. Soc. Amer.*, vol. 51, pp. 1194-1197, 1972.

[49] F. B. Hildebrand, *Introduction to Numerical Values.* New York: McGraw-Hill, 1956.

[50] F. Itakura and S. Saito, "Analysis synthesis telephony based on the maximum likelihood method," in *Rep. 6th Int. Congr. Acoustics*, Y. Kohasi, Ed., pp. C17-C20, Paper C-5-5, Aug. 1968.

[51] —, "A statistical method for estimation of speech spectral density and formant frequencies," *Electron. Commun. Japan*, vol. 53-A, no. 1, pp. 36-43, 1970.

[52] —, "Digital filtering techniques for speech analysis and synthesis," in *Conf. Rec., 7th Int. Congr. Acoustics*, Paper 25 C 1, 1971.

[53] —, "On the optimum quantization of feature parameters in the Parcor speech synthesizer," in *IEEE Conf. Rec., 1972 Conf. Speech Communication and Processing*, pp. 434-437, Apr. 1972.

[54] G. M. Jenkins and D. G. Watts, *Spectral Analysis and Its Applications.* San Francisco, Calif.: Holden-Day, 1968.

[55] T. Kailath, "A view of three decades of linear filtering theory," *IEEE Trans. Inform. Theory*, vol. IT-20, pp. 146-181, Mar. 1974.

[56] R. E. Kalman, "A new approach to linear filtering and prediction problems," *Trans. ASME, J. Basic Eng.*, Series D82, pp. 35-45, 1960.

[57] H. Kobayashi and L. R. Bahl, "Image data compression by predictive coding I: Prediction algorithms," *IBM J. Res. Dev.*, pp. 164-171, Mar. 1974.

[58] A. Kolmogorov, "Interpolation und Extrapolation von stationären zufälligen Folgen," *Bull. Acad. Sci., U.S.S.R., Ser. Math.*, vol. 5, pp. 3-14, 1941.

[59] K. S. Kunz, *Numerical Analysis.* New York: McGraw-Hill, 1957.

[60] R. C. K. Lee, *Optimal Estimation, Identification, and Control*, Research Monograph no. 28. Cambridge, Mass.: M.I.T. Press, 1964.

[61] N. Levinson, "The Wiener RMS (root mean square) error criterion in filter design and prediction," *J. Math. Phys.*, vol. 25, no. 4, pp. 261-278, 1947. Also Appendix B, in N. Wiener, *Extrapolation, Interpolation and Smoothing of Stationary Time Series.* Cambridge, Mass.: M.I.T. Press, 1949.

[62] J. Makhoul, "Spectral analysis of speech by linear prediction," *IEEE Trans. Audio Electroacoust.*, vol. AU-21, pp. 140-148, June 1973.

[63] —, "Selective linear prediction and analysis-by-synthesis in speech analysis," Bolt Beranek and Newman Inc., Cambridge, Mass., Rep. 2578, Apr. 1974.

[64] J. Makhoul and J. J. Wolf, "Linear prediction and the spectral analysis of speech," Bolt Beranek and Newman Inc., Cambridge, Mass., NTIS AD-749066, Rep. 2304, Aug. 1972.

[65] J. Makhoul and R. Viswanathan, "Quantization properties of transmission parameters in linear predictive systems," Bolt Beranek and Newman Inc., Cambridge, Mass., Rep. 2800, Apr. 1974.

[66] H. B. Mann and A. Wald, "On the statistical treatment of linear stochastic difference equations," *Econometrica*, vol. 11, nos. 3 and 4, pp. 173-220, 1943.

[67] W. D. Mark, "Spectral analysis of the convolution and filtering of non-stationary stochastic processes," *J. Sound Vib.*, no. 11, pp. 19-63, 1970.

[68] J. D. Markel, "Digital inverse filtering–A new tool for formant trajectory estimation," *IEEE Trans. Audio Electroacoust.*, vol. AU-20, pp. 129-137, June 1972.

[69] J. D. Markel and A. H. Gray, "On autocorrelation equations as applied to speech analysis," *IEEE Trans. Audio Electroacoust.*, vol. AU-20, pp. 69–79, Apr. 1973.

[70] E. Matsui *et al.*, "An adaptive method for speech analysis based on Kalman filtering theory," *Bull Electrotech. Lab*, vol. 36, no. 3, pp. 42–51, 1972 (in Japanese).

[71] R. N. McDonough and W. H. Huggins, "Best least-squares representation of signals by exponentials," *IEEE Trans. Automat. Contr.*, vol. AC-13, pp. 408–412, Aug. 1968.

[72] S. K. Mitra and R. J. Sherwood, "Estimation of pole-zero displacements of a digital filter due to coefficient quantization," *IEEE Trans. Circuits Syst.*, vol. CAS-21, pp. 116–124, Jan. 1974.

[73] R. E. Nieman, D. G. Fisher, and D. E. Seborg, "A review of process identification and parameter estimation techniques," *Int. J. Control*, vol. 13, no. 2, pp. 209–264, 1971.

[74] R. H. Norden, "A survey of maximum likelihood estimation," *Int. Statist. Rev.*, vol. 40, no. 3, pp. 329–354, 1972.

[75] ——, "A survey of maximum likelihood estimation, Part 2," *Int. Statist. Rev.*, vol. 41, no. 1, pp. 39–58, 1973.

[76] A. Oppenheim and J. M. Tribolet, "Pole-zero modeling using cepstral prediction," Res. Lab. Electronics, M.I.T., Cambridge, Mass., QPR 111, pp. 157–159, 1973.

[77] A. V. Oppenheim, R. W. Schafer and T. G. Stockham, "Nonlinear filtering of multiplied and convolved signals," *Proc. IEEE*, vol. 56, pp. 1264–1291, Aug. 1968.

[78] M. Pagano, "An algorithm for fitting autoregressive schemes," *J. Royal Statist. Soc., Series C (Applied Statistics)*, vol. 21, no. 3, pp. 274–281, 1972.

[79] A. Papoulis, *Probability, Random Variables, and Stochastic Processes.* New York: McGraw-Hill, 1965.

[80] L. R. Rabiner *et al.*, "Terminology in digital signal processing," *IEEE Trans. Audio Electroacoust.*, vol. AU-20, pp. 322–337, Dec. 1972.

[81] A. Ralston, *A First Course in Numerical Analysis.* New York: McGraw-Hill, 1965.

[82] A. Ralston and H. Wilf, *Mathematical Methods for Digital Computers*, vol. II. New York: Wiley, 1967.

[83] M. P. Ristenbatt, "Alternatives in digital communications," *Proc. IEEE*, vol. 61, pp. 703–721, June 1973.

[84] E. A. Robinson, *Multichannel Time Series Analysis with Digital Computer Programs.* San Francisco, Calif.: Holden-Day, 1967.

[85] ——, *Statistical Communication and Detection.* New York: Hafner, 1967.

[86] ——, "Predictive decomposition of time series with application to seismic exploration," *Geophysics*, vol. 32, no. 3, pp. 418–484, June 1967.

[87] E. A. Robinson and S. Treitel, *The Robinson-Treitel Reader*, 3rd ed. Tulsa, Okla.: Seismograph Service Corp., 1973.

[88] A. P. Sage and J. L. Melsa, *System Identification.* New York: Academic Press, 1971.

[89] R. W. Schafer and L. R. Rabiner, "Digital representations of speech signals," this issue, pp. 662–677.

[90] F. C. Schweppe, *Uncertain Dynamic Systems.* Englewood Cliffs, N.J.: Prentice-Hall, 1973.

[91] J. L. Shanks, "Recursion filters for digital processing," *Geophysics*, vol. XXXII, no. 1, pp. 33–51, 1967.

[92] R. A. Silverman, "Locally stationary random processes," *IRE Trans. Inform. Theory*, vol. IT-3, pp. 182–187, Sept. 1957.

[93] G. Szegö, *Orthogonal Polynomials.* New York: Amer. Math. Soc. Colloquium Publ., vol. XXIII, N.Y., 1959.

[94] S. A. Tretter and K. Steiglitz, "Power-spectrum identification in terms of rational models," *IEEE Trans. Automat. Control*, vol. AC-12, pp. 185–188, Apr. 1967.

[95] J. M. Tribolet, "Identification of linear discrete systems with applications to speech processing," Master's thesis, Dep. Elec. Eng., M.I.T., Cambridge, Mass., Jan. 1974.

[96] A. Van den Bos, "Alternative interpretation of maximum entropy spectral analysis," *IEEE Trans. Inform. Theory*, vol. IT-17, pp. 493–494, July 1971.

[97] H. Wakita, "Direct estimation of the vocal tract shape by inverse filtering of acoustic speech waveforms," *IEEE Trans. Audio Electroacoust.*, vol. AU-21, pp. 417–427, Oct. 1973.

[98] G. Walker, "On periodicity in series of related terms," *Proc. Royal Soc.*, vol. 131-A, p. 518, 1931.

[99] R. J. Wang, "Optimum window length for the measurement of time-varying power spectra," *J. Acoust. Soc. Amer.*, vol. 52, no. 1 (part 1), pp. 33–38, 1971.

[100] R. J. Wang and S. Treitel, "The determination of digital Wiener filters by means of gradient methods," *Geophysics*, vol. 38, no. 2, pp. 310–326, Apr. 1973.

[101] P. D. Welch, "A direct digital method of power spectrum estimation," *IBM J. Res. Dev.*, vol. 5, pp. 141–156, Apr. 1961.

[102] A. Wennberg and L. H. Zetterberg, "Application of a computer-based model for EEG analysis," *Electroencephalogr. Clin. Neurophys.*, vol. 31, no. 5, pp. 457–468, 1971.

[103] P. Whittle, "Some recent contributions to the theory of stationary processes," Appendix 2, in H. Wold, *A Study in the Analysis of Stationary Time Series.* Stockholm, Sweden: Almqvist and Wiksell, 1954.

[104] ——, "On the fitting of multivariate autoregressions, and the approximate canonical factorization of a spectral density matrix," *Biometrika*, vol. 50, nos. 1 and 2, pp. 129–134, 1963.

[105] ——, *Prediction and Regulation by Linear Least Square Methods.* London, England: English Universities Press, 1963.

[106] B. Widrow, P. E. Mantey, L. J. Griffiths, and B. B. Goode, "Adaptive antenna systems," *Proc. IEEE*, vol. 55, pp. 2143–2159, Dec. 1967.

[107] N. Wiener, *Extrapolation, Interpolation and Smoothing of Stationary Time Series With Engineering Applications.* Cambridge, Mass.: M.I.T. Press, 1949.

[108] D. J. Wilde, *Optimum Seeking Methods.* Englewood Cliffs, N.J.: Prentice-Hall, 1964.

[109] J. H. Wilkinson, "Error analysis of direct methods of matrix inversion," *J. Ass. Comput. Mach.*, vol. 8, no. 3, pp. 281–330, 1961.

[110] ——, "The solution of ill-conditioned linear equations," in Ralston and Wilf, *Mathematical Methods for Digital Computers*, pp. 65–93, 1967.

[111] J. H. Wilkinson and C. Reinsch, *Linear Algebra*, vol. II. New York: Springer-Verlag, 1971.

[112] H. Wold, *A Study in the Analysis of Stationary Time Series.* Stockholm, Sweden: Almqvist and Wiksell, 1954.

[113] H. O. A. Wold, *Bibliography on Time Series and Stochastic Processes.* Cambridge, Mass.: M.I.T. Press, 1965.

[114] L. C. Wood and S. Treitel, "Seismic signal processing," this issue, pp. 649–661.

[115] G. U. Yule, "On a method of investigating periodicities in disturbed series, with special reference to Wolfer's sunspot numbers," *Phil. Trans. Roy. Soc.*, vol. 226-A, pp. 267–298, 1927.

# Part 2
# One-Dimensional Digital Filtering

## A. Finite Impulse Response Filter Design

The field of finite impulse response (FIR) filter design has grown rapidly since the introduction of the window design method several years ago. Most of the activity has been in the area of designing optimal (in a minimax or Chebyshev sense) filters. However, some important work has also been done in the area of quantifying the design relationships for the Kaiser or $I_0$-sinh window.

The original work in the area of optimal FIR filter design was done by Herrmann and Schuessler in Germany [1], [2] and Hofstetter *et al.* at M.I.T. [3], [4]. These researchers devised methods for designing a subset of the class of optimal FIR filters, the so-called maximal ripple or extraripple filters. Techniques capable of designing any extraripple filter were subsequently proposed by Parks and McClellan [5] and Rabiner [6]. The Parks and McClellan design procedure was based on a Remez exchange algorithm approach, whereas Rabiner used the slower, but less constraining, method of linear programming to design an arbitrary optimal FIR filter. Because of its inherent speed and capabilities, the Parks and McClellan method has become the accepted design method of optimal FIR filters.

Following the introduction of the design methods for optimal FIR filters, much effort went into studies of the properties of the resulting designs [7], [8] and the relationships between the various filter parameters for several classes of filters [9]–[13].

The three papers on optimal FIR filters included in this section were chosen to give a good global coverage of optimal filter design. The tutorial paper by Rabiner *et al.* provides a general introduction to the problem of FIR filter design using weighted Chebyshev approximation. The computer program in the paper by McClellan *et al.* is a realization of the design methods presented in the tutorial paper. The paper on design relationships for low-pass filters by Rabiner provides a set of formulas for choosing filter design parameters for the low-pass filter case.

The paper by Kaiser on the $I_0$-sinh window provides a comprehensive summary of the window design method and gives valuable analytic formulas for choosing filter parameters for both the Kaiser window and for the Chebyshev designs previously discussed.

### References


[1] O. Herrmann, "Design of nonrecursive digital filters with linear phase," *Electron. Lett.*, vol. 6, pp. 328–329, May 1970.

[2] O. Herrmann and H. W. Schuessler, "Design of nonrecursive digital filters with minimum phase," *Electron. Lett.*, vol. 6, pp. 329–330, May 1970.

[3] E. Hofstetter, A. V. Oppenheim, and J. Siegel, "A new technique for the design of nonrecursive digital filters," in *Proc. 5th Annu. Conf. Information Sciences and Systems*, 1970, pp. 64–72.

[4] —— "On optimum nonrecursive digital filters," in *Proc. 9th Allerton Conf.*, 1971, pp. 789–798.

[5] T. Parks and J. H. McClellan, "Chebyshev approximation for nonrecursive digital filters with linear phase," *IEEE Trans. Circuit Theory*, vol. CT-19, pp. 189–194, Mar. 1972.

[6] L. R. Rabiner, "Linear program design of finite impulse response (FIR) digital filters," *IEEE Trans. Audio Electroacoust.*, vol. AU-20, pp. 280–288, Oct. 1972.

[7] T. W. Parks, L. R. Rabiner, and J. H. McClellan, "On the transition width of finite imdulse response digital filters," *IEEE Trans. Audio Electroacoust.*, vol. AU-21, pp. 1–4, Feb. 1973.

[8] L. R. Rabiner and O. Herrmann, "The predictability of certain optimum finite-impulse-response digital filters," *IEEE Trans. Circuit Theory*, vol. CT-20, pp. 401–408, July 1973.

[9] O. Herrmann, L. Rabiner, and D. Chan, "Practical design rules for optimum FIR digital filters," *Bell Syst. Tech. J.*, vol. 52, pp. 769–799, July–Aug. 1973.

[10] L. R. Rabiner and O. Herrmann, "On the design of optimum FIR low-pass filters with even impulse response duration," *IEEE Trans. Audio Electroacoust.*, vol. AU-21, pp. 329–336, Aug. 1973.

[11] L. Rabiner and R. Schafer, "On the behavior of minimax relative error FIR digital differentiators," *Bell Syst. Tech. J.*, vol. 53, pp. 333–361, Feb. 1974.

[12] ——, "On the behavior of minimax FIR digital Hilbert transformers," *Bell Syst. Tech. J.*, vol. 53, pp. 363–390, Feb. 1974.

[13] L. R. Rabiner, J. F. Kaiser, and R. W. Schafer, "Some considerations in the design of multiband finite-impulse-response digital filters," *IEEE Trans. Acoust., Speech, Signal Processing*, vol. ASSP-22, pp. 462–472, Dec. 1974.


## B. Infinite Impulse Response Filter Design

The papers included in this section may be classified as relating to either analytical or algorithmic design procedures. In the first reprint volume, *Digital Signal Processing*, there were several papers which discussed classical infinite impulse response (IIR) design procedures. Included among these papers were discussions of the various filter transformation procedures and the classical filter design methods. In this section the analytical design procedures are extended further in the papers by Thiran, Fettweis, and Brophy and Salazar. Thiran discusses the problem of designing IIR digital filters with equal-ripple delay characteristics. Fettweis presents a simple, but elegant method for designing maximally flat group delay digital filters. An earlier paper by Thiran [1] also presents results on this problem, but, for reasons of space, the editors could not include both papers. The paper by Brophy and Salazar discusses the Padé approximant technique in which a simple approximation to a desired (smooth) frequency response is obtained in a fairly straightforward manner. In cases where this first approximation is not adequate, it can be used as an initial guess for an appropriate algorithmic design procedure.

The final two papers in this section deal with algorithmic design of IIR digital filters. The paper by Deczky on minimum $p$-error approximation is a natural extension of the work by Steiglitz [2] presented in the first reprint volume. The paper by Rabiner *et al.* presents a linear programming approach to approximating an arbitrary magnitude function for an IIR filter. The important advantage of this technique is its guaranteed mathematical convergence to the globally minimax error. A difficulty of using this method is the computational accuracy required to obtain accurate and reliable solutions, since a squared-magnitude function is used in the approximation.

### References

[1] J. P. Thiran, "Recursive digital filters with maximally flat group delay," *IEEE Trans. Circuit Theory*, vol. CT-18, pp. 659–664, Nov. 1971.

[2] K. Steiglitz, "Computer-aided design of recursive digital filters," *IEEE Trans. Audio Electroacoust.*, vol. AU-18, pp. 123–129, June 1970.

## C. Digital Filter Theory

In the first two sections, the papers have primarily been concerned with design methods for filters. In this section the papers deal with several related issues in digital filter theory, namely, analysis of the interpolation problem, the effects of jitter in time sampling on the resulting frequency response of a filter, and a comparison between FIR and IIR filters.

The process of interpolation has long been of interest to mathematicians, scientists, and engineers and has classically been approached from a numerical analysis point of view. The paper by Schafer and Rabiner presents a new approach to this problem. Here they treat the process of interpolation (and decimation) as a signal processing problem with well-defined filter specifications. Further, they show that FIR filters have strong advantages over IIR filters for the actual implementation of interpolators and decimators. Finally, they describe a method for designing optimal (in a Chebyshev sense) FIR interpolating filters using linear programming techniques. The paper by Oetken *et al.* extends the results of the Schafer and Rabiner paper by discussing an analytical design method for minimum mean-square-error interpolating FIR filters.

The paper by Kan and Aggarwal discusses the effects of jitter in the sampling period on the transfer function of the signal and gives bounds on the variance of the error introduced by the jitter.

The last paper in this section, by Rabiner *et al.* provides one of the first comparisons between FIR and IIR digital filters designed to meet the same magnitude specifications. This paper also provides a comprehensive summary of the design specifications and formulas for each class of filters on the basis of magnitude specifications alone. In addition, this paper looks at how a phase equalizer for the IIR filter affects the overall comparisons.

# FIR Digital Filter Design Techniques Using Weighted Chebyshev Approximation

LAWRENCE R. RABINER, MEMBER, IEEE, JAMES H. McCLELLAN, MEMBER, IEEE, AND THOMAS W. PARKS, MEMBER, IEEE

*Invited Paper*

*Abstract*—**This paper discusses the various approaches to designing FIR digital filters using the theory of weighted Chebyshev approximation. The different design techniques are explained and compared on the basis of their capabilities and limitations. The relationships between filter parameters are briefly discussed for the case of low-pass filters. Extensions of the theory to the problems of magnitude and complex approximation are also included, as are some recent results on the design of two-dimensional FIR filters by transformation.**

## I. Introduction

In THE PAST few years, powerful computer optimization algorithms have been developed to solve the design problem for finite-duration impulse response (FIR) digital filters. It is the purpose of this paper to review these techniques in the light of Chebyshev approximation theory and to describe some of the extensions of this theory.

Manuscript received June 10, 1974; revised September 25, 1974. This work was supported in part by NSF Grant GK-23697.

L. R. Rabiner is with Bell Laboratories, Murray Hill, N.J. 07974.
J. H. McClellan is with the M.I.T. Lincoln Laboratory, Lexington, Mass. 02173.
T. W. Parks is with the Department of Electrical Engineering, Rice University, Houston, Tex. 77001.

FIR digital filters possess certain desirable properties which make them attractive for digital signal processing applications. Among these are the ability to have exactly linear phase and the absence of stability problems in nonrecursive realizations. While long sequences are sometimes necessary to achieve sharp cutoff filters, use of the fast Fourier transform (FFT) can make the realization of such filters computationally competitive even with sharp cutoff infinite-duration impulse response (IIR) elliptic filters.

The process of designing and realizing a digital filter to meet some desired specifications consists of five basic steps.

1) Choose a design technique and convert the desired specifications into a precise mathematical formulation in order to approximate the ideal filter shape.

2) Solve the approximation problem to determine the filter coefficients which minimize a performance measure.

3) Choose a specific structure in which the filter will be realized and quantize the resulting filter coefficients to a fixed word length.

4) Quantize the digital filter variables, i.e., the input, output, and intermediate variable word lengths.

Reprinted from *Proc. IEEE*, vol. 63, pp. 595–610, Apr. 1975.

5) Verify by simulation that the resulting design meets given performance specifications.

The results of step 5) generally lead to revisions in steps 1), 2), 3), or 4) in order to meet specifications.

Although it would be desirable to be able to perform steps 2)–4) simultaneously, i.e., to be able to solve the approximation problem for arbitrary structures, with arbitrary word lengths, it is not likely that such a design procedure will be available in the foreseeable future. Thus for the time being we must be content to solve each of these problems independently.

It is quite easy to overlook the importance of step 1) both in the choice of design methods and in the ways that the desired specifications are converted for input to the different design methods. For example, some design algorithms specify the cutoff frequencies and minimize passband and/or stopband deviation whereas other algorithms specify the passband and stopband deviations and obtain the cutoff frequencies only after performing step 2). Algorithms of both types will be described later in this paper.

This paper will be devoted exclusively to a discussion of optimal techniques for solving the approximation problem of step 2). The word "optimal" will denote weighted Chebyshev approximation of a desired frequency response by an FIR digital filter. The presentation will concentrate on linear phase filters where the design techniques are highly developed because the approximation problem is real. We will also discuss the interrelationships between the various design parameters for several filter types. Such relationships are helpful in performing step 1), where the designer must specify parameters such as the filter length, cutoff frequencies, and passband ripple or stopband attenuation. Often these parameters are only implicit in the desired specifications.

Several extensions of the Chebyshev approximation techniques will also be discussed. These include approximation of the magnitude response without regard to the phase, and two-dimensional approximation. Before discussing specific design methods we review various alternative solutions to the FIR filter approximation problem.

## II. History of the FIR Filter Approximation Problem

One of the earliest attempts at deriving the coefficients of an FIR digital filter in order to approximate an ideal desired frequency response was the method of windowing in which the desired frequency response is expanded in a Fourier series and truncated to the desired filter length [1]–[7]. The resulting filter minimizes the least-squares error between the desired response and the filter response. However, the Chebyshev error (the maximum absolute value of the error) from this approach is rather large, due to the Gibbs phenomenon which occurs at discontinuities of the desired frequency response.

Instead of simply truncating the infinite Fourier series, the technique of windowing seeks to reduce the Gibbs phenomenon by multiplying the coefficients of the Fourier series by a smooth time-limited window. Among the more popular windows are the Kaiser window [2], the Hamming window [3], the hanning window [3], and the Dolph–Chebyshev window [4]–[7]. One of the attributes of windowing is that it is an analytical technique, whereas, most other FIR design techniques are iterative in nature.

A second FIR design technique [8]–[14] (the frequency sampling method) was originally proposed by Gold and Jordan [10] and further developed by Rabiner *et al.* [11]. The basic idea behind this method is that one can approximate a specified frequency response by fixing most of its discrete Fourier transform (DFT) coefficients (the frequency samples) and leaving unspecified those DFT coefficients which lie in transition bands. An optimization algorithm is used to choose values for the unspecified coefficients so as to minimize a weighted approximation error over the frequency range of interest. The problem can be shown to be a linear programming problem with very few independent variables, but a large number of constraints.

Herrmann [16] was the first to develop a method for designing optimal (in a Chebyshev sense) FIR filters. By assuming that the frequency response of the optimal low-pass filter was equiripple in both the passband and the stopband, and by fixing the number of ripples in each band, Herrmann was able to write down a set of nonlinear equations which completely described the filter. He then proceeded to solve these equations directly, using an iterative descent method. The length of filters designed in this manner was limited to about 40.

Hofstetter *et al.* [19], [20] removed the restriction on length in Herrmann's approach by developing an algorithm which was "reminiscent of the Remez exchange algorithm" in order to solve the nonlinear equations. In view of later results, it is possible to show that the filters of Herrmann and Hofstetter are a restricted subset of optimal min-max filters, the so-called extraripple or maximal ripple filters. One drawback of the approach of both Herrmann and Hofstetter *et al.* is that it is not possible to specify *a priori* the locations of the passband and stopband cutoff frequencies.

Parks and McClellan [21] formulated the lowpass approximation of the desired response on two disjoint intervals, the passband and the stopband with a transition band left unspecified. Necessary and sufficient conditions for the best Chebyshev approximation were obtained from the classical alternation theorem, and the Remez exchange algorithm was demonstrated to be an effective tool for the computation of these optimal filters. Subsequently, this formulation was extended to include all types of linear phase FIR filters [31].

Rabiner [22], [23] showed that linear programming offered an alternative method for computing the best Chebyshev approximation. Although linear programming is very flexible and can be used to approximate a wide variety of desired filter shapes, it is comparatively slow and hence, the length of the filters it can design is limited.

This paper will describe the last four design methods. Each of these methods yields optimal filters and the theory of Chebyshev approximation provides the underlying mathematical explanation of this behavior. In all cases the filter is restricted to have linear phase so that the approximation problem will be real. In the next section, the linear phase condition for FIR filters is reviewed and the characteristics of the various types of FIR filters are derived.

## III. Characteristics of FIR Filters with Linear Phase

Let $\{h(n)\}$ be a causal finite duration sequence defined over the interval $0 \leqslant n \leqslant N-1$.

The $z$ transform of $\{h(n)\}$ is

$$H(z) = \sum_{n=0}^{N-1} h(n)\, z^{-n} = h(0) + h(1)\, z^{-1} + \cdots + h(N-1)\, z^{-(N-1)} \tag{1}$$

TABLE I

| | $\alpha$ | $\beta$ |
|---|---|---|
| Case 1–$N$ odd, Symmetric impulse response | $\frac{N-1}{2}$ | 0 |
| Case 2–$N$ even, Symmetric impulse response | $\frac{N-1}{2}$ | 0 |
| Case 3–$N$ odd, Antisymmetric impulse response | $\frac{N-1}{2}$ | $\frac{\pi}{2}$ |
| Case 4–$N$ even, Antisymmetric impulse response | $\frac{N-1}{2}$ | $\frac{\pi}{2}$ |

and the Fourier transform of $\{h(n)\}$ is

$$H(e^{j\omega}) = \sum_{n=0}^{N-1} h(n)\, e^{-j\omega n} \tag{2}$$

We define a linear phase filter as one whose frequency response can be written in the form

$$H(e^{j\omega}) = G(e^{j\omega})\, e^{j(\beta - \alpha\omega)} \tag{3}$$

where $G(e^{j\omega})$ is real valued. Notice that $G(e^{j\omega})$ is not the magnitude of the frequency response,[1] since $G(e^{j\omega})$ can be negative. Also, (3) only requires the filter to have a constant group delay.

It can be shown that the only solutions for $\alpha$ and $\beta$ are $\alpha = (N-1)/2$ and $\beta = 0$ or $\beta = \pi/2$. When $\beta = 0$ the impulse response is symmetric, i.e., $h(n) = h(N-1-n)$ $n = 0, 1, \cdots, N-1$, and when $\beta = \pi/2$ the impulse response is antisymmetric, i.e., $h(n) = -h(N-1-n)$, $n = 0, 1, \cdots, N-1$. The class of linear phase FIR filters can be divided into four cases according to whether $N$ is odd or even and whether $\beta = 0$ or $\pi/2$. Table I lists these four cases as we will refer to them throughout this paper. It should be noted that when $N$ is even (i.e., cases 2 and 4) the filter group delay $\alpha = (N-1)/2$ is not an integer number of samples. Thus for $N = 10$, $\alpha = 4\frac{1}{2}$ samples delay. This extra "half sample" delay is of importance in some applications, but in most cases it has little effect on the overall processing.

The function $G(e^{j\omega})$ of (3) may be expressed in terms of the impulse response coefficients for each of the four cases of a linear phase filter. Such formulas are derived now for later use in describing various techniques for designing FIR filters to match prescribed frequency response characteristics.

*Case 1:* $N$ odd, symmetric impulse response

$$G(e^{j\omega}) = \sum_{n=0}^{M} a(n) \cos(\omega n) \tag{4}$$

where $M = (N-1)/2$, $a(0) = h(M)$ and $a(n) = 2h(M-n)$ for $n = 1, 2, \cdots, M$.

*Case 2:* $N$ even, symmetric impulse response

$$G(e^{j\omega}) = \sum_{n=1}^{M} b(n) \cos[\omega(n - \tfrac{1}{2})] \tag{5}$$

where $M = N/2$ and $b(n) = 2h(M-n)$ for $n = 1, \cdots, M$.

[1] The terms which are used to refer to frequency are $\omega$ and $f$, with $\omega = 2\pi f$. Throughout this paper the terms $\omega$, and $f$, are used interchangeably.

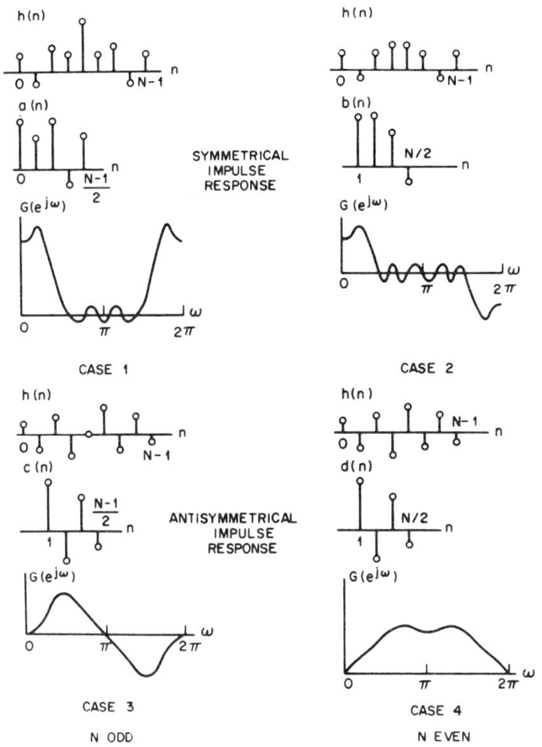


Fig. 1. The four cases of linear phase FIR filters.

*Case 3:* $N$ odd, antisymmetric impulse response

$$G(e^{j\omega}) = \sum_{n=1}^{M} c(n) \sin(\omega n) \tag{6}$$

where $M = (N-1)/2$ and $c(n) = 2h(M-n)$ for $n = 1, 2, \cdots, M$ and $h(M) = 0$.

*Case 4:* $N$ even, antisymmetric impulse response

$$G(e^{j\omega}) = \sum_{n=1}^{M} d(n) \sin[\omega(n - \tfrac{1}{2})] \tag{7}$$

where $M = N/2$ and $d(n) = 2h(M-n)$ for $n = 1, \cdots, M$.

Fig. 1 presents a comprehensive summary of the results of this section. Shown in this figure are typical impulse response sequences $h(n)$, the resulting shifted sequence ($a(n)$ through $d(n)$, depending on the case) and typical frequency response functions $G(e^{j\omega})$, for each of the four cases of linear phase FIR filter.

## IV. Chebyshev Approximation

The formulation of the linear phase FIR design problem as a Chebyshev approximation problem provides a set of conditions (the alternation theorem) which completely characterize the optimal filter. The alternation theorem is the basis of the Remez exchange algorithm which is a very efficient method for calculating the optimal filter coefficients. Furthermore, this characterization has shown that other methods yield a restricted class of optimal filters. In this section we will show how the linear phase FIR design problem can be formulated as a Chebyshev approximation problem and in succeeding sections it will become clear that this formulation is the key to understanding the various FIR design procedures.

The linear phase FIR design problem is an approximation problem in the sense that one is trying to match some ideal amplitude response with the function $G(e^{j\omega})$. For each of

TABLE II

| | $L$ | $G(e^{j\omega})$ |
|---|---|---|
| Case 1—$N$ odd, Symmetric impulse response | 0 | $\sum_{n=0}^{(N-1)/2} a(n)\cos(\omega n)$ |
| Case 2—$N$ even, Symmetric impulse response | 0 | $\sum_{n=1}^{N/2} b(n)\cos[\omega(n-1/2)]$ |
| Case 3—$N$ odd, Antisymmetric impulse response | 1 | $\sum_{n=1}^{(N-1)/2} c(n)\sin(\omega n)$ |
| Case 4—$N$ even, Antisymmetric impulse response | 1 | $\sum_{n=1}^{N/2} d(n)\sin[\omega(n-1/2)]$ |

TABLE III

| | $Q(e^{j\omega})$ | $P(e^{j\omega})$ |
|---|---|---|
| Case 1 | 1 | $\sum_{n=0}^{M} a(n)\cos(\omega n)$ |
| Case 2 | $\cos(\omega/2)$ | $\sum_{n=0}^{M-1} \tilde{b}(n)\cos(\omega n)$ |
| Case 3 | $\sin(\omega)$ | $\sum_{n=0}^{M-1} \tilde{c}(n)\cos(\omega n)$ |
| Case 4 | $\sin(\omega/2)$ | $\sum_{n=0}^{M-1} \tilde{d}(n)\cos(\omega n)$ |

the four cases the function $G(e^{j\omega})$ is a linear combination of a different set of basis functions. It is convenient to reduce all cases to a common one with cosine basis functions.

In the previous section on linear phase FIR filters, it was shown that the frequency response of the four cases of linear phase filters could be written in the form

$$H(e^{j\omega}) = e^{-j\omega[(N-1)/2]}\, e^{j(\pi/2)L}\, G(e^{j\omega}). \tag{8}$$

Values for $L$ and the form for $G(e^{j\omega})$ are given in Table II for each of the four cases of linear phase.

Using simple trigonometric identities, each of the expressions for $G(e^{j\omega})$ in Table II can be written as a product of a fixed function of $\omega$ (call this $Q(e^{j\omega})$), and a term which is a sum of cosines (call this $P(e^{j\omega})$). Thus the expressions for $G(e^{j\omega})$ in Table II become [31] as follows.

*Case 1:*

No change.

*Case 2:*

$$\sum_{n=1}^{M} b(n)\cos[\omega(n-1/2)] = \cos(\omega/2)\sum_{n=0}^{M-1}\tilde{b}(n)\cos(\omega n). \tag{9}$$

*Case 3:*

$$\sum_{n=1}^{M} c(n)\sin(\omega n) = \sin(\omega)\sum_{n=0}^{M-1}\tilde{c}(n)\cos(\omega n). \tag{10}$$

*Case 4:*

$$\sum_{n=1}^{M} d(n)\sin[\omega(n-1/2)] = \sin(\omega/2)\sum_{n=0}^{M-1}\tilde{d}(n)\cos(\omega n). \tag{11}$$

The coefficients $\tilde{b}(n)$, $\tilde{c}(n)$, and $\tilde{d}(n)$ in (9)–(11) are simply related to $b(n)$, $c(n)$, and $d(n)$ of Table II as shown in [31]. Table III shows a summary of the functions $Q(e^{j\omega})$ and $P(e^{j\omega})$ for each of the four cases of linear phase filters.

For Cases 2-4, $Q(e^{j\omega})$ is constrained to be zero at either $\omega = 0$ or $\omega = \pi$, or both.

To formulate the optimal linear phase FIR filter design problem as a Chebyshev approximation problem, it is necessary to define $D(e^{j\omega})$, the desired (real) frequency response of the filter, and $W(e^{j\omega})$, a weighting function on the approximation error which enables the designer to choose the relative size of the error in different frequency bands. The weighted error of approximation $E(e^{j\omega})$ is, by definition,

$$E(e^{j\omega}) = W(e^{j\omega})\,[D(e^{j\omega}) - G(e^{j\omega})]. \tag{12}$$

By writing $G(e^{j\omega})$ as a product of $P(e^{j\omega})$ and $Q(e^{j\omega})$, $E(e^{j\omega})$ can be rewritten as

$$E(e^{j\omega}) = W(e^{j\omega})\,[D(e^{j\omega}) - P(e^{j\omega})\,Q(e^{j\omega})]. \tag{13}$$

Since $Q(e^{j\omega})$ is a fixed function of frequency, it can be factored out of (13), giving

$$E(e^{j\omega}) = W(e^{j\omega})\,Q(e^{j\omega})\left[\frac{D(e^{j\omega})}{Q(e^{j\omega})} - P(e^{j\omega})\right]. \tag{14}$$

Equation (14) is valid everywhere except possibly at $\omega = 0$ and/or $\omega = \pi$. Defining $\hat{W}(e^{j\omega})$ and $\hat{D}(e^{j\omega})$ as

$$\hat{W}(e^{j\omega}) = W(e^{j\omega})\,Q(e^{j\omega}) \tag{15}$$

and

$$\hat{D}(e^{j\omega}) = \frac{D(e^{j\omega})}{Q(e^{j\omega})} \tag{16}$$

the error function may be written as

$$E(e^{j\omega}) = \hat{W}(e^{j\omega})\,[\hat{D}(e^{j\omega}) - P(e^{j\omega})]. \tag{17}$$

The Chebyshev approximation problem may now be stated as finding the set of coefficients ($a(n)$, $\tilde{b}(n)$, $\tilde{c}(n)$, or $\tilde{d}(n)$) to minimize the maximum absolute value of $E(e^{j\omega})$ over the frequency bands in which the approximation is being performed. Using the notation $\|E(e^{j\omega})\|$ to denote this minimum value (i.e., the $L_\infty$-norm of $E(e^{j\omega})$), the Chebyshev approximation problem may be stated mathematically as

$$\|E(e^{j\omega})\| = \min_{\{\text{coefficients}\}}\left[\max_{\omega\in A} |E(e^{j\omega})|\right] \tag{18}$$

where $A$ represents the disjoint union of all the frequency bands of interest.

A well-known property of this class of Chebyshev approximation problems may be used to obtain a characterization of the solution to (18). This is the so-called alternation theorem which may be stated as follows [41].

*Theorem:* If $P(e^{j\omega})$ is a linear combination of $r$ cosine functions (i.e., $P(e^{j\omega}) = \sum_{n=0}^{r-1}\alpha(n)\cos(\omega n)$) then a necessary and sufficient condition that $P(e^{j\omega})$ be the unique, best weighted Chebyshev approximation to a continuous function $\hat{D}(e^{j\omega})$ on $A$, a compact subset of $[0, \pi]$, is that the weighted error function $E(e^{j\omega})$ exhibit at least $r+1$ extremal frequencies in $A$, i.e., there must exist $r+1$ points $\omega_i$ in $A$ such that $\omega_1 < \omega_2 < \cdots < \omega_r < \omega_{r+1}$ and such that $E(e^{j\omega_i}) = -E(e^{j\omega_{i+1}})$, $i = 1, 2, \cdots, r$, and $|E(e^{j\omega_i})| = \max_{\omega \in A} |E(e^{j\omega})|$.

The preceding alternation theorem is extremely powerful in that it expresses a necessary and sufficient set of conditions for obtaining the optimal Chebyshev solution. A number of techniques have been devised for obtaining this optimal solution, depending on the interpretation of this theorem.

It is worth noting that the alternation theorem depends very strongly on the fact that the basis functions satisfy the Haar condition [41]. When one attempts to do constrained Chebyshev approximation or two-dimensional approximation, it turns out that the basis functions do not satisfy the Haar condition. Thus, a characterization of the optimal solution in the form of an alternation theorem is no longer possible. Before discussing any specific algorithm for designing optimal filters, the next section presents an important result on the maximum number of extrema of a linear phase FIR filter.

*Constraint on the Number of Extrema of the Frequency Response of a Linear Phase Filter*

The alternation theorem states that for the optimal linear phase FIR filter, the error function has at least $r+1$ extrema where $r$ is the number of cosine functions being used in the approximation. Since for many cases of interest the extrema of $G(e^{j\omega})$ are also the extrema of $E(e^{j\omega})$, (i.e., both $dW(e^{j\omega})/d\omega$ and $dD(e^{j\omega})/d\omega$ will be zero when $dG(e^{j\omega})/d\omega$ is zero), it is important to know the maximum number of extrema of $G(e^{j\omega})$. By adding to this number the number of extrema of $E(e^{j\omega})$ which are not extrema of $G(e^{j\omega})$ the total maximum number of extrema of $E(e^{j\omega})$ can be found.

By differentiating $G(e^{j\omega})$ with respect to $\omega$, it can be shown that $N_e$, the number of extrema of $G(e^{j\omega})$, obeys

$$\begin{aligned} N_e &\leq (N+1)/2 && \text{case 1} \\ N_e &\leq N/2 && \text{case 2} \\ N_e &\leq (N-1)/2 && \text{case 3} \\ N_e &\leq N/2 && \text{case 4.} \end{aligned} \tag{19}$$

Equation (19) only constrains the number of extrema of $G(e^{j\omega})$. It is readily seen that if the approximation problem is being solved over a union of disjoint frequency bands, the error function can obtain an extremum at *each* band edge, whereas these points will generally *not* be extrema of $G(e^{j\omega})$ [21]. The exception to this rule is when the band edges are at either $\omega = 0$ or $\omega = \pi$ where $G(e^{j\omega})$ will often have an extremum. Thus, for example, the error function for Case 1 lowpass filter (a two-band approximation problem) can have a maximum of $(N+5)/2$ extrema, i.e., $(N+1)/2$ extrema of $G(e^{j\omega})$ and 2 extra extrema for the passband and stopband edges. The error function for a Case 1 bandpass filter (a three-band approximation problem) can have a maximum of $(N+9)/2$ extrema, i.e., $(N+1)/2$ extrema of $G(e^{j\omega})$ and 4 extra extrema for the passband and stopband edges.

Foreknowledge of the maximum number of extrema of $E(e^{j\omega})$ is important because it relates to the exact ways in which design techniques have been devised to design optimal filters. For example, two of the optimal design techniques are only capable of designing optimal filters with the *maximum* possible number of extrema. These design techniques are of limited utility in that the alternation theorem shows that, in general, filters with the maximum number of extrema in their error functions are special cases of the theorem, and hence are only a subset of the larger class of optimal filters. In the following sections a discussion of the various optimal filter design algorithms is given. Both for historical reasons, and for development purposes, we describe first the two algorithms which only are capable of designing optimal filters with the maximum possible number of extrema in their error functions. Then a discussion is given of a Remez type algorithm and finally a linear programming method for designing any optimal, linear phase, FIR filter.

## V. Maximal Ripple FIR Filters

In the preceding section, it was shown that the number of frequencies at which $G(e^{j\omega})$ could attain an extremum is strictly a function of the case of linear phase filter under investigation. At each extremum, the value of $G(e^{j\omega})$ is predetermined by a combination of the weighting function $W(e^{j\omega})$, the desired frequency response $D(e^{j\omega})$, and the quantity $\delta$ which represents the peak error of approximation. By distributing the frequencies at which $G(e^{j\omega})$ attained an extremal value among the different frequency bands over which a desired response was being approximated, and by requiring the resulting filter to have the maximum number of extremal frequencies, a unique optimal filter can be obtained. Since these filters have the maximum number of alternations, or ripples, in their error of approximation curve, they have been called maximal ripple filters. For the case of lowpass filters these maximal ripple filters have also been called extraripple filters [21] because only a single extra ripple above the minimum number required for optimality is present.

The manner in which a set of nonlinear equations is obtained for describing the maximal ripple filter is as follows. At each of the $N_e$ unknown extremal frequencies, $E(e^{j\omega})$ attains the maximum value of $\pm\delta$, and $E(e^{j\omega})$, or equivalently, $G(e^{j\omega})$ has zero derivative. Thus $2N_e$ equations of the form

$$G(e^{j\omega_i}) = \frac{\pm\delta}{W(e^{j\omega_i})} + D(e^{j\omega_i}), \qquad i = 1, 2, \cdots, N_e$$

$$\frac{d}{d\omega}[G(e^{j\omega})]\Big|_{\omega=\omega_i} = 0, \qquad i = 1, 2, \cdots, N_e$$

for $i = 1, 2, \cdots, N_e$ are obtained. These equations represent a set of $2N_e$ nonlinear equations in $2N_e$ unknowns ($N_e$ impulse response coefficients, and $N_e$ frequencies at which $G(e^{j\omega})$ obtains the extremal value). The set of $2N_e$ equations may be solved iteratively using a nonlinear optimization procedure such as the well-known Fletcher–Powell algorithm.

Two facts should be noted about this procedure. First the quantity $\delta$ (i.e., the peak error) is a fixed quantity and is not minimized by the optimization scheme. Thus the shape of $G(e^{j\omega})$ is postulated *a priori*, and only the frequencies at which $G(e^{j\omega})$ attains the extremal values are unknown. Second, the design procedure has no way of specifying band edges for the different frequency bands of the filter. Thus the optimization algorithm does not work on given frequency bands, but instead is free to select exactly where the bands will lie. This lack of control over frequency band edges diminishes the utility of this and the next algorithm to be discussed.

To illustrate a specific set of equations for optimization we consider the design of a case 1 low-pass filter with $N = 15$, a

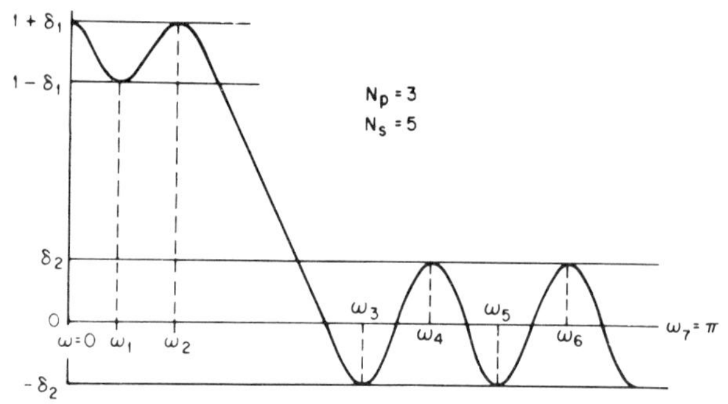


Fig. 2. The frequency response of a maximal ripple low-pass filter.

peak ripple of $\delta = \delta_2$, a weighting function defined as

$$W(e^{j\omega}) = \begin{cases} \delta_2/\delta_1, & \omega \text{ in the passband} \\ 1, & \omega \text{ in the stopband} \end{cases}$$

and a desired response of

$$D(e^{j\omega}) = \begin{cases} 1, & \omega \text{ in the passband} \\ 0, & \omega \text{ in the stopband.} \end{cases}$$

Fig. 2 illustrates $G(e^{j\omega})$ for this example. The extremal frequencies are the set $\omega = 0, \omega_1, \omega_2, \omega_3, \omega_4, \omega_5, \omega_6$, and $\omega_7 = \pi$. At $\omega = 0$ and $\omega = \pi$, $G(e^{j\omega})$ has zero derivative, independent of the impulse response coefficients for a case 1 design. For this example, the $N_e = 8$ extremal frequencies are divided so that $N_p = 3$ occur in the passband, and $N_s = 5$ occur in the stopband. Thus for this set of conditions, the following equations are obtained:

| Function Constraints | Derivative Constraints |
|---|---|
| $G(e^{j0}) = 1 + \delta_1$ | |
| $G(e^{j\omega_1}) = 1 - \delta_1$ | $\left.\frac{d}{d\omega} G(e^{j\omega})\right\rvert_{\omega=\omega_1} = 0$ |
| $G(e^{j\omega_2}) = 1 + \delta_1$ | $\left.\frac{d}{d\omega} G(e^{j\omega})\right\rvert_{\omega=\omega_2} = 0$ |
| $G(e^{j\omega_3}) = -\delta_2$ | $\left.\frac{d}{d\omega} G(e^{j\omega})\right\rvert_{\omega=\omega_3} = 0$ |
| $G(e^{j\omega_4}) = +\delta_2$ | $\left.\frac{d}{d\omega} G(e^{j\omega})\right\rvert_{\omega=\omega_4} = 0$ |
| $G(e^{j\omega_5}) = -\delta_2$ | $\left.\frac{d}{d\omega} G(e^{j\omega})\right\rvert_{\omega=\omega_5} = 0$ |
| $G(e^{j\omega_6}) = +\delta_2$ | $\left.\frac{d}{d\omega} G(e^{j\omega})\right\rvert_{\omega=\omega_6} = 0.$ |
| $G(e^{j\pi}) = -\delta_2.$ | |

Once this set of equations has been solved for the unknown frequencies and the impulse-response coefficients, the passband and stopband edges may be solved for by searching for the frequency beyond $\omega_2$ where $G(e^{j\omega})$ exactly equals $1 - \delta_1$ (passband edge), and the frequency before $\omega_3$ where $G(e^{j\omega})$ exactly equals $+\delta_2$ (stopband edge).

The preceding optimization procedure has been used by Herrmann [16] to design low-pass and bandpass filters for values of $N$ up to about 61. The next section discusses another technique for designing maximal ripple filters where much larger filters can be designed efficiently.

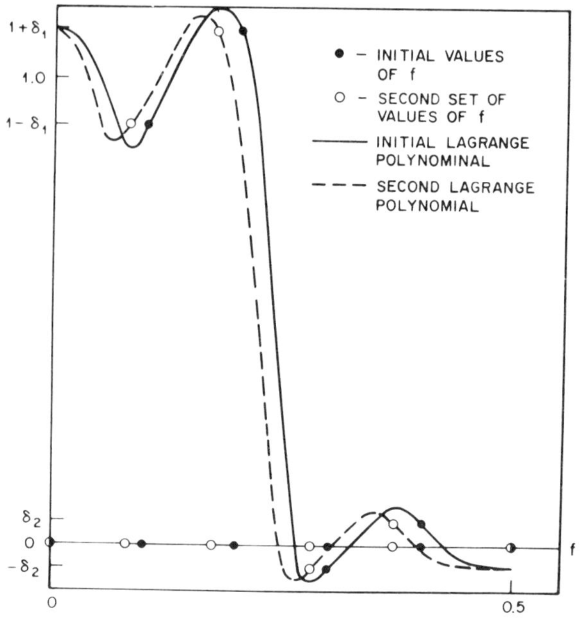


Fig. 3. Polynomial interpolation iterative solution for maximal ripple low-pass filter.

*Polynomial Interpolation Solution for Maximal Ripple FIR Filters*

A second, more efficient, method has been proposed for designing maximal ripple filters. This algorithm is basically an iterative technique for producing a polynomial ($G(e^{j\omega})$) that has extrema of desired values. The algorithm begins by making an initial estimate of the frequencies at which the extrema in $G(e^{j\omega})$ will occur, and then uses the well-known Lagrange interpolation formula to obtain a polynomial that alternatingly goes through the maximum allowable ripple values at these frequencies. It has been experimentally found that the initial guess of extremal frequencies does not affect the ultimate convergence of the algorithm, but instead affects the number of iterations required to achieve the desired result.

Rather than consider the general filter design problem, it is instructive to consider the design of a case 1 low-pass filter as an example of how the algorithm works. Fig. 3 shows the frequency response of a low-pass filter with $N = 11$, peak ripple $\delta = \delta_2$, weighting function $W(e^{j\omega})$, and desired frequency response $D(e^{j\omega})$, as defined in the preceding section. The number of extremal frequencies $N_e$ is 6 for this example, and they are divided into $N_p = 3$ passband extrema, and $N_s = 3$ stopband extrema. The filled dots along the frequency axis of Fig. 3 show the initial guess as to the extremal frequencies of $G(e^{j\omega})$. The solid line shows the initial Lagrange polynomial obtained by choosing polynomial coefficients so that the values of the polynomial at the guessed set of frequencies are identical to the assigned extreme values. As seen in Fig. 3, the polynomial associated with the initial guess does not have extrema that achieve the maximum allowable errors, but rather it has extrema that exceed these values. The next stage of the algorithm is to locate the frequencies at which the extrema of the first Lagrange interpolation occur. These frequencies are used as an updated improved guess of the frequencies for which the extrema of the filter response will achieve the desired ripple values. This second set of frequencies is indicated by the open dots in Fig. 3. The algorithm uses these new frequencies to construct another Lagrange polynomial (shown by the dotted line in Fig. 3) that achieves the desired values at these frequencies. At this point the iterative nature of the

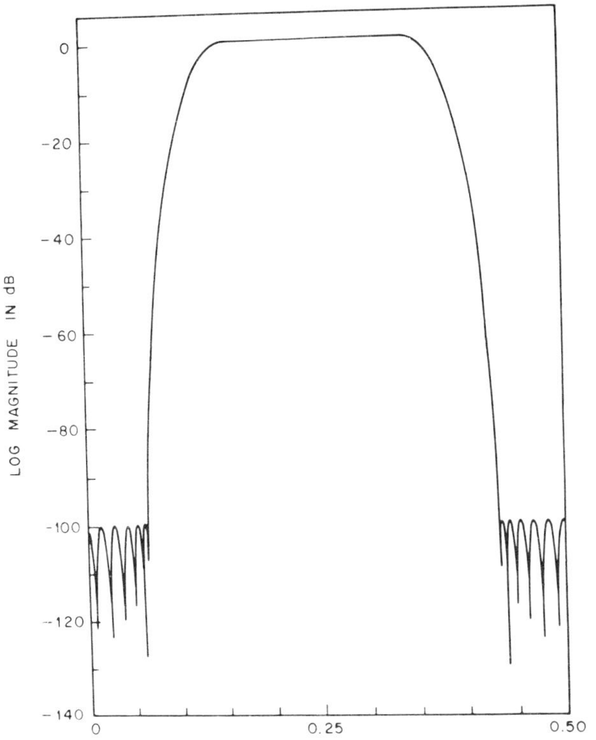


Fig. 4. The frequency response of a maximal ripple bandpass filter.

algorithm has emerged. By locating the extrema of the new polynomial, another iteration of the algorithm is begun. This algorithm is quite similar to the well-known Remez multiple exchange algorithm of Chebyshev approximation theory.

Two typical maximal ripple filters designed by Hofstetter *et al.* [19], [20] using this algorithm are shown in Figs. 4 and 5. Figure 4 shows the log magnitude response of a case 1 bandpass filter with $N = 41$, (i.e., $N_e = 21$) with 6 extrema of $G(e^{j\omega})$ in each stopband, and 9 extrema in the passband. The peak ripple in the stopbands is $\delta_2 = 0.00001$ (or $-100$ dB), whereas the peak ripple in the passband is 0.005. Fig. 5 shows the log magnitude response of a case 1 low-pass filter with $N = 251$ and with 33 extrema of $G(e^{j\omega})$ in the passband and 94 extrema in the stopband. The peak ripple in the passband is $\delta_1 = 0.01$ and the peak ripple in the stopband is $\delta_2 = 0.00004$ (or $-88$ dB).

Although this improved algorithm has essentially eliminated the difficulty of designing filters with large values of $N$, the inherent problem still remains that the filter band edge frequencies cannot be specified *a priori*, i.e., they must be calculated from the final solution. Furthermore, both techniques are only capable of designing maximal ripple filters, which, as discussed earlier, are a subclass of the class of optimal filters. In the next sections design techniques are presented which are capable of designing any optimal filter.

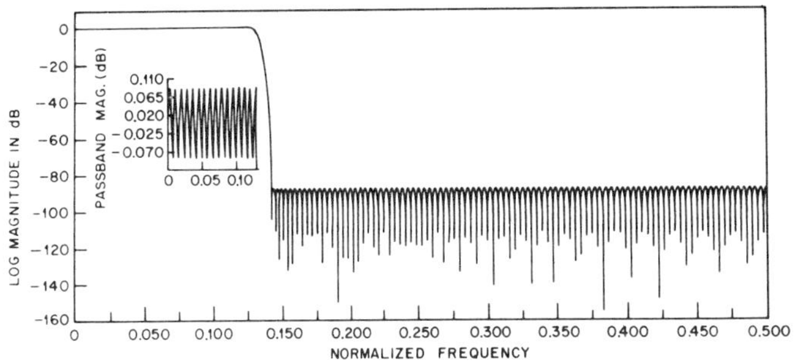


Fig. 5. The frequency response of a maximal ripple low-pass filter.

## VI. Remez Exchange Algorithm Design of Optimal FIR Filters

As shown earlier, the optimal linear phase FIR filter design problem is a Chebyshev approximation problem where the approximating function $P(e^{j\omega})$ in (17) is a sum of $r$ independent cosine functions. The alternation theorem gave a set of necessary and sufficient conditions on the weighted error function $E(e^{j\omega})$ (see (17)) such that the solution was the unique best approximation to the desired frequency response $\hat{D}(e^{j\omega})$. The Remez exchange algorithm is an algorithm which solves the Chebyshev approximation problem by searching for the extremal frequencies of the best approximation. This is accomplished as follows. At the beginning of each iteration one has a set of $r + 1$ extremal frequencies $\{\omega_k\}$. Equation (20) gives the set of equations which must be solved for the generalized polynomial approximating function $P(e^{j\omega})$ whose weighted error function has magnitude $\delta$ with alternating signs on the set $\{\omega_k\}$

$$\hat{W}(e^{j\omega_k})[\hat{D}(e^{j\omega_k}) - P(e^{j\omega_k})] = (-1)^k \delta, \quad k = 0, 1, \cdots, r \tag{20}$$

where $P(e^{j\omega})$ is of the form

$$P(e^{j\omega}) = \sum_{n=0}^{r-1} \alpha(n) \cos(\omega n).$$

Equation (20) can be rewritten in matrix form as shown in (21). The invertibility of this matrix is guaranteed by the Haar condition on the basis functions.

$$\begin{bmatrix} 1 & \cos\omega_0 & \cos 2\omega_0 & \cdots & \cos[(r-1)\omega_0] & \dfrac{1}{\hat{W}(e^{j\omega_0})} \\ 1 & \cos\omega_1 & & & & \\ \vdots & & & & & \\ 1 & \cos\omega_r & & & & \dfrac{(-1)^r}{\hat{W}(e^{j\omega_r})} \end{bmatrix} \begin{bmatrix} \alpha(0) \\ \alpha(1) \\ \vdots \\ \alpha(r-1) \\ \delta \end{bmatrix} = \begin{bmatrix} \hat{D}(e^{j\omega_0}) \\ \hat{D}(e^{j\omega_1}) \\ \vdots \\ \hat{D}(e^{j\omega_r}) \end{bmatrix} \tag{21}$$

Since direct solution of (21) is both difficult, and slow, it is more efficient to calculate $\delta$ analytically as

$$\delta = \frac{a_0\hat{D}(e^{j\omega_0}) + a_1\hat{D}(e^{j\omega_1}) + \cdots + a_r\hat{D}(e^{j\omega_r})}{a_0/\hat{W}(e^{j\omega_0}) - a_1/\hat{W}(e^{j\omega_1}) + \cdots + (-1)^r a_r/\hat{W}(e^{j\omega_r})} \tag{22}$$

where

$$a_k = \prod_{\substack{i=0 \\ i \neq k}}^{r} \frac{1}{(x_k - x_i)} \tag{23}$$

$$x_i = \cos\omega_i. \tag{24}$$

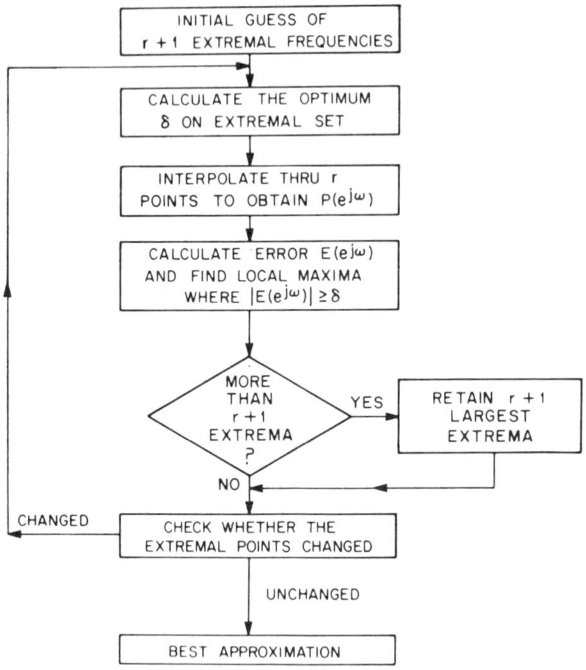


Fig. 6. Flow chart of the Remez exchange algorithm.

After calculating $\delta$, the Lagrange interpolation formula in the barycentric form[2] is used to interpolate $P(e^{j\omega})$ on the $r$ points $\omega_0, \omega_1, \cdots, \omega_{r-1}$ to the values

$$C_k = \hat{D}(e^{j\omega_k}) - (-1)^k \frac{\delta}{\hat{W}(e^{j\omega_k})}, \qquad k = 0, 1, \cdots, r-1 \tag{25}$$

$$P(e^{j\omega}) = \frac{\sum_{k=0}^{r-1} \left[\frac{\beta_k}{x - x_k}\right] C_k}{\sum_{k=0}^{r-1} \left[\frac{\beta_k}{x - x_k}\right]} \tag{26}$$

where

$$\beta_k = \prod_{\substack{i=0 \\ i \neq k}}^{r-1} \frac{1}{(x_k - x_i)}. \tag{27}$$

Note that $P(e^{j\omega})$ will also interpolate to $D(e^{j\omega_r}) - [(-1)^r \delta / \hat{W}(e^{j\omega_r})]$ since it satisfies (20). The next step in the process is to evaluate $E(e^{j\omega})$ on a dense set of the frequency axis. If $|E(e^{j\omega})| \leqslant \delta$ for all frequencies in the dense set, then the optimal approximation has been found. If $|E(e^{j\omega})| > \delta$ for some frequencies in the dense set, then a new set of $r + 1$ frequencies must be chosen as candidates for the extremal frequencies. The new points are chosen as the peaks of the resulting error curve, thereby forcing $\delta$ to increase and ultimately converge to its upper bound which corresponds to the solution to the problem. In the event that there are more than $r + 1$ extrema in $E(e^{j\omega})$ at any iteration, the $r + 1$ frequencies at which $|E(e^{j\omega})|$ is largest are retained as the guessed set of extremal frequencies for the next iteration. Fig. 6 summarizes the exchange algorithm in a block diagram form.

[2]See R. W. Hamming, *Numerical Methods for Scientists and Engineers*, 1st ed. New York: McGraw-Hill, for a discussion of the barycentric form.

The filter impulse response is obtained by evaluating $P(e^{j\omega})$ at $N$ equally spaced frequencies and using the DFT to get the sequence $\{\alpha(n)\}$, from which the impulse response coefficients may be derived. Depending on which case linear phase filter is derived, a unique formula can be written for obtaining $h(n)$ from $\alpha(n)$.

A general purpose computer program has been written to implement this algorithm and has found widespread use in filter design applications [34].

## VII. Linear Programming Design of Optimal FIR Filters

The optimal linear phase FIR filter is the one for which the maximum error $E(e^{j\omega})$ is minimized over all $\omega$. Letting $\delta$ represent the maximum error, a set of linear inequalities can be written to describe this minimax problem, i.e.,

$$-\delta \leqslant \hat{W}(e^{j\omega_i}) [\hat{D}(e^{j\omega_i}) - P(e^{j\omega_i})] \leqslant \delta, \qquad \omega_i \in F \tag{28}$$

where $F$ is a dense grid of frequencies in the bands over which the approximation is being made. Equation (28) can formally be written as a linear program, i.e.,

$$-\hat{W}(e^{j\omega_i}) \sum_{m=0}^{r-1} \alpha(m) \cos(m\omega_i) - \delta \leqslant -\hat{W}(e^{j\omega_i}) \hat{D}(e^{j\omega_i})$$

$$\hat{W}(e^{j\omega_i}) \sum_{m=0}^{r-1} \alpha(m) \cos(m\omega_i) - \delta \leqslant \hat{W}(e^{j\omega_i}) \hat{D}(e^{j\omega_i})$$

minimize $\delta$.

Linear programming techniques can be used to solve the preceding set of equations [22], [23]. However, since linear programming is basically a single exchange method, it is significantly slower than the Remez method, and hence is avoided for this class of problems. In a later section, however, it will be shown how when time response constraints are added to the design problem, linear programming is perhaps the only simple method of solving the problem.

## VIII. Characteristics of Optimal Case 1 Low-Pass Filters

For a low-pass filter the optimal design problem consists of specifying the filter length $N$, the passband cutoff frequency $F_p$, the stopband cutoff frequency $F_s$, and the ripple ratio $K = \delta_1/\delta_2$ which describes the desired weighting function $W(e^{j\omega})$ as

$$W(e^{j\omega}) = \begin{cases} 1/K = \delta_2/\delta_1, & 0 \leqslant \omega \leqslant 2\pi F_p \\ 1, & 2\pi F_s \leqslant \omega \leqslant \pi \end{cases} \tag{29}$$

where $\delta_1$ is the passband ripple, and $\delta_2$ is the stopband ripple. Fig. 7 shows the frequency response of a Case 1 low-pass filter. The auxiliary parameter $\Delta F$ is defined as

$$\Delta F = F_s - F_p \tag{30}$$

and serves as a measure of the width of the transition band of the filter.

It was shown earlier that the error curve for the optimal lowpass filter could have either $r + 1$ or $r + 2$ extrema where $r = (N + 1)/2$ for case 1, and $r = N/2$ for case 2. It is important to understand the nature of the optimum lowpass filter to see under what conditions the number of ripple extrema attains the maximum value. It has been found experimentally that a reasonably straightforward and informative way of summariz-

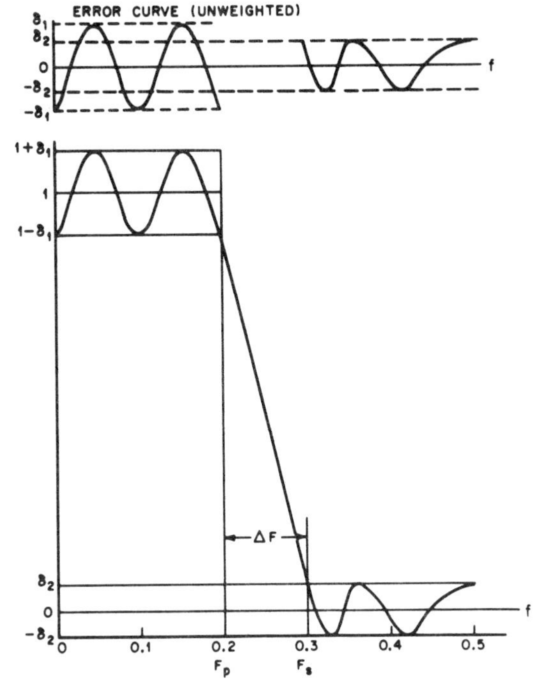


Fig. 7. Frequency response and error curve of optimal low-pass filter.

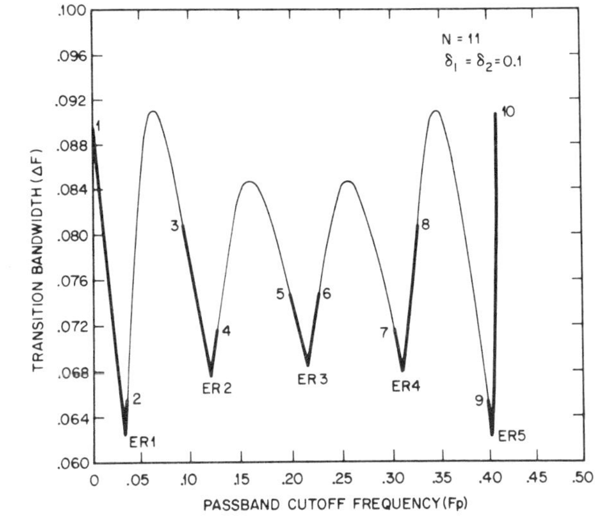


Fig. 8. The curve of transition width versus passband cutoff frequency for optimal low-pass filters.

ing the behavior of the optimum filter is to plot the transition width of the filter ($\Delta F$) versus passband cutoff frequency $F_p$, for fixed values of $N$, $\delta_1$, and $\delta_2$. Fig. 8 shows such a plot for case 1 data for $N = 11$, $\delta_1 = \delta_2 = 0.1$. As seen in this figure the curve of $\Delta F$ versus $F_p$ has an oscillatory behavior, alternating between sharp minima, and flat-topped maxima. The local minima of the curves (labelled $ER1$ to $ER5$) have been found to be the maximal ripple (extraripple) filters for the particular choice of $N$, $\delta_1$, and $\delta_2$. (Recall that extraripple filters have $(N+5)/2$ equal amplitude extrema in their error curves.) There are exactly $(N-1)/2$ of these extraripple filters. In between the extraripple solutions, it has been found that there are two types of optimum filters—scaled extraripple filters and equiripple filters with exactly $(N+3)/2$ equal amplitude extrema in their error curves.

The scaled extraripple filters (shown in Fig. 8 by the heavy lines on the curve) have the property that their error curves

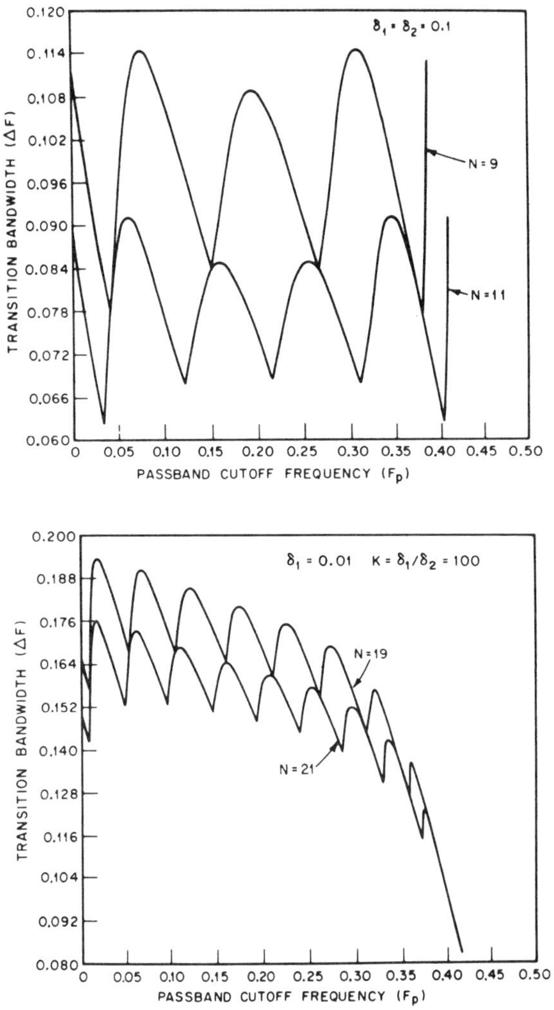


Fig. 9. The curves of transition width versus passband cutoff frequency for several values of low-pass filter parameters.

have $(N+3)/2$ equal amplitude extrema, as well as one smaller amplitude extremum which is either at $f = 0$ or $f = 0.5$. These filters can be derived from their neighboring extraripple filter by a simple scaling procedure [25], [26].

The optimum filters between scaled extraripple solutions all have exactly $(N+3)/2$ equal amplitude extrema in their error curves. No simple linear scaling procedure has been found to account for their presence.

However, an extraripple filter of length $N-2$ has $(N+3)/2$ equal amplitude extrema in its error curves and thus it is an optimal length $N$ filter (with the highest order coefficient zero). This point is evident in Fig. 9 which shows the curves of transition width versus passband cutoff frequency for $N = 9$ and 11, with $\delta_1 = \delta_2 = 0.1$ as well as for $N = 19$ and 21 with $\delta_1 = 0.001$ and $\delta_2 = 0.0001$. The curves touch at the local minima of the $N = 9$ curve at the top and at the local minima of the $N = 19$ curve at the bottom.

Fig. 10 presents a summary of the types of optimal filters which may be obtained by varying the filter cutoff frequencies. The first filter shown is an extraripple solution with $N = 25$, $\delta_1 = \delta_2 = 0.05$. Below it are two different scaled solutions where the frequency response is 0 at $f = 0.5$ for the first filter and then 0.03 at $f = 0.5$ for the second case. The last filter in the first column represents the maximum possible scaling, i.e., the frequency response is 0.05 at $f = 0.5$ with $(N+3)/2$ equal

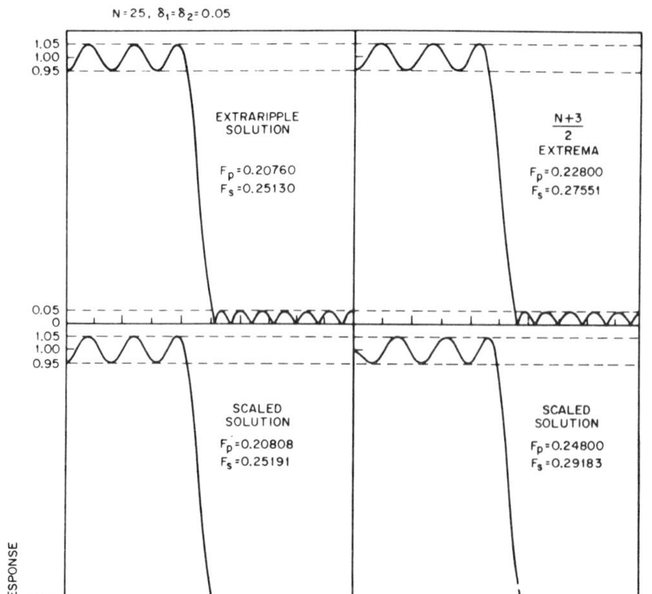


Fig. 10. Summary of the types of optimal low-pass filters.

amplitude extrema in the error curve. In the second column, the first filter is a point approximately midway between extraripple solutions. The next two filters represent scaled extraripple solutions where the error at $f = 0$ is not of the same value as the other error extrema. For the first of these filters the error is about -0.005 at $f = 0$, whereas for the second it is about 0.015 at $f = 0$. The last filter in the second column corresponds to the next extraripple solution.

## IX. Relations Between Optimal Low-Pass Filter Parameters

A great deal has been learned about the relationships between the parameters of optimal low-pass filters. In this section we summarize some of the key results.

### *A. Chebyshev Solutions [27]*

An analytical solution to the optimal filter design problem exists for the case of extraripple designs with either one passband, or one stopband ripple. Since these cases are either very wide-band, or very narrow-band designs, they are not generally of much interest, except for the insights they provide into analytical relations between the various filter parameters.

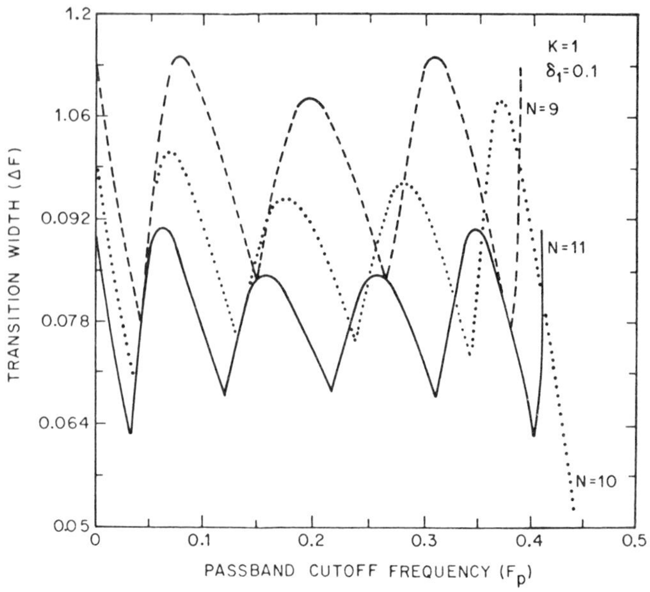


Fig. 11. A comparison of the curves of transition width versus passband cutoff frequency for low-pass filters with even and odd values of $N$.

### *B. Symmetry Relations [27]*

A symmetry exists in the design parameters in the following sense. If $H(e^{j\omega})$ is the frequency response of an optimal low-pass filter with parameters $N$, $F_p$, $F_s$, $\delta_1$, and $\delta_2$, then $G(e^{j\omega}) = 1 - H(e^{j(\pi-\omega)})$ is the frequency response of another optimal low-pass filter with parameters $N$, $F_p' = 0.5 - F_s$, $F_s' = 0.5 - F_p$, $\delta_1' = \delta_2$, $\delta_2' = \delta_1$. This symmetry explains the behavior of the curve of $\Delta F$ versus $F_p$ of Fig. 8 since, in this case, $\delta_1 = \delta_2$; therefore, any filter with parameter $F_p$ has a symmetrical partner with parameter $0.5 - F_s$.

### *C. Case 2 Low-Pass Filters—N Even [29]*

An interesting design relation exists when comparing case 1 and case 2 low-pass filters. Fig. 11 shows a plot of $\Delta F$ versus $F_p$ for $\delta_1 = \delta_2 = 0.1$ and $N = 9$, 10, and 11. Notice that for certain values of $F_p$, the transition width is smaller for $N = 10$ designs than for $N = 9$; whereas for other values of $F_p$, the transition width is smaller for $N = 9$ designs than for $N = 10$. Thus monotonicity of transition width as a function of $N$ is not preserved across both even and odd values but instead holds only for comparing either $N$ odd designs, or $N$ even designs.

### *D. Design Formulas [27], [30]*

Although exact analytical relations do not exist between the 5 low-pass filter parameters, a set of approximate relations can be given which is valid to within some reasonable bounds. Kaiser has proposed the particularly simple formula

$$N \approx \frac{-20 \log_{10}(\sqrt{\delta_1 \delta_2}) - 13}{14.6\, \Delta F} + 1 \tag{31}$$

for predicting the filter length $N$ from the ripple specifications and the band edge frequencies. A somewhat more accurate formula due to Herrmann *et al.* [27] is

$$N \approx \frac{D_\infty(\delta_1, \delta_2) - f(\delta_1, \delta_2)(\Delta F)^2}{\Delta F} + 1 \tag{32}$$

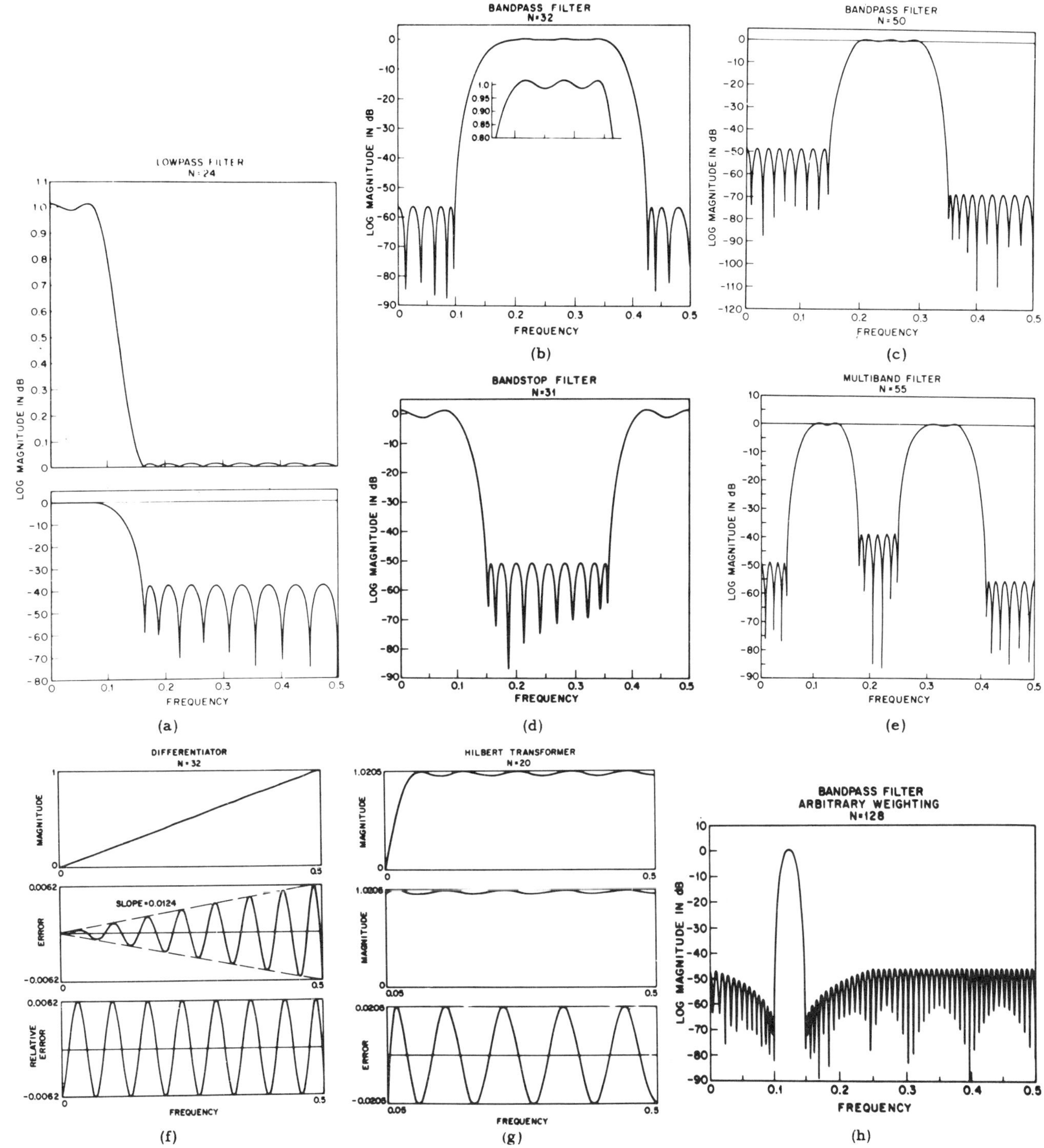


Fig. 12. A summary of several optimal FIR filters.

where

$$D_\infty = [0.005309 (\log_{10} \delta_1)^2 + 0.07114 (\log_{10} \delta_1) - 0.4761] \log_{10} \delta_2 - [0.00266 (\log_{10} \delta_1)^2 + 0.5941 \log_{10} \delta_1 + 0.4278] \quad (33)$$

and

$$f(\delta_1, \delta_2) = 11.012 + 0.51244 (\log_{10} \delta_1 - \log_{10} \delta_2). \quad (34)$$

Such design formulas have proved exceedingly useful for getting a good estimate of the required filter length.

## X. Examples of Optimal Filters

Fig. 12 illustrates some typical optimal filters which have been designed using the iterative approximation method of Section VI, and the program implementation of McClellan *et al.* These examples are meant to illustrate the power of the design method in approximating a wide range of filter types. Fig. 12(a) shows an $N = 25$ low-pass filter; Fig. 12(b) shows an $N = 32$ bandpass filter with equal sidelobe levels; Fig. 12(c) shows an $N = 50$ bandpass filter with unequal sidelobe levels; Fig. 12(d) shows an $N = 31$ bandstop filter; Fig. 12(e) shows an $N = 55$ multiband filter with 3 stopbands and 2 passbands; Fig. 12(f) shows an $N = 32$ equiripple relative error differentiator; Fig. 12(g) shows an $N = 20$ Hilbert transformer; and Fig. 12(h) shows an $N = 128$ bandpass filter with arbitrary weighting near the edges of the stopbands.

## XI. Design of Filters with Time and Frequency Domain Constraints

We have discussed the design of digital filters which approximate characteristics of a specified frequency response only. Quite often one would like to impose simultaneous restrictions on both the time and frequency response of the filter. For example, in the design of lowpass filters, one would often like to limit the step response overshoot or ripple; at the same time maintaining some reasonable control over the frequency response of the filter. Since the step response is a linear function of the impulse response coefficients, a linear program is capable of setting up constraints of the type discussed above. By way of example, we consider the design of a case 1 lowpass filter with the following specifications.

*Passband:*

$$1 - \delta_1 \leqslant H^*(e^{j\omega}) \leqslant 1 + \delta_1. \tag{35}$$

*Stopband:*

$$-\delta_2 \leqslant H^*(e^{j\omega}) \leqslant \delta_2. \tag{36}$$

*Step Response:*

$$-\delta_3 \leqslant g(n) \leqslant \delta_3 \tag{37}$$

where $g(n)$ is the step response of the filter, and is defined as

$$g(n) = \sum_{m=0}^{n} h(m). \tag{38}$$

Clearly $g(n)$ is a linear combination of the impulse response coefficients; hence (33)–(35) can be solved using linear programming techniques on the deltas. For example, one could fix any one or two of the parameters $\delta_1$, $\delta_2$, or $\delta_3$ and minimize the other(s). Alternatively one could set $\delta_1 = \alpha_1 \delta$, $\delta_2 = \alpha_2 \delta$, and $\delta_3 = \alpha_3 \delta$ where $\alpha_1$, $\alpha_2$, and $\alpha_3$ are constants, and simultaneously minimize all three parameters by minimizing $\delta$.

Another application is in the design of interpolation filters [28] where some of the impulse response coefficients must be constrained to be zero. In this case the Alternation Theorem no longer applies because the basis functions of the approximation do not satisfy the Haar condition. Thus, the Remez exchange algorithm cannot be used to calculate the best approximation. But since linear programming does not depend on the Haar condition for its convergence, it can and has been applied to this problem.

## XII. Filters with Optimum Magnitude and Minimum Phase

There may be applications where the linear phase characteristic may not be necessary and one is only interested in the shortest possible filter length for a given magnitude response. In such cases a natural question is: how much can the filter length be reduced for a given $F_p$, $F_s$, $\delta_1$, and $\delta_2$ by dropping the requirement for linear phase? Since half of the filter coefficients are constrained by the symmetry required for linear phase as shown in Fig. 1, one might at first guess that by dropping the linear phase requirement the required filter length might be cut in half. This is generally not the case. In fact the length reduction to be expected depends very much on the type of filter under consideration as will be described in this section.

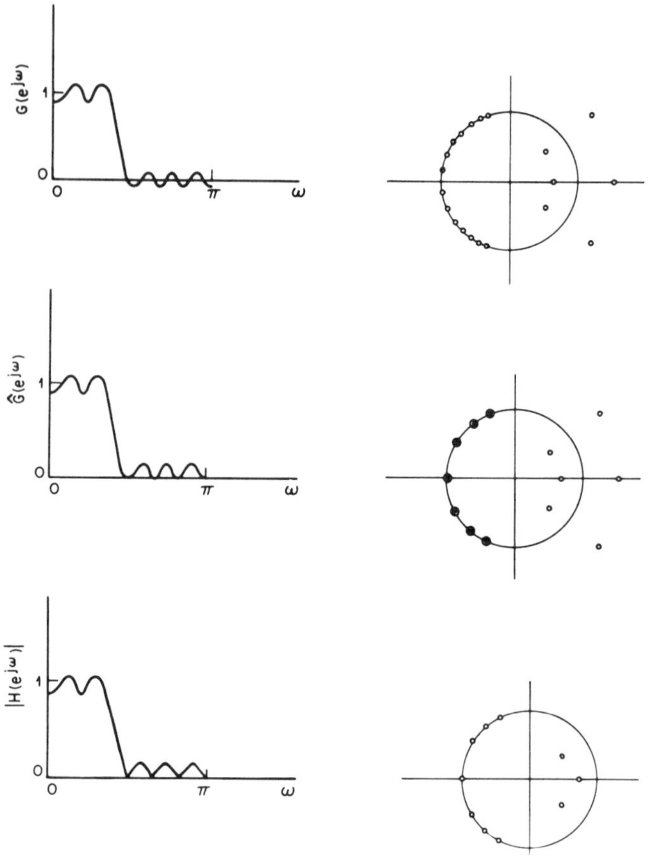


Fig. 13. The procedure for obtaining the optimal magnitude low-pass filter from an optimal linear phase design.

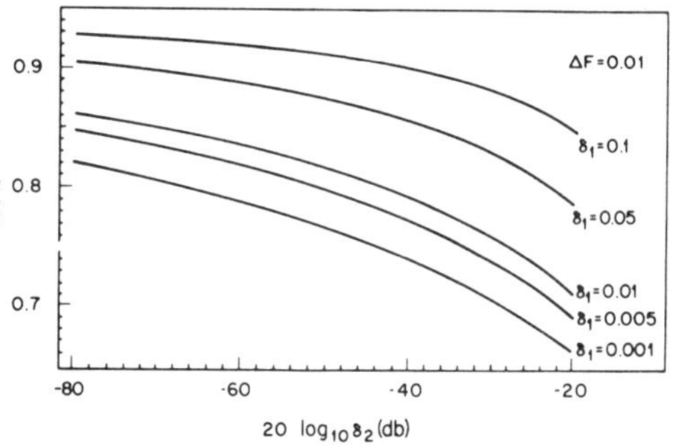


Fig. 14. Ratio of $N'$, the length of the optimum magnitude filter, to $N$, the length of the optimum linear phase filter, with identical low-pass parameters.

First, let us define what we mean by an optimum magnitude approximation. For a given desired magnitude $D(e^{j\omega})$ and weight function $W(e^{j\omega})$, the optimum magnitude approximation is the filter which minimizes

$$\max_{\omega \in \Omega} W(e^{j\omega}) \, |D(e^{j\omega}) - |H(e^{j\omega})||$$

where $\Omega$ is the union of the frequency bands of interest.

For the case where $D(e^{j\omega})$ is piecewise constant over the frequency bands of interest (e.g., a low-pass or bandpass filter), a procedure suggested by Herrmann and Schussler [17] will yield the optimum magnitude filter. Briefly, this procedure works as follows for the lowpass case. A weighted Chebyshev approximation problem is solved for a linear combination of $N$ cosines by using the Remez algorithm described

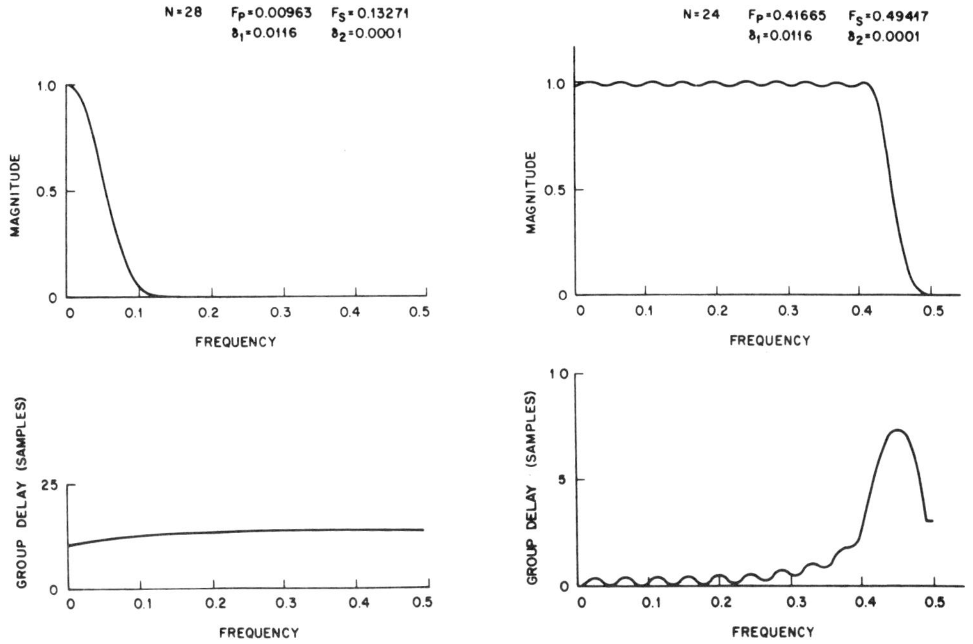


Fig. 15. The magnitude and group delay responses for a narrow-band and a wide-band optimum magnitude filter.

earlier to design a length $2N - 1$ linear phase filter (see Fig. 13(a)). The response $G(e^{j\omega})$ is then scaled (by addition of the constant $\delta_2$ to the frequency response) to give a $\hat{G}(e^{j\omega})$ which is positive with double zeros in the stopband as shown in Fig. 13(b). The resulting order $2N - 2$ mirror image polynomial is factored retaining one each of the double zeros on the unit circle and all zeros inside the unit circle. Finally, the filter is scaled so that in the passband the filter response approximates one as shown in Fig. 13(c). The resulting length $N$ filter is the optimum magnitude approximation and is also a minimum phase transfer function [17].

To see how much the filter length can be reduced from the linear phase design, while retaining the same magnitude characteristics, the design formula (30) can be used to calculate $N'/N$, the ratio of lengths for optimum squared magnitude and linear phase filters.

Fig. 14 shows how this calculated improvement depends on the other filter parameters. Since (30) does not apply to either very narrow or very wide-band filters, a separate analysis is required. In fact it can be shown that for very narrow-band filters where all zeros are on the unit circle of the $z$ plane the optimum squared magnitude filter is a linear phase filter and $N'/N = 1$. For the other extreme of very wide-band filters one should expect $N'/N = 1/2$.

The actual realization of these optimum magnitude filters will be less efficient (in terms of the number of multiplications per sample in a direct form realization) than a linear phase filter with similar performance. This is due to the fact that the symmetry of the linear phase impulse response will allow a 2 to 1 reduction in the number of multiplies and this will offset the 10- or 20-percent reduction in length of the optimum magnitude filters. In cases where one is interested in minimizing the required number of delays, optimum magnitude filters become more attractive.

The phase characteristics of the optimum magnitude filters also depend very much on the filter bandwidth. As the bandwidth is increased from the very narrow-band optimum magnitude filter with linear phase (constant group delay), the group delay characteristic deteriorates until for very wide band filters the group delay is reminiscent of that for an elliptic filter with a sharp peak near the band edge. To illustrate this point, Fig. 15 shows examples of the magnitude and group delay response for a narrow-band and a wide-band low-pass filter.

While this section has focused on low-pass filters for simplicity, the procedure for designing optimum magnitude filters can also be applied to the general bandpass filter case with several pass and stop bands. The improvement to be expected by dropping the linear phase requirement is very small for filters with narrow passbands but much larger for filters with narrow stopbands.

A further generalization of the design of linear phase filters would allow the simultaneous approximation of both magnitude and phase. Such filters could find applications as phase equalizers. However, at this time extensions of the Chebyshev theory to this case have met with little success [35]. One promising algorithm which has been used for the design of filters with both magnitude and phase specifications is the Lawson algorithm [44].

## XIII. The Two-Dimensional FIR Approximation Problem

The approximation problem for two-dimensional FIR digital filters is a much more difficult problem than the corresponding one-dimensional design problem. Some of the one-dimensional filter design techniques have been extended to two dimensions [37], [38], but for other techniques such a generalization appears unlikely. In particular, the iterative design methods based on the Remez exchange algorithm have not been extended to the two-dimensional case, and so at this time there is no efficient procedure available in two dimensions for designing optimal Chebyshev filters.[3]

There are two reasons why the optimal algorithm based on the alternation theorem cannot be extended to two dimen-

[3] Recently, work on exchange algorithms in two dimensions has been carried out by Kamp and Thiran [41], and Hersey and Mersereau [42].

sions. First, it is impossible for any set of functions defined on a two-dimensional domain to satisfy the Haar condition. Thus, the alternation theorem applies in a weaker form. Secondly, there is no possibility of ordering the extremal frequencies as in the one-dimensional case, where increasing ordering guarantees that the error changes sign from one point to the next. Even if the method could be extended, the size of the problem is a handicap. For example, the design of an optimum $31 \times 31$ FIR linear phase filter involves optimization over $16 \times 16 = 256$ parameters.

Since linear programming does not depend on the Haar condition, it can be applied to the two-dimensional approximation problem. While convergence of the linear program is guaranteed, the size of the problem (proportional to $N^2$) and the inefficiency of the linear programming technique has limited this technique to the design of low-order filters, e.g., $9 \times 9$ is the largest reported in Hu and Rabiner [38].

The suboptimal techniques of windowing and frequency sampling have also been extended to two dimensions and are capable of designing higher order filters.

In view of the computational difficulties of linear programming and the lack of theory for an exchange algorithm, two-dimensional approximation techniques have concentrated on suboptimal techniques. Included in the class of suboptimal methods are the windowing technique, the frequency sampling technique and an approach based on transformations of one-dimensional filters. This transformation technique in some cases may yield an optimal filter. In this section we outline the basic idea for the transformation method of design.

A two-dimensional digital filter with impulse response matrix $\{h(k,p)\}, k = 0, 1, \cdots, N_1 - 1; p = 0, 1, \cdots, N_2 - 1$ has a frequency response defined by the two-dimensional Fourier transform

$$H(e^{j\omega_1}, e^{j\omega_2}) = \sum_{k=0}^{N_1-1} \sum_{p=0}^{N_2-1} h(k,p) \exp\left[-j(k\omega_1 + p\omega_2)\right]. \tag{39}$$

If the impulse response is constrained to be symmetric,

$$\begin{aligned} h(N_1 - m - 1, p) &= h(m,p), \qquad m = 0, 1, \cdots, n_1 = (N_1 - 1)/2 \\ h(k, N_2 - m - 1) &= h(k,m), \qquad m = 0, 1, \cdots, n_2 = (N_2 - 1)/2 \end{aligned} \tag{40}$$

then the frequency response can be rewritten as

$$H(e^{j\omega_1}, e^{j\omega_2}) = \exp\left[-j(n_1\omega_1 + n_2\omega_2)\right] \hat{H}(\omega_1, \omega_2) \tag{41}$$

where

$$\hat{H}(e^{j\omega_1}, e^{j\omega_2}) = \sum_{k=0}^{n_1} \sum_{p=0}^{n_2} a(k,p) \cos k\omega_1 \cos p\omega_2. \tag{42}$$

Since $\exp[-j(n_1\omega_1 + n_2\omega_2)]$ has magnitude one, $|\hat{H}(e^{j\omega_1}, e^{j\omega_2})|$ is the magnitude of the frequency response. The folding frequency along each axis is $\pi$, so the magnitude response is completely specified on the square $[0, \pi] \times [0, \pi]$.

Recall the type-1 one-dimensional filter where $G(e^{j\omega})$ can be written in the form

$$G(e^{j\omega}) = \sum_{n=0}^{M} b(n) \cos(n\omega) \tag{43}$$

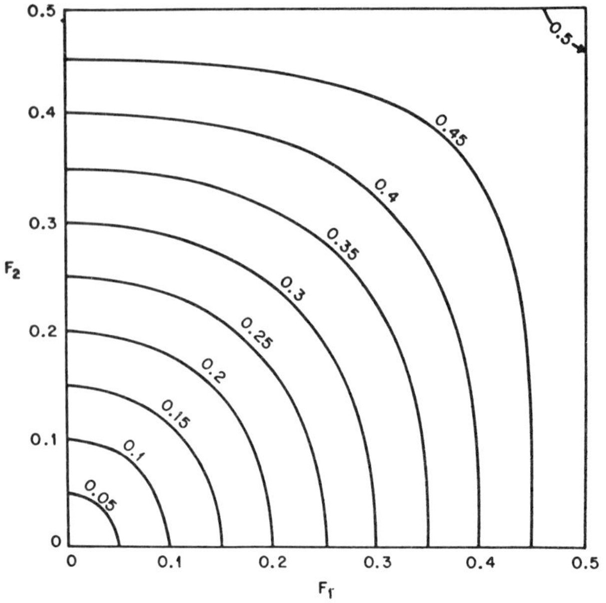


Fig. 16. Contours of the mapping in frequency from one to two dimensions.

where $M = (N-1)/2$, $b(0) = h(M)$, and $b(n) = 2h(M-n)$, $n = 1, 2, \cdots, M$. The magnitude of the frequency response is $|G(e^{j\omega})|$.

If the change of variables

$$\cos\omega = A\cos\omega_1 + B\cos\omega_2 + C\cos\omega_1\cos\omega_2 + D \tag{44}$$

is made in $G(e^{j\omega})$ then a function of the form $\hat{H}(e^{j\omega_1}, e^{j\omega_2})$ results. Thus the one-dimensional response $G(e^{j\omega})$ is mapped to a two-dimensional response $\hat{H}(e^{j\omega_1}, e^{j\omega_2})$. Given a fixed value of $\omega \in [0, \pi]$ there corresponds a curve in the $(\omega_1, \omega_2)$ plane, and along this curve the transformed two-dimensional frequency response is a constant equal to the value of the one-dimensional frequency response at $\omega$. As $\omega$ varies, a family of contours is generated which completely describes the transformed two-dimensional frequency response. For the choice of parameters $A = B = C = -D = \frac{1}{2}$, the contours are shown in Fig. 16. Thus a low-pass filter will be mapped by this particular change of variables to a low-pass circularly symmetric two-dimensional filter.

An important feature of this new method is that it is not limited to the design of small filters. With a larger filter it is possible to obtain smaller deviations and/or a sharper cutoff. Fig. 17 shows the magnitude of the frequency response for a $31 \times 31$ circularly symmetric lowpass filter with a transition region of width $0.10\pi$. The design time for this filter was approximately 5 s on an IBM 370/155 computer.

In addition, it can be shown that the filter of Fig. 17 is an optimal filter, if one is willing to accept the approximately circular contours shown in Fig. 16.

The possibilities for new design algorithms for two-dimensional filters are still wide open but for the present, suboptimal techniques are the only ways to get large filters.

## XIV. Summary

The mathematical theory of Chebyshev approximation has been used as the unifying theme in the presentation of recent design algorithms for FIR digital filters. Special emphasis was placed on the linear phase design problem where the methods are highly developed due to the direct applicability of the

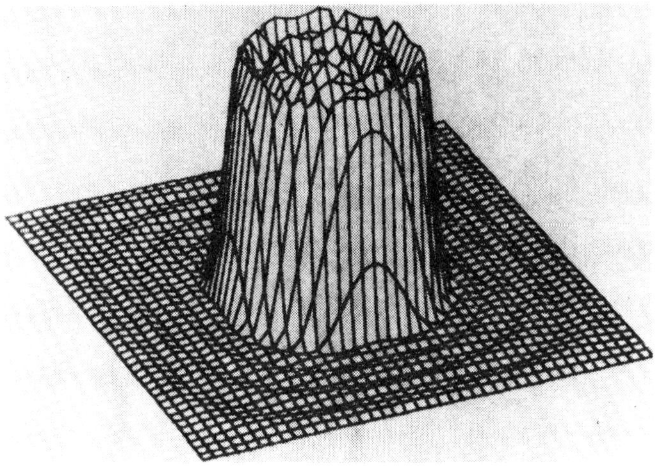

Fig. 17. Frequency response of a 31 X 31 low-pass filter.

Chebyshev theory. In the more general cases of magnitude approximation and two-dimensional approximation, it was shown that special techniques can be used to obtain optimal low-pass or bandpass filters.

Within the realm of Chebyshev approximation of FIR digital filters several research problems remain. These include the simultaneous approximation of magnitude and phase, general two-dimensional approximation, and the optimization of the filter response under the constraint of finite wordlength coefficients.

## REFERENCES


*General Reference*

L. R. Rabiner and B. Gold, "Theory and implementation of FIR digital filters," ch. 3 in *Theory and Application of Digital Signal Processing.* Englewood Cliffs, N.J.: Prentice-Hall, 1975.

*Windows*

[1] J. F. Kaiser, "Design methods for sampled data filters," in *Proc. 1st Allerton Conf. Circuit and System Theory*, pp. 221-236, Nov. 1963.

[2] J. F. Kaiser, "Digital filters," ch. 7 in *System Analysis by Digital Computer*, F. F. Kuo and J. F. Kaiser, Eds. New York: Wiley, 1966.

[3] R. B. Blackman and J. W. Tukey, *The Measurement of Power Spectra.* New York: Dover, 1958.

[4] G. E. Heyliger, "The scanning function approach to the design of numerical filters," Martin Co., Denver, Colo., Rep. R-63-2, Apr. 1963.

[5] —, "Design of numerical filters: Scanning functions and equal ripple approximation," in *Proc. 4th Allerton Conf. Circuit and System Theory*, pp. 175-185, 1966.

[6] G. E. Heyliger and C. A. Halijak, "Topics in the design of moving average numerical filters," in *Proc. 2nd Asilomar Conf. Circuits and Systems*, pp. 214-220, 1968.

[7] H. D. Helms, "Nonrecursive digital filters: Design methods for achieving specifications on frequency response," *IEEE Trans. Audio Electroacoust.*, vol. AU-16, pp. 336-342, Sept. 1968.

*Frequency Sampling Filters*

[8] B. Gold and C. M. Rader, *Digital Processing of Signals.* New York: McGraw-Hill, 1969.

[9] B. Gold and K. L. Jordan, Jr., "A note on digital filter synthesis," *Proc. IEEE* (Lett.), vol. 56, pp. 1717-1718, Oct. 1968.

[10] —, "A direct search procedure for designing finite duration impulse response filters," *IEEE Trans. Audio Electroacoust.*, vol. AU-17, pp. 33-36, Mar. 1969.

[11] L. R. Rabiner, B. Gold, and C. A. McGonegal, "An approach to the approximation problem for nonrecursive digital filters," *IEEE Trans. Audio Electroacoust.*, vol. AU-18, pp. 83-106, June 1970.

[12] L. R. Rabiner and K. Steiglitz, "The design of wide-band recursive and nonrecursive digital differentiators," *IEEE Trans. Audio Electroacoust.*, vol. AU-18, pp. 204-209, June 1970.

[13] L. R. Rabiner and R. W. Schafer, "Recursive and nonrecursive realizations of digital filters designed by frequency sampling techniques," *IEEE Trans. Audio Electroacoust.*, vol. AU-19, pp. 200-207, Sept. 1971.

[14] —, "Correction to 'Recursive and nonrecursive realization of digital filters designed by frequency sampling techniques,'" *IEEE Trans. Audio Electroacoust.* (Corresp.), vol. AU-20, pp. 104-105, Mar. 1972.

*Optimal Filters*

[15] L. R. Rabiner, "Techniques for designing finite-duration impulse response digital filters," *IEEE Trans. Commun. Technol.*, vol. COM-19, pp. 188-195, Mar. 1971.

[16] O. Herrmann, "Design of nonrecursive digital filters with linear phase," *Electron. Lett.*, vol. 6, no. 11, pp. 328-329, 1970.

[17] O. Herrmann and H. W. Schuessler, "Design of nonrecursive digital filters with minimum phase," *Electron. Lett.*, vol. 6, no. 11, pp. 329-330, 1970.

[18] H. D. Helms, "Digital filters with equiripple or minimax responses," *IEEE Trans. Audio Electroacoust.*, vol. AU-19, pp. 87-93, Mar. 1971.

[19] E. Hofstetter, A. Oppenheim, and J. Siegel, "A new technique for the design of nonrecursive digital filters," in *Proc. 5th Annu. Princeton Conf. Information Sciences and Systems*, pp. 64-72, 1971.

[20] —, "On optimum nonrecursive digital filters," in *Proc. 9th Allerton Conf. Circuit and System Theory*, Oct. 1971.

[21] T. W. Parks and J. H. McClellan, "Chebyshev approximation for nonrecursive digital filters with linear phase," *IEEE Trans. Circuit Theory*, vol. CT-19, pp. 189-194, Mar. 1972.

[22] L. R. Rabiner, "The design of finite impulse response digital filters using linear programming techniques," *Bell Syst. Tech. J.*, vol. 51, pp. 1117-1198, July-Aug. 1972.

[23] —, "Linear program design of finite impulse response (FIR) digital filters," *IEEE Trans. Audio Electroacoust.*, vol. AU-20, pp. 280-288, Oct. 1972.

[24] T. W. Parks and J. H. McClellan, "A program for the design of linear phase finite impulse response digital filters," *IEEE Trans. Audio Electroacoust.*, vol. AU-20, pp. 195-199, Aug. 1972.

[25] T. W. Parks, L. R. Rabiner, and J. H. McClellan, "On the transition width of finite impulse response digital filters," *IEEE Trans. Audio Electroacoust.*, vol. AU-21, pp. 1-4, Feb. 1973.

[26] L. R. Rabiner and O. Herrmann, "The predictability of certain optimum finite impulse response digital filters," *IEEE Trans. Circuit Theory*, vol. CT-20, pp. 401-406, July 1973.

[27] O. Herrmann, L. R. Rabiner, and D. S. K. Chan, "Practical design rules for optimum finite impulse response lowpass digital filters," *Bell Syst. Tech. J.*, vol. 52, pp. 769-799, July-Aug. 1973.

[28] R. W. Schafer and L. R. Rabiner, "A digital signal processing approach to interpolation," *Proc. IEEE*, vol. 61, pp. 692-702, June 1973.

[29] L. R. Rabiner and O. Herrmann, "On the design of optimum FIR low-pass filters with even impulse response duration," *IEEE Trans. Audio Electroacoust.*, vol. AU-21, pp. 329-336, Aug. 1973.

[30] L. R. Rabiner, "Approximate design relationships for low-pass FIR digital filters," *IEEE Trans. Audio Electroacoust.*, vol. AU-21, pp. 456-460, Oct. 1973.

[31] J. H. McClellan and T. W. Parks, "A unified approach to the design of optimum FIR linear phase digital filters," *IEEE Trans. Circuit Theory*, vol. CT-20, pp. 697-701, Nov. 1973.

[32] L. R. Rabiner and R. W. Schafer, "On the behavior of minimax FIR digital differentiators," *Bell Syst. Tech. J.*, Feb. 1974.

[33] —, "On the behavior of minimax FIR digital Hilbert transformers," *Bell Syst. Tech. J.*, Feb. 1974.

[34] J. H. McClellan, T. W. Parks, and L. R. Rabiner, "A computer program for designing optimum FIR linear phase digital filters," *IEEE Trans. Audio Electroacoust.*, vol. AU-21, pp. 506-526, Dec. 1973.

[35] J. Fisher, "Design of finite impulse response digital filters," Ph.D. dissertation, Rice Univ., Houston, Tex., 1973.

*Two-Dimensional Filters*

[36] J. H. McClellan, "On the design of one-dimensional and two-dimensional FIR digital filters," Ph.D. dissertation, Rice Univ., Houston, Tex., Apr. 1973.

[37] T. S. Huang, "Two-dimensional windows," *IEEE Trans. Audio Electroacoust.* (Corresp.), vol. AU-20, pp. 88-89, Mar. 1972.

[38] J. V. Hu and L. R. Rabiner, "Design techniques for two-dimensional digital filters," *IEEE Trans. Audio Electroacoust.*, vol. AU-20, pp. 249-257, Oct. 1972.

[39] J. H. McClellan and T. W. Parks, "Equiripple approximation of fan filters," *Geophysics*, vol. 37, pp. 573-583, Aug. 1972.

[40] J. H. McClellan, "The design of two-dimensional digital filters by transformations," in *Proc. 7th Annu. Princeton Conf. Information Sciences and Systems*, pp. 247-251, 1973.

[41] Y. Kamp and J. P. Thiran, "Chebyshev approximation for two-dimensional nonrecursive digital filters," *IEEE Trans. Circuits Syst.*, to be published.
[42] H. Hersey and R. Mersereau, "Design techniques for two-dimensional finite impulse response circularly symmetric digital filters," Res. Lab. Electron., M.I.T., Cambridge, Mass., QPR 114, pp. 160-171, July 1974.

*Optimization Methods*

[43] E. W. Cheney, *Introduction to Approximation Theory.* New York: McGraw-Hill, 1966.
[44] J. R. Rice, *The Approximation of Functions*, vols. 1 and 2. Reading, Mass.: Addison-Wesley, 1964.
[45] G. Meinardus, *Approximation of Functions: Theory and Numerical Methods.* New York: Springer-Verlag, 1967.
[46] M. Aoki, *Introduction to Optimization Techniques.* New York: Macmillan, 1971.
[47] E. Ya. Remez, *General Computational Methods of Chebyshev Approximation.* Kiev, USSR: Atomic Energy Translation 4491, 1957, pp. 1-85.

*Linear Programming*

[48] G. Dantzig, *Linear Programming and Extensions.* Princeton, N.J.: Princeton Univ Press, 1963.
[49] G. Hadley, *Linear Programming.* Reading, Mass.: Addison-Wesley, 1963.
[50] T. C. Hu, *Integer Programming and Network Flows.* Reading, Mass.: Addison-Wesley, 1969.
[51] W. A. Spivey and R. M. Thrall, *Linear Optimization.* New York: Holt, Rinehart and Winston, 1970.
[52] S. I. Gass, *Linear Programming.* New York: McGraw-Hill, 1969.
[53] M. Simonnand, *Linear Programming.* Englewood Cliffs, N.J.: Prentice-Hall, 1966.

# A Computer Program for Designing Optimum FIR Linear Phase Digital Filters

JAMES H. McCLELLAN, Student Member, IEEE,
THOMAS W. PARKS, Member, IEEE, and
LAWRENCE R. RABINER, Member, IEEE

*Abstract*—This paper presents a general-purpose computer program which is capable of designing a large class of optimum (in the minimax sense) FIR linear phase digital filters. The program has options for designing such standard filters as low-pass, high-pass, bandpass, and bandstop filters, as well as multipassband-stopband filters, differentiators, and Hilbert transformers. The program can also be used to design filters which approximate arbitrary frequency specifications which are provided by the user. The program is written in Fortran, and is carefully documented both by comments and by detailed flowcharts. The filter design algorithm is shown to be exceedingly efficient, e.g., it is capable of designing a filter with a 100-point impulse response in about 20 s.

## I. Introduction

This paper presents a general algorithm for the design of a large class of finite impulse response (FIR) linear phase digital filters. Emphasis is placed on a description of how the algorithm works, and several examples are included which illustrate specific applications. A unified treatment of the theory behind this approach is available in [1].

The algorithm uses the Remez exchange method [2], [3] to design filters with minimum weighted Chebyshev error in approximating a desired ideal frequency response $D(f)$. Several authors have studied the FIR design problem for special filter types using several different algorithms [4]-[13]. The advantage of the present approach is that it combines the speed of the Remez procedure with a capability for designing a large class of general filter types. While the algorithm to be described has a special section for the more common filter types (e.g., bandpass filters with multiple bands, Hilbert transform filters, and differentiators), an arbitrary frequency response can also be approximated.

Manuscript received August 6, 1973. The work of J. H. McClellan and T. W. Parks was supported by NSF Grant GK-23697.
J. H. McClellan and T. W. Parks are with Rice University, Houston, Tex.
L. R. Rabiner is with Bell Laboratories, Murray Hill, N.J. 07974.

## II. Formulation of the Approximation Problem

The frequency response of an FIR digital filter with an $N$-point impulse response $\{h(k)\}$ is the $z$-transform of the sequence evaluated on the unit circle, i.e.,

$$H(f)^1 = H(z)\Big|_{z=e^{j2\pi f}} = \sum_{k=0}^{N-1} h(k)\, e^{-j2\pi kf}. \quad (1)$$

The frequency response of a linear phase filter can be written as

$$H(f) = G(f)\, e^{j\left(\frac{L\pi}{2} - \left(\frac{N-1}{2}\right) 2\pi f\right)} \quad (2)$$

where $G(f)$ is a real valued function and $L = 0$ or 1. It is possible to show that there are exactly four cases of linear phase FIR filters to consider [1]. These four cases differ in the length of the impulse response (even or odd) and the symmetry of the impulse response [positive ($L = 0$) or negative ($L = 1$)]. By positive symmetry we mean $h(k) = h(N - 1 - k)$, and by negative symmetry $h(k) = -h(N - 1 - k)$.

In all cases, the real function $G(f)$ will be used to approximate the desired ideal magnitude specifications since the linear phase term in (2) has no effect on the magnitude response of the filter. The form of $G(f)$ depends on which of the four cases is being used. Using the appropriate symmetry relations, $G(f)$ can be expressed as follows.

*Case 1:* Positive symmetry, odd length:

$$G(f) = \sum_{k=0}^{n} a(k) \cos(2\pi kf) \quad (3)$$

where $n = (N - 1)/2$, $a(0) = h(n)$, and $a(k) = 2h(n - k)$ for $k = 1, 2, \cdots, n$.

*Case 2:* Positive symmetry, even length:

$$G(f) = \sum_{k=1}^{n} b(k) \cos[2\pi(k - \tfrac{1}{2})f] \quad (4)$$

where $n = N/2$ and $b(k) = 2h(n - k)$ for $k = 1, \cdots, n$.

*Case 3:* Negative symmetry, odd length:

$$G(f) = \sum_{k=1}^{n} c(k) \sin(2\pi kf) \quad (5)$$

where $n = (N - 1)/2$ and $c(k) = 2h(n - k)$ for $k = 1, 2, \cdots, n$ and $h(n) = 0$.

*Case 4:* Negative symmetry, even length:

$$G(f) = \sum_{k=1}^{n} d(k) \sin[2\pi(k - \tfrac{1}{2})f] \quad (6)$$

where $n = N/2$ and $d(k) = 2h(n - k)$ for $k = 1, \cdots, n$.

Earlier efforts at designing FIR filters concentrated on Case 1 designs, but it is now possible to combine

[1] For convenience, throughout this paper the notation $H(f)$ rather than $H(e^{j2\pi f})$ is used to denote the frequency response of the digital filter.

Reprinted from *IEEE Trans. Audio Electroacoust.*, vol. AU-21, pp. 506-526, Dec. 1973.

all four cases into one algorithm. This is accomplished by noting that $G(f)$ can be rewritten as $G(f) = Q(f)P(f)$ where $P(f)$ is a linear combination of cosine functions. Thus, results that have been worked out for Case 1 can be applied to the other three cases as well. For these purposes, it is convenient to express the summations in (4)-(6) as a sum of cosines directly. Simple manipulations of (4)-(6) yield the expressions.

*Case 2:*

$$\sum_{k=1}^{n} b(k) \cos [2\pi(k - \tfrac{1}{2})f] = \cos (\pi f) \sum_{k=0}^{n-1} \tilde{b}(k) \cos (2\pi k f). \quad (7)$$

*Case 3:*

$$\sum_{k=1}^{n} c(k) \sin (2\pi k f) = \sin (2\pi f) \sum_{k=0}^{n-1} \tilde{c}(k) \cos (2\pi k f). \quad (8)$$

*Case 4:*

$$\sum_{k=1}^{n} d(k) \sin [2\pi(k - \tfrac{1}{2})f] = \sin (\pi f) \sum_{k=0}^{n-1} \tilde{d}(k) \cos (2\pi k f) \quad (9)$$

where

$$\text{Case 2:} \begin{cases} b(1) = \tilde{b}(0) + \frac{1}{2}\tilde{b}(1) \\ b(k) = \frac{1}{2}[\tilde{b}(k-1) + \tilde{b}(k)], \quad k = 2, 3, \cdots, n-1 \\ b(n) = \frac{1}{2}\tilde{b}(n-1) \end{cases} \quad (10)$$

$$\text{Case 3:} \begin{cases} c(1) = \tilde{c}(0) - \frac{1}{2}\tilde{c}(2) \\ c(k) = \frac{1}{2}[\tilde{c}(k-1) - \tilde{c}(k+1)], \quad k = 2, 3, \cdots, n-2 \\ c(n-1) = \frac{1}{2}\tilde{c}(n-2) \\ c(n) = \frac{1}{2}\tilde{c}(n-1) \end{cases} \quad (11)$$

$$\text{Case 4:} \begin{cases} d(1) = \tilde{d}(0) - \frac{1}{2}\tilde{d}(1) \\ d(k) = \frac{1}{2}[\tilde{d}(k-1) - \tilde{d}(k)], \quad k = 2, 3, \cdots, n-1 \\ d(n) = \frac{1}{2}\tilde{d}(n-1). \end{cases} \quad (12)$$

The motivation for rewriting the four cases in a common form is that a single central computation routine (based on the Remez exchange method) can be used to calculate the best approximation in each of the four cases. This is accomplished by modifying both the desired magnitude function and the weighting function to formulate a new equivalent approximation problem.

The original approximation problem can be stated as follows: given a desired magnitude response $D(f)$ and a positive weight function $W(f)$, both continuous on a compact subset $F \subset [0, \frac{1}{2}]$ (note that the sampling rate is 1.0) and one of the four cases of linear phase filters [i.e., the forms of $G(f)$], then one wishes to minimize the maximum absolute weighted error, defined as

$$\| E(f) \| = \max_{f \in F} W(f) \,| D(f) - G(f) | \quad (13)$$

over the set of coefficients of $G(f)$.

The error function $E(f)$ can be rewritten in the form

$$E(f) = W(f)\,[D(f) - G(f)] = W(f)\,Q(f) \left[ \frac{D(f)}{Q(f)} - P(f) \right] \quad (14)$$

if one is careful to omit those endpoint(s) where $Q(f) = 0$. Letting $\hat{D}(f) = D(f)/Q(f)$ and $\hat{W}(f) = W(f)\,Q(f)$, then an equivalent approximation problem would be to minimize the quantity

$$\| E(f) \| = \max_{f \in F'} \hat{W}(f) \,| \hat{D}(f) - P(f) | \quad (15)$$

by choice of the coefficients of $P(f)$. The set $F$ is replaced by $F' = F - \{\text{endpoints where } Q(f) = 0\}$.

The net effect of this reformulation of the problem is a unification of the four cases of linear phase FIR filters from the point of view of the approximation problem. Furthermore, (15) provides a simplified viewpoint from which it is easy to see the necessary and sufficient conditions which are satisfied by the best approximation. Finally, (15) shows how to calculate this best approximation using an algorithm which can do only cosine approximations. The set of necessary and sufficient conditions for this best approximation is given in the following alternation theorem [2].

*Alternation theorem:* If $P(f)$ is a linear combination of $r$ cosine functions i.e.,

$$P(f) = \sum_{k=0}^{r-1} \alpha(k) \cos 2\pi k f,$$

then a necessary and sufficient condition that $P(f)$ be the unique best weighted Chebyshev approximation to a continuous function $\hat{D}(f)$ on $F'$ is that the weighted error function $E(f) = \hat{W}(f)\ [\hat{D}(f) - P(f)]$ exhibit *at least* $r + 1$ extremal frequencies in $F'$.

These extremal frequencies are a set of points $\{F_i\}$, $i = 1, 2, \cdots, r + 1$ such that $F_1 < F_2 < \cdots < F_r < F_{r+1}$, with $E(F_i) = -E(F_{i+1})$, $i = 1, 2, \cdots, r$ and $|E(F_i)| = \max_{f \in F'} E(f)$.

An algorithm can now be designed to make the

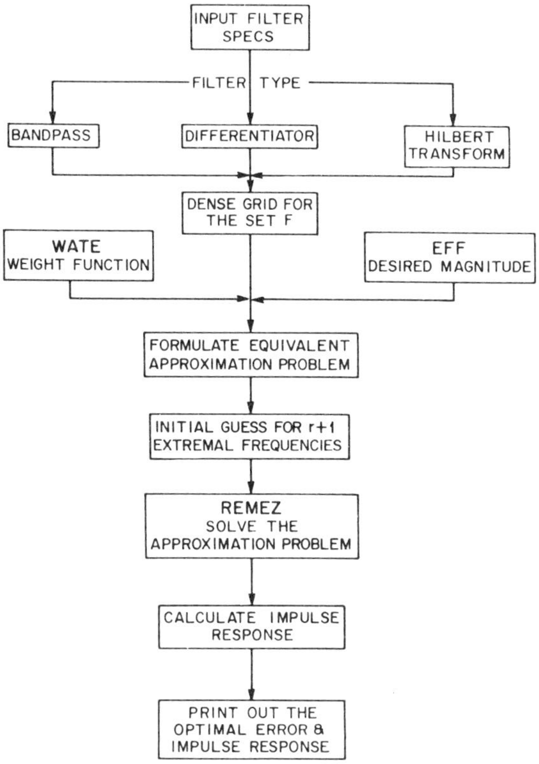


**Fig. 1. Overall flowchart of filter design algorithm.**

error function of the filter satisfy the set of necessary and sufficient conditions for optimality as stated in the alternation theorem. The next section describes such an algorithm along with details as to its implementation.

### III. Description of the Design Algorithm

As seen in Fig. 1, the design algorithm consists of an input section, formulation of the appropriate equivalent approximation problem, solution of the approximation problem using the Remez exchange method, and calculation of the filter impulse response. The flowcharts of Figs. 2–5 give details of the exact structure of the computer program.

The input which describes the filter specifications consists of the following.

1) The filter length, $3 \leqslant$ NFILT $\leqslant$ NFMAX (the upper limit set by the programmer).

2) The type of filter (JTYPE):

a) Multiple passband/stopband (JTYPE=1)

b) Differentiator (JTYPE=2)

c) Hilbert transformer (JTYPE=3).

3) The frequency bands, specified by upper and lower cutoff frequencies (EDGE array) up to a maximum of 10 bands.

4) The desired frequency response (FX array) in each band.

5) A positive weight function (WTX array) in each band.

6) The grid density (LGRID), assumed to be 16 unless specified otherwise.

7) Impulse response punch option (JPUNCH).

Part 3) specifies the set $F$ to be of the form $F = \cup B_i$ where each frequency band $B_i$ is a closed subinterval of $[0, \frac{1}{2}]$. The inputs 4) and 5) are interpreted differently by the program for a differentiator than for the other two types of filters (see the EFF and WATE subroutines in Figs. 3 and 4). The weight specification in the case of a differentiator results in a *relative* error tolerance as is used in all other cases.

The set $F$ must be replaced by a finite set of points for implementation on a computer. A dense grid of points is used with the spacing between points being $0.5/(\text{LGRID} \times r)$ where $r$ is the number of cosine basis functions. Both $D(f)$ and $W(f)$ are evaluated on this grid by the subroutines EFF and WATE, respectively. Then the auxiliary approximation problem is set up by forming $\hat{D}(f)$ and $\hat{W}(f)$ as above, and an initial guess of the extremal frequencies is made by taking $r + 1$ equally spaced frequency values. The subroutine REMEZ (Fig. 5) is called to perform the calculation of the best approximation for the equivalent problem. The mechanics of the Remez algorithm will not be discussed here since they are treated elsewhere for the particular case of low-pass filters [9]. (The flowchart of Fig. 5 gives details about the mechanics of the Remez algorithm as implemented in this design program.)

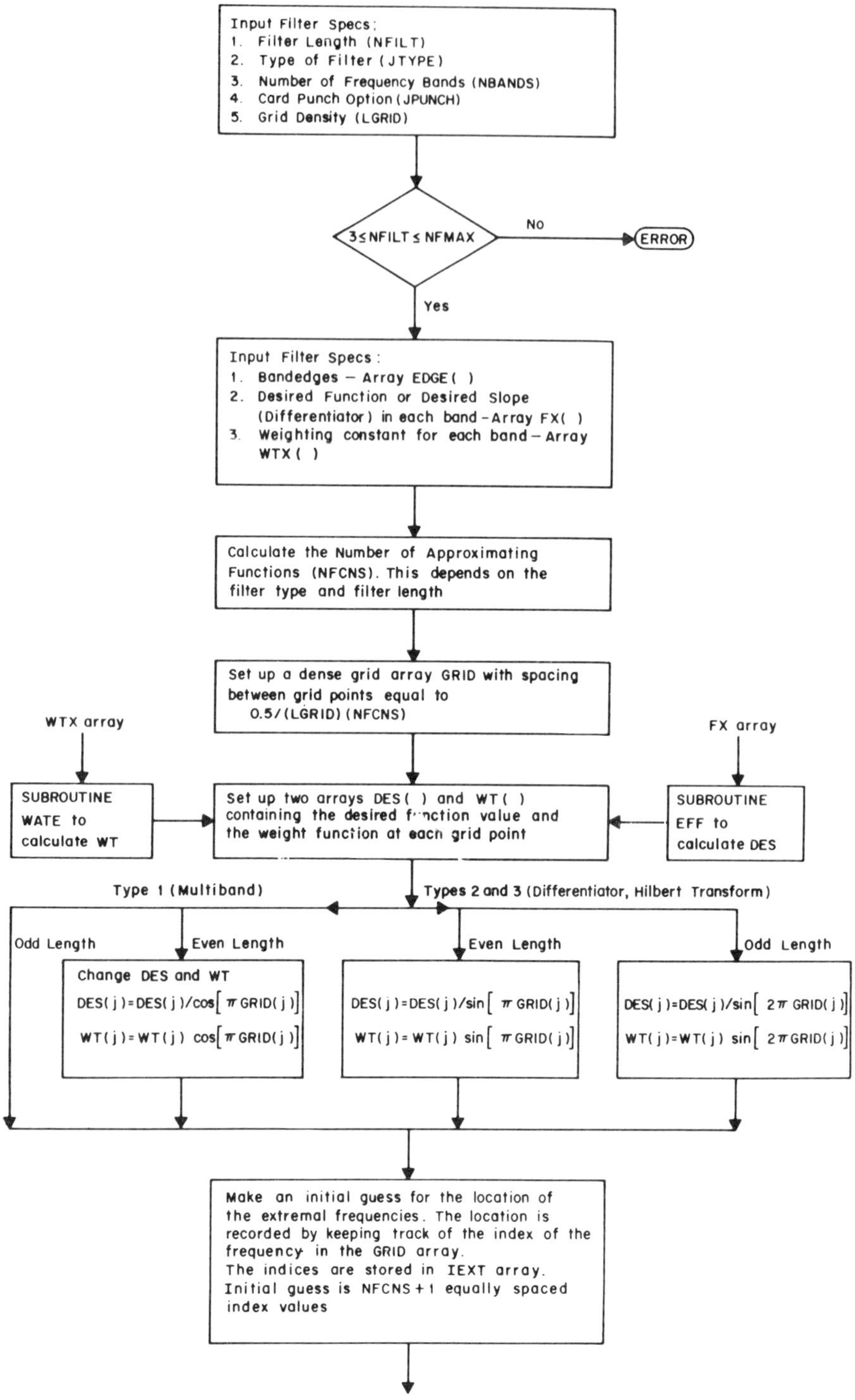


**Fig. 2. Detailed flowchart for filter design algorithm.**

The appropriate equations (3)–(12) are used to recover the impulse response from the coefficients of the best cosine approximation obtained in the REMEZ subroutine. The outputs of the program are the impulse response, the optimal error (min $\|E(f)\|$), and the $r + 1$ extremal frequencies where $E(f) = \pm\|E(f)\|$.

It is possible that one might want to design a filter to approximate a magnitude specification which is not included in the scheme given above, or change the weight function to get a desired tolerance scheme. A flowchart of such a program is given in Fig. 6. In such cases, the user must code the subroutines EFF and WATE to calculate $D(f)$ and $W(f)$. The input is the same as before, except that there are only two types of filters, depending on whether the impulse symmetry is positive or negative.

A detailed program listing of the generalized design program is given in the Appendix. Representative

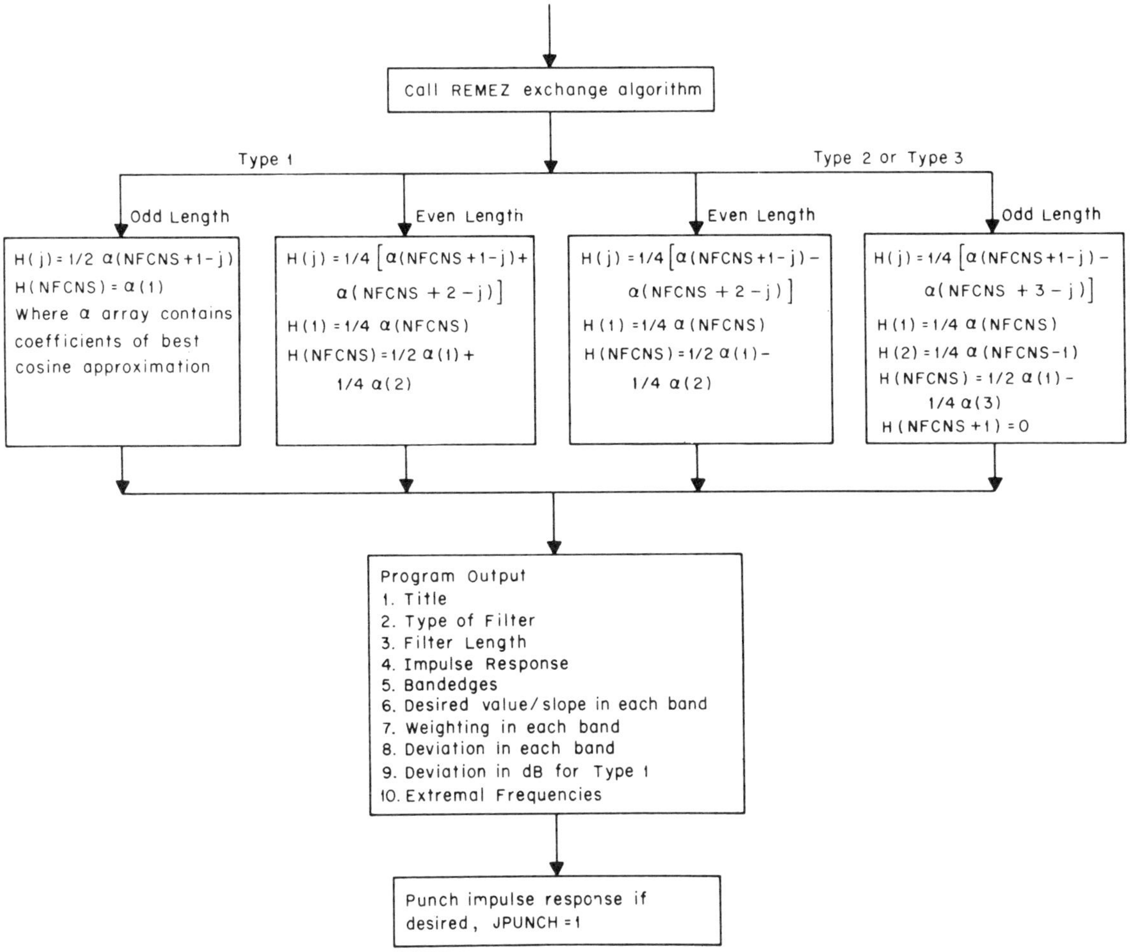


Fig. 2. (Continued.)

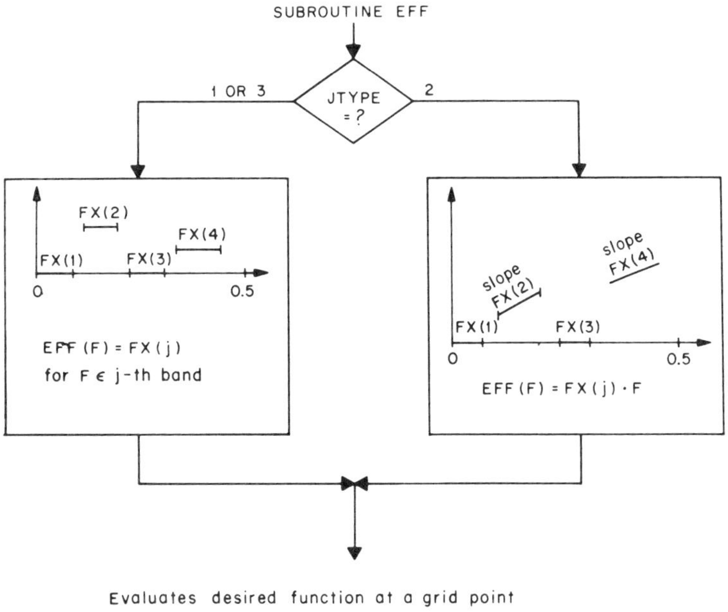


Fig. 3. Flowchart for subroutine EFF.

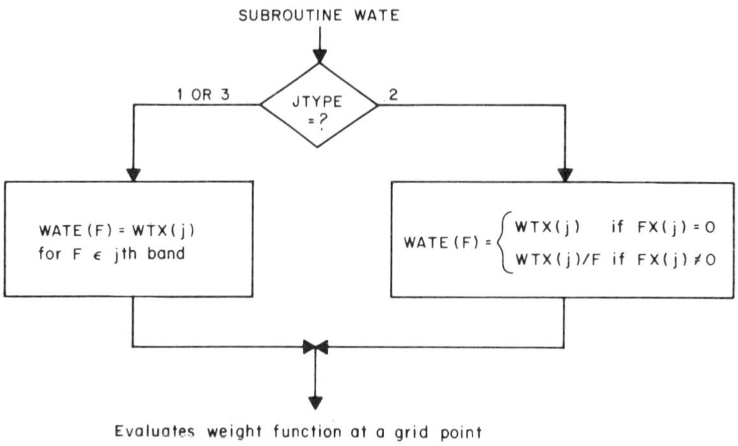


**Fig. 4. Flowchart for subroutine WATE.**

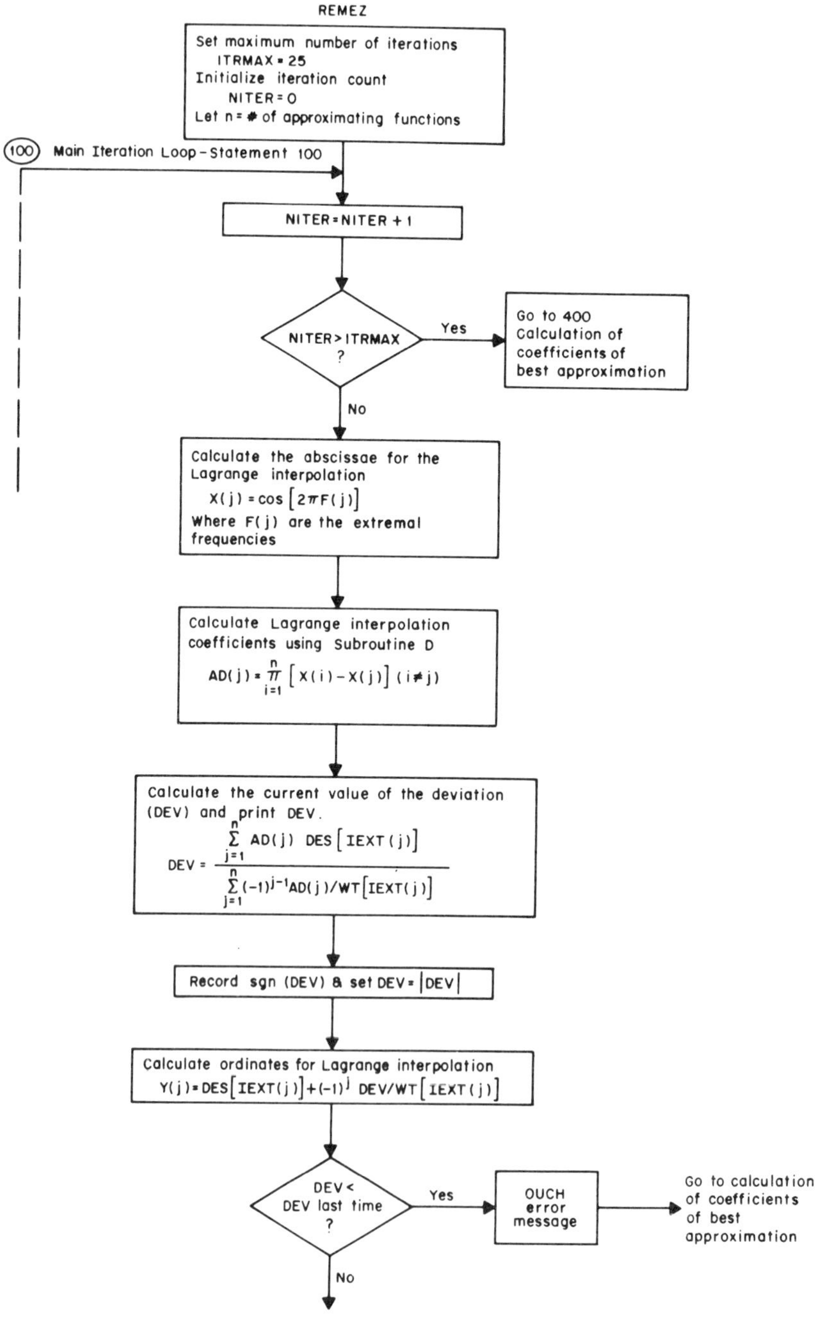


**Fig. 5. Detailed flowchart for subroutine REMEZ.**

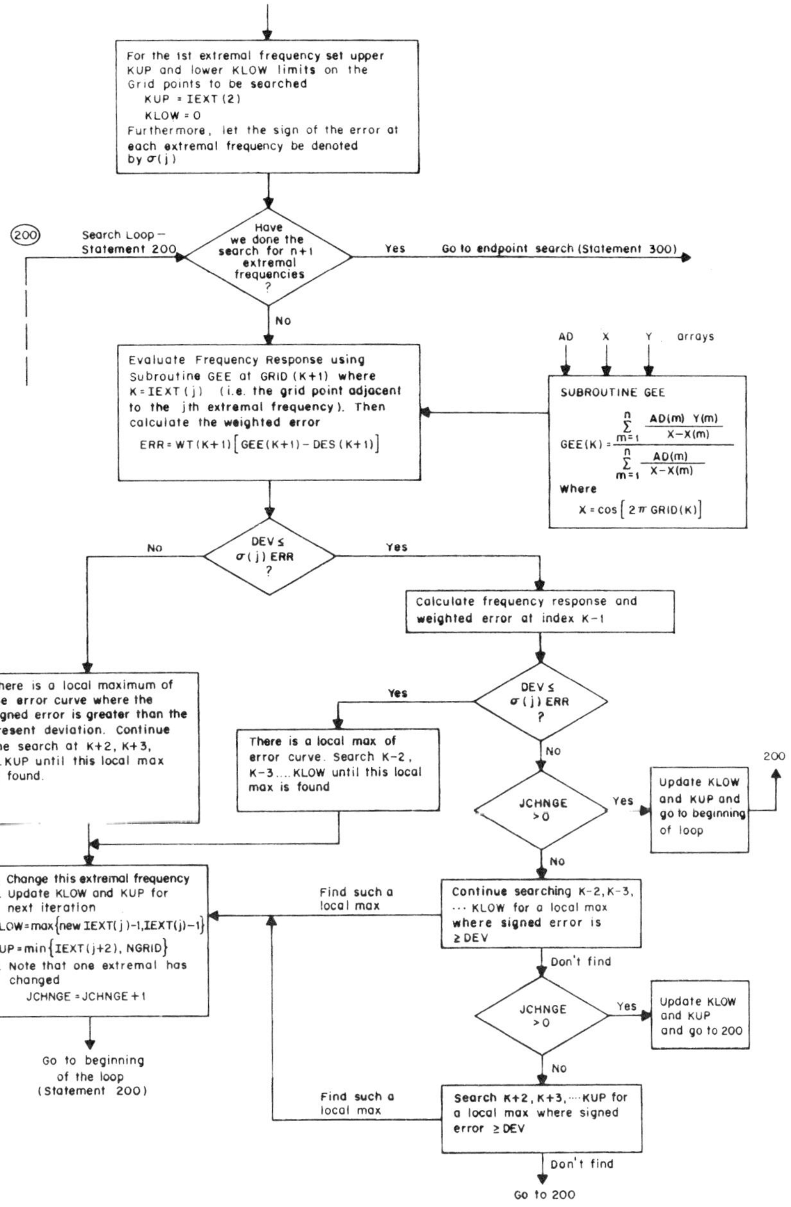


Fig. 5. (Continued.)

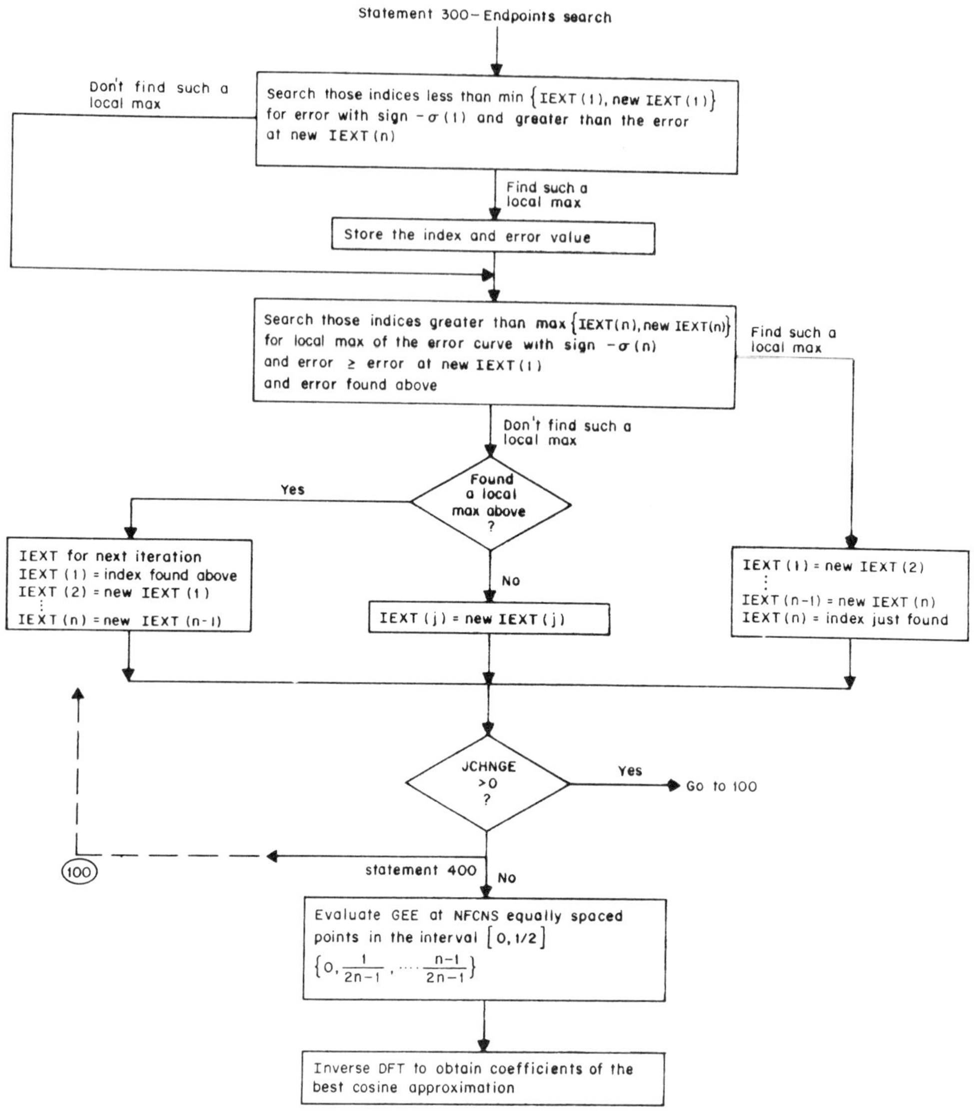


**Fig. 5. (Continued.)**

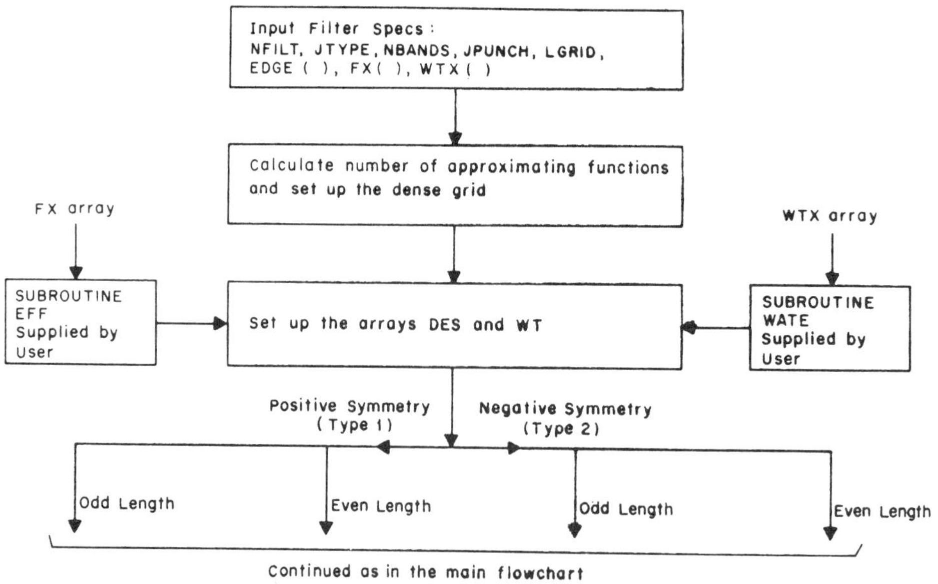


**Fig. 6. Flowchart for arbitrary magnitude filter design algorithm.**

```
************************************************************************

                    FINITE IMPULSE RESPONSE (FIR)
                    LINEAR PHASE DIGITAL FILTER DESIGN
                    REMEZ EXCHANGE ALGORITHM
                    BANDPASS FILTER
            FILTER LENGTH =  24
            ***** IMPULSE RESPONSE *****
                 H(  1) =  0.33740917E-02 = H(  24)
                 H(  2) =  0.14938299E-01 = H(  23)
                 H(  3) =  0.10569360E-01 = H(  22)
                 H(  4) =  0.25415067E-02 = H(  21)
                 H(  5) = -0.15929392E-01 = H(  20)
                 H(  6) = -0.34085343E-01 = H(  19)
                 H(  7) = -0.38112177E-01 = H(  18)
                 H(  8) = -0.14629169E-01 = H(  17)
                 H(  9) =  0.40089541E-01 = H(  16)
                 H( 10) =  0.11540713E 00 = H(  15)
                 H( 11) =  0.13850752E 00 = H(  14)
                 H( 12) =  0.23354606E 00 = H(  13)

                    BAND 1             BAND 2            BAND
LOWER BAND EDGE     0.                 0.16000000
UPPER BAND EDGE     0.08000000         0.50000000
DESIRED VALUE       1.00000000         0.
WEIGHTING           1.00000000         1.00000000
DEVIATION           0.01243364         0.01243364
DEVIATION IN DB   -38.10803413       -38.10803413

EXTREMAL FREQUENCIES
   0.             0.0364583     0.0677083     0.0800000     0.1600000
   0.1730208      0.2068750     0.2459375     0.2876042     0.3318750
   0.3787500      0.4256251     0.4751043

************************************************************************
TIME=        0.7694063 SECONDS
```

**Fig. 7. Output listing for an $N$ = 24 low-pass filter.**

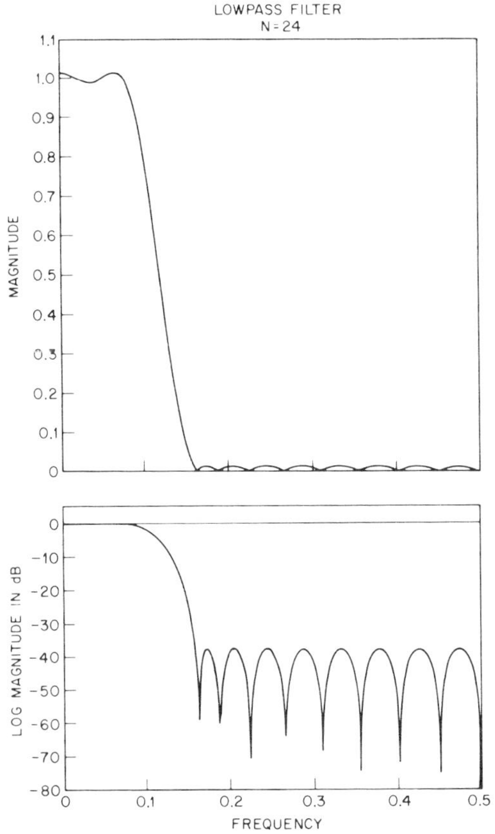


**Fig. 8. Magnitude responses, on linear and log scales, for an $N$ = 24 low-pass filter.**

input card sequences are given for the design of a bandpass filter and a differentiator. To approximate an arbitrary magnitude response and/or an arbitrary weighting function, all the user has to do is change the subroutines EFF and WATE and use the program in the Appendix. In the next section, representative filters designed using these algorithms are presented.

## IV. Design Examples

Figs. 7-22 show specific examples of use of the design program for several typical filters of interest. For each of these filters, one figure shows the computer output listing (including the run time on a Honeywell 6000 computer), and the other figure shows a plot of the filter frequency response on either a linear or a log magnitude scale (or sometimes both). Figs. 7 and 8 are for an $N$ = 24 low-pass filter. For this example, the run time was 0.77 s. Figs. 9 and 10 are for an $N$ = 32 bandpass filter. This example is the first example listed in the prologue to the program in the Appendix. The run time for this example was 0.82 s. Figs. 11 and 12 are for an $N$ = 50 bandpass filter in which unequal weighting was used in the two stopbands. Thus the peak error in the upper stopband is ten times smaller than the peak error in the lower stopband. A total of 2.96 s was required to design this filter. Figs. 13 and 14 are for an $N$ = 31 bandstop filter with equal weighting in both passbands. For the design of this filter 1.61 s were required.

To illustrate the multiband capability of the pro-

```
*********************************************************************

                 FINITE IMPULSE RESPONSE (FIR)
                 LINEAR PHASE DIGITAL FILTER DESIGN
                 REMEZ EXCHANGE ALGORITHM
                 BANDPASS FILTER
          FILTER LENGTH =  32
          ***** IMPULSE RESPONSE *****
               H(  1) = -0.57534121E-02 = H(  32)
               H(  2) =  0.39027198E-03 = H(  31)
               H(  3) =  0.75733545E-02 = H(  30)
               H(  4) = -0.65141192E-02 = H(  29)
               H(  5) =  0.13960525E-01 = H(  28)
               H(  6) =  0.22351469E-02 = H(  27)
               H(  7) = -0.19994067E-01 = H(  26)
               H(  8) =  0.71369560E-02 = H(  25)
               H(  9) = -0.39657363E-01 = H(  24)
               H( 10) =  0.11260114E-01 = H(  23)
               H( 11) =  0.66233643E-01 = H(  22)
               H( 12) = -0.10497223E-01 = H(  21)
               H( 13) =  0.85136133E-01 = H(  20)
               H( 14) = -0.12024393E 00 = H(  19)
               H( 15) = -0.29678577E 00 = H(  18)
               H( 16) =  0.30410917E 00 = H(  17)

                        BAND 1           BAND 2          BAND 3          BAND
LOWER BAND EDGE      0.               0.20000000      0.42500000
UPPER BAND EDGE      0.10000000       0.35000000      0.50000000
DESIRED VALUE        0.               1.00000000      0.
WEIGHTING           10.00000000       1.00000000     10.00000000
DEVIATION            0.00151312       0.01513118      0.00151312
DEVIATION IN DB    -56.40254641     -36.40254641    -56.40254641

EXTREMAL FREQUENCIES
   0.            0.0273437    0.0527344    0.0761719    0.0937500
   0.1000000     0.2000000    0.2195312    0.2527344    0.2839844
   0.3132812     0.3386719    0.3500000    0.4250000    0.4328125
   0.4503906     0.4796875

*********************************************************************
TIME=       0.8245625 SECONDS
```

**Fig. 9. Output listing for an $N$ = 32 bandpass filter.**

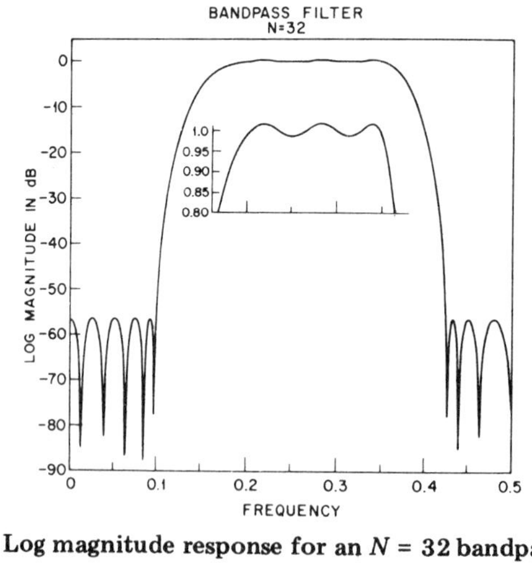


**Fig. 10. Log magnitude response for an $N$ = 32 bandpass filter.**

```
***********************************************************************
                    FINITE IMPULSE RESPONSE (FIR)
                    LINEAR PHASE DIGITAL FILTER DESIGN
                    REMEZ EXCHANGE ALGORITHM
                    BANDPASS FILTER
          FILTER LENGTH =  50
          ***** IMPULSE RESPONSE *****
               H(  1) =  0.15648409E-02 = H(  50)
               H(  2) =  0.50816298E-02 = H(  49)
               H(  3) = -0.31745254E-02 = H(  48)
               H(  4) = -0.61380034E-02 = H(  47)
               H(  5) =  0.74350685E-02 = H(  46)
               H(  6) =  0.38368361E-02 = H(  45)
               H(  7) = -0.11103735E-01 = H(  44)
               H(  8) = -0.10101927E-01 = H(  43)
               H(  9) =  0.89349206E-02 = H(  42)
               H( 10) =  0.28980188E-02 = H(  41)
               H( 11) =  0.26632392E-02 = H(  40)
               H( 12) =  0.12021958E-01 = H(  39)
               H( 13) = -0.20657142E-01 = H(  38)
               H( 14) = -0.27189007E-01 = H(  37)
               H( 15) =  0.32337130E-01 = H(  36)
               H( 16) =  0.28305613E-01 = H(  35)
               H( 17) = -0.20922041E-01 = H(  34)
               H( 18) = -0.18761153E-02 = H(  33)
               H( 19) = -0.22823357E-01 = H(  32)
               H( 20) = -0.53926217E-01 = H(  31)
               H( 21) =  0.90472539E-01 = H(  30)
               H( 22) =  0.12315772E 00 = H(  29)
               H( 23) = -0.15639221E 00 = H(  28)
               H( 24) = -0.17733448E 00 = H(  27)
               H( 25) =  0.13078165E 00 = H(  26)

                         BAND 1          BAND 2          BAND 3          BAND
LOWER BAND EDGE      0.              0.20000000      0.35000000
UPPER BAND EDGE      0.15000000      0.30000000      0.50000000
DESIRED VALUE        0.              1.00000000      0.
WEIGHTING           10.00000000      1.00000000    100.00000000
DEVIATION            0.00370505      0.03705048      0.00037050
DEVIATION IN DB    -48.62412214    -28.62412214    -68.62412262

EXTREMAL FREQUENCIES
   0.           0.0200000    0.0400000    0.0612500    0.0812500
   0.1012500    0.1225000    0.1412500    0.1500000    0.2000000
   0.2100000    0.2287500    0.2500000    0.2712500    0.2900000
   0.3000000    0.3500000    0.3537500    0.3637500    0.3762500
   0.3925000    0.4099999    0.4287499    0.4487499    0.4687499
   0.4899999

***********************************************************************
TIME=        2.9562969 SECONDS
```

**Fig. 11.** Output listing for an $N = 50$ bandpass filter with unequal weighting in the stopbands.

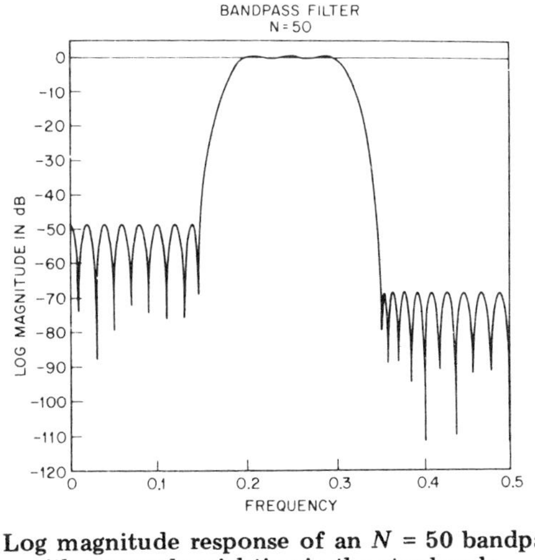


**Fig. 12.** Log magnitude response of an $N = 50$ bandpass filter with unequal weighting in the stopbands.

```
***********************************************************************

                    FINITE IMPULSE RESPONSE (FIR)
                    LINEAR PHASE DIGITAL FILTER DESIGN
                    REMEZ EXCHANGE ALGORITHM
                    BANDPASS FILTER
               FILTER LENGTH =  31
               ***** IMPULSE RESPONSE *****
                  H(  1) = -0.43725800E-02 = H(  31)
                  H(  2) =  0.19295933E-01 = H(  30)
                  H(  3) = -0.56382901E-02 = H(  29)
                  H(  4) =  0.52360280E-01 = H(  28)
                  H(  5) =  0.31550241E-02 = H(  27)
                  H(  6) =  0.43481227E-01 = H(  26)
                  H(  7) =  0.11696224E-01 = H(  25)
                  H(  8) = -0.37915416E-01 = H(  24)
                  H(  9) =  0.34844146E-02 = H(  23)
                  H( 10) = -0.87599027E-01 = H(  22)
                  H( 11) = -0.10393063E-01 = H(  21)
                  H( 12) =  0.44455166E-01 = H(  20)
                  H( 13) = -0.69347169E-02 = H(  19)
                  H( 14) =  0.31144824E 00 = H(  18)
                  H( 15) =  0.96629834E-02 = H(  17)
                  H( 16) =  0.45296733E 00 = H(  16)

                       BAND 1          BAND 2          BAND 3          BAND
 LOWER BAND EDGE      0.              0.15000000      0.42000000
 UPPER BAND EDGE      0.10000000      0.35000000      0.50000000
 DESIRED VALUE        1.00000000      0.              1.00000000
 WEIGHTING            1.00000000     50.00000000      1.00000000
 DEVIATION            0.14402014      0.00288040      0.14402014
 DEVIATION IN DB    -16.83153510    -50.81093502    -16.83153510

 EXTREMAL FREQUENCIES
    0.           0.0390625    0.0781250    0.1000000    0.1500000
    0.1578125    0.1753906    0.2007813    0.2261719    0.2554688
    0.2828125    0.3082031    0.3335938    0.3500000    0.4258594
    0.4629688    0.5000000

***********************************************************************
TIME=        1.6110156 SECONDS
```

Fig. 13. Output listing for an $N = 31$ bandstop filter.

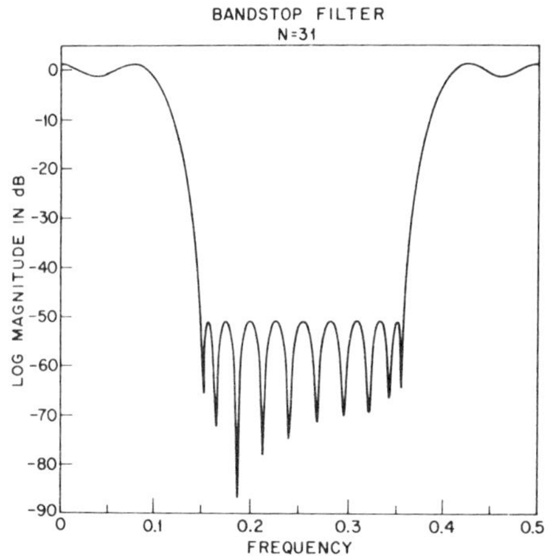


Fig. 14. Log magnitude response for an $N = 31$ bandstop filter.

gram, Figs. 15 and 16 show results for an $N = 55$ five-band filter with three stopbands and two passbands. The weighting in each of the stopbands is different, making the peak approximation error differ in each of these bands. A total of 3.81 s was required to design this filter.

Figs. 17–20 show typical examples of single band approximations to a differentiator and a Hilbert transformer. Figs. 17 and 18 show results for an $N = 32$ full band differentiator (this filter is the second example listed in the prologue to the Appendix), whereas Figs. 19 and 20 show results for an $N = 20$ Hilbert transformer where the upper cutoff frequency is 0.5 and the lower cutoff frequency is 0.05. The peak (relative) approximation error is 0.0062 for the differentiator and 0.02 for the Hilbert transformer. The design times for these two examples are 1.11 s for the differentiator and 0.48 s for the Hilbert transformer.

Finally, Figs. 21 and 22 show an example of an $N = 128$ bandpass filter with an arbitrary weighting function of the form

```
***************************************************************************

                     FINITE IMPULSE RESPONSE (FIR)
                     LINEAR PHASE DIGITAL FILTER DESIGN
                     REMEZ EXCHANGE ALGORITHM
                     BANDPASS FILTER
 FILTER LENGTH =  55
 ***** IMPULSE RESPONSE *****
         H(  1) =  0.10662552E-02 = H(  55)
         H(  2) =  0.63777615E-02 = H(  54)
         H(  3) =  0.35755610E-02 = H(  53)
         H(  4) = -0.30677355E-02 = H(  52)
         H(  5) = -0.90906379E-02 = H(  51)
         H(  6) =  0.23155629E-02 = H(  50)
         H(  7) =  0.39637965E-02 = H(  49)
         H(  8) =  0.11172050E-01 = H(  48)
         H(  9) =  0.11646759E-01 = H(  47)
         H( 10) = -0.33630784E-02 = H(  46)
         H( 11) = -0.92384245E-02 = H(  45)
         H( 12) = -0.20406392E-01 = H(  44)
         H( 13) = -0.19460483E-01 = H(  43)
         H( 14) =  0.31243013E-01 = H(  42)
         H( 15) =  0.63045567E-02 = H(  41)
         H( 16) = -0.20482303E-01 = H(  40)
         H( 17) =  0.65740513E-02 = H(  39)
         H( 18) = -0.11202127E-02 = H(  38)
         H( 19) =  0.41956985E-01 = H(  37)
         H( 20) =  0.35784266E-01 = H(  36)
         H( 21) =  0.34744802E-01 = H(  35)
         H( 22) =  0.71496359E-01 = H(  34)
         H( 23) = -0.17138831E 00 = H(  33)
         H( 24) = -0.18255044E 00 = H(  32)
         H( 25) =  0.74059024E-01 = H(  31)
         H( 26) = -0.10317421E 00 = H(  30)
         H( 27) =  0.25716721E-01 = H(  29)
         H( 28) =  0.37813547E 00 = H(  28)

                      BAND 1          BAND 2          BAND 3          BAND 4
LOWER BAND EDGE    0.              0.10000000      0.18000000      0.30000000
UPPER BAND EDGE    0.05000000      0.15000000      0.25000000      0.36000000
DESIRED VALUE      0.              1.00000000      0.              1.00000000
WEIGHTING         10.00000000      1.00000000      3.00000000      1.00000000
DEVIATION          0.00344486      0.03444859      0.01148286      0.03444859
DEVIATION IN DB  -49.25657034    -29.25657034    -38.79839549    -29.25657034

                      BAND 5          BAND
LOWER BAND EDGE    0.41000000
UPPER BAND EDGE    0.50000000
DESIRED VALUE      0.
WEIGHTING         20.00000000
DEVIATION          0.00172243
DEVIATION IN DB  -55.27717018

EXTREMAL FREQUENCIES
   0.            0.0167411    0.0323661    0.0446429    0.0500000
   0.1000000     0.1089286    0.1267857    0.1424107    0.1500000
   0.1800000     0.1855804    0.1978571    0.2134821    0.2302232
   0.2436160     0.2500000    0.3000000    0.3122768    0.3323661
   0.3502232     0.3600000    0.4100000    0.4155804    0.4289732
   0.4457143     0.4635714    0.4814285    0.5000000

***************************************************************************
TIME=      3.8164219 SECONDS
```

Fig. 15. Output listing for an $N = 55$ multiband filter.

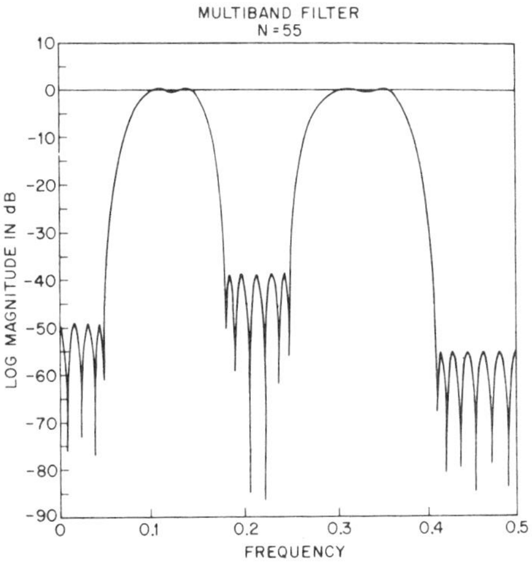


Fig. 16. Log magnitude response for an $N = 55$ multiband filter.

```
**********************************************************************
                 FINITE IMPULSE RESPONSE (FIR)
                 LINEAR PHASE DIGITAL FILTER DESIGN
                 REMEZ EXCHANGE ALGORITHM
                 DIFFERENTIATOR
         FILTER LENGTH =  32
         ***** IMPULSE RESPONSE *****
              H(  1) = -0.62713091E-03 = -H(  32)
              H(  2) =  0.85633433E-03 = -H(  31)
              H(  3) = -0.42418549E-03 = -H(  30)
              H(  4) =  0.39901518E-03 = -H(  29)
              H(  5) = -0.43437273E-03 = -H(  28)
              H(  6) =  0.49969450E-03 = -H(  27)
              H(  7) = -0.59634961E-03 = -H(  26)
              H(  8) =  0.73277031E-03 = -H(  25)
              H(  9) = -0.93002681E-03 = -H(  24)
              H( 10) =  0.12270042E-02 = -H(  23)
              H( 11) = -0.17012820E-02 = -H(  22)
              H( 12) =  0.25272341E-02 = -H(  21)
              H( 13) = -0.41601159E-02 = -H(  20)
              H( 14) =  0.81294555E-02 = -H(  19)
              H( 15) = -0.22539097E-01 = -H(  18)
              H( 16) =  0.20266535E 00 = -H(  17)

                      BAND 1          BAND
LOWER BAND EDGE       0.
UPPER BAND EDGE       0.50000000
DESIRED SLOPE         1.00000000
WEIGHTING             1.00000000
DEVIATION             0.00620231

EXTREMAL FREQUENCIES
    0.0019531    0.0332031    0.0664062    0.0996094    0.1328125
    0.1640625    0.1972656    0.2304687    0.2636719    0.2968750
    0.3300781    0.3632812    0.3945312    0.4277344    0.4589844
    0.4863281    0.5000000

**********************************************************************
TIME=        1.1072656 SECONDS
```

**Fig. 17. Output listing for an $N = 32$ differentiator.**

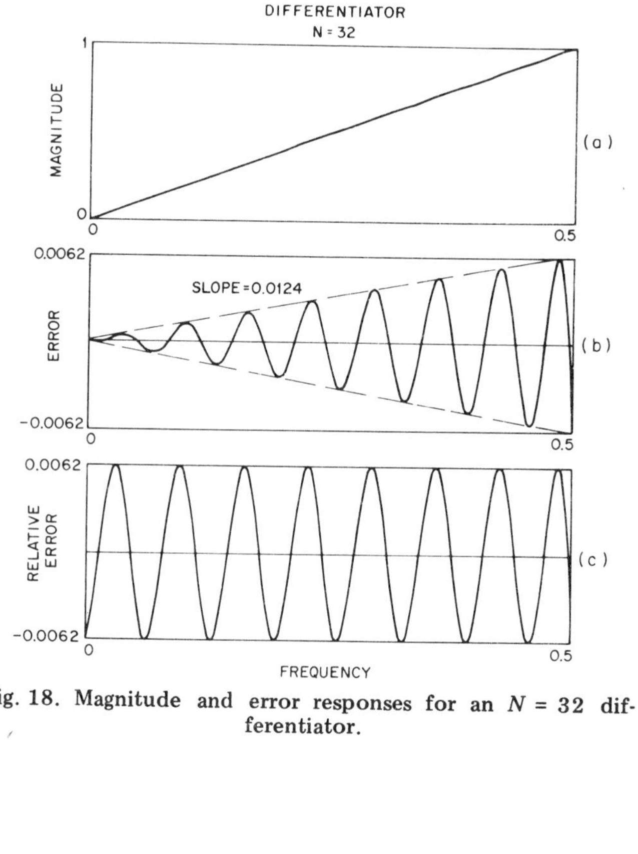


**Fig. 18. Magnitude and error responses for an $N = 32$ differentiator.**

```
**********************************************************************

                    FINITE IMPULSE RESPONSE (FIR)
                    LINEAR PHASE DIGITAL FILTER DESIGN
                    REMEZ EXCHANGE ALGORITHM
                    HILBERT TRANSFORMER
          FILTER LENGTH =  20
          ***** IMPULSE RESPONSE *****
              H(  1) =  0.10026189E-01 = -H(  20)
              H(  2) =  0.14173287E-01 = -H(  19)
              H(  3) =  0.20452437E-01 = -H(  18)
              H(  4) =  0.28736882E-01 = -H(  17)
              H(  5) =  0.39852580E-01 = -H(  16)
              H(  6) =  0.55333300E-01 = -H(  15)
              H(  7) =  0.78542752E-01 = -H(  14)
              H(  8) =  0.11823755E 00 = -H(  13)
              H(  9) =  0.20664125E 00 = -H(  12)
              H( 10) =  0.63475619E 00 = -H(  11)

                    BAND 1           BAND
 LOWER BAND EDGE    0.05000000
 UPPER BAND EDGE    0.50000000
 DESIRED VALUE      1.00000000
 WEIGHTING          1.00000000
 DEVIATION          0.02055604

 EXTREMAL FREQUENCIES
    0.0500000   0.0656250   0.1031250   0.1468750   0.1937500
    0.2437500   0.2937500   0.3468750   0.3968751   0.4500001
    0.5000000

**********************************************************************
TIME=        0.4825937 SECONDS
```

**Fig. 19. Output listing for an $N$ = 20 Hilbert transformer.**

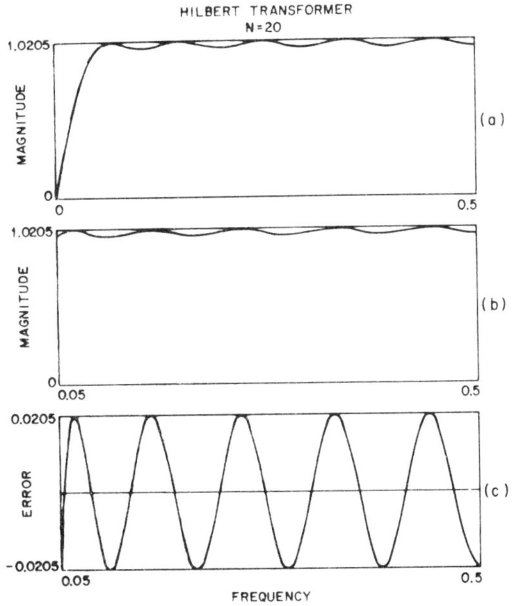


**Fig. 20. Magnitude and error responses for an $N$ = 20 Hilbert transformer.**

```
***********************************************************************

                 FINITE IMPULSE RESPONSE (FIR)
                 LINEAR PHASE DIGITAL FILTER DESIGN
                 REMEZ EXCHANGE ALGORITHM
                 BANDPASS FILTER
 FILTER LENGTH = 128
 ***** IMPULSE RESPONSE *****
      H(  1) = -0.20662533E-02 = H( 128)
      H(  2) =  0.68616867E-03 = H( 127)
      H(  3) =  0.48620260E-03 = H( 126)
      H(  4) =  0.67255029E-03 = H( 125)
      H(  5) =  0.93764407E-03 = H( 124)
      H(  6) =  0.68975566E-03 = H( 123)
      H(  7) = -0.15411817E-03 = H( 122)
      H(  8) = -0.97912618E-03 = H( 121)
      H(  9) = -0.10910514E-02 = H( 120)
      H( 10) = -0.35942515E-03 = H( 119)
      H( 11) =  0.74040653E-03 = H( 118)
      H( 12) =  0.15280502E-02 = H( 117)
      H( 13) =  0.15296961E-02 = H( 116)
      H( 14) =  0.72595198E-03 = H( 115)
      H( 15) = -0.40411612E-03 = H( 114)
      H( 16) = -0.11471390E-02 = H( 113)
      H( 17) = -0.10636777E-02 = H( 112)
      H( 18) = -0.33855668E-03 = H( 111)
      H( 19) =  0.39066916E-03 = H( 110)
      H( 20) =  0.59175241E-03 = H( 109)
      H( 21) =  0.29497579E-03 = H( 108)
      H( 22) =  0.14988618E-04 = H( 107)
      H( 23) =  0.28683373E-03 = H( 106)
      H( 24) =  0.10619103E-02 = H( 105)
      H( 25) =  0.15913830E-02 = H( 104)
      H( 26) =  0.93133046E-03 = H( 103)
      H( 27) = -0.11501053E-02 = H( 102)
      H( 28) = -0.35816184E-02 = H( 101)
      H( 29) = -0.44120705E-02 = H( 100)
      H( 30) = -0.21895665E-02 = H(  99)
      H( 31) =  0.26225995E-02 = H(  98)
      H( 32) =  0.73509685E-02 = H(  97)
      H( 33) =  0.84703702E-02 = H(  96)
      H( 34) =  0.40213218E-02 = H(  95)
      H( 35) = -0.44717526E-02 = H(  94)
      H( 36) = -0.12166497E-01 = H(  93)
      H( 37) = -0.13554154E-01 = H(  92)
      H( 38) = -0.62183138E-02 = H(  91)
      H( 39) =  0.67914337E-02 = H(  90)
      H( 40) =  0.17947235E-01 = H(  89)
      H( 41) =  0.19518513E-01 = H(  88)
      H( 42) =  0.87717158E-02 = H(  87)
      H( 43) = -0.93473921E-02 = H(  86)
      H( 44) = -0.24259875E-01 = H(  85)
      H( 45) = -0.25905771E-01 = H(  84)
      H( 46) = -0.11425838E-01 = H(  83)
      H( 47) =  0.12021431E-01 = H(  82)
      H( 48) =  0.30664841E-01 = H(  81)
      H( 49) =  0.32239440E-01 = H(  80)
      H( 50) =  0.14010583E-01 = H(  79)
      H( 51) = -0.14528283E-01 = H(  78)
      H( 52) = -0.36550244E-01 = H(  77)
      H( 53) = -0.37904738E-01 = H(  76)
      H( 54) = -0.16241339E-01 = H(  75)

      H( 55) =  0.16669269E-01 = H(  74)
      H( 56) =  0.41368275E-01 = H(  73)
      H( 57) =  0.42361176E-01 = H(  72)
      H( 58) =  0.17925027E-01 = H(  71)
      H( 59) = -0.18188843E-01 = H(  70)
      H( 60) = -0.44592980E-01 = H(  69)
      H( 61) = -0.45118319E-01 = H(  68)
      H( 62) = -0.18854116E-01 = H(  67)
      H( 63) =  0.18954474E-01 = H(  66)
      H( 64) =  0.45897002E-01 = H(  65)
```

**Fig. 21.** Output listing for an $N = 128$ bandpass filter with arbitrary weighting characteristics.

```
                     BAND 1          BAND 2          BAND 3          BAND 4
LOWER BAND EDGE     0.              0.12000000      0.15000000      0.25000000
UPPER BAND EDGE     0.10000000      0.13000000      0.25000000      0.50000000
DESIRED VALUE       0.              1.00000000      0.              0.
WEIGHTING          10.00000000      1.00000000     10.00000000     10.00000000
DEVIATION           0.00500134      0.05001341      0.00500134      0.00500134
DEVIATION IN DB   -46.01827145    -26.01827145    -46.01827145    -46.01827145

EXTREMAL FREQUENCIES
   0.            0.0102539    0.0200195    0.0288086    0.0371094
   0.0454102     0.0532227    0.0610352    0.0688477    0.0761719
   0.0834961     0.0898437    0.0952148    0.0986328    0.1000000
   0.1200000     0.1248828    0.1300000    0.1500000    0.1514648
   0.1548828     0.1602539    0.1666016    0.1734375    0.1807617
   0.1880859     0.1958984    0.2037109    0.2115234    0.2193359
   0.2271484     0.2349609    0.2427734    0.2509766    0.2592773
   0.2670898     0.2749023    0.2827148    0.2910156    0.2988281
   0.3066406     0.3144531    0.3222656    0.3305664    0.3383789
   0.3461914     0.3540039    0.3618164    0.3696289    0.3779297
   0.3857422     0.3935547    0.4013672    0.4091797    0.4169922
   0.4252930     0.4331055    0.4409180    0.4487305    0.4565430
   0.4643555     0.4721680    0.4804687    0.4882812    0.4960937

*****************************************************************************
TIME=      23.7952969 SECONDS
```

Fig. 21. (Continued.)

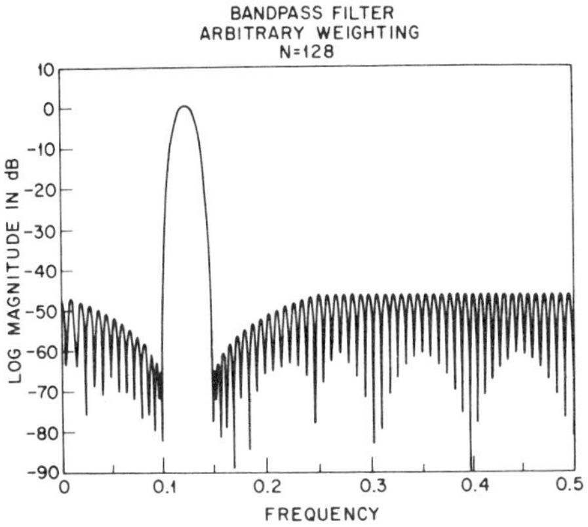


**Fig. 22. Log magnitude response for an $N = 128$ bandpass filter with arbitrary weighting characteristics.**

$$W(f) = \begin{cases} \dfrac{10}{1 - 9f} & 0 \leqslant f \leqslant 0.1 \\ 1 & 0.12 \leqslant f \leqslant 0.13 \\ \dfrac{10}{9f - 1.25} & 0.15 \leqslant f \leqslant 0.25 \\ 10 & 0.25 \leqslant f \leqslant 0.5. \end{cases}$$

Thus the tolerance scheme is linear in the intervals $0 \leqslant f \leqslant 0.1$ and $0.15 \leqslant f \leqslant 0.25$. The error at the stopband edges is 0.0005 (-66 dB), and the peak error increases linearly to 0.005 (-46 dB). The time required to design this filter was 23.8 s.

## Summary

A general-purpose linear phase FIR filter design program is presented which is capable of designing a wide variety of standard filters as well as any desired magnitude response which can be specified by the user. The speed of the algorithm, as well as its generality, make this program an attractive one for a wide variety of design applications.

## Appendix

```
C   PROGRAM FOR THE DESIGN OF LINEAR PHASE FINITE IMPULSE
C   RESPONSE (FIR) FILTERS USING THE REMEZ EXCHANGE ALGORITHM
C   JIM MCCLELLAN, RICE UNIVERSITY, APRIL 13, 1973
C   THREE TYPES OF FILTERS ARE INCLUDED--BANDPASS FILTERS
C   DIFFERENTIATORS, AND HILBERT TRANSFORM FILTERS
C
C   THE INPUT DATA CONSISTS OF 5 CARDS
C
C   CARD 1--FILTER LENGTH, TYPE OF FILTER.  1-MULTIPLE
C   PASSBAND/STOPBAND, 2-DIFFERENTIATOR, 3-HILBERT TRANSFORM
C   FILTER.  NUMBER OF BANDS, CARD PUNCH DESIRED, AND GRID
C   DENSITY.
C
C   CARD 2--BANDEDGES, LOWER AND UPPER EDGES FOR EACH BAND
C   WITH A MAXIMUM OF 10 BANDS.
C
C   CARD 3--DESIRED FUNCTION (OR DESIRED SLOPE IF A
C   DIFFERENTIATOR) FOR EACH BAND.
C
C   CARD 4--WEIGHT FUNCTION IN EACH BAND.  FOR A
C   DIFFERENTIATOR, THE WEIGHT FUNCTION IS INVERSELY
C   PROPORTIONAL TO F.
C
C   THE FOLLOWING INPUT DATA SPECIFIES A LENGTH 32 BANDPASS
C   FILTER WITH STOPBANDS 0 TO 0.1 AND 0.425 TO 0.5, AND
C   PASSBAND FROM 0.2 TO 0.35 WITH WEIGHTING OF 10 IN THE
C   STOPBANDS AND 1 IN THE PASSBAND.  THE IMPULSE RESPONSE
C   WILL BE PUNCHED AND THE GRID DENSITY IS 32.  THIS IS THE
C   FILTER IN FIGURES 9 AND 10 IN THE TEXT.
C   SAMPLE INPUT DATA SETUP
C   32,1,3,1,32
C   0,0.1,0.2,0.35,0.425,0.5
C   0,1,0
C   10,1,10
C
C   THE FOLLOWING INPUT DATA SPECIFIES A LENGTH 32 WIDEBAND
C   DIFFERENTIATOR WITH SLOPE 1 AND WEIGHTING OF 1/F.  THE
C   IMPULSE RESPONSE WILL NOT BE PUNCHED AND THE GRID
C   DENSITY IS ASSUMED TO BE 16.  THIS IS THE FILTER IN
C   FIGURES 17 AND 18 IN THE TEXT.
C   32,2,1,0,0
C   0,0.5
C   1.0
C   1.0
C
C
      COMMON PI2,AD,DEV,X,Y,GRID,DES,WT,ALPHA,IEXT,NFCNS,NGRID
      DIMENSION IEXT(66),AD(66),ALPHA(66),X(66),Y(66)
      DIMENSION H(66)
      DIMENSION DES(1045),GRID(1045),WT(1045)
      DIMENSION EDGE(20),FX(10),WTX(10),DEVIAT(10)
      DOUBLE PRECISION PI2,PI
      DOUBLE PRECISION AD,DEV,X,Y
      PI2=6.283185307179586
      PI=3.141592653589793
```

```
C
C  THE PROGRAM IS SET UP FOR A MAXIMUM LENGTH OF 128, BUT
C  THIS UPPER LIMIT CAN BE CHANGED BY REDIMENSIONING THE
C  ARRAYS IEXT, AD, ALPHA, X, Y, H TO BE NFMAX/2 + 2.
C  THE ARRAYS DES, GRID, AND WT MUST DIMENSIONED
C  16(NFMAX/2 + 2).
C
      NFMAX=128
  100 CONTINUE
      JTYPE=0
C
C  PROGRAM INPUT SECTION
C
      READ *,NFILT,JTYPE,NBANDS,JPUNCH,LGRID
      IF(NFILT.GT.NFMAX.OR.NFILT.LT.3) CALL ERROR
      IF(NBANDS.LE.0) NBANDS=1
C
C  GRID DENSITY IS ASSUMED TO BE 16 UNLESS SPECIFIED
C  OTHERWISE
C
      IF(LGRID.LE.0) LGRID=16
      JB=2*NBANDS
      READ *,(EDGE(J),J=1,JB)
      READ *,(FX(J),J=1,NBANDS)
      READ *,(WTX(J),J=1,NBANDS)
      IF(JTYPE.EQ.0) CALL ERROR
      NEG=1
      IF(JTYPE.EQ.1) NEG=0
      NODD=NFILT/2
      NODD=NFILT-2*NODD
      NFCNS=NFILT/2
      IF(NODD.EQ.1.AND.NEG.EQ.0) NFCNS=NFCNS+1
C
C  SET UP THE DENSE GRID.  THE NUMBER OF POINTS IN THE GRID
C  IS (FILTER LENGTH + 1)*GRID DENSITY/2
C
      GRID(1)=EDGE(1)
      DELF=LGRID*NFCNS
      DELF=0.5/DELF
      IF(NEG.EQ.0) GO TO 135
      IF(EDGE(1).LT.DELF) GRID(1)=DELF
  135 CONTINUE
      J=1
      L=1
      LBAND=1
  140 FUP=EDGE(L+1)
  145 TEMP=GRID(J)
C
C  CALCULATE THE DESIRED MAGNITUDE RESPONSE AND THE WEIGHT
C  FUNCTION ON THE GRID
C
      DES(J)=EFF(TEMP,FX,WTX,LBAND,JTYPE)
      WT(J)=WATE(TEMP,FX,WTX,LBAND,JTYPE)
      J=J+1
      GRID(J)=TEMP+DELF
      IF(GRID(J).GT.FUP) GO TO 150
      GO TO 145
  150 GRID(J-1)=FUP
      DES(J-1)=EFF(FUP,FX,WTX,LBAND,JTYPE)
      WT(J-1)=WATE(FUP,FX,WTX,LBAND,JTYPE)
      LBAND=LBAND+1
      L=L+2
      IF(LBAND.GT.NBANDS) GO TO 160
      GRID(J)=EDGE(L)
      GO TO 140
  160 NGRID=J-1
      IF(NEG.NE.NODD) GO TO 165
      IF(GRID(NGRID).GT.(0.5-DELF)) NGRID=NGRID-1
  165 CONTINUE
C
C  SET UP A NEW APPROXIMATION PROBLEM WHICH IS EQUIVALENT
C  TO THE ORIGINAL PROBLEM
C
      IF(NEG) 170,170,180
  170 IF(NODD.EQ.1) GO TO 200
      DO 175 J=1,NGRID
      CHANGE=DCOS(PI*GRID(J))
      DES(J)=DES(J)/CHANGE
  175 WT(J)=WT(J)*CHANGE
      GO TO 200
  180 IF(NODD.EQ.1) GO TO 190
      DO 185 J=1,NGRID
      CHANGE=DSIN(PI*GRID(J))
      DES(J)=DES(J)/CHANGE
  185 WT(J)=WT(J)*CHANGE
      GO TO 200
  190 DO 195 J=1,NGRID
      CHANGE=DSIN(PI2*GRID(J))
      DES(J)=DES(J)/CHANGE
  195 WT(J)=WT(J)*CHANGE
C
C  INITIAL GUESS FOR THE EXTREMAL FREQUENCIES--EQUALLY
C  SPACED ALONG THE GRID
C
  200 TEMP=FLOAT(NGRID-1)/FLOAT(NFCNS)
      DO 210 J=1,NFCNS
  210 IEXT(J)=(J-1)*TEMP+1
      IEXT(NFCNS+1)=NGRID
      NM1=NFCNS-1
      NZ=NFCNS+1
C
C  CALL THE REMEZ EXCHANGE ALGORITHM TO DO THE APPROXIMATION
C  PROBLEM
C
      CALL REMEZ(EDGE,NBANDS)
C
C  CALCULATE THE IMPULSE RESPONSE.
C
      IF(NEG) 300,300,320
  300 IF(NODD.EQ.0) GO TO 310
      DO 305 J=1,NM1
  305 H(J)=0.5*ALPHA(NZ-J)
      H(NFCNS)=ALPHA(1)
      GO TO 350
  310 H(1)=0.25*ALPHA(NFCNS)
      DO 315 J=2,NM1
  315 H(J)=0.25*(ALPHA(NZ-J)+ALPHA(NFCNS+2-J))
      H(NFCNS)=0.5*ALPHA(1)+0.25*ALPHA(2)
      GO TO 350
  320 IF(NODD.EQ.0) GO TO 330
      H(1)=0.25*ALPHA(NFCNS)
      H(2)=0.25*ALPHA(NM1)
      DO 325 J=3,NM1
  325 H(J)=0.25*(ALPHA(NZ-J)-ALPHA(NFCNS+3-J))
      H(NFCNS)=0.5*ALPHA(1)-0.25*ALPHA(3)
      H(NZ)=0.0
      GO TO 350
  330 H(1)=0.25*ALPHA(NFCNS)
      DO 335 J=2,NM1
  335 H(J)=0.25*(ALPHA(NZ-J)-ALPHA(NFCNS+2-J))
      H(NFCNS)=0.5*ALPHA(1)-0.25*ALPHA(2)
C
C  PROGRAM OUTPUT SECTION.
C
  350 PRINT 360
  360 FORMAT(1H1, 70(1H*)//25X,'FINITE IMPULSE RESPONSE (FIR)'/
     125X,'LINEAR PHASE DIGITAL FILTER DESIGN'/
     225X,'REMEZ EXCHANGE ALGORITHM'/)
      IF(JTYPE.EQ.1) PRINT 365
  365 FORMAT(25X,'BANDPASS FILTER'/)
      IF(JTYPE.EQ.2) PRINT 370
  370 FORMAT(25X,'DIFFERENTIATOR'/)
      IF(JTYPE.EQ.3) PRINT 375
  375 FORMAT(25X,'HILBERT TRANSFORMER'/)
      PRINT 378,NFILT
  378 FORMAT(15X,'FILTER LENGTH = ',I3/)
      PRINT 380
  380 FORMAT(15X,'***** IMPULSE RESPONSE *****')
      DO 381 J=1,NFCNS
      K=NFILT+1-J
      IF(NEG.EQ.0) PRINT 382,J,H(J),K
      IF(NEG.EQ.1) PRINT 383,J,H(J),K
  381 CONTINUE
  382 FORMAT(20X,'H(',I3,') = ',E15.8,' = H(',I4,')')
  383 FORMAT(20X,'H(',I3,') = ',E15.8,' = -H(',I4,')')
      IF(NEG.EQ.1.AND.NODD.EQ.1) PRINT 384,NZ
  384 FORMAT(20X,'H(',I3,') =  0.0')
      DO 450 K=1,NBANDS,4
      KUP=K+3
      IF(KUP.GT.NBANDS) KUP=NBANDS
      PRINT 385,(J,J=K,KUP)
  385 FORMAT(/24X,4('BAND',I3,8X))
      PRINT 390,(EDGE(2*J-1),J=K,KUP)
  390 FORMAT(2X,'LOWER BAND EDGE',5F15.9)
      PRINT 395,(EDGE(2*J),J=K,KUP)
  395 FORMAT(2X,'UPPER BAND EDGE',5F15.9)
      IF(JTYPE.NE.2) PRINT 400,(FX(J),J=K,KUP)
  400 FORMAT(2X,'DESIRED VALUE',2X,5F15.9)
      IF(JTYPE.EQ.2) PRINT 405,(FX(J),J=K,KUP)
  405 FORMAT(2X,'DESIRED SLOPE',2X,5F15.9)
      PRINT 410,(WTX(J),J=K,KUP)
  410 FORMAT(2X,'WEIGHTING',6X,5F15.9)
      DO 420 J=K,KUP
  420 DEVIAT(J)=DEV/WTX(J)
      PRINT 425,(DEVIAT(J),J=K,KUP)
  425 FORMAT(2X,'DEVIATION',6X,5F15.9)
      IF(JTYPE.NE.1) GO TO 450
      DO 430 J=K,KUP
  430 DEVIAT(J)=20.0*ALOG10(DEVIAT(J))
      PRINT 435,(DEVIAT(J),J=K,KUP)
  435 FORMAT(2X,'DEVIATION IN DB',5F15.9)
  450 CONTINUE
      PRINT 455,(GRID(IEXT(J)),J=1,NZ)
  455 FORMAT(/2X,'EXTREMAL FREQUENCIES'/(2X,5F12.7))
      PRINT 460
  460 FORMAT(/1X,70(1H*)/1H1)
      IF(JPUNCH.NE.0) PUNCH *,(H(J),J=1,NFCNS)
      IF(NFILT.NE.0) GO TO 100
      RETURN
      END

      FUNCTION EFF(TEMP,FX,WTX,LBAND,JTYPE)
C
C  FUNCTION TO CALCULATE THE DESIRED MAGNITUDE RESPONSE
C  AS A FUNCTION OF FREQUENCY.
C
      DIMENSION FX(5),WTX(5)
      IF(JTYPE.EQ.2) GO TO 1
      EFF=FX(LBAND)
      RETURN
```

```
    1 EFF=FX(LBAND)*TEMP
      RETURN
      END



      FUNCTION WATE(TEMP,FX,WTX,LBAND,JTYPE)
C
C  FUNCTION TO CALCULATE THE WEIGHT FUNCTION AS A FUNCTION
C  OF FREQUENCY.
C
      DIMENSION FX(5),WTX(5)
      IF(JTYPE.EQ.2) GO TO 1
      WATE=WTX(LBAND)
      RETURN
    1 IF(FX(LBAND).LT.0.0001) GO TO 2
      WATE=WTX(LBAND)/TEMP
      RETURN
    2 WATE=WTX(LBAND)
      RETURN
      END



      SUBROUTINE ERROR
      PRINT 1
    1 FORMAT(' ************ ERROR IN INPUT DATA **********')
      STOP
      END



      SUBROUTINE REMEZ(EDGE,NBANDS)
C
C  THIS SUBROUTINE IMPLEMENTS THE REMEZ EXCHANGE ALGORITHM
C  FOR THE WEIGHTED CHEBYCHEV APPROXIMATION OF A CONTINUOUS
C  FUNCTION WITH A SUM OF COSINES.  INPUTS TO THE SUBROUTINE
C  ARE A DENSE GRID WHICH REPLACES THE FREQUENCY AXIS, THE
C  DESIRED FUNCTION ON THIS GRID, THE WEIGHT FUNCTION ON THE
C  GRID, THE NUMBER OF COSINES, AND AN INITIAL GUESS OF THE
C  EXTREMAL FREQUENCIES.  THE PROGRAM MINIMIZES THE CHEBYCHEV
C  ERROR BY DETERMINING THE BEST LOCATION OF THE EXTREMAL
C  FREQUENCIES (POINTS OF MAXIMUM ERROR) AND THEN CALCULATES
C  THE COEFFICIENTS OF THE BEST APPROXIMATION.
C
      COMMON PI2,AD,DEV,X,Y,GRID,DES,WT,ALPHA,IEXT,NFCNS,NGRID
      DIMENSION EDGE(20)
      DIMENSION IEXT(66),AD(66),ALPHA(66),X(66),Y(66)
      DIMENSION DES(1045),GRID(1045),WT(1045)
      DIMENSION A(66),P(65),Q(65)
      DOUBLE PRECISION PI2,DNUM,DDEN,DTEMP,A,P,Q
      DOUBLE PRECISION AD,DEV,X,Y
C
C  THE PROGRAM ALLOWS A MAXIMUM NUMBER OF ITERATIONS OF 25
C
      ITRMAX=25
      DEVL=-1.0
      NZ=NFCNS+1
      NZZ=NFCNS+2
      NITER=0
  100 CONTINUE
      IEXT(NZZ)=NGRID+1
      NITER=NITER+1
      IF(NITER.GT.ITRMAX) GO TO 400
      DO 110 J=1,NZ
      DTEMP=GRID(IEXT(J))
      DTEMP=DCOS(DTEMP*PI2)
  110 X(J)=DTEMP
      JET=(NFCNS-1)/15+1
      DO 120 J=1,NZ
  120 AD(J)=D(J,NZ,JET)
      DNUM=0.0
      DDEN=0.0
      K=1
      DO 130 J=1,NZ
      L=IEXT(J)
      DTEMP=AD(J)*DES(L)
      DNUM=DNUM+DTEMP
      DTEMP=K*AD(J)/WT(L)
      DDEN=DDEN+DTEMP
  130 K=-K
      DEV=DNUM/DDEN
      NU=1
      IF(DEV.GT.0.0) NU=-1
      DEV=-NU*DEV
      K=NU
      DO 140 J=1,NZ
      L=IEXT(J)
      DTEMP=K*DEV/WT(L)
      Y(J)=DES(L)+DTEMP
  140 K=-K
      IF(DEV.GE.DEVL) GO TO 150
      CALL OUCH
      GO TO 400
  150 DEVL=DEV
      JCHNGE=0
      K1=IEXT(1)
      KNZ=IEXT(NZ)
      KLOW=0
      NUT=-NU
      J=1
C
C  SEARCH FOR THE EXTREMAL FREQUENCIES OF THE BEST
C  APPROXIMATION
C
  200 IF(J.EQ.NZZ) YNZ=COMP
      IF(J.GE.NZZ) GO TO 300
      KUP=IEXT(J+1)
      L=IEXT(J)+1
      NUT=-NUT
      IF(J.EQ.2) Y1=COMP
      COMP=DEV
      IF(L.GE.KUP) GO TO 220
      ERR=GEE(L,NZ)
      ERR=(ERR-DES(L))*WT(L)
      DTEMP=NUT*ERR-COMP
      IF(DTEMP.LE.0.0) GO TO 220
      COMP=NUT*ERR
  210 L=L+1
      IF(L.GE.KUP) GO TO 215
      ERR=GEE(L,NZ)
      ERR=(ERR-DES(L))*WT(L)
      DTEMP=NUT*ERR-COMP
      IF(DTEMP.LE.0.0) GO TO 215
      COMP=NUT*ERR
      GO TO 210
  215 IEXT(J)=L-1
      J=J+1
      KLOW=L-1
      JCHNGE=JCHNGE+1
      GO TO 200
  220 L=L-1
  225 L=L-1
      IF(L.LE.KLOW) GO TO 250
      ERR=GEE(L,NZ)
      ERR=(ERR-DES(L))*WT(L)
      DTEMP=NUT*ERR-COMP
      IF(DTEMP.GT.0.0) GO TO 230
      IF(JCHNGE.LE.0) GO TO 225
      GO TO 260
  230 COMP=NUT*ERR
  235 L=L-1
      IF(L.LE.KLOW) GO TO 240
      ERR=GEE(L,NZ)
      ERR=(ERR-DES(L))*WT(L)
      DTEMP=NUT*ERR-COMP
      IF(DTEMP.LE.0.0) GO TO 240
      COMP=NUT*ERR
      GO TO 235
  240 KLOW=IEXT(J)
      IEXT(J)=L+1
      J=J+1
      JCHNGE=JCHNGE+1
      GO TO 200
  250 L=IEXT(J)+1
      IF(JCHNGE.GT.0) GO TO 215
  255 L=L+1
      IF(L.GE.KUP) GO TO 260
      ERR=GEE(L,NZ)
      ERR=(ERR-DES(L))*WT(L)
      DTEMP=NUT*ERR-COMP
      IF(DTEMP.LE.0.0) GO TO 255
      COMP=NUT*ERR
      GO TO 210
  260 KLOW=IEXT(J)
      J=J+1
      GO TO 200
  300 IF(J.GT.NZZ) GO TO 320
      IF(K1.GT.IEXT(1)) K1=IEXT(1)
      IF(KNZ.LT.IEXT(NZ)) KNZ=IEXT(NZ)
      NUT1=NUT
      NUT=-NU
      L=0
      KUP=K1
      COMP=YNZ*(1.00001)
      LUCK=1
  310 L=L+1
      IF(L.GE.KUP) GO TO 315
      ERR=GEE(L,NZ)
      ERR=(ERR-DES(L))*WT(L)
      DTEMP=NUT*ERR-COMP
      IF(DTEMP.LE.0.0) GO TO 310
      COMP=NUT*ERR
      J=NZZ
      GO TO 210
  315 LUCK=6
      GO TO 325
  320 IF(LUCK.GT.9) GO TO 350
      IF(COMP.GT.Y1) Y1=COMP
      K1=IEXT(NZZ)
  325 L=NGRID+1
      KLOW=KNZ
      NUT=-NUT1
      COMP=Y1*(1.00001)
  330 L=L-1
      IF(L.LE.KLOW) GO TO 340
      ERR=GEE(L,NZ)
      ERR=(ERR-DES(L))*WT(L)
      DTEMP=NUT*ERR-COMP
      IF(DTEMP.LE.0.0) GO TO 330
      J=NZZ
      COMP=NUT*ERR
      LUCK=LUCK+10
      GO TO 235
```

```
  340 IF(LUCK.EQ.6) GO TO 370
      DO 345 J=1,NFCNS
  345 IEXT(NZZ-J)=IEXT(NZ-J)
      IEXT(1)=K1
      GO TO 100
  350 KN=IEXT(NZZ)
      DO 360 J=1,NFCNS
  360 IEXT(J)=IEXT(J+1)
      IEXT(NZ)=KN
      GO TO 100
  370 IF(JCHNGE.GT.0) GO TO 100


C  CALCULATION OF THE COEFFICIENTS OF THE BEST APPROXIMATION
C  USING THE INVERSE DISCRETE FOURIER TRANSFORM
C
  400 CONTINUE
      NM1=NFCNS-1
      FSH=1.0E-06
      GTEMP=GRID(1)
      X(NZZ)=-2.0
      CN=2*NFCNS-1
      DELF=1.0/CN
      L=1
      KKK=0
      IF(EDGE(1).EQ.0.0.AND.EDGE(2*NBANDS).EQ.0.5) KKK=1
      IF(NFCNS.LE.3) KKK=1
      IF(KKK.EQ.1) GO TO 405
      DTEMP=DCOS(PI2*GRID(1))
      DNUM=DCOS(PI2*GRID(NGRID))
      AA=2.0/(DTEMP-DNUM)
      BB=-(DTEMP+DNUM)/(DTEMP-DNUM)
  405 CONTINUE
      DO 430 J=1,NFCNS
      FT=(J-1)*DELF
      XT=DCOS(PI2*FT)
      IF(KKK.EQ.1) GO TO 410
      XT=(XT-BB)/AA
      FT=ARCOS(XT)/PI2
  410 XE=X(L)
      IF(XT.GT.XE) GO TO 420
      IF((XE-XT).LT.FSH) GO TO 415
      L=L+1
      GO TO 410
  415 A(J)=Y(L)
      GO TO 425
  420 IF((XT-XE).LT.FSH) GO TO 415
      GRID(1)=FT
      A(J)=GEE(1,NZ)
  425 CONTINUE
      IF(L.GT.1) L=L-1
  430 CONTINUE
      GRID(1)=GTEMP
      DDEN=PI2/CN
      DO 510 J=1,NFCNS
      DTEMP=0.0
      DNUM=(J-1)*DDEN
      IF(NM1.LT.1) GO TO 505
      DO 500 K=1,NM1
  500 DTEMP=DTEMP+A(K+1)*DCOS(DNUM*K)
  505 DTEMP=2.0*DTEMP+A(1)
  510 ALPHA(J)=DTEMP
      DO 550 J=2,NFCNS
  550 ALPHA(J)=2*ALPHA(J)/CN
      ALPHA(1)=ALPHA(1)/CN
      IF(KKK.EQ.1) GO TO 545
      P(1)=2.0*ALPHA(NFCNS)*BB+ALPHA(NM1)
      P(2)=2.0*AA*ALPHA(NFCNS)
      Q(1)=ALPHA(NFCNS-2)-ALPHA(NFCNS)
      DO 540 J=2,NM1
      IF(J.LT.NM1) GO TO 515
      AA=0.5*AA
      BB=0.5*BB
  515 CONTINUE
      P(J+1)=0.0
      DO 520 K=1,J
      A(K)=P(K)
  520 P(K)=2.0*BB*A(K)
      P(2)=P(2)+A(1)*2.0*AA
      JM1=J-1
      DO 525 K=1,JM1
  525 P(K)=P(K)+Q(K)+AA*A(K+1)
      JP1=J+1
      DO 530 K=3,JP1
  530 P(K)=P(K)+AA*A(K-1)
      IF(J.EQ.NM1) GO TO 540
      DO 535 K=1,J
  535 Q(K)=-A(K)
      Q(1)=Q(1)+ALPHA(NFCNS-1-J)
  540 CONTINUE
      DO 543 J=1,NFCNS
  543 ALPHA(J)=P(J)
  545 CONTINUE
      IF(NFCNS.GT.3) RETURN
      ALPHA(NFCNS+1)=0.0
      ALPHA(NFCNS+2)=0.0
      RETURN
      END


      DOUBLE PRECISION FUNCTION D(K,N,M)
C
C  FUNCTION TO CALCULATE THE LAGRANGE INTERPOLATION
C  COEFFICIENTS FOR USE IN THE FUNCTION GEE.
C
      COMMON PI2,AD,DEV,X,Y,GRID,DES,WT,ALPHA,IEXT,NFCNS,NGRID
      DIMENSION IEXT(66),AD(66),ALPHA(66),X(66),Y(66)
      DIMENSION DES(1045),GRID(1045),WT(1045)
      DOUBLE PRECISION AD,DEV,X,Y
      DOUBLE PRECISION Q
      DOUBLE PRECISION PI2
      D=1.0
      Q=X(K)
      DO 3 L=1,M
      DO 2 J=L,N,M
      IF(J-K)1,2,1
    1 D=2.0*D*(Q-X(J))
    2 CONTINUE
    3 CONTINUE
      D=1.0/D
      RETURN
      END


      DOUBLE PRECISION FUNCTION GEE(K,N)
C
C  FUNCTION TO EVALUATE THE FREQUENCY RESPONSE USING THE
C  LAGRANGE INTERPOLATION FORMULA IN THE BARYCENTRIC FORM
C
      COMMON PI2,AD,DEV,X,Y,GRID,DES,WT,ALPHA,IEXT,NFCNS,NGRID
      DIMENSION IEXT(66),AD(66),ALPHA(66),X(66),Y(66)
      DIMENSION DES(1045),GRID(1045),WT(1045)
      DOUBLE PRECISION P,C,D,XF
      DOUBLE PRECISION PI2
      DOUBLE PRECISION AD,DEV,X,Y
      P=0.0
      XF=GRID(K)
      XF=DCOS(PI2*XF)
      D=0.0
      DO 1 J=1,N
      C=XF-X(J)
      C=AD(J)/C
      D=D+C
    1 P=P+C*Y(J)
      GEE=P/D
      RETURN
      END


      SUBROUTINE OUCH
      PRINT 1
    1 FORMAT(' ************ FAILURE TO CONVERGE **********'/
     1'0PROBABLE CAUSE IS MACHINE ROUNDING ERROR'/
     2'0THE IMPULSE RESPONSE MAY BE CORRECT'/
     3'0CHECK WITH A FREQUENCY RESPONSE')
      RETURN
      END
```

## References


[1] J. H. McClellan and T. W. Parks, "A unified approach to the design of optimum FIR linear phase digital filters," *IEEE Trans. Circuit Theory*, vol. CT-20, pp. 697–701, Nov. 1973.

[2] E. W. Cheney, *Introduction to Approximation Theory*. New York: McGraw-Hill, 1966, pp. 72–100.

[3] E. Ya. Remez, "General computational methods of Tchebycheff approximation," Kiev, 1957 (Atomic Energy Translation 4491, pp. 1–85).

[4] B. Gold and K. L. Jordan, "A direct search procedure for designing finite duration impulse response filters," *IEEE Trans. Audio Electroacoust.*, vol. AU-17, pp. 33–36, Mar. 1969.

[5] L. R. Rabiner, B. Gold, and C. McGonegal, "An approach to the approximation problem for nonrecursive digital filters," *IEEE Trans. Audio Electroacoust.*, vol. AU-18, pp. 83–106, June 1970.

[6] L. R. Rabiner and K. Steiglitz, "The design of wideband recursive and nonrecursive digital differentiators," *IEEE Trans. Audio Electroacoust.*, vol. AU-18, pp. 204–209, June 1970.

[7] O. Herrmann, "Design of nonrecursive digital filters with linear phase," *Electron. Lett.*, pp. 328–329, 1970.

[8] E. Hofstetter, A. V. Oppenheim, and J. Siegel, "A new technique for the design of nonrecursive digital filters," in *Proc. 5th Annu. Princeton Conf. Inform. Sci. and Syst.*, Mar. 1971, pp. 64–72.

[9] T. W. Parks and J. H. McClellan, "Chebyshev approximation for nonrecursive digital filters with linear phase," *IEEE Trans. Circuit Theory*, vol. CT-19, pp. 189–194, Mar. 1972.

[10] L. R. Rabiner, "The design of finite impulse response

digital filters using linear programming techniques," *Bell Syst. Tech. J.*, vol. 51, pp. 1177-1198, July-Aug. 1972.

[11] T. W. Parks and J. H. McClellan, "A program for the design of linear phase finite impulse response digital filters," *IEEE Trans. Audio Electroacoust.*, vol. AU-20, pp. 195-199, Aug. 1972.

[12] O. Herrmann, "Transversal filters for the Hilbert Transformation," *Arch. Elek. Übertragung.*, vol. 23, pp. 581-587, 1969.

[13] J. H. McClellan, "On the design of one-dimensional and two-dimensional FIR digital filters," Ph.D. dissertation, Rice Univ., Houston, Tex., Apr. 1973.

# Approximate Design Relationships for Low-Pass FIR Digital Filters

LAWRENCE R. RABINER

*Abstract*—In this paper, a set of simple, approximate relationships between FIR, linear phase, low-pass filter parameters is presented. Based on these relationships, it is shown how an existing, readily available, filter design program can be used to efficiently design low-pass filters that meet or exceed arbitrary input specifications.

## Introduction

Although a great deal has been learned about how to design linear phase FIR low-pass digital filters [1]–[5] the relationships between filter parameters is not yet fully understood quantitatively. Fig. 1 shows the frequency response of a low-pass filter with passband cutoff frequency $F_p$; stopband cutoff frequency $F_s$; passband deviation $\delta_1$; and stopband deviation $\delta_2$. Along with the preceding four parameters, an FIR low-pass filter is specified by $N$, the filter impulse response duration in samples. Additional useful low-pass filter parameters are transition width $\Delta F$ and deviation ratio $K$, defined as

$$\Delta F = F_s - F_p \tag{1}$$

$$K = \delta_1/\delta_2. \tag{2}$$

An optimal low-pass filter (in the Chebyshev sense) is one for which $N$, $F_p$, $F_s$, and $K$ are specified, and $\delta_1$ and $\delta_2$ are minimized. Parks and McClellan [4] have given the necessary and sufficient conditions for such an optimal filter to exist and be unique. Furthermore, they have presented in great detail a design program [6] which on input accepts values of $N$, $F_p$, $F_s$, and $K$, and then designs an optimum filter with the minimum values of $\delta_1$ and $\delta_2$. There is little flexibility in the algorithm in that the user cannot arbitrarily choose *any* four of the five filter parameters, but must instead choose the ones listed previously. This constraint on the design program led Herrmann *et al.* [7] to investigate the relationships between filter parameters so as to be able to predict the required value of $N$ to meet given values of $F_p$, $F_s$, $\delta_1$, and $\delta_2$. In this paper the design relationships are extended so as to interface with the Parks-McClellan algorithm in an iterative, but straight forward manner so as to allow the filter designer the option of choosing any four of the five low-pass parameters and having the resulting filter meet or exceed the given specifications.

Manuscript received April 26, 1973; revised June 4, 1973.
The author is with Bell Laboratories, Murray Hill, N.J. 07974.

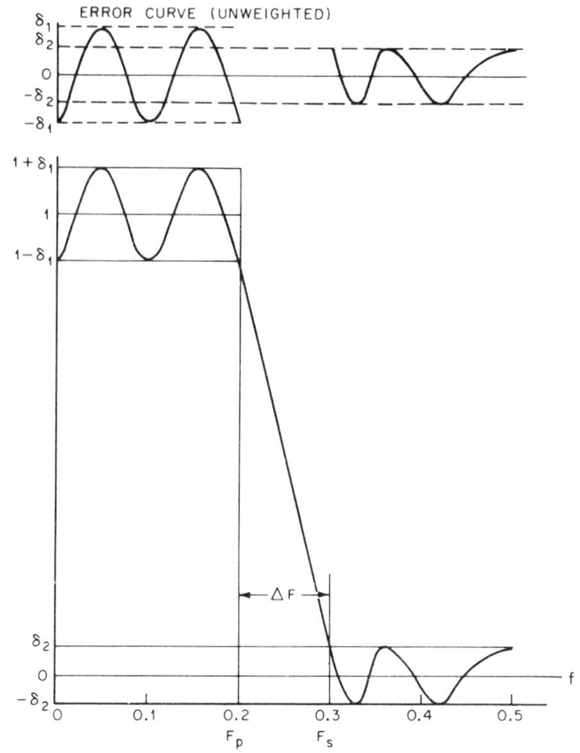


**Fig. 1.** The frequency response and error curve of an FIR low-pass filter defining, filter parameters $F_p$, $F_s$, $\Delta F$, $\delta_1$, and $\delta_2$.

## FIR Design Relationships

Based on measurements on an extensive set of optimal, linear phase, low-pass filters, Herrmann *et al.* empirically determined the relationship

$$D = D_\infty(\delta_1, \delta_2) - f(\delta_1, \delta_2)(\Delta F)^2 \tag{3}$$

where

$$D = (N - 1)\Delta F \tag{4}$$

and

$$D_\infty(\delta_1, \delta_2) = [a_1(\log_{10}\delta_1)^2 + a_2 \log_{10}\delta_1 + a_3] \cdot \log_{10}\delta_2 + [a_4(\log_{10}\delta_1)^2 + a_5 \log_{10}\delta_1 + a_6] \tag{5}$$

with

$$a_1 = 5.309 \times 10^{-3}$$
$$a_2 = 7.114 \times 10^{-2}$$
$$a_3 = -4.761 \times 10^{-1}$$
$$a_4 = -2.66 \times 10^{-3}$$
$$a_5 = -5.941 \times 10^{-1}$$
$$a_6 = -4.278 \times 10^{-1}$$

and

$$f(\delta_1, \delta_2) = b_1 + b_2 \log_{10}\delta_1 - b_2 \log_{10}\delta_2 \tag{6}$$

Reprinted from *IEEE Trans. Audio Electroacoust.*, vol. AU-21, pp. 456–460, Oct. 1973.

with

$$b_1 = 11.01217$$
$$b_2 = 0.51244.$$

The coefficients in (5) and (6) were determined by a minimum mean-square error fitting procedure to the data, whereas the forms for (5) and (6) were suggested by some simple data fitting procedures.

We now consider the original problem—that is, given any four of the five parameters $N, F_p, F_s, \delta_1, \delta_2$, how can the unspecified parameter be estimated and then adjusted using feedback from the Parks-McClellan algorithm, until specifications on all parameters are met or exceeded. We now consider the five possibilities for the unspecified parameter.

*Case 1:* $F_p, F_s, \delta_1, \delta_2$ specified—$N$ unspecified. In this case (3) and (4) may be used to give $\hat{N}$, the estimate of $N$, as

$$\hat{N} = \frac{D_\infty(\delta_1, \delta_2)}{\Delta F} - f(\delta_1, \delta_2)(\Delta F)^2 + 1. \tag{7}$$

Fig. 2 shows the logic required to obtain the actual value of $N$ that is required. After estimating $\hat{N}$ from (7), the direction parameter *JD* is initialized to 0. The parameters $\hat{N}$, $F_p$, $F_s$, and $K = \delta_1/\delta_2$ are used as input to the optimal design algorithm that returns the value $\hat{\delta}_2$ as the actual deviation in the stopband. This value is compared with $\delta_2$ and if they are equal (to within some tolerance) the algorithm is done. If $\hat{\delta}_2 > \delta_2$ then $\hat{N}$ is incremented by 2 (i.e., one filter order) and a check is made to see if the direction parameter *JD* was -1, indicating that $\hat{N}$ had previously been decreasing. If so, the new value of $\hat{N}$ is the smallest $N$ that meets or exceeds specifications and the algorithm is done. If not, the value of *JD* is set to 1 and the updated value of $\hat{N}$ is used as input to the optimal design algorithm. A similar path is taken if $\hat{\delta}_2 < \delta_2$ whereby if *JD* was 1, the current value of $\hat{N}$ is the minimum value of $N$. Otherwise $N$ is decreased by 2 and *JD* set to -1 and the algorithm repeats.

It has been found that two to three iterations suffice for a large range of filter parameters.

*Case 2:* $N, F_s, \delta_1$, and $\delta_2$ specified—$F_p$ unspecified. In this case (3) and (4) are used to give $\hat{F}_p$, the estimate of $F_p$, as

$$\Delta\hat{F} = F_s - \hat{F}_p = \frac{(N-1)}{2f(\delta_1, \delta_2)}, \left(\sqrt{1 + \frac{4f(\delta_1,\delta_2)D_\infty(\delta_1,\delta_2)}{(N-1)^2}} - 1\right) \tag{8}$$

or

$$\hat{F}_p = F_s - \Delta\hat{F}. \tag{9}$$

Fig. 3 shows a flow chart of the logic required to obtain the correct value of $F_p$. First $\hat{F}_p$ is calculated and the frequency step size (FST) is set to $\Delta\hat{F}/10$, and the direction parameter *JD* is set to 0. The initial value $\hat{F}_p$ along with $N$, $F_s$, and $K$ are the inputs to the design algorithm that returns the value $\hat{\delta}_2$ as the actual stopband deviation. If $\hat{\delta}_2 = \delta_2$ to within some specified tolerance (typically 1 percent or 0.1 percent), the algorithm is done. If $\hat{\delta}_2 > \delta_2$, and the direction parameter *JD* is -1, the step size is divided by 3 to

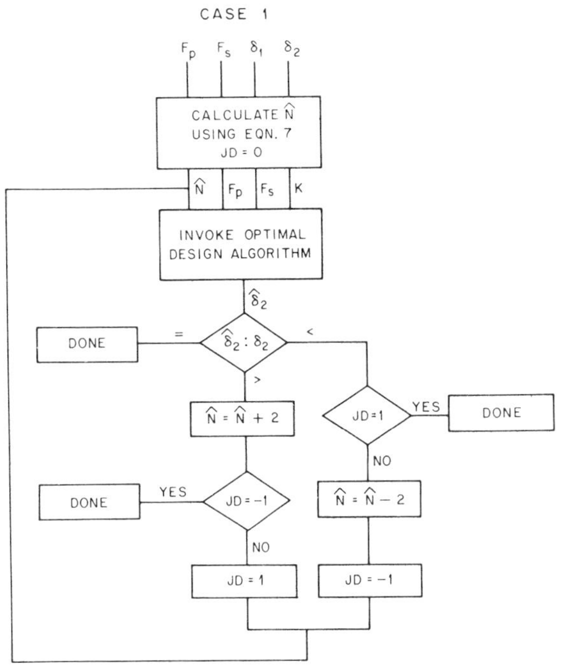


**Fig. 2. Algorithm for choosing smallest $N$ for FIR low-pass filter to meet specifications on $F_p$, $F_s$, $\delta_1$, and $\delta_2$.**

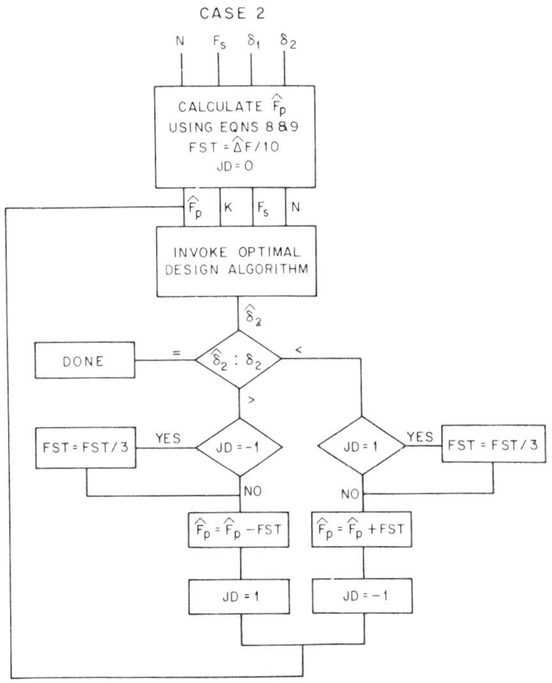


**Fig. 3. Algorithm for choosing value of $F_p$ for FIR low-pass filter to meet specifications on $N$, $F_s$, $\delta_1$, and $\delta_2$.**

take into account the fact that the search has changed direction, and hence significantly narrowed down the interval under investigation. The value of $\hat{F}_p$ is decremented by the step size FST (giving a larger transition width, hence a smaller value of $\hat{\delta}_2$), and the direction parameter is set to 1. If $\hat{\delta}_2 < \delta_2$, a similar path is executed whereby $\hat{F}_p$ is incremented by the current step size.

*Case 3:* $N, F_p, \delta_1$, and $\delta_2$ specified—$F_s$ unspecified. This case is almost identical to Case 2. Equation (8) is used to estimate $\Delta F$, and $\hat{F}_s$ is obtained as $F_p + \Delta\hat{F}$. The flow chart of Fig. 3 can be used for this case with two simple modifications. First $\hat{F}_p$ and $\hat{F}_s$ are interchanged, and second, for the box where $\hat{F}_p$ was decremented, $\hat{F}_s$ is incremented, whereas for the box where $\hat{F}_p$ was incremented, $\hat{F}_s$ is decremented. The similarity between Cases 2 and 3 should be clear.

*Case 4:* $N, F_p, F_s, \delta_1$ specified—$\delta_2$ unspecified. A formula for estimating $\delta_2$ may be obtained by rewriting (5) and (6) in the form

$$D_\infty(\delta_1, \delta_2) = c_1 \log_{10} \delta_2 + c_2 \quad (10)$$

$$f(\delta_1, \delta_2) = d_1 - b_2 \log_{10} \delta_2 \quad (11)$$

where

$$c_1 = a_1 (\log_{10} \delta_1)^2 + a_2 \log_{10} \delta_1 + a_3 \quad (12)$$

$$c_2 = a_4 (\log_{10} \delta_1)^2 + a_5 \log_{10} \delta_1 + a_6 \quad (13)$$

$$d_1 = b_1 + b_2 \log_{10} \delta_1 \quad (14)$$

and then solving for $\log_{10} \hat{\delta}_2$, the estimate of $\log_{10} \delta_2$ from (3) as

$$\log_{10} \hat{\delta}_2 = \frac{(N-1)\Delta F + d_1 (\Delta F)^2 - c_2}{c_1 + b_2 (\Delta F)^2} = \alpha \quad (15)$$

or

$$\hat{\delta}_2 = 10^\alpha. \quad (16)$$

The initial estimate $\hat{\delta}_2$ serves to give an initial estimate $\hat{K}$ since $\delta_2$ is not explicitly used as an input parameter. The parameter $\hat{K}$ is then varied until the value $\hat{\delta}_1$ returned by the design algorithm is within a specified tolerance of $\delta_1$, the input specification. Fig. 4 shows a flow chart of how $\hat{K}$ is varied to achieve the desired specifications on $\delta_1$.

First $\hat{\delta}_2$ and $\hat{K} = \delta_1/\hat{\delta}_2$ are computed using (15) and (16); the step size multiplier on $K$ ($KM$) is set to 2.0; and the direction indicator is initialized to 0. The initial parameters are used in the design algorithm that returns the value $\hat{\delta}_1 = \hat{K}\hat{\delta}_2$ as the actual passband deviation. If $\hat{\delta}_1 = \delta_1$ to within the prescribed tolerance, the algorithm terminates. If $\hat{\delta}_1 > \delta_1$, the estimate $K$ is decreased by the factor $(1 + KM)$ where $KM$ is the current step size multiplier. If a change in direction was detected, the step size multiplier is halved.

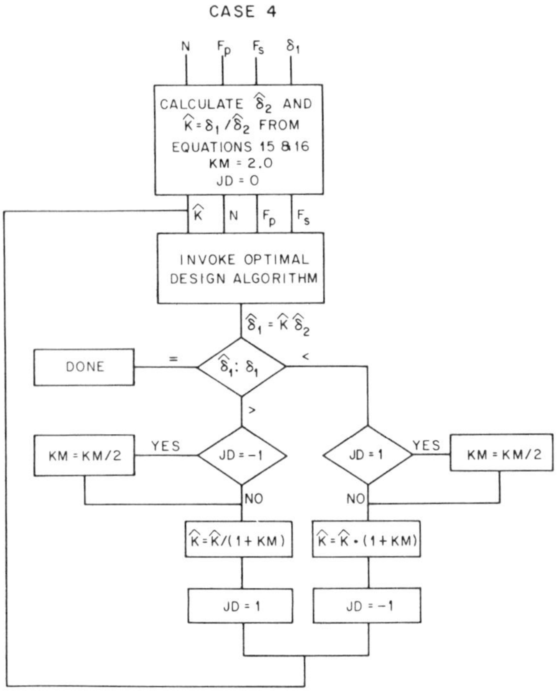


Fig. 4. Algorithm for choosing value of $K$ for FIR low-pass filter to meet specifications on $N$, $F_p$, $F_s$, and $\delta_1$.

The direction indicator is then set to 1. If $\hat{\delta}_1 < \delta_1$, a similar path is executed with $\hat{K}$ being increased by the factor $(1 + KM)$. The new estimate $\hat{K}$ is used as updated information for the design algorithm.

*Case 5:* $N, F_p, F_s, \delta_2$ specified—$\delta_1$ unspecified. A formula for estimating $\delta_1$ may be obtained by rewriting (5) and (6) as

$$D_\infty(\delta_1, \delta_2) = e_1 (\log_{10} \delta_1)^2 + e_2 (\log_{10} \delta_1) + e_3 \quad (17)$$

$$f(\delta_1, \delta_2) = g_1 + b_2 \log_{10} \delta_1 \quad (18)$$

where

$$e_1 = a_1 \log_{10} \delta_2 + a_4 \quad (19)$$

$$e_2 = a_2 \log_{10} \delta_2 + a_5 \quad (20)$$

$$e_3 = a_3 \log_{10} \delta_2 + a_6 \quad (21)$$

$$g_1 = b_1 - b_2 \log_{10} \delta_2. \quad (22)$$

Equations (3) and (4) can be used to give a quadratic equation in $\log_{10} \delta_1$, i.e.,

$$(\log_{10} \delta_1)^2 + g_2 \log_{10} \delta_1 + g_3 = 0 \quad (23)$$

where $g_2$ and $g_3$ are defined as

$$g_2 = \left(\frac{e_2 - b_2 (\Delta F)^2}{e_1}\right) \quad (24)$$

$$g_3 = \frac{e_3 - g_1 (\Delta F)^2 - (N-1)\Delta F}{e_1}. \quad (25)$$

Solving for $\log_{10} \hat{\delta}_1$, the estimate of $\log_{10} \delta_1$, we get

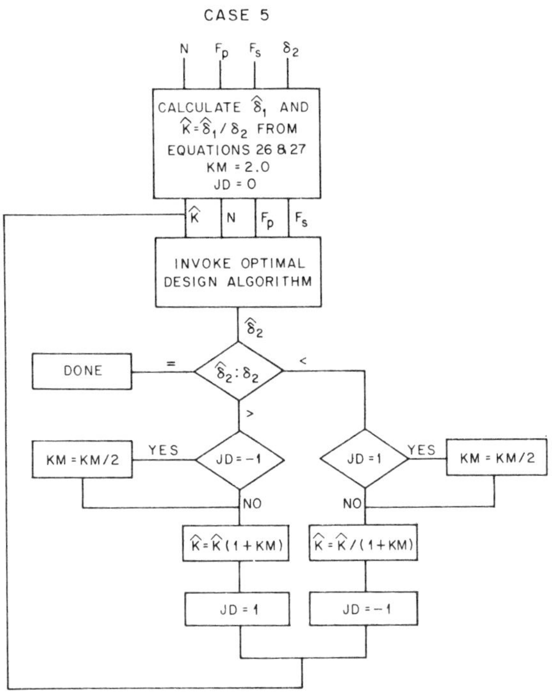


**Fig. 5. Algorithm for choosing value of $K$ for FIR low-pass filter to meet specifications on $N$, $F_p$, $F_s$, and $\delta_2$.**

$$\log_{10} \hat{\delta}_1 = \frac{-g_2}{2} + \sqrt{\frac{g_2^2}{4} - g_3} = \beta \tag{26}$$

or

$$\hat{\delta}_1 = 10^{\beta}. \tag{27}$$

(We have used the positive sign for the square root because the negative value leads to unrealistically small values of $\hat{\delta}_1$.)

From the estimate $\hat{\delta}_1$, an estimate $\hat{K} = \hat{\delta}_1/\delta_2$ is obtained. The flow diagram of Fig. 5 shows how $\hat{K}$ is adjusted until the filter specifications are achieved. Since this flow chart is so similar to the one of Fig. 4, we will not discuss the details.

### Example of Use of Design Formulas

To illustrate how the above formulas are used, the parameters $N = 19$, $F_p = 0.14$, $F_s = 0.3182422$, $\delta_1 = 0.01$, and $\delta_2 = 0.0001$ were used as input to the preceding design rules with one of the parameters left unspecified. (These parameters are those for an optimal filter which had previously been designed.)

For Case 1, $N$ was unspecified. The initial estimate $\hat{N}$ was 17. Table I shows values for $\hat{N}$, $\delta_2$, and $JD$ for the two iterations that were required. In this case, for a 1.0 percent accuracy criterion the value $N = 19$ was selected. The total run time on a Honeywell 6000 computer was 0.20 s.

For Case 2, $F_p$ was unspecified. The initial estimate $\hat{F}_p$ was 0.1605246. Table II shows values for $\hat{F}_p$, $\hat{\delta}_2$, FST, and $JD$ for the five iterations that were required to get $\hat{\delta}_2$ to within 1.0 percent tolerance. The total run time was 0.47 s. The final value of $F_p$ was 0.1394956 as opposed to the exact value of 0.140 for which $\delta_2 = 0.0001$.

For Case 3, $F_s$ was unspecified. The initial estimate $\hat{F}_s$ was 0.2977176. In this case ten iterations were required to get $\hat{\delta}_2$ to within 1.0 percent tolerance. The run time here was 0.95 s. The final value of $F_s$ was 0.3181625 for which $\delta_2$ was 0.000099.

For Case 4, $\delta_2$ was unspecified. The initial estimate $\hat{\delta}_2$ was 0.0000166 or $\hat{K}$ was 601. Seven iterations were required to get $\delta_1$ to within 1.0 percent of $\delta_1$. Table III shows values of $\hat{\delta}_1$, $\hat{K}$, $KM$, and $JD$ for the seven iterations. The run time was approximately 0.63 s.

TABLE I

| Iteration | $\hat{N}$ | $\hat{\delta}_2$ | $JD$ |
|---|---|---|---|
| 1 | 17 | 0.0001931 | 0 |
| 2 | 19 | 0.0001001 | 1 |

TABLE II

| Iteration | $\hat{F}_p$ | $\delta_2$ | FST | $JD$ |
|---|---|---|---|---|
| 1 | 0.1605246 | 0.000151 | 0.0157718 | 0 |
| 2 | 0.1447529 | 0.000110 | 0.0157718 | 1 |
| 3 | 0.1289811 | 0.000078 | 0.0157718 | 1 |
| 4 | 0.1342384 | 0.000088 | 0.0052573 | -1 |
| 5 | 0.1394956 | 0.000099 | 0.0052573 | -1 |

TABLE III

| Iteration | $\hat{K}$ | $\hat{\delta}_1$ | $KM$ | $JD$ |
|---|---|---|---|---|
| 1 | 601.0 | 0.025003 | 2.0 | 0 |
| 2 | 200.4 | 0.015567 | 2.0 | 1 |
| 3 | 66.8 | 0.007426 | 2.0 | 1 |
| 4 | 133.6 | 0.012181 | 1.0 | -1 |
| 5 | 89.0 | 0.009207 | 0.5 | 1 |
| 6 | 111.3 | 0.010786 | 0.25 | -1 |
| 7 | 98.9 | 0.009933 | 0.125 | 1 |

Finally, for Case 5, $\delta_1$ was unspecified. The initial estimate $\hat{\delta}_1$ was 0.002655 or $\hat{K} = 26.55$. It required ten iterations for $\hat{\delta}_2$ to be within 1 percent of $\delta_2$. The run time for this case was about 1.0 s.

## Summary

An optimal FIR low-pass filter can now be designed where any four of the five parameters $N$, $F_p$, $F_s$, $\delta_1$, and $\delta_2$ are specified, and the remaining parameter is chosen so as to meet or exceed specifications on all given parameters. A set of simple, approximate formulas was given for obtaining initial estimates of the unspecified parameter. Finally, simple iterative rules were given for varying the unspecified parameter from its initial estimate so as to meet input specifications to within a given tolerance.

## References


[1] J. F. Kaiser, "Design methods for sampled data filters," in *Proc. 1st Allerton Conf. Circuit and System Theory*, 1963, pp. 221-236.

[2] L. R. Rabiner, "Techniques for designing finite-duration impulse-response digital filters," *IEEE Trans. Commun. Technol.*, vol. COM-19, pp. 188-195, Apr. 1971.

[3] O. Herrmann, "Design of nonrecursive digital filters with linear phase," *Electron. Lett.*, vol. 6, no. 11, pp. 328-329, 1970.

[4] T. W. Parks and J. H. McClellan, "Chebyshev approximation for nonrecursive digital filters with linear phase," *IEEE Trans. Circuit Theory*, vol. CT-19, pp. 189-194, Mar. 1972.

[5] L. R. Rabiner, "The design of finite impulse response digital filters using linear programming techniques," *Bell Syst. Tech. J.*, vol. 51, pp. 1177-1198, July-Aug. 1972.

[6] T. W. Parks and J. H. McClellan, "A program for the design of linear phase finite impulse response digital filters," *IEEE Trans. Audio Electroacoust.*, vol. AU-20, pp. 195-199, Aug. 1972.

[7] O. Herrmann, L. R. Rabiner, and D. S. K. Chan, "Practical design rules for optimum finite impulse response low-pass digital filters," *Bell Syst. Tech. J.*, vol. 52, no. 6, pp. 769-799, July-Aug. 1973.

# NONRECURSIVE DIGITAL FILTER DESIGN USING THE $I_0$-SINH WINDOW FUNCTION

J. F. Kaiser
Bell Telephone Laboratories, Incorporated
Murray Hill, New Jersey 07974

**Abstract**

The design of nonrecursive digital filters of the lowpass, symmetric bandpass, and highpass types is shown to be especially simple when the $I_0$-sinh window function is used. A simple design expression for filter order in terms of filter specifications is given and compared to another more exact expression.

## 1. INTRODUCTION

The design of nonrecursive, (FIR), digital filters of the lowpass type has been shown to be especially simple when the window function method is used with the direct Fourier series approach.[1-4] The window functions are utilized to improve the convergence of the Fourier series in the vicinity of discontinuities where the well-known Gibbs' phenomenon is present[5].

The first windows used for this purpose were the raised cosine window and those of Hamming, Blackman, Tukey, and Bartlett[6]; these windows represented only isolated solutions to the problem each with its own particular properties, an ad hoc approach at best. The introduction of a family of window functions with an adjustable parameter gave the designer the flexibility he needed for effective use of the technique.

In this paper the properties of the $I_0$-sinh family of window functions are explored in some detail and a simple design expression advanced relating filter order to magnitude ripple and transition band width. This design expression is then compared to a more comprehensive one advanced by Herrmann et al for equiripple FIR filter design. The paper concludes with an outline of the design procedure for digital filters with a single passband.

## 2. THE $I_0$-SINH WINDOW FUNCTIONS

A desirable property of a function to be used as a window function is that the function be of limited duration in the time domain and also have most of its energy concentrated at low frequencies in its Fourier transform. The prolate spheroidal wave functions have this property in an optimal way but are rather difficult to compute. A very good approximation to the zeroth-order prolates are the $I_0$-sinh functions defined by

$$w(t) = \begin{cases} \dfrac{I_0\left[\alpha\sqrt{1-(t/\tau)^2}\right]}{I_0(\alpha)}, & |t| \le \tau \\ 0 & |t| > \tau \end{cases} \tag{1}$$

whose Fourier transform is given by

$$W(\omega) = \frac{2\tau \sin\left[\alpha\sqrt{(\omega/\omega_a)^2 - 1}\right]}{\alpha I_0(\alpha)\sqrt{(\omega/\omega_a)^2 - 1}} \tag{2}$$

where

$$\alpha \triangleq \omega_a \tau \tag{3}$$

and $I_0(x)$ is the modified Bessel function of the first kind and zeroth order. $I_0(x)$ is most easily computed from its power series expansion given by

$$I_0(x) = 1 + \sum_{k=1}^{\infty}\left[\frac{(x/2)^k}{k!}\right]^2 \tag{4}$$

A simple FORTRAN subroutine which realizes (4) is

```
      SUBROUTINE INO(X,S)
      S=1.
      DS=1.
      D=0.
    1 D=D+2.
      DS=DS*X*X/(D*D)
      S=S+DS
      IF(DS-.2E-8*S)2,1,1
    2 RETURN
      END
```

Reprinted from *Proc. 1974 IEEE Int. Symp. on Circuits and Syst.*, Apr. 22–25, 1974, pp. 20–23.

In using this family of window functions in design a knowledge of the first integral of $W(\omega)$ is required. From Fig.1, the

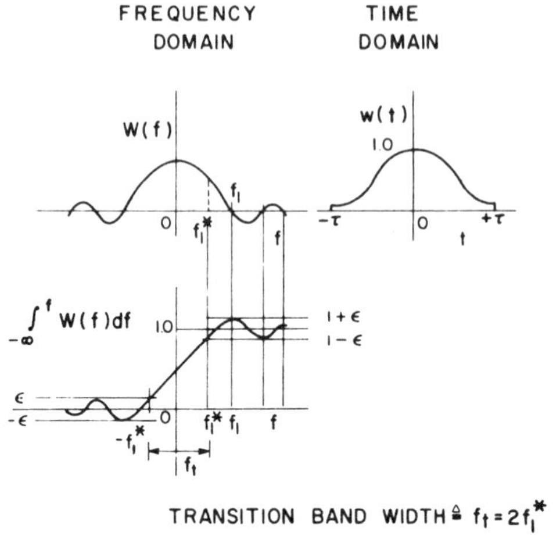


Figure 1 Window Function Transform Pair

amount of overshoot, $\epsilon$ , translates directly to both the amount of passband ripple and stopband loss in the digital filter

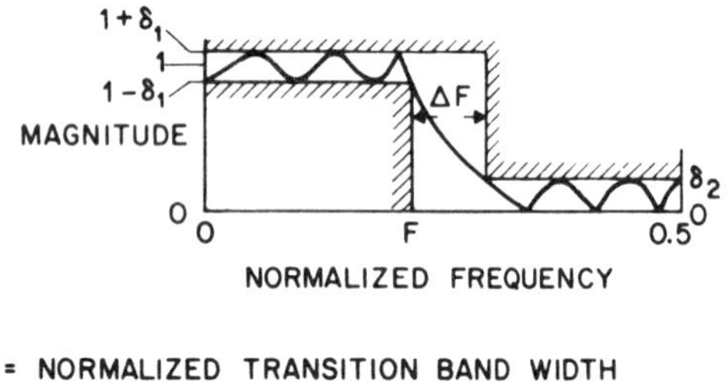


Figure 2 Lowpass Filter Specification

design, as shown in Figure 2. Thus defining

$$\epsilon = \delta = \delta_1 = \delta_2 \tag{5}$$

and

$$ATT = -20 \log_{10}\delta \text{ (dB)} \tag{6}$$

there results the following useful empirical design relations for the window shape parameter, $\alpha$ , and the normalized window width, D, in terms of the desired ripple, $\epsilon$ , expressed in decibels (ATT)

$$\alpha = \begin{cases} 0.1102*(ATT-8.7) & ATT>50 \\ 0.5842*(ATT-21)^{0.4}+0.07886*(ATT-21), & 21<ATT<50 \\ 0.0 & ATT<21 \end{cases} \tag{7}$$

$$D = \begin{cases} (ATT - 7.95)/14.36 & ATT>21 \\ 0.9222 & \end{cases} \tag{8}$$

These expressions yield values of $\alpha$ and D to better than 1% accuracy over the useful range of values of ATT. Note that $\alpha$ =0 corresponds to the straight truncation window with a transform of the form sin x/x. Tables I and II present some of the data from which these empirical relations were obtained.

TABLE I. Performance of $I_0$-sinh Window, Integer Values of $\alpha$

| $\alpha$ | Overshoot % | ATT dB | D ($4f_1^*\tau$) |
|---|---|---|---|
| 0 | 8.95 | 20.96 | 0.922 |
| 2.0 | 3.49 | 29.15 | 1.479 |
| 3.0 | 1.44 | 36.81 | 2.009 |
| 4.0 | 0.545 | 45.27 | 2.597 |
| 5.0 | 0.197 | 54.10 | 3.211 |
| 5.4414 | 0.125 | 58.05 | 3.486 |
| 6.0 | 0.0701 | 63.08 | 3.836 |
| 7.0 | 0.0247 | 72.14 | 4.467 |
| 8.0 | 0.0087 | 81.23 | 5.100 |
| 9.0 | 0.0030 | 90.37 | 5.736 |
| 10.0 | 0.0011 | 99.49 | 6.373 |

The overshoot values in Table I correct the factor of 2 error present in Table 7.1, page 235 of [2] and reprinted in [3].

TABLE II. Performance of $I_0$-sinh Window Integer Attenuation Values

| ATT dB | $\alpha$ | D ($4f_1^*\tau$) | In-band Ripple-dB |
|---|---|---|---|
| 30 | 2.210 | 1.536 | ±0.270 |
| 40 | 3.384 | 2.228 | ±0.0864 |
| 50 | 4.538 | 2.926 | ±0.0274 |
| 60 | 5.658 | 3.621 | ±0.00868 |
| 70 | 6.764 | 4.317 | ±0.00275 |
| 80 | 7.865 | 5.015 | ±0.00089 |
| 90 | 8.960 | 5.712 | ±0.00027 |
| 100 | 10.056 | 6.408 | ±0.00009 |

From the normalized window width parameter, D, the desired design equation can be obtained. First write

$$D = 4f_1^*\tau = (2f_1^*)\cdot(2\tau)$$

D=(transition band width)(window span)

If the filter is to have N terms i.e. of $(N-1)^{th}$ order, then the window span is simply (N-1)T where $T = 1/f_{samp}$ is the sampling interval. The transition band width is defined as $\Delta f$. Hence

$$D = \Delta f(N-1)/f_S$$

Solving for N gives

$$N = \frac{D}{\Delta f/f_S} + 1 \qquad (9)$$

Defining the normalized transition band width as

$$\Delta F = \Delta f/f_S$$

and using equations (6) and (8) yields the desired design equation

$$N = \frac{-20 \log_{10}\delta - 7.95}{14.36\ \Delta F} + 1 \qquad (10)$$

or

$$N - 1 \approx \frac{ATT - 8}{14.36\ \Delta F} \qquad (11)$$

This design equation is quite accurate for nonrecursive filter design using the $I_o$-sinh window function. In the resulting designs the passband ripple and stopband ripple are approximately equal in size with maximum ripple occurring in the immediate vicinity of the bandedge or transition region. Thus one can expect the overall ripple to be somewhat less (see Reference 8, page 108). In filter designs with very narrow passbands or stopbands relative to the transition band width one can expect the ripple amplitude to be in error by at most a factor of 2 (6 dB) because of interference effects of the tails of the integrals of $W(\omega)$.

If designs are desired with $\delta_1 \neq \delta_2$ then the window function design method can be used to obtain a good starting design whose coefficients can then be perturbed to obtain the desired result. In this situation the design equation (10) is modified by replacing $\delta$ with the geometric mean of $\delta_1$ and $\delta_2$ as

$$N \approx \frac{-20 \log_{10}\sqrt{\delta_1\delta_2} - 8}{14.36\ \Delta F} + 1 \qquad (12)$$

In fact the nearness of designs obtained by the window function method to those equiripple designs of Herrmann and Rabiner[7] is made apparent if a fit of the form of (12) is made to their design data. The result is

$$N \approx \frac{-20 \log_{10}\sqrt{\delta_1\delta_2} - 13}{14.6\ \Delta F} + 1 \qquad (13)$$

or

$$D \approx \frac{-20 \log_{10}\sqrt{\delta_1\delta_2} - 13}{14.6} \qquad (14)$$

Figure 3 shows a plot of D as a function of $20 * \log_{10}\delta_2$. This figure is to be compared with Figure 8 of Reference 7; the resemblence is very close as one would expect.

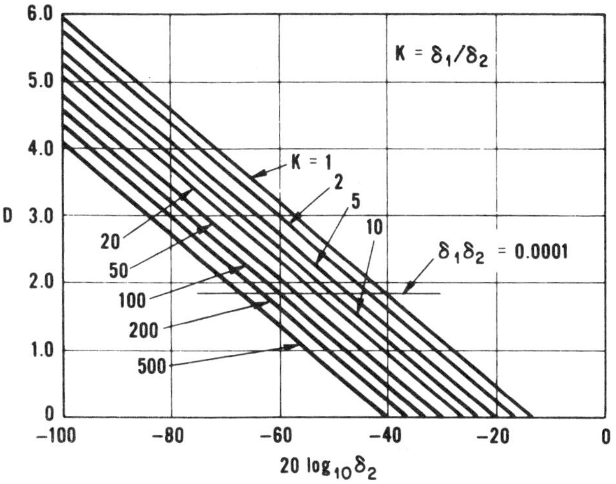


Figure 3 Approximate Design Relation for FIR Filters

Marked discrepancies occur only for large $\delta_1$ and $\delta_2$ values, i.e. small D values, and for designs with passbands and stopbands very narrow in width relative to the transition bands. Thus for approximate estimation of number of terms required in an equiripple FIR filter design, equation (13) can be applied directly.

The primary advantages of the window function method of nonrecursive filter design are that only Fourier series expansions of the simple ideal filter characteristics are required and that full control is held over passband and stopband ripples and transition band widths with the nature of the tradeoff between filter order and these parameters directly apparent. The disadvantages are that pass and stopband ripple amplitudes must be identical and the widths of all transition regions must be equal. The penalty in filter order for equal band ripples ($\delta_1 = \delta_2$) rather than unequal ($\delta_1 \neq \delta_2$) is easily obtained from (13).

As will be shown in the next section the filter design procedure is sufficiently simple so that a subroutine can easily be coded to do the entire filter design directly from the filter performance specifications.

## 3. BANDPASS FILTER DESIGN

Consider the bandpass filter specifications as as shown in Figure 4.

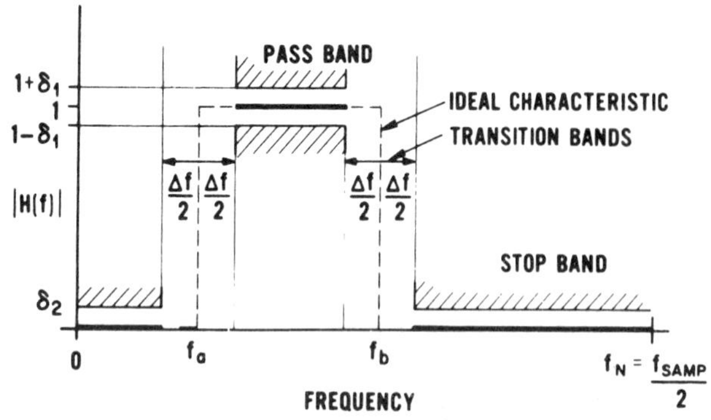


Figure 4 Bandpass Filter Specifications

A nonrecursive filter H(z) of the form

$$H(z) = c_0 + \sum_{n=1}^{N_p} c_n(z^n+z^{-n}) \qquad (16)$$

is desired. First a Fourier series expansion is made of $|H(f)|$ as

$$H(f) = a_0 + \sum_{n=1}^{\infty} 2a_n \cos(n\pi f/f_N) \qquad (17)$$

which yields

$$a_0 = (f_b - f_a)/f_N$$

$$a_n = \frac{1}{n\pi}\left[\sin(n\pi f_b/f_N) - \sin(n\pi\, f_a/f_N)\right] \qquad (18)$$

For lowpass filters $f_a = 0$ and for highpass filters $f_b = 0$.

Using equation (7) the window parameter $\alpha$ is determined; from (10) the number of terms, N, is determined and rounded up to the nearest odd integer. The number of pairs of terms $N_p$ is then $N_p = (N-1)/2$.

The actual filter coefficients are then given by

$$c_n = a_n \cdot \frac{I_0\left[\alpha\sqrt{1-(n/N_p)^2}\right]}{I_0(\alpha)} \qquad (19)$$

$$n = 0,1,2\ldots N_p$$

This completes the design of the symmetric single passband digital filter of the nonrecursive type using the window function method.

## References


1. Kaiser, J. F., "Design Methods for Sampled Data Filters", First Allerton Conference on Circuit and System Theory, Monticello, Illinois, 1963, pp 221-236; reprinted in IEEE Press Monograph on Digital Signal Processing 1972, pp 20-34.

2. Kaiser, J.F., "Digital Filters", Chapter 7 in book "System Analysis by Digital Computer", Kuo and Kaiser, John Wiley 1966, pp 228-243.

3. Huelsman, L. P., "Active Filters* Lumped, Distributed, Integrated, Digital, and Parametric", McGraw Hill, 1970, pp 237-245.

4. Rabiner, L. R., "Techniques for Designing Finite-Duration Impulse-Response Digital Filters", IEEE Trans. Commun. Technol. Vol COM-19, April 1971, pp 188-195; reprinted in IEEE Press Monograph on Digital Signal Processing, 1972, pp 150-157.

5. Hamming, R. W., "Numerical Methods for Scientists and Engineers", Second edition, McGraw Hill, N.Y., 1973, pp 527-538.

6. Blackman, R. B. and J. W. Tukey, "The Measurement of Power Spectra", Dover, 1958.

7. Herrmann, O., L. R. Rabiner, D. S. K. Chan, "Practical Design Rules for Optimum Finite Impulse Response Low-Pass Digital Filters", BSTJ, Vol 52, No. 6, July-August 1973, pp 769-799.

8. Schüssler, H. W., "Digitale Systeme zur Signalverarbeitung", Springer-Verlag, 1973, pp 105-108.

# Equal-Ripple Delay Recursive Digital Filters

JEAN-PIERRE THIRAN, MEMBER, IEEE

**Abstract**—All-pole recursive digital low-pass filters approximating a constant group delay in an equal-ripple sense are described. The equal-ripple conditions are given by a system of nonlinear equations in the poles of the transfer function. The closed-form solution obtained for a particular value of the cutoff frequency yields good first approximations for the iterative procedure required in the general case. The set of approximated delays and ripples leading to realizable filters is discussed. Numerical examples illustrating the procedure are included.

## I. Introduction

Among the problems which cannot be satisfactorily solved by the classical method of designing recursive digital filters, i.e., digitalization of continuous transmittances, one finds the approximation of a constant group delay. The choice of the maximally flat criterion for the direct procedure thus required leads to the use of a Gauss hypergeometric function [1]. Obviously, this error norm is not the best one from the approximation standpoint and the present paper is intended to give the equal-ripple solution to the same problem.

For the continuous case, two methods have been proposed, one by Ulbrich and Piloty [2] and the other by Abele [3], to get the equal-ripple delay filters. Both methods are characterized by a system of nonlinear equations to be solved by iterative techniques. While, for the first solution, the only unknowns of the system are the poles of the transfer function, the second one computes, in addition, the abscissas of the extrema of the delay response: the order of the system equals approximately twice the degree of the filter. However, contrary to the Ulbrich–Piloty method, the Bessel filter is explicitly the solution of the Abele system and can be used as an excellent first guess for the iterative process. To take advantage of the lower order of the system, the method developed in this paper for digital filters parallels the Ulbrich–Piloty procedure, and as an explicit solution is obtained in a particular case, the lack of first approximations can be overcome. The allowable approximated delays and ripples for a given cutoff frequency are discussed and some amplitude and delay characteristics are shown.

## II. Statement of the Problem

In an approximation problem for prescribed group delay, it is natural to determine only the poles of the transfer function. Indeed, it is well known that a "mirror-image polynomial" used for the numerator makes it possible to meet additional requirements on attenuation [4] without affecting

Manuscript received March 8, 1971.
The author is with the MBLE Research Laboratory, Brussels, Belgium.

Reprinted from *IEEE Trans. Circuit Theory*, vol. CT-18, pp. 664–669, Nov. 1971.

group delay. Therefore the investigated recursive transmittances are of the all-pole type

$$H(z) = \frac{K}{p(z^{-1})}.$$

The constant $K$ is a scaling factor in amplitude which will be fixed to have zero loss at dc, and the polynomial $p(z^{-1})$ can be defined either by its coefficients:

$$p(z^{-1}) = \sum_{i=0}^{n} a_i z^{-i}, \qquad \text{with } a_n = 1 \tag{1}$$

or by its roots:

$$p(z^{-1}) = \prod_{i=1}^{n} (z^{-1} - z_i^{-1}) \tag{2}$$

with $|z_i| < 1$ to guarantee stability.

The phase shift of $H(z)$ is obviously half that of the all-pass function $G(z)=p(z)/p(z^{-1})$ and is thus given by ln $[G(e^{j\omega})]/2j$ if the sampling frequency is normalized to unity. Hence the analytic continuation of the group delay is

$$t(z) = -\frac{1}{2}\frac{d \ln G(z)}{d \ln z}$$

or

$$t(z) = -\frac{1}{2}\left[\frac{zp'(z)}{p(z)} + \frac{z^{-1}p'(z^{-1})}{p(z^{-1})}\right]$$

where $p'$ denotes the derivative of $p$.

Using, respectively, (1) and (2), one gets

$$t(z) = -\frac{1}{2}\left\{\frac{\sum_{i=0}^{n} a_i i z^i}{\sum_{i=0}^{n} a_i z^i} + \frac{\sum_{i=0}^{n} a_i i z^{-i}}{\sum_{i=0}^{n} a_i z^{-i}}\right\} \tag{3}$$

and

$$t(z) = -\frac{1}{2}\sum_{i=1}^{n}\left(\frac{z}{z - z_i^{-1}} + \frac{z^{-1}}{z^{-1} - z_i^{-1}}\right).$$

This last expression can be rewritten in the form

$$t(z) = -\frac{n}{2} + \frac{1}{2}\sum_{i=1}^{n}\frac{(z_i - z_i^{-1})/2}{(z + z^{-1})/2 - (z_i + z_i^{-1})/2} \tag{4}$$

which suggests [5] the use of the transformation of variable $w=(z+z^{-1})/2$ mapping the unit circle of the $z$ plane onto the interval $[-1, +1]$ of the real axis of the $w$ plane. If one denotes by $w_i$ the transformed value of $z_i$, the group delay becomes in terms of the new variable

$$T(w) = -\frac{n}{2} + \frac{1}{2}\sum_{i=1}^{n}\frac{\sqrt{w_i^2 - 1}}{w - w_i}$$

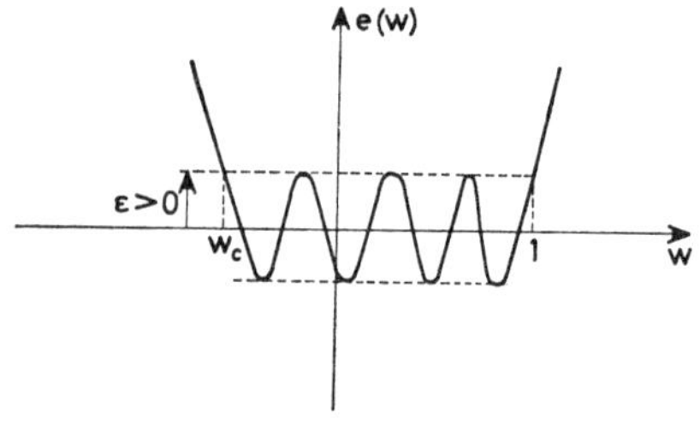


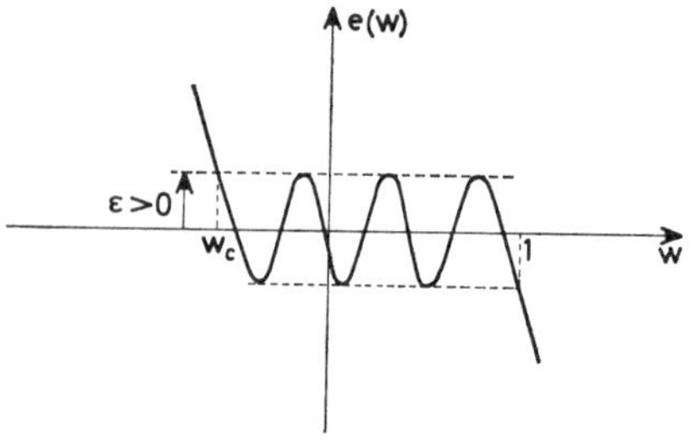


Fig. 1. Optimal error characteristics. (a) $n$ even. (b) $n$ odd.

where the sign of the square root is uniquely determined by the stability condition

$$|z_i| = |w_i + \sqrt{w_i^2 - 1}| < 1. \tag{5}$$

The problem can now be stated in the following terms: in the low-pass interval $[0, \omega_c]$, a constant delay $\tau$ must be approximated by $T(w)$ according to the equal-ripple norm or, in other words, so that the error

$$e(w) = \tau + \frac{n}{2} - \frac{1}{2}\sum_{i=1}^{n}\frac{\sqrt{w_i^2 - 1}}{w - w_i}$$

alternates [6, p. 54] in $[1, w_c = \cos \omega_c]$ the maximum number of times, i.e., $n$ since $e(w)$ is a rational fraction of degree $n$ in $w$. The optimal error is sketched on Fig. 1 for positive ripples, if it is agreed that the ripple $\epsilon$ is positive with the deviation at $w_c$.

## III. Equal-Ripple Conditions

As is well known [7], [8], given $g(p)$ any strictly Hurwitzian polynomial of degree $n$, the rational fraction of degree $n$ in $p^2$

$$R(p^2) = \frac{1}{2}\left[\frac{g(p)}{g(-p)} + \frac{g(-p)}{g(p)}\right]$$

alternates $n$ times with a unit ripple on the whole negative real axis. In fact, the same result holds for $g(p)$ anti-Hurwitzian which amounts to a mere permutation between $g(p)$ and $g(-p)$, but if $g(p)$ had roots in both left and right half-planes, the number of alternations would be reduced.

Now, by using the transformation [9]

$$p^2 = \frac{w - w_c}{w - 1} \tag{6}$$

which maps the whole negative real axis of $p^2$ onto the low-pass interval $[w_c, 1]$, one gets the desired equal-ripple error in the $w$ plane by

$$e(w) \equiv \epsilon R\left(\frac{w - w_c}{w - 1}\right). \tag{7}$$

Since $R(0)$ is equal to $+1$, $e(w_c)$ is positive with $\epsilon$ in accordance with Fig. 1.

Identifying the residues of both members of (7) at the poles $w_i$ gives

$$-\frac{1}{2}\sqrt{w_i^2 - 1} = \epsilon \operatorname*{res}_{w=w_i}\left[R\left(\frac{w - w_c}{w - 1}\right)\right], \qquad i = 1, 2, \cdots, n. \tag{8}$$

As is easily seen, the roots of $g(p)$, denoted by $p_i$ $(i=1, 2, \cdots, n)$, correspond to the poles $w_i$ so that one has

$$\sqrt{w_i^2 - 1} = \sqrt{2(1 - w_c)}\,\frac{\sqrt{p_i^2 - (w_c + 1)/2}}{p_i^2 - 1} \tag{9}$$

and, from the expression $\{d[1/f(t)]/dt\}_{t=t_0}^{-1}$ of the residue of a function $f(t)$ at a simple pole $t_0$,

$$\operatorname*{res}_{w=w_i}\left[R\left(\frac{w - w_c}{w - 1}\right)\right] = \operatorname*{res}_{p^2=p_i^2}\,[R(p^2)]\left(\frac{dw}{dp^2}\right)_{p^2=p_i^2}. \tag{10}$$

By means of (9) and (10) together with

$$\operatorname*{res}_{p^2=p_i^2}\,[R(p^2)] = -2p_i^2 \prod_{\substack{j=1\\ \neq i}}^{n} \frac{p_j + p_i}{p_j - p_i}$$

$$\left(\frac{dw}{dp^2}\right)_{p^2=p_i^2} = -\frac{1 - w_c}{(p_i^2 - 1)^2}$$

(8) can be rewritten in the form

$$\sqrt{p_i^2-(w_c+1)/2}+2\epsilon\sqrt{2(1-w_c)}\,\frac{p_i^2}{p_i^2-1}\prod_{\substack{j=1\\ \neq 1}}^{n}\frac{p_j+p_i}{p_j-p_i}=0, \qquad i = 1, 2, \cdots, n. \tag{11}$$

The sign of $[p_i^2-(\omega_c+1)/2]^{1/2}$ is fixed according to (5) transformed in the new plane:

$$\left| p_i^2 - w_c + \sqrt{2(1 - w_c)}\,\sqrt{p_i^2 - (w_c + 1)/2}\right| < \left| p_i^2 - 1\right|. \tag{12}$$

Finally, the identification of both members of (7) is completed by equating their values at $w=\infty$:

$$\tau + \frac{n}{2} = \frac{\epsilon}{2}\left(\prod_{i=1}^{n} \frac{p_i - 1}{p_i + 1} + \prod_{i=1}^{n} \frac{p_i + 1}{p_i - 1}\right). \tag{13}$$

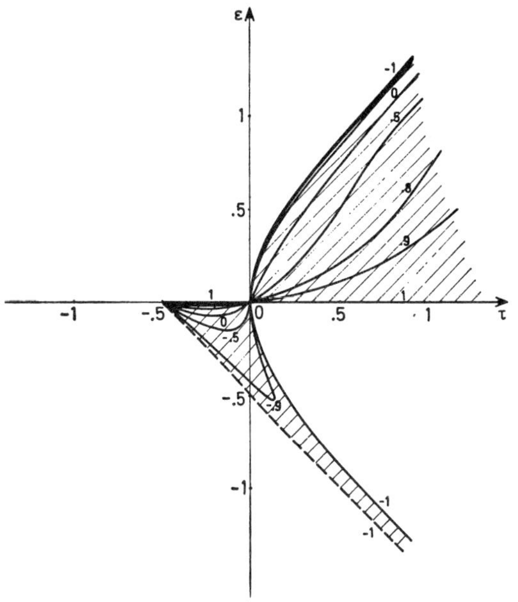


Fig. 2. Admissible values of $\epsilon$ and $\tau$ for $n=1$ and $w_c$ ranging from $-1$ to $+1$.

Equations (11) and (13) together with conditions (12) form a system of $(n+1)$ equations in the roots $p_i$ $(i=1, 2, \cdots, n)$ and one of the three parameters $(\epsilon, \tau, w_c)$, the two others being freely chosen. For $|w_c| \leq 1$, any strictly Hurwitzian solution $g(p)$, whose roots are simple and do not belong to the transformed stopband $[-\sqrt{(1+w_c)/2}, 0]$, will yield a realizable equal-ripple delay filter.

## IV. Explicit Solution for $n=1$

For $n=1$, the equal-ripple conditions are readily deduced from (4)

$$\tau + \epsilon = -\frac{1}{2} + \frac{1}{2}\,\frac{(z_1 - z_1^{-1})/2}{1 - (z_1 + z_1^{-1})/2}$$

$$\tau - \epsilon = -\frac{1}{2} + \frac{1}{2}\,\frac{(z_1 - z_1^{-1})/2}{w_c - (z_1 + z_1^{-1})/2}.$$

From the former condition it follows that $z_1=(\tau+\epsilon)/(\tau+\epsilon+1)$, and inserting this value into the latter one gets

$$w_c = 1 - \frac{\epsilon}{(\tau + \epsilon)(\tau + \epsilon + 1)(\tau - \epsilon + 1/2)}.$$

Obviously, the case $\epsilon=0$ corresponds to the maximally flat delay filter; i.e., $w_c=1$ or $\omega_c=0$ and $z_1=\tau/(\tau+1)$ [1]. The stability condition $|z_1|<1$ implies that $\tau+\epsilon+1/2$ must be strictly positive. On the other hand, the locus $\omega_c=\pi$ or $w_c=-1$ given by

$$(\tau + \epsilon)(\tau + \epsilon + 1)(\tau - \epsilon + 1/2) - \epsilon/2 = 0$$

can be factorized into

$$(\tau + \epsilon + 1/2)(\tau^2 + \tau - \epsilon^2) = 0.$$

Therefore, one easily verifies that the values of $\tau$ and $\epsilon$ yielding stable solutions ($|z_1|<1$) and admissible cutoff frequencies ($|w_c| \leq 1$) must lie in the shaded areas of Fig. 2.

## V. Admissible Values of $\epsilon$ and $\tau$ for Arbitrary $n$

In order to generalize Fig. 2 for arbitrary degrees, one needs the locus corresponding to the extreme values of $\omega_c$, i.e., 0 and $\pi$.

On the one hand, for $\omega_c=0$ or $\epsilon=0$, it has been proved in [1] that for any $n$ the range of values of $\tau$ leading to stable maximally flat delay filters is the interval $]-1/2, +\infty[$.

On the other hand, for $\omega_c=\pi$, (11) reduces to

$$p_i\left(1 + 4\epsilon \frac{p_i}{p_i^2 - 1} \prod_{\substack{j=1 \\ \neq i}}^{n} \frac{p_j + p_i}{p_j - p_i}\right) = 0, \qquad i = 1, 2, \cdots, n$$

where the left member is the product of two factors. If $\omega_c=\pi$, $p_i=0$ corresponds to a root $z_i$ at $1-$, and for a stable solution one has to choose the second factor which can be rewritten as

$$-\frac{1}{\epsilon} = \frac{4p_i}{p_i^2 - 1} \prod_{\substack{j=1 \\ \neq i}}^{n} \frac{p_j + p_i}{p_j - p_i}, \qquad i = 1, 2, \cdots, n \quad (14)$$

and after summation over all $i$ becomes

$$-\frac{n}{\epsilon} = \sum_{i=1}^{n} \frac{4p_i}{p_i^2 - 1} \prod_{\substack{j=1 \\ \neq i}}^{n} \frac{p_j + p_i}{p_j - p_i}. \quad (15)$$

From (13), one easily deduces

$$2\tau + n + 2\epsilon = \epsilon \frac{\left[\prod_{i=1}^{n} (p_i + 1) + \prod_{i=1}^{n} (p_i - 1)\right]^2}{\prod_{i=1}^{n} (p_i^2 - 1)}$$

$$2\tau + n - 2\epsilon = \epsilon \frac{\left[\prod_{i=1}^{n} (p_i + 1) - \prod_{i=1}^{n} (p_i - 1)\right]^2}{\prod_{i=1}^{n} (p_i^2 - 1)}$$

so that one has

$$(2\tau + n + 2\epsilon)(2\tau + n - 2\epsilon) = \left[\epsilon\left(\prod_{i=1}^{n} \frac{p_i + 1}{p_i - 1} - \prod_{i=1}^{n} \frac{p_i - 1}{p_i + 1}\right)\right]^2. \quad (16)$$

Consider now the function

$$f(p) = \prod_{i=1}^{n} \frac{p + p_i}{p - p_i} - \prod_{i=1}^{n} \frac{p - p_i}{p + p_i}.$$

The partial fraction expansion of each term gives

$$f(p) = \sum_{i=1}^{n} 2p_i \prod_{\substack{j=1 \\ \neq i}}^{n} \frac{p_i + p_j}{p_i - p_j}\left(\frac{1}{p - p_i} + \frac{1}{p + p_i}\right)$$

and by evaluating both expressions of $f(p)$ at $p=1$, one gets the identity

$$\prod_{i=1}^{n} \frac{1 + p_i}{1 - p_i} - \prod_{i=1}^{n} \frac{1 - p_i}{1 + p_i} = \sum_{i=1}^{n} \frac{4p_i}{1 - p_i^2} \prod_{\substack{j=1 \\ \neq i}}^{n} \frac{p_i + p_j}{p_i - p_j}. \quad (17)$$

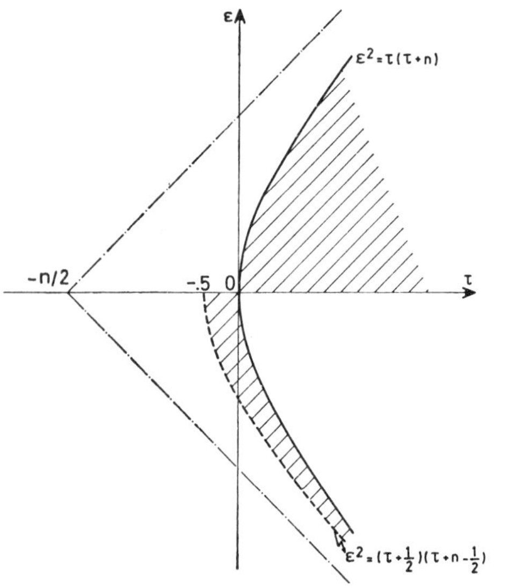


Fig. 3. Admissible values of $\epsilon$ and $\tau$ for arbitrary $n$.

Hence by means of (15)–(17) the locus is

$$(2\tau + n + 2\epsilon)(2\tau + n - 2\epsilon) = n^2$$

which, after simplification, turns out to be the hyperbola $\epsilon^2=\tau(\tau+n)$.

The second locus, which corresponds to the stability boundary drawn in the dashed line in Fig. 2, is obtained by taking one root, e.g., $p_n$, equal to zero while keeping the other $(n-1)$ equations (14). The equal-ripple conditions then become

$$\frac{1}{\epsilon} = \frac{4p_i}{p_i^2 - 1} \prod_{\substack{j=1 \\ \neq i}}^{n-1} \frac{p_j + p_i}{p_j - p_i}, \qquad i = 1, 2, \cdots, n - 1$$

$$\tau + \frac{n}{2} = -\frac{\epsilon}{2}\left(\prod_{i=1}^{n-1} \frac{p_i - 1}{p_i + 1} + \prod_{i=1}^{n-1} \frac{p_i + 1}{p_i - 1}\right).$$

Consequently, the second locus is readily deduced from the first one by replacing $n$, $\tau$, and $\epsilon$, respectively, by $n-1$, $\tau+1/2$, and $-\epsilon$, thereby yielding $\epsilon^2=(\tau+1/2)(\tau+n-1/2)$. To summarize, the allowable values of $\epsilon$ and $\tau$ lie in the shaded regions indicated in Fig. 3.

## VI. Explicit Solution for $\omega_c=\pi$

For $\omega_c=\pi$, the passband interval corresponds to the whole unit circle of the $z$ plane and, as could be expected, the $z$ and $p$ variables are related by the bilinear transformation $p=(1+z^{-1})/(1-z^{-1})$. Therefore, from the expression (3) of the group delay, the equal-ripple condition (7) becomes

$$\tau + \frac{1}{2}\left\{\frac{\sum_{i=0}^{n} a_i i z^i}{\sum_{i=0}^{n} a_i z^i} + \frac{\sum_{i=0}^{n} a_i i z^{-i}}{\sum_{i=0}^{n} a_i z^{-i}}\right\} \equiv \frac{\epsilon}{2}\left[\frac{g(p)}{g(-p)} + \frac{g(-p)}{g(p)}\right]_{p=(1+z^{-1})/(1-z^{-1})}$$

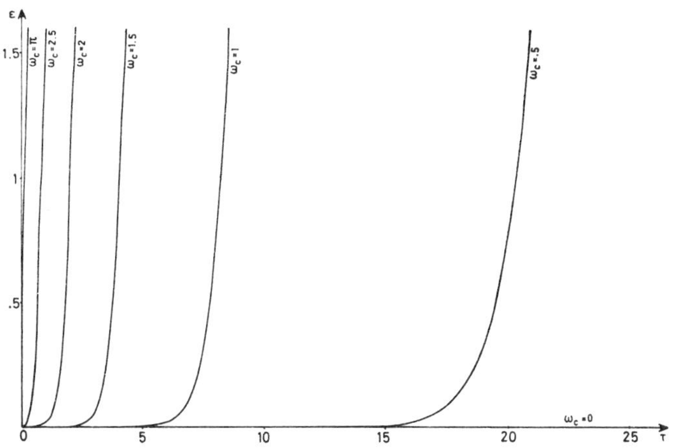


Fig. 4. Values of $\tau$ versus positive $\epsilon$ for $n=10$ and $\omega_c$ ranging from 0 to $\pi$.

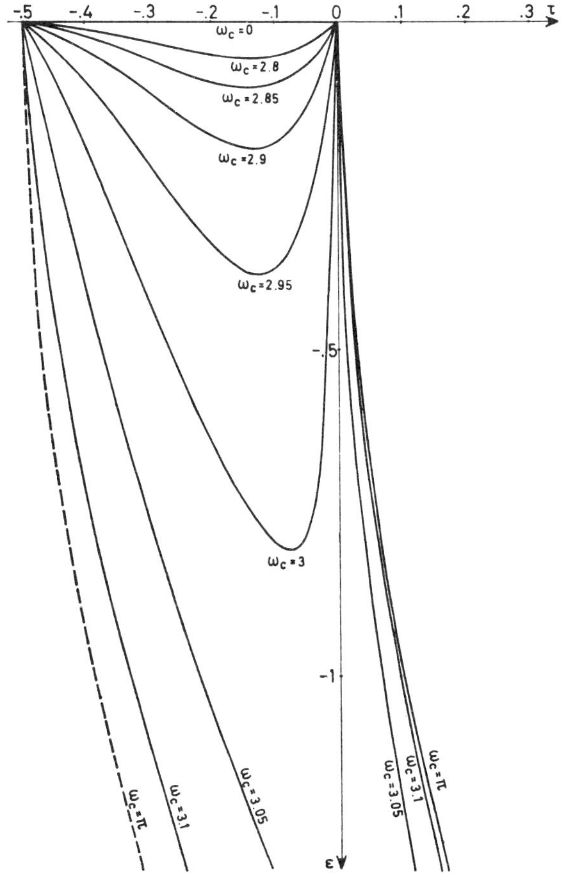


Fig. 5. Values of $\tau$ versus negative $\epsilon$ for $n=10$ and $\omega_c$ ranging from 0 to $\pi$.

which can be rewritten in the form

$$\frac{\sum_{i=0}^{n}[a_i(\tau+i)/\epsilon]z^i}{\sum_{i=0}^{n}a_iz^i}+\frac{\sum_{i=0}^{n}[a_i(\tau+i)/\epsilon]z^{-i}}{\sum_{i=0}^{n}a_iz^{-i}}$$

$$\equiv\frac{(-z)^nG(z^{-1})}{G(z)}+\frac{(-z^{-1})^nG(z)}{G(z^{-1})}$$

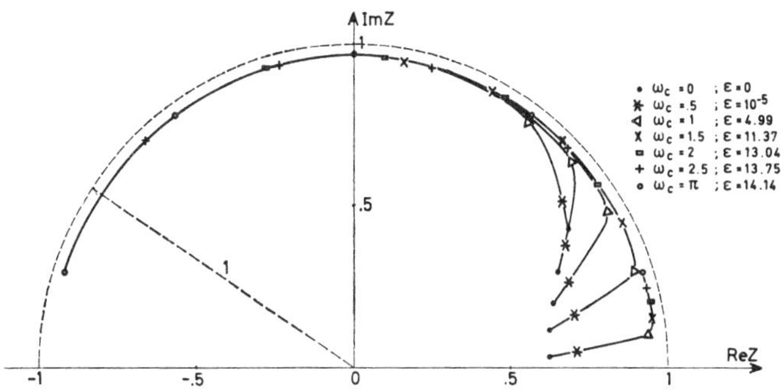


Fig. 6. Pole locus of the tenth-degree filter for $\tau=10$ and $\omega_c$ ranging from 0 to $\pi$.

where $G(z^{-1})=(1-z^{-1})^n g[(1+z^{-1})/(1-z^{-1})]$ has all its roots outside the unit circle if $g(p)$ is strictly Hurwitzian. The only way of obtaining a stable solution is thus to identify $G(z^{-1})$ with $\sum_{i=0}^{n} a_i z^{-i}$ so that one has

$$\sum_{i=0}^{n}[a_i(\tau+i)/\epsilon]z^{-i}\equiv(-z^{-1})^nG(z)$$

or

$$a_i(\tau+i)/\epsilon=(-1)^n a_{n-i}, \qquad i=0,1,\cdots,n.$$

Since $a_n$ is normalized to unity, one gets $a_0=(-1)^n\epsilon/\tau$ and $(-1)^n(\tau+n)/\epsilon$ for $i=0$ and $n$, respectively, and the locus of stable solutions for $\omega_c=\pi$ is accordingly the hyperbola $\epsilon^2=\tau(\tau+n)$ as shown in Section V. All other coefficients are obviously zero, and the denominator of the equal-ripple delay transmittance turns out to be

$$p(z^{-1})=z^{-n}+(-1)^n\epsilon/\tau.$$

## VII. Iterative Solution in the General Case and Numerical Results

For $n\neq 1$ and $\omega_c\neq 0$ or $\pi$, a closed-form expression of the equal-ripple delay filters could not be obtained. Therefore, a generalized Newton–Raphson procedure was implemented for solving the nonlinear equations (11) and (13). In order to reduce the system to only equation (11), it is preferable to choose $\epsilon$ and $w_c$ as free parameters so that one can deduce $\tau$ from (13) after solving system (11). Obviously, the maximally flat delay filter is not an explicit solution of (11) since transformation (6) becomes meaningless for $\omega_c=0$ or $w_c=1$. Yet, the solution given in Section VI for $\omega_c=\pi$ is a starting point for generating the whole equal-ripple family when $\omega_c$ is decreased.

As an illustration for $n=10$, Figs. 4 and 5 give some details of the general Fig. 3 for positive and negative ripples, respectively. In particular, one notices in Fig. 4 that for any cutoff frequency it is possible to achieve a small error tolerance provided the approximated delay is less than some upper limit: the narrower the bandwidth, the larger this upper bound. On the other hand, Fig. 5 shows how the range $]-1/2,\ 0[$ of negative delays yielding stable maximally flat delay filters [1] is transformed when $\omega_c$ is increased from 0 to $\pi$. The pole locus of the tenth-degree filter is plotted in Fig. 6 for $\tau=10$ and $\omega_c$ ranging from 0 to $\pi$: the

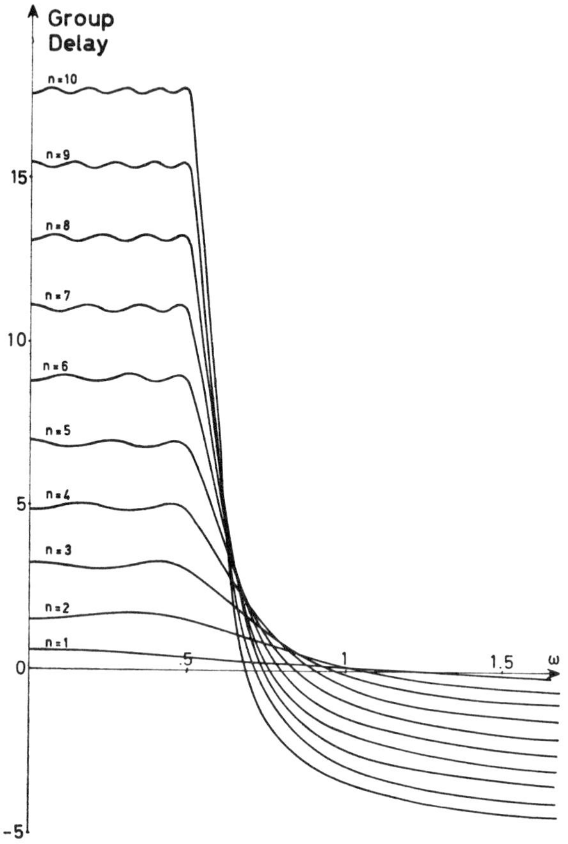


Fig. 7. Delay characteristics for $\omega_c=0.5$, $\epsilon=0.1$ and $n$ ranging from 1 to $n$.

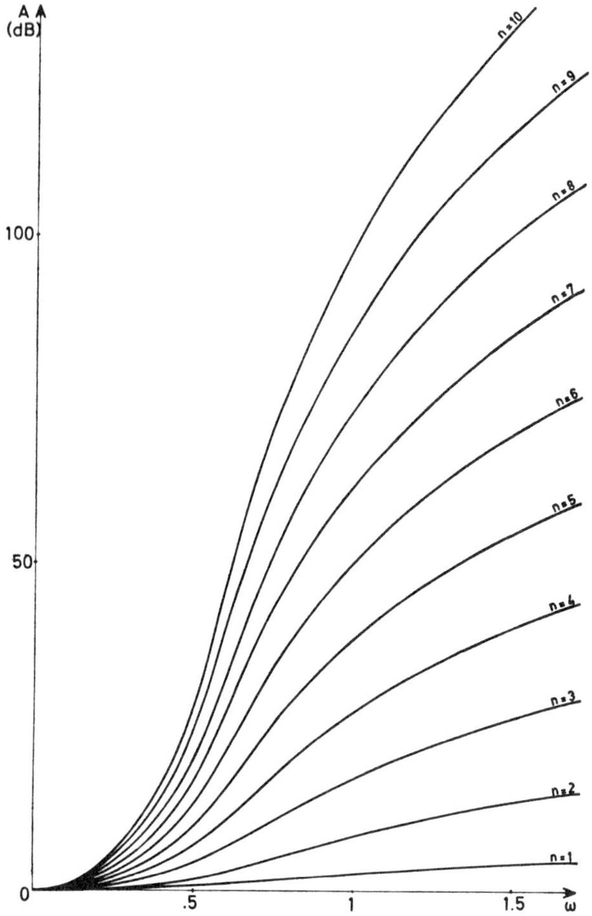


Fig. 8. Attenuation characteristics for $\omega_c=0.5$, $\epsilon=0.1$ and $n$ ranging from 1 to $n$.

boundaries of the locus correspond to the solution given in [1] for $\omega_c=0$ and in Section VI for $\omega_c=\pi$. Finally, in the case of $\omega_c=0.5$ and $\epsilon=0.1$, Figs. 7 and 8 give the delay and attenuation characteristics of filters of degrees 1 to 10.

## Acknowledgment

The author wishes to thank M. J. Praet for carrying out the numerical computations.

## References

[1] J. P. Thiran, "Recursive digital filters with maximally flat group delay," this issue, pp. 659–664.

[2] E. Ulbrich and H. Piloty, "Über den Entwurf von Allpässen, Tiefpässen, und Bandpässen mit einer in Tschebyscheffschen Sinne approximierten konstanten Gruppenlaufzeit," *Arch. Elek. Übertragung*, vol. 14, Oct. 1960, pp. 451–467.

[3] T. A. Abele, "Übertragungsfaktoren mit Tschebyscheffscher Approximation konstanter Gruppenlaufzeit," *Arch. Elek. Übertragung*, vol. 16, Jan. 1962, pp. 9–18.

[4] M. Sablatash, "Approximation theory for digital filters," in *Dig. IEEE Int. Symp. Circuit Theory* (Atlanta, Ga.), Dec. 1970, pp. 141–142.

[5] R. Unbehauen, "Zur Synthese digitaler Filter," *Arch. Elek. Übertragung*, vol. 24, July–Aug. 1970, pp. 305–313.

[6] J. R. Rice, *The Approximation of Functions*, vol. 1. Reading, Mass.: Addison-Wesley, 1964.

[7] B. J. Bennett, "Synthesis of electric filters with arbitrary phase characteristics," *IRE Conv. Rec.*, pt. 5, 1953, pp. 19–26.

[8] G. Szentirmai, "A filter synthesis program," in *System Analysis by Digital Computer*, F. F. Kuo and J. F. Kaiser, Eds. New York: Wiley, 1966.

[9] A. G. Constantinides, "Digital filters with equiripple passbands," *IEEE Trans. Circuit Theory* (Corresp.), vol. CT-16, Nov. 1969, pp. 535–538.

# A Simple Design of Maximally Flat Delay Digital Filters

ALFRED FETTWEIS, Senior Member, IEEE
Dep. Elec. Eng.
Univ. Bochum
Bochum, Germany

Abstract

**The problem of maximally flat delay design of recursive digital filters has been solved by Thiran, using a direct $z$-domain approach. In this paper, the solution is obtained in a much simpler way by making use of the familiar bilinear transform $s$ of the variable $z$. A suitable continued fraction expansion available in the mathematical literature is shown to lead immediately to the required solution. The approach corresponds to the one used by Abele in transmission line filter design.**

## I. Introduction

Thiran [1]–[3] has recently described an explicit solution for the problem of determining for recursive digital filters a transfer function that possesses a maximally flat delay characteristic. Thiran's method is based on a $z$-domain analysis. The purpose of this paper is to show that the same result can be obtained in a much simpler way by operating in the $s$ domain, where $s$ is the familiar bilinear transform of the variable $z$.

The present method is related to the one used by Storch [4] for solving the similar problem in classical LC filter design, and it corresponds to the solution described by Abele [5] for microwave unit element filters. It should be pointed out, however, that the lengthy proof given by Abele for verifying that the desired property has actually been obtained is not required. As we shall see, a simpler proof follows easily from known properties of continued fraction expansions of the type to be used.

Only the basic aspects of the method will be given, and the application to transfer functions of particular practical importance will be discussed. Further properties can be derived from the corresponding properties described by Abele [5]. In particular, the identity of the present result with the one obtained by Thiran can easily be established by comparing Abele's and Thiran's expressions in terms of hypergeometric functions.

Manuscript received October 15, 1971.

## II. Direct Derivation of the Desired Transfer Function

Thiran bases his analysis on a transfer function having the general form

$$H = 1 \Big/ \left( \sum_{i=0}^{n} a_i z^{-i} \right), \qquad z = e^{pT} \tag{1}$$

where $p$ is the complex frequency and $T$ the sampling period. Using the definition

$$s = (z - 1)/(z + 1) = \tanh (pT/2), \tag{2}$$

we can thus write

$$H = \frac{(1 + s)^n}{d} = \frac{(1 + s)^n}{(1 + F_n)d_e} \tag{3}$$

where $d = d(s)$ is a polynomial in $s$ of degree $n$ and where

$$F_n = d_o/d_e, \qquad d_o + d_e = d, \tag{4}$$

$d_o$ and $d_e$ being the odd and even part of $d$, respectively. Stability requires that the polynomial $d$ is Hurwitz. According to (3), we henceforth consider $H$ to be a function of $s$, i.e., $H = H(s)$.

We define a function $B_n = B_n(s)$ by means of

$$e^{j2B_n} = H_*/H \tag{5}$$

Reprinted from *IEEE Trans. Audio Electroacoust*, vol. AU-20, pp. 112–114, June 1972.

where $H_*$ stands, in the usual notation, for $H(-s)$. The function $B_n(s)$ is analytic in $s$ and thus also in $p$; for $p=j\omega$, $B_n$ is equal to the phase corresponding to the transfer function $H$. Comparing (5) with (3), we can write

$$jB_n = -pnT/2 + \tanh^{-1} F_n. \quad (6)$$

In order to obtain the ideal phase characteristic, the function $jB_n$ would have to be equal to the function

$$jB = p\tau \quad (7)$$

where $\tau$ is the delay to be approximated. Thus, replacing $B_n$ in (6) by $B$ and writing $F$ instead of $F_n$, we see that the ideal function $F$ to be realized is

$$F = \tanh(\mu pT/2) \quad (8)$$

where

$$\mu = n + 2\tau/T. \quad (9)$$

Due to (2), (8) can also be written

$$F = \tanh(\mu \tanh^{-1} s). \quad (10)$$

Our approximation problem now consists in determining $B_n$ in such a way that it coincides with $B$ up to the $n$th order in $p$. Due to the general properties of Taylor series expansions, this is equivalent to approximating $F$ by $F_n$ up to the $n$th order in $p$. Since $s=0$ for $p=0$, however, this latter problem is equivalent to approximating $F$ up to the $n$th order in $s$; this must be done in such a way that the resulting function $F_n(s)$ is the ratio of an odd to an even polynomial in $s$ whose sum $d$ is Hurwitz.

The solution to the problem thus defined is easily obtained by making use of a suitable continued fraction expansion. By appropriately rewriting the expansion of tan ($\mu\tan^{-1}y$) given in the standard textbooks of Perron [6] and Wall [7], we obtain the following expression for the function $F$ defined by (10), using Pringheim's notation for representing a continued fraction:

$$F = \frac{\mu|}{|1/s} + \frac{(\mu^2-1)|}{|3/s} \quad \frac{(\mu^2-4)|}{|5/s} + \cdots + \frac{(\mu^2-m^2)|}{|(2m+1)/s} + \cdots. \quad (11)$$

Let us choose for $F_n$ the $n$th convergent of the continued fraction (11), i.e.,

$$F_n = \frac{\mu|}{|1/s} + \frac{(\mu^2-1)|}{|3/s} + \cdots + \frac{[\mu^2-(n-1)^2]|}{|(2n-1)/s}, \quad (12)$$

which, clearly, is a rational function of degree $n$ in $s$ and which consists of a ratio of an odd to an even polynomial in $s$. By well-known properties of continued fraction expansions of the type (11), the function $F_n(s)$ defined by (12) coincides with the function $F(s)$ defined by (10) precisely up to the $n$th order.

Even without regard to the Hurwitz property, still to be discussed, the function $F_n$ thus obtained is certainly the only one that satisfies the approximation requirement. This can be seen by the fact that the approximation problem is also equivalent to finding a certain Padé approximant in $s^2$ of the even function $F(s)/s$, and this problem is known to have a unique solution [6].

The Routh–Hurwitz criterion for testing the Hurwitz character of a polynomial is known to amount to verifying that the ratio of its odd and even parts is a reactance function, which in turn amounts to requiring that the coefficients appearing in a continued fraction expansion of the type of (12) all are positive. Hence, due to (4) and (12), $d(s)$ is Hurwitz if $\mu>n-1$, i.e., because of (9) if

$$\tau > -T/2. \quad (13)$$

This condition is always fulfilled for positive values of the delay $\tau$ to be approximated.

### III. Related Approximation Problems

For practical applications, transfer functions of the form

$$H = \frac{f}{d} = \frac{f}{(1+F_n)d_e}, \quad (14)$$

where $d$ is still a Hurwitz polynomial in $s$ of degree $n$, with $F_n$, $d_e$, and $d_o$ defined as before, but where $f$ is an even or odd polynomial in $s$, are of greater importance than those of the form (3). The degree $m$ of the polynomial $f$ then satisfies the condition $m \leq n$, since otherwise $H$ would be infinite for points located on the imaginary $p$ axis. In fact, the theory to be given hereafter remains valid even if $f$, although being either even or odd, contains, e.g., square root expressions, as may be the case for wave digital filters [8]–[11].

We now replace the definition (5) by

$$e^{j2B_n} = \pm H_*/H \quad (15)$$

where the plus sign has to be chosen for $f$ even and the minus sign for $f$ odd. The function $B_n$ thus defined is, for $p=j\omega$, still equal to the phase corresponding to $H$, expect for an additive term equal to a multiple of $\pi/2$ appearing in case of $f$ odd. This additional term does not affect the group delay and is thus of no importance to our problem. Due to (4) and (14), we derive from (15) the relation

$$jB_n = \tanh^{-1} F_n. \quad (16)$$

The ideal value for $B_n$ is again the value of $B$ given by (7). As can be seen from (16), the ideal value for $F_n$ is thus again the value of $F$ given by (8), and thus by (10), but with $\mu$ now defined by

$$\mu = 2\tau/T. \quad (17)$$

Using this definition rather than (9), the function $F_n$ defined by (12) is still the desired approximation to be used, together with (4), in (14). The condition for positiveness of all coefficients in (12), i.e., $\mu>n-1$, however, is now given by

$$\tau > (n-1)T/2. \quad (18)$$

It is thus no longer sufficient to require $\tau$ to be positive in order to satisfy the stability requirement.

The transfer functions studied by Abele [5] are special cases of those given by (14), and the results described by this author coincide with those obtained in this section. Furthermore, the results of Section II can be derived easily from the present ones by writing (3) in the form

$$H = \frac{f}{d} e^{npT/2}, \tag{19}$$

which can easily be obtained by making use of (2) and by defining $f$ by means of

$$f = (1 - s^2)^{n/2}.$$

Thus, the problem of choosing the function $H$ defined by (19) in such a way that the corresponding delay approximates $\tau$ is equivalent to choosing $f/d$ such that the corresponding delay approximates $\tau + nT/2$. This shows in particular why (18) reduces to (13) if $\tau$ is replaced by $\tau + nT/2$.

It may be surprising to find that condition (13) actually is compatible with a certain range of negative values of $\tau$. This paradox can be explained by noticing that even for the extreme value $\tau = -T/2$, the value of $|\tau|$ is only equal to half a sampling period, which is too small a value to allow proper interpretation as the precise value of the time delay of a signal. If a discrete-time system approaches a continuous-time system, we have $T \to 0$, in which case (13) approaches the inequality $\tau > 0$.

## References


[1] J.-P. Thiran, "Recursive digital filters with maximally flat group delay," in *1970 IEEE Int. Symp. Circuit Theory, Dig. Tech. Papers*, pp. 139–140.

[2] ——, "Recursive digital filters with maximally flat group delay," *IEEE Trans. Circuit Theory*, vol. CT-18, pp. 659–664, Nov. 1971.

[3] ——, "L'approximation des caractéristiques idéales par les filtres digitaux récursifs," Ph.D. dissertation, Catholic Univ. Louvain, Louvain, Belgium, 1971.

[4] L. Storch, "Synthesis of constant-time-delay networks using Bessel polynomials," *Proc. IRE*, vol. 42, pp. 1666–1675, Nov. 1954.

[5] T. A. Abele, "Transmission line filters approximating a constant delay in a maximally flat sense," *IEEE Trans. Circuit Theory*, vol. CT-14, pp. 298–306, Sept. 1967.

[6] O. Perron, *Die Lehre von den Kettenbrüchen*, vol. II. Stuttgart: Teubner, 1957, p. 154.

[7] H. S. Wall, *Analytic Theory of Continued Fractions*. New York: Van Nostrand, 1958.

[8] A. Fettweis, "Entwurf von Digitalfiltern in Anlehnung an Verfahren der klassischen Netzwerktheorie," NTZ Report in *NTG-Fachtagung "Analyse und Synthese von Netzwerken"* (Stuttgart), Oct. 1970, pp. 17–20.

[9] J. A. C. Bingham, "A new type of digital filter with a reciprocal ladder configuration," in *1970 IEEE Int. Symp. Circuit Theory, Dig. Tech. Papers*, pp. 129–130.

[10] A. Fettweis, "Digital filter structures related to classical filter networks," *Arch. Elek. Übertragung*, vol. 25, pp. 79–89, Feb. 1971.

[11] ——, "Some principles of designing digital filters imitating classical filter structures," *IEEE Trans. Circuit Theory* (Corresp.), vol. CT-18, pp. 314–316, Mar. 1971.

# Considerations of the Padé Approximant Technique in the Synthesis of Recursive Digital Filters

FRANCIS BROPHY and ANDRES C. SALAZAR,
Member, IEEE

*Abstract*—The Padé approximant technique provides a quick design of recursive digital filters. An added advantage of the technique lies in that spectrum shaping requirements as well as linear phase constraints can be handled easily, even for higher order filters. This is important in supplying initial guesses of the filter parameters to iterative routines that would then seek a locally optimal design solution. These advantages are among those discussed in a partly tutorial presentation of the technique that relates to filter needs found in data transmission systems. In addition, the question of stability is treated and a new criterion is presented. The criterion provides sufficient conditions in establishing stability for a filter designed by using the Padé approximant technique.

## I. Introduction

The design of spectrum-shaping recursive digital filters in the $z$-plane often requires the use of a routine that calculates the extremum of an object function of several variables. The function is generally nonlinear and positive definite and indicates the "closeness" of the designed spectrum to the desired spectrum. In some cases, depending on the complexity of the function, the number of iterations or even convergence to an extremum is dependent on the initial guess for the $\alpha$ and $\beta$ (feedforward and feedback) parameters.

By working in the time domain the degrees of freedom available can be used to match a set of time samples exactly, thus reducing the design to the solution of a linear system of equations. While this approach, call the Padé approximant,[1] does not lead to a locally optimal solution as an iterative technique would, it nevertheless provides a viable solution in a fraction of the time.

In the following we show how the Padé approximant technique can yield a simple digital filter design for spectrum shaping networks with linear (or nonlinear, if so desired) phase constraints often required in data transmission systems. The problem of stability is discussed and sufficient conditions are given to ensure that the design procedure will not lead to an unstable filter.

## II. Padé Approximate in Digital Filter Design

Let $H(\omega)$ denote in the interval $[-2\pi W, 2\pi W]$ the bounded filter amplitude characteristic that is to be synthesized. Since $H(\omega) \in L_p[-2\pi W, 2\pi W]$, $p \geqslant 1$ it has a Fourier series expansion

$$H(\omega) = \sum h_n e^{-jn\omega/2W} \tag{1}$$

Manuscript received December 28, 1972; revised June 29, 1973.

The authors are with Bell Laboratories, Holmdel, N.J. 07733.

[1] "Prony's method" is related to the Padé approximant technique through a transformation of variables (see [1] for more details).

Reprinted from *IEEE Trans. Audio Electroacoust*, vol. AU-21, pp. 500–505, Dec. 1973.

where $h_n$ are the samples of the impulse response of $H(\omega)$ at the appropriate sampling period $1/2W$ s.

Also, given $\epsilon > 0$, there exists an integer $K$ such that

$$\frac{1}{2W} \sum_{-K}^{K} h_n^2 \geqslant \frac{1}{2\pi} \int_{-2\pi W}^{2\pi W} |H(\omega)|^2 \, d\omega - \epsilon.$$

In fact, the associated Fourier expansion of $H(\omega)$, viz.,

$$\sum_{-K}^{K} h_n e^{-jn\omega/2W},$$

is the best approximation in the mean ($L_2$) to $H(\omega)$ on the interval $[-2\pi W, 2\pi W]$ using only $2K + 1$ coefficients.

By adding the constant delay $K/2W$ to $H(\omega)$ we can reindex the samples $\{h_n\}_{-K}^{\infty}$ to $\{h_n\}_0^{\infty}$. Also, normalization on $H(\omega)$ can be induced by dividing $\{h_n\}_0^{\infty}$ by $h_0 \neq 0$ so that $\{h_n\}_0^{\infty}$ now represents the samples of a time function whose Fourier transform closely approximates a delayed and normalized version of $H(\omega)$.

A realizable digital filter transfer function has the form of a rational function[2] in $z^{-1}$:

$$G(z^{-1}) = \frac{1 + \alpha_1 z^{-1} + \alpha_2 z^{-2} + \cdots + \alpha_M z^{-M}}{1 - \beta_1 z^{-1} - \beta_2 z^{-2} - \cdots - \beta_N z^{-N}}. \tag{2}$$

If $z^{-1} = 0$ is not a point of singularity then

$$G(z^{-1}) = \sum_{n=0}^{\infty} g_n(\vec{\alpha}, \vec{\beta}) z^{-n} \tag{3}$$

is the Maclaurin series where $\vec{\alpha} = (\alpha_1, \alpha_2, \cdots, \alpha_M)$ and $\vec{\beta} = (\beta_1, \beta_2, \cdots, \beta_N)$ for a positive radius of convergence $0 < |z^{-1}| < r_0$. If stability constraints[3] are placed on the $\beta$'s we then have that $r_0 \geqslant 1$. The term $g_n(\vec{\alpha}, \vec{\beta})$ is easily found recursively from

$$g_n(\vec{\alpha}, \vec{\beta}) = \beta_1 g_{n-1}(\vec{\alpha}, \vec{\beta}) + \cdots + \beta_N g_{n-N}(\vec{\alpha}, \vec{\beta}) + \alpha_n \tag{4}$$

where $g_0(\vec{\alpha}, \vec{\beta}) = 1$ and $\alpha_n = 0$ if $n > M$.

If we represent $h_n$ as the response samples of the digital filter we wish to synthesize we can equate (1), $(z = e^{jn\omega/2W})$, and (4) and find equations involving the $\alpha$ and $\beta$ parameters:

$$h_n = \beta_1 g_{n-1} + \beta_2 g_{n-2} + \cdots + \beta_N g_{n-N} + \alpha_n, \qquad 1 < n < M \tag{5}$$

$$h_n = \beta_1 g_{n-1} + \beta_2 g_{n-2} + \cdots + \beta_N g_{n-N}, \qquad n > M \tag{6}$$

where the arguments of $g_n$ have been dropped. Clearly there are not enough degrees of freedom in $G(z^{-1})$ upon fixing $M$ and $N$ to generate any response $\{h_n\}_{n=0}^{\infty}$ that satisfies $\Sigma |h_n| < \infty$. However, there is a possibility of minimizing $\mathcal{E}$, the weighted squared difference between the two responses through the $T$th sample, $T \geqslant M + N$:

$$\mathcal{E} = \sum_{n=1}^{T} W_n (h_n - g_n)^2, \; W_n \geqslant 0, \qquad n = 1, 2, \cdots T.$$

Since $g_n$ is a nonlinear function of the $\alpha$ and $\beta$ parameters the minimization of $\mathcal{E}$ for $T \geqslant M + N$ can be achieved only through iterative means. In one case, however, viz., $T = M + N$, the $\alpha$ and $\beta$ parameters which minimize $\mathcal{E}$ can be found by solving $M + N$ linear equations.[4] By assuming $g_n = h_n$, $n = 1, \cdots M$, we can then solve (6) for the $\beta$'s that produce $g_n = h_n$, $n = M + 1, M + 2, \cdots, M + N$. Further we note that for a given set of $\{\beta_k\}_{k=1}^{N}$ we can solve (5) for $\alpha$'s so that $g_n = h_n$, $n = 1, 2, \cdots, M$. This procedure amounts to equating the truncated power series of (1), i.e.,

$$\sum_{n=0}^{M+N} h_n z^{-n},$$

to the first $M + N + 1$ terms of the MacLaurin series (3). This approximation of a power series by a rational function is commonly referred to as the Padé approximate procedure. This technique generates the $M + N + 1$ samples $\{h_n\}_0^{M+N}$ exactly but produces a tail $\{g_n\}_{M+N+1}^{\infty}$ that can serve to yield a better approximation to $H(\omega)$.

Time domain synthesis of *analog* filters (i.e., s plane) utilizing this method is well documented [3]. By approximating the desired digital filter response by reproducing the first $M + N + 1$ samples ($h_0$ is given free by initial normalization) it is hoped that the overall time or frequency response from the approximating filter will not deviate unreasonably from the desired one. There is, however, no guarantee that the $\beta$'s compose a stable filter. In Section III we will discuss stability considerations of the Padé technique.

## III. Stability of the Padé Approximant

Since the Padé approximant produces a rational function approximation, we know how the stability of the approximation can be related to the moduli of the poles of the approximant. We would like to know, however, if we can relate the stability of the proposed Padé approximant to some measure other than the poles.

We can state the matrix equation of the Padé approximant in the following form by reindexing $\{g_n\}_0^{\infty}$ and $\{h_n\}_0^{\infty}$ to $\{g_n\}_{-N+1}^{\infty}$ and $\{h_n\}_{-N+1}^{\infty}$, respectively.

[2] For notational convenience $G$ is normalized to 1 at $z^{-1} = 0$.

[3] A filter is stable if bounded output results from bounded input.

[4] This special case has been discussed in [2] and [6] for synthesis of a "time" sequence $\{h_n\}$.

$$(\beta_1 \cdots \beta_N)\begin{bmatrix} h_0 & h_1 & \cdots & h_{N-1} \\ h_{-1} & h_0 & & h_{N-2} \\ \vdots & h_{-1} & & \vdots \\ \vdots & & & \\ h_{-N+1} & h_{-N+2} \cdots & h_{-1} & h_0 \end{bmatrix} = \begin{bmatrix} h_1 \\ \vdots \\ h_N \end{bmatrix}^t.$$

We define

$$Q_i = \begin{bmatrix} q_i & q_{i+1} & \cdots & q_{i+N-1} \\ q_{i-1} & q_i & & \vdots \\ \vdots & & & \\ q_{i-N+1} & & \cdots & q_i \end{bmatrix} \tag{9}$$

where $q_i = h_i$ or $g_i$ and $Q_i = H_i$ or $G_i$. Note that for the Toeplitz forms $G_i$ there is a recurrence relation:

$$G_{i+2} = BG_{i+1} \text{ or } G_{i+2} = B^{i+2} H_0$$

since $G_0 = H_0$ and $G_1 = H_1$, where $B$ is the $N \times N$ companion[5] matrix, $B = H_1 H_0^{-1}$.

$$B = \begin{bmatrix} \beta_1 & \beta_2 & & & \beta_N \\ 1 & 0 & & & 0 \\ & 1 & & & 0 \\ & & 1 & & \cdot \\ & & & \ddots & \vdots \\ & & & 1 & 0 \end{bmatrix}. \tag{10}$$

We note that

$$H_1 = BH_0 \tag{11}$$

where $B$ has eigenvalues $\{\lambda_1, \lambda_2, \cdots, \lambda_N\}$.[6] We know, in addition, that the eigenvector corresponding to $\lambda_i$ is $(1, \lambda_i^{-1}, \lambda_i^{-2}, \cdots, \lambda_i^{-N+2}, \lambda_i^{-N+1})$. We can decompose $B$ in terms of this eigensystem to get

$$B = P \Lambda P^{-1} \quad \text{and} \quad G_n = P \Lambda^n P^{-1} H_0 \tag{12}$$

where $\Lambda$ is a diagonal eigenvalue matrix and $P$ is an $N \times N$ matrix whose columns are the eigenvectors of $B$.

From (12) we can see that a bounded impulse response $\{g_n\}$ is possible only if $|\lambda_k| \leqslant 1$, $k = 1, 2, \cdots N$. Since $g_n \to 0$ necessarily for a stable filter, this means $|\lambda_k| < 1$. The converse is also true, i.e., $|\lambda_k| < 1$ implies filter stability. This can be most easily seen by noting that the eigenvalues of $B$ are precisely the roots of the denominator polynomial of $G(z^{-1})$ since $B$ is its companion matrix. This development, of course, is well known in linear system theory. The point we make here, however, is that the expression for $B$, namely $B = H_1 H_0^{-1}$, involves only the first $2N + 1$ samples of the target sequence $\{h_n\}$. Hence, we can test $H_1 H_0^{-1}$ for its eigenvalue of maximum modulus. If it is less than one, the Padé procedure yields a stable filter. Hence, a valuable condition for stability of the synthesized Padé filter is easily obtained in this fashion.

To see this, let us index the column vectors of $H_0$ by $\{\vec{h}_0, \vec{h}_1, \cdots, \vec{h}_{N-1}\}$ where $\vec{h}_k^t = (h_{k0}, h_{k1}, \cdots, h_{kN-1})$ for $k = 0, 1, 2, \cdots, N-1$. Since $H_0$ was assumed invertible we have then that $\{\vec{h}_k\}_{k=0}^{N-1}$ form a linearly independent set of vectors. Let $\vec{x}$ be a nonzero vector in $R^N$ (Euclidean $N$ space). Then there exist constants $\gamma_0, \gamma_1, \cdots, \gamma_{N-1}$ (not all zero) such that $\vec{x} = \sum_{k=0}^{N-1} \gamma_k \vec{h}_k$. Hence, the squared Euclidean norm of $\vec{x}$ is

$$\|\vec{x}\|^2 = \left\| \sum_{k=0}^{N-1} \gamma_k \vec{h}_k \right\|^2 = \sum_{l=0}^{N-1} \left\{ \sum_{k=0}^{N-1} \gamma_k h_{kl} \right\}^2$$

$$\|\vec{x}\|^2 = \sum_{k=0}^{N-1} \sum_{m=0}^{N-1} \gamma_k \gamma_m \sum_{l=0}^{N-1} h_{kl} h_{ml}. \tag{13}$$

Now define

$$c_{km} = \sum_{l=0}^{N-1} h_{kl} h_{ml},$$

which forms a strictly positive definite matrix, $C_0$, Similarly, we can form

$$\|B\vec{x}\|^2 = \left\| \sum_{k=0}^{N-1} \gamma_k B\vec{h}_k \right\|^2. \tag{14}$$

But we recall from (11) that $B\vec{h}_k = \vec{h}_{k+1}$, $k = 0, 1, \cdots, N-1$, [where $\vec{h}_N^t = (h_{N0}, h_{N1}, \cdots, h_{NN-1})$]. Thus, we can now write

$$\|B\vec{x}\|^2 = \sum_{l=0}^{N-1} \left\{ \sum_{k=0}^{N-1} \gamma_k h_{k+1,l} \right\}^2$$

$$= \sum_{k=0}^{N-1} \sum_{m=0}^{N-1} \gamma_k \gamma_m \sum_{l=0}^{N-1} h_{k+1,l} h_{m+1,l}. \tag{15}$$

Form

$$c_{k+1,m+1} = \sum_{l=0}^{N-1} h_{k+1,l} h_{m+1,l},$$

$$k, m = 0, 1, \cdots, N-1, \tag{16}$$

a positive definite matrix, $C_1$. We are now ready to formulate the following.

*Stability Criterion:* The digital filter formed by the

[5] $H_0$ is invertible since it is nonsingular for cases of interest, (e.g., $H(z^{-1})$ is not a rational function). Fast techniques for inverting $H_0$ are available [4].

[6] It does not seem overly restrictive to assume simple roots here.

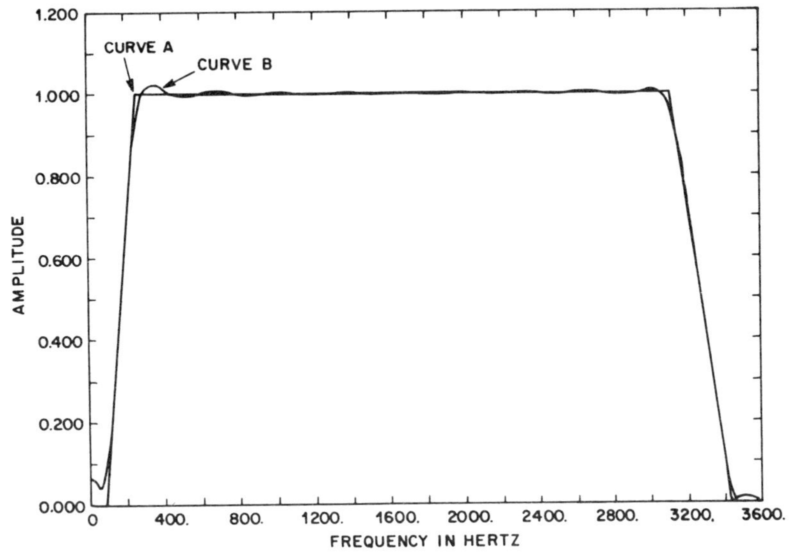


**Fig. 1. Amplitude spectrum for bandpass filter (24th order).**

Padé approximant procedure outlined in Section II for $M = N$ and invertible $H_0$ is stable if $C_0 - C_1 > 0$, that is, the matrix is a strictly positive definite matrix.

*Proof:* Let us begin by first defining $\vec{\gamma} = (\gamma_0, \gamma_1, \cdots, \gamma_{N-1})$. Then $(C_0\vec{\gamma}, \vec{\gamma}) = \|\vec{x}\|^2$ from (13) and the definition of $C_0$ while $(C_1\vec{\gamma}, \vec{\gamma}) = \|B\vec{x}\|^2$ from (15) and (16). Hence, $((C_0 - C_1)\vec{\gamma}, \vec{\gamma}) > 0$, for nontrivial $\vec{x}$. So $\|\vec{x}\|^2 - \|B\vec{x}\|^2 > 0$ or $\|B\vec{x}\| < \|\vec{x}\|$ for arbitrary $\vec{x}$ so conclude $\|B\| < 1$. But we know that the eigenvalue of maximum modulus of $B$, namely $|\lambda_{\max}(B)|$ is bounded above by $\|B\|$ so we have the desired conclusion.

Armed with this stability criterion we check for the possibility of instability in a Padé designed filter by simply following three steps.

*Step 1:* Obtain the time sequence $\{h_k\}_{-N+1}^{N-1}$.

*Step 2:* Form the matrices $C_0$ and $C_1$, (total of $N(N+3)/2$ inner products are necessary).

*Step 3:* Calculate the principal minors of $C_0 - C_1$ and check for positivity.

The last step is the well-known test for positive definiteness of a matrix [5, p. 59]. The usefulness of following the procedure is appreciated when the alternatives for checking stability are considered. For example, stability can be determined by simply solving for the set of $\beta$'s and then applying a root finding routine to the resulting polynomial. Equivalently, we can compute $H_1 H_0^{-1} = B$ and then apply an eigenvalue-eigenvector routine to the matrix. Last, we consider solving for the $\beta$'s and then computing the response $\{g_n\}_{N+1}^{T}$ of (4). If the response is seen to decrease the filter is stable. In all three cases, a sizable amount of computation is involved when compared to the three step stability algorithm discussed above.

To avoid misrepresentation, it must be pointed out that stability is still possible even if $C_0 - C_1 \not> 0$. In fact, if $h_N$ of $\{h_n\}_0^\infty$ is chosen to be the peak of the impulse response of $H(\omega)$, then $C_0 - C_1$ can not be expected to be strictly positive definite. However, we improve the chance that $C_0 - C_1 > 0$ if we choose the index of the peak closer to $N/2$.[7]

Last, we mention that the condition that $H_0$ be invertible has always been met easily in all the filter designs we have attempted using the Padé technique.[8]

## V. Examples

Realistic examples have been chosen to illustrate the development in the previous sections. They constitute two filter designs that are typical of data transmission systems employing digital signal processing.

As the first example[9] consider the bandpass shape depicted in Fig. 1 and labeled by curve $A$. Suppose that design criteria for this spectrum specify 3-dB loss points at 200 Hz and 3200 Hz and no more than a 0.25-dB peak to peak ripple over the passband. Linear phase over the passband as well as slope loss greater than 12 dB/octave in the rejection bands are other requirements. The classical approach to designing a

[7] The intention of our work on the stability question has been to employ the full symmetry of $B = H_1 H_0^{-1}$. The classical approach to determining conditions on $B$ so that $B^n \to 0$ has not been especially useful in our case in spite of this symmetry. (See [7] and [8] for a summary of results in this area.)

[8] We offer this statement in response to the warnings given by a theorem (commonly found in studies of Toeplitz forms) that says that the minimum eigenvalue of $H_0$, $|\lambda_{\min}(H_0)|$, monotonically approaches $\inf_\omega |H(\omega)|$ (as $N \to \infty$) which, of course, is zero for most filters of interest—low pass, bandpass; etc. Hence, $H_0$ would not be invertible if $|\lambda_{\min}(H_0)| = 0$. More importantly, machine computation of $H_0$ inverse becomes more difficult as $N$ increases even though $|\lambda_{\min}(H_0)| \neq 0$. (See [9, p. 147] and [10, pp. 201-202] for a precise statement of this theorem.)

[9] In each of the examples the sampling rate is 7200 samples/s.

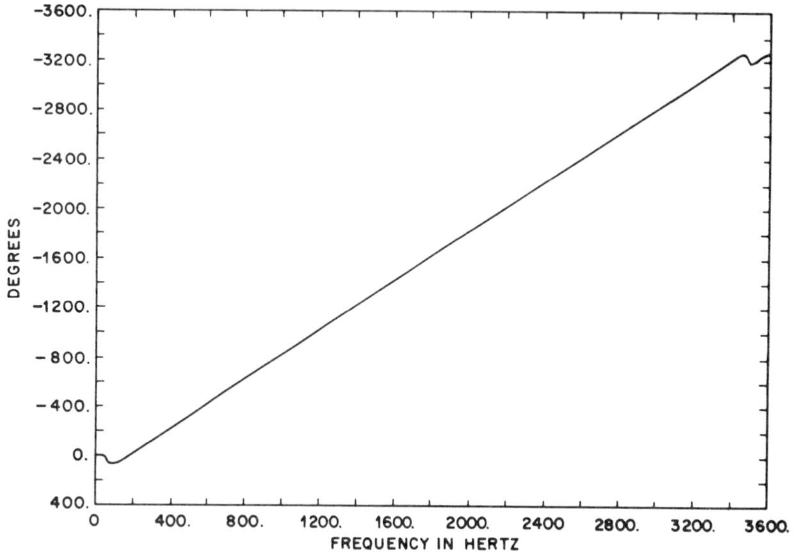


Fig. 2. Phase for bandpass filter (24th order).

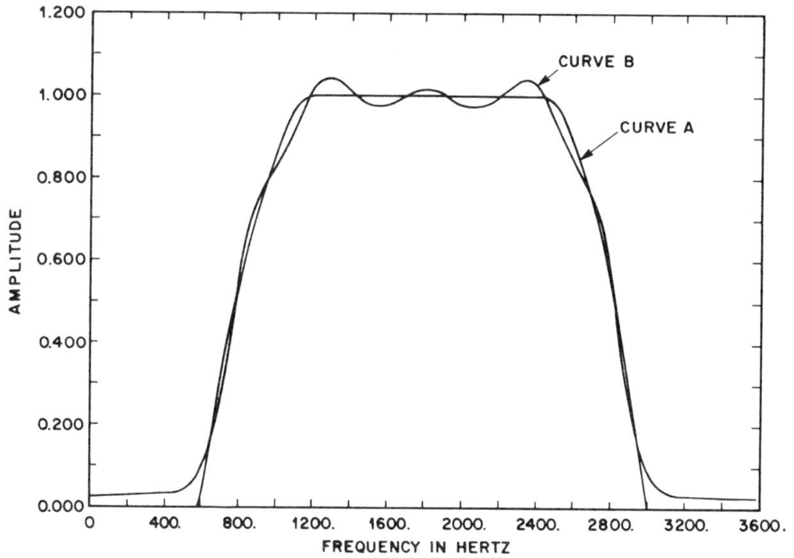


Fig. 3. Amplitude spectrum for cosine rolloff filter (10th order).

filter of this type would be to use several degrees of freedom to satisfy the amplitude spectrum requirements and then attempt phase equalization with the remaining degrees of freedom. Curve $B$ represents the amplitude spectrum of a Padé realization of the desired spectrum in 12 second-order sections (24th order filter). The resultant phase is shown in Fig. 2. The largest absolute error for the time samples in this realization is 0.0018 where $\max |h_n| = 1$.

Similarly, a tenth ($N = 10$) order recursive digital filter has been synthesized to match the spectral shape of the bandpass function with cosine rolloff labeled $A$ in Fig. 3. (This receiving filter shape can be used for baseband data transmission on channels with no frequency offset.) A time sequence comparison reveals $\max_n |h_n - g_n| = 0.024$ where $\max_n |h_n| = 1$. Its approximately linear phase is shown in Fig. 4. This example provides us with an opportunity to discuss the stopband behavior of Padé-realized filters. At times a minimum loss of 40, 50, or even 80 dB is specified for a filter with a stopband. This type of requirement cannot be easily worked into the time domain terms of the Padé synthesis procedure. Hence, only iterative routines, working in the frequency domain, from the Padé initial $\alpha$ and $\beta$ values, could work these requirements into the final design. From our experience in obtaining initial designs by the Padé

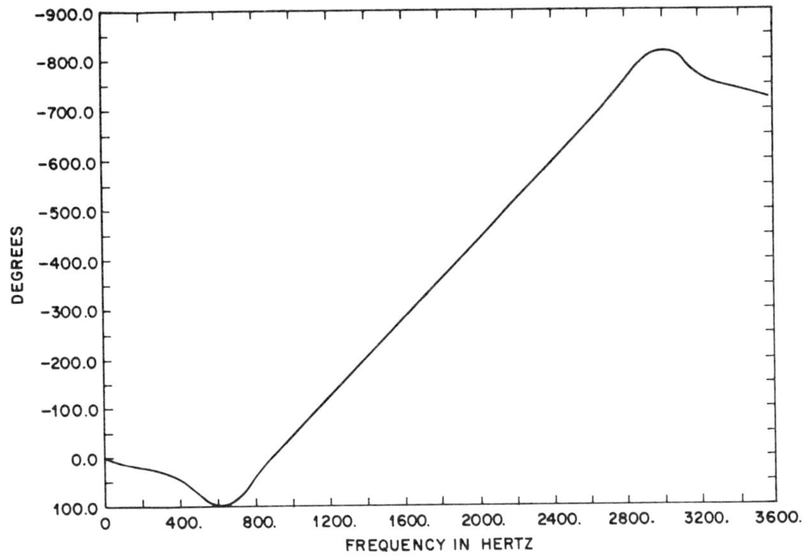


**Fig. 4. Phase for cosine rolloff filter (10th order).**

technique we have observed that the stopband loss ranges from 20 to 40 dB, which in many cases may be satisfactory.

## References

[1] L. Weiss and R. N. McDonough, "Prony's method, z-transforms, and Padé approximation," *SIAM Rev.*, vol. 5, no. 2, p. 145, 1963.

[2] C. S. Burrus and T. W. Parks, "Time domain design of recursive digital filters," *IEEE Trans. Audio Electroacoust.*, vol. AU-18, pp. 137-141, June 1970.

[3] K. L. Su, *Time Domain Synthesis of Linear Networks.* Englewood Cliffs, N.J.: Prentice-Hall, 1971, ch. 2.

[4] W. F. Trench, "An algorithm for the inversion of finite toeplitz matrices," *J. Appl. Math.*, vol. 12, pp. 515-522, Sept. 1964.

[5] E. F. Beckenbach and R. Bellman, *Inequalities.* New York: Springer Verlag, 1965.

[6] J. L. Shanks, "Recursion filters for digital processing," *Geophysics*, vol. 32, pp. 33-51, Feb. 1967.

[7] P. Stein, "Some general theorems on iterants," *J. Res. Nat. Bur. Stand.*, vol. 48, pp. 82-83, Jan. 1952.

[8] O. Taussky (Todd), "Matrices $C$ with $C^n \to 0$," *J. Algebra*, vol. 1, pp. 5-10, 1964.

[9] U. Grenander and G. Szego, *Toeplitz Forms and Their Applications.* Berkeley, Calif: Univ. California Press, 1958.

[10] H. Widom, "Toeplitz matrices," in *Studies in Mathematics*, vol. 3, I. I. Hirschman, Jr., Ed. Math. Assoc. of Amer. Englewood Cliffs, N.J.: Prentice-Hall, 1965, pp. 179-209.

# Synthesis of Recursive Digital Filters Using the Minimum $p$-Error Criterion

ANDREW G. DECZKY

*Abstract*—The problem of designing a stable recursive digital filter to have an arbitrarily prescribed frequency response may be considered as an approximation problem. Using the minimum $p$-error criterion, a new problem of minimizing a function of $n$ variables results, which is successfully solved using the Fletcher–Powell algorithm. An important theorem guaranteeing the existence of a stable optimum for a large class of synthesis problems is stated, and necessary modifications to the Fletcher–Powell algorithm to assure stability are considered. Finally a number of results of the application of this method are given.

## I. The Approximation Problem

The problem of designing a recursive digital filter to have an arbitrarily prescribed frequency response may be regarded as a classical approximation problem. The advantage of this approach is that many methods of solution for such problems exist.

The classical approximation problem may be stated as follows [14]. Let $f(x)$ be a given real-valued function defined on a set $X$, and let $F(A, x)$ be a real-valued approximating function depending continuously on $x \in X$ and on $n$ parameters $A$. Given a distance function $\rho$, determine the $n$ parameters $A^* \in P$ such that

$$\rho[F(A^*, x), f(x)] \leq \rho[F(A, x), f(x)], \quad \text{for all } A \in P. \quad (1)$$

The solution to this problem is said to be a best approximation with respect to the distance function $\rho$ chosen.

For recursive digital filter synthesis we have the following identification. Let the transfer function $H(z)$ be a function of $n$ parameters (e.g., the filter coefficients) and order these in a vector $A$. The independent variable $x$ is now the digital frequency $\phi = \omega T$, and the set $X$ is $(\phi: 0 \leq \phi \leq \pi)$. Then the frequency response of the filter (such as magnitude or group delay) may be expressed as a real-valued function of $A$ and $\phi$, i.e., $F(A, \phi)$, while the desired frequency response becomes $f(\phi)$.

Having defined the problem in these general terms, we must now make specific choices for the distance function $\rho$, the form of the filter transfer function $H(z)$, the parameter vector $A$, and the method of solution (this depending on the distance function $\rho$ chosen).

## II. The Distance Function

The distance function we have chosen is one of the most often used, namely the weighted $L_p$ norm

$$\|L(A)\|_p = \left\{ \int_0^\pi w(\phi) \,|\, F(A, \phi) - f(\phi) \,|^p d\phi \right\}^{1/p} \quad (2)$$

where $w(\phi)$ is a positive weighting function. The reason for this choice is that $L_p$ approximations have been extensively studied, and their properties are well known. Furthermore, the problem of finding the best $L_p$ approximation numerically is transformed into the minimization of a function of $n$ variables, for which very powerful algorithms exist. Finally the two most often used error criteria, namely the minimum square and the minimax are included as special cases of this distance

Manuscript received March 7, 1972. This work was supported by the Swiss National Science Foundation. This paper is part of a dissertation submitted to the Institute of Applied Physics, Swiss Federal Institute of Technology, Zurich, in partial fulfillment of the requirements for the Ph.D. degree. This paper was presented at the IEEE Arden House Workshop on Digital Filtering, Harriman, N. Y., January 1972.

The author is with the Institute of Applied Physics, Swiss Federal Institute of Technology, Zurich 8049, Switzerland.

Reprinted from *IEEE Trans. Audio Electroacoust.*, vol. AU-20, pp. 257–263, Oct. 1972.

function. The former follows from the definition with $p=2$, while the latter is a consequence of the following theorem [14].

*Polya Algorithm:* If there is a unique best Chebyshev (or minimax) approximation $F(A^*, x)$ to $f(x)$ on (0, 1) and $F(A_p^*, x)$ is the best $L_p$ approximation to $f(x)$, then

$$\lim_{p\to\infty} A_p^* = A^*. \tag{3}$$

The best minimax approximation for a continuous function is defined as

$$\|L(A^*)\|_\infty = \min_{A\in P} \max_{x\in X} | F(A, x) - f(x) | . \tag{4}$$

(Note that this result may be transposed to the interval $0\leq\phi\leq\pi$ by using the transformation $x=\sin \phi/2$.)

In numerical work it is convenient to use the simpler distance function

$$L_{2p}(A) = \sum_{\kappa=1}^{K} w(\phi_\kappa)[(F(A, \phi_\kappa) - f(\phi_\kappa)]^{2p} \tag{5}$$

since the minimum of $L_{2p}(A)=\|L(A)\|_{2p}{}^{2p}$ is the same as the minimum of $\|L(A)\|_{2p}$. The positive index $2p$ is used here for ease of computation of the derivatives. We also note that the points $\phi_\kappa$ need not be equally spaced. In fact they should be so distributed that the compressed error function exhibits a sinusoidal character. The number of points $K$ taken should then be about 10 per period.

Finally for a minimax approximation the index $2p$ is taken as a sequence such as 2, 4, 10, 40, the distance function being minimized each time. This defines a descent algorithm that converges to the minimax solution if the latter is unique. The largest index we tried was 40, which gave satisfactory results for most of the problems attempted.

### III. The Form of the Transfer Function *H(z)*

The form of the digital filter transfer function is important for several reasons. The physical structure of the filter has a large influence on the error caused by quantization effects and finite coefficient size [2], [8], [13]. Thus the direct form is the worst in this respect, while the cascade and parallel forms each have distinct advantages. If we want to have sharply defined zeros on the unit circle (well-defined stopbands), then the cascade form is superior and was chosen for this reason.

The second reason for this choice is that the stability of a cascade filter is readily tested, whereas for a direct form this requires lengthy calculations which are impractical. Since during the process of optimization we have to assure in some way that the filter stays stable, a cascade structure is again advantageous.

Finally the frequency response of the cascade filter takes on a particularly simple functional form, permitting easy calculation of the first derivatives and providing insight into the way the poles and zeros influence the distance function.

Thus we have chosen the decompose $H(z)$ into a cascade of second-order sections

$$H(z) = k_0 \prod_{i=1}^{N} H_i(z) = k_0 \prod_{i=1}^{N} \frac{z^2 + a_{1i}z + a_{2i}}{z^2 + b_{1i}z + b_{2i}} \tag{6}$$

where $k_0$ is a positive normalizing constant.

### IV. The Parameter Vector A

The most obvious choice for the parameter vector $A$ is to take the multiplier coefficients of the cascade structure in (6) directly. Thus

$$A = (a_{11}, a_{12}, b_{11}, b_{12}, \cdots, a_{1i}, a_{2i}, b_{1i}, b_{2i}, \cdots, k_0). \tag{7}$$

In this case we have the following expressions for the amplitude and group delay of the overall filter. (For simplicity we have assumed that the numerator and denominator have the same degree, though of course this is not necessary.)

$$\alpha(A, \phi) = H(e^{j\phi}) = k_0 \prod_{i=1}^{N} \frac{(1+a_{2i}) \cos \phi + a_{1i} + j(1-a_{2i}) \sin \phi}{(1+b_{2i}) \cos \phi + b_{1i} + j(1-b_{2i}) \sin \phi} \tag{8}$$

$$\tau(A, \phi) = -\frac{d}{d\phi} [\measuredangle H(z)]_{z=e^{j\phi}} = \sum_{i=1}^{N} \left[ \mathrm{Re}\left( \frac{2 \cos \phi + b_{1i} + j2 \sin \phi}{(1+b_{2i}) \cos \phi + b_{1i} + j(1-b_{2i}) \sin \phi} \right) - \mathrm{Re}\left( \frac{2 \cos \phi + a_{1i} + j2 \sin \phi}{(1+a_{2i}) \cos \phi + a_{1i} + j(1-a_{2i}) \sin \phi} \right) \right] \tag{9}$$

where $j=\sqrt{-1}$. Note that these are fairly complicated complex expressions where the variation of $\alpha$ or $\tau$ as a function of the $a$'s and $b$'s is not evident. The derivatives of these expressions with respect to the parameters are even more complicated, though readily derived.

Let us now consider another choice for the parameters, namely the polar coordinates of the poles and zeros. We then have

$$A = (r_{01}, \phi_{01}, r_{p1}, \phi_{p1}, \cdots, r_{0i}, \phi_{0i}, r_{pi}, \phi_{pi}, \cdots, k_0). \tag{10}$$

In this case we get the following real expressions for the magnitude and group delay:

$$\alpha(A, \phi) = k_0 \prod_{i=1}^{N} \frac{\{1 - 2r_{0i} \cos (\phi - \phi_{0i}) + r_{0i}^2\}^{1/2}\{1 - 2r_{0i} \cos (\phi + \phi_{0i}) + r_{0i}^2\}^{1/2}}{\{1 - 2r_{pi} \cos (\phi - \phi_{pi}) + r_{pi}^2\}^{1/2}\{1 - 2r_{pi} \cos (\phi + \phi_{pi}) + r_{pi}^2\}^{1/2}} \tag{11}$$

$$\tau(A, \phi) = \sum_{i=1}^{N} \left[ \frac{1 - r_{pi} \cos (\phi - \phi_{pi})}{1 - 2r_{pi} \cos (\phi - \phi_{pi}) + r_{pi}^2} + \frac{1 - r_{pi} \cos (\phi + \phi_{pi})}{1 - 2r_{pi} \cos (\phi + \phi_{pi}) + r_{pi}^2} - \frac{1 - r_{0i} \cos (\phi - \phi_{0i})}{1 - 2r_{0i} \cos (\phi - \phi_{0i}) + r_{0i}^2} - \frac{1 - r_{0i} \cos (\phi + \phi_{0i})}{1 - 2r_{0i} \cos (\phi + \phi_{0i}) + r_{0i}^2} \right]. \tag{12}$$

The influence of a single pole (or zero) on these expressions is readily visualized if we plot the group delay or magnitude of a single pole (or zero) located at $z=r_0e^{j\phi_0}$ as a function of $(\phi-\phi_0)/\pi$ with parameter $r_0$. (See Figs. 1 and 2.)

Finally, for completeness we give here the various derivatives with respect to the parameters. Thus we have

$$\frac{\partial\alpha}{\partial r_{0i}} = \alpha\left[\frac{r_{0i} - \cos(\phi - \phi_{0i})}{1 - 2r_{0i}\cos(\phi - \phi_{0i}) + r_{0i}^2} + \frac{r_{0i} - \cos(\phi + \phi_{0i})}{1 - 2r_{0i}\cos(\phi + \phi_{0i}) + r_{0i}^2}\right] \quad (13)$$

$$\frac{\partial\alpha}{\partial \phi_{0i}} = \alpha\left[\frac{r_{0i}\sin(\phi - \phi_{0i})}{1 - 2r_{0i}\cos(\phi - \phi_{0i}) + r_{0i}^2} - \frac{r_{0i}\sin(\phi + \phi_{0i})}{1 - 2r_{0i}\cos(\phi + \phi_{0i}) + r_{0i}^2}\right] \quad (14)$$

$$\frac{\partial\tau}{\partial r_{pi}} = \frac{(1 + r_{pi}^2)\cos(\phi - \phi_{pi}) - 2r_{pi}}{(1 - 2r_{pi}\cos(\phi - \phi_{pi}) + r_{pi}^2)^2} + \frac{(1 + r_{pi}^2)\cos(\phi + \phi_{pi}) - 2r_{pi}}{(1 - 2r_{pi}\cos(\phi + \phi_{pi}) + r_{pi}^2)^2} \quad (15)$$

$$\frac{\partial\tau}{\partial \phi_{pi}} = \frac{r_{pi}(1 - r_{pi}^2)\sin(\phi - \phi_{pi})}{(1 - 2r_{pi}\cos(\phi - \phi_{pi}) + r_{pi}^2)^2} - \frac{r_{pi}(1 - r_{pi}^2)\sin(\phi + \phi_{pi})}{(1 - 2r_{pi}\cos(\phi + \phi_{pi}) + r_{pi}^2)^2} \quad (16)$$

and similarly for $\partial\alpha/\partial r_{pi}$, $\partial\alpha/\partial\phi_{pi}$, $\partial\tau/\partial r_{0i}$, and $\partial\tau/\partial\phi_{0i}$, but with changed signs.

Using these expressions, the partial derivatives of the distance functions $L_{2p}{}^{\alpha}$ and $L_{2p}{}^{\tau}$ are given by

$$\frac{\partial L_{2p}{}^{\alpha}(A)}{\partial r_{0i}} = \sum_{k=1}^{K} p w_{\alpha}(\phi_k)\frac{\partial\alpha}{\partial r_{0i}}(\alpha(A, \phi_k) - f(\phi_k))^{2p-1} \quad (17)$$

$$\frac{\partial L_{2p}{}^{\tau}(A)}{\partial r_{0i}} = \sum_{k=1}^{K} p w_{\tau}(\phi_k)\frac{\partial\tau}{\partial r_{0i}}(\tau(A, \phi_k) - g(\phi_k))^{2p-1} \quad (18)$$

and similarly for $\partial L_{2p}{}^{\alpha}(A)/\partial\phi_{0i}$ and $\partial L_{2p}{}^{\tau}(A)/\partial\phi_{0i}$, etc.

## V. The Fletcher–Powell Algorithm

As we have seen above, our problem reduces to the minimization of a nonlinear function of $n$ variables. Although many methods for the solution of this class of problems exist [3], [17], probably the most powerful method known to date is the one derived by Fletcher and Powell [6].

This algorithm has several advantages. Only first derivative information is required, which in our case is readily obtained. A programmed version of the algorithm exists in the IBM Scientific Subroutine Package, making it almost universally available (though as we shall see slight modifications were made for our problem). Finally, it has been used with success over the last several years by many people [3], [16], [17], demonstrating its good convergence properties.

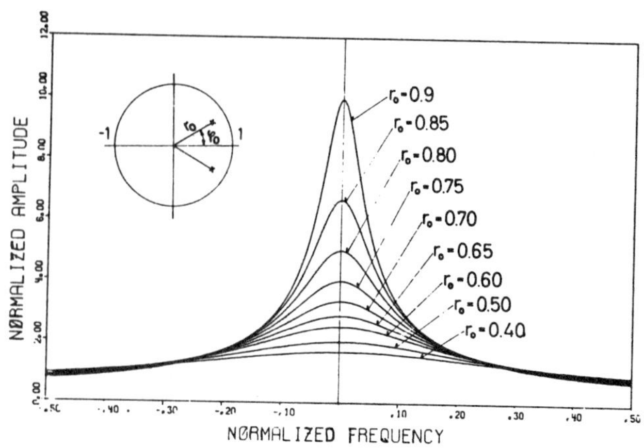


**Fig. 1.** Magnitude of a single pole at $z=z_0e^{j\phi_0}$ as a function of $(\phi-\phi_0)/\pi$ and $r_0$.

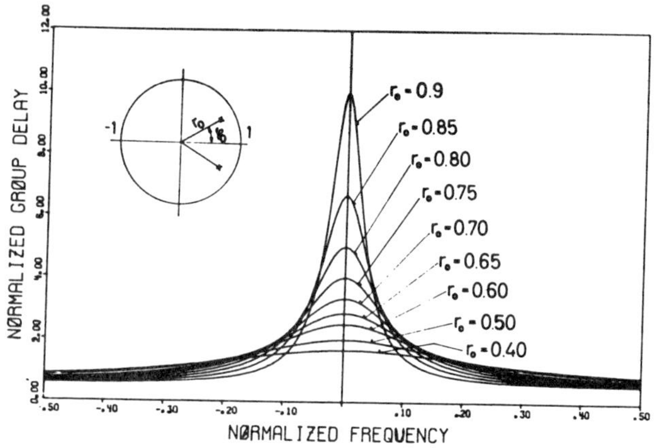


**Fig. 2.** Group delay of a single pole at $z=r_0e^{j\phi_0}$ as a function of $(\phi-\phi_0)/\pi$ and $r_0$.

The method is described in the original article by Fletcher and Powell and will not be treated in detail here. We merely mention that the method is based on the fact that for a quadratic function of $n$ variables a set of $n$ conjugate directions and a convergent set of matrices may be generated, such that after $n$ steps the matrices converge to the inverse Hessian matrix of second derivatives, evaluated at the minimum. Since $n$ conjugate directions span a space of $n$ dimensions, it follows that if we minimize the function successively in these directions, the minimum is attained in exactly $n$ steps.

Clearly, for a nonquadratic function this no longer holds. However, if the function is quadratic near the minimum, the method generates information about the local Hessian at that point and will converge quite rapidly.

A few remarks concerning the application of this algorithm to our problem are in order. First we notice that the method depends on finding local minima along certain directions. It turns out that the accuracy with which these minima are determined plays an important role in the rate of convergence. Thus it is advantageous to spend more time determining these minima accurately rather then taking a larger number of iterations.

A second remark concerns the region of search in the $n$-dimensional parameter space $P$. It is clear that we are

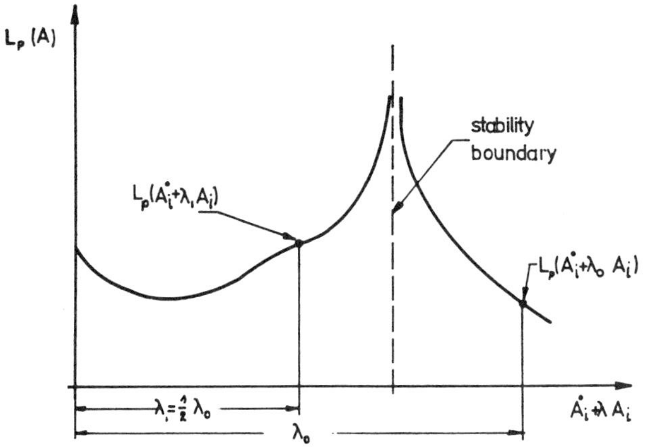


**Fig. 3.** The general nature of the distance function $L_p(A)$ and the choice of the step size.

interested in only stable filters. Thus it is logical to restrict the search to inside the stable region in $P$, that is, to the subspace $(A \in P_s : r_{pi} < 1$, for all $i)$. This is achieved by restricting the step size in any one direction such that the new point is inside the stable region $P_s$. Clearly this algorithm will converge if a (local) minimum of the distance function exists inside $P_s$, away from the boundary. That this is so is confirmed by the following important theorem.

*Theorem:* Starting with a stable approximating function $\alpha(A, \phi)$ or $\tau(A, \phi)$, a bounded function $f(\phi)$ or $g(\phi)$, and positive weighting functions $w_\alpha(\phi)$ and $w_\tau(\phi)$, a (local) minimum of the distance functions

$$L_p^\alpha(A) = \int_0^\pi w_\alpha(\phi) \mid \alpha(A, \phi) - f(\phi) \mid^p d\phi$$

$$L_p^\tau(A) = \int_0^\pi w_\tau(\phi) \mid \tau(A, \phi) - g(\phi) \mid^p d\phi \qquad (19)$$

exists for $p \geq 2$ such that the optimum parameters $A$ fulfill the stability constraints $r_{pi} < 1$ for all $i$.

The proof of this theorem is based on the fact that the amplitude or group delay of a digital filter becomes very large $(\rightarrow \infty)$ as any one of the poles approaches the stability boundary from within. Thus we see that the general nature of the distance function is as shown in Fig. 3. Should a chosen step size $\lambda_0$ be greater than the distance to the stability boundary in the given direction, it is successively halved until the new point lies inside the stable region (see Fig. 3). It has been found that with the inclusion of this condition a stable optimum was always attained.

## VI. Conclusions and Results

The method outlined in this paper for the synthesis of recursive digital filters using a weighted minimum $p$-error criterion has been applied to a large class of problems. Thus filters having arbitrary magnitude, arbitrary group delay using all-pass sections, and combined magnitude and group delay specifications, as well as all-pass group delay equalizers, were synthesized using this method. Although the running time of these programs is somewhat long (typically a few minutes for a complex ten-order filter on the CDC 6500) their great generality and flexibility make them some of the most powerful known to date for the synthesis of completely arbitrary frequency responses.

Finally we shall give a few examples of synthesis problems handled.

### Example 1

This example treats the synthesis of all-pass group delay equalizers for a low-pass elliptic filter of order 4. The original group delay distortion of this filter was about $10T$ seconds and its specifications were as follows:

| | | |
|---|---|---|
| maximum passband ripple | 0.5 dB | $0 \leq \phi \leq 0.5 f_n$ |
| minimum stopband attenuation | 32.0 dB | $0.6 f_n \leq \phi \leq f_n$ |

where $f_n = 1/2T$, the Nyquist frequency coefficients:

$$k_0 = 1.47295E - 01$$

$$a_{11} = 1.62178E + 00 \quad b_{11} = -4.03133E - 01$$
$$a_{12} = 7.18956E - 01 \quad b_{12} = 5.14010E - 02$$
$$a_{21} = 1.0 \quad b_{21} = 2.33280E - 01$$
$$a_{22} = 1.0 \quad b_{22} = 7.97295E - 01.$$

The loss characteristic of this filter is shown in Fig. 4.

The passband group delay of this filter was next equalized using all-pass sections of the form

$$H_e(z) = \prod_{i=1}^{M} \frac{(c_{2i}z^2 + c_{1i}z + 1)}{(z^2 + c_{1i}z + c_{2i})}. \qquad (20)$$

The distance function used in this case was

$$L_{2p}^\tau(A, \tau_0) = \sum_{k=1}^{K} (\tau_e(A, \phi_k) + \tau_s(\phi_k) - \tau_0)^{2p} \qquad (21)$$

where $\tau_s$ is the group delay of the original filter, $\tau_e$ is that of the equalizer, and $\tau_0$ is a constant delay whose exact value is to be determined by minimizing $L_{2p}(A, \tau_0)$ with respect to $\tau_0$.

Using a one-section equalizer $(M=1)$, an index $2p = 2$, and the initial set of poles (in polar coordinates using normalized angles)

$$r_{p1} = 0.8 \quad \phi_{p1} = 0.18 \quad \tau_0 = 0.$$

the program converged after 24 function evaluations (with an error criterion $\epsilon = 10^{-3}$ in the Fletcher–Powell subroutine) to

$$L_2^\tau(A, \tau_0) = 5.09035E + 00$$
$$c_{11} = -1.12708E + 00 \quad c_{12} = 4.36150E - 01$$
$$\tau_0 = 3.57216E - 01.$$

The group delay of this filter before and after equalization is shown in Fig. 5.

Next the index $2p$ was increased in the sequence 2, 4, 10, using the same initial poles as above. In this case the solution found after 128 function evaluations was $(\epsilon = 10^{-3})$

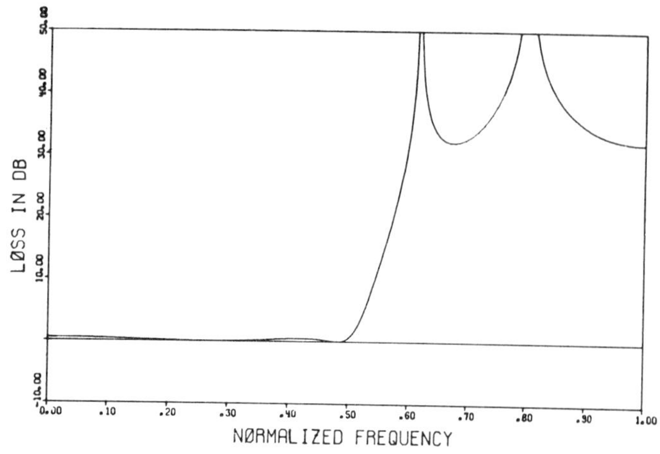


Fig. 4. Loss characteristics of a fourth-order elliptic digital low-pass filter.

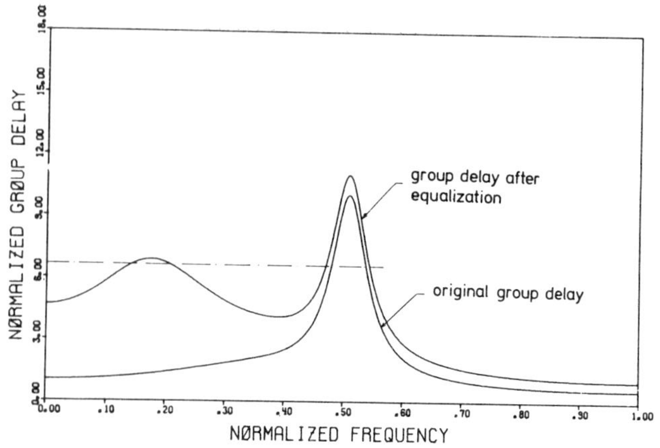


Fig. 5. Group delay of the fourth-order low-pass filter of Fig. 4 before and after equalization by a one-section all-pass with index $2p=2$.

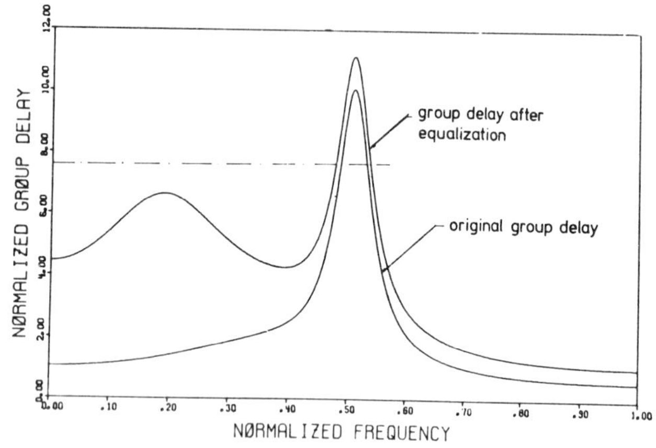


Fig. 6. Group delay of the fourth-order low-pass filter of Fig. 4 before and after equalization by a one-section all-pass with index $2p=10$.

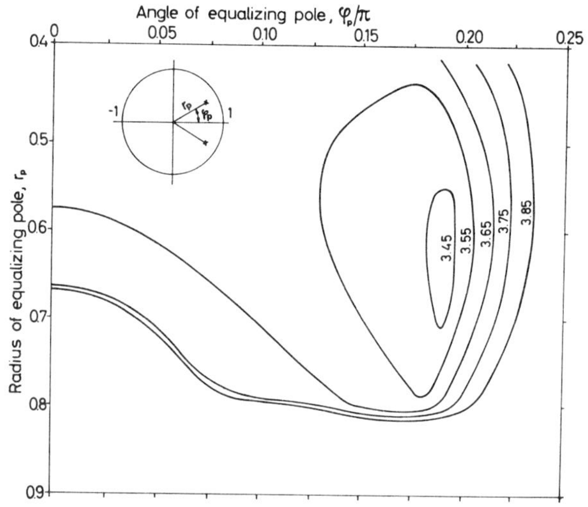


Fig. 7. Contours of the function $L_\infty(A, \tau_0)$ for the one-section group delay equalizer of Fig. 6.

$$L_{10}^{\tau}(A, \tau_0) = 6.61108E+04$$
$$c_{11} = -1.07596E+00 \quad c_{12} = 4.17425E-01$$
$$\tau_0 = 1.50940E+00.$$

This solution is shown in Fig. 6. We note here an interesting phenomenon. This solution, although quite close to the minimax one, is not equiripple, but is degenerate in the sense that the number of peaks where the maximum error is attained does not correspond to the number of degrees of freedom in the problem. To confirm this and to make sure that the optimum is unique, the exact minimax error contours were plotted in Fig. 7. (Notice here that the error surface becomes very steep as $r \to 1$, confirming the theorem in Section V.)

Finally using four second-order sections, indices 2, 4, and 10, and the initial poles

$$r_{p1} = 0.75 \quad \phi_{p1} = 0.05 \quad r_{p2} = 0.75 \quad \phi_{p2} = 0.2$$
$$r_{p3} = 0.79 \quad \phi_{p3} = 0.3 \quad r_{p4} = 0.8 \quad \phi_{p4} = 0.4$$

the following solution was obtained after 224 function evaluations ($\epsilon = 10^{-3}$)

$$L_{10}^{\tau}(A, \tau_0) = 2.58756E-03$$
$$\tau_0 = 9.90211E+00$$
$$c_{11} = -1.52790E+00 \quad c_{12} = 6.05715E-01$$
$$c_{21} = -1.30459E+00 \quad c_{22} = 6.06295E-01$$
$$c_{31} = -8.88274E-01 \quad c_{32} = 6.11083E-01$$
$$c_{41} = -3.65353E-01 \quad c_{42} = 6.46197E-01.$$

The group delay distortion in this case was reduced to $1.3T$, and the degeneracy has disappeared. The solution is shown in Fig. 8.

### Example 2

In this example we consider the synthesis of a digital all-pass filter with prescribed group delay

$$\tau_s(\phi) = 15\phi/f_n, \qquad 0.1f_n \le \phi \le 0.9f_n$$
$$f_n = 1/2T$$

where $\phi$ is a normalized frequency variable.

The distance function used in this case is

$$L_{2p}^{\tau}(A, \tau_0) = \sum_{k=1}^{K} w(\phi_k)(\tau_e(A, \phi_k) - \tau_s(\phi_k) - \tau_0)^{2p}$$
$$w(\phi_k) = 1/\tau_s(\phi_k) \qquad (22)$$

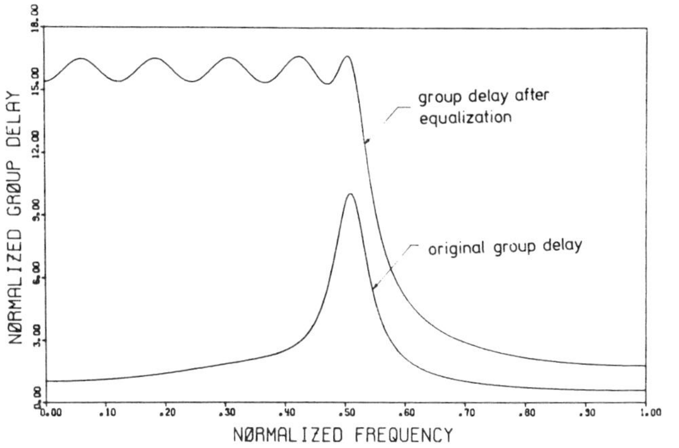


Fig. 8. Group delay of the fourth-order low-pass filter of Fig. 4 before and after equalization by a four-section all-pass with index $2p=10$.

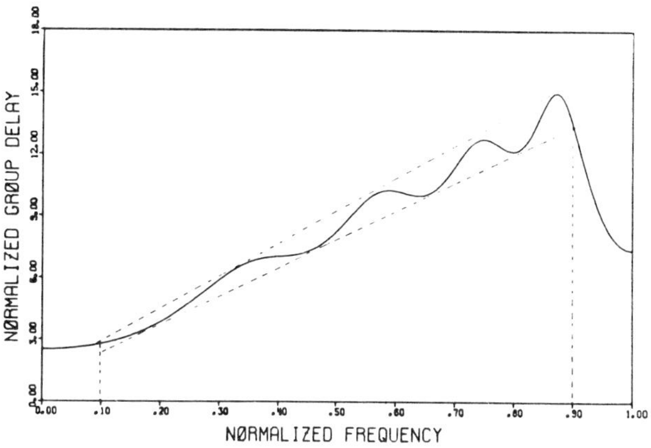


Fig. 9. All-pass digital filter having linearly increasing group delay with index $2p=4$.

where $\tau_0$ is again a variable contant delay to be determined.

Using a four-section filter, indices $2p=2$, 4, and the initial pole positions (using normalized angles):

$$r_{p1} = 0.83 \quad \phi_{p1} = 0.28 \quad r_{p2} = 0.86 \quad \phi_{p2} = 0.49$$

$$r_{p3} = 0.89 \quad \phi_{p3} = 0.68 \quad r_{p4} = 0.93 \quad \phi_{p4} = 0.78$$

and

$$\tau_0 = -0.6$$

gave the following solution after 147 function evaluations ($\epsilon=10^{-4}$):

$$L_4^\tau(A, \tau_0) = 2.69357E-05$$

$$\tau_0 = 1.12145E+00$$

$$c_{11} = -5.52893E-01 \quad c_{12} = 3.90022E-01$$

$$c_{21} = 3.30090E-01 \quad c_{22} = 5.09065E-01$$

$$c_{31} = 1.05511E+00 \quad c_{32} = 5.82468E-01$$

$$c_{41} = 1.51618E+00 \quad c_{42} = 6.69592E-01.$$

The solution is shown in Fig. 9.

### Example 3

Finally the simultaneous specification of magnitude and group delay was attempted. Here we have used the method of Kwan and Bach [10], which in the case of digital filters, means that the numerator of the transfer function is a mirror-image polynomial, i.e.,

$$H(z) = k_0 z^{-R} \frac{\prod_{i=1}^{N} (z^2 + a_{1i}z + a_{2i})(a_{2i}z^2 + a_{1i}z + 1) \prod_{i=1}^{L} (z^2 + c_{1i}z + 1)}{\prod_{i=1}^{M} (z^2 + b_{1i}z + b_{2i})} \tag{23}$$

where $R=4N+2L-2M$.

Thus we see that the group delay of this filter is determined solely by the poles of $H(z)$. The specifications were as follows:

$$\begin{aligned} \text{magnitude} \quad \alpha_s(\phi) &= 1, \quad 0 \le \phi \le 0.3f_n \\ &= 0, \quad 0.6f_n \le \phi \le 1.0f_n \\ \text{group delay} \quad \tau_s(\phi) &= 4, \quad 0 \le \phi \le 0.5f_n. \end{aligned}$$

The distance function used in this case was

$$L_{2p,2q}{}^{\alpha,\tau}(A, \tau_0) = \delta \sum_{k=1}^{K} (\alpha(A, \phi_k) - \alpha_s(\phi_k))^{2p} + (1-\delta) \sum_{r=1}^{S} (\tau(A, \phi_r) - \tau_s(\phi_r) - \tau_0)^{2q} \tag{24}$$

where $\tau_0$ is again a constant variable delay.

Using the indices $2p=2q=4$, $N=1$, $L=3$, $M=3$, and the initial poles and zeros (with normalized angles)

$$r_{p1} = 0.619 \quad \phi_{p1} = 0.101 \quad r_{p2} = 0.629$$

$$\phi_{p2} = 0.274 \quad r_{p3} = 0.732 \quad \phi_{p3} = 0.458$$

$$r_{01} = 0.42 \quad \phi_{01} = 0 \quad r_{02} = 1.0 \quad \phi_{02} = 0.632$$

$$r_{03} = 1.0 \quad \phi_{03} = 0.851 \quad r_{04} = 0.507 \quad \phi_{04} = 0.238$$

and

$$k_0 = 1.0 \quad \tau_0 = 0 \quad \delta = 0.9.$$

The following solution was obtained after 306 function evaluations:

$$L_{2p,2q}{}^{\alpha,\tau}(A, \tau_0) = 1.54936E-06$$

$$\tau_0 = 3.24856E+00$$

$$k_0 = 5.50165E-03$$

$$a_{11} = -4.86871E-01 \quad a_{21} = 1.08827E-01$$

$$c_{11} = -3.01720E+00$$

$$c_{12} = 7.69136E-01$$

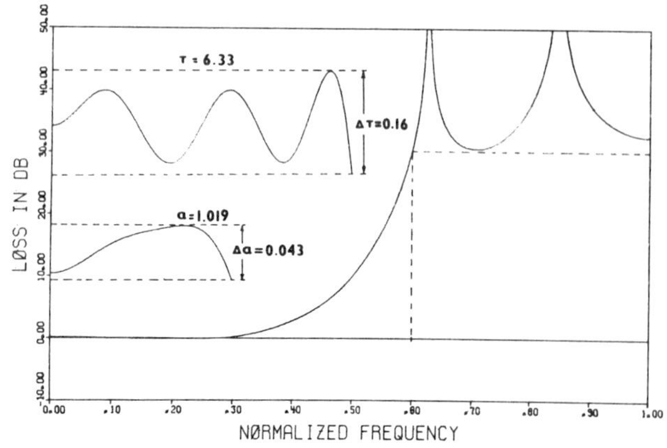


**Fig. 10.** Digital low-pass filter of order 10 having prescribed amplitude and group delay with index $2p=2q=4$.

$$c_{13} = 1.77316E+00$$

$$b_{11} = -1.11099E+00 \quad b_{21} = 3.39431E-01$$

$$b_{12} = -7.12005E-01 \quad b_{22} = 3.68174E-01$$

$$b_{13} = -4.36870E-02 \quad b_{23} = 4.57282E-01.$$

The solution for this case is shown in Fig. 10.

## Acknowledgment

The author would like to thank Prof. E. Baumann for his encouragement and support, and Prof. T. S. Huang and Dr. F. Pellandini for their help and suggestions in correcting the manuscript.

## References

[1] J. A. Athanassopoulos and J. F. Kaiser, "Constrained optimization problems in digital filter design," presented at the Arden House Workshop on Digital Filtering, Jan. 1970.
[2] F. Bonzanigo and F. Pellandini, "Problèmes de réalisation des filtres digitaux," AGEN 9, pp. 56–72, July 1969.
[3] G. C. Bown, "Design and optimization of circuits by computer," *Proc. Inst. Elec. Eng.*, vol. 118, pp. 649–661, May 1971.
[4] A. G. Deczky, "General expression for the group delay of digital filters," *Electron. Lett.*, vol. 5, pp. 663–665, Dec. 1969.
[5] A. V. Fiacco and G. P. McCormick, *Nonlinear Programming.* Reading, Mass.: Addison-Wesley, 1964.
[6] R. Fletcher and M. J. D. Powell, "A rapidly convergent descent method for minimization," *Comput. J.*, vol. 6, pp. 163–168, 1963.
[7] H. D. Helms, "Digital filters with equiripple or minimax responses," *IEEE Trans. Audio Electroacoust.*, vol. AU-19, pp. 87–93, Mar. 1971.
[8] J. B. Knowles and E. M. Olcayto, "Coefficient accuracy and digital filter response," *IEEE Trans. Circuit Theory*, vol. CT-15, pp. 31–41, Mar. 1968.
[9] F. F. Kuo and W. Magnuson, Eds., *Computer Oriented Circuit Design.* Englewood Cliffs, N. J.: Prentice-Hall, 1968.
[10] R. K. Kwan and G. G. Bach, "Simultaneous approximation in filter design," *IEEE Trans. Circuit Theory* (Corresp.), vol. CT-16, pp. 117–121, Feb. 1969.
[11] J. B. Naylor and J. M. Townsend, "Lowpass digital filters using least-squared-error design," *Electron. Lett.*, vol. 5, pp. 647–657, 1969.
[12] D. A. Pierre, *Optimization Theory with Applications.* New York: Wiley, 1969.
[13] C. M. Rader and B. Gold, "Digital filter design techniques in the frequency domain," *Proc. IEEE*, vol. 55, pp. 149–171, Feb. 1967.
[14] J. R. Rice, *The Approximation of Functions.* Reading, Mass.: Addison-Wesley, 1964.
[15] R. A. Rohrer, "Explicit step-size determination in gradient exploitation minimization techniques," *IEEE Trans. Circuit Theory* (Corresp.), vol. CT-17, pp. 154–155, Feb. 1970.
[16] K. Steiglitz, "Computer-aided design of recursive digital filters," *IEEE Trans. Audio Electroacoust.* (*Special Issue on Digital Filtering*), vol. AU-18, pp. 123–129, June 1970.
[17] G. C. Temes and D. A. Calahan, "Computer-aided network optimization—The state-of-the-art," *Proc. IEEE*, vol. 55, pp. 1832–1863, Nov. 1967.

# Linear Programming Design of IIR Digital Filters with Arbitrary Magnitude Function

LAWRENCE R. RABINER, MEMBER, IEEE, NANCY Y. GRAHAM, AND HOWARD D. HELMS, SENIOR MEMBER, IEEE

*Abstract*—**This paper discusses the use of linear programming techniques for the design of infinite impulse response (IIR) digital filters. In particular, it is shown that, in theory, a weighted equiripple approximation to an arbitrary magnitude function can be obtained in a predictable number of applications of the simplex algorithm of linear programming. When one implements the design algorithm, certain practical difficulties (e.g., coefficient sensitivity) limit the range of filters which can be designed using this technique. However, a fairly large number of IIR filters have been successfully designed and several examples will be presented to illustrate the range of problems for which we found this technique to be useful.**

## INTRODUCTION

A LARGE NUMBER of techniques are available for designing infinite impulse response (IIR) digital filters [1], [2]. The techniques of impulse invariant design, bilinear transformation, and matching poles and zeros [3], for transforming a given analog filter to an "equivalent" digital filter are well known and widely used. These techniques, however, are limited in that they are generally applied only to the case of transforming standard analog filters—e.g., low-pass, bandpass, bandstop, or high-pass filters. When one is interested in designing a digital filter with a nonstandard frequency response, i.e., one which has not been exhaustively studied, then some algorithmic (as opposed to closed form) design procedure is generally used. Several frequency domain (e.g., [4]–[7]) and one time domain algorithmic design procedure [8] have been recently proposed for designing IIR filters. One difficulty with almost all of these procedures is that convergence of the optimization procedure that is used to design the filter is not guaranteed, and even when the procedure converges, the optimality of the resulting filter is also not guaranteed. In this paper a frequency domain IIR filter design procedure is discussed which uses linear programming techniques (the simplex algorithm) to choose filter coefficients to approximate an arbitrary magnitude characteristic. Theoretical convergence of the optimization algorithm is guaranteed, and the resulting filter can be shown to be optimal in the given design sense (e.g., minimum absolute weighted error). The optimization algorithm itself has been designed to minimize the number of applications of the simplex algorithm.

The next section presents the design procedure with a discussion of the practical aspects of implementing the method. Following this, several representative filter designs are given. Finally, some discussion is given as to practical limitations in using the method.

## THEORY

Let $H(z)$ be the transfer function of an IIR digital filter. Assume $H(z)$ has the form

$$H(z) = \frac{N(z)}{D(z)} = \sum_{i=0}^{m} b_i z^{-i} \Big/ \sum_{i=0}^{n} a_i z^{-i} \qquad (1)$$

where the numerator polynomial $N(z)$ is of $m$th degree, and the denominator polynomial $D(z)$ is of $n$th degree. The $a_0$ term in (1) can be set to 1.0 without any loss in generality. The magnitude response of the filter is ob-

Manuscript received December 13, 1973.
L. R. Rabiner and N. Y. Graham are with Bell Laboratories, Murray Hill, N. J. 07974.
H. D. Helms is with Bell Laboratories, Holmdel, N. J. 07733.

Reprinted from *IEEE Trans. Acoust., Speech, Signal Processing,* vol. ASSP-22, pp. 117–123, Apr. 1974.

tained by evaluating (1) on the unit circle (i.e., for $z = \exp[j\omega]$),[1] and taking its magnitude, thus giving

$$|H(\exp[j\omega])| = \left|\frac{N(\exp[j\omega])}{D(\exp[j\omega])}\right|$$

$$= \left|\sum_{i=0}^{m} b_i \exp[-j\omega i] / \sum_{i=0}^{n} a_i \exp[-j\omega i]\right|. \quad (2)$$

In many frequency domain filter design problems it is desired that the magnitude of the resulting filter approximate a given magnitude function, $M(\exp[j\omega])$, to within a tolerance $G(\omega,\delta)$, where $G(\omega,\delta)$ is a monotically increasing function of $\delta$ for fixed $\omega$.[2] Thus, the resultant approximation problem is to choose the filter coefficients (the $a_i$'s and $b_i$'s) to minimize the quantity $\delta$ consistent with that constraint inequality

$$\left|\left|\frac{N(\exp[j\omega])}{D(\exp[j\omega])}\right| - M(\exp[j\omega])\right| \le G(\omega,\delta). \quad (3)$$

Inequality (3) is generally evaluated over a union of disjoint subintervals of the band $0 \le \omega \le \pi$.

The above approximation problem is a nonlinear one in that the filter coefficients enter into the constraint equation nonlinearity. Although various techniques have been proposed for solving this nonlinear problem (e.g., [4] and [5]), a linear approximation problem can be defined by considering the magnitude squared function of the filter. From (1) we get the relation

$$H(z)H(z^{-1}) = \frac{N(z)N(z^{-i})}{D(z)D(z^{-1})} \quad (4)$$

$$= \left(\sum_{i=0}^{m} b_i z^{-i}\right)\left(\sum_{j=0}^{m} b_j z^{+j}\right) / \left(\sum_{i=0}^{n} a_i z^{-1}\right)\left(\sum_{j=0}^{n} a_j z^{+j}\right) \quad (5)$$

$$= \sum_{i=-m}^{m} c_i z^{-i} / \sum_{i=-n}^{n} d_i z^{-i} \quad (6)$$

where

$$c_i = c_{-i} \qquad i = 1,2,\cdots,m \quad (7)$$

$$d_i = d_{-i} \qquad i = 1,2,\cdots,n. \quad (8)$$

The magnitude squared function of the filter is obtained by evaluating (6) on the unit circle giving

$$|H(\exp[j\omega])|^2$$

$$= H(z)H(z^{-1})\,|_{z=\exp[j\omega]} = \frac{\hat{N}(\omega)}{\hat{D}(\omega)} \quad (9)$$

$$= \left[c_0 + \sum_{i=1}^{m} 2c_i \cos(\omega i)\right] / \left[d_0 + \sum_{i=1}^{n} 2d_i \cos(\omega i)\right]. \quad (10)$$

(Again the $d_0$ term in (10) can be set to 1.0 without loss of generality.) Equation (10) shows that the magnitude squared function of the filter is a ratio of trigonometric polynomials. It is also seen that both $\hat{N}(\omega)$, the numerator polynomial, and $\hat{D}(\omega)$, the denominator polynomial, are linear in the unknown filter coefficients $\{c_i\}$ and $\{d_i\}$. It is now shown how linear programming techniques can be used to determine the $c_i$'s and $d_i$'s such that $|H(\exp[j\omega])|^2$ approximates a given magnitude squared characteristic $F(\omega)$ where the peak weighted error of approximation is minimized—i.e., the weighted error is an equiripple function.

If we let $F(\omega)$ be the desired magnitude squared characteristic, then the approximation problem consists of finding the filter coefficients such that

$$-\epsilon(\omega) \le \frac{\hat{N}(\omega)}{\hat{D}(\omega)} - F(\omega) \le \epsilon(\omega) \quad (11)$$

where $\epsilon(\omega)$ is a tolerance function on the error which allows for unequal weighting of errors as a function of frequency. Since $F(\omega)$ and $\epsilon(\omega)$ are generally specified functions of frequency, (or depend on some parameter in a manner explained below), (11) can be expressed as a set of linear inequalities in the $c_i$'s and $d_i$'s by writing it in the form

$$\hat{N}(\omega) - \hat{D}(\omega)F(\omega) \le \epsilon(\omega)\hat{D}(\omega)$$
$$-\hat{N}(\omega) + \hat{D}(\omega)F(\omega) \le \epsilon(\omega)\hat{D}(\omega) \quad (12)$$

or

$$\hat{N}(\omega) - \hat{D}(\omega)[F(\omega) + \epsilon(\omega)] \le 0 \quad (13)$$

$$-\hat{N}(\omega) + \hat{D}(\omega)[F(\omega) - \epsilon(\omega)] < 0. \quad (14)$$

The additional linear inequalities

$$-\hat{N}(\omega) \le 0 \quad (15)$$

$$-\hat{D}(\omega) \le 0 \quad (16)$$

completely define the approximation problem.

Thus, the question of whether or not there exists a digital filter with magnitude squared characteristic $F(\omega)$ and tolerance function $\epsilon(\omega)$ is equivalent to the question of whether or not there exists a set of filter coefficients satisfying the system of constraints defined by (13)–(16). The question can be answered by using linear programming techniques [9]. First, an auxiliary variable $\nu$ is subtracted from the left side of each constraint, forming the new set of constraints

$$\hat{N}(\omega) - \hat{D}(\omega)[F(\omega) + \epsilon(\omega)] - \nu \le 0 \quad (17)$$

$$-\hat{N}(\omega) + \hat{D}(\omega)[F(\omega) - \epsilon(\omega)] - \nu \le 0 \quad (18)$$

$$-\hat{N}(\omega) - \nu \le 0 \quad (19)$$

$$-\hat{D}(\omega) - \nu \le 0. \quad (20)$$

[1] The quantity $\omega$ is the normalized frequency variable (i.e., the sampling period, $T$, is assumed to be 1.0).

[2] It should be noted that the function $G(\omega,\delta)$ is generally determined as soon as $M(\exp[j\omega])$ is specified by the designer, as will be seen in the examples later in the paper.

The objective function

$$z = \nu \tag{21}$$

is chosen to be minimized under the constraints of (17)–(20).[3] Clearly a solution to constraints (13)–(16) exists if and only if the minimum value of $z$ under constraints (17)–(20) is zero. If the minimum value of $\nu$ is 0, then a solution exists to the approximation problem and the filter coefficients may be obtained directly as the output of the linear programming routine. If $\nu > 0$, then no solution to the approximation problem exists, and either $F(\omega)$, or $\epsilon(\omega)$, or both must be modified in order to obtain a solution.

To illustrate the above procedure, consider the design of a low-pass filter. If we let $\delta$ be the peak approximation error in the stopband, and $K\delta$ ($K$ is a constant expressing the ratio of passband to stopband ripple) be the peak approximation error in the passband, then the magnitude function for an equiripple error approximation is as shown in Fig. 1 (a). The quantity $\delta$ is unknown and is to be minimized in the ultimate design program. (Of course in this case the resulting filter is an elliptic filter which can be readily designed in closed form, but we are only using this as an example of how to apply the design technique.) The passband cutoff frequency is $\omega_p = 2\pi F_p$ and the stopband cutoff frequency is $\omega_s = 2\pi F_s$. The magnitude squared function of the filter is the square of the response in Fig. 1 (a) and is shown in Fig. 1 (b). This magnitude squared function can be viewed as a weighted equiripple approximation of the function $F(\omega)$ [shown in Fig. 1 (c)], with peak approximation error $\epsilon(\omega)$ [shown in Fig. 1 (d)], defined by

$$\begin{aligned} F(\omega) &= 1 + K^2\delta^2 & 0 \le \omega \le \omega_p \\ &= \delta^2/2 & \omega_s \le \omega \le \pi \end{aligned} \tag{22}$$

$$\begin{aligned} \epsilon(\omega) &= 2K\delta & 0 \le \omega \le \omega_p \\ &= \delta^2/2 & \omega_s \le \omega \le \pi. \end{aligned} \tag{23}$$

It is easily verified that $|H(\exp[j\omega])|^2$ of Fig. 1 (b) is less than or equal to $F(\omega) + \epsilon(\omega)$ and greater than or equal to $F(\omega) - \epsilon(\omega)$ in both the passband and the stopband.

To determine the smallest value of $\delta$ such that the filter approximation problem has a solution (i.e., to find the $\delta$ of the elliptic filter) an iterative procedure must be used since $\delta$ enters into the design constraints in a nonlinear

[3] The linear programming problem defined above (by constraints (17)–(20) and objective function $z = \nu$) may be solved by a straightforward application of the simplex (or revised simplex) algorithm. However, since the number of constraints is generally much larger than the number of filter coefficients, it is of course much more efficient to apply the simplex algorithm to the dual linear programming problem. Furthermore, since the range of values of filter coefficients, for the examples of interest here, is between $-1$ and $+1$, a change of variables to $c_i' = c_i + 1$, $d_i' = d_i + 1$ is performed before application of the simplex algorithm, which does not allow variables to assume negative values.

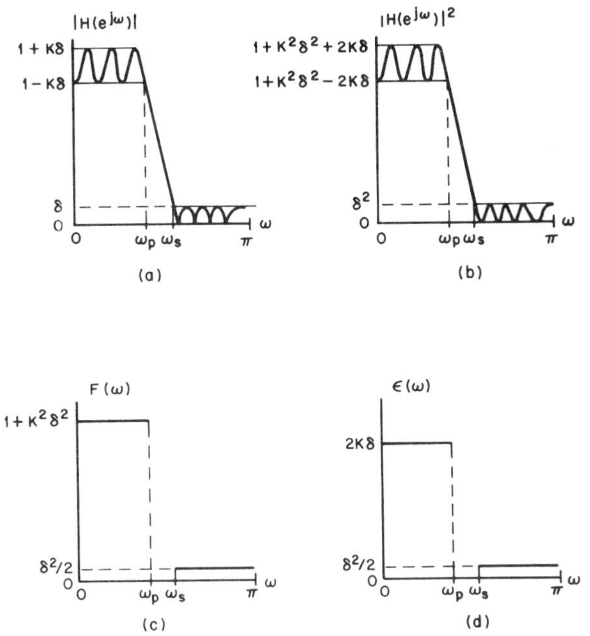


Fig. 1. Specifications for designing an equiripple error low-pass filter. (a) Bounds on the approximation. (b) Bounds on the square of the approximation. (c) Function obtained by averaging the upper and lower bounds on the square of the approximation. (d) Error bounds obtained by subtracting Fig. 1(c) from Fig. 1(b).

manner. If we let $\delta^*$ denote the minimum value of $\delta$ for which the approximation problem has a solution, then $\delta^*$ satisfies the inequality

$$0 < \delta^* \le \frac{1}{K+1} \tag{24}$$

since the sum of passband ripple ($K\delta^*$) and stopband ripple ($\delta^*$) must be less than or equal to 1.0 because otherwise the passband and stopband would no longer be well defined. Based on (24), a binary search may be used to locate $\delta^*$, as in the following procedure. (This procedure is different from the usual binary search in that the search is performed on the log of $\delta$ instead of $\delta$ itself.)

*Step* 1: Let $\delta_+{}^i$ and $\delta_-{}^i$ denote initial upper and lower bounds for $\delta^*$. For example, choose $\delta_+{}^i = (K+1)^{-1}$, $\delta_-{}^i = 10^{-8}$ (since $\delta_-{}^i < 10^{-8}$ is unrealistic). Initialize $\delta$ at $(\delta_+{}^i\delta_-{}^i)^{1/2}$, the geometric mean of the *initial* upper and lower bounds. (Note that the geometric mean of two quantities is equal to the arithmetic mean of the logs of these quantities.)

*Step* 2: Solve the linear programming problem (17)–(21) with this value of $\delta$. If $z = 0$ a solution to the approximation problem exists and $\delta^* < \delta$. In this case set $\delta_+ = \delta$. Otherwise no solution exists for this value of $\delta$ and $\delta < \delta^*$; in which case set $\delta_- = \delta$.

*Step* 3: Set $\delta = (\delta_+\delta_-)^{1/2}$ and repeat Step 2. This procedure is iterated until a predetermined accuracy criterion in locating $\delta^*$ is satisfied.

It should be noted that, when $\delta^*$ is small ($<10^{-2}$) (as in practical filter design problems), choosing $\delta$ to be the geometric mean of the upper and lower bounds for $\delta^*$ will result in a smaller number of iterations required to achieve relative accuracy in locating $\delta^*$ than required by the usual

TABLE I
RESULTS ON THE DESIGN OF SEVERAL LOW-PASS FILTERS

| Filter No. | $n$ | $m$ | $F_p$ | $F_s$ | $K$ | $-20 \log_{10} \delta^*$ | $-20 \log_{10} \delta_+$ | No. of Iterations | Total Run Time (s) |
|---|---|---|---|---|---|---|---|---|---|
| 1 | 4 | 4 | 0.30 | 0.35 | 5.8 | 37.9 | 37.8 | 11 | 40.6 |
| 2 | 4 | 4 | 0.15 | 0.18 | 2.0 | 29.8 | 29.9 | 11 | 43.5 |
| 3 | 4 | 4 | 0.10 | 0.15 | 12.0 | 43.1 | 42.9 | 11 | 44.2 |
| 4 | 4 | 4 | 0.10 | 0.14 | 6.5 | 38.7 | 38.6 | 11 | 46.9 |
| 5 | 4 | 4 | 0.10 | 0.13 | 3.4 | 34.0 | 33.9 | 11 | 48.2 |
| 6 | 4 | 4 | 0.10 | 0.12 | 1.7 | 28.5 | 28.4 | 11 | 50.6 |
| 7 | 6 | 6 | 0.20 | 0.23 | 8.5 | 48.3 | 48.4 | 11 | 134.3 |
| 8 | 6 | 6 | 0.20 | 0.25 | 71.9 | 61.6 | 61.8 | 11 | 141.7 |

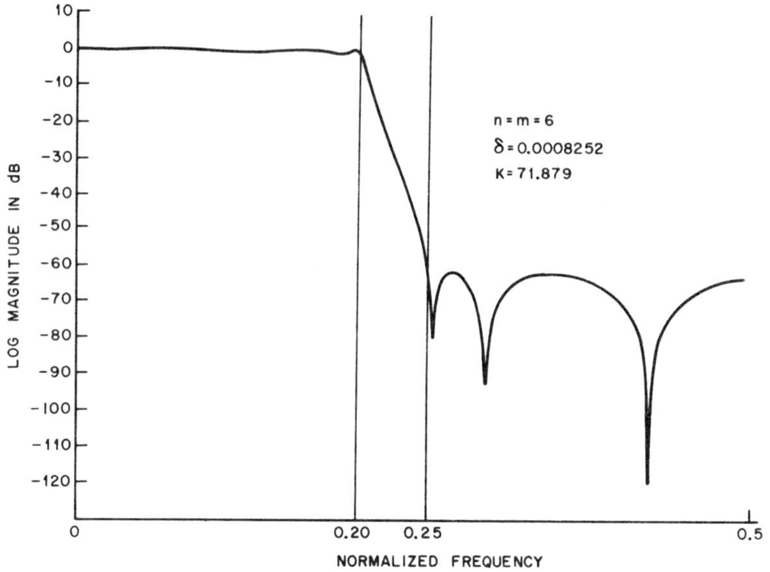


Fig. 2. Log magnitude response of a sixth order low-pass filter designed using linear programming methods.

binary search in which $\delta$ is chosen to be the arithmetic mean of the upper and lower bounds. In particular, the number $n$ of iterations required for a maximum relative error $p$ (where $p$ is defined as

$$p = \frac{\delta_+{}^f - \delta_-{}^f}{\delta_-{}^f} \tag{25}$$

with $\delta_+{}^f$ and $\delta_-{}^f$ being the final upper and lower bounds determined in step 3 above) can be shown to be of the form

$$n = \left[\log_2\left(\frac{\log_{10}(\delta_+{}^i) - \log_{10}(\delta_-{}^i)}{\log_{10}(\delta_+{}^f) - \log_{10}(\delta_-{}^f)}\right)\right] + 1 \tag{26}$$

$$= \left[\log_2\left(\frac{\log_{10}(\delta_+{}^i) - \log_{10}(\delta_-{}^i)}{\log_{10}(1+p)}\right)\right] + 1 \tag{27}$$

where $[y]$ denotes the largest integer $\leq y$.

Theoretically the value of $\delta^*$ may be bounded as tightly as desired by using a large number of iterations thereby reducing the tolerance (difference between $\delta_+$ and $\delta_-$) at much as desired. In practice, a relative error of $p = 0.01 =$ 1 percent on the deltas is sufficient for most problems.

The values of $c_i$ and $d_i$ associated with the final value of $\delta_+$ are used in the polynomials in $z$ (6), which are then factored. A $z$-transform, which is stable, and minimum phase is obtained by retaining only zeros and poles which lie inside or on the unit circle. (If zeros lie on the unit circle, only half of the pairs are retained.)

The extension of the above procedure to other types of filters other than low-pass filters is straightforward and will not be discussed here. In the next section we present examples of several filters which were designed using the above iterative procedure.

## FILTER EXAMPLES

In order to test out the procedure a number of filters were designed. Table I gives the results for a set of 8 low-pass filters. The data in this table correspond to the 8 low-pass filters designed by Swanton [10] using linear programming methods in a sequence of individual numerator and denominator optimization iterations. These data are for low-order filters (4th or 6th order). The values for $F_p$ and $F_s$ are the corresponding filter cutoff frequencies where $F_p = \omega_p/2\pi$ and $F_s = \omega_s/2\pi$. The quantity $20 \log_{10} \delta^*$ is the theoretical stopband attenuation for the elliptic filter and $20 \log_{10} \delta_+$ is the actual attenuation for the filter designed using the linear programming technique described above. Table I also gives the number of iterations on the delta's (for a 1 percent tolerance on $\delta$) and the overall run time on a Honeywell 6000 Computer. It is seen from Table I that the resulting filters meet approximately the same specifications as the equivalent elliptic filter. To further illustrate these results, Fig. 2 shows the log magni-

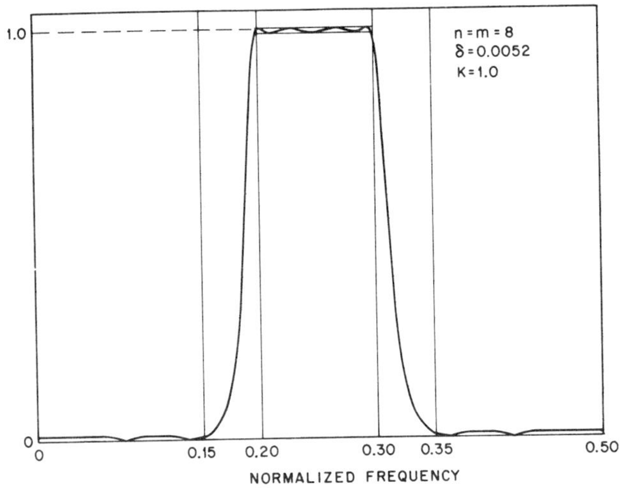


Fig. 3. Magnitude response of an eighth order bandpass filter designed using linear programming techniques.

tude response of the eighth filter in the table which achieved a $\delta = \delta_+$ of 0.0008144 or a stopband attenuation of 61.8 dB.

As seen in Table I the run time for the simple 4th order examples was about 45 s whereas for the two 6th order examples it was about 138 s. These times could be significantly reduced by an improved initial guess of $\delta$, or by relaxing the convergence criterion (the 1 percent tolerance on the deltas). These were not done for the examples in Table I in order to verify that the procedure would converge without any starting information.

In addition to low-pass filters, several 3, 4 and 5 band filters, and wide-band differentiators have been designed with this procedure. Figs. 3–7 show the frequency responses of some typical filters which were designed. Fig. 3 gives an example of a standard bandpass filter of eighth order (i.e., $n = m = 8$). This filter took 11 iterations to achieve the desired 1 percent accuracy, and required 187 s of processor time. The value of $K$ was 1.0 giving a value of $\delta$ of 0.0051. The two filter stopbands were from 0.0 to 0.15 and from 0.35 to 0.5. Fig. 3 shows the magnitude response of the filter including the tolerances in each of the bands and the filter band edges.

Fig. 4 shows the frequency response of a 4 band filter with a stopband followed by 2 disjoint passbands, followed by another stopband. The filter is of eighth order and the tolerances in all the bands were the same. It required 11 iterations for convergence of the delta to within 1 percent and took 238 s of computation. Fig. 4(a) shows the log magnitude response of the filter; Fig. 4 (b) shows the positions of the poles and zeros in the $z$-plane; and Fig. 4(c) shows the group delay response of this filter. The pole and zero positions of the filter are obtained by factoring the denominator and numerator polynomials of the magnitude squared function of the filter and assigning poles inside the unit circle to the resulting denominator, and zeros inside or on the unit circle to the resulting numerator. (Generally the zeros of the magnitude squared function will be on the unit circle in pairs. Thus one of each pair of zeros on the unit circle is assigned to the resulting filter.)

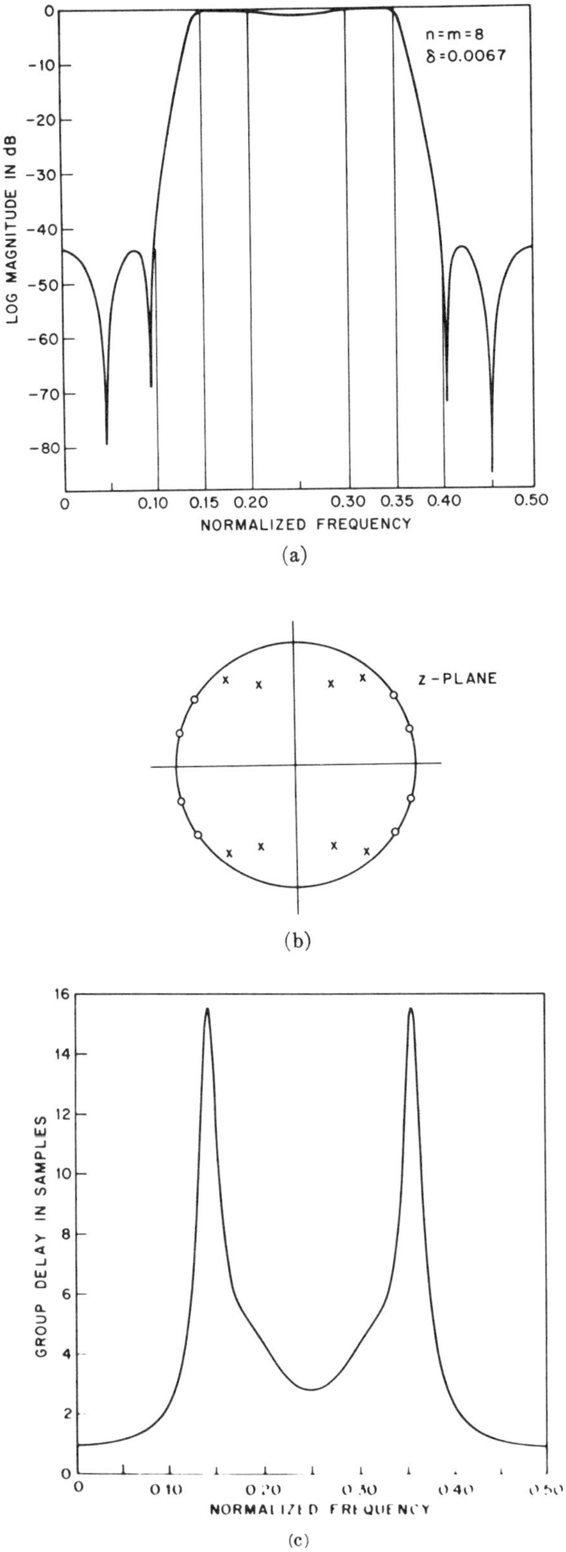


Fig. 4. Log magnitude response, $z$-plane pole–zero diagram, and group delay response of an eight order-four band filter.

Fig. 5 shows an example of a 5 band filter with 2 passbands and 3 stopbands. An eight order filter gave a delta of 0.035 with equal weighing in each of the five bands. The design procedure took 11 iterations and required 224 s to design.

Fig. 6 shows another 5 band filter with the arbitrary specifications:

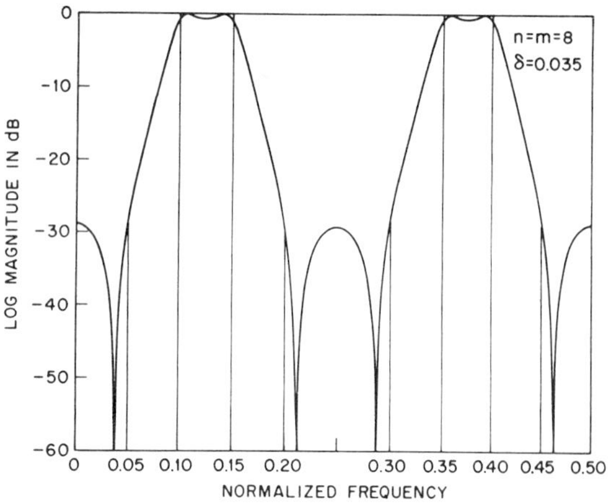


Fig. 5. Log magnitude response of an eight order filter with 5 alternating stopbands and passbands.

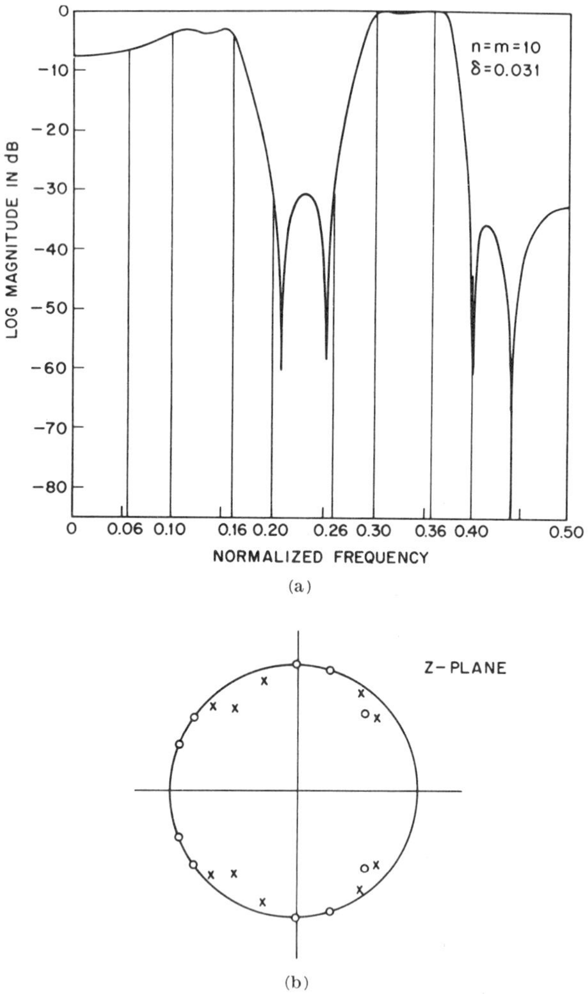


Fig. 6. Log magnitude response and $z$-plane pole–zero positions of a tenth order five band filter with arbitrary magnitude specifications.

$$
\begin{aligned}
|H(\exp[j2\pi f])| &= 0.5 & 0.00 \le f \le 0.06\\
&= 0.75 & 0.10 \le f \le 0.16\\
&= 0.0 & 0.20 \le f \le 0.26\\
&= 1.0 & 0.30 \le f \le 0.36\\
&= 0.0 & 0.40 \le f \le 0.50
\end{aligned}
$$

and with equal tolerances on the error in each band. The resulting filter was of 10th order and had a $\delta$ of 0.0303. Even though the filter met specifications in all the bands, the magnitude squared function blew up between the fourth and fifth bands because of a pole which was on the unit circle. The simple expedient of moving the pole slightly inside the unit circle gave a stable filter whose log magnitude response and pole–zero positions are plotted in Fig. 5. It should be noted that the presence of a pole on the unit circle in one of the unconstrained regions of the frequency scale is not a violation of the design procedure and is perfectly acceptable to the optimization procedure. Of course the resulting filter cannot tolerate such a situation.

Finally, Fig. 7 shows the error response of a differentiator of fourth degree designed using the technique. In this case the error criterion was a minimum relative error criterion. The resulting value of $\delta$ was 0.00000763 when the desired band for differentiation was from 0 to 0.45. (In this case $\delta$ represents the peak relative error for the differentiator.) This example required 11 iterations and took 51 s of computation time.

## DISCUSSION

The preceding examples have shown that the linear programming method does, in many cases, give reasonable solutions for IIR filters which approximate arbitrary magnitude specifications with arbitrary weighing of the error function. In this section we discuss what we believe are the practical limitations of this technique.

One of the major difficulties with the proposed method is that one is forced to work with magnitude squared characteristics to solve for the filter coefficients. Kaiser [11] has shown that an extreme coefficient sensitivity problem exists for sharp cutoff filters when implemented in the direct form. This coefficient sensitivity is aggravated by using magnitude squared functions rather than the magnitude function itself. Thus the results derived by Kaiser, along with our own practical experience indicate the following.

1) Sharp cutoff filters are difficult to design with this procedure. Thus, if the width of a transition band is small, the coefficient sensitivity will make the procedure unstable.

2) High-order filters are difficult to design. Filters with order greater than about 12 cannot readily be designed with double precision arithmetic on a 36 bit word length computer since the high order polynomial coefficients are extremely sensitive to small changes in the filter specifications.

3) Filters with deltas on the order of $10^{-5}$ or less cannot generally be designed even with double precision arithmetic since, a tolerance of the magnitude function on the order of $\le 10^{-5}$ in a band implies a tolerance of the magnitude squared function on the order of $\le 10^{-10}$ in that band. The attainment of such a high degree of numerical precision is of course limited by the precision capabilities of the computer. Experience has shown that the linear programming routine required to solve for the filter coefficients must be implemented in double precision

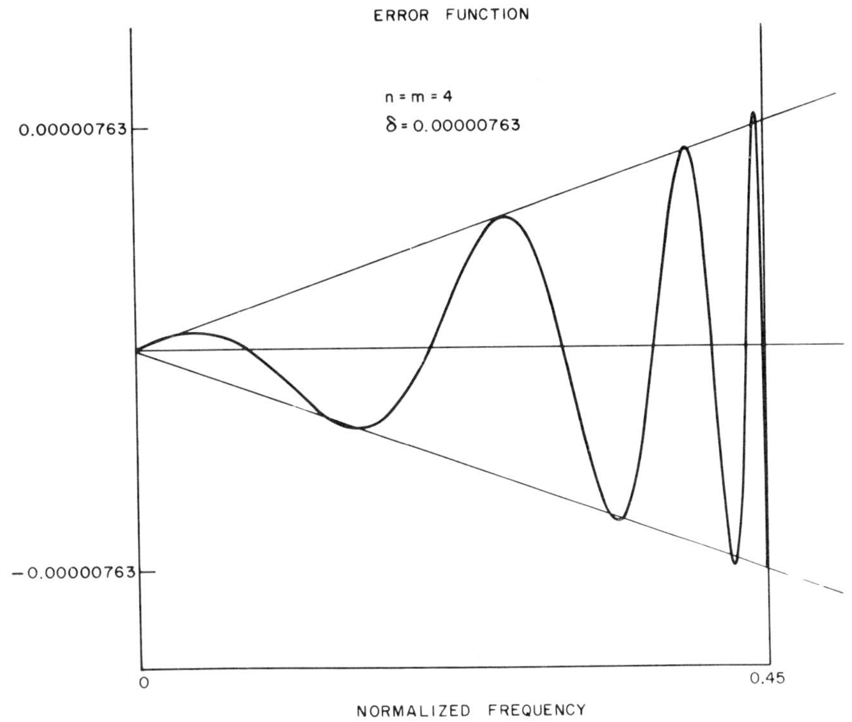


Fig. 7. Error function of a fourth order differentiator with minimum relative error over the band $0 \leq f \leq 0.45$.

arithmetic in order to obtain sufficient accuracy for deltas in the range $10^{-1} > \delta \geq 10^{-5}$, whereas for deltas $< 10^{-5}$, sufficient accuracy is almost impossible to consistently obtain even with double precision arithmetic on a 36-bit computer.

4) Experience has also shown that for filter design problems with $\geq 8$ degrees of freedom in numerator and denominator polynomials, the number of pivots within a given application of the simplex algorithm may be exceedingly large, say $> 400$.

Fortunately, despite the above practical limitations, there is a large class of problems where the linear programming technique can be used to advantage. One of the key properties here is the guaranteed theoretical convergence and the clear statement of optimality of the resulting approximation. Also, compilers which permit arbitrary precision arithmetic may circumvent the above limitations, although at the cost of greatly increased computation time.

## SUMMARY

A new technique for designing IIR filters which can approximate arbitrary magnitude specifications was presented. The technique sets up a linear programming problem which is solved iteratively for the best approximation to the given specifications. Several examples of filters designed using this technique were given and the ultimate limitations of the procedure were discussed.


## ACKNOWLEDGMENT

The authors wish to thank J. E. Heller of Bell Laboratories, Whippany, N. J. and W. L. Roach of Bell Laboratories, Holmdel, N. J. for their comments and suggestions.


## REFERENCES


[1] J. F. Kaiser, "Digital filters," in *System Analysis by Digital Computer*, F. F. Kuo and J. F. Kaiser, Ed. New York: Wiley, 1966.

[2] B. Gold and C. M. Rader, *Digital Processing of Signals*. New York: McGraw-Hill, 1969.

[3] R. M. Golden, "Digital filter synthesis by sampled-data transformation," *IEEE Trans. Audio Electroacoust.*, vol. AU-16, pp. 321–329, Sept. 1968.

[4] K. Steiglitz, "Computer-aided design of recursive digital filters," *IEEE Trans. Audio Electroacoust.*, vol. AU-18, pp. 123–129, June 1970.

[5] A. G. Deczky, "Synthesis of recursive digital filters using the minimum *p*-error criterion," *IEEE Trans. Audio Electroacoust.*, vol. AU-20, pp. 257–263, Oct. 1972.

[6] P. Thajchayapong and P. J. Rayner, "Recursive digital filter design by linear programming," *IEEE Trans. Audio Electroacoust.*, vol. AU-21, pp. 107–112, Apr. 1973.

[7] F. X. Brophy and A. C. Salazar, "Recursive digital filter synthesis in the time domain," *IEEE Trans. Acoust., Speech, and Signal Processing*, vol. ASSP-22, pp. 45–56, Feb. 1974.

[8] C. S. Burrus and T. W. Parks, "Time domain design of recursive digital filters," *IEEE Trans. Audio Electroacoust.*, vol. AU-18, pp. 137–141, June 1970.

[9] H. L. Loeb, "Algorithms for Chebyshev approximations using the ratios of linear forms," *J. Soc. Ind. Appl. Math.*, vol. 8, pp. 458–465, Sept. 1960.

[10] D. Swanton, "Linear programming design of recursive digital filters," M.Sc. Thesis, McGill Univ., Mar. 1973.

[11] J. F. Kaiser, "Some practical considerations in the realization of linear digital filters," in *Proc. 3rd Annu. Allerton Conf. on Circuit and System Theory*, 1965, pp. 621–623.

# A Digital Signal Processing Approach to Interpolation

RONALD W. SCHAFER AND LAWRENCE R. RABINER

***Abstract*—In many digital signal processing systems, e.g., vocoders, modulation systems, and digital waveform coding systems, it is necessary to alter the sampling rate of a digital signal. Thus it is of considerable interest to examine the problem of interpolation of bandlimited signals from the viewpoint of digital signal processing. A frequency domain interpretation of the interpolation process, through which it is clear that interpolation is fundamentally a linear filtering process, is presented.**

**An examination of the relative merits of finite duration impulse response (FIR) and infinite duration impulse response (IIR) digital filters as interpolation filters indicates that FIR filters are generally to be preferred for interpolation. It is shown that linear interpolation and classical polynomial interpolation correspond to the use of the FIR interpolation filter. The use of classical interpolation methods in signal processing applications is illustrated by a discussion of FIR interpolation filters derived from the Lagrange interpolation formula. The limitations of these filters lead us to a consideration of optimum FIR filters for interpolation that can be designed using linear programming techniques. Examples are presented to illustrate the significant improvements that are obtained using the optimum filters**

## I. Introduction

THE PROCESS of interpolation is familiar to anyone who has had occasion to "read between the lines" in a table of mathematical functions. In digital signal processing, interpolation is required whenever it is necessary to change from one sampling rate to another. For example, in speech processing systems, estimates of speech parameters are often computed at a low sampling rate for low bit-rate storage or transmission; however, for constructing a synthetic speech signal from the low bit-rate representation, the speech parameters are normally required at much higher sampling rates [1], [2]. In such cases, the sampling rate must be increased by a digital interpolation process. As another example, an efficient digital realization of a frequency-multiplexed single-sideband system has been obtained [3] by performing complicated filtering functions at a low sampling rate and simpler functions at the high sampling rate required for grouping several channels into a frequency-multiplexed format. In this process, there is a need for both increasing and decreasing the sampling rate. Another example where sampling rate reduction is required is in converting a delta modulation representation of a waveform to a pulse-code modulation (PCM) representation [4].

In these and other examples, it is important to thoroughly understand the process of interpolation from the point of view of digital signal processing rather than from a numerical analysis viewpoint. For example, tables of mathematical functions are generally constructed so that linear interpolation produces sufficiently accurate results, and for cases where linear interpolation is inadequate, there exists a great variety of higher order polynomial interpolation formulas. In signal processing applications there is a great temptation to try to get along with linear interpolation because it is a simple technique. In this paper we present a frequency-domain interpretation of the interpolation process in which it is clear that interpolation is fundamentally a linear filtering process. This discussion makes it abundantly clear that linear interpolation is generally not appropriate for digital signal processing applications. We discuss the advantages of finite duration impulse response (FIR) over infinite duration impulse response (IIR) digital filters for use as interpolation filters and we discuss the application of recently developed design techniques for FIR filters to the design of optimum interpolation filters. These filters are compared to filters derived from classical polynomial interpolation formulas to illustrate the improvement that can be achieved.

## II. Digital Sampling Rate Alteration

### *A. Sampling Continuous-Time Signals*

Consider a continuous-time signal $\hat{x}(t)$ with Fourier transform

$$\hat{X}(\omega) = \int_{-\infty}^{\infty} \hat{x}(t)e^{-j\omega t}dt.$$

The signal $\hat{x}(t)$ is sampled to produce the sequence

$$x(n) = \hat{x}(nT), \qquad -\infty < n < \infty$$

where $T$ is the sampling period. The $z$ transform of the sequence $x(n)$ is defined as

$$X(z) = \sum_{n=-\infty}^{\infty} x(n)z^{-n}.$$

The $z$ transform evaluated on the unit circle $X(e^{j\omega T})$ will be called the ***Fourier transform*** of the sequence $x(n)$. It is well known that the Fourier transform of the sequence $x(n)$ is related to the Fourier transform of $\hat{x}(t)$ by [5]

$$X(e^{j\omega T}) = \frac{1}{T}\sum_{k=-\infty}^{\infty} \hat{X}\left(\omega + k\frac{2\pi}{T}\right). \tag{1}$$

If $\hat{x}(t)$ is bandlimited, i.e., $\hat{X}(\omega)=0$ for $|\omega| \geq \Omega$, and if $T \leq \pi/\Omega$, then it can be seen from (1) that

$$X(e^{j\omega T}) = \frac{1}{T}\hat{X}(\omega), \qquad -\frac{\pi}{T} \leq \omega \leq \frac{\pi}{T}$$

as depicted in Fig. 1 where $T=\pi/\Omega$.

Assuming that $\hat{x}(t)$ is bandlimited, the original continuous-time signal can be obtained uniquely from the samples $x(n)$ through the interpolation formula

$$\hat{x}(t) = \sum_{k=-\infty}^{\infty} x(k)\frac{\sin\left[\frac{\pi}{T}(t-kT)\right]}{\frac{\pi}{T}(t-kT)}. \tag{2}$$

In many digital signal processing problems, we are given a sequence $x(n)$, corresponding to sampling period $T$, and we must obtain from the sequence $x(n)$ a sequence $y(n)=\hat{x}(nT')$; i.e., the sequence $y(n)$ corresponds to sampling $\hat{x}(t)$ at a different sampling rate. If we evaluate (2) for $t=nT'$, we obtain

Manuscript received December 6, 1972; revised March 7, 1973.
The authors are with Bell Laboratories, Murray Hill, N. J. 07974.

Reprinted from *Proc. IEEE*, vol. 61, pp. 692–702, June 1973.

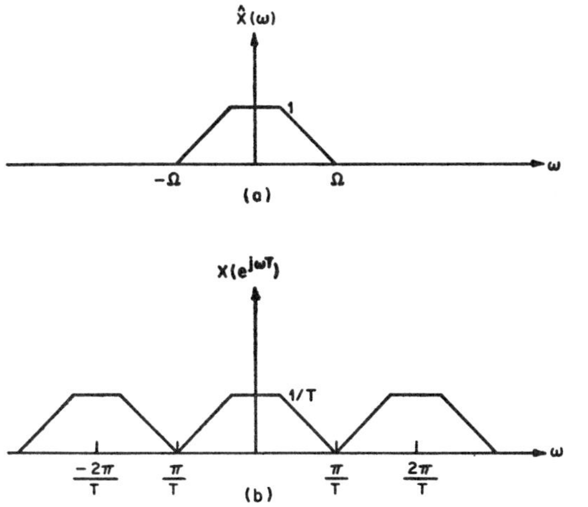


Fig. 1. Illustration of the relationship between the Fourier transform of a continuous-time signal and the Fourier transform of the sequence obtained by sampling with period $T$.

a direct relationship between $y(n)$ and $x(n)$, but it is clear that such an equation is impossible to evaluate because the functions $\sin\left[(\pi/T)(t-kT)\right]/\left[(\pi/T)(t-kT)\right]$ are of infinite duration. Rather than simply truncate these functions, it is more reasonable to design finite duration interpolators. To understand how such interpolators can be designed and to understand the limitations of classical interpolators, it is useful to consider the frequency-domain representation of the process of changing the sampling rate.

### *B. Sampling Rate Reduction—Integer Factors*

Suppose that the desired sampling period is $T'=MT$. If $M$ is an integer, this simply implies that the new sequence is

$$\begin{aligned} y(n) &= \hat{x}(nT') = \hat{x}(nMT) \\ &= x(Mn). \end{aligned}$$

That is, the sequence $x(n)$ is "sampled" by retaining only one out of each group of $M$ consecutive samples. The values of the sequence $y$ are samples of the original waveform $\hat{x}(t)$; however, these samples will uniquely determine $\hat{x}(t)$ if and only if $T' \leq \pi/\Omega$. This is clearly just a consequence of the sampling theorem as expressed by (1).

Since we are interested in direct relationships between the sequences $y(n)$ and $x(n)$, it is instructive to derive an equation similar to (1) that relates the Fourier transforms of the two sequences. The derivation of the equation is facilitated by the definition of a new sequence $w(n)$ which is nonzero only at integer multiples of $M$; that is

$$\begin{aligned} w(n) &= x(n), \qquad n = 0, \ \pm M, \ \pm 2M, \cdots \\ &= 0, \qquad \text{elsewhere} \end{aligned}$$

where the sampling period is assumed to be $T$ for both sequences. A convenient representation of $w(n)$ is

$$w(n) = x(n)\left\{\frac{1}{M}\sum_{l=0}^{M-1} e^{j(2\pi/Mln)}\right\}, \qquad -\infty < n < \infty$$

where the term in brackets may be recognized as a discrete Fourier series representation of a periodic sequence that is one at integer multiples of $M$ and zero otherwise. The sequence $y(n)$, corresponding to sampling period $T'=MT$, is

$$y(n) = w(Mn), \qquad -\infty < n < \infty.$$

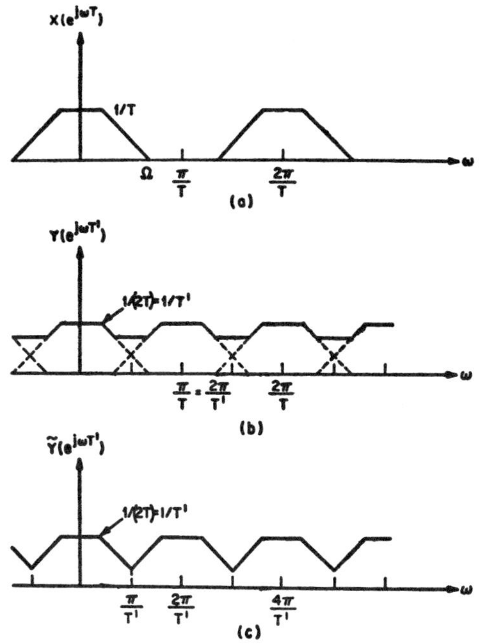


Fig. 2. Sampling rate reduction ($T'=2T$). (a) Fourier transform of original sequence $x(n)$. (b) Fourier transform of sequence $y(n)=x(2n)$ showing aliasing. (c) Fourier transform obtained after sampling rate reduction of a low-pass filtered version of $x(n)$.

Therefore

$$Y(z) = \sum_{n=-\infty}^{\infty} w(Mn)z^{-n}.$$

Since $w(n)$ is zero except at integer multiples of $M$, we obtain

$$\begin{aligned} Y(z) &= \sum_{n=-\infty}^{\infty} w(n)z^{-n/M} \\ &= \sum_{n=-\infty}^{\infty} x(n)\frac{1}{M}\sum_{l=0}^{M-1} e^{j(2\pi/M)ln}\, z^{-n/M} \\ &= \frac{1}{M}\sum_{l=0}^{M-1}\sum_{n=-\infty}^{\infty} x(n)e^{j(2\pi/M)ln}\, z^{-n/M} \end{aligned}$$

$$Y(z) = \frac{1}{M}\sum_{l=0}^{M-1} X(e^{-j(2\pi/M)l}z^{1/M}). \tag{3}$$

If we evaluate $Y(z)$ on the unit circle, with normalization appropriate for the new sampling rate, we obtain

$$Y(e^{j\omega T'}) = \frac{1}{M}\sum_{l=0}^{M-1} X(e^{j(\omega T'-2\pi l)/M}). \tag{4}$$

There is a clear similarity between (4) and (1).

An example of sampling rate reduction by a factor of 2 is shown in Fig. 2. The Fourier transform of $x(n)$ is depicted in Fig. 2(a) for the case when $\pi/2\Omega < T < \pi/\Omega$ so that

$$X(e^{j\omega T}) = \frac{1}{T}\hat{X}(\omega), \qquad -\frac{\pi}{T} < \omega \leq \frac{\pi}{T}.$$

Fig. 2(b) shows $Y(e^{j\omega T'})$ for $T'=2T$. In this case, aliasing occurs and it is clear that, in general, aliasing will occur in the process of digital sampling rate reduction unless the original sampling period satisfies

$$T \leq \frac{\pi}{M\Omega} \tag{5}$$

where $\Omega$ is the Nyquist frequency of $\hat{x}(t)$. If this inequality is satisfied, however, then

$$Y(e^{j\omega T'}) = \frac{1}{M} X(e^{j\omega T})$$

$$= \frac{1}{MT} \hat{X}(\omega)$$

$$= \frac{1}{T'} \hat{X}(\omega), \qquad -\frac{\pi}{T'} < \omega < \frac{\pi}{T'}.$$

If the original sampling period does not satisfy (5), aliasing distortion can be avoided only by passing the sequence $x(n)$ through an ideal low-pass digital filter with cutoff frequency $\pi/T'$. It can be seen that the filter must have unit gain, since (4) provides the factor $1/M$ needed to correct the amplitude for the new sampling rate. This, of course, results in a sequence $y(n)$ corresponding to a continuous time signal $\tilde{y}(t)$ which is a low-pass filtered version of the original signal $\hat{x}(t)$.

### *C. Sampling Rate Increase—Integer Factors*

If the sampling rate is increased by an integer factor $L$, then the new sampling period will be $T'=T/L$. Since the sequence $x(n)$ provides samples of the desired sequence only at intervals of $L$ samples at the new sampling rate, the remaining samples must be filled in by interpolation. To see how this can be done using a digital filter, consider the sequence

$$v(n) = x(n/L), \qquad n = 0, \quad \pm L, \quad \pm 2L, \cdots$$

$$= 0, \qquad \text{otherwise}.$$

The $z$ transform of $v(n)$ is

$$V(z) = \sum_{n=-\infty}^{\infty} x(n/L)z^{-n}$$

$$= \sum_{n=-\infty}^{\infty} x(n)z^{-Ln} = X(z^L). \tag{6}$$

The Fourier transform of this sequence is

$$V(e^{j\omega T'}) = X(e^{j\omega T'L})$$

$$= X(e^{j\omega T}).$$

Thus $V(e^{j\omega T'})$ is periodic with period $2\pi/T = 2\pi/LT'$, rather than $2\pi/T'$ as is the case in general for sequences associated with a sampling period $T'$. Fig. 3(a) shows $V(e^{j\omega T'})$ for the case $T'=T/3$. If we wish to obtain a sequence $y(n)$ such that

$$y(n) = \hat{x}(nT')$$

then we must insure that

$$Y(e^{j\omega T'}) = \frac{1}{T'} \hat{X}(\omega), \quad -\frac{\pi}{T'} \leq \omega \leq \frac{\pi}{T'}. \tag{7}$$

Assuming that

$$X(e^{j\omega T}) = \frac{1}{T} \hat{X}(\omega), \qquad -\frac{\pi}{T} \leq \omega \leq \frac{\pi}{T}$$

then it is clear from Fig. 3(a), that the images of $(1/T)\hat{X}(\omega)$

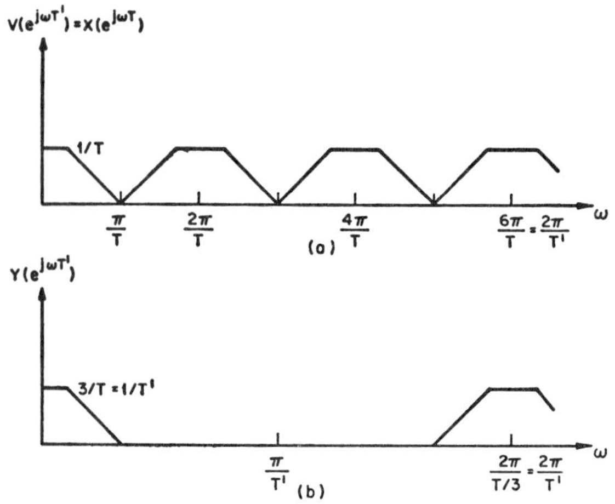


Fig. 3. Sampling rate increase ($T'=T/3$). (a) Fourier transform of sequences $x(n)$ and $v(n)$. (b) Fourier transform of desired output of interpolation process.

n $V(e^{j\omega T'})$ that are centered at $\omega = 2\pi/T$ and $4\pi/T$ must be removed by a digital low-pass filter that rejects all frequencies in the range $\pi/T < |\omega| < \pi/T'$. Furthermore, to insure that the amplitude is correct for sampling interval $T'$, the gain of the filter must be $L$. That is,

$$Y(e^{j\omega T'}) = H(e^{j\omega T'})V(e^{j\omega T'}) = H(e^{j\omega T'})X(e^{j\omega T})$$

$$= \frac{1}{T} H(e^{j\omega T'})\hat{X}(\omega)$$

where $H(e^{j\omega T'})$ is periodic with period $2\pi/T'$ and

$$H(e^{j\omega T'}) = L, \qquad |\omega| \leq \frac{\pi}{T}$$

$$= 0, \qquad \frac{\pi}{T} < |\omega| < \frac{\pi}{T'}. \tag{8}$$

Thus the ideal interpolation scheme for increasing the sampling rate requires the creation of a sequence of $L-1$ zero-valued samples between each value of the original sequence, which is then filtered with an ideal low-pass filter as in (8).

### *D. Changing by Noninteger Factors*

In the previous two subsections we have discussed methods for increasing or decreasing the sampling rate using a linear time-invariant digital filter. However, because of the fact that we required that the process be entirely a discrete-time process, the new sampling period $T'$ was restricted to be either an integer multiple or submultiple of the original sampling period $T$. This restriction can be eased somewhat by a two-step process involving a sampling rate increase followed by a decrease.

Suppose the desired sampling period is $T'=(M/L)T$, where $M$ and $L$ are integers. This does not seem to be a significant limitation in practice since any factor can be approximated as closely as desired by proper choice of $M$ and $L$. The two-step process for interpolating to a sampling period $T'=(M/L)T$ is as follows:

1) Increase the sampling rate by a factor of $L$ by interpolation as in Section II-C. Let the resulting sequence be denoted $y_1(n)$ with sampling period $T_1 = T/L$.

2) Decrease the sampling rate by a factor of $M$ as in Section II-B to obtain the desired sequence $y(n)$ with sampling period $T' = MT_1 = (M/L)T$.

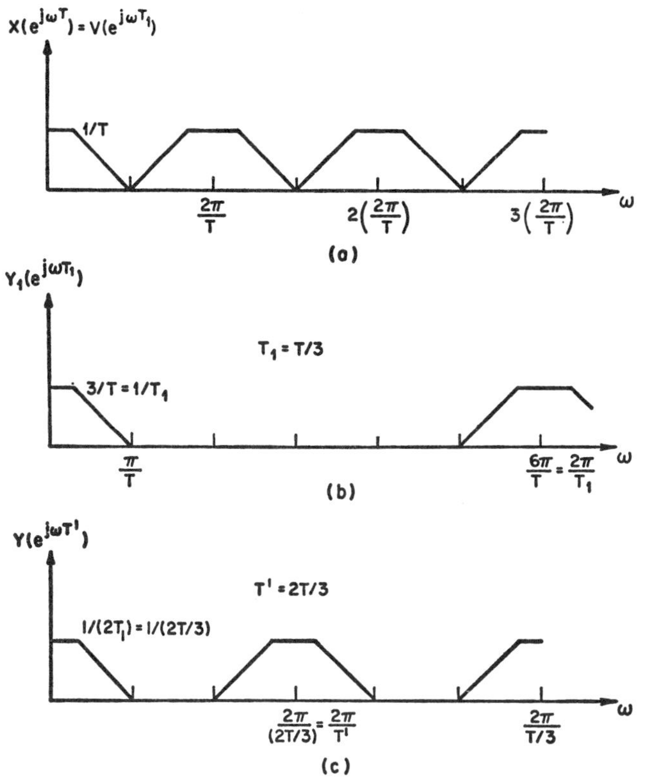


Fig. 4. Sampling rate change by noninteger factor ($T' = \frac{2}{3}T$). (a) Fourier transform of $x(n)$ and $v(n)$. (b) Fourier transform for increase of sampling rate by a factor of 3. (c) Fourier transform after sampling rate reduction by factor of 2.

This process is illustrated in Fig. 4 for $T' = \frac{2}{3}T$; i.e., for a net increase in sampling rate. Fig. 4(a) shows the Fourier transform of the original sequence $x(n)$. Fig. 4(b) shows the Fourier transform of the intermediate sequence $y_1(n)$ which results from filtering the sequence

$$\begin{aligned} v(n) &= x(n), \qquad n = 0, \pm 3, \pm 6, \cdots \\ &= 0, \qquad \text{elsewhere} \end{aligned}$$

with an ideal low-pass filter having gain 3 and cutoff frequency $\pi/T$. The result of reducing the sampling rate of $y_1(n)$ by a factor of 2 is shown in Fig. 4(c).

If $M < L$, i.e., there is a net increase in sampling rate as in Fig. 4, no aliasing can occur, and the interpolation filter can have a cutoff frequency $\pi/T$. However if $M > L$, there is a possibility of aliasing. In this case aliasing can be avoided by making the cutoff frequency of the interpolation filter equal to $\pi/T'$. As in Section II-B, the resulting output sequence $y(n)$ will correspond to samples of a low-pass filtered version of the original continuous time signal.

## III. Interpolation Using FIR Digital Filters

The previous section makes it abundantly clear that the process of changing the sampling rate requires a low-pass filter. Since it is impossible to realize the ideal low-pass filter that is required for exact results, we must consider digital filters that approximate this ideal behavior. As in all filtering problems, there are many important considerations. A basic consideration is the choice between filters from the class of FIR filters and from the class of IIR filters. Given the type of filter to be used, the problem of approximating the ideal low-pass filter must be solved. Finally, there are important considerations in how the filter is realized as software or hardware. All of these facets of the problem are interrelated—resulting in arbitrary tradeoffs between accuracy of interpolation and efficiency of realization. In this section we present some observations on filter design and realization that seem to imply that FIR filters are the proper choice for most interpolation problems.

### *A. Phase Distortion*

The ideal interpolator has zero phase or at most a linear phase corresponding to an integer number of samples of delay. IIR filters cannot have precisely linear phase [6]. In contrast, there currently exist several techniques for designing optimal FIR digital filters with precisely linear phase. These filters are optimal in the sense that the width of transition band between passband and stopband is minimum for given values of passband and stopband ripple and specified passband and stopband cutoff frequencies [7]–[11]. These filters can be designed with arbitrarily small values for passband ripple, stopband ripple, and transition bandwidth, at the cost of increased impulse response duration. Thus with FIR filters, the interpolation error due to phase nonlinearity can be zero and the error due to amplitude distortion can be made arbitrarily small. In the case of IIR filters, although extremely good amplitude characteristics can be achieved, there will always be an interpolation error due to phase nonlinearity.

### *B. Filter Realization*

IIR filters have recursive realizations that are very economical in terms of computational complexity. Leaving phase considerations aside, FIR filters in general require more computation to achieve a given accuracy of approximation to the desired amplitude response than do IIR filters. However, the particular nature of the interpolation problem makes FIR filters computationally competitive with IIR filters.

Consider for convenience a zero-phase FIR filter with impulse response $h(n)$ which is nonzero in the interval

$$-\left(\frac{N-1}{2}\right) \le n \le \left(\frac{N-1}{2}\right)$$

where $N$ is an odd integer.[1] In reducing the sampling rate by an integer factor, i.e., $T' = MT$, we may need a unity gain low-pass filter to insure that no aliasing occurs in retaining only every $M$th sample of the sequence $x(n)$. In this case computation is reduced because of the nature of the desired output sequence. The filtered output sequence at the original sampling rate, defined as $\hat{y}(n)$, is

$$\hat{y}(n) = \sum_{k=-((N-1)/2)}^{((N-1)/2)} h(k)x(n-k) \tag{9}$$

where all sequences in (9) are associated with sampling period $T$. Clearly, all values of the sequence $\hat{y}(n)$ need not be computed since the desired output is

$$y(n) = \hat{y}(nM) = \sum_{k=-((N-1)/2)}^{((N-1)/2)} h(k)x(nM-k) \tag{10a}$$

where $y(n)$ is associated with a sampling period $T' = MT$. This is in contrast to a comparable IIR filter where the computations required to realize the poles of the system function would have to be carried out at the original sampling rate even

[1] See Section III-C for a comment on why $N$ should be odd.

though $M-1$ out of $M$ samples of $\hat{y}(n)$ would be discarded. A further simplification results from the fact that zero-phase FIR filters have the property

$$h(n) = h(-n).$$

Thus (10a) becomes

$$y(n) = \sum_{k=1}^{((N-1)/2)} h(k)[x(nM - k) + x(nM + k)] + h(0)x(nM). \quad (10b)$$

In the case of sampling rate increase, the interpolated output is obtained with sampling period $T' = T/L$ by filtering the sequence

$$\begin{aligned} v(n) &= x(n/L), \qquad n = 0, \pm L, \pm 2L, \cdots \\ &= 0, \qquad \text{otherwise.} \end{aligned}$$

In this case it is convenient to write

$$y(n) = \sum_{k=n-((N-1)/2)}^{n+((N-1)/2)} v(k)h(n-k). \quad (11)$$

Substituting for $v(k)$ results in

$$\begin{aligned} y(n) &= \sum_{k=n-((N-1)/2)}^{n+((N-1)/2)} x(k/L)h(n-k), \qquad k/L \text{ an integer} \\ &= \sum_{k=[(n/L)-((N-1)/2L)]}^{[(n/L)+((N-1)/2L)]} x(k)h(n-kL) \end{aligned} \quad (12)$$

where $[a]$ means "the largest integer contained in $a$." Although (11) suggests that computation required for each output sample is proportional to $N$, we note that only one out of every $L$ samples of $v(n)$ is nonzero. Thus we see from (12) that the actual computation required is proportional to $N/L$. Note that in this case the symmetry of the impulse response cannot be exploited to reduce computation.

If an IIR filter were used, relatively little saving could be achieved. In fact, in a cascade realization, almost no computational saving could be attained by considering the zeroes in the input sequence.

Changing the sampling period according to $T' = (M/L)T$, requires that we first increase the sampling rate by a factor $L$ and then reduce it by a factor $M$. Clearly, the savings previously discussed could be incorporated into both steps of this process.

### C. Impulse Response Constraints

The previous discussion has presented compelling reasons for the use of linear-phase FIR filters in interpolation. To conclude this section we discuss some constraints on the impulse response that are specific to the problem of interpolation. Recall that the output $y(n)$ is given by (12). A reasonable requirement on the interpolation filter is that the values of the output at the original sampling times be the same as the original samples.[2] That is, for $r$ an integer,

$$y(rL) = x(rL/L) = x(r), \qquad -\infty < r < \infty.$$

Equation (12) implies that

$$y(rL) = x(r) = \sum_{k=[r-(N-1)/2L]}^{[r+(N-1)/2L]} x(k)h(rL - kL) \quad (13)$$

[2] This is also a property of (2), as can be easily verified.

for all integer values of $r$. From this equation, we see that

$$\begin{aligned} h(0) &= 1 \\ h(n) &= 0, \qquad n = \pm L, \pm 2L, \cdots, \left[\frac{N-1}{2L}\right]L. \end{aligned} \quad (14)$$

Constraints of this type are rather difficult to impose on an IIR filter design procedure.

Some final comments on the choice of the length of the impulse response $N$ are in order. First, we have assumed that $N$ is an odd integer; however, in general $N$ can be either even or odd for the FIR filter. It can be shown [6], [12] that for $N$ even, a linear-phase FIR filter must have a delay of at least one-half sample. This one-half sample delay itself corresponds to interpolation between samples; thus such a filter could not preserve the samples of the original sequence. Although there may be instances where the half-sample delay may not be objectionable, or may even be desirable, odd values of $N$ appear to be appropriate for most applications.

A second comment regarding the choice of $N$ concerns the fact that we have asserted that by increasing $N$, we can achieve a better approximation in the frequency domain to the ideal interpolating filter. In the time domain, increasing $N$ implies that more values of the original sequence are involved in the computation of a given interpolated sample and thus we should expect increasingly better results as $N$ increases. We have observed that because the sequence $v(n)$ has mostly zero samples, the computation is proportional to $N/L$, since the impulse response always spans approximately $N/L$ nonzero samples. Suppose that we wish to always have $Q$ samples of the original input sequence involved in the computation of each interpolated sample. Using (12) it is easily shown that the length of the impulse response should be

$$\begin{aligned} N &= QL, \qquad \text{if } Q \text{ and } L \text{ are odd} \\ &= QL - 1, \qquad \text{if either } Q \text{ or } L \text{ are even.} \end{aligned} \quad (15)$$

If $N$ is slightly larger than this value, the computation of some interpolated samples will involve $Q+1$ original samples while the rest will involve only $Q$. If $N$ is slightly smaller than the value given by (15), then the interpolation will involve either $Q$ or $Q-1$ samples. Thus to insure that $Q$ samples of the original sequence are always involved, $N$ should satisfy (15).

This constraint is satisfied, as we will see, for filters derived from classical interpolation formulas. It is also important in hardware and software realizations of interpolation filters where it allows the computations to be structured so that one does not have to check for zero samples in order to eliminate multiplications by zero.

## IV. Classical Polynomial Interpolation

In this section we apply the previous discussion of the interpolation process to a study of classical polynomial interpolation methods. Our aim is to give a frequency domain interpretation of these formulas that will shed some light on their applicability in interpolation problems arising in digital signal processing. In the course of this discussion, we shall indicate how to derive linear phase interpolation filters from tables of Lagrange coefficients. We begin with a discussion of linear interpolation.

### A. Linear Interpolation

Linear interpolation involves only two consecutive samples of the original sequence $x(n)$ in the computation of an interpolated sample. Specifically, the values interpolated between two samples $x(0)$ and $x(1)$ lie on a straight line connecting the two original samples,[3] with the original samples of course being preserved. Thus the equation relating the output $y(n)$, having sampling period $T'=T/L$, to the input sequence $x(n)$, having sampling period $T$, is

$$y(n) = x(0) + \frac{x(1)-x(0)}{LT'-0}(nT'-0)$$
$$= x(0)\left(1-\frac{n}{L}\right) + x(1)\left(\frac{n}{L}\right), \quad 0 \le n < L. \tag{16}$$

In order to interpret linear interpolation as a linear filtering process, we must derive an impulse response for the linear interpolation filter. This can be done by comparing (16) to (12). To begin, it is clear that the length of the impulse response must be

$$N = 2L - 1$$

as discussed in Section III-C. If $N$ were larger, more than 2 samples of $x(n)$ would be involved in the computation of some of the interpolated values. Likewise, if $N$ were smaller, only one sample of the input would enter into the computation of some of the interpolated values. Using this information, we can write (12) as

$$y(n) = x(0)h(n) + x(1)h(n-L), \quad 0 \le n < L. \tag{17}$$

Comparing (16) and (17) we see that

$$h(n) = 1 - \frac{n}{L}, \qquad 0 \le n < L$$
$$h(n-L) = \frac{n}{L}, \qquad 0 \le n < L.$$

Thus $h(n)$ is seen to be

$$h(n) = 1 - |n|/L, \qquad |n| < L$$
$$= 0, \qquad \text{otherwise}. \tag{18}$$

Clearly $h(n)$ satisfies the requirements of (14) since $h(0)=1$ and $h(n)=0$ for $|n|>L$. The length of the impulse response is

$$N = 2L - 1$$

consistent with (15) for $Q=2$.

Fig. 5 depicts linear interpolation as a convolution process. The sequence $v(k)$ and the triangular envelope of the impulse response $h(n-k)$ are shown for the case of $T'=T/5$. It is clear that because the impulse response has duration $N=5(2)-1=9$, only two nonzero samples of $v(k)$ are ever coincident with $h(n-k)$. Also it is clear that

$$y(n) = x(n/L), \qquad n = 0, \pm L, \pm 2L, \cdots.$$

The system function corresponding to linear interpolation is

$$H(e^{j\omega T'}) = \frac{1}{L}\left\{\frac{\sin[\omega LT'/2]}{\sin[\omega T'/2]}\right\}^2. \tag{19}$$

This system function is plotted in Fig. 6 for $L=5$. (Curve labelled $Q=2$.) We recall that the purpose of the interpolation filter is to remove the images of the signal spectrum that are centered at integer multiples of $2\pi/T$, while leaving the frequencies below $\pi/T$ unaltered. It can be seen from Fig. 6, that linear interpolation achieves significant attenuation in only a very small region around each integer multiple of $2\pi/T$. Specifically, attenuation is 40 dB or greater in a band of width $0.35\pi/T$ centered at $2\pi/T$, $4\pi/T$, etc. Thus it seems reasonable to note that linear interpolation is appropriate only if the original sampling rate is many times the Nyquist rate.

[3] We can choose samples $x(0)$ and $x(1)$ without loss of generality since interpolation is a time-invariant process.

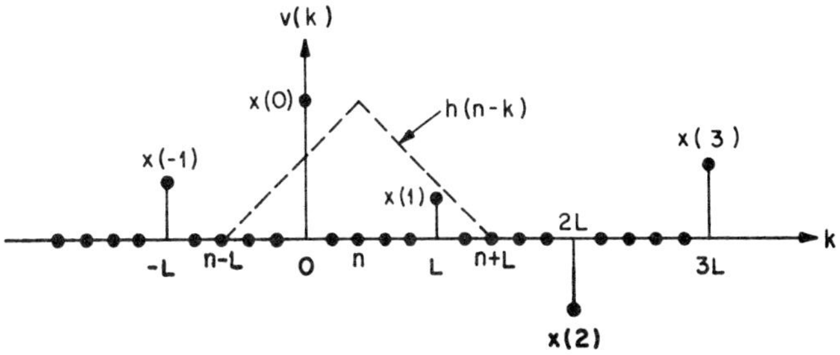


Fig. 5. Linear filtering interpretation of linear interpolation.

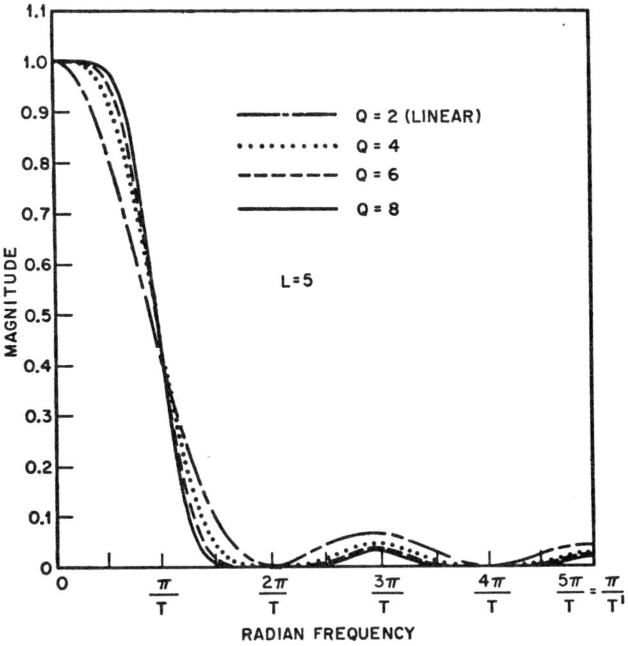


Fig. 6. Comparison of Lagrange interpolation frequency responses for $L=5$.

*B. Lagrange Interpolation*

Clearly linear interpolation will not be satisfactory in many digital signal processing applications. In classical numerical analysis, the inadequacies of linear interpolation lead to the use of higher order polynomials; i.e., in contrast to connecting two points by a straight line, one finds a polynomial of degree $Q-1$ that passes through $Q$ original samples. The interpolated values then, are samples of this polynomial. A variety of formulas have been derived for obtaining samples of the polynomial directly from the samples $x(n)$, however since we are only interested in the interpolation filter corresponding to polynomial interpolation, we shall use the most convenient form; namely, the Lagrange interpolation formula [13], [14]. The form corresponding to interpolation with equal spacing from sampling period $T$ to $T'=T/L$ is

$$y(n) = \sum_{k=-((Q-2)/2)}^{Q/2} A_k^Q(n/L)x(k), \qquad Q \text{ even} \qquad (20a)$$

$$y(n) = \sum_{k=-((Q-1)/2)}^{((Q-1)/2)} A_k^Q(n/L)x(k), \qquad Q \text{ odd} \qquad (20b)$$

where we have again chosen to consider interpolation around some arbitrary time labelled $O$. In (20), the quantities $A_k^Q(\cdot)$ are called the Lagrange coefficients and are given by the equations

$$A_k^Q(t) = \frac{(-1)^{k+Q/2}}{\left(\left(\frac{Q-2}{2}\right)+k\right)!\left(\frac{Q}{2}-k\right)!(t-k)} \cdot \prod_{i=1}^{Q}\left(t+\frac{Q}{2}-i\right), \qquad Q \text{ even} \qquad (21a)$$

$$A_k^Q(t) = \frac{(-1)^{k+(Q-1)/2}}{\left(\left(\frac{Q-1}{2}\right)+k\right)!\left(\left(\frac{Q-1}{2}\right)-k\right)!(t-k)} \cdot \prod_{i=0}^{Q-1}\left(t+\left(\frac{Q-1}{2}\right)-i\right), \qquad Q \text{ odd.} \qquad (21b)$$

Extensive tabulations of these coefficients are available in tables of mathematical functions [15]. It is interesting to note that

$$\begin{aligned} A_k^Q(t) &= 0, \qquad t \text{ an integer, and } \neq k. \\ &= 1, \qquad t = k. \end{aligned}$$

This is a result of the fact that the interpolation polynomial passes through the $Q$ original data samples. That is,

$$y(kL) = x(k), \qquad \text{for } k \text{ an integer.}$$

Thus the condition of (14) is satisfied for Lagrange interpolation filters.

In general, the formulas in (20) may be evaluated for any value of $n$ such that

$$-\left(\frac{Q-2}{2}\right)L < n < \frac{Q}{2}L, \qquad Q \text{ even}$$

$$-\left(\frac{Q-1}{2}\right)L < n < \left(\frac{Q-1}{2}\right)L, \qquad Q \text{ odd.}$$

In this case the *same* $Q$ original samples are involved in the computation of all the interpolated samples in the time interval spanned by the original samples. If we perform the interpolation with a linear filter, we can choose the length of the filter impulse response so that $Q$ original samples are always involved in the computation; however, a given set of $Q$ original samples is used only to compute $L-1$ interpolated samples. Thus we can interpret the Lagrange formulas in (20) in terms of $Q-1$ different impulse responses—corresponding to the $Q-1$ different interpolation intervals between the $Q$ original samples. As an example, consider the case for $Q=3$. In comparing (12) and (20b), we note that the three samples $x(-1)$, $x(0)$, and $x(1)$ can enter into the interpolation in two

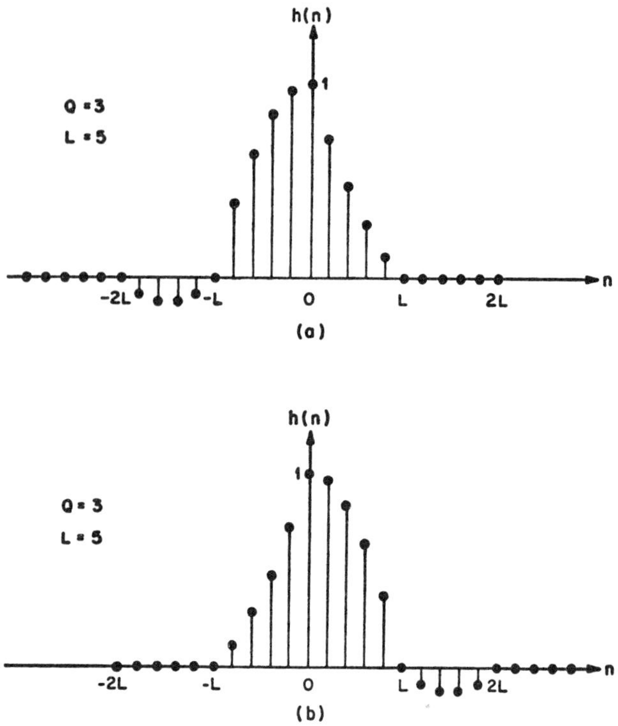


Fig. 7. Impulse responses for three-point ($Q=3$) Lagrange interpolation. (a) Interpolation in the interval $-L<n\leq 0$. (b) Interpolation in the interval $0\leq n<L$.

intervals. Thus

$$\begin{aligned} y(n) &= \sum_{k=-1}^{1} A_k^Q(n/L)x(k) \\ &= \sum_{k=-1}^{1} x(k)h(n-kL) \end{aligned}$$

for values of $n$

$$\begin{aligned} -L < n &\leq 0 \\ 0 \leq n &< L. \end{aligned}$$

If we compare the above two equations, we obtain

$$h(n+L) = A_{-1}^3(n/L)$$

$$h(n) = A_0^3(n/L)$$

$$h(n-L) = A_1^3(n/L).$$

If these equations are evaluated for the first interval we obtain the impulse response of Fig. 7(a) where $L=5$. Likewise, Fig. 7(b) shows the impulse response corresponding to Lagrange interpolation in the second interval. Clearly neither of these impulse responses has linear phase, since they do not satisfy the symmetry condition $h(n)=h(-n)$. Indeed, it is easy to show that whenever $Q$ is odd, none of the impulse responses corresponding to Lagrange interpolation can have linear phase. However, if $Q$ is even, one of the $Q-1$ impulse responses does have linear phase

As a second example suppose $Q=4$. Using (20a) we obtain

$$y(n) = \sum_{k=-1}^{2} A_k^4(n/L)\,x(k). \qquad (22)$$

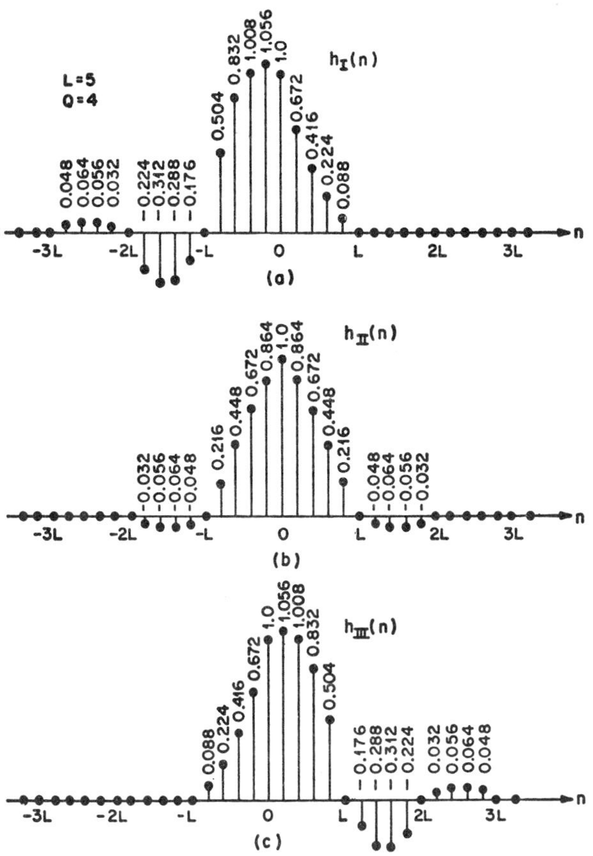


Fig. 8. Impulse responses for four-point ($Q=4$) Lagrange interpolation.

As in the previous example we can evaluate (12) for $N=QL-1$ and we obtain the equation

$$y(n) = \sum_{k=-1}^{2} x(k)h(n-kL)$$

from which by comparison to (22) we obtain

$$\begin{aligned} h(n+L) &= A_{-1}^{4}(n/L) \\ h(n) &= A_{0}^{4}(n/L) \\ h(n-L) &= A_{1}^{4}(n/L) \\ h(n-2L) &= A_{2}^{4}(n/L). \end{aligned}$$

These equations can be evaluated in the 3 intervals

$$\begin{aligned} \text{I:} &\quad -L < n \le 0 \\ \text{II:} &\quad 0 < n \le L \\ \text{III:} &\quad L < n \le 2L \end{aligned}$$

to obtain the three impulse responses shown in Fig. 8 for the condition $L=5$. The exact values for the three impulse response coefficients are also given in Fig. 8. We note that the impulse response corresponding to interpolation in the central interval $0 \le n < L$, is symmetric and thus has zero phase. The two other impulse responses clearly do not have linear phase. Indeed, it is clear that, in general, Lagrange interpolation has phase distortion except in the $Q$ even case and interpolation in the central interval. Thus if we wish to use impulse responses derived from the Lagrange interpolation formula for interpolation in signal processing applications we should use the conditions that yield zero phase. In general the linear phase impulse response derived from a $Q$ point Lagrange interpolation formula is obtained from the equations

$$\begin{aligned} h\left(n+\left(\frac{Q-2}{2}\right)L\right) &= A_{-(Q-2)/2}^{Q}(n/L) \\ &\cdots \\ h(n+L) &= A_{-1}^{Q}(n/L) \\ h(n) &= A_{0}^{Q}(n/L) \\ h(n-L) &= A_{1}^{Q}(n/L) \\ &\cdots \\ h\left(n-\frac{Q}{2}L\right) &= A_{Q/2}^{Q}(n/L) \end{aligned} \tag{23}$$

where $0 \le n < L$.

In Fig. 6 we have plotted the frequency response of the zero-phase interpolators derived in this manner for $Q=2, 4, 6, 8$, and $L=5$. Since the impulse response duration is $N=QL-1$, $N=9$, 19, 29, 39 for these cases. Clearly, the effect of increasing $Q$ is to improve the frequency response of the interpolator. Whereas the linear interpolator has 40-dB attenuation in a bandwidth of $0.35\pi/T$ centered around each integer multiple of $2\pi/T$, for the 4-point and 6-point interpolators, the bandwidths are $0.7\pi/T$ and $0.9\pi/T$, respectively, for at least 40-dB attenuation. Thus at the expense of increased computation, we can achieve a significantly better interpolation by using a filter derived using (23).

The interpolation filters derived from the Lagrange interpolation formula achieve high attenuation in a narrow band around integer multiples of $2\pi/T$ because the zeroes of the system function tend to be clustered about those frequencies. For example, in the case of linear interpolation ($Q=2$), by looking at (19) we see that $H(z)$ has a double zero on the unit circle at integer multiples of $2\pi/T$. For $Q=4$, we have found that there are clusters of 4 zeroes not precisely on the unit circle but close in the vicinity of $\omega=2\pi/T$, $4\pi/T$, etc. As a result, the attenuation close to frequencies $2\pi/T$ and $4\pi/T$ is very high. This is clear from Fig. 9, where the system function for the case $Q=6$ is plotted on a log scale. However, it is also clear that between $2\pi/T$ and $4\pi/T$, the response of the Lagrange interpolation filter leaves much to be desired. Clearly, as $Q$ gets larger, the impulse response gets longer and there are more zeroes to distribute so as to increase the attenuation and broaden the attenuation bands. This raises the question as to how we might design FIR digital filters so as to make the best use of the filter zeroes.

## V. Optimum FIR Interpolation Filters

In practical situations, signals are often sampled at a rate that is only slightly higher than twice the Nyquist frequency in order to minimize the computation required for digital filtering and other signal processing procedures. In this case, the ideal interpolation filter for increasing the sampling rate has constant gain in the frequency range $0 \le |\omega| < \pi/T$, and zero gain elsewhere. For such signals we are clearly interested in the best possible approximations to the ideal low-pass filter.

On the other hand, in situations where the original sequence $x(n)$ is derived by sampling at a rate considerably higher than twice the Nyquist frequency, we require a relatively narrow passband of constant gain and a number of stop-

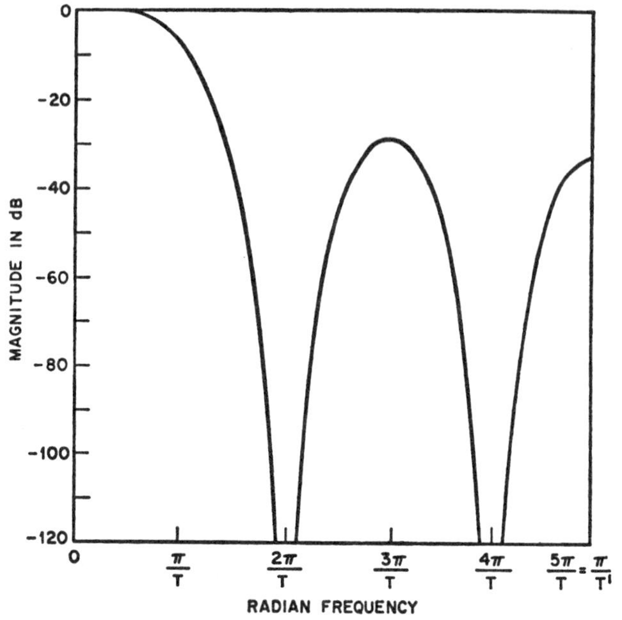


Fig. 9. Frequency response of six-point ($Q=6$) linear phase Lagrange interpolation filter. $L=5$, $N=29$.

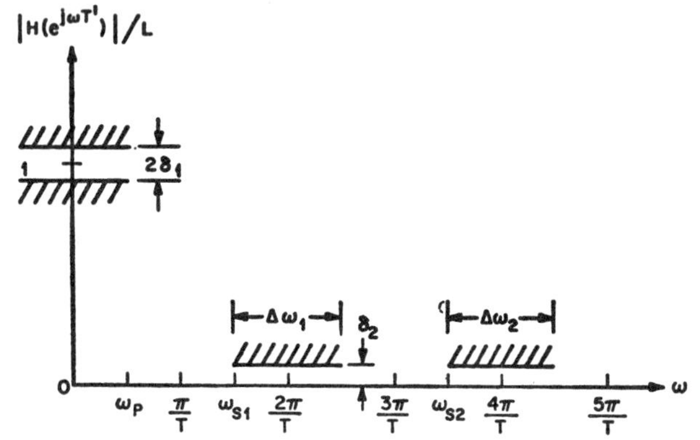


Fig. 10. General tolerance scheme for digital interpolation.

bands of zero gain, with the frequency response being somewhat arbitrary elsewhere. This means that high-order polynomial interpolation filters may be quite satisfactory for signals that are sufficiently oversampled. However, it is generally possible to achieve significantly better interpolation filters using optimization techniques.

### A. Design Specifications

A scheme for approximating the ideal interpolator is shown in Fig. 10. There is a band of frequencies, $0 \leq \omega \leq \omega_p$, called the passband in which the frequency response should be close to 1.[4] In practice we allow for an error of $\pm\delta_1$. In addition, there are one or more stopbands in which the frequency response is required to be within $\pm\delta_2$ of zero. The choice of the parameters in Fig. 10 depends upon the desired accuracy of interpolation and to some extent upon the nature of the input sequence. The choice of passband and stopbands depends upon the Nyquist frequency of the original signal. If the original sampling period is such that $\pi/T \approx \Omega$, then $\omega_p$ must be very close to $\pi/T$, and the first stopband, beginning at $\omega_{s1}$, must also be close to $\pi/T$. Thus the transition band $\omega_p \leq \omega \leq \omega_{s1}$, must be very narrow, which implies that a large value of $N$ will be required.

[4] If the sampling rate is increased by a factor of $L$, the filter gain must approximate $L$ in the passband instead of 1.0. This is easily accomplished by multiplying the impulse response samples by $L$.

In such cases, it is reasonable to define only one stopband, $\omega_{s1} \leq \omega \leq \pi/T'$. However, in cases where $\pi/T$ is significantly higher than the Nyquist frequency, the transition band between passband and stopband can be wider, and it makes sense to define stopbands around each integer multiple of $2\pi/T$, with transition bands $\omega_{s_1}+\Delta\omega_1 \leq \omega \leq \omega_{s_2}$, etc., in which the frequency response is unconstrained. As can be seen from Figs. 6 and 9, this is the type of frequency response that characterizes the Lagrange interpolation filters.

### B. Design Techniques

The frequency response of a zero-phase FIR filter can be expressed as

$$H(e^{j\omega T'}) = h(0) + 2 \sum_{n=1}^{(N-1)/2} h(n) \cos(\omega n T'). \quad (24)$$

The tolerance scheme depicted in Fig. 10 can be expressed as the following set of inequalities:

$$1-K\delta_2 \leq H(e^{j\omega T'}) \leq 1+K\delta_2, \qquad 0 \leq \omega \leq \omega_p \quad (25)$$

$$-\delta_2 \leq H(e^{j\omega T'}) \leq \delta_2, \qquad \omega_{s_1} \leq \omega \leq \omega_{s_1}+\Delta\omega_1$$

$$\omega_{s_2} \leq \omega \leq \omega_{s_2}+\Delta\omega_2 \quad (26)$$

where we have defined $K=\delta_1/\delta_2$ as the ratio of the passband to stopband error. For a given $N$ and $K$, these equations may be evaluated on a dense set of frequencies in the specified passband and stopbands and may be solved for $\{h(n)\}$ and $\delta_2$ using either linear programming techniques [11] or discrete Chebyshev approximation techniques [10]. It has been shown [10] that the filters designed by these techniques are optimum in the sense of having the narrowest transition bands for given $\delta_1$ and $\delta_2$. For these filters, the error in the passband and stopbands exhibits an equiripple behavior. An example of a low-pass filter designed by linear programming is shown in Fig. 11. In this case $L=5$, $N=29$ ($Q=6$), $\omega_p=0.6\pi/T$, $\omega_{s_1}=1.4\pi/T$, $K=1.0$. The resulting filter has $\delta_1=\delta_2=0.00586$. (The filter gain has been normalized to 0 dB for convenience in plotting.) The equiripple behavior of the stopband is readily apparent.

The filter in this example does not satisfy the constraints on $h(n)$ given in (14); however, these constraints are linear and easily can be added to the constraints of (25) and (26). Thus in the optimization procedure $h(0)$ is constrained to 1.0 and $h(\pm L)$, $h(\pm 2L)$, etc., are set to 0. The filter performance is only slightly degraded by the addition of these constraints. Fig. 11 shows an example where all the fixed-design parameters were the same as in the previous example. In this case the resulting value of $\delta_1=\delta_2$ was 0.00599. It can be seen from Fig. 12 that the equiripple nature of the frequency response is destroyed, but with little sacrifice in performance.

If the original signal bandwidth is much less than $\pi/T$, a bandstop filter may provide superior performance to a comparable low-pass filter. Fig. 13 shows an example for $L=5$ where $N=29$, $\omega_p=0.3\pi/T$, $\omega_{s_1}=1.7\pi/T$, $\Delta\omega_1=\Delta\omega_2=0.6\pi/T$, $\omega_{s_2}=3.7\pi/T$, and $K=1.0$. In this case $\delta_1=\delta_2=0.0000681$. Thus this filter does a much better job of attenuating the images of the signal spectrum around the frequencies $2\pi/T$ and $4\pi/T$, but only if the original signal bandwidth was much less than $\pi/T$. It can be seen that the extra attenuation around $2\pi/T$ and $4\pi/T$ is obtained at the expense of the regions around

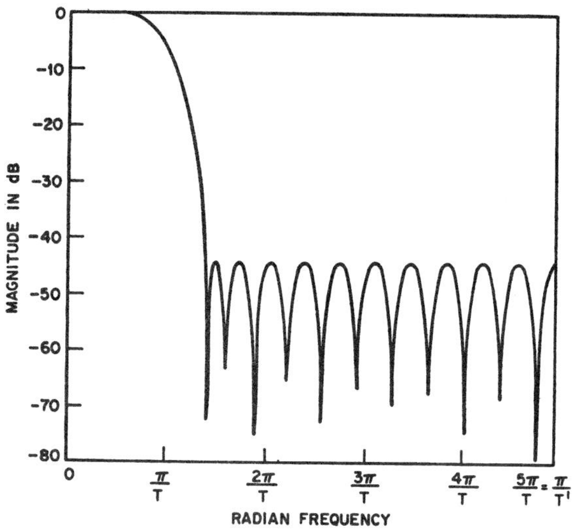


Fig. 11. Frequency response of a typical low-pass interpolation filter designed by linear programming. $L=5$, $Q=6$, $N=29$, $\omega_p=0.6\pi/T$, $\omega_{s_1}=1.4\pi/T$, $\delta_1=\delta_2=0.00586$. (No impulse response constraints.)

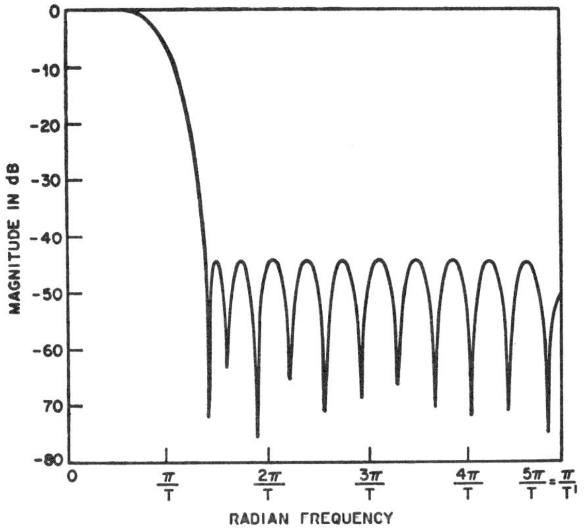


Fig. 12. Frequency response for the example of Fig. 11, with impulse response constraints. In this case $\delta_1=\delta_2=0.00599$.

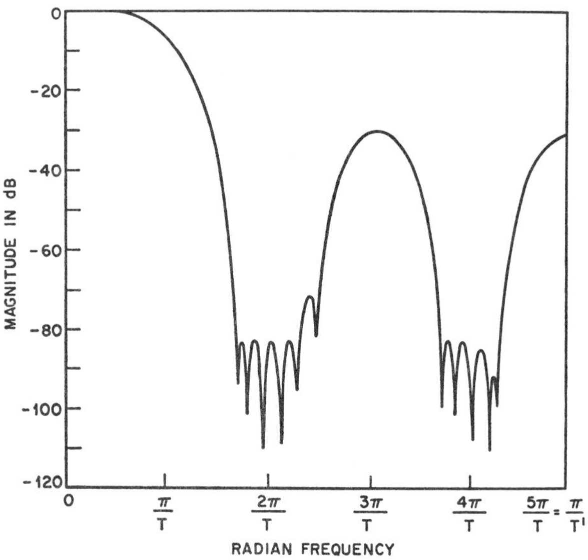


Fig. 13. Frequency response of a typical bandstop interpolation filter designed by linear programming. $L=5$, $Q=6$, $N=29$, $\omega_p=0.3\pi/T$, $\omega_{s_1}=1.7\pi/T$, $\omega_{s_2}=3.7\pi/T$, $\Delta\omega_1=\Delta\omega_2=0.6\pi/T$, and $\delta_1=\delta_2=0.0000681$.

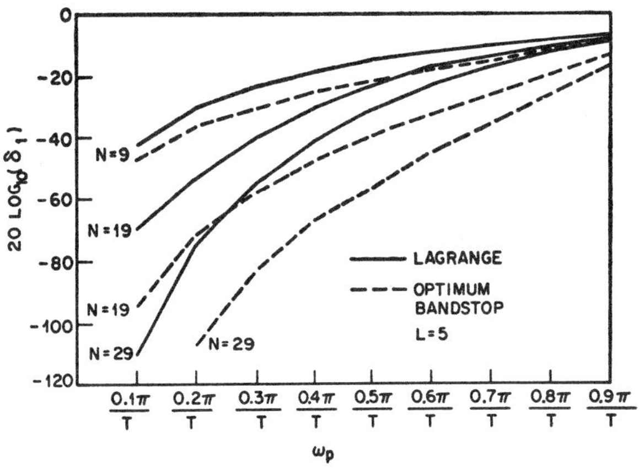


Fig. 14. Comparison of linear-phase Lagrange interpolators and optimum bandstop filters ($L=5$). Plots show minimum stopband attenuation as a function of half the stopband width. For the optimum bandstop filter ($N=29$), the curve is off scale for $\omega_p<0.2\pi/T$.

$3\pi/T$ and $5\pi/T$ where the attenuation is much less. Thus optimum filters of this type (and classical filters) should only be used if one is certain of the bandwidth of the input sequence.

### *C. Comparison to Classical Interpolators*

It is of considerable interest to compare the filters derived from classical interpolation formulas and those designed by linear programming. In the design of the optimum filters, the parameters of the tolerance scheme of Fig. 10 were set so that $\delta_1=\delta_2$ with symmetrical stopbands of width $\Delta\omega_1=\Delta\omega_2=\cdots=2\omega_p$ centered at integer multiples of $2\pi/T$. This is reasonable since in interpolation, preservation of the passband is generally as important as rejection of stopbands. In order to compare the Lagrange filters to the optimal filters, values of $\delta_1$ and $\delta_2$ were measured at the edges of the passband and the first stopband, respectively. Since for the Lagrange filters, $\delta_1$ is always greater than $\delta_2$ for the above definition of passband and stopbands, the comparison has been made on the basis of passband error.

Fig. 14 shows such a comparison for the case $L=5$.[5] The solid curves are for linear-phase filters derived from the Lagrange interpolation formula where $N=QL-1$ and $Q=2$, 4, and 6. These curves show the error in decibels at the edge of the passband, i.e., $20 \log_{10} \delta_1$, as a function of $\omega_p$. As an example, for linear interpolation ($Q=2$ and $N=9$) we see that the passband error is $-42$ dB for $\omega_p=0.1\pi/T$. Furthermore, since $\delta_1>\delta_2$ we can say that the attenuation is at least 42 dB in the region $(2\pi/T-0.1\pi/T)\leq\omega\leq(2\pi/T+0.1\pi/T)$. For wider bandwidths, the performance is worse; however, for higher order interpolators the performance becomes appreciably better as is expected.

The dotted curves in Fig. 14 show passband error for bandstop filters designed by linear programming. By comparing corresponding curves, it can be seen that the optimum designs are always significantly better than the corresponding classical interpolator, with the improvement being most striking for narrow bandwidths and for the higher order filters.

Clearly, there are a variety of optimum designs corresponding to situations in which passband and stopband approximation errors are not treated as being of equal importance. For example if $\delta_1=K\delta_2$, then the case $K>1$ corresponds to placing more importance on stopband attenuation than on passband

[5] The comparisons are similar for larger values of $L$.

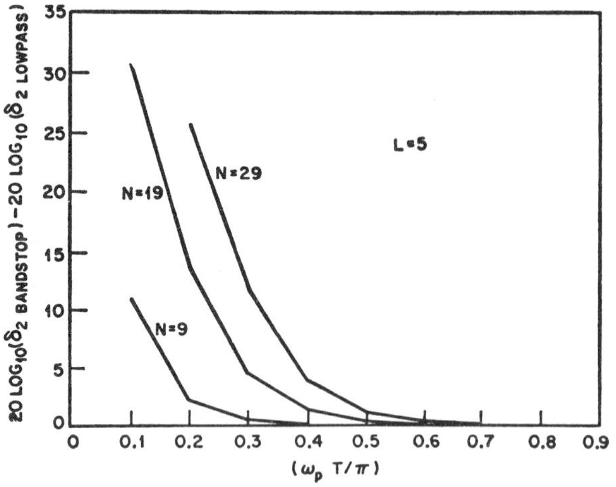


Fig. 15. Comparison of optimum bandstop and optimum low-pass interpolation filters ($L=5$). The curve for $N=29$ is off scale for $\omega_p < 0.2\pi/T$.

error. This situation would be a more favorable situation for the classical filters, although it is always possible to design a better filter using the optimum design procedures.

Another interesting comparison is between optimum low-pass filters and optimum bandstop filters. To compare these filters, we set

$$\omega_p = \frac{2\pi}{T} - \omega_{s_1} = \frac{4\pi}{T} - \omega_{s_2}$$

and $\Delta\omega_1 = \Delta\omega_2 = 2\omega_p$ for the bandstop filters and

$$\omega_p = \frac{2\pi}{T} - \omega_{s_1}$$

and $\Delta\omega = \pi/T' - \omega_{s_1}$ for the low-pass filters. That is, both filters were designed to accommodate the same input signal bandwidth. Fig. 15 shows the difference between stopband attenuations for the bandstop filters and the low-pass filters as a function of bandwidth. From these curves we see that for narrow bandwidths the bandstop filters have significantly greater attenuation; however as the bandwidth approaches half the original sampling frequency, there is no difference between the two types of filters.

## VI. Conclusions

In this paper we have discussed the process of interpolation as a problem in digital filtering. Most of our discussion has involved frequency-domain representations of the interpolation process and design criteria for digital interpolation filters. We have taken this approach because it is the most reasonable for digital signal processing applications where it is necessary to either raise or lower the sampling rate of a signal. This point of view is in contrast to that of interpolation in tables where one is concerned primarily with minimizing the error in a particular interpolated sample. Because of the variety of factors involved in the design of an interpolation filter, we have not tried to give design formulas and error bounds that would have limited value, but rather we have chosen to attempt to illuminate the important factors involved in the interpolation process and to discuss general design procedures that can be adapted to a variety of situations.

In particular, we have argued that linear-phase FIR filters have many attractive features for discrete-time interpolation and have shown how they may be efficiently utilized. Classical polynomial interpolation has been discussed in the context of digital signal processing. Interpolation filters derived from polynomial interpolation formulas are attractive because the impulse response can be easily computed or looked up in a table. However, we have seen that the frequency response of such systems leaves much to be desired in digital signal processing applications where the original sampling rate may be only slightly above the Nyquist rate.

As an alternative to filters based on classical interpolation formulas, we discussed optimum low-pass and bandstop FIR filters that were designed by linear programming. The bandstop filters have frequency responses that are very similar to the classical interpolators, but are always superior. The bandstop designs appear to be most important for cases when the original sampling rate is several times the Nyquist rate, while the low-pass designs are appropriate when the original sampling rate is close to the Nyquist rate.

## Acknowledgment

The authors would like to acknowledge the helpful comments and criticisms provided by Dr. O. Herrmann and Dr. J. F. Kaiser.

## References


[1] A. E. Rosenberg, R. W. Schafer, and L. R. Rabiner, "Effects of smoothing and quantizing the parameters of formant-coded voiced speech," *J. Acoust. Soc. Amer.*, vol. 50, pp. 1532–1538, Dec. 1971.

[2] R. W. Schafer and L. R. Rabiner, "Design and simulation of a speech analysis-synthesis system based on short-time Fourier analysis," *IEEE Trans. Audio Electroacoust.*, vol. AU-21, pp. 165–174, June 1973.

[3] S. L. Freeney, R. B. Kieburtz, K. V. Mina, and S. K. Tewksbury, "Design of digital filters for an all digital frequency division multiplex–time division multiplex translator," *IEEE Trans. Circuit Theory*, vol. CT-18, pp. 702–711, Nov. 1971.

[4] D. J. Goodman, "The application of delta modulation to analog-to-PCM encoding," *Bell Syst. Tech. J.*, vol. 48, pp. 321–344, Feb. 1969.

[5] H. Freeman, *Discrete-Time Systems.* New York: Wiley, 1965.

[6] A. J. Gibbs, "On the frequency-domain responses of causal digital filters," Ph.D. dissertation, University of Wisconsin, Madison, 1969.

[7] L. R. Rabiner, "Techniques for designing finite-duration impulse-response digital filters," *IEEE Trans. Commun. Technol.*, vol. COM-19, pp. 188–195, Apr. 1971.

[8] O. Herrmann, "On the design of nonrecursive digital filters with linear phase," *Electron. Lett.*, vol. 6, pp. 328–329, 1970.

[9] E. Hofstetter, A. V. Oppenheim, and J. Siegel, "A new technique for the design of nonrecursive digital filters," in *Proc. 5th Ann. Princeton Conf. on Information Sciences and Systems*, pp. 64–72, 1971.

[10] T. W. Parks and J. H. McClellan, "Chebyshev approximation for nonrecursive digital filters with linear phase," *IEEE Trans. Circuit Theory*, vol. CT-19, pp. 189–194, Mar. 1972.

[11] L. R. Rabiner, "The design of finite impulse response digital filters using linear programming techniques," *Bell Syst. Tech. J.*, vol. 51, pp. 1177–1198, July–Aug. 1972.

[12] L. R. Rabiner and R. W. Schafer, "Recursive and nonrecursive realizations of digital filters designed by frequency sampling techniques," *IEEE Trans. Audio Electroacoust.*, vol. AU-19, pp. 200–207, Sept. 1971.

[13] R. W. Hamming, *Numerical Methods for Scientists and Engineers.* New York: McGraw-Hill, 1962.

[14] F. B. Hildebrand, *Introduction to Numerical Analysis.* New York: McGraw-Hill, 1956.

[15] *Handbook of Mathematical Functions*, M. Abramowitz and I. A. Stegun, Eds. (Nat. Bur. Stand. Appl. Math. Ser.–55.) Washington, D. C.: U. S. Gov. Printing Office, 1964.

# New Results in the Design of Digital Interpolators

GEERD OETKEN, THOMAS W. PARKS, MEMBER, IEEE, AND HANS W. SCHÜSSLER, MEMBER, IEEE

*Abstract*—The design of optimum interpolators for band-limited sequences is considered. Newly found properties of these interpolators are discussed and used to simplify the design.

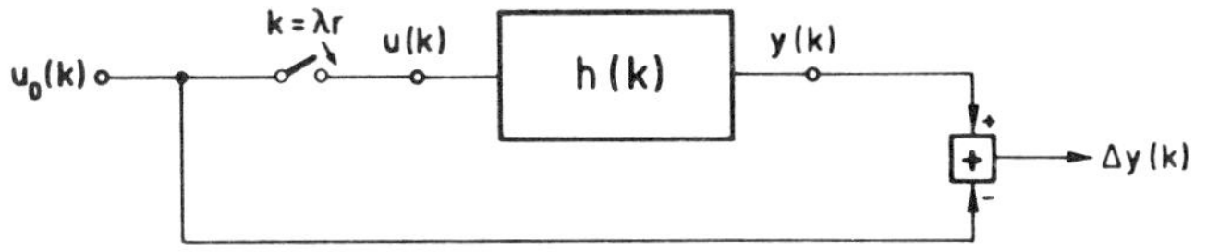


Fig. 1. General block diagram of the system to be investigated.

## I. Introduction

In an earlier paper the problem of designing a digital filter for interpolation has been considered [1]. Further investigation of the solution previously found has led to a more general formulation and to a simplification of the design. Therefore it seems justified to treat the subject again.

For convenience we state the problem again briefly, using a somewhat different notation (see Fig. 1).

We are given a sequence $\{u(k)\}$, the values of which are possibly unequal to zero only at $k = \lambda \cdot r$, where $\lambda = \cdots -1, 0, +1 \cdots$ and $r$ is an integer greater than 1. The $u(k)$ will be considered as being a sampled version of an unknown, but band-limited sequence $u_0(k)$

$$u(k) = \begin{cases} u_0(k) & \forall\, k = \lambda r \\ 0 & \forall\, k \neq \lambda r. \end{cases} \tag{1}$$

The interpolation filter should be designed such that its output sequence $y(k)$ approximates $u_0(k)$ as closely as possible [1], [2].

Stated in this way the interpolation problem is simply the discrete time version of the classical problem of recovering a band-limited signal from its samples. Previous work on this problem for continuous time output functions considered interpolators having infinite [3], [4] as well as finite duration impulse responses [5]. In [6] it was assumed that the input sequence as well as the impulse response was of finite duration. Beside the relation to [1] this paper is closely related to the work by Kellogg referred to in [5].

## II. Formulation of the Problem

If we assume the degree of the filter to be designed is $n = 2rL$ the output of the filter will be

$$y(k) = \sum_{\mu=-rL}^{+rL} h(\mu)u(k-\mu), \tag{2}$$

where $h(\mu)$ is the impulse response of the filter, which has been assumed for the time being to be noncausal for a convenient notation. Since $u(k) = 0$ for $k \neq \lambda r$ according to (1) we get for $k = \lambda r$

$$y(\lambda r) = \sum_{\lambda_1=-L}^{+L} h(\lambda_1 r)u((\lambda - \lambda_1)r) \tag{3a}$$

and for $k = \lambda r + \rho$ with $\rho = 1, \cdots, r-1$

$$y(\lambda r + \rho) = \sum_{\lambda_1=-L}^{+L-1} h(\lambda_1 r + \rho)u((\lambda - \lambda_1)r). \tag{3b}$$

If $k = \lambda r$, $y(k)$ will be calculated using only $2L + 1$ samples of $u(k)$ and $h_0(k)$ while $2L$ samples of both sequences are used if $k \neq \lambda r$. For simplification we introduce the following notation (see Figs. 2 and 3, where $r = 4$ has been assumed):

$$x_\rho(k) = \begin{cases} u_0(k) & k = \lambda r + \rho \\ 0 & k \neq \lambda r + \rho \quad \lambda = \cdots -1, 0, 1, \cdots \end{cases} \tag{4a}$$

$$y_\rho(k) = \begin{cases} y(k) & k = \lambda r + \rho \quad \rho = 0, 1, \cdots, r-1 \\ 0 & k \neq \lambda r + \rho. \end{cases} \tag{4b}$$

Obviously we have

$$u_0(k) = \sum_{\rho=0}^{r-1} x_\rho(k)$$

and

$$y(k) = \sum_{\rho=0}^{r-1} y_\rho(k).$$

Now we define $r$ different error sequences

$$\Delta y_\rho(k) = y_\rho(k) - x_\rho(k). \tag{5}$$

According to (4) we have

Manuscript received February 1, 1974; revised July 18, 1974.
G. Oetken and H. W. Schüssler are with the Institut für Nachrichtentechnik, Universität Erlangen-Nürnberg, Germany.
T. W. Parks is with the Institut für Nachrichtentechnik, Universität Erlangen-Nürnberg, Germany, on leave from Rice University, Houston, Tex. 77001.

Reprinted from *IEEE Trans. Acoust., Speech, Signal Processing*, vol. ASSP-23, pp. 301–309, June 1975.

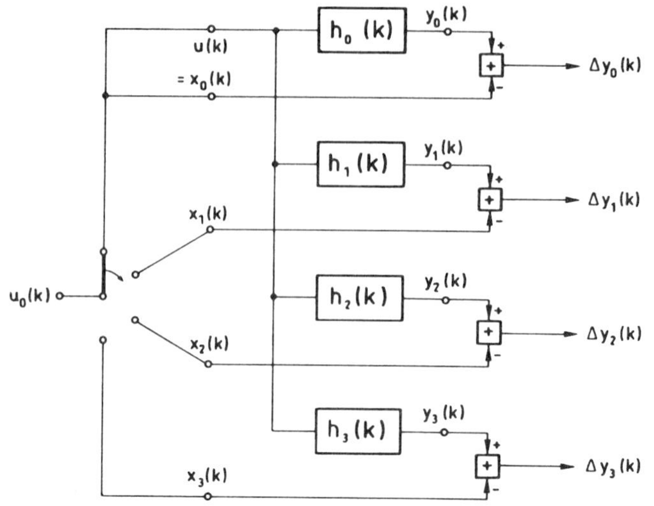


Fig. 2. Model of the interpolator consisting of $r = 4$ subsystems having $r$ distinct impulse responses $h_\rho(k)$.

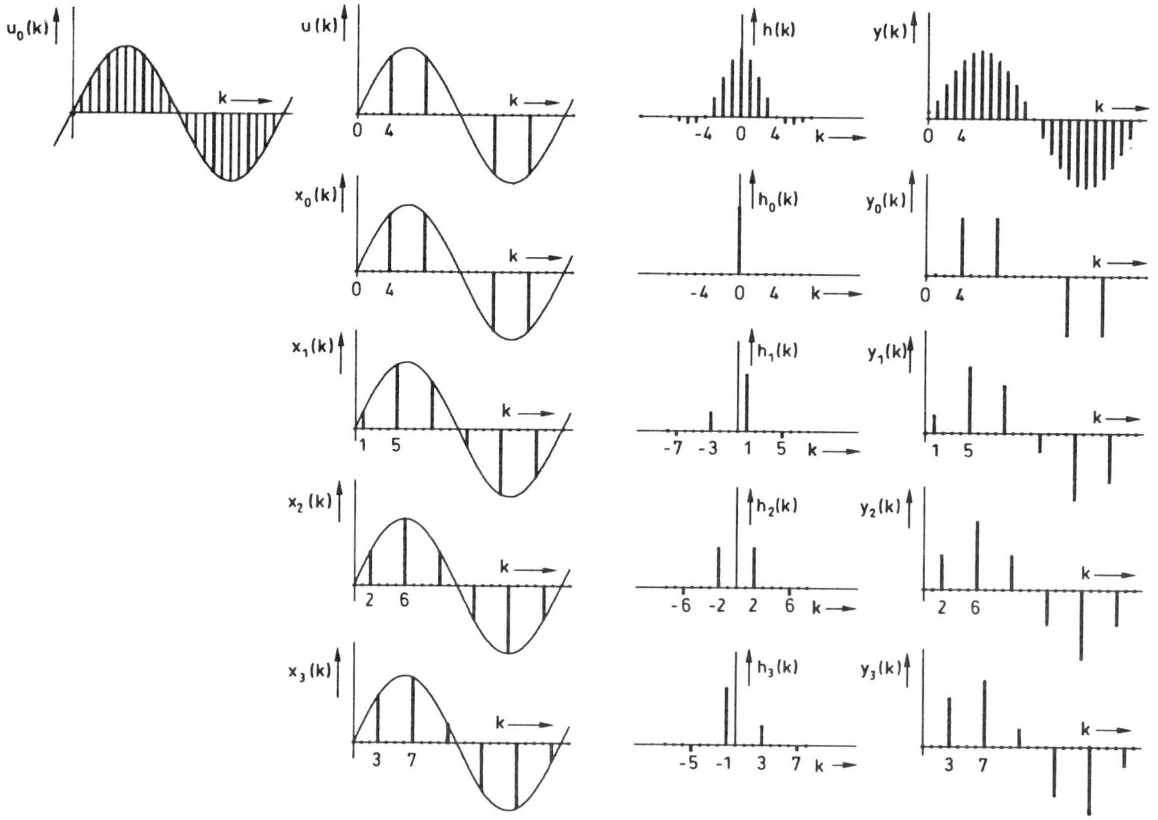


Fig. 3. Impulse responses, input and output sequences for the interpolator and the subsystems, if $r = 4$, $L = 2$.

$$\Delta y(k) = \sum_{\rho=0}^{r-1} \Delta y_\rho(k)$$

and as a measure for the total error

$$||\Delta y||^2 = \sum_{\rho=0}^{r-1} ||\Delta y_\rho||^2. \tag{6}$$

Each of the error sequences $\Delta y_\rho(k)$ depends on a specific subset of the $\{h(k)\}$ only. Therefore the minimization of $||\Delta y(k)||^2$ can be done by minimizing the $||\Delta y_\rho||^2$ separately. With

$$h_\rho(k) = \begin{cases} h(k) & k = \lambda r + \rho \quad \lambda = -L, \cdots, 0, 1, \cdots L-1 \\ 0 & k \neq \lambda r + \rho \quad \rho = 0, 1, \cdots, r-1 \end{cases} \tag{7}$$

(see Fig. 3) we introduce the impulse response of a subsystem to be designed such that $||\Delta y_\rho||^2$ becomes a minimum. The transfer function of this subsystem will be

$$H_\rho(z) = \sum_{k=-n}^{+n} h_\rho(k) z^{-k}$$

$$= \sum_{\lambda=-L}^{L-1} h_\rho(\lambda r + \rho)\,(z^{-r})^\lambda z^{-\rho}. \quad (8)$$

Now the problem of designing an interpolator of degree $n = 2Lr$ with the impulse response $h(k)$ can be split up into the design of $r$ subfilters with transfer functions of degree $2L$ in $z^{-r}$, having the impulse responses $h_\rho(k)$. Each of these subsystems calculates an estimate for $u_0(\lambda r + \rho)$, using only the samples $u(k) = u_0(\lambda r)$ according to (1).

## III. Solution

In order to minimize $||\Delta y_\rho||^2$, we first write the sequence $\Delta y_\rho(k)$ more explicitly for those values of $k$ where the samples might be $\neq 0$. According to (3) and (4) we get

$$\Delta y_\rho(\lambda r + \rho) = \sum_{\lambda_1=-L}^{L-1} h_\rho(\lambda_1 r + \rho) u((\lambda - \lambda_1) r) - x_\rho(\lambda r + \rho)$$

$$= \sum_{\lambda_1=-L}^{L-1} h_\rho(\lambda_1 r + \rho) u_0((\lambda - \lambda_1) r) - u_0(\lambda r + \rho).$$

We see immediately, that we get $\Delta y_0(k) = 0$, if

$$h_0(k) = \begin{cases} 1 & k = 0 \\ 0 & k \neq 0. \end{cases} \quad (9)$$

The minimization of the other $||\Delta y_\rho||^2$ has to be treated for different classes of input sequences $u_0(k)$ separately.

### *A. Sinusoidal Input*

The case $u_0(k) = U_0(\Omega) e^{j\Omega k}$ with $U_0(\Omega) = 0, \Omega \geqslant \Omega_g = \alpha\pi/r$, $0 < \alpha < 1$ has been treated in [1]. It will be considered again in a different way.

First we specialize (4a) for this input sequence. With

$$x_\rho(k) = u_0(k) p_\rho(k),$$

where

$$p_\rho(k) = \frac{1}{r} \sum_{\nu=0}^{r-1} e^{-j(k-\rho)\nu(2\pi/r)}$$

$$= \begin{cases} 1 & k = \lambda r + \rho \\ 0 & k \neq \lambda r + \rho \end{cases} \quad (10)$$

we get

$$x_\rho(k) = \frac{1}{r} \sum_{\nu=0}^{r-1} U_0(\Omega) e^{j\rho\nu(2\pi/r)}\, e^{j[(\Omega - \nu(2\pi/r)]k} \quad (11a)$$

and especially

$$u(k) = x_0(k) = \frac{1}{r} \sum_{\nu=0}^{r-1} U_0(\Omega) e^{j[\Omega - \nu(2\pi/r)]k}. \quad (11b)$$

With the simplified notation $H_\rho(\Omega)$ instead of $H_\rho(e^{j\Omega})$ for the frequency response of the subsystem described by (8), we get for the error sequence

$$\Delta y_\rho(k, \Omega) = \frac{1}{r} \sum_{\nu=0}^{r-1} \left( H_\rho\left(\Omega - \nu \frac{2\pi}{r}\right) - e^{j\rho\nu(2\pi/r)} \right) \cdot U_0(\Omega) e^{j[\Omega - \nu(2\pi/r)]k}. \quad (12)$$

If we define

$$||\Delta y_\rho||^2 = \lim_{K \to \infty} \frac{1}{2K+1} \sum_{k=-K}^{+K} |\Delta y_\rho(k)|^2$$

we get

$$||\Delta y_\rho(\Omega)||^2 = \frac{1}{r^2} \sum_{\nu=0}^{r-1} \left| H_\rho\left(\Omega - \nu \frac{2\pi}{r}\right) - e^{j\rho\nu(2\pi/r)} \right|^2 |U_0(\Omega)|^2. \quad (13)$$

A reasonable measure for the error of the $\rho$th subsystem if excited by an input sequence with a power spectrum $|U_0(\Omega)|^2$ seems to be

$$\epsilon_\rho^2 = \int_{-\Omega_g}^{+\Omega_g} ||\Delta y_\rho(\Omega)||^2\, d\Omega$$

$$= \frac{1}{r^2} \sum_{\nu=0}^{r-1} \int_{-\alpha\pi/r}^{\alpha\pi/r} \left| H_\rho\left(\Omega - \nu \frac{2\pi}{r}\right) - e^{j\rho\nu(2\pi/r)} \right|^2 |U_0(\Omega)|^2 d\Omega \quad (14a)$$

and for the total error according to (6)

$$\epsilon^2 = \sum_{\rho=0}^{r-1} \epsilon_\rho^2. \quad (14b)$$

In [1] this measure was used to define a figure of merit for interpolators as

$$P = \frac{\Omega_g}{\epsilon}.$$

The optimization of the interpolator will be treated again for fixed $\Omega_g$, this time by minimizing the $\epsilon_\rho^2$'s separately. As is to be seen by (14a), minimization of $\epsilon_\rho^2$ means the approximation of a transfer function $e^{j\rho\nu(2\pi/r)}$ for $|\Omega - \nu(2\pi/r)| \leqslant \Omega_g = \alpha\pi/r$, taking into account a weighting factor $|U_0(\Omega)|^2$. This ideal frequency response is illustrated in Fig. 4 for the example $r = 8$, $\rho = 1$, $\alpha = 0{,}75$. It should be mentioned that the necessary satisfaction of the causality condition requires an additional delay.

Minimization of $\epsilon_\rho^2$ is a straightforward problem [7]. With

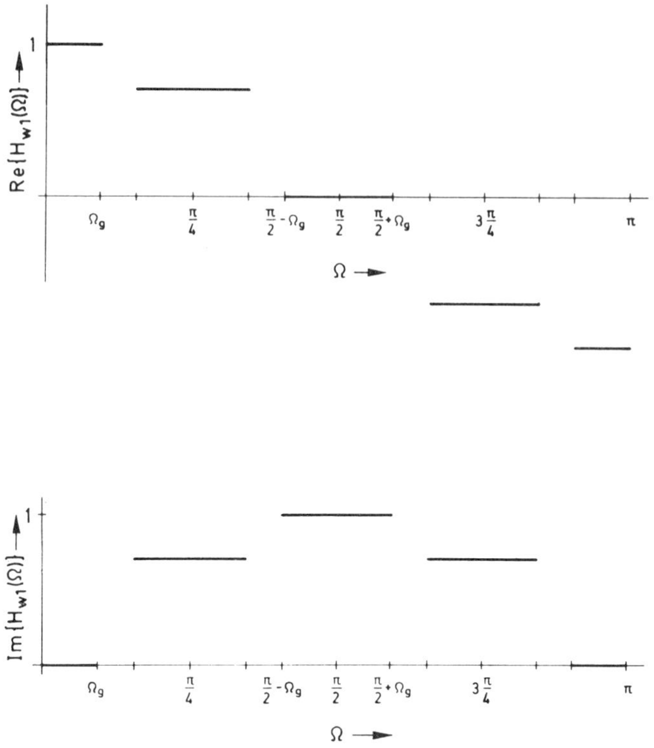


Fig. 4. Frequency response of an ideal subsystem for $\rho = 1$, $r = 8$, $\alpha = 0.75$.

$$H_\rho(\Omega) = \sum_{\lambda=-L}^{L-1} h(\lambda r + \rho)\, e^{-j(\lambda r+\rho)\Omega},$$

where according to (7) $h(\lambda r + \rho) = h_\rho(\lambda r + \rho)$, $\lambda = -L, \cdots, 0, 1, \cdots L - 1$, $\rho = 0, 1, \cdots, r - 1$ and with

$$\phi(k) = \int_{-\alpha^{\pi}/r}^{\alpha^{\pi}/r} |U_0(\Omega)|^2 \cos k\Omega d\Omega \tag{15}$$

we get

$$\epsilon_\rho^2 = \frac{1}{r}\Bigg\{\phi(0) - 2\sum_{\lambda=-L}^{L-1} h(\lambda r + \rho)\phi(\lambda r + \rho) + \sum_{\lambda=-L}^{L-1}\sum_{\lambda_1=-L}^{L-1} h(\lambda r + \rho)h(\lambda_1 r + \rho) \cdot \phi((\lambda - \lambda_1)r)\Bigg\}. \tag{16}$$

The solution of the minimization problem is a set of $2L$ linear equations for the $h(\lambda r + \rho)$

$$\sum_{\lambda=-L}^{L-1} h(\lambda r + \rho)\,\phi((\lambda - \lambda_1)r) = \phi(\lambda_1 r + \rho)$$

$$\lambda_1 = -L, \cdots, 0, 1, \cdots L - 1 \tag{17a}$$

or

$$\phi_T h_\rho = \phi_\rho. \tag{17b}$$

Here

$$\phi_T = \begin{bmatrix} \phi(0) & \phi(r) & \cdots & \phi((2L-1)r) \\ \phi(r) & \phi(0) & & \phi((2L-2)r) \\ \phi(2r) & \phi(r) & \ddots & \phi((2L-3)r) \\ \vdots & & \ddots & \\ \phi((2L-1)r) & & \phi(r) & \phi(0) \end{bmatrix} \tag{18a}$$

is a Töplitz matrix, which does not depend on $\rho$. Furthermore

$$h_\rho = (h(-Lr + \rho), h[-(L-1)r + \rho], \cdots, h(\rho), \cdots h[(L-1)r + \rho])^T \tag{18b}$$

and

$$\phi_\rho = (\phi(-Lr + \rho), \phi(-(L-1)r + \rho), \cdots, \phi(\rho), \cdots \phi((L-1)r + \rho))^T \tag{18c}$$

are samples of the unknown impulse response $h(k)$ and the sequence $\phi(k)$, respectively, taken in intervals of length $r$.

Equation (18) leads to

$$h_\rho = \phi_T^{-1}\phi_\rho. \tag{19}$$

Therefore the calculation of the $h(k)$ requires the inversion of one $2L \times 2L$-matrix and $(r - 1)$ multiplications of this inverted matrix by the vector (18c). Further simplifications and properties of the solution will be discussed later.

### *B. Other Input Sequences*

Next we consider a band-limited input sequence $\{u_0(k)\}$ of finite energy having the properties

$$||u_0||^2 = \sum_{-\infty}^{+\infty} u_0^2(k) < \infty \tag{20a}$$

and

$$U_0(\Omega) = 0 \qquad |\Omega| \geqslant \Omega_g = \alpha\frac{\pi}{r} \tag{20b}$$

where

$$U_0(\Omega) = \sum_{k=-\infty}^{+\infty} u_0(k)e^{-j\Omega k}.$$

Using (10) we can calculate the spectrum at the input of the $\rho$th subsystem as

$$X_\rho(\Omega) = Z\{x_\rho(k)\}_{z=e^{j\Omega}} = Z\{u_0(k)p_\rho(k)\}_{z=e^{j\Omega}} = \frac{1}{r}\sum_{\nu=0}^{r-1} e^{j\rho\nu(2\pi/r)}\, U_0\left(\Omega + \nu\frac{2\pi}{r}\right), \tag{21}$$

which corresponds to (11a). If we define an error measure as

$$\epsilon_\rho^2 = ||\Delta y_\rho||^2 = \sum_{k=-\infty}^{+\infty} \Delta y_\rho^2(k), \tag{22}$$

Parseval's theorem leads directly to (17a), where $\phi(k)$ are samples of the autocorrelation function of $u_0(k)$. Thus the filters designed in Section III-A are optimal for sequences of finite energy as well.

Now we assume that $u_0(k)$ is a stationary random process with zero mean and autocorrelation function $\phi(k)$. In this case the error measure $E\{\Delta y_\rho^2\}$ is minimized by the orthogonality principle [8]

$$E\left\{\left[\sum_{\lambda_1=-L}^{L-1} h(\lambda_1 r + \rho) u_0((\lambda - \lambda_1)r) - u_0(\lambda r + \rho)\right] u_0((\lambda - \lambda_2)r)\right\} = 0 \tag{23}$$

for $\lambda_2 = -L, \cdots, 0, 1, \cdots, L-1$, yielding exactly the same equations (17a) for $h(k)$. Therefore the filters are optimum also for this class of signals.

## IV. Properties of Optimum Filters

The optimum interpolators have some interesting symmetry properties. First we compare $h_\rho$ and $h_{r-\rho}$. We get according to (19)

$$h_{r-\rho} = \phi_T^{-1} \phi_{r-\rho}.$$

Since $\phi(-k) = \phi(k)$, the elements of $\phi_{r-\rho}$ are the same as those of $\phi_\rho$, taken in reversed order. Therefore the same statement holds for $h_{r-\rho}$, and $h_\rho$. For the impulse responses of the subsystems we get

$$h_\rho(k) = h_{r-\rho}(-k) \tag{24}$$

and for the transfer functions

$$H_\rho(z) = H_{r-\rho}(z^{-1}). \tag{25}$$

With

$$h(k) = \sum_{\rho=0}^{r-1} h_\rho(k),$$

we get

$$h(k) = h(-k), \tag{26}$$

i.e., the optimal interpolator has linear phase. This symmetry property reduces by a factor of 2 the number of vectors $h_\rho$ which must be calculated.

By making further use of symmetries (17b) can be solved by inverting two different $L \times L$ matrices rather than the $2L \times 2L$ matrix $\phi_T$. This possibility will be explained using the case $L = 2$ as an example. Equation (17b) becomes

$$\begin{bmatrix} \phi(0) & \phi(r) & \phi(2r) & \phi(3r) \\ \phi(r) & \phi(0) & \phi(r) & \phi(2r) \\ \phi(2r) & \phi(r) & \phi(0) & \phi(r) \\ \phi(3r) & \phi(2r) & \phi(r) & \phi(0) \end{bmatrix} \begin{bmatrix} h(-2r+\rho) \\ h(-r+\rho) \\ h(\rho) \\ h(r+\rho) \end{bmatrix} = \begin{bmatrix} \phi(-2r+\rho) \\ \phi(-r+\rho) \\ \phi(\rho) \\ \phi(r+\rho) \end{bmatrix}.$$

By adding rows 1 and 4 and rows 2 and 3 we get

$$2 \begin{bmatrix} \phi(0)+\phi(3r) & \phi(r)+\phi(2r) \\ \phi(r)+\phi(2r) & \phi(0)+\phi(r) \end{bmatrix} \begin{bmatrix} h(-2r+\rho)+h(r+\rho) \\ h(-r+\rho)+h(\rho) \end{bmatrix} = \begin{bmatrix} \phi(-2r+\rho)+\phi(r+\rho) \\ \phi(-r+\rho)+\phi(\rho) \end{bmatrix}.$$

In the same way we get by subtraction

$$2 \begin{bmatrix} \phi(0)-\phi(3r) & \phi(r)-\phi(2r) \\ \phi(r)-\phi(2r) & \phi(0)-\phi(r) \end{bmatrix} \cdot \begin{bmatrix} h(-2r+\rho)-h(r+\rho) \\ h(-r+\rho)-h(\rho) \end{bmatrix} = \begin{bmatrix} \phi(-2r+\rho)-\phi(r+\rho) \\ \phi(-r+\rho)-\phi(\rho) \end{bmatrix}.$$

Thus the design of an optimum interpolator can be done by inverting two different $L \times L$ matrices and then multiplying the inverses by $r/2$ ($(r-1)/2$ for odd $r$) vectors.

Fig. 5 shows as an example the frequency response of an optimum subsystem for $\rho = 1$, $r = 8$, $L = 4$, $|U_0(\Omega)| = 1$ for $|\Omega| \leqslant \alpha\pi/r$ and $\alpha = 0{,}75$, which can be compared with that of an idealized system as drawn in Fig. 4. Examples for frequency responses of the whole system for different $L$, $r$ and $\alpha$ are given in [1].

Another property will be discussed, which simplifies the design of filters for various values of $r$. It will be derived as follows. We write the $\phi(k)$ differently by using a frequency normalization. With $\Omega = \Omega'/r$ (15) becomes

$$\phi(k) = \frac{1}{r} \int_{-\alpha\pi}^{+\alpha\pi} |U_0(\Omega'/r)|^2 \cos(k\Omega'/r)\, d\Omega'.$$

If we introduce $k = (\lambda - \lambda_1)r$ we get for the elements of the matrix $\phi_T$

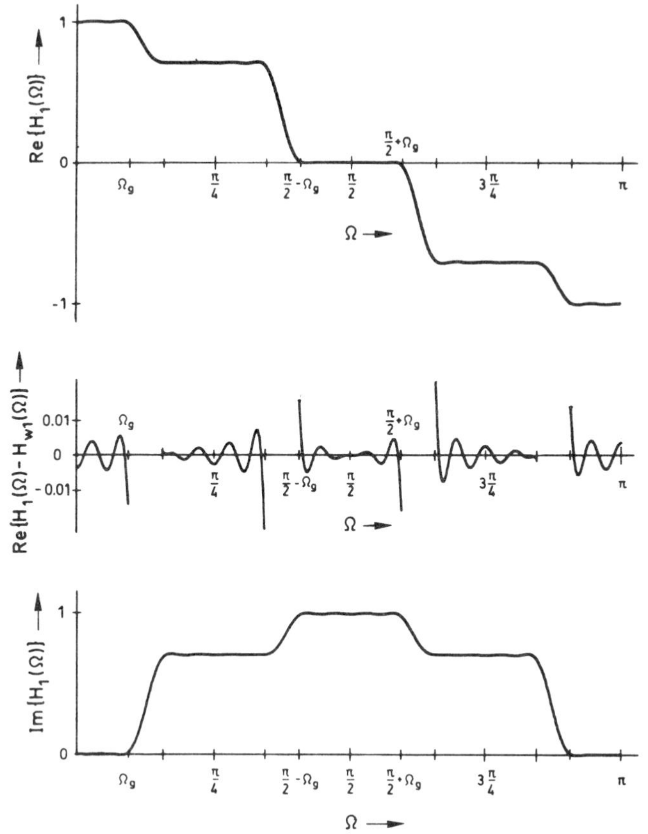


Fig. 5. Frequency response of an optimum subsystem for $\rho = 1, r = 8, L = 4, |U_0(\Omega)| = 1$ for $|\Omega| \leqslant \alpha\pi/r$ and $\alpha = 0.75$.

$$\phi[(\lambda - \lambda_1)r] = \frac{1}{r}\int_{-\alpha\pi}^{\alpha\pi} |U_0(\Omega'/r)|^2 \cdot \cos[(\lambda - \lambda_1)\Omega']\, d\Omega'$$

and with $k = \lambda_1 r + \rho = (\lambda_1 + \rho')r$, where $0 \leqslant \rho' < 1$ for the elements of the vector $\phi_\rho$

$$\phi[(\lambda_1 + \rho')r] = \frac{1}{r}\int_{-\alpha\pi}^{\alpha\pi} |U_0(\Omega'/r)|^2 \cdot \cos[(\lambda_1 + \rho')\Omega']\, d\Omega'.$$

It is to be seen that the matrix $\phi_T' = r\phi_T$ becomes independent of $r$ and can be calculated and inverted once for given $L$, $\alpha$ and $|U_0|$. The vector $\phi_{\rho'}' = r\phi_\rho$ depends on the normalized distance $\rho'$ only. Thus the $\rho$th subsystem only depends on this distance as well.

The result is that the design of an interpolator for a large $r$ delivers all interpolators for those values of $r$, which are fractions of $r$. A continuous impulse response $h(\tau = k/r)$ can be calculated being the envelope of all interpolators for the same design parameters $L$, $\alpha$ and $|U_0|$. As an example Fig. 6 shows this function for $L = 4$, $\alpha = 0,5$ and $|U_0(\Omega)| = 1$ for $|\Omega| \leqslant \alpha \cdot \pi/r$.

Another symmetry can be observed by investigating the error measure. We calculate $\epsilon_\rho^2$ for the optimum interpolator. With (17a) and (16) we get

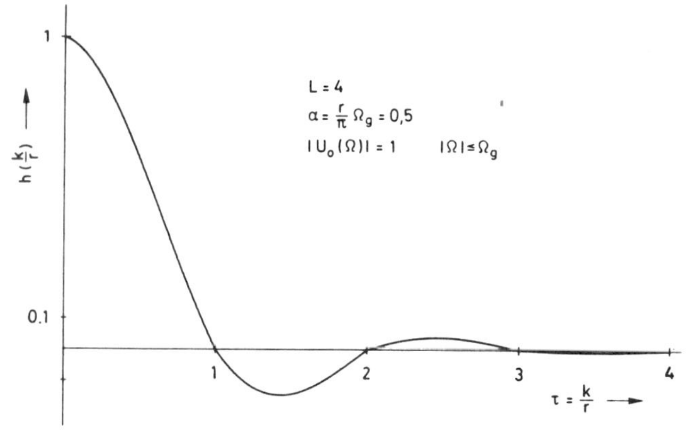


Fig. 6. Envelope $h(\tau = k/r)$ for all impulse responses with $L = 4$, $|U_0(\Omega)| = 1$ for $|\Omega| \leqslant \alpha\pi/r$ and $\alpha = 0.5$ for different $r$.

$$\min \epsilon_\rho^2 = \phi(0) - \sum_{\lambda=-L}^{L-1} h(\lambda r + \rho)\phi(\lambda r + \rho).$$

According to the symmetry of $h(k)$ and $\phi(k)$ we find

$$\min \epsilon_\rho^2 = \min \epsilon_{r-\rho}^2.$$

This property means that each individual error term depends only on the distance $\rho$ between the sample to be estimated and the nearest data sample. Fig. 7 shows the normalized error measure as a function of the normalized distance $\rho'$ for $L = 4$, $|U_0(\Omega)| = 1$ for $|\Omega| \leqslant \alpha\pi/r$ and $\alpha = 0, 5$. Finally in Fig. 8 the total error of optimum interpolators has been drawn for different values of $L$ and $\alpha$. As could be expected $\epsilon^2$ decreases rapidly with increasing $L$ and increases if $\alpha$ becomes larger.

## V. Conclusion

In this paper the numerical difficulties of designing long interpolating filters have been greatly reduced by breaking up the problem of designing a length $2rL + 1$ filter into $r$ separate subfilter design problems.

First two $L \times L$ matrices must be inverted. These same inverse matrices are then used to calculate all $r$ subfilters. Thus only two $L \times L$ matrix inversions are required for the entire design, where $L$ is less than 10 for most applications.

Once an impulse response is calculated using a large value of $r$, filters for smaller values, $r' = r/k$, are obtained by simply using every $k$th sample of the long impulse response.

The decomposition into $r$ subfilters can also be used to simplify hardware required to realize interpolators. The symmetries presented allow further simplification. Detailed error properties are discussed giving insight into individual interpolation errors. A listing of a Fortran program for the design of these filters is included in the Appendix.

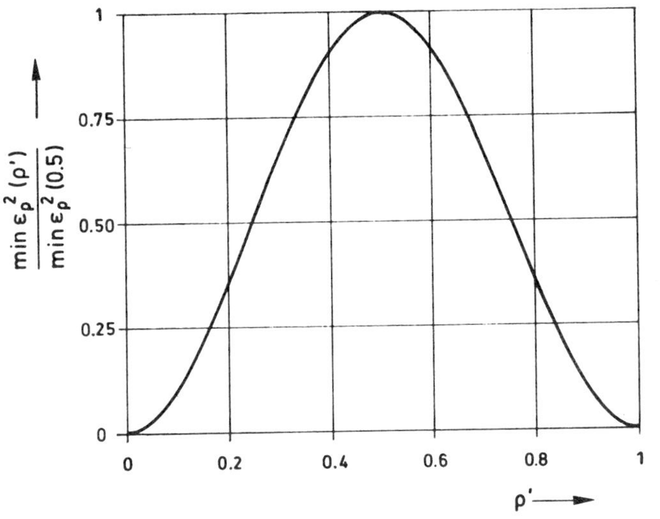


Fig. 7. Envelope of the normalized error measures $\min \epsilon_\rho^2(\rho')/\min \epsilon_\rho^2(0.5)$ for optimal subsystems. $L = 4$, $|U_0(\Omega)| = 1$ for $|\Omega| \leqslant \alpha\pi/r$, $\alpha = 0.5$.

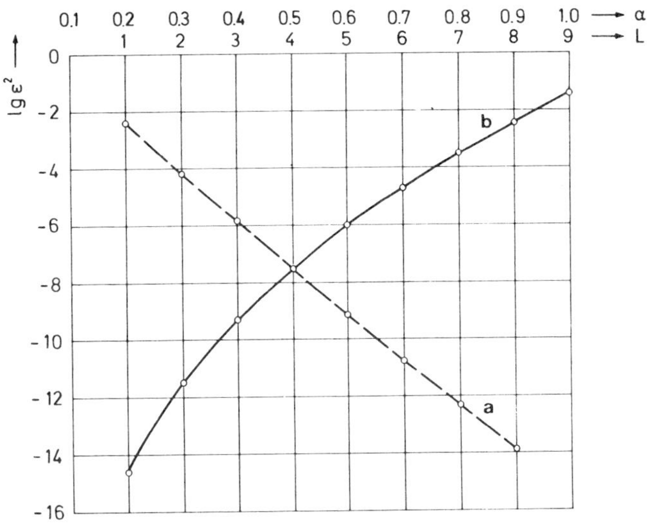


Fig. 8. Total error of optimum interpolators. Curve a) $lg\epsilon^2(L)$ for $\alpha = 0.5$, $r = 4$. Curve b) $lg\epsilon^2(\alpha)$ for $L = 4$, $r = 4$.

## APPENDIX

```
C TITLE/ DODIF      24.07.74
C PURPOSE/
C       SUBROUTINE TO CALCULATE THE OPTIMAL DIGITAL INTERPOLATING
C       FILTER
C AUTHOR/
C       GEERD OETKEN - INSTITUT FUER NACHRICHTENTECHNIK
C                      UNIVESITAET ERLANGEN
C FORTRAN 4 USAGE/
C       CALL DODIF(L,R,ALPHA,DIF)
C DESCIPTION OF PARAMETERS/
C       R       - WANTED INCREASE IN SAMPLING RATE,I.E. THE FILTER
C                 CALCULATES R-1 SAMPLES BETWEEN EACH PAIR OF
C                 INPUT SAMPLES. RANGE FOR R/ ANY INTEGER ; 1
C       L       - INTEGER WHICH DETERMINES THE DEGREE OF THE FILTER.
C                 THE DEGREE IS 2*L*R.
C                 RANGE FOR L/ 1(1)10, MAY BE INCREASED AT THE
C                 RISK OF NUMERICAL INSTABILITY (CHANGE LMAX AND
C                 ARRAY BOUNDS OF AP,AM,D FOR L ; 10).
C       ALPHA   - CUTOFF FREQUENCY (RADIANS) OF THE FILTER INPUT DATA
C                 DIVIDED BY PI/R.
C                 RANGE FOR ALPHA/ 0.DO ; ALPHA ;= 1.DO (DOUBLE
C                 PRECISION).
C       DIF     - DOUBLE PRECISION ARRAY IN WHICH THE FIRST HALF (L*R+1
C                 VALUES) OF THE UNIT SAMPLE RESPONSE IS RETURNED
C                 THE OPTIMAL FILTER HAS LINEAR PHASE).
C
C REMARKS/
C       BIG VALUES OF L AND SMALL VALUAS OF ALPHA INCREASE THE
C       POSSIBILITY OF PROBLEMS WITH MATRIX INVERSION (DEPENDING UPON
C       THE WORDLENGTH OF THE COMPUTER ). A VERY GOOD POSSIBILITY TO
C       CONTROL THE QUALITY OF THE MATRIX INVERSION IS TO CHECK THE
C       MAGNITUDE VALUES OFTHE UNIT SAMPLE RESPONSE DIF(K) AT K=M*R+1,
C       M=0(1)L-1, WHICH IDEALLY HAVE TO BE ZERO.
C
C SUBROUTINES AND FUNCTIONS REQUIRED/
C       DSINV,DMFSD (BOTH IBM/SSP SUBROUTINES)
C       DPHI
C
C
C
      SUBROUTINE DODIF(L,R,ALPHA,DIF)
C
      DOUBLE PRECISION AP(100),AM(100),D(200)
C
      DOUBLE PRECISION DBLE,DFLOAT,S1,S2,DPHI,ALPHA,H,X,Y,DIF(1)
      INTEGER R
C
      DATA LMAX /10/
C (MAXIMUM VALUE OF L SPECIFIED BY ARRAY BOUNDS)
C
      DATA IODEV/6/
C       (DEVICE NO. ON WHICH ERROR MESSAGES ARE TO HAPPEN)
C
      DATA EPS/1.E-6/
C (RELATIVE TOLERANCE FOR LOSS OF SIGNIFICANCE TEST IN MATRIX)
C
      DFLOAT(I)=DBLE(FLOAT(I))
```

```
C
C TEST IF L WITHIN CURRENT BOUNDS
      IF(L.GT.LMAX) GOTO 10
C
C CALCULATE THE TWO L*L MATRICES
      LR=0
      DO 2 M=1,L
      DO 1 K=1,M
      K1=K+LR
      S1=K-M
      S1=DABS(S1)
      S2=DFLOAT(2*L+1-M-K)
      S1=DPHI(S1,ALPHA)
      S2=DPHI(S2,ALPHA)
      AP(K1)=S1+S2
1     AM(K1)=S1-S2
2     LR=LR+M
C
C INVERT BOTH MATRICES USING IBM/SSP SUBROUTINES
      CALL DSINV(AP,L,EPS,IER1)
      CALL DSINV(AM,L,EPS,IER2)
      IF(IER1.EQ.-1.OR.IER2.EQ.-1) GOTO 12
C
C CALCULATE THE L*2L MATRIX D
      DO 3 M=1,L
      M1=(M-1)*L
      DO 3 K=1,L
      K1=2*(M1+K)-1
      K2=(K-1)*K/2+M
      IF(M.GT.K) K2=(M-1)*M/2+K
      S1=AP(K2) + AM(K2)
      S2=AP(K2) - AM(K2)
      D(K1)=S1
3     D(K1+1)=S2
C
C CALCULATE UNIT SAMPLE RESPONSE FROM D AND DPHI
      J=0
      DO 7 I=1,L
      I1=(I-1)*2*L
      DO 7 M=1,R
      M1=M-1
      GOTO 5
4     M1=R
5     X=DFLOAT(M1)/DFLOAT(R)
      H=0.D0
      DO 6 K=1,L
      Y=DFLOAT(L-(K-1))-X
      H=H+D(I1+2*K-1)*DPHI(Y,ALPHA)
      Y=DFLOAT(L-K)+X
6     H=H+D(I1+2*K)*DPHI(Y,ALPHA)
      J=J+1
C
C FIRST HALF OF UNIT SAMPLE RESPONSE IS STORED IN DIF(J)
      DIF(J)=.5D0*H
      IF(I.EQ.L.AND.M1.EQ.R-1) GOTO 4
7     CONTINUE
      RETURN
C
C
C ERROR MESSAGES
10    WRITE(IODEV,11)  LMAX
11    FORMAT(18H L IS GREATER THAN,I4)
      GOTO 14
12    WRITE(IODEV,13)
13    FORMAT(53HERROR DURING MATRIX INVERSION, ILL CONDITIONED SYSTEM)
14    RETURN
      END
```

```
C TITLE/ DPHI
C
C PURPOSE/
C          FUNCTION REQUIRED BY DODIF
C
C REMARKS/
C          SPECIAL VERSION FOR LOWPASS INTERPOLATORS
C          FURTHER REMARKS SEE DODIF
C
      DOUBLE PRECISION FUNCTION DPHI(X,ALPHA)
      DOUBLE PRECISION X,ALPHA,PI
      DATA IBR/1/
      GOTO (1,2),IBR
1     PI=4.D0*DATAN(1.D0)
      IBR=2
2     IF(DABS(X).LT.1.D-10) GOTO 3
      X=ALPHA*PI*X
      DPHI=DSIN(X)/X
      RETURN
3     CONTINUE
      DPHI=1.D0
      RETURN
      END
```

## References

[1] G. Oetken and H. W. Schüssler, "On the design of digital filters for interpolation," *Arch. Elek. Übertragung.*, vol. 27, pp. 471–476, 1973.
[2] R. W. Schafer and L. R. Rabiner, "A digital signal processing approach to interpolation," *Proc. IEEE*, vol. 61, pp. 692–702, June 1973.
[3] S. P. Lloyd and B. McMillan, "Linear least squares filtering and prediction of sampled signals," in *Proc. Symp. Modern Network Synthesis*, Symposia Ser. vol. V, pp. 221–247.
[4] J. Jess, "Eine neue Klasse von Filtern zur Rekonstruktion abgetasteter Signale," *Nachrichtentech. Z.*, vol. 20, pp. 658–662, 1967.
[5] W. C. Kellogg, "Information rates in sampling and quantization," *IEEE Trans. Inform. Theory*, vol. IT-13, pp. 506–511, July 1967.
[6] D. W. Tufts and N. Johnson, "Methods of recovering a random waveform from a finite number of samples," *IEEE Trans. Circuit Theory*, vol. CT-12, pp. 32–39, Mar. 1965.
[7] E. W. Cheney, *Introduction to Approximation Theory.* New York: McGraw-Hill, 1966.
[8] A. Papoulis, *Probability, Random Variables, and Stochastic Processes.* New York: McGraw-Hill, 1965.

# Randomly Sampled Digital Filters

EDWIN P. F. KAN, Member, IEEE
J. K. AGGARWAL, Member, IEEE
Dep. Elec. Eng. and Electron. Res. Cen.
Univ. Texas at Austin
Austin, Tex. 78712

Abstract

The concept of a randomly sampled digital filter is introduced and formulated in this paper. An equivalent problem, which is of utmost importance in digital filtering, is also considered, namely, the problem of digital filtering with faulty samplers. Error analysis is carried out in two different approaches: 1) frequency response method, and 2) worst-case analysis. With the confirmation of experimental verification, very simple formulas are derived that are applicable under different realistic assumptions, and are most appealing to practical design purposes.

Manuscript received March 15, 1971.

This research was supported by NSF Grant GK-1879 and JESP Grant AF-AFOSR-766-67.

## I. What is a Randomly Sampled Digital Filter?

The term digital filter has been well understood [1] to be a system that processes an input $\{x_k\}_{k=0}^{\infty}$ and transforms it to an output $\{w_k\}_{k=0}^{\infty}$ according to the following linear difference equation:

$$w_n = \sum_{i=0}^{M} b_i x_{n-i} - \sum_{i=1}^{N} a_i w_{n-i}, \qquad \forall n. \tag{1}$$

There are two interpretations of the operation of this system: a) the two sequences $\{x_k\}$ and $\{w_k\}$ may be considered as just any sequences of numbers without any implication of time ordering, in which case the process of digital filtering is just the execution of the algorithm (1) [see Fig. 1(A)]; and 2) these sequences $\{x_k\}$ and $\{w_k\}$ may be considered as samples, as in sampled-data control theory, of continuous signals $x(\cdot)$ and $w(\cdot)$, in which case, a time ordering is implied; and this time ordering is usually uniform, with samples spaced at $T$ s apart [see Fig. 1(B)]. With this latter interpretation, the idea of a randomly sampled digital filter can be introduced as follows.

Consider then a continuous input signal $x(\cdot)$, randomly sampled by a sampler which closes once in every sampling period of duration $T$. (See Fig. 2.) A sequence of numbers $\{\hat{x}_k\}_{k=0}^{\infty}$ is thus evolved, where $\hat{x}_k = x(t_k)$, for all $k$, $t_k$ being the sampling instant in the $k$th sampling period. This sequence of numbers can be processed directly algorithmically as in Fig. 1(A). Otherwise, in order to introduce a time ordering, the sequence $\{\hat{x}_k\}$ can be stored (in a computer) and

Reprinted from *IEEE Trans. Audio Electroacoust.*, vol. AU-20, pp. 52–57, Mar. 1972.

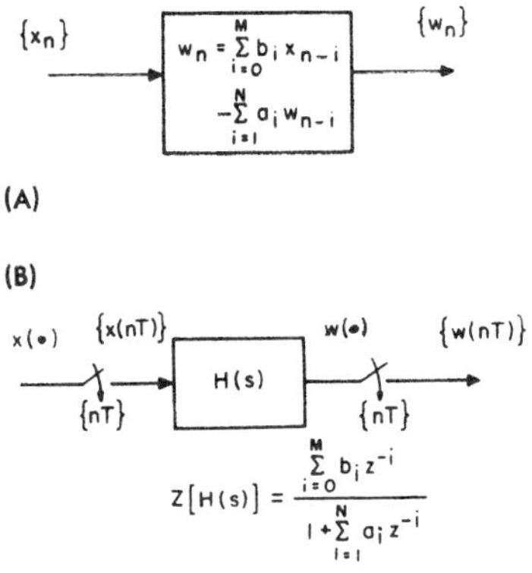

Fig. 1. Two interpretations of digital filtering. (A) As an algorithm. (B) As a sampled-data system.

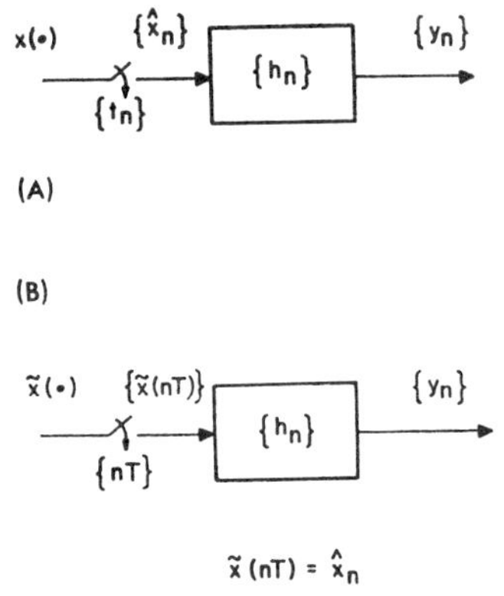

Fig. 2. A randomly sampled digital filter. (A) The system. (B) An equivalent system with the hypothetical input $\tilde{x}(\cdot)$.

periodically retrieved sequentially, say with a period of $T$, giving the time series $\{\tilde{x}(mT)\}_{m=m_0}^{\infty}$ (for an arbitrary $m_0$). In this case, $\tilde{x}(\overline{k+m_0T})=\hat{x}_k$, $\forall k$. And, $\{\tilde{x}(mT)\}$ can be processed as in Fig. 2(B), as if $\{\tilde{x}(mT)\}$ had been uniformly sampled from a perturbed continuous input signal $\tilde{x}(\cdot)$.

This device of a randomly sampled digital filter is by no means a contrived one. Actually, it is not difficult to conceive that it is only natural to introduce this idea of random sampling when, say, digital filters are time shared. And of course, for this time sharing to operate rationally, each channel (input to the filter) must share its use of the computer (i.e., the filter) once in every $T$-s interval, a feature imposed on the randomly sampled digital filter in the previous paragraph. Also, such randomness does take place, say, in radar tracking of a rapidly moving evader, also in which case a sample of the evader position is recorded once every sweep cycle (as long as the object remains in screen). In fact, the ideas of random sampling in feedback control systems have been introduced earlier and found to be realistic models of certain physiological systems. (See [2], [3].) This reveals then, that random sampling should be regarded as an integral part of the digital filtering technology.

## II. An Equivalent Problem: Previous Works and Present Investigation

Instead of viewing this randomly sampled digital filter as a new device, the following equivalent problem is recognized: the randomness of the sampling process may be considered as an unintentional process as opposed to an intentional process, i.e. as a natural phenomenon versus a contrived feature. In other words, when a faulty sampler is present, the designer would like to find out how much this faultiness affects the system performance. Error analysis of this faulty system is then naturally the first step towards the study of this system.

It is only timely here to point out that Balakrishnan [4] performed an intensive analysis of the problem of "time jittering in sampling" back in 1962. In that paper, the spectrum of the sampled signal (i.e. the randomly sampled signal) is computed for the cases of deterministic and of stationary stochastic signals. The analysis of the latter case is relatively straightforward, since the derived (i.e., sampled) signal is again stationary. As to the former case of deterministic input, it was noted that the sampled signal is not stationary. To overcome this difficulty of nonstationarity, an average spectrum was defined, which then brought the analysis of this case into parallelism with that of the other case. As digital filtering is fundamentally a linear process, the output of the filter is thus well defined through a spectrum shaping of the sampled signal. Thus, Balakrishnan's study essentially spelled out the whole error analysis of the randomly sampled digital filter, save the complex nature of his formulas.

The purpose of this paper is to use a different approach, essentially an engineer's practical approach, to tackle the problem of error analysis of such filters in the case of deterministic input. And, instead of defining the average spectrum and propagating this spectrum through the filter, the following two approaches are taken.

*1) Frequency Response Method:* A modified transfer function is derived and its expected value computed, which is then plotted against frequency. Also the average steady state noise-to-signal ratio is computed and displayed in frequency-plots.

*2) Worst-Case Analysis:* Bounds on the expected values of the error and its energy are expressed in terms of the maximum input "velocity" and "acceleration."

Formulas derived on these two analyses are found to be very simple and most easily computed; and the validity of these are checked by experimental verifications. Furthermore, the derived formulas depend on some practically accessible quantities such as the input's harmonic contents, or the bounds on the input's derivatives. And thus, the present analyses are seen to be most attractive to practical design purposes.

## III. Formulation

Referring back to Fig. 2(A), the deviation of the random sampling instants from the uniform instants is defined as follows:

$$t_n = n + Z_n, \qquad \forall n, \tag{2}$$

where $T$, the sampling period, has been normalized to unity, and $Z_n$ is a random variable taking values in $[-\Delta, \Delta]$, with $\Delta \equiv \alpha T/2$, $0 \le \alpha < 1$. $Z_n$ is assumed to be evenly distributed, with the result that all the odd moments vanish. Typical distributions are 1) rectangular, 2) triangular, and 3) trapezoidal with parameter $p$, $0 \le p \le 1$. These distributions are shown in Fig. 3. Their first four even moments are tabulated in Table I. Formally, the system assumptions are as follows.

### A. System Assumptions

*Assumption 1:* $x(\cdot)$ is a deterministic signal, assumed to be infinitely differentiable.

*Assumption 2:* $\{h_n\}_{n=0}^{\infty}$ is the impulse response of the digital filter, assumed to be strictly stable.

*Assumption 3:* $T$, the sampling period, is normalized to unity.

*Assumption 4:* $Z_n$ is a random variable evenly distributed in $[-\Delta, \Delta]$, with

$$\Delta \equiv \alpha T/2, \qquad 0 \le \alpha < 1, \qquad \forall n.$$

$$Z_n \text{ is independent of } Z_m, \qquad \forall n \ne m.^1$$

*Assumption 5:* $\{Z_n\}_{n=0}^{\infty}$ is a stationary stochastic process, with $\nu^2$ and $\eta^4$ denoting $E(Z_n^2)$ and $E(Z_n^4)$, respectively.

### B. Basic Analysis

By Assumption 1, $x(t_n) = \hat{x}_n$ is expanded as

$$\begin{aligned}\hat{x}_n &= x(n + Z_n) \\ &= x_n + \dot{x}_n Z_n + \tfrac{1}{2}\ddot{x}_n Z_n^2 + \cdots .\end{aligned} \tag{3}$$

Define the input error $\{\mathcal{E}_n\}$

$$\begin{aligned}\mathcal{E}_n &\triangleq \hat{x}_n - x_n \\ &= \dot{x}_n Z_n + \tfrac{1}{2}\ddot{x}_n Z_n^2 + \cdots .\end{aligned} \tag{4}$$

Since the digital filter is a linear system, the output error $\{e_n\}$ is

$$\begin{aligned}e_n &\triangleq y_n - w_n \\ &= \sum_{i=0}^{n} h_{n-i}\, \mathcal{E}_i\end{aligned} \tag{5}$$

where $\{w_n\}$ and $\{y_n\}$ denote the outputs of the uniformly sampled system and the randomly sampled system respectively. (See Figs. 1 and 2.) With the assumption on $\{Z_n\}$, the following expressions are readily obtained.

[1] By the terminology used in [4], $\{Z_n\}$ is said to be "pure white."

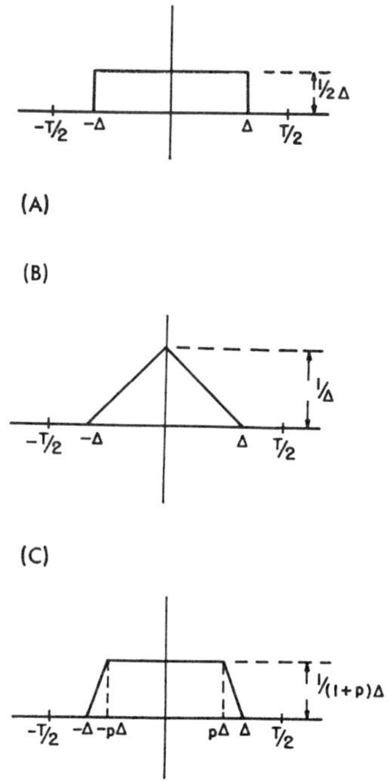


Fig. 3. Three typical distributions of $Z_n$. (A) Rectangular. (B) Triangular. (C) Trapezoidal with parameter $p$.

TABLE

The First Four Even Moments of $Z_n$ Having Rectangular, Triangular, and Trapezoidal with $p = \frac{1}{2}$ Distribution over $[-\Delta, \Delta]$

| | Rectangular | Triangular | Trapezoidal with $p = \frac{1}{2}$ |
|---|---|---|---|
| $E(Z_n^2)$ | $\frac{1}{3}\Delta^2$ | $\frac{1}{6}\Delta^2$ | $\frac{5}{18}\Delta^2$ |
| $E(Z_n^4)$ | $\frac{1}{5}\Delta^4$ | $\frac{1}{15}\Delta^4$ | $\frac{19}{240}\Delta^4$ |
| $E(Z_n^6)$ | $\frac{1}{7}\Delta^6$ | $\frac{1}{28}\Delta^6$ | $\frac{85}{1792}\Delta^6$ |
| $E(Z_n^8)$ | $\frac{1}{9}\Delta^8$ | $\frac{1}{45}\Delta^8$ | $\frac{1023}{34560}\Delta^8$ |

$$E(\mathcal{E}_n) = \tfrac{1}{2}\ddot{x}_n \nu^2 + \cdots \tag{6a}$$

$$E(\mathcal{E}_n \mathcal{E}_m) = E(\mathcal{E}_n) E(\mathcal{E}_m), \qquad n \ne m \tag{6b}$$

$$E(\mathcal{E}_n^2) = \dot{x}_n^2 \nu^2 + (\tfrac{1}{4}\ddot{x}_n^2 + \tfrac{1}{3}\dot{x}_n \dddot{x}_n)\eta^4 + \cdots \tag{6c}$$

$$E(e_n) = \sum_{i=0}^{n} h_{n-i} E(\mathcal{E}_i) \tag{6d}$$

$$E(e_n^2) = E^2(e_n) + \sum_{i=0}^{n} h_{n-i}^2 \operatorname{var}(\mathcal{E}_i). \tag{6e}$$

## IV. Error Analysis (I): Frequency Response

In this section, the input is assumed to be a cosinusoid, i.e.,

$$x(t) = \cos \omega t \tag{7}$$

with $0 \leq \omega \leq N_f$. Here $N_f$ denotes the Nyquist frequency, which equals $\pi/T = \pi$ rad/s. This range $[0, N_f]$ is the largest frequency band in which the digital filter, while uniformly sampled, will operate meaningfully, and thus indicates the plausibly largest operating range of frequencies for the randomly sampled digital filter. In fact, the following will show that, depending on the distribution of $Z_n$ on $[-\Delta, \Delta]$, only a portion of $[0, N_f]$ is operational, in the sense of tolerable noise-to-signal ratio (NSR). Denote

$$\beta \equiv \omega/N_f. \tag{8}$$

Noting that $\dot{x}(t) = (-\omega) \sin \omega t$, $\ddot{x}(t) = (-\omega^2) \cos \omega t, \cdots$, (6a) and (c) can be rewritten as

$$E(\mathcal{E}_n) = (-\tfrac{1}{2}\omega^2\nu^2) \cos n\omega + \cdots \tag{9a}$$

$$E(\mathcal{E}_n^2) = (\omega^2\nu^2) \sin^2 n\omega + (\omega^4\eta^4)(\tfrac{1}{4} \cos^2 n\omega - \tfrac{1}{3} \sin^2 n\omega) + \cdots. \tag{9b}$$

In order to see whether these infinite series can be truncated at the terms indicated, it is necessary to study the convergence of the terms $\omega^2E(Z_n^2)$, $\omega^4E(Z_n^4)$, $\omega^6E(Z_n^6)$, $\omega^8E(Z_n^8) \cdots$. Taking the example of rectangular distribution, $\omega^2E(Z_n^2)$ can be rewritten

$$\begin{aligned}\omega^2E(Z_n^2) &= (\beta N_f)^2(\tfrac{1}{3}\Delta^2) \\ &= (\beta\pi/T)^2 \cdot \tfrac{1}{3}(\alpha T/2)^2 \\ &= \frac{\pi^2}{12} (\beta\alpha)^2.\end{aligned} \tag{10}$$

Note the independence of $T$ and the dependence on the product parameter $\beta\alpha$. A list of these quantities for the three distributions is tabulated in Table II. Reasonably fast convergence is easily seen to be obtainable when: 1) $\beta\alpha \leq 0.5$ for the rectangular distribution, 2) $\beta\alpha \leq 0.75$ for the triangular distribution, and 3) intermediate ranges for $\beta\alpha$ for the trapezoidal distribution depending on the parameter $p$. It is further revealed by a comment at the end of this section that useful operation of the randomly sampled digital filter usually requires that the frequency ranges be restricted such that these bounds on $\beta\alpha$ are satisfied. Thus, approximation of (9) by the indicated truncated series are valid in the operational frequency ranges. Rewriting then,

$$E(\mathcal{E}_n) = (-\tfrac{1}{2}\omega^2\nu^2) \cos n\omega \tag{11a}$$

$$\begin{aligned}E(\mathcal{E}_n^2) &= (\omega^2\nu^2) \sin^2 n\omega + (\omega^4\eta^4)(\tfrac{1}{4} \cos^2 n\omega - \tfrac{1}{3} \sin^2 n\omega) \\ &= (\tfrac{1}{2}\omega^2\nu^2 - \tfrac{1}{24}\omega^4\eta^4) \\ &\quad + (-\tfrac{1}{2}\omega^2\nu^2 + \tfrac{7}{24}\omega^4\eta^4) \cos n(2\omega).\end{aligned} \tag{11b}$$

And, the output to error $e_n$ has its first two moments

TABLE II

Tabulating $\omega^{2k}E(Z_n^{2k})$, $k = 1, 2, 3, 4$ for the Three Distributions of $Z_n$ as a Function of $\beta\alpha$ where $\beta \equiv \omega/N_f$ and $\alpha \equiv 2\Delta/T$

| | Rectangular | Triangular | Trapezoidal with $p = \frac{1}{2}$ |
|---|---|---|---|
| $\omega^2E(Z_n^2)$ | $\frac{\pi^2}{12}(\beta\alpha)^2$ | $\frac{\pi^2}{24}(\beta\alpha)^2$ | $\frac{5}{72}\pi^2(\beta\alpha)^2$ |
| $\omega^4E(Z_n^4)$ | $\frac{\pi^4}{80}(\beta\alpha)^4$ | $\frac{\pi^4}{240}(\beta\alpha)^4$ | $\frac{19}{3840}\pi^4(\beta\alpha)^4$ |
| $\omega^6E(Z_n^6)$ | $\frac{\pi^6}{448}(\beta\alpha)^6$ | $\frac{\pi^6}{1792}(\beta\alpha)^6$ | $\frac{85}{114688}\pi^6(\beta\alpha)^6$ |
| $\omega^8E(Z_n^8)$ | $\frac{\pi^8}{2304}(\beta\alpha)^8$ | $\frac{\pi^8}{11520}(\beta\alpha)^8$ | $\frac{1023}{8847360}\pi^8(\beta\alpha)^8$ |

$$E(e_n) = (-\tfrac{1}{2}\omega^2\nu^2) \sum_{i=0}^{n} h_{n-i} \cos i\omega \tag{12a}$$

$$\begin{aligned}E(e_n^2) &= E^2(e_n) + (\tfrac{1}{2}\omega^2\nu^2 - \tfrac{1}{24}\omega^4\eta^4) \sum_{i=0}^{n} h_{n-i}^2 \\ &\quad + (-\tfrac{1}{2}\omega^2\nu^2 + \tfrac{7}{24}\omega^4\eta^4) \sum_{i=0}^{n} h_{n-i}^2 \cos i(2\omega) \\ &\quad - (\tfrac{1}{4}\omega^4\nu^4) \sum_{i=0}^{n} h_{n-i}^2 \cos^2 i\omega \\ &= E^2(e_n) + (\tfrac{1}{2}\omega^2\nu^2 - \tfrac{1}{24}\omega^4\eta^4 - \tfrac{1}{8}\omega^4\nu^4) \sum_{i=0}^{n} h_i^2 \\ &\quad + (-\tfrac{1}{2}\omega^2\nu^2 + \tfrac{7}{24}\omega^4\eta^4 - \tfrac{1}{8}\omega^4\nu^4) \\ &\quad \cdot \sum_{i=0}^{n} h_{n-i}^2 \cos i(2\omega).\end{aligned} \tag{12b}$$

The steady-state expressions are recognized to be

$$E(e_n)_{ss} = (-\tfrac{1}{2}\omega^2\nu^2)H(z \mid z = e^{j\omega}) \cos n\omega \tag{13a}$$

$$\begin{aligned}E(e_n^2)_{ss} &= E^2(e_n)_{ss} + \|h\|_2^2(\tfrac{1}{2}\omega^2\nu^2 - \tfrac{1}{24}\omega^4\eta^4 - \tfrac{1}{8}\omega^4\nu^4) \\ &\quad + (-\tfrac{1}{2}\omega^2\nu^2 + \tfrac{7}{24}\omega^4\eta^4 - \tfrac{1}{8}\omega^4\nu^4)\hat{H}(z \mid z = e^{j2\omega}) \\ &\quad \cdot \cos n(2\omega)\end{aligned} \tag{13b}$$

where $\|h\|_2^2 \equiv \sum_{i=0}^{\infty} h_i^2$, and $\hat{H}(z) \equiv \sum_{i=0}^{\infty} h_i^2 z^{-i}$. From (13a), a "modified function" $H_m(z)$ can be defined describing the input-output relation of the randomly sampled digital filter excited by a cosine input, as if the input were still uniformly sampled but the transfer function of the filter were modified. And, the first moment of $H_m(z)$ is given by

$$E(H_m(z)) = (1 - \tfrac{1}{2}\omega^2\nu^2)H(z), \qquad z = e^{j\omega}. \tag{14}$$

Notice that only the expected amplitude response is distorted, while its expected phase response remains unchanged. As to the second moment, i.e., the energy contained in the error signal, a meaningful measure is to find the NSR:

$$\mathrm{NSR}(\omega) \triangleq 10 \log_{10} (P_e(\omega)/P_w(\omega)) \text{ dB} \tag{15}$$

at steady state, where $P_e(\omega)$, $P_w(\omega)$ denote the average expected power contained in $\{e_n\}$ and $\{w_n\}$ at their steady states, corresponding to an input of frequency $\omega$, and with the average taken over a sufficiently large number of samples. For example,

$$\begin{aligned}(w_n)_{ss} &= H(z \mid z = e^{j\omega}) \cos n\omega \\ &= \mid H(z \mid z = e^{j\omega}) \mid \cos (n\omega + \phi), \qquad (16)\end{aligned}$$

with $\phi \triangleq \angle H(z \mid z = e^{j\omega})$; and

$$(w_n)_{ss}^2 = \mid H(z \mid z = e^{j\omega}) \mid^2 \cos^2 (n\omega + \phi). \qquad (17)$$

Averaging $(w_n^2)_{ss}$ over a "cycle,"[2]

$$P_w(\omega) = \mid H(z \mid z = e^{j\omega}) \mid^2 \cdot \tfrac{1}{2}. \qquad (18)$$

In the same way, averaging $E(e_n^2)_{ss}$ over the same cycle,

$$\begin{aligned}P_e(\omega) &= (-\tfrac{1}{2}\omega^2\nu^2)^2 P_w(\omega) \\ &\quad + \|h\|_2^2(\tfrac{1}{2}\omega^2\nu^2 - \tfrac{1}{24}\omega^4\eta^4 - \tfrac{1}{8}\omega^4\nu^4). \qquad (19)\end{aligned}$$

Thus, by (15), the NSR is given by

$$\mathrm{NSR}(\omega) = 10 \log_{10} \left\{ \tfrac{1}{4}\omega^4\nu^4 + \frac{2\|h\|_2^2(\tfrac{1}{2}\omega^2\nu^2 - \tfrac{1}{24}\omega^4\eta^4 - \tfrac{1}{8}\omega^4\nu^4)}{\mid H(z \mid z = e^{j\omega}) \mid^2} \right\} \text{ dB}. \qquad (20)$$

Simplification of (20) is possible when $\tfrac{1}{4}\omega^4\nu^4 \ll \|h\|_2^2\omega^2\nu^2 / \mid H(e^{j\omega})\mid^2$ and $\omega^4\nu^4$, $\omega^4\eta^4 \ll \omega^2\nu^2$. This leads to

$$\mathrm{NSR}(\omega) = 10 \log_{10} \{\|h\|_2^2\omega^2\nu^2 / \mid H(e^{j\omega}) \mid^2\} \text{ dB}. \qquad (21)$$

This simplification can easily be seen to offer good approximation in the operational frequency range (see comment following the example, concerning the truncation of infinite series) and when the filter does not have excessive high gains. As a matter of fact (21) is used instead of (20) in the following example.

*Example:* Take the filter having the following transfer function $H(z) = 0.1/(1-0.9z^{-1})$. By (14) and (21), $\mid H(e^{j\omega})\mid^2$, $\mid E[H_m(e^{j\omega})]\mid^2$ and NSR are plotted for: 1) rectangular distribution, 2) triangular distribution, and 3) trapezoidal distribution with $p = \tfrac{1}{2}$ of $Z_n$ spread over $[-\Delta, \Delta]$, where $\Delta \equiv \alpha T/2$, with $\alpha = 1$, 0.5 and 0.1. Experimental values of NSR for the rectangular distribution are also obtained by a brute-force calculation, in which the library function RANF in the CDC 6600 computer system was used to generate $\{Z_n\}$. These values are also plotted in Figs. 4, 5, and 6. Agreement with the theoretical results is evident.

## Comment

With regard to the operational frequency range and the truncation of the infinite series in (6), here is the following observation. Taking the case of rectangular distribution and

[2] When the sampling rate is large compared to the input frequency, $\{\cos n\omega T\}_n$ behaves like $\cos \omega(\cdot)$. This cycle, then, can be taken over $2\pi/\omega$. Otherwise, a larger interval must be chosen accordingly.

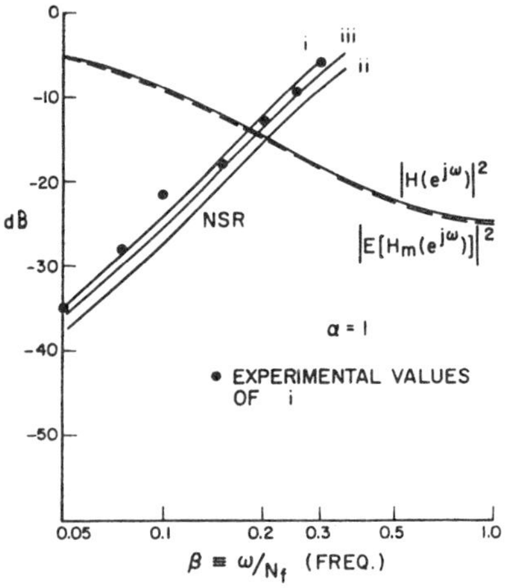


Fig. 4. Frequency plots of $H(z)$, $E(H_m(z))$, and NSR for $H(z) = 0.1/(1-0.9z^{-1})$, with $\alpha = 1$, for *i* rectangular, *ii* triangular, and *iii* trapezoidal with $p = \tfrac{1}{2}$ distribution of $Z_n$.

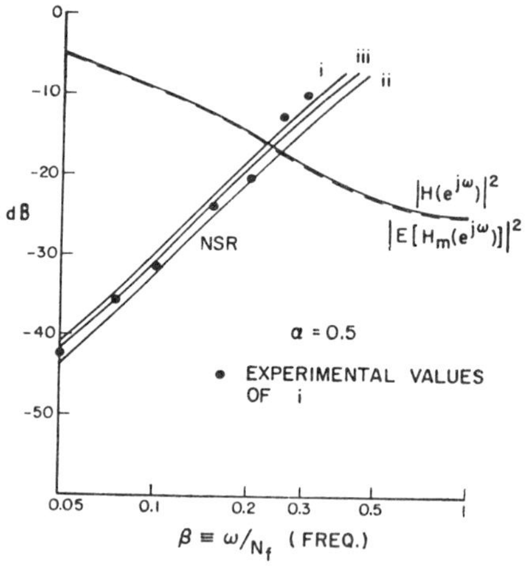


Fig. 5. Frequency plots for the same example for $\alpha = 0.5$.

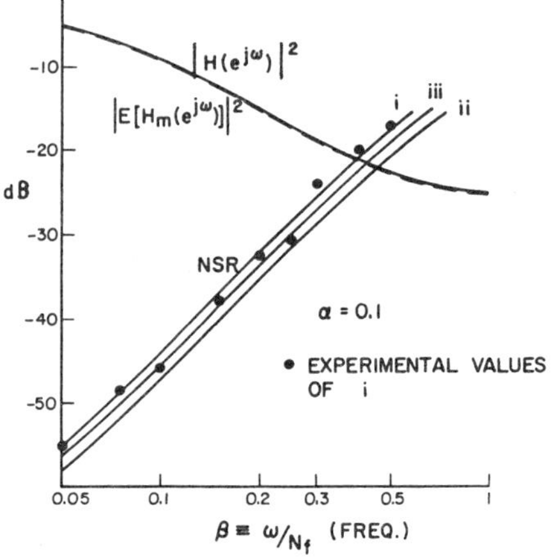


Fig. 6. Frequency plots for the same example for $\alpha = 0.1$.

$\alpha=1$ in the above example (refer to curve $i$ in Fig. 4), the NSR becomes intolerable ($>-5$ dB) for $\beta\equiv\omega/N_f>0.3$. (The degree of tolerance depends on the application of the filter, say for transmission of voices or transmission of pictures, in which cases the former can tolerate much heavier distortion than the latter. Typical figures are $-10$ dB for the former and $-30$ dB for the latter.) That means that successful operation of the randomly sampled digital filter is not available beyond $\beta>0.3$ in the present illustration. (This is certainly plausible, in regard of the extreme wildsess of the random sampling when $\alpha=1$.) By this observation, and recalling an earlier assertion that the truncated expressions (11) and (12) are valid approximations of (6) in the range $0\leq\beta\alpha=\beta\leq0.5$ in the present case, it is then seen that these truncations are indeed legal in the operational frequency range $0\leq\beta\leq0.3$. Similar agreements for other cases thus justify completely the truncations of the infinite series in (6).

## V. Error Analysis (II): Worst-Case Analysis

In this section, upper bounds on the expected error and its variance are obtained on the assumption of knowledge of the bounds on the input velocity and acceleration. It is to be noted that these input quantities are easily accessible and practically controllable. In addition, it is assumed here that $x(\cdot)$ is smooth enough and $Z_n$ small enough such that $x_n^{(2k)}E(Z_n^{2k})$ goes to zero with $k$ reasonably fast, where $x^{(2k)}$ denotes the $(2k)$th derivative of $x(\cdot)$. This assumption would be easily satisfied in common practice when $T$ is small (of the order of milliseconds). Formally, here are the extra assumptions.

*Assumption 6:* $|\dot{x}(t)|\leq c_1$, $|\ddot{x}(t)|\leq c_2$, $\forall t$.

*Assumption 7:* $x_n^{(2k)}E(Z_n^{2k})\underset{k}{\rightarrow}0$ reasonably fast, $\forall n$.

Then, (4) can be truncated and written as

$$\mathcal{E}_n = \dot{x}_n Z_n + \tfrac{1}{2}\ddot{x}_n Z_n^2. \tag{22}$$

Thus,

$$E(\mathcal{E}_n) = \tfrac{1}{2}\ddot{x}_n \nu^2 \tag{23a}$$

$$E(\mathcal{E}_n^2) = \dot{x}_n^2\nu^2 + \tfrac{1}{4}\ddot{x}_n^2\eta^4 \tag{23b}$$

$$\text{var}\,(\mathcal{E}_n) = \dot{x}_n^2\nu^2 + \tfrac{1}{4}\ddot{x}_n^2(\eta^4 - \nu^4). \tag{23c}$$

Note that $\eta^4-\nu^4\geq0$ because

$$\text{var}\,(Z_n^2) = E(Z_n^4) - E^2(Z_n^2) \geq 0.$$

And, the first two moments of the output error $e_n=\sum_{i=0}^n h_{n-i}\mathcal{E}_i$ are given by the following expressions:

$$E(e_n) = \tfrac{1}{2}\left(\sum_{i=0}^{n} h_{n-i}\ddot{x}_i\right)\nu^2 \tag{24a}$$

$$E(e_n^2) = E^2(e_n) + \left(\sum_{i=0}^{n} h_{n-i}^2\dot{x}_i^2\right)\nu^2 + \tfrac{1}{4}\left(\sum_{i=0}^{n} h_{n-i}^2\ddot{x}_i^2\right)(\eta^4-\nu^4). \tag{24b}$$

These expressions are bounded by

$$|E(e_n)| \leq \tfrac{1}{2}c_2\|h\|_1\nu^2 \tag{25a}$$

$$E(e_n^2)\leq c_1^2\|h\|_2^2\nu^2+\tfrac{1}{4}c_2^2\{\|h\|_1^2\nu^2+\|h\|_2^2(\eta^4-\nu^4)\}. \tag{25b}$$

where $\|h\|_1\equiv\sum_{i=0}^{\infty}|h_i|$.

## VI. Conclusion

The idea of randomly sampled digital filters has been introduced and formulated in this paper. Its importance and natural occurrence are pointed out. Also, it is unveiled that an equivalent problem is the very important aspect of digital filtering with faulty samplers. It is recognized that the latter problem has been very rigorously analyzed by Balakrishnan in the context of time jittering in sampling, who arrived at exact analytic expressions of the spectral densities of the randomly sampled signals in the case of deterministic input as well as stochastic input. However, the present paper aims at and successfully arrives at relatively much simpler expressions for the error analysis of such filters when excited by a smooth deterministic signal. The derived expressions are approximations, which are proved to be valid under various realistic assumptions. As a matter of fact, two approaches are proposed here: 1) frequency response method, and 2) worst-case analysis. These are indeed familiar tools to the design engineers. The first approach arrives at frequency plots of the expected modified transfer function and the output noise-to-signal ratio as a function of input frequency. The second approach assumes a knowledge of the bounds on the input velocity and acceleration—also practically accessible quantities—and arrives at upper bounds on the possible expected error and its variance. Evidence of experimental verification also confirms the theory. All these then lead to the conclusion that the present analyses are most suitable and appealing to practical engineering design purposes.

## References


[1] B. Gold and C. M. Rader, *Digital Processing of Signals.* New York: McGraw-Hill, 1969.

[2] R. G. Agniel and E. I. Jury, "Stability of nonlinear randomly sampled systems," in *Proc. 9th Annu. Allerton Conf. Circuits Syst. Theory*, 1969.

[3] G. A. Bekey, J. M. Biddle, and A. J. Jacobson, "The effect of a random sampling interval on a sampled-data model of a human operator," in *Proc. 3rd Conf. Man. Contr.*, Rep. NASA-SP-144, 1967.

[4] A. V. Balakrishnan, "On the problem of time jitter in sampling," *IEEE Trans. Inform. Theory*, vol. IT-18, pp. 226–236, Apr. 1962.

# Some Comparisons Between FIR and IIR Digital Filters

By L. R. RABINER, J. F. KAISER, O. HERRMANN,
and M. T. DOLAN

(Manuscript received July 27, 1973)

*The purpose of this paper is to make comparisons between optimum, linear phase, finite impulse response (FIR) digital filters and infinite impulse response (IIR) digital filters which meet equivalent frequency domain specifications. The basis of comparison is, for the most part, the number of multiplications per sample required in the usual realizations of these filters—i.e., the cascade form for IIR filters, and the direct form for FIR filters. Comparisons are also made between group-delay equalized filters and linear phase FIR filters. Considerations dealing with finite word-length effects are discussed for both these filter types. A set of design charts is also presented for determining the minimum filter order required to meet given low-pass filter specifications for both digital and analog filters.*

## I. INTRODUCTION

Although a great deal is known about the properties of different types of digital filters, very little has been done to relate the various designs as to performance and complexity of realization. Thus the filter designer must learn the details of several design procedures before being able to make a wise decision on a suitable filter for his specific application. It is the purpose of this paper to add insight into some of the problems that have been encountered by filter designers by: (*i*) presenting new and useful design curves for digital and analog low-pass filters, and (*ii*) making several comparisons between optimum (quasi-equiripple), linear phase, FIR low-pass filters and equiripple (elliptic) IIR filters which meet equivalent frequency domain specifications. Although these results are presented for low-pass filter designs, they are easily extended to the case of bandpass, bandstop, and high-pass filters by the well-known frequency band transformations.

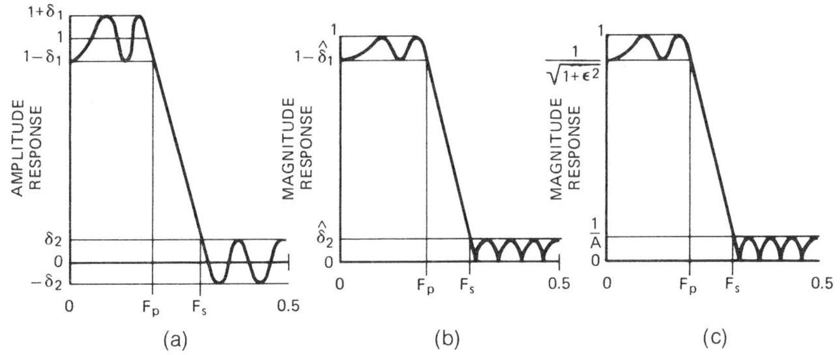


Fig. 1—Terminology used to describe low-pass filter characteristics.

The organization of the paper is as follows. After defining the terminology to be used, the design relationships between the FIR filter parameters are reviewed. The design relationships for the IIR filter parameters are developed and a novel graphical interpretation of these relations is presented in the form of useful filter design charts (applicable to both digital and analog filters). Using the design relationships, the filter orders required to achieve equivalent performance are compared for different ranges of filter parameters.

## II. TERMINOLOGY

The design procedures for the two general classes of digital filters, FIR and IIR, have essentially progressed along independent paths. As a result, the terminology used in specifying the filter performance is generally not quite the same. Thus it is instructive to define the most commonly accepted definitions of the filter parameters for these two classes of filters. The relations between these parameter sets for equivalence are then established. Figure 1a shows the amplitude response of a typical optimum FIR low-pass filter and Fig. 1b shows the magnitude response of a typical elliptic low-pass filter. For the FIR case the amplitude response in the passband ($0 \leqq f \leqq F_p$)* generally oscillates between $1 + \delta_1$ and $1 - \delta_1$, where $\delta_1$ is the passband ripple. In the stopband ($F_s \leqq f \leqq 0.5$) the amplitude response oscillates between $+\delta_2$ and $-\delta_2$ where $\delta_2$ is the stopband ripple. For the elliptic case the magnitude response is constrained to always be less

* Throughout this paper the frequency scale has been normalized with respect to the sampling frequency. Thus the normalized sampling frequency is 1.0 and the frequency range graphed is $0 \leqq f \leqq 0.5$ or, equivalently, $0 \leqq \omega \leqq \pi$.

than 1.0. Thus, in the passband ($0 \leqq f \leqq F_p$) the magnitude response oscillates between 1 and $1 - \hat{\delta}_1$. In the stopband the magnitude response oscillates between $\hat{\delta}_2$ and 0. It is straightforward to relate $\delta_1$, $\delta_2$, $\hat{\delta}_1$, and $\hat{\delta}_2$ so the resulting magnitude characteristics are equivalent. If the FIR amplitude characteristic is scaled by $1/(1 + \delta_1)$ and the magnitude of the resulting amplitude response is taken, then the following relationships are obtained:

$$\hat{\delta}_1 = \frac{2\delta_1}{1 + \delta_1} \tag{1}$$

$$\hat{\delta}_2 = \frac{\delta_2}{1 + \delta_1} \tag{2}$$

$$\delta_1 = \frac{\hat{\delta}_1}{2 - \hat{\delta}_1} \tag{3}$$

$$\delta_2 = \frac{2\hat{\delta}_2}{2 - \hat{\delta}_1} \cdot \tag{4}$$

Thus, given either $(\delta_1, \delta_2)$ or $(\hat{\delta}_1, \hat{\delta}_2)$, eqs. (1) through (4) can be used to find the equivalent specifications for the other type of filter.

Although the notation of Fig. 1b is acceptable for the magnitude response of an elliptic filter, it is not the most widely used form for these filters. Figure 1c shows the same magnitude response described in terms of passband parameter $\epsilon$ and stopband parameter $A$. Comparing Figs. 1b and 1c it is easy to relate $\epsilon$ and $A$ to $\hat{\delta}_1$ and $\hat{\delta}_2$ as

$$\epsilon = \frac{\sqrt{2 - \hat{\delta}_1}\,\sqrt{\hat{\delta}_1}}{(1 - \hat{\delta}_1)} \tag{5}$$

$$A = \frac{1}{\hat{\delta}_2} \cdot \tag{6}$$

At this point it is convenient to define the additional filter terms $E$, ATT, and $\eta$ as

$$E = \text{(in-band) ripple} = 20 \log_{10} \sqrt{(1 + \epsilon^2)} \tag{7}$$

$$\text{ATT} = \text{stopband attenuation} = 20 \log_{10} A \tag{8}$$

$$\eta = \frac{\epsilon}{\sqrt{A^2 - 1}} = \frac{\hat{\delta}_2\sqrt{\hat{\delta}_1}\sqrt{2 - \hat{\delta}_1}}{(1 - \hat{\delta}_1)\sqrt{1 - \hat{\delta}_2^2}} = \frac{2\delta_2(\sqrt{\delta_1})}{(1 - \delta_1)\sqrt{(1 + \delta_1)^2 - \delta_2^2}} \cdot \tag{9}$$

Thus, parameters $E$ and ATT are a third set of parameters which describe the characteristics of the magnitude response of the elliptic filter. The parameter $\eta$ has been shown to be a basic analog filter parameter[1] which will be used in the filter design curves given in a later section.

## III. FIR DESIGN RELATIONS

The five basic FIR filter parameters are $F_p$, $F_s$, $\delta_1$, $\delta_2$, and $N$, the duration of the filter impulse response in samples. For the general case of optimum, linear phase, low-pass FIR filters, there exist no simple analytical relationships between these five filter parameters, except in special cases, e.g., one passband or one stopband ripple. However, an approximate empirical relationship between the filter parameters has recently been obtained[2] which accurately satisfies known design results for a wide range of values of the filter parameters. The relationship is of the form:

$$N = 1 + \frac{D_\infty(\delta_1, \delta_2)}{\Delta F} - f(\delta_1, \delta_2)\Delta F, \tag{10}$$

where

$$\Delta F = F_s - F_p = \text{relative transition width}, \tag{11}$$

$$\begin{aligned} D_\infty(\delta_1, \delta_2) = {} & [0.005309 (\log_{10} \delta_1)^2 + 0.07114 \log_{10} \delta_1 - 0.4761] \log_{10} \delta_2 \\ & - [0.00266 (\log_{10} \delta_1)^2 + 0.5941 \log_{10} \delta_1 + 0.4278], \end{aligned} \tag{12}$$

and

$$f(\delta_1, \delta_2) = 0.51244 \log_{10} (\delta_1/\delta_2) + 11.01. \tag{13}$$

Equation (10) can generally predict the value of $N$ required to meet specifications on $\delta_1$, $\delta_2$, $F_p$, and $F_s$ to within $\pm 2$. In the cases where $F_p$ is very close to 0, or $F_s$ is very close to 0.5, eq. (10) tends to overestimate the required $N$. It should be noted that eq. (10) shows the estimate of $N$ to be independent of specific values of $F_s$ or $F_p$, but instead is dependent only on the transition width, $(F_s - F_p)$.

A simpler expression giving a less accurate estimate of $N$ is

$$N = \frac{-10 \log_{10} (\delta_1 \cdot \delta_2) - 15}{14\Delta F} + 1. \tag{14}$$

This expression is a modification of the design relationship for FIR filters designed by windowing techniques (where $\delta_1 = \delta_2$). See Ref. 3, pp. 237–238.

## IV. IIR DESIGN RELATIONS

One of the most general procedures for designing IIR digital filters is through the bilinear transformation of an appropriate continuous filter. There are two equivalent techniques for obtaining the desired digital filter using the bilinear transformation and these are illustrated in Fig. 2. The technique of Fig. 2a begins with an analog low-pass filter with normalized passband cutoff frequency of 1 radian per second, and analog stopband cutoff frequency of $\Omega_s$ radians per second. This filter is bilinearly transformed[4] to give a "normalized" digital filter with passband cutoff frequency $\pi/2$ radians per second (or $f = 0.25$ on the normalized scale) and stopband cutoff frequency $\hat{\omega}_s$. The relation between the frequency variables $\Omega$ and $\hat{\omega}$ is given by

$$\Omega = \tan(\hat{\omega}/2). \tag{15}$$

Thus $\Omega_s$ and $\hat{\omega}_s$ are simply related by eq. (15) with $\Omega = \Omega_s$, and $\hat{\omega} = \hat{\omega}_s$. Finally, a digital all-pass transformation[5] is used to give the desired digital low-pass filter with passband cutoff frequency $\omega_p$ and stopband cutoff frequency $\omega_s$. The relation between the frequency variables $\omega$ and $\hat{\omega}$ is given by

$$\tan \omega = \frac{(1 - \alpha^2) \sin \hat{\omega}}{(1 + \alpha^2) \cos \hat{\omega} - 2\alpha}, \tag{16}$$

where

$$\alpha = \frac{\sin(\omega_p/2) - \cos(\omega_p/2)}{\sin(\omega_p/2) + \cos(\omega_p/2)}. \tag{17}$$

The second technique (shown in Fig. 2b) begins with the identical analog low-pass filter as in the first technique and immediately performs a low-pass–to–low-pass transformation to give an analog filter with passband cutoff frequency $\hat{\Omega}_p$ and stopband cutoff frequency $\hat{\Omega}_s$. The relation between the frequency variables $\Omega$ and $\hat{\Omega}$ is

$$\Omega = \hat{\Omega}/\hat{\Omega}_p. \tag{18}$$

The resulting low-pass filter is then transformed to a digital filter using the bilinear transformation, giving the same end result as in the first technique (Fig. 2a). The relation between frequency variables $\omega$ and $\hat{\Omega}$ is

$$\hat{\Omega} = \tan(\omega/2). \tag{19}$$

In the case where the prototype normalized analog filter is an elliptic filter, it is relatively easy to derive a design formula relating the various filter parameters, both in the analog and digital cases. For either the analog or digital case the order, $n$, of the elliptic filter is related to the remaining filter parameters by the equation

$$n = \frac{K(k)K(\sqrt{1 - k_1^2})}{K(k_1)K(\sqrt{1 - k^2})}, \tag{20}$$

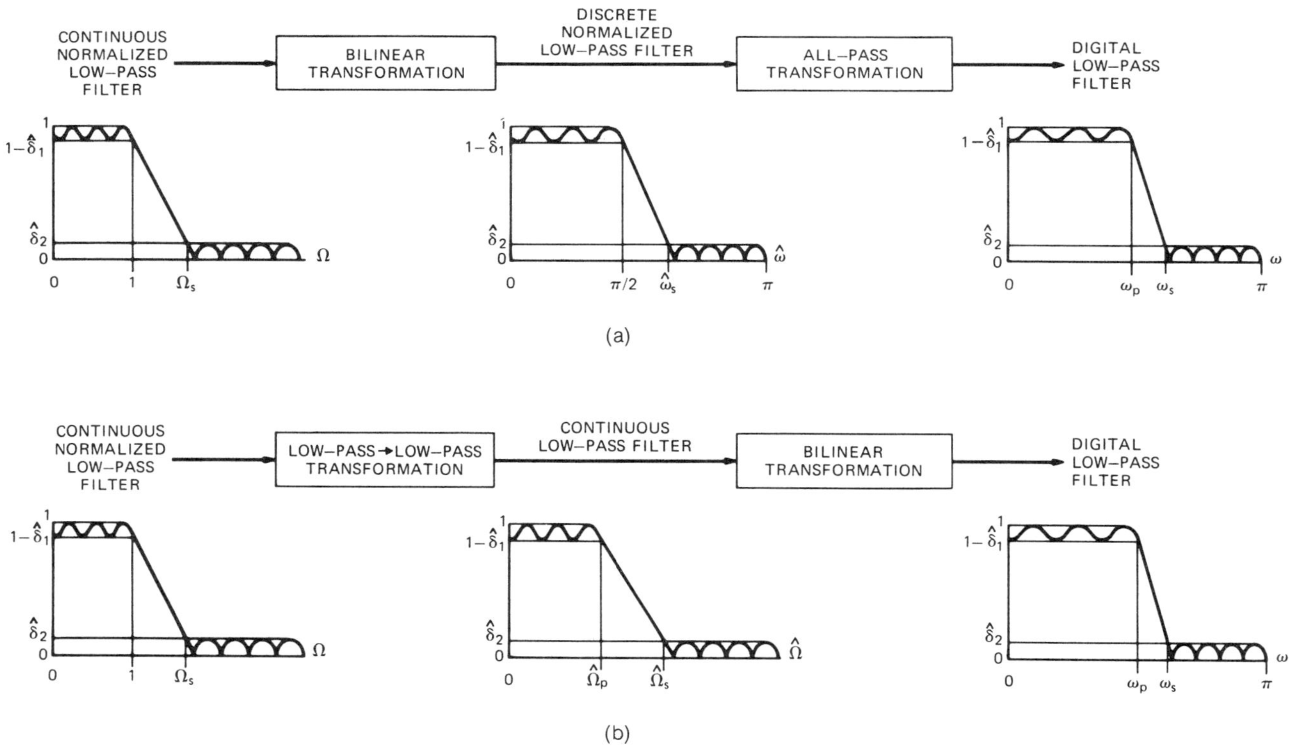


Fig. 2—Two techniques for transforming a continuous normalized low-pass filter to a digital low-pass filter.

where $K(\cdot)$ is the complete elliptic integral of the first kind and

$$k = \text{transition ratio} = \frac{1}{\Omega_s} = \frac{\tan(\omega_p/2)}{\tan(\omega_s/2)} \tag{21}$$

and

$$k_1 = \eta = \frac{\epsilon}{\sqrt{A^2 - 1}} = \frac{2\delta_2\sqrt{\delta_1}}{(1 - \delta_1)\sqrt{(1 + \delta_1)^2 - \delta_2^2}}. \tag{22}$$

Thus eq. (20) relates filter order, $n$, to the parameters $F_p$, $F_s$ [through eq. (21)] and $\delta_1$ and $\delta_2$ [through eq. (22)].

For the case when the prototype filter is a Chebyshev filter (either type I—equiripple passband, monotone stopband, or type II—maximally flat passband, equiripple stopband) the design equation becomes

$$n = \frac{\cosh^{-1}(1/\eta)}{\ln \beta}, \tag{23}$$

where

$$\beta = \frac{1 + \sqrt{1 - k^2}}{k} \tag{24}$$

and $\eta$ and $k$ are defined as eqs. (21) and (22). Finally for a prototype Butterworth filter (maximally-flat magnitude, all pole) the design equation is

$$n = \frac{\ln \eta}{\ln k}. \tag{25}$$

Although eqs. (20) through (25) completely describe the design curves for both analog and digital filters, it is generally quite helpful to see the relationships between filter parameters displayed in a meaningful way. Since, in general, there are five filter parameters there is no simple way of presenting these relationships on a single plot, even in terms of well-known nomograph procedures.[6] There is, however, a simple and straightforward way of including all design relations for both digital and analog filters, for any prototype filter, using a sequence of three charts.

The first chart(s) relates the filter design parameter $\eta$ to the passband and stopband ripple specifications $\delta_1$ and $\delta_2$ or their equivalents. The second chart(s) graphs the filter design equation relating filter order $n$, design parameter $\eta$, and transition ratio $k$. The third chart(s) relates transition ratio $k$ to passband cutoff frequency $F_p$ and transition bandwidth $\nu$.

Figures 3a through 3d show four possibilities for Chart No. 1. The graphs of Figs. 3a and 3b correspond to digital filters with $\delta_1$ as a param-

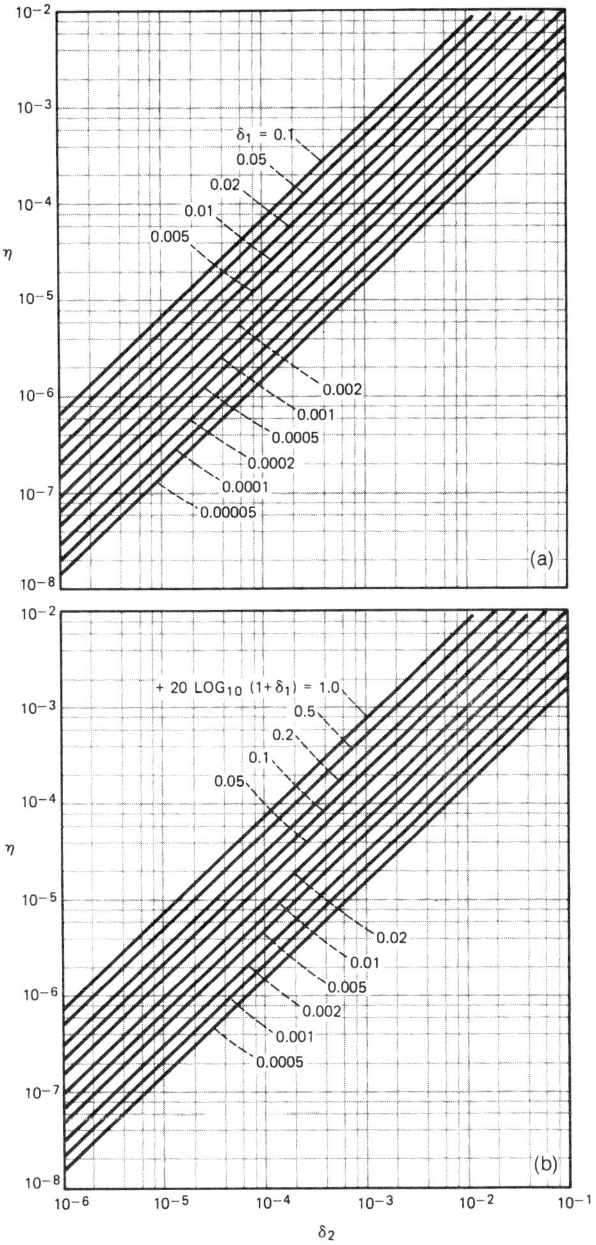


Fig. 3—Plots of $\eta$ versus stopband specification, with parameter passband specification for low-pass filters.

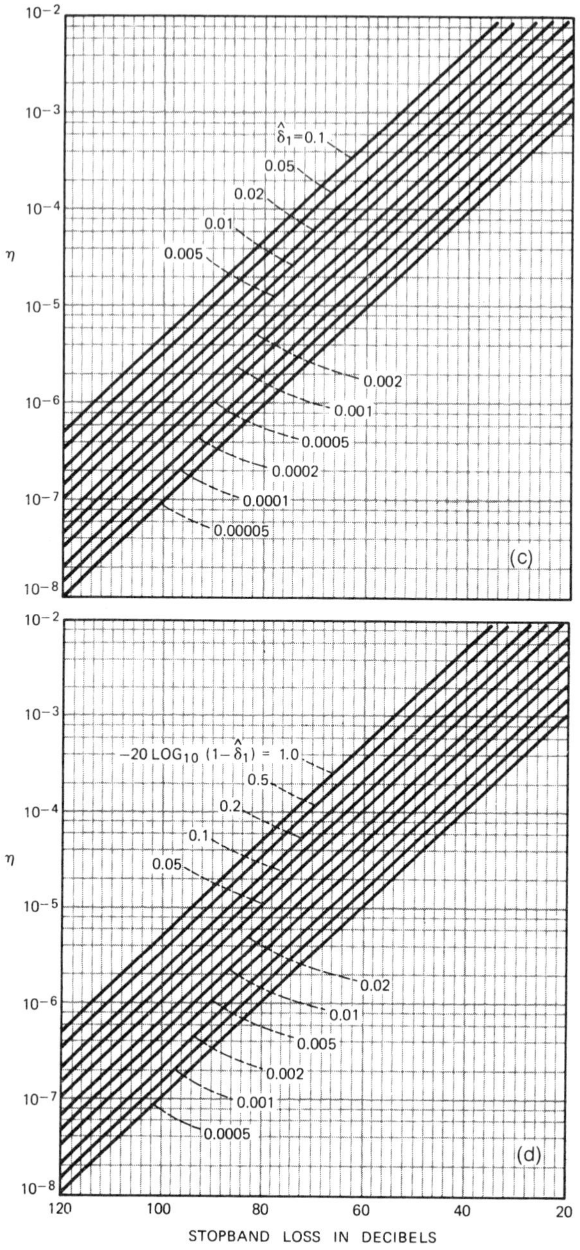


Fig. 3 (continued).

eter (Fig. 3a) or $20 \log_{10} (1 + \delta_1)$(dB) as a parameter (Fig. 3b). The graphs of Figs. 3c and 3d correspond to analog filters with absolute ripple $\hat{\delta}_1$ as a parameter (Fig. 3c) or total ripple $20 \log_{10} [1/(1 - \hat{\delta}_1)]$(dB) as a parameter (Fig. 3d).

Chart No. 2 represents the design relations particular to the prototype filters, i.e., eq. (20) for elliptic filters, eq. (23) for Chebyshev filters, and eq. (25) for Butterworth filters. For these graphs the parameter $\eta$ is plotted versus transition ratio, $k$, with filter order, $n$, as the parameter. Figures 4a through 4c show the resulting graphs for elliptic filters, Chebyshev filters, and Butterworth filters, respectively. The horizontal scale on each of these graphs is a nonuniform scale which was chosen to provide a reasonably good spacing of the curves for the various values of $n$. The actual nonlinear scale used is represented by the equation

$$x = \frac{k + k^8}{2}, \tag{26}$$

where $x$ is the $x$-axis coordinate $(0 \leqq x \leqq 1)$ and $k$ is the transition width. Thus, the scale is linear for small values of $k$ and highly nonlinear near $k = 1.0$.

Chart No. 3 represents the relation between the transition ratio and the filter cutoff frequencies [eq. (21)]. For these graphs the passband cutoff frequency, $F_p$, is plotted versus transition ratio, $k$, for various values of normalized transition width, $\nu$, defined as

$$\nu = F_s - F_p = \frac{\omega_s - \omega_p}{2\pi}. \tag{27}$$

Figures 5a and 5b show the resulting graphs for digital and analog filters. The scale for transition ratio is identical to the scale used for Chart No. 2.

## V. USE OF CHARTS

To illustrate how to use the set of charts of Figs. 3 through 5, consider the determination of filter order $n$ required to meet the following specifications:

$\delta_1 = 0.01$ ($\approx \pm 0.086$-dB passband ripple)
$\delta_2 = 0.0001$ (80-dB stopband loss)
passband cutoff frequency = 480 Hz
stopband edge frequency = 520 Hz
sampling frequency = 8000 Hz.

Normalizing the band-edge frequencies gives

$$F_p = \frac{480}{8000} = 0.06$$

$$F_s = \frac{520}{8000} = 0.065.$$

For the determination of filter order $n$ for a digital filter of the elliptic type, the charts of Figs. 3a, 4a, and 5a are used (Fig. 4a specializes the design to the elliptic type). To obtain the value of $\eta$ on Fig. 3a, we use the curve $\delta_1 = 0.01$ and find its intersection with the line $\delta_2 = 0.0001$ which yields a value of $\eta$ approximately equal to $2 \times 10^{-5}$. To obtain the transition ratio, we use Fig. 5a by finding the intersection of the curve $\nu = F_s - F_p = 0.005$ with line $F_p = 0.06$; this yields a value of 0.923 for the transition ratio (this agrees nicely with $F_p/F_s = 0.06/0.065 = 0.923$, an alternate way of arriving at the same result). Finally, the filter order, $n$, can now be determined from Fig. 4a by finding the intersection of the lines $\eta = 2 \times 10^{-5}$ and transition ratio = 0.923; thus the required theoretical elliptic filter order is $\approx 11.5$. In order to meet specifications on all four parameters, a 12th-order filter must be used.

However, there are several tradeoffs possible for the final filter specifications. For example, if $\eta$ is held fixed at $2 \times 10^{-5}$ and the transition ratio is changed to approximately 0.94 to lie on the $n = 12$ curve, then either $F_s$ or $F_p$ can be varied to match this new value of transition ratio. The tradeoffs here are obtained from Fig. 5a. If the transition ratio is held fixed, then for $n = 12$ we find $\eta$ is $\approx 1.0 \times 10^{-5}$; from Chart No. 1 (Fig. 3a) we can observe the tradeoff as $\delta_1$ and $\delta_2$ are varied for this new value of $\eta$. Finally, both transition ratio and $\eta$ can be varied, e.g., to 0.93 for transition ratio and $1.5 \times 10^{-5}$ for $\eta$, so as to make their intersection remain on the $n = 12$ curve; now all four filter parameters can be varied to match the new values of $\eta$ and transition ratio.

It is interesting to note that if a Chebyshev or Butterworth filter type is specified in place of the elliptic, the designer need only substitute Figs. 4b or 4c for Fig. 4a as Chart No. 2 and proceed as before. In both cases of the example given, the required filter order considerably exceeds the maximum limit of 20 of the curves; thus the "efficiency" of the elliptic design is clearly seen.

Clearly, this design procedure presents a tremendous amount of flexibility to the designer—more so than is generally available in most

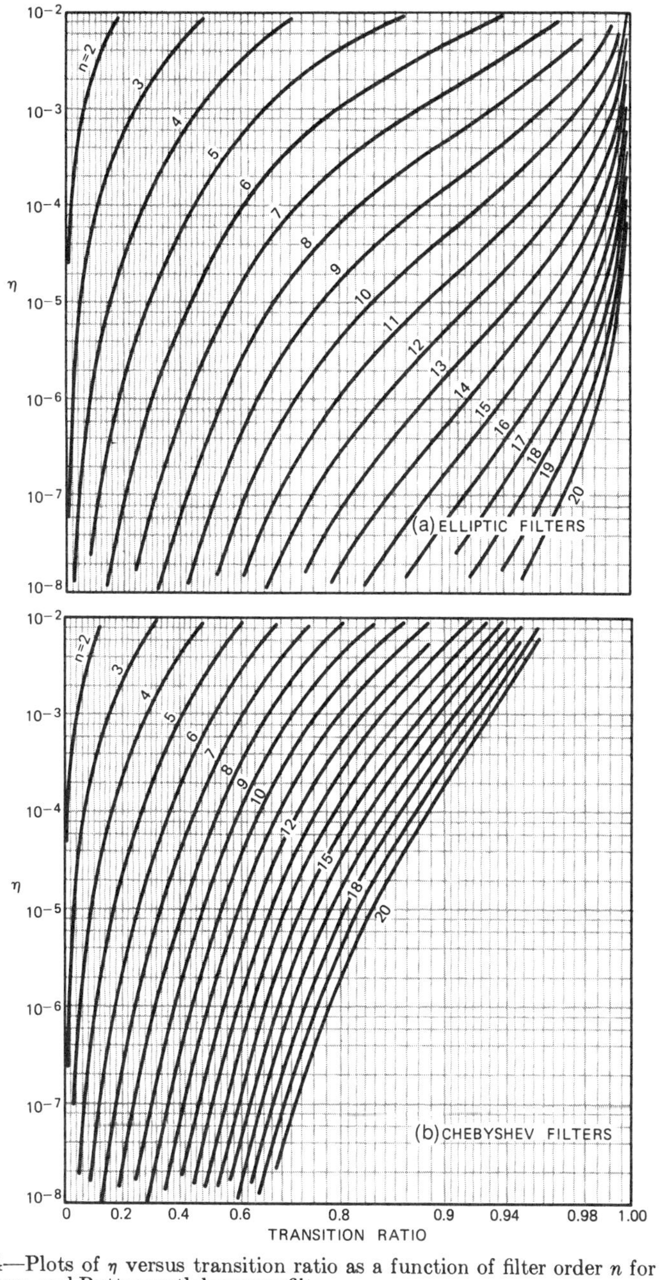


Fig. 4—Plots of $\eta$ versus transition ratio as a function of filter order $n$ for elliptic, Chebyshev, and Butterworth low-pass filters.

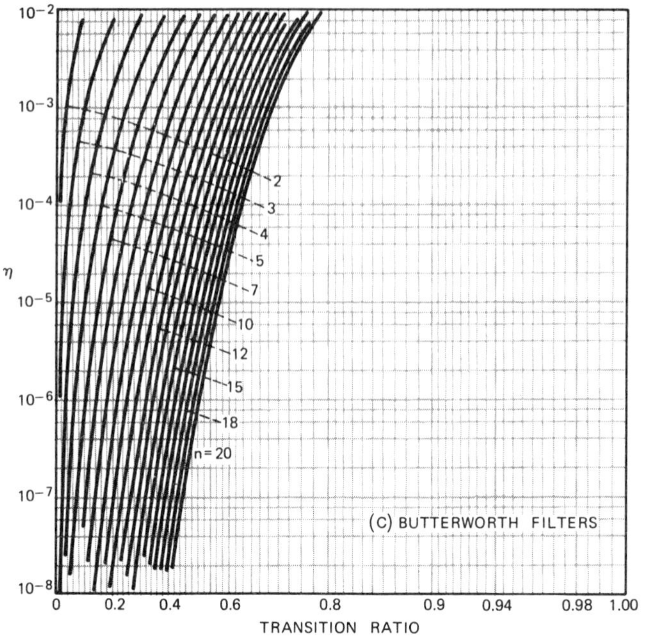


Fig. 4 (continued).

programs for filter order determination. *Furthermore, the insight into the design problem afforded by this graphical technique allows the designer to get a feeling for the way in which small changes in filter specification affect the required filter order.* Quite often the designer is willing to changes his ideas on "required" specifications, especially if it reduces the filter order necessary to meet his specifications.

## VI. COMPARISONS BETWEEN OPTIMUM FIR AND ELLIPTIC DIGITAL FILTERS

Based on the design formulas of the preceding sections, it is possible to make some comparisons between optimum FIR low-pass filters and equivalent elliptic filters. The main basis of comparison will be the number of multiplications per input sample* required in the most standard realization of each filter type, i.e., the direct form for the FIR case and the cascade form for the elliptic case.[7] Direct realization of an $N$-point impulse response filter with linear phase requires

* The number of multiplications per input sample is a useful measure of the computational complexity of the filtering operations as it represents the number of multiply-add operations required for a software implementation of the algorithm as well as for a general hardware implementation.

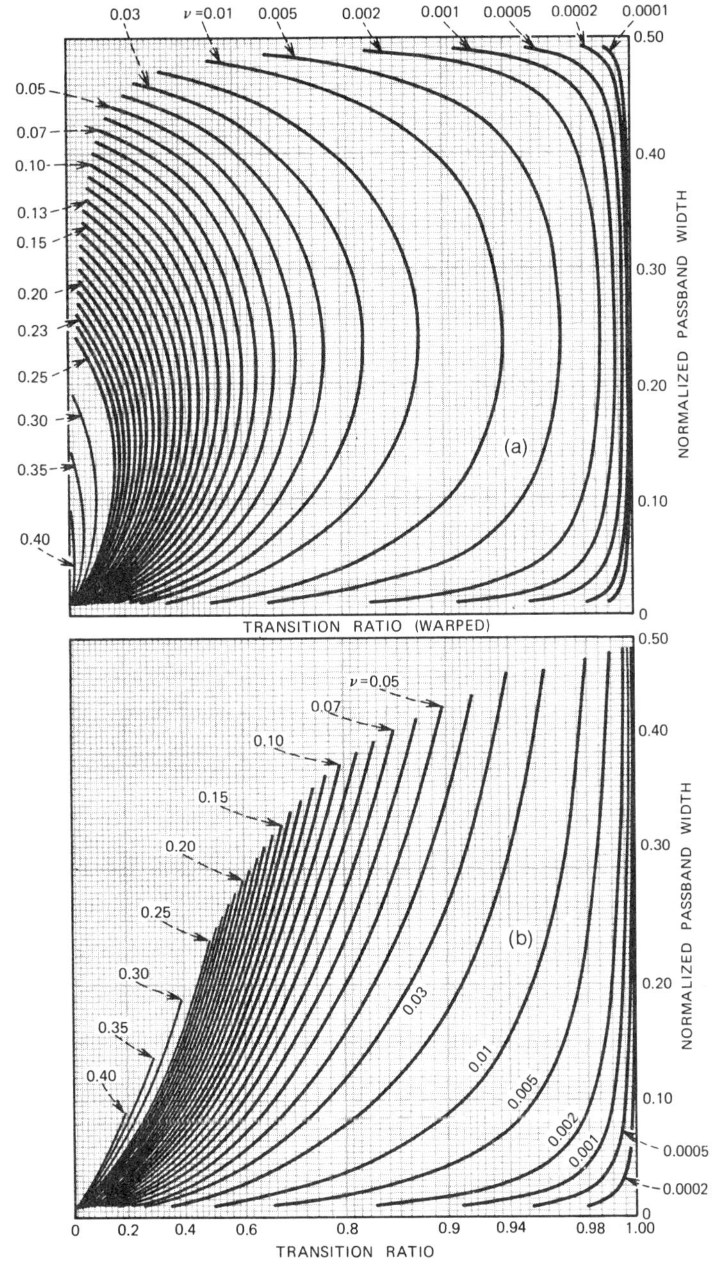


Fig. 5—Plots of passband cutoff frequency versus transition ratio as a function of transition width for discrete and continuous low-pass filters.

$[(N+1)/2]$ multiplications per sample, whereas cascade realization of an $n$th-order elliptic filter (all zeros on the unit circle) requires $[(3n+3)/2]$* multiplications per sample where $[\cdot]$ denotes "integer part of."

Thus, one basis of comparison between equivalent filter designs (i.e., both meeting the same specifications on $\delta_1$, $\delta_2$, $F_p$, and $F_s$) is in terms of the efficiency of the respective realizations, i.e., which structure requires fewer multiplications per sample. Equivalence between structures is attained when the condition

$$\left[\frac{3n+3}{2}\right] = \left[\frac{(N+1)}{2}\right] \tag{28}$$

or equivalently

$$\frac{N}{n} \approx 3 + \frac{1}{n}. \tag{29}$$

Using the appropriate filter design formulas, we have measured the quantity $N/n$ as a function of $n$ for a large range of values of $F_p$, $\delta_1$, and $\delta_2$. Figure 6 shows two typical sets of curves which were obtained. Figure 6a shows data for the case $F_p = 0.15$, $\delta_1 = 0.1$, $\delta_2 = 0.1, 0.01, 0.001, 0.0001$, and Fig. 6b shows data for $F_p = 0.35$, $\delta_1 = 0.00001$, and the same range of $\delta_2$ as in Fig. 6a. Also shown in these plots is the line $N/n = 3$ for showing where the data lie with respect to the fixed portion of eq. (29). As seen in this figure, for certain values of $F_p$, $\delta_1$, and $\delta_2$, the ratio of $N/n$ falls below the equivalence level of eq. (29), i.e., the FIR filter is more efficient than the elliptic filter. However, in general, the elliptic filter is more efficient than the optimum FIR filter, and, in the case of high-order elliptic designs, the ratio of $N/n$ is often in the hundreds or thousands.

Based on our examination of large amounts of data, the following general observation can be made: the most favorable conditions for the FIR design are large values of $\delta_1$, small values of $\delta_2$, and large transition widths (i.e., small transition ratios). One also observes the following behavior:

(*i*) For values of $F_p \geqq 0.3$, the ratio $N/n$ always exceeded $3 + 1/n$ for all values of $\delta_1$, $\delta_2$, and $n$.

(*ii*) For values of $n \geqq 7$, the ratio $N/n$ always exceeded $3 + 1/n$ for all values of $\delta_1$, $\delta_2$, and $F_p$.

---

* This number of multiplications per sample for the IIR filter assumes that any scaling between sections is an integer power of 2 and is performed entirely by shifts of the data. If finer scaling multipliers are included between each cascade section, the realization requires $[(4n+3)/2]$ multiplications per sample.

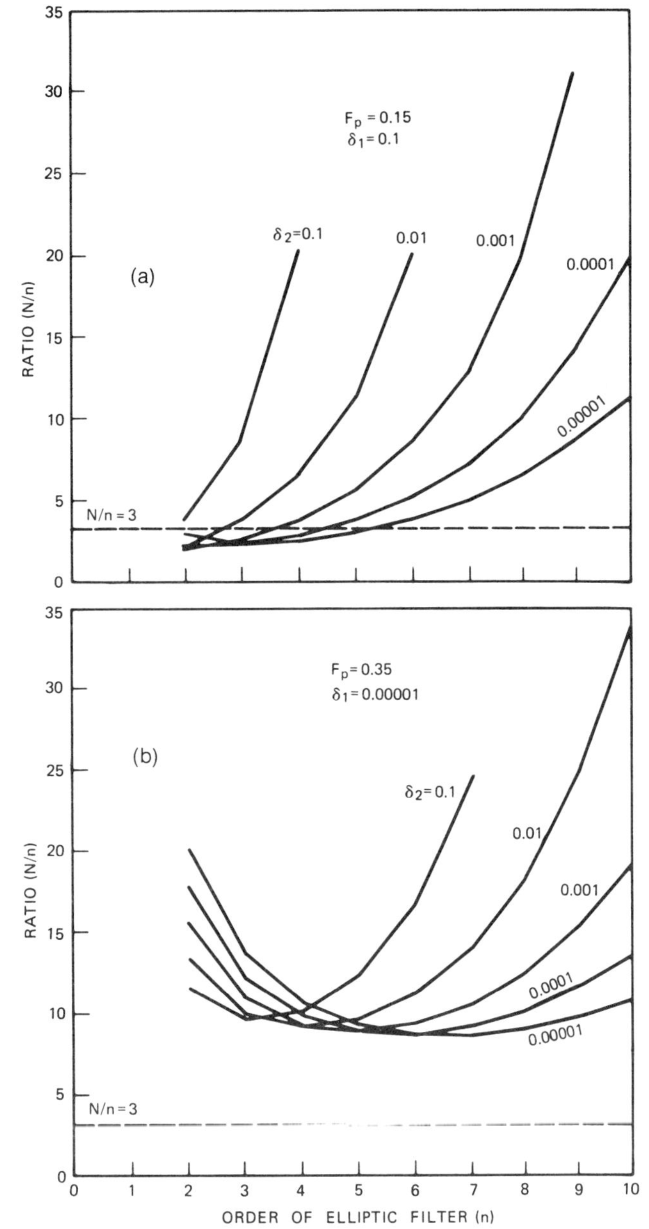


Fig. 6—Plots of the ratio $N/n$ as a function of $n$ for optimum FIR filters and elliptic filters meeting identical specifications on $\delta_1$, $\delta_2$, $F_p$, and $F_s$.

(*iii*) The smaller the value of $F_p$, the larger the range of $\delta_1$, $\delta_2$, and $n$ for which $N/n$ was less than $3 + 1/n$.

Since the design formula for $N$ for the optimum FIR case is not exact but only an estimate, measurements were also made of the required theoretical value of $n$ (elliptical filter order) to meet the specifications of optimum FIR filters which had already been designed. Typical results of these measurements are shown in Fig. 7. Figure 7a shows the theoretical order $n$ ($n$ need not be an integer) required to match specifications on $F_p$, $F_s$, for $\delta_1 = 0.1$, $\delta_2 = 0.1$, 0.01, 0.001, 0.0001, and 0.00001, as a function of $F_p$ for a set of optimum FIR filters with $N = 21$. (It should be noted that as $F_p$ varies, $F_s$ also varies so as to achieve the desired specifications on $\delta_1$ and $\delta_2$.) Figure 7b shows similar measurements for $N = 41$. In Fig. 7a the theoretical point of equivalence is $n = 6.3$, whereas in Fig. 7b it is $n = 13$. From this figure it is seen that for these cases the elliptic filter is always more efficient than the equivalent FIR filter, as anticipated by the discussion in the preceding paragraphs.

In summary, elliptic filters are generally more efficient in achieving given specifications on the frequency response than optimum FIR filters. However, the FIR filters have the additional useful property that their phase is exactly linear, i.e., there is no group delay distortion. For the elliptic filter, however, there is generally a large amount of group delay distortion (concentrated primarily near the band edge). A question of both theoretical and practical importance is whether, in cases when the additional requirement of a flat delay is specified, it is more desirable to equalize the delay of an elliptic filter or to use the equivalent optimum FIR filter (with its constant group delay). In the next section we discuss various aspects of this question. It should be noted that the above alternatives are not the only possibilities for obtaining a digital filter which meets frequency domain specifications on both magnitude and group delay responses. For example, a filter can be designed, using modern optimization procedures, where the number of poles and zeros are unequal. In such cases, the comparisons between FIR and IIR filters are quite distinct from those to be discussed in the next section.

## VII. COMPARISONS OF OPTIMUM FIR FILTERS AND DELAY-EQUALIZED ELLIPTIC FILTERS

Recently developed optimization procedures[8] make it possible to design an all-pass equalizer which can equalize the group delay of any

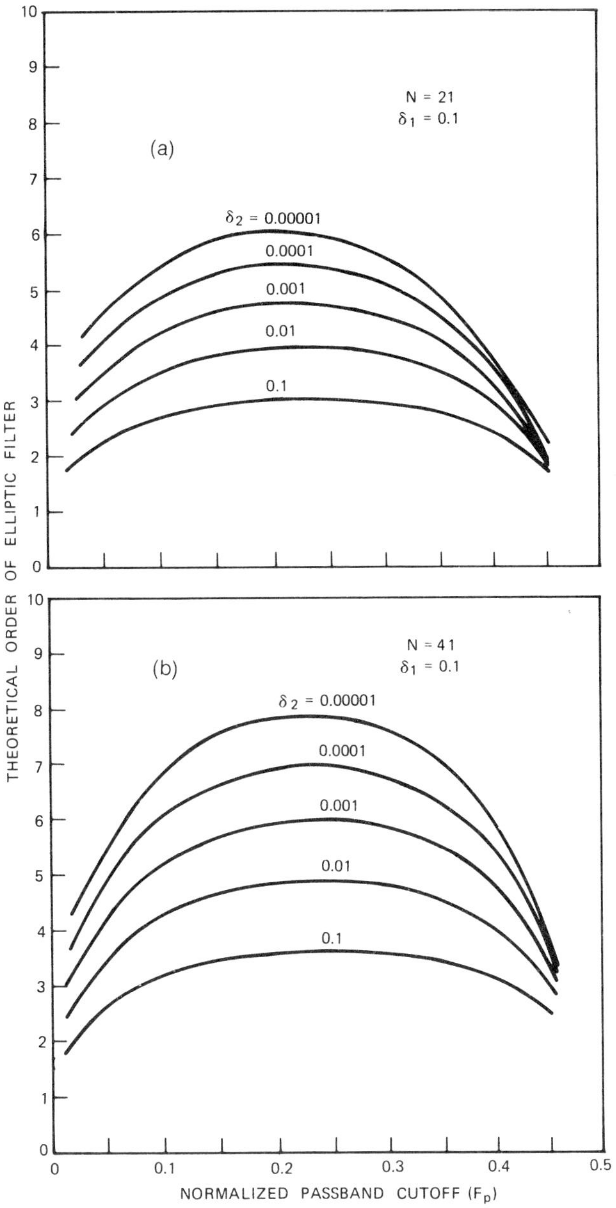


Fig. 7—Theoretical order of elliptic filters required to meet given specification on $\delta_1$, $\delta_2$, $F_p$, and $F_s$ as a function of $F_p$ for various values of $\delta_2$. Optimum FIR filters with $N = 21$ meet the specifications for all filters of (a), whereas $N = 41$ is required for all filters of (b).

digital filter to any desired accuracy over a restricted band of frequencies. As an example of the use of this procedure, Fig. 8 shows plots of the group delay of a 6th-order (unequalized) elliptic filter (with parameters $\delta_1 = 0.01$, $\delta_2 = 0.0001$, $F_p = 0.24163$, $F_s = 0.34842$) and the equalized group delay using a 10th-order all-pass filter. The relative error in the equalized delay curve is 3.6 percent of the average delay in the passband. In this case the equalized elliptic filter requires 20 multiplications per sample, whereas an optimum FIR filter which achieves the same specifications requires only 11 multiplications per sample.

The difficulty with trying to equalize the group delay of a filter lies in the fact that the equalized filter must have a total delay greater than the largest delay in the unequalized filter which always occurs near the passband cutoff frequency. Thus, in the example of Fig. 8, even though the delay throughout most of the passband is between 2 and 6 samples, the delay at the edge of the band is about 15 samples. It can be shown that an all-pass equalizer of degree $n_e$ has the

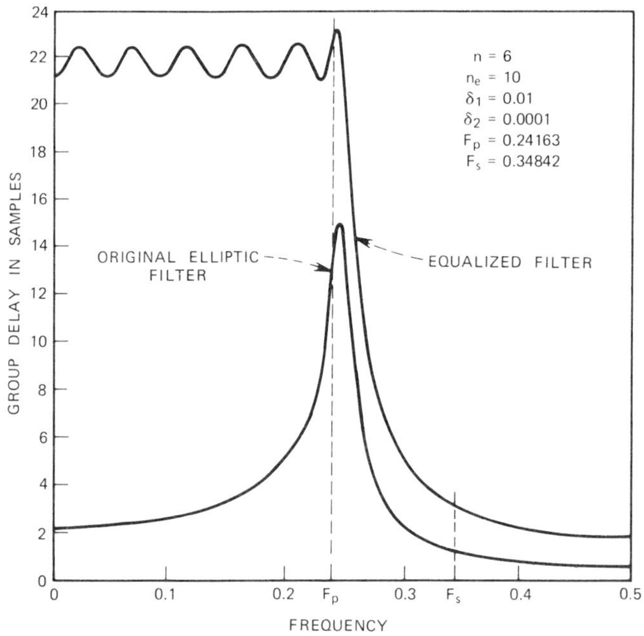


Fig. 8—The group delay of an unequalized and an equalized elliptic filter. The equalizer is of 10th degree and the elliptic filter is of 6th degree.

property

$$\frac{1}{2\pi}\int_0^{\pi} \tau_g(\omega)d\omega = 0.5n_e, \tag{30}$$

where $\tau_g(\omega)$ is the equalizer group delay and the integral is taken over half the sampling interval $(0 \leqq \omega \leqq \pi)$. Since $\tau_g(\omega) \geqq 0$, i.e., group delays add, to justify eq. (30) it is sufficient to show that a first-degree all-pass equalizer has the required property. The $z$-transform of a first-degree all-pass equalizer is

$$H(z) = \frac{1 - z^{-1}/a}{1 - az^{-1}}, \tag{31}$$

where $a$ is the pole position and $1/a$ is the zero position in the $z$-plane. The group delay is commonly defined as

$$\tau_g(\omega) = -\frac{d[\measuredangle H(e^{j\omega})]}{d\omega}, \tag{32}$$

where $\measuredangle H(e^{j\omega})$ is the phase of the transfer function. Using eqs. (31) and (32) we obtain

$$\tau_g(\omega) = \frac{1 - a^2}{1 + a^2 - 2a\cos\omega} \tag{33}$$

for the first-degree equalizer. Integrating eq. (33) from 0 to $\pi$ and normalizing by $2\pi$ gives

$$\frac{1}{2\pi}\int_0^{\pi} \frac{1 - a^2}{1 + a^2 - 2a\cos\omega}\,d\omega = \frac{1}{\pi}\tan^{-1}\left[\frac{(1 - a^2)\tan(\omega/2)}{(1 - a)^2}\right]\Bigg|_0^{\pi} = \frac{\pi}{2\pi} = 0.5.$$

The significance of eq. (30) is that one can estimate the minimum-order equalizer required to equalize a given group delay characteristic by determining the area between the line $\tau = \tau_{\max}$ and the curve $\tau_g(\omega)$ and dividing by $\pi$, where $\tau_{\max}$ is the maximum value of $\tau_g(\omega)$ in the passband. Thus in the example of Fig. 8, the estimated order of the equalizer is approximately $(13 \times \pi/2)/\pi = 6.5$. Of course, the required order of the equalizer must be greater than the estimate given above, since this estimate assumes the delay of the equalizer exactly compensates the delay of the unequalized filter. As the degree of the equalizer is increased over the estimate, the peak error of approximation decreases monotonically.

We have used the above algorithm, along with initial estimates of equalizer order, to equalize three sets of elliptic filters. The data for

Table I — Comparisons between optimum FIR and equalized elliptic digital filters
(Set 1: $\delta_1 = 0.01$, $\delta_2 = 0.0001$)

| $F_p$ | $F_s$ | $n$ | $N$ | $n_e$ | $\bar{\tau}_g$ | $r$ | $N_1$* | $N_2$* |
|---|---|---|---|---|---|---|---|---|
| 0.0502 | 0.13999 | 5 | 21 | 2 | 28.7 | 12.1 | 11 | 11 |
| | | | | 4 | 42.7 | 3.4 | | 13 |
| 0.09846 | 0.24111 | 5 | 21 | 2 | 14.5 | 11.6 | 11 | 11 |
| | | | | 4 | 22.2 | 4.1 | | 13 |
| | | | | 6 | 29.4 | 0.8 | | 15 |
| 0.14722 | 0.25297 | 6 | 21 | 4 | 17.6 | 13.1 | 11 | 14 |
| | | | | 6 | 23.0 | 6.3 | | 16 |
| | | | | 8 | 28.5 | 2.6 | | 18 |
| 0.19507 | 0.30647 | 6 | 21 | 4 | 13.8 | 16.0 | 11 | 14 |
| | | | | 6 | 17.8 | 8.7 | | 16 |
| | | | | 8 | 22.0 | 4.2 | | 18 |
| 0.24163 | 0.34842 | 6 | 21 | 6 | 14.5 | 11.1 | 11 | 16 |
| | | | | 8 | 18.3 | 7.0 | | 18 |
| | | | | 10 | 21.8 | 3.6 | | 20 |
| 0.28664 | 0.41668 | 5 | 21 | 6 | 11.6 | 8.4 | 11 | 15 |
| | | | | 8 | 14.5 | 3.8 | | 17 |
| | | | | 10 | 17.3 | 1.6 | | 19 |
| 0.33014 | 0.43727 | 5 | 21 | 6 | 10.7 | 14.7 | 11 | 15 |
| | | | | 8 | 13.1 | 8.3 | | 17 |
| | | | | 10 | 15.7 | 4.5 | | 19 |
| 0.37254 | 0.47479 | 4 | 21 | 6 | 8.7 | 19.1 | 11 | 13 |
| | | | | 8 | 11.1 | 6.5 | | 15 |
| | | | | 10 | 13.4 | 3.2 | | 17 |
| 0.41665 | 0.49417 | 3 | 21 | 8 | 9.6 | 6.3 | 11 | 14 |
| | | | | 10 | 11.8 | 3.2 | | 16 |

* $N_1$ is the number of multiplications per sample for the optimum FIR filter; $N_2$ is the number of multiplications per sample for the equalized elliptic filter.

these three sets of filters are given in Tables I through III. Included in the table are the filter specifications ($\delta_1$, $\delta_2$, $F_p$, $F_s$); the required elliptic order $n$; the required FIR filter duration $N$; the equalizer order $n_e$; the average passband delay, $\bar{\tau}_g$ (in samples), of the equalized filter; the percentage ripple, $r$, in the passband group delay of the equalized filter; and a comparison between the number of multiplications per sample required in both the optimum FIR filter and the equalized elliptic filters. The data in these tables indicate that to achieve equalization to within about a 3-percent error requires on the order of 30 percent more multiplications per sample for the equalized filter than for the optimum FIR design, although in most cases the unequalized elliptic filter was more efficient than the optimum FIR designs. Thus it would appear that, at least for these restricted results, if constant group delay is required in addition to the equiripple magni-

Table II — Comparisons between optimum FIR and equalized elliptic digital filters
(Set 2: $F_p = 0.25$, $\delta_1 = 0.02$, $\delta_2 = 0.001$)

| $F_s$ | $n$ | $N$ | $n_e$ | $\bar{\tau}_g$ | $r$ | $N_1$* | $N_2$* |
|---|---|---|---|---|---|---|---|
| 0.4893 | 2 | 11 | 2 | 3.3 | 1.2 | 6 | 6 |
| | | | 4 | 5.6 | 0.1 | | 8 |
| 0.44816 | 3 | 13 | 2 | 4.5 | 9.4 | 7 | 8 |
| | | | 4 | 7.3 | 1.0 | | 10 |
| 0.39146 | 4 | 19 | 2 | 5.9 | 25.1 | 10 | 9 |
| | | | 4 | 8.8 | 8.0 | | 11 |
| | | | 6 | 11.9 | 2.2 | | 13 |
| 0.34153 | 5 | 29 | 2 | 8.4 | 37.4 | 15 | 11 |
| | | | 4 | 10.6 | 21.6 | | 13 |
| | | | 6 | 13.7 | 11.6 | | 15 |
| | | | 8 | 16.9 | 5.6 | | 17 |
| | | | 10 | 20.3 | 2.4 | | 19 |
| 0.30639 | 6 | 45 | 4 | 13.8 | 34.7 | 23 | 14 |
| | | | 6 | 16.0 | 25.0 | | 16 |
| | | | 8 | 18.7 | 16.9 | | 18 |
| | | | 10 | 22.0 | 11.7 | | 20 |
| | | | 12 | 25.5 | 7.9 | | 22 |
| | | | 14 | 29.4 | 5.2 | | 24 |
| | | | 16 | 32.8 | 3.2 | | 26 |
| | | | 18 | 36.3 | 1.8 | | 28 |

* $N_1$ is the number of multiplications per sample for the optimum FIR filter; $N_2$ is the number of multiplications per sample for the equalized elliptic filter.

tude characteristics, then the optimum FIR filter is always more efficient than an equalized elliptic filter. It should also be noted that the delay of the optimum FIR filter [$(N - 1)/2$ samples] was *always* less than the delay of the equalized elliptic filter.

The examples of Tables I through III considered filters where the order of the unequalized elliptic filter was six or less. It can be argued that, for higher-order elliptic designs, the relative efficiency of the elliptic filter over the optimum FIR filter is far greater than for lower-order designs; hence in these cases perhaps the equalized filter may still be more efficient than the optimum FIR design. This conjecture turns out to be untestable because high-order elliptic filters have a peak passband delay $\tau_{\max}$ which is much larger than for low-order filters, hence the order required for the equalizer becomes extremely large and thus is not even practical to consider if equalization over the entire passband is required. To illustrate this point, Fig. 9 shows the group delay of a 10th-order elliptic low-pass filter with $F_p = 0.25$. Using eq. (30) to get an estimate of $n_e$ we arrive at a value of $n_e = 45$. Since this value of $n_e$ is only an underbound on the actual order of the

Table III — Comparisons between optimum FIR and equalized elliptic digital filters
(Set 3: $F_p = 0.25$, $\delta_1 = 0.02$, $\delta_2 = 0.0001$)

| $F_s$ | $n$ | $N$ | $n_e$ | $\bar{\tau}_g$ | $r$ | $N_1$* | $N_2$* |
|---|---|---|---|---|---|---|---|
| 0.49661 | 2 | 11 | 2 | 3.3 | 1.2 | 6 | 6 |
| | | | 4 | 5.6 | 0.1 | | 8 |
| 0.47564 | 3 | 11 | 2 | 4.5 | 9.1 | 6 | 8 |
| | | | 4 | 7.3 | 1.0 | | 10 |
| 0.43591 | 4 | 17 | 2 | 5.8 | 23.3 | 9 | 9 |
| | | | 4 | 8.8 | 7.0 | | 11 |
| | | | 6 | 11.8 | 1.7 | | 13 |
| 0.38983 | 5 | 21 | 2 | 8.0 | 33.4 | 11 | 11 |
| | | | 4 | 10.3 | 18.0 | | 13 |
| | | | 6 | 13.5 | 8.7 | | 15 |
| | | | 8 | 16.7 | 3.7 | | 17 |
| | | | 10 | 20.0 | 1.4 | | 19 |
| 0.34878 | 6 | 31 | 4 | 12.8 | 28.9 | 16 | 14 |
| | | | 6 | 15.5 | 19.2 | | 16 |
| | | | 8 | 18.2 | 11.8 | | 18 |
| | | | 10 | 22.0 | 7.7 | | 20 |
| | | | 12 | 25.3 | 4.3 | | 22 |
| | | | 14 | 28.8 | 2.2 | | 24 |

* $N_1$ is the number of multiplications per sample for the optimum FIR filter; $N_2$ is the number of multiplications per sample for the equalized elliptic filter.

equalizer, it is clear that it is not practical to try to obtain such a high-degree equalizer.

Another interesting question which arises when one considers the idea of equalizing an IIR filter is how does the cascade combination of an elliptic filter and an all-pass equalizer compare to the optimum IIR filter which best approximates both the desired magnitude and group delay characteristics? It is clear that the optimum IIR filter can be no worse than the cascade; the question remains as to how much

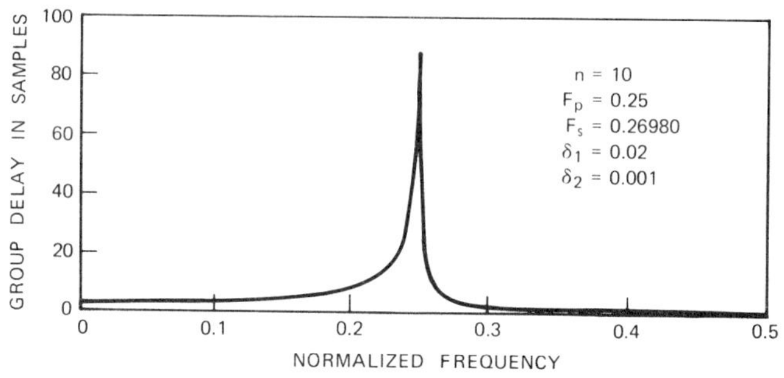


Fig. 9—The group delay of a 10th-order elliptic filter with $F_p = 0.25$.

better it can be. There is no clear-cut answer to this question. However, based on our experience with equalized elliptic filters, several observations can be made. (We shall use the $z$-plane pole-zero plot of a typical equalizer filter, shown in Fig. 10, to aid in understanding the nature of the equalized filter.)

(*i*) The zeros of the elliptic filter lie on the unit circle to give good stopband attenuation.
(*ii*) The zeros of the equalizer lie outside the unit circle to give positive delay.
(*iii*) The poles of the elliptic filter are constrained by the transition width requirements of the low-pass filter.
(*iv*) The poles of the equalizer lie approximately on a circle of fixed radius, and are approximately equally spaced in the passband.

If the zeros of the optimum filter are not constrained to lie on the unit circle, then each second-order section will require four multiplications per sample, rather than the three multiplications for each second-order section of the elliptic design and the two multiplications for each second-order section of the all-pass equalizer. Based on the

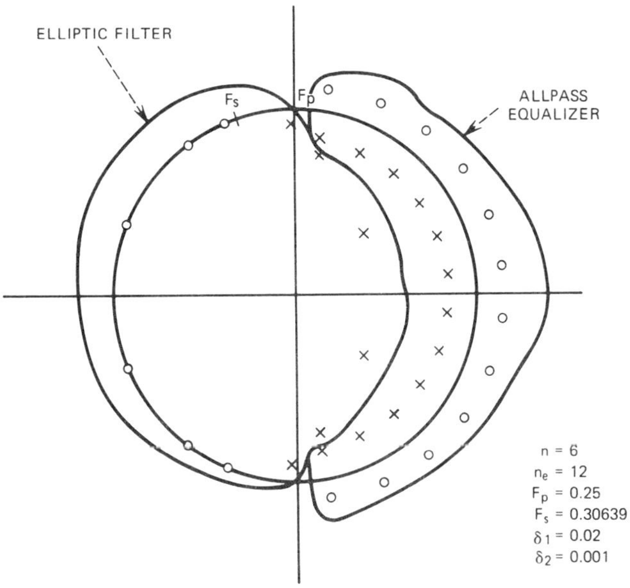


Fig. 10—The pole-zero positions of an equalized elliptic filter.

above observations, it seems unlikely that there is much to gain by using the optimum IIR filter over the equalized filter.

### VIII. GENERAL DISCUSSION

In this paper we have considered only one basis for comparison between optimum FIR filters and equivalent IIR designs, that measure being the number of multiplications per sample required in the standard method of realization for each of these filter types. The justification for this measure is that in hardware (and generally in software) the number of multiplications per sample is an excellent measure of the complexity required in the implementation as well as the factor which determines the maximum throughput rate of the system.[9] However, there are many other ways for comparing these filter types when one takes into consideration the various finite word-length effects which occur in a practical design situation. In this section we review several of these design issues.

Among the various finite word-length effects are roundoff noise, both uncorrelated and correlated (e.g., limit cycles), and coefficient quantization sensitivity. For direct-form FIR realization, the peak roundoff noise can easily be made to be less than $\frac{1}{2}$ of the least significant bit by accumulating partial sums in an extended length register and then rounding the final result. For cascade IIR filters realized with fixed-point arithmetic, the roundoff noise problem is inherently related to the dynamic range problem,[7] and involves the concepts of pole-zero pairing and section ordering. Jackson[10] has shown that with reasonable pairing and ordering the uncorrelated roundoff noise variance can be minimized. However, even in the best of situations, the roundoff noise is equivalent to several bits. In terms of correlated roundoff noise, i.e., limit cycles, the direct-form FIR realization has no zero-input limit cycles (because no feedback is present), whereas the cascade IIR realization will generally exhibit zero-input limit cycles. Kaiser[11] has extensively studied these limit cycles and has developed bounds and estimates for their amplitude and frequency.

The coefficient quantization problem is one of the most difficult finite word length effects to treat analytically. Rounding of infinite precision filter coefficients to a fixed number of bits alters the overall frequency response of the filter in a complicated manner. Avenhaus[12] has shown that straight rounding of the infinite precision filter coefficients is generally inferior to optimizing the filter performance over the finite set of fixed precision filter coefficients. However, there are no general procedures for performing this optimization, nor are there any guarantees of convergence of the existing methods. Furthermore, in many cases the advantage of optimizing finite precision coefficients over straight rounding of the infinite precision coefficients is small. Thus for the case of coefficient quantization neither direct-form realization of FIR filters nor cascade realization of IIR filters seems to offer a relative advantage here.

Thus it is difficult, if not impossible, to be quantitative in comparing FIR and IIR filters based on anything other than number of multiplications per sample. This is why we have used this measure throughout this paper.

## IX. SUMMARY

In this paper some comparisons were made between equivalent FIR and IIR digital filters based on the number of multiplications per sample required to realize these filters. In the case of low-pass filters with quasi-equiripple magnitude characteristics, IIR elliptic filters could generally be realized more efficiently than equivalent linear phase FIR filters. When the additional requirement of constant group delay in the passband was added to the specifications, comparisons showed the linear phase FIR filters to be more efficient than group-delay-equalized elliptic IIR filters.

Additionally, a novel set of design charts for determining the minimum filter order required to meet given filter specifications for both digital and analog elliptic, Chebyshev, and Butterworth low-pass filters was presented. Explanation of how to use these charts to gain insight into the various filter parameter tradeoffs was also given.

## REFERENCES


1. J. E. Storer, *Passive Network Synthesis*, New York: McGraw-Hill Book Co., 1957.
2. O. Herrmann, L. R. Rabiner, and D. S. K. Chan, "Practical Design Rules for Optimum Finite Impulse Response Low-Pass Digital Filters," B.S.T.J., *52*, No. 6 (July-August 1973), pp. 769–799.
3. J. F. Kaiser, "Digital Filters," in *System Analysis by Digital Computer*, edited by F. F. Kuo and J. F. Kaiser, New York: John Wiley and Sons, 1966.
4. R. M. Golden and J. F. Kaiser, "Design of Wideband Sampled Data Filters," B.S.T.J., *43*, No. 4, Pt. 2 (July 1964), pp. 1533–1546.
5. A. G. Constantinides, "Spectral Transformations for Digital Filters," Proc. IEE, *117*, No. 8, 1970, pp. 1585–1590.
6. E. Christian and E. Eisenmann, *Filter Design Tables and Graphs*, New York: John Wiley and Sons, 1966.
7. L. B. Jackson, "On the Interaction of Roundoff Noise and Dynamic Range in Digital Filters," B.S.T.J., *49*, No. 2 (February 1970), pp. 159–184.
8. A. G. Deczky, "Synthesis of Recursive Digital Filters Using the Minimum p-Error Criterion," IEEE Trans. Audio and Electroacoustics, *AU-20*, No. 4 (October 1972), pp. 257–263.

9. L. B. Jackson, J. F. Kaiser, and H. S. McDonald, "An Approach to the Implementation of Digital Filters," IEEE Trans. Audio and Electroacoustics, *AU-16*, No. 3 (September 1968), pp. 413–421.
10. L. B. Jackson, "Roundoff-Noise Analysis for Fixed-Point Digital Filters Realized in Cascade or Parallel Form," IEEE Trans. Audio and Electroacoustics, *AU-18*, No. 2 (June 1970), pp. 107–122.
11. J. F. Kaiser, "An Overview on Digital Filters," Newsletter of IEEE Circuit Theory Group, *6*, No. 1 (March 1972).
12. E. Avenhaus, "On the Design of Digital Filters with Coefficients of Limited Word Length," IEEE Trans. Audio and Electroacoustics, *AU-20*, No. 3 (August 1972), pp. 206–212.

---

FIR AND IIR DIGITAL FILTERS

# Part 3
# Two-Dimensional Signal Processing

In the predecessor volume, *Digital Signal Processing*, there were no papers from the field of two-dimensional digital signal processing. A large part of this volume, however, is devoted to this area, which is indicative that this field has come into its own in the past few years. The papers in this part fall into three general areas: the theory of two-dimensional processing, stability criteria for infinite impulse response (IIR) filters, and design methods for two-dimensional finite impulse response (FIR) and IIR digital filters.

The first paper, by Mersereau and Oppenheim, discusses algorithms for digitally reconstructing multidimensional objects from their projections. It represents the application of digital signal processing concepts to a uniquely multidimensional problem. The next paper, by Mersereau and Dudgeon, presents an overview of the problems encountered in two-dimensional processing and a discussion of how these problems differ from the corresponding one-dimensional problems.

The use of cepstral analysis has been important in digital signal processing. In two dimensions the definition of a complex cepstrum has long been thought to be mathematically difficult, if not impossible. In the next paper, Dudgeon discusses the problem of obtaining the cepstrum for a two-dimensional rational function and produces a necessary and sufficient condition for its existence.

Two-dimensional problems pose data handling issues not encountered in one dimension. This is due both to the size of the arrays involved and to the fact that the data must be accessed both row by row and column by column in many cases. In the final paper in Part 3-A, Eklundh presents a fast method for matrix transposition, which is required in the computation of the two-dimensional fast Fourier transform (FFT).

In the area of two-dimensional FIR filter design, the paper by Huang extends the window method to two dimensions in a simple and natural way; in the paper by Hu and Rabiner, the frequency sampling and linear programming design procedures are also extended to two dimensions. In the next paper, Kamp and Thiran present an algorithm for performing weighted Chebyshev approximations in two dimensions. This generalization from the corresponding one-dimensional problem is a real accomplishment.

McClellan, in the next paper presents a novel technique for converting one-dimensional FIR linear phase filters into two-dimensional FIR linear phase filters. This technique is probably the most powerful method available today for designing certain large FIR filters.

The problems of testing an IIR digital system for stability and stabilizing an unstable design are basic, and in two dimensions they represent difficult problems. Two key papers on stability criteria have thus been included. The paper by Shanks *et al.* presents a stability criterion for two-dimensional IIR filters, and the paper following it, by Huang, extends the criterion and presents a scheme for implementing it. Another paper [1], which was not included in this selection of papers because of limited space, presents an improved implementation of Huang's test.

In two dimensions, the problems of filter design and filter implementation are not independent. The structure is implied by the design procedure. Three papers on IIR design have been included, and each in turn implies a unique implementation. The first of these papers, by Treitel and Shanks, presents a technique for either IIR or FIR design based on a decomposition of the filter into a series of separable filters. The paper by Maria and Fahmy presents a technique for finding the coefficients for a cascade realization using the minimum $p$th-error criterion. The paper by Costa and Venetsanopoulos presents a method for designing circularly symmetric filters using rotated one-dimensional filters.

## References

[1] B. D. O. Anderson and E. I. Jury, "Stability test for two-dimensional recursive filters," *IEEE Trans. Audio Electroacoust.*, vol. AU-21, pp. 366–372, Aug. 1973.

# Digital Reconstruction of Multidimensional Signals from Their Projections

RUSSELL M. MERSEREAU, MEMBER, IEEE, AND ALAN V. OPPENHEIM, SENIOR MEMBER, IEEE

*Invited Paper*

*Abstract*—In a wide variety of applications it is necessary to infer the structure of a multidimensional object from a set of its projections. There has been a long-standing interest in this problem and a number of different techniques have been proposed. In this paper, we present a tutorial review of the reconstruction problem and some of the algorithms which have been proposed for its solution. In addition, we present a number of new algorithms that appear to have some advantages over previous algorithms. Some comparisons of these algorithms applied to reconstructions of two-dimensional pictures are given. Furthermore, a number of new theoretical results are presented relating to the minimum number of projections necessary for exact reconstruction.

## I. Introduction

In a wide variety of applications, it is necessary to infer the structure of a multidimensional object from a set of its projections. X-ray photographs, for example, represent two-dimensional projections of the irradiated three-dimensional structures. Details can be obtained from the projections considered collectively that are not apparent in any one of them alone. Similarly, transmission electron micrographs represent two-dimensional projections of the three-dimensional structures being examined. Other examples arise in the inversion of fan-beam radio-telescope scans and in the recovery of a point-spread function from a set of line spreads.

Because of its potential benefits to so many different fields, particularly medicine and molecular biology, there has been a long-standing interest in this problem, and a number of different techniques have been proposed. While some analog techniques have been considered, specifically optical methods, most of the techniques proposed are in part numerical and thus require the use of a digital computer. Typically, the computations involved are sufficiently complex that off-line processing is required. A part of the present interest in the problem appears to stem from an awareness that with the continuing trend toward faster, cheaper, digital computers and hardware it may soon be, and to a limited extent already is, possible to implement systems on-line for reconstruction from projections.

The techniques that exist for reconstruction fall into two basic classes—one in which the reconstruction is performed in normal signal space and one in which it is performed in Fourier space. While any technique can, of course, be interpreted and analyzed in either space or in both together, most techniques are more easily implemented in one space than the other. Whether implemented in signal space or Fourier space, the reconstruction algorithms can be conveniently interpreted by means of a straightforward and interesting theorem which we refer to as the projection-slice theorem. In essence this theorem states that the Fourier transform of a projection is a slice of the Fourier transform of the projected object. This theorem appears to have first been applied to the reconstruction problem in 1956 by Bracewell [1]. His interest involved the inversion of fan-beam radio-telescope scans. The algorithm proposed at that time was implemented in Fourier space. Although still based on the projection-slice theorem, a modified algorithm was later proposed by Bracewell and Riddle [2] which could be implemented entirely in signal space.

In 1968 DeRosier and Klug [3] proposed a Fourier space technique again based on the projection-slice theorem which was similar to, but apparently derived independently of, Bracewell's results. DeRosier and Klug were interested in the reconstruction problem primarily within the framework of molecular biology, where the projections obtained corresponded to

*This invited paper is one of a series planned on topics of general interest—The Editor.*

Manuscript received January 31, 1974; revised May 10, 1974. This work was supported in part by the Advanced Research Projects Agency monitored by ONR under Contract N-00014-67-A-0204-0064 and in part by the National Science Foundation under Grant GK-31353.

The authors are with the Department of Electrical Engineering and Research Laboratory of Electronics, Massachusetts Institute of Technology, Cambridge, Mass. 02139.

Reprinted from *Proc. IEEE*, vol. 62, pp. 1319–1338, Oct. 1974.

electron micrographs. They subsequently applied their method to a variety of virus structures and in many of the examples were able to take advantage of considerable symmetry in the structures studied. In 1971 Vainshtein [4] and Ramachandran and Lakshminarayanan [5], following DeRosier and Klug, formulated a signal-space algorithm similar to that proposed by Bracewell and Riddle.

An optical technique was used by Garrison *et al.* [6] for making reconstructions from medical X-rays. It implemented a variation of the Bracewell and Riddle algorithm optically. Fourier domain reconstructions from medical X-rays were also investigated by Tretiak, Ozonoff, Klopping, and Eden [7].

Another interesting set of approaches to the problem are the iterative algebraic techniques, implemented in signal space, by Herman [8], Gordon, Bender, and Herman [9], and Gilbert [10]. A new system for on-line radiographic analysis of the brain using reconstructions made by an iterative algebraic technique has recently been marketed [11]. It has proven useful for locating tumors and blood clots in the brain.

In this paper, we present a tutorial review of the reconstruction problem and some of the algorithms which have been proposed for its solution. In addition, we shall present a number of new algorithms that appear to have some advantages over previous algorithms. Furthermore, a number of new theoretical results will be presented relating to the minimum number of projections necessary for exact reconstruction. Mersereau has shown, in particular, that, taken at the right orientation, exact reconstruction can theoretically be carried out from a single projection [12], [13]. While, as will be discussed, this is not a practical algorithm for reconstruction because of the high sensitivity involved, it appears to be an important theoretical result. In particular, it provides one means for mapping multidimensional functions to one-dimensional functions and may have potential application to bandwidth compression and multidimensional digital filter design [14].

In Section II we discuss the theoretical background for understanding the reconstruction problem including the projection-slice theorem and in Section III we consider the question of the number of projections theoretically required for exact reconstruction. Section IV is concerned with a review of a variety of practical reconstruction algorithms which are conveniently interpreted, although not necessarily best implemented, in Fourier space. Section V considers some new modifications of these algorithms. In Section VI we discuss briefly several other algorithms including the iterative algebraic reconstruction techniques referred to earlier.

## II. Theoretical Background

Although sometimes implemented in signal space, most of the reconstruction algorithms which will be discussed in subsequent sections are conveniently interpreted in terms of Fourier space and the projection-slice theorem. Thus in this section, we summarize briefly the $N$-dimensional continuous Fourier transform (CFT) for continuous variables and the $N$-dimensional discrete Fourier transform (DFT) for discrete variables. We shall then define projections and prove the projection-slice theorem.

### A. The N-Dimensional Continuous Fourier Transform (CFT)

We consider here a function $f(x_1, x_2, \cdots, x_N)$ of $N$ continuous variables $x_1, x_2, \cdots, x_N$. We will generally find it convenient to express the $N$-tuple $(x_1, x_2, \cdots, x_N)$ as a vector $\boldsymbol{x}$ and refer to the function as $f(\boldsymbol{x})$. The $N$-dimensional Fourier transform of $f(\boldsymbol{x})$ is denoted by $F(\omega_1, \omega_2, \cdots, \omega_N)$ or $F(\boldsymbol{\omega})$. The domain of $f(\boldsymbol{x})$ will be referred to as signal space and the domain of $F(\boldsymbol{\omega})$ as Fourier space. The $N$-dimensional function $f(\boldsymbol{x})$ and its Fourier transform are related by

$$F(\omega_1, \omega_2, \cdots, \omega_N) = \int_{-\infty}^{+\infty} \cdots \int_{-\infty}^{+\infty} f(x_1, x_2, \cdots, x_N) \cdot \exp\left[-j(\omega_1 x_1 + \omega_2 x_2 + \cdots + \omega_N x_N)\right] dx_1 dx_2 \cdots dx_N \quad (1)$$

$$f(x_1, x_2, \cdots, x_N) = \frac{1}{(2\pi)^N} \int_{-\infty}^{+\infty} \cdots \int_{-\infty}^{+\infty} F(\omega_1, \omega_2, \cdots, \omega_N) \cdot \exp\left[j(\omega_1 x_1 + \omega_2 x_2 + \cdots + \omega_N x_N)\right] d\omega_1 d\omega_2 \cdots d\omega_N \quad (2)$$

or, expressed in vector notation,

$$F(\boldsymbol{\omega}) = \int_{-\infty}^{+\infty} f(\boldsymbol{x}) \exp\left[-j(\boldsymbol{x} \cdot \boldsymbol{\omega})\right] d\boldsymbol{x} \quad (3)$$

and

$$f(\boldsymbol{x}) = \frac{1}{(2\pi)^N} \int_{-\infty}^{+\infty} F(\boldsymbol{\omega}) \exp\left[j(\boldsymbol{x} \cdot \boldsymbol{\omega})\right] d\boldsymbol{\omega} \quad (4)$$

where $\boldsymbol{x} \cdot \boldsymbol{\omega}$ denotes the dot product of the vectors $\boldsymbol{x}$ and $\boldsymbol{\omega}$ or equivalently with $\boldsymbol{x}$ and $\boldsymbol{\omega}$ interpreted as row matrices, $\boldsymbol{x} \cdot \boldsymbol{\omega} = \boldsymbol{x}\boldsymbol{\omega}^t$.

A useful property of the $N$-dimensional Fourier transform pair which we will want to use later is the fact that if $f(\boldsymbol{x})$ and $F(\boldsymbol{\omega})$ form a Fourier transform pair, then $f(\boldsymbol{x}A)$ and $F(\boldsymbol{\omega}A)$ form a Fourier transform pair if $A$ is an orthogonal matrix, i.e., $A^t = A^{-1}$.

This property is easily verified by direct substitution into (3). Thus an orthogonal transformation or equivalently an orthogonal change of coordinates in signal space results in the same change of coordinates in Fourier space. For example, for $N = 2$, if

$$A = \begin{bmatrix} \cos\theta & \sin\theta \\ -\sin\theta & \cos\theta \end{bmatrix} \quad (5)$$

so that $f(\boldsymbol{x})$ is rotated by an angle $\theta$, then its Fourier transform will be rotated by the same angle $\theta$.

### B. The N-Dimensional Discrete Fourier Transform (DFT)

We shall generally be interested in functions which can be processed by a digital computer and consequently can be represented by their samples. Thus we consider the class of band-limited functions. Specifically, $f(\boldsymbol{x})$ is said to be band-limited if there exists an $N$-tuple $(W_1, W_2, \cdots, W_N)$ such that $F(\boldsymbol{\omega})$ is zero for $|\omega_i| > W_i$, $i = 1,2, \cdots, N$. The $N$-tuple $\boldsymbol{W}$ will be referred to as the vector bandwidth. In some cases it is convenient to set $W = \max(W_1, W_2, \cdots, W_N)$ and refer to the scalar $W$ as the bandwidth of $f(\boldsymbol{x})$.

The $N$-dimensional sampling theorem states that if $f(\boldsymbol{x})$ is sampled in signal space on a rectangular lattice with the sample spacing in dimension $x_i$ less than $\pi/W_i$, then $f(\boldsymbol{x})$ can be recovered from its samples. Sampling on a rectangular lattice will be referred to as periodic Cartesian sampling.

Let us denote by $g(\boldsymbol{n})$ the $N$-dimensional sequence corresponding to sampling $f(\boldsymbol{x})$ with a sample spacing of $\pi/V_i$ in the dimension $x_i$ where $V_i > W_i$ so that

$$g(n_1, n_2, \cdots, n_N) = f\left\{\frac{\pi n_1}{V_1}, \frac{\pi n_2}{V_2}, \cdots, \frac{\pi n_N}{V_N}\right\}. \tag{6}$$

From the sampling theorem the Fourier transform $F(\boldsymbol{\omega})$ of $f(\boldsymbol{x})$ can be obtained from $g(\boldsymbol{n})$ by the relation

$$F(\boldsymbol{\omega}) = \frac{\pi^N}{V_1 V_2 \cdots V_N}\left\{\sum_{\boldsymbol{n}=-\infty}^{+\infty} g(\boldsymbol{n}) \exp\left[-j\pi(\boldsymbol{n}\cdot\boldsymbol{\omega}_V)\right]\right\}\cdot b_V(\boldsymbol{\omega}) \tag{7}$$

where $\boldsymbol{\omega}_V$ denotes the vector $\left(\frac{\omega_1}{V_1}, \frac{\omega_2}{V_2}, \cdots, \frac{\omega_N}{V_N}\right)$ and

$$b_V(\boldsymbol{\omega}) = \begin{cases} 1, & |\omega_i| < V_i, \quad i = 1, 2, \cdots, N \\ 0, & \text{otherwise.} \end{cases}$$

Likewise, the sequence $g(\boldsymbol{n})$ can be obtained from $F(\boldsymbol{\omega})$ by the relation

$$g(\boldsymbol{n}) = \frac{1}{(2\pi)^N}\int_{-V}^{V} F(\boldsymbol{\omega}) \exp\left\{j\pi(\boldsymbol{n}\cdot\boldsymbol{\omega}_V)\right\} d\boldsymbol{\omega}. \tag{8}$$

The original $N$-dimensional function $f(\boldsymbol{x})$ can be obtained from the sequence $g(\boldsymbol{n})$ by means of the interpolation formula

$$f(\boldsymbol{x}) = \sum_{\boldsymbol{n}=-\infty}^{+\infty} g(\boldsymbol{n})\, \phi(\boldsymbol{n}, \boldsymbol{x}) \tag{9}$$

where

$$\phi(\boldsymbol{n}, \boldsymbol{x}) = \prod_{i=1}^{N} \frac{\sin V_i\left(x_i - \frac{n_i\pi}{V_i}\right)}{V_i\left(x_i - \frac{n_i\pi}{V_i}\right)}. \tag{10}$$

When only a finite number of the samples of $f(\boldsymbol{x})$ are nonzero, the Fourier transform $F(\boldsymbol{\omega})$ can be represented by a finite set of Cartesian samples. The relationship between the Cartesian samples of $F(\boldsymbol{\omega})$ and the Cartesian samples of $f(\boldsymbol{x})$ is the $N$-dimensional DFT. Specifically, let us assume that

$$g(\boldsymbol{n}) = 0, \qquad \text{if } n_i \geqslant M_i \text{ or } n_i < 0, \quad i = 1, 2, \cdots, N.$$

We now consider the Cartesian samples of $F(\boldsymbol{\omega})$, which we denote by $G(\boldsymbol{k})$ given by

$$G(k_1, k_2, \cdots, k_N) = \frac{\pi^N}{V_1 V_2 \cdots V_N} \cdot F\left(\frac{2V_1}{M_1}k_1, \frac{2V_2}{M_2}k_2, \cdots, \frac{2V_N}{M_N}k_N\right)$$

where $k_i$ is an integer such that

$$-\frac{M_i}{2} + 1 \leqslant k_i \leqslant \frac{M_i}{2}, \qquad \text{if } M_i \text{ is even}$$

$$-\frac{M_i - 1}{2} \leqslant k_i \leqslant \frac{M_i - 1}{2}, \qquad \text{if } M_i \text{ is odd.}$$

Then

$$G(k_1, k_2, \cdots, k_N) = \sum_{n_1=0}^{M_1-1} \cdots \sum_{n_N=0}^{M_N-1} g(n_1, n_2, \cdots, n_N) \cdot \exp\left[-j2\pi\left(\frac{k_1 n_1}{M_1} + \frac{k_2 n_2}{M_2} + \cdots + \frac{k_N n_N}{M_N}\right)\right] \tag{11}$$

and

$$g(n_1, n_2, \cdots, n_N) = \frac{1}{M_1 \cdot M_2 \cdots M_N} \cdot \sum_{k_1} \cdots \sum_{k_N} G(k_1, k_2, \cdots, k_N) \cdot \exp\left[j2\pi\left(\frac{k_1 n_1}{M_1} + \cdots + \frac{k_N n_N}{M_N}\right)\right]. \tag{12}$$

Since $G(k_1, k_2, \cdots, k_N)$ as defined in (11) is periodic in $k_i$ it is frequently convenient to use the values of $k_i$ in the range $0 \leqslant k_i \leqslant M_i - 1$ rather than over the ranges given above. Adopting this convention and defining the vector $\boldsymbol{k}_M$ as the $N$-tuple

$$\left(\frac{k_1}{M_1}, \frac{k_2}{M_2}, \cdots, \frac{k_N}{M_N}\right)$$

we can express (11) and (12) as

$$G(\boldsymbol{k}) = \sum_{\boldsymbol{n}=0}^{M-1} g(\boldsymbol{n}) \exp\left\{-j2\pi(\boldsymbol{n}\cdot\boldsymbol{k}_M)\right\} \tag{13}$$

and

$$g(\boldsymbol{n}) = \frac{1}{M_1 M_2 \cdots M_N}\sum_{\boldsymbol{k}=0}^{M-1} G(\boldsymbol{k}) \exp\left\{j2\pi(\boldsymbol{n}\cdot\boldsymbol{k}_M)\right\}. \tag{14}$$

Equations (13) and (14) are referred to as the $N$-dimensional DFT pair. The $N$-dimensional DFT can be computed efficiently by using the one-dimensional fast Fourier transform (FFT) algorithm, since the summations in (13) and (14) can each be decomposed as a cascade of one-dimensional transforms.

The class of functions $f(\boldsymbol{x})$ which can be represented by a finite number of samples will be referred to as band-limited functions of finite order $M$ where

$$M = \max\{M_1, M_2, \cdots, M_N\}.$$

### C. Projections

A projection is a mapping of an $N$-dimensional function to an $(N-1)$-dimensional function obtained by integrating the function in a particular direction. For example, $p_{x_2}(x_1)$ given by

$$p_{x_2}(x_1) = \int_{-\infty}^{+\infty} f(x_1, x_2)\, dx_2 \tag{15}$$

is an example of a projection of the two-dimensional function $f(x_1, x_2)$ onto one dimension.

For the general case, we define a projection as follows: Let $f(\boldsymbol{x})$ denote an $N$-dimensional function and let $\boldsymbol{u}$ denote a new set of coordinates where

$$\boldsymbol{x} = \boldsymbol{u}\boldsymbol{A}$$

and $\boldsymbol{A}$ is an orthogonal transformation. Then a projection onto the hyperplane $(u_1, u_2, \cdots, u_{i-1}, u_{i+1}, \cdots, u_N)$ is defined as

$$p_{u_i}(u_1, u_2, \cdots, u_{i-1}, u_{i+1}, \cdots, u_N) = \int_{-\infty}^{\infty} f(\boldsymbol{u}A)\, du_i. \tag{16}$$

The coordinate axis $u_i$, which is normal to the hyperplane onto which $f(\boldsymbol{x})$ is projected, will be referred to as the projection axis.

For $N = 2$, the matrix $A$ is given by

$$A = \begin{bmatrix} \cos\theta & \sin\theta \\ -\sin\theta & \cos\theta \end{bmatrix}.$$

In this case, the $u_1$, $u_2$ coordinate axes are offset from the $(x_1, x_2)$ axes by an angle of $\theta$. For two-dimensional functions it will generally be convenient to refer to a projection by its angle $\theta$. A projection at angle $\theta$ will be interpreted to mean a projection onto the coordinate $u_1$, which is at an angle $\theta$ with $x_1$. Equivalently, then, the *projection axis* $u_2$ is at an angle $\theta$ to the coordinate axis $x_2$. Equation (15) corresponds to a projection at an angle $\theta = 0$ or equivalently with $x_2$ as the projection axis.

### D. The Projection-Slice Theorem

The projection-slice theorem relates the $(N-1)$-dimensional Fourier transforms of the projections to the $N$-dimensional Fourier transform of the original function. Basically, the theorem states that the $(N-1)$-dimensional Fourier transform of a projection is a "slice" through the $N$-dimensional Fourier transform of $f(\boldsymbol{x})$.

First, let us consider a projection for which the projection axis is one of the coordinate axes of $f(\boldsymbol{x})$, for example, $x_1$. Then $p_{x_1}(x_2, \cdots, x_N)$ is given by

$$p_{x_1}(x_2, \cdots, x_N) = \int_{-\infty}^{+\infty} f(\boldsymbol{x})\, dx_1 \tag{17}$$

and its $(N-1)$-dimensional Fourier transform is given by

$$P_{x_1}(\omega_2, \cdots, \omega_N) = \int_{-\infty}^{+\infty} \cdots \int_{-\infty}^{+\infty} p_{x_1}(x_2, \cdots, x_N) \cdot \exp\left[-j(\omega_2 x_2 + \cdots + \omega_N x_N)\right]. \tag{18}$$

Comparing (18) and (1), we see that

$$P_{x_1}(\omega_2, \cdots, \omega_N) = F(\omega_1, \omega_2, \cdots, \omega_N)\big|_{\omega_1 = 0}. \tag{19}$$

In other words, $P_{x_1}(\omega_2, \cdots, \omega_N)$ is a "slice" of $F(\omega_1, \omega_2, \cdots, \omega_N)$ defined by $\omega_1 = 0$. Clearly, a projection whose axis is any coordinate axis $x_i$ has a Fourier transform that is a slice of $F(\omega_1, \omega_2, \cdots, \omega_N)$ defined by $\omega_i = 0$.

A general projection was defined in (16) where $A$ is an orthogonal transformation. It was argued previously that if $F(\boldsymbol{\omega})$ is the Fourier transform of $f(\boldsymbol{x})$ then $F(\boldsymbol{\Omega})$ is the Fourier transform of $f(\boldsymbol{u})$ where

$$\boldsymbol{x} = \boldsymbol{u}A$$

$$\boldsymbol{\omega} = \boldsymbol{\Omega}A. \tag{20}$$

From this, (19) is easily generalized to state that a projection for which the projection axis is the transformed coordinate $u_i$ has an $(N-1)$-dimensional Fourier transform which is a slice of $F(\boldsymbol{\Omega})$ for $\Omega_i = 0$ where the coordinate systems $\boldsymbol{u}$ and $\boldsymbol{\Omega}$ are related to the coordinate systems $\boldsymbol{x}$ and $\boldsymbol{\omega}$ by the same orthogonal transformation. In two dimensions, for example, the projection-slice theorem states that the one-dimensional Fourier transform of a projection at an angle $\theta$ is a slice at the same angle of the two-dimensional Fourier transform of the original object. This relationship is depicted in Fig. 1.

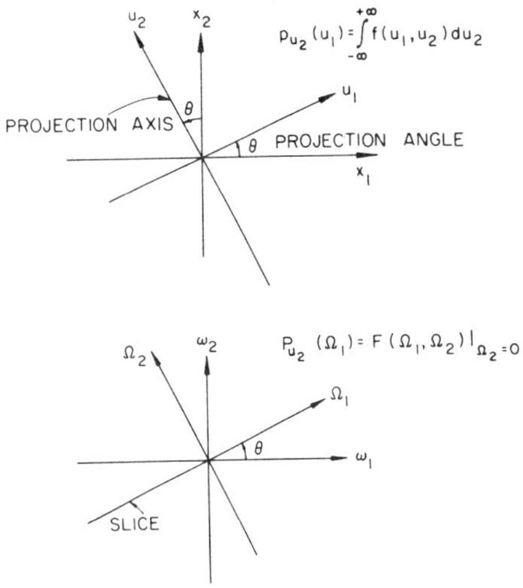


Fig. 1. The relationship between the projection of a two-dimensional function and the slice of its Fourier transform.

### E. The Basis for Reconstruction from Projections

From the projection-slice theorem, we see that specification of a projection in signal space corresponds to the specification of a slice in Fourier space and thus represents a partial specification of the signal itself. In principle, then, if an unlimited number of projections at different orientations are available, the Fourier transform of $f(\boldsymbol{x})$ can be obtained and therefore so can $f(\boldsymbol{x})$ itself. Generally, in any practical context, we are restricted to a finite number of projections.

Under certain assumptions, it is possible to carry out exact reconstruction from a finite number of projections. If the structure is highly symmetrical, for example, a finite number of projections might suffice for exact reconstruction. For example, for a two-dimensional circularly symmetric function all of its projections are identical and consequently such a function can be represented exactly by a single projection. Similarly, in three dimensions, for an object which is cylindrically symmetric all of the projections for which the projection axis is normal to the longitudinal axis are identical and consequently, in this case also, a single projection is sufficient. Helical objects can also be often reconstructed from a single projection.

In Section III we shall consider another class of functions that can be represented by a finite number of projections.

In utilizing projections for reconstruction, many of the algorithms involve computing the Fourier transforms of the projections. The Fourier transform of each projection is a function of a set of continuous variables, but only a finite number of points from each Fourier transform can be computed and stored. Thus from the projections, only samples in the Fourier domain are available, in part because of the limited number of projections and in part because only samples of the Fourier transform on each slice can be obtained. The essence of the reconstruction problem, then, is to approximate all of Fourier space from its values on a discrete point set.

### *F. Reduction of the Dimensionality of the Reconstruction Problem*

As we saw in Section II-E, the underlying basis for reconstruction is to obtain samples in the Fourier plane by transforming projections. Intuitively it seems reasonable that projections need not be taken in all orientations. For three-dimensional objects, for example, we could imagine using only projections in the spatial domain on planes parallel to one of the coordinate axes, say $x_1$. Slices of these projections at $x_1 = A$ are then projections of the two-dimensional function $f(A, x_2, x_3)$ and, consequently, a two-dimensional reconstruction algorithm can be applied to reconstructing this two-dimensional slice of the three-dimensional object. In this way, the three-dimensional object can be built up slice by slice and, consequently, the three-dimensional problem can be reduced to a series of two-dimensional problems. In the general case, we can apply a similar argument to reduce an $N$-dimensional problem to a set of $(N-1)$-dimensional problems each of which can in principle be reduced to an $(N-1)$-dimensional problem, etc.

Thus in principle, an $N$-dimensional problem can be reduced to a set of two-dimensional problems. Often this procedure requires considerably less storage and is simpler computationally than attacking the $N$-dimensional problem directly. Furthermore, in many cases, we may be content with very coarse sampling in one or several dimensions. For the three-dimensional problem, for example, reconstruction of only a few slices may be sufficient.

In the next sections, we discuss a number of reconstruction theorems and algorithms. The discussion will be phrased in terms of reconstructing two-dimensional functions from their one-dimensional projections. This is motivated by a number of considerations. Notationally and mathematically, the two-dimensional problem is easier to express and all of the results presented easily generalize to the $N$-dimensional case. Furthermore, as we discussed above, an $N$-dimensional problem can be reduced to a set of two-dimensional reconstruction problems and for most practical examples, this is an efficient procedure computationally.

## III. Exact Reconstruction from a Finite Number of Projections

The exact reconstruction of an arbitrary signal requires an infinite number of projections, since an infinite number of slices are required to encompass all of Fourier space. If, however, the unknown signal is known to have some underlying mathematical form, then exact reconstructions can often be performed from a limited set of projections. For example, as indicated previously, if the unknown is circularly symmetric then all of its projections are necessarily identical and $f(x_1, x_2)$ can be recovered from any one of them. Similarly, if $f(x_1, x_2)$ is separable so that it is of the form $f(x_1, x_2) = g(x_1)h(x_2)$ then it can be reconstructed from the two projections taken normal to the coordinate axes. This follows from the fact that in this case $p_{x_1}(x_2)$ is proportional to $h(x_2)$ and $p_{x_2}(x_1)$ is proportional to $g(x_1)$. The classes of circularly symmetric or separable functions do not represent very general classes of signals. There is, however, a class of functions that are somewhat more general and that can, in principle, be reconstructed from a finite number of projections. This is the class of band-limited functions of finite order, i.e., the class of functions that can be specified by a finite number of Cartesian samples. This class is very general, as evidenced by the fact that most signals processed digitally are assumed to be of this class since they are represented digitally by a finite number of samples. In this section, we consider the possibility of exact reconstruction from a finite number of projections for this class of signals. While the results do not lead to practical reconstruction algorithms they emphasize the underlying structure in this class of problems.

Throughout the discussion we will assume that the unknown signal can be represented by a square raster of samples $M$ points on a side. We could, of course, assume with more generality that we have a rectangular raster of $M_1$ samples in one dimension and $M_2$ samples in the orthogonal dimension. None of our results would be substantially changed although the introduction of an additional parameter would have made the expressions slightly more complex. Seeing no particular reason (theoretical or practical) for introducing this additional parameter, we have chosen not to do so.

### *A. The One-Projection Theorem [12]*

An important property of band-limited functions of finite order is that they can be represented by a single projection, as discussed below. This theorem will be referred to as the one-projection theorem.

*Theorem (One-Projection Theorem):* Let $A$ and $B$ denote two positive or negative integers and $m$, $m'$, $n$, and $n'$ integers in the range $0 \leq m, n, m', n' \leq M-1$. A band-limited function $f(x_1, x_2)$ of bandwidth $W$ and order $M$ can be represented exactly by a single projection at angle $\theta = \tan^{-1}(A/B)$ if $A$ and $B$ have no common factors and if $Bm + An = Bm' + An'$ only for $m = m'$ and $n = n'$.

The angle $\theta$ for such a projection will be referred to as a critical angle and the corresponding slice of $F(\omega_1, \omega_2)$ will be referred to as a critical slice.

*Proof:* Since the function is band-limited, it is sufficient to determine its samples on a Cartesian raster, i.e., to determine the samples $f((m\pi/W), (n\pi/W))$ for $0 \leq (m, n) \leq M-1$. From (7), the Fourier transform $F(\omega_1, \omega_2)$ of $f(x_1, x_2)$ is expressed in terms of these samples as

$$F(\omega_1, \omega_2) = \frac{\pi^2}{W^2} \sum_{m=0}^{M-1} \sum_{n=0}^{M-1} f\left(\frac{m\pi}{W}, \frac{n\pi}{W}\right) \cdot \exp\left[-j\frac{\pi}{W}(m\omega_1 + n\omega_2)\right] b_W(\omega) \quad (21)$$

where

$$b_W(\omega) = \begin{cases} 1, & |\omega| \leq \dfrac{W\sqrt{A^2 + B^2}}{\max[|A|, |B|]} \\ 0, & \text{otherwise.} \end{cases} \quad (22)$$

Now consider a slice of $F(\omega_1, \omega_2)$ denoted by $S_\theta(\omega)$, at angle $\theta$, corresponding to the one-dimensional Fourier transform of the projection at that angle. With $\omega$ representing frequency along the slice, the slice is specified by

$$\omega_1 = \omega \cos\theta$$

$$\omega_2 = \omega \sin\theta$$

so that

$$S_\theta(\omega) = F(\omega\cos\theta, \omega\sin\theta). \quad (23)$$

Thus with

$$\theta = \tan^{-1} \frac{A}{B}$$

$$S_\theta(\omega) = \frac{\pi^2}{W^2} \sum_{m=0}^{M-1} \sum_{n=0}^{M-1} f\left(\frac{m\pi}{W}, \frac{n\pi}{W}\right)$$

$$\cdot \exp\left[-j \frac{\pi}{W} \frac{\omega}{\sqrt{A^2+B^2}} (Bm + An)\right] b_W(\omega). \quad (24)$$

We observe that for

$$|\omega| \leqslant \frac{W\sqrt{A^2+B^2}}{\max[|A|, |B|]}$$

(24) corresponds to a one-dimensional polynomial of the form

$$p(z) = a_Q z^Q + a_{Q-1} z^{Q-1} + \cdots + a_R z^R \quad (25)$$

where

$$z = \exp\left[-j \frac{\pi\omega}{W\sqrt{A^2+B^2}}\right] \quad (26)$$

$$Q = \max(Bm + An), \qquad 0 \leqslant m, n \leqslant M - 1$$

and

$$R = \min(Bm + An), \qquad 0 \leqslant m, n \leqslant M - 1.$$

$R$ can, and in general will, be negative. The polynomial $p(z)$ is specified by $[(|A| + |B|)(M - 1) + 1]$ coefficients or equivalently by $[(|A| + |B|)(M - 1) + 1]$ samples. Thus it will be convenient to refer to $p(z)$ as a polynomial of degree $[(|A| + |B|)(M - 1)]$.

It is clear from (24) that $p(z)$, or equivalently $S_\theta(\omega)$, has at most $M^2$ nonzero coefficients. Under the stated conditions on $A$ and $B$, we note that the coefficients of the polynomial $p(z)$ are the function samples $f((m\pi/W), (n\pi/W))$. If the slice $S_\theta(\omega)$ is completely specified then all of its coefficients, and hence the function samples, can be found. Consequently, the stated conditions represent sufficient conditions for $f(x_1, x_2)$ to be represented exactly by a single projection. If the stated conditions are not met then at least one of the coefficients in $p(z)$ will consist of a sum of two or more function samples and the number of nonzero coefficients in $p(z)$ will be reduced accordingly. In this case then, the function samples cannot be uniquely recovered from the polynomial coefficients. This completes the proof of the theorem.

At this point, a comment should be made on the degrees of the polynomials corresponding to the different critical slices. The preceding theorem has shown that there are an infinite number of critical slices. Each corresponds to a polynomial of degree at least $M^2 - 1$, since there are $M^2$ nonzero function samples and thus $M^2$ nonzero coefficients that each such polynomial must contain. There is no maximum degree for these polynomials. For example, consider the slice at $\theta = \tan^{-1} 1/2M$. From our theorem this is seen to be a critical slice and it corresponds to a polynomial of degree $2M^2 - M - 1$. Of the $2M^2 - M$ coefficients that it contains, however, only $M^2$ are nonzero; $M^2 - M$ of the coefficients are identically zero. On the other hand, the slice at angle $\theta = \tan^{-1} 1/M$ corresponds to a polynomial of degree $M^2 - 1$. Since there are $M^2$ function samples, no critical slice can be a polynomial of degree less than $M^2 - 1$ and thus $\theta = \tan^{-1} 1/M$ represents an optimum in the sense that of all the critical slices this one is of minimum degree.

To obtain the function samples from the slice polynomial we must first obtain a sufficient number of samples of $S_\theta(\omega)$, the number of samples required being one plus the degree of the polynomial. The samples can be obtained by sampling the projection above its Nyquist rate and performing the appropriate DFT computation. For example, when $\theta = \tan^{-1} 1/M$, $M^2$ such samples are needed. When the slice samples are obtained from the DFT, they will be equally spaced along the slice and they will extend along the entire nonzero frequency band. Since for this example the width of this band is $2W\sqrt{M^2+1}/M$, the frequency spacing between adjacent samples is

$$\Delta\omega = \frac{2W}{M^3} \sqrt{M^2 + 1}. \quad (27)$$

Thus substituting in (24), the values of the $M^2$ samples are

$$S_{\theta_0}(k\Delta\omega) = \frac{\pi^2}{W^2} \sum_{m=0}^{M-1} \sum_{n=0}^{M-1} f\left(\frac{m\pi}{W}, \frac{n\pi}{W}\right)$$

$$\cdot \exp\left[-j \frac{2\pi k}{M^3}(Mm + n)\right], \qquad k = -\frac{M^2}{2} + 1, \cdots, 0, 1, \cdots, \frac{M^2}{2}. \quad (28)$$

For notational simplicity, let us define

$$g(mM + n) = f\left(\frac{m\pi}{W}, \frac{n\pi}{W}\right).$$

This mapping of a two-dimensional sequence to a one-dimensional sequence is unambiguous because of the limited ranges of $m$ and $n$. Then setting $p = mM + n$, (28) can be expressed as

$$S_{\theta_0}(k\Delta\omega) = \frac{\pi^2}{W^2} \sum_{p=0}^{M^2-1} g(p) \exp\left[-j \frac{2\pi k p}{M^3}\right],$$

$$k = -\frac{M^2}{2} + 1, \cdots, \frac{M^2}{2}. \quad (29)$$

Equation (28) or equivalently (29) is in principle invertible so that the $M^2$ function samples $g(p)$ can be computed from $M^2$ samples of the critical slice. In practice, however, this set of equations is nearly impossible to invert. In particular, assume that the slice samples $S_{\theta_0}(k\Delta\omega)$ are not known exactly due to measurement errors, quantization effects, model inaccuracies, etc. If the observed slice samples are $S_{\theta_0}(k\Delta\omega) + \epsilon(k)$ where $\epsilon(k)$ is a random sequence with zero mean and for which

$$E[\epsilon(k)\epsilon(l)] = 0, \qquad k \neq l \quad (30)$$

$$E[\epsilon^2(k)] = \sigma^2 \quad (31)$$

where $E[\cdot]$ denotes the expected value, then with $\delta(p)$ denoting the error in $g(p)$ due to the error $\epsilon(k)$, it has been shown [12] that $\delta(p)$ has zero mean and

$$E[\delta^2(p)] \simeq \frac{\sigma^2}{M^4} \left(\frac{2Me}{\pi}\right)^{2M^2-2}. \quad (32)$$

For $M = 256$, $\log[\delta^2/\sigma^2]$ is more than $10^6$ bits, indicating that the inversion of (29) is extremely ill-conditioned for any but very small values of $M$.

The problems involved in reconstructing band-limited functions of order $M$ are computational ones and are due to the high orders of the polynomials involved. This is unavoidable, since to recover all $M^2$ samples from a single projection, it is necessary to work with a polynomial of degree greater than $M^2 - 1$.[1] If, however, we are willing to use more than one projection, then we can work with lower degree polynomials. This procedure can perhaps be best explained in terms of an example.

Let us consider a band-limited picture of order 4 which we wish to reconstruct from the two projections at $\theta_1 = \tan^{-1} \frac{1}{2}$ and $\theta_2 = \tan^{-1} -\frac{1}{2}$. Since each of these slices has a rational slope, each corresponds to a polynomial of finite degree. In this case the degree is 9 and thus

$$\begin{aligned} S_{\theta_1}(k\Delta\omega) &= f(0,0) + f(0,1)E_k + [f(0,2) + f(1,0)]\,E_k^2 \\ &\quad + [f(0,3) + f(1,1)]\,E_k^3 + [f(1,2) \\ &\quad + f(2,0)]\,E_k^4 + [f(1,3) + f(2,1)]\,E_k^5 \\ &\quad + [f(2,2) + f(3,0)]\,E_k^6 + [f(2,3) \\ &\quad + f(3,1)]\,E_k^7 + f(3,2)E_k^8 + f(3,3)E_k^9 \\ &= \sum_{i=0}^{9} a_i E_k^i \end{aligned} \tag{33}$$

where

$$E_k = \exp\left[-j\,\frac{2\pi k}{20}\right].$$

Similarly we can write

$$\begin{aligned} S_{\theta_2}(k\Delta\omega) &= f(0,3)E_k^{-3} + f(0,2)E_k^{-2} + f[(0,1) + f(1,3)]\,E_k^{-1} \\ &\quad + [f(0,0) + f(1,2)] + [f(2,3) + f(1,1)]\,E_k \\ &\quad + [f(2,2) + f(1,0)]\,E_k^2 + [f(2,1) + f(3,3)]\,E_k^3 \\ &\quad + [f(2,0) + f(3,2)]\,E_k^4 + f(3,1)E_k^5 + f(3,0)E_k^6 \\ &= \sum_{i=0}^{9} b_i E_k^{i-3}. \end{aligned} \tag{34}$$

If $S_{\theta_1}(k\Delta\omega)$ and $S_{\theta_2}(k\Delta\omega)$ are each available for 10 different evenly spaced values of $k$ then we can obtain the two sequences $a_i$ and $b_i$. From these we can compute the function samples by

$$\begin{array}{llll} f(0,0) = a_0 & f(1,0) = a_2 - b_1 & f(2,0) = b_7 - a_8 & f(3,0) = b_9 \\ f(0,1) = a_1 & f(1,1) = a_3 - b_0 & f(2,1) = b_6 - a_9 & f(3,1) = b_8 \\ f(0,2) = b_1 & f(1,2) = b_3 - a_0 & f(2,2) = a_6 - b_9 & f(3,2) = a_8 \\ f(0,3) = b_0 & f(1,3) = b_2 - a_1 & f(2,3) = a_7 - b_8 & f(3,3) = a_9. \end{array}$$

The technique for performing a reconstruction from more than one projection, following this approach, can be summarized in the following steps:

1) find a sufficient set of slice angles;
2) sample the slices at $d + 1$ points, where $d$ is the degree of the polynomial along each slice, and compute the coefficients corresponding to these polynomials;
3) set up a system of linear equations relating the polynomial coefficients to the function samples;
4) solve this set of equations to obtain the Nyquist samples of the unknown;
5) use the band-limited interpolation formula (9) and (10) to recover $f(x, y)$ from its Nyquist samples.

None of these steps is particularly simple. In our example, we reconstructed from the slices at $\theta_1 = \tan^{-1} \frac{1}{2}$ and $\theta_2 = \tan^{-1} -\frac{1}{2}$. Had we instead chosen to work with the slices at $\theta_1 = \tan^{-1} \frac{1}{2}$ and $\theta_2 = \tan^{-1} 2$, which are of the same degrees, we would not have been able to solve the resulting linear equations. We must, therefore, be careful when we choose a set of slice angles to insure that the resulting equations are solvable. This is difficult to do for large $M$ since the number of equations and the number of variables grow as $M^2$. It can be shown that the projections at $\theta_1 = \tan^{-1} 2/M$ and $\theta_2 = \tan^{-1} -2/M$ are sufficient to reconstruct a function of order $M$, although from (32) we see that for $M > 4$ it is difficult to determine the coefficients of the resulting polynomials. It can also be shown that an $8 \times 8$ array can be reconstructed from the four projections $\theta_1 = \tan^{-1} \frac{1}{2}$, $\theta_2 = \tan^{-1} 2$, $\theta_3 = \tan^{-1} -\frac{1}{2}$, $\theta_4 = \tan^{-1} -2$. In all of these cases the resulting linear equations are straightforward to compute.

From the orders of the polynomials involved and (32), we can perhaps infer that to reconstruct an $M \times M$ picture will require on the order of $M/2$ projections by these techniques, due to the sensitivity problems, alluded to earlier, associated with the coefficient calculations. Since for large values of $M$ this requires formidable matrix manipulations, it is perhaps reasonable to consider alternative reconstruction procedures which, using simpler algorithms, will produce good quality reconstructions from roughly the same number of projections. This we will do in the next section. Before doing so, however, we wish to close this section with a discussion of a somewhat different procedure for reconstructing exactly a band-limited function of finite order $M$ from a set of $M + 1$ projections. While this procedure is also somewhat difficult to implement exactly, it forms the basis for an approximate procedure to be discussed in Section V that appears to be among the most successful.

To develop the exact reconstruction procedure we again consider $f(x,y)$ to be band-limited with bandwidth $W$ and of finite order $M$. Then

$$F(\omega_1, \omega_2) = \frac{\pi}{W^2} \sum_{m=0}^{M-1} \sum_{n=0}^{M-1} f\left(\frac{m\pi}{W}, \frac{n\pi}{W}\right) z_1^m z_2^n\, b_W(\omega_1, \omega_2)$$

where

$$z_1 = \exp\left(-j\,\frac{\pi}{W}\,\omega_1\right) \quad \text{and} \quad z_2 = \exp\left(-j\,\frac{\pi}{W}\,\omega_2\right).$$

Thus over the region of support ($|\omega_1| \leqslant W$, $|\omega_2| \leqslant W$), $F(\omega_1, \omega_2)$ corresponds to a two-dimensional polynomial of degree $M - 1$ in each of the complex variables $z_1$ and $z_2$. Along horizontal ($\omega_1$ constant) or vertical ($\omega_2$ constant) lines in the Fourier plane, $F(\omega_1, \omega_2)$ varies as a one-dimensional polynomial of degree $M - 1$ in either $z_1$ or $z_2$. Therefore, if $F(\omega_1, \omega_2)$ is known at any $M$ points along any horizontal or vertical line in the Fourier plane (subject to the constraint that the points all must lie within the region of support) it is known at all points on that line. Now consider a slice at an arbitrary angle $\theta$ in the Fourier plane. Since $f(x_1, x_2)$ is constrained to be band-limited, $S_\theta(\omega) = 0$ for $|\omega| \geqslant W_\theta$ where

$$W_\theta = \frac{W}{\max\{|\cos\theta|, |\sin\theta|\}} \tag{35}$$

[1] A similar result was obtained by Smith *et al.* [15] for a slightly different class of functions.

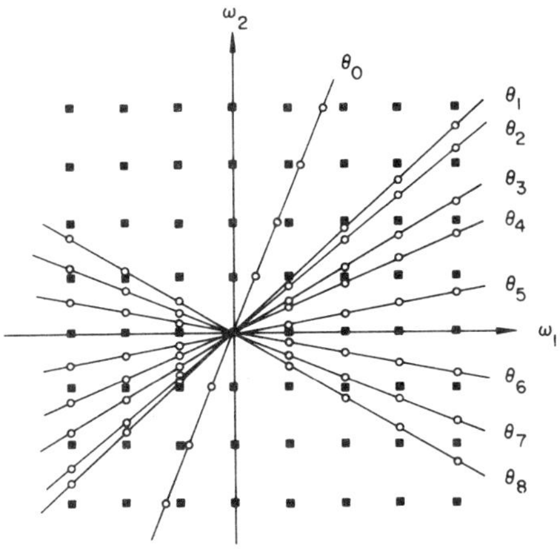


Fig. 2. The set of Fourier plane samples from which an 8 by 8 picture can be reconstructed exactly under the assumption that the picture is band-limited and of order 8.

is the bandwidth of the projection at this particular angle. Thus we can expand the projection in terms of its samples according to (36).

$$p_\theta(u) = \sum_{n=-\infty}^{\infty} p_\theta\left(\frac{n\pi}{W_\theta}\right) \frac{\sin \frac{W_\theta}{\pi}\left(u - \frac{n\pi}{W_\theta}\right)}{\frac{W_\theta}{\pi}\left(u - \frac{n\pi}{W_\theta}\right)}. \tag{36}$$

Now suppose that we limit $\theta$ to fall in the range $-(\pi/4) \leqslant \theta \leqslant (\pi/4)$ and we evaluate the value of the slice function at $M$ equally spaced points from the sampled projection. Thus (36) becomes

$$S_\theta\left(\frac{2Wk}{M\cos\theta}\right) = \frac{\pi\cos\theta}{W} \sum_{n=-\infty}^{\infty} p_\theta\left(\frac{n\pi\cos\theta}{W}\right) \cdot \exp\left[-j\frac{2\pi}{M}nk\right], \quad k = -\frac{M}{2}+1, \cdots, \frac{M}{2}. \tag{37}$$

If this is done for $M$ different angles in the range $-(\pi/4) \leqslant \theta_i \leqslant (\pi/4)$ then $F(\omega_1, \omega_2)$ will be known at an array of sample locations such as that illustrated in Fig. 2 for $M = 8$. Included in the same figure are locations for the $M^2$ DFT samples of $f(x_1, x_2)$. It should be noted that $M$ vertical lines can be drawn through the slice samples and that the DFT samples lie along these same vertical lines. Therefore, from our earlier statement, it is seen that all of the DFT samples, except for those which lie along the line $\omega_x = 0$, are completely determined from the raster of slice samples. If, however, in addition we know $M$ slice samples of a slice whose angle falls outside the range $[-(\pi/4), (\pi/4)]$, denoted by $\theta_0$ in Fig. 2, then by interpolating along horizontal lines in the Fourier plane we can determine the DFT samples along the line $\omega_1 = 0$, since along each horizontal line we now know $F(\omega_1, \omega_2)$ at $M$ points: $M - 1$ from the previously computed DFT points and 1 from this last projection. Thus all $M^2$ samples of $F(\omega_1, \omega_2)$ that correspond to its DFT can be determined from $M + 1$ projections and these samples, in turn, completely specify $F(x_1, x_2)$. This procedure for exact reconstruction basically involves one-dimensional polynomial interpolation. In Section V we shall consider an approximation of this interpolation procedure. First, however, we consider in the next section approximate reconstruction from a polar raster.

## IV. Approximate Reconstructions from a Polar Raster

Since specification of a projection in signal space corresponds to specification of a slice in Fourier space, if all the projections for a continuous range of angles $0 \leqslant \theta \leqslant \pi$ are known, the entire Fourier space is swept out and consequently the function is known exactly. Conversely, if only a finite number of projections are known, an arbitrary function cannot be specified exactly, since its Fourier transform is constrained only on a finite set of slice lines and is elsewhere merely constrained to be infinitely differentiable. If $F(\omega_1, \omega_2)$ is expressed in polar coordinates as $\mathcal{F}(\omega, \theta)$ then, substituting into the inverse Fourier transform integral,

$$f(x_1, x_2) = \frac{1}{(2\pi)^2} \int_{-\infty}^{+\infty} \int_0^\pi \mathcal{F}(\omega, \theta) \cdot \exp\left[j(x_1\omega\cos\theta + x_2\omega\sin\theta)\right] |\omega|\, d\omega d\theta \tag{38}$$

or, since $\mathcal{F}(\omega, \theta)$ for fixed $\theta$ is the slice of $F(\omega_1, \omega_2)$ at angle $\theta$,

$$f(x_1, x_2) = \frac{1}{(2\pi)^2} \int_{-\infty}^{+\infty} \int_0^\pi S_\theta(\omega) \cdot \exp\left[j(x_1\omega\cos\theta + x_2\omega\sin\theta)\right] |\omega|\, d\omega d\theta. \tag{39}$$

Equation (39) tells us how $f(x_1, x_2)$ can be reconstructed from its projections. In general, however, to apply (39), an infinite number of slices are needed, while in practice only a finite number will be available. Thus in general, (39) must be approximated from the values of $S_\theta(\omega)$ at only a finite number of discrete angles. Intuitively, one would expect that as the number of projections used for the reconstruction increases, the accuracy of the reconstruction, or more specifically, the resolution of detail in the reconstructed image will increase. This is basically a consequence of the fact that the slices $S_\theta(\omega)$ sample the two-dimensional Fourier transform at only a finite number of discrete angles. Some attempts to quantify the relationship between the number of projections and the resulting resolution available in the reconstruction are discussed in [2], [16]–[18].

In addition to requiring an infinite number of slices, (39) also requires that each slice be specified for all values of $\omega$. This is, in general, impossible to do because the Fourier transforms which must be computed to determine the slices of $F(\omega_1, \omega_2)$ must be computed digitally. If the projections are all band-limited, which is the case if $f(x_1, x_2)$ is band-limited because of the projection-slice theorem, then samples of $S_\theta(\omega)$ can be computed by applying the DFT to samples of the projection $p_\theta(u)$. With a finite set of equally spaced values of $\theta$, and with samples of $S_\theta(\omega)$ equally spaced in $\omega$ computed from the projections, we obtain a set of samples of $\mathcal{F}(\omega, \theta)$ on a regular polar raster as indicated in Fig. 3. The first set of approximate algorithms to be considered will be those utilizing such a set of samples, although they are not all implemented in the Fourier plane.

Before addressing the problem of reconstructing from this raster, however, let us briefly consider some issues associated

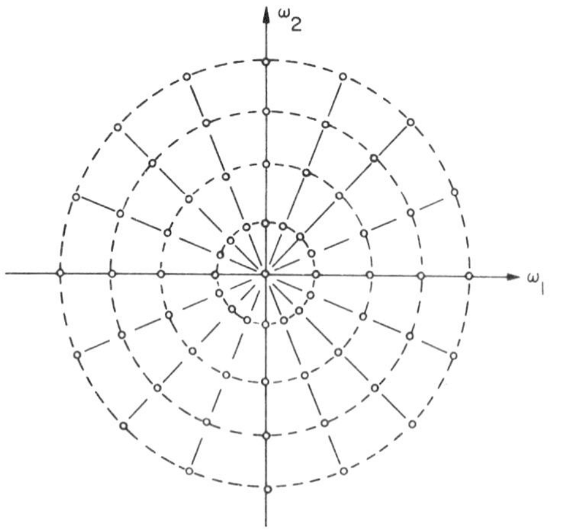


Fig. 3. Polar raster indicating the set of points at which a two-dimensional Fourier transform will be known from 8 projections with the Fourier transforms computed digitally.

with obtaining these polar samples. The radius of the polar raster is, of necessity, finite and thus only a limited portion of the Fourier plane can be directly sampled. It thus follows that if $f(x_1, x_2)$ can be reconstructed from such a raster then the remainder of the Fourier plane either is unimportant in specifying $f(x_1, x_2)$ or is completely determined by the samples in the raster. We shall assume the former situation—that $f(x_1, x_2)$ is either band-limited or very nearly band-limited and that the polar raster samples the entire nonzero frequency region. Therefore, we shall assume that $F(\omega_1, \omega_2)$ is nonzero only over a bounded region of the Fourier plane that can be enclosed by a circle of radius $\sqrt{2}\,W$. Each projection then corresponds to a band-limited one-dimensional function with bandwidth less than $\sqrt{2}\,W$ so that if each is periodically sampled with an intersample distance less than $\pi/\sqrt{2}\,W$ (meters) no information will be lost.

Each sampled projection represents a sequence of numbers of infinite duration. We can obtain $M$ samples of the slice corresponding to a particular projection by converting that infinitely long sequence to an $M$ point sequence by aliasing. Thus if

$$\left\{p_\theta\left(\frac{\pi}{\sqrt{2}\,W}\,n\right),\quad n = -\infty, \cdots, -1, 0, 1, \cdots, \infty\right\}$$

represents the infinitely long sequence and we define

$$\tilde{p}_\theta(n) = \sum_{m=-\infty}^{\infty} p_\theta\left(\frac{\pi}{\sqrt{2}\,W}(Mm + n)\right), \qquad n = 0, 1, \cdots, M-1 \tag{40}$$

then $\tilde{p}_\theta(n)$ is a sequence of length $M$ whose DFT corresponds to evenly spaced samples of the Fourier transform of $p_\theta(u)$. In practice, of course, the sum in (40) must be performed using finite limits. If this procedure is followed for each of the $N$ projections that are evenly spaced in angle over the range 0 to $\pi$, then we will know the Fourier transform $F(\omega_1, \omega_2)$ on a regular polar raster such as that in Fig. 3.

If $f(x_1, x_2)$ is band-limited and of order $M$ then it is completely specified by its $M \times M$ point DFT. If it is not of finite order, then it can hopefully be approximately determined by its DFT. Since the DFT corresponds to samples of $F(\omega_1, \omega_2)$ on a Cartesian raster, one possible reconstruction algorithm is to interpolate from the known samples on a polar raster to estimate the samples on a Cartesian raster, perform an inverse DFT, and use the result as an estimate of samples on a Cartesian raster of the unknown. This is the basis for the first set of approximate algorithms that we shall consider.

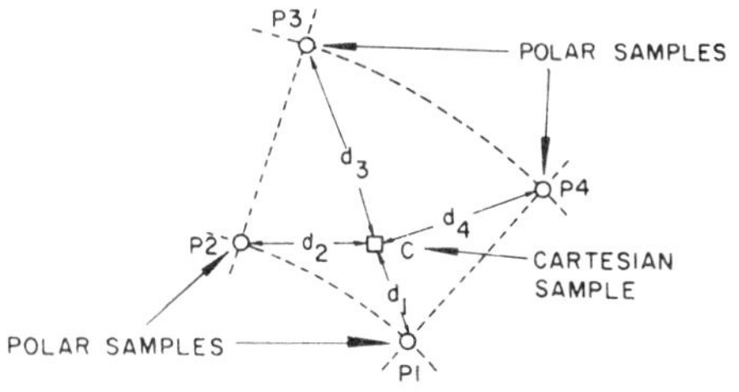


Fig. 4. Parameters for the definition of zeroth-order and linear interpolation.

To perform the necessary interpolation we may consider the use of a simple form of polynomial interpolation. Two interpolation techniques have been used—zeroth-order and linear interpolation. Each of these approximations is shown in Fig. 4. With zeroth-order interpolation, each Cartesian (DFT) sample is assigned the value of the nearest polar sample, whereas with linear interpolation it is assigned a weighted average of the four nearest polar values, the weighting varying inversely with the Euclidean distance between the Cartesian sample and the polar sample in question. In particular, for the Cartesian sample $c$ indicated in Fig. 4, with $d_1 = \min\,[d_1, d_2, d_3, d_4]$, for zeroth-order interpolation we choose $F(c) = F(P1)$. With linear interpolation, $F(c)$ is computed as

$$F(c) = \frac{\dfrac{1}{d_1}F(P1) + \dfrac{1}{d_2}F(P2) + \dfrac{1}{d_3}F(P3) + \dfrac{1}{d_4}F(P4)}{\dfrac{1}{d_1} + \dfrac{1}{d_2} + \dfrac{1}{d_3} + \dfrac{1}{d_4}}. \tag{41}$$

The reconstruction algorithm can thus be summarized in the following steps:

1) Compute the Fourier transform of the sampled projections using a one-dimensional DFT to obtain a collection of sampled slices of $F(\omega_1, \omega_2)$ which represents a polar sampling of $F(\omega_1, \omega_2)$.

2) Use zeroth-order or linear interpolation to estimate the Cartesian samples of $F(\omega_1, \omega_2)$ which correspond to the DFT of a two-dimensional sequence

$$\left\{f\left(\frac{m\pi}{W}, \frac{n\pi}{W}\right), \qquad 0 \leqslant m, n \leqslant M-1\right\}.$$

3) Estimate

$$\left\{f\left(\frac{m\pi}{W}, \frac{n\pi}{W}\right), \qquad 0 \leqslant m, n \leqslant M-1\right\}$$

using a two-dimensional inverse DFT.

4) Approximate $f(x_1, x_2)$ as

$$\hat{f}(x_1, x_2) = \sum_{m=0}^{M-1}\sum_{n=0}^{M-1} f\left(\frac{m\pi}{W}, \frac{n\pi}{W}\right) \cdot \frac{\sin\dfrac{W}{\pi}\left(x - \dfrac{m\pi}{W}\right)\sin\dfrac{W}{\pi}\left(y - \dfrac{n\pi}{W}\right)}{\dfrac{W^2}{\pi^2}\left(x - \dfrac{m\pi}{W}\right)\left(y - \dfrac{n\pi}{W}\right)}. \tag{42}$$

Fig. 5. Reconstruction of the pictures of Fig. 7 from their projections using zeroth-order interpolation from a polar raster.

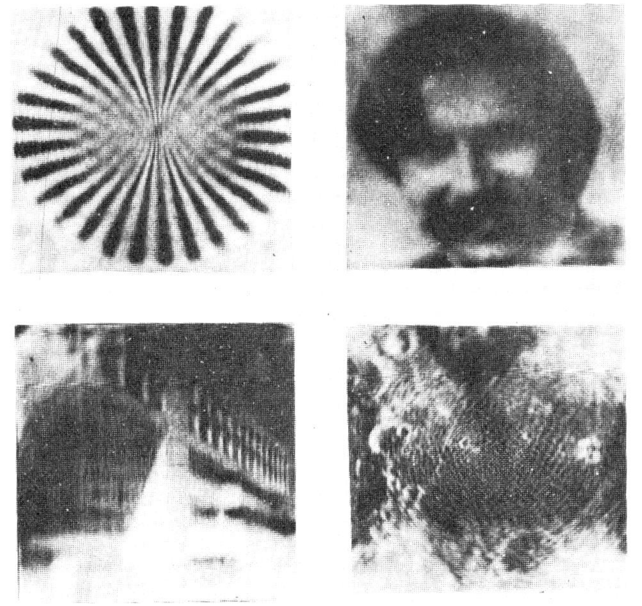

Fig. 6. Reconstruction of the pictures of Fig. 7 from their projections using linear interpolation from a polar raster.

For the interpolation to be effective, the Fourier transform must be reasonably smooth. Thus each projection should be considered as a two-sided sequence with its origin at the center. Choosing the origin at one of the end points of the projection introduces a linear phase component into the Fourier transform which will seriously impair the quality of the resulting reconstructions.

Figs. 5 and 6 present some reconstructions obtained by using the above technique with zeroth-order and linear interpolation, respectively.[2] Each of these reconstructions was made from 64 computer-generated projections equally spaced in angle, with each projection represented by 256 samples. The original pictures from which these projections were computed are included for comparison in Fig. 7. Discrepancies between the reconstructions and the originals are due to two distinct factors—errors made in computing the projections themselves, and errors introduced in performing the reconstructions. All of the reconstructions performed in this section were performed from the same set of projections and thus any comparison between different algorithms which reconstruct from a polar raster is dependent on only the latter type of error. In computing the projections it was assumed that the original photographs represented two-dimensional band-limited functions of order $M$, where the DFT computed in the reconstruction algorithms was of dimension $M \times M$.

[2] All of the reconstructions presented in this paper were performed on a Digital Equipment Corporation PDP-9 computer with 24 000 words of core memory using 18-bit fixed-point arithmetic.

Fig. 7. Four original pictures for which the reconstruction from projections is to be carried out using a variety of algorithms.

A cursory examination of Figs. 5 and 6 reveals that the quality of the final reconstruction is very much dependent upon the method of interpolation which was used, although neither set of reconstructions is as good, perhaps, as one might wish. The reconstructions performed using linear interpolation seem preferable to those performed using zeroth-order interpolation. This is certainly reasonable since if $F(\omega_1, \omega_2)$ is continuous and slowly varying, one would expect the error from a piecewise planar approximation to be less than that from a piecewise constant approximation.

One of the possible explanations for the errors that were introduced by these algorithms is that insufficient projections were taken for linear interpolation to approximate accurately the entire Fourier plane. Indeed, it was noted experimentally, as expected, that as the number of projections was increased, the quality of the resulting reconstructions improved. Taking an increasing number of projections is impossible, however, in many practical situations because this requires more X-rays, more electron micrographs (a specimen can be destroyed by too much electron bombardment), etc. Another possibility to improve reconstruction quality is to use higher order interpolation algorithms, but this involves a major computational effort and would significantly add to the total computation time required. A third alternative, which is closely related, would be to augment the set of projections with additional projections computed from the given ones by some computationally efficient scheme, and then from this larger set of projections the linear interpolation algorithm can then be applied as before. Were these extra projections computed from the original set by linear or zeroth-order interpolation, then our reconstructions would be identical to those of the previous section, but by using some other interpolation scheme different and (it is hoped) improved reconstructions might be obtained. This is the central idea behind an algorithm which we refer to as the polar expansion algorithm.

Let us assume that $F(\omega_1, \omega_2)$ can be expanded in the form

$$F(\omega \cos \theta, \omega \sin \theta) = \mathfrak{F}(\omega, \theta) = \sum_{m=-\infty}^{\infty} \sum_{n=-\infty}^{\infty} f_{mn} \Psi_{mn}(\omega, \theta) \tag{43}$$

where $\omega$ and $\theta$ are polar coordinate variables and where

$$\Psi_{mn}(\omega, \theta) = \exp\left[j\left(\frac{2\pi m \omega}{\sqrt{2}\, W} + n\theta\right)\right] b_W(\omega) \tag{44}$$

$$b_W(\omega) = \begin{cases} 1, & 0 \leqslant \omega < \sqrt{2}\, W \\ 0, & \text{otherwise.} \end{cases} \tag{45}$$

This choice is made because the set $\{\Psi_{mn}(\omega, \theta)\}$ is complete over the set of band-limited functions with bandwidth $\sqrt{2}\, W$ and a finite subset of the set should be capable of producing a good approximation to most unknowns.

In order to compute the coefficients $f_{mn}$ of (43) we require an infinite number of projections. If, however, we approximate $\mathfrak{F}(\omega, \theta)$ by a truncated expansion

$$\mathfrak{F}(\omega, \theta) \simeq \sum_{m=-M_1/2+1}^{M_1/2} \sum_{n=-M_2/2+1}^{M_2/2} f_{mn} \cdot \exp\left[j\left(\frac{2\pi m \omega}{\sqrt{2}\, W} + n\theta\right)\right] b_W(\omega) \tag{46}$$

the coefficients $f_{mn}$ can be computed efficiently from a finite number of projections using the DFT. In particular, we recognize that with $f_{mn}$ viewed as a two-dimensional Cartesian sequence, (46) corresponds to its two-dimensional Fourier transform. By inserting finite limits in (46) we assumed that $f_{mn}$ is of finite order. Consequently, the two-dimensional DFT of $f_{mn}$ corresponds to samples of $\mathfrak{F}(\omega, \theta)$ equally spaced in $\omega$ and equally spaced in $\theta$. Hence $f_{mn}$ can be obtained from these samples by means of the inverse DFT relation, i.e.,

$$f_{mn} = \frac{1}{M_1 M_2} \sum_{a=0}^{M_1-1} \sum_{b=0}^{M_2-1} \mathfrak{F}\left(\frac{\sqrt{2}\, W}{M_1} a, \frac{2\pi}{M_1} b\right) \cdot \exp\left[-j2\pi\left(\frac{ma}{M_1} + \frac{nb}{M_2}\right)\right] \tag{47}$$

where

$$\mathfrak{F}\left(\frac{\sqrt{2}\, W}{M_1} a, \frac{2\pi}{M_2} b\right), \quad a = 0, 1, \cdots, M - 1 \text{ and } b = 0, 1, \cdots, M_2 - 1$$

correspond to samples of $F(\omega_1, \omega_2)$ on a regular polar raster formed by taking the $2M_1 - 1$ point DFT's of $M_2/2$ evenly spaced projections. From the coefficients $\{f_{mn}\}$ we can calculate the values of $F(\omega_1, \omega_2)$ along any other slices we wish using the expansion of (46). In particular, to obtain samples of $F(\omega_1, \omega_2)$ on a polar raster with $M_4$ rays of $M_3$ samples each, (46) becomes

$$\mathfrak{F}\left(\frac{\sqrt{2}\, W}{M_3} k, \frac{2\pi}{M_4} l\right) = \sum_{m=-M_3/2+1}^{M_3/2} \sum_{n=-M_4/2+1}^{M_4/2} f_{mn} \cdot \exp\left[j\left(\frac{2\pi mk}{M_3} + \frac{2\pi nl}{M_4}\right)\right]. \tag{48}$$

Incorporating (47) we can thus arrive at a formula for interpolating the values of these extra slices. Specifically

$$\begin{aligned} \mathfrak{F}\left(\frac{\sqrt{2}\, W}{M_3} k, \frac{2\pi}{M_4} l\right) &= \sum_{a=0}^{M_1-1} \sum_{b=0}^{M_2-1} \mathfrak{F}\left(\frac{\sqrt{2}\, W}{M_1} a, \frac{2\pi}{M_2} b\right) \cdot \exp\left[-j\pi\left(\frac{k}{M_3} - \frac{a}{M_1} + \frac{l}{M_4} - \frac{b}{M_2}\right)\right] \\ &\cdot \frac{\sin \pi M_1 \left(\frac{k}{M_3} - \frac{a}{M_1}\right) \sin \pi M_2 \left(\frac{l}{M_4} - \frac{b}{M_2}\right)}{M_1 M_2 \sin \pi\left(\frac{k}{M_3} - \frac{a}{M_1}\right) \sin \pi\left(\frac{l}{M_4} - \frac{b}{M_2}\right)} PW(\omega), \\ &-\frac{M_3}{2} + 1 \leqslant k \leqslant \frac{M_3}{2} \quad -\frac{M_4}{2} + 1 \leqslant l \leqslant \frac{M_4}{2}. \end{aligned} \tag{49}$$

Fig. 8. Reconstruction of the pictures of Fig. 7 from their projections using the polar expansion algorithm.

Fig. 8 shows some reconstructions that were made from 64 original projections of 256 points each. Using the above techniques, 256 projections of 256 points apiece were then computed according to (48). Linear interpolation was then used to estimate the 256 X 256 DFT of the unknown, and an inverse 256 X 256 point DFT was performed. On the basis of these restricted examples it might be inferred that this algorithm yields slightly improved reconstructions compared with simple linear interpolation. This algorithm is quite similar to one presented by DeRosier and Moore [19] using Hankel transformations, as discussed in Section VI of this paper. Their algorithm as used by themselves and others has been quite successful at reconstructing viruses, which do exhibit such symmetries.

Both algorithms which have been presented thus far have been based on interpolating from a polar raster to a Cartesian raster and then using the DFT to obtain samples of the original function. An alternative class of algorithms is based on a direct approximation of (39) using a two-dimensional extension of the trapezoidal rule. In particular, let us approximate

$$S_\theta(\omega) \exp\{j\omega(x\cos\theta + y\sin\theta)\} = g(\theta, \omega)$$

by a piecewise planar function that is planar over the "rectangular" region bounded by two radii and two circular sections. Assume that slices are known at angles $\theta_0, \theta_1, \cdots, \theta_{N-1}$ (not necessarily evenly spaced) and that along each slice the values of $S_{\theta_i}(\omega)$ are known at the $R_1$ values ($R_1$ even) designated by $\Omega[-(R_1/2)+1], \cdots, \Omega_0, \Omega_1, \cdots, \Omega_{R_1}/2$. Since these samples will presumably be computed from sampled projections using a DFT, we shall assume that the radial samples are evenly spaced and that

$$\Omega_{i+1} - \Omega_i = \Delta\omega = \frac{2\sqrt{2}\,W}{R_1}. \tag{50}$$

The angular spacing shall be designated by

$$\Delta\theta_i = \begin{cases} \theta_{i+1} - \theta_i, & i = 0, 1, \cdots, R_2 - 2 \\ \theta_0 - \theta_{R_2-1} + \pi, & i = R_2 - 1. \end{cases}$$

It should be noted that $\sum_{i=0}^{R_2-1} \Delta\theta_i = \pi$ and that if the projections are evenly spaced $\Delta\theta_i = \pi/R_2$. Referring then to Fig. 9, we can approximate $g(\theta, \omega)$ for $\theta_i \leqslant \theta \leqslant \theta_{i+1}, \Omega_k \leqslant \omega \leqslant \Omega_{k+1}$ by the planar section

$$g(\theta, \omega) \simeq A + B(\omega - \Omega_k) + C(\theta - \theta_i) + D(\omega - \Omega_k)(\theta - \theta_1) \tag{51}$$

where

$$A = g(\theta_i, \Omega_k)$$

$$B = \frac{g(\theta_i, \Omega_{k+1}) - g(\theta_i, \Omega_k)}{\Delta\omega}$$

$$C = \frac{g(\theta_{i+1}, \Omega_k) - g(\theta_i, \Omega_k)}{\Delta\theta_i}$$

$$D = \frac{g(\theta_{i+1}, \Omega_{k+1}) - g(\theta_{i+1}, \Omega_k) - g(\theta_i, \Omega_{k+1}) + g(\theta_i, \Omega_k)}{\Delta\omega\Delta\theta_i}.$$

By doing this for all sections, (39) becomes

$$f(x_1, x_2) \simeq \frac{2W^2}{3R_1^2\pi} S_{\theta_0}(0) + \frac{2W^2}{R_1^2\pi^2} \sum_{i=0}^{R_2-1} \Delta\theta_i \sum_{k=-R_1/2+1}^{R_1/2} |k| \cdot S_{\theta_i}\left(\frac{2\sqrt{2}\,kW}{R_1}\right) \exp\left\{j\frac{2\sqrt{2}\,kW}{R_1}(x_1\cos\theta_i + x_2\sin\theta_i)\right\}. \tag{52}$$

Equation (52) represents a reconstruction algorithm since it expresses $f(x_1, x_2)$ in terms of $S_\theta(k\Delta\omega)$, the slice samples. The direct computation of (52) is difficult as that equation is written. However, we can reinterpret it and use the DFT to make it easier to compute. Reinterpreting (52) has multiple advantages. It results in a straightforward computational procedure for implementing the algorithm, provides us with some insight into how the algorithm works, and shows us that this algorithm is equivalent to another algorithm—the so-called smearing algorithm which can be implemented in the space domain. With the smearing algorithm, each projection is back-projected or smeared in the direction of the original projection and the smeared projections are then weighted and summed. Let us define what is meant by smearing. If $p_{\theta_i}(u_i)$ represents the projection of $f(x, y)$ at angle $\theta_i$ where

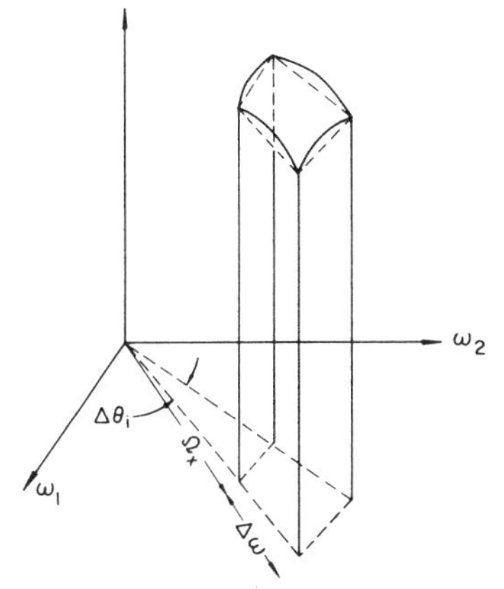


Fig. 9. Approximation of $g(\theta, \omega)$ by a planar section.

$$u_i = x_1\cos\theta_i + x_2\sin\theta_i$$
$$v_i = -x_1\sin\theta_i + x_2\cos\theta_i \tag{53}$$

then the smeared projection at angle $\theta_i$, $p_{\theta_i}(u_i, v_i)$, is defined by

$$p_{\theta_i}(u_i, v_i) = p_{\theta_i}(u_i). \tag{54}$$

To express (52) in terms of smeared projections, we can first rewrite it as

$$f(x_1, x_2) \simeq \hat{f}(x_1, x_2) = \frac{2W^2}{3\pi R_1^2} S_{\theta_0}(0) + \frac{2W^2}{\pi^2 R_1^2} \sum_{i=0}^{R_2-1} \Delta\theta_i g_i(x_1, x_2) \tag{55}$$

where

$$g_i(x_1, x_2) = \sum_{k=-R_1/2+1}^{R_1/2} |k| S_{\theta_i}(k\Delta\omega) \cdot \exp\left[jk\Delta\omega(x_1\cos\theta_i + x_2\sin\theta_i)\right]. \tag{56}$$

If we define $\hat{g}_i(u_i, v_i) = g_i(x, y)$ with a change of coordinates, then

$$\hat{g}_i(u_i, v_i) = \sum_{k=-R_1/2+1}^{R_1/2} |k| S_{\theta_i}(k\Delta\omega) \exp\left[jku_i\Delta\omega\right]. \tag{57}$$

From (57) we see that $\hat{g}(u_i, v_i)$ is a function only of $u_i$ and that it is thus a smeared quantity. Furthermore, the quantity that is smeared is a filtered projection. This filtering can be accomplished by computing the DFT of the projection, multiplying the resulting sequence by $|k|$, and then inverse transforming. According to (56), these filtered smeared projections $\{g_i(x_1, x_2)\}$ are then weighted, summed, and added to the dc bias (constant) function.

Instead of computing $\hat{g}_i(u_i)$ from $p_{\theta_i}(u_i)$ by computing transforms, we could confine ourselves totally to the space domain and express $\hat{g}_i(u_i)$ as a (circular) convolution of $p_{\theta_i}(u_i)$ and the inverse DFT of $|k|$. This computation is cumbersome, however, and the modification of the projections is more easily carried out in the frequency domain than by performing the convolution in the space domain. Ramachandran and Lakshminarayanan [5] and Shepp and Logan [21] have shown, however, that by modifying the weighting function slightly, the convolution is more conveniently done directly than by using Fourier transforms.

This algorithm has been used by Bracewell [1], Vainshtein [22], Ramachandran [23], Herman [8], Gilbert [24], and perhaps others, with minor variations. There are a number of nice features to it. First, it imposes no requirements on $f(x_1, x_2)$ except that it be nearly band-limited and that $F(\omega_1, \omega_2)$ be "slowly varying" in $\omega_1$ and $\omega_2$ so that our original approximation of an integral by a sum is valid. These are not severe restrictions to impose. Another advantage of the algorithm is that the entire procedure can be implemented in the space domain (although, as we have seen, it may be more efficient to filter the projections in the frequency domain). Omitting the modification of the projections, this system can, in fact, be implemented optically and such an optical system has been built by Garrison, Grant *et al.* [6], [25]. Another feature of this algorithm, actually the nicest feature of all, is that it yields very good reconstructions.

There are, on the other hand, several problems inherent in use of the smearing algorithm. One of the most serious is that the reconstructions are very sensitive to inaccuracies in the input projections. Any error or "noise" in the projection is amplified in the reconstructions and, as a result, the final reconstructions appear noisy; that is, the reconstructions have a mottled or speckled appearance.

Some reconstructions formed by using this algorithm are shown in Figs. 10 and 11. In Fig. 10 several photographs are reconstructed from 64 projections—the projections that were used for the preceding algorithms. In Fig. 11 reconstructions are shown for 16, 32, 64, and 128 projections. It will be noted that for the example chosen the degradation with a decrease in the number of projections is serious. Such a degradation is to be expected but it is particularly serious for this example. It is informative to compare these reconstructions with those produced by linear interpolation. The smearing reconstructions clearly display more resolved details, although they do have a grainy or speckled appearance due to the noise mentioned earlier. A possible explanation is due to the fact that with the linear interpolation algorithm, the values of Cartesian samples of $F(\omega_1, \omega_2)$ were estimated from neighboring polar samples prior to performing an inverse DFT. As a result, polar samples which were not close to Cartesian samples were ignored. For large values of $\omega$ (the radial variable in a polar coordinate system) this effect is negligible since almost all of these polar samples are used, but for polar samples close to the origin this effect becomes significant, especially since for photographs, $|F(\omega_1, \omega_2)|$ is largest for those values of $\omega_1$ and $\omega_2$ near the origin. On the other hand, the smearing algorithm uses all of the information contained in the polar samples since all of the projections are treated identically.

The noise which is introduced in the smearing reconstructions is due to the high-frequency emphasis filter that is applied to the projections and it introduces computational errors similar to those introduced by numerical differentiation.

We close this section with the discussion of an algorithm similar to the polar expansion algorithm. This technique has not been used by the authors, but it has been used successfully

Fig. 10. Reconstruction of the pictures of Fig. 7 from 64 projections using the smearing algorithm.

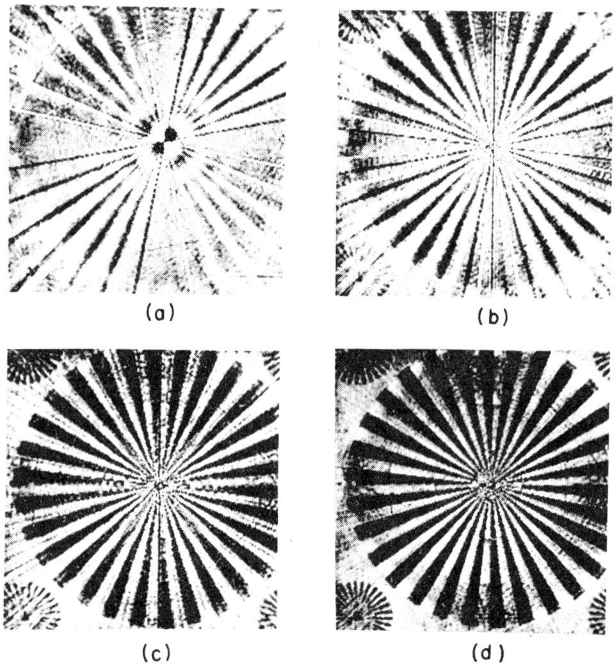


Fig. 11. Reconstruction of the test pattern of Fig. 7 using the smearing algorithm for different numbers of projections. (a) 16 projections. (b) 32 projections. (c) 64 projections. (d) 128 projections.

by DeRosier and Klug and their colleagues [3], [14], [26], [27] to reconstruct from electron micrographs.

Let $f(r, \phi)$ represent the picture function in polar coordinates and let $F(\omega, \theta)$ represent its Fourier transform. Then, since $f(r, \phi)$ is necessarily a periodic function in $\phi$ with period $2\pi$, we can express $f(r, \phi)$ in a Fourier series as

$$f(r, \phi) = \sum_{n=-\infty}^{\infty} f_n(r) \exp[jn\phi] \qquad (58)$$

where

$$f_n(r) = \frac{1}{2\pi} \int_{-\pi}^{\pi} f(r, \phi) \exp[-jn\phi]\, d\phi. \qquad (59)$$

However, since $f(r, \phi)$ and $F(\omega, \theta)$ are a Fourier transform pair it is true that

$$f(r, \phi) = \frac{1}{4\pi^2} \int_0^{\infty} \int_{-\pi}^{\pi} F(\omega, \theta) \exp\left[j r\omega \cos(\theta - \phi)\right] \omega \, d\theta d\omega. \tag{60}$$

Thus

$$f_n(r) = \frac{1}{2\pi} \int_{-\pi}^{\pi} \left\{ \frac{1}{4\pi^2} \int_0^{\infty} \int_{-\pi}^{\pi} F(\omega, \theta) \cdot \exp\left[j r\omega \cos(\theta - \phi)\right] \omega \, d\theta d\omega \right\} \exp\left[-jn\phi\right] d\phi. \tag{61}$$

Interchanging the order of the integrations,

$$f_n(r) = \frac{1}{8\pi^3} \int_0^{\infty} \int_{-\pi}^{\pi} F(\omega, \theta) \, \omega \, d\theta d\omega \cdot \int_{-\pi}^{\pi} \exp\left[j(r\omega \cos(\phi - \theta) - n\phi)\right] d\phi \tag{62}$$

$$= \frac{1}{8\pi^3} \int_0^{\infty} \int_{-\pi}^{\pi} F(\omega, \theta) \exp\left[jn\left(\theta + \frac{\pi}{2}\right)\right] \omega \, d\theta d\omega \cdot \int_{-\pi+\theta}^{\pi+\theta} \exp\left[j r\omega \sin\alpha - n\alpha\right] d\alpha \tag{63}$$

$$= \frac{1}{4\pi^2} \int_0^{\infty} \int_{-\pi}^{\pi} F(\omega, \theta) \exp\left[jn\left(\theta + \frac{\pi}{2}\right)\right] J_n(r\omega) \, d\theta d\omega \tag{64}$$

where $J_n(x)$ is the $n$th-order Bessel function of the first kind. Since $F(\omega, \theta)$ is a periodic function in $\theta$ with period $2\pi$ we can express it in a Fourier series as well, as

$$F(\omega, \theta) = \sum_{n=-\infty}^{\infty} F_n(\omega) \exp\left[+jn\left(\theta + \frac{\pi}{2}\right)\right] d\theta. \tag{65}$$

The factor of $\pi/2$ has been included in the exponent of (65) so that by comparing (65) and (64) we can see that

$$f_n(r) = \frac{1}{2\pi} \int_0^{\infty} F_n(\omega) J_n(r\omega) \, d\omega \tag{66}$$

$$F_n(\omega) = \int_0^{\infty} f_n(r) J_n(r\omega) \, dr. \tag{67}$$

Equations (66) and (67) say that $f_n(r)$ and $F_n(\omega)$ form a Fourier–Bessel or Hankel transform pair.

From this derivation we can formulate a reconstruction algorithm. From the slices we compute the sequences $F_n(\omega)$ using the inverse of (65). If the slice angles are evenly spaced, this can be done using a DFT. Then we can take the $n$th-order Hankel transform of each of the $F_n(\omega)$ to obtain the $f_n(r)$. Then, using another DFT calculation, $f(r, \phi)$ can be approximated on any desired number of radii and angles.

The major difficulties with this technique are computational. Each of the Hankel transforms which must be taken is different since each uses a different Bessel function. The amount of computation can, therefore, be large. DeRosier and Klug [3], in their original work, reconstructed helices. The unique properties of helices allowed them to obtain several of the $F_n(\omega)$ from a single projection.

## V. Approximate Reconstruction from a Concentric-Squares Raster

In Section III we considered a procedure for exact reconstruction of a band-limited function of finite order based on using the projections to obtain samples of the Fourier transform on a concentric-squares raster. An exact reconstruction procedure then consists of interpolating or extrapolating from this concentric-squares raster to a Cartesian raster. Since the vertical lines of the concentric-squares and Cartesian rasters are identical, interpolation from a concentric-squares raster to a Cartesian raster involves only one-dimensional interpolation rather than two-dimensional interpolation as was required for the polar raster.

One difficulty with the concentric-squares raster is that with the slices confined to the range $-(\pi/4) \leqslant \theta \leqslant +(\pi/4)$, as was done in Section III, for small radial values the samples are closely spaced over a small frequency range and consequently extrapolation far away from these points is necessary.

In this section we consider a modification of this procedure. In particular, by generalizing the raster of Fig. 2 to include additional projections, it is possible to avoid the extrapolation problem previously mentioned. In addition, for approximate reconstruction, we will use low-order interpolation.

In deriving the polar raster it was assumed that the Fourier transform of the unknown was confined to a circular region in the Fourier plane of radius $\sqrt{2}\,W$. Here we assume that it is instead confined to a square region of half-side $W$, as illustrated in Fig. 12. It should be noted that such a function is also confined to a disk of radius $\sqrt{2}\,W$.

For such a function, by the projection-slice theorem it is seen that each projection can be represented by a band-limited function whose bandwidth is a function of the projection angle $\theta$. Thus we can represent the projection at angle $\theta$, $p_\theta(u)$ by

$$p_\theta(u) = \sum_{n=-\infty}^{\infty} p_\theta\left(\frac{n\pi}{W_\theta}\right) \frac{\sin W_\theta\left(u - \frac{n\pi}{W_\theta}\right)}{W_\theta\left(u - \frac{n\pi}{W_\theta}\right)} \tag{68}$$

where

$$W_\theta \geqslant \frac{W}{\max\{|\cos\theta|, |\sin\theta|\}}. \tag{69}$$

The right-hand side of (69) specifies the minimum sampling rate that can be used to sample the projection of a band-limited function of angle $\theta$. If a lower sampling rate is used, information will be lost through the sampling process. If each projection is sampled at this rate and the resulting sequences are transformed by using an $N$ point DFT to compute the sum

$$S_\theta(k\Delta\omega) = \sum_{n=-\infty}^{\infty} p_\theta\left(\frac{n\pi}{W_\theta}\right) \exp\left[-j\frac{2\pi}{N}nk\right], \quad k = -\frac{N}{2} + 1, \cdots, 0, \cdots, \frac{N}{2}$$

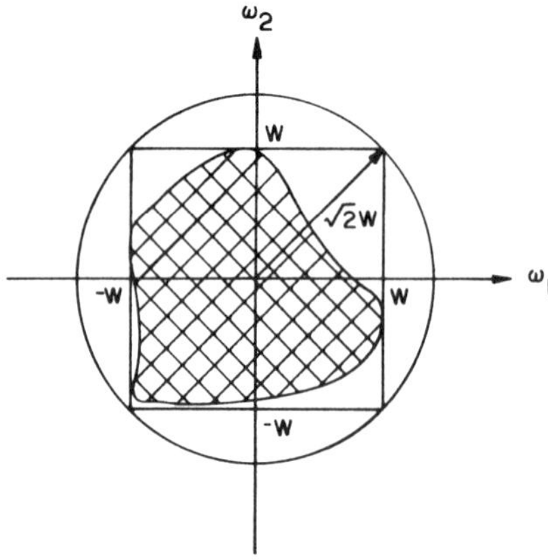


**Fig. 12. Illustration of the Fourier transform confined to a square region of half side $W$.**

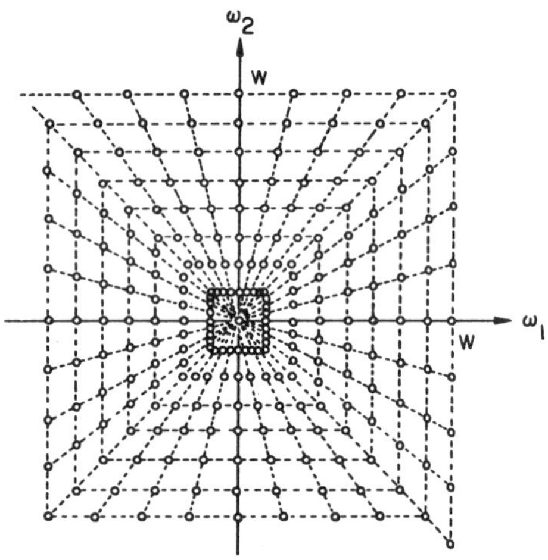


**Fig. 13. A concentric-squares raster.**

we will obtain a set of slice samples that are different from the polar case. In fact, we can in general control where the samples of $S_\theta(\omega)$ are located by controlling the sampling interval used to sample the projection $p_\theta(u)$. We shall define a concentric-square raster to be a raster of sample locations formed by sampling a finite number of projections at their respective Nyquist rates and then taking $N$ point transforms of the results. Such a raster is illustrated in Fig. 13.

As a comment, it should be noted that if the projections are made by a source of collimated radiation such as an X-ray source then concentric-squares samples can be obtained with a hardware scanner which does not have an adjustable sampling interval. This technique is illustrated in Fig. 14. Assume that the object that we wish to identify is irradiated by a collimated beam and that it lies between the beam source and a recording surface such as a photographic plate. Instead of rotating the unknown to produce different projections, the beam and the photographic plate will be moved simultaneously. In particular, assume that we begin with the beam perpendicular to the $x$ axis and the plate parallel to the $x$ axis as indicated in Fig. 14(a). As the beam is rotated, the center of the plate is shifted as indicated in Fig. 14(b). When the projection angle exceeds $\pi/4$, the plate is placed parallel to the $y$ axis as indicated in Fig. 14(c), and as the projection angle increases further, the center of the plate shifts accordingly, as indicated in Fig. 14(d). Thus the photographic plate is moved in such a way that its center follows a square. In this way the

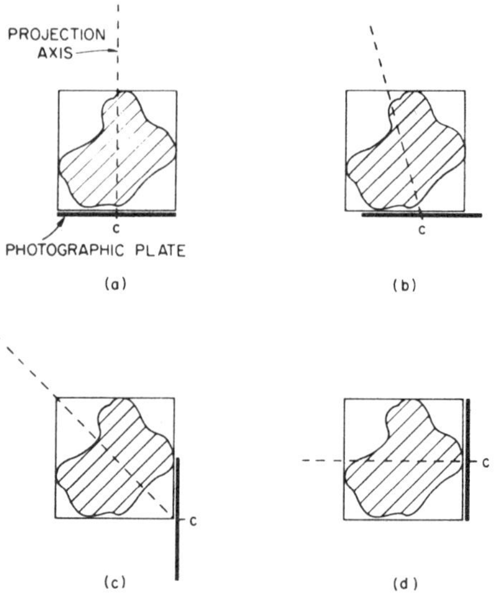


**Fig. 14. Illustration of a technique for obtaining projections for concentric-squares samples.**

**Fig. 15. Reconstructions of the pictures of Fig. 7 using linear interpolation from a concentric-squares raster of 64 evenly spaced slices.**

recorded picture is seen to be expanded by a factor of

$$\frac{1}{\max\{|\cos\theta|, |\sin\theta|\}}$$

by the obliqueness of the recording surface. If all projections are now sampled with the same sampling interval, $\pi/W$, the resulting sequences correspond to a set of sampled projections on a concentric-squares raster.

As with the polar raster, we can use linear interpolation to estimate the values of Cartesian samples of $F(\omega_1, \omega_2)$ from the known samples on the concentric-squares raster.

Fig. 15 shows some reconstructions made by using linear interpolation from a concentric-squares raster of 64 evenly spaced slices. In Fig. 16 we have used a similar algorithm with a different set of slice (projection) angles. Instead of choosing the angles to be evenly spaced, we have chosen them

Fig. 16. Reconstruction of the pictures of Fig. 7 using the same algorithm as for Fig. 15 but with a different set of projection angles.

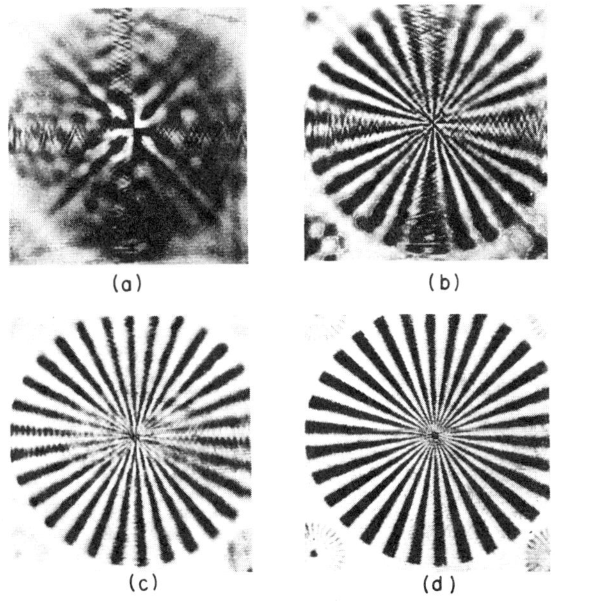


Fig. 17. Reconstruction of the test pattern of Fig. 7 using the identical algorithm as for Fig. 16 but varying the number of projections. (a) 16 projections. (b) 32 projections. (c) 64 projections. (d) 128 projections.

so that the intersections of the slices with the concentric squares are evenly spaced on each square. This set of angles yields a computationally simpler algorithm and it results in a higher density of samples in the "corners" of the nonzero region of $F(\omega_1, \omega_2)$. It should be noted that these two different sets of projection angles yield virtually identical reconstructions. In Fig. 17 we have taken this latter strategy for choosing projection angles and examined reconstructions made with 16, 32, 64, and 128 projections to see what improvement in reconstruction quality could be gained by adding more projections.

From these examples we note that the reconstructions from linear interpolation from a concentric-squares raster seem to be of better quality than those from a polar raster. This is probably due to two factors. The interpolation is one-

Fig. 18. Reconstruction based on high-order polynomial interpolation along the sides of a series of concentric squares.

dimensional and the Fourier transforms of the photographs used have regions of support in the Fourier plane which are more nearly square than circular. From Fig. 17 we note a steady improvement in reconstruction quality with increased projections. This result is certainly not surprising. The improvement is particularly striking when we go from 32 to 64 projections, although this is somewhat dependent upon the example chosen which has 30 circularly arranged black spokes.

As the preceding algorithm represented a straightforward extension of the polar linear interpolation algorithm, so also can we extend the polar expansion algorithm and compute extra concentric-squares slices from a limited number which are available from projections. Instead of performing an expansion in polar coordinates, however, we must perform it in "concentric-squares coordinates," i.e., along the slice lines and along the squares. Actual interpolation along the slices is unnecessary since all of the concentric-squares samples already lie on the squares which contain all of the DFT samples—this, of course, is the advantage of using concentric-squares coordinates. Thus our algorithm reduces to the implementation of a high-order polynomial interpolation (in $\exp[j\omega_1]$ or $\exp[j\omega_2]$) along the sides of a series of concentric squares, a task that can be accomplished using a DFT.

Suppose, for example, that $N_v$ slices intersect the vertical sides of an array of concentric squares where $N_v \geq N$. (If $N_v \leq N$, then by the arguments of Section III, exact reconstruction can be performed.) We can approximate $F(\omega_1, \omega_2)$ along each vertical side by a polynomial of degree $N_v - 1$ in the variable $\exp[-j(\pi/W)\,\omega_2]$. Similarly, if each horizontal side is intersected by $N_h$ slices ($N_h < N$), we can approximate $F(\omega_1, \omega_2)$ along these lines by a polynomial of degree $N_h - 1$ in the variable $\exp[j(\pi/W)\,\omega_1]$. We know, however, that $f(x_1, x_2)$ is of finite order and that in fact $F(\omega_1, \omega_2)$ varies as a polynomial of degree $N - 1$ in the appropriate variable along all of these lines. Thus to the extent that the higher order polynomial can be closely approximated by the lower one, we can expect an accurate reconstruction.

Fig. 18 illustrates four reconstructions performed by this technique. Whereas these reconstructions represent an improvement over the corresponding polar reconstructions, they represent very little improvement over the straightforward linear interpolation from a concentric-squares raster.

As a final step in our development of parallel polar and concentric-squares reconstruction algorithms, it remains to implement the smearing algorithm in concentric-squares fashion. This is straightforward. First, we must express $F(\omega_1, \omega_2)$ in concentric-squares coordinates. Defining $\omega$ to be a variable that designates the square of interest, i.e., $\omega$ remains constant on each square, and $\lambda$ as the normalized arc length, we obtain

$$\begin{aligned} f(x_1, x_2) &= \frac{1}{4\pi^2}\int_0^W \int_0^1 \omega\hat{F}(\omega,\lambda) \exp\left[j\omega(x_1+\lambda x_2)\right] d\lambda d\omega \\ &- \frac{1}{4\pi^2}\int_0^W \int_1^3 \omega\hat{F}(\omega,\lambda) \exp\left[j\omega((2-\lambda)x_1+x_2)\right] d\lambda d\omega \\ &- \frac{1}{4\pi^2}\int_0^W \int_3^5 \omega\hat{F}(\omega,\lambda) \exp\left[j\omega(-x_1+(4-\lambda)x_2)\right] d\lambda d\omega \\ &+ \frac{1}{4\pi^2}\int_0^W \int_5^7 \omega\hat{F}(\omega,\lambda) \exp\left[j\omega((\lambda-6)x_1-x_2)\right] d\lambda d\omega \\ &+ \frac{1}{4\pi^2}\int_0^W \int_7^8 \omega\hat{F}(\omega,\lambda) \exp\left[j\omega(x_1+(\lambda-8)x_2)\right] d\lambda d\omega. \end{aligned} \tag{70}$$

Now we can apply the trapezoidal rule to the samples in concentric-squares coordinates. We shall assume that the samples are located on a regular concentric-squares raster, i.e., evenly spaced in $\omega$ and evenly spaced in $\lambda$. We thus obtain a summation that very much resembles the similar result in polar coordinates:

$$\begin{aligned} f(x_1, x_2) &= \frac{4W^2}{8M^2\pi^2} S_{\theta_0}(0) \\ &+ \frac{4W^2}{NM^2\pi^2}\sum_{i=0}^{N/4-1}\sum_{k=-M/2+1}^{M/2} |k|\hat{F}\left(\frac{2Wk}{M}, \frac{4i}{N}\right) \cdot \exp\left[j\frac{2Wk}{M}\left(x_1 - \frac{4i}{N}x_2\right)\right] \\ &+ \frac{4W^2}{NM^2\pi^2}\sum_{i=N/4}^{3N/4-1}\sum_{k=-M/2+1}^{M/2} |k|\hat{F}\left(\frac{2Wk}{M}, \frac{4i}{N}\right) \cdot \exp\left[j\frac{2Wk}{M}\frac{2N-4i}{N}(x_1+x_2)\right] \\ &+ \frac{4W^2}{NM^2\pi^2}\sum_{i=3N/4}^{N-1}\sum_{k=-M/2+1}^{M/2} |k|\hat{F}\left(\frac{2Wk}{M}, \frac{4i}{N}\right) \cdot \exp\left[j\frac{2Wk}{M}\left(-x_1 + \frac{4N-4i}{N}x_2\right)\right]. \end{aligned} \tag{71}$$

Equation (71) can be shown to have an interpretation in terms of smearing exactly as in the polar case. Each projection is sampled, and transformed to obtain concentric-squares samples of $F(\omega_1, \omega_2)$. They are then multiplied by the weighting function $|\omega|$, inverse-transformed, and back-projected as before. As in the polar case, the mean gray levels of the original and of the reconstruction are adjusted to be equal.

Fig. 19. Reconstructions obtained by using concentric-squares smearing.

(a) (b) (c) (d)

Fig. 20. Reconstruction of the test pattern of Fig. 7 using concentric-squares smearing for several choices of the number of projections. (a) 16 projections. (b) 32 projections. (c) 64 projections. (d) 128 projections.

Fig. 19 presents some reconstructions generated by using concentric-squares smearing and in Fig. 20 we present some reconstructions of one of these photographs for 16, 32, 64, and 128 projections. It should be noted that the quality of the reconstructions, at least when more than 64 projections are used, is very good—in fact, it is better than with any other algorithm used for this paper.

One objection to the use of the concentric-squares raster might be that it assumes a particular orientation for the object being reconstructed and that the projections are not all being treated equally in some sense. There are two important cases, however, in which this can be a distinct advantage. One is

(a)

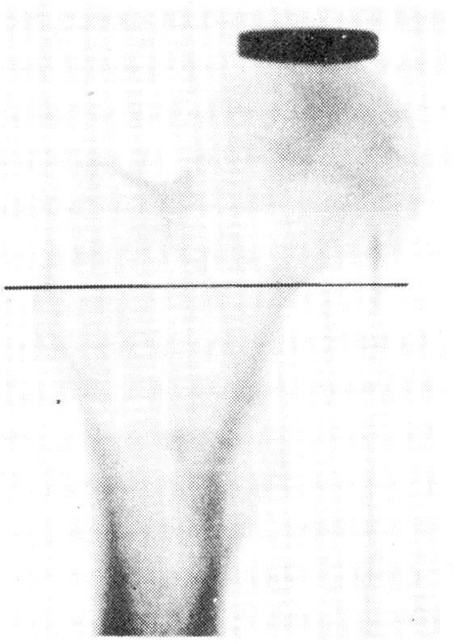

(b)

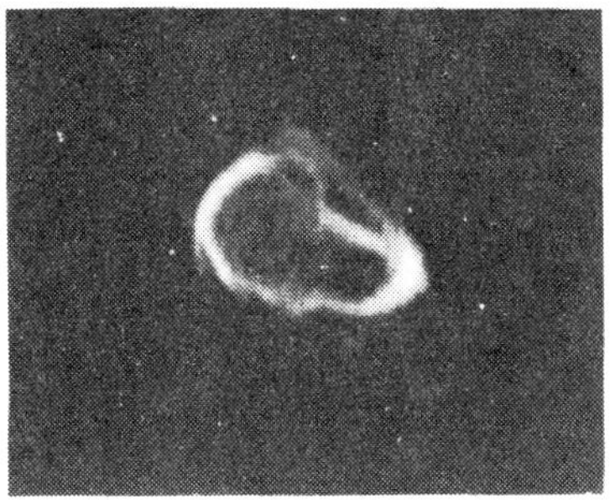

Fig. 21. Reconstruction from a series of X-rays of an excised human femur using the concentric-squares linear interpolation algorithm. (a) One of the set of X-rays. (b) Reconstruction of cross section at the level of the dark line indicated in Fig. 21(a).

when the unknown itself has a preferred orientation, i.e., when it tends to have a square region of support. The second is when the reconstruction algorithm itself does not treat all projections identically. This class of algorithms includes all of the interpolation algorithms which perform an inverse two-dimensional DFT, but does not include the smearing algorithm or the algorithm in Section IV which uses Hankel transforms.

In an effort to see how well one of these algorithms would perform on real data, a series of X-rays of an excised human femur were obtained. The complete series consisted of 36 X-rays which were taken normal to the long axis of the bone at 5° intervals. Since the algorithms were all designed to perform two-dimensional reconstructions from one-dimensional projections, each of the X-rays was sampled logarithmically along a single line normal to the long axis and a single cross section of the bone was reconstructed. Each projection was sampled at 256 points and the sampling interval was the same for each projection. These samples were then used to compute both polar and concentric-squares samples of $F(\omega_1, \omega_2)$. The concentric-squares linear interpolation reconstruction is presented in Fig. 21(b) and one of the X-rays from which it is produced is in Fig. 21(a). The hollow tubular structure of the bone is evident from this reconstruction. In addition, material of lighter density is seen inside the bone. The reconstruction from a polar grid was virtually identical.

## VI. Algebraic Reconstruction Techniques

In the previous sections we have concentrated on a number of reconstruction algorithms that are conveniently interpreted in Fourier space. In this section we wish to briefly discuss a

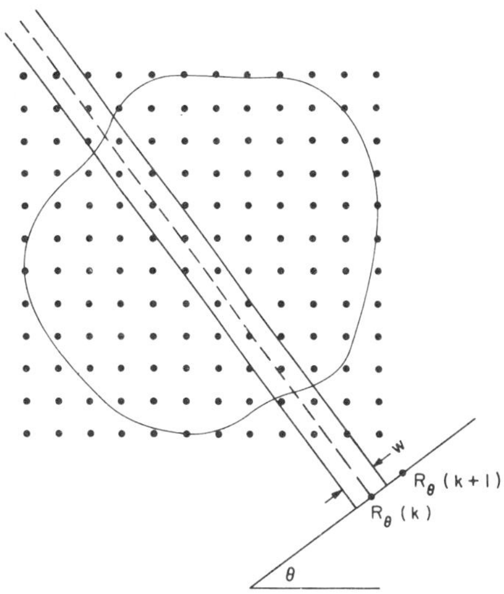


Fig. 22. Procedure for obtaining the $k$th value of the pseudo-projection at angle $\theta$.

class of algebraic reconstruction techniques which are implemented directly in signal space. A variety of such approaches have been proposed by Gordon, Bender, and Herman [9]; Gilbert [10]; Goitein [28]; and Hounsfield [11]. The latter has resulted in a marketable device for performing brain scans. All of these techniques begin with the assumption that the object to be reconstructed is a sampled array, i.e., it is represented by samples of its density on a Cartesian grid. A pseudo-projection at angle $\theta$ for such an object is defined by considering the sum of the densities contained within a set of rays of width $w$ parallel to the projection axis. For a two-dimensional object, for example, we consider the object as representable by the densities $\rho(i, j)$ on a Cartesian grid. The $k$th value of pseudo-projection at angle $\theta$, denoted by $R_\theta(k)$, is obtained as indicated in Fig. 22. Specifically, $R_\theta(k)$ is equal to the sum of the densities associated with the points contained within the indicated ray.

If the projections from which the object is to be reconstructed corresponded to the pseudo-projections, then, in principle, the reconstruction could be carried out algebraically. Specifically, if the unknown can be represented by an $N \times N$ grid then it is defined by $N^2$ density values. Each pseudo-projection value $R_\theta(k)$ represents a linear equation in the $N^2$ unknown density values. Thus with a sufficient number of projections a set of $N^2$ linear equations can be obtained which can then be solved for the unknown densities $\rho(i, j)$. For any reasonable value of $N$ this would result in a large number of equations—too many equations in fact to solve algebraically in a practical system. As an additional consideration, the approximation of pseudo-projections by projections and the effect of measurement errors are likely to make the set of $N^2$ equations ill-conditioned or inconsistent. Therefore, the equations are solved approximately in an iterative fashion. This removes the constraint that the number of equations equal $N^2$. In fact, Hounsfield uses $4N^2$ equations and Gordon, Bender, and Herman use a much smaller number, typically $N^2/4$ or so.

In the class of algebraic reconstruction techniques, an initial set of densities for the object are assumed. These initial densities are then changed iteratively in such a manner that the calculated pseudo-projections converge to the available projection data. A variety of methods have been proposed for iteratively changing the densities and the reader is referred to the literature for specific discussions of the various methods. A typical approach is the following adapted from Herman, Lent, and Rowland [29]. Let $P_\theta(k)$ denote samples of the projection at angle $\theta$ and $R_\theta^q(k)$ the pseudo-projection at angle $\theta$ on the $q$th iteration of the algorithm. Furthermore, let $\rho^q(i,j)$ denote the density at the $i,j$th point on the $q$th iteration. The algorithm then sets

$$\rho^{q+1}(i,j) = \rho^q(i,j) + \delta(i,j,k,\theta)\frac{R_\theta(k) - R_\theta^q(k)}{N_{\theta,k}} \qquad (72)$$

where $\delta(i,j,k,\theta)$ is an indicator variable which is 1 when $\rho(i,j)$ is included in the ray corresponding to $R_\theta^q(k)$ and 0 when it is not and $N_{\theta,k}$ is the number of elements in that ray.

The family of algebraic reconstruction techniques has generated a certain amount of controversy [30], [31], which centers primarily on the comparison with Fourier methods. The development and modification of these techniques and the comparison with other methods appear to still be an active area of discussion. These techniques being signal space techniques have the important advantage that signal space constraints, such as the fact that densities are of necessity positive, are relatively easy to impose, whereas they suffer from the drawback that they become computationally inefficient for very large values of $N$. These techniques, like the Fourier space techniques, impose certain constraints upon the object to be reconstructed. Which set of constraints is the more appropriate will probably vary with the object under investigation and the required resolution of the finished reconstruction.

## Acknowledgment

The authors were first introduced to this problem by Dr. B. Gold of the M.I.T. Lincoln Laboratories, who also suggested the possibility that it might benefit from digital signal processing techniques. They would like to express their appreciation for this suggestion and for his continuing interest. The authors would also like to thank Dr. O. Tretiak for providing the X-ray data used in the reconstruction of Fig. 21, and the Cognitive Information Processing Group in the M.I.T. Research Laboratory of Electronics for the use of their facilities for A/D and D/A conversion of the photographs.

## References

[1] R. N. Bracewell, "Strip integration in radio astronomy," *Aust. J. Phys.*, vol. 9, pp. 198–217, 1956.

[2] R. N. Bracewell and A. C. Riddle, "Inversion of fan-beam scans in radio astronomy," *The Astrophys. J.*, vol. 150, pp. 427–434, 1967.

[3] D. J. DeRosier and A. Klug, "Reconstruction of three-dimensional structures from electron micrographs," *Nature*, vol. 217, pp. 130–134, 1968.

[4] B. K. Vainshtein, "Finding the structure of objects from projections," *Sov. Phys.–Crystall.*, vol. 15, no. 5, pp. 781–787.

[5] G. N. Ramachandran and A. V. Lakshminarayanan, "Three-dimensional reconstruction from radiographs and electron micrographs: II. Application of convolutions instead of Fourier transforms," *Proc. Nat. Acad. Sci.*, vol. 68, no. 9, pp. 2236–2240.

[6] J. B. Garrison, D. G. Grant, W. H. Guier, and R. J. Johns, "Three-dimensional roentgenography," *Amer. J. Roentgenol., Rad. Therapy, Nucl. Med.*, vol. 60, no. 4, pp. 903–908, 1969.

[7] O. Tretiak, D. Ozonoff, J. Klopping, and M. Eden, "Calculation of internal structure from multiple radiograms," in *Proc. Two-Dimensional Digital Signal Processing Conf.* (Univ. of Missouri-Columbia), pp. 6-2-1–6-2-3.

[8] G. T. Herman, "Two direct methods for reconstructing pictures from their projections: A comparative study," *Comp. Graph. Image Proc.*, vol. 1, pp. 123–144, 1973.

[9] R. Gordon, R. Bender, and G. T. Herman, "Algebraic reconstruction techniques (ART) for three-dimensional electron microscopy and X-ray photography," *J. Theor. Biol.*, vol. 29, pp. 471–481, 1970.

[10] P. Gilbert, "Iterative methods for the three-dimensional reconstruction of an object from projections," *J. Theor. Biol.*, vol. 36, pp. 105–117.

[11] G. N. Hounsfield, "Method and apparatus for measuring X or γ-radiation absorption or transmission at plural angles and analyzing the data," U.S. Patent 3 778 614, Dec. 11, 1973.

[12] R. M. Mersereau, "The digital reconstruction of multi-dimensional signals from their projections," Sc.D. dissertation, M.I.T., Dept. of Elec. Eng., 1973.

[13] ——, "Recovering multi-dimensional signals from their projections," *Comp. Graph. Image Proc.*, vol. 1, no. 6, pp. 179–195, Oct. 1973.

[14] R. M. Mersereau and D. E. Dudgeon, "The representation of two-dimensional sequences as one-dimensional sequences," *IEEE Trans. Acoust., Speech, Signal Processing*, vol. ASSP-22, pp. 320–325, Oct. 1974.

[15] P. R. Smith, T. M. Peters, and R. H. T. Bates, "Image reconstruction from finite numbers of projections," *J. Phys. A: Math, Nucl. Gen.*, vol. 6, pp. 361–382, 1973.

[16] R. A. Crowther, D. J. DeRosier, and A. Klug, "The reconstruction of a three-dimensional structure from projections and its application to electron microscopy," *Proc. Roy. Soc. London, Ser. A*, vol. 317, pp. 319–340, 1970.

[17] O. Tretiak, M. Eden, and W. Simon, "Internal structure from X-ray images," presented at the Int. Conf. on Medicine Biology in Engineering, 1969.

[18] A. Klug and R. A. Crowther, "Three-dimensional image reconstruction from the viewpoint of information theory," *Nature*, vol. 238, pp. 435–440, Aug. 25, 1972.

[19] D. J. DeRosier and P. B. Moore, "Reconstruction of three-dimensional images from electron micrographs of structures with helical symmetry," *J. Mol. Biol.*, vol. 52, pp. 355–369, 1970.

[20] D. J. DeRosier, "Three-dimensional image reconstruction of helical structures," *Ber. Bensen-Gesellschaft*, vol. 74, no. 11, pp. 1127–1128, 1970.

[21] L. A. Shepp and B. F. Logan, "The Fourier reconstruction of a head section," *IEEE Trans. Nucl. Sci.*, vol. NS-21, pp. 21–43, June 1974.

[22] B. K. Vainshtein, "Synthesis of projecting functions," *Sov. Phys.–Dokl.*, vol. 16, no. 2, pp. 66–69.

[23] G. N. Ramachandran, "Reconstruction of substance from shadow: I. Mathematical theory with application to three-dimensional radiography and electron," *Proc. Indian Acad. Sci.*, vol. 74, pp. 14–24.

[24] P. F. C. Gilbert, "The reconstruction of a three-dimensional structure from projections and its application to electron microscopy: II. Direct methods," *Proc. Roy. Soc. London, Ser. B*, vol. 182, pp. 89–102.

[25] D. G. Grant, J. B. Garrison, and R. J. Johns, "Three-dimensional radiography," *APL Tech. Dig.*, pp. 10–15, Jan. 1970.

[26] R. A. Crowther, L. A. Amos, J. T. Finch, D. J. DeRosier, and A. Klug, "Three-dimensional reconstructions of spherical viruses by Fourier synthesis from electron micrographs," *Nature*, vol. 226, pp. 421–425, 1970.

[27] R. A. Crowther, "Three-dimensional reconstruction and the architecture of spherical viruses," *Endeavor*, vol. 30, pp. 124–129.

[28] M. Goitein, "Three-dimensional density reconstruction from a series of two-dimensional projections," *Nucl. Instrum. Methods*, vol. 101, pp. 509–518, 1972.

[29] G. T. Herman, A. Lent, and S. Rowland, "ART: Mathematics and applications (a report on the mathematical foundations and on the applicability to real data of the algebraic reconstruction techniques)," *J. Theor. Biol.*, vol. 32, pp. 1–32, 1971.

[30] R. A. Crowther and A. Klug, "Art and science or conditions for three-dimensional reconstruction from electron micrograph images," *J. Theor. Biol.*, vol. 32, pp. 199–203, 1971.

[31] S. H. Bellman, R. Bender, R. Gordon, and J. E. Rowe, "ART is science, being a defense of algebraic reconstruction techniques for three-dimensional electron microscopy," *J. Theor. Biol.*, vol. 32, pp. 205–216, 1971.

## Additional Bibliography

*Fourier Space Techniques*

In addition to [2], [3], [7], [12], [13], [19], [20], [26], [27]:

[32] R. A. Crowther, "Procedures for three-dimensional reconstruction of spherical viruses by Fourier synthesis from electron micrographs," *Phil. Trans. Roy. Soc. London, Ser. B*, vol. 261, pp. 221–230, 1971.

[33] R. A. Crowther and L. Amos, "Three-dimensional image reconstructions of some small spherical viruses," in *Proc. Cold Spring Harbor Symp.*, vol. 36, pp. 489–494, 1971.

[34] ——, "Harmonic analysis of electron microscope images with rotational symmetry," *J. Mol. Biol.*, vol. 60, pp. 123–130, 1971.

[35] D. J. DeRosier, "The reconstruction of three-dimensional images from electron micrographs," *Contemp. Phys.*, vol. 12, no. 5, pp. 437-452.
[36] A. Klug, "III. Applications of image analysis techniques in electron microscopy: Optical diffraction and filtering and three-dimensional reconstructions from electron micrographs," *Phil. Trans. Roy. Soc. London, Ser. A*, vol. 261, pp. 173-179.
[37] J. A. Lake, "Reconstruction of three-dimensional structures from sectioned helices by deconvolution of partial data," *J. Mol. Biol.*, vol. 66, pp. 255-269, 1972.
[38] R. M. Mersereau, "The digital reconstruction of multi-dimensional signals from their projections," in *Proc. 10th Annu. Allerton Conf. on Circuit and System Theory* (Oct. 4-6, 1972), pp. 326-334.
[39] T. M. Peters, P. R. Smith, and R. D. Gibson, "Computer aided transverse body-section radiography," *Brit. J. Radiol.*, vol. 46, pp. 314-317, 1973.

*Direct Signal Space Techniques*

In addition to [1], [4], [5], [8], [21]-[24]:
[40] A. M. Cormack, "Representation of a function by its line integrals, with some radiological applications," *J. Appl. Phys.*, vol. 34, pp. 2722-2727, 1963.
[41] ——, "Representation of a function by its line integrals with some radiological applications—II," *J. Appl. Phys.*, vol. 35, pp. 2908-2913, 1964.
[42] ——, "Reconstruction of densities from their projections, with applications in radiological physics," *Phys. Med. Biol.*, vol. 18, pp. 195-207, 1973.
[43] A. M. Mikhailov and B. K. Vainshtein, "Electron microscope determination of the three-dimensional structure of the extended tail of the P6 bacteriophage," *Sov. Phys.—Crystall.*, vol. 16, pp. 428-436, 1971.
[44] G. Muehllehner and R. A. Wetzel, "Section imaging by computer calculation," *J. Nucl. Med.*, vol. 12, pp. 76-84, 1972.
[45] G. N. Ramachandran and A. V. Lakshminarayanan, "Three-dimensional reconstruction from radiographs and electron micrographs: III. Description and application of the convolution method," *Indian J. Pure Appl. Phys.*, vol. 9, pp. 997-1003, 1971.
[46] B. K. Vainshtein and A. M. Mikhailov, "Some properties of the synthesis of projecting functions," *Sov. Phys.—Crystall.*, vol. 17, pp. 217-222, 1972.
[47] B. K. Vainshtein and S. S. Orlov, "Theory of the recovery of functions from their projections," *Sov. Phys.—Crystall.*, vol. 17, pp. 213-216, 1972.

*Iterative Techniques*

In addition to [8]-[11], [21], [28]-[31]:
[48] R. Bender, S. H. Bellman, and R. Gordon, "ART and the ribosome: A preliminary report on the three-dimensional structure of individual ribosomes determined by an algebraic reconstruction technique," *J. Theor. Biol.*, vol. 29, pp. 483-487, 1970.
[49] G. Frieder and G. T. Herman, "Resolution in reconstructing objects from electron micrographs," *J. Theor. Biol.*, vol. 33, pp. 189-211, 1971.
[50] N. T. Gaarder and G. T. Herman, "Algorithms for reproducing objects from their X-rays," *Comp. Graph. Image Proc.*, vol. 1, pp. 97-106.
[51] R. Gordon and G. T. Herman, "Reconstruction of pictures from their projections," *Commun. Ass. Comput. Mach.*, vol. 14, no. 12, pp. 759-768, Dec. 1971.
[52] G. T. Herman and S. Rowland, "Resolution in art: An experimental investigation of the resolving power of an algebraic picture reconstruction technique," *J. Theor. Biol.*, vol. 33, pp. 213-223, 1971.
[53] G. N. Hounsfield, "A method of and apparatus for examination of a body by radiation such as X or gamma radiation," Patent Office, London, England, Patent Specification 1 283 915, 1972.
[54] S. Krishnan, S. S. Prabhu, and E. V. Krishnamurthy, "Probabilistic reinforcement algorithms for the reconstruction of pictures from their projections," *Int. J. Syst. Sci.*, vol. 4, pp. 661-670, 1973.

*Other Reconstruction Techniques*

In addition to [6], [25]:
[55] S. K. Chang, "The reconstruction of binary patterns from their projections," *Commun. Ass. Comput. Mach.*, vol. 14, no. 1, pp. 21-25, Jan. 1973.
[56] S. K. Chang and C. K. Chow, "The reconstruction of three-dimensional objects from two orthogonal projections and its applications to cardiac cineangiography," *IEEE Trans. Comput.*, vol. C-22, pp. 18-28, Jan. 1973.
[57] S. K. Chang and G. L. Shelton, "Two algorithms for multiple-view binary pattern reconstruction," *IEEE Trans. Syst., Man, Cybern.* (Corresp.), vol. SMC-1, pp. 90-94, Jan. 1971.
[58] D. G. Grant, "TOMOSYNTHESIS: A three-dimensional radiographic imaging technique," *IEEE Trans. Biomed. Eng.*, vol. BME-19, pp. 20-28, Jan. 1972.
[59] D. E. Kuhl and R. Q. Edwards, "Reorganizing data from transverse section scans of the brain using digital processing," *Radiology*, vol. 91, pp. 975-983, 1968.
[60] E. R. Miller, E. M. McCurry, and B. Hruska, "An infinite number of laminagrams from a finite number of radiographs," *Radiology*, vol. 98, pp. 249-255, 1971.
[61] S. S. Sandler, "Direct three-dimensional analysis of electron micrograph pictures," *Pattern Recogn.*, vol. 4, pp. 353-359, 1972.

*Others*

[62] M. V. Berry and D. F. Gibbs, "The interpretation of optical projections," *Proc. Roy. Soc. London, Ser. A*, vol. 314, pp. 143-152, 1970.
[63] R. P. Kruger, E. C. Mahen, A. J. Carlson, S. J. Dwyer, and G. S. Lodwick, "Image analysis of radiograms," in *21st SWIEECO Rec.*, pp. 147-151.
[64] D. Ludwig, "The Radon transform on Euclidean space," *Commun. Pure Appl. Math.*, vol. 19, pp. 49-81, 1966.
[65] A. Papoulis, *Systems and Transforms with Applications in Optics*. New York: McGraw-Hill, 1968.
[66] J. Radon, "Ueber die Bestimmung von Funktionen durch Ihre Interralwerte Laengs Gewisser Mannigfaltigkeiten" (On the determination of functions from their integrals along certain manifolds), *Berichte Saechsische Akad. Wissenschaft.* (*Leipzig*), *Math. Phys. Klass 69*, pp. 262-271.

# Two-Dimensional Digital Filtering

RUSSELL M. MERSEREAU, MEMBER, IEEE, AND DAN E. DUDGEON, MEMBER, IEEE

*Invited Paper*

*Abstract*—The problems of designing and implementing LSI systems for the processing of 2-*D* digital data, such as images or geophone arrays, are reviewed and discussed. This discussion encompasses both FIR and IIR digital filters and with respect to the latter, the issues of stability testing and filter stabilization are also considered. Techniques are also presented whereby such filtering can be accomplished using either 1 or 2-*D* LSI systems.

## I. Introduction

OUR OBJECTIVE in this paper is to review the mathematical framework underlying the two-dimensional (2-*D*) digital filtering problem and to explore available techniques for the design and implementation of 2-*D* linear shift-invariant (LSI) digital filters. In the course of this discussion, the many similarities to and differences from the one-dimensional (1-*D*) digital filtering problem will hopefully become apparent. Restricting our attention to studying LSI systems allows us to use the powerful techniques of Fourier analysis which have proved their value in numerous problems of practical interest in both 1 and 2 dimensions.

Other papers in this special issue [1]–[3] have touched upon the need for 2-*D* digital filters within the larger context of 2-*D* digital signal processing. Such filters are central to many image and array processing applications—such as X-ray enhancement, image deblurring, scene analysis, weather predictions, seismic analysis, and the processing of radar and sonar arrays, to name just a few. The discussions in this paper, while applicable to any of these problems, will not be specifically directed toward any one of them.

1-*D* LSI systems represent a special case of two and higher dimensional systems. Therefore, many 2-*D* concepts will look vaguely familiar. For this reason, those 2-*D* results which are straightforward extensions of 1-*D* results will be presented without extensive discussions. On the other hand, there are many properties of 1-*D* LSI systems which cannot be easily generalized, which is why 2-*D* digital signal processing remains a challenging and interesting field of study. These difficulties are almost always related to the fact that there is no fundamental theorem of algebra for polynomials in two independent variables [4]. The reader will recall that it is this theorem which allows us to factor a 1-*D* polynomial of degree $n$ into a product of $n$ polynomial factors of first degree, thereby allowing us to find the roots of polynomials, to check the stability of a filter by finding the locations of the poles of its system function, and to realize digital filters as cascade or parallel structures [5].

Because there are many operations which are more easily performed using 1-*D* mathematics, it is perhaps worthwhile to use 1-*D* filters to perform 2-*D* tasks. Interestingly this can be done, and some techniques will be presented later in this paper. While the ultimate value of such an approach remains to be established, it is the authors' belief that this approach will prove to be quite useful both conceptually as well as in practice. Inasmuch as this material is fairly new, it has received a rather heavy emphasis in this paper.

The remainder of this paper is divided into four parts. In Section II, a number of definitions are presented, primarily to establish notation and to serve as a reference for later sections. The representation of 2-*D* arrays by 1-*D* sequences is also discussed here. In Section III, we discuss the related issues of stability testing and the stabilization of unstable

Manuscript received August 14, 1974; revised October 7, 1974.
This work was supported in part by the Advanced Research Projects Agency monitored by ONR under Contract N-00014-67-A-0204-0064 and in part by the National Science Foundation under Grant GK-31353.
R. M. Mersereau is with the Department of Electrical Engineering and Research Laboratory of Electronics, Massachusetts Institute of Technology, Cambridge, Mass. 02139.
D. E. Dudgeon was with the Department of Electrical Engineering and Research Laboratory of Electronics, Massachusetts Institute of Technology, Cambridge. He is now with Bolt, Beranek, and Newman, Inc., Cambridge, Mass. 02138.

Reprinted from *Proc. IEEE*, vol. 63, pp. 610–623, Apr. 1975.

filters while in Section IV, we discuss the implementation of 2-*D* digital systems, FIR and IIR, using both 1-*D* and 2-*D* digital filters. In Section V, we shall review design techniques for 2-*D* digital filters.

## II. Definitions and Generalizations from One Dimension

In this section, we will examine the underlying mathematical basis for 2-*D* LSI digital filtering. For the most part, these results are straightforward extensions of familiar 1-*D* definitions, but there are some important differences which will be discussed as they arise.

The signals with which we shall concern ourselves are discrete, that is they are functions of integer variables. 1-*D* signals are functions of a single integer variable and 2-*D* signals are functions of two integer variables. Consider the signal $x(m, n)$, shown in Fig. 1. One interpretation of $x$ is as a function which assigns a (generally complex) number to each integer ordered pair $(m, n)$. The function $x$ is not defined when either or both of its arguments $(m, n)$ is not an integer. Such a signal will be interchangeably referred to as either a 2-*D* array, a 2-*D* sequence or simply as a 2-*D* signal. In considering the sequence $x(m, n)$ as an array, referring to Fig. 1, it will be noticed that $m$ thus corresponds to a column index and $n$ corresponds to a row index. This is the opposite of the convention generally used in linear algebra for describing 2-*D* arrays.

An important issue which we will not consider in this paper is the problem of extracting a discrete 2-*D* array from a continuous 2-*D* signal. Clearly, in any practical digital signal processing application this must be done. For completeness, let us, therefore, state that 2-*D* sampling theorems do exist which are conceptually similar to the familiar Nyquist sampling ideas of 1-*D* bandlimited signals. The interested reader is referred to Goodman [6], to the classic paper by Peterson and Middleton [7], or to the accompanying paper by Hunt [2]. In addition, there are discrete representations of continuous signals which are not bandlimited [8]. These ideas can be readily extended to two dimensions and may be helpful for particular 2-*D* digital signal processing problems.

A system which is characterized by the operator $T[\cdot]$ is said to be *linear* if, for arbitrary inputs $x_1(m, n)$ and $x_2(m, n)$ and for an arbitrary complex constant $c$, the following relation is true:

$$T[x_1(m,n) + cx_2(m,n)] = T[x_1(m,n)] + cT[x_2(m,n)]. \quad (1)$$

If $y(m, n) = T[x(m, n)]$, then the system characterized by $T[\cdot]$ is said to be *shift-invariant* if and only if

$$y(m - m_0, n - n_0) = T[x(m - m_0, n - n_0)] \quad (2)$$

for all $x$ and for all arbitrary integers $m_0$ and $n_0$. Linearity and shift-invariance are independent properties of a system in that neither property implies the other. Thus the system

$$T[x(m,n)] = h(m,n)\, x(m,n)$$

is linear but not shift invariant and the system

$$T[x(m,n)] = x^2(m,n)$$

is shift-invariant but not linear.

Any LSI system is easily shown to be completely specified by its impulse response $h(m, n)$. The impulse response (called in image processing the point spread function) is the output

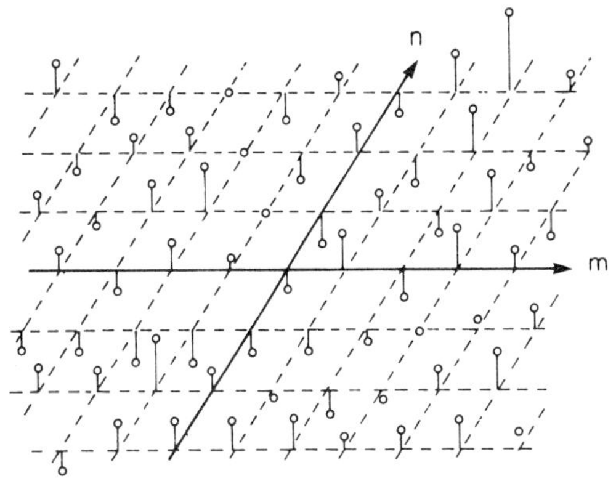


Fig. 1. A pictorial representation for a 2-*D* discrete array.

of the system when the input is a 2-*D* unit sample $\delta(m, n)$ where

$$\delta(m,n) = \begin{cases} 1, & m = n = 0 \\ 0, & \text{otherwise.} \end{cases} \quad (3)$$

Thus

$$h(m,n) = T[\delta(m,n)].$$

As with 1-*D* LSI systems, the output $y(m, n)$ of a 2-*D* LSI system is the *convolution* of the input sequence $x(m, n)$ with the impulse response $h(m, n)$. That is,

$$y(m,n) = \sum_{k=-\infty}^{\infty} \sum_{l=-\infty}^{\infty} x(k,l)h(m-k, n-l)$$

$$= x(m,n) ** h(m,n) \quad (4)$$

where the asterisks are a shorthand representation of the convolution sum of (4).

It is straightforward to verify that complex exponentials of the form $x(m, n) = w^m z^n$ ($w$ and $z$ are arbitrary complex numbers) are eigenfunctions of LSI systems. The associated eigenvalue is

$$H(w,z) = \sum_{k=-\infty}^{\infty} \sum_{l=-\infty}^{\infty} h(k,l) w^{-k} z^{-l} \quad (5)$$

where $h$ is the impulse response. For such an exponential input the output would be

$$y(m,n) = w^m z^n H(w,z).$$

$H(w, z)$ is called the *system function* for the LSI system. Like the impulse response $h$, it too completely characterizes the system.

Let us formally define the 2-*D* *z transform* of a discrete array $x$ as

$$X(w,z) = \sum_{m=-\infty}^{\infty} \sum_{n=-\infty}^{\infty} x(m,n) w^{-m} z^{-n}. \quad (6)$$

The system function is thus seen to be the $z$ transform of the impulse response. We shall adopt the convention of using upper case letters to denote $z$ transforms of arrays which are denoted by the corresponding lower case letter. The sum in (6) may not converge for all (or any) values of the complex variables $w$ and $z$. Those values of $w$ and $z$ for which the sum does converge constitute a *region of convergence* in the $(w, z)$ hyperplane.

A 2-*D* *z* transform can be inverted using the inversion formula

$$x(m, n) = \left(\frac{1}{2\pi j}\right)^2 \oint_{c_1} \oint_{c_2(z)} X(w, z) w^{m-1} z^{n-1} \, dw \, dz. \quad (7)$$

The contours for both integrals must lie completely within the region of convergence in the $(w, z)$ hyperplane; they must be closed and they must encircle the origin counterclockwise in the plane of their respective variables.

At this point, an example might be beneficial. Consider the 2-*D* *z* transform

$$H(w, z) = \frac{1}{1 - aw^{-1} - bz^{-1}}$$

where $|a| + |b| < 1$ and the region of convergence includes the surface $|w| = |z| = 1$. Then

$$h(m, n) = \left(\frac{1}{2\pi j}\right)^2 \oint\oint \frac{w^m z^n}{wz - az - bw} \, dw \, dz$$

$$= \left(\frac{1}{2\pi j}\right)^2 \oint\oint \frac{\frac{1}{(z - b)} w^m z^n}{w - \frac{az}{z - b}} \, dw \, dz.$$

Considering $z$ to be a parameter, the integrand of the $w$ integration considered as a function of $w$ has a simple pole at $az/(z - b)$. If we take the contours of integration to be $|z| = 1$ and $|w| = 1$, then it can be shown that $|az/(z - b)| < 1$ and hence that the pole is inside the contour $|w| = 1$. Applying Cauchy's residue theorem to perform the $w$ integration, we get

$$h(m, n) = \begin{cases} \frac{1}{2\pi j} \oint \frac{a^m z^{m+n}}{(z - b)^{m+1}} \, dz, & \text{for } m \geqslant 0 \\ 0, & \text{for } m < 0. \end{cases}$$

Now we have an $(m + 1)$st-order pole at $z = b$, which lies inside the contour $|z| = 1$. Applying the residue theorem once again yields the final result

$$h(m, n) = \begin{cases} a^m b^n \frac{m!n!}{(m + n)!}, & \text{for } m \geqslant 0, n \geqslant 0 \\ 0, & \text{otherwise.} \end{cases}$$

If a 2-*D* sequence is absolutely summable, it can be represented by its 2-*D* Fourier transform which is a representation of the sequence as a linear combination of sinusoids

$$X(e^{j\mu}, e^{j\nu}) = \sum_{m=-\infty}^{\infty} \sum_{n=-\infty}^{\infty} x(m, n) \exp\left[-j(\mu m + \nu m)\right]. \quad (8)$$

The sequence can be recovered from $X(e^{j\mu}, e^{j\nu})$ by the inverse Fourier transform

$$x(m, n) = \frac{1}{4\pi^2} \int_{-\pi}^{\pi} \int_{-\pi}^{\pi} X(e^{j\mu}, e^{j\nu}) \exp\left[j(\mu m, \nu n)\right] d\mu \, d\nu. \quad (9)$$

The notation for the Fourier transform, although cumbersome, is used to emphasize the fact that the Fourier transform is equal to the $z$ transform evaluated on the surface $w = e^{j\mu}$, $z = e^{j\nu}$. It also serves to remind us that $X$ is periodic in both $\mu$ and $\nu$ with period $2\pi$ in each.

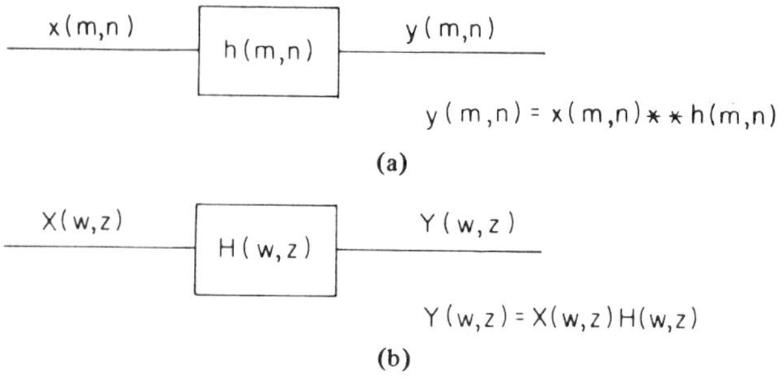


Fig. 2. Two representations of a 2-*D* LSI digital filter. (a) Signal space representation. (b) Transform space representation.

Using (4), (6), and (8), it is straightforward to show that if

$$y(m, n) = x(m, n) * h(m, n)$$

then

$$Y(w, z) = X(w, z) \, H(w, z) \quad (10)$$

and

$$Y(e^{j\mu}, e^{j\nu}) = X(e^{j\mu}, e^{j\nu}) \, H(e^{j\mu}, e^{j\nu}). \quad (11)$$

An LSI digital filter is often drawn in input–output fashion, as in Fig. 2. The variables can be specified as sequences or as transforms. In the former case it is understood that $y$ is the convolution of $x$ and $h$, and in the latter, $Y$ is the product of $X$ and $H$.

The interpretation of the Fourier transform is the same for both 1-*D* and 2-*D* signals. It is a measure of the frequency content of a signal. Slowly varying signals will have Fourier transforms which are mostly concentrated near the origin of the $(\mu, \nu)$ plane and rapidly varying sequences have more frequency content for $\mu$ or $\nu$ near $\pi$.

The $z$ transform and the Fourier transform both suffer from the computational disadvantage that they are functions of continuous variables. As in 1-*D*, the *discrete Fourier transform* (DFT) has thus proven to be an extremely useful tool for examining spectra and for implementing digital filters, particularly since the discovery of a particularly efficient algorithm (the fast Fourier transform (FFT) algorithm) for its evaluation [9].

A 2-*D* DFT may be defined by a straightforward extension of the 1-*D* definition. If we assume that the sequence $x(m, n)$ is nonzero only for $0 \leqslant m \leqslant M - 1$ and $0 \leqslant n \leqslant N - 1$, then we can define $X(k, l)$, the 2-*D* DFT of $x(m, n)$, to be

$$X(k, l) = \sum_{m=0}^{M-1} \sum_{n=0}^{N-1} x(m, n) \exp\left[-j\left[\frac{2\pi km}{M} + \frac{2\pi ln}{N}\right]\right]. \quad (12)$$

As a notational convenience, the substitution

$$W_M = \exp\left[-j\frac{2\pi}{M}\right]$$

is often made yielding

$$X(k, l) = \sum_{m=0}^{M-1} \sum_{n=0}^{N-1} x(m, n) \, W_M^{km} \, W_N^{ln}. \quad (13)$$

The inverse DFT is similar in form

$$x(m, n) = \frac{1}{MN} \sum_{k=0}^{M-1} \sum_{l=0}^{N-1} X(k, l) \, W_M^{-km} \, W_N^{-ln}. \quad (14)$$

The careful reader will notice that the DFT is equal to the Fourier transform $X(e^{j\mu}, e^{j\nu})$ evaluated at $\mu = 2\pi k/M, \nu = 2\pi l/N$ for $0 \leq k \leq M-1$, $0 \leq l \leq N-1$. Thus our notation of $X(k, l)$ for the DFT is ambiguous, although this double usage is quite common. Generally, it will be clear from the context whether we are referring to the $z$ transform or the DFT. If the arguments are restricted to integer values we will always be referring to the latter.

If $M$ and $N$ are highly composite numbers, a 1-*D* FFT algorithm can be used efficiently to evaluate the 2-*D* DFT. This is due to the separability of the DFT kernel $W_M^{km} W_N^{ln}$. (A function $x$ is *separable* if $x(m, n) = x_1(m)x_2(n)$). This allows us to evaluate (13) as follows: for each column of the array $x$ ($m$ = constant) perform an $N$-point DFT on $x(m, n)$ to get an intermediate array $r(m, l)$. Then perform an $M$-point DFT separately on each row ($l$ = constant) of $r(m, l)$. The result is $X(k, l)$. If $M$ and $N$ are powers of two, the total number of real multiplications required is

$$2MN \log_2 N + 2MN \log_2 M = 2MN \log_2 MN.$$

Note that this is the same number of multiplies which would have to be performed to find the 1-*D* DFT of an $MN$ point sequence.

The normal objective of digital filtering is to alter the frequency spectrum $X(e^{j\mu}, e^{j\nu})$ of some signal by multiplying it by some frequency response function $H(e^{j\mu}, e^{j\nu})$. The digital filtering problem consists of designing impulse responses $h(m, n)$, which simultaneously have desirable frequency responses and which can be efficiently implemented. In general, the latter constraint implies that ideal frequency responses can only by approximated by realizable systems. One approach to the approximation problem is to assume that the system function $H(w, z)$ must be a polynomial of finite degree in $w$ and $z$ (or $w^{-1}$ and $z^{-1}$). From the inverse $z$ transform relationship (7) we see that such an impulse response $h(m, n)$ must have only a finite number of nonzero samples. Such filters are referred to as finite-extent impulse response (FIR) filters. With FIR filters we can compute the output directly using the convolution sum (4) or by using the DFT (see Section IV).

Another approach is to assume that the system function is the ratio of finite-degree polynomials in $w$ and $z$.

$$H(w, z) = \sum_{m=-\infty}^{\infty} \sum_{n=-\infty}^{\infty} h(m, n) w^{-m} z^{-n} = \frac{A(w, z)}{B(w, z)} \tag{15a}$$

where

$$A(w, z) = \sum_{k=0}^{M_1} \sum_{l=0}^{N_1} a(k, l) w^{-k} z^{-l} \tag{15b}$$

$$B(w, z) = \sum_{k=0}^{M_2} \sum_{l=0}^{N_2} b(k, l) w^{-k} z^{-l}.$$

If $X(e^{j\mu}, e^{j\nu})$ is the Fourier transform of the input and $Y(e^{j\mu}, e^{j\nu})$ is the Fourier transform of the output of the filter, then

$$Y(e^{j\mu}, e^{j\nu}) = H(e^{j\mu}, e^{j\nu}) X(e^{j\mu}, e^{j\nu}) = \frac{A(e^{j\mu}, e^{j\nu})}{B(e^{j\mu}, e^{j\nu})} X(e^{j\mu}, e^{j\nu})$$

or

$$B(e^{j\mu}, e^{j\nu})\, Y(e^{j\mu}, e^{j\nu}) = A(e^{j\mu}, e^{j\nu})\, X(e^{j\mu}, e^{j\nu}). \tag{16}$$

Equation (16) implies

$$\sum_{k, l} b(k, l) y(m-k, n-l) = \sum_{k, l} a(k, l) x(m-k, n-l). \tag{17}$$

Equation (17) is an example of a 2-*D difference equation* which relates weighted sums of shifted versions of the input and output. The output $y$ can be computed recursively from the input $x$ and a set of initial conditions. This can be done by rewriting (17) as

$$y(m-i, n-j) = \frac{1}{b(i, j)} \sum_{k, l} a(k, l) x(m-k, n-l) - \frac{1}{b(i, j)} \sum_{k, l}{}' b(k, l) y(m-k, n-l). \tag{18}$$

The prime by the second summation indicates that the sum is for values of the integer ordered pair $(k, l)$ excluding the value $(i, j)$. Such a filter is referred to as recursive or infinite-duration impulse response (IIR). The latter term is somewhat of a misnomer since not all infinite-length impulse responses can be realized using difference equations, only those with rational transfer functions.

It is assumed in (18) that all values of $y(m-k, n-l)$ needed on the right-hand side of that equation are known either from previous calculations or from the initial conditions. If it is possible to increment the values of the indices $(m, n)$ in such a fashion that all of the values of $y$ can be computed in turn from a given set of initial conditions, the difference equation will be called *recursible*. In general, the recursibility of (18) will depend upon which term $y(m-i, n-j)$ is deleted from the right-hand side summation, i.e., it depends upon the value of $(i, j)$. Recursibility will be studied in greater detail in Section IV.

In several interesting cases, 2-*D* arrays can be represented exactly by 1-*D* sequences [10]. This approach has great potential for 2-*D* digital signal processing since the resulting sequences can be processed by familiar 1-*D* techniques and the results can then be remapped to two dimensions.

A 2-*D* array of finite duration

$$\{0 \leq m \leq M-1, 0 \leq n \leq N-1\}$$

can be mapped to a 1-*D* sequence $g(p)$ by means of the mapping

$$g(Nm + n) = f(m, n), \qquad 0 \leq m \leq M-1, \quad 0 \leq n \leq N-1. \tag{19}$$

The samples of $g(p)$ are thus chosen by concatenating the columns of $f$. This is sometimes referred to as a lexicographic ordering of the 2-*D* sequence into a 1-*D* sequence. Due to the limited extent of $f$, the mapping of (19) is readily seen to be invertible. If this mapping is considered in the $z$ plane, then

$$G(z) = F(z^N, z). \tag{20}$$

Thus the $z$ transform of $g$ corresponds to the $z$ transform of $f$ evaluated along the contour $w = z^N$. The function $G(z)$ is referred to as a *slice* of $F(w, z)$ and $g(p)$ is called a *projection* of $f(m, n)$.

Restricting our attention to the Fourier plane, (20) becomes

$$G(e^{j\omega}) = F(e^{jN\omega}, e^{j\omega}). \tag{21}$$

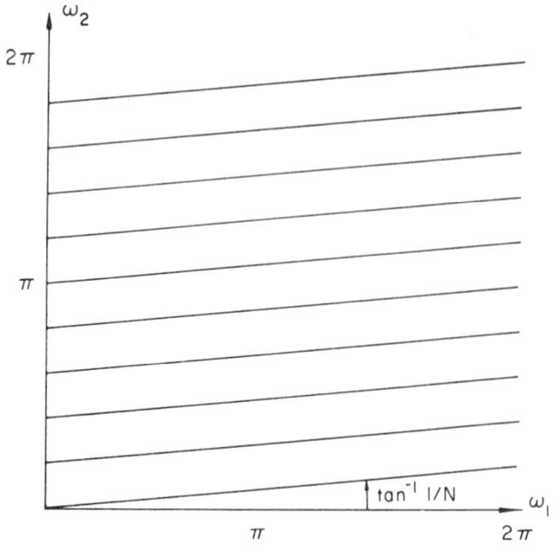


Fig. 3. Lines in the 2-*D* Fourier plane along which $F(\exp(j\mu), \exp(j\nu))$ and $G \exp(j\omega)$ are numerically equal for $N = 10$.

Since $F(e^{j\mu}, e^{j\nu})$ is doubly periodic in $\mu$ and $\nu$ with period $2\pi$ in each, $G(e^{j\omega})$ corresponds to the evaluation of $F(e^{j\mu}, e^{j\nu})$ along a series of $N$ parallel lines in the 2-*D* Fourier plane. Each of these $N$ lines forms an angle $\theta = \tan^{-1} 1/N$ with the $\mu$ axis, as illustrated in Fig. 3. $F(e^{j\mu}, e^{j\nu})$ can be recovered from $G(e^{j\omega})$ using the interpolation formula

$$F(e^{j\mu}, e^{j\nu}) = \sum_{k=0}^{N-1} G \exp\left(j\left(\frac{\mu + 2\pi k}{N}\right)\right) \frac{\sin \frac{N}{2}\left(\frac{\mu + 2\pi k}{N} - \nu\right)}{\sin \frac{1}{2}\left(\frac{\mu + 2\pi k}{N} - \nu\right)}$$
$$\times \exp\left[-j \frac{N-1}{2}\left(\frac{\mu + 2\pi k}{N} - \nu\right)\right]. \quad (22)$$

If we are given a rational 2-*D* frequency response

$$H(e^{j\mu}, e^{j\nu}) = \frac{A(e^{j\mu}, e^{j\nu})}{B(e^{j\mu}, e^{j\nu})} \quad (23)$$

there exist slices from which $H(e^{j\mu}, e^{j\nu})$ can be recovered [10]. If $A$ is a 2-*D* polynomial of degree $N_1$ in $\exp[j\nu]$ and if $B$ is a polynomial of degree $N_2$ in $\exp[j\nu]$, then letting $N = \max[N_1, N_2]$, $H$ can be recovered from the slice at angle $\tan^{-1} 1/N$. If $\hat{h}(p)$ denotes the projection associated with this slice, then

$$\hat{h}(p) = \sum_{m=-\infty}^{\infty} h(m, p - Nm). \quad (24)$$

The sequences $\hat{a}(p)$ and $\hat{b}(p)$, which are the inverse transforms of $A(z^N, z)$ and $B(z^N, z)$, respectively, can be recovered from $\hat{h}(p)$ by Prony's method, Shanks' method, or any similar technique which estimates the poles and zeros of a 1-*D* rational system function from samples of its impulse response [11], [12]. Then $\hat{a}(p)$ and $\hat{b}(p)$ can be separately back-projected, as in the finite duration sequence case, to give the coefficients of the numerator and denominator polynomials $a(m, n)$ and $b(m, n)$ from which $H(e^{j\mu}, e^{j\nu})$ can be reconstructed or $h(m, n)$ determined.

## III. Stability of IIR Filters

An important issue in the design of digital filters is their stability. If a filter is unstable, any input, including computational noise, can cause the filter output to grow without bound, obliterating the desired output of the filter. Thus the condition is generally imposed that the output of the digital filter must remain bounded for any bounded input. This is sometimes referred to as bounded-input–bounded output (BIBO) stability. It can be easily shown [13] that BIBO stability is equivalent to requiring that the impulse response of the filter $h(m, n)$ be absolutely summable

$$\sum_{m=-\infty}^{\infty} \sum_{n=-\infty}^{\infty} |h(m, n)| < \infty. \quad (25)$$

For FIR filters, this condition is always satisfied, since $h(m, n)$ has only a finite number of nonzero terms. For IIR filters, on the other hand, stability is a very important design constraint.

In the subsequent discussion, we shall restrict ourselves to impulse responses which are nonzero only for nonnegative values of $m$ and $n$ (first quadrant sequences) and to system functions whose numerator and denominator polynomials are the $z$ transforms of first quadrant sequences. (That is, the numerator and denominator of the system function are polynomials in $w^{-1}$ and $z^{-1}$.) This restriction is commonly imposed upon IIR filters so that they have recursible implementations, although there is a slightly more general class of 2-*D* IIR filters which are also recursible [14].

The basic theorem for guaranteeing BIBO stability is due to Shanks [13], [15] and has been generalized by Farmer and Bednar [16]. Restated in our notation, Shanks' theorem states the following.

*Theorem (Shanks):* A recursive (IIR) filter

$$H(w, z) = \frac{A(w, z)}{B(w, z)} = \frac{\sum_{m=0}^{M} \sum_{n=0}^{N} a(m, n) w^{-m} z^{-n}}{\sum_{k=0}^{K} \sum_{l=0}^{L} b(k, l) w^{-k} z^{-l}}$$

is BIBO stable if and only if there are no values of $w$ and $z$ such that $B(w, z) = 0$ for $|w| \geq 1$ and $|z| \geq 1$ simultaneously.

This is analogous in 1-*D* to requiring that the system function have no poles outside the unit circle. One important difference is that 2-*D* singularities defined by $B(w, z) = 0$ are not isolated points as poles are, but are rather multidimensional surfaces or manifolds.

To apply the Shanks test is conceptually straightforward but computationally involved. The region of the $z$ plane, $|z| \geq 1$ must be mapped into the $w$ plane by the algebraic mapping $B(w, z) = 0$. If the image of that map lies completely within the circle $|w| = 1$ the filter is stable, otherwise, it is not.

A simpler stability test to apply is based on a theorem due to Huang [17].

*Theorem (Huang):* A recursive filter

$$H(w, z) = \frac{A(w, z)}{B(w, z)}$$

is stable if and only if the image of the unit circle $|z| = 1$ mapped into the $w$ plane by $B(w, z) = 0$ does not intersect the region $|w| \geq 1$ and in addition, no point in the region $|w| \geq 1$ maps into the point $z = \infty$.

To apply Huang's test, the $z$ unit circle must be mapped into the $w$ plane by $B(w, z) = 0$. For the filter to be stable, its image must not intersect $|w| \leq 1$. In addition, we must solve $B(w, z) = 0$ for $w$, setting $z^{-1} = 0$. The magnitude of all of the roots of this equation must be less than one. Both this test and the preceding one require the mapping of a continuous region of $(w, z)$ hyperspace and thus, in theory, both require an infinite amount of computation. A fair amount of effort has thus been expended to find implementations of Huang's test which require only a finite amount of computation. The first such attempt was by Huang himself in the same paper. He showed that his test could be reformulated in a form which only required a finite number of computational steps using a test of Ansell's [18] for continuous systems. Anderson and Jury [19] and Maria and Fahmy [20] have also proposed more efficient implementations of Huang's test. The former showed that the main component of Huang's test could be accomplished using a Schur–Cohn matrix test. This test has also been generalized for checking the stability of $N$-dimensional systems [21], [22]. Although the Anderson–Jury test is probably the most efficient stability testing algorithm, the testing of stability remains a major computational effort.

Any stability test, however simple, is not sufficient for designing 2-*D* recursive filters. What is required is a procedure to stabilize an unstable filter without perturbing the magnitude of its frequency response, or a design technique which will guarantee a stable filter. This is easily accomplished in one dimension since the poles of the system function which are located outside the unit circle, if any, which account for the instability, can be replaced by poles in the conjugate reciprocal locations, which are necessarily within the unit circle. This is equivalent to cascading the unstable filter with a digital all-pass filter which has no effect upon the magnitude of the frequency response. This procedure can either be performed algebraically or using discrete Hilbert transform (DHT) techniques. In 2-*D* there are difficulties with both approaches. The algebraic approach fails due to the inability to factor 2-*D* polynomials. There are also difficulties associated with the DHT.

Read and Treitel [23] have defined a 2-*D* DHT to be used for the stabilization of recursive filters. Their idea was to obtain the log-magnitude function of the denominator polynomial and use the 2-*D* DHT to calculate the minimum phase function associated with that log-magnitude function. A new denominator polynomial is then constructed by a complex exponentiation. This technique was used on examples with denominator polynomials up to degree $5 \times 5$ with moderate success. Their difficulties arose from nonexact computations of the DHT and the fact that in many cases the reconstructed minimum phase denominator polynomial is not of finite degree.

A similar attempt at stabilization utilizing the notion of a 2-*D* complex cepstrum was performed by Pistor [24] and Dudgeon [25]. As in the 1-*D* case a *complex cepstrum* can be defined as the inverse Fourier transform of the complex logarithm of the Fourier transform of a sequence. Thus, if two sequences are convolved, their cepstra add. Dudgeon [14], [26] has shown that cepstra exist for 2-*D* rational functions

$$H(w, z) = \frac{A(w, z)}{B(w, z)}$$

provided that:

1) $A(w, z)$ and $B(w, z)$ are never zero at any point $(w, z)$ such that $|w| = 1$ and $|z| = 1$;
2) any linear phase components are removed from the Fourier transform by an appropriate linear shift of the sequence $h(m, n)$.

As an aside, it might be mentioned that we can also talk about the 1-*D* cepstrum of an image by considering the cepstrum of the projection of the image. This is also the projection of the cepstrum and like the standard 2-*D* cepstrum, it has the property that the 1-*D* cepstra add when images are convolved. The 1-*D* cepstra of 2-*D* rational functions are invertible and are defined whenever the standard 2-*D* cepstrum is defined. More importantly, they can be defined in many cases when standard cepstra cannot!

Both Pistor and Dudgeon used 2-*D* cepstra to decompose magnitude-squared frequency responses to get stable, recursible 2-*D* filters. Pistor decomposed the magnitude-squared function into a cascade of four stable recursive one-quadrant filters, each of whose impulse responses was defined on a different quadrant. Dudgeon's approach was similar in spirit. He decomposed the magnitude-squared frequency response into a cascade of two stable, recursible, half-plane filters. (The half-plane filter is one for which the impulse response is required to be zero over half the $(m, n)$ plane rather than over three quadrants, as in the one-quadrant filter case). The filters from both techniques, having rational transfer functions, could then be implemented using difference equations [14], [27].

Both Pistor's and Dudgeon's techniques suffer from the serious difficulty also encountered by Read and Treitel [23] that an exact decomposition yields a denominator polynomial which is generally not of finite degree, in spite of the fact that the denominator of the original magnitude-squared function is of finite degree. Both Pistor and Dudgeon were forced to approximate their infinite extent results by finite extent results in order to get a practical implementation. This procedure alters the frequency response of the filter. In some cases the effect is tolerable, and in others it is quite serious.

A completely unrelated stabilization procedure is based on a conjecture due to Shanks [15]. His conjecture states that the planar least squares inverse (PLSI) of an array is a minimum phase array. To illustrate his procedure, assume that we have a possibly unstable filter

$$H(w, z) = \frac{A(w, z)}{B(w, z)}$$

which we want to stabilize. Letting $b(m, n)$ denote the coefficients of the denominator polynomial, we seek an inverse to $b$, denoted by $p$ such that

$$\delta(m, n) \simeq b(m, n) ** p(m, n) \tag{26}$$

where $**$ denotes 2-*D* convolution. In general, the equality in (26) will not be exact and $p$ must be chosen to minimize the error between the left and right-hand sides according to some norm. If $p$ is chosen to minimize the mean-squared error

$$E = \sum_{m,n} (\delta(m, n) - b(m, n) ** p(m, n))^2$$

then it is referred to as the PLSI of $b(m, n)$. By the conjecture, $p(m, n)$ is a minimum phase array. Shanks further states that if the array of denominator coefficients of a rational transfer function is a minimum phase array, then the filter is stable. (An array $x(m, n)$ is minimum phase if the 1-*D* Fourier transform $X(e^{j\mu_0}, e^{j\nu})$ is minimum phase for all fixed $\mu_0$ and all real $\nu$ and if $X(e^{j\mu}, e^{j\nu_0})$ is a 1-*D* minimum phase function for fixed $\nu_0$ and all real $\mu$.)

To stabilize an unstable filter, Shanks *et al.* [15] proposed taking the double PLSI of the denominator array. This double inversion, by the hypothesis, will yield a stable filter and the frequency response of the final result will hopefully approximate the original frequency response. The final frequency response will only be an approximation, however, and should be measured to see that the approximation is sufficiently good. Although the hypothesis has never been proven, it has been proven in the 1-*D* case and has been used in the design of several hundred 2-*D* filters. A counterexample has never appeared. It is perhaps reasonable, however, to test the stability of a filter obtained by this procedure.

## IV. Digital Filter Implementation

As in the 1-*D* case, 2-*D* digital filters fall into two classes—those whose impulse responses contain a finite number of samples and whose transfer functions are polynomials, the FIR filters, and those whose transfer functions are rational functions, the IIR filters. Both questions of design and questions of implementation are quite different for the two classes of filters and for this reason in this section and the next, the two classes will be treated separately. FIR filters can either be implemented directly or they can be implemented using high speed convolution techniques (using either DFT or number theoretic transforms [28]). 2-*D* IIR filters are invariably implemented recursively using difference equations.

The choice between using an FIR filter or an IIR filter depends upon the application, although most applications are probably better served by FIR designs. They are always stable and they can be designed to have either zero or linear phase, which is an important consideration in image processing applications. 1-*D* design methods are generally extendable and filters of high order can be designed relatively quickly. The primary advantage of IIR filters is that fewer arithmetic operations can often be employed than by using comparable FIR filters, however these filters can be unstable. IIR filters are particularly useful in applications such as seismic data processing where volumes of data must be processed very quickly and where the extra time required to design the filter is easily justified.

### FIR Filters

Referring to Fig. 2, let $x(m, n)$ and $y(m, n)$ denote the input and output respectively of a digital filter characterized by the impulse response $h(m, n)$. Since $h(m, n)$ has an impulse response which is of finite extent, we can say with no loss of generality that

$$h(m, n) = 0, \qquad (m, n) \notin I_{MN}$$

$$I_{MN} = \{(m, n): 0 \leqslant m \leqslant M - 1, 0 \leqslant n \leqslant N - 1\}.$$

$M$ is thus the horizontal extent of the array and $N$ is its vertical extent. The output $y$ is then the convolution of $x$ and $h$

$$y(m, n) = \sum_{k=0}^{M-1} \sum_{l=0}^{N-1} x(k, l) h(m - k, n - l). \tag{27}$$

Equation (27) represents one realization of the filter.

As a tutorial device it may be instructive to define the concept of an input mask as in Fig. 4. To produce the value $y(m, n)$ the input mask is placed over the appropriate samples of the input array. Each sample is then multiplied by the impulse response coefficient associated with that position in the mask and the products are summed to give $y(m, n)$. This is repeated for all $(m, n)$. The $(m, n)$ can be incremented by rows, by columns, or by any other ordering since the values of the output sequence do not depend on the order in which those values are computed.

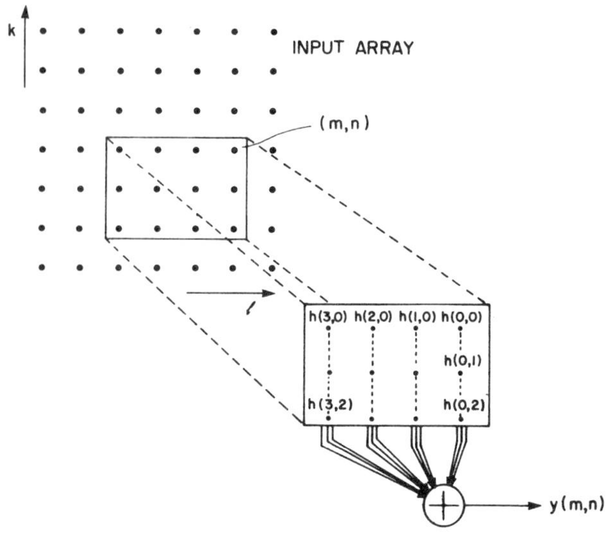


Fig. 4. An example of an input mask for a 3 × 3 FIR filter.

The realization of (27) has the additional advantage shared by the corresponding realization for 1-*D* FIR digital filters that there is little arithmetic quantization error. If the input samples are quantized to $b_1$ bits and the impulse response samples are quantized to $b_2$ bits, the full $b_1 + b_2$ bit products can be accumulated and only the final sum needs to be quantized.

This realization also allows savings for those filters where the impulse response possesses symmetries. If several samples of the impulse response have the same value then by applying the arithmetic distributive law and adding the appropriate input samples before multiplying by their common coefficient value, a savings in the number of multiplications can be realized. As an example, consider an impulse response which is $N_H \times N_H$ points which is used to filter an $N_X \times N_X$ input. In the absence of any exploitable symmetries in the impulse response, using (27) approximately $N_H^2$ multiplies and adds must be performed for each sample of the output. This results in a total of $N_X^2 N_H^2$ multiplies. In the case of octagonal symmetry

$$h(m, n) = h(-m, n) = h(m, -n) = h(n, m)$$

which is common for zero phase filters which approximate circularly symmetric frequency responses, the number of distinct coefficients is $[(N_H + 1)/2][(N_H + 3)/2]/2$, where $[\cdot]$ denotes the greatest integer function. Filtering the same image with such a filter requires $N_X^2[(N_H + 1)/2][(N_H + 3)/2]/2$ multiplies and $N_X^2 N_H^2$ adds. This is a reduction in the number of multiplies by roughly a factor of eight.

FIR filters can also be implemented using the FFT. It can be readily shown by paralleling the 1-*D* derivation that if

$$x(m, n) ** h(m, n) = y(m, n)$$

then

$$X(k, l) H(k, l) = Y(k, l) \tag{28}$$

where $X(k, l)$, $H(k, l)$, and $Y(k, l)$ are the 2-*D* DFT's of $x$, $h$, and $y$, respectively. Equation (28) is true for discrete Fourier transforms of any size, although to be useful for implementing filters $y(m, n)$ must be recoverable from $Y(k, l)$. Thus, if

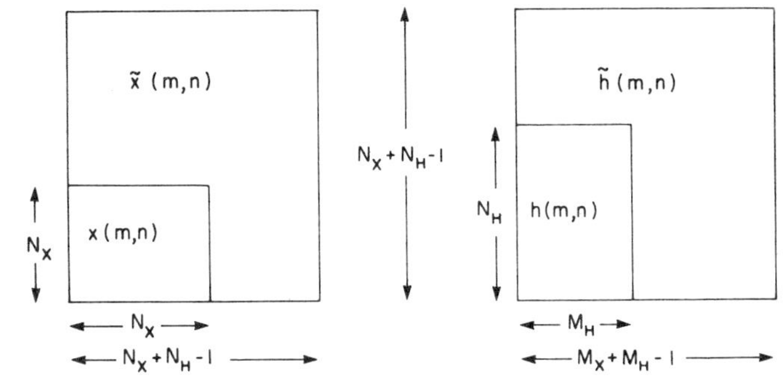


**Fig. 5.** This figure illustrates padding with zeros. The inverse DFT of the product of the DFT's of $\tilde{x}$ and $\tilde{h}$ corresponds to the linear convolution of $x$ and $h$.

$x(m, n)$ is an $M_X \times N_X$ array and $h(m, n)$ is an $M_H \times N_H$ array so that $y(m, n)$ is an $M_Y \times N_Y$ array where

$$M_Y = M_X + M_H - 1$$
$$N_Y = N_X + N_H - 1$$

then the 2-*D* DFT's must all contain at least $M_Y \times N_Y$ points. (They can, of course, be larger in either or both dimensions.) To take the $M_Y \times N_Y$ point DFT's of the arrays $x$ and $h$ we can simply replace $x$ and $h$ by $\tilde{x}$ and $\tilde{h}$, respectively, where

$$\tilde{x}(m, n) = \begin{cases} x(m, n), & (m, n) \in I_{M_X N_X} \\ 0, & (m, n) \in I_{M_Y N_Y}, \quad (m, n) \notin I_{M_X N_X} \end{cases} \tag{29}$$

$$\tilde{h}(m, n) = \begin{cases} h(m, n), & (m, n) \in I_{M_H N_H} \\ 0, & (m, n) \in I_{M_Y N_Y}, \quad \notin I_{M_H N_H}. \end{cases} \tag{30}$$

Such an operation, commonly referred to as padding with zeros, is represented in Fig. 5.

If such an operation is performed on real sequences $x$ and $h$ (and the DFT of $h$ has been precomputed and stored), then the convolution $y$ can be performed using approximately

$$2\, M_Y N_Y \log_2 M_Y N_Y + 2\, M_Y N_Y$$

real multiplications and real additions if $M_Y$ and $N_Y$ are both powers of two. If $h(m, n)$ has octagonal symmetry, there are essentially no additional savings since this fact can only be used to compute $H(k, l)$, which by assumption, was precomputed.

For filtering very large inputs when machine storage is a limiting factor or when the size of the input is considerably larger than the size of the impulse response, the ideas of block convolution using either an overlap-add or overlap-save approach [29] can be applied as in the 1-*D* case. The extension to 2-*D* is straightforward. The idea is to partition the input into blocks and to filter each of the blocks separately and then appropriately combine the results. The size of the blocks used is generally limited by machine considerations and the efficiency of the whole procedure varies greatly with the block-size chosen. Hunt [30] has analyzed this problem in some depth.

For short impulse responses, the direct convolution results in a faster implementation than the one which uses the FFT, whereas for longer impulse responses the FFT approach is faster. To get a feeling for where the crossover point lies, let us assume that the total number of multiplications involved is a figure of merit. Then the FFT approach will be faster if

$$M_X N_X M_H N_H > 2\, M_Y N_Y \log_2 M_Y N_Y + 2\, M_Y N_Y. \tag{31}$$

Assuming that $M_H = N_H = M_X = N_X$, (31) is true for $N > 8$! If the filter possesses octagonal symmetry the FFT would be favored for impulse responses larger than about $24 \times 24$. If $M_X = N_X = 256$ and $M_H = N_H$ then (31) is true for $N_H > 6$ without octagonal symmetry and for $N_H > 18$ with it. Thus for FIR filters of reasonable size the FFT should nearly always be used in implementation.

Instead of using the Fourier transform to perform the convolution, the number theoretic transform [28] could be used, with perhaps even greater computational efficiency. In fact, it has been suggested that due to the lengths of the transforms involved, the number of multiplications involved etc., that these transforms are ideally suited for implementing 2-*D* FIR filters.

2-*D* FIR filters can also be implemented as 1-*D* FIR filters using projections. To see how this can be done, if

$$y(m, n) = x(m, n) ** h(m, n)$$

then from the properties of $z$ transforms, we know that

$$Y(w, z) = X(w, z) H(w, z)$$

and thus

$$Y(z^N, z) = X(z^N, z) H(z^N, z). \tag{32}$$

From the relations between projections and their $z$ transforms which we developed in Section II, it follows that

$$g_y(p) = g_x(p) * g_h(p) \tag{33}$$

where $g_y$, $g_x$, and $g_h$ are the projections of $y$, $x$, and $h$, respectively, and the convolution is 1-*D*. In order for (33) to be useful in a filter implementation, $y(m, n)$ must be recoverable from $g_y(p)$. This implies that the parameter $N$ which appears in (32), which defines the projection, must be greater than or equal to the vertical extent of $y(m, n)$. To obtain projections of $x$ and $h$, based on this same parameter, we merely lexicographically order the arrays $x$ and $h$ which were defined in (29) and (30). The final filter implementation appears in Fig. 6. The filter $g_h(p)$ is a 1-*D* FIR filter and it can be realized in any of the classical forms for these filters, including using the FFT. It can be readily seen that the number of multiplications and the amount of storage involved are identical to the corresponding 2-*D* realizations discussed here. This type of an implementation has also been treated by MacAdam [31] and is the reverse of the problem treated by Agarwal and Burrus [32] who used higher dimensional convolutions to implement 1-*D* convolutions.

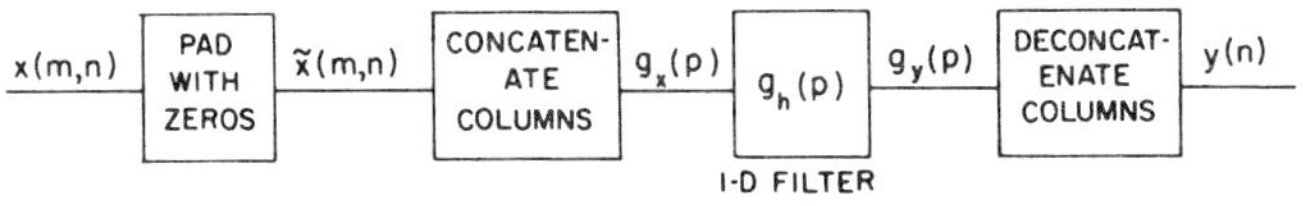


**Fig. 6.** A representation of a 2-*D* filter using a single 1-*D* filter using projections.

Another implementation using 1-*D* filters is due to Treitel and Shanks [33] and Tretiak [34]. They have shown that any $N_H \times N_H$ point filter can be realized as a sum of separable

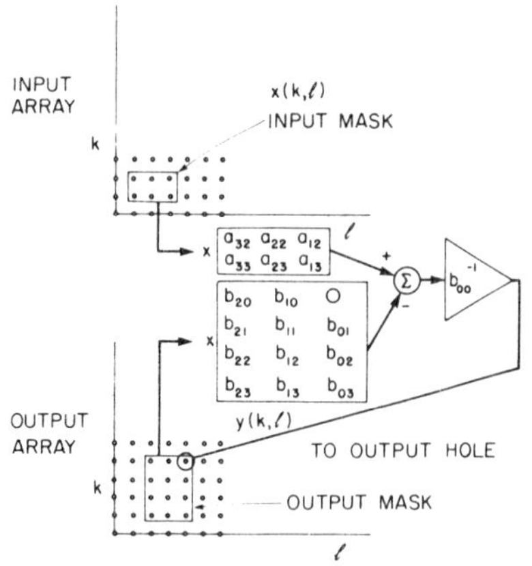


Fig. 7. An example of input and output masks to calculate $y(4, 4)$.

filters

$$h(m, n) = \sum_{i=1}^{N_H} \lambda_i f_i(m) g_i(n). \tag{34}$$

Separable filters are particularly simple to implement. Each filter in the sum can be realized by filtering each row by the 1-*D* filter $f_i(m)$ (which requires $N_H N_X^2$ multiplications) and then filtering each column of the result by $g_i(n)$ (which requires $N_X N_H (N_X + N_H - 1)$ multiplications). The complete filter thus requires $N_X N_H^2 (2N_X + N_H - 1)$ multiplications, which is roughly two times the number needed for a direct implementation. On the other hand, if a number of the eigenvalues $\{\lambda_i\}$ are sufficiently small so that only $k$ stages of the filter are needed to approximate $h$, then computational speed advantages can still be realized if

$$k(2N_X + N_H)(N_X N_H) < N_X^2 N_H^2$$

or if

$$k < \frac{N_X N_H}{2N_X + N_H}.$$

For details on how the $\lambda_i$, $f_i(m)$, and $g_i(n)$ are computed, the reader is referred to [33]. These results can be generalized by using the singular value decomposition of an $M_H \times N_H$ matrix to matrices which are not square.

## IIR Filters

2-*D* IIR digital filters are generally specified by 2-*D* difference equations of the general form

$$\sum_{k=0}^{M_1} \sum_{l=0}^{N_1} b(k, l) y(m-k, n-l) = \sum_{k=0}^{M_2} \sum_{l=0}^{N_2} a(k, l) x(m-k, n-l). \tag{35}$$

If $b(0, 0) \neq 0$, this can be rewritten as

$$y(m, n) = \frac{1}{b(0,0)} \left\{ - \sum_{\substack{k=0 \\ k \cdot l \neq 0}}^{M_1} \sum_{l=0}^{N_1} b(k, l) y(m-k, n-l) + \sum_{k=0}^{M_2} \sum_{l=0}^{N_2} a(k, l) x(m-k, n-l) \right\} \tag{36}$$

which expresses $y(m, n)$ as a linear combination of input samples and previously computed values of the output $y$. Generalizing the concept of an input mask which we used before, we can define two masks, an input mask and an output mask, as depicted in Fig. 7. The input mask is the same as in the FIR case. It sweeps over the input array $x(m, n)$ and multiplies the appropriate values of $x$ by the $(M_2 + 1) \times (N_2 + 1)$ array of coefficients $a(k, l)$. The output mask is similar to the input mask except that it possesses a hole at the location of the output value presently being computed. The mask corresponding to (37) thus has a hole location in its upper right (NE) hand corner. Other masks can be chosen corresponding to the difference equation of (35). In particular, masks could be developed with holes in the NW, SW, or SE corners in addition to the NE. The impulse responses corresponding to these different masks will be different for each of the four and they will not all be necessarily stable. In fact, it has been shown [24] that at most one of the four masks can correspond to a stable filter for a particular set of denominator coefficients $b(k, l)$. Generally, the choice of a mask, given a set of coefficients $b(k, l)$ is thus determined by stability considerations. An IIR filter is said to be *recursible* if for every output point $(m, n)$ the output mask covers only points which have been previously computed. Thus the output mask in Fig. 7 corresponds to a recursible realization of (36) since if the values of the output are computed in the order $\cdots (m, n), (m, n+1), (m, n+2), \cdots, (m+1, n-1), (m+1, n), (m+1, n+1), \cdots$ the output mask only covers points which have been previously computed. Dudgeon [27] has shown that the output mask in Fig. 7 is not the most general form for a recursible output mask but is rather a special case of the mask in Fig. 8 which is columnwise recursible. It is common practice in the field, however, to deal with the masks similar to that in Fig. 7, and this we shall continue to do. This is the form, for example, assumed by most stability tests.

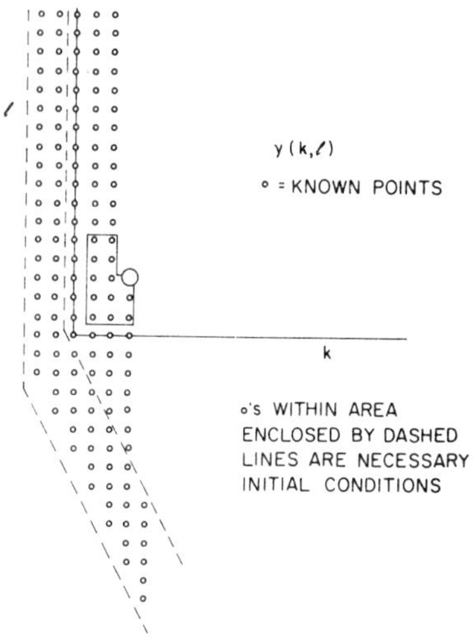


Fig. 8. A more general form for a recursible output mask. This filter is columnwise recursible if the appropriate initial conditions are given.

Implicit in the notion of recursive implementations of digital filters is the requirement that a set of initial conditions be specified. It can be shown that these initial conditions must be zero in order that the resulting system be linear, although for many problems strict linearity may not be a requirement. For the realization of (36), which recurses in the NE direction, a sufficient set of initial conditions is shown in Fig. 9. Here we are assumed to know the initial values of $y$ in an L-shaped band which contains $M_1$ columns and $N_1$ rows. Other sets of sufficient initial conditions can be generated by linearly displacing this set of locations in either the $m$ or $n$ directions or both.

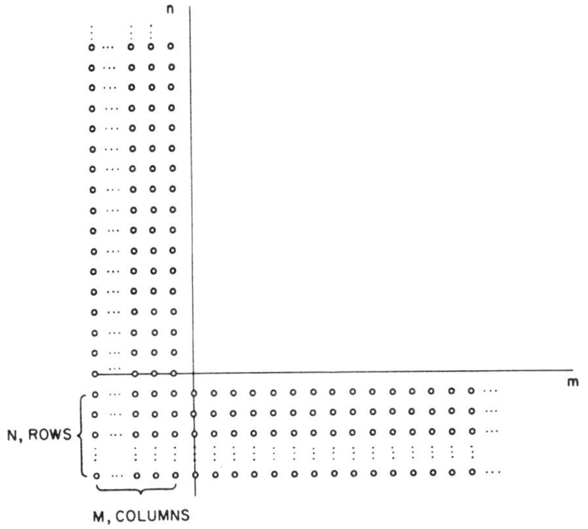


Fig. 9. A sufficient set of initial conditions for the filter of (36).

As in 1-*D*, other realizations of the filter in (35) can be obtained by rearranging the computations involved. This is most often easily done by manipulating the *z* transform. The realization which we have been discussing, the direct form, combines together the terms corresponding to delayed values of the input and delayed values of the output. Alternatively by writing the *z* transform in the form

$$H(w,z) = \{ \sum\sum b(k,l)\, w^{-k} z^{-l} \}^{-1} \{ \sum\sum a(k,l)\, w^{-k} z^{-l} \} \tag{37}$$

we can consider the same filter as a cascade of two systems—one which uses only past values of its output and one which has a finite length impulse response. This is the so-called canonic form and it corresponds to rewriting (36) as

$$u(m,n) = \frac{1}{b(0,0)} \left\{ - \sum_{\substack{k=0 \\ k \cdot l \neq 0}}^{M_1} \sum_{l=0}^{N_1} b(k,l)\, u(m-k, n-l) + x(m,n) \right\}$$

$$y(m,n) = \sum_{k=0}^{M_2} \sum_{l=0}^{N_2} a(k,l)\, u(m-k, n-l). \tag{38}$$

Although the two realizations use the same coefficients and have the same frequency response, the two realizations will perform differently with respect to the effect of the noise which is generated as a result of arithmetic roundoff using finite precision arithmetic. Ni and Aggarwal [35] have considered the arithmetic roundoff question and while they have shown that the structures will perform differently, no definite conclusions could be made concerning which structure to prefer.

Other 1-*D* structures such as the cascade, parallel, and continued fraction structures do not generalize to two dimensions in the general case, due to the inability to factor 2-*D* polynomials. Rational *z* transforms which are factorable can be profitably implemented using a cascade realization, however. Such a realization is also attractive for purposes of guaranteeing stability. However, since in general the factorability property will not be present it must be imposed into the design algorithm for the filter. This complicates the design algorithm.

IIR filters can also be implemented using projections much as FIR filters can, as seen using the system in Fig. 10. The difficulty with the system is that the back projection operator is complicated. In fact, the only known technique for performing this operation is to use the technique mentioned in Section II. This is sufficiently complicated as to make this type of implementation impractical. Dudgeon [14] has suggested a simpler scheme where the backprojection operation is replaced by the simple deconcatenator of Fig. 6. The idea is as follows. A 1-*D* input sequence is formed by concatenating the length *N* columns of the input array. This sequence is then processed by a 1-*D* IIR filter, which can be implemented in any standard configuration. The resulting 1-*D* output sequence is then used to form a 2-*D* array by taking the first *N* points as the first column, the next *N* points as the second column, etc. The overall system is linear but shift varying with the result that only periodic impulse responses of the form $h(m, n) = h(m + 1, n - N)$ can be realized exactly by this technique. However, it is possible to approximate other impulse responses in this manner.

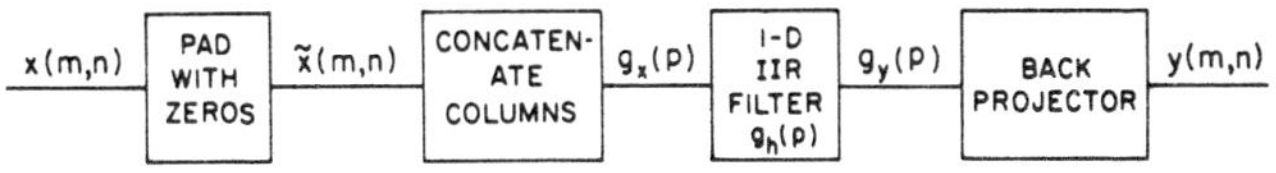


Fig. 10. A system for implementing 2-*D* IIR filters using a 1-*D* IIR filter.

This technique can be refined by considering a 2-*D* array to be a superposition of subarrays each of which contains a single row, padded with appropriate zeros. Each subarray can be filtered separately by the preceding technique and the outputs superimposed to give a final output array. Now the net operation corresponds to a LSI operation which requires doing *N* 1-*D* filtering operations. However, the *N* filtering operations can each be done with different 1-*D* filters which will give an overall system which is shift-varying in a controlled manner. Manry and Aggarwal [36] have also explored a similar approach to realizing 2-*D* IIR filters with 1-*D* IIR filters. By exploring mappings of the 2-*D* index $(m, n)$ to the integers, they develop realizations which approximate the exact 2-*D* filter very closely (in terms of perturbation of the output for the same noise-like input) and offer considerable savings in storage. Their work offers an intriguing first step in trying to simplify realizations of 2-*D* digital recursive filters.

With FIR filters it is necessary to store a number of input rows (columns). With IIR filters it is also necessary to store a number of previously computed rows (columns) of output as well as a number of initial conditions. This is complicated by the fact that in general the output rows are infinitely long. Fortunately, in most applications we only require a limited number of samples of the output, but the fact remains that the number of rows of output that we desire effects the amount of storage required by the filter. This is a problem not encountered in 1-*D* implementations.

## V. Two-Dimensional Digital Filter Design

In this section, we will discuss various methods by which 2-*D* digital filters may be designed to approximate an ideal 2-*D* frequency response. The FIR and IIR design problems will be considered separately. Our objective is to give an overview of the various design techniques. References will be provided so that the interested reader can consult the original sources if he is interested in the details of any particular method.

### *A. FIR Filters*

The first FIR design technique, which is also the simplest, is windowing which is easily extendable from 1-*D* [29]. Assume that we have an ideal frequency response $I(e^{j\mu}, e^{j\nu})$ which we wish to approximate by a FIR filter whose impulse response is

$h(m, n)$. We can calculate the ideal impulse response $i(m, n)$ by performing an inverse 2-*D* Fourier transform either exactly using analytical techniques or approximately by taking a large inverse 2-*D* DFT. (The size of the DFT should be several times as large as the ultimate impulse response $h(m, n)$ so as to minimize aliasing effects in the space domain.) In general $i(m, n)$ will be of infinite extent. We can get a finite extent impulse response by multiplying the ideal by a window function $w(m, n)$ which is of finite extent

$$h(m, n) = i(m, n)\, w(m, n). \qquad (39)$$

$H(e^{j\mu}, e^{j\nu})$ is then related to $I(e^{j\mu}, e^{j\nu})$ by a convolution in the frequency domain

$$H(e^{j\mu}, e^{j\nu}) = \frac{1}{(2\pi)^2} \int_{-\pi}^{\pi} \int_{-\pi}^{\pi} I(e^{j\alpha}, e^{j\beta}) \cdot W(e^{j(\mu-\alpha)}, e^{j(\nu-\beta)})\, d\alpha\, d\beta. \qquad (40)$$

For $H$ to approximate $I$, $W(e^{j\mu}, e^{j\nu})$ should approximate an impulse function. A number of finite extent windows have been proposed which approximate an impulse in their frequency response. Huang [37] has studied the use of continuous 2-*D* circularly symmetric windows taken from 1-*D* continuous windows and Rabiner and Gold [38] give examples of digital filters designed with discrete versions of these windows. Separable windows of the form

$$w(m, n) = w_1(m) w_2(n) \qquad (41)$$

where $w_1(m)$ and $w_2(n)$ are good 1-*D* windows, are also frequently used. Designs based on the window method can be obtained very quickly but for a filter of a given size the approximation to the ideal frequency response will not be as good in general as with other techniques.

FIR filters can also be designed using frequency sampling as in 1-*D*. Here the filter is specified by determining its DFT. The impulse response of the filter can be determined at the end by performing a simple inverse DFT. The values of the DFT at selected frequencies is fixed and at other frequencies it is left variable. (Typically samples in the passband of a filter will be set equal to one, values in the stopband will be set to zero and values in a transition band, which is located between the two, will be left unspecified.) The error in the resulting approximation over the entire frequency plane is then minimized by varying the values of the unspecified DFT samples. If the filter being designed is symmetric through the origin so that its frequency response is real, linear programming can be used to perform the minimization. Hu and Rabiner [39] have used this technique for designing octagonally symmetric filters as large as 25 X 25. The primary drawback with the method is that it is computationally quite expensive.

Both Hu and Rabiner [40] and Fiasconaro [41] have designed filters by linear programming. This can be considered as an extension of frequency sampling where all of the frequency samples are considered as variables. Since the frequency response $H(e^{j\mu}, e^{j\nu})$ is linear in both its DFT samples and also its impulse response coefficients, the linear program can be applied to manipulate either. If $I(e^{j\mu}, e^{j\nu})$ is the ideal frequency response, then the constraints

$$H(e^{j\mu}, e^{j\nu}) \leqslant I(e^{j\mu}, e^{j\nu}) + \delta \qquad (42)$$

$$H(e^{j\mu}, e^{j\nu}) \geqslant I(e^{j\mu}, e^{j\nu}) - \delta \qquad (43)$$

can be applied to a dense grid of points in the $(\mu, \nu)$ frequency plane. Linear programming then adjusts $h(m, n)$ while minimizing $\delta$ and satisfying the above constraints much as in 1-*D* [1]. The resulting filters are optimal in the minimax or Chebyshev sense.

Unfortunately this brute strength approach is even more computationally expensive than frequency sampling. Design times of an hour on high-powered computers are not uncommon. Fiasconaro [40] has designed a more efficient algorithm for designing zero-phase FIR filters. His algorithm is in the same spirit as the 1-*D* Remez second exchange algorithm used for 1-*D* filter design [1], in that it is iterative: at each iteration a linear programming minimization is performed on a small number of constraint points (typically 100) and this set of points varies from iteration to iteration. The constraint points for a given iteration include those located at the extrema of the error function from the previous iteration.

Recently, Kamp and Thiran [41] and Hersey and Mersereau [42], [43] have proposed similar algorithms which used an exchange algorithm instead of the linear programming of Fiasconaro's algorithm. Some speedup in computational time was obtained but the results were not particularly dramatic since the bulk of the time is consumed, not in the minimization but rather in the evaluation of the error function over the entire Fourier plane which all of these algorithms must perform many times. The largest optimal filter which has been designed by any of these methods was 17 X 17 and that filter possessed octagonal symmetry. The use of optimal techniques is thus confined to relatively small impulse responses.

Several methods exist for designing 2-*D* FIR filters from 1-*D* FIR designs. These methods offer the advantage of using relatively little computer time but in general the resulting filters are not optimal, although they can be very good.

The first technique is for approximating ideal frequency responses which are separable

$$I(e^{j\mu}, e^{j\nu}) = I_1(e^{j\mu}) I_2(e^{j\nu}). \qquad (44)$$

Here $I_1$ and $I_2$ can be designed using any 1-*D* algorithm. If the 1-*D* filters are optimal, the 2-*D* filter will be optimal. This sort of an approach can also be used for filters which are sums of separable filters

$$I(e^{j\mu}, e^{j\nu}) = \sum_{i=1}^{N} I_{1i}(e^{j\mu}) I_{2i}(e^{j\nu}). \qquad (45)$$

McClellan (see [1]) has proposed an ingenious approach for converting 1-*D* linear phase designs into 2-*D* linear phase designs. His algorithm converts a 1-*D* design $H(e^{j\omega})$ into a 2-*D* design $H(e^{j\mu}, e^{j\nu})$ by means of the substitution

$$\cos\omega = A\cos\mu + B\cos\nu + C\cos\mu\cos\nu + D. \qquad (46)$$

This technique will not be elaborated here since it is discussed elsewhere in this issue [1] but it is an extremely useful approach for designing many common types of 2-*D* filters. (Circularly symmetric filters can be approximated by $A = B = C = -D = \frac{1}{2}$.) High order filters (e.g., 63 X 63 or 127 X 127) can be designed in seconds.

Projections may also be useful for some 2-*D* FIR filter design problems. To design a filter which approximates an ideal 2-*D* function, we can first evaluate the ideal on the appropriate slice to obtain a 1-*D* ideal function. Then using any 1-*D* FIR filter design algorithm, we may obtain a 1-*D* ideal. The unit sample response of this filter can then be backprojected to give a 2-*D* unit sample response, which hopefully

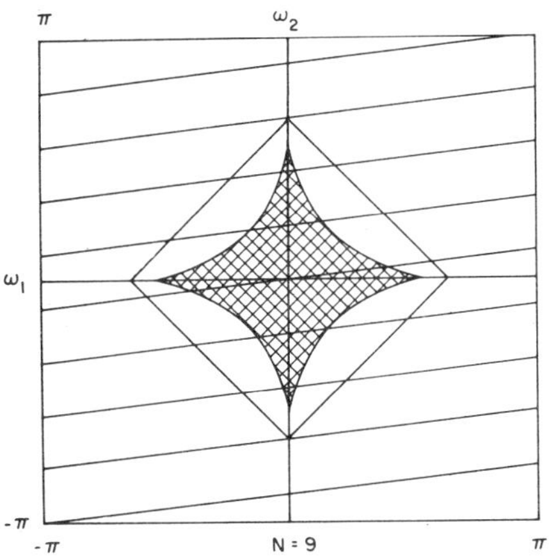


Fig. 11. An ideal filter characteristic which we want to approximate by a 9 X 9 point FIR filter. The cross hatched region represents a passband, the area outside the diamond is the stopband, and the area between the two is the transition band. The $N = 9$ slice is superimposed.

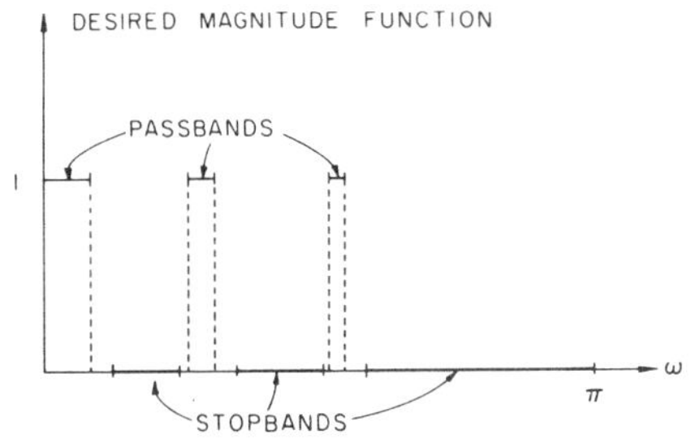


Fig. 12. The frequency response of an 81 point one-dimensional filter which approximates the $N = 9$ slice of the ideal.

Fig. 13. An intensity plot of the log-magnitude of the resulting 2-*D* filter.

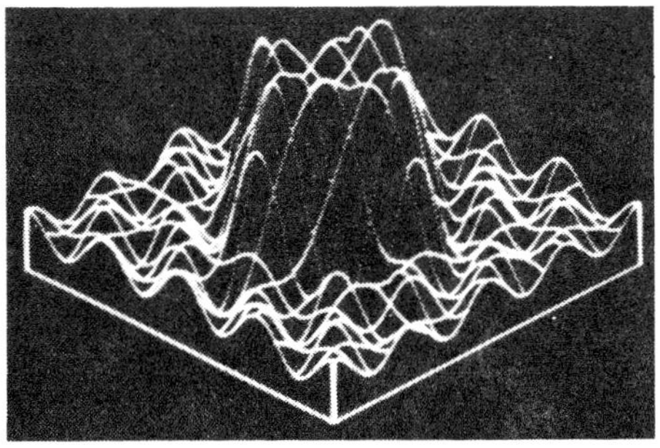

Fig. 14. A surface drawing of the filter of Fig. 13, which is an approximation to the filter of Fig. 11.

approximates the 2-*D* ideal. This technique is outlined in Figs. 11–14. In Fig. 11 is the 2-*D* ideal. It is of the form of a lowpass filter with a cusp-shaped passband and the stopband is the exterior of a diamond. The region between the pass and stop bands is a transition band where the ideal response is not specified.

In Fig. 12 is shown an approximation to the 1-*D* ideal. The approximation was obtained by using a Parks–McClellan equiripple design [21] with an 81 point impulse response. The frequency response of the 9 X 9 2-*D* filter, which results from back projecting the filter of Fig. 12, is shown in Figs. 13 and 14. Fig. 13 is an intensity plot of the log magnitude of the frequency response, and Fig. 14 is a surface drawing of the frequency response. Note that the approximation does not possess all of the symmetry of the ideal.

Recently, Hersey and Mersereau [43] have been working on ways to impose symmetry for FIR designs on a slice. This is easily done by requiring that the 2-*D* approximating function have certain coefficients equal to each other. However, this means the set of basis functions no longer satisfies the Haar condition, and degeneracies may occur if Remez' second exchange algorithm is used (as in a Parks–McClellan design). Hersey and Mersereau avoid this problem by using Remez' first exchange algorithm, which does not require the Haar condition, to perform the necessary minimization. They have successfully designed slice filters with 8-fold symmetry which closely approximate the optimal minimax designs.

Of all the design techniques, these authors feel that the most valuable are windowing, the transformation approach of McClellan and for very small filters, the optimum minimax designs of Kamp and Thiran or Hersey and Mersereau. The former two techniques are quite efficient computationally. The method of windows is the only one which can be used to approximate a completely arbitrary complex frequency response.

### B. *IIR Filters*

In 2-*D* filter design, much less work has been done on recursive filters than on FIR filters. Recursive filters have the potential of being able to realize a given frequency response in fewer multiplications than in IIR filter. However, the recursive filter's frequency response is not linear in the design parameters, so the linear minimization techniques so successful for designing FIR filters are not directly applicable to the recursive case.

Shanks *et al.* [15] have given us two methods for designing recursive filters. The first one we will consider is the space domain approximation. Assume we are given an ideal impulse response $i(m, n)$ which we wish to approximate. (It may have come from an ideal frequency response by inverse transforming.) If we design a recursive filter with impulse response $h(m, n)$, a numerator polynomial with coefficients $a(m, n)$ and a denominator polynomial with coefficients $b(m, n)$, we know that

$$b(m, n) ** h(m, n) = a(m, n). \tag{47}$$

Both $b$ and $a$ are finite extent arrays. So for certain values of $(m, n)$ in a region $R$ we will have

$$b(m, n) ** h(m, n) = 0, \quad (m, n) \in R. \tag{48}$$

If $h(m, n)$ is a good approximation to $i(m, n)$ we can write

$$b(m, n) ** i(m, n) = e(m, n) \simeq 0, \quad (m, n) \in R \tag{49}$$

where $e(m, n)$ is an error term. If we minimize

$$\sum_{m,n} |e(m, n)|^2, \quad (m, n) \in R \tag{50}$$

by differentiating with respect to all the coefficients $b(m, n)$ and setting the derivatives equal to zero, we will get a set of linear equations for the $b(m, n)$ which are easily solved. The coefficients $a(m, n)$ may be computed by minimizing the mean-square error between $h(m, n)$ and $i(m, n)$, or less accurately, by convolving

$$b(m, n) ** i(m, n) \simeq a(m, n).$$

Burrus and Parks [44] give an excellent discussion of this technique for 1-*D* recursive filter design, and most of their results may be extended to two dimensions.

Shanks also gives a conceptually simple method of filter design in the frequency domain. In 1-*D*, recursive filters may be designed by applying the bilinear transformation to an analog filter (e.g., Butterworth, Chebyshev, or elliptic). This is not as easily done in 2-*D*, since very little design work has been done for 2-*D* analog filters. However, for certain applications, we can use 1-*D* analog filters as 2-*D* analog filters. Let $H(s)$ be a classical analog filter's Laplace transform. Then define a 2-*D* analog filter by

$$H(s_1, s_2) = H(s_1). \tag{51}$$

That is, it is constant as a function of $s_2$, and some function of $s_1$. By applying the transformation

$$s_1 = s_1' \cos\theta + s_2' \sin\theta$$

$$s_2 = s_2' \cos\theta - s_1' \sin\theta$$

we can rotate the filter response in the 2-*D* Fourier plane by the angle $\theta$. If we now apply the bilinear transformation to each variable

$$s_1' = \frac{w-1}{w+1} \qquad s_2' = \frac{z-1}{z+1}$$

we will have a 2-*D* recursive filter. By cascading filters of different rotations and different 1-*D* prototypes, many filters can be synthesized. They gave an example of a 2-*D* filter with a pass region in the middle of the first quadrant by cascading a low-pass filter rotated at 45° with a bandpass filter rotated at -45°.

The primary drawback to this method is that a trial and error method may be needed to synthesize a desired filter. In addition, if the number of sections which must be cascaded is high, it may prove more efficient (in terms of a realization) to use an optimum design technique.

Costa and Venetsanopoulos [45] have used this design method to design good approximations to circularly symmetric filters by cascading rotated filters. In addition they have proved that if the angle $\theta$ lies within the region $270° < \theta < 360°$ the rotated filter will be stable. To implement filters with angles of rotation not in this range, the input data is simply rotated by the appropriate multiple of 90°.

Recently several workers [13], [47], [48] began exploring optimum design techniques for 2-*D* recursive filters. These methods are based on the 2-*D* differential correction method for rational approximation [46] which has been used successfully for 1-*D* recursive filter design [49], [50].

Basically the algorithm is as follows. We are given an ideal magnitude-squared frequency response $I(e^{j\mu}, e^{j\nu})$ which we wish to approximate with a ratio of 2-*D* polynomials.

$$I(e^{j\mu}, e^{j\nu}) \simeq R(e^{j\mu}, e^{j\nu}) = \frac{P(\mu, \nu)}{Q(\mu, \nu)}$$

$$= |H(e^{j\mu}, e^{j\nu})|^2 = \frac{|A(e^{j\mu}, e^{j\nu})|^2}{|B(e^{j\mu}, e^{j\nu})|^2}.$$

$P$ and $Q$ are 2-*D* cosine polynomials of the form

$$P(\mu, \nu) = p(0,0) + 2 \sum_{m,n} 2p(m, n) \cos(m\mu + n\nu).$$

For the error criterion we will use the maximum difference between the ideal and the design functions

$$\Delta = \max_{\mu,\nu} |I(e^{j\mu}, e^{j\nu}) - R(e^{j\mu}, e^{j\nu})|.$$

Thus we seek to minimize $\Delta$. The algorithm is iterative and at the $k$th iteration we have an approximation $R_k(e^{j\mu}, e^{j\nu})$:

1) Calculate the error $\Delta_k$ for $R_k(e^{j\mu}, e^{j\nu})$.
2) Next form the functional
$$\delta_k(R) = \max_{\mu,\nu} \left\{ \frac{|IQ - P| - \Delta_k Q}{Q_k} \right\}$$
where $R = P/Q$ and $Q_k$ is the denominator of $R_k = P_k/Q_k$.
3) Find $R_{k+1}$ which minimized $\delta_k$ by varying the coefficients of $P$ and $Q$.
4) If min $\delta_k \geq 0$ stop. $R_k$ is a best approximation to $I(\mu, \nu)$. If not, go to step 1 and start over with $R_{k+1}$.

The minimization of $\delta_k$ can be structured into a linear programming problem, thus using linear programming in an iterative fashion to solve a nonlinear optimization problem.

There are two serious drawbacks to this method. Computation time is large. (Times on the order of 1 hour are not uncommon for filters with 42 coefficients and 544 constraint points [14].) Also when the design is finished one has the cosine polynomials $P(\mu, \nu)$ and $Q(\mu, \nu)$. However, it remains to solve for $A(e^{j\mu}, e^{j\nu})$ and $B(e^{j\mu}, e^{j\nu})$. This can be done by using a 2-*D* DHT as indicated in section III, but an infinite extent answer may result which then may have to be truncated in some fashion.

Bednar and Farmer [47] took a slightly different approach. They set up the minimization of $\Delta$ in terms of $A$ and $B$ directly, using them to approximate an ideal magnitude (rather than magnitude-square) function. In order to do this, however, the minimization of $\Delta$ becomes a nonlinear optimization problem. This leads to the usual numerical problems with nonlinear optimization, namely very large amounts of computation for minimizations with many variables, sensitivity to starting points, and the possibility of converging to a local rather than global minimum. In addition, it is necessary to check the stability of the designed filter at each iteration of the algorithm.

## VI. Summary

We have tried to present a number of recently developed results for the design and implementation of 2-*D* digital filters. Although the rapidly changing nature of the subject is such that no such compendium could ever be complete, we have hopefully presented the flavor of what types of techniques are available and perhaps more importantly, which are still needed.

## References

[1] L. R. Rabiner, J. H. McClellan, and T. W. Parks, "FIR digital filter design techniques using weighted Chebyshev approximation," this issue, pp. 595-610.
[2] B. R. Hunt, "Digital image processing," this issue, pp. 693-708.
[3] L. C. Wood and S. Treitel, "Seismic signal processing," this issue, pp. 649-661.
[4] T. S. Huang, W. F. Schreiber, and O. Tretiak, "Image processing," *Proc. IEEE*, vol. 59, pp. 1586-1609, Nov. 1971.

[5] R. E. Crochiere and A. V. Oppenheim, "Analysis of linear digital networks," this issue, pp. 581–595.

[6] J. W. Goodman, *Introduction to Fourier Optics*. New York: McGraw-Hill, 1968.

[7] D. P. Peterson and D. Middleton, "Sampling and reconstruction of wave number limited functions in $N$-dimensional Euclidean spaces," *Inform. Contr.*, vol. 5, pp. 279–323, 1962.

[8] A. V. Oppenheim and D. H. Johnson, "Discrete representation of signals," *Proc. IEEE*, vol. 60, pp. 681–691, June 1972.

[9] G. D. Bergland, "A guided tour of the fast Fourier transform," *IEEE Spectrum*, vol. 6, pp. 41–52, July 1969.

[10] R. M. Mersereau and D. E. Dudgeon, "The representation of two-dimensional sequences as one-dimensional sequences," *IEEE Trans. Acoust., Speech, and Sig. Proc.*, vol. ASSP-22, No. 5, pp. 320–325, Oct. 1974.

[11] J. L. Shanks, "Recursive filters for digital processing," *Geophysics*, vol. 32, pp. 33–51, 1967.

[12] C. S. Burrus and T. W. Parks, "Time domain design of recursive digital filters," *IEEE Trans. Audio Electroacoust.*, vol. AU-18, pp. 137–141, June 1970.

[13] J. L. Shanks, "Two-dimensional recursive filters," in *1969 SWIEEECO Rec.*, pp. 19E1–19E8.

[14] D. E. Dudgeon, "Two-dimensional recursive filtering," Sc.D. dissertation Dep. Elec. Eng., Massachusetts Inst. Technol., Cambridge, 1974.

[15] J. L. Shanks, S. Treitel, and J. H. Justice, "Stability and synthesis of two-dimensional recursive filters," *IEEE Trans. Audio Electroacoust.*, vol. AU-20, pp. 115–128, June 1972.

[16] C. Farmer and J. B. Bednar, "Stability of spatial digital filters," *Math. Biosci*, vol. 14, pp. 113–119, 1972.

[17] T. S. Huang, "Stability of two-dimensional recursive filters," *IEEE Trans. Audio Electroacoust.*, vol. AU-20, pp. 158–163, June 1972.

[18] H. G. Ansell, "On certain two-variable generalizations of circuit theory, with applications to networks of transmission lines and lumped reactances," *IEEE Trans. Circuit Theory*, vol. CT-11, pp. 214–223, June 1964.

[19] B. D. O. Anderson and E. I. Jury, "Stability test for two-dimensional recursive filters," *IEEE Trans. Audio Electroacoust.*, vol. AU-21, pp. 366–372, Aug. 1973.

[20] G. A. Maria and M. M. Fahmy, "On the stability of two-dimensional digital filters," *IEEE Trans. Audio Electroacoust.*, vol. AU-21, pp. 470–472, Oct. 1973.

[21] J. H. Justice and J. L. Shanks, "Stability criterion for $N$-dimensional digital filters," *IEEE Trans. Autom. Contr.*, vol. AC-18, pp. 284–286, June 1973.

[22] B. D. O. Anderson and E. I. Jury, "Stability of multidimensional digital filters," *IEEE Trans. Circuit and Syst.*, vol. CAS-21, pp. 300–304, Mar. 1974.

[23] R. R. Read and S. Treitel, "The stabilization of two-dimensional recursive filters via the discrete Hilbert transform," *IEEE Trans. Geoscience Elec.*, vol. GE-11, pp. 153–160, July 1973.

[24] P. Pistor, "Stability criterion for recursive filters," *IBM J. Res. Dev.*, vol. 18, pp. 59–71, Jan. 1974.

[25] D. E. Dudgeon, "Using a two-dimensional discrete Hilbert transform to factor two-dimensional polynomials," M.I.T. Res. Lab. Electron., Quart. Progr. Rep. 113, pp. 143–150, Apr. 1974.

[26] D. E. Dudgeon, "Existence of Cepstra for Two-Dimensional Rational Polynomials," M.I.T. Res. Lab. Electron., Quart. Progr. Rep. 114, pp. 153–157, July 15, 1974.

[27] D. E. Dudgeon, "Recursibility of two-dimensional difference equations," M.I.T. Res. Lab. of Electron., Quart. Progr. Rep. 113, pp. 151–158, Apr. 1974.

[28] R. C. Agarwal and C. S. Burrus, "Number theoretic transforms to implement fast digital convolution, this issue, pp. 550–560.

[29] A. V. Oppenheim and R. W. Schafer, *Digital Signal Processing*. Englewood Cliffs, N.J.: Prentice-Hall, 1974.

[30] B. R. Hunt, "Minimizing the computation time for using the technique of sectioning for digital filtering of pictures," *IEEE Trans. Comp.*, vol. C-21, pp. 1219–1222, Nov. 1972.

[31] D. P. MacAdam, "Digital image restoration by constrained deconvolution," *J. Opt. Soc. Amer.*, vol. 60, pp. 1617–1627, Oct. 1970.

[32] R. C. Agarwal and C. S. Burrus, "Fast one-dimensional digital convolution by multidimensional techniques," *IEEE Trans. Acoust., Speech, Sig. Proc.*, vol. ASSP-22, pp. 1–10, Feb. 1974.

[33] S. Treitel and J. L. Shanks, "The design of multistage separable planar filters," *IEEE Trans. Geosci. Elec.*, vol. GE-9, pp. 10–27, 1971.

[34] O. J. Tretiak, "Approximation of a matrix by a sum of products," M.I.T. Res. Lab. of Electron., Quart. Progr. Rep. 98, pp. 157–169, July 1970.

[35] M. D. Ni and J. K. Aggarwal, "Two-dimensional digital filtering and its error analysis," Electron. Res. Cent., Univ. of Texas, Austin, *Inform. Syst. Research Labor. Tech. Rep. 157*, Mar. 1974. Also in *IEEE Trans. Comp.*, vol. C-23, pp. 942–953, Sept. 1974.

[36] M. T. Manry and J. K. Aggarwal, "Picture processing using one-dimensional implementations of discrete planar filters," *IEEE Trans. Acoust. Speech, Sig. Proc.*, vol. ASSP-22, pp. 164–173, June 1974.

[37] T. S. Huang, "Two-dimensional windows," *IEEE Trans. Audio Electroacoust.*, vol. AU-20, pp. 88–90, Mar. 1972.

[38] L. R. Rabiner and B. Gold, *Theory and Applications of Digital Signal Processing*. Englewood Cliffs, N.J.: Prentice-Hall, Inc., New York, 1974.

[39] J. V. Hu and L. R. Rabiner, "Design techniques for two-dimensional digital filters," *IEEE Trans. Audio Electroacoust.*, vol. AU-20, pp. 249–257, Oct. 1972.

[40] J. G. Fiasconaro, "Two-Dimensional Nonrecursive Digital Filters," M.I.T. Lincoln Lab., Lexington, Mass., Tech. Rep. 505.

[41] Y. Kamp and J. P. Thiran, "Chebyshev approximation for two-dimensional nonrecursive digital filters," MBLE Res. Lab. Rep. R245, Mar. 1974. Also to be published in *IEEE Trans. Circuits and Syst.*

[42] H. S. Hersey and R. M. Mersereau, "An algorithm to perform minimax approximation in the absence of the Haar condition," M.I.T. Res. Lab. Electron., Quart. Progr. Rep. 114, Cambridge, Mass., pp. 157–160, July 15, 1974.

[43] ——, "Design techniques for two-dimensional FIR circularly symmetric digital filters," M.I.T. Res. Lab. Electron., Quart. Progr. Rep. 114, pp. 160–171, July 15, 1974.

[44] C. S. Burrus and T. W. Parks, "Time domain design of recursive digital filters," *IEEE Trans. Audio Electroacoust.*, vol. AU-18, pp. 137–141, June 1970.

[45] J. M. Costa and A. N. Venetsanopoulos, "Design of circularly symmetric two-dimensional recursive filters," *IEEE Trans. Acoust., Speech, Sig. Proc.*, vol. ASSP-22, pp. 432–443, Dec. 1974.

[46] I. Barrodale, M. J. D. Powell, and F. D. K. Roberts, "The differential correction algorithm for rational $l_\infty$ approximation," *SIAM J. Numer. Anal.*, vol. 9, pp. 493–504, Sept. 1972.

[47] J. B. Bednar and C. Farmer, "Implementation of Chebyshev digital filter design," to be published in *IEEE Trans. Circuits and Syst.*

[48] E. H. Kaufman, Jr., and C. D. Taylor, "An application of linear programming to rational approximation," to be published in *Rocky Mountain J. Math.*

[49] D. E. Dudgeon, "Recursive filter design using differential correction," *IEEE Trans. Acoust., Speech, Sig. Proc.*, vol. ASSP-22, pp. 443–448, Dec. 1974.

[50] F. Brophy and A. Salazar, "Synthesis of spectrum shaping digital filters of recursive design," in *Rec Int. Symp. Circuits and Systems*, Apr. 1974. Also to be published in *IEEE Trans. Circuits and Syst.*

# The Existence of Cepstra for Two-Dimensional Rational Polynomials

DAN E. DUDGEON

***Abstract*—The use of cepstral analysis is helpful for some problems where two one-dimensional signals are combined by convolution [1]. In such problems it is important to ensure that the phase function associated with the resultant signal may be defined so that it is a continuous, odd, and periodic function of frequency [2], [3]. One class of one-dimensional signals which have this property is the class whose $z$-transforms are rational polynomials [2]. In this correspondence, we shall show that these results are extendible to two dimensions, and that 2-$D$ cepstra can be defined for 2-$D$ rational polynomials.**

## INTRODUCTION

In signal processing problems where two signals are combined by convolution, homomorphic filtering techniques [1] may be useful. For example, such a technique may be useful for inverse filtering problems or problems where a signal and noise are combined by convolution. To make the definition of the complex cepstrum mathematically tractable, it is important to ensure that the phase function associated with a signal is continuous, odd, and periodic. Defining the phase by using the imaginary part of the principal part of the complex logarithm or the arctangent is usually not satisfactory [3]. Therefore phase must be defined as an integral [1] to ensure continuity. If the phase so defined is not periodic, it may be made so by subtracting off a linear phase component. This simply corresponds to a shift in the time origin of the original signal.

At this juncture, it should be noted that we must restrict the set of signals for which cepstra can be defined to exclude signals whose Fourier transforms go to zero or infinity at some frequency. This corresponds to their $z$-transforms having a pole or a zero on the unit circle, resulting in a phase discontinuity which cannot be removed by the integral formulation. Equivalently the log of their Fourier transforms become infinite at such frequencies.

One class of signals for which cepstra may be defined is the class whose $z$-transforms are rational polynomials and are analytic and nonzero on the unit circle. For these signals the cepstra also have $z$-transforms which are analytic on the unit circle [1].

Manuscript received June 10, 1974; revised October 25, 1974. This work was supported in part by the Advanced Research Projects Agency of the Department of Defense under Contract N0014-67-A-0204-0064 at the M.I.T. Research Laboratory of Electronics, and in part by the M.I.T. Lincoln Laboratory (sponsored by the U. S. Air Force).

The author was with the M.I.T. Research Laboratory of Electronics, Digital Signal Processing Group, Cambridge, Mass. He is now with the Computer Systems Division, BBN Inc., Cambridge, Mass. 02138.

In an exactly analogous manner, we can demonstrate that any 2-$D$ array having a rational $z$-transform will also have a well-defined 2-$D$ complex cepstrum provided 1) the Fourier transform is not equal to zero or infinity at any frequency, and 2) we are careful to eliminate any linear phase components by an appropriate shift of the original array.

This is most easily seen by first considering finite extent 2-$D$ arrays. These arrays have Fourier transforms which are 2-$D$ polynomials in $\exp[j\mu]$ and $\exp[j\nu]$. We shall now show that if a finite extent array $b(m,n)$ has a Fourier transform $B(\mu,\nu)$ which is nonzero for all $\mu$ and $\nu$, then the phase function associated with $B(\mu,\nu)$ will be the sum of a linear component plus a continuous, odd, and periodic component. First, let $z = \exp[j\nu]$, and consider

$$B_\mu(z) = \sum_n [\sum_m b(m,n) \exp(-j\mu m)]z^{-n}$$

as a one-dimensional polynomial in $z$ with a parameter $\mu$. Since $b(m,n)$ is of finite extent, $B_\mu(z)$ will have poles inside the unit circle only at $z = 0$. Now let us define the phase function $\phi(\mu,\nu)$ as a contour integral

$$\phi(\mu,\nu) = \operatorname{Im}\left\{\oint \frac{B_\mu'(z)}{B_\mu(z)}\,dz\right\} + \phi(\mu,0) \tag{1}$$

where the prime denotes differentation with respect to $z$. The contour of integration starts at $z = 1$ and proceeds around the unit circle to $z = \exp[j\nu]$. It is necessary to define the constant $\phi(\mu,0)$. This is done by considering $\phi(\mu,0)$ to be the phase as a function of $\mu$ for $\nu = 0$, and may be derived as in Oppenheim and Schafer [3].

By constructing $\phi(\mu,\nu)$ in this manner we are assured that $\phi(\mu,\nu)$ is continuous and odd; that is, $\phi(\mu,\nu) = -\phi(-\mu,-\nu)$.

When $\nu = 2\pi$, the contour of integration in (1) is a closed curve. Using Cauchy's residue theorem and [4, theorems 1 and 2] we have

$$\phi(\mu,2\pi) = 2\pi r - 2\pi N$$

where $r$ is the number of roots (including multiplicities) of $B_\mu(z)$ inside the unit circle and $N$ is the number of poles at $z = 0$. If we let $k_\nu = r - N$, then it is clear that

$$\phi(\mu,\nu + 2\pi) = \phi(\mu,\nu) + 2\pi k_\nu.$$

Similarly we can derive the relation

$$\phi(\mu + 2\pi,\nu) = \phi(\mu,\nu) + 2\pi k_\mu.$$

What remains to be shown is that $k_\nu$ is not a function of $\mu$, and that $k_\mu$ is not a function of $\nu$. If we examine the roots of $B_\mu(z)$ as we continuously vary the parameter $\mu$ from zero to $2\pi$, we discover that the roots move about in a continuous manner [4, p. 3]. Thus, for a root to move from inside to outside the unit circle (or vice versa), it must lie on the unit circle for some value of $\mu$. This, however, violates the hypothesis that $B(\mu,\nu) \neq 0$. Therefore the number of roots $r$ inside the unit circle is not a function of $\mu$, and hence $k_\nu$ is not a function of $\mu$. A similar argument can be made to show that $k_\mu$ is not a function of $\nu$.

Given a continuous, odd phase function $\phi(\mu,\nu)$ we can subtract the linear phase component

$$\phi_L(\mu,\nu) = k_\nu\nu + k_\mu\mu$$

to give a term $\phi_A(\mu,\nu)$ which is easily shown to be continuous, odd, and periodic.

This result is easily extended to arrays with rational Fourier transforms. The phase function for such an array can be defined as the phase function of the numerator polynomial minus the phase function of the denominator polynomial. Clearly this difference can also be written as a linear phase component plus a continuous, odd, and periodic component.

Reprinted from *IEEE Trans. Acoust., Speech, Signal Processing*, vol. ASSP-23, pp. 242–243, Apr. 1975.

It is easy to verify that subtracting off the linear phase component is equivalent to a shift of the array. Therefore, if we are given an array $h(m,n)$ whose Fourier transform is a ratio of polynomials, we can form a new sequence

$$g(m,n) = h(m - k_\mu, n - k_\nu)$$

whose phase function will be $\phi_A(\mu,\nu)$ which is continuous, odd, and periodic. We may then form the function

$$\hat{G}(\mu,\nu) = \ln |H(\mu,\nu)| + j\phi_A(\mu,\nu)$$

which has a real inverse Fourier transform denoted by $\hat{g}(m,n)$ and called the cepstrum of $g(m,n)$. Since $H(\mu,\nu)$ is a ratio of polynomials, the properties of $\hat{g}(\mu,\nu)$ imply that the $z$-transform of $\hat{g}(m,n)$ is an analytic function for $|w| = |z| = 1$. This may be seen by considering $\hat{G}(\mu,\nu)$ to be a one-dimensional Fourier transform in $\nu$ with a parameter $\mu$. Then the $z$-transform is analytic for all values of $\mu$ and $|z| = 1$. Similarly the $z$-transform is analytic for all values of $\nu$ and $|w| = 1$. Therefore it is analytic for $|w| = |z| = 1$.

## SUMMARY

From this derivation two conclusions can be drawn. First, any real 2-$D$ array whose Fourier transform has a log-magnitude which is continuous, even, and periodic, and whose phase is continuous, odd, and periodic over the frequency plane will have a well-defined 2-$D$ cepstrum which is real. Second, one class of 2-$D$ arrays with well-defined cepstra is the class whose $z$-transforms are the ratio of two 2-$D$ polynomials (with the restriction that the $z$-transforms are not zero or infinite for $|w| = |z| = 1$), care having been taken first to shift the arrays to eliminate any linear phase component. In addition, for this class of 2-$D$ arrays we can argue that the logarithm of the $z$-transform is analytic on the surface $|w| = |z| = 1$.

## ACKNOWLEDGMENT

The author would like to express his thanks to Prof. A. V. Oppenheim of M.I.T. for his suggestions and encouragement during the course of this work.

## REFERENCES

[1] A. V. Oppenheim, R. W. Schafer, and T. G. Stockham, Jr., "Nonlinear filtering of multiplied and convolved signals," *Proc. IEEE*, vol. 56, pp. 1264–1291, Aug. 1968.
[2] B. Gold and C. M. Rader, *Digital Processing of Signals.* New York: McGraw-Hill, 1969.
[3] A. V. Oppenheim and R. W. Schafer, *Digital Signal Processing.* Englewood Cliffs, N. J.: Prentice-Hall, 1975.
[4] M. Marden, *The Geometry of the Zeros of a Polynomial in a Complex Variable.* New York: Amer. Math. Soc., 1949.

# A Fast Computer Method for Matrix Transposing

J. O. EKLUNDH

*Abstract*—A method is given for transposition of $2^n \times 2^n$ data matrices, larger than available high-speed storage. The data should be stored on an external storage device, allowing direct access. The performance of the algorithm depends on the size of the main storage, which at least should hold $2^{n+1}$ points. In that case the matrix has to be read in and written out $n$ times.

*Index Terms*—Externally stored matrices, fast Fourier transform, fast Hadamard transform, matrix transposition, two-dimensional transformations.

## INTRODUCTION

The purpose of this correspondence is to present a fast computer method for matrix transposing. This method makes it possible to apply fast transforms such as the fast Fourier transform (FFT) and the fast Hadamard transform (FHT) on matrices of size exceeding available main storage. The algorithm presented is useful provided that each row (or column) of the matrix can be addressed, e.g., by use of direct access. When the data records have to be accessed sequentially, the computational and input–output time required will be excessive. In fact, the number of backspace and read operations will then be of the same magnitude as the number of matrix elements.

A characteristic feature of the FFT, as presented by Cooley and Tukey [1], is the assumption that all the data points in the array to be transformed are kept in the main storage simultaneously. This may be a serious restriction considering two-dimensional applications, such as image processing.

To state the problem explicitly, let

$$f = (f(m, n)), \qquad m = 0, 1, \cdots, M-1, \; n = 0, 1, \cdots, N-1 \quad (1)$$

be the matrix to be transformed. The transform is defined by

$$F(p, q) = \sum_{m=0}^{M-1} \sum_{n=0}^{N-1} f(m, n) \exp\left(-2\pi j\left(\frac{mp}{M} + \frac{nq}{N}\right)\right),$$
$$p = 0, 1, \cdots, M-1, \; q = 0, 1, \cdots, N-1. \quad (2)$$

Suppose that $f$ is stored on an external unit, e.g., a disk or a tape, row by row, that is in the order $f(0, 0), f(0, 1), \cdots, f(0, N-1), f(1, 0), f(1, 1), \cdots, f(M-1, N-1)$. We may then read $f$ row by row and calculate the sums

Manuscript received December 9, 1971; revised March 1, 1972.
The author is with the Research Institute of National Defence, Stockholm, Sweden.

Reprinted from *IEEE Trans. Comput.*, vol. C-21, pp. 801–803, July 1972.

$$G(m, q) = \sum_{n=0}^{N-1} f(m, n) \exp\left(-2\pi j \frac{nq}{N}\right), \qquad q = 0, 1, \cdots, N-1. \qquad (3)$$

To minimize required storage, $G(m, q)$ is stored in the location that earlier contained $f(m, q)$. If we write the resulting arrays on an external data set—that may or may not be identical with the one containing $f$—we still have a matrix stored row by row. Now the next operation is to calculate

$$F(p, q) = \sum_{m=0}^{M-1} G(m, q) \exp\left(-2\pi j \frac{mp}{M}\right), \qquad p = 0, 1, \cdots, M-1, \qquad (4)$$

that is, to form the sum of the elements in the matrix $G$ multiplied by their corresponding exponential factor *columnwise*. But since $G$ is stored row by row, we cannot obtain the elements in one column without reading through the whole matrix, picking out for each row the proper element. Doing this straightforwardly, we shall have to go through the matrix $M$ times, which will result in excessive input–output time, even in the case of moderate matrix sizes. One way to avoid this is evidently to transpose the matrix $G$, which, however, by no means is trivial. The problem can be formulated as the one of how to construct a (quite specific) permutation of a given sequence when only part of the sequence can be kept in the main storage and operated on at the same time, minimizing the number of times the matrix is "worked through," and each element is taken into the core.

Singleton [2] and Brenner [3] have presented procedures for computation of the Fourier transform of externally stored one-dimensional data. Their methods may be modified to apply also in two dimensions (see [3]). In both cases, matrices of a size exceeding the available main storage have to be transposed.

In this correspondence we shall demonstrate a fast method for transposition of matrices of the order $2^n \times 2^n$, that is, matrices appearing in the two-dimensional application of the FFT and FHT. Simple generalizations to arbitrary matrices will be discussed later.

Theoretically, the algorithm to be described requires that the matrix—that we suppose is stored on an external data set unit—is brought into the main storage and returned again $n$ times. However, this number can be reduced substantially by efficient programming. In fact, two times suffice in most implementations. This implies that we have to work through the matrix only one extra time, since any linear transformation requires that the data are taken into the main storage and that the result is written out on some unit at least once.

## The Algorithm

The idea of the method is that a $2^n \times 2^n$ matrix can be transposed, taking two rows at the same time into the main storage, interchanging certain elements in the two rows, and finally returning the resulting two rows to the very same locations.

Consider the matrix

$$A = (\alpha_{IK}), \qquad I = 0, 1, \cdots, 2^n - 1, \; K = 0, 1, \cdots, 2^n - 1. \qquad (5)$$

We can write $I$ and $K$ in the form

$$I = \sum_{p=0}^{n-1} i_p 2^p = b(i_{n-1}, i_{n-2}, \cdots, i_1, i_0) \qquad (6)$$

$$K = \sum_{p=0}^{n-1} k_p 2^p = b(k_{n-1}, k_{n-2}, \cdots, k_1, k_0) \qquad (7)$$

where $i_p$ and $k_p$ are binary integers.

Let us also write $\alpha(I, K)$ instead of $\alpha_{IK}$.

We transpose $A$ in $n$ steps, forming pairs of rows (or columns) and interchanging elements according to the following rules.

In step $j$, $0 \le j \le n-1$,

row $I'$ and row $I''$ form a pair iff

$$i_j' \ne i_j'' \text{ and } i_p' = i_p'', \qquad \text{for all values of } p \ne j. \qquad (8)$$

In each pair of rows $I'$ and $I''$, the elements $\alpha(I', K')$ and $\alpha(I'', K'')$ change places iff

$$k_j' \ne k_j'', \; k_j' \ne i_j', \text{ and } k_p' = k_p'', \qquad \text{for all values of } p \ne j. \qquad (9)$$

It follows from (8) and (9) that

$$k_j'' = i_j' \quad \text{and} \quad k_j' = i_j''. \qquad (10)$$

Hence if $\alpha(I', K')$ is the element in row $I$ and column $K$ before step $j$ is performed, the new element in this place will be $\alpha(I'', K'')$ where

$$I'' = b(i_{n-1}', \cdots, i_{j+1}', k_j', i_{j-1}', \cdots, i_0') \qquad (11)$$

$$K'' = b(k_{n-1}', \cdots, k_{j+1}', i_j', k_{j-1}', \cdots, k_0'), \qquad (12)$$

the primes referring to $I'$ and $K'$.

Consider an arbitrary (but fixed) element $\alpha(I, K)$ of $A$. Let $I_j$ and $K_j$, respectively, be the indices of the row and the column containing $\alpha(I, K)$ when step $j$ is performed. Using the notation of (6) and (7), we get from (11) and (12) that

$$\begin{aligned} (I_0, K_0) &= (b(i_{n-1}, \cdots, i_2, i_1, k_0), b(k_{n-1}, \cdots, k_2, k_1, i_0)) \\ (I_1, K_1) &= (b(i_{n-1}, \cdots, i_2, k_1, k_0), b(k_{n-1}, \cdots, k_2, i_1, i_0)) \\ &\vdots \\ (I_{n-1}, K_{n-1}) &= (b(k_{n-1}, \cdots, k_2, k_1, k_0), b(i_{n-1}, \cdots, i_2, i_1, i_0)). \end{aligned} \qquad (13)$$

(Incidentally, these formulas also apply to the case when $k_j' = i_j'$ in (9).)

We conclude that after $n$ steps the element $\alpha(I, K)$ is found in row $K$ and column $I$, which means that the matrix is transposed.

A simple example will illustrate how the algorithm works.

*Example:* Suppose that the $4 \times 4$ matrix

$$A = \begin{pmatrix} \alpha_{11} & \alpha_{12} & \alpha_{13} & \alpha_{14} \\ \alpha_{21} & \alpha_{22} & \alpha_{23} & \alpha_{24} \\ \alpha_{31} & \alpha_{32} & \alpha_{33} & \alpha_{34} \\ \alpha_{41} & \alpha_{42} & \alpha_{43} & \alpha_{44} \end{pmatrix} \qquad (14)$$

is stored row by row. We transpose it in the following two steps.

*Step 1:* First read row 1 and row 2 and let $\alpha_{12}$ and $\alpha_{14}$ change places with $\alpha_{21}$ and $\alpha_{23}$, respectively. This means that the elements 2 and 4 in row 1 change places with the elements 1 and 3, respectively, in row 2. Then write row 1 and row 2 in their former locations and repeat the procedure with rows 3 and 4. We get

$$A_1 = \begin{pmatrix} \alpha_{11} & \alpha_{21} & \alpha_{13} & \alpha_{23} \\ \alpha_{12} & \alpha_{22} & \alpha_{14} & \alpha_{24} \\ \alpha_{31} & \alpha_{41} & \alpha_{33} & \alpha_{43} \\ \alpha_{32} & \alpha_{42} & \alpha_{34} & \alpha_{44} \end{pmatrix}. \qquad (15)$$

*Step 2:* Read row 1 and row 3 (of $A_1$), interchange element 3 in row 1 and element 1 in row 3, element 4 in row 1 and element 2 in row 3, and finally write row 1 and row 3 in their proper locations. Repeat the procedure with row 2 and row 4. We obtain

$$A_2 = \begin{pmatrix} \alpha_{11} & \alpha_{21} & \alpha_{31} & \alpha_{41} \\ \alpha_{12} & \alpha_{22} & \alpha_{32} & \alpha_{42} \\ \alpha_{13} & \alpha_{23} & \alpha_{33} & \alpha_{43} \\ \alpha_{14} & \alpha_{24} & \alpha_{34} & \alpha_{44} \end{pmatrix} = A^t. \qquad (16)$$

Hence $A$ being a $2^2 \times 2^2$ matrix, we get $A^t$, the transpose of $A$, by twice reading through the matrix, swapping the elements, and writing it out again.

It may be noted that the rows are put together in couples in the same manner as the elements are interchanged.

## Programming Considerations

Evidently, the optimal way to implement the algorithm is to take more than two rows into the main storage simultaneously. For instance, during the first three steps, the rows numbered 0, 1, 2, $\cdots$, 7 form a group in which the elements are interchanged independently of the other rows. Suppose that $2^j$ rows can be held in the high-speed storage. Then, if $n = j_1 + \cdots + j_k$, where $j_p \le j$ for all

TABLE I

NUMBER OF EXTRA I/O PASSES WHEN $2^j$ ROWS OUT OF $2^n$ CAN BE HELD IN THE MAIN STORAGE

| $j$ \ $n$ | 10 | 9 | 8 | 7 |
|---|---|---|---|---|
| 5 | 1 | 1 | 1 | 1 |
| 4 | 2 | 2 | 1 | 1 |
| 3 | 3 | 2 | 2 | 2 |
| 2 | 4 | 4 | 3 | 3 |

TABLE II

CENTRAL PROCESSOR (LEFT) AND ESTIMATED INPUT–OUTPUT (RIGHT) TIME IN SECONDS FOR TRANSPOSITION OF $2^n \times 2^n$ COMPLEX MATRIX WHEN $2^j$ ROWS FIT INTO CORE

| $j$ \ $n$ | 10 | 9 | 8 | 7 | $j$ \ $n$ | 10 | 9 | 8 | 7 |
|---|---|---|---|---|---|---|---|---|---|
| 5 | 23.0 | 5.8 | 2.2 | 0.8 | 5 | 134 | 54 | 24 | 12 |
| 4 | 25.2 | 6.9 | 2.2 | 0.8 | 4 | 268 | 108 | 24 | 12 |
| 3 | 27.4 | 6.9 | 2.8 | 1.0 | 3 | 402 | 108 | 48 | 24 |
| 2 | 29.6 | 9.1 | 3.3 | 1.3 | 2 | 536 | 216 | 71 | 34 |

values of $p$, $k$ input–output passes are needed, e.g., the matrix has to be taken into the main storage and returned again $k$ times. Hence the number of input–output passes caused by the transposition algorithm is $k-1$. Table I lists $k-1$ as a function of $n$ and $j$.

Required computing and input–output times can be expressed as

$$T_c = (k-1)(2^n T_{\text{read}} + 2^n T_{\text{write}}) + n 2^{2n-2} T_{\text{change}} \tag{17}$$

and

$$T_{I/O} = (k-1)(2^{n+1} T_{\text{acc}} + 2^{2n+1} T_{\text{tran}}). \tag{18}$$

Here $T_{\text{read}}$ and $T_{\text{write}}$ are the central processor times to read and write one data record, and $T_{\text{change}}$ is the time to perform a permutation in the core of two complex elements. Moreover, $T_{\text{acc}}$ and $T_{\text{tran}}$ are the input–output times to access one data record and read or write one complex number, respectively.

The algorithm has been implemented on an IBM 360/75, with 2314 disk packs. The routine is written in Fortran and utilizes the so-called BSAM routines for input and output. Its performance as a function of $n$ and $j$ is exemplified in Table II.

It should be pointed out that $T_c$ is dominated by the second term, which is independent of the working storage area. (Typical values for the time constants are $T_{\text{read}} = T_{\text{write}} = 1$ ms, $T_{\text{change}} = 8$ μs.) Consequently, even a drastic reduction of high-speed storage will affect the computing time but little. This may be compared with the method suggested by Brenner [3], which for transposition requires $2^{2n-j}$ records to be read and written. Computing time, then, is dominated by the time to read and write, and hence is approximately proportional to working storage.

Generalization to arbitrary square matrices seems feasible if we use the approach of Bergland [4] for the FFT case. Obviously, there exists no generalization to nonsquare matrices, as row and column length must coincide when the elements are changed. However, if enough external storage is accessible, an arbitrary matrix could be imbedded in a square matrix, filling out the latter one with zero elements. Of course, this trick can also be used to obtain a $2^n \times 2^n$ matrix from an arbitrary one.

REFERENCES

[1] J. W. Cooley and J. W. Tukey, "An algorithm for the machine calculation of complex Fourier series," *Math. Comput.*, vol. 19, pp. 297–301, Apr. 1965.
[2] R. C. Singleton, "A method for computing the fast Fourier transform with auxiliary memory and limited high-speed storage," *IEEE Trans. Audio Electroacoust.*, vol. AU-15, pp. 91–98, June 1967.
[3] N. M. Brenner, "Fast Fourier transform of externally stored data," *IEEE Trans. Audio Electroacoust.*, vol. AU-17, pp. 128–132, June 1969.
[4] G. D. Bergland, "The fast Fourier transform recursive equations for arbitrary length records," *Math. Comput.*, vol. 21, pp. 236–238, Apr. 1967.

## Two-Dimensional Windows

### I. Introduction

Two-dimensional windows find applications in many diverse fields, such as the spectral estimation of random fields, the design of two-dimensional digital filters, optical apodization, and antenna array design.

Many good one-dimensional windows have been devised [1]–[4]; however, relatively few two-dimensional windows have been investigated [5], [6]. In this correspondence, we establish a result which enables us to get good two-dimensional windows from good one-dimensional windows.

### II. The Problem

We first review briefly the one-dimensional problem. Let the Fourier transform of a function $f(x)$ be $F(u)$. For some reason, we want to truncate $f(x)$:

$$g(x) = f(x)w(x) \tag{1}$$

$$w(x) = 0, \quad \text{for } |x| > A \tag{2}$$

where $A$ is a constant.

The Fourier transform of $g(x)$ is

$$G(u) = F(u) \otimes W(u) \tag{3}$$

Manuscript received June 22, 1971.
This work was performed at M.I.T. Lincoln Laboratory, Lexington, Mass., under Air Force Contract F19628-70-C-0230.

where $W(u)$ is the Fourier transform of $w(x)$, and $\otimes$ denotes convolution. Our problem is to choose an appropriate shape for the window function $w(x)$ such that $G(u)$ is close to $F(u)$ and in any region surrounding a discontinuity of $F(u)$, $G(u)$ will not contain excessive ripples. It is well known that the Fourier transform $W(u)$ of a good window $w(x)$ should have a big central peak and small sidelobes.

In two dimensions, the problem is entirely similar. Let the two-dimensional Fourier transform of a function $f_2(x, y)$ be $F_2(u, v)$, and let

$$g_2(x, y) = f_2(x, y)w_2(x, y) \tag{4}$$

where

$$w_2(x, y) = 0, \quad \text{for } |x^2 + y^2| > A^2. \tag{5}$$

The Fourier transform of $g_2(x, y)$ is

$$G_2(u, v) = F_2(u, v) \otimes W_2(u, v) \tag{6}$$

where $W_2(u, v)$ is the Fourier transform of $w_2(x, y)$. The problem is to choose an appropriate shape for the two-dimensional window function $w_2(x, y)$ such that $G_2(u, v)$ is close to $F_2(u, v)$ and in the neighborhood of a discontinuity of $F_2(u, v)$, $G_2(u, v)$ does not contain excessive ripples.

### III. The Result

Intuitively, we feel that if $w(x)$ is a good symmetrical one-dimensional window, then

$$w_2(x, y) = w(\sqrt{x^2 + y^2}) \tag{7}$$

will be a good two-dimensional window. This is indeed partially verified by the following two examples. The first example is

$$w(x) = \begin{cases} 1, & \text{for } |x| \le 1 \\ 0, & \text{for } |x| > 1. \end{cases} \tag{8}$$

Then the Fourier transform is

$$W(u) = \frac{2 \sin u}{u} \tag{9}$$

whose first sidelobe peak is about 20 percent of the peak at $u=0$. The corresponding two-dimensional window

$$w_2(x, y) = \begin{cases} 1, & \text{for } |x^2 + y^2| \le 1 \\ 0, & \text{for } |x^2 + y^2| > 1 \end{cases} \tag{10}$$

has the two-dimensional Fourier transform

$$W_2(u, v) = \frac{2\pi J_1(\sqrt{u^2 + v^2})}{\sqrt{u^2 + v^2}} \tag{11}$$

whose first sidelobe peak is only about 12 percent of the peak at $u=0=v$.

The second example is

$$w(x) = \begin{cases} 1 - x, & \text{for } |x| \le 1 \\ 0, & \text{for } |x| > 1. \end{cases} \tag{12}$$

The Fourier transform is

$$W(u) = \left[ \frac{\sin \frac{u}{2}}{\frac{u}{2}} \right]^2 \tag{13}$$

whose sidelobe peak is about 4 percent of the peak at $u=0$. The corresponding two-dimensional window

Reprinted from *IEEE Trans. Audio Electroacoust.*, vol. AU-20, pp. 88–89, Mar. 1972.

$$w_2(x, y) = \begin{cases} 1-\sqrt{x^2+y^2}, & \text{for } |x^2+y^2| \le 1 \\ 0, & \text{for } |x^2+y^2| > 1 \end{cases} \quad (14)$$

has a two-dimensional Fourier transform

$$W_2(u, v) = 2\pi \left[ \rho^{-3} \int_0^\rho J_0(t)\, dt - \rho^{-2} J_0(\rho) \right] \quad (15)$$

where $\rho = \sqrt{u^2+v^2}$, whose first sidelobe peak is only about 2 percent of the peak at $u=0=v$.

The above comparisons are, however, unfair. Because when we convolve a window with a discontinuity, what count in the one-dimensional case are the areas under the sidelobes, while in the two-dimensional case what count are the volumes under the sidelobes. A fair comparison would be to look at the result of the convolution. One thing we can say along this line is contained in the following theorem which is the main result of this correspondence.

*Theorem:* If a symmetrical one-dimensional window $w(x)$ and a two-dimensional window $w_2(x, y)$ are related by

$$w_2(x, y) = w(\sqrt{x^2 + y^2}) \quad (16)$$

then their Fourier transforms $W(u)$ and $W_2(u, v)$ satisfy the relation

$$\frac{1}{2\pi} W_2(u, v) \otimes H_2(u, v) = W(u) \otimes H(u) \quad (17)$$

where

$$H(u) = \begin{cases} 1, & \text{for } u \ge 0 \\ 0, & \text{for } u < 0 \end{cases} \quad (18)$$

$$H_2(u, v) = \begin{cases} 1, & \text{for } u \ge 0 \text{ and all } v \\ 0, & \text{for } u < 0 \text{ and all } v \end{cases} \quad (19)$$

and $\otimes$ denotes convolution.

*Proof:* We first show that

$$W(u) = \frac{1}{2\pi} \int_{-\infty}^{\infty} dv W_2(u, v). \quad (20)$$

By definition,

$$w_2(x, y) = \frac{1}{4\pi^2} \int_{-\infty}^{\infty} du\, dv W_2(u, v) e^{j(xu+yv)}.$$

Whence

$$w_2(x, 0) = \frac{1}{2\pi} \int_{-\infty}^{\infty} du\, e^{jxu} \cdot \left[ \frac{1}{2\pi} \int_{-\infty}^{\infty} dv W_2(u, v) \right]. \quad (21)$$

But from (16),

$$w_2(x, 0) = w(x).$$

Therefore from (21), $1/2\pi \int_{-\infty}^{\infty} dv W_2(u, v)$ is the Fourier transform of $w(x)$. This established (20). Now,

$$W(u) \otimes H(u) = \int_{-\infty}^{\infty} dt W(t) H(u-t) = \int_{-\infty}^{u} dt W(t) \quad (22)$$

and

$$W_2(u, v) \otimes H_2(u, v) = \iint_{-\infty}^{\infty} dt\, ds W(t, s) H_2(u-t, v-s) = \int_{-\infty}^{u} dt \int_{-\infty}^{\infty} ds W(t, s). \quad (23)$$

From (22) and (20), we have

$$W(u) \otimes H(u) = \frac{1}{2\pi} \int_{-\infty}^{u} dt \int_{-\infty}^{\infty} ds W(t, s) = \frac{1}{2\pi} W_2(u, v) \otimes H_2(u, v)$$

by virtue of (23).

## IV. The Design of Two-Dimensional Nonrecursive Ideal Low-Pass Filters

One way of designing one-dimensional nonrecursive digital filters is the so-called window method [7]. To fix ideas, let us consider the design of an ideal low-pass filter. The ideal frequency response is

$$F(u) = \begin{cases} 1, & \text{for } |u| \le B \\ 0, & \text{for } |u| > B \end{cases} \quad (24)$$

where $B$ is a constant. (This frequency response is actually repeated periodically because the impulse response is sampled.) The inverse Fourier transform $f(x)$ of $F(u)$, which is the impulse response, has infinite duration, but in reality we have to use a finite-duration impulse response. So we truncate $f(x)$ by using a window:

$$g(x) = f(x) w(x) \quad (25)$$

$$w(x) = 0, \quad \text{for } |x| > A \quad (26)$$

where $A$ is a constant. The actual frequency response we are getting is then

$$G(u) = F(u) \otimes W(u). \quad (27)$$

Suppose now we wish to design a two-dimensional ideal low-pass filter with an ideal frequency response

$$F_2(u, v) = F(\sqrt{u^2 + v^2}) = \begin{cases} 1, & \text{for } |u^2 + v^2| \le B^2 \\ 0, & \text{for } |u^2 + v^2| > B^2. \end{cases} \quad (28)$$

And we truncate the two-dimensional impulse response $f_2(x, y)$ by a two-dimensional window $w_2(x, y)$. Then the actual frequency response we are getting is

$$G_2(u, v) = F_2(u, v) \otimes W_2(u, v). \quad (29)$$

Let us assume that the widths of $W(u)$ and $W_2(u, v)$ are much smaller than $B$, then near the discontinuities of $F(u)$, viz., $u = \pm B$, $G(u)$ is essentially equal to the convolution of $W(u)$ and a one-dimensional step function, and similarly, nears the discontinuities of $F_2(u, v)$, viz., $u^2+v^2=B^2$, $G_2(u, v)$ is essentially equal to the convolution of $W_2(u, v)$ and a two-dimensional step function. It therefore follows from our theorem that if

$$w_2(x, y) = w(\sqrt{x^2 + y^2}) \quad (30)$$

then

$$\frac{1}{2\pi} G_2(u, v) \approx G(\sqrt{u^2 + v^2}). \quad (31)$$

This means that we can design a good two-dimensional low-pass filter by using the window $w_2(x, y)$ as given by (30), if $w(x)$ is a good window to use in designing a good one-dimensional low-pass filter.

## V. Summary

We have shown that if $w(x)$ is a good symmetrical one-dimensional window, then $w_2(x, y) = w(\sqrt{x^2+y^2})$ is a good circularly symmetrical two-dimensional window.

T. S. HUANG[1]
Dep. Elec. Eng.
M.I.T. Lincoln Lab.[2]
Lexington, Mass.

## References

[1] R. B. Blackman and J. W. Tukey, *The Measurement of Power Spectra.* New York: Dover, 1959.

[2] R. B. Blackman, *Linear Data-Smoothing and Prediction in Theory and Practice.* Reading, Mass.: Addison-Wesley, 1965.

[3] J. F. Kaiser, "Digital filters," in *System Analysis by Digital Computer*, F. F. Kuo and J. F. Kaiser, Eds. New York: Wiley, 1966, ch. 7.

[4] H. D. Helms, "Nonrecursive digital filters: Design methods for achieving specifications on frequency response," *IEEE Trans. Audio Electroacoust.*, vol. AU-16, pp. 336–342, Sept. 1968.

[5] A. Papoulis, *Systems and Transforms with Applications in Optics.* New York: McGraw-Hill, 1969.

[6] J. L. Allen, "The theory of array antennas (with emphasis on radar applications)," M.I.T. Lincoln Lab., Lexington, Mass., Tech. Rep. 323, Memo DDC AD-422945, July 1963.

[7] L. R. Rabiner, B. Gold, and C. A. McGonegal, "An approach to the approximation problem for nonrecursive digital filters," *IEEE Trans. Audio Electroacoust.*, vol. AU-18, pp. 83–106, June 1970.

[1] Now with the Department of Electrical Engineering, Massachusetts Institute of Technology, Cambridge, Mass.

[2] Operated with support from the U. S Air Force.

# Design Techniques for Two-Dimensional Digital Filters

JAMES V. HU and LAWRENCE R. RABINER

*Abstract*—The theory for designing finite-duration impulse response (FIR) digital filters can readily be extended to two or more dimensions. Using linear programming techniques, both frequency sampling and optimal (in the sense of Chebyshev approximation over closed compact sets) two-dimensional filters have been successfully designed. Computational considerations have limited the filter impulse response durations (in samples) to 25 by 25 in the frequency sampling case, and to 9 by 9 in the optimal design case. However, within these restrictions, a large number of filters have been investigated. Several of the issues involved in designing two-dimensional digital filters are discussed.

## Introduction

Techniques for designing one-dimensional digital filters have been investigated for several years, and are fairly well established in the literature [1]–[7]. Very few of these techniques, however, have been extended to two or more dimensions. Since there is no lack of interest in two-dimensional signal processing, it is worth investigating those design procedures that are readily extendable to two dimensions. In this paper, the frequency sampling technique and optimal filter design procedures are discussed in the case of the design of circularly symmetric two-dimensional low-pass filters.

Before discussing the results obtained, it is worth reviewing the state of the art in designing two-dimensional digital filters. Two of the most common classes of one-dimensional digital filters are usually called recursive and nonrecursive filters. (The alternate terminology infinite impulse response (IIR) and finite impulse response (FIR) has recently been proposed [8] and will be used throughout this paper.) IIR filters can be efficiently realized by factoring their $z$-transforms into a cascade of real or complex pole-zero networks and realizing the filter using these simple sections. In two dimensions, however, the two-dimensional $z$-transforms cannot generally be factored into lower order systems [9], [10]. This leads to many design problems, such as the difficulty in determining the stability of the filter, as well as in finding an efficient realization where the coefficients may be truncated to a reasonable number of bits. Because of these difficulties, very little work has been done on designing IIR two-dimensional digital filters.

For FIR digital filters, the problems of stability do not exist in one dimension as the $z$-transform is a finite polynomial. Thus design techniques in one dimension are often directly extendable to two or more dimensions by appropriate modifications to the design procedure. Huang [11] has already shown how the windowing technique may be extended to two dimensions by forming two-dimensional circularly symmetric windows from one-dimensional windows. The remainder of this paper discusses the design of circularly symmetric frequency sampling filters and optimal (equiripple) filters.

Manuscript received May 2, 1972. This paper was presented at the IEEE Arden House Workshop on Digital Filtering, Harriman, N. Y., January 1972.

The authors are with Bell Telephone Laboratories, Inc., Murray Hill, N. J. 07974.

## Theory of Two-Dimensional FIR Filters

Let $h(n_1, n_2)$ be the impulse response of a two-dimensional filter where $-\infty < n_1, n_2 < \infty$. The two-dimensional $z$ transform of the filter is defined as

$$H(z_1, z_2) = \sum_{n_1=-\infty}^{\infty} \sum_{n_2=-\infty}^{\infty} h(n_1, n_2) z_1^{-n_1} z_2^{-n_2}. \qquad (1)$$

If the resulting filter is to be stable, the impulse response must satisfy the condition

$$\sum_{n_1=-\infty}^{\infty} \sum_{n_2=-\infty}^{\infty} |h(n_1, n_2)| < \infty. \qquad (2)$$

For IIR filters (when any one of the four limits on the summation is infinite), it is generally quite difficult to determine whether or not a particular filter is stable, i.e., whether (2) is satisfied. For FIR filters, on the other hand, stability is always guaranteed since all four limits on the summation are finite and $|h(n_1, n_2)| < \infty$.

If we restrict ourselves to the case when $h(n_1, n_2)$ is a finite sequence defined from $0 \le n_1 \le N_1 - 1$, $0 \le n_2 \le N_2 - 1$, then (1) becomes

$$H(z_1, z_2) = \sum_{n_1=0}^{N_1-1} \sum_{n_2=0}^{N_2-1} h(n_1, n_2) z_1^{-n_1} z_2^{-n_2}. \qquad (3)$$

The "frequency response" of the filter is obtained by evaluating (3) for $z_1 = e^{j\omega_1}$ and $z_2 = e^{j\omega_2}$, giving

$$H(e^{j\omega_1}, e^{j\omega_2}) = \sum_{n_1=0}^{N_1-1} \sum_{n_2=0}^{N_2-1} h(n_1, n_2) \exp\left[-j(\omega_1 n_1 + \omega_2 n_2)\right]. \quad (4)$$

The two-dimensional FIR filter may be realized in several ways. The simplest technique is a direct-convolution realization using the relation

$$y(n_1, n_2) = \sum_{m_1=0}^{N_1-1} \sum_{m_2=0}^{N_2-1} h(m_1, m_2) x(n_1 - m_1, n_2 - m_2) \qquad (5)$$

Reprinted from *IEEE Trans. Audio Electroacoust.*, vol. AU-20, pp. 249–257, Oct. 1972.

where $x(n_1, n_2)$ is the input to the filter and $y(n_1, n_2)$ is the filter output.

The discrete Fourier transform (DFT) relations for the filter may be obtained by evaluating (4) at the discrete set of values

$$\omega_{k_1} = k_1\left(\frac{2\pi}{N_1}\right), \qquad k_1 = 0, 1, \cdots, N_1 - 1 \tag{6}$$

$$\omega_{k_2} = k_2\left(\frac{2\pi}{N_2}\right), \qquad k_2 = 0, 1, \cdots, N_2 - 1 \tag{7}$$

giving[1]

$$H(k_1, k_2) = \sum_{n_1=0}^{N_1-1} \sum_{n_2=0}^{N_2-1} h(n_1, n_2) \cdot \exp\left[-j2\pi\left(\frac{k_1 n_1}{N_1} + \frac{k_2 n_2}{N_1}\right)\right]. \tag{8}$$

The inverse DFT is readily obtained as

$$h(n_1, n_2) = \frac{1}{N_1 N_2} \sum_{k_1=0}^{N_1-1} \sum_{k_2=0}^{N_2-1} H(k_1, k_2) \cdot \exp\left[j2\pi\left(\frac{k_1 n_1}{N_1} + \frac{k_2 n_2}{N_2}\right)\right]. \tag{9}$$

It is interesting to note that two-dimensional DFT's, in (8), may be evaluated using one-dimensional DFT's by partitioning (8) as

$$H(k_1, k_2) = \sum_{n_1=0}^{N_1-1} \underbrace{\left[\sum_{n_2=0}^{N_2-1} h(n_1, n_2) e^{-j2\pi(n_2 k_2/N_2)}\right]}_{g(n_1, k_2)} e^{-j2\pi(n_1 k_1/N_1)} \tag{10}$$

and realizing that the inner summation is a one-dimensional DFT where $n_1$ is held fixed; once all the inner summation DFT's have been calculated (calling the results $g(n_1, k_2)$, the outer summation is a DFT where $k_2$ is held fixed.

A frequency interpolation formula from the DFT coefficients may be derived by inserting (9) into (4), giving

$$H(e^{j\omega_1}, e^{j\omega_2}) = \sum_{n_1=0}^{N_1-1} \sum_{n_2=0}^{N_2-1} \left[\frac{1}{N_1 N_2} \sum_{k_1=0}^{N_1-1} \sum_{k_2=0}^{N_2-1} H(k_1, k_2) \cdot \exp\left[j2\pi\left(\frac{k_1 n_1}{N_1} + \frac{k_2 n_2}{N_2}\right)\right]\right] \cdot \exp\left[-j(\omega_1 n_1 + \omega_2 n_2)\right]. \tag{11}$$

Interchanging orders of summation, and summing over the $n_1$ and $n_2$ indices, gives

$$H(e^{j\omega_1}, e^{j\omega_2}) = \sum_{k_1=0}^{N_1-1} \sum_{k_2=0}^{N_2-1} H(k_1, k_2) A(k_1, k_2, \omega_1, \omega_2) \tag{12}$$

where

$$A(k_1, k_2, \omega_1, \omega_2) = \frac{1}{N_1 N_2}\left[\frac{1-\exp(-jN_1\omega_1)}{1-\exp\left[j\left(2\pi\frac{k_1}{N_1} - \omega_1\right)\right]}\right] \cdot \left[\frac{1-\exp(-jN_2\omega_2)}{1-\exp\left[j\left(2\pi\frac{k_2}{N_2} - \omega_2\right)\right]}\right]. \tag{13}$$

Equation (12) serves as the basis for designing frequency sampling two-dimensional filters. As seen from (12), the continuous frequency response of the filter is a linear combination of shifted interpolating functions $(A(k_1, k_2, \omega_1, \omega_2))$ weighted by the DFT coefficients $H(k_1, k_2)$. The DFT coefficients are called frequency samples as they exactly specify the value of the frequency response at uniformly spaced frequencies. For designing frequency sampling filters, the majority of the frequency samples are given specific values depending on the frequency response being approximated. The remaining unspecified frequency samples are left as free variables to be optimized according to some minimization criterion. The details of frequency sampling design of two-dimensional low-pass filters will be left for the next section.

For designing optimal (equiripple) filters whose peak error of approximation, over several closed areas in the $\omega_1$, $\omega_2$ plane, is minimized, (4) can be applied directly.[2] In this case, the filter coefficients are the impulse response samples that are chosen to minimize the approximation error using an optimization procedure. Since the filter frequency response of (4) is still linear in the filter coefficients $(h(n_1, n_2))$, optimization procedures similar to those used to design frequency sampling filters may be applied. The details of this method are discussed in a later section.

Finally, it can be observed from (4) that if certain symmetries are maintained in the filter impulse response coefficients (similar statements can be made about the DFT coefficients), the filter's frequency response becomes purely real, to within a linear phase shift in each of the two dimensions. The necessary symmetries are

$$h(n_1, n_2) = h(N_1 - 1 - n_1, n_2) = h(n_1, N_2 - 1 - n_2). \tag{14}$$

The application of (14) to (4) gives (assuming $N_1$, $N_2$ even)

[1] The terminology $H(k_1, k_2)$ rather than $H(e^{j\omega_{k_1}}, e^{j\omega_{k_2}})$ is used for notational convenience.

[2] Equivalently, by making all the DFT coefficients variable, (12) can be applied directly for designing optimal (equiripple) filters.

$$H(e^{j\omega_1}, e^{j\omega_2}) = e^{-j\omega_1((N_1-1)/2)}e^{-j\omega_2((N_2-1)/2)} \cdot \sum_{n_1=0}^{(N_1/2)-1} \sum_{n_2=0}^{(N_2/2)-1} h(n_1, n_2) \cdot \cos\left[\left(\frac{N_1-1}{2}\right) - n_1\right] \cdot \cos\left[\left(\frac{N_2-1}{2}\right) - n_2\right] \quad (15)$$

which is purely real, except for the linear phase terms outside the summation.

### Design of Frequency Sampling Filters

For the specific case of linear phase frequency sampling filters, (12) and (13) can be modified through use of the symmetry relations on the DFT coefficients which are of the form

$$H(k_1, k_2) = |H(k_1, k_2)|\, e^{j\theta(k_1,k_2)} \quad (16)$$

where

$$|H(k_1, k_2)| = |H(k_1, N_2 - k_2)| = |H(N_1 - k_1, k_2)| \quad (17)$$

$$\theta(k_1, k_2) = \theta(k_1) + \theta(k_2) \quad (18)$$

$$\theta(k_1) = \begin{cases} -\dfrac{2\pi}{N_1} k_1\left(\dfrac{N_1-1}{2}\right), & k_1 = 0, 1, \cdots, NU \\ \dfrac{2\pi}{N_1}(N_1 - k_1)\left(\dfrac{N_1-1}{2}\right), & k_1 = NU+1, \cdots, N_1 - 1 \end{cases} \quad (19)$$

$$\theta(k_2) = \begin{cases} -\dfrac{2\pi}{N_2} k_2\left(\dfrac{N_2-1}{2}\right), & k_2 = 0, 1, \cdots, NV \\ \dfrac{2\pi}{N_2}(N_2 - k_2)\left(\dfrac{N_2-1}{2}\right), & k_2 = NV+1, \cdots, N_2 - 1 \end{cases} \quad (20)$$

and

$$NU = \begin{cases} N_1/2, & N_1 \text{ even} \\ \dfrac{N_1-1}{2}, & N_1 \text{ odd} \end{cases} \quad (21)$$

$$NV = \begin{cases} N_2/2, & N_2 \text{ even} \\ \dfrac{N_2-1}{2}, & N_2 \text{ odd.} \end{cases} \quad (22)$$

Whenever $N_1$ or $N_2$ is even, the additional constraints

$$\theta(k_1, N_2/2) = \theta(N_1/2, k_2) = 0 \quad (23)$$

$$H(k_1, N_2/2) = H(N_1/2, k_2) = 0 \quad (24)$$

must be maintained.

Applying these constaints into (12) and (13) yields the following equations (after considerable arithmetic manipulations):

$$H(e^{j\omega_1}, e^{j\omega_2}) = \exp\left[-j\left[\omega_1\left(\frac{N_1-1}{2}\right) + \omega_2\left(\frac{N_2-1}{2}\right)\right]\right] \cdot \frac{1}{N_1N_2} \cdot \Bigg[H(0,0)\alpha(\omega_1, N_1)\alpha(\omega_2, N_2) + \sum_{k_1=1}^{NU} |H(k_1, 0)|\, \alpha(\omega_2, N_2)\beta(\omega_1, k_1, N_1) + \sum_{k_2=1}^{NV} |H(0, k_2)|\, \alpha(\omega_1, N_1)\beta(\omega_2, k_2, N_2) + \sum_{k_1=1}^{NU}\sum_{k_2=1}^{NV} |H(k_1, k_2)|\, \beta(\omega_1, k_1, N_1)\beta(\omega_2, k_2, N_2)\Bigg] \quad (25)$$

where

$$\alpha(\omega, N) = \frac{\sin \dfrac{\omega N}{2}}{\sin \dfrac{\omega}{2}} \quad (26)$$

$$\beta(\omega, k, N) = \frac{\sin\left[\left(\dfrac{\omega}{2} - \dfrac{\pi k}{N}\right)N\right]}{\sin\left(\dfrac{\omega}{2} - \dfrac{\pi k}{N}\right)} + \frac{\sin\left[\left(\dfrac{\omega}{2} + \dfrac{\pi k}{N}\right)N\right]}{\sin\left(\dfrac{\omega}{2} + \dfrac{\pi k}{N}\right)} \quad (27)$$

$$NU = \begin{cases} N_1/2 - 1, & N_1 \text{ even} \\ (N_1 - 1)/2, & N_1 \text{ odd} \end{cases} \quad (28)$$

$$NV = \begin{cases} N_2/2 - 1, & N_2 \text{ even} \\ (N_2 - 1)/2, & N_2 \text{ odd.} \end{cases} \quad (29)$$

Although cumbersome in appearance, (25) is seen to consist basically of a sum of simple interpolating functions that are shifted (up or down) in frequency depending on $k_1$ and $k_2$. As such, (25)—without the leading linear phase terms—is the basic design equation for frequency sampling filters.

To apply (25) to the design of a circularly symmetric low-pass filter, consider Fig. 1 in which the filter response regions are plotted in the $\omega_1$, $\omega_2$ plane. The passband is the region where

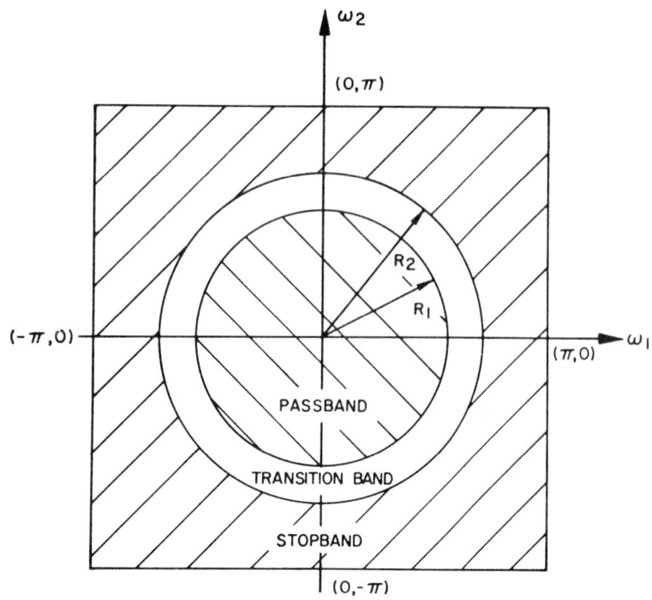


Fig. 1. Regions in the ($\omega_1$, $\omega_2$) plane over which the filter response is specified for a low-pass filter.

$$\rho(\omega_1, \omega_2) = (\omega_1^2 + \omega_2^2)^{1/2} \leq R_1. \tag{30}$$

The transition band is the region where

$$R_1 < \rho(\omega_1, \omega_2) < R_2 \tag{31}$$

and the stopband is the region where

$$\rho(\omega_1, \omega_2) \geq R_2. \tag{32}$$

With respect to Fig. 1, the DFT points, or frequency samples, are located on an $N_1$ by $N_2$ grid of points in the $\omega_1$, $\omega_2$ plane as shown in Fig. 2 for $N_1=9$, $N_2=9$. To design a frequency sampling approximation to the ideal circularly symmetric response

$$\hat{H}(e^{j\omega_1}, e^{j\omega_2}) = \begin{cases} 1, & \rho(\omega_1, \omega_2) \leq R_1 \\ 0, & \rho(\omega_1, \omega_2) \geq R_2 \end{cases} \tag{33}$$

the DFT values that occur inside the passband are given a value of 1.0; those that occur in the stopband are given a value 0.0; those that lie in the transition band are free variables whose value will be chosen by a minimization criterion.

To simplify both the equations and the computation involved in performing the desired minimizations, the following assumptions will be used in the remainder of this paper. First, we assume $N_1=N_2=$odd integer. Next, we assume

$$H(k_1, k_2) = H(k_2, k_1) \tag{34}$$

i.e., there is symmetry across the 45° diagonals in the $\omega_1$, $\omega_2$ plane. This is a reasonable assumption for designing circularly symmetric filters as (34) is valid in the ideal case of zero approximation error. Finally, we assume the linear phase term in (25) can be disregarded in terms of the design procedure.

With these assumptions in mind, (25) can be simplified to the form

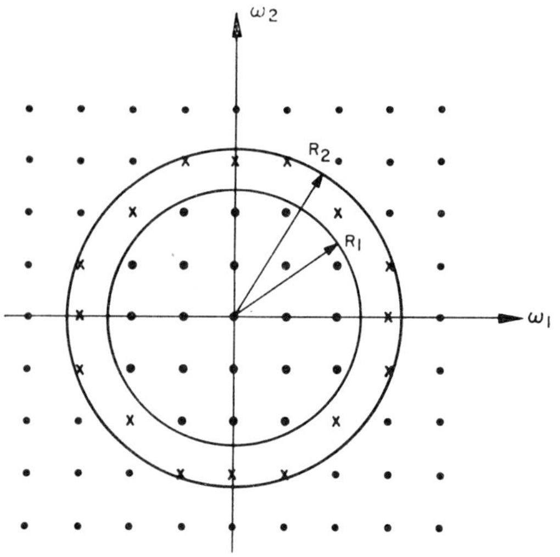


Fig. 2. Locations of frequency samples (DFT points) for a 9 by 9 grid.

$$H(e^{j\omega_1}, e^{j\omega_2}) = H_F(\omega_1, \omega_2) + \sum_{m=1}^{M} T_m H_m(\omega_1, \omega_2) \tag{35}$$

where $H_F(\omega_1, \omega_2)$ represents the contribution of the fixed DFT coefficients (the 1.0's), $T_m$ represents the amplitude of the $m$th DFT coefficient in the transition band, $H_m(\omega_1, \omega_2)$ represents the interpolation function appropriate to the $m$th transition coefficient, and $M$ is the total number of transition coefficients. The pertinent design equations can now be put in the form

$$1 - \alpha\delta \leq H(e^{j\omega_1}, e^{j\omega_2}) \leq 1 + \alpha\delta, \quad \text{for } \rho(\omega_1, \omega_2) \leq R_1 \tag{36}$$

$$-\delta \leq H(e^{j\omega_1}, e^{j\omega_2}) \leq \delta, \quad \text{for } \rho(\omega_1, \omega_2) \geq R_2 \tag{37}$$

where $\delta$ represents the peak approximation error in the stopband, and $\alpha\delta$ ($\alpha$ any constant) represents the peak approximation error in the passband. The design goal is to choose the transition samples so as to minimize $\delta$.

The above problem is readily seen to be a linear programming [7] problem by evaluating (36) and (37) at a dense set of points in the passband and stopband and by writing the equations explicitly as linear inequalities in the form

$$\left.\begin{aligned} \sum_{m=1}^{M} T_m H_m(\omega_1, \omega_2) - \alpha\delta &\leq 1 - H_F(\omega_1, \omega_2) \\ -\sum_{m=1}^{M} T_m H_m(\omega_1, \omega_2) - \alpha\delta &\leq -1 + H_F(\omega_1, \omega_2) \end{aligned}\right\} \quad \text{for all points with } \rho(\omega_1, \omega_2) \leq R_1 \tag{38}$$

$$\left.\begin{aligned} \sum_{m=1}^{M} T_m H_m(\omega_1, \omega_2) - \delta &\leq -H_F(\omega_1, \omega_2) \\ -\sum_{m=1}^{M} T_m H_m(\omega_1, \omega_2) - \delta &\leq H_F(\omega_1, \omega_2) \end{aligned}\right\} \quad \text{for all points with } \rho(\omega_1, \omega_2) \geq R_2 \tag{39}$$

choosing ($T_1$, $T_2$, $\cdots$, $T_M$) to minimize $\delta$.

**TABLE I**

**Stopband Attenuation for Low-Pass Filters (Filter Attenuation in Decibels)**

| | $(R_2-R_1)\cdot(12.5/\pi)$ | | | | |
|---|---|---|---|---|---|
| $R_1\cdot(12.5/\pi)$ | 1 | 1.5 | 2 | 2.5 | 3 |
| 0 | | | −42.8 | | −69.28 |
| 1 | | | −37.37 | | −65.08 |
| 2 | | | −42.06 | | |
| 3 | | | AF | | −67.92 |
| 4 | | | −38.56 | | |
| 5 | | | −41.83 | | NC |
| 6 | | | AF | | |
| 7 | | | AF | | |
| 8 | | | NC | | |
| 8.5 | | | | NC | |
| 9 | | | −44.68 | | |
| 9.5 | | −31.85 | | NC | |
| 10 | −14.42 | NC | −48.35 | | |
| 10.5 | −21.87 | −33.72 | | | |
| 11 | −11.25 | | | | |

AF = algorithm failed for undetermined reasons.
NC = convergence was not attained within a specified number of iterations.

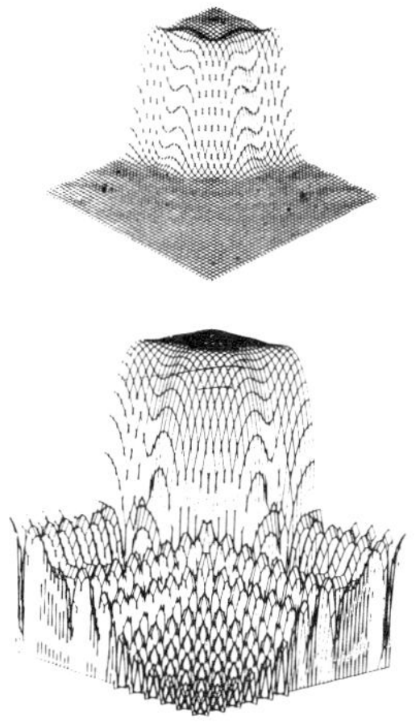


**Fig. 3. Amplitude and log magnitude response of a typical low-pass filter for a transition width of $R_2-R_1=2\pi/12.5$.**

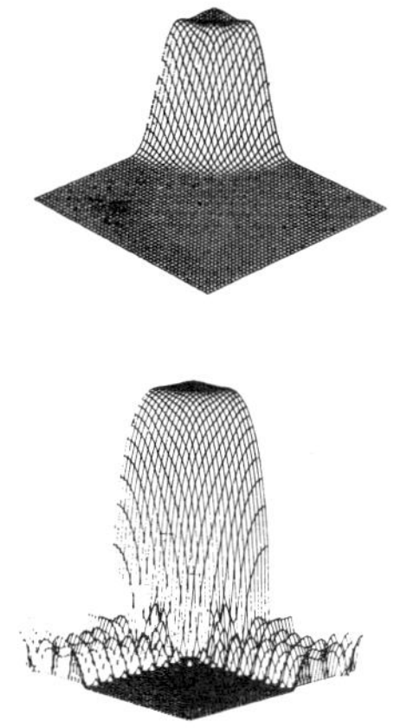


**Fig. 4. Amplitude and log magnitude response of a typical low-pass filter for a transition width of $R_2-R_1=3\pi/12.5$.**

## Results of Low-Pass Filter Design

The linear programming problem of (38) and (39) was solved on a large central computer (GE-635) using the scientific subroutine APMM [12]. A 64-1 interpolation was used throughout the stopband and passband to evaluate $H(e^{j\omega_1}, e^{j\omega_2})$ over a fine grid. Table I presents a summary of results[3] obtained for the case where $N_1=N_2=N=25$ and $\alpha\gg 1$, i.e., in-band ripple was not constrained. The entries in this table give values for $20\log_{10}\delta$ as a function of $R_1$, the in-band radius, and $R_2-R_1$, the width of the transition band. For many of the entries in the table, the linear program subroutine did not converge within a specified number of iterations (on the order of 200) and the run was terminated. Each filter that was designed required on the order of 6-min computation time to determine the transition coefficients. This was due to the large number of constraint equations (on the order of several thousand), even though the number of variable coefficients was small (on the order of 10).

If the passband ripple is unconstrained, the results of Table I indicate that a transition width of $2(\pi/12.5)$ gives about 40-dB stopband attenuation, whereas a width of $3(\pi/12.5)$ gives about 60-dB attenuation. Figs. 3 and 4 show two-dimensional perspectives [14] of the frequency response of two of the filters in Table I. Fig. 3 shows linear and log magnitude responses for a filter with $R_1=5\pi/12.5$, $R_2-R_1=2\pi/12.5$. Fig. 4 shows linear and log magnitude responses for a filter with $R_1=3\pi/12.5$, $R_2-R_1=3\pi/12.5$. These figures clearly show that the peak in-band ripple occurs at the edges of the passband, and the peak out-of-band ripple occurs at the edges of the stopband. (The filter response only in the region $0\le\omega_1, \omega_2\le\pi$ is shown in these figures as the response in the rest of the $\omega_1, \omega_2$ plane is determined from symmetry considerations.)

[3] The exact filter coefficients corresponding to the entries in Table I are available in [13].

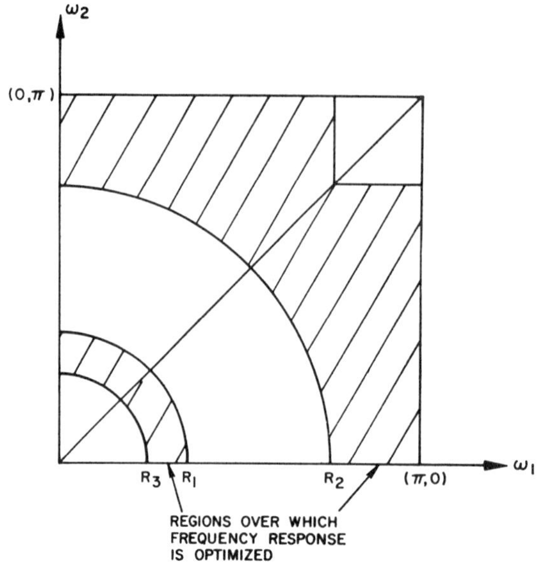


**Fig. 5. Modified regions in the ($\omega_1$, $\omega_2$) plane over which the filter response is optimized for a circularly symmetric low-pass filter.**

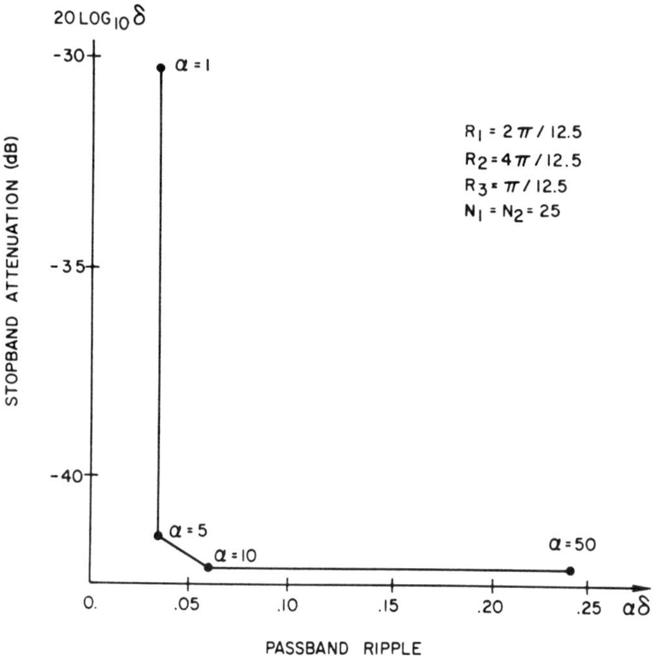


**Fig. 6. Plot of tradeoffs obtainable between passband ripple and peak stopband attenuation for a fixed transition width.**

Since the worst ripples occurred near the edges of the transition band, a new set of filters was designed using the constraint on passband ripple. In an effort to save computation, the filter response was optimized over the region shown in Fig. 5. The regions $0 \leq \rho(\omega_1, \omega_2) \leq R_3$ and $\omega_1 > \omega_2 > R_2$ were unconstrained because the ripple in the passband and stopband becomes quite small far from the band edges. A filter was designed with $R_3 = \pi/12.5$, $R_1 = 2\pi/12.5$, $R_2 = 4\pi/12.5$, and with $N_1 = N_2 = 25$. Values of $\alpha$ of 50, 10, 5, and 1 were used. Fig. 6 gives a plot of the measured tradeoffs between the stopband ripple 20 $\log_{10} \delta$ and the passband ripple $\alpha\delta$. This curve shows that considerable improvement can be made in passband ripple (a reduction from 0.24 to 0.04) with hardly any change in stopband attenuation (0.8 dB). On the basis of these results, it would seem worth the computational effort to constrain the in-band ripple at the edges of the passband. Figs. 7–10 show the frequency responses for the cases $\alpha = 50$ and $\alpha = 1$, as well as magnitude contour plots (magnitude $\geq 0.75$) for these two cases. In the contour plots, a dotted contour is drawn each 0.05. The passband and the transition region for the $\alpha = 1$ case are significantly more circularly symmetric than for the $\alpha = 50$ case, because the $\alpha = 50$

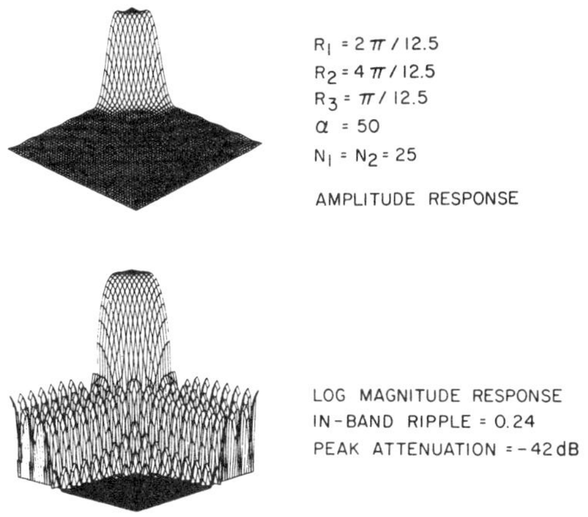


Fig. 7. Amplitude and log magnitude response of a low-pass filter with $\alpha = 50$.

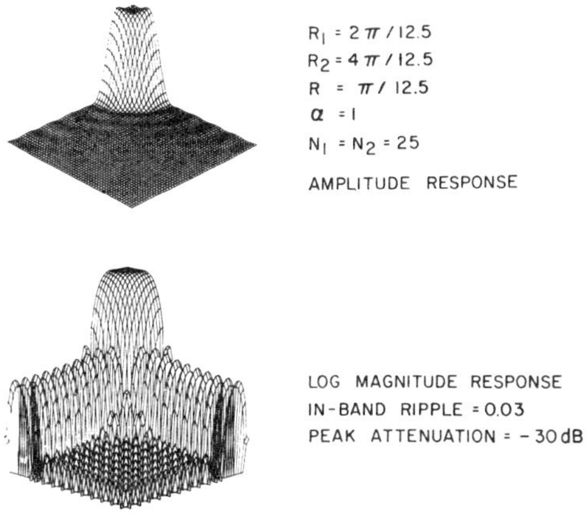


Fig. 8. Amplitude and log magnitude response of a low-pass filter with $\alpha = 1$.

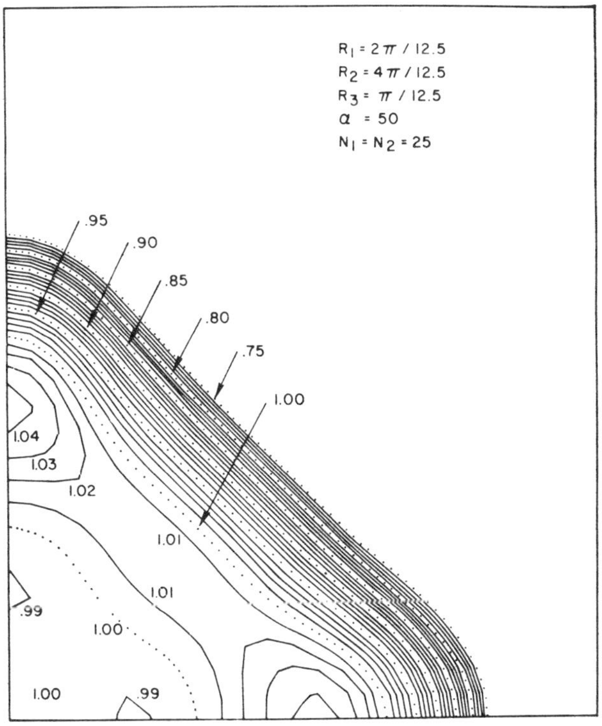


Fig. 9. Contour plot of passband and transition band response for the filter of Fig. 7.

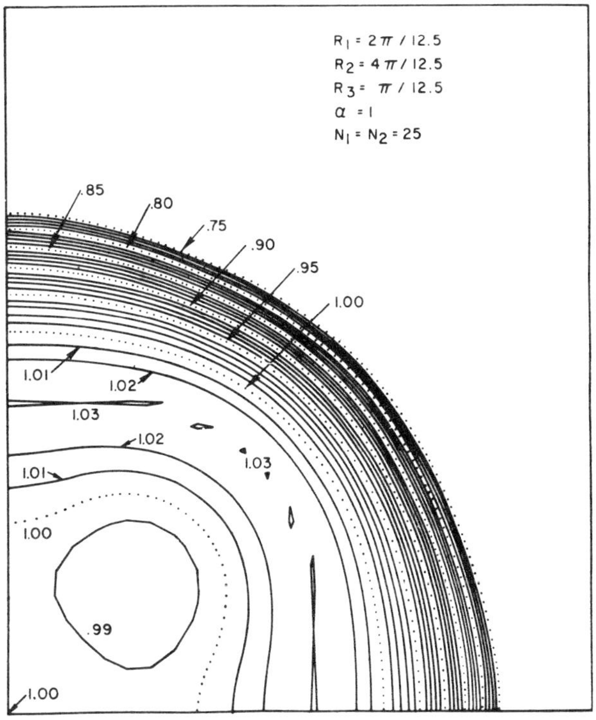


Fig. 10. Contour plot of passband and transition band response for the filters of Fig. 8.

case essentially has an unconstrained passband and thus the requirements for maintaining a good approximation to circular symmetry in the passband are not met.

### Design of Optimal (Equiripple) Filters

By allowing all the impulse response coefficients to vary, or equivalently all the DFT coefficients, filters can be designed which are optimal in the sense of Chebyshev approximation over closed regions. For the case when all the DFT coefficients are varied, as was done here, the filter frequency response is evaluated using (25) and treating all the filter coefficients $H(k_1, k_2)$ as unknowns. Optimal approximations to circularly symmetric filters may be designed (using linear programming techniques) by evaluating (25) at a dense set of points in both the passband and the stopband and minimizing both the stopband ripple $\delta$ and the passband ripple $\alpha\delta$.

Because of the increased amount of computation, necessitated both by the need to evaluate $H(e^{j\omega_1}, e^{j\omega_2})$ over the entire $\omega_1$, $\omega_2$ plane (except, of course, the transition band) and by the increased number of variables, the size of the impulse responses considered was reduced to $N_1 = N_2 = 9$. Within these limitations, four optimal (equiripple) filters were designed. A summary of the results for these four cases is given in Table II. Figs. 11 and 12 show the frequency response of the

TABLE II
Optimal Filters

| | |
|---|---|
| Filter 1 | $R_1=2\pi/4.5$, $R_2=3\pi/4.5$, $\alpha=1$<br>in-band ripple = 0.0867<br>out-of-band attenuation = −21.24 dB |
| Filter 2 | $R_1=\pi/4.5$, $R_2=2\pi/4.5$, $\alpha=10$<br>in-band ripple = 0.287<br>out-of-band attenuation = −30.84 dB |
| Filter 3 | $R_1=1.5\pi/4.5$, $R_2=3\pi/4.5$, $\alpha=10$<br>in-band ripple = 0.079<br>out-of-band attenuation = −41.97 dB |
| Filter 4 | $R_1=2\pi/4.5$, $R_2=3\pi/4.5$, $\alpha=10$<br>in-band ripple = 0.235<br>out-of-band attenuation = −32.51 dB |

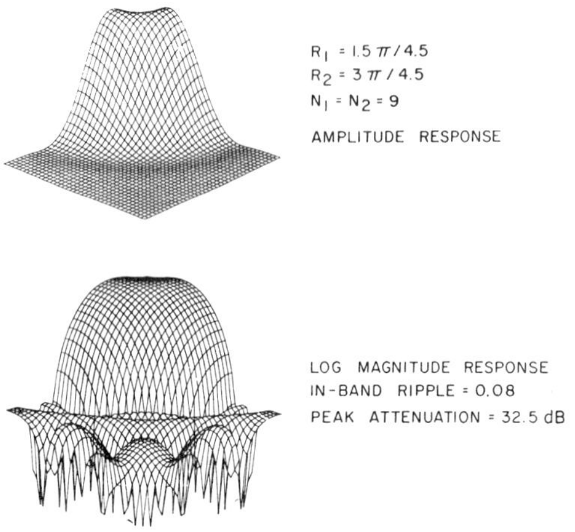


Fig. 11. Amplitude and log magnitude response of an optimal low-pass filter.

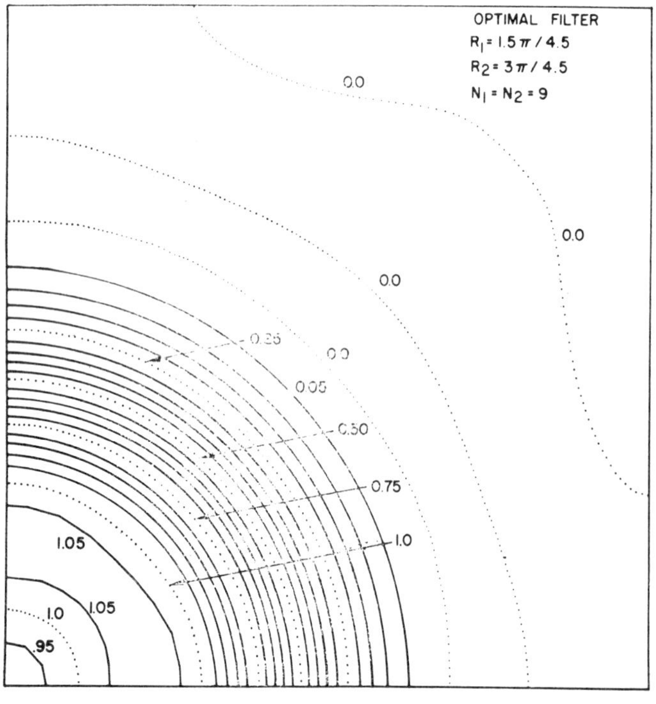


Fig. 12. Contour plot of the response of the filter of Fig. 11.

third filter in Table II and a contour plot of the log magnitude response, respectively. As the contour lines in the passband and transition region show, the passband response is a good approximation to a circularly symmetric response. Fig. 11 also shows the interesting phenomenon that there are equiripple ridges, as well as peaks. Thus the ripple of the frequency response may attain a maximum or minimum in a continuous region of the $\omega_1$, $\omega_2$ plane (corresponding to the equiripple ridge), as well as at isolated unconnected values. This is an added complexity to the one-dimensional design case where, because of the well-ordered frequency scale, the ripple peaks alternate between maxima and minima. This phenomenon suggests that simply counting points at which $H(e^{j\omega_1}, e^{j\omega_2})$ attains a maximum value can lead to significant errors in designing equiripple filters. For this reason several of the one-dimensional design algorithms [3], [5], [6] are not readily extended to two dimensions.

## Conclusions

This paper represents preliminary work on techniques for designing two-dimensional digital filters. It shows that the frequency sampling and optimal (equiripple) design methods can be carried over to the two-dimensional design problem directly, although at considerable computational cost. Future efforts along these lines should be aimed at lowering these costs.

## Acknowledgment

The authors wish to thank Mrs. N. Graham for her work in devising a novel algorithm for making the perspective plots shown.

## References

[1] J. F. Kaiser, "Digital filters," in *Systems Analysis by Digital Computer*, F. Kuo and J. F. Kaiser, Eds. New York: Wiley, 1966, ch. 7.
[2] B. Gold and C. M. Rader, *Digital Processing of Signals.* New York: McGraw-Hill, 1969.

[3] H. W. Schuessler, "On the approximation problem in the design of digital filters," in *Proc. 5th Annual Princeton Conf. Information Sciences and Systems*, Mar. 1971, pp. 54–63.
[4] L. R. Rabiner, B. Gold, and C. A. McGonegal, "An approach to the approximation problem for nonrecursive digital filters," *IEEE Trans. Audio Electroacoust.*, vol. AU-18, pp. 83–106, June 1970.
[5] E. Hofstetter, A. V. Oppenheim, and J. Siegel, "A new technique for the design of nonrecursive digital filters," in *Proc. 5th Annual Princeton Conf. Information Sciences and Systems*, Mar. 1971, pp. 64–72.
[6] T. W. Parks and J. H. McClellan, "Chebyshev approximation for nonrecursive digital filters with linear phase," *IEEE Trans. Circuit Theory*, vol. CT-19, pp. 189–194, Mar. 1972.
[7] L. R. Rabiner, "The design of finite impulse response digital filters using linear programming techniques," *Bell Syst. Tech. J.*, Aug. 1972.
[8] ——, "Techniques for designing finite-duration impulse-response digital filters," *IEEE Trans. Commun. Technol.*, vol. COM-19, pp. 188–195, Apr. 1971.
[9] J. L. Shanks, "The design of stable two-dimensional recursive filters," in *Proc. M. J. Kelly Communications Conf.*, Univ. of Missouri, Columbia, 1970.
[10] ——, "Two-dimensional recursive filters," in *SWIEECO Rec.*, 1969, pp. 19E1–19E8.
[11] T. S. Huang, "Two-dimensional windows," *IEEE Trans. Audio Electroacoust.* (Corresp.), vol. AU-20, pp. 88–89, Mar. 1972.
[12] *IBM System/360 Programming Manual*, Scientific Subroutine Package (360 A-CM-03 X), Fortran Program APMM, pp. 283–288.
[13] J. Hu, "Frequency sampling design of two-dimensional finite impulse response digital filters," M.S. thesis, Mass. Inst. Tech., Cambridge, June 1972.
[14] N. Graham, "Perspective drawing of surfaces with hidden line elimination," *Bell Syst. Tech. J.*, vol. 51, pp. 843–861, Apr. 1972.

# Chebyshev Approximation for Two-Dimensional Nonrecursive Digital Filters

Y. KAMP AND J. P. THIRAN, MEMBER, IEEE

*Abstract*—**The Remez exchange algorithm is extended for the design of two-dimensional nonrecursive digital filters approximating circularly symmetrical low-pass specifications according to a weighted Chebyshev error norm. Since the approximating function does not satisfy the Haar condition, the optimal solution is not necessarily unique and a straightforward extension of the one-dimensional exchange method may fail to converge. It is shown how the algorithm has to be complemented with a perturbation technique in order to force convergence under all circumstances. In the case of nonuniqueness the solution provided by the algorithm is a vertex of the polyhedron containing all optimal solutions, and a method is given which allows one to compute an adjacent vertex located on the same edge, thereby allowing the successive determination of all the vertices defining the polyhedron. For this case also, a procedure is described which selects, among all optimal solutions, the best one according to some additional criterion. Finally, the efficiency and accuracy aspects of the algorithm are considered and practical conclusions are drawn as an aid for the designer.**

## I. Introduction

WORK IN THE DOMAIN of two-dimensional digital filters has been up to now mainly concerned with nonrecursive designs because they escape the stability problem which is inherent in recursive filters [1] and because they can achieve a strictly linear phase which is a desirable feature for the distortionless processing of images.

Two-dimensional nonrecursive digital filters can be obtained by extending some of the methods already available for the design of one-dimensional filters. This was done for the windowing technique [2], linear programming [3], and for designs according to a maximally flat criterion [4]. Recently [5], an ingenious solution was proposed to transform a one-dimensional Chebyshev design into a two-dimensional design through a change of variables. The method is limited by the fact that the optimality of the result is tied to special classes of cutoff boundaries and, in particular, optimal approximations of specifications with exact circular symmetry cannot be achieved. This demonstrates the need for a method in which the two-dimensional nature of the problem is effectively taken into account. In this line the present paper investigates the problems connected with the Chebyshev approximation by functions of two variables. As a result, an extension of the Remez exchange algorithm [6] is presented for the design of two-dimensional nonrecursive filters which approximate low-pass specifications having exactly circular pass- and stop-band limits. Moreover, most of the results obtained here also apply to the design of a much wider class of filters.

Manuscript received March 11, 1974; revised July 5, 1974.

The authors are with the MBLE Research Laboratory, Brussells, Belgium.

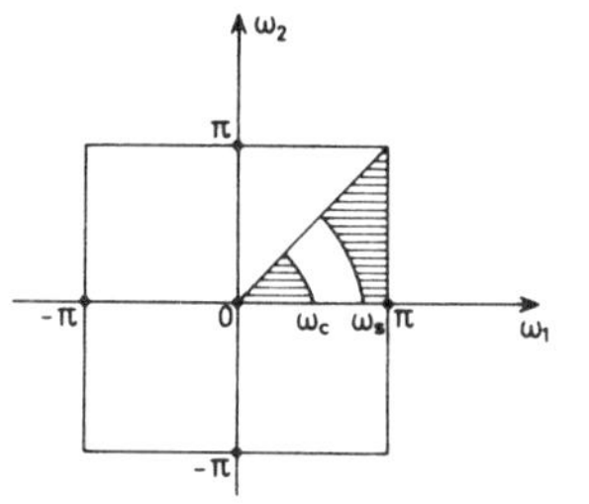


Fig. 1. Approximation domain.

## II. The Approximation Problem and its Solution

### *A. Statement of the Problem*

It has been shown in [4] that the requirement of circular symmetry imposes the following form for the frequency response of a nonrecursive two-dimensional digital filter

$$G(e^{j\omega_1},e^{j\omega_2}) = \sum_{I=0}^{N} \sum_{J=0}^{I} B_{IJ}(\cos I\omega_1 \cos J\omega_2 + \cos J\omega_1 \cos I\omega_2). \quad (1)$$

For the discussion of the approximation problem, it will prove more convenient to rewrite it as

$$f(x) = \sum_{i=1}^{m} a_i f_i(x) \quad (2)$$

where

$$\begin{aligned} f_i(x) &= \cos I\omega_1 \cos J\omega_2 + \cos J\omega_1 \cos I\omega_2, \\ x &= (\omega_1,\omega_2), \\ a_i &= B_{IJ}, \\ i &= I(I+1)/2 + J + 1, \\ m &= (N+1)(N+2)/2. \end{aligned} \quad (3)$$

The purpose is to design a circularly symmetric low-pass filter and the function to be approximated is thus

$$F(x) = \begin{cases} 1 & \text{if } \omega_1^2 + \omega_2^2 \le \omega_c^2 \\ 0 & \text{if } \omega_1^2 + \omega_2^2 \ge \omega_s^2 \end{cases}$$

where $\omega_c$ and $\omega_s$ are, respectively, the pass- and stopband limits. Since the frequency response (1) is already symmetrical with respect to both axes and both diagonals of the frequency plane, it will be sufficient to control its behavior in the shaded region of Fig. 1. In fact, this continuous domain will be replaced by a discrete finite point set $X = \{x_1,x_2,\cdots,x_n\}$ with $n > m$.

If $V$ denotes the linear space of approximating functions (2) when $\{a_i\} \in \boldsymbol{R}^m$, the following linear Chebyshev approx-

Reprinted from *IEEE Trans. Circuits Syst.*, vol. CAS-22, pp. 208–218, Mar. 1975.

imation problem has to be solved: find $f^* \in V$ which minimizes the norm

$$\|f - F\| = \max_{x \in X} w(x)|f(x) - F(x)| \quad (4)$$

where the weight function $w(x)$ is strictly positive on $X$. The minimum of this norm is the Chebyshev deviation

$$H = \min_{f \in V} \|f - F\|. \quad (5)$$

The solution lies in a generalization of the Remez algorithm which will be summarized in the following two sections by using the approach of Stiefel [6].

### B. Reference and Reference Function

The theory of the exchange algorithm relies on the following definitions. For the approximating function (2) the *characteristic vector* $\phi_j$ relative to a point $x_j$ is defined by $\phi_j = [f_1(x_j), f_2(x_j), \cdots, f_m(x_j)]^T$.

A *reference* $R \subset X$ is a set of $(m + 1)$ distinct points $R = \{x_1, x_2, \cdots, x_{m+1}\}$ such that the corresponding matrix $[\phi_1, \phi_2, \cdots, \phi_{m+1}]$ has rank $m$. The existence of at least one reference in $X$ is equivalent to the claim that the base functions $f_i(x)$ are linearly independent on $X$.

A *basis* is a set of $m$ linearly independent characteristic vectors. By definition, a reference contains at least one basis. Let then $B = [\phi_1, \phi_2, \cdots, \phi_m]$ be a basis for $R$. The *characteristic relation of a reference* expresses the characteristic vector $\phi_{m+1}$ of the reference as a linear combination of the other vectors

$$\phi_{m+1} = \sum_{j=1}^{m} \lambda_j \phi_j \quad (6)$$

where

$$\Lambda = [\lambda_1, \lambda_2, \cdots, \lambda_m]^T = B^{-1}\phi_{m+1}. \quad (7)$$

By putting

$$\lambda_{m+1} = -1 \quad (8)$$

one derives from (6)

$$\sum_{j=1}^{m+1} \lambda_j f(x_j) = 0, \qquad \forall f \in V \quad (9)$$

and $[{}^{\Lambda}_{-1}]$ is then called the *vector of characteristic coefficients.*

The *error* $h(x)$ associated with an approximating function is defined by

$$h(x) = w(x)[f(x) - F(x)]. \quad (10)$$

A function $f(x)$ is a *reference function for R* if

$$\operatorname{sgn} h(x_j) = \eta \operatorname{sgn} \lambda_j, \qquad 1 \le j \le m + 1 \quad (11)$$

where $\eta = \pm 1$ and where sgn $x$ is arbitrary for $x = 0$. By (8) the latter definition is equivalent to

$$h(x_j)h(x_{m+1})\lambda_j \le 0, \qquad 1 \le j \le m. \quad (12)$$

In the one-dimensional problem the characteristic coefficients have alternating signs and (11) becomes the well-known alternation property of the reference function. A *leveled reference function for R* satisfies (11) and in addition $|h(x_j)| = |h|$ for $1 \le j \le m + 1$. Owing to (9), (10), and (11), the *reference deviation* $|h|$ of a reference can be computed as

$$|h| = \left|\sum_{j=1}^{m+1} \lambda_j F(x_j)\right| \Bigg/ \left|\sum_{j=1}^{m+1} |\lambda_j| w^{-1}(x_j)\right| \quad (13)$$

and the *coefficient vector* $A = [a_1, a_2, \cdots, a_m]^T$ of the leveled reference function by

$$A^T = \left[F(x_1) + \frac{h(x_1)}{w(x_1)}, \cdots, F(x_m) + \frac{h(x_m)}{w(x_m)}\right] B^{-1}. \quad (14)$$

From (9), (10), (11), and (13), Theorem 1 follows.

*Theorem 1* [6]: If $f(x)$ is a reference function for $R$ with associated error $h(x)$, then the reference deviation of $R$ is given by

$$|h| = \sum_{j=1}^{m+1} |\lambda_j| |h(x_j)| w^{-1}(x_j) \Bigg/ \sum_{j=1}^{m+1} |\lambda_j| w^{-1}(x_j).$$

### C. Exchange Algorithm

*Theorem 2* [6]: Let $f(x)$ be a reference function for $R$, $h(x)$ the corresponding error and $x_k \notin R$ an arbitrary point in $X$. There exists a point $x_l \in R$ such that by exchanging $x_l$ with $x_k$ one obtains a new reference $R'$ for which $f(x)$ is still a reference function. The point to be exchanged is found according to the following rule. For the reference $\{x_1, x_2, \cdots, x_m, x_k\}$, let

$$\phi_k = \sum_{j=1}^{m} \mu_j \phi_j \quad (15)$$

be the characteristic relation with

$$M = [\mu_1, \mu_2, \cdots, \mu_m]^T = B^{-1}\phi_k. \quad (16)$$

*Case 1:* If

$$h(x_j)h(x_k)\mu_j \le 0, \qquad 1 \le j \le m \quad (17)$$

then $x_l = x_{m+1}$.

*Case 2:* If

$$E = \{x_j, 1 \le j \le m \mid h(x_j)h(x_k)\mu_j > 0\} \ne — \quad (18)$$

the index $l$ of the point to be exchanged is such that

$$|\lambda_l/\mu_l| = \min_{j \in E} |\lambda_j/\mu_j|. \quad (19)$$

It is indeed easily shown that for the new reference $R'$ a condition of the type (12) is satisfied. The new characteristic relation is (15) in Case 1 and results from the elimination of $\phi_l$ between (6) and (15) in Case 2, giving

$$\phi_{m+1} = \frac{\lambda_l}{\mu_l}\phi_k + \sum_{\substack{j=1 \\ \ne l}}^{m} \lambda_j' \phi_j \quad (20)$$

with

$$\lambda_j' = \lambda_j - \lambda_l \mu_j / \mu_l, \qquad 1 \le j \ne l \le m. \quad (21)$$

Let it be noted also that for Case 2 it is not necessary to compute the inverse of the new basis $B'$ by matrix inversion.

Since $B$ and $B'$ only differ by one characteristic vector, $(B')^{-1}$ can directly be derived from $B^{-1}$ [7]. This may, however, lead to important error accumulation, and it is therefore suggested to periodically restart the algorithm with a fresh inverse generated by matrix inversion in order to keep the reference well behaved.

*Method of Ascent:* Using the preceding theorems a monotonically increasing sequence of reference deviations is constructed as follows.

Let $f(x)$ be the leveled reference function for $R$ and $h(x)$ the associated error. If

$$\max_X |h(x)| = |h| \tag{22}$$

$f(x)$ is an *optimal solution* for the Chebyshev approximation $F(x)$ on $X$ [6]. Else let $x_k \notin R$ be a point of $X$ such that

$$|h(x_k)| > |h|. \tag{23}$$

By the above exchange Theorem 2, a new reference $R'$ is constructed whose characteristic relation is either (15) or (20). The new reference deviation $|h'|$ results then from Theorem 2.

*Case 1:*

$$|h'| = \frac{|h(x_k)|w^{-1}(x_k) + |h| \sum_{j=1}^{m} |\mu_j| w^{-1}(x_j)}{w^{-1}(x_k) + \sum_{j=1}^{m} |\mu_j| w^{-1}(x_j)} \tag{24}$$

and by (23), $|h'| > |h|$.

*Case 2:*

$$|h'| = \frac{|h(x_k)\lambda_l \mu_l^{-1}| w^{-1}(x_k) + |h| \left[ w^{-1}(x_{m+1}) + \sum_{\substack{j=1 \\ \neq l}}^{m} |\lambda_j'| w^{-1}(x_j) \right]}{|\lambda_l \mu_l^{-1}| w^{-1}(x_k) + w^{-1}(x_{m+1}) + \sum_{\substack{j=1 \\ \neq l}}^{m} |\lambda_j'| w^{-1}(x_j)} \tag{25}$$

and (23) again implies $|h'| > |h|$ except in the case $\lambda_l = 0$ which will be considered in the next section. The fact that the reference deviation is strictly increasing is of prime importance for the convergence proof of the algorithm because in a finite number of iterations a final reference is obtained for which (22) holds.

*Multiple Exchange:* In order to reduce the number of iterations, it is advantageous to increase as much as possible the new reference deviation $|h'|$. To this end, it would be preferable to introduce in the new reference not only a single point but several points where the absolute value of the error exceeds the reference deviation $|h|$. For the one-dimensional case [8], this is easy to achieve because the sign alternation of the $\lambda_j$ allows the simultaneous exchange of all the reference points. For two-variable approximations there is no such simple rule for the sign of the characteristic coefficients but one can try to perform a multiple exchange in the following way.

In the neighborhood of each reference point, the local maximum of the absolute value of the error will be retained if its value is larger than the reference deviation. Let then $(x_k^{(1)}, x_k^{(2)}, \cdots, x_k^{(r)})(1 \leq r \leq m + 1)$ be the set of the distinct local maxima. One calculates the vector $M_1 = B^{-1}\phi_k^{(1)}$ where $\phi_k^{(1)}$ denotes the characteristic vector of $x_k^{(1)}$ and the exchange Theorem 2 gives then the point $x_l$ of the reference $R$ which has to be exchanged. Let $R_1$ be the new reference, $B_1$ the corresponding basis. The same sequence of operations is repeated for the second maximum $x_k^{(2)}$ by computing $M_2 = B_1^{-1}\phi_k^{(2)}$ but this second exchange is effectively performed only if it does not eliminate $x_k^{(1)}$. One proceeds in this way for all local maxima with the rule that an exchange will not be carried out if it would eliminate one of the previously introduced maxima.

### D. Perturbation Technique for Degenerate Problems

An approximation problem becomes *degenerate for a reference R* if the corresponding characteristic relation contains at least one vanishing $\lambda_j$. The occurrence of degeneracies is a direct consequence of the fact that practically no two-variable functions satisfy the Haar condition [9]. According to the methods of ascent described in the preceding section, one has Theorem 3.

*Theorem 3:* A *static exchange* $|h'| = |h|$ occurs if and only if the problem is degenerate with respect to $R$ and if among the points $x_j$ of $R$ for which $\lambda_j = 0$, one at least satisfies the inequality $h(x_j)h(x_k)\,\mu_j > 0$ where $\mu_j$ is defined by (16).

The possibility of static exchanges for degenerate problems entails that the same reference can be repeated periodically. In these cases of cycling, the above exchange algorithm does not converge and one has to resort to a special perturbation technique. This technique was originally developed in linear programming [7] to cope with similar convergence problems and was later on adapted to suit the exchange algorithm. The adaptation presented here is similar to but more efficient than that of Bittner [10] and Descloux [11]. Like in linear programming, the perturbation technique is only of theoretical importance because it ensures convergence without exception, but in practice no degenerate cases have been known to cycle for this problem.

The perturbation technique consists in replacing the degenerate problem $P$ by a perturbed problem $P(\varepsilon)$ satisfying the twofold condition that it never becomes degenerate and that it coincides with the original problem when $\varepsilon$ tends to zero. In this way, the convergence of the perturbed problem also insures that of the degenerate problem. For convenience it will be assumed that the points of $R$ have been ordered such that $\lambda_j = 0\ (1 \leq j \leq p)$ and $\lambda_j \neq 0\ (p + 1 \leq j \leq m)$. According to Theorem 3, it is also assumed that at least one of the points $x_j\ (1 \leq j \leq p)$ is an element of the set $E$ defined in (18). To construct the perturbed problem, the vector $\phi_{m+1}$ of the degenerate problem is replaced by $\phi_{m+1}(\varepsilon) = \phi_{m+1} + \sum_{j=1}^{p} \varepsilon^j \phi_j$

where $\varepsilon$ is a small positive quantity. The characteristic relation of the perturbed problem is then

$$\phi_{m+1}(\varepsilon) = \sum_{j=1}^{m} \lambda_j(\varepsilon)\phi_j \tag{26}$$

where, by (7),

$$\begin{cases} \lambda_j(\varepsilon) = \varepsilon^j, & 1 \le j \le p \\ \lambda_j(\varepsilon) = \lambda_j \ne 0, & p+1 \le j \le m \end{cases} \tag{27}$$

thus showing that the perturbed problem is nondegenerate. Since by hypothesis some indices $j$ $(1 \le j \le p)$ belong to $E$, the exchange Theorem 2 will give a point $x_l$ with $1 \le l \le p$, whence

$$\lambda_l(\varepsilon) = \varepsilon^l. \tag{28}$$

The characteristic relation (20) of the new reference $R'$ becomes here

$$\phi_{m+1} = \lambda_k'(\varepsilon)\phi_k + \sum_{\substack{j=1 \\ \ne l}}^{m} \lambda_j'(\varepsilon)\phi_j$$

with

$$\begin{cases} \lambda_j'(\varepsilon) = \lambda_j(\varepsilon) - \lambda_l(\varepsilon)\mu_j/\mu_l, & (1 \le j \ne l \le m) \\ \lambda_k'(\varepsilon) = \lambda_l(\varepsilon)/\mu_l. \end{cases} \tag{29}$$

The new reference deviation $|h'(\varepsilon)|$ is still given by (25) provided that the characteristic coefficients are replaced by (29). Owing to (28),

$$\lim_{\varepsilon \to 0} |h'(\varepsilon)| = |h|$$

and the exchange of the perturbed problem corresponds to a static exchange for the original problem. In addition, it is seen that the $\lambda_j'(\varepsilon)$ $(j = k,\ 1 \le j \ne l \le m)$ are polynomials without constant term and which are linearly independent because $\lambda_j'(\varepsilon)$ is the only polynomial having a term in $\varepsilon^j$. In particular, $\lambda_j'(\varepsilon) \not\equiv 0$ and the perturbed problem is still nondegenerate.

In order to compute the leveled reference function for $R'$ at the next iteration, one should take

$$\operatorname{sgn} h'(x_j) = \eta \lim_{\varepsilon \to 0} [\operatorname{sgn} \lambda_j'(\varepsilon)], \qquad j = k, \quad 1 \le j \ne l \le p$$

which means that the sign is determined by the coefficient of the lowest power of $\varepsilon$ in $\lambda_j'(\varepsilon)$. For this iteration, a new set $E'$ has to be defined analogously to (18) and the perturbed problem continues if among the $\lambda_j'(\varepsilon)$ $(j = k, 1 \le j \ne l \le p)$ some have still indices belonging to $E'$. During the subsequent iterations, the perturbed problem remains nondegenerate because linear combinations of the linearly independent polynomials $\lambda_j'(\varepsilon)$ $(j = k, 1 \le j \ne l \le p)$ are again linearly independent. All the exchanges of the perturbed problem correspond to static exchanges for the original one.

If among the $\lambda_j'(\varepsilon)$ $(j = k, 1 \le j \ne l \le p)$ no index belongs to $E'$, then the perturbed problem stops. Indeed, a point $x_q$ $(k \ne q \ge p + 1)$ is then exchanged such that by (29) and (27) $\lambda_q'(0) = \lambda_q \ne 0$. A relation similar to (25) for the new reference $R''$ shows then that

$$\lim_{\varepsilon \to 0} |h''(\varepsilon)| > |h|$$

which corresponds to a nonstatic exchange for the original problem.

Bittner [10] makes the perturbed problem unnecessarily complicated by defining $\phi_{m+1}(\varepsilon) = \phi_{m+1} + \sum_{j=1}^{n} \varepsilon^j\phi_j$ instead of (26). The resulting computations are much more time-consuming since normally $p < m \ll n$. Descloux [11], on the other hand, oversimplifies the perturbed problem by taking $\phi_{m+1}(\varepsilon) = \phi_{m+1} + \varepsilon \sum_{j=1}^{p} \alpha_j\phi_j$ where the $\alpha_j$ are arbitrary constants. Indeed, (27) becomes then $\lambda_j(\varepsilon) = \varepsilon\alpha_j$ $(1 \le j \le p)$ and it is no longer possible to guarantee that the coefficients $\lambda_j'(\varepsilon)$ $(1 \le j \ne l \le p)$ in (29) are nonzero. This means that the perturbed problem can be degenerate and its convergence is thus no longer insured.

For the problem defined in Section II-A several cases of degeneracy can theoretically exist. However, experience has shown that in practice only one type of degeneracy actually occurs which is characterized by the fact that the reference contains exactly $(N + 2)$ points on the frequency axis $\omega_1$, in short, degeneracy with respect to the $\omega_1$-axis. At these points the base functions $f_i(x)$ in (3) reduce to linear combinations of $\cos I\omega_1$ $(0 \le I \le N)$ and therefore, like in the one-variable case, the characteristic vector of an arbitrary point on $\omega_1$ is a linear combination of the characteristic vectors corresponding to the remaining $(N + 1)$ points on $\omega_1$. By the definitions of Section II-B, a basis for such a reference contains exactly $(N + 1)$ points on $\omega_1$ while the $(N + 2)$th point plays the role of $x_{m+1}$ in $R$. Consequently, in the vector $\Lambda$ only the characteristic coefficients corresponding to the points on $\omega_1$ are nonzero.

## III. Nonuniqueness

Since the approximating function (2) does not satisfy the Haar condition on $X$, the Chebyshev approximation problem stated in Section II-A has not necessarily a unique solution.

### A. The Set of Best Chebyshev Approximations

In order to compute all best approximations, the approach used in [10] will now be extended. It is based on the concept of *extreme optimal solution* which is an optimal solution such that 1) the error has at least $m + 1$ *extremal points*, i.e., points where it attains the Chebyshev deviation, and 2) the approximating function is a reference function for $m + 1$ extremal points.

By the assumption of independent base functions on $X$, the set of best Chebyshev approximations are characterized by the following theorem [11].

*Theorem 4:* The set $C$ of best Chebyshev approximations is bounded. In the $m$-dimensional space of the coefficient vector $A$, it is a convex polyhedron whose vertices are the extreme optimal solutions.

Since a convex polyhedron is the convex hull of its vertices, the determination of $C$ amounts to computing all extreme optimal solutions. Starting from the extreme optimal solution obtained by the exchange algorithm, we will first determine an *adjacent extreme optimal solution* which belongs to the same *edge* of $C$, i.e., the set of optimal solutions corresponding to the same $m - 1$ extremal points with independent characteristic vectors.

In order to express the dependence of $h(x)$ on the coefficient vector $A$, one rewrites the error function (10) at the point $x_i$ as

$$h_i(A) = w_i(\phi_i^T A - F_i) \tag{30}$$

with $w_i = w(x_i)$ and $F_i = F(x_i)$. One can then define $C$ as the admissible solutions $A$ of the set of inequalities

$$\begin{cases} h_i(A) \leq H \\ -h_i(A) \leq H, & (1 \leq i \leq n) \end{cases} \tag{31}$$

where $H$ is the Chebyshev deviation.

Let $A$ be the extreme optimal solution obtained by the exchange algorithm with no more than $m + 1$ extremal points $\{x_j; 1 \leq j \leq m + 1\}$ such that one has (12) with (6), and suppose we want to compute the adjacent extreme optimal solution $A^*$ located on the edge $h_j(A^*) = h_j(A)$ $(1 \leq j \leq m - 1)$. From (30), this edge is still defined by

$$\phi_j^T(A^* - A) = 0, \qquad (1 \leq j \leq m - 1). \tag{32}$$

The extreme optimal solution $A^*$ will be different from $A$ if sgn $h_m(A)h_m(A^*) < H$ or, by (30), if

$$h_m(A)\phi_m^T(A^* - A) < 0. \tag{33}$$

The infinite number of $A^*$ satisfying (32) together with (33) must now be restricted by the remaining constraints of (31). First, one has sgn $h_{m+1}(A)h_{m+1}(A^*) \leq H$ which, by means of (30), (6), (32), and (33), can be rewritten as $h_{m+1}(A) \cdot h_m(A)\lambda_m \geq 0$. Obviously, this inequality is compatible with (12) only if $\lambda_m = 0$ which yields

$$h_{m+1}(A^*) = h_{m+1}(A). \tag{34}$$

Therefore, the number of extreme optimal solutions which are adjacent to the solution obtained by the exchange method is equal to the number of zero characteristic coefficients corresponding to the final reference of the exchange algorithm. In particular, if every characteristic coefficient is different from zero, the Chebyshev approximation is unique. On the other hand, all best approximations have the same error at the extremal points relative to the nonzero characteristic coefficients: according to the terminology used in [12], they form a *critical point set.*

In addition to (32)–(34), the admissible values of $A^*$ are finally restricted by the set of inequalities

$$\begin{cases} -\text{sgn } h_m(A)h_m(A^*) \leq H \\ \pm h_i(A^*) \qquad\qquad \leq H, & (m + 1 < i \leq n) \end{cases}$$

which, by means of (30), (32), (33), and the relations $\phi_i = \sum_{j=1}^{m} \mu_{ij}\phi_j$ $(m + 1 < i \leq n)$, are equivalent to

$$\begin{cases} -\text{sgn } h_m(A)\phi_m^T(A^* - A) \leq \dfrac{2H}{w_m} \\ -\text{sgn } h_m(A)\phi_m^T(A^* - A) \leq \dfrac{H \mp h_i(A)}{\mp w_i\mu_{im} \text{ sgn } h_m(A)}, \\ \qquad\qquad \text{for } \{m + 1 < i \leq n \mid \mp\mu_{im}h_m(A) > 0\} \end{cases}$$

where the three upper and lower signs correspond to each other. Consequently, one has Theorem 5.

*Theorem 5:* The extreme optimal solution $A^*$ adjacent to $A$ and located on the edge (32) with $\lambda_m = 0$, is characterized by the extremal point set $\{x_1,\cdots,x_{m-1},x_m^*,x_{m+1}\}$ where $x_m^*$ corresponds to the minimum of the following expressions

$$\begin{cases} \dfrac{2H}{w_m} \\ \dfrac{H \mp h_i(A)}{w_i|\mu_{im}|}, & \text{for } \{m + 1 < i \leq n \mid \mp\mu_{im}h_m(A) > 0\}. \end{cases}$$

The errors at the points $\{x_1,\cdots,x_{m-1},x_{m+1}\}$ of the old reference are unchanged whereas the error sign at $x_m^*$ is given by sgn $[-h_m(A)]$ if $x_m^* = x_m$ and by sgn $[-\mu_{im}h_m(A)]$ if $x_m^* = x_i$ $(m + 1 < i \leq n)$.

When there are $t$ zero characteristic coefficients in (6), the above theorem allows one to compute $t$ adjacent extreme optimal solutions. If the $t$ points $x_m^*$ together with the error sign are identical, it is readily seen that the number of vertices of $C$ is equal to $t + 1$: the set of best approximations is thus completely defined. Otherwise, the process must be repeated for each new extreme optimal solution and the number of vertices may be very high.

Finally, it should be pointed out that, if the optimal solution obtained by the exchange algorithm has more than $m + 1$ extremal points, the number of adjacent extreme optimal solutions can be smaller than $t$: in fact, the best approximation may be unique even if some coefficients of the characteristic relation (6) are zero.

### B. Strict Approximation

In case of nonuniqueness of the best approximation, one is faced with the problem of choosing one of the infinite number of solutions belonging to $C$. One way to overcome this difficulty is the strict approximation as proposed by Rice [12]. Suppose that in the characteristic relation corresponding to the extremal points of the optimal solution obtained by the exchange method, one has $\lambda_i \neq 0$ $(1 \leq i \leq t)$ and $\lambda_i = 0$ $(t < i \leq m)$. As mentioned above, all elements of $C$ have the same error at the critical point set $\{x_1,x_2,\cdots, x_t,x_{m+1}\}$ and, consequently, at all points $x_k$ depending on this set, i.e., such that $\phi_k = \sum_{i=1}^{t} \mu_{ki}\phi_i$. If $X^*$ denotes this set of points where all best approximations have identical errors, the strict approximation is the element of $C$ whose error has the minimum deviation in $X - X^*$.

Since the matrix of the characteristic vectors corresponding to the points $x^* \in X^*$ has rank $t$, each base function $f_i(x^*)$ can be expressed in terms of $t$ independent functions $\{g_1(x^*),\cdots,g_t(x^*)\}$ as

$$f_i(x^*) = \sum_{j=1}^{t} b_{ij}g_j(x^*), \qquad (1 \leq i \leq m) \tag{35}$$

where the matrix $[b_{ij}]$ has rank $t$. By suitable ordering of the base functions $f_i$, it is always possible to make the first $t$ rows of $[b_{ij}]$ linearly independent. Let $\{g_1,\cdots,g_m\}$ be a set of new base functions such that the first $t$ functions are solution of the nonsingular system of equations

$$f_i(x) = \sum_{j=1}^{t} b_{ij}g_j(x), \qquad (1 \leq i \leq t)$$

whereas the $m - t$ last ones are given by

$$g_k(x) = f_k(x) - \sum_{j=1}^{t} b_{kj} g_j(x), \qquad (t < k \leq m). \quad (36)$$

The approximating function (2) then becomes

$$f(x) = G_1(x) + G_2(x) \quad (37)$$

in which

$$\begin{cases} G_1(x) = \sum_{i=1}^{t} \alpha_i g_i(x) \\ G_2(x) = \sum_{i=t+1}^{m} a_i g_i(x) \end{cases} \quad (38)$$

with $\alpha_i = \sum_{j=1}^{m} a_j b_{ij}$. For $x = x^*$, by (36) and (35), $G_2(x^*) = 0$ and $f(x^*) = G_1(x^*)$. Therefore, since the functions $\{g_1, \cdots, g_t\}$ are independent on $X^*$, all optimal approximating functions have the same coefficients $\alpha_i$. The strict approximation will then be obtained by approximating $F(x) - G_1(x)$ by $G_2(x)$ on the set $X - X^*$. It may be possible that this new problem has no unique solution either in which case a further decomposition of the approximating function must be carried out.

As mentioned in Section II-D, exhaustive computations have yielded only degeneracies with respect to the $\omega_1$-axis, in which case $t = N + 1$ and $X^* = \{x^* = (\omega_1, 0)\}$. Applying the above procedure to (1), one gets

$$G_1(x) = \sum_{I=0}^{N} \alpha_I \cos I\omega_1 \cos I\omega_2$$

with

$$\alpha_I = 2B_{II} + \sum_{J=0}^{I-1} B_{IJ} + \sum_{J=I+1}^{N} B_{JI} \quad (39)$$

and

$$G_2(x) = -\sum_{I=1}^{N} \sum_{J=0}^{I-1} B_{IJ}(\cos I\omega_1 - \cos J\omega_1) \cdot (\cos I\omega_2 - \cos J\omega_2). \quad (40)$$

In fact, $G_1(x^*)$ is the best approximation on the $\omega_1$-axis and, as is well known [8], approximation by one-dimensional filters is very fast and accurate. In order to take advantage of this property, one can directly compute the strict approximation by first performing the approximation on the $\omega_1$-axis which yields the optimal solution $\{\alpha_I; 0 \leq I \leq N\}$ with Chebyshev deviation $H^*$. Once the coefficients $B_{IJ}$ $(1 \leq I \leq N; 0 \leq J \leq I - 1)$ have been determined by the approximation outside the $\omega_1$-axis, it remains to deduce $B_{II}$ $(0 \leq I \leq N)$ from relation (39). Obviously, one does not know from the outset whether the problem is degenerate with respect to the final reference of the exchange algorithm but, thanks to the following theorem, the method turns out to be advantageous even in case of nondegeneracy.

*Theorem 6:* Let $R^* = \{x_i^*; 1 \leq i \leq t + 1\}$ be the final reference of the exchange algorithm on $X^*$ with Chebyshev deviation $H^*$. If $R' = \{x_i; 1 \leq i \leq m - t + 1\}$ denotes a reference on $X - X^*$ with reference deviation $|h|$ greater than $H^*$, it is possible to build from $R^*$ and $R'$ a reference $R$ on the whole region $X$, whose reference deviation is greater than $H^*$.

The two references $R^*$ and $R'$ correspond to the characteristic relations

$$G_1(x_{t+1}^*) = \sum_{i=1}^{t} \lambda_i^* G_1(x_i^*), \qquad (\lambda_i^* \neq 0)$$

and

$$G_2(x_{m-t+1}) = \sum_{i=1}^{m-t} \mu_i G_2(x_i) \quad (41)$$

with

$$h(x_{m-t+1}) h(x_i) \mu_i \leq 0, \qquad (1 \leq i \leq m - t). \quad (42)$$

As $G_1(x^*) = f(x^*)$, $f(x)$ is also a reference function for $\bar{R} = \{x_1^*, \cdots, x_t^*, x_1, \cdots, x_{m-t}, x_{t+1}^*\}$ with reference deviation $H^*$ and characteristic relation

$$f(x_{t+1}^*) = \sum_{i=1}^{t} \lambda_i^* f(x_i^*) + \sum_{i=1}^{m-t} \lambda_i f(x_i) \quad (43)$$

in which all $\lambda_i$ are zero. By (37), (41) becomes

$$f(x_{m-t+1}) = \sum_{i=1}^{m-t} \mu_i f(x_i) + G_1(x_{m-t+1}) - \sum_{i=1}^{m-t} \mu_i G_1(x_i). \quad (44)$$

Since $G_1(x)$ is given in (38) in terms of $t$ independent functions, any $G_1(x_j)$ for $1 \leq j \leq m - t + 1$ can be expressed as a linear combination of $t$ independent values $G_1(x_i^*) = f(x_i^*)$ $(1 \leq i \leq t)$ and (44) becomes

$$f(x_{m-t+1}) = \sum_{i=1}^{t} \mu_i^* f(x_i^*) + \sum_{i=1}^{m-t} \mu_i f(x_i). \quad (45)$$

By applying the exchange theorem to the characteristic relations (43) and (45), it is now possible to introduce the point $x_{m-t+1}$ with deviation $|h| > H^*$ in $\bar{R}$ in order to form $R$. This exchange cannot be a static one because condition (42) holds for all zero $\lambda_i$ in (43) and the reference deviation of $R$ will thus be larger than $H^*$.

In consequence of this theorem, as soon as the reference deviation $|h|$ of the exchange method for $X - X^*$ becomes greater than $H^*$, one can restart the exchange algorithm for the whole region $X$ with a reference deviation greater than $H^*$, thus avoiding any further degeneracy with respect to the $\omega_1$-axis.

## IV. Numerical Results

In order to save computation time, it is advantageous to use a Cartesian grid for the discrete point set $X$ which is to replace the continuous approximation domain of Fig. 1. Indeed, evaluating (1) along a vertical or horizontal line reduces to univariable computation. Nevertheless, an accurate representation of the continuous domain requires to include the intersections of the grid lines with the passband and stopband boundary circles. If one characterizes the quality of approximation by the precision factor $(h_{max} - H)/H$ where $h_{max}$ is the maximum of the absolute value of the error function on the continuous approximation domain, experience shows that a Cartesian grid of about $25\ N$ by $25\ N$ points is sufficient for getting a precision factor on the order of $10^{-3}$, which is quite acceptable for practical purposes.

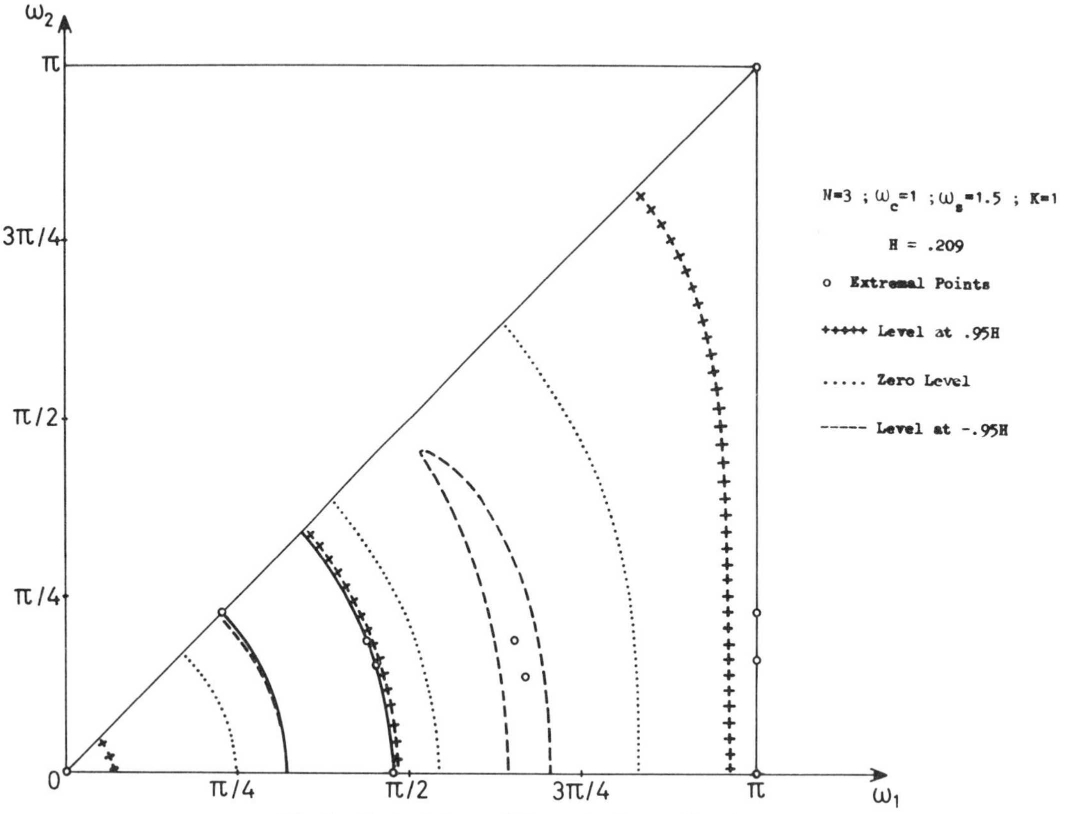


Fig. 2. Typical shape of the optimal error function.

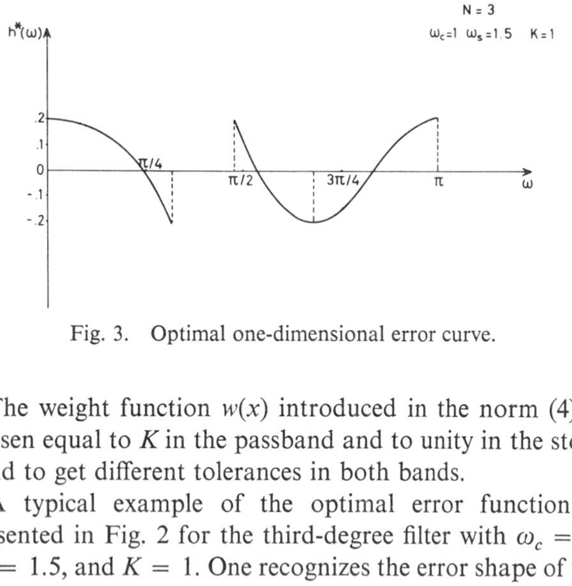


Fig. 3. Optimal one-dimensional error curve.

The weight function $w(x)$ introduced in the norm (4) is chosen equal to $K$ in the passband and to unity in the stopband to get different tolerances in both bands.

A typical example of the optimal error function is presented in Fig. 2 for the third-degree filter with $\omega_c = 1$, $\omega_s = 1.5$, and $K = 1$. One recognizes the error shape of the optimal one-dimensional filter drawn in Fig. 3, rotated in all directions with the final reference points located approximately on circular ridges whose radii correspond to the abscissas of the extremal points of Fig. 3. In fact, on a circle ($\omega_1 = R \cos t$, $\omega_2 = R \sin t$), each base function (3) can be expanded in terms of Bessel functions of the first kind as

$$\cos I\omega_1 \cos J\omega_2 + \cos J\omega_1 \cos I\omega_2$$
$$= 2J_0[R(\sqrt{I^2 + J^2})]$$
$$+ 4 \sum_{k=1}^{\infty} J_{4k}[R(\sqrt{I^2 + J^2})] \cos 4k\alpha_{IJ} \cos 4kt$$

where $\alpha_{IJ} = \sin^{-1}[I/(I^2 + J^2)^{1/2}]$. Since the Bessel function $J_n(x)$ has a zero of multiplicity $n$ at the origin, it has negligible values when its argument $x$ is small with respect to the order $n$. For the above-mentioned third-degree filter, on the boundary circle $\omega_s = 1.5$, the approximating function can be represented with an error on the order of $10^{-3}$, as a second-degree trigonometric polynomial in $4t$. Hence, since there are three extremal points on $\omega_1$, the error function attains approximately the Chebyshev deviation at all points of the circle. The circular symmetry of the error function is less pronounced for circles with larger radii and, in particular, the ridge corresponding to the last extremal point of Fig. 3, i.e., the Nyquist frequency, is the straight line $\omega_1 = \pi$. As, for $\omega_1 = \pi$, the approximating function is a third-degree polynomial in $\omega_2$, the four extremal points located on $\omega_1 = \pi$ produce a ridge of strictly constant height for all $\omega_2$. In order to take advantage of this property during the exchange algorithm, one can split the approximating function into two parts according to (37) such that

$$G_1(x) = \sum_{I=0}^{N} \alpha_I(-1)^I \cos I\omega_1 \cos I\omega_2$$

$$G_2(x) = -\sum_{I=1}^{N} \sum_{J=0}^{I-1} B_{IJ}[\cos I\omega_1 - (-1)^{I+J} \cos J\omega_1] \cdot [(-1)^{I+J} \cos I\omega_2 - \cos J\omega_2]$$

with

$$\alpha_I = (-1)^I 2B_{II} + \sum_{J=0}^{I-1} (-1)^J B_{IJ} + \sum_{J=I+1}^{N} (-1)^J B_{JI}.$$

TABLE I

$\omega_c = 1$ ; $\omega_s = 2$ ; K=1

Precision factor = $10^{-3}$

| N | Single Exchange | | Multiple Exchange | |
|---|---|---|---|---|
| | Iteration Number | Computation Time (Sec) | Iteration Number | Computation Time (Sec) |
| 1 | 1 | 9 | 1 | 9 |
| 2 | 3 | 9 | 3 | 9 |
| 3 | 16 | 46 | 18 | 54 |
| 4 | 25 | 114 | 15 | 118 |
| 5 | 85 | 564 | 30 | 401 |
| 6 | 184 | 2276 | 51 | 1276 |

TABLE II

$\omega_c = 1$ ; $\omega_s = 2$ ; K=1

Precision factor = $10^{-2}$

| N | Single Exchange | | Multiple Exchange | |
|---|---|---|---|---|
| | Iteration Number | Computation Time (Sec) | Iteration Number | Computation Time (Sec) |
| 1 | 1 | 9 | 1 | 9 |
| 2 | 2 | 8 | 2 | 8 |
| 3 | 9 | 29 | 9 | 41 |
| 4 | 15 | 85 | 9 | 77 |
| 5 | 68 | 534 | 21 | 360 |
| 6 | 128 | 1714 | 16 | 463 |

Since $G_2(x) = 0$ for $\omega_1 = \pi$, imposing a constant value of $f(x)$ on $\omega_1 = \pi$ amounts to canceling all coefficients $\alpha_I$ except $\alpha_0$. It turns out that the exchange algorithm modified this way yields very good results, at least within the precision factor of $10^{-3}$, if $\pi$ belongs to the extremal point set of the best univariable approximation. On the other hand, since nonuniqueness generally occurs only when $\pi$ is not an extremal point of the optimal unidimensional error, the following rule of thumb was found to be very efficient for solving the approximation problem. It consists in performing the approximation in two steps:

1) approximation along the $\omega_1$-axis by a univariable approximating function which yields the extremal point set $R^* = \{x_k^*; 1 \leq k \leq N + 2\}$;

2) if $x_{N+2}^* = \pi$, approximation in the whole region by maintaining a constant value of the error on $\omega_1 = \pi$; and

2′) if $x_{N+2}^* \neq \pi$, approximation by means of the procedure of Section III-B which allows a direct computation of the strict approximation.

In both cases, the points of the initial reference for starting the exchange algorithm are equally spaced on circles of radius $x_k^*$ and their number is proportional to the index $k$.

As regards the rate of convergence of the iterative process, a necessary condition [13] for the Remez algorithm, applied to univariable functions consisting of $m$ base functions, to converge quadratically, is that the optimal error function has exactly $m + 1$ extremal points: on the contrary, the convergence is slower if their number exceeds $m + 1$. In particular, this property has been verified for one-dimensional filters [14]. Although such a theorem has not been proved for bivariable functions, it is reasonable to expect a similar phenomenon for two-dimensional filters. Since the optimal error exhibits ridges of extrema, the convergence is slow as evidenced by Table I which gives the iteration number and the computation time (on the Electrologica X8 computer: cycle time = 1 $\mu$s) of the single and multiple exchange algorithms in the case of $\omega_c = 1$, $\omega_s = 2$, $K = 1$ and for various degrees. If both algorithms are comparable for $N$ ranging from 1 to 4, the multiple exchange method becomes quite advantageous when the degree is higher. In order to stress the importance of the precision factor on the computation time, Table II indicates how the figures of Table I vary when the precision factor is relaxed from $10^{-3}$ to $10^{-2}$: from these results, it appears that the difficulty for designing the filters, especially the highest degree filter, strongly depends on the quality of approximation which is required by the practical problem.

TABLE III

N=1 ; $\omega_c = 1.5$ ; $\omega_s = 2$ ; K=10

| Filter Coefficients | Extreme Optimal Solution $A_1$ | Extreme Optimal Solution $A_2$ | Strict Approximation |
|---|---|---|---|
| $A_{00}$ | .793917 | .790117 | .790699 |
| $A_{10} = A_{01}$ | .0552223 | .0571223 | .0568314 |
| $A_{11}$ | .017049 | .016099 | .0162445 |

To serve as an illustration for nonuniqueness, Table III presents several optimal solutions relative to the parameters $N = 1$, $\omega_c = 1.5$, $\omega_s = 2$, and $K = 10$. When degeneracy with respect to the $\omega_1$-axis occurs for a first-degree polynomial, there is only one vanishing characteristic coefficient, and the number of extreme optimal solutions is exactly equal to 2. Therefore, the two extreme optimal solutions $A_1$ and $A_2$ given in the first two columns of Table III completely define the set of best approximations and one can verify that the strict approximation of the third column is the convex combination $\delta A_1 + (1 - \delta)A_2$ with $\delta = 0.153124$.

Another example of nonuniqueness results from the data set $N = 3$, $\omega_c = 1.5$, $\omega_s = 2$, and $K = 1$. In that case, there are tens of different extreme optimal solutions and it is no longer possible to mention all of them. As regards the strict approximation, it is worth pointing out that performing the approximation outside the $\omega_1$-axis with the function (40) yields four extremal points located on the parallel of the Cartesian grid closest to the $\omega_1$-axis, which corresponds to a new degenerate problem. To remove this difficulty, it is thus necessary to apply the procedure of Section III-B to the reduced approximation problem. Fig. 4 shows the optimal error function of the strict approximation together with the three critical point sets.

If the exchange algorithm is used to compute an extreme optimal solution, since the problem is degenerate with respect to the final reference, the last iterations consist of static exchanges. In order to ensure unconditional convergence, one has to resort to the perturbation technique of Section I-D. For the example at hand, the number of iterations of the unperturbed and perturbed exchange algorithms is equal to 36 and 41, respectively, while the corresponding computation times are 92 s and 137 s. Therefore, although both iteration numbers are comparable,

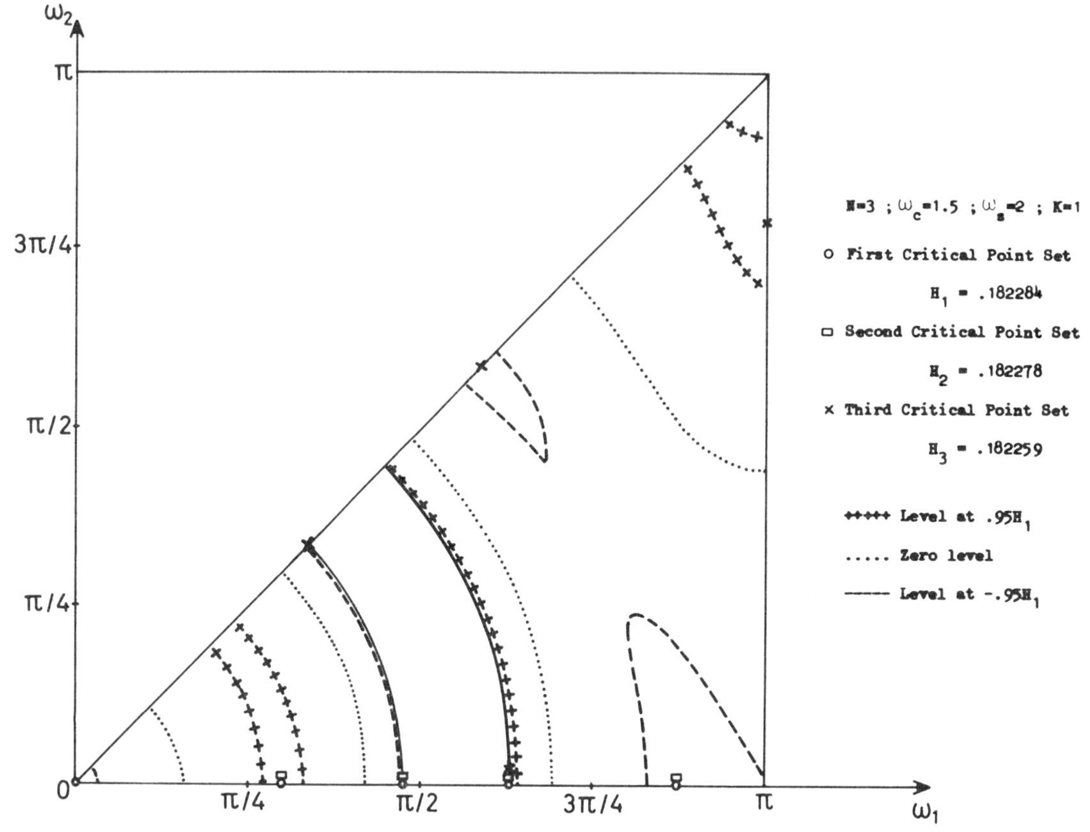


Fig. 4. Optimal error function and critical point sets of a strict approximation.

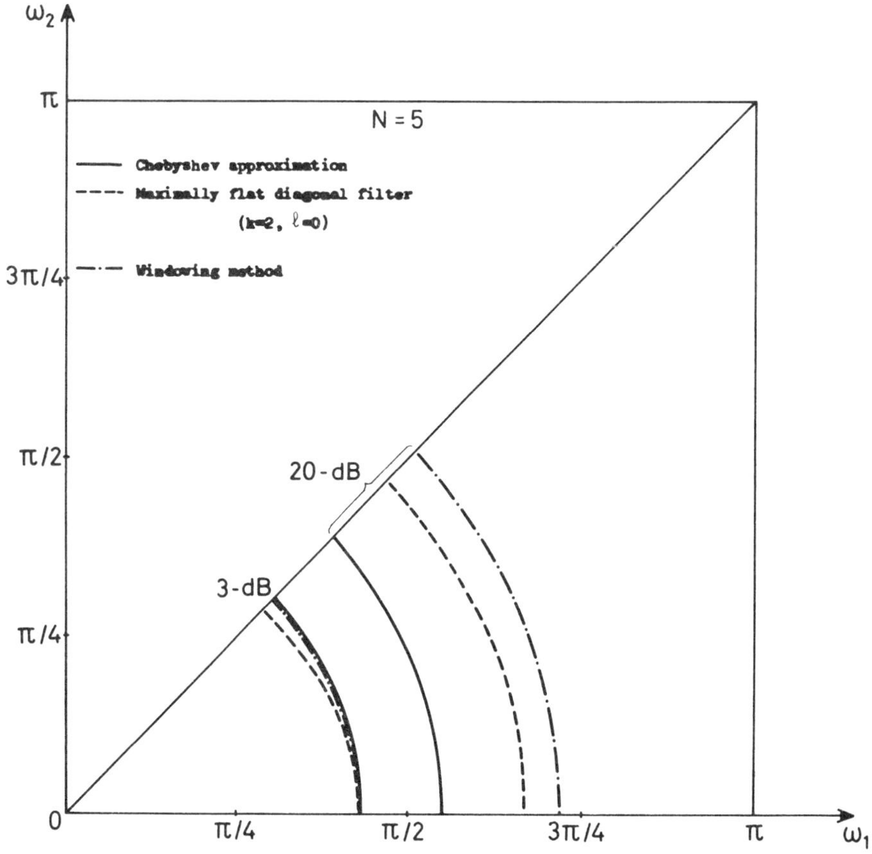


Fig. 5. Comparison with other design methods.

the computation time of the perturbed algorithm is higher owing to the more involved computations at each iteration step. On the other hand, experience has proven that the unperturbed exchange algorithm never failed to converge. Hence, as pointed out in Section I-D, the perturbation method is more important from the theoretical standpoint than from the practical one.

Hu and Rabiner [3] have used linear programming to design frequency sampling and optimal (in the sense of Chebyshev approximation) two-dimensional filters. Results for the latter design have been reported only for a few filters of the fourth degree and were found to be comparable with those obtained by the exchange algorithm of this paper. By lack of a detailed description of their methods, it seems likely that the small discrepancies are due to different discretization schemes of the continuous approximation domain. As regards the frequency sampling design, the comparison was made on a few samples of the filters of degree 12 published in [3]. For these examples, design by Chebyshev approximation showed that a degree 7 was already sufficient to achieve the same filtering performance.

To visualize the superiority of the Chebyshev approximation on other design methods, a fifth-degree filter was computed having a 3-dB passband and a 20-dB stopband (Fig. 5). For comparison, the 3- and 20-dB contour lines have been plotted for a maximally flat diagonal filter [4] and for a filter designed with a windowing technique [2], both having the same degree and the same passband limit along the $\omega_1$-axis. The circular symmetry of these designs is comparable but the selectivity given by the Chebyshev approximation is significantly better, i.e., smaller transition bandwidth for substantially the same passband boundary.

## V. Conclusion

The preceding analysis has clearly shown that the Chebyshev approximation of two-dimensional nonrecursive filters is fundamentally more difficult and inherently slower than in the one-dimensional case. The reasons behind this deserve some further comments.

First, two-variable approximating functions do generally not satisfy the Haar condition and this is reflected by the existence of degeneracies. If degeneracy occurs before the end of the algorithm it may fail to converge because the same reference is repeated periodically. To force convergence even in this case, it was shown how to complement the algorithm with a suitable perturbation technique. This technique is only of theoretical interest, however, because it is computationally very costly and because in practice cycling of the algorithm has never been observed.

A second difference is that the optimal solution is not characterized by a simple alternation of the error like in the one-dimensional problem and, as a consequence, the rule for exchanging one point of the reference becomes more complicated. This fact makes the one-by-one exchange more time-consuming and even prevents a simultaneous exchange, a procedure which precisely makes the algorithm so efficient in the single variable case. An improvement in this direction was made, however, by proposing a method which allows to exchange several points simultaneously, and the experiments have demonstrated that it leads to important reductions of the run time when the degree is high.

Further, the maximum detection over a two-dimensional region is much longer than the corresponding work over an interval and is made even slower by the ridges in the error function whose existence was already mentioned in [3]. Finally, the computation time is also increased by the fact that in two dimensions there exists no counterpart of the Lagrange interpolation formula for the leveled reference function and by the fact that the number of coefficients is proportional to the square of the degree.

These considerations show that the nonfulfilled Haar condition is not the only reason for the slowness of the two-dimensional approximation. Several measures were taken to improve this situation. Besides the multiple exchange already mentioned, a heuristic procedure was proposed where the approximation is first performed on the $\omega_1$-axis or where the value of the approximating function is kept constant over $\omega_1 = \pi$. Its advantage lies in a significant saving of computation time and although it does not always give the optimal solution, the accuracy of the result is in most cases satisfactory. It is clear that future efforts should be aimed at further reducing the run time.

Comparison with other design methods (frequency sampling, maximally flat criterion, and windowing technique) has made evident the superiority of the Chebyshev design for what regards selectivity. For this norm, the one-by-one exchange is known to be more economical than linear programming [15] and this speed advantage is increased by the possibility of performing a multiple exchange.

This paper has solved the typical problem of designing a two-dimensional nonrecursive filter which approximates circularly symmetrical low-pass specifications. Most of the results, however, are not restricted to circular symmetry or low-pass specifications but apply to the design of any filter whose frequency response can be put under the form (2) within a linear phase factor.

## Acknowledgment

The authors wish to thank J. Praet for his programming of the exchange algorithm.

## References

[1] T. S. Huang, W. F. Schreiber, and O. J. Tretiak, "Image processing," *Proc. IEEE*, vol. 59, pp. 1586–1609, Nov. 1971.

[2] T. S. Huang, "Two-dimensional windows," *IEEE Trans. Audio Electroacoust.*, vol. AU-20, pp. 88–89, Mar. 1972.

[3] J. V. Hu and L. R. Rabiner, "Design techniques for two-dimensional digital filters," *IEEE Trans. Audio Electroacoust.*, vol. AU-20, pp. 249–257, Oct. 1972.

[4] Y. Kamp and J. P. Thiran, "Maximally flat nonrecursive two-dimensional digital filters," *IEEE Trans. Circuits Syst.*, vol. CAS-21, pp. 437–449, May 1974.

[5] J. H. McClellan, "On the design of one-dimensional and two-dimensional FIR digital filters," Ph.D. dissertation, Rice University, Houston, Tex., Apr. 1973.

[6] E. Stiefel, "Über diskrete und lineare Tschebyscheff-Approx-

imationen," *Numer. Math.*, vol. 1, pp. 1–28, 1959.
[7] S. I. Gass, *Linear Programming*. New York: McGraw-Hill, 1958.
[8] T. W. Parks and J. H. McClellan, "Chebyshev approximation for nonrecursive digital filters with linear phase," *IEEE Trans. Circuit Theory*, vol. CT-19, pp. 189–194, Mar. 1972.
[9] L. Collatz, "Approximation von Functionen bei einer und bei mehreren unabhängigen Veränderlichen," *Z. Angew. Math. Mech.*, vol. 36, pp. 198–211, May/June 1956.
[10] L. Bittner, "Das Austauschverfahren der linearen Tschebyscheff-Approximation bei nicht erfüllter Haarscher Bedingung," *Z. Angew. Math. Mech.*, vol. 41, pp. 238–256, June 1961.
[11] J. Descloux, "Dégénérescence dans les approximations de Tschebyscheff linéaires et discrètes," *Numer. Math.*, vol. 3, pp. 180–187, 1961.
[12] J. R. Rice, *The Approximation of Functions*, vol. 2, *Nonlinear and Multivariate Theory*. Reading, Mass.: Addison-Wesley, 1969.
[13] G. Meinardus, *Approximation von Funktionen und ihre numerische Behandlung*. Berlin: Springer, 1964.
[14] A. N. Godard, "Design of optimum FIR bandpass filters through frequency transformations," *Philips Res. Rep.*, vol. 29, pp. 243–252, June 1974.
[15] E. Stiefel, "Note on Jordan elimination, linear programming and Tchebycheff approximation," *Numer. Math.*, vol. 2, pp. 1–17, 1960.

# THE DESIGN OF TWO-DIMENSIONAL DIGITAL FILTERS BY TRANSFORMATIONS

James H. McClellan
Rice University


## Abstract

This paper describes a technique for designing two-dimensional (2-D) Finite Impulse Response (FIR) digital filters by application of a change of variables to an optimal min-max one-dimensional (1-D) filter. The advantages of such an approach are that very large filter arrays can be designed in an amount of time comparable to that required to do the 1-D design, and in many cases the filters which are obtained are optimal.

Among the types of filters which can be treated by such an approach are circularly symmetric lowpass and bandpass filters. Several examples are presented to illustrate the method and to compare it with existing designs.


## I. Introduction

Two-dimensional (2-D) digital signal processing is very important in such areas as geophysical seismic signal processing and image enhancement for picture processing. However, many of the techniques for designing one-dimensional (1-D) digital filters have not been extended to 2-D digital filter design. In particular, in the area of finite impulse response (FIR) linear phase digital filter design the iterative design methods based on the Remez exchange algorithm have not been extended to the 2-D case, and so at this time there is no efficient design procedure for optimum Chebychev filters available in two dimensions. Since the problem of 2-D design in the FIR linear phase case is a linear approximation problem, linear programming can be applied[1], but this approach has met with only limited success. Linear programming is slow and has been limited to the design of 9x9 filters. This is because of the size of the problem. For example, the design of an optimum 31x31 FIR filter involves optimization over 16x16=256 parameters. Therefore, it is important to consider methods of design which may yield suboptimum filters but which are computationally more efficient. Within such a class of methods are the windowing technique[2] and the frequency sampling technique[1]. This paper describes a new approach based on transformations of 1-D filters, and shows in addition that many of the new designs are optimal min-max 2-D filters.

The technique consists of transforming a 1-D filter into a 2-D filter by a change of variables. The class of filters which can be approximated by this method is limited, but included in the class are many important filters, such as those filters whose magnitude is constant over regions of the 2-D frequency plane (e.g., lowpass and bandpass circularly symmetric filters). The main advantage of this technique is that it is very fast because the computing time is devoted almost entirely to the 1-D design which is known to be efficient[3]. Furthermore, it is possible to prove that many filters designed by this method are optimal in the Chebychev sense.

## II. Problem Formulation

A two-dimensional digital filter with impulse response matrix $h(k,p)$ $k=0,1,\ldots N_1-1$; $p=0,1,\ldots N_2-1$ has a frequency response defined by the 2-D Discrete Fourier transform

$$H(F_1,F_2) = \sum_{k=0}^{N_1-1} \sum_{p=0}^{N_2-1} h(k,p)\exp(-j2\pi(F_1K+F_2p))$$

If the impulse response is constrained so that it is $N_1 \times N_2$, where $N_1$ and $N_2$ are odd integers, and the symmetry conditions

$$h(N_1-m-1,p) = h(m,p) \quad m = 0,1,\ldots n_1 = \tfrac{1}{2}(N_1-1)$$
$$h(k,N_2-m-1) = h(k,m) \quad m = 0,1,\ldots n_2 = \tfrac{1}{2}(N_2-1)$$

are imposed, then the frequency response can be rewritten

$$H(F_1,F_2)=\exp(-j2\pi(n_1F_1 + n_2F_2)) \sum_{k=0}^{n_1} \sum_{p=0}^{n_2} a(k,p)\cos 2\pi kF_1 \cos 2\pi pF_2 \quad (1)$$

where

$$a(0,0) = h(n_1,n_2)$$
$$a(0,p) = 2h(n_1,n_2-p) \quad p = 1,\ldots n_2$$
$$a(k,0) = 2h(n_1-k,n_2) \quad k = 1,\ldots n_1$$
$$a(k,p) = 4h(n_1-k,n_2-p) \quad k = 1,\ldots n_1;\ p = 1,\ldots n_2$$

Let

$$\hat{H}(F_1,F_2) = \sum_{k=0}^{n_1} \sum_{p=0}^{n_2} a(k,p)\cos 2\pi kF_1 \cos 2\pi pF_2$$

Since $\exp(-j2\pi(n_1F_1 + n_2F_2))$ has magnitude one, $|\hat{H}(F_1,F_2)|$ is the magnitude of the frequency response.

A one dimensional filter with impulse response $h(k)$, $k = 0,1,\ldots N-1$ has a frequency response defined by the Fourier transform

$$G(F) = \sum_{k=0}^{N-1} h(k)\exp(-j2\pi kF)$$

If the length of the impulse response is odd $N = 2n+1$ and symmetric, $h(k) = h(N-1-k)$, $k=0,1,\ldots n$ then

$$G(F) = \exp(-j2\pi nF) \sum_{k=0}^{n} b(k)\cos 2\pi kF \quad (2)$$

where $b(0) = h(n)$ and $b(k) = 2h(n-k), k=1,\ldots n$

Letting

$$\hat{G}(F) = \sum_{k=0}^{n} b(k)\cos 2\pi kF$$

the magnitude of the frequency response is $|\hat{G}(F)|$.

Reprinted with permission from *Proc. 7th Annu. Princeton Conf. on Inform. Sci. and Syst.*, Mar. 1973, pp. 247–251.

The change of variables to be described depends on the fact that both $\hat{G}(F)$ and $\hat{H}(F_1,F_2)$ are sums of cosine functions. $\hat{G}(F)$ can be rewritten as

$$\hat{G}(F) = \sum_{k=0}^{n} \hat{b}(k) \, (\cos 2\pi F)^k$$

by the choice of suitable coefficients $\hat{b}(k)$. This can be seen by letting $x = \cos 2\pi F$. Then $\cos 2\pi k F = \cos k(\arccos x) = T_k(x)$, where $T_k$ is the k-th Chebychev polynomial. Thus,

$$\sum_{k=0}^{n} b(k) T_k(x) = \sum_{k=0}^{n} \hat{b}(k) \, x^k$$

where both sides of the equation are now polynomials in x.

The following change of variables is proposed

$$\cos 2\pi F = A \cos 2\pi F_1 + B \cos 2\pi F_2 + C \cos 2\pi F_1 \cos 2\pi F_2 + D \qquad (3)$$

Substitution of (3) into G(F) yields a bivariate polynomial in $\cos 2\pi F_1$ and $\cos 2\pi F_2$

$$\sum_{k=0}^{n} \sum_{p=0}^{n} \hat{a}(k,p) (\cos 2\pi F_1)^k \, (\cos 2\pi F_2)^p$$

which can be rewritten uniquely in the same form as $H(F_1,F_2)$ namely,

$$\sum_{k=0}^{n} \sum_{p=0}^{n} a(k,p) \cos 2\pi k F_1 \cos 2\pi p F_2$$

Thus, it has been shown that the proposed change of variables will preserve the proper form for a 2-D filter and the resulting filter will be a square array (i.e., $N_1 = N_2$).

## III. Change of Variables

The change of variables (3) defines a mapping from the interval [0,0.5] of the 1-D frequency axis to the square [0,0.5] × [0,0.5] in the 2-D frequency plane. In order to be able to use the change of variables to produce a desired frequency response (i.e., pick A,B,C, and D), it is necessary to describe the mechanics of the transformation (3).

Solving (3) for $F_2$ as a function of $F_1$,

$$F_2 = \frac{1}{2\pi} \operatorname{Arccos} \left\{ \frac{\cos 2\pi F - D - A\cos 2\pi F_1}{B + C\cos 2\pi F_1} \right\} \qquad (4)$$

it is easy to see that for a fixed F we get a curve in the $(F_1,F_2)$ plane, and along this curve the transformed 2-D frequency response is a constant equal to the value of the 1-D frequency response at F. As F varies we get a family of

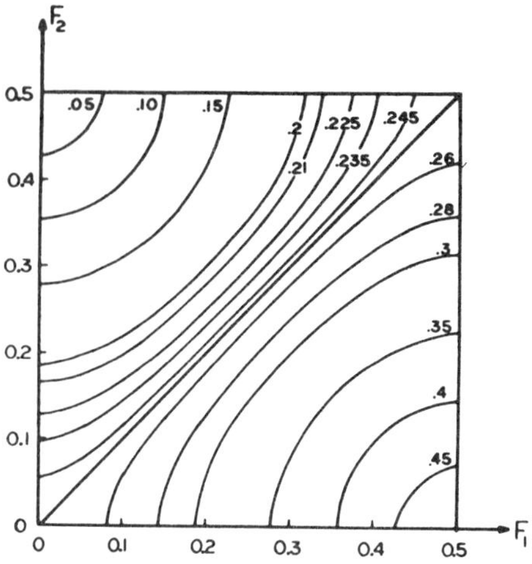


Figure 1. Contour lines for equiripple fan filter as F varies from 0 to 0.5 in steps of 0.05.

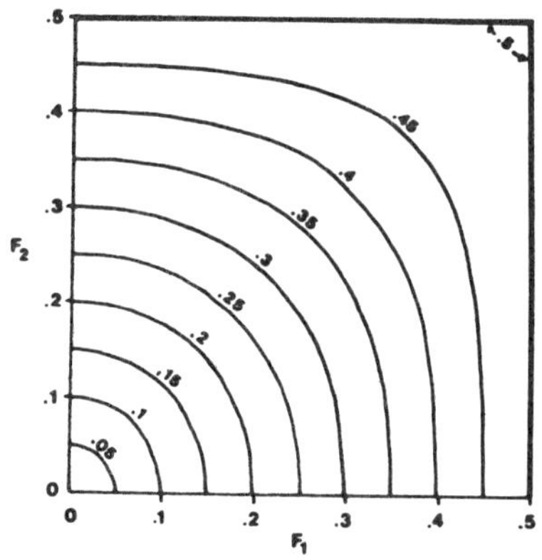


Figure 2. Contour lines for circularly symmetric filter as F varies from 0 to 0.5 in steps of 0.05.

contours which completely describe the transformed 2-D frequency response. For example, A = -B = 0.5, C = D = 0 gives the contours of Figure 1; whereas, A = B = C = -D = 0.5 gives the contours of Figure 2. The former has been used in the design of equiripple fan filters[4], and the latter can be used to design circularly symmetric filters[1].

Letting $u = \cos 2\pi F_1$, $v = \cos 2\pi F_2$, and $x = \cos 2\pi F$, (4) is rewritten as

$$v = \frac{x - D - Au}{B + Cu} \qquad (5)$$

As x varies from -1 to +1, v as a function of u is a family of curves in the square [-1,1] × [-1,1]. Fixing x and calculating the derivative,

$$\frac{dv}{du} = (-AB + CD - Cx)(B + Cu)^{-2}$$

shows that v is either a monotone increasing or decreasing function of u, because the sign of the derivative doesn't change as u varies from -1 to +1. Thus, the same thing is true for $F_2$ vs. $F_1$; namely, $F_2$ is monotone increasing or decreasing as a function of $F_1$. If the derivative is zero, it is identically zero, meaning that v is a constant. In order to avoid such horizontal contours, it is sufficient to assume that $|C| \leq |A|$. Similarly, to exclude vertical contours one considers u as a function of v and concludes that $|C| \leq |B|$ is a sufficient condition. Note that the equal sign will permit horizontal or vertical contours at the boundaries of square $[0,0.5] \times [0,0.5]$ as in the contour line for $F = 0.5$ in Figure 2.

In order to insure that the mapping from F to $(F_1,F_2)$ is well defined, it is necessary that A,B,C, and D be chosen so that the right hand side of (3) is between +1 and -1 for all $F \in [0,0.5]$. With this constraint there are two cases: all contours increasing or all contours decreasing.

If all the contours are increasing, as in Figure 1, then the point $F = 0$ is mapped to the point $(F_1 = 0,\ F_2 = 0.5)$ and the point $F = 0.5$ to $(0.5,0)$, or vice-versa. Thus,

$$\pm 1 = A - B - C + D$$

$$\mp 1 = -A + B - C + D$$

which simplifies to $A = \pm 1 - B$ and $C = D$. If all the contours are decreasing, as in Figure 2, $F = 0$ is mapped to $(0,0)$ and $F = 0.5$ to $(0.5,\ 0.5)$, or vice-versa. Thus,

$$\pm 1 = A + B + C + D$$

$$\mp 1 = -A - B + C + D$$

which simplifies to $A = \pm 1 - B$ and $C = -D$. In both cases, the number of parameters to choose has been reduced from four to two.

## IV. Design Procedure

Figure 3 shows the desired specification of a 2-D circularly symmetric lowpass filter. Between the pass region and the stop region we have allowed for a transition region, which is bounded by the arcs of two different circles. Since the arcs are strictly decreasing functions, $C = -D$ and $A = 1 - B$. The object of the design procedure is to choose A and C such that for some $F = F_p$ the resulting contour line closely approximates the inner circular arc and for some $F = F_s$ the resulting contour line approximates the outer arc. Then one designs an optimal Chebychev 1-D lowpass filter with passband cutoff frequency $F_p$ and stopband cutoff frequency $F_s$, using the Remez exchange algorithm and applies the change of variables to get the desired 2-D filter.

It is well known[5] that the transition width of a 1-D lowpass filter affects the attenuation

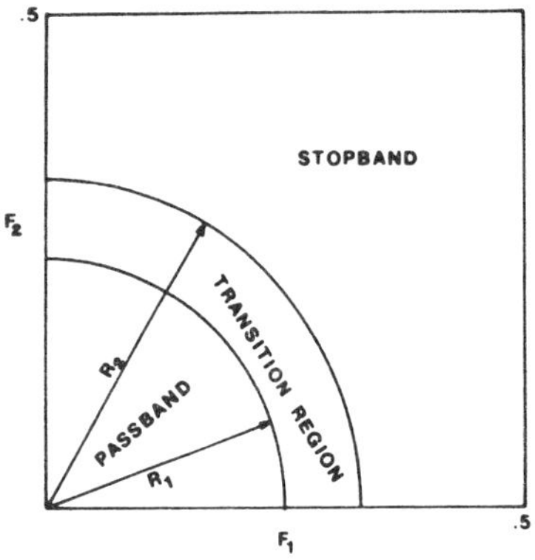


Figure 3. Desired specifications for a circularly symmetric lowpass filter.

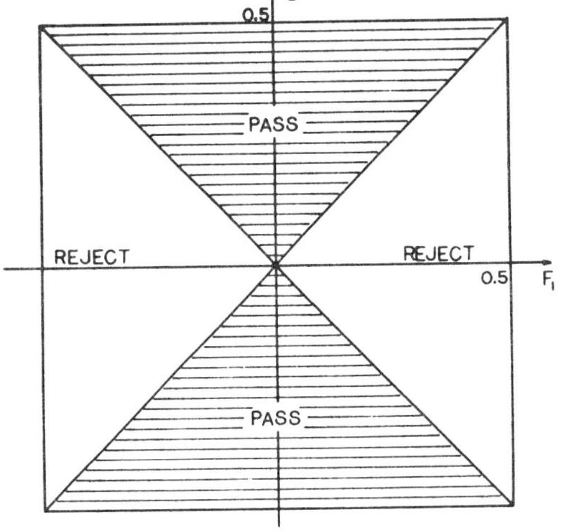


Figure 4. Desired specifications for a 2-D fan filter.

which can be obtained from such a filter. Therefore, in the above procedure, one would like to be able to design a 1-D filter whose transition width is as large as possible. This can be done by picking A and C such that the contours are bunched together closely in the transition region. The best choice for A and C is 0.5 in order to approximate circular symmetry (Fig. 2).

The design procedure for fan filters is similar. Figure 4 shows the desired response. Since the cutoff boundary is an increasing function, $C = D$ and $A = 1 + B$. The best choices for A,B,C and D are $A = -B = 0.5$ and $C = D = 0$ (Fig. 1). The details of fan filter design have been described elsewhere[4].

## V. Optimality of the 2-D Filters

When using the change of variables described above it is often possible to show that the 2-D filter which is obtained is an optimal Chebychev approximation to the desired frequency response. The argument used to prove this assertion is to recognize that along a line in the 2-D frequency square we are trying to approximate a desired 1-D frequency response. Hence, the best 1-D approximation to this desired 1-D frequency response yields a lower bound for the error in approximating the desired 2-D frequency response over the entire frequency square. If this lower bound can be achieved then the resulting 2-D filter is a best Chebychev approximation. Let's illustrate the argument for the case of circularly symmetric filters.

Circularly symmetric 2-D filters are obtained from 1-D filters by using the change of variables (3) with A = B = C = -D = 0.5. Restricted to the line $F_2 = 0$, the 2-D response is a 1-D response in the variable $F_1$. Since $\cos 2\pi F =$

$0.5 \cos 2\pi F_1 + 0.5 + 0.5 \cos 2\pi F_1 - 0.5 = \cos 2\pi F_1$, the 2-D response along $F_2 = 0$ is exactly the same as the optimal 1-D response used to generate the 2-D filter. Thus the deviation of the 1-D filter is a lower bound for the deviation of the 2-D filter. However, since the change of variables preserves the equiripple nature of the error, the deviation of the 2-D response is equal to this lower bound and the filter is optimal in the Chebychev sense.

A similar argument can be made to prove that fan filters designed with the change of variables A = 0.5, B = -0.5, and C = D = 0 are optimal Chebychev filters. In this case the line to be used is $F_2 = 0.5 - F_1$. We remark that in an earlier paper[4] on fan filter design it was felt that this technique might not give an optimum Chebychev filter, but now we have proved the optimality of those designs.

## VI. Results

The above methods have been applied to several different types of filters to illustrate the flexibility of the method considered here. Figure 5 shows a lowpass circularly symmetric 9 X 9 filter (with A = B = C = -D = 0.5) for comparison with an optimal filter designed by linear programming (see filter 3 in Table II of reference 1). Using the above argument, it is possible to show that both filters have exactly the same error in the passband ($\delta_p = 0.079$) and the stopband ($\delta_s = 0.0079$), since along the line $F_2 = 0$ they are exactly the same filter. Both filters are optimal min-max filters, the difference being that the passband edges and stopband edges are slightly different. Figure 6 shows the difference in cutoff boundaries for the two filters. The dashed lines are circular arcs which represent the linear programming design, the solid lines are the edges for the design by the change of variables.

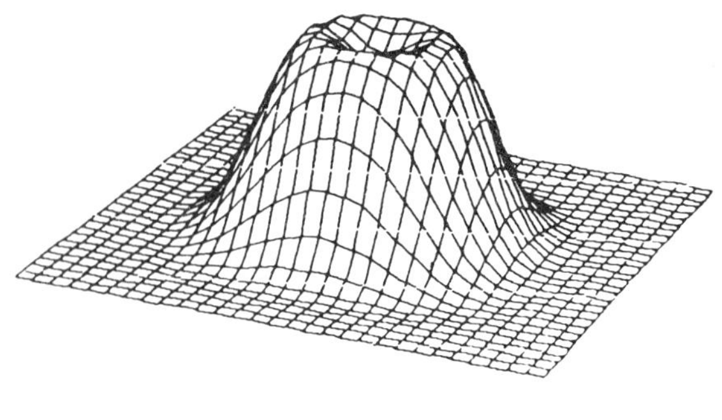

Figure 5. 9 X 9 circularly symmetric lowpass filter $R_1 = 0.1666...$, $R_2 = 0.333...$, $\delta_p = 0.079$, and $\delta_s = 0.0079$ .

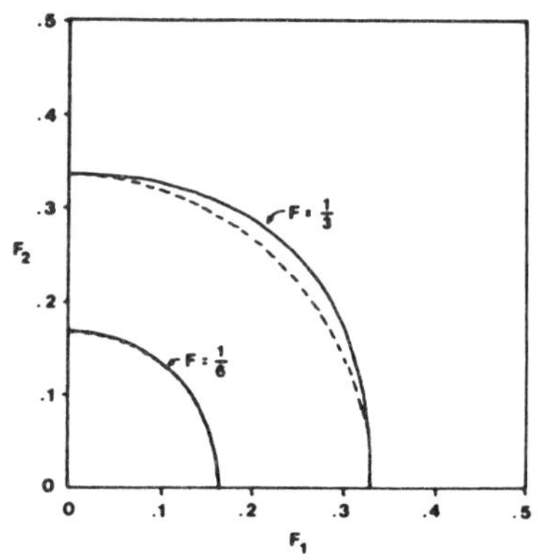


Figure 6. Comparison of cutoff contours for 9 X 9 circularly symmetric lowpass filter. Solid lines denote change of variables, dashed lines denote linear programming.

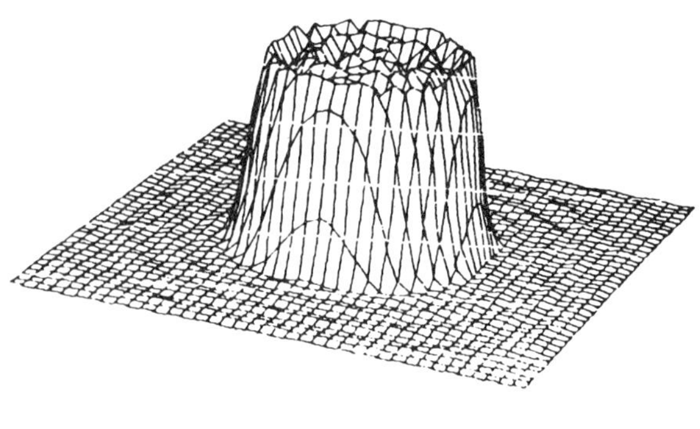

Figure 7. 31 X 31 circularly symmetric lowpass filter. $R_1 = 0.2$, $R_2 = 0.25$, $\delta_p = 0.0755$, and $\delta_s = -42.45$db.

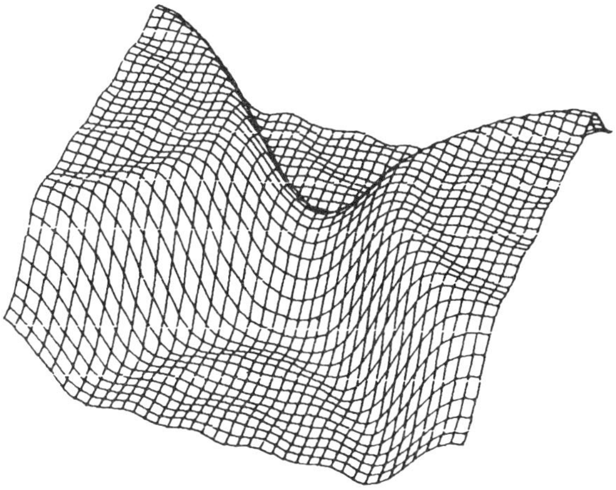

Figure 8. 21 x 21 fan filter. $\delta_p = \delta_s = 0.0229$.

The important feature of this new method is that it is not restricted to the design of such small filters as the length 9 x 9 filter above. With a larger filter it is possible to obtain smaller deviations and/or a sharper cutoff. Figure 7 shows the magnitude of the frequency response for a 31 x 31 circularly symmetric lowpass filter with a transition region of width 0.05. The design time for this filter was approximately 5 seconds on an IBM 370/155 computer.

Figure 8 shows the magnitude response of a 21 x 21 fan filter. The 1-D lowpass filter used to generate the fan filter has a passband cutoff at $F_p = 0.21$ and a stopband cutoff at $F_s = 0.29$. The resulting deviation is $\delta_p = \delta_s = 0.0229$.

## VII. Summary

A method for transforming 1-D digital filters into 2-D digital filters has been described. This method has the advantage that it is fast, because only a 1-D filter must be designed, and that often the resulting 2-D filter is optimal in the Chebychev sense. Thus, several important types of 2-D filters can be designed using this technique.

## References


1. J. V. Hu and L. R. Rabiner, "Design Techniques for Two-Dimensional Digital Filters", IEEE Trans. Audio Electroacoustics, vol. AU-20, pp. 249-257, Oct. 1972.
2. T. S. Huang, "Two-Dimensional Windows", IEEE Trans. Audio Electroacoustics (Corresp.), vol. AU-20, pp. 88-89, March 1972.
3. T. W. Parks and J. H. McClellan, "Chebyshev Approximation for Non-Recursive Digital Filters with Linear Phase", IEEE Trans. Circuit Theory, vol. CT-19,pp. 189-194, March 1972.
4. J. H. McClellan and T. W. Parks, "Equiripple Approximation of Fan Filters", Geophysics, vol. 37, pp. 573-583, August 1972.
5. O. Herrmann, L. R. Rabiner, and D. S. K. Chan, "Practical Design Rules for Optimum Finite Impulse Response Lowpass Digital Filters", submitted for publication.

# Stability and Synthesis of Two-Dimensional Recursive Filters

JOHN L. SHANKS, Member, IEEE
SVEN TREITEL
Res. Cen.
Amoco Production Co.
Tulsa, Okla. 74102

JAMES H. JUSTICE
Dep. Math.
Univ. Tulsa
Tulsa, Okla.

**Abstract**

**Two-dimensional recursive filters are conveniently described in terms of two-dimensional $z$ transforms. The designer of these filters faces two fundamental problems, their stability and their synthesis. Stability is determined by the location of the zero-valued region of the filter's denominator polynomial. A conjecture based on a one-dimensional stability theorem leads to a useful empirical stabilization procedure. Two-dimensional recursive filters can be synthesized to approximate large varieties of desired two-dimensional pulse responses. A conformal transformation yields two-dimensional recursive bandpass filters from appropriately specified one-dimensional filters.**

## I. Introduction

There are many problems that call for digital filtering of sampled two-dimensional data. For example, the geophysical industry uses two-dimensional filtering for processing seismic records [1], [2] and gravity and magnetic data [3]. Two-dimensional filtering can also be used for the enhancement of photographic data such as weather photos, air photos, and medical X-rays [4].

Conventional implementation has made use of the two-dimensional convolution algorithm and, more recently, of the two-dimensional fast Fourier transform [28]. It is possible to extend another one-dimensional technique into two dimensions, namely recursive filtering [5], [6]. In this technique the filter algorithm uses previously computed output values as well as the input values. One-dimensional recursive filtering has often proven more efficient than either pure convolution in the time domain, or, alternatively, filtering in the frequency domain. Thus it should be interesting to study the properties of two-dimensional recursive filters.

Two problem areas of two-dimensional recursive filtering are stability and synthesis. Since a portion of the output values are used by the recursive algorithm, it is possible for the output values to become arbitrarily large independent of the size of the input values. A filter of this type is said to be unstable, a condition which is undesirable. Thus we need to know what constraints to put on the recursive filter coefficients so that the filtering operation will be stable. This problem is similar to the stability problem for one-dimensional filters, except that the added dimension materially increases the complexity of the analysis. However, we can construct stability diagrams for any particular two-dimensional recursive filter. By inspecting these diagrams we can determine whether or not the filter is stable. We shall now introduce a technique for stabilizing an unstable two-dimensional recursive filter.

In order to stabilize an unstable filter, we shall use a conjecture stating that the planar least squares (Wiener) inverse of an array is minimum phase. A rigorous proof for this property exists only for the one-dimensional case. However, we present a number of examples which uphold this conjecture.

Synthesis is concerned with the design of two-dimensional recursive filters that perform some desirable filtering operations. In this paper we show two ways to synthesize recursive filters. In the first of these we assume that the impulse response for some desired filtering operation is known. We then find the coefficients of a two-dimensional recursive filter that approximates that desired impulse response in a least mean square error sense. The second synthesis method concerns the design of two-dimensional recursive filters that have some desirable two-dimensional frequency response. In the technique presented we show how to take one-dimensional filters and convert them into two-dimensional filters with arbitrary directivity in a two-dimensional frequency response plane. These techniques have been presented briefly in the past [7]–[9]. In this paper we hope to present the complete arguments, with as much background as is practical.

### The Two-Dimensional $z$ Transform

Just as in one-dimensional digital filtering, the $z$ transform can be used to characterize two-dimensional digital filters and two-dimensional digital filtering operations. Consider an array of data values

$$D = \begin{bmatrix} d_{11} & d_{12} & d_{13} & d_{14} \\ d_{21} & d_{22} & d_{23} & d_{24} \\ d_{31} & d_{32} & d_{33} & d_{34} \end{bmatrix}.$$

We can write $D$ as a polynomial in two variables $z_1$ and $z_2$

$$D(z_1, z_2) = \begin{bmatrix} d_{11} + d_{12}z_2 + d_{13}z_2^2 + d_{14}z_2^3 \\ + d_{21}z_1 + d_{22}z_1z_2 + d_{23}z_1z_2^2 + d_{24}z_1z_2^3 \\ + d_{31}z_1^2 + d_{32}z_1^2z_2 + d_{33}z_1^2z_2^2 + d_{34}z_1^2z_2^3 \end{bmatrix}.$$

The power of the $z_1$ and $z_2$ variables indicates the position of the associated filter coefficient in the array.

Manuscript received November 19, 1971.

Reprinted from *IEEE Trans. Audio Electroacoust.*, vol. AU-20, pp. 115–128, June 1972.

In general, the $z$ transform for an $M \times N$ array is

$$D(z_1, z_2) = \sum_{i=1}^{M} \sum_{j=1}^{N} d_{ij} z_1^{i-1} z_2^{j-1}.$$

## II. Two-Dimensional Recursive Filters

It is possible to describe a two-dimensional filter as a ratio of 2 two-dimensional polynomials in $(z_1, z_2)$. Consider the filter

$$F(z_1, z_2) = \frac{A(z_1, z_2)}{B(z_1, z_2)} = \frac{\sum_{i,j=1}^{M_a, N_a} \sum a_{ij} z_1^{i-1} z_2^{j-1}}{\sum_{k,l=1}^{M_b, N_b} \sum b_{kl} z_1^{k-1} z_2^{l-1}}. \tag{1}$$

This filter is applied to a data array by use of the two-dimensional recursive algorithm. Given an input data array $D(z_1, z_2)$, the output of the filter $F(z_1, z_2)$ is

$$R(z_1, z_2) = F(z_1, z_2) \cdot D(z_1, z_2).$$

Using (1) and clearing fractions, we get

$$B(z_1, z_2) R(z_1, z_2) = A(z_1, z_2) D(z_1, z_2) \tag{2}$$

or

$$\Bigg(1 + \sum_{\substack{k=1 \\ k \cdot l \neq 1}}^{M_b} \sum_{l=1}^{N_b} b_{kl}' z_1^{k-1} z_2^{l-1}\Bigg) R(z_1, z_2) = \Bigg(\sum_{i=1}^{M_a} \sum_{j=1}^{N_a} a_{ij}' z_1^{i-1} z_2^{j-1}\Bigg) D(z_1, z_2) \tag{3}$$

where

$$b_{kl}' = b_{k,l}/b_{11}$$
$$a_{kl}' = a_{k,l}/b_{11}$$
$$b_{11} \neq 0.$$

Solving (3) for $R(z_1, z_2)$ we obtain

$$R(z_1, z_2) = \Bigg[\sum_{i=1}^{M_a} \sum_{j=1}^{N_a} a_{ij}' z_1^{i-1} z_2^{j-1}\Bigg] D(z_1, z_2) - \Bigg[\sum_{\substack{k=1 \\ k \cdot l \neq 1}}^{M_b} \sum_{l=1}^{N_b} b_{kl}' z_1^{k-1} z_2^{l-1}\Bigg] R(z_1, z_2). \tag{4}$$

From (4) we deduce the recursive algorithm for computing $r_{mn}$, the output value at the array grid point $(m, n)$

$$r_{mn} = \sum_{i=1}^{M_a} \sum_{j=1}^{N_a} a_{ij}' d_{m-i+1, n-j+1} - \sum_{\substack{k=1 \\ k \cdot l \neq 1}}^{M_b} \sum_{l=1}^{N_b} b_{kl}' r_{m-k+1, n-l+1}. \tag{5}$$

Here we assume either that all output values $r_{m-k+1, n-l+1}$ have been computed previously or are equal to zero.

Let us consider a particular filter

$$F(z_1, z_2) = \frac{a_{11} + a_{12} z_2 + a_{21} z_1 + a_{22} z_1 z_2}{1.0 + b_{12} z_2 + b_{13} z_2^2 + b_{21} z_1 + b_{22} z_1 z_2 + b_{23} z_1 z_2^2}.$$

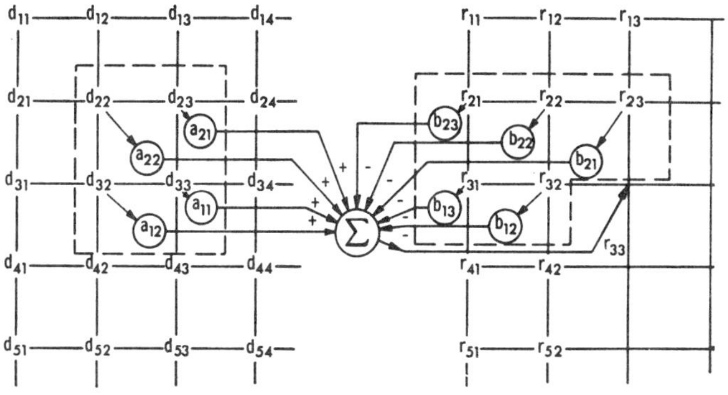


Fig. 1. Recursive filtering of an input array $D$ to produce the output array $R$.

From (5), the recursive algorithm is

$$r_{mn} = a_{11} d_{m,n} + a_{12} d_{m,n-1} + a_{21} d_{m-1,n} + a_{22} d_{m-1,n-1} - b_{12} r_{m,n-1} - b_{13} r_{m,n-2} - b_{21} r_{m-1,n} - b_{22} r_{m-1,n-1} - b_{23} r_{m-1,n-2}. \tag{6}$$

The algorithm is illustrated in Fig. 1. The input array $D$ is being filtered to produce the output array $R$. The recursive algorithm is used in Fig. 1 to compute the output value $r_{33}$. The input values $d_{22}$, $d_{23}$, $d_{32}$, and $d_{33}$ and the previously computed output values $r_{21}$, $r_{22}$, $r_{23}$, $r_{31}$, and $r_{32}$ are used to compute $r_{33}$. After $r_{33}$ has been computed, $r_{43}$ can be computed, assuming that $r_{41}$ and $r_{42}$ have been previously computed. A convenient implementation of this algorithm would be to compute the output array one column at a time. Alternately, the output could be computed one row at a time.

## III. Stability

Just as in the case of one-dimensional digital feedback filters, it is possible for a two-dimensional recursive filter to be unstable. That is, given a bounded input array, the output array can be unbounded. We would like to be able to place constraints on the coefficients of the two-dimensional recursive filter to assure its stability.

The stability of the two-dimensional recursive filter is determined by the coefficients of the denominator $B(z_1, z_2)$ [see (1)]. The general criterion for stability is given by the following theorem. The proof of this theorem is being prepared for publication [26].

*Theorem 1:* Given that $B(z_1, z_2)$ is a polynomial in $(z_1, z_2)$, for the coefficients of expansion of $1/B(z_1, z_2)$ in a positive powers of $z_1$ and $z_2$ to converge absolutely, it is necessary and sufficient that $B(z_1, z_2)$ not be zero for $|z_1|$ and $|z_2|$ simultaneously less than or equal to one.

Stated another way, the theorem says that if there are any values (real or complex) of $z_1$ and $z_2$ for which $B(z_1, z_2)$ is zero, and for which $z_1$ and $z_2$ are simultaneously less than or equal to one in magnitude, then the filter $1/B(z_1, z_2)$ will be unstable. If there are not, then $1/B(z_1, z_2)$ will be stable.

It is much more difficult to determine the stability of two-dimensional filters than of one-dimensional filters. In one dimension, it is only necessary to locate a finite set of roots in the $z$ plane [10], [11]. In order to mechanize the theorem, we must in general find the continuum of $(z_1, z_2)$ values for which $B(z_1, z_2)=0$. The authors know of no techniques for locating the continuum of zeros for a general two-dimensional polynomial in $(z_1, z_2)$. However, for a low-order denominator, we can estimate the region of zeros. We can do this by assigning values to the variable $z_1$, and finding the roots in terms of $z_2$. For stability, all the roots of $z_2$ must be greater than one in magnitude if $|z_1|$ is less than one in magnitude.[1]

For example, consider the filter $F(z_1, z_2)=1/B(z_1, z_2)$, where

$$B(z_1, z_2) = 1 - 1.5z_1 + 0.6z_1^2 - 1.2z_2 + 1.8z_1z_2 - 0.72z_1^2z_2 + 0.5z_2^2 - 0.75z_1z_2^2 + 0.25z_1^2z_2^2.$$

Letting $z_1=\hat{z}_1$, we get

$$B(\hat{z}_1, z_2) = (1 - 1.5\hat{z}_1 + 0.6\hat{z}_1^2) + (-1.2 + 1.8\hat{z}_1 - 0.72\hat{z}_1^2)z_2 + (0.5 - 0.75\hat{z}_1 + 0.25\hat{z}_1^2)z_2^2. \tag{7}$$

Equation (7) is a quadratic in $z_2$ with complex coefficients. Thus we can compute the roots of (7) and plot them as a function of their location in the $z_1$ plane. Fig. 2 shows a contour map of the magnitude of one of the $z_2$ roots as a function of complex $z_1=u+iv$. The $z_2$ root chosen is the one with the minimum magnitude, since we are interested in finding those roots for which $|z_2| \leq 1.0$. The contour interval is 0.1, and the darkest contour corresponds to $|z_2| = 1.0$. The shaded area is for $|z_2| < 1.0$ and is therefore the mapping of the interior of the $z_2$ unit circle onto the $z_1$ plane for $B(z_1, z_2)=0$. In this case the filter is unstable, since the shaded area intersects the $z_1$ unit circle.

Fig. 3 shows the mapping of the $z_2$ unit circle into the $z_1$ plane for the polynomial

$$B(z_1, z_2) = 1 - 1.2z_2 + 0.5z_2^2 - 1.5z_1 + 1.8z_1z_2 - 0.75z_1z_2^2 + 0.6z_1^2 - 0.72z_1^2z_2 + 0.29z_1^2z_2^2 = 0.$$

As before, the shaded area represents the mapping of the $z_2$ unit circle onto the $z_1$ plane. In this case the $z_2$ unit circle does not intersect the $z_1$ unit circle and the filter $F(z_1, z_2) = 1/B(z_1, z_2)$ is stable.[2]

In the special case that the denominator polynomial is first order in both directions, the $z_1$ plane$\leftrightarrow z_2$ plane transformation becomes bilinear [15], [16] and we can set some

[1] Weinberg [12] suggests a stability test involving an adaptation of a two-variable Hurwitz test as described by Ansell [13]. Huang [14] has suggested a test involving only the unit circle in one $z$ plane and the origin of the other $z$ plane.

[2] Figs. 2 and 3 illustrate the fact that the root maps are symmetric with respect to the $u$ axis. The remaining root maps show only the upper half $z$ plane.

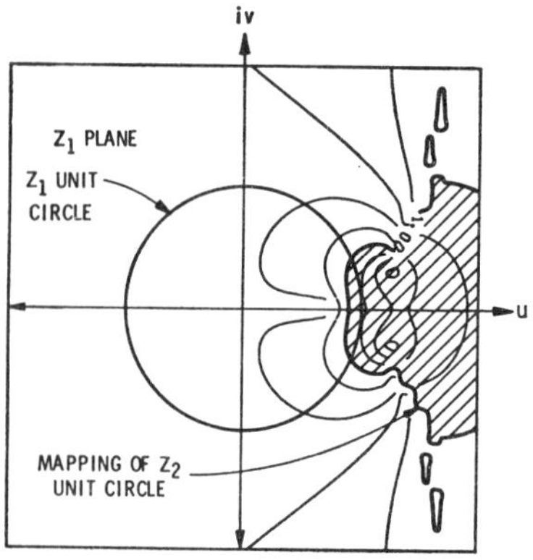


Fig. 2. Mapping of $z_2$ unit circle onto $z_1$ plane for unstable filter.

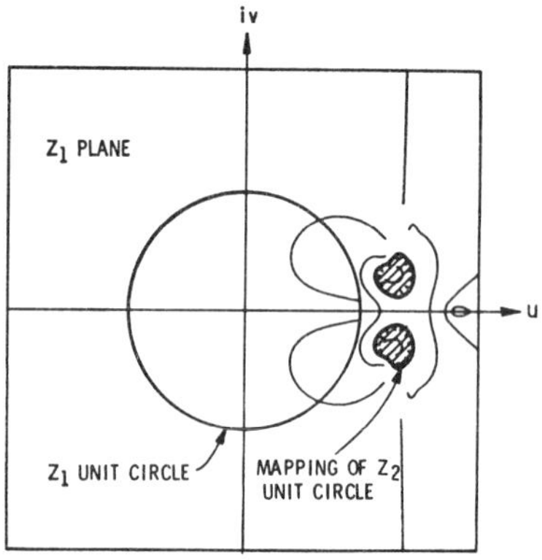


Fig. 3. Mapping of $z_2$ unit circle onto $z_1$ plane for stable filter.

specific limits on the filter coefficients. Let $F(z_1, z_2)=1/B(z_1, z_2)$, where $B(z_1,z_2)=1+b_{21}z_1+b_{12}z_2+b_{22}z_1z_2$. The filter $F(z_1, z_2)$ will be stable if and only if the following conditions hold.

*Condition 1:*

$$\left|\frac{1}{b_{12}}\right| > 1.$$

*Condition 2:*

$$\left|\frac{1 + b_{12}}{b_{21} + b_{22}}\right| > 1.$$

*Condition 3:*

$$\left|\frac{1 + b_{12}}{b_{21} - b_{22}}\right| > 1. \tag{8}$$

The derivation of these conditions is being prepared for future publication.

For example, let us consider the filter

$$F(z_1, z_2) = 1/(1 - 0.7z_1 - 0.5z_2 + 0.3z_1, z_2). \tag{9}$$

**TABLE I**

Unit Pulse Response
$F(z_1, z_2) = 1/(1-0.7z_1-0.5z_2+0.3z_1z_2)$

| m \ n | 1 | 2 | 3 | 4 | 5 | 6 | 7 | 8 | 9 | 10 | 11 |
|---|---|---|---|---|---|---|---|---|---|---|---|
| 1 | 1.0 | 0.5 | 0.25 | 0.125 | 0.063 | 0.031 | 0.016 | 0.008 | 0.004 | 0.002 | 0.001 |
| 2 | 0.7 | 0.4 | 0.225 | 0.125 | 0.069 | 0.038 | 0.02 | 0.011 | 0.006 | 0.003 | 0.002 |
| 3 | 0.49 | 0.315 | 0.195 | 0.117 | 0.069 | 0.04 | 0.023 | 0.013 | 0.007 | 0.004 | 0.002 |
| 4 | 0.343 | 0.245 | 0.165 | 0.106 | 0.066 | 0.041 | 0.024 | 0.014 | 0.008 | 0.005 | 0.003 |
| 5 | 0.24 | 0.189 | 0.136 | 0.093 | 0.061 | 0.039 | 0.024 | 0.015 | 0.009 | 0.005 | 0.003 |
| 6 | 0.168 | 0.144 | 0.111 | 0.08 | 0.055 | 0.036 | 0.024 | 0.015 | 0.009 | 0.006 | 0.004 |
| 7 | 0.117 | 0.109 | 0.090 | 0.067 | 0.048 | 0.033 | 0.022 | 0.014 | 0.009 | 0.006 | 0.004 |
| 8 | 0.082 | 0.082 | 0.082 | 0.055 | 0.041 | 0.029 | 0.02 | 0.014 | 0.009 | 0.006 | 0.004 |
| 9 | 0.058 | 0.062 | 0.062 | 0.045 | 0.035 | 0.026 | 0.018 | 0.013 | 0.008 | 0.005 | 0.004 |
| 10 | 0.04 | 0.046 | 0.046 | 0.037 | 0.029 | 0.022 | 0.016 | 0.011 | 0.008 | 0.005 | 0.004 |
| 11 | 0.028 | 0.034 | 0.034 | 0.03 | 0.024 | 0.019 | 0.014 | 0.01 | 0.007 | 0.004 | 0.003 |

**TABLE II**

Unit Pulse Response
$F(z_1, z_2) = 1/(1-0.95z_1-0.9z_2+0.5z_1z_2)$

| m \ n | 1 | 2 | 3 | 4 | 5 | 6 | 7 | 8 | 9 | 10 | 11 |
|---|---|---|---|---|---|---|---|---|---|---|---|
| 1 | 1.0 | 0.9 | 0.81 | 0.73 | 0.66 | 0.59 | 0.53 | 0.48 | 0.43 | 0.39 | 0.35 |
| 2 | 0.95 | 1.2 | 1.4 | 1.5 | 1.6 | 1.7 | 1.7 | 1.8 | 1.8 | 1.7 | 1.7 |
| 3 | 0.90 | 1.5 | 2.1 | 2.6 | 3.1 | 3.7 | 4.2 | 4.5 | 4.8 | 5.1 | 5.4 |
| 4 | 0.86 | 1.7 | 2.8 | 4.0 | 5.3 | 6.6 | 8.0 | 9.5 | 10. | 12. | 14. |
| 5 | 0.81 | 1.9 | 3.5 | 5.6 | 8.0 | 11. | 14. | 18. | 22. | 26. | 30. |
| 6 | 0.77 | 2.1 | 4.3 | 7.4 | 12. | 17. | 23. | 30. | 39. | 49. | 59. |
| 7 | 0.74 | 2.3 | 5.1 | 9.5 | 15. | 24. | 35. | 49. | 66. | 86. | 110. |
| 8 | 0.70 | 2.5 | 5.9 | 12. | 21. | 34. | 52. | 76. | 106. | 145. | 191. |
| 9 | 0.66 | 2.6 | 6.7 | 14. | 27. | 46. | 74. | 112. | 164. | 232. | 318. |
| 10 | 0.63 | 2.7 | 7.4 | 17. | 33. | 60. | 101. | 161. | 245. | 359. | 509. |
| 11 | 0.60 | 2.8 | 8.3 | 19. | 41. | 78. | 136. | 225. | 355. | 536. | 788. |

The parameters given in (9) meet all of the conditions of (8). Therefore, the filter of (9) is stable.

It would be interesting to look at the unit pulse response of the filter of (9). The unit pulse response is the output of the filter when the input is

$$U = \begin{bmatrix} 1.0 & 0 & 0 & 0 & \cdots \\ 0 & 0 & 0 & 0 & \cdots \\ 0 & 0 & 0 & 0 & \cdots \\ 0 & 0 & 0 & 0 & \cdots \\ \cdot & \cdot & \cdot & \cdots \\ \cdot & \cdot & \cdot & \cdots \\ \cdot & \cdot & \cdot & \cdots \end{bmatrix}. \tag{10}$$

The size of the unit pulse array is arbitrary and depends on how many output values are desired from the filter.

The first (11×11) unit pulse response values for the filter in (9) are shown in Table I. It is characteristic of a stable two-dimensional recursive filter that its unit pulse response values tend to decrease at points away from the origin (upper left-hand corner). As can be seen in Table I, the unit pulse response values tend to decrease as $m$ and $n$ become large.

As an example of an unstable filter, consider the filter

$$F(z_1, z_2) = 1/(1 - 0.95z_2 - 0.95z_1 + 0.5z_1z_2). \tag{11}$$

This filter satisfies Conditions 1 and 3 of (8), but does not meet Condition 2. We can see in Table II that the first 11×11 values of the unit pulse response tend to get larger as $m$ and $n$ increase. This is characteristic of an unstable filter.[3]

## IV. Minimum Phase in Two Dimensions

It would be useful to be able to distinguish between those denominators which produce unstable filters and those which produce stable filters. For this purpose, it is useful to introduce the concept of a two-dimensional minimum phase array.

*Definition:* We shall denote a minimum phase array $A$ as one that satisfies the following two conditions.

[3] The stability of a filter cannot be determined by just looking at a finite section of the unit pulse response. However, these two examples do illustrate the decaying and the expanding characteristics of stable and unstable filters.

*Condition 1:* The two-dimensional array $A$, with the spectrum $A(f_1, f_2)$ evaluated at any real frequency $\hat{f}_1$, shall be a one-dimensionally minimum phase function of frequency $f_2$.

*Condition 2:* The two-dimensional array $A$, with the spectrum $A(f_1, f_2)$ evaluated at any real frequency $\hat{f}_2$, shall be a one-dimensionally minimum phase function of $f_1$.

The spectrum $A(f_1, f_2)$ is the two-dimensional Fourier transform of the array $A$. The function $A(f_1, f_2)$ can also be thought of as the $z$ transform polynomial $A(z_1, z_2)$ evaluated along the $z_1$ and $z_2$ unit circles. Any particular frequency point $(f_1, f_2)$ corresponds to the point in the $z_1$ and $z_2$ planes, $(z_1, z_2) = (\exp[-i\pi f_1/f_{N1}], \exp[-i\pi f_2/f_{N2}])$, where $f_{N1}$ and $f_{N2}$ are the Nyquist frequencies and $i = \sqrt{-1}$.[4]

The statement that a discrete sequence is one-dimensionally minimum phase is equivalent to the statement that its $z$ transform has no zeros inside the $z$ plane unit circle [17]. Therefore, since we can relate the frequencies $f_1$ and $f_2$ to unit circle points $z_1$ and $z_2$, we may also state Conditions 1 and 2 in the following way.

1) The two-dimensional $z$ polynomial $A(\hat{z}_1, z_2) = 0$, evaluated at any $|\hat{z}_1| = 1.0$, has no zeros inside or on the $z_2$ unit circle.

2) The two-dimensional $z$ polynomial $A(z_1, \hat{z}_2) = 0$ evaluated at any $|\hat{z}_2| = 1.0$, has no zeros inside or on the $z_1$ unit circle.

Having defined two-dimensional minimum phase in this way, we are ready to state a second important theorem.

*Theorem 2:* An array $A$ is two-dimensionally minimum phase if and only if the polynomial $A(z_1, z_2) = 0$ has no zeros for which $|z_1| \leq 1$ and $|z_2| \leq 1$, simultaneously. The proof of this theorem is being prepared for future publication.

From Theorems 1 and 2 we see that the filter $F(z_1, z_2) = 1/A(z_1, z_2)$ is stable if $A(z_1, z_2)$ is minimum phase. Conversely, if $A(z_1, z_2)$ is not minimum phase, then $F(z_1, z_2) = 1/A(z_1, z_2)$ will not be stable. For example, the array from (9)

$$B_1 = \begin{bmatrix} 1 & -0.5 \\ -0.7 & 0.3 \end{bmatrix}$$

is a minimum phase array, since the filter $F(z_1, z_2) = 1/B_1(z_1, z_2)$ is stable. On the other hand, the array from (11)

$$B_2 = \begin{bmatrix} 1 & -0.9 \\ -0.95 & 0.5 \end{bmatrix} \tag{12}$$

is a nonminimum phase array since it produces an unstable filter.

## V. The Planar Least Squares Inverse Filter

The design of a stable filter is facilitated by an understanding of the properties of the planar least squares inverse filter. Let us suppose that we have some given array $C$. We would like to find an array $P$ such that $C$ convolved with $P$ approximates the unit pulse array $U$. That is

$$C * P \approx U \tag{13}$$

where the symbol $*$ denotes two-dimensional convolution. In general it will not be possible to make $C * P$ exactly equal to $U$. In actuality, $C * P$ will be equal to some other array, $G$. If we choose $P$ such that the sum of the squares of the elements of $U - G$ is minimized, we call $P$ a *planar least squares inverse* (PLSI) of $C$.

The size of the $P$ array is arbitrary. The $U$ and $G$ arrays must have the same sizes, which depend on the size of the $C$ and $P$ arrays. If $C$ is a $(m \times n)$ array, and $P$ is a $(k \times l)$ array, then $G$ and $U$ must be $(m+k-1)$ by $(n+l-1)$ arrays. Since the size of $P$ is arbitrary, there are many PLSI's of the array $C$, one for each possible set of dimensions of $P$. However, once the dimensions $k$ and $l$ of the matrix $P$ are fixed, there is one and only one array $P$ that minimizes the mean square difference between $G$ and $U$. This array $P$ is determined from the matrices $U$ and $C$ by a two-dimensional Wiener technique described by Wiggins [18].

As a group, PLSI's have interesting properties. One property in particular which we shall use is described by the following conjecture.

*Conjecture 1:* Given an arbitrary real finite array $C$, any planar least squares inverse of $C$ is minimum phase. This is an important conjecture because it implies that the filter $F(z_1, z_2) = 1/P(z_1, z_2)$ must be stable if $P$ is a PLSI.

Unfortunately we do not have a proof for this conjecture, which is an extension of a one-dimensional theorem [25]. However, we have tested the conjecture on a large number of numerical cases, and have failed to find any counterexamples.

### Numerical Examples

In order to test Conjecture 1, we chose a number of arbitrary arrays for which PLSI's were determined. We restricted the size of the PLSI's to $(2 \times 2)$ and $(3 \times 3)$, so that they could be easily tested for the minimum phase property. The PLSI's of the various test arrays were obtained with a planar Wiener filter program written for the Univac 1108 computer.

Table III shows a set of three examples for which $2 \times 2$ PLSI's were computed. The first column gives the input array $C$. Using the planar Wiener inverse filter program a $2 \times 2$ PLSI for $C$ was computed. This result is shown in the second column. The $P$ arrays were all normalized so that the upper left coefficient $p_{11}$ is equal to 1.0. Finally, each $P$ array was examined for stability by use of the three tests given in (8). All three tests must result in magnitudes greater than 1.0 in order for $P$ to be minimum phase. As can be seen in Table III, every test was satisfied for each of the three samples. Therefore, each of the three $P$ arrays are minimum phase, and the filter $F(z_1, z_2) = 1/P(z_1, z_2)$ would be stable in each example.

We also computed a number of $(3 \times 3)$ PLSI's for various arrays. We tested the stability of the $(3 \times 3)$ PLSI's by plotting a root map for each PLSI. In these root maps we show the mapping of the $z_2$ unit circle and its interior (unit

[4] Following the usual definition of the $z$ transform variable, we have that $z_1 = \exp[-i2\pi f_1 \Delta x] = \exp[-i\pi f_1/f_{N1}]$, where $f_{N1} = 1/2\Delta x$ is the Nyquist frequency, and $\Delta x$ is the uniform sampling increment.

Table III

Stability Tests on 2×2 PLSI's

| Example | $C$ | $P$ Normalized so that $p_{11}=1.0$ | Stability Tests on $P$ |
|---|---|---|---|
| 1 | $\begin{bmatrix} 1. & 3. \\ 2. & 4. \end{bmatrix}$ | $\begin{bmatrix} 1.0 & -0.3719 \\ -0.4703 & 0.2127 \end{bmatrix}$ | I. 2.69<br>II. 2.44<br>III. 2.01 |
| 2 | $\begin{bmatrix} 1. & 4. & 7. \\ 2. & 5. & 8. \\ 3. & 6. & 9. \end{bmatrix}$ | $\begin{bmatrix} 1.0 & -0.5437 \\ -0.5862 & 0.3793 \end{bmatrix}$ | I. 1.84<br>II. 2.21<br>III. 1.60 |
| 3 | $\begin{bmatrix} 4. & 1. & 9. \\ 2. & 4. & 1. \\ 3. & 1. & 0. \\ 6. & 2. & 1. \\ 7. & 6. & 2. \end{bmatrix}$ | $\begin{bmatrix} 1.0 & -0.2633 \\ -0.3254 & -0.2515 \end{bmatrix}$ | I. 3.80<br>II. 2.10<br>III. 4.21 |

$$\begin{bmatrix} 1. & 4. & 7. \\ 2. & 5. & 8. \\ 3. & 6. & 9. \end{bmatrix}$$

(A)

$$\begin{bmatrix} 1.0 & -.63405 & .16389 \\ -.78183 & .52113 & -.13153 \\ .19267 & -.13399 & .03399 \end{bmatrix}$$

(B)

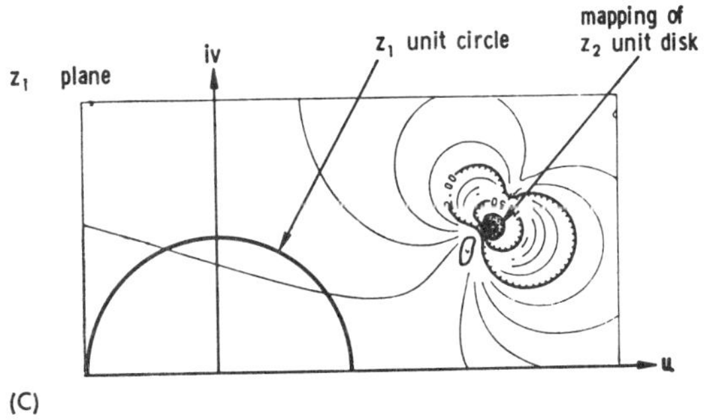


(C)

Fig. 4. Minimum phase test of (3×3) PLSI. (A) Input matrix *A*. (B) (3×3) PLSI of *A*. (C) Root map of PLSI.

disk) onto the $z_1$ plane. If the mapping of the $z_2$ unit disk does not intersect the $z_1$ unit disk, the array is minimum phase. For example, Figs. 4–6 show the input arrays, their (3×3) PLSI's, and the $z$ plane root maps for the PLSI's. The symbols $u$ and $v$ denote the real and imaginary parts of the complex variable $z_1$, i.e., $z_1=u+iv$. The PLSI's have been normalized so that the upper-left coefficient is 1.0. The dark area on each root map is the mapping of the $z_2$ unit disk onto the $z_1$ plane. These root maps show that all of the PLSI's are minimum phase.

$$\begin{bmatrix} 1.0 & -.75 & .9 \\ 1.5 & -1.2 & 1.3 \\ 1.2 & .9 & .5 \end{bmatrix}$$

(A)

$$\begin{bmatrix} 1.0 & .20189 & -.17283 \\ -.47681 & -.09545 & .0036 \\ .11856 & -.00605 & -.02144 \end{bmatrix}$$

(B)

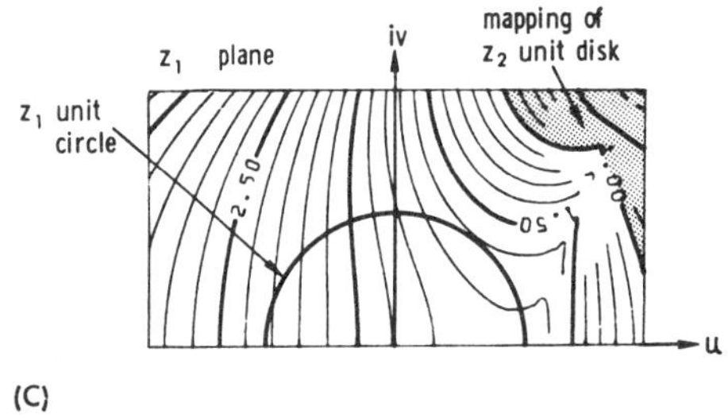


(C)

Fig. 5. Minimum phase test of (3×3) PLSI. (A) Input matrix *A*. (B) (3×3) PLSI of *A*. (C) Root map of (3×3) PLSI.

$$\begin{bmatrix} 1.0 & .5 \\ -.5 & 2. \end{bmatrix}$$

(A)

$$\begin{bmatrix} 1.0 & .17151 & .00427 \\ -.17151 & -.48877 & -.11991 \\ .00427 & .11991 & .19953 \end{bmatrix}$$

(B)

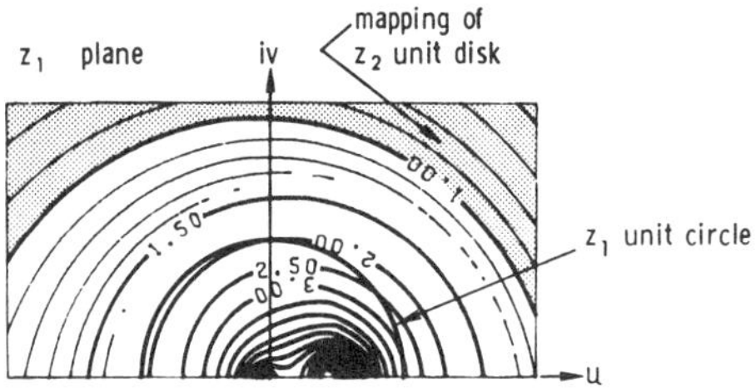


(C)

Fig. 6. Minimum phase test of (3×3) PLSI. (A) Input matrix *A*. (B) (3×3) PLSI of *A*. (C) Root map of PLSI.

Notice in particular Fig. 4. This shows that the (3×3) inverse of the array

$$A=\begin{bmatrix} 1 & 4 & 7 \\ 2 & 5 & 8 \\ 3 & 6 & 9 \end{bmatrix}$$

is minimum phase. It was also shown in Table III, Example 2, that the (2×2) PLSI of this same array is minimum phase. According to Conjecture 1 all PLSI's of an array should be minimum phase. It would be interesting to plot root maps for PLSI's larger than (3×3), but our current root map program is limited to (3×3) arrays. The examples in this section illustrate, but do not prove Conjecture 1. However, we have yet to find a counterexample, and present indications are that Conjecture 1 is true.

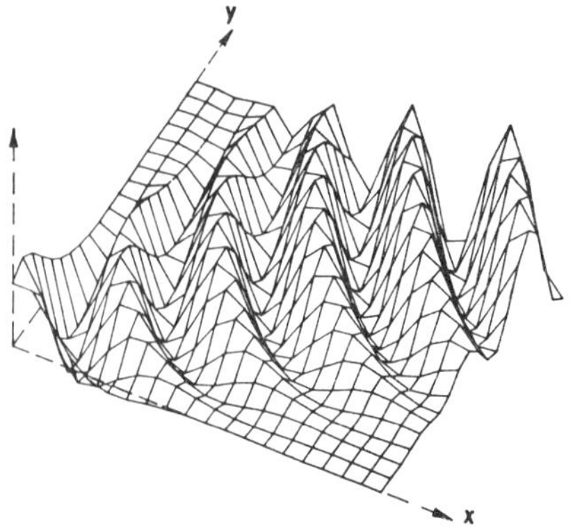


Fig. 7. Pulse response of $F(z_1, z_2) = 1/B(z_1, z_2)$.

## VI. The Stabilization of an Unstable Two-Dimensional Recursive Filter

Suppose that we have designed a two-dimensional recursive filter $F(z_1, z_2) = 1/B(z_1, z_2)$, only to find that it is unstable. How can we alter the coefficients in order to produce a stable filter?

Let us denote the denominator array of this unstable filter by $\boldsymbol{B}$. We can compute a planar least squares inverse $\boldsymbol{B}'$ of $\boldsymbol{B}$. Suppose that we compute a PLSI of $\boldsymbol{B}'$, and call it $\hat{\boldsymbol{B}}$. Now $\hat{\boldsymbol{B}}$ is the inverse of the inverse, or the "double inverse" of $\boldsymbol{B}$. One intuitively feels that $\hat{\boldsymbol{B}}$ and $\boldsymbol{B}$ have some characteristics in common. Moreover, $\hat{\boldsymbol{B}}$ is itself a PLSI. Hence it is minimum phase, while $\boldsymbol{B}$ is not. Therefore, the filter $\hat{F}(z_1, z_2) = 1/\hat{B}(z_1, z_2)$ is stable. For example, consider the filter with the denominator array

$$B = \begin{bmatrix} 1.0 & -1.20002759 & 0.40002239 \\ -1.00003018 & 1.70007079 & -0.65005088 \\ 0.40002035 & -0.7005488 & 0.25004387 \end{bmatrix}. \tag{14}$$

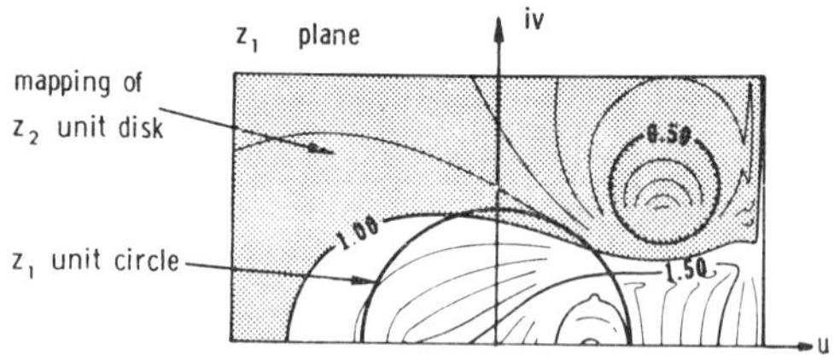


Fig. 8. Root map of $\boldsymbol{B}$ (14).

Fig. 7 shows a perspective view of the unit pulse response of $F(z_1, z_2)$. Note that the unit pulse response tends to grow larger and larger with increasing values of the spatial coordinates $x$ and $y$. We can verify this instability by plotting the root map of $\boldsymbol{B}$, which is shown in Fig. 8. Since a portion of the dark area (mapping of the $z_2$ unit disk) falls inside the $z_1$ unit circle, $\boldsymbol{B}$ is not minimum phase.

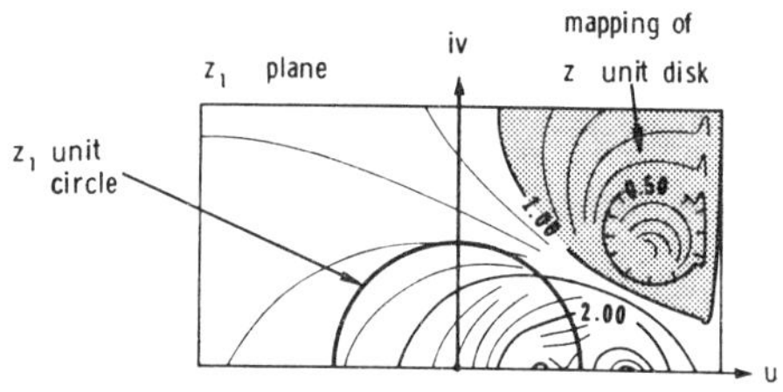


Fig. 9. Root map of $\hat{B}$ (15).

In order to produce a stable filter, we first computed an (8×8) PLSI, $\boldsymbol{B}'$, of $\boldsymbol{B}$. We then computed a (3×3) PLSI of $\boldsymbol{B}'$, which we shall call $\hat{\boldsymbol{B}}$. The evaluation of $\hat{\boldsymbol{B}}$ yielded

$$\hat{B} = \begin{bmatrix} 1.11567 & -1.23469 & 0.38431 \\ -1.0932 & 1.68765 & -0.66821 \\ 0.37577 & -0.65178 & 0.29767 \end{bmatrix}. \tag{15}$$

The root map of $\hat{\boldsymbol{B}}$ is shown in Fig. 9, which shows that $\hat{\boldsymbol{B}}$ is minimum phase. Hence $\hat{F}(z_1, z_2) = 1/\hat{B}(z_1, z_2)$ should be stable. Fig. 10 shows the unit pulse response of $\hat{F}(z_1, z_2)$, which is clearly seen to decay as $x$ and $y$ increase.

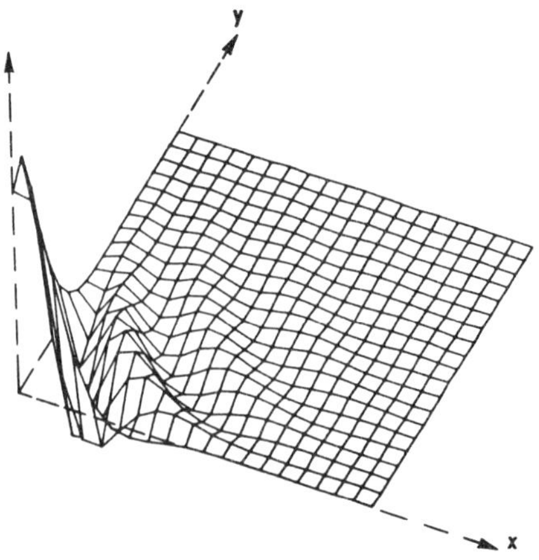


Fig. 10. Pulse response of $\hat{F}(z_1, z_2) = 1/\hat{B}(z_1, z_2)$.

In what sense are $\boldsymbol{B}$ and $\hat{\boldsymbol{B}}$ similar? We have already seen that $\hat{\boldsymbol{B}}$ is minimum phase, while $\boldsymbol{B}$ does not necessarily have this property. On the other hand, $\hat{\boldsymbol{B}}$ is the double PLSI of $\boldsymbol{B}$. Hence we conjecture that $\hat{\boldsymbol{B}}$ must in fact be an approximate

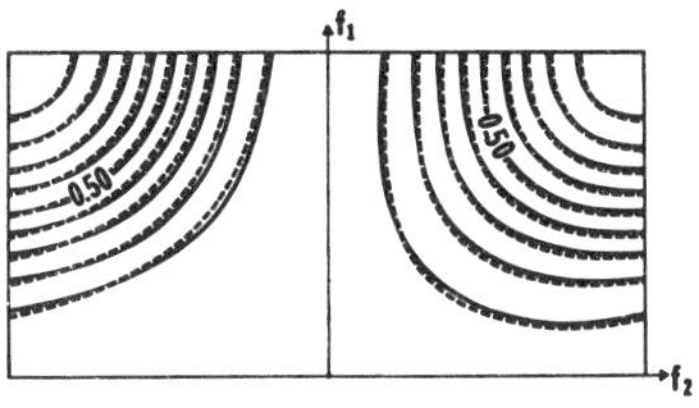


Fig. 11. Amplitude spectra of $\hat{B}$ (——) and $B$ (- - - - - -).

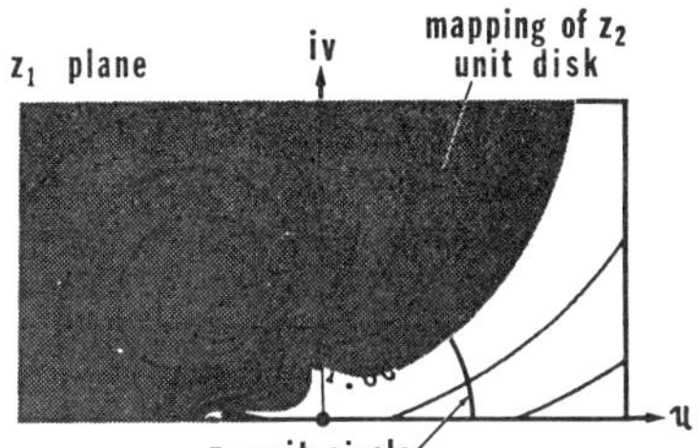


Fig. 13. Root map for array $A$ (Fig. 5).

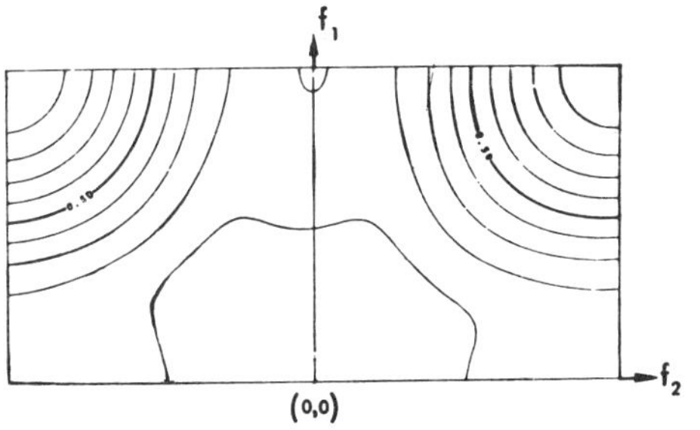


Fig. 12. Amplitude spectrum of $\hat{B}$ from (3×3) $B'$.

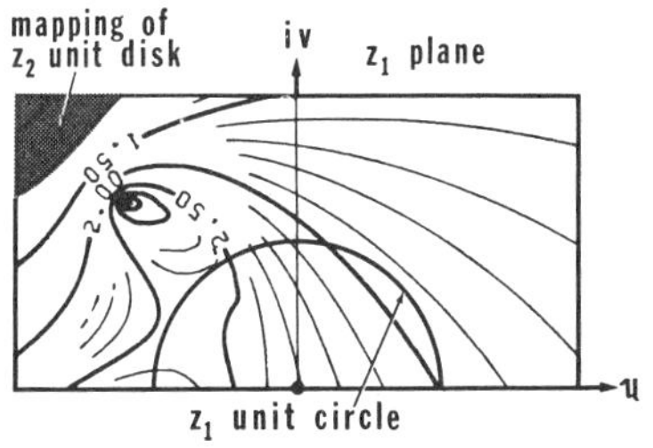


Fig. 14. Root map for array $\hat{A}$ (16) from (3×3) $A'$.

minimum phase version of $B$. In other words, we would expect the amplitude spectra of $B$ and $\hat{B}$ to be roughly equal. Fig. 11 shows the two-dimensional amplitude spectra of $B$ and $\hat{B}$ in the form of contour plots.[5] We observe that these spectra are similar, but not identical.

One of the factors that appears to govern the quality of the approximation is the size of the intermediate array $B'$. The larger the size of $B'$, the better is the resemblance of $\hat{B}$ to the minimum phase version array $B$. In order to test this assertion, we computed a $B'$ array of size (3×3) as compared to the (8×8) $B'$ array of the previous example. We then obtained the (3×3) PLSI of this smaller $B'$ and evaluated the amplitude spectrum of the resulting new $\hat{B}$. This spectrum is shown in Fig. 12; as expected, the agreement with the spectrum of $B$ depicted in Fig. 11 is now much poorer.

As a second example, let us consider the filter $G(z_1, z_2) = 1/A(z_1, z_2)$, where $A$ is the array shown in Fig. 5. The root map for $A$ is shown in Fig. 13. As can be seen, $A$ is not minimum phase, and the filter $F(z_1, z_2) = 1/A(z_1, z_2)$ will be unstable. We computed $A'$, a (3×3) PLSI of $A$. The array $A'$ and its root map are shown in Fig. 5. We then computed $\hat{A}$, the (3×3) PLSI of $A'$ to get

$$\hat{A} = \begin{bmatrix} 6.761 & -1.248 & 1.258 \\ 3.156 & -0.602 & 1.027 \\ 0.706 & 0.209 & 0.368 \end{bmatrix}. \tag{16}$$

[5] The two-dimensional amplitude spectrum must be radially symmetric about the (0,0)-frequency point. Therefore, we show only two quadrants of the two-dimensional spectrum.

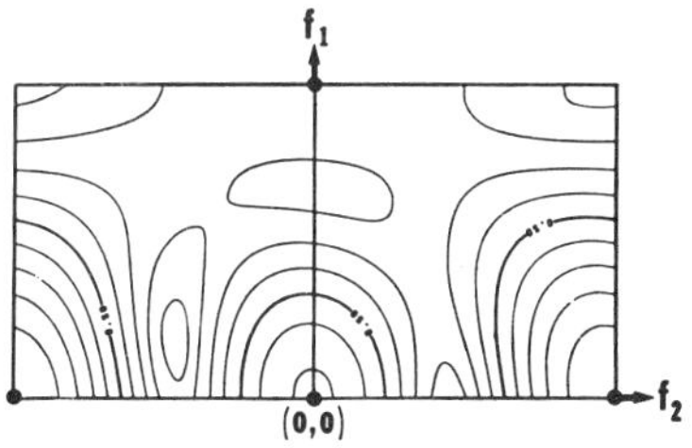


Fig. 15. Amplitude spectrum of array $A$ (Fig. 5).

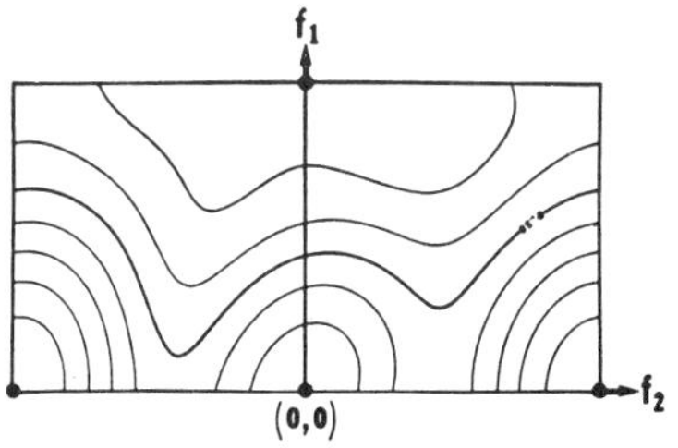


Fig. 16. Amplitude spectrum at array $\hat{A}$ from (3×3) $A'$.

$\hat{A}$ is minimum phase, as shown by its root map in Fig. 14. We can compare the amplitude spectra of $A$ and $\hat{A}$ in Figs. 15 and 16. The two spectra are generally similar, but there are differences in detail.

To get a better approximation, we computed an (8×8) $A'$ and a (15×15) $A'$. From these arrays the corresponding

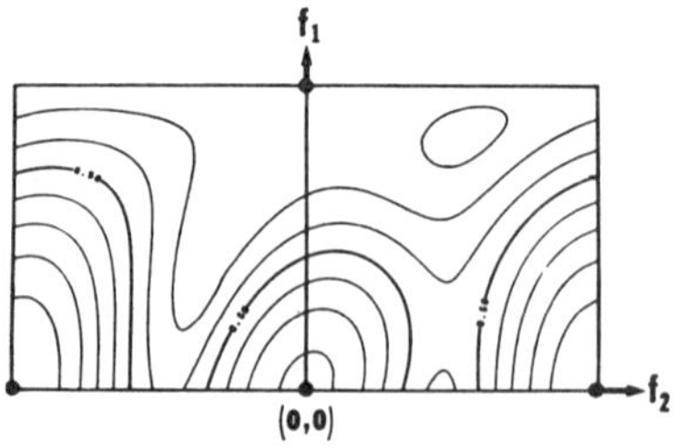


Fig. 17. Amplitude spectrum of array $\hat{A}$ from (8×8) $A'$.

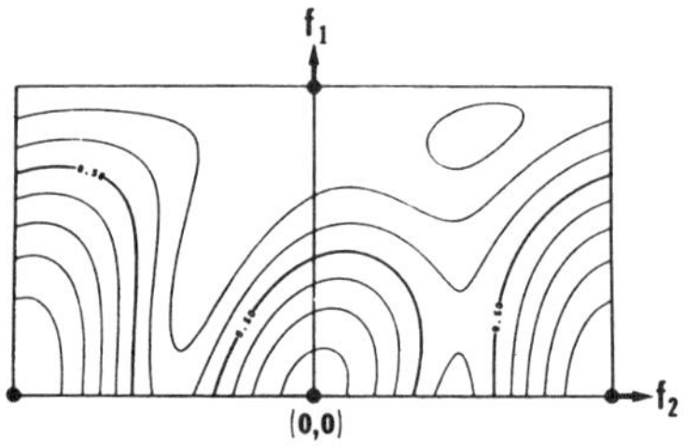


Fig. 18. Amplitude spectrum of array $\hat{A}$ from (15×15) $A'$.

(3×3) $\hat{A}$ arrays were obtained. Figs. 17 and 18 show the amplitude spectra of the $\hat{A}$ arrays. The amplitude spectrum of $\hat{A}$ obtained via the (8×8) $A'$ array (Fig. 17) is in better agreement with the true amplitude spectrum of $A$ (Fig. 15) than is the case for the (3×3) $A'$ array employed previously (Fig. 16). However, Fig. 18 shows that very little improvement is obtained by increasing the $A'$ array size from (8×8) to (15×15).

## VII. Synthesis

We would like to be able to synthesize a two-dimensional recursive filter to perform some desired filtering operation. In this section we shall present two synthesis techniques, one in the space (measurement) domain, and one in the frequency domain.

In the special case that the filter is separable, then the synthesis becomes one-dimensional. In that case, we can express $F(z_1, z_2)$ as

$$F(z_1, z_2) = F_1(z_1)F_2(z_2)$$

and the two-dimensional filter becomes two one-dimensional filters in cascade [19]. If $F(z_1, z_2)$ is a convolutional filter, nonseparable filters can be realized by sums of separable filters [20]. In the remainder of this paper, we shall concern ourselves with the design of nonseparable two-dimensional recursive filters.

### Synthesis in the Space Domain

Suppose we know the two-dimensional unit pulse response of the discrete filter which we wish to use on the data. We could use two-dimensional convolution or the two-dimensional Fourier transform to apply this filter. However, it might also be possible to approximate the impulse response with a two-dimensional recursive filter. In some cases, the recursive filter can be more efficient than straight convolution or filtering in the frequency domain.

The following is an extension of a synthesis technique in one dimension [6], [21], [22]. Suppose the desired impulse response is $d_{ij}$, for $i=1, 2, 3, \cdots, M_d$, and $j=1, 2, 3, \cdots, N_d$. That is,

$$D(z_1, z_2) = \sum_{i=1}^{M_d} \sum_{j=1}^{N_d} d_{ij} z_1^{i-1} z_2^{j-1}. \tag{17}$$

Let the approximating two-dimensional recursive filter be

$$F(z_1, z_2) = \frac{A(z_1, z_2)}{B(z_1, z_2)} \tag{18}$$

where

$$A(z_1, z_2) = \sum_{i=1}^{M_a} \sum_{j=1}^{N_a} a_{ij} z_1^{i-1} z_2^{j-1}$$

$$B(z_1, z_2) = \sum_{i=1}^{M_b} \sum_{j=1}^{N_b} b_{ij} z_1^{i-1} z_2^{j-1}$$

$$b_{11} = 1.0$$

$$M_a, N_a, M_b, N_b = \text{arbitrary (but fixed) parameters.}$$

From (18), $F(z_1, z_2)B(z_1, z_2) = A(z_1, z_2)$. Since multiplication of the $z$ polynomials is equivalent to convolution of the arrays

$$a_{mn} = \sum_{i=1}^{M_b} \sum_{j=1}^{N_b} b_{ij} f_{m-i+1, n-j+1}. \tag{19}$$

The coefficients $a_{mn}$ are defined by (19) over the integers $m=1, 2, \cdots, M_a$, $n=1, 2, \cdots, N_a$. Outside this region, the $a_{mn}$ are zero. Let us define the set of integer pairs

$$I_a = \{(i, j) : 1 \le i \le M_a,\ 1 \le j \le N_a\}.$$

Furthermore, let us define $\hat{I}_a$ as the set of all other values of $(m, n)$ *greater than zero.*

$$\hat{I}_a = \{(m, n) : m > 0,\ n > 0,\ (m, n) \notin I_a\}.$$

Thus, for $(m, n) \in I_a$, the values of $a_{mn}$ are yet to be determined. For $(m, n) \in \hat{I}_a$, $a_{mn} \equiv 0$. Therefore, using (19) and the fact that $b_{11} = 1.0$, we get

$$f_{mn} = -\sum_{\substack{i=1 \\ i \cdot j \ne 1}}^{M_b} \sum_{j=1}^{N_b} b_{ij} f_{m-i+1, n-j+1}, \qquad \text{for } (m, n) \in \hat{I}_a.$$

If we judiciously choose the $b_{ij}$, the $f_{mn}$ will approximate the desired response $d_{mn}$. Thus we may write

$$d_{mn} \approx -\sum_{\substack{i=1 \\ i \cdot j \ne 1}}^{M_b} \sum_{j=1}^{N_b} b_{ij} d_{m-i+1, n-j+1}, \quad \text{for } (m, n) \in \{\hat{I}_a \cap I_d\} \tag{20}$$

where $(m, n)$ must be a member of the intersection of the set $\hat{I}_a$ and the set

$$I_d = \{(m, n): 1 \leq m \leq M_d, 1 \leq n \leq N_d\}.$$

Let us define an error $e_{mn}$ that can be added to the right side of (20) to produce equality.

$$d_{mn} = e_{mn} - \sum_{\substack{i=1 \\ i\cdot j\neq 1}}^{M_b} \sum_{j=1}^{N_b} b_{ij} d_{m-i+1,n-j+1}, \quad \text{for } (m, n) \in \{\hat{I}_a \cap I_d\}.$$

Therefore, we can put $d_{mn}$ inside the summation to get

$$e_{mn} = \sum_{i=1}^{M_b} \sum_{j=1}^{N_b} b_{ij} d_{m-i+1,n-j+1}, \quad \text{for } (m, n) \in \{\hat{I}_a \cap I_d\}.$$

Now that we have defined the error, let us choose the $b_{ij}$ such as to minimize the mean square error,

$$\overline{e^2} = \sum_{(m,n)} \sum \left[ \sum_{(i,j)} b_{ij} d_{m-i+1,n-j+1} \right]^2,$$

$$\text{for } (i, j) \in I_b, \qquad (m, n) \in \{\hat{I}_a \cap I_d\}. \quad (21)$$

Except for $b_{11}$, which is 1.0, the $b_{ij}$ are not yet specified. By differentiating (21) with respect to the $b_{ij}$ and setting the resultant equations equal to zero, we can find those $b_{ij}$ that minimize the mean square error. This will give us $(M_b \cdot N_b - 1)$ equations of the type

$$\sum_{\substack{i=1 \\ i\cdot j\neq 1}}^{M_b} \sum_{j=1}^{N_b} b_{ij} \Phi_{klij} = \phi_{kl}, \qquad \text{for } k = 1, 2, 3, \cdots, M_b, \quad (22)$$
$$l = 1, 2, 3, \cdots, N_b,$$
$$\text{but } k \cdot l \neq 1$$

where

$$\Phi_{klij} = \sum_{(m,n)} d_{m-i+1,n-j+1} d_{m-k+1,n-l+1} \quad (23)$$

$$\phi_{kl} = - \sum_{(m,n)} d_{mn} d_{m-k+1,n-j+1}, \qquad (m, n) \in \{\hat{I}_a \cap I_d\}. \quad (24)$$

Thus, (22) describes a set of $(M_b \cdot N_b - 1)$ simultaneous linear equations with $(M_b \cdot N_b - 1)$ unknowns. The solutions to (22) gives us the $b_{ij}$ that minimizes the mean square error.

Having computed the denominator $B(z_1, z_2)$ of the filter, we now compute the numerator $A(z_1, z_2)$. One way to compute $A(z_1, z_2)$ is to compute those coefficients that minimize the mean square difference between the coefficients of $F(z_1, z_2) = A(z_1, z_2)/B(z_1, z_2)$ and the coefficients of the desired response $D(z_1, z_2)$. This is a discrete Wiener filtering problem in two dimensions [18]. It consists of finding the optimum filter $A(z_1, z_2)$ given an input $1/B(z_1, z_2)$ and a desired output $D(z_1, z_2)$.

In a simpler but less accurate method we can convolve the $B$ array with the $D$ array to get the $A$ array. Since $A(z_1, z_2)/B(z_1, z_2) \approx D(z_1, z_2)$, we compute the coefficients $a_{mn}$ from $A(z_1, z_2) = B(z_1, z_2) \cdot D(z_1, z_2)$ for $(m, n) \in I_a$. If $B(z_1, z_2)$ has been chosen well, the coefficients of $B(z_1, z_2) \cdot D(z_1, z_2)$ will be comparatively small for $(m, n) \in \hat{I}_a$.

A computer program was written to obtain a two-dimensional recursive filter using the above technique. In order to test the program, we computed the impulse response of a two-dimensional recursive filter, $D(z_1, z_2) = A'(z_1, z_2)/B'(z_1, z_2)$. Both $A'$ and $B'$ were $3\times3$ arrays. The coefficients of the $A'$ and $B'$ arrays are shown in Table IV, under columns $A'$ and $B'$. The first $20\times20$ values of the impulse response were used as inputs to the computer program that finds the two-dimensional recursive filter that approximates that response. $M_a$, $M_b$, $N_a$, and $N_b$ were set at three. On a CDC 3300 computer, the program yielded the coefficients given by (23) and (24), and solved (22) to find the denominator coefficients. The numerator coefficients were found by two-dimensional convolution of $B(z_1, z_2)$ with $D(z_1, z_2)$. The recursive coefficients computed by the program are shown in Table IV under Columns $A$ and $B$.

TABLE IV

Test of Space Domain Synthesis

| Term | $A'$ | $A$ | $B'$ | $B$ |
|---|---|---|---|---|
| Constant | 1.0 | 1.0 | 1.0 | 1.0 |
| $z_1$ | 3.0 | 3.0000005 | −1.2 | −1.1999995 |
| $z_1^2$ | 2.0 | 2.0000016 | 0.5 | 0.4999996 |
| $z_2$ | 2.0 | 2.0000002 | −1.5 | −1.4999998 |
| $z_1z_2$ | 4.0 | 4.0000014 | 1.8 | 1.7999989 |
| $z_1^2z_2$ | −1.0 | −0.9999976 | −0.75 | −0.7499992 |
| $z_2^2$ | −1.0 | −0.9999995 | 0.6 | 0.5999998 |
| $z_1z_2^2$ | 2.0 | 2.0000002 | −0.72 | −0.7199994 |
| $z_1^2z_2^2$ | 1.0 | 0.9999998 | 0.29 | 0.2899996 |

*Note:* $A'$, $B'$ are the actual coefficients. $A$, $B$ are the estimated coefficients.

Thus we see that this technique is fairly good if the desired response can be approximated by a recursive type filter.

### Synthesis in the Frequency Domain

Many frequency domain recursive filter design techniques are by now available for the one-dimensional case [5], [23]. Much of the work in one-dimensional frequency domain synthesis has been in the design of the low-pass and band-pass type filter. We shall discuss a technique for designing such filters in two dimensions.

In one dimension, the design parameters are often the cutoff frequencies, or the points of transition between the "passband" and the "stopband." In two dimensions the transition is a boundary separating the pass and reject regions. The shape of the boundary depends on the designer's filter requirements and the number of coefficients he is willing to have in the filter.

Suppose we design a low-pass or bandpass filter in one dimension. Such a filter can also be viewed as a two-dimensional filter that varies in one dimension only. In terms of the two-dimensional Laplace variables [27] $(s_1, s_2)$ such a filter would be written

$$F(s_1, s_2) = F_2(s_2).$$

The frequency response of this filter can be found by setting $s_1 = i2\pi f_1$ and $s_2 = i2\pi f_2$. Thus the response varies only in the $f_2$ frequency direction. However, suppose we were to perform a transformation, or mapping, of the $(s_1, s_2)$ plane into an $(s_1', s_2')$ such that the passband of $F(s_1, s_2)$ maps into a suitable region in the $(s_1', s_2')$ plane. In particular, let us consider rotating the axes of the $(s_1, s_2)$ plane.

We can rotate the $(s_1, s_2)$ axes by an angle $\beta$ by means of the transformation

$$s_1 = s_1' \cos\beta + s_2' \sin\beta \tag{25}$$

$$s_2 = s_2' \cos\beta - s_1' \sin\beta. \tag{26}$$

Rotating the $(s_1, s_2)$ axes also causes the two-dimensional frequency axes $(f_1, f_2)$ to be rotated. Consider a filter with the transfer function $F(s_1, s_2)$. Using (25) and (26), we can rotate the $(f_1, f_2)$ axes by an angle $\beta$ to a new position $(f_1', f_2')$. Conversely, this has the effect of rotating the "pass" region by an angle $(-\beta)$ with respect to the new frequency axes.

For example, consider the filter

$$F(s_1, s_2) = \frac{1}{s_2 + a}\cdot \tag{27}$$

The amplitude response of this varies in the $f_2$ direction only. Applying the transformation equations (25) and (26) we get

$$F'(s_1', s_2') = \frac{1}{s_2' \cos\beta - s_1' \sin\beta + a}\cdot \tag{28}$$

This filter varies in both frequency directions.

We can design two-dimensional bandpass filters whose response pattern has a particular angle with respect to the $(f_1, f_2)$ frequency axes. First we would design a suitable one-dimensional low-pass or bandpass filter in the one-dimensional $s$ plane. Expressing this filter as a function of either $s_1$ or $s_2$, we would use either (25) or (26) to rotate this filter some desired angle into the $(s_1', s_2')$ plane. This resultant form $F'(s_1', s_2')$ describes a continuous two-dimensional filter.

It is still necessary to produce the equivalent two-dimensional discrete filter. For this purpose we can use the bilinear $z$ transform [23]. We have

$$s_1' = \frac{2}{\Delta_1}\,\frac{1 - z_1}{1 + z_1} \tag{29}$$

$$s_2' = \frac{2}{\Delta_2}\,\frac{1 - z_2}{1 + z_2} \tag{30}$$

where $\Delta_1$ and $\Delta_2$ are the sample intervals. Applying the bilinear equations to the $F'(s_1', s_2')$ given in (28), we get

$$F(z_1, z_2) = \frac{\Delta_1\Delta_2}{2}\,\frac{(1 + z_1)(1 + z_2)}{1 + b_{21}z_1 + b_{12}z_2 + b_{22}z_1z_2}, \tag{31}$$

where

$$b_{21} = \left(\frac{a\Delta_1\Delta_2}{2} + \Delta_1\cos\beta + \Delta_2\sin\beta\right)\Big/ D \tag{32}$$

$$b_{12} = \left(\frac{a\Delta_1\Delta_2}{2} - \Delta_1\cos\beta - \Delta_2\sin\beta\right)\Big/ D \tag{33}$$

$$b_{22} = \left(\frac{a\Delta_1\Delta_2}{2} - \Delta_1\cos\beta + \Delta_2\sin\beta\right)\Big/ D \tag{34}$$

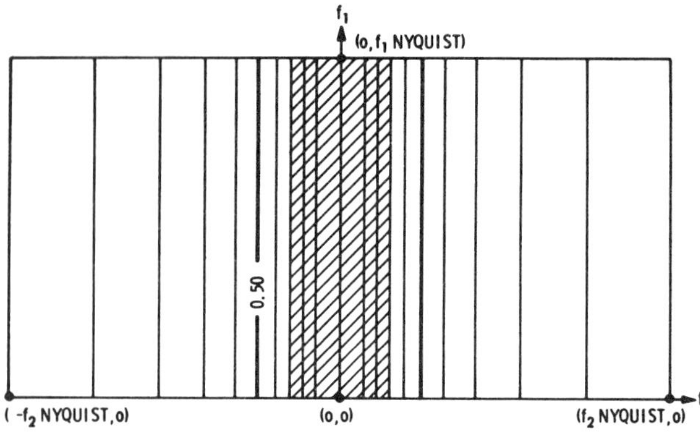


Fig. 19. Two-dimensional amplitude response of $(1+z_2)/(1-0.6186\,z_2)$.

and

$$D = \frac{a\Delta_1\Delta_2}{2} + \Delta_1\cos\beta - \Delta_2\sin\beta.$$

These equations can be simplified by letting $\Delta_1 = \Delta_2 = \Delta$. In equation (27) the constant $a$ represents the cutoff frequency (3 dB down) in rad/unit[6] [24]. Let us set

$$a = 2\pi f_c f_N$$

where $f_c$ is cutoff frequency as a fraction of the Nyquist frequency, and $f_N$ is the Nyquist frequency in cycles per unit. Since $f_N = 1/2\Delta$,

$$a = \frac{\pi f_c}{\Delta}\cdot$$

Thus we get

$$b_{21} = \{\pi f_c + 2(\cos\beta + \sin\beta)\}/D$$
$$b_{12} = \{\pi f_c - 2(\cos\beta + \sin\beta)\}/D$$
$$b_{22} = \{\pi f_c - 2(\cos\beta - \sin\beta)\}/D$$
$$D = \{\pi f_c + 2(\cos\beta - \sin\beta)\}.$$

For example, let us consider a filter $F(s) = 1/(s+a)$, where $a$ is chosen to give a cutoff frequency at 0.15 Nyquist. Thus, $f_c = 0.15$ and for $\beta = 0°$, we get

$$F(z_1, z_2) = \frac{\Delta^2}{2}\,\frac{(1 + z_1)(1 + z_2)}{\begin{bmatrix} 1 - 0.6186z_2 \\ + z_1 - 0.6186z_1z_2 \end{bmatrix}} = \frac{\Delta^2}{2}\,\frac{1 + z_2}{1 - 0.6186z_2}\cdot$$

Fig. 19 shows a contour map of the amplitude response of this filter, normalized to a peak response of 1.0. The contour interval is 0.1. The pass region, for which the amplitude response is greater than 0.7, is shaded. Since $\beta = 0$, this response does not vary along the $f_1$ frequency axis. Along

[6] The unit depends on the problem. In the case of seismic records, the unit in one direction would be feet, and in the other direction, time. In the case of a photograph, both units would be spatial.

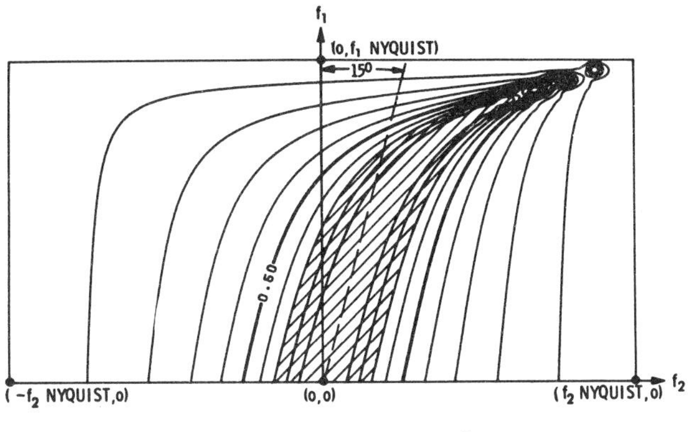


Fig. 20. Amplitude response of filter rotated 15°.

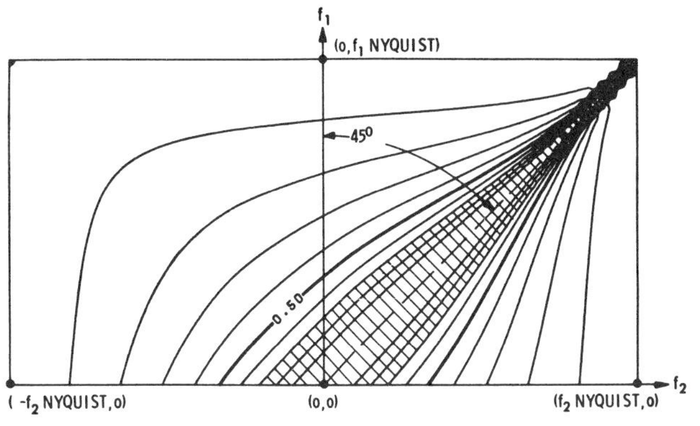


Fig. 21. Amplitude response of filter rotated 45°.

the $f_2$ frequency axis, the filter is a low-pass filter. Note that the cutoff frequency is 15 percent of the Nyquist frequency, which agrees with the design parameter $f_c = 0.15$.

For $\beta = 15°$, we get the filter

$$F(z_1, z_2) = \frac{\Delta^2}{2} \frac{(1 + z_1)(1 + z_2)}{\begin{bmatrix} 1 - 0.3229 z_2 \\ +0.6455 z_1 - 0.6773 z_1 z_2 \end{bmatrix}}.$$

Fig. 20 shows a contour map of the amplitude response of this filter. The contour interval is again 0.1. As required, the amplitude response has been rotated by 15° from the $f_1$ axis. However, the contour lines are no longer straight, and tend to be distorted at the higher $(f_1, f_2)$ frequencies. This effect is caused by the bilinear $(z_1, z_2)$ transform. When the bilinear equations (29) and (30) are applied to (28), the numerator of $F(z_1, z_2)$ will contain the factors $(1+z_1)(1+z_2)$. The factor $(1+z_1)$ causes $F(z_1, z_2)$ to be zero for all frequencies for which $f_1$ is equal to the Nyquist. Similarly, the factor $(1+z_2)$ causes the $F(z_1, z_2)$ to go to zero at the $f_2$ Nyquist frequencies.

Fig. 21 shows the amplitude response of a filter rotated to 45°. The filter equation is

$$F(z_1, z_2) = \frac{(1 + z_1)(1 + z_2)}{\begin{bmatrix} 1 + 0.1428 z_2 \\ +0.1428 z_1 - 0.7144 z_1 z_2 \end{bmatrix}}.$$

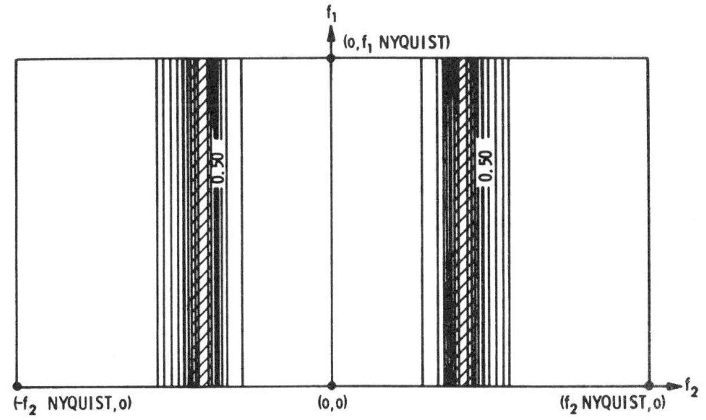


Fig. 22. Amplitude response of bandpass filter.

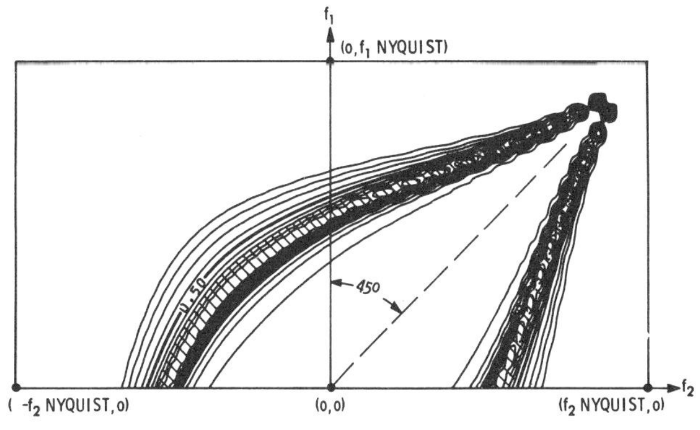


Fig. 23. Amplitude response of bandpass filter rotated 45°.

We can also "rotate" the bandpass type of filter. Fig. 22 shows the amplitude response of a bandpass filter. This is a bilinear transform of an eight-order Butterworth filter [6, pp. 37–41]. Fig. 23 shows the response of the same filter rotated 45°, again using the bilinear transform.

In order to obtain a rotation through a negative angle, we can filter the data in a different manner. Suppose $D(z_1, z_2)$ is an $M \times N$ input and $F(z_1, z_2)$ is the filter which produces the response shown in Fig. 23. Instead of applying $F(z_1, z_2)$ in the normal way, as described by equation (5), let us start filtering at the input point $d_{1,N}$, and filter backwards in the $y$ direction. This is equivalent to applying the filter

$$G(z_1, z_2) = F(z_1, 1/z_2).$$

This filter has the amplitude response curve shown in Fig. 24, with the bandpass characteristic rotated by −45°.

We can also combine these rotated filters to obtain other filter shapes. Consider the filter with an amplitude response shown in Fig. 25. This is a low-pass filter rotated 45°. Suppose we filter the data with this filter in cascade with the filter described by Fig. 24. The overall amplitude response will be

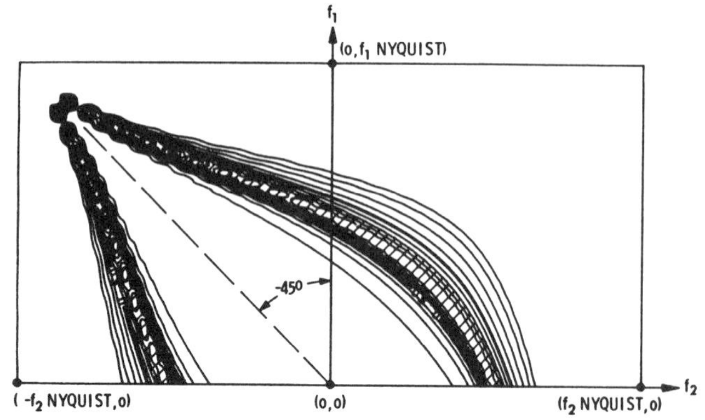


Fig. 24. Amplitude response of bandpass filter rotated $-45°$.

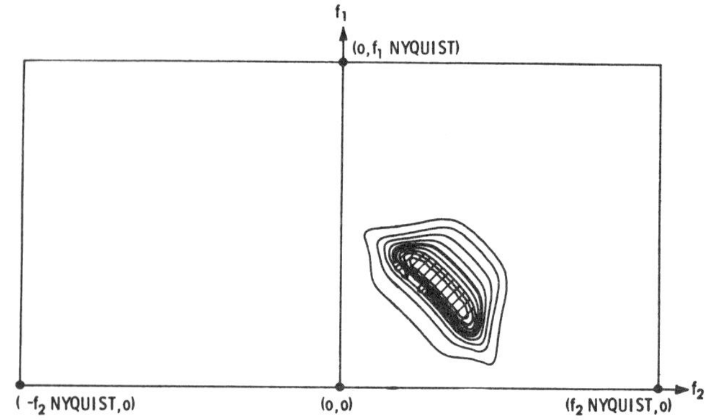


Fig. 26. Amplitude response of cascade response of filters from Figs. 24 and 25.

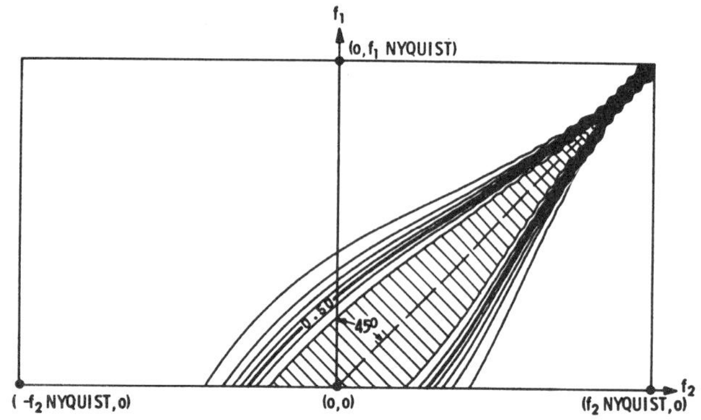


Fig. 25. Amplitude response of low-pass filter rotated 45°.

the product of the two amplitude responses. The amplitude response of the cascaded filters is shown in Fig. 26.

### Phase Response of Two-Dimensional Recursive Filters

The recursive filtering algorithm of (5) produces a filter with a "one-quadrant" impulse response. If the input to the filter is a unit pulse at some grid point, the filter responds in the positive $m$ and $n$ directions only. Due to its one-quadrant response, the recursive filter cannot have a zero-phase response. However, by the proper combination of two or more recursive filters, we can produce a filtering operation which has zero phase or other useful symmetries.

A zero-phase two-dimensional filter has the particular property that its impulse response is symmetric about the $(x, y)$ origin along any radius through the origin. We can perform zero-phase filtering with recursive filters by first filtering an input array with a recursive filter, starting at one corner of the input array. We then take the resultant array and, starting at the diagonally opposite corner, filter the result in the opposite direction with the same recursive filter.[7] This is equivalent to filtering with the filter

$$G(z_1, z_2) = F(z_1, z_2) \cdot F(1/z_1, 1/z_2) \tag{35}$$

where $F(z_1, z_2)$ is the one-quadrant recursive filter.

Setting $z_1 = e^{i\omega_1}$ and $z_2 = e^{i\omega_2}$ in (35) gives us the two-dimensional frequency response of $G(z_1, z_2)$. Since $F(e^{i\omega_1}, e^{i\omega_2})$ is the complex conjugate of $F(e^{-i\omega_1}, e^{-i\omega_2})$, $G(e^{i\omega_1}, e^{i\omega_2})$ will be real for any value of $(\omega_1, \omega_2)$. Therefore, the filter $G(z_1, z_2)$ has zero-phase response. The amplitude response of $G(z_1, z_2)$ will be the square of the amplitude response of $F(z_1, z_2)$. That is, $|G(\omega_1, \omega_2)| = |F(\omega_1, \omega_2)|^2$.

We can also get zero-phase response by filtering an input array forward and backward with the same recursive filter, and adding the results. This is equivalent to filtering with

$$H(z_1, z_2) = F(z_1, z_2) + F(1/z_1, 1/z_2).$$

Like $G(z_1, z_2)$, $H(z_1, z_2)$ has zero-phase response. However, the amplitude responses are different. The amplitude response of $H(z_1, z_2)$ is

$$|H(\omega_1, \omega_2)| = 2 \operatorname{Re} \{F(\omega_1, \omega_2)\}.$$

A filtering operation can be made symmetric with respect to the $x$ axis (or $y$ axis) by filtering forward and reverse in the $x$ (or $y$) direction. For example, if we relate $z_1$ with the $x$ axis and $z_2$ with the $y$ axis, $G(z_1, z_2) = F(z_1, z_2) \cdot F(1/z_1, z_2)$ and $H(z_1, z_2) = F(z_1, z_2) + F(1/z_1, z_2)$ are symmetric about the $x$ axis.

Finally, the filtering operations

$$G(z_1, z_2) = F(z_1, z_2) \cdot F(1/z_1, z_2) \cdot F(z_1, 1/z_2) \cdot F(1/z_1, 1/z_2)$$

and

$$H(z_1, z_2) = F(z_1, z_2) + F(1/z_1, z_2) + F(z_1, 1/z_2) + F(1/z_1, 1/z_2)$$

are symmetric with respect to both the $x$ and $y$ axes, as well as having zero-phase response.

## VIII. Summary

We can perform two-dimensional filtering with the two-dimensional recursive filter algorithm. The primary prob-

[7] Use of this technique requires that the computer output array be sufficiently large to allow the output values to decay to reasonably small numbers.

lems are stability and synthesis. We have given the general theorem for stability and shown that the stability can be computed for a two by two array. For larger denominator arrays, the stability question is more difficult. We can estimate the stability by using the mapping technique as illustrated by two examples.

We have demonstrated that the denominator array of a two-dimensional recursive filter must be minimum phase if this recursive filter is to be stable. A given nonminimum phase denominator array can be stabilized by computing two successive planar least squares inverse filters. This method is based on a conjecture which states that the planar least squares inverse of any matrix is minimum phase. This conjecture appears to be a reasonable extrapolation from a theorem in the one-dimensional case. We have also presented a number of examples, all of which uphold the validity of our conjecture.

We have shown two techniques for designing or synthesizing two-dimensional recursive filters. There are several problems where the space domain synthesis might be useful. More work needs to be done on frequency domain synthesis to give the designer more control of the pass-reject boundaries. Some mappings other than the simple rotation might be useful.

We need to study further the theory and characteristics of two-dimensional recursive filters, and to find applications that will make efficient use of these techniques.

## Acknowledgment

The authors wish to express their appreciation to Amoco Production Company for allowing them to publish this material.

## References

[1] J. P. Fail and G. Grau, "Les filtres en éventail," *Geophys. Prospect.*, vol. 11, pp. 131–163, June 1963.
[2] S. Treitel, J. L. Shanks, and C. W. Frasier, "Some aspects of fan filtering," *Geophysics*, vol. 32, pp. 789–800, Oct. 1967.
[3] E. G. Zurflueh, "Applications of two-dimensional linear wavelength filtering," *Geophysics*, vol. 32, pp. 1015–1035, Dec. 1967.
[4] R. H. Selzer, "Improving biomedical image quality with computers," NASA Tech. Rep. 32-1336, 1968.
[5] C. M. Rader and B. Gold, "Digital filter design techniques in the frequency domain," *Proc. IEEE*, vol. 55, pp. 149–171, Feb. 1967.
[6] J. L. Shanks, "Recursion filters for digital processing," *Geophysics*, vol. 32, pp. 33–51, Feb. 1967.
[7] ——, "Two-dimensional recursive filters," in *SWIEEECO Rec.*, Apr. 1969, pp. 19E1–19E8.
[8] ——, "The design of stable two-dimensional recursive filters," in *Proc. Univ. Missouri—Mervin J. Kelly Commun. Conf.*, Oct. 1970, pp. 15-2-1–15-2-5.
[9] ——, "Two planar digital filtering algorithms," presented at the 5th Princeton Conf. Inform. Sci. Syst., Mar. 1971.
[10] J. T. Tou, *Digital and Sampled-Data Control Systems.* New York: McGraw-Hill, 1959, pp. 230–249.
[11] S. Treitel and E. A. Robinson, "The stability of digital filters," *IEEE Trans. Geosci. Electron.*, vol. GE-2, pp. 6–18, Nov. 1964.
[12] L. Weinberg, "Approximation and stability—Test procedures suitable for digital filters," presented at the Arden House Workshop on Digital Filtering, Harriman, N. Y., Jan. 1970.
[13] H. G. Ansell, "On certain two-variable generalizations of circuit theory, with applications to networks of transmission lines and lumped reactances," *IEEE Trans. Circuit Theory*, vol. CT-11, pp. 214–223, June 1964.
[14] T. S. Huang, private communication, Mar. 1971.
[15] H. Kober, *Dictionary of Conformal Representations.* New York: Dover, 1952, pp. 3–8.
[16] R. V. Churchill, *Introduction to Complex Variables and Applications.* New York: McGraw-Hill, 1948, pp. 57–63.
[17] S. Treitel and E. A. Robinson, "The stability of digital filters," *IEEE Trans. Geosci. Electron.*, vol. GE-2, pp. 15–16, Nov. 1964.
[18] R. A. Wiggins, "On factoring the correlations of discrete multivariable stochastic processes," Mass. Inst. Tech., Cambridge, Sci. Rep. 9 of Contract AF 19 (604) 7378, pp. 127–152, 1965.
[19] W. D. Fryer and G. E. Richmond, "Two-dimensional spatial filtering and computers," in *Proc. Nat. Electron. Conf.*, vol. 18, 1962, pp. 529–534.
[20] S. Treitel and J. L. Shanks, "The design of multistage separable planar filters," *IEEE Trans. Geosci. Electron.*, vol. GE-9, pp. 10–27, Jan. 1971.
[21] C. S. Burris and T. W. Parks, "Time domain design of recursive digital filters," *IEEE Trans. Audio Electroacoust.*, vol. AU-18, pp. 137–141, June 1970.
[22] T. W. Parks and C. S. Burris, "Applications and extensions of Prony's method to parameter identification and digital filtering," presented at the 5th Princeton Conf. Inform. Sci. Syst., Mar. 1971.
[23] R. M. Golden and J. F. Kaiser, "Design of wideband sampled-data filters," *Bell Syst. Tech. J.*, pp. 1533–1546, July 1964.
[24] R. G. Brown and J. W. Nilsson, *Introduction to Linear Systems Analysis.* New York: Wiley, 1962, pp. 121–125.
[25] E. A. Robinson, *Statistical Communication and Detection.* New York: Hafner, 1967, pp. 173–174.
[26] J. H. Justice and J. L. Shanks, "Stability criterion for *N*-dimensional digital filters," unpublished.
[27] V. A. Ditkin and A. P. Prudnikov, *Operational Calculus in Two Variables* (English translation by D. M. G. Wishart). New York: Pergamon, 1962.
[28] W. T. Cochran *et al.*, "What is the fast Fourier transform?" *Proc. IEEE*, vol. 55, pp. 1667–1674, Oct. 1967.

# Stability of Two-Dimensional Recursive Filters

THOMAS S. HUANG, Member, IEEE
Dep. Elec. Eng.
Res. Lab. Electron.
and Lincoln Lab.
Mass. Inst. Tech.
Cambridge, Mass. 02173

**Abstract**

**We discuss some aspects of the stability problem in two-dimensional recursive filtering. In particular, we derive a simplified version of a stability theorem due to Shanks and show that it is equivalent to some results of Ansell. We also give several examples of stability tests and pose a few unsolved problems.**

Manuscript received November 23, 1971.

The work reported in this paper was performed at Lincoln Laboratory, a center for research operated by the Massachusetts Institute of Technology, Cambridge, Mass., with the support of the Department of the Air Force.

## I. Introduction

In digital image processing, we often want to do linear filtering. The use of recursive instead of nonrecursive filters has the potential of saving computation time. A two-dimensional digital recursive filter is characterized by the two-dimensional $z$ transform

$$H(z_1, z_2) = \frac{\sum_{m=0}^{n} \sum_{n=0}^{q} a_{mn} z_1^m z_2^n}{\sum_{m=0}^{p} \sum_{n=0}^{q} b_{mn} z_1^m z_2^n} \tag{1}$$

where $a_{mn}$ and $b_{mn}$ are constants, and without loss of generality we can set $b_{oo} = 1$. The degrees of the numerator and the denominator polynomials do not have to be equal, since some of the coefficients $a_{mn}$ and $b_{mn}$ can be zero. The variables $z_1$ and $z_2$ are defined as

$$\begin{aligned} z_1 &= e^{-s_1 A} \\ z_2 &= e^{-s_2 B} \end{aligned} \tag{2}$$

where $s_1$ and $s_2$ are, respectively, the horizontal and vertical complex spatial frequency variables and $A$ and $B$ are constants (sampling periods in the horizontal and vertical directions, respectively).

Let $f(m, n)$ and $g(m, n)$ be the input and output, respectively, of the filter. Then the spatial-domain difference equation corresponding to (1) is

Reprinted from *IEEE Trans. Audio Electroacoust.*, vol. AU-20, pp. 158–163, June 1972.

$$\sum_{m=0}^{p} \sum_{n=0}^{q} b_{mn} g(M-m, N-n) = \sum_{m=0}^{p} \sum_{n=0}^{q} a_{mn} f(M-m, N-n). \quad (3)$$

We can express $g(M, N)$ in terms of the rest, and thus obtain a recursive filter that recurses in the $(+m, +n)$ direction. We can also express $g(M-p, N-q)$ in terms of the rest, and obtain a recursive filter that recurses in the $(-m, -n)$ direction. Similarly, we can get from (3) recursive filters recursing in the $(+m, -n)$ and the $(-m, +n)$ directions. (See Fig. 1.) We shall call the recursive filter that recurses in the $(+m, +n)$ direction *causal*. Thus, the $z$ transform of (1) can be associated with four different recursive filters. However, unless otherwise specified, we shall associate it with the causal one.

To start recursing, we need initial conditions. For the causal filter, we need to know the values of the output $g$ in the shaded region in Fig. 2.

There are two major problems in the design of a recursive filter: approximation and stability. The approximation problem consists of determining the coefficients $a_{mn}$ and $b_{mn}$ in (1) so that $H(e^{-juA}, e^{-jvB})$ approximates a given frequency responses $H_1(ju, jv)$, where $u$ and $v$ are, respectively, the horizontal and vertical spatial frequencies. In order for the resulting filter $H(z_1, z_2)$ to be useful, we have to require that it be stable, i.e., if we expand $H(z_1, z_2)$ into a power series

$$H(z_1, z_2) = \sum_{m=0}^{\infty} \sum_{n=0}^{\infty} h_{mn} z_1^m z_2^m \quad (4)$$

the coefficients $h_{mn}$, which represent the impulse response of the filter, should satisfy

$$\sum_{m=0}^{\infty} \sum_{n=0}^{\infty} |h_{mn}| < \infty. \quad (5)$$

If the filter is unstable, any noise (including the roundoff errors in computation) will propagate through the output and possibly be amplified. Furthermore, in many important applications, such as inverse filtering for image restoration, the initial conditions are usually not known. We, therefore, desire that the result of the recursive filtering be roughly independent of the initial conditions. This will be true if the impulse response $|h_{mn}| \approx 0$ for $m, n > A$, where $A$ is a constant that is much smaller than the linear dimensions of the input and output images. Therefore, in some cases, stability itself is not sufficient; we require in addition that the impulse response die out fast enough.

In this paper we shall discuss some aspects of the stability problem of two-dimensional recursive filters. In particular, we shall derive a simplified version of the Shanks stability theorem [1], [2] and show that it is equivalent to the Ansell stability theorem [3], [4].

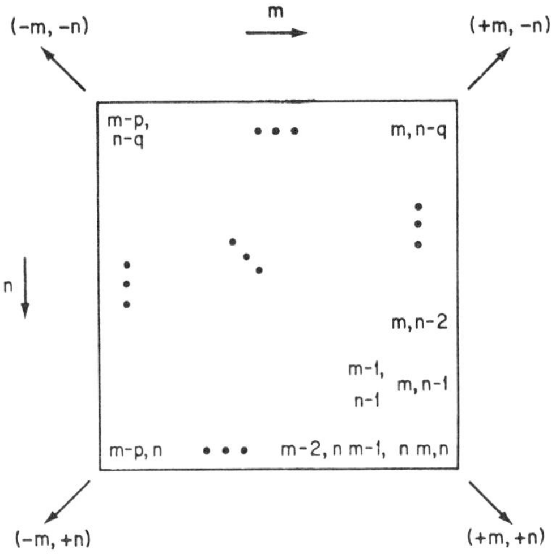


Fig. 1. Four directions of recursion.

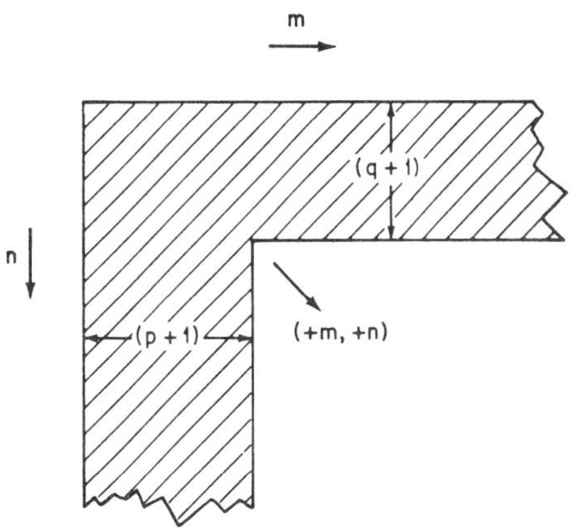


Fig. 2. Initial conditions for a causal filter.

## II. Shanks Stability Theorem

Both the theorems of Shanks and Ansell tell us stability conditions in the frequency domain. In this section we shall state and prove the Shanks stability theorem and discuss its implications.

### A. Statement of the Theorem

*Theorem 1 (Shanks):* A causal recursive filter with the $z$ transform $H(z_1, z_2) = A(z_1, z_2)/B(z_1, z_2)$, where $A$ and $B$ are polynomials in $z_1$ and $z_2$, is stable if and only if there are no values of $z_1$ and $z_2$ such that $B(z_1, z_2) = 0$ and $|z_1| \leq 1$ and $|z_2| \leq 1$.

*Proof:* Let

$$H(z_1, z_2) = \frac{A(z_1, z_2)}{B(z_1, z_2)} = \sum_{m=0}^{\infty} \sum_{n=0}^{\infty} h_{mn} z_1^m z_2^n. \quad (6)$$

What we want to show is that

$$\sum_{m=0}^{\infty} \sum_{n=0}^{\infty} |h_{mn}| < \infty$$

if and only if $H(z_1, z_2)$ is analytic in the region $D = \{(z_1, z_2); |z_1| \leq 1 \cap |z_2| \leq 1\}$.

*The "if" part:* If $H(z_1, z_2)$ is analytic in $D$, we can find $\epsilon > 0$ such that $H(z_1, z_2)$ is analytic in $D_1 = \{(z_1, z_2); |z_1| < 1+\epsilon \cap |z_2| < 1+\epsilon\}$, which implies that

$$\sum_m \sum_n h_{mn} z_1^m z_2^n$$

is absolutely convergent in $D_1$. Therefore

$$\sum_m \sum_n |h_{mn}| < \infty.$$

*The "only if" part:* If

$$\sum_m \sum_n |h_{mn}| < \infty,$$

then by the $M$ test,

$$\sum_m \sum_n h_{mn} z_1^m z_2^n$$

is absolutely convergent in $D$, which implies in turn that $H(z_1, z_2)$ is analytic in $D$. Q.E.D.

Independent proofs of this theorem have been given by Shanks *et al.* [5] and by Farmer [6].

### B. Test Procedure

Testing stability using the Shanks theorem is hard work. One way is to map the unit disk $d_1 \equiv (z_1, |z_1| \leq 1)$ in the $z_1$ plane into the $z_2$ plane by the implicit mapping relation $B(z_1, z_2) = 0$. The filter is stable if and only if the image of $d_1$ in the $z_2$ plane does not overlap the unit disk $d_2 \equiv (z_2, |z_2| \leq 1)$. This is hard work because for each particular value of $z_1 = z_1^0 \epsilon d_1$, we have to solve the equation $B(z_1^0, z_2) = 0$ for $z_2$. And we have to do that for all (in practice, a large number of points) in $d_1$.

### C. An Implication

We emphasize that the Shanks theorem was stated for causal filters, that is, filters that recurse in the $(+m, +n)$ direction. Let us consider the filter associated with $H(z_1, z_2) = A(z_1, z_2)/B(z_1, z_2)$ that recurses in the $(-m, -n)$ direction. We assume that the orders of $A$ in both $z_1$ and $z_2$ are smaller than the corresponding orders of $B$. Then the stability condition becomes: the image of $(z_1, |z_1| \geq 1)$ in the $z_2$ plane, according to $B(z_1, z_2) = 0$, does not overlap $(z_2; |z_2| \geq 1.)$

This can be shown by expanding $H(z_1, z_2)$ in terms of $z_1^{-1}$ and $z_2^{-1}$:

$$H(z_1, z_2) = \frac{A(z_1, z_2)}{B(z_1, z_2)} = \sum_{m=0}^{\infty} \sum_{n=0}^{\infty} h_{mn}{}^1 z_1^{-m} z_2^{-n}. \tag{7}$$

Making the change of variables

$$w_1 = z_1^{-1}$$
$$w_2 = z_2^{-1} \tag{8}$$

we get

$$H(w_1^{-1}, w_2^{-1}) = \frac{A(w_1^{-1}, w_2^{-1})}{B(w_1^{-1}, w_2^{-1})} = \sum_m \sum_n h_{mn}{}^1 w_1^m w_2^n. \tag{9}$$

Now

$$\frac{A(w_1^{-1}, w_2^{-1})}{B(w_1^{-1}, w_2^{-1})} = w_1^a w_2^b \frac{A_1(w_1, w_2)}{B_1(w_1, w_2)} \tag{10}$$

where $a$ and $b$ are nonnegative integers, and $B_1(w_1, w_2) = 0$ if and only if $B(w_1^{-1}, w_2^{-1}) = 0$. Applying the Shanks theorem to

$$\frac{w_1^a w_2^b A_1(w_1, w_2)}{B_1(w_1, w_2)} = \sum_m \sum_n h_{mn}{}^1 w_1^m w_2^n \tag{11}$$

we know that

$$\sum_m \sum_n |h_{mn}{}^1| < \infty$$

if and only if the image of $(w_1, |w_1| \leq 1)$ in the $w_2$ plane, according to $B_1(w_1, w_2) = 0$, does not overlap $(w_2; |w_2| \leq 1)$. But $w_1 = z_1^{-1}$, and $w_2 = z_2^{-1}$. So our assertion has been proved.

Similarly, we can get the stability conditions for the filters associated with $H(z_1, z_2)$ that recurse in the $(+m, -n)$ and the $(-m, +n)$ directions. We can easily convince ourselves that at most one of the four stability conditions, corresponding to the four recursing directions, can be satisfied. Therefore, we have the following.

*Theorem 2:* Among the four recursive filters we can associate with the $z$ transform of (1), or equivalently the difference equation of (3), at most one can be stable.

## III. A Simplified Version of the Shanks Theorem

Shanks' result as stated in Theorem 1 is very tedious to apply. We now show that his stability conditions can be simplified considerably.

### A. The Result

*Theorem 3:* A causal filter with a $z$ transform $H(z_1, z_2) = A(z_1, z_2)/B(z_1, z_2)$, where $A$ and $B$ are polynomials, is stable if and only if: 1) the map of $\partial d_1 \equiv (z_1; |z_1| = 1)$ in the $z_2$ plane, according to $B(z_1, z_2) = 0$, lies outside $d_2 \equiv (z_2; |z_2| \leq 1)$; and 2) no point in $d_1 = (z_1; |z_1| \leq 1)$ maps into the point $z_2 = 0$ by the relation $B(z_1, z_2) = 0$.

*Proof:* We want to establish that the stability conditions of Theorems 1 and 3 are equivalent. It is obvious that the stability conditions of Theorem 1 imply those of Theorem 3. So we proceed to show the implication in the other direction.

The two-variable polynomial $B(z_1, z_2) = 0$ defines an algebraic function [7] $z_2 = f(z_1)$. We first modify the unit-circle contour in the $z_1$ plane to exclude any singular points of $f$ inside the contour, resulting in a modified contour $\partial d_1'$, as shown in Fig. 3. We use $d_1'$ to denote the closed region enclosed by $\partial d_1'$. A point $z_1 = z_1^0$ is called a singular point of $z_2 = f(z_1)$, if $B(z_1^0, z_2) = 0$, considered as an equation in $z_2$, has multiple (finite or infinite) roots.

According to the theory of algebraic functions, in $d_1'$ the function $z_2 = f(z_1)$ has a number of branches, each of which is holomorphic. Therefore, from the maximum-modulus theorem, the maximum of $|f(z_1)|$ over $d_1'$ occurs on $\partial d_1'$, and the minimum of $|f(z_1)|$ over $d_1'$ can occur in the interior only if the minimum is zero. However, condition 2) of Theorem

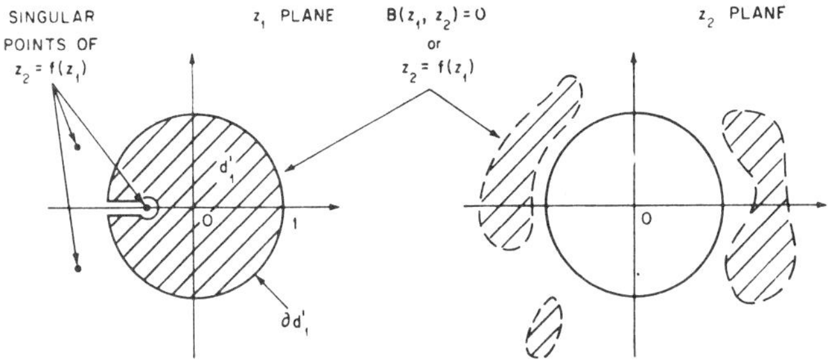


Fig. 3. Mapping $z_2 = f(z_1)$.

3 says $f(z_1)$ is never zero in $d_1'$. Therefore, the minimum of $f(z_1)$ occurs on $\partial d_1'$, i.e.,

$$|f(d_1')| \geq \min |f(\partial d_1')| \tag{12}$$

which implies that if $|f(\partial d_1')| > 1$, then $|f(d_1')| > 1$, i.e., to ensure that $f(d_1')$ lies outside the unit disk $d_2 = (z_2, |z_2| \leq 1)$, it is sufficient to ensure that $f(\partial d_1')$ lies outside $d_2$.

We are almost there, but not quite. What we really want to show is that if $|f(\partial d_1)| > 1$, then $|f(d_1)| > 1$, where $d_1 \equiv (z_1, |z_1| \leq 1)$ and $\partial d_1 \equiv (z_1; |z_1| = 1)$. Since the detour in $\partial d_1'$ can be any path leading from $\partial d_1$ to the singular point, what is left to show is simply that $|f(s)| > 1$, where $s$ is the singular point. But since each branch of $z_2 = f(z_1)$ is continuous at $z_1 = s$, and since $|f(s + \epsilon e^{j\theta})| > 1$ for arbitrarily small $\epsilon$ and any $\theta$, we have $|f(s)| > 1$. Q.E.D.

### B. Test Procedure

To test stability using Theorem 3, we have to map $\partial d_1 = (z_1; |z_1| = 1)$ into the $z_2$ plane according to $B(z_1, z_2) = 0$, and see whether the image lies outside $d_2 \equiv (z_2; |z_2| \leq 1)$. Also, we have to solve $B(z_1, 0) = 0$ to see whether there is any root with magnitude smaller than 1.

## IV. Ansell Stability Theorem

To test the stability of a two-dimensional recursive filter using Theorem 3 is much simpler than using Shanks' original Theorem 1. However, the procedure is still infinite in the sense that in principle we have to map every point on the continuum $\partial d$ into the $z_2$ plane. In this section we show that Theorem 3 can be reduced to a result due to Ansell. We shall see that Ansell's result enables us to test stability in a finite number of steps, which, unfortunately, can still be very tedious.

Make the change of variables

$$p_1 = \frac{1 - z_1}{1 + z_1} \tag{13a}$$

and

$$p_2 = \frac{1 - z_2}{1 + z_2} \tag{13b}$$

and let

$$H(z_1, z_2) = \frac{E(p_1, p_2)}{F(p_1, p_2)} \tag{14}$$

where $E$ and $F$ are polynomials in $p_1$ and $p_2$. We can restate Theorem 3 as follows.

*Theorem 4 (Ansell):* The causal recursive filter $H(z_1, z_2)$ is stable if and only if: 1) for all real finite $w$, the complex polynomial in $p_2$, $F(jw, p_2)$ has no zeros in Re $p_2 \geq 0$; and 2) the real polynomial in $p_1$, $F(p_1, 1)$ has no zeros in Re $p_1 \geq 0$.

This theorem is essentially the same as [3, Lemma 6]. The nice part about this theorem is that condition 1) can be tested using standard techniques of circuit theory [8].

*Theorem 5 (Ansell):* Condition 1) of Theorem 4 is equivalent to the following. Let us express $F(jw, j\Omega)$, where $w$ and $\Omega$ are real, as

$$F(jw, j\Omega) = b_0(w)\Omega^n + b_1(w)\Omega^{n-1} + \cdots + b_n(w) + j[a_0(w)\Omega^n + a_1(w)\Omega^{n-1} + \cdots + a_n(w)] \tag{15}$$

where $a_i(w)$ and $b_i(w)$ are real polynomials in $w$ and neither $a_0(w)$ nor $b_0(w)$ is identically zero. Let $C_{r,s}(w)$ be defined as

$$C_{r,s} = a_r b_s - a_s b_r \tag{16}$$

for $0 \leq r, s \leq n$ [using zero for $a$'s and $b$'s not present in $F(jw, j\Omega)$]. Let $D(w)$ denote the $n \times n$ symmetrical polynomial matrix whose typical element $D_{ij}(w)(1 \leq i, j \leq n)$ is the sum of all those $C_{r,s}(w)(0 \leq r, s \leq n)$ for which both

$$s + r = i + j - 1 \tag{17a}$$

and

$$s - r > |i - j| \tag{17b}$$

are satisfied. Then, the $n$ successive principal minors of $D(w)$ must be positive for all real $w$.

This is part of Theorem 2 of Ansell [3], [9]. We note that Sturm's method [8] can be used to test whether each minor of $D(w)$ is positive for all real $w$.

## V. Examples

We now give some examples illustrating the use of Theorems 3, 4, and 5. Ideally, we would like to determine what relations the coefficients $a_{ij}$ and $b_{ij}$ in (1) must satisfy to ensure the stability of a causal recursive filter $H(z_1, z_2)$. Unfortunately, this does not seem possible except for very simple filters.

### A. First-Order Filters

Let us consider the first-order filter

$$H(z_1, z_2) = \frac{1}{1 + az_1 + bz_2} \tag{18}$$

where $a$ and $b$ are constants. In this simple case, we can easily determine the stability conditions in terms of $a$ and $b$ using Theorem 1.

Let

$$B(z_1, z_2) = 1 + az_1 + bz_2 = 0.$$

We have

$$z_2 = -\frac{1}{b} - \frac{a}{b} z_1. \tag{19}$$

Therefore, $B=0$ maps the unit disk in the $z_1$ plane into a disk in the $z_2$ plane with center at $z_2=-1/b$ and radius $|a/b|$. This image will not overlap the unit disk in the $z_2$ plane if and only if

$$\left|\frac{1}{b}\right| - \left|\frac{a}{b}\right| > 1 \tag{20}$$

or equivalently

$$|a| + |b| < 1. \tag{21}$$

We now try to establish the same result by using Theorems 4 and 5. Making the change of variables according to (13), we get

$$H(z_1, z_2) = \frac{E(p_1, p_2)}{F(p_1, p_2)} = \frac{1 + p_1 + p_2 + p_1p_2}{(1 + a + b) + (1 - a + b)p_1 + (1 + a - b)p_2 + (1 - a - b)p_1p_2}. \tag{22}$$

Condition 2) of Theorem 4 dictates that $F(p_1, 1)=2(1+a)+2(1-a)p_1=0$ has no root in Re $p_1 \geq 0$. Therefore

$$|a| < 1. \tag{23}$$

Now we tackle condition 1) of Theorem 4 with the help of Theorem 5. We have

$$F(jw, j\Omega) = (1 + a + b) - (1 - a - b)w\Omega + j[(1 - a + b)w + (1 + a - b)\Omega]$$

and we want

$$C_{0,1} = (1+a-b)(1+a+b) + (1-a+b)(1-a-b)w^2 > 0, \qquad \text{for all real } w. \tag{24}$$

Let $A_1=1+a-b$, $A_2=1+a+b$, $A_3=1-a+b$, and $A_4=1-a-b$. Then inequality (24) is valid if either of the following four conditions is satisfied: 1) $A_1>0$, $A_2>0$, $A_3>0$, $A_4>0$; 2) $A_1>0$, $A_2>0$, $A_3<0$, $A_4<0$; 3) $A_1<0$, $A_2<0$, $A_3>0$, $A_4>0$; 4) $A_1<0$, $A_2<0$, $A_3<0$, $A_4<0$. It is easy to show that 1) is equivalent to inequality (21), while 2)–4) are incompatible with inequality (23).

### B. A Special Class of Second-Order Filters

We next consider the filter

$$H(z_1, z_2) = \frac{1}{1 + az_1 + bz_2 + cz_1z_2} \tag{25}$$

which we shall call a bilinear filter. We first establish stability conditions in terms of the coefficients $a$, $b$, and $c$, using Theorem 3.

Let

$$B(z_1, z_2) = 1 + az_1 + bz_2 + cz_1z_2 = 0.$$

We have

$$z_2 = -\frac{1 + az_1}{b + cz_1} \tag{26}$$

which is a bilinear transformation mapping circles into circles. The image of the unit circle $\partial d_1 \equiv (z_1;\ |z_1| = 1)$ in the $z_2$ plane is then a circle. From (26), the center of this image circle is on the real axis, and it intersects the real axis at

$$z_2 = -\frac{1 - a}{b - c} \quad \text{and} \quad z_2 = -\frac{1 + a}{b + c}.$$

It is easy to see that condition 1) of Theorem 3 is satisfied if

$$\left|\frac{1 - a}{b - c}\right| > 1 \tag{27a}$$

and

$$\left|\frac{1 + a}{b + c}\right| > 1. \tag{27b}$$

Finally, condition 2) of Theorem 3 is equivalent to

$$|a| < 1. \tag{27c}$$

Conditions (27) agree with those of Shanks [1].

We now try to establish the stability conditions using Theorems 4 and 5. Making the change of variables according to (13), we get

$$H(z_1, z_2) = \frac{E(p_1, p_2)}{F(p_1, p_2)} = \frac{1 + p_1 + p_2 + p_1p_2}{B_1 + B_2p_1 + B_3p_2 + B_4p_1p_2} \tag{28}$$

where

$$\begin{aligned} B_1 &= 1 + a + b + c \\ B_2 &= 1 - a + b - c \\ B_3 &= 1 + a - b - c \\ B_4 &= 1 - a - b + c. \end{aligned} \tag{29}$$

For condition 2) of Theorem 4, we form

$$F(p_1, 1) = 2(1 + a) + 2(1 - a)p_1 = 0$$

where Re $p_1 > 0$ means

$$|a| < 1. \tag{30}$$

Now we tackle condition 1) of Theorem 4. We have

$$F(jw, j\Omega) = B_1 - B_4w\Omega + j[(B_2w + B_3\Omega)].$$

According to Theorem 5, we want

$$C_{0,1} = B_3B_1 + B_2B_4w^2 > 0, \qquad \text{for all real } w \tag{31}$$

which is valid if either of the following four conditions is satisfied: 1) $B_1>0$, $B_3>0$, $B_2>0$, $B_4>0$; 2) $B_1>0$, $B_3>0$, $B_2<0$, $B_4<0$; 3) $B_1<0$, $B_3<0$, $B_2>0$, $B_4>0$; 4) $B_1<0$, $B_3<0$, $B_2<0$, $B_4<0$.

It is easy to show that 1) is equivalent to

$$|a+b| - 1 < c < 1 - |a-b| \tag{32}$$

and 2)–4) are ruled out by inequality (30). It can be verified readily that (32) is equivalent to (27).

### C. A Numerical Example

Unfortunately, filters described by (18) and (25) are the only types for which we were able to derive stability conditions in terms of the coefficients in the $z$ transforms. For more complicated filters, the application of Theorem 3 becomes unwieldy. However, for any filter with numerical coefficients, we can always apply Theorems 4 and 5 and determine whether or not it is stable in a finite number of steps. We illustrate by an example.

Consider the filter

$$H(z_1, z_2) = \frac{1}{1 + \frac{1}{2}z_1 + \frac{1}{2}z_2 + \frac{1}{4}z_1z_2 + \frac{1}{4}z_1^2 + \frac{1}{4}z_2^2}. \tag{33}$$

Making the change of variables according to (13), and proceeding as before, we find that

$$F(p_1, 1) = 3p_1^2 + 6p_1 + 7$$

whose zeros have negative real parts. Therefore, condition 2) of Theorem 4 is satisfied.

For condition 1) of Theorem 4, we form

$$C_{0,1}(w) = \frac{3}{4}w^4 - \frac{3}{2}w^2 - \frac{15}{4}$$

which certainly can be negative for some real values of $w$. Therefore, condition 1) is violated. The filter of (33) is unstable.

## VI. Concluding Remarks

We have discussed in this paper how to test the stability of two-dimensional recursive filters in the frequency domain. In conclusion, we would like to point out that what we have discussed is only part of the stability problem. Furthermore, it is not the most important part. The more important questions we want to ask are: How do we design filters that are guaranteed to be stable? And, if a given filter is unstable, how do we stabilize it without changing its frequency response?

One approach to attacking the first question is to study what types of frequency responses we can get from classes of simple filters whose stability we know how to control.

With respect to the second question, we might mention a conjecture of Shanks [2]. Let $H(z_1, z_2) = 1/B(z_1, z_2)$ be an unstable filter. We first determine a least mean-square inverse of $B(z_1, z_2)$, which we call

$$C(z_1, z_2) = \sum_{m=0}^{a} \sum_{n=0}^{b} c_{mn} z_1^m z_2^n \tag{34}$$

where $a$ and $b$ are positive integers.

Let

$$B(z_1, z_2)C(z_1, z_2) = 1 + d_{10}z_1 + d_{01}z_2 + d_{11}z_1z_2 \cdots.$$

Then among all polynomials of order $(a, b)$, $C$ is the least mean-square inverse of $B$ if the coefficients of $C$ are such that $(d_{10}^2 + d_{01}^2 + d_{11}^2 + \cdots)$ is minimized. We next determine a least mean-square inverse of $C(z_1, z_2)$, which we call $\hat{B}(z_1, z_2)$. Then Shanks' conjecture states that the filter $\hat{H}(z_1, z_2) = 1/\hat{B}(z_1, z_2)$ is stable and that the magnitude of the frequency response of $\hat{H}$ is approximately equal to that of $H$. The approximation becomes better when we use larger values for $a$ and $b$ in (34). It was proven by Robinson [10], [11] that a one-dimensional version of Shanks' procedure does yield stable (one-dimensional) filters. However, whether the procedure yields stable filters in two dimensions is still an open question. Also, no analysis is available on how close the magnitude of the frequency response of $\hat{H}$ is to that of $H$.

We have yet another open question with respect to question 2. In the one-dimensional case, if we are doing non-real-time filtering, then we can always decompose an unstable filter with no poles on the unit circle into two stable ones recursing in opposite directions. We associate the poles of the original filter outside the unit circle with a filter recursing in the positive direction, and the poles inside the unit circle with a filter recursing in the negative direction. Is there an analogous procedure in two dimension?

### References


[1] J. L. Shanks, "Two-dimensional recursive filters," in *1969 SWIEECO Rec.*, pp. 19E1–19E8.

[2] ——, "The design of stable two-dimensional recursive filters," in *Proc. UMR—M. J. Kelley Commun. Conf.*, Univ. Missouri, Rolla, Oct. 1970.

[3] H. G. Ansell, "On certain two-variable generalizations of circuit theory, with applications to networks of transmission lines and lumped reactances," *IEEE Trans. Circuit Theory*, vol. CT-11, pp. 214–223, June 1964.

[4] L. Weinberg, "Approximation and stability-test procedures suitable for digital filters," presented at the Arden House Workshop on Digital Filtering, Harriman, N. Y., Jan. 1970.

[5] J. L. Shanks, S. Treitel, and J. H. Justice, "Stability and synthesis of two-dimensional recursive filters," this issue, pp. 115–128.

[6] C. Farmer, Tulsa Univ., Tulsa, Okla., private communication, May 1971.

[7] G. A. Bliss, *Algebraic Functions.* New York: Amer. Math. Soc., 1933.

[8] E. A. Guillemin, *Synthesis of Passive Networks.* New York: Wiley, 1957.

[9] F. R. Gantmacher, *Applications of the Theory of Matrices.* New York: Interscience, 1959, pp. 295–296.

[10] E. A. Robinson, *Statistical Communication and Detection.* New York: Hafner, 1967.

[11] ——, "Structural properties of stationary stochastic processes with applications," ch. 12 in *Time Series Analysis*, M. Rosenblatt, Ed. New York: Wiley, 1963.

# The Design of Multistage Separable Planar Filters

SVEN TREITEL AND JOHN L. SHANKS, MEMBER, IEEE

*Abstract*—A two-dimensional, or planar, digital filter can be described in terms of its planar response function, which is in the form of a matrix of weighting coefficients, or filter array. In many instances the dimensions of these matrices are so large that their implementation as ordinary planar convolutional filters becomes computationally inefficient. It is possible to expand the given coefficient matrix into a finite and convergent sum of matrix-valued stages. Each stage can be separated with no error into the product of an $m$-length column vector multiplied into an $n$-length row vector, where $m$ is the number of rows and $n$ is the number of columns of the original filter array. Substantial savings in computer storage and speed result if the given filter array can be represented with a tolerably small error by the first few stages of the expansion. Since each constituent stage consists of two vector-valued factors, further computational economies accrue if the one-dimensional sequences described by these vectors are in turn approximated by one-dimensional recursive filters. Two geophysical examples have been selected to illustrate how the present design techniques may be reduced to practice.

## I. Introduction

A SUBSTANTIAL amount of geophysical data is recorded in the form of two-dimensional, or planar, arrays of discrete observations. A seismogram consists of a multiplicity of one-dimensional time sequences, or traces, each of which is registered at a different point in space. If the receivers are spaced along a straight line at uniform distances, and if the individual traces are sampled at a uniform temporal sampling increment, the resulting discrete seismogram is conveniently visualized as a planar array of observations, or weighting coefficients. Filters that are designed to operate explicitly on such planar input data are characterized as planar, or two-dimensional, in contradistinction to filters that act only on one-dimensional sequences, such as a single seismic trace. The velocity, or fan filter, is a familiar illustration of a planar digital operator that has found widespread use in the reduction of seismic data. Other geophysical examples also abound. For instance, potential field surveys are often recorded on an observation plane, rather than merely along a profile. If this is not done at uniform sampling increments, the data are reducible to this form by the application of two-dimensional interpolation techniques. More recently, the field implementation of the acoustical holography method has required the recording of ground motion over a plane at very fine sampling rates [5].

Manuscript received July 16, 1970; revised August 30, 1970. This paper was presented at the IEEE Workshop on Digital Filtering, Harriman, N. Y., January 13, 1970.

The authors are with the Research Center, Pan American Petroleum Corporation, Tulsa, Okla. 74102.

Planar digital filters can be obtained to satisfy a wide variety of optimization criteria [17]. The response functions of such operators must often be described by large arrays of weighting coefficients, so that computing times grow rapidly and computer-storage allocation problems become increasingly complex. These difficulties have stimulated the search for methods that reduce the total computational labor necessary for the implementation of a given planar digital filter. In the present treatment we assume that the two-dimensional response function of the filter is already known in the form of a uniformly sampled array, or matrix, of weighting coefficients. The problem of reducing the computational labor associated with the application of planar filters to planar input data can be attacked in several ways. One method that shows promise involves the approximation of the given planar response function by a two-dimensional recursive filter [14]. The current approach, on the other hand, is based on the expansion of the given coefficient matrix into a finite and converging sum of matrices having the same dimensions as the original filter array. Each constituent matrix of this sum is separable; that is, each matrix in the expansion can be expressed as the product of a column vector multiplied into a row vector. In many cases the given matrix can be represented with small error by only the first few terms of the expansion, and thus substantial gains in computing efficiency arise. Since each term, or filter stage, is expressible as a product of vectors, which are one-dimensional arrays, further speed and storage advantages result if each vector is itself approximated by an appropriate one-dimensional recursive filter.

The matrix expansion method presented here is based on the so-called "spectral representation" of an arbitrary rectangular matrix in terms of the eigenvalues and eigenvectors of two particular matrices associated with the given coefficient array. The salient features of this expansion technique are described in the main text, while several lengthy proofs are relegated to three appendices. A somewhat related approach has been given by Sanathanan [12], but the expansion he proposes does not in general converge in a finite number of matrix-valued stages. In addition, his method requires *a priori*, knowledge of one of the two vectors that comprise each stage of the expansion, and such information is not generally available in most applications. The present approach, on the other hand, determines these vectors uniquely in terms of the given planar-filter coefficient array, and no preliminary knowledge of their particular form is necessary.

Reprinted from *IEEE Trans. Geosci. Electron.*, vol. GE-9, pp. 10–27, Jan. 1971.

Throughout this treatment we make extensive use of eigenvalue theory. Readers unfamiliar with this subject are referred to standard reference books such as Frazer *et al.* [7] and Hildebrand [8, ch. 1].

## II. Expansion of a Planar Filter into a Sum of Individually Separable Planar Stages

### A. Theoretical Background

Consider an arbitrary real and nonnull planar response function $\boldsymbol{D}$[1] which is described by the $(m \times n)$ rectangular matrix $\boldsymbol{D}$

$$\boldsymbol{D} = \begin{bmatrix} d_{11} & d_{12} & \cdots & d_{1n} \\ d_{21} & & & \cdot \\ \cdot & & & \cdot \\ \cdot & & & \cdot \\ \cdot & & & \\ d_{m1} & d_{m2} & \cdots & d_{mn} \end{bmatrix}. \tag{1}$$

We say that $\boldsymbol{D}$ is separable when it can be split into the product of an $m$-length column vector $\mathbf{c}$ and an $n$-length row vector $\boldsymbol{r}$

$$\boldsymbol{D} = \mathbf{c}\boldsymbol{r} = \begin{bmatrix} c_1 \\ c_2 \\ \cdot \\ \cdot \\ \cdot \\ c_m \end{bmatrix} [r_1 r_2 \cdots r_n] \tag{2}$$

or

$$\boldsymbol{D} = \mathbf{c}\boldsymbol{r} = \begin{bmatrix} c_1r_1 & c_1r_2 & \cdots & c_1r_n \\ c_2r_1 & c_2r_2 & \cdots & c_2r_n \\ \cdot & & & \\ \cdot & & & \\ c_mr_1 & c_mr_2 & \cdots & c_mr_n \end{bmatrix}. \tag{3}$$

From the last of the preceding relations it is evident that a separable matrix has proportional rows and proportional columns; that is, all rows of $\boldsymbol{D}$ are of the form $c_1\boldsymbol{r}, c_2\boldsymbol{r}, \cdots, c_m\boldsymbol{r}$, and all columns of $\boldsymbol{D}$ are of the form $r_1\mathbf{c}, r_2\mathbf{c}, \cdots, r_n\mathbf{c}$.

The array $\boldsymbol{D}$ of (1) is described by $mn$ elements, or weighting coefficients, but if $\boldsymbol{D}$ is a separable matrix, we see from (2) that it is equally well describable by the $(m+n)$ elements of its constituent vectors $\mathbf{c}$ and $\boldsymbol{r}$. As $m$ and $n$ become large, one sees that substantial savings in computer time and computer storage will be achieved if the planar-filter $\boldsymbol{D}$ is so separable.

Unfortunately, we cannot generally expect that an arbitrary planar-filter matrix satisfies such stringent conditions. Suppose, however, that we require only that $\boldsymbol{D}$ be approximately separable; that is, we seek two vectors $\mathbf{c}$ and $\boldsymbol{r}$ such that

$$\boldsymbol{D} = \mathbf{c}\boldsymbol{r} + \boldsymbol{E} \tag{4}$$

where $\boldsymbol{E}$ is an $(m \times n)$ error matrix, whose elements we would like to be as small as possible in relation to the elements of $\boldsymbol{D}$. If this is not the case, we might then attempt to generalize the separation (4) in the form

$$\begin{aligned} \boldsymbol{D} &= \mathbf{c}_1\boldsymbol{r}_1 + \mathbf{c}_2\boldsymbol{r}_2 + \cdots + \mathbf{c}_i\boldsymbol{r}_i + \cdots + \mathbf{c}_k\boldsymbol{r}_k + \boldsymbol{E}_k \\ &= \boldsymbol{\Gamma}_1 + \boldsymbol{\Gamma}_2 + \cdots + \boldsymbol{\Gamma}_i + \cdots + \boldsymbol{\Gamma}_k + \boldsymbol{E}_k \end{aligned} \tag{5}$$

where each term,

$$\boldsymbol{\Gamma}_i = \mathbf{c}_i\boldsymbol{r}_i, \qquad i = 1, 2, \cdots, k \tag{6}$$

is thus an exactly separable $(m \times n)$ matrix, and where $\boldsymbol{E}_k$ is the $(m \times n)$ error matrix associated with a $k$-term expansion of $\boldsymbol{D}$ in the form (5). In keeping with nomenclature from filter theory, we call each term $\boldsymbol{\Gamma}_i$ a stage, and we say that (5) represents the "multistage separable" expansion of $\boldsymbol{D}$ into a sum of individually separable stages $\boldsymbol{\Gamma}_i$ plus a not necessarily separable error matrix $\boldsymbol{E}_k$. We demonstrate later that each stage is implementable as a member of a parallel configuration of individually separable planar filters.

Assuming that $\boldsymbol{D}$ is well approximated by $k$ stages, we require $k(m+n)$ elements, or weighting coefficients, to describe the filter in this manner. Substantial savings in computer time and storage will result as long as $k(m+n) \ll mn$. Further advantages of this expansion will be discussed in later sections of the present treatment.

Up to this point we have merely postulated that $\boldsymbol{D}$ is representable as a sum of individually separable stages. We demonstrate in Appendix I that such an expansion can always be obtained and that in fact an exact representation of $\boldsymbol{D}$ can be guaranteed with a finite number of stages. Here we shall only state the salient features of the derivation—for detailed proofs and references the reader is referred to Appendix I.

Without loss of generality we may assume that $m$, the number of rows of $\boldsymbol{D}$, is equal to or greater than $n$, the number of columns of $\boldsymbol{D}$. We also assume first that the $n$ columns of $\boldsymbol{D}$ are linearly independent.[2]

Then a solution to our problem is given by the spectral representation of $\boldsymbol{D}$ [see (54)],

$$\begin{aligned} \boldsymbol{D} &= \sqrt{\lambda_1}\boldsymbol{\kappa}_1\boldsymbol{\varrho}_1 + \sqrt{\lambda_2}\boldsymbol{\kappa}_2\boldsymbol{\varrho}_2 + \cdots + \sqrt{\lambda_n}\boldsymbol{\kappa}_n\boldsymbol{\varrho}_n \\ &= \sum_{i=1}^{n} \sqrt{\lambda_i}\boldsymbol{\kappa}_i\boldsymbol{\varrho}_i \end{aligned} \tag{7}$$

where

$\lambda_1, \lambda_2, \cdots, \lambda_n$ real and positive eigenvalues of the $n$-order, symmetric and positive definite matrix $\boldsymbol{S}=\boldsymbol{D}^T\boldsymbol{D}$ (the superscript $T$ denotes the matrix transpose);

$\boldsymbol{\varrho}_1, \boldsymbol{\varrho}_2, \cdots, \boldsymbol{\varrho}_n$ normalized row eigenvectors, or eigenrows, associated with the corresponding eigenvalues $\lambda_1, \lambda_2, \cdots, \lambda_n$ of $\boldsymbol{S}$,

[1] A symbol in boldface type indicates that a matrix or a vector is implied.

[2] This restriction is relaxed at the end of the present section.

$\boldsymbol{\kappa}_1, \boldsymbol{\kappa}_2, \cdots, \boldsymbol{\kappa}_n$ normalized column eigenvectors, or eigencolumns associated with the corresponding nonvanishing eigenvalues $\lambda_1, \lambda_2, \cdots, \lambda_n$ of the $m$-order, symmetric, and positive semidefinite matrix $\boldsymbol{Q} = \boldsymbol{D}\boldsymbol{D}^T$.

Each term of the spectral representation (7) is the separable $(m \times n)$ matrix resulting from the multiplication of the $m$-length column vector $\sqrt{\lambda_i}\boldsymbol{\kappa}_i$ into the $n$-length row vector $\boldsymbol{\varrho}_i$. Furthermore, this representation is *exact*, provided that all $n$ terms associated with the $n$ eigenvalues $\lambda_1, \lambda_2, \cdots, \lambda_n$ are retained in the expansion. The matrices $\boldsymbol{Q} = \boldsymbol{D}\boldsymbol{D}^T$ and $\boldsymbol{S} = \boldsymbol{D}^T\boldsymbol{D}$ have an *identical* set of nonzero eigenvalues, namely, $\lambda_1, \lambda_2, \cdots, \lambda_n$.

We accordingly *identify* the spectral representation (7) with our desired multistage expansion of $\boldsymbol{D}$, and hence,

$$\boldsymbol{D} = \sum_{i=1}^{n} \sqrt{\lambda_i}\boldsymbol{\kappa}_i\boldsymbol{\varrho}_i = \sum_{i=1}^{n} \boldsymbol{\Gamma}_i \tag{8}$$

where the $i$th stage $\boldsymbol{\Gamma}_i$ is given by

$$\boldsymbol{\Gamma}_i = \sqrt{\lambda_i}\boldsymbol{\kappa}_i\boldsymbol{\varrho}_i. \tag{9}$$

Comparison of (6) and (7) yields the explicit forms of the desired column vectors $\boldsymbol{c}_i$ and of the desired row vectors $\boldsymbol{r}_i$

$$\boldsymbol{c}_i = \sqrt{\lambda_i}\boldsymbol{\kappa}_i$$

$$\boldsymbol{r}_i = \boldsymbol{\varrho}_i$$

so that the $\boldsymbol{c}_i$'s are scale-factored versions of the normalized eigen columns $\boldsymbol{\kappa}_i$, while the $\boldsymbol{r}_i$'s are identically equal to the normalized eigenrows $\boldsymbol{\varrho}_i$.

The identification of the multistage expansion of $\boldsymbol{D}$ with the spectral representation of $\boldsymbol{D}$ does not constitute the only possible solution of our problem. We show in Appendix II that a least squares approach also leads to an expansion in terms of separable stages, but in general only an infinite number of stages computed in this manner will represent $\boldsymbol{D}$ with no error. It is further demonstrated in Appendix II that the spectral representation of $\boldsymbol{D}$ can be obtained as a special case of the least squares formulation.

The exact representation of $\boldsymbol{D}$ requires $(m+n)$ coefficients per stage, or a total of $n(m+n) = nm + n^2$ coefficients for the $n$ separable-stages $\boldsymbol{\Gamma}_i$. Since the original filter matrix $\boldsymbol{D}$ is described by only $mn$ coefficients, one obviously does not obtain any direct computing speed advantages in this way.[3] On the other hand, an approximate multistage representation of $\boldsymbol{D}$ is always possible by truncating the exact expansion (8) to only $k$ terms. Computing speed advantages will then still result as long as $k(m+n) < mn$.

In general the $n$ real and positive eigenvalues $\lambda_i$ of the symmetric positive definite $n$-order matrix

$$\boldsymbol{S} = \boldsymbol{D}^T\boldsymbol{D} \tag{10}$$

will have distinct magnitudes, so that some will be larger than others. One may therefore suspect that the $n$ terms of (8) do not contribute equally to the entire expansion. If this should turn out to be the case, one will naturally wish to retain only the $k$ *dominant* terms.

Accordingly, we suppose that the eigenvalues $\lambda_i$ are listed in order of decreasing magnitude,

$$\lambda_1, \lambda_2, \cdots, \lambda_{n-1}, \lambda_n \tag{11}$$

where $\lambda_1$ is the largest and $\lambda_n$ the smallest eigenvalue. We suppose further that any repeated eigenvalues are labeled separately.

Let us note that the expansion (8) can be written in any one of the $n$ forms,

$$\boldsymbol{D} = \sum_{i=1}^{k} \boldsymbol{\Gamma}_i + \boldsymbol{E}_k, \qquad k = 1, 2, \cdots, n \tag{12}$$

where the $(m \times n)$ array $\boldsymbol{E}_k$ is the truncation error matrix arising from the retention of the first $k$ terms in the expansion (8). Since the expansion is exact for $k = n$, we have

$$\boldsymbol{E}_n = \boldsymbol{0}$$

where $\boldsymbol{0}$ is the $(m \times n)$ null matrix.

A convenient scalar-valued measure of the truncation error is given by the normalized error energy $\epsilon_k$, which we define to be

$$\epsilon_k = \frac{J_k}{J_0} \cdot \tag{13}$$

Here $J_k$ is the sum of the squares of the elements of the truncation error matrix $\boldsymbol{E}_k$, while $J_0$ is the sum of the squares of the elements of the given matrix $\boldsymbol{D}$.

It is shown in Appendix I (66) that $\epsilon_k$ is uniquely determined by the eigenvalues $\lambda_i$ of $\boldsymbol{S}$,

$$\epsilon_k = 1 - \frac{\lambda_1 + \lambda_2 + \cdots + \lambda_k}{\lambda_1 + \lambda_2 + \cdots + \lambda_n}, \qquad k = 1, 2, \cdots, n. \tag{14}$$

The normalized error energy $\epsilon_k$ decreases as $k$ increases, for the term

$$\frac{\lambda_1 + \lambda_2 + \cdots + \lambda_k}{\lambda_1 + \lambda_2 + \cdots + \lambda_n}$$

is less than unity for $k < n$ and becomes unity when $k = n$. The bounds on $\epsilon_k$ are then

$$0 \leq \epsilon_k < 1.$$

Let us see how rapidly the normalized error energy approaches zero as the number of retained stages $k$ in-

[3] Other advantages do, however, arise from the exact multistage representation of $\boldsymbol{D}$. These are described in Section IV.

crease. From (14) we obtain an expression for the difference between successive values of $\epsilon_k$,

$$\epsilon_k - \epsilon_{k+1} = \frac{\lambda_{k+1}}{\lambda_1 + \lambda_2 + \cdots + \lambda_n}, \qquad k = 1, 2, \cdots, n-1. \tag{15}$$

Suppose for the moment that all eigenvalues $\lambda_i$ are distinct. Then we have that $\lambda_1 > \lambda_2 > \cdots > \lambda_n$, and $\epsilon_k$ diminishes rapidly for the smaller values of $k$. As $k$ increases further, the difference $\epsilon_k - \epsilon_{k+1}$ becomes smaller because $\lambda_{k+1}$ becomes progressively smaller. In other words, the dominant terms of the expansion (8) are associated with the larger, or dominant, eigenvalues of the matrix $\boldsymbol{S} = \boldsymbol{D}^T\boldsymbol{D}$. The ordering (11) then guarantees that the leading terms of (8) are also the dominant ones. Knowledge of the eigenvalues of $\boldsymbol{S}$ is sufficient to establish the number of stages $k$ required to reduce $\epsilon_k$ equal to or below some desired value in the range $0 \leq \epsilon_k < 1$.

If all eigenvalues $\lambda_i$ are equal, we deduce from (15) that no term of the expansion (8) is dominant; if some of the $\lambda_i$ are equal, corresponding terms of (8) contribute equally to the expansion.

Thus far it has been assumed that the $n$ columns of $\boldsymbol{D}$ are linearly independent; that is, no set of constants $a_1, a_2, \cdots, a_n$, at least one of which is not zero, exists such that

$$a_1\boldsymbol{d}_1 + a_2\boldsymbol{d}_2 + \cdots + a_i\boldsymbol{d}_i + \cdots + a_n\boldsymbol{d}_n = 0$$

where $\boldsymbol{d}_i$ $(i = 1, 2, \cdots, n)$ is the $i$th $m$-length column of $\boldsymbol{D}$ [8, p. 25]. If this is not the case, and say a certain number $p(<n)$ of the columns $\boldsymbol{d}_i$ are linearly dependent, the spectral representation for $\boldsymbol{D}$ given by (7) terminates with no error for $n-p$ stages. Then the expansion (12) becomes

$$\boldsymbol{D} = \sum_{i=1}^{k} \boldsymbol{\Gamma}_i + \boldsymbol{E}_k, \qquad k = 1, 2, \cdots, n-p \tag{16}$$

where

$$\boldsymbol{E}_{n-p} = \boldsymbol{0}.$$

The proof for this result is given in the discussion leading to (67) of Appendix I. In particular, if $p = n-1$ so that only one column of $\boldsymbol{D}$ is linearly independent, the expansion (16) terminates with no error for $(n - [n-1]) = 1$ stage,

$$\boldsymbol{D} = \boldsymbol{\Gamma}_1$$

where we have by (9) that

$$\boldsymbol{\Gamma}_1 = \sqrt{\lambda_1}\,\boldsymbol{\kappa}_1\boldsymbol{\varrho}_1 = \boldsymbol{c}_1\boldsymbol{r}_1.$$

In this instance the given array $\boldsymbol{D}$ is thus *itself* separable.

The numerical problem of computing the eigenvalues $\lambda_i$ and their associated eigenvectors $\boldsymbol{\kappa}_i$ and $\boldsymbol{\varrho}_i$ is the subject of the subsequent section.

*B. The Numerical Solution*

We have previously seen that the most general form of the multistage expansion of $\boldsymbol{D}$ is given by (16),

$$\boldsymbol{D} = \sum_{i=1}^{k} \boldsymbol{\Gamma}_i + \boldsymbol{E}_k, \qquad k = 1, 2, \cdots, n-p \tag{17}$$

and where, in particular,

$$\boldsymbol{E}_{n-p} = \boldsymbol{0}.$$

The $i$th separable stage $\boldsymbol{\Gamma}_i$ is given by (9)

$$\boldsymbol{\Gamma}_i = \sqrt{\lambda_k}\,\boldsymbol{\kappa}_i\boldsymbol{\varrho}_i. \tag{18}$$

It may be deduced from the derivations of Appendix I that the $m$-length column vectors $\boldsymbol{\kappa}_i$ constitute the set of $(n-p)$ orthonormal eigencolumns of the symmetric, positive semidefinite, $m$-order matrix $\boldsymbol{Q}$,

$$\boldsymbol{Q} = \boldsymbol{D}\boldsymbol{D}^T.$$

Similarly, the $n$-length row vectors $\boldsymbol{\varrho}_i$ constitute the set of $(n-p)$ orthonormal eigenrows of the symmetric $n$th order matrix $\boldsymbol{S}$,

$$\boldsymbol{S} = \boldsymbol{D}^T\boldsymbol{D}$$

where $\boldsymbol{S}$ is positive definite when $p = 0$ and positive semidefinite otherwise. Furthermore, both $\boldsymbol{Q}$ and $\boldsymbol{S}$ have the identical set of nonvanishing eigenvalues,

$$\lambda_1, \lambda_2, \cdots, \lambda_{n-p}$$

where $\lambda$ is the largest and $\lambda_{n-p}$ the smallest in magnitude.

Throughout this treatment we assume that $m \geq n$, and it is shown in Appendix I that no loss of generality results from such a restriction. If $m > n$, it is obviously better to extract the $(n-p)$ nonvanishing $\lambda_i$'s from $\boldsymbol{S}$, which is only of order $n$, rather than from $\boldsymbol{Q}$, which is of order $m$. If $m = n$, the eigenvalues may be extracted equally well either from $\boldsymbol{Q}$ or from $\boldsymbol{S}$, but for the sake of convenience we shall always extract these eigenvalues from $\boldsymbol{S}$.

Once the $(n-p)$ nonvanishing eigenvalues of $\boldsymbol{S}$ are known, we can compute the normalized error energy $\epsilon_k$ from the formula

$$\epsilon_k = 1 - \frac{\lambda_1 + \lambda_2 + \cdots + \lambda_k}{\lambda_1 + \lambda_2 + \cdots + \lambda_{n-p}}, \qquad k = 1, 2, \cdots, n-p \tag{19}$$

which is the generalization of (14) for the case $p \neq 0$. It is convenient to compute the full set of $(n-p)$ normalized error energies at this point, because one may thus determine the number of stages $k$ which are required to reduce $\epsilon_k$ to or below some desired value in the range $0 \leq \epsilon_k < 1$. In this manner one avoids the calculation of unnecessary eigenvectors.

Next one computes the first $k$ orthonormal eigenrows $\boldsymbol{\varrho}_i$ of $\boldsymbol{S}$, which are associated with the eigenvalues

$\lambda_1, \cdots, \lambda_k$, $(k \leq n-p)$. These vectors constitute the set of desired $k$ row vectors of the multistage expansion (16).

We could now calculate the required first $k$ orthonormal eigencolumns $\boldsymbol{\kappa}_i$ directely from $\boldsymbol{Q}$. The extraction of an eigenvector from a matrix is generally a laborious numerical problem. Although this must be done for the eigenrows $\boldsymbol{\varrho}_i$ of $\boldsymbol{S}=\boldsymbol{D}^T\boldsymbol{D}$, we may make use of the expression

$$\sqrt{\lambda_i}\,\boldsymbol{\kappa}_i = \boldsymbol{D}\boldsymbol{\varrho}_i^T, \qquad i = 1, 2, \cdots, k \tag{20}$$

which is (53a) of Appendix I. Indeed, it is shown there that this is one of the two relations that provide the coupling between the eigenvalue problems for $\boldsymbol{S}=\boldsymbol{D}^T\boldsymbol{D}$ and $\boldsymbol{Q}=\boldsymbol{D}\boldsymbol{D}^T$.

From (18) we see that the desired column vectors are actually of the scale-factored form

$$\boldsymbol{c}_i = \sqrt{\lambda_i}\,\boldsymbol{\kappa}_i. \tag{21}$$

Combination of (20) and (21) yields

$$\boldsymbol{c}_i = \boldsymbol{D}\boldsymbol{\varrho}_i^T.$$

In summary, if we express the $i$th stage $\boldsymbol{\Gamma}_i$ as the product

$$\boldsymbol{\Gamma}_i = \boldsymbol{c}_i\boldsymbol{r}_i \tag{22}$$

the vectors comprising this stage are

$$\left.\begin{aligned} \boldsymbol{r}_i &= \boldsymbol{\varrho}_i, \\ \boldsymbol{c}_i &= \boldsymbol{D}\boldsymbol{\varrho}_i^T, \end{aligned}\right\} \quad i = 1, 2, \cdots, k \tag{23}$$

and we see that the eigencolumns $\boldsymbol{\kappa}_i$ need not be computed explicitly. On the other hand, it will always be necessary to solve the eigenvalue problem associated with $\boldsymbol{S}=\boldsymbol{D}^T\boldsymbol{D}$; that is, one will have to make use of one of the many numerical schemes available for the extraction of the eigenvalues $\lambda_1, \cdots, \lambda_n$ and the desired associated first $k$ eigenrows $\boldsymbol{\varrho}_1, \cdots, \boldsymbol{\varrho}_k$.

The choice of a particular eigenvalue algorithm will be governed by many factors, including desired computational speed and accuracy and the idiosyncracies of one's digital computer installation. The numerical examples described in the present treatment were calculated with the aid of a Fortran IV subroutine[4] based on the method of Jacobi. Readers interested in an exhaustive analysis of the numerical eigenvalue problem are referred to the excellent book by Wilkinson [18].

A small illustrative example will be helpful. Let $\boldsymbol{D}$ be the matrix

$$\boldsymbol{D} = \begin{bmatrix} 1 & 2 \\ 2 & 1 \\ 1 & 3 \end{bmatrix}$$

[4] Subroutine EIGEN, IBM System 360 Scientific Subroutine Package, H20-0205-3, 4th ed., IBM Corporation, New York, 1968.

so that $m=3$ and $n=2$. Then

$$\boldsymbol{S} = \boldsymbol{D}^T\boldsymbol{D} = \begin{bmatrix} 1 & 2 & 1 \\ 2 & 1 & 3 \end{bmatrix}\begin{bmatrix} 1 & 2 \\ 2 & 1 \\ 1 & 3 \end{bmatrix} = \begin{bmatrix} 6 & 7 \\ 7 & 14 \end{bmatrix}.$$

The eigenvalues of $\boldsymbol{S}$ are

$$\lambda_1 = 18.06226$$
$$\lambda_2 = \phantom{0}1.93774.$$

Since neither of these eigenvalues is zero, $p=0$; i.e., the columns of $\boldsymbol{D}$ are linearly independent, and therefore we expect $\boldsymbol{D}$ to be representable exactly with $(n-p)=2$ stages. From (19) we obtain the normalized error energies

$$\epsilon_1 = 1 - \frac{\lambda_1}{\lambda_1 + \lambda_2} = 0.09689 \approx 0.1$$

$$\epsilon_2 = 1 - \frac{\lambda_1 + \lambda_2}{\lambda_1 + \lambda_2} = 0.$$

The associated orthonormal eigenrows are

$$\boldsymbol{\varrho}_1 = [0.50193 \quad 0.86491] = \boldsymbol{r}_1$$
$$\boldsymbol{\varrho}_2 = [0.86491 \quad -0.50193] = \boldsymbol{r}_2.$$

We compute the column vectors $\boldsymbol{c}_1$ and $\boldsymbol{c}_2$ from (23),

$$\boldsymbol{c}_1 = \begin{bmatrix} 1 & 2 \\ 2 & 1 \\ 1 & 3 \end{bmatrix}\begin{bmatrix} 0.50193 \\ 0.86491 \end{bmatrix} = \begin{bmatrix} 2.23175 \\ 1.86876 \\ 3.09666 \end{bmatrix}$$

$$\boldsymbol{c}_2 = \begin{bmatrix} 1 & 2 \\ 2 & 1 \\ 1 & 3 \end{bmatrix}\begin{bmatrix} 0.86491 \\ -0.50193 \end{bmatrix} = \begin{bmatrix} -0.13894 \\ 1.22789 \\ -0.64087 \end{bmatrix}.$$

The desired two-stage expansion of $\boldsymbol{D}$ is thus

$$\boldsymbol{D} = \boldsymbol{\Gamma}_1 + \boldsymbol{\Gamma}_2$$

where

$$\boldsymbol{\Gamma}_1 = \begin{bmatrix} 2.23175 \\ 1.86876 \\ 3.09666 \end{bmatrix}[0.50193 \quad 0.86491]$$

$$\boldsymbol{\Gamma}_2 = \begin{bmatrix} -0.13894 \\ 1.22789 \\ -0.64087 \end{bmatrix}[0.86491 \quad -0.50193].$$

Retention of the first and dominant stage $\boldsymbol{\Gamma}_1$ produces the truncation error matrix

$$\boldsymbol{E}_1 = \boldsymbol{D} - \boldsymbol{\Gamma}_1 = \begin{bmatrix} -0.12017 & 0.06974 \\ 1.06202 & -0.61631 \\ -0.55430 & 0.32167 \end{bmatrix}.$$

The elements of $\boldsymbol{E}_1$ are certainly not small in comparison to the elements of $\boldsymbol{D}$, and hence one would use both stages in order to obtain a satisfactory (in this instance, perfect) agreement between $\boldsymbol{D}$ and its multistage expansion. The single-stage normalized error energy $\epsilon_1$ is as we have seen, of the order of 0.1. The preceding example illustrates that 0.1 is not "small" because the single-stage $\boldsymbol{\Gamma}_1$ does not represent $\boldsymbol{D}$ adequately. In a subsequent section we shall find that a good multistage expansion is achievable whenever $k$ is large enough to reduce $\epsilon_k$ to the order of 0.01 or less.

The skeptical reader has probably noticed that even the single-stage expansion requires $(m+n)=(3+2)=5$ coefficients in comparison to the $mn=6$ coefficients that describe $\boldsymbol{D}$. The advantages of the multistage expansion method only become apparent for larger values of $m$ and $n$, and the above example is merely meant to illustrate the principles of the technique.

## III. The Implementation of Multistage Separable Planar Filters

In the preceding section we have demonstrated how an arbitrary planar response function can be expanded into a series of individually separable stages. Here we shall discuss how such filters may be reduced to practice.

It is convenient to represent the planar filtering operation in terms of the two-dimensional, or planar, $z$ transform [9, p. 73].[5] Suppose that we have an $(M\times N)$ input matrix $\boldsymbol{X}$ that is to be filtered with an $(m\times n)$ filter matrix $\boldsymbol{F}$. Both $\boldsymbol{X}$ and $\boldsymbol{F}$ are representable as two-dimensional, or planar, polynomials in the variables $z_1$ and $z_2$.

$$X(z_1, z_2) = \sum_{i=1}^{M}\sum_{j=1}^{N} x_{ij} z_1^{i-1} z_2^{j-1} \tag{24}$$

and

$$F(z_1, z_2) = \sum_{i=1}^{m}\sum_{j=1}^{n} f_{ij} z_1^{i-1} z_2^{j-1} \tag{25}$$

where $x_{ij}$ is an element of $\boldsymbol{X}$ and $f_{ij}$ is an element of $\boldsymbol{F}$.

Let $\boldsymbol{Y}$ be the output matrix resulting from the action of the filter $\boldsymbol{F}$ on the input $\boldsymbol{X}$. Then this operation can be represented by the planar polynomial product

$$Y(z_1, z_2) = X(z_1, z_2)F(z_1, z_2) \tag{26}$$

[14]. Alternatively, this filtering operation is expressible as the planar convolution

$$y_{kl} = \sum_{i=1}^{m}\sum_{j=1}^{n} f_{ij} x_{k-i+1, l-j+1} \tag{27}$$

where $y_{kl}$ is an element of $\boldsymbol{Y}$, and where we assume that the input elements $x_{ij}$ vanish outside the range

$$i, j: \quad i = 1, 2, \cdots, M$$

and

$$j = 1, 2, \cdots, N.$$

The convolution of an $m$-length one-dimensional filter with an $M$-length one-dimensional input sequence results in a one-dimensional output sequence of length $M+m-1$. Similarly, the planar convolution of an $(m\times n)$ filter $\boldsymbol{F}$ with an $(M\times N)$ input $\boldsymbol{X}$ yields an $(M+m-1)$ by $(N+n-1)$ output array $\boldsymbol{Y}$, so that (27) holds for the subscript ranges

$$k = 1, 2, \cdots, \quad M + m - 1$$
$$l = 1, 2, \cdots, \quad N + n - 1.$$

Suppose that the filter $\boldsymbol{F}$ is exactly separable. Then its planar response function is given by (2).[6]

$$\boldsymbol{F} = \boldsymbol{c}\boldsymbol{r} = \begin{bmatrix} c_1 \\ c_2 \\ \cdot \\ \cdot \\ \cdot \\ c_m \end{bmatrix} [r_1 r_2 \cdots r_n] \tag{28}$$

where $\boldsymbol{c}$ is an $m$-length column vector and $\boldsymbol{r}$ is an $n$-length row vector. In this case the elements of $\boldsymbol{F}$ are simply

$$f_{ij} = c_i r_j$$

and substitution of this relation into (27) yields

$$y_{kl} = \sum_{i=1}^{m}\sum_{j=1}^{n} c_i r_j x_{k-i+1, l-j+1}.$$

In terms of planar $z$ transforms, the separation (28) is

$$F(z_1, z_2) = C(z_1)R(z_2) \tag{29}$$

where

$$c(z_1) = c_1 + c_2 z_1 + c_3 z_1^2 + \cdots + c_m z_1^{m-1}$$
$$R(z_2) = r_1 + r_2 z_2 + r_3 z_2^2 + \cdots + r_n z_2^{n-1}.$$

The planar $z$ transform of the output $Y(z_1,z_2)$ is obtained by inserting (29) into (26),

$$Y(z_1, z_2) = X(z_1, z_2)C(z_1)R(z_2). \tag{30}$$

Since polynomial multiplication is commutative, $Y(z_1, z_2)$ may also be expressed in the form

$$Y(z_1, z_2) = X(z_1, z_2)R(z_2)C(z_1). \tag{31}$$

Let us define the intermediate outputs

$$W(z_1, z_2) = X(z_1, z_2)C(z_1)$$

and

$$V(z_1, z_2) = X(z_1, z_2)R(z_2).$$

Then (30) and (31) become

$$Y(z_1, z_2) = W(z_1, z_2)R(z_2) \tag{32a}$$

and

$$Y(z_1, z_2) = V(z_1, z_2)C(z_1). \tag{32b}$$

[5] Our $z$-transfom definition differs from Jury's [9] in that his $z$ is our $z_1^{-1}$ and his $w$ is our $z_2^{-1}$.

[6] In the sequel we use the symbol $\boldsymbol{F}$ to represent the filter; the symbol $\boldsymbol{D}$ was used in Section II to denote the fact that either the filter or the input is expressible in terms of a series of separable stages.

The formulations (32a) and (32b) both produce the same output $Y(z_1,z_2)$, but they do not necessarily have the same computing efficiency. This efficiency can be established by determining the number of "multiply and add" operations (MADS) required by each formulation. Let us call the number of MADS required by the algorithms (32a) and (32b) MADS $(a)$ and MADS $(b)$, respectively. We show in Appendix III that

$$\begin{aligned} \text{MADS}(a) &= MN(m+n) + n(m-1)N \\ \text{MADS}(b) &= MN(m+n) + m(n-1)M. \end{aligned} \tag{33}$$

Since the first terms of these two expressions are identical, we will wish to pick the algorithm for which the second term is smallest. If we make the generally reasonable assumption that $n(m-1)$ is of the order of $m(n-1)$, the scheme (32a) will be more efficient when $N \ll M$, while the scheme (32a) will be more efficient when $N \gg M$. If $M \approx N$, both schemes have roughly similar efficiency. Furthermore, if $mn \ll MN$, so that the filter array is of much smaller size than the input array, the second terms of the two expressions (33) can be neglected. Then

$$\text{MADS}(a) \approx \text{MADS}(b) \approx MN(m+n) \tag{34}$$

and again both schemes exhibit comparable computational efficiency.

If the filter $F(z_1, z_2)$ is not separable, the original formulation (26) would have to be used. It is shown in Appendix III that the number of MADS required for the complete convolution of the $(m \times n)$ filter array $\boldsymbol{F}$ with the $(M \times N)$ input array $\boldsymbol{X}$ are

$$MNmn. \tag{35}$$

The computing efficiencies of the various filtering algorithms can be conveniently compared by the introduction of a quantity that we call the "effective length" $L_{\text{eff}}$ of a given algorithm. We define the effective length to be the number of MADS required by a given filtering algorithm divided by the size $MN$ of the input array. For standard planar convolution this quantity follows directly from (35) and is

$$L_{\text{eff}} = \frac{MNmn}{MN} = mn. \tag{36}$$

If the filter $F$ is separable and if $mn \ll MN$, the appropriate effective length is by (34)

$$L_{\text{eff}} = \frac{MN(m+n)}{MN} = m + n. \tag{37}$$

Thus the effective length for a nonseparable filter grows as the product of the filter dimensions, whereas it grows only as their sum if the filter is separable.

In the previous section we showed how an arbitrary planar response function can be expanded into a sum of $k$ separable stages $\boldsymbol{\Gamma}_i$, $(i = 1, 2, \cdots, k)$. Assuming that $m \geq n$, we also established that this expansion is exact in, at most, $k = n$ stages, and that good approximations of the given filter response are often achievable when $k < n$. Let us represent the $k$-stage approximation of the planar filter $\boldsymbol{F}$ by the symbol $\boldsymbol{F}^{(k)}$, where

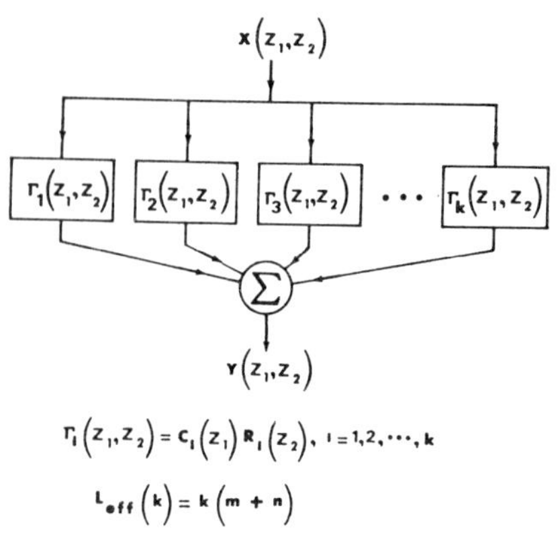


Fig. 1. Block diagram showing implementation of a $k$-stage planar filter [see (38)].

$$\boldsymbol{F}^{(k)} = \boldsymbol{\Gamma}_1 + \boldsymbol{\Gamma}_2 + \cdots + \boldsymbol{\Gamma}_k = \sum_{i=1}^{k} \boldsymbol{\Gamma}_i.$$

For every filter $\boldsymbol{F}^{(k)}$ there exists an associated normalized error energy $\epsilon_k$ which vanishes for some value of $k \leq n$, and whose explicit form is given by (14).

The planar $z$ transform of the filter $\boldsymbol{F}^{(k)}$ is

$$F^{(k)}(z_1, z_2) = \sum_{i=1}^{k} \Gamma_i(z_1, z_2).$$

If we replace $F(z_1,z_2)$ by $F^{(k)}(z_1,z_2)$, in (26) and use the preceding expansion, we obtain

$$Y(z_1, z_2) = X(z_1, z_2) \sum_{i=1}^{k} \Gamma_i(z_1, z_2). \tag{38}$$

But every stage $\Gamma_i$ is separable, and thus (29) generalizes to

$$\Gamma_i(z_1, z_2) = C_i(z_1) R_i(z_2) \tag{39}$$

where $C_i(z_1)$ and $R_i(z_2)$ are the one-dimensional $z$ transforms of the $m$-length column vectors $\boldsymbol{c}_i$ and the $n$-length row vectors $\boldsymbol{r}_i$, respectively. Substitution of (39) into (38) yields

$$Y(z_1, z_2) = X(z_1, z_2) \sum_{i=1}^{k} C_i(z_1) R_i(z_2). \tag{40}$$

We thus obtain a sum of the $k$ planar $z$ transform products of the type (30). Equation (40) naturally reduces to (30) when $k = 1$ if we make the identifications $C_1(z_1) = C(z_1)$ and $R_1(z_2) = R(z_2)$. But the effective length $L_{\text{eff}}$ for a single stage of the filter $\boldsymbol{F}^{(k)}$ is given by (37), and hence the effective length $L_{\text{eff}}(k)$ for the $k$-stage filtering operation represented by (38) or (40) is

$$L_{\text{eff}}(k) = k(m+n). \tag{41}$$

The block diagram of Fig. 1 illustrates the implementation of the filter equation (38). We observe that each of the $k$ separable stages $\mathbf{\Gamma}_i$ operates individually on the common input $\boldsymbol{X}$, so that $k$ partial output arrays are produced. The sum of these output matrices then results in the overall output matrix $\boldsymbol{Y}$.

If the filter matrix $\boldsymbol{F}$ can be suitably approximated by a sum of $k$ stages $\mathbf{\Gamma}_i$ such that

$$k(m+n) < mn \tag{42}$$

the expanded filter $\boldsymbol{F}^{(k)}$ is more efficient than the given filter $\boldsymbol{F}$. The inequality (42) suggests the introduction of the efficiency parameter $\alpha$,

$$\alpha = \frac{mn}{k(m+n)} = \frac{mn}{L_{\text{eff}}(k)} \cdot \tag{43}$$

If $\alpha$ is greater than unity, the algorithm (38) is more efficient, while the standard filtering algorithm (26) is more efficient whenever $\alpha$ is less than unity. If $\alpha=1$, both algorithms are equally efficient. Of course the smaller the acceptable value of the number of stages $k$ the greater is the magnitude of $\alpha$ for a given set of filter dimensions $m$ and $n$. The price one pays as $k$ decreases is an increasing value of the normalized error energy $\epsilon_k$ (see Section II), so that one will want to pick a value of $k$ large enough to render $\epsilon_k$ acceptably small. For example, suppose that $m=n=100$. If five stages constitute a suitable approximation of $\boldsymbol{F}$, $k=5$ and then

$$\alpha = \frac{10\,000}{1000} = 10$$

so that the expansion scheme (38) is ten times as efficient as the standard scheme (26). On the other hand, if as many as ten stages are required to yield a tolerably small value of $\epsilon_k$, $k=10$ so that

$$\alpha = \frac{10\,000}{2000} = 5$$

and the scheme (38) is only five times as efficient as the scheme (26).

We remark also that the efficiency parameter $\alpha$ always has a lower limit, since the multistage expansion of $\boldsymbol{F}^{(k)}$ is exact when $k$ is at most equal to $n$. The minimum value of $\alpha$, $\alpha_{\min}$, is obtainable from (43) by setting $k=n$

$$\alpha_{\min} = \frac{m}{m+n} \cdot$$

If $m=n$ (so that $\boldsymbol{F}$ is a square filter matrix), $\alpha_{\min}=1/2$. Such a value of $\alpha$ does not warrant use of the expansion scheme unless further refinements of the method can be made.

One such refinement becomes possible once we recognize that the columns $C_i(z_1)$ and the rows $R_i(z_2)$ of all stages $\Gamma_i(z_1,z_2)$ are themselves one-dimensional convolutional filters. In many instances these filters can be well approximated by one-dimensional recursive filters [13], and thus further gains in computing efficiency result that are over and above those achievable by the multistage separation technique alone. All these matters will be illustrated by example in the following section.

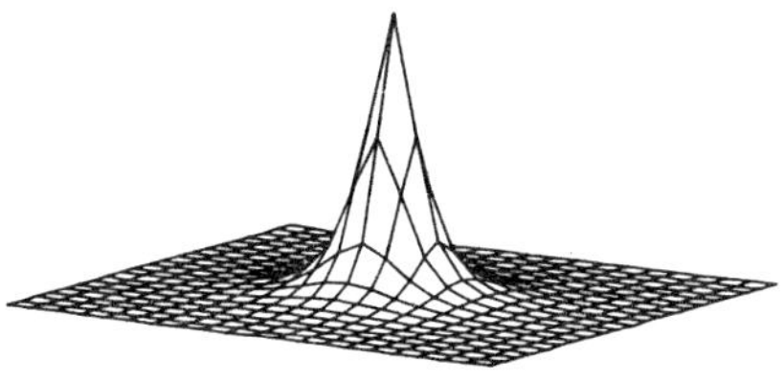

Fig. 2. Perspective view of discrete upward continuation operator. $L_{\text{eff}}=441$.

## IV. Illustrations of the Design Technique

The previously developed expansion method will here be illustrated with two examples. The first application is to a filter array that can be expanded with good accuracy into only a small number of individually separable stages $\Gamma_i$, while the second involves a response function requiring a greater number of stages for a satisfactory approximation.

### *A. Upward Continuation Operator*

The upward continuation operator is a planar filter that acts on gridded potential field data of one level and yields as output the field that would be measured at a distance $h$ above the original recording plane. The continuous form of this filter in the space domain is

$$f(x,y) = \frac{1}{2\pi}\,\frac{h}{(h^2+x^2+y^2)^{3/2}}$$

where $x$ and $y$ are spatial coordinates [2]. If this response function is truncated in both $x$ and $y$ beyond a point for which the magnitude of $f(x, y)$ is negligibly small, and if the resulting function is then sampled in both directions at a uniform increment $\Delta s$, a planar array of weighting coefficients is obtained. Here we have

$$h = 1.5\Delta s$$

chosen to make[7] and to truncate $f(x, y)$ in such a manner that a square array $\boldsymbol{F}$ of size $(21)\times(21)$ is produced, and hence $m=n=21$.

Fig. 2 shows a perspective view of the discrete upward continuation operator, where the magnitudes of the elements $f_{ij}$ of $\boldsymbol{F}$ are plotted against the vertical axis. Use of the standard planar convolution algorithm (26) leads to the effective length $L_{\text{eff}}$ given by (36), namely,

$$L_{\text{eff}} = (21)^2 = 441.$$

The matrix $\boldsymbol{F}$ was expanded into a single stage $\mathbf{\Gamma}_1=\mathbf{c}_1\boldsymbol{r}_1$, with an associated normalized error energy (NEE)

$$\epsilon_1 = 0.01136.$$

The effective length for this single-stage ($k=1$) expan-

[7] It can be shown that the aliasing error is negligible as long as $h/\Delta s \geq 1$. S. Treitel, unpublished note.

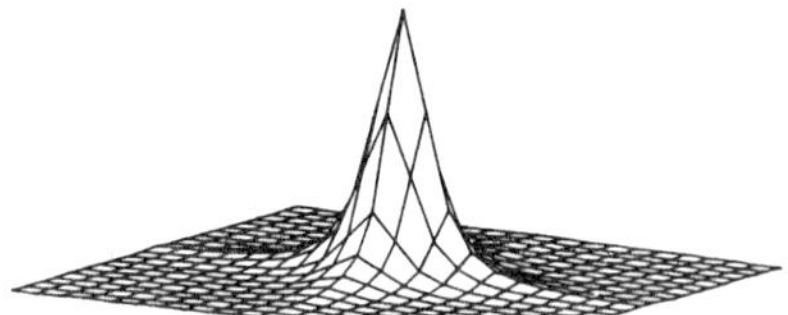

Fig. 3. Perspective view of single-stage expansion of discrete upward continuation operator. $\epsilon_k = 0.01136$ and $L_{eff} = 42$.

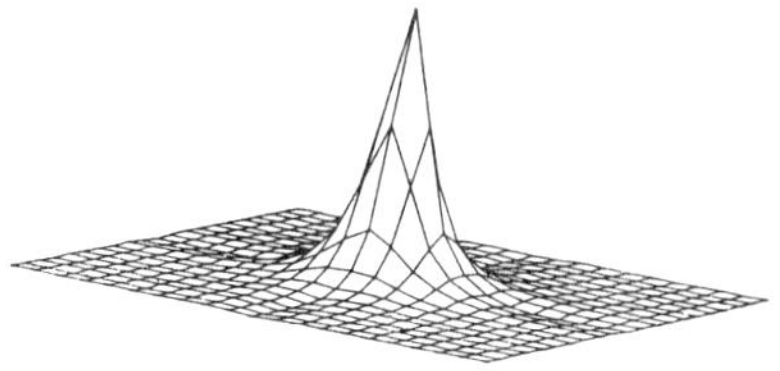

Fig. 4. Perspective view of two-stage expansion of discrete upward continuation operator. $\epsilon_k = 0.00057$ and $L_{eff} = 84$.

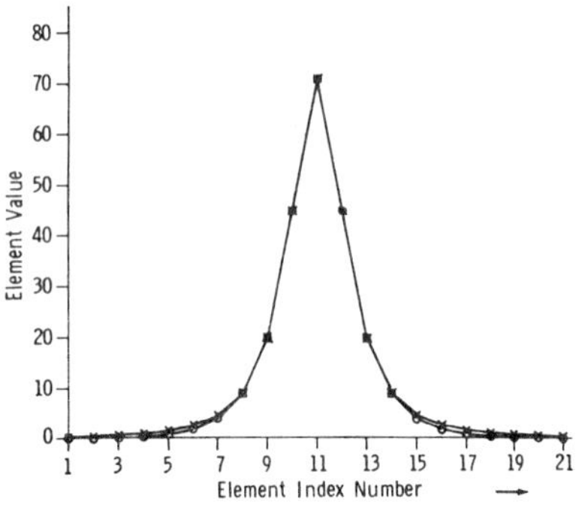


Fig. 5. The column vector $C_1$ (X) and its recursive approximation (O).

sion of $F$ is then by (41)

$$L_{eff}(1) = 21 + 21 = 42.$$

The array $\Gamma_1$ is shown in perspective view in Fig. 3. Comparison with Fig. 2 clearly reveals that even the single separable stage $\Gamma_1$ constitutes a satisfactory though not perfect approximation of the given array $F$.

The array

$$F^{(2)} = \Gamma_1 + \Gamma_2$$

that results from the two-stage ($k=2$) expansion of $F$ is depicted in Fig. 4. The value of $\epsilon_k$ has now decreased to

$$\epsilon_2 = 0.00057$$

while the effective length is by (41)

$$L_{eff}(2) = 2(21 + 21) = 84.$$

The two-stage expansion $F^{(2)} = \Gamma_1 + \Gamma_2$ is seen to be in excellent agreement with the given upward continuation filter array $F$. The efficiency parameter $\alpha$ is by (43) for $k=2$

$$\alpha = \frac{441}{84} = 5.25.$$

Therefore use of the filter $F^{(2)}$ in the expanded convolution scheme (38) is more than five times as efficient as use of the given filter $F$ in the standard convolution scheme (26).

The planar filter $F^{(2)}$ consists of four individual one-dimensional filters, so that in terms of $z$ transforms we have

$$\begin{aligned} F^{(2)}(z_1, z_2) &= \Gamma_1(z_1, z_2) + \Gamma_2(z_1, z_2) \\ &= C_1(z_1)R_1(z_2) + C_2(z_1)R_2(z_2). \end{aligned}$$

The computing efficiency of the present expansion method can be further increased by replacing the four one-dimensional convolutional filters $C_1(z_1)$, $C_2(z_1)$, $R_1(z_2)$, and $R_2(z_2)$ by corresponding one-dimensional recursive filters [13].

Fig. 5 shows a plot of the 21-length operator $C_1(z_1)$. We notice that this weighting coefficient sequence is symmetric with respect to the central, or eleventh coefficient. It was therefore decided to approximate $C_1(z_1)$ as a sum of two individually recursive stages that act in the forward and reverse directions, respectively [15]. In other words, given the sequence

$$C_1 = (c_1, c_2, \cdots, c_{11}, \cdots, c_{20}, c_{21})$$

we first design a "forward" recursive filter $C_1'(z_1)$ that approximates the sequence

$$0.5c_{11}, c_{12}, \cdots, c_{21}.$$

Then, since the sequence $C_1$ is symmetric about $c_{11}$, the "reverse" recursive filter is merely $C_1'(1/z_1)$; that is, the reverse filter is obtained by replacing $z_1$ by $1/z_1$ in the expression for $C_1'(z_1)$. The reverse filter thus approximates the sequence

$$c_1, c_2, \cdots, 0.5c_{11}$$

and evidently the sum

$$C_1(z_1) \approx C'(z_1) + C'(1/z_1) \qquad (44)$$

will then constitute an approximation to the given filter $C_1(z_1)$. The method for computing the coefficients of a recursive filter as well as the algorithms required to apply them to digitized data are described by Shanks [13]. Fig. 5 also shows a plot of the approximation (44), which is seen to be quite satisfactory. In this particular case $C_1'(z_1)$ is given by

$$C_1'(z_1) = \frac{0.35447 + 0.29128z_1}{1.0 - 0.44040z_1}$$

so that three MADS per output value are required. Thus, whereas use of the convolutional filter $C_1(z_1)$ requires 21 MADS per output value, use of the recursive form (44) leads to only $2 \times 3$, or six MADS per output value.

The remaining one-dimensional convolution filters $R_1(z_2)$, $C_2(z_1)$, and $R_2(z_2)$ can likewise be approximated by sums of recursive filters of the form (44). For the present

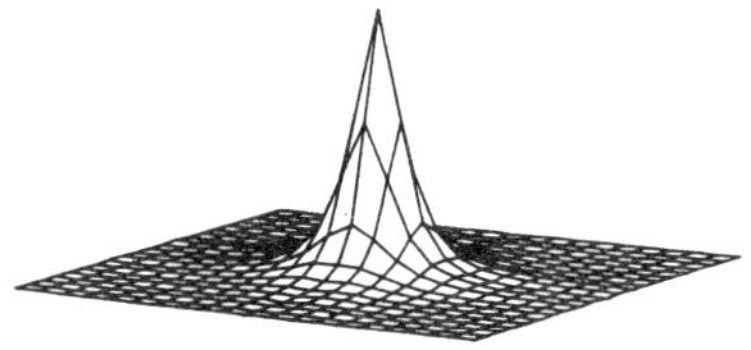

Fig. 6. Perspective of two-stage expansion of the discrete upward continuation operator. The constituent vectors have been approximated by one-dimensional recursive filters. $\epsilon_k = 0.002$ and $L_{\text{eff}} = 28$.

example satisfactory fits have been obtained by approximating $C_1(z_1)$ and $R_1(z_2)$ by sums of two recursive stages that each require three MADS per output value, while the stages corresponding to $C_2(z_1)$ and $R_2(z_2)$ each require four MADS per output value. As a result the effective length of the recursive approximation to the filter $\boldsymbol{F}^{(2)}$ is

$$L_{\text{eff}} = (4 \times 3) + (4 \times 4) = 28.$$

This effective length must be compared with $L_{\text{eff}} = 84$ for the original two-stage expansion of $\boldsymbol{F}$, and with $L_{\text{eff}} = 441$ for use of $\boldsymbol{F}$ in its original nonsuspended form. Fig. 6 shows a perspective view of the two-stage filter $\boldsymbol{F}^{(2)}$ after its constituent one-dimensional convolutional filters $C_i(z_1)$ and $R_i(z_2)$, $(i = 1, 2)$ have been approximated by one-dimensional recursive filters. The value of the normalized error energy $\epsilon_2$, is now

$$\epsilon_2 = 0.002$$

which compares with $\epsilon_2 = 0.00057$ for the original two-stage expansion. Thus a price for this further increase in efficiency is paid, but comparison of Fig. 6 with Fig. 2 reveals good agreement beween the original filter $\boldsymbol{F}$ and its recursively approximated two-stage expansion.

Quite evidently the use of one-dimensional recursive filters can produce savings in computer time and storage beyond those achievable by the planar expansion method alone.

### B. Fan-Pass Filter

The fan-pass operator is a planar filter that is designed to pass events on a seismogram whose apparent velocities fall within a given wedge-shaped region in the frequency–wavenumber, or $(f, \kappa)$, plane [4], [6], and [15]. Here we have chosen a 22-trace fan filter, each of whose rows consists of 21 sampled values. This filter is therefore described by a matrix of weighting coefficients having $m = 22$ rows and $n = 21$ columns. Fig. 7 shows a contour map of the two-dimensional amplitude spectrum of the operator.[8] Also shown is a perspective view of the shaded portion of the contour map. The performance of the separation technique for the upward continuation operator was illustrated by successive views of the appropriate approximations to the given space-domain response function. A fan filter is more easily visualized in terms of its $(f, \kappa)$-domain transfer function. Accordingly, we illustrate the performance of the expansion technique by displaying the $(f, \kappa)$ amplitude spectra of the successive approximations. These are, of course, computed by the expansion method from the exact representation of the fan filter in the space–time, or $(x, t)$-domain.

[8] The first and second quadrants of this spectrum are shown. The quadrants are bounded by the respective temporal and spatial Nyquist frequencies.

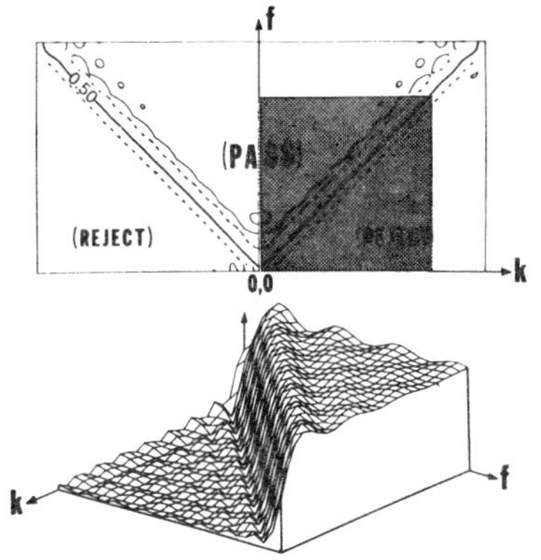


Fig. 7. Amplitude spectrum of the 22-trace 21 sampled values per trace fan filter. $L_{\text{eff}} = 462$.

TABLE I

EFFECTIVE LENGTHS AND NORMALIZED ERROR ENERGIES FOR THE EXPANSION OF THE FAN FILTER $(k = 1, 2, \cdots, 5)$

| Number of Stages $(k)$ | Effective Length $L_{\text{eff}}(k)$ | Normalized Error Energy $(\epsilon_k)$ |
|---|---|---|
| 1 | 43 | 0.173 |
| 2 | 86 | 0.081 |
| 3 | 129 | 0.048 |
| 4 | 172 | 0.031 |
| 5 | 215 | 0.021 |

The effective length of the given $(22 \times 21)$ operator $\boldsymbol{F}$ is

$$L_{\text{eff}} = (22) \times (21) = 462.$$

This operator was expanded into its first five dominant separable stages $\boldsymbol{\Gamma}_k (k = 1, 2, \cdots, 5)$. The effective lengths and associated normalized error energies for the approximations $\boldsymbol{F}^{(1)}, \boldsymbol{F}^{(2)}, \cdots, \boldsymbol{F}^{(5)}$ are given in Table I. The large value of the error energy for $k = 1$ indicates that the given filter is not readily separable, and thus the single-stage approximation $\boldsymbol{F}^{(1)}$ does not constitute an acceptable representation of $\boldsymbol{F}$.

Figs. 8–12 show contour maps of the two-dimensional amplitude spectra of the approximations $\boldsymbol{F}^{(k)}$, $(k = 1, 2, \cdots, 5)$. In all these figures perspective views of the shaded portions of the contour maps are also displayed. Comparison of Fig. 8 with Fig. 7 confirms that the single-stage approximation of $\boldsymbol{F}$ is inadequate. However, progressive comparison of Figs. 9–12 with Fig. 7 clearly reveals that the approximation improves in quality as $k$ increases. For $k = 5$ the agreement is quite good; the efficiency parameter for the filter $\boldsymbol{F}^{(5)}$ is by (43)

$$\alpha = \frac{462}{215} \approx 2.15$$

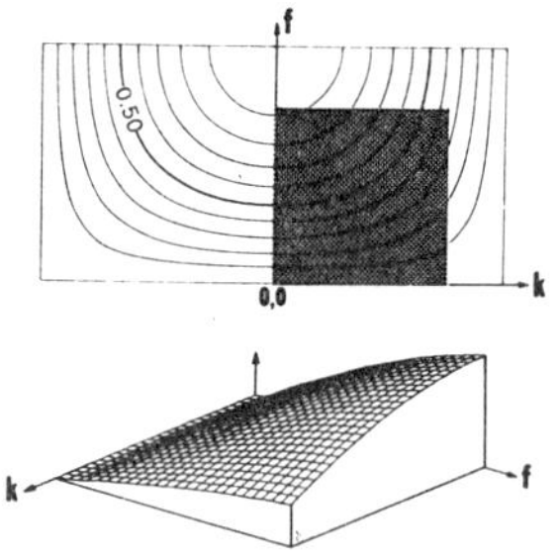


Fig. 8. Amplitude spectrum of single-stage approximation to fan-pass filter. $\epsilon_k = 0.173$ and $L_{eff} = 43$.

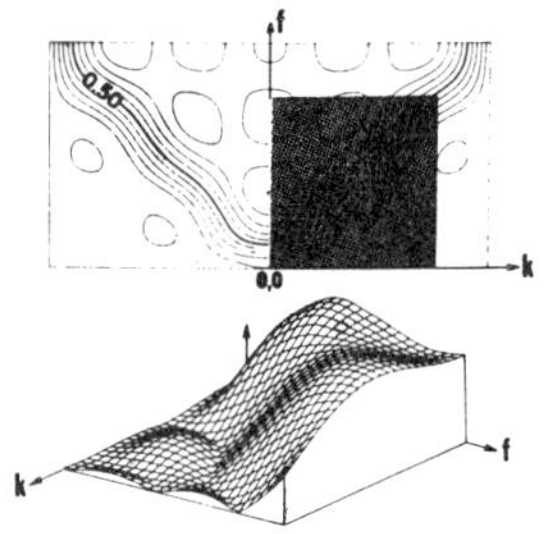


Fig. 10. Amplitude spectrum of three-stage approximation to fan-pass filter. $\epsilon_k = 0.048$ and $L_{eff} = 129$.

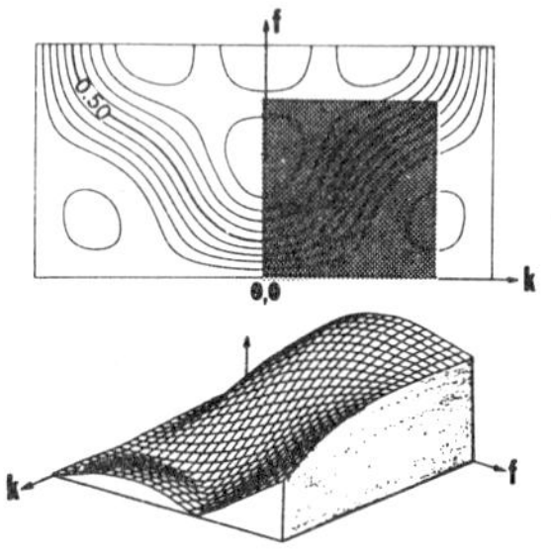


Fig. 9. Amplitude spectrum of two-stage approximation to fan-pass filter. $\epsilon_k = 0.081$ and $L_{eff} = 86$.

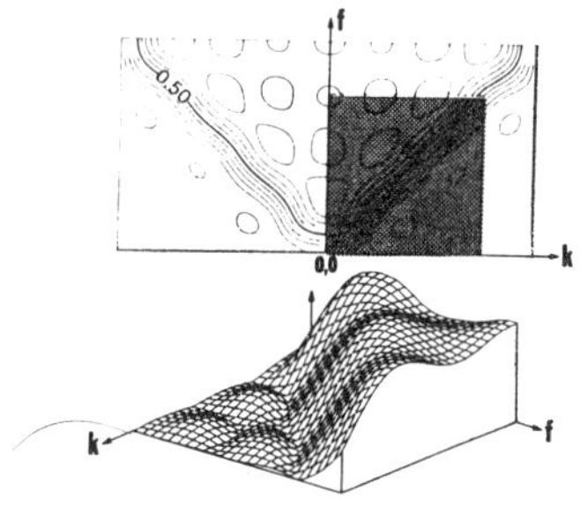


Fig. 11. Amplitude spectrum of four-stage approximation to fan-pass filter. $\epsilon_k = 0.031$ and $L_{eff} = 172$.

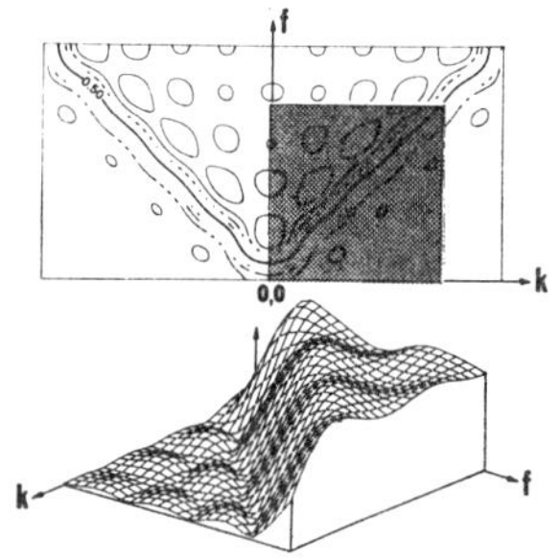


Fig. 12. Amplitude spectrum of five-stage approximation to fan-pass filter. $\epsilon_k = 0.021$ and $L_{eff} = 215$.

so that use of the filter $\boldsymbol{F}^{(5)}$ in the expanded convolution scheme (38) is more than twice as efficient as use of the given filter $\boldsymbol{F}$ in the standard convolution scheme (26). Further gains in computing efficiency could again be achieved if the convolutional components of each separable stage $\boldsymbol{\Gamma}_k$ were approximated by recursive filters. This refinement will be omitted here because its usefulness has already been demonstrated for the previous example.

## V. Concluding Remarks

We have presented a method that permits the expansion of an arbitrary planar filter response function into a converging sum of individually separable planar stages. Judicious use of this technique can result in substantial savings in computer time and storage. These savings are further increased if the one-dimensional convolutional filters of each separable stage are approximated by appropriate one-dimensional recursive filters.

The advent of digital computers with parallel processing units renders the present scheme attractive even when the total number of weighting coefficients in the expanded form of the filter exceeds the number of such coefficients in the original filter response function array. One computer of this type is the ILLIAC IV [1], which will have the capacity to execute simultaneously a large number of identical algorithms. Thus each of the $k$ planar stages of the expanded filter could operate *simultaneously* on the common input array $\boldsymbol{X}$, and in this manner the output is obtainable much more rapidly than if the original filter array is convolved directly with the input. Such a scheme will be particularly useful if the same planar filter is to be applied to a large number of different input arrays.

The expanded filter is expressible as a sum of products of column vectors and row vectors. The one-dimensional fast Fourier transform algorithm (FFT) can then be used to evaluate the Fourier transforms of each constituent vector, and the resulting complex-valued vector product sum will constitute an approximation to the two-dimensional Fourier transform of the original array. In many cases, particularly when the given array is expressible with small error by only a few separable stages, this scheme could result in faster computing speeds than if the two-dimensional FFT algorithm is applied directly to the original array. Such an approach to planar Fourier analysis certainly merits closer examination.

## Appendix I

### Eigenvector Expansion of a Planar Array

We desire to find an expansion of the given $(m \times n)$ matrix $D$ into a series of individually separable $(m \times n)$ matrix stages $\Gamma_i$ such that the sum

$$\Gamma_1 + \Gamma_2 + \cdots + \Gamma_i + \cdots$$

gives a good approximation to $D$ with a small number of stages. Each stage $\Gamma_i$ can be written

$$\Gamma_i = c_i r_i \tag{45}$$

where $c_i$ is an $m$-length column vector and $r_i$ is an $n$-length row vector.

Let us consider the two square matrices

$$Q = DD^T$$
$$S = D^T D$$

where the superscript $(^T)$ denotes the matrix transpose. We note that $Q$ and $S$ are real, symmetric, and of order $m$ and $n$, respectively [7, p. 26]. Such matrices are expressible in terms of the finite matrix sums

$$Q = \sum_{i=1}^{m} \mu_i \kappa_i \kappa_i^T \tag{46}$$

$$S = \sum_{i=1}^{n} \lambda_i \varrho_i^T \varrho_i. \tag{47}$$

See [8, p. 38]. The numbers $\mu_i (i=1, 2, \cdots, m)$ are the eigenvalues of $Q$, while the $m$-length vectors $\kappa_i$ are the associated normalized column eigenvectors, or eigencolumns of $Q$. Similarly, the numbers $\lambda_i (i=1, 2, \cdots, n)$ are the eigenvalues of $S$, while the $n$-length vectors $\varrho_i$ are the associated normalized row eigenvectors, or eigenrows of $S$. The eigenvalues $\mu_i$ and $\lambda_i$ are real because the matrices $Q$ and $S$ are symmetric [8, p. 32].

We observe that the term

$$\mu_i \kappa_i \kappa_i^T$$

of (46) is a separable matrix, since it results from the multiplication of the column vector $\mu_i \kappa_i$ into the row vector $\kappa_i^T$. Similarly, the term

$$\lambda_i \varrho_i^T \varrho_i$$

of (47) is a separable matrix because it results from the multiplication of the column vector $\lambda_i \varrho_i^T$ into the row vector $\varrho_i$.

The expansions for $Q$ and $S$ as given by (46) and (47) are called "spectral representations" [16, p. 49]. Quite evidently these representations at once enable one to split $Q$ and $S$ into a sum of separble matrix stages. The problem at hand, however, is to determine whether the matrix $D$ can be split into such a sum of separable matrices or, equivalently, to determine whether there exists a spectral representation for $D$. We shall show that such a representation can indeed be found from knowledge of the spectral representations for the associated matrices $Q = DD^T$ and $S = D^T D$.

We now assume that $m \geq n$ so that the number of rows of $D$ are always equal to or greater than the number of columns of $D$. This assumption does not result in a loss of generality, since any matrix $\tilde{D}$ for which $m < n$ can always be transposed, and then the resulting matrix $\tilde{D}^T = D$ does satisfy the postulated condition $m \geq n$. Let us assume further that the $n$ columns of $D$ are linearly independent.[9] The matrices $Q = DD^T$ and $S = D^T D$ are both products resulting from the multiplication of a rectangular matrix by its transpose. In view of these considerations, the $m$th-order matrix $Q$ is positive semidefinite and of deficiency $(m-n)$; that is, $Q$ is only of rank $n$. On the other hand, the $n$-order matrix $S$ is positive definite and thus of full rank $n$ [7, p. 20] and [3, p. 268].

Since $Q$ is positive semidefinite and only of rank $n$, $(m-n)$ of its eigenvalues $\mu_i$ vanish, while its remaining $n$ eigenvalues are real and positive. Since $S$ is positive definite and of full rank $n$, none of its eigenvalues $\lambda_i$ vanish, and these eigenvalues are all real and positive. Thus both $Q$ and $S$ have the same number $n$ of real and positive eigenvalues.

An even closer relationship exists between the eigenvalue problems associated with $Q$ and $S$. This matter has been treated with great elegance by Lanczos [10, pp. 115–117], whose approach we use here with some modifications. Consider the eigenvalue problem

$$Uw = \sqrt{\lambda}\, w \tag{48}$$

where the $(m+n)$ by $(m+n)$ matrix $U$ and the $(m+n)$-length eigenvector $w$ are formed as shown in Fig. 13. Here $0_m$ and $0_n$ denote the null matrices of order $m$ and $n$, respectively. The reason for writing an eigenvalue of $U$ as the square root of a number $\lambda$ will become apparent in a moment. Clearly, the system (48) can be written in the coupled form

$$D\varrho^T = \sqrt{\lambda}\, \kappa \tag{49a}$$

$$D^T \kappa = \sqrt{\lambda}\, \varrho^T. \tag{49b}$$

Premultiplying the first of the relations by $D^T$ and the second by $D$, we obtain

$$D^T D \varrho^T = \sqrt{\lambda}\, D^T \kappa \tag{50a}$$

$$DD^T \kappa = \sqrt{\lambda}\, D\varrho^T. \tag{50b}$$

[9] This restriction will be relaxed in the sequel.

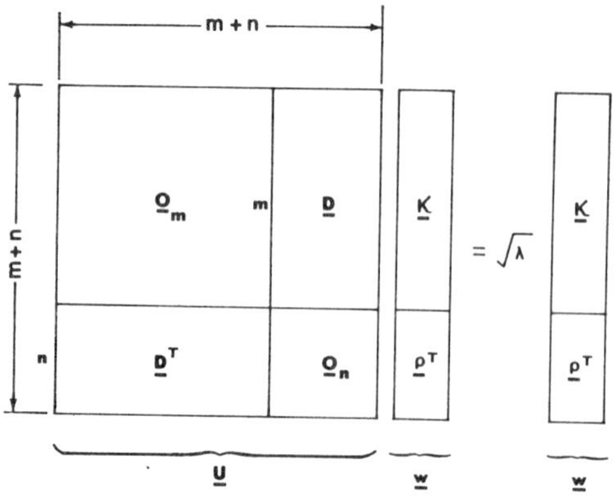


Fig. 13. The eigenvalue problem $Uw = \sqrt{\lambda}\, w$ in terms of $D$, $\mathfrak{u}$, and $\varrho$.

Substitution of (49b) into (50a) and of (49a) into (50b) yields

$$D^T D \varrho^T = \lambda \varrho^T$$

$$DD^T \kappa = \lambda \kappa$$

and we see why we have have written the eigenvalue as $\sqrt{\lambda}$ in (48). Since $Q = DD^T$ and $S = D^T D$ these relations can be written

$$S\varrho^T = \lambda \varrho^T \tag{51a}$$

$$Q\kappa = \lambda \kappa. \tag{51b}$$

Now we have already established that $(m-n)$ of the eigenvalues of $Q$ vanish, and that the remaining $n$ eigenvalues of $Q$ are real and positive. From (51a) and (51b) we then deduce that the $n$ nonvanishing eigenvalues of $Q$ coincide with the $n$ eigenvalues of $S$, all of which are real and positive since $S$ is positive definite. In other words, we have

$$\mu_i = \lambda_i, \qquad i = 1, 2, \cdots, n$$

$$\mu_i = 0, \qquad i = n+1, n+2, \cdots, m.$$

This means that the spectral representations for $Q$ and $S$ given by (46) and (47) become

$$Q = \sum_{i=1}^{n} \lambda_i \kappa_i \kappa_i^T \tag{52a}$$

$$S = \sum_{i=1}^{n} \lambda_i \varrho_i^T \varrho_i. \tag{52b}$$

Let us next realize that the coupling relations (49a) and (49b) can equally well be written in the subscripted form

$$D\varrho_i^T = \sqrt{\lambda_i}\, \kappa_i \tag{53a}$$

$$D^T \kappa_i = \sqrt{\lambda_i}\, \varrho_i^T \tag{53b}$$

for $i = 1, 2, \cdots, n$. If we replace $S$ by $D^T D$ and express the eigenvalue $\lambda_i$ as $\sqrt{\lambda_i}\,\sqrt{\lambda_i}$, (52b) can be rewritten in the form

$$D^T D = \sum_{i=1}^{n} \sqrt{\lambda_i}\, \varrho_i^T \sqrt{\lambda_i}\, \varrho_i.$$

Substitution of the coupling relation (53b) into the right-hand side of this equation produces

$$D^T D = \sum_{i=1}^{n} D^T \kappa_i \sqrt{\lambda_i}\, \varrho_i$$

$$= D^T \sum_{i=1}^{n} \sqrt{\lambda_i}\, \kappa_i \varrho_i.$$

From this matrix identity we deduce that

$$D = \sum_{i=1}^{n} \sqrt{\lambda_i}\, \kappa_i \varrho_i \tag{54}$$

which is the spectral representation for the given matrix $D$. An identical result is obtained if we replace $Q$ by $DD^T$ in (52a) and substitute into the resulting equation the transpose of (53a).

Consider the $i$th term of the spectral representation for $D$,

$$\sqrt{\lambda_i}\, \kappa_i \varrho_i.$$

This term can be viewed as the *separable* $(m \times n)$-matrix resulting from the multiplication of the $m$-length column vector $\sqrt{\lambda_i}\kappa_i$ into the $n$-length row vector $\varrho_i$. We also see that the sum of $n$ such separable matrices represents the given $(m \times n)$ matrix $D$ with no error. But these are the very criteria that we ourselves had set for the implementation of the multistage expansion of $D$, and hence we *identify* the spectral representation (54) with this desired multistage expansion:

$$D = \Gamma_1 + \Gamma_2 + \cdots + \Gamma_i + \cdots + \Gamma_n$$

$$= \sum_{i=1}^{n} \Gamma_i$$

$$= \sum_{i=1}^{n} \sqrt{\lambda_i}\, \kappa_i \varrho_i. \tag{55}$$

The $i$th stage, or term, of this expansion is thus

$$\Gamma_i = \sqrt{\lambda_i}\, \kappa_i \varrho_i, \qquad i = 1, 2, \cdots, n. \tag{56}$$

Comparing (56) with (45), we see further that the desired series of column vectors $c_i$ can be written

$$c_i = \sqrt{\lambda_i}\, \kappa_i$$

and that the desired series of row vectors $r_i$ is then

$$r_i = \varrho_i.$$

It is to be particularly noted that the multistage expansion (55) is exact when the number of individually separable stages $\Gamma_i$ is $n$, and we recall that $n \leq m$. We remark that the spectral representation of $D$ constitutes only one possible solution of the problem we have posed originally (see Appendix II). However, it has the distinct advantage of yielding an exact representation of $D$ in a finite number of stages. Other advantages will become evident later.

In many applications (see Section IV) we will be satisfied with a multistage separation that will merely approximate $\boldsymbol{D}$. One way to do this is to retain only $k$ stages in the expansion (55). Quite obviously we will wish to retain the dominant stages and neglect those that contribute little to the expansion. We accordingly suppose that the real and positive eigenvalues $\lambda_i$ are listed in order of decreasing magnitude,

$$\lambda_1, \lambda_2, \cdots, \lambda_{n-1}, \lambda_n \tag{57}$$

where $\lambda_1$ is the largest and $\lambda_n$ the smallest eigenvalue of the positive definite matrix $\boldsymbol{S}$.[10] We suppose also that any repeated eigenvalues are labeled separately.

Let us then assume that the sequence of terms of the multistage expansion (55) conforms with the ordering (57), so that $\boldsymbol{\Gamma}_1$ is the stage associated with the largest eigenvalue $\lambda_1$, $\boldsymbol{\Gamma}_2$ is the stage associated with the second largest eigenvalue $\lambda_2$, and so on. If we retain the first $k$ terms in the expansion, we may write

$$\boldsymbol{D} = \sum_{i=1}^{k} \sqrt{\lambda_i}\, \boldsymbol{\kappa}_i \boldsymbol{\varrho}_i + \boldsymbol{E}_k, \qquad k = 1, 2, \cdots, n \tag{58}$$

where the $(m \times n)$ matrix $\boldsymbol{E}_k$ is the truncation error matrix resulting from the neglect of all terms beyond the $k$th. Since the expansion is exact for $k = n$, we have that

$$\boldsymbol{E}_n = \boldsymbol{0}$$

where 0 is the $(m \times n)$ null matrix. Furthermore, the truncation error matrix must be equal to the sum of all neglected terms in the expansion (55).

$$\boldsymbol{E}_k = \sum_{i=k+1}^{n} \sqrt{\lambda_i}\, \boldsymbol{\kappa}_i \boldsymbol{\varrho}_i. \tag{59}$$

Although we could inspect the individual magnitudes of the $mn$ elements of $\boldsymbol{E}_k$ in order to determine how well the sum of the first $k$ separable stages approximate $\boldsymbol{D}$, it is more practical to measure this performance with a single number. A convenient scalar-valued error indicator is simply the sum of the squares of the elements in the matrix $\boldsymbol{E}_k$, which we call the "error energy" in $\boldsymbol{E}_k$, and which we identify with the symbol $J_k$,

$$J_k = \sum_{i=1}^{m} \sum_{j=1}^{n} e_{k,ij}^2$$

where $e_{k,ij}$ is an element of $\boldsymbol{E}_k$. But $J_k$ is also equal to the trace[11] of the $n$th-order matrix $\boldsymbol{S}_k$,

$$\boldsymbol{S}_k = \boldsymbol{E}_k^T \boldsymbol{E}_k$$

so that

$$J_k = \operatorname{Tr}(\boldsymbol{S}_k)$$

where the notation $\operatorname{Tr}(\cdot)$ denotes the trace of the matrix with parentheses.

[10] The usual convention is to label $\lambda_n$ the largest eigenvalue and $\lambda_1$ the smallest eigenvalue, but here we will find it convenient to reverse standard practice.

[11] The trace of a square matrix is equal to the sum of the elements on its principal diagonal. It is easily verified that the energy in any rectangular matrix $\boldsymbol{X}$ is equal to the trace of the square matrices $\boldsymbol{X}^T\boldsymbol{X}$ or $\boldsymbol{X}\boldsymbol{X}^T$.

It will be desirable to express $J_k$ as a function of the neglected terms in (55). Using (59) we have

$$\begin{aligned}\boldsymbol{E}_k^T \boldsymbol{E}_k &= \left[\sum_{i=k+1}^{n} \sqrt{\lambda_i}\, \boldsymbol{\kappa}_i \boldsymbol{\varrho}_i\right]^T \left[\sum_{j=k+1}^{n} \sqrt{\lambda_j}\, \boldsymbol{\kappa}_j \boldsymbol{\varrho}_j\right] \\ &= \sum_{i=k+1}^{n} \sum_{j=k+1}^{n} \sqrt{\lambda_i}\sqrt{\lambda_j}\, \boldsymbol{\varrho}_i^T \boldsymbol{\kappa}_i^T \boldsymbol{\kappa}_j \boldsymbol{\varrho}_j \end{aligned}\tag{60}$$

where $j$ is a convenient dummy summation index. From (46) and (47) we recall that $\boldsymbol{\kappa}_i$ and $\boldsymbol{\varrho}_i$ are the normalized eigencolumns and eigenrows of the symmetric matrices $\boldsymbol{Q}$ and $\boldsymbol{S}$, respectively. But the normalized eigenvectors of a symmetric matrix are mutally orthonormal, that is,

$$\boldsymbol{\kappa}_i^T \boldsymbol{\kappa}_j = \delta_{ij} \tag{61}$$

$$\boldsymbol{\varrho}_i \boldsymbol{\varrho}_j^T = \delta_{ij} \tag{62}$$

where

$$\delta_{ij} = \begin{cases} 1, & i = j \\ 0, & i \neq j \end{cases}$$

[8, pp. 32–33]. Substitution of the orthonormality relation (61) into (60) produces

$$\begin{aligned}\boldsymbol{E}_k^T \boldsymbol{E}_k &= \sum_{i=k+1}^{n} \sum_{j=k+1}^{n} \sqrt{\lambda_i}\sqrt{\lambda_j}\, \boldsymbol{\varrho}_i^T \delta_{ij} \boldsymbol{\varrho}_j \\ &= \sum_{i=k+1}^{n} \sqrt{\lambda_i}\, \boldsymbol{\varrho}_i^T \sum_{j=k+1}^{n} \sqrt{\lambda_j}\, \delta_{ij} \boldsymbol{\varrho}_j \\ &= \sum_{i=k+1}^{n} \sqrt{\lambda_i}\, \boldsymbol{\varrho}_i^T \big[\sqrt{\lambda_{k+1}}\, \delta_{i,k+1} \boldsymbol{\varrho}_{k+1} \\ &\qquad + \sqrt{\lambda_{k+2}}\, \delta_{i,k+2} \boldsymbol{\varrho}_{k+2} + \cdots + \sqrt{\lambda_n}\, \delta_{i,n} \boldsymbol{\varrho}_n\big].\end{aligned}$$

Since $\delta_{ij} = 1$ only when $i = j$ and vanishes otherwise, we obtain

$$\boldsymbol{S}_k = \boldsymbol{E}_k^T \boldsymbol{E}_k = \sum_{i=k+1}^{n} \lambda_i \boldsymbol{\varrho}_i^T \boldsymbol{\varrho}_i \tag{63}$$

which is the spectral representation of $\boldsymbol{S}_k$. Finally,

$$\begin{aligned} J_k &= \operatorname{Tr}(\boldsymbol{S}_k) \\ &= \operatorname{Tr}(\lambda_{k+1} \boldsymbol{\varrho}_{k+1}^T \boldsymbol{\varrho}_{k+1} + \cdots + \lambda_n \boldsymbol{\varrho}_n^T \boldsymbol{\varrho}_n).\end{aligned}$$

Now the trace of a sum of equal-order matrices is equal to the sum of the traces of the component matrices. Furthermore, each of these component matrices is the product obtained by multiplying the column vector $\lambda_i \boldsymbol{\varrho}_i^T$ into the row vector $\boldsymbol{\varrho}_i$.

For such matrices one easily verifies that

$$\operatorname{Tr}(\lambda_i \boldsymbol{\varrho}_i^T \boldsymbol{\varrho}_i) = \lambda_i \boldsymbol{\varrho}_i \boldsymbol{\varrho}_i^T = \lambda_i$$

where the last equality follows from the orthonormality relation (62). These considerations lead to the interesting result

$$\begin{aligned} J_k &= \lambda_{k+1} + \lambda_{k+2} + \cdots + \lambda_n \\ &= \sum_{i=k+1}^{n} \lambda_i \end{aligned}\tag{64}$$

and we see that the error energy resulting from the retention of the first $k$ terms of the expansion (55) is equal to the sum of the "neglected" eigenvalues $\lambda_{k+1}, \lambda_{k+2}, \cdots \lambda_n$. In other words, the energy in the truncation error matrix $\boldsymbol{E}_k$ of (58) is equal to the sum of the eigenvalues associated with the neglected stages of this approximate multistage expansion.

The particular magnitude of the error energy $J_k$ will be a function of the particular magnitudes of the elements of $\boldsymbol{D}$, and hence one finds it convenient to define the normalized energy $\epsilon_k$ in the form

$$\epsilon_k = J_k/J_0 \tag{65}$$

where $J_0$ is the energy in the matrix $\boldsymbol{D}$. The energy in $\boldsymbol{D}$ is equal to the trace of the matrix $\boldsymbol{S}=\boldsymbol{D}^T\boldsymbol{D}$,

$$J_0 = \sum_{i=1}^{m}\sum_{j=1}^{n} d_{ij}^2 = \text{Tr}\,(\boldsymbol{S}).$$

Then, using the spectral represention (52b) for $\boldsymbol{S}$ and proceeding as before, we find

$$J_0 = \sum_{i=1}^{n} \lambda_i.$$

Thus the energy in the given matrix $\boldsymbol{D}$ is equal to the sum of the eigenvalues of the matrix $\boldsymbol{S}=\boldsymbol{D}^T\boldsymbol{D}$. The normalized error energy $\epsilon_k$ now becomes

$$\epsilon_k = \frac{\lambda_{k+1} + \lambda_{k+2} + \cdots + \lambda_n}{\lambda_1 + \lambda_2 + \cdots + \lambda_n}$$

or

$$\epsilon_k = 1 - \frac{\lambda_1 + \lambda_2 + \cdots + \lambda_k}{\lambda_1 + \lambda_2 + \cdots + \lambda_n}, \quad k = 1, 2, \cdots, n. \tag{66}$$

In particular,

$$\epsilon_1 = 1 - \frac{\lambda_1}{\lambda_1 + \lambda_2 + \cdots + \lambda_n}$$

$$\epsilon_2 = 1 - \frac{\lambda_1 + \lambda_2}{\lambda_1 + \lambda_2 + \cdots + \lambda_n}$$

$$\cdots\cdots\cdots\cdots\cdots\cdots$$

$$\epsilon_n = 1 - 1 = 0.$$

Since the $\lambda_1$ are all real and positive, the bounds on $\epsilon_i$ are

$$1 > \epsilon_1 > \epsilon_2 > \cdots > \epsilon_{n-1}, \qquad \epsilon_n = 0$$

and thus the normalized error energy is always less than unity in magnitude no matter what the particular values taken on by the elements $d_{ij}$ of the given matrix $\boldsymbol{D}$ may be.

Up to this point we have assumed that the $n$ columns of $\boldsymbol{D}$ are linearly independent. This means that $\boldsymbol{S}=\boldsymbol{D}^T\boldsymbol{D}$ is positive definite, and hence has no vanishing eigenvalues. Suppose, however, that $p$ of the columns of $\boldsymbol{D}$ are not linearly independent. Then $\boldsymbol{S}$ is positive semidefinite, and $p$ of its $n$ eigenvalues $\lambda_i$ vanish [7, p. 29]. This means that the ordering (57) now becomes

$$\lambda_1, \lambda_2, \cdots, \lambda_{n-p}$$

so that $\lambda_{n-p}$ is the smallest nonvanishing eigenvalue of $\boldsymbol{S}$. As a result the multistage expansion (55) is exact for only $(n-p)$ stages. Equation (58) is rewritten in the form

$$\boldsymbol{D} = \sum_{i=1}^{k} \sqrt{\lambda_i}\,\boldsymbol{\kappa}_i\boldsymbol{\varrho}_i + \boldsymbol{E}_k, \qquad k = 1, 2, \cdots, n-p \tag{67}$$

where

$$\boldsymbol{E}_{n-p} = \boldsymbol{0}.$$

If $p=n-1$, the expansion (67) terminates with no error after the first stage. Therefore only one of the $n$ columns of $\boldsymbol{D}$ is linearly independent, so that all the columns (and rows) of $\boldsymbol{D}$ are proportional to each other [7]. This is just the condition that $\boldsymbol{D}$ must satisfy in order to be exactly separable with a single stage,

$$\boldsymbol{D} = \boldsymbol{\Gamma}_1 = \boldsymbol{c}_1\boldsymbol{r}_1 = \sqrt{\lambda_1}\,\boldsymbol{\kappa}_1\boldsymbol{\varrho}_1$$

for then the rows and columns of $\boldsymbol{\Gamma}_1$ and hence of $\boldsymbol{D}$ are proportional (compare with Section II). The present method yields the gratifying result of an error-free single-stage separation when $\boldsymbol{D}$ is so separable by construction.

Finally, if $p$ of the columns of $\boldsymbol{D}$ are not linearly independent, the normalized error energy (66) becomes

$$\epsilon_k = 1 - \frac{\lambda_1 + \lambda_2 + \cdots + \lambda_k}{\lambda_1 + \lambda_2 + \cdots + \lambda_{n-p}}, \qquad k = 1, 2, \cdots, n-p. \tag{68}$$

## Appendix II

### Least Squares Expansion of a Planar Array

Suppose that we are given the $(m\times n)$ matrix $\boldsymbol{D}$ and any nonnull *a priori* estimate of the $n$-length row vector $\boldsymbol{r}_1$. We first seek that $m$-length column vector $\boldsymbol{c}_1$ that minimizes the error energy $J_1$ in the error matrix $\boldsymbol{E}_1$

$$\boldsymbol{E}_1 = \boldsymbol{D} - \boldsymbol{c}_1\boldsymbol{r}_1$$

where

$$J_1 = \sum_{i=1}^{m}\sum_{j=1}^{n} e_{1,ij}^2 \tag{69}$$

and $e_{1,ij}$ is an element of $\boldsymbol{E}_1$. Since we can also write

$$\boldsymbol{D} = \boldsymbol{c}_1\boldsymbol{r}_1 + \boldsymbol{E}_1 \tag{70}$$

we see that the least squares solution $\boldsymbol{c}_1$ and the *a priori* estimate $\boldsymbol{r}_1$ will constitute a possible single-stage separation of $\boldsymbol{D}$ in the least squares sense.

Equation (70) may be written in the alternate form

$$d_{ij} = c_i r_j + e_{1,ij} \tag{71}$$

where $d_{ij}$, $c_i$, and $r_j$ are elements of $\boldsymbol{c}_1$, $\boldsymbol{r}_1$, and $\boldsymbol{E}_1$, respectively.

Substitution of (71) into (69) allows us to express the error energy $J_1$ in the form

$$J_1 = \sum_{i=1}^{m} \sum_{j=1}^{n} (d_{ij} - c_i r_j)^2. \quad (72)$$

In order to find the elements $c_i$ that minimize $J_1$ we differentiate this expression with respect to each of the elements $c_l (l=1, 2, \cdots, m)$ and set the resulting system of $m$ normal equations equal to zero,

$$\begin{aligned} \frac{\partial J_1}{\partial c_l} &= \frac{\partial}{\partial c_l} \sum_{i=1}^{m} \sum_{j=1}^{n} (d_{ij}^2 - 2d_{ij}c_i r_j + c_i^2 r_j^2) \\ &= -2 \sum_{j=1}^{n} d_{ij} r_j \frac{\partial}{\partial c_l} \sum_{i=1}^{m} c_i + \sum_{j=1}^{n} r_j^2 \frac{\partial}{\partial c_l} \sum_{i=1}^{m} c_i^2 \\ &= -2 \sum_{j=1}^{n} d_{lj} r_j + 2c_l \sum_{j=1}^{n} r_j^2 \\ &= 0, \qquad l = 1, 2, \cdots, m. \end{aligned} \quad (73)$$

This system is linear in the unknown elements $c_l$, and its solution can be found directly from the last of the preceding relations,

$$c_l = \sum_{j=1}^{n} d_{lj} r_j \bigg/ \sum_{j=1}^{n} r_j^2, \qquad l = 1, 2, \cdots, m \quad (74)$$

or using matrix notation,

$$\mathbf{c}_1 = \frac{\boldsymbol{D}\boldsymbol{r}_1^T}{\boldsymbol{r}_1\boldsymbol{r}_1^T}. \quad (75)$$

This column vector thus minimizes the error energy in the single-stage expansion of $\boldsymbol{D}$ for a given known *a priori* estimate of the row vector $\boldsymbol{r}_1$.

If the elements of $\boldsymbol{E}_1$ are not small in comparison to the elements of $\boldsymbol{D}$, we may apply this technique once more, but this time to $\boldsymbol{E}_1$ rather than to $\boldsymbol{D}$ itself. Let $\boldsymbol{r}_2$ be a nonnull *a priori* estimate in the separation

$$\boldsymbol{E}_1 = \mathbf{c}_2\boldsymbol{r}_2 + \boldsymbol{E}_2 \quad (76)$$

where $\mathbf{c}_2$ will be computed such that the error energy $J_2$ in the error matrix $\boldsymbol{E}_2$ is minimized. If we compare (76) with (70), we see that the least squares problems connected with these relations have identical structure, and hence the solution for $\mathbf{c}_2$ may be written down by appropriately extending the solution for $\mathbf{c}_1$ given by (75),

$$\mathbf{c}_2 = \frac{\boldsymbol{E}_1\boldsymbol{r}_2^T}{\boldsymbol{r}_2\boldsymbol{r}_2^T}. \quad (77)$$

Finally, if we desire a $k$-stage expansion in this sense, we deduce inductively that

$$\mathbf{c}_i = \frac{\boldsymbol{E}_{i-1}\boldsymbol{r}_i^T}{\boldsymbol{r}_i\boldsymbol{r}_i^T}, \qquad i = 1, 2, \cdots, k. \quad (78)$$

For $i=1$, this solution reduces to (75) if we set $\boldsymbol{E}_0 = \boldsymbol{D}$. We are thus able to expand $\boldsymbol{D}$ in the form

$$\boldsymbol{D} = \mathbf{c}_1\boldsymbol{r}_1 + \mathbf{c}_2\boldsymbol{r}_2 + \cdots + \mathbf{c}_k\boldsymbol{r}_k + \boldsymbol{E}_k \quad (79)$$

where $\boldsymbol{E}_k$ is the error matrix resulting from this $k$-stage least squares expansion of $\boldsymbol{D}$.

The minimum error energy $\hat{J}_1$ for the single-stage expansion of $\boldsymbol{D}$ is obtained by substituting the solution (74) into the expression for the error energy (72). After some algebra this gives

$$\hat{J}_1 = J_0 - \frac{\boldsymbol{r}_1\boldsymbol{D}^T\boldsymbol{D}\boldsymbol{r}_1^T}{\boldsymbol{r}_1\boldsymbol{r}_1^T}$$

where $J_0 = \sum_{i=1}^{m} \sum_{j=1}^{n} d_{ij}^2$ = energy in the given matrix $\boldsymbol{D}$ [11]. It is convenient to deal with the normalized minimum square error (NMSE), which we define to be

$$\begin{aligned} \epsilon^{(1)} &= \hat{J}_1/J_0 \\ &= 1 - \frac{1}{J_0} \frac{\boldsymbol{r}_1\boldsymbol{D}^T\boldsymbol{D}\boldsymbol{r}_1^T}{\boldsymbol{r}_1\boldsymbol{r}_1^T}. \end{aligned} \quad (80)$$

But $\epsilon^{(1)}$ is a sum of squares and cannot hence be negative. The matrix $\boldsymbol{D}$ and the *a priori* vector estimates $\boldsymbol{r}_i$ are nonnull by hypothesis. Therefore the bounds on $\epsilon^{(1)}$ are

$$0 \leq \epsilon^{(1)} \leq 1$$

[11].

It can be similarly shown that the NMSE for the least squares separation of $\boldsymbol{E}_1$ (76), namely $\epsilon^{(2)}$, also satisfies these bounds, and in general

$$0 \leq \epsilon^{(i)} \leq 1, \qquad (i = 1, 2, \cdots, k) \quad (81)$$

where $\epsilon^{(i)}$ is the NMSE for the least squares separation of $\boldsymbol{E}_{i-1}$,

$$\boldsymbol{E}_{i-1} = \mathbf{c}_i\boldsymbol{r}_i + \boldsymbol{E}_i. \quad (82)$$

Moreover, (80) can be generalized in the form

$$\epsilon^{(i)} = \hat{J}_i/\hat{J}_{i-1}, \qquad i = 1, 2, \cdots, k \quad (83)$$

where $\hat{J}_i$ is the energy in the error matrix $\boldsymbol{E}_1$, which is minimum for a given *a priori* choice of the vector $\boldsymbol{r}_i$ and where for $i=1$, we set $\hat{J}_0 = \hat{J}_0$, the energy in $\boldsymbol{D}$.

But (83) can be written

$$\hat{J}_i = \epsilon^{(i)}\hat{J}_{i-1}$$

and in particular,

$$\begin{aligned} \hat{J}_1 &= \epsilon^{(1)}J_0 \\ \hat{J}_2 &= \epsilon^{(2)}\hat{J}_1 = \epsilon^{(2)}\epsilon^{(1)}J_0 \\ &\cdots\cdots\cdots\cdots \\ \hat{J}_k &= \epsilon^{(k)}\hat{J}_{k-1} = \epsilon^{(k)}\epsilon^{(k-1)} \cdots \epsilon^{(1)}J_0. \end{aligned}$$

The last of these relations can be written,

$$\hat{J}_k = J_0 \prod_{i=1}^{k} \epsilon^{(i)}.$$

Now $\hat{J}_k/J_0$ is the ratio of the energy in $\boldsymbol{E}_k$ to the energy in $\boldsymbol{D}$, and hence we have

$$\epsilon_{(k)} = \hat{J}_k/J_0 = \prod_{i=1}^{k} \epsilon^{(i)} \quad (84)$$

where $\epsilon_{(k)}$ is the *cumulative* NMSE for a $k$-stage expansion of $\boldsymbol{D}$. But the bounds (81) guarantee that $\epsilon_{(k)}<1$ unless all $\epsilon^{(i)}$ is of unit magnitude.[12] If no $\epsilon^{(i)}$ is of unit magnitude for $i=1, 2, \cdots, k$, we conclude from (84) that the least squares multistage expansion (79) converges uniformly to $\boldsymbol{D}$ as $k$ increases. If for any stage $\epsilon^{(i)}=0$, the least squares expansion represents $\boldsymbol{D}$ exactly with a finite number of stages.

Although this result is gratifying at first thought, it becomes less so when one reflects that an inordinately large number of stages may be required to bring $\epsilon_{(k)}$ down to an acceptable small level. Up to this point we have assumed that the least squares computation of the vector $\mathbf{c}_i$ is always carried out with an arbitrary, nonnull estimate of the row vector $\boldsymbol{r}_i$ (82).

Instead, let us now go back to the solution for $\mathbf{c}_1$ given by (75), and let us there *identify* the vector $\boldsymbol{r}_1$ with $\boldsymbol{\varrho}_1$, where $\boldsymbol{\varrho}_1$ is the normalized eigenrow associated with the largest eigenvalue $\lambda_1$ of the matrix $\boldsymbol{S}=\boldsymbol{D}^T\boldsymbol{D}$. Then (75) becomes

$$\mathbf{c}_i = \boldsymbol{D}\boldsymbol{\varrho}_1^T \tag{85}$$

since $\boldsymbol{r}_1\boldsymbol{r}_1^T=\boldsymbol{\varrho}_1\boldsymbol{\varrho}_1^T=1$. If we compare (85) with (23), we obtain the remarkable result that the single-stage least squares solution for $\mathbf{c}_1$ is now identical to the solution for $\mathbf{c}_1$ obtained via the spectral representation of $\boldsymbol{D}$.

Next, let us identify the vector $\boldsymbol{r}_2$ of (77) with $\boldsymbol{\varrho}_2$, where $\boldsymbol{\varrho}_2$ is the normalized eigenrow associated with the second-largest eigenvalue, $\lambda_2$, of $\boldsymbol{S}$. Then (77) becomes

$$\mathbf{c}_2 = \boldsymbol{E}_1\boldsymbol{\varrho}_2^T. \tag{86}$$

But we have previously seen that $\boldsymbol{E}_1$ can be expressed as the difference,

$$\boldsymbol{E}_1 = \boldsymbol{D} - \mathbf{c}_1\boldsymbol{r}_1$$

which by the result just obtained is

$$\boldsymbol{E}_1 = \boldsymbol{D} - \mathbf{c}_1\boldsymbol{\varrho}_1.$$

Substituting this expression in (86) we obtain

$$\begin{aligned}\mathbf{c}_2 &= (\boldsymbol{D} - \mathbf{c}_1\boldsymbol{\varrho}_1)\boldsymbol{\varrho}_2^T \\ &= \boldsymbol{D}\boldsymbol{\varrho}_2^T - \mathbf{c}_1\boldsymbol{\varrho}_1\boldsymbol{\varrho}_2^T. \qquad (87)\end{aligned}$$

It has been previously state (62) that the $n$ normalized eigenrows $\boldsymbol{\varrho}_i$ $(i=1, 2, \cdots, n)$ of the $n$th order symmetric matrix $\boldsymbol{S}$ form an orthonormal set[13]

$$\boldsymbol{\varrho}_i\boldsymbol{\varrho}_j^T = \delta_{ij}, \qquad \delta_{ij} = \begin{cases}1, & i=j\\ 0, & i\neq j.\end{cases}$$

[12] From (80) we deduce that $\epsilon^{(1)}=1$ if and only if $\boldsymbol{r}_1$ is orthogonal to all rows of $\boldsymbol{D}$. Similarly, $\epsilon^{(2)}=1$ if and only if $\boldsymbol{r}_2$ is orthogonal to all rows of $\boldsymbol{E}_1$, and so on. One could at each stage avoid a unit value for the NMSE by chosing an $\boldsymbol{r}_i$ that is not orthogonal to all rows of $\boldsymbol{E}_{i-1}$. This entire question is obviated in view of our subsequent results.

[13] We assume for the sake of simplicity that $\boldsymbol{S}$ has no vanishing eigenvalues, i.e., that $\boldsymbol{S}$ is positive definite. The present results are easily generalized for the case when $\lambda_i$ is a $p$-fold null eigenvalue of $\boldsymbol{S}$ (see also Appendix I).

In particular, this implies

$$\boldsymbol{\varrho}_i\boldsymbol{\varrho}_2^T = 0$$

and thus the second term of (87) vanishes so that

$$\mathbf{c}_2 = \boldsymbol{D}\boldsymbol{\varrho}_2^T. \tag{88}$$

If we compare (88) with (23) we find that the second stage of the least squares expansion of $\boldsymbol{D}$ is now identical to the second stage as determined from the spectral representation of $\boldsymbol{D}$. This result, together with the previous result (85), then implies that the two-stage least squares expansion of $\boldsymbol{D}$ is equivalent term by term to the expansion obtained via its spectral representation.

It is evident that if in the general least squares solution (78) we identify $\boldsymbol{r}_i$ with $\boldsymbol{\varrho}_i$, there results

$$\begin{aligned}\mathbf{c}_i &= \boldsymbol{E}_{i-1}\boldsymbol{\varrho}_i^T\\ &= (\boldsymbol{D} - \mathbf{c}_1\boldsymbol{\varrho}_1 - \mathbf{c}_2\boldsymbol{\varrho}_2 - \cdots - \mathbf{c}_{i-1}\boldsymbol{\varrho}_{i-1})\boldsymbol{\varrho}_i^T\\ &= \boldsymbol{D}\boldsymbol{\varrho}_i^T, \qquad i = 1, 2, \cdots, n\end{aligned}$$

where we have made use of (79) to express $\boldsymbol{E}_{i-1}$ in terms of $\boldsymbol{D}$ and the previously computed separable stages, and where we have again used the fact that the $\boldsymbol{\varrho}_i$ are orthonormal.

Therefore the identification $\boldsymbol{r}_i=\boldsymbol{\varrho}_i$ for $i=1, 2, \cdots, n$ renders the least squares expansion for $\boldsymbol{D}$ term by term equivalent to the spectral representation for $\boldsymbol{D}$. Moreover, the least squares expansion of $\boldsymbol{D}$ then terminates with no error for a total of $k=n$ stages (see Section II-A and Appendix I).

It may be of interest to note that the authors arrived at the spectral representation of $\boldsymbol{D}$ only after having obtained the least squares expansion of $\boldsymbol{D}$ in an arbitrary, not necessarily finite number of stages [Compare with (79) and (84))]. If for no other reason, this Appendix has value because it illustrates that research rarely proceeds in the chronological sequence in which it is finally presented!

Finally, we remark that an identical result will be achieved if we pose the original problem in such a way that the column vectors $\mathbf{c}_i$ constitute *a priori* estimates; the row vectors $\boldsymbol{r}_i$ are then the solutions to appropriate least squares problems.

## Appendix III

## Number of "Multiply and Add" Operations Required for the Implementation of Planar Filtering Algorithms

We begin by computing the number of "multiply and adds" MADS required for the one-dimensional convolution of an $m$-length sequence $a_i$ with an $M$-length sequence $b_j$. In terms of one-dimensional $z$-transforms, we have

$$\begin{aligned}C(z) &= A(z)B(z)\\ &= \left\{\sum_{i=1}^{m} a_i z^{i-1}\right\}\left\{\sum_{j=1}^{M} b_j z^{j-1}\right\}\end{aligned} \tag{89}$$

where $C(z)$ is the $z$ transform of the $(m+M-1)$-length output sequence $c_k$. When we carry out the preceding polynomial multiplication, we observe that each of the $m$ coefficients $a_i$ is multiplied by each of the $M$ coefficients $b_j$. Thus the number of MADS required to convolve the two sequences is simply $mM$, and the total length of the output sequence is $m+M-1$.

Let us next determine the number of MADS required for the implementation of the algorithm given by (32a)

$$Y(z_1, z_2) = W(z_1, z_2)R(z_2) \tag{90}$$

where $W(z_1, z_2)$ is the intermediate output,

$$W(z_1, z_2) = X(z_1, z_2)C(z_1). \tag{91}$$

We first carry out the operations associated with (91). The $m$-length column vector $\boldsymbol{C}$ is convolved with each of the $N$ $M$-length columns of $\boldsymbol{X}$. Since each such one-dimensional convolution requires $mM$ MADS, and since $\boldsymbol{X}$ has $N$ columns, the total number of MADS needed for the implementation of (91) is $mMN$. In order to carry out the operations associated with (90), we notice that the intermediate output array $\boldsymbol{W}$ contains $m+M-1$ rows, since every column of $\boldsymbol{W}$ resulted from the one-dimensional convolution of the $m$-length column vector $\boldsymbol{C}(z_1)$ with each $M$-length column of $\boldsymbol{X}$. Each $N$-length row of $W$ is convolved with the $n$-length row vector $\boldsymbol{R}$, so that $nN$ MADS are required. But since $W$ contains $(m+M-1)$ rows, the total number of MADS needed for the implementation of (89) is $(m+M-1)$ $(nN)$. Therefore the number of MADS necessary for the combined evaluation of (90) and (91) is

$$mMN + (m + M - 1)(nN)$$

or after rearrangement,

$$MN(m + n) + n(m - 1)N$$

which is the first of the relations (33).

The number of MADS required for the implementation of the filtering algorithm (32b) can be obtained in a similar manner, and this analysis then leads to the second of the relations (33).

Finally, we determine the number of MADS needed for the implementation of the standard planar convolution algorithm (26),

$$Y(z_1, z_2) = X(z_1, z_2)F(z_1, z_2). \tag{92}$$

The planar $z$ transform of the filter $F(z_1, z_2)$ can be written

$$F(z_1, z_2) = \sum_{j=1}^{n} F_j(z_1)z_2^{j-1}$$

where $F_j(z_2)$ is the one-dimensional $z$ transform of the $j$th column of $\boldsymbol{F}$. Similarly, $X(z_1, z_2)$ can be written

$$X(z_1, z_2) = \sum_{k=1}^{N} X_k(z_1)z_2^{k-1}$$

where $X_k(z_1)$ is the one-dimensional $z$ transform of the $k$th column of $\boldsymbol{X}$. Thus (92) becomes

$$Y(z_1, z_2) = \left\{\sum_{k=1}^{N} X_k(z_1)z_2^{k-1}\right\} \left\{\sum_{j=1}^{n} F_j(z_1)z_2^{j-1}\right\}. \tag{93}$$

We observe that (93) is structurally similar to (89), except that now the "coefficients" $X_k$ and $F_j$ are polynomials in $z_1$, rather than mere constants. Then by analogy with our analysis of (89) we find that the evaluation of (93) requires the computation of $nN$ polynomial products of the form

$$X_k(z_1)F_j(z_1).$$

But each such product represents the one-dimensional convolution of the $M$-length column vector $\boldsymbol{X}_k$ with the $m$-length column vector $\boldsymbol{F}_j$, for whose evaluation $mM$ MADS are needed. Since there are $nN$ convolutions of this kind, we deduce that the standard planar convolution algorithm (26) requires a total of $mn$ $MN$ MADS, which is the result (35).

## References


[1] G. M. Ackins and D. J. Kuck, "Seismic signal processing via the ILLIAC IV computer," *IEEE Trans. Geosci. Electron.*, vol. GE-7, Jan. 1969, pp. 34–41.

[2] W. C. Dean, "Frequency analysis for gravity and magnetic interpretation," *Geophysics*, vol. 23, 1958, pp. 97–127.

[3] P. M. De Russo, R. J. Roy, and C. M. Close, *State Variables for Engineers.* New York: Wiley, 1965.

[4] J. P. Fail and G. Grau, "Les filtres en eventail," *Geophys. Prospect.*, vol. 11, 1963, pp. 131–163.

[5] J. B. Farr, "Acoustical holography experiments using digital processing," in *Acoustical Holography*, vol. 2. New York: Plenum, 1970, pp. 225–242.

[6] M. R. Foster, R. L. Sengbush, and R. J. Watson, "Suboptimum filter systems for seismic data processing," *Geophys. Prospect.*, vol. 12, 1964, pp. 173–191.

[7] R. A. Frazer, W. J. Duncan, and A. R. Collar, *Elementary Matrices.* New York: Cambridge, 1938.

[8] F. B. Hildebrand, *Methods of Applied Mathematics*, 2nd ed. Englewood Cliffs, N. J.: Prentice-Hall, 1965.

[9] E. I. Jury, *Theory and Application of the Z-Transform Method.* New York: Wiley, 1964.

[10] C. Lanczos, *Linear Differential Operators.* New York: Van Nostrand-Reinhold, 1961.

[11] E. A. Robinson and S. Treitel, "Principles of digital Wiener filtering," *Geophys. Prospect.*, vol. 15, 1967, pp. 311–333.

[12] C. K. Sanathanan, "On the synthesis of space dependent transfer functions," *IEEE Trans. Automat. Contr.*, vol. AC-11, Oct. 1966, pp. 724–729.

[13] J. L. Shanks, "Recursion filters for digital processing," *Geophysics*, vol. 32, 1967, pp. 33–51.

[14] J. L. Shanks, "Two-dimensional recursive filters," *SWIEEECO Rec.*, 1969, pp. 19E1–19 E8.

[15] S. Treitel, J. L. Shanks, and C. W. Frasier, "Some aspects of fan filtering," *Geophysics*, vol. 32, 1967, pp. 789–800.

[16] P. Whittle, *Prediction and Regulation.* London: The English Universities Press, 1963.

[17] R. A. Wiggins, "ω-k filter design," *Geophys. Prospect.*, vol. 14, 1966, pp. 427–440.

[18] J. H. Wilkinson, *The Algebraic Eigenvalue Problem.* New York: Oxford, 1965.

# An $l_p$ Design Technique for Two-Dimensional Digital Recursive Filters

GAMAL A. MARIA, Student Member, IEEE, and
MOUSTAFA M. FAHMY, Member, IEEE

*Abstract*—In this paper a design technique for the two-dimensional filters is proposed. An optimization algorithm is developed to minimize the $p$-error criterion under the constraint that the resulting filter be stable. Design of one-dimensional filter may be considered as a special case to which the proposed algorithm is applicable. Several examples are solved to illustrate the technique.

## I. Introduction

Recently there has been increasing interest in using the two-dimensional digital filters in processing a variety of two-dimensional signals. Some of the design techniques available for the one-dimensional filters have been extended to the two-dimensional case [1], [2]. In this paper an approximation technique is proposed for the design of two-dimensional recursive filters consisting of first and second order sections connected in cascade. This cascade form has the following advantages over the direct form.[1]

1) The word length that guarantees the stability of the filter after the rounding of its coefficients is shorter. This is due to the fact that the sensitivity of the mapping of the unit circle of the $Z_2^{-1}$-plane into the $Z_1^{-1}$-plane under the denominator of the transfer function is lower [3].

2) The error caused by quantization effect and finite coefficient size is lower [4].

3) In two-dimensional filters, the number of multiplications needed for each output using the cascade form is less than that needed when using the direct form.

4) It is easier to check the stability of the cascade filter during the optimization process and make any necessary changes in the coefficients of the unstable section to restore the overall stability. This guarantees that the resulting filter will be stable.

Unlike the one-dimensional filters, it is not always possible, in the two-dimensional case, to factorize the transfer function

$$F(Z_1^{-1}, Z_2^{-1}) = \frac{\sum_{j=0}^{K_1^2} \sum_{i=0}^{K_1^1} a_{ij} Z_1^{-(i-1)} Z_2^{-(j-1)}}{1 + \sum_{\substack{j=0 \\ i,j \neq 0}}^{K_2^2} \sum_{i=0}^{K_2^1} d_{ij} Z_1^{-(i-1)} Z_2^{-(j-1)}}$$

as a product of first and second order sections. Therefore, in this paper the approximating function is chosen, from the start, to be in the following form:

Manuscript received January 31, 1973; revised May 1, 1973 and July 27, 1973.

The authors are with the Department of Electrical Engineering, Queen's University, Kingston, Ont., Canada.

[1] Note that these advantages, except (3), are valid for both one-dimensional and two-dimensional filters.

Reprinted from *IEEE Trans. Acoust., Speech, Signal Processing,* vol. ASSP-22, pp. 15–21, Feb. 1974.

$$F(Z_1^{-1}, Z_2^{-1}) = A \frac{\prod_{l=1}^{K_1} N^{(l)}(Z_1^{-1}, Z_2^{-1})}{\prod_{l=1}^{K_2} D^{(l)}(Z_1^{-1}, Z_2^{-1})} \tag{1}$$

where

$$N^{(l)}(Z_1^{-1}, Z_2^{-1}) = [1\ Z_1^{-1} \cdots Z_1^{-(n_l^1-1)}] \cdot \begin{bmatrix} 1 & a_{21}^l & \cdots & a_{1,m_l^1}^l \\ a_{21}^l & & & \\ \vdots & & & \\ a_{n_l^1,1}^l & & & a_{n_l^1,m_l^1}^l \end{bmatrix} \begin{bmatrix} 1 \\ Z_2^{-1} \\ \vdots \\ Z_2^{-(m_l^1-1)} \end{bmatrix},$$

$$D^{(l)}(Z_1^{-1}, Z_2^{-1}) = [1\ Z_1^{-1} \cdots Z_1^{-(n_l^2-1)}] \cdot \begin{bmatrix} 1 & d_{12}^l & \cdots & d_{1,m_l^2}^l \\ d_{21}^l & & & \\ \vdots & & & \\ d_{n_l^2,1}^l & & & d_{n_l^2,m_l^2}^l \end{bmatrix} \begin{bmatrix} 1 \\ Z_2^{-1} \\ \vdots \\ Z_2^{-(m_l^2-1)} \end{bmatrix}$$

and $n_l^1$, $n_l^2$, $m_l^1$, $m_l^2$ can only assume the numerical value two and three.

In this paper, the problem is first mathematically formulated, and then the optimization technique proposed is developed and illustrated by several examples.

## II. Statement of the Problem

The problem considered here can be stated as follows. Let the magnitude characteristic $Y = [Y_{mn}]$ defined on the discrete set of frequency pairs $(\omega_m^{(1)}, \omega_n^{(2)})$, $m = 1, \cdots, M$; $n = 1, \cdots, N$, be given. Let

$$\Phi = [\Phi_1^T, \Phi_2^T, A]^T$$

where

$$\Phi_1^T = [a_{21}^1, a_{31}^1, \cdots, a_{n_1^1,1}^1, a_{12}^1, a_{22}^1, \cdots, a_{n_1^1,2}^1, \cdots, a_{n_1^1,m_1^1}^1, \cdots, a_{n_{K_1}^1, m_{K_1}^1}^{K_1}]$$

$$\Phi_2^T = [d_{21}^1, d_{31}^1, \cdots, d_{n_1^2,1}^1, d_{12}^1, d_{22}^1, \cdots, d_{n_1^2,2}^1, \cdots, d_{n_1^2,m_1^2}^1, \cdots, d_{n_{K_2}^2, m_{K_2}^2}^{K_2}]$$

and the $a$'s, $d$'s, and $A$ are the parameters of the transfer function

$$F(Z_1^{-1}, Z_2^{-1}) = A \frac{\prod_{l=1}^{K_1} N^{(l)}(Z_1^{-1}, Z_2^{-1})}{\prod_{l=1}^{K_2} D^{(l)}(Z_1^{-1}, Z_2^{-1})}$$

defined in Section I. Here $(\omega^{(1)}, \omega^{(2)})$ are related to the variables $(Z_1, Z_2)$ by

$$(Z_{1m}, Z_{2m}) = (e^{j\omega_m^{(1)}\Pi}, e^{j\omega_n^{(2)}\Pi}).$$

Let the performance index $J(\Phi)$ be defined as

$$J(\Phi) = \sum_{m=1}^{M} \sum_{n=1}^{N} [|F_{mn}| - Y_{mn}]^p \tag{2}$$

where $p$ is an even positive integer.[2] It is required to select the elements of the vector $\Phi$ that minimize the performance index $J(\Phi)$ under the constraint that the resulting filter $F(Z_1^{-1}, Z_2^{-1})$ be stable.

## III. The Optimization Procedure

Here we will use the same optimization procedure used in [5]. The algorithm starts by assuming initial values for the elements of the parameter vector $\Phi$. After each iteration the value of these elements are changed by the increment $\Delta\Phi$ to be chosen such that

$$\frac{dJ(\Phi + \Delta\Phi)}{d\Delta\Phi} \cong \frac{d(J(\Phi) + \nabla J^T \cdot \Delta\Phi + \frac{1}{2}(\Delta\Phi)^T \cdot [H] \cdot \Delta\Phi)}{d\Delta\Phi} = 0.$$

From the above, we get the set of linear equations

$$H\Delta\Phi = -\nabla J \tag{3}$$

where $H$ is the Hessian matrix with the element

$$h(f, s) = \frac{\partial^2 J}{\partial\phi_f\, \partial\phi_s},$$

and $\nabla J$ is the gradient vector with the element

$$\nabla J_i = \frac{\partial J}{\partial \phi_i}.$$

This optimization procedure calls for the computation of the gradient vector $\nabla J$ and the Hessian matrix $H$ and the solution of the system of linear equations (3) in the unknown $\Delta\Phi$.

The elements of the gradient vector can be calculated as follows:

$$\nabla_k J = \frac{\partial J}{\partial \phi_k} = \sum_{m=1}^{M} \sum_{n=1}^{N} p[|F_{mn}| - Y_{mn}]^{p-1} \frac{\partial |F_{mn}|}{\partial \phi_k}. \tag{4}$$

[2] Here $|F_{mn}| = |F(Z_{1m}^{-1}, Z_{2n}^{-1})|$.

It can be easily shown that

$$\frac{\partial |F_{mn}|}{\partial \phi_k} = \frac{1}{|F_{mn}|} \operatorname{Re}\left(\overline{F}_{mn} \cdot \frac{\partial F_{mn}}{\partial \phi_k}\right) = |F_{mn}| \operatorname{Re}(X_k(m,n)) \tag{5}$$

where $X_k(m, n)$ are the components of $\overline{X}$ defined by

$$\overline{X}^T = (X_1(m,n), \cdots, X_k(m,n), \cdots)$$
$$= \left(\alpha_{21}^1, \alpha_{31}^1, \cdots, \alpha_{ij}^l, \cdots, \alpha_{n_{K_1}^1, m_{K_1}^1}^{K_1}, \beta_{21}^1, \beta_{31}^1, \cdots, \beta_{ij}^l, \cdots, \beta_{n_{K_2}^2, m_{K_2}^2}^{K_2}, \frac{1}{A}\right)$$

and

$$\alpha_{ij}^l = \frac{Z_{1m}^{-(i-1)} Z_{2m}^{-(j-1)}}{N^{(l)}(Z_{1m}^{-1} Z_{2n}^{-1})}, \quad l = 1, \cdots, K_1, \quad i = 1, \cdots, n_l^1, \quad j = 1, \cdots, m_l^1, \quad i, j \neq 1$$

$$\beta_{ij}^l = -\frac{Z_{1m}^{-(i-1)} Z_{2n}^{-(j-1)}}{D^{(l)}(Z_{1m}^{-1} Z_{2n}^{-1})}, \quad l = 1, \cdots, K_2, \quad i = 1, \cdots, n_l^2, \quad j = 1, \cdots, m_l^2, \quad i, j \neq 1.$$

Substituting (5) in (4), we get

$$\nabla_k J = \sum_{m=1}^{M} \sum_{n=1}^{N} p\, \epsilon_{m,n}^{p-1} |F_{mn}| \operatorname{Re}[X_k(m,n)] \tag{6}$$

where $\epsilon_{m,n} = (|F_{mn}| - Y_{mn})$. Equation (6) is used to compute the gradient vector.

Turning now to the problem of computing the elements of the Hessian matrix defined in equation (3), we proceed as follows:

$$h(f,s) = \frac{\partial^2 J}{\partial \phi_f \partial \phi_s}$$
$$= \sum_{m=1}^{M} \sum_{n=1}^{N} p\, \epsilon_{m,n}^{p-1} \frac{\partial^2 |F_{mn}|}{\partial \phi_f \partial \phi_s} + p(p-1)\, \epsilon_{m,n}^{p-2} \frac{\partial |F_{mn}|}{\partial \phi_f} \frac{\partial |F_{mn}|}{\partial \phi_s}$$
$$= \sum_{m=1}^{M} \sum_{n=1}^{N} p\, \epsilon_{m,n}^{p-1} \frac{\partial^2 |F_{mn}|}{\partial \phi_f \partial \phi_s} + p(p-1)\, \epsilon_{m,n}^{p-2} \cdot |F_{mn}|^2 \operatorname{Re}(X_f(m,n)) \operatorname{Re}(X_s(m,n)). \tag{7}$$

To evaluate the first term, we must get the value of $(\partial^2 |F_{mn}|)/(\partial \phi_f \partial \phi_s)$. It can be easily shown that

$$\frac{\partial^2 |F_{mn}|}{\partial \phi_f \partial \phi_s} = \frac{1}{|F_{mn}|} \operatorname{Re}\left(\overline{F}_{mn} \cdot \frac{\partial^2 F_{mn}}{\partial \phi_f \partial \phi_s}\right) + |F_{mn}| \operatorname{Im}(X_f(m,n)) \operatorname{Im}(X_s(m,n)). \tag{8}$$

Noting that

$$\frac{\partial^2 F_{mn}}{\partial a_{ij}^l \partial a_{ce}^l} = 0 \tag{9a}$$

$$\frac{\partial^2 F_{mn}}{\partial A^2} = 0 \tag{9b}$$

$$\frac{\partial^2 F_{mn}}{\partial d_{ij}^l \partial d_{ce}^l} = 2F_{mn} \beta_{ij}^l \beta_{ce}^l \tag{9c}$$

and

$$\frac{\partial^2 F_{mn}}{\partial \phi_f \partial \phi_s} = F_{mn} X_f(m,n) \cdot X_s(m,n) \tag{9d}$$

for all other combination of $f$, $s$, it follows that $(\partial^2 |F_{mn}|)/(\partial \phi_f \partial \phi_s)$ takes the following form:

$$\frac{\partial^2 |F_{mn}|}{\partial a_{ij}^l \partial a_{ce}^l} = |F_{mn}| \operatorname{Im}(\alpha_{ij}^l) \operatorname{Im}(\alpha_{ce}^l) \tag{10a}$$

$$\frac{\partial^2 |F_{mn}|}{\partial A^2} = 0 \tag{10b}$$

$$\frac{\partial^2 |F_{mn}|}{\partial d_{ij}^l \partial d_{ce}^l} = |F_{mn}| [2 \operatorname{Re}(\beta_{ij}^l) \operatorname{Re}(\beta_{ce}^l) - \operatorname{Im}(\beta_{ij}^l) \operatorname{Im}(\beta_{ce}^l) \tag{10c}$$

and

$$\frac{\partial^2 |F_{mn}|}{\partial \phi_f \partial \phi_s} = |F_{mn}| \operatorname{Re}(X_f(m,n)) \operatorname{Re}(X_s(m,n)) \tag{10d}$$

for all other combinations of $f$, $s$.

Substituting these values in (7) we can compute all the elements of the Hessian matrix.

### IV. Steps of the Computer Program Used for Optimization

1) Assume initial values for the elements of the parameter vector $\Phi$.

2) Compute $(X_k(m, n))$ for all $k$. Start by computing

$$\gamma_{1_l}(m,n) = \frac{1}{N^{(l)}(Z_{1m}^{-1}, Z_{2n}^{-1})}, \quad l = 1, \cdots, K_1$$

and

$$\gamma_{2_l}(m,n) = \frac{1}{D^{(l)}(Z_{1m}^{-1}, Z_{2n}^{-1})}, \quad l = 1, \cdots, K_2.$$

Store these values to be used in computing $|F_{mn}|$.

3) Compute $|F_{mn}|$ using the formula

$$|F_{mn}| = A \frac{\prod_{l=1}^{K_2} |\gamma_{2_l}(m,n)|}{\prod_{l=1}^{K_1} |\gamma_{1_l}(m,n)|}.$$

4) Compute the gradient vector $\nabla J$ using (6).

5) Compute the Hessian matrix $H$ using (7), (10).

6) Solve the linear equations (3).

7) Put $\Phi = \Phi + \Delta\Phi$.

8) Examine the stability of the filter [8]. If unstable, successively reduce by half the incremental change in the coefficients of the unstable sections until the filter becomes stable.

9) Repeat these steps (2–8) until the approximation is satisfactory (i.e., change in the performance index is less than a specified number) or a certain number of iteration has been completed.

### V. Concluding Remarks

1) If the function $Y$ is symmetrical with respect to $\omega^{(1)}$ and $\omega^{(2)}$, we can put a constraint on the coefficients of $F(Z_1^{-1}, Z_2^{-1})$ such that $d_{ij}^l = d_{ji}^l$, $a_{ij}^l = a_{ji}^l$ for all $i$, $j$, $l$. This reduces the number of parameters to be optimized. The elements of the vector $\overline{X}$ defined in (5) will be redefined as follows:

$$\alpha_{ij}^l = \frac{Z_{1m}^{-(i-1)} Z_{2n}^{-(j-1)} + S Z_{1m}^{-(j-1)} Z_{2n}^{-(i-1)}}{N^{(l)}(Z_{1m}^{-1}, Z_{2n}^{-1})}, \quad j \leqslant i, \quad i,j \neq 1$$

and

$$\beta_{ij}^l = \frac{-(Z_{1m}^{-(i-1)} Z_{2n}^{-(j-1)} + S Z_{1m}^{-(j-1)} Z_{2n}^{-(i-1)})}{D^{(l)}(Z_{1m}^{-1}, Z_{2n}^{-1})}, \quad j \leqslant i, \quad i,j \neq 1$$

with

$$S = 1 \text{ for } i \neq j$$

$$S = 0 \text{ for } i = j.$$

Also it is sufficient to compute $|F_{mn}|$ only for $m = 1, \cdots, M$ and $n$ such that $|\omega_n^{(2)}| \leqslant |\omega_m^{(1)}|$, thus saving considerable amount of computation.

2) This procedure can be used for the design of one-dimensional filters by putting $N = 1$ and $m_l^1 = m_l^2 = 1$ for all $l$. Here, the proposed algorithm has the advantage of not restricting the search to the class of filters with complex poles and zeros, as is the case with the algorithm suggested by Deczky [9]. Deczky's algorithm excludes the class of filters with real poles and/or zeros within which the optimal filter may lie (as, for example, in the case of wide-band differentiators [6]).

As an illustration of using the proposed algorithm to design a one-dimensional filter, the first example in [6] is solved here using the proposed method. Starting from the same initial conditions, exactly the same results have been obtained after 32 iterations. For this example, it was found that the computation time needed for each iteration of the proposed algorithm is almost twice the computation time needed for each function evaluation in Fletcher–Powell algorithm. So these 32 iterations are equivalent to almost 64 function evaluations in Fletcher–Powell algorithm compared to 155 reported in [6].

3) It was shown by Temes and Zai [7] that $J$ will decrease if $[H]$ is positive definite and $\Delta\phi$ small. To ensure that $[H]$ be so, they suggested a positive definite approximation of the Hessian matrix to be used. An alternative method to ensure the positive definiteness of $[H]$, and which converges faster in the neighborhood of the minimum, is to add a sufficiently large positive constant to the diagonal elements of the Hessian matrix. It was shown [7] that the addition of a positive constant to the diagonal elements of $[H]$ will keep $\Delta\phi$ small and thus $J$ will decrease. This positive constant is decreased in steps and is put to zero near the extremal point. If the extremal point reached is a minimum, $[H]$ will be positive definite and $J$ will continue on decreasing until the minimum is reached. If the extremal point is not a minimum, $J$ will start to increase. In such a case, the search for the minimum should be reinitiated by selecting another set of initial conditions. Of course, the proposed method, as any other nonlinear optimization technique (including Fletcher–Powell technique), does not guarantee that the minimum point reached is the global one.

It is to be noted that, since $J$ goes to infinity at the stability boundary, there exists, in all directions where $J$ decreases, a certain magnitude of $\Delta\phi$ below which the filter will remain in the stability region. This has been discussed in detail in [9]. Also, the fact that a stable minimum does exist has been shown by Deczky [9].

4) For $p = 2$, the coefficient $A$ can be optimized separately using the equation

$$A = \frac{\sum_{m=1}^{M} \sum_{n=1}^{N} Y_{mn} H_{mn}}{\sum_{m=1}^{M} \sum_{n=1}^{N} |H_{mn}|^2}$$

where

$$H_{mn} = \frac{\prod_{l=1}^{K_1} N^{(l)}(Z_{1m}^{-1}, Z_{2n}^{-1})}{\prod_{l=1}^{K_2} D^{(l)}(Z_{1m}^{-1}, Z_{2n}^{-1})}.$$

The above equation can also be used to generate the initial value of $A$.

### VI. Examples

*Example 1:* In this example we will consider one- and two-dimensional low-pass filters with cutoff frequency equal to one tenth of the Nyquist frequency. The specifications of this filter are as follows:

A) The one-dimensional filter:

$$Y_m = 1 \qquad \omega_m = 0, 0.01, \cdots, 0.09$$

$$\phantom{Y_m} = 0.5 \qquad \omega_m = 0.1$$

$$= 0.0 \qquad \omega_m = 0.1, \cdots, 0.2, 0.3, \cdots, 1.0.$$

B) The two-dimensional filter:

$$Y_{mn} = 1 \qquad [(\omega_m^{(2)})^2 + (\omega_n^{(2)})^2]^{1/2} \leqslant 0.08$$

$$= 0.5 \qquad 0.08 < [(\omega_m^{(1)})^2 + (\omega_n^{(2)})^2]^{1/2} < 0.12$$

$$= 0 \qquad [(\omega_m^{(1)})^2 + (\omega_n^{(2)})^2]^{1/2} \geqslant 0.12$$

$$\omega_m^{(1)} = 0, 0.02, \cdots, 0.2, 0.4, \cdots, 1.0$$

and

$$\omega_n^{(2)} = 0, 0.02, \cdots, 0.2, 0.4, \cdots, 1.$$

This filter was approximated by one second order section in both the one- and two-dimensional case. The optimization starts with the one-dimensional filter with $p = 2$. After 32 iterations the algorithm converges to the filter

$$F(Z^{-1}) = 0.11024 \frac{1 - 1.64382 Z^{-1} + Z^{-2}}{1 - 1.79353 Z^{-1} + 0.84098 Z^{-2}}$$

with $J(\Phi) = 0.56732$.

Initial conditions for the two-dimensional low-pass filter is taken to be the coefficients of the filter

$$F_0(Z_1^{-1}, Z_2^{-1}) = F(Z_1^{-1}) F(Z_2^{-1}).$$

After 20 more iterations the algorithm converges to

$$F(Z_1^{-1}, Z_2^{-1}) = 0.00895$$

$$\frac{(1\ Z_1^{-1}\ Z_1^{-2}) \begin{bmatrix} 1 & -1.62151 & 0.99994 \\ -1.62151 & 2.63704 & 1.62129 \\ 0.99994 & 1.62129 & 1.00203 \end{bmatrix} \begin{bmatrix} 1 \\ Z_2^{-1} \\ Z_2^{-2} \end{bmatrix}}{(1\ Z_1^{-1}\ Z_1^{-2}) \begin{bmatrix} 1 & -1.78813 & 0.82930 \\ -1.78813 & 3.20640 & 1.49271 \\ 0.82930 & 1.49271 & 0.69823 \end{bmatrix} \begin{bmatrix} 1 \\ Z_2^{-1} \\ Z_2^{-2} \end{bmatrix}}$$

with $J(\Phi) = 1.6361$.

The resulting one- and two-dimensional filters are shown in Fig. 1 and Fig. 2. Three sections of the two-dimensional low-pass filter obtained here connected in series will give peak attenuation of $-45$ dB and in-band ripple of 0.2 dB (better attenuation performance can still be obtained by optimizing the coefficients of the additional two sections). In [1], the same attenuation has been obtained by $25 \times 25$ nonrecursive filter with the same passband and twice the transient band. This means that for each output of the recursive filter designed using the proposed method, we need 31 multiplications compared to 625 multiplications when using the nonrecursive filter designed in [1].

*Example 2:* Here we will consider the wide-band differentiator. This filter will be approximated by: 1) one first order section and 2) two first order sections. To show the effect of $p$ in the resulting filter, the example was solved with $p = 2$, $p = 6$, $p = 10$. The specifications of this filter are as follows:

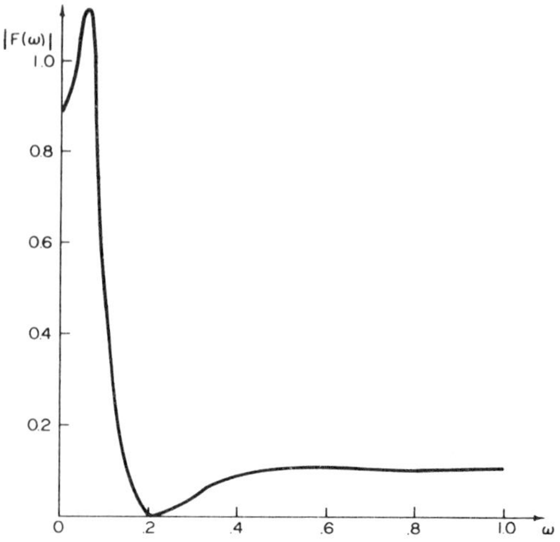


**Fig. 1. Example 1: amplitude response of the one-dimensional optimal low-pass filter.**

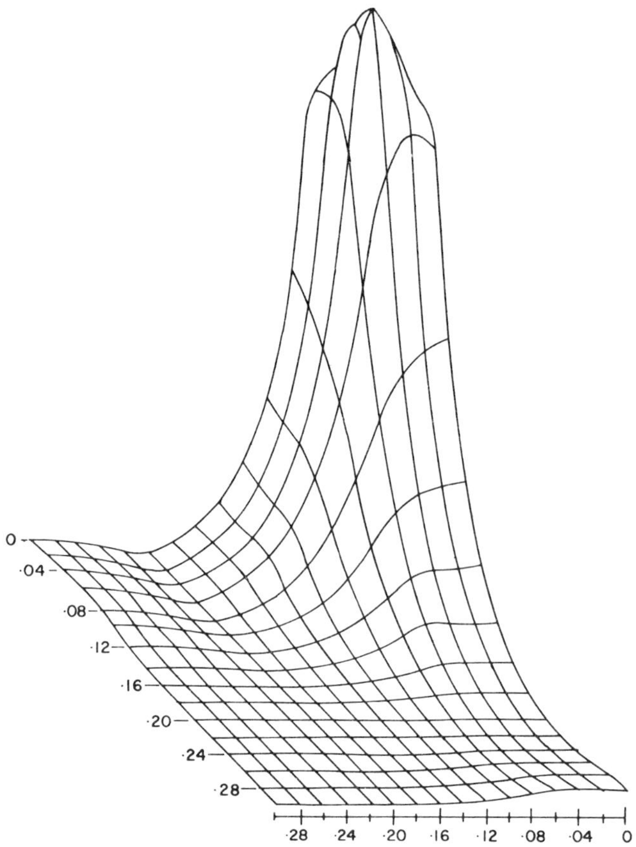


**Fig. 2. Example 1: amplitude response of the two-dimensional optimal low-pass filter.**

$$Y_{mn} = |\omega_m^{(1)} \omega_n^{(2)}|$$

$$\omega_m^{(1)} = 0, 0.1, \cdots, 1$$

$$\omega_n^{(1)} = 0, 0.1, \cdots, 1.$$

The results of the approximation by one first order section are shown in Table I.

The results of the approximation by two first order sections are as follows.

**TABLE I**

| | Initial Conditions | $p = 2$ | $p = 6$ | $p = 10$ |
|---|---|---|---|---|
| $A$ | 0.11 | 0.12785 | 0.12188 | 0.11580 |
| $a_{21}$ | -0.95 | -1.00050 | -0.96536 | -1.00475 |
| $a_{22}$ | 0.95 | 1.00094 | 0.95619 | 1.01491 |
| $d_{21}$ | 0.4 | 0.24616 | 0.28993 | 0.29738 |
| $d_{22}$ | 0.16 | 0.06059 | 0.08637 | 0.09064 |
| $J(\Phi)$ | — | $4.7312 \times 10^{-2}$ | $2.8324 \times 10^{-7}$ | $2.4619 \times 10^{-12}$ |
| Bounds for the deviation | — | -10 percent | -6.5 percent | -6 percent |
| From the specifications | — | +5.9 percent | +5.7 percent | +5.4 percent |
| Number of iterations | — | 15 | 20 | 20 |

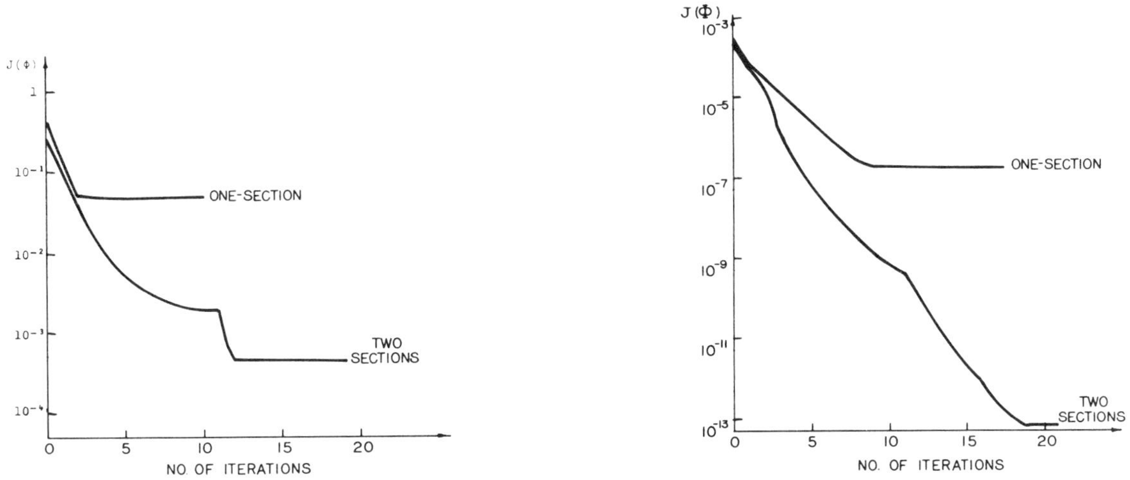


(a) (b)

**Fig. 3. Example 2. (a) Error versus iteration for $p = 2$. (b) Plot of error versus iteration for $p = 6$.**

For $p = 2$, the algorithm converges after 14 iterations to

$$F_2(Z_1^{-1}, Z_2^{-1}) = 0.13452 \frac{(1\ Z_1^{-1}) \begin{bmatrix} 1 & -1 \\ -1 & 1 \end{bmatrix} \begin{bmatrix} 1 \\ Z_2^{-1} \end{bmatrix} (1\ Z_1^{-1}) \begin{bmatrix} 1 & 0.66411 \\ 0.66411 & 0.44104 \end{bmatrix} \begin{bmatrix} 1 \\ Z_2^{-1} \end{bmatrix}}{(1\ Z_1^{-1}) \begin{bmatrix} 1 & 0.13842 \\ 0.13842 & 0.01916 \end{bmatrix} \begin{bmatrix} 1 \\ Z_2^{-1} \end{bmatrix} (1\ Z_1^{-1}) \begin{bmatrix} 1 & 0.71390 \\ 0.71390 & 0.50966 \end{bmatrix} \begin{bmatrix} 1 \\ Z_2^{-1} \end{bmatrix}}$$

with $J(\Phi) = 4.1608 \times 10^{-4}$. The deviation from the specifications is bounded by -0.7 percent and +0.4 percent.

For $p = 6$, the algorithm converges after 21 iterations to

$$F_6(Z_1^{-1}, Z_2^{-1}) = 0.13327 \frac{(1\ Z_1^{-1}) \begin{bmatrix} 1 & -1.00316 \\ -1.00316 & 1.01024 \end{bmatrix} \begin{bmatrix} 1 \\ Z_2^{-1} \end{bmatrix} (1\ Z_1^{-1}) \begin{bmatrix} 1 & 0.67876 \\ 0.67876 & 0.46209 \end{bmatrix} \begin{bmatrix} 1 \\ Z_2^{-1} \end{bmatrix}}{(1\ Z_1^{-1}) \begin{bmatrix} 1 & 0.14762 \\ 0.14762 & 0.02314 \end{bmatrix} \begin{bmatrix} 1 \\ Z_2^{-1} \end{bmatrix} (1\ Z_1^{-1}) \begin{bmatrix} 1 & 0.72434 \\ 0.72434 & 0.52571 \end{bmatrix} \begin{bmatrix} 1 \\ Z_2^{-1} \end{bmatrix}}$$

with $J(\Phi) = 0.6395 \times 10^{-13}$. The deviation from the specifications is bounded by -0.5 percent and +0.4 percent.

The rates of convergence in both cases for $p = 2$ and $p = 6$ are shown in Fig. 3.

*Example 3:* In this example we will consider the linear high-pass filter with the following specifications:

$$Y_{mn} = [(\omega_m^{(1)})^2 + (\omega_n^{(2)})^2]^{1/2}$$

$$\omega_m^{(1)} = 0, \frac{1}{14}, \cdots, 1.0$$

and

$$\omega_n^{(2)} = 0, \frac{1}{14}, \cdots, 1.0.$$

This filter was approximated by one first order section. The optimization starts with the initial conditions

$$F_0(Z_1^{-1}, Z_2^{-1}) = 0.7 \frac{(1\ Z_1^{-1}) \begin{bmatrix} 1 & -0.67 \\ -0.67 & 0.35 \end{bmatrix} \begin{bmatrix} 1 \\ Z_2^{-1} \end{bmatrix}}{(1\ Z_1^{-1}) \begin{bmatrix} 1 & -0.35 \\ -0.35 & 0.1 \end{bmatrix} \begin{bmatrix} 1 \\ Z_2^{-1} \end{bmatrix}}$$

and $p = 2$. After 27 iterations the algorithm converges to

$$F(Z_1^{-1}, Z_2^{-1}) = 0.59944 \cdot \frac{(1\ Z_1^{-1}) \begin{bmatrix} 1 & -0.04668 \\ -0.04668 & -0.46761 \end{bmatrix} \begin{bmatrix} 1 \\ Z_2^{-1} \end{bmatrix}}{(1\ Z_1^{-1}) \begin{bmatrix} 1 & -0.42602 \\ -0.42602 & 0.10692 \end{bmatrix} \begin{bmatrix} 1 \\ Z_2^{-1} \end{bmatrix}}$$

with $J(\Phi) = 0.28110$.

The amplitude response of the resulting filter is shown in Fig. 4.

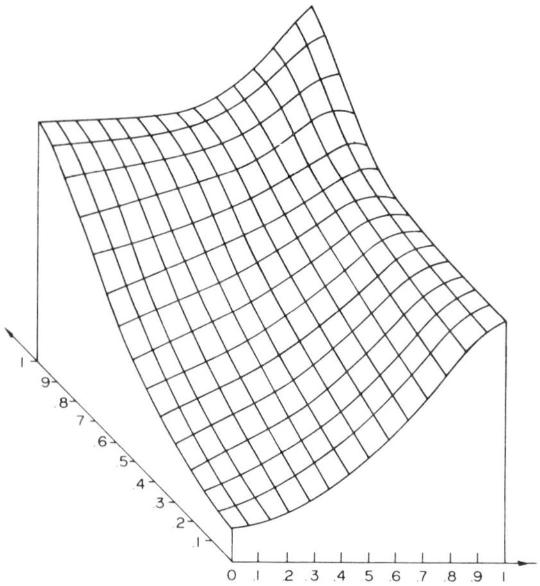


**Fig. 4. Example 3: amplitude response of the optimal two-dimensional linear high-pass filter.**

### References

[1] J. V. Hu and L. R. Rabiner, "Design techniques for two-dimensional digital filters," *IEEE Trans. Audio Electroacoust.*, vol. AU-20, pp. 249-257, Oct. 1972.

[2] E. L. Hall, "Digital filtering of images," Ph.D. dissertation, Univ. of Missouri, Columbia, 1971.

[3] G. A. Maria and M. M. Fahmy, "Effect of rounding on the stability of the two-dimensional digital filters," to be pubiished in *IEEE Trans. Circuits Syst.*, July 1974.

[4] J. B. Knowles and E. M. Olcayto, "Coefficient accuracy and digital filters response," *IEEE Trans. Circuit Theory*, vol. CT-14, pp. 31-41, Mar. 1968.

[5] G. A. Maria and M. M. Fahmy, "$l_p^m$-approximation by exponentials," *IEEE Trans. Circuit Theory*, vol. CT-20, pp. 71-74, Jan. 1973.

[6] K. Steiglitz, "Computer-aided design of recursive digital filters," *IEEE Trans. Audio Electroacoust.*, vol. AU-18, pp. 123-129, June 1970.

[7] G. C. Temes and D. Y. Zai, "Least *P*th approximation," *IEEE Trans. Circuit Theory*, vol. CT-16, pp. 235-237, May 1969.

[8] T. S. Huang, "Stability of two dimensional recursive filters," *IEEE Trans. Audio Electroacoust.*, vol. AU-20, pp. 158-163, June 1972.

[9] A. G. Deczky, "Synthesis of recursive digital filters using the minimum *p*-error criterion," *IEEE Trans. Audio Electroacoust.*, vol. AU-20, pp. 257-263, Oct. 1972.

# Design of Circularly Symmetric Two-Dimensional Recursive Filters

JOSÉ M. COSTA, STUDENT MEMBER, IEEE, AND ANASTASIOS N. VENETSANOPOULOS, MEMBER, IEEE

***Abstract*—The digital filtering of two-dimensional signals offers the many advantages characteristic of digital computers, such as flexibility and accuracy. Applications exist in the processing of images and geophysical data. A technique is presented for designing stable two-dimensional recursive filters whose magnitude response is approximately circularly symmetric. This is achieved by cascading a number of elementary filters which are called rotated filters because they are designed by rotating one-dimensional continuous filters and using the two-dimensional $z$-transform to obtain the corresponding digital filter. Stability of these filters is considered in detail and the results obtained are stated in two corollaries. In particular it is proved that rotated filters are stable if the angle of rotation is between 270° and 360°. Finally, methods of analysis and design of the shape, circular symmetry, and cutoff frequency of two-dimensional recursive filters are discussed.**

## I. INTRODUCTION

THE digital filtering of two-dimensional signals offers the many advantages characteristic of digital computers, such as flexibility and accuracy. Applications exist in the processing of images and geophysical data. In image processing there is no preferred spatial frequency axis. It is therefore desirable to process images with filters whose frequency response is approximately circularly symmetric.

The design techniques for nonrecursive filters in one dimension can often be extended to two dimensions in order to obtain filters with circularly-symmetric magnitude response [1]–[3]. On the other hand, the design of recursive filters in two dimensions becomes difficult due to the fact that a polynomial in two variables $P(z_1,z_2)$ cannot in general be factored into first- or second-order polynomials. This implies that many one-dimensional design techniques cannot be readily extended to two dimensions and that a high-order two-dimensional filter cannot in general be realized in parallel or cascade form to reduce the effect of quantization noise. It is also difficult to test the stability of two-dimensional recursive filters except for simple filters. Because of these difficulties, little work has been done on designing two-dimensional recursive filters, even though the use of recursive instead of nonrecursive filters has the potential of saving both computer time and computer memory [4]. A technique exists for mapping one-dimensional into two-dimensional filters with arbitrary directivity in a two-dimensional frequency response plane [5], [6]. These filters are called rotated filters because they are obtained by rotating one-dimensional filters.

In this paper we show how to cascade a number of rotated filters in order to obtain a stable circularly-symmetric filter. This approach approximates the circular-shaped contour levels of the magnitude response by a polygonal shape.

In Section II we review some concepts on two-dimensional recursive filters that will be used later in the paper. Section III shows a generalization of the technique of Shanks [6] to design rotated filters and in Section IV we solve the problem of determining the stability of rotated filters beforehand. The technique for designing circularly symmetric recursive filters is described in Section V. Finally, in Section VI we show briefly some methods of analysis and design of the shape, circular symmetry, and cutoff frequency of two-dimensional recursive filters.

Manuscript received December 28, 1973; revised May 29, 1974. This work was supported by the National Research Council of Canada under Grant No. A-7397.

The authors are with the Department of Electrical Engineering, University of Toronto, Toronto, Ont., Canada.

## II. TWO-DIMENSIONAL RECURSIVE FILTERS

A two-dimensional recursive filter can be described by a linear difference equation. The general form is

$$g(m,n) = \sum_{i=1}^{M_a}\sum_{j=1}^{N_a} a_{ij} f(m-i+1,n-j+1) - \sum_{i=1}^{M_b}\sum_{j=1}^{N_b} b_{ij} g(m-i+1,n-j+1), \quad (1)$$

$i,j \neq 1$ simultaneously.
Here it is assumed either that all output values $g(m-i+1,n-j+1)$ have been computed previously or are equal to zero (boundary conditions).

Equation (1) can be rewritten in the form

$$\sum_{i=1}^{M_b}\sum_{j=1}^{N_b} b_{ij} g(m-i+1,n-j+1) = \sum_{i=1}^{M_a}\sum_{j=1}^{N_a} a_{ij} f(m-i+1,n-j+1) \quad (2)$$

where $b_{11} = 1$.

Equation (2) can be solved for $g(m,n)$ [cf. (1)], $g(m-M_b+1,n)$, $g(m,n-N_b+1)$, or $g(m-M_b+1, n-N_b+1)$ and in each case the difference equation obtained corresponds to a two-dimensional recursive filter recursing in the $(+m,+n)$, $(-m,+n)$, $(+m,-n)$, or $(-m,-n)$ directions, respectively [7].

A recursive filter is said to be causal if it recurses in the $(+m,+n)$ direction [7]. A causal filter is realizable if its impulse response satisfies the property

$$h(m,n) = 0, \qquad m,n < 0.$$

A useful representation of (2) is obtained by the two-dimensional $z$ transform or double $z$ transform. This transform is defined as follows

Reprinted from *IEEE Trans. Acoust., Speech, Signal Processing*, vol. ASSP-22, pp. 432–443, Dec. 1974.

$$X(z_1,z_2) \triangleq \sum_{m=-\infty}^{\infty} \sum_{n=-\infty}^{\infty} x(m,n) z_1^m z_2^n \tag{3}$$

where $x(m,n)$ is a two-dimensional sequence and $X(z_1,z_2)$ its two-dimensional $z$ transform.[1] The domain of definition for $X(z_1,z_2)$ is its region of convergence in the $z_1,z_2$ plane. The properties of the two-dimensional $z$ transform are similar to the properties of the $z$ transform in one dimension.

If the sequence $x(m,n)$ is of finite duration and is bounded then (3) has finite limits in all the summations and $X(z_1,z_2)$ converges everywhere in the $z_1z_2$ plane.

The two-dimensional $z$ transform of (2) can be shown [8] to be

$$A(z_1,z_2)F(z_1,z_2) = B(z_1,z_2)G(z_1,z_2)$$

or

$$G(z_1,z_2) = H(z_1,z_2)F(z_1,z_2)$$

where $F(z_1,z_2)$ and $G(z_1,z_2)$ are the two-dimensional $z$ transforms of the input and the output sequences of the filter, respectively, and

$$H(z_1,z_2) = \frac{A(z_1,z_2)}{B(z_1,z_2)} \triangleq \frac{\sum_{i=1}^{M_a}\sum_{j=1}^{N_a} a_{ij} z_1^{i-1} z_2^{j-1}}{\sum_{i=1}^{M_b}\sum_{j=1}^{N_b} b_{ij} z_1^{i-1} z_2^{j-1}} \tag{4}$$

is the transfer function of the filter. The frequency response of the filter is obtained by evaluating (4) for $z_1 = \exp(-j\omega_1 X)$ and $z_2 = \exp(-j\omega_2 Y)$, where $X$ and $Y$ are the sample intervals in the $x$ and $y$ directions, respectively.

The transfer function $H(z_1,z_2)$ given in (4) can be associated with the recursive filters recursing in the directions $(-m,n)$, $(m,-n)$, and $(-m,-n)$ by changing $z_1$ by $z_1^{-1}$, $z_2$ by $z_2^{-1}$, or both, respectively.

## III. DESIGN OF ROTATED FILTERS

Suppose a one-dimensional continuous filter whose impulse response is real, is given in its factored form

$$H_1(s) = H_0\left[\prod_{i=1}^{m}(s-q_i)\Big/\prod_{i=1}^{n}(s-p_i)\right] \tag{5}$$

where $H_0$ is a scalar gain constant. The zero locations $q_i$ and the pole locations $p_i$ may be complex, in which case their conjugates are also present in the corresponding product. Assume that this filter has unity cutoff angular frequency.

The filter given in (5) can also be viewed as a two-dimensional filter that varies in one dimension only and could be written as follows:

[1] Some authors use an equivalent definition by replacing the unit delay operators $z_1$ and $z_2$ in (3) by $z_1^{-1}$ and $z_2^{-1}$, respectively. We use the definition in (3) because it is more common in the two-dimensional recursive filter literature [6], [7].

$$H_2(s_1,s_2) = H_1(s_2) = H_0\left[\prod_{i=1}^{m}(s_2-q_i)\Big/\prod_{i=1}^{n}(s_2-p_i)\right]. \tag{6}$$

Rotating clockwise the $(s_1,s_2)$ axes through an angle $\beta$ by means of the transformation (7) we obtain the filter of (8), whose frequency response is rotated by an angle $-\beta$ with respect to the frequency response of (6)

$$s_1 = s_1' \cos\beta + s_2' \sin\beta \tag{7a}$$

$$s_2 = -s_1' \sin\beta + s_2' \cos\beta \tag{7b}$$

$$H_2(s_1',s_2') = H_0 \frac{\prod_{i=1}^{m}[(s_2'\cos\beta - s_1'\sin\beta) - q_i]}{\prod_{i=1}^{n}[(s_2'\cos\beta - s_1'\sin\beta) - p_i]}. \tag{8}$$

$H_2(s_1',s_2')$ describes a continuous two-dimensional filter in the new coordinate system of $s_1'$ and $s_2'$. To produce the corresponding two-dimensional discrete filter we use the two-dimensional bilinear $z$ transform [6] defined by the following two equations:

$$s_1' = \frac{2}{T}\frac{1-z_1}{1+z_1} \tag{9a}$$

$$s_2' = \frac{2}{T}\frac{1-z_2}{1+z_2}. \tag{9b}$$

It is assumed throughout the study that the sample interval $T$ is the same in both directions. Substituting (9) into (8) we obtain

$$H(z_1,z_2) = A \prod_{i=1}^{M} \frac{a_{11}^i + a_{21}^i z_1 + a_{12}^i z_2 + a_{22}^i z_1 z_2}{b_{11}^i + b_{21}^i z_1 + b_{12}^i z_2 + b_{22}^i z_1 z_2} \tag{10}$$

where

$$\left.\begin{aligned}
A &= H_0(\tfrac{1}{2}T)^{n-m} \\
M &= \max(m,n) \\
&\left.\begin{aligned}
a_{11}^i &= \cos\beta - \sin\beta - \tfrac{1}{2}Tq_i \\
a_{21}^i &= \cos\beta + \sin\beta - \tfrac{1}{2}Tq_i \\
a_{12}^i &= -\cos\beta - \sin\beta - \tfrac{1}{2}Tq_i \\
a_{22}^i &= -\cos\beta + \sin\beta - \tfrac{1}{2}Tq_i
\end{aligned}\right\} \text{for } 1 \le i \le m \\
a_{11}^i &= a_{21}^i = a_{12}^i = a_{22}^i = 1, \quad \text{for } m < i \le M \\
&\left.\begin{aligned}
b_{11}^i &= \cos\beta - \sin\beta - \tfrac{1}{2}Tp_i \\
b_{21}^i &= \cos\beta + \sin\beta - \tfrac{1}{2}Tp_i \\
b_{12}^i &= -\cos\beta - \sin\beta - \tfrac{1}{2}Tp_i \\
b_{22}^i &= -\cos\beta + \sin\beta - \tfrac{1}{2}Tp_i
\end{aligned}\right\} \text{for } 1 \le i \le n \\
b_{11}^i &= b_{21}^i = b_{12}^i = b_{22}^i = 1, \quad \text{for } n < i \le M.
\end{aligned}\right\} \tag{11}$$

Equation (10) can be physically interpreted as a cascade of bilinear second-order systems and can be readily realized by complex cascade programming [8]. The operations are programmed using real arithmetic and are simplified using the fact that the impulse response of each pair of conjugate filters is real. The cascade form of two-dimensional recursive filters has some advantage over the direct form [9].

The desired cutoff frequency of the filter in (10) can be obtained by substituting in (11) $T$ by $\pi f_r$ or by $2\tan(\pi/2)f_r$ if the frequency axis has to be prewarped [10, cf. eq. (9)] where $f_r$ is the desired cutoff frequency expressed as a fraction of the Nyquist frequency, that is

$$f_r = f/f_N$$

where $f_N \triangleq 1/2T$ is the Nyquist frequency.

Examples of rotated filters are given in Figs. 1 and 2. These figures show contour maps of the magnitude responses of a second-order Butterworth filter rotated through 0° and 285°, respectively. Frequencies are shown as fractions of the Nyquist frequency and in every case the magnitude response has been normalized to a peak response of 1.0, the contour interval is 0.1, and the cutoff frequency is 0.2.

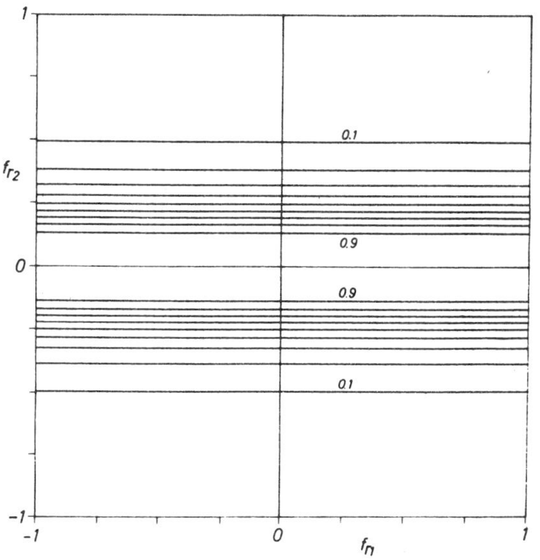


Fig. 1. Contour plot of the two-dimensional magnitude response of a second-order Butterworth filter rotated 0°.

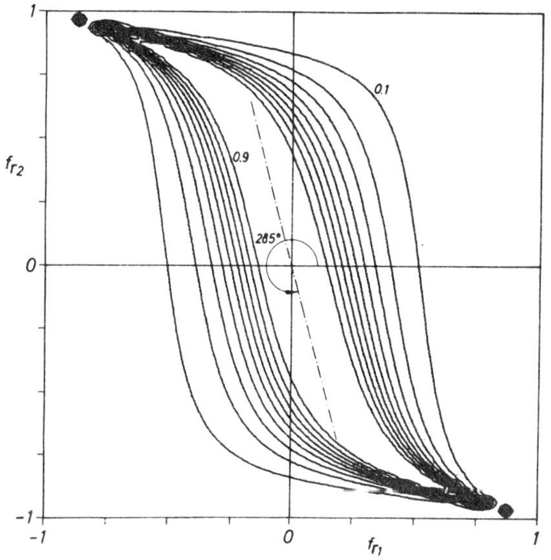


Fig 2. Contour plot of the two-dimensional magnitude response of second-order Butterworth filter rotated 285°.

## IV. STABILITY OF ROTATED FILTERS

Some basic theorems concerning stability of two-dimensional recursive filters are reviewed in Appendix A. Using Theorem 3 the conditions that the coefficients in (10) must satisfy to ensure the stability of the filter were derived. The results are given in the following two corollaries.

*Corollary 1*

A causal recursive filter with transfer function[2]

$$H(z_1,z_2) = \prod_{i=1}^{N} H_i(z_1,z_2)H_i^*(z_1^*,z_2^*) \tag{12}$$

where

$$H_i(z_1,z_2) = A_i(z_1,z_2)/B_i(z_1,z_2),$$

$A_i(z_1,z_2)$, and $B_i(z_1,z_2)$ $i = 1,2,\cdots,N$, are complex polynomials in $z_1$ and $z_2$ (for those $A_i$ and $B_i$ with real coefficients the complex conjugate factor $H_i^*(z_1^*,z_2^*)$ is set to unity), and $B_i(z_1,z_2)$ is of the form

$$B_i(z_1,z_2) = b_{11}{}^i + b_{21}{}^i z_1 + b_{12}{}^i z_2 + b_{22}{}^i z_1 z_2$$

is stable if

$$\big|\,|b_{21}{}^i b_{22}{}^{i*} - b_{11}{}^i b_{12}{}^{i*}| - |b_{11}{}^i b_{22}{}^i - b_{21}{}^i b_{12}{}^i|\,\big| > \big|\,|b_{12}{}^i|^2 - |b_{22}{}^i|^2\,\big| \tag{13}$$

and

$$|b_{11}{}^i| > |b_{21}{}^i| \tag{14}$$

for $i = 1,2,\cdots,N$.

*Proof:* We want to show that for the class of filters specified in this corollary, (13) and (14) are equivalent to the "if" part of Theorem 3 in Appendix A.

Setting $B_i(z_1,z_2) = 0$ we obtain

$$z_2 = -\frac{b_{11}{}^i + b_{21}{}^i z_1}{b_{12}{}^i + b_{22}{}^i z_1} \triangleq f_i(z_1) \tag{15}$$

which is a bilinear transformation mapping circles into circles.[3]

Clearly, for the filter $H_i(z_1,z_2)$ condition 1) of Theorem 3 is satisfied if and only if

$$\big|\,|c_i| - r_i\,\big| > 1 \tag{16}$$

where $c_i$ is the center and $r_i$ is the radius of the circle image of the unit circle by the transformation $z_2 = f_i(z_1)$.

[2] The * denotes complex conjugate.

[3] Here we include straight lines as circles with infinite radii.

We have derived expressions for $c_i$ and $r_i$ as a function of the filter coefficients in Lemma 1 of Appendix B.

But (16) is equivalent to (13) by substituting (B2) and (B3) into (16), that is

$$\left| \left| \frac{b_{21}{}^i b_{22}{}^{i*} - b_{11}{}^i b_{12}{}^{i*}}{|b_{12}{}^i|^2 - |b_{22}{}^i|^2} \right| - \left| \frac{b_{11}{}^i b_{22}{}^i - b_{21}{}^i b_{12}{}^i}{|b_{12}{}^i|^2 - |b_{22}{}^i|^2} \right| \right| > 1$$

which is equivalent to (13).

For condition 2) consider the inverse transformation

$$z_1 = f_i^{-1}(z_2) = -\frac{b_{11}{}^i + b_{12}{}^i z_2}{b_{21}{}^i + b_{22}{}^i z_2}$$

$$\hat{z}_1 \triangleq f_i^{-1}(z_2)\,|_{z_2=0} = -b_{11}{}^i/b_{21}{}^i. \qquad (17)$$

Therefore, for the filter $H_i(z_1,z_2)$ the second condition of Theorem 3 is verified if and only if $|\hat{z}_1| > 1$, that is

$$|-b_{11}{}^i/b_{21}{}^i| > 1$$

which is equivalent to (14).

Since (13) and (14) compare magnitudes only, the same conditions apply to $H_i(z_1,z_2)$ and $H_i^*(z_1^*,z_2^*)$; therefore, in general (13) and (14) have to be checked for $i = 1,2,\cdots,N = (M + N_r)/2 \leq M$ only, where $N_r$ is the number of filters $H_i(z_1,z_2)$ with real coefficients and $M$ is the total number of filters $H_i(z_1,z_2)$ in the cascade.

If every filter $H_i(z_1,z_2)$ in the cascade is stable, then clearly the filter $H(z_1,z_2)$ is stable. Q.E.D.

Specializing Corollary 1 to rotated filters with their coefficients as defined in (11) we obtain Corollary 2.

*Corollary 2*

Rotating a stable one-dimensional continuous filter by an angle $\beta$ in the $(s_1,s_2)$ plane and applying the two-dimensional bilinear $z$ transform, the resulting two-dimensional digital filter [cf. (10)] is stable if $270° \leq \beta \leq 360°$.

*Proof:* Substituting (11) into (13) and (14) we obtain (18) and (19) as sufficient stability conditions for rotated filters

$$||a_i| - |\cos\beta|| > |a_i + \cos\beta| \qquad (18)$$

$$|\cos\beta - \sin\beta - a_i| > |\cos\beta + \sin\beta - a_i|, \quad \text{for } i = 1,2,\cdots,M \qquad (19)$$

where $a_i = \mathrm{Re}\,[(T/2)p_i]$ and $p_i$ represents the location of a pole.

There are no values of $a_i$ and $\beta$ satisfying (18); at best, both sides of (18) are equal. This is because the transformation $B_i(z_1,z_2) = 0$ has a fixed point at $z_1 = z_2 = -1$ (the other fixed point of the transformation is $-b_{11}/b_{22}$). This pole is cancelled by a zero because the transformation $A_i(z_1,z_2) = 0$ has also a fixed point at $z_1 = z_2 = -1$. However, the pole and zero might not exactly cancel each other because of quantization. To avoid this, since the center of the circle image lies on the real axis (see the corollary in Appendix B), we can always shift the circle image to the left by a small distance $\epsilon$. This is illustrated in Fig. 3. This displacement eliminates the overlapping of the two circles at the point $z_1 = z_2 = -1$. However, in all the filters that we have designed using the rotation procedure there was never a need for this shift because the impulse responses decayed sufficiently quickly.

The set of values that satisfy (18) with equality is[4]

$$S_i = \{a_i,\beta: (a_i < 0 \cap \cos\beta > 0) \cup (a_i > 0 \cap \cos\beta < 0)\}.$$

The set of values that satisfy (19) is

$$S_2 = \{a_i,\beta: (\sin\beta > 0 \cap a_i > \cos\beta) \cup (\sin\beta < 0 \cap a_i < \cos\beta)\}.$$

The set of values of $a_i$ and $\beta$ for which rotated filters are stable is given by the intersection of set $S_1$ and set $S_2$.

$$S_1 \cap S_2 = \{a_i,\beta: (a_i < 0 \cap 270° < \beta < 360°) \cup (a_i > 0 \cap 90° < \beta < 180°)\}.$$

The regions of stability are illustrated in Fig. 4.

For $\beta = 270°$ and $\beta = 360°$ the bilinear transformation in (15) is degenerate. However, in these cases the two-dimensional filter varies in one dimension only and it can easily be shown that it is stable provided that the one-dimensional filter has all its poles in the LHP (left-hand plane).

Therefore, for stability $a_i < 0$, $i = 1,2,\cdots,M$, and $270° \leq \beta \leq 360°$. Q.E.D.

Theorem 4 in Appendix A indicates that it might be possible to obtain stable filters for other angles of rotation by changing the direction of recursion. That is, by substituting either $z_1$ by $z_1^{-1}$, or $z_2$ by $z_2^{-1}$, or both, in the filter transfer function $H(z_1,z_2)$.

The application of Corollary 1 to all cases of interest results in Table I. Nevertheless, all the possibilities indicated by Table I result in a filter with the same magnitude response, that is, the effective angle of rotation always satisfies $270° < \beta_e < 360°$. To obtain a rotation through a different angle the data can be filtered in a distinct manner. Consider for example the following.

In the frequency domain the filtering operation can be represented as

$$G[\exp(-j\omega_1 T),\exp(-j\omega_2 T)] = H[\exp(-j\omega_1 T),\exp(-j\omega_2 T)] \cdot F[\exp(-j\omega_1 T),\exp(-j\omega_2 T)]. \qquad (20)$$

It is easy to see that a rotation of the filter transfer function $H[\exp(-j\omega_1 T),\exp(-j\omega_2 T)]$ by an angle $\theta$ is equivalent to a rotation of the two-dimensional Fourier transform of the two-dimensional input signal $F[\exp(-j\omega_1 T),\exp(-j\omega_2 T)]$ by an angle $-\theta$ and a rotation of the output by an angle $\theta$. Since the rotation of a function of two variables in the space domain causes an equal rotation of its Fourier transform in the frequency domain [8], to obtain a filter rotated by an angle $\beta$, when $0° < \beta < 270°$, we may rotate the matrix contain-

[4] ∩ means "and" and ∪ means "or."

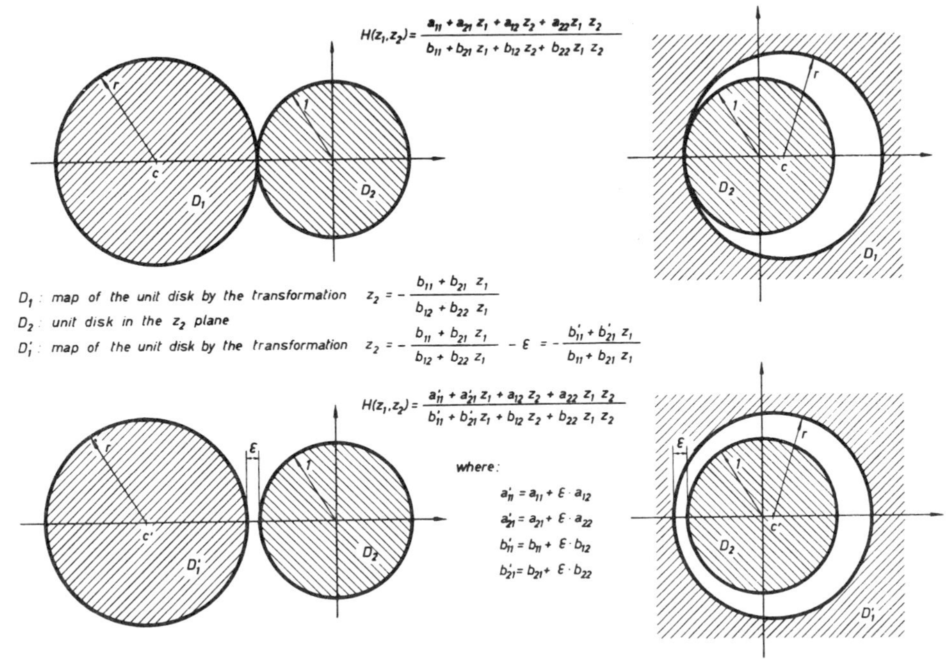


Fig. 3. Stabilization of marginally unstable rotated filters.

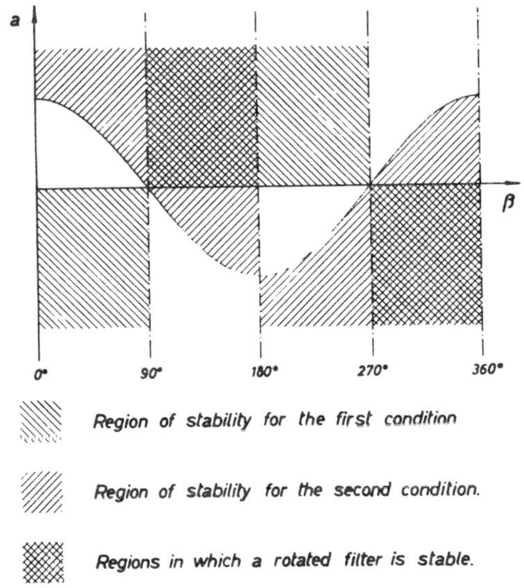


Fig. 4. Regions of stability for rotated filters with transfer function $H(z_1, z_2)$.

TABLE I
STABILITY REGIONS FOR ROTATED FILTERS

| Transfer function of the two-dimensional filter | Location of the poles of the original one-dimensional filter H(s) | |
|---|---|---|
| | L.H.P. | R.H.P |
| $H(z_1,z_2)$ | $270^\circ < \beta < 360^\circ$ | $90^\circ < \beta < 180^\circ$ |
| $H(z_1^{-1},z_2)$ | $0^\circ < \beta < 90^\circ$ | $180^\circ < \beta < 270^\circ$ |
| $H(z_1,z_2^{-1})$ | $180^\circ < \beta < 270^\circ$ | $0^\circ < \beta < 90^\circ$ |
| $H(z_1^{-1},z_2^{-1})$ | $90^\circ < \beta < 180^\circ$ | $270^\circ < \beta < 360^\circ$ |

ing the input data by $-90°$, $-180°$, or $-270°$, rotate the filter by $\beta_f$, where $270° \leq \beta_f \leq 360°$ for stability, perform the filtering operation and finally rotate the output matrix to the original position. Since the data are usually given in the form of a matrix, which allows us to perform the digital filtering operation, it can be easily rotated by angles which are multiples of 90°.

When choosing the angles of rotation for the data and for the filter the following equation has to be satisfied

$$\beta_e = \beta_f - \beta_d \tag{21}$$

where

- $\beta_e$ effective angle of rotation of the filter with respect to the original data matrix;
- $\beta_f$ angle of rotation of the filter ($270° \leq \beta_f \leq 360°$ for stability);
- $\beta_d$ $k90°$ = angle of rotation of the input data matrix ($k$ is an integer).

Thus, given $\beta_e$, the appropriate value of $k$ has to be chosen such that $\beta_f = \beta_e + k90°$ and $270° \leq \beta_f \leq 360°$ are satisfied. Note that all the angles are measured counter-clockwise and modulo 360°, so $k \geq 0$ because $\beta_f \geq \beta_e$ always. After filtering, the output matrix has to be rotated by an angle $-\beta_d$. A block diagram of this technique is given in Fig. 5(a), where $R$ denotes a rotation by $\beta_d$ and $H$ denotes a $\beta_f$-rotated filter. The system is equivalent to a $\beta_e$-rotated filter with $\beta_e = \beta_f - \beta_d$.

The rotation of the data is not the only possible solution. There are other transformations of the data which change the direction of recursion. These transformations are the matrix transposition with respect to each diagonal

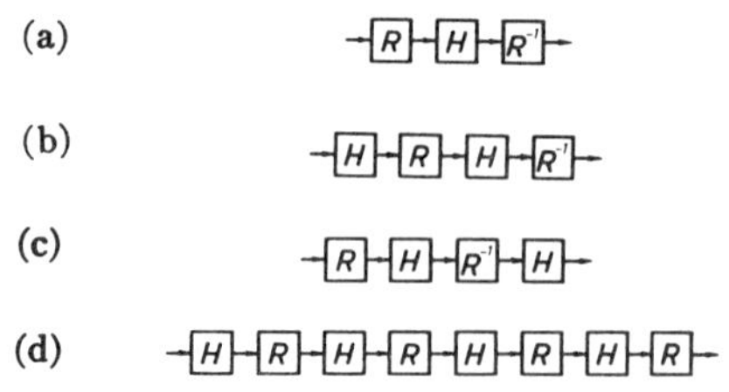


Fig. 5. Realization of noncausal filters by combining data rotations with a causal filter.

and the horizontal and vertical mirror images. An alternative solution to the transformation of the data may be obtained by using algorithms recursing in other directions (noncausal filters).

## V. DESIGN OF FILTERS WITH CIRCULAR SYMMETRY

We have shown that stable rotated filters can be obtained for any effective angle of rotation ($\beta_e$). Therefore, cascading a number of rotated filters whose angles of rotation are uniformly distributed over 180° results in a magnitude response which approximates a circularly symmetric function by a polygonal. This polygonal has an even number of sides because with each rotated filter we obtain two sides of the polygonal. The more filters that are cascaded, the more sides the polygonal has and the better the circular symmetry.

Suppose that the angles of rotation are between 180° and 360°. When the angle of rotation is $\beta$, $180° < \beta < 270°$, we have to transform the data matrix according to what was said in Section IV, then filter with a $\beta + 90°$ rotated filter, and inverse-transform the output matrix. If $\beta$ is distributed between 180° and 360°, these filters rotated by $\beta + 90°$, $180° < \beta < 270°$, coincide with those rotated by $\beta$, $270° < \beta < 360°$. Therefore, we need a filter consisting of rotated filters whose angles of rotation are distributed between 270° and 360°. The contour levels of the magnitude response of that filter have approximately elliptical shapes with major axis oriented 315°. We refer to this class of filters as elliptically shaped filters. Fig. 6 shows an example of an elliptically shaped filter obtained by cascading three second-order Butterworth filters rotated by 285°, 315°, and 345° and cutoff frequency 0.2. To obtain a circularly symmetric magnitude response the sequence of operations indicated in Fig. 5(b) is used, where $H$ denotes the elliptically shaped filter and $R$ denotes a data rotation by 90°. Here we assume that the impulse response of the elliptically shaped filter becomes negligible fast enough with respect to the dimensions of the data. An alternative solution is shown in Fig. 5(c).

The combination shown in Fig. 5(a) with $R$ a rotation by 90° and $H$ an elliptical filter is equivalent to an elliptically shaped filter with major axis oriented 225°. The magnitude response of this filter is shown in Fig. 7. Cascading the filters in Figs. 6 and 7 we obtain the filter

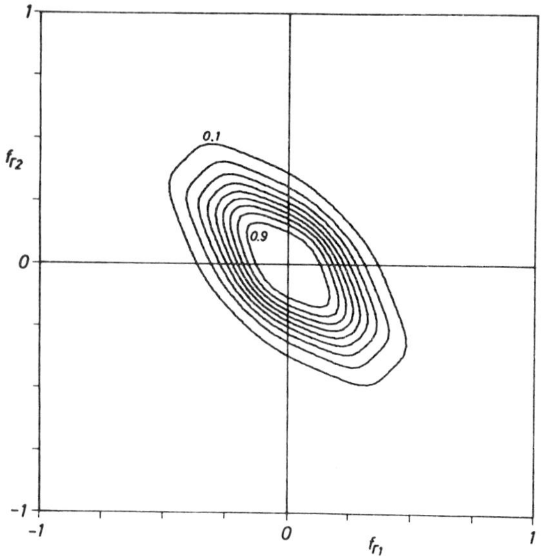


Fig. 6. Contour plot of the two-dimensional magnitude response of a cascade of three second-order Butterworth filters rotated 285°, 315°, and 345°.

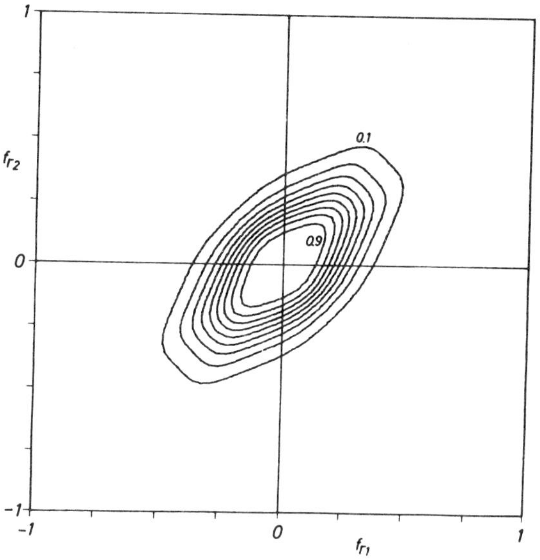


Fig. 7. Contour plot of the two-dimensional magnitude response of a cascade of three second-order Butterworth filters rotated 195°, 225°, and 255°.

shown in Fig. 8. A fine detail in one quadrant of the same magnitude response is shown in Fig. 11.

Figs. 9 through 11 show the magnitude responses obtained by cascading 2, 4, and 6 rotated filters, respectively. We observe that cascading more than four rotated filters the polygonal shape of the contour levels cannot be appreciated. Note also that as the number of rotated filters which are cascaded increases, the attenuation also increases and the cutoff region becomes steeper.

Two-dimensional filters with zero-phase response can be obtained [6]. Suppose, for example, that the linear operation indicated in Fig. 5(d) is performed, where $H$ denotes an elliptically shaped filter with major axis

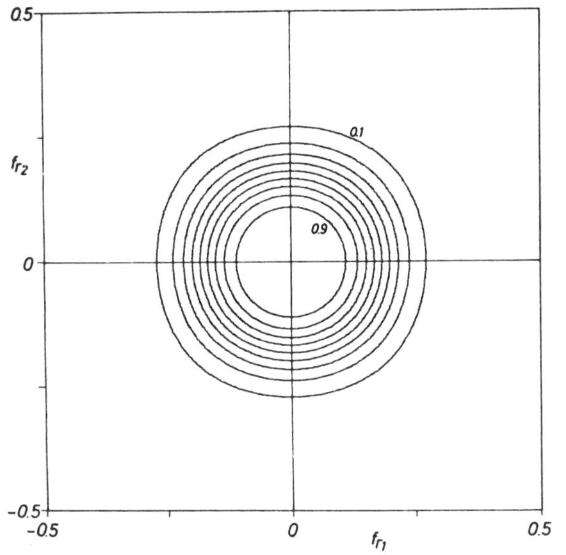


Fig. 8. Contour map of the two-dimensional magnitude response of the composite filter formed by cascading the two filters whose magnitude responses are shown in Figs. 6 and 7.

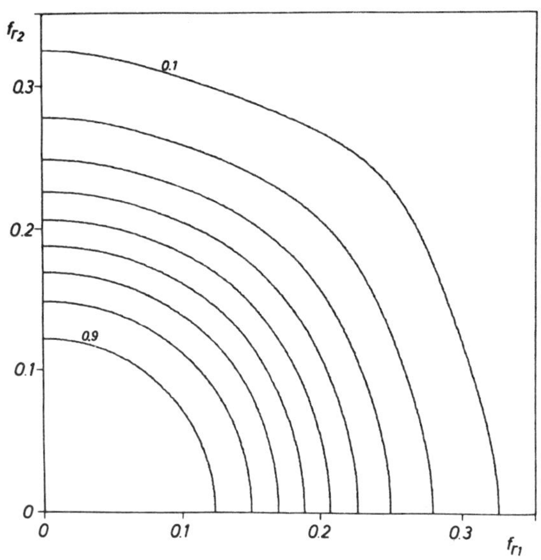


Fig. 10. Contour plot of the two-dimensional magnitude response of a cascade of four second-order Butterworth filters rotated by multiples of 45°.

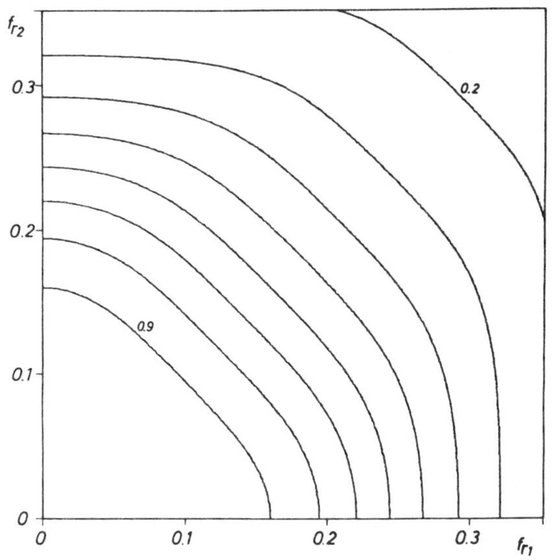


Fig. 9. Contour plot of the two-dimensional magnitude response of a cascade of two second-order Butterworth filters rotated 225° and 315°.

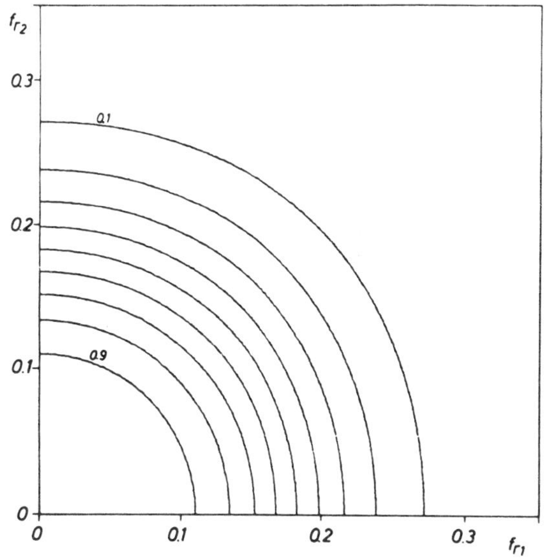


Fig. 11. Contour plot of the two-dimensional magnitude response of a cascade of six second-order Butterworth filters rotated by multiples of 30°.

oriented 315° and $R$ denotes a rotation by 90°. This is equivalent to a cascade of an even number of rotated filters whose angles of rotation are uniformly distributed between 0° and 360°. This filter has zero-phase response. Indeed, its transfer function can be written as follows:

$$\hat{H}(z_1,z_2) = H(z_1,z_2)H(z_1^{-1},z_2^{-1})$$

where $H(z_1,z_2)$ is the transfer function of a cascade of rotated filters whose angles of rotation are distributed over 180°. Clearly, the filter $\hat{H}(z_1,z_2)$ has zero-phase response and its magnitude response is the square of the magnitude response of $H(z_1,z_2)$.

## VI. CHOICE OF FILTER PARAMETERS

In the previous section it was pointed out that the cutoff frequency and the shape of the magnitude response of a two-dimensional filter depend on the number of rotated filters being cascaded. In this section we show briefly how these parameters can be controlled.

### *A. Cutoff Frequency*

The technique used to obtain a two-dimensional filter with a specified cutoff frequency in a given direction consists of an iteration that modifies the cutoff frequency of the original one-dimensional continuous filter in the proper amount and sense until the desired cutoff frequency of the two-dimensional digital filter is reached.

This method was implemented by writing a program in Fortran IV featuring the following steps.

1) Read in the coefficients of the one-dimensional continuous filter (with cutoff angular frequency normalized to unity), the number of rotations desired, and the de-

sired cutoff frequency $f_u$ of the two-dimensional digital filter given as a fraction of the Nyquist frequency.

2) Determine the coefficients of a two-dimensional digital filter [using (11)] derived from a one-dimensional filter with cutoff frequency $f_u$.

3) Find the cutoff frequency $f_c$ of the filter obtained in a given direction.

4) Determine the coefficients of a two-dimensional digital filter derived from a one-dimensional filter with cutoff frequency $f_d = 2f_u - f_c$.

5) Repeat steps 3) and 4) until $|f_u - f_c| < \epsilon$ where $\epsilon$ is the specified maximum error of the cutoff frequency in that direction.

6) Determine the shape factors (see Section VI-B).

7) Check for stability.

8) Evaluate the magnitude response and plot a contour map.

The program described previously was used to design many filters of different shapes and cutoff frequencies. The convergence was obtained quickly. The number of iterations required increases with the cutoff frequency and decreases with the order of the original one-dimensional filter.

## B. Shape Factors

Shape factors will be used here as a measurement of how good a filter is. Rather than giving large tables of shape factors for various classes of filters, we describe some different kinds of shape factors that are suitable for two-dimensional digital filters. We also give some tables containing shape factors of two-dimensional recursive filters derived from Butterworth filters of different orders. Since the shape of the magnitude response of a filter depends on the cutoff frequency, all these tables are given for filters normalized to a cutoff frequency of 0.1. Different kinds of shape factors are shown in Fig. 12.

The first shape factor is the ratio in decibels of the magnitude response at the cutoff frequency and the magnitude response at a frequency at a distance $d$ beyond the cutoff frequency in a specified direction. This shape factor, that we call shape factor number 1, is given in Table II for $d = 0.05$ in the directions 0° and 45°.

The second shape factor (No. 2) is the distance between the cutoff frequency and the frequency corresponding to an attenuation of $A$ dB. Table III gives this shape factor for $A = 40$ in the directions 0° and 45°.

The third shape factor (No. 3) concerns the circular symmetry only and gives the area between the $-3$-dB contour and a circle of radius the cutoff frequency. These areas are given in Table IV.

The fourth shape factor (No. 4) is both a measure of the steepness of the magnitude response and a measure of the circular symmetry. It gives the volume (see Table V) between the magnitude response of the filter and the magnitude response of an ideal filter (low-pass). The magnitude response of an ideal low-pass two-dimensional filter in polar coordinates is

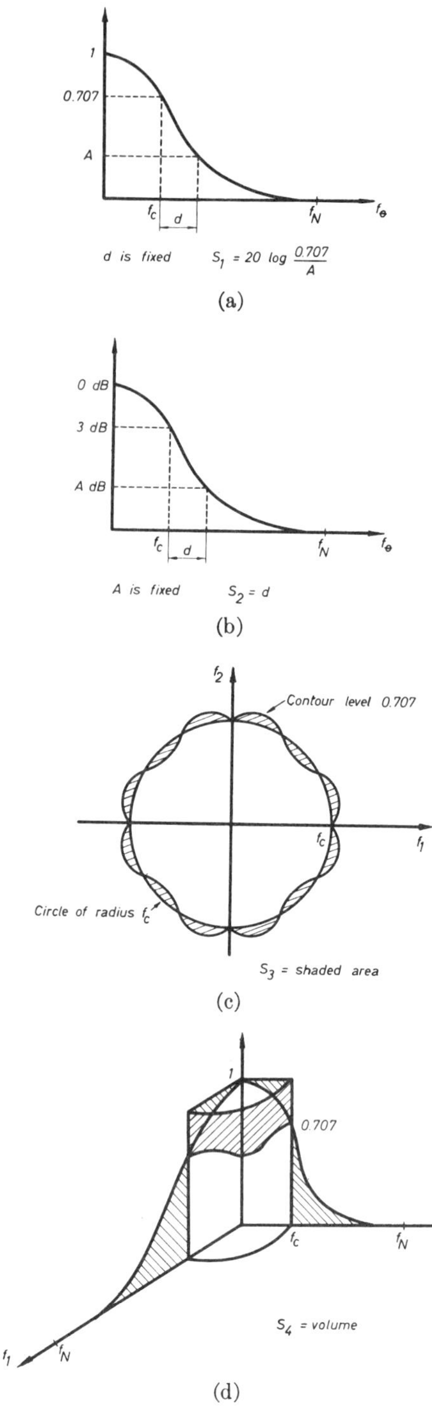


Fig. 12. Shape factors.

$$H(\nu) = \begin{cases} 1 & \text{for } \nu < f_c \\ 0 & \text{otherwise.} \end{cases}$$

From the observation of the tabulated results, the following general conclusions can be made.

The steepness of the filter is highly dependent on the number of rotated filters being cascaded, the type of one dimensional filters used, and the order of the filters. Shape factors 1 and 2 indicate the steepness of the filter, usually considered the main criterion for low-pass filters. In one dimension, the steepness of the filter is well known if given

TABLE II
SHAPE FACTOR NUMBER 1 (IN DECIBELS)

| Order of Butterworth filter | Number of rotated filters being cascaded | | | | | |
|---|---|---|---|---|---|---|
| | 2 | 4 | 6 | 8 | 10 | 12 |
| | d=0.05 θ=0° | | | | | |
| 1 | 2.80 | 3.05 | 3.28 | 3.41 | 3.50 | 3.56 |
| 2 | 7.06 | 7.34 | 8.56 | 9.26 | 9.77 | 10.15 |
| 3 | 12.59 | 12.75 | 15.56 | 17.54 | 19.10 | 20.36 |
| 4 | 18.96 | 19.04 | 23.48 | 27.53 | 30.80 | 33.63 |
| 5 | 25.78 | 25.82 | 31.75 | 38.51 | 43.92 | 48.86 |
| 6 | 32.83 | 32.84 | 40.15 | 49.98 | 57.72 | 65.15 |
| 7 | 39.98 | 39.98 | 48.60 | 61.67 | 71.82 | 81.99 |
| 8 | 47.17 | 47.17 | 57.10 | 73.43 | 86.05 | 99.07 |
| 9 | 54.38 | 54.38 | 65.65 | 85.20 | 100.30 | 116.27 |
| 10 | 61.59 | 61.60 | 74.26 | 96.94 | 114.53 | 133.49 |
| | d=0.05 θ=45° | | | | | |
| 1 | 1.95 | 2.99 | 3.21 | 3.34 | 3.32 | 3.48 |
| 2 | 5.50 | 7.17 | 8.34 | 9.02 | 9.51 | 9.87 |
| 3 | 9.71 | 12.46 | 15.12 | 17.05 | 18.54 | 19.74 |
| 4 | 14.19 | 18.61 | 23.05 | 26.76 | 29.91 | 32.60 |
| 5 | 18.79 | 25.26 | 31.62 | 37.46 | 42.70 | 47.41 |
| 6 | 23.42 | 32.16 | 40.48 | 48.66 | 56.26 | 63.27 |
| 7 | 28.06 | 39.18 | 49.46 | 60.09 | 70.20 | 79.70 |
| 8 | 32.72 | 46.25 | 58.44 | 71.60 | 84.32 | 96.39 |
| 9 | 37.37 | 53.34 | 67.37 | 83.13 | 98.50 | 113.20 |
| 10 | 42.02 | 60.44 | 76.25 | 94.64 | 112.70 | 130.04 |

TABLE III
SHAPE FACTOR NUMBER 2

| Order of Butterworth filter | Number of rotated filters being cascaded | | | | | |
|---|---|---|---|---|---|---|
| | 2 | 4 | 6 | 8 | 10 | 12 |
| | A=40 θ=0° | | | | | |
| 1 | 0.6532 | 0.4694 | 0.3967 | 0.3545 | 0.3288 | 0.3120 |
| 2 | 0.2546 | 0.2007 | 0.1565 | 0.1371 | 0.1264 | 0.1194 |
| 3 | 0.1392 | 0.1283 | 0.0951 | 0.0843 | 0.0775 | 0.0731 |
| 4 | 0.0937 | 0.0920 | 0.0694 | 0.0614 | 0.0564 | 0.0531 |
| 5 | 0.0702 | 0.0699 | 0.0558 | 0.0486 | 0.0447 | 0.0420 |
| 6 | 0.0559 | 0.0559 | 0.0471 | 0.0405 | 0.0372 | 0.0349 |
| 7 | 0.0465 | 0.0465 | 0.0412 | 0.0349 | 0.0320 | 0.0300 |
| 8 | 0.0397 | 0.0397 | 0.0366 | 0.0308 | 0.0281 | 0.0263 |
| 9 | 0.0346 | 0.0346 | 0.0329 | 0.0277 | 0.0252 | 0.0236 |
| 10 | 0.0307 | 0.0307 | 0.0298 | 0.0253 | 0.0228 | 0.0214 |
| | A=40 θ=45° | | | | | |
| 1 | 0.8898 | 0.5498 | 0.4541 | 0.3944 | 0.3622 | 0.3414 |
| 2 | 0.6127 | 0.2117 | 0.1619 | 0.1422 | 0.1308 | 0.1234 |
| 3 | 0.2800 | 0.1329 | 0.0992 | 0.0865 | 0.0794 | 0.0748 |
| 4 | 0.1661 | 0.0945 | 0.0720 | 0.0628 | 0.0575 | 0.0542 |
| 5 | 0.1149 | 0.0716 | 0.0565 | 0.0496 | 0.0455 | 0.0427 |
| 6 | 0.0869 | 0.0571 | 0.0466 | 0.0412 | 0.0378 | 0.0355 |
| 7 | 0.0696 | 0.0473 | 0.0399 | 0.0355 | 0.0325 | 0.0305 |
| 8 | 0.0578 | 0.0404 | 0.0351 | 0.0313 | 0.0286 | 0.0268 |
| 9 | 0.0494 | 0.0352 | 0.0315 | 0.0282 | 0.0257 | 0.0239 |
| 10 | 0.0431 | 0.0312 | 0.0288 | 0.0257 | 0.0233 | 0.0217 |

TABLE IV
SHAPE FACTOR NUMBER 3 ($\times 10^{-6}$)

| Order of Butterworth filter | Number of rotated filters being cascaded | | | | | |
|---|---|---|---|---|---|---|
| | 2 | 4 | 6 | 8 | 10 | 12 |
| 1 | 4.38 | 0.19 | 0.14 | 0.23 | 0.21 | 0.20 |
| 2 | 6.02 | 0.32 | 0.14 | 0.14 | 0.16 | 0.16 |
| 3 | 16.14 | 1.94 | 0.34 | 0.16 | 0.21 | 0.16 |
| 4 | 27.15 | 1.09 | 0.40 | 0.12 | 0.32 | 0.97 |
| 5 | 38.08 | 0.19 | 0.67 | 0.13 | 0.16 | 0.21 |
| 6 | 48.90 | 1.73 | 1.05 | 0.40 | 0.22 | 0.37 |
| 7 | 59.69 | 3.72 | 1.24 | 0.44 | 0.16 | 0.19 |
| 8 | 70.15 | 5.47 | 1.02 | 0.56 | 0.37 | 0.84 |
| 9 | 81.00 | 7.24 | 1.03 | 0.95 | 0.83 | 0.37 |
| 10 | 90.93 | 9.25 | 0.20 | 1.02 | 0.78 | 0.85 |

TABLE V
SHAPE FACTOR NUMBER 4 ($\times 10^{-2}$)

| Order of Butterworth filter | Number of rotated filters being cascaded | | | | | |
|---|---|---|---|---|---|---|
| | 2 | 4 | 6 | 8 | 10 | 12 |
| 1 | 17.65 | 12.09 | 9.98 | 8.92 | 8.31 | 7.93 |
| 2 | 4.53 | 3.90 | 3.16 | 2.88 | 2.72 | 2.62 |
| 3 | 2.34 | 2.24 | 1.86 | 1.71 | 1.62 | 1.56 |
| 4 | 1.55 | 1.53 | 1.33 | 1.21 | 1.15 | 1.11 |
| 5 | 1.15 | 1.14 | 1.03 | 0.94 | 0.89 | 0.86 |
| 6 | 0.90 | 0.90 | 0.84 | 0.76 | 0.72 | 0.69 |
| 7 | 0.73 | 0.73 | 0.70 | 0.64 | 0.60 | 0.57 |
| 8 | 0.61 | 0.61 | 0.59 | 0.54 | 0.50 | 0.48 |
| 9 | 0.51 | 0.51 | 0.50 | 0.46 | 0.42 | 0.40 |
| 10 | 0.43 | 0.43 | 0.42 | 0.39 | 0.36 | 0.33 |

the type of filter, such as Butterworth or Chebyshev, and the order of the filter. However, in two dimensions the number of rotated filters being cascaded also affects the steepness. Tables II and III show that the steepness of the filters is monotonically increasing with respect to both the order of the filter and the number of filter rotations. As a design criterion, the steepness of a filter increases more rapidly with the change of order than with the increase of the number of filters being cascaded. The effect of the number of filter rotations on steepness decreases as the number of rotations increases (the gain in steepness is the greatest when the filter is realized by a lower number of rotations).

In contrast to the monotonic relation of shape factors 1 and 2, shape factor 3, which indicates the exactness of the realized filter symmetry to the ideal circular symmetry, shows no simple relationship with respect to the two parameters considered. This shape factor varies in an erratic manner, although some general pattern can be noticed. The reason for this variation is that shape factor 3 is highly dependent on the direction chosen for the exact cutoff frequency. A fairer comparison between filters could be done if shape factors 3 and 4 were determined with filters whose cutoff frequency were exact in the optimum direction, that is in the direction in which shape factors 3 and 4 were minimized.

## VII. CONCLUSIONS

The work described previously concerns the design of stable two-dimensional recursive filters whose magnitude response approximates a circularly symmetric function. The principal advantages of the method are the following.

1) Stability can be guaranteed.

2) The method leads to a recursive realization which is more efficient than a nonrecursive one.

3) The filters are factorable into lower order filters, so that they can be efficiently realized. In general this factorization is not possible for two-dimensional recursive filters.

4) The calculation of the coefficients of a filter using a digital computer is fast.

Finally, we have shown how to obtain the desired cutoff frequency in a given direction with a specified maximum error in that direction. We have also given tables of shape factors to help the designer choosing the order of the original filter and the number of rotations for a particular application.

Future work could be done on frequency domain synthesis in order to have more control of the error from the ideal characteristics of the filter. In particular, a problem of interest is to choose a direction for the exact cutoff frequency such that the deviation from the cutoff frequency in all other directions is minimized (i.e., minimize shape factor number 3, see Section VI-B). The suitability of the filters discussed here for filtering images (or other practical applications) could be tested.

## APPENDIX A

The following theorems concern stability of two-dimensional recursive filters.

*Theorem 1* [11]

A two-dimensional filter is said to be stable (in the sense that a bounded input produces a bounded output) if and only if its impulse response satisfies the constraint

$$\sum_{m=-\infty}^{\infty} \sum_{n=-\infty}^{\infty} | h(m,n) | < \infty. \tag{A1}$$

*Theorem 2* [6], [7]

A causal recursive filter with transfer function

$$H(z_1,z_2) = A(z_1,z_2)/B(z_1,z_2)$$

where $A$ and $B$ are polynomials in $z_1$ and $z_2$, is stable if and only if there are no values of $z_1$ and $z_2$ such that $B(z_1,z_2) = 0$, $| z_1 | \leq 1$ and $| z_2 | \leq 1$.

*Theorem 3* [7]

A causal recursive filter with transfer function $H(z_1,z_2) = A(z_1,z_2)/B(z_1,z_2)$, where $A$ and $B$ are polynomials is stable if and only if the following.

1) The map of the unit circle of the $z_1$ plane, $\{z_1: | z_1 | = 1\}$ into the $z_2$ plane, according to the equation $B(z_1,z_2) = 0$, lies outside the unit disk in the $z_2$ plane, $\{z_2: | z_2 | \leq 1\}$.

2) No point in the unit disk of the $z_1$ plane $\{z_1: | z_1 | \leq 1\}$ maps into the point $z_2 = 0$ by the relation $B(z_1,z_2) = 0$.

*Theorem 4* [7]

Among the four recursive filters which can be associated with $H(z_1,z_2) = A(z_1,z_2)/B(z_1,z_2)$, that is, $H(z_1,z_2)$ itself, $H(z_1^{-1},z_2)$, $H(z_1,z_2^{-1})$, and $H(z_1^{-1},z_2^{-1})$, at most one is stable.

## APPENDIX B

*Lemma 1*

The image of the unit circle $\{z_1: | z_1 | = 1\}$ by a bilinear transformation of the type

$$z_2 = - \frac{b_{11} + b_{21}z_1}{b_{12} + b_{22}z_1} \tag{B1}$$

where $b_{11}$, $b_{21}$, $b_{12}$, and $b_{22}$ are any complex constants, is another circle in the $z_2$ plane of center $c$ and radius $r$ given by

$$c = \frac{b_{21}b_{22}^* - b_{11}b_{12}^*}{| b_{12} |^2 - | b_{22} |^2} \tag{B2}$$

$$r = \left| \frac{b_{11}b_{22} - b_{21}b_{12}}{| b_{12} |^2 - | b_{22} |^2} \right| . \tag{B3}$$

*Proof:* The proof that (B1) maps circles into circles can be found in any text on complex variables (see for example [12]) and will not be included here.

To find the center and the radius of the circle image we have to determine two points $z_{1A}$ and $z_{1B}$ over the unit circle $\{z_1: | z_1 | = 1\}$ such that their images $z_{2A}$ and $z_{2B}$ are diametrically opposite (i.e., maximally far apart) over the circle image. Under these circumstances

$$c = \frac{z_{2A} + z_{2B}}{2} \tag{B4}$$

$$r = \left| \frac{z_{2A} - z_{2B}}{2} \right| . \tag{B5}$$

In order to see how we choose $z_{1A}$ and $z_{1B}$ we write (B1) as follows:

$$z_2 = - \frac{b_{11} + b_{21}z_1}{b_{12} + b_{22}z_1} = - \frac{b_{21}}{b_{22}} - \frac{b_{11} - (b_{21}/b_{22})b_{12}}{b_{12} + b_{22}z_1}$$

$$= \mu + \frac{\nu}{b_{12} + b_{22}z_1} \tag{B6}$$

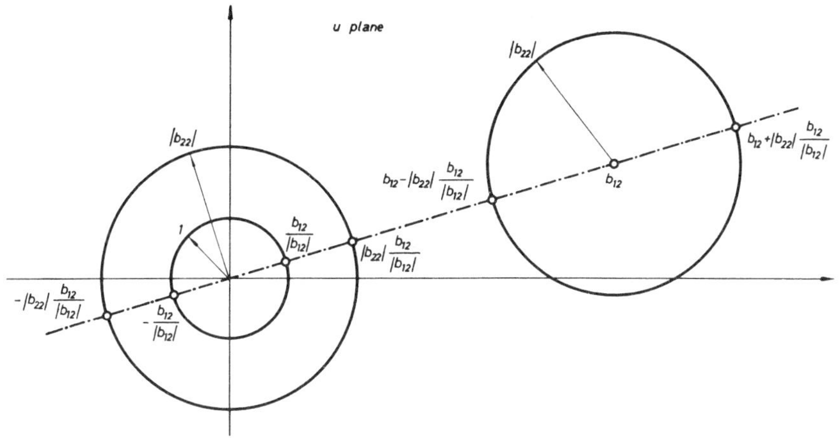


Fig. 13. Mapping the unit circle by the transformation $u = b_{12} + b_{22}z_1$.

where

$$\mu = -b_{21}/b_{22}$$

$$\nu = -b_{11} + (b_{21}/b_{22})b_{12}.$$

The transformation (B6) can be considered as a combination of the following transformations:

$$u = b_{12} + b_{22}z_1$$

$$v = 1/u$$

$$z_2 = \mu + \nu v$$

namely, the product of a similarity, an inversion, and another similarity.

The similarity has the property of mapping straight lines into straight lines and so does the inversion $v = 1/u$ for those straight lines crossing the origin in the $u$ plane. Since our domain of interest is $\{z_1: |z_1| = 1\}$, the points in the $u$ plane lying on a straight line crossing the origin, and that, according to the transformation $u = b_{12} + b_{22}z_1$, are images of some $z_1$ such that $|z_1| = 1$, are the following (see Fig. 13):

$$u_A = b_{12} - |b_{22}|(b_{12}/|b_{12}|) \tag{B7}$$

$$u_B = b_{12} + |b_{22}|(b_{12}/|b_{12}|). \tag{B8}$$

Therefore, by the transformation

$$z_2 = -b_{21}/b_{22} - [b_{11} - (b_{21}/b_{22})b_{12}](1/u)$$

the points $z_{2A}$ and $z_{2B}$ are

$$z_{2A} \triangleq z_2|_{u=u_A} = -\frac{b_{21}}{b_{22}} - \frac{b_{11}b_{22} - b_{21}b_{12}}{b_{22}} \frac{|b_{12}|}{b_{12}(|b_{12}| - |b_{22}|)} \tag{B9}$$

$$z_{2B} \triangleq z_2|_{u=u_B} = -\frac{b_{21}}{b_{22}} - \frac{b_{11}b_{22} - b_{21}b_{12}}{b_{22}} \frac{|b_{12}|}{b_{12}(|b_{12}| + |b_{22}|)}. \tag{B10}$$

Substituting (B9) and (B10) into (B4) and (B5) the center and the radius are

$$c = \frac{z_{2A} + z_{2B}}{2} = \frac{b_{21}b_{22}^* - b_{11}b_{12}^*}{|b_{12}|^2 - |b_{22}^2|}$$

$$r = \frac{z_{2A} - z_{2B}}{2} = \frac{b_{11}b_{22} - b_{21}b_{12}}{|b_{12}|^2 - |b_{22}|^2}. \qquad \text{Q.E.D.}$$

*Corollary*

If the coefficients $b_{11}$, $b_{21}$, $b_{12}$, and $b_{22}$ of (B1) are defined by (11) the center $c$ and the radius $r$ of the circle image are given by

$$c_i = -\frac{a_i}{a_i + \cos\beta} \tag{B11}$$

$$r_i = \frac{\cos\beta}{a_i + \cos\beta} \tag{B12}$$

where $a_i = \text{Re}\,[(T/2)p_i]$.

*Remarks*

Note that the center given by (B11) indicates that it will always lie on the real axis. If $a_i + \cos\beta = 0$ the circle image becomes a straight line perpendicular to the real axis at the point $(-1,0)$.

## REFERENCES

[1] T. S. Huang, "Two-dimensional windows," *IEEE Trans. Audio Electroacoust.* (Corresp.), vol. AU-20, pp. 88–89, Mar. 1972.
[2] J. V. Hu and L. R. Rabiner, "Design techniques for two-dimensional digital filters," *IEEE Trans. Audio Electroacoust.* (*Special Issue on Digital Filtering*), vol. AU-20, pp. 249–257, Oct. 1972.
[3] J. H. McClellan, "The design of two-dimensional digital filters by transformations," in *Proc. Seventh Annu. Princeton Conf. on Information Sciences and Systems*, Mar. 1973, pp. 247–251.
[4] E. L. Hall, "A comparison of computations for spatial frequency filtering," *Proc. IEEE* (*Special Issue on Digital Picture Processing*), vol. 60, pp. 887–891, July 1972.
[5] C. H. Farmer and D. S. Gooden, "Rotation and stability of a recursive digital filter," in *Proc. Two-Dimensional Digital Signal Processing Conf.*, Columbia, Mo., Oct. 1971, pp. 1-2-1–1-2-12.
[6] J. L. Shanks, S. Treitel, and J. H. Justice, "Stability and synthesis of two-dimensional recursive filters," *IEEE Trans. Audio Electroacoust.*, vol. AU-20, pp. 115–128, June 1972.
[7] T. S. Huang, "Stability of two-dimensional recursive filters," *IEEE Trans. Audio Electroacoust.*, vol. AU-20, pp. 158–163, June 1972.

[8] J. M. Costa, "Design and realization of stable two-dimensional recursive filters," M.A.Sc. thesis, Univ. of Toronto, Toronto, Ont., Canada, Sept. 1973.

[9] G. A. Maria and M. M. Fahmy, "An $l_p$ design technique for two-dimensional digital recursive filters," *IEEE Trans. Acoust., Speech, Signal Processing*, vol. ASSP-22, pp. 15–21, Feb. 1974.

[10] R. M. Golden, "Digital filters synthesis by sampled-data transformation," *IEEE Trans. Audio Electroacoust.* (*Special Issue on Digital Filters: The Promise of LSI Applied to Signal Processing*), vol. AU-16, pp. 321–329, Sept. 1968.

[11] L. R. Rabiner and B. Gold, "Processing of two-dimensional signals," unpublished notes, May 1972.

[12] W. Rudin, *Real and Complex Analysis.* New York: McGraw-Hill, 1966, p. 270.

# Part 4
# Finite Word Length Effects

When digital signal processing algorithms are implemented on a digital computer or with special-purpose hardware, account must be taken of effects due to finite word length in signal values, in coefficient values, and in the arithmetic operations performed. The paper by Oppenheim and Weinstein, which constitutes the first section, gives a detailed overview of the many different finite arithmetic effects that can occur in real system implementations. All other papers are divided into two groups, one pertaining to coefficient sensitivity and the other to limit cycles.

## Coefficient Sensitivity

The required coefficient word length of a digital filter has a strong effect on both the cost and speed of the filter; if the word length is large, then the cost of the filter will generally be high and the operating speed low. Therefore it is of interest to keep the coefficient word length as small as possible. This can be accomplished by two approaches: by choosing a structure that has low coefficient sensitivity (i.e., low accuracy requirements) and by appropriately choosing the quantized binary representation of the coefficients such that the coefficient word length is as small as possible for a specific structure without exceeding the design specifications.

The paper by Knowles and Olcayto represents one of the earliest attempts at the first approach by comparing structures on the basis of a coefficient sensitivity norm. The paper by Avenhaus presents an improvement on this approach and establishes the concept of a statistical measure of coefficient word length (sometimes referred to as a "statistical" word length) as a basis of comparison. This paper also addresses the problem of minimizing word length for a specific structure and shows that it can be treated as a nonlinear integer optimization problem. The paper by Chan and Rabiner presents a comprehensive analysis of the important class of direct form finite impulse response (FIR) structures on the basis of a similar statistical word length measure.

The paper by Crochiere is directed at the second approach, i.e., that of reducing the coefficient word length for a specific structure. It is shown that the word length can often be reduced by an optimization procdure which minimizes the statistical word length for a specific set of design specifications. The papers by Steiglitz, and Charalambous and Best, treat the minimum word length problem for a specific structure as a nonlinear integer optimization problem and suggest various ways of performing this optimization, such as the Hooke and Jeeves method and the branch-and-bound algorithm.

## Limit Cycles

In stable recursive implementations of digital filters, limit cycles or periodic output sequences can appear as a result of the rounding or truncation in the successive finite-precision arithmetic operations. The phenomenon is a nonlinear one and is dependent on the filter configuration, the type of arithmetic, the rounding rule, and the actual filter coefficent values. The limit cycle phenomenon may be used to advantage as in the case of a periodic sequence generator; however, it is usually considered to be a type of unwanted noise or error. Thus effort has been spent on understanding the nature of the limit cycle phenomena, conditions for their existence, bounds on their amplitude, and estimates of their frequency and length. While some effort has been spent on studying the general $n$th-order system, most interest and effort has been centered on the second-order system. These linear second-order digital systems form the bulk of the implementations, both hardware and software, of digital filters.

The early work [1], [2] concentrated on understanding the basic mechanism of the limit cycle phenomena and on developing ways to both analyze and illustrate the obtained results. The wide range of problems covered is well illustrated in the introductory work by Parker and Hess [2] that developed bounds on amplitudes, introduced the successive-value phase plane, noted symmetries in the limit cycle sequences, and described the expected behavior of limit cycles under forced conditions. These problems and others form the principal topics of the papers included in this section.

In the paper by Sandberg and Kaiser, a bound on the root-mean-square value of limit cycle amplitude is developed for $n$th-order systems implemented with fixed-point arithmetic. The results are then specialized to the second-order digital filter. In the paper by Long and Trick, absolute bounds on limit cycle amplitude are derived and compared with other known bounds. The important concept of accessibility of limit cycles and further remarks on classification and symmetry are the subjects of the paper by Claasen *et al.* Especially notable here is that Jackson's estimate of limit cycle amplitude is also the amplitude of the smallest accessible limit cycle for the second-order section with two rounding nonlinearities.

In the second paper by Claasen *et al.*, it is shown that

by using a single magnitude truncation quantizer in the second-order digital filter the possible existence of limit cycles is confined to those filters with very low damping and critical frequencies near zero or near half the sampling frequency. In the third paper by Claasen *et al.*, a more general approach to the question of existence of limit cycles is formulated in the frequency domain for *n*th-order systems with general nonlinear characteristics. The results of this analysis are then specialized to the case of a second-order digital filter with either one or two quantization nonlinearities.

An interesting and illuminating account of limit cycle phenomena in a tenth-order digital filter is given in the paper by Kieburtz. The results illustrate many of the ideas developed in the previous papers and thus give the reader some insight into how limit cycles arise, how they propagate in the system, how large they can become, and how they may be controlled. The final paper, by Kaneko, shows that the limit cycle phenomenon is not confined only to those filters employing fixed-point arithmetic, but can occur also in filters using floating-point arithmetic. The conditions for existence of floating-point limit cycles in first- and second-order digital filters is derived and amplitude estimates are given indicating the magnitude of the problem.

While these papers cover only a fraction of the work that has been done on various aspects of the limit cycle problem, they do represent the range of research in this currently active area.

## References


[1] L. B. Jackson, "An analysis of limit cycles due to multiplication rounding in recursive digital (sub) filters," in *Proc. 7th Allerton Conf. Circuit and System Theory*, Monticello, Ill., 1969, pp. 69–78; also in *Digital Signal Processing.* New York: IEEE Press, 1972, pp. 439–447.

[2] S. R. Parker and S. F. Hess, "Limit-cycle oscillations in digital filters," *IEEE Trans. Circuit Theory*, vol. CT-18, pp. 687–697, Nov. 1971.

# Effects of Finite Register Length in Digital Filtering and the Fast Fourier Transform

ALAN V. OPPENHEIM, SENIOR MEMBER, IEEE, AND
CLIFFORD J. WEINSTEIN, MEMBER, IEEE

*Invited Paper*

*Abstract*—When digital signal processing operations are implemented on a computer or with special-purpose hardware, errors and constraints due to finite word length are unavoidable. The main categories of finite register length effects are errors due to A/D conversion, errors due to roundoffs in the arithmetic, constraints on signal levels imposed by the need to prevent overflow, and quantization of system coefficients. The effects of finite register length on implementations of linear recursive difference equation digital filters, and the fast Fourier transform (FFT), are discussed in some detail. For these algorithms, the differing quantization effects of fixed point, floating point, and block floating point arithmetic are examined and compared.

The paper is intended primarily as a tutorial review of a subject which has received considerable attention over the past few years. The groundwork is set through a discussion of the relationship between the binary representation of numbers and truncation or rounding, and a formulation of a statistical model for arithmetic roundoff. The analyses presented here are intended to illustrate techniques of working with particular models. Results of previous work are discussed and summarized when appropriate. Some examples are presented to indicate how the results developed for simple digital filters and the FFT can be applied to the analysis of more complicated systems which use these algorithms as building blocks.

## I. Introduction

In PRACTICE, digital signal processing requires the representation of sequence values in a binary format with a finite register length. The effect of the finite word-length constraint manifests itself in several different ways. If a sequence to be processed is derived by sampling an analog waveform, then the finite word-length constraint requires that the analog-to-digital conversion produce only a finite number of values. This represents quantization of the input waveform. Even when we start with data representable with a finite word length, the result of processing will naturally lead to values requiring additional bits for their representation. For example, a $b$-bit data sample multiplied by a $b$-bit coefficient results in a product which is $2b$ bits long. If in a recursive digital filter we do not quantize the result of arithmetic operations, the number of bits required will increase indefinitely, since after the first iteration $2b$ bits are required, after the second iteration $3b$ bits are required, etc. The effect of quantization in such a context depends on such factors as whether we are considering fixed-point or floating-point arithmetic, and whether for fixed-point arithmetic we are using a representation of numbers in terms of fractions or integers, or perhaps a mixture. We will be treating the case of fixed-point arithmetic and floating-point arithmetic separately. For fixed-point arithmetic, it is natural in a signal processing context to consider a register as representing a fixed-point fraction. In this way the product of two numbers remains a fraction and the limited register length can be maintained by truncating or rounding the least significant bits. With this type of representation the result of addition on fixed-point fractions need not be truncated or rounded but it can increase in magnitude so that the sum eventually is not a fraction. This effect is commonly referred to as overflow, and can be handled by requiring that the input data be sufficiently small so that the possibility of overflow is avoided. In considering floating-point arithmetic, dynamic range considerations generally can be neglected due to the large range of representable numbers, but quantization is introduced both for multiplication and for addition.

A third effect of finite word length is inaccuracies in parameter values. While generally signal processing parameters are initially specified with unlimited accuracy, they can only be utilized with finite word length. This effect is similar to the effect which arises in implementing analog processing using inaccurate circuit elements. There are two possible approaches to handling the inaccuracies in parameter values. One possibility is to develop design procedures which inherently are insensitive to parameter inaccuracies. An alternate is to choose specifications which are consistent with the limited register length. There is a certain amount that is understood about the effect of inaccuracies in parameter values, but for the most part present results lead to guidelines rather than hard design or analytical strategies.

In the following discussion the relationship between the binary representation of numbers and truncation or rounding is discussed and a statistical model for arithmetic roundoff is presented. This statistical model is then applied to the analysis of fixed-point and floating-point rounding errors in digital filters. The analysis includes a consideration of the effect of dynamic range in developing and comparing signal-to-noise ratios for fixed-point and floating-point filters. It is not always possible to treat the effects of arithmetic roundoff in terms of a simple statistical model. Some approaches and results are available in the literature on the limit cycle behavior of digital filters due to arithmetic roundoff, and a discussion of some of these results is included.

For the analysis of arithmetic roundoff in computation of the discrete Fourier transform using the fast Fourier trans-

Manuscript received May 11, 1972. A. V. Oppenheim was supported in part by the National Science Foundation under Grant GK-31353 and in part by the Advanced Research Project Agency of the Department of Defense, monitored by ONR under Contract N00014-67-A-0204-0064; C. J. Weinstein was supported in part by the U. S. Air Force. *This invited paper is one of a series planned on topics of general interest—The Editor.*

A. V. Oppenheim is with the Department of Electrical Engineering and the Research Laboratory of Electronics, Massachusetts Institute of Technology, Cambridge, Mass. 02139.

C. J. Weinstein is with Lincoln Laboratory, Massachusetts Institute of Technology, Lexington, Mass. 02173.

Reprinted from *Proc. IEEE*, vol. 60, pp. 957–976, Aug. 1972.

form (FFT) algorithm a statistical model is used. With this model the signal-to-noise ratio is developed and compared for fixed-point and floating-point arithmetic.

While for any given filter configuration or spectral analysis problem it can be difficult to carry out a detailed analysis of the effects of finite register length there are a number of general guidelines that can be distilled from the results presented here. In Section IV some examples and guidelines are presented for filters implemented with fixed-point arithmetic and with floating-point arithmetic as well as for filters implemented with the FFT.

This paper is intended primarily as a tutorial review of a subject which has received considerable attention over the past few years. The analyses which are presented here are selected to illustrate techniques of working with particular models. Previous work is freely referenced, discussed, and borrowed from.

## II. Number Representation and Its Effect on Quantization

### A. *Fixed-Point and Floating-Point Numbers*

The manner in which finite word-length effects are manifested is closely tied to the way in which numbers are represented.

Digital computers and special purpose digital hardware for the most part use a number representation with a radix of 2, i.e., a binary representation. Therefore, a number is represented by a sequence of binary digits which are either zero or unity. Just as a decimal number is represented as a string of decimal digits with a decimal point dividing the integer part from the fractional part, the sequence of binary digits is divided by a binary point into those representing the integer part of the number and those representing the fractional part. Thus if $\Delta$ denotes the location of the binary point, the binary number $1001_\Delta 0110$ has the decimal value of $(1\times 2^3 + 0\times 2^2 + 0\times 2^1 + 1\times 2^0) + (0\times 2^{-1} + 1\times 2^{-2} + 1\times 2^{-3} + 0\times 2^{-4})$. This representation always corresponds to a positive number.

The manner in which arithmetic is implemented in a digital computer or in a special purpose hardware depends on where in the register the binary point is located. For fixed-point arithmetic, the implementation is based on the assumption that the location of the binary point is fixed. The manner in which addition is carried out will not depend on the location of the binary point for fixed-point arithmetic as long as it is the same for every register. For multiplication, however, the location of the binary point must be known. For example, consider the product of the two 4-bit numbers $1001_\Delta$ and $0011_\Delta$. In general, of course, the product of two $b$-bit numbers will be $2b$ bits long. The 8-bit product of the above number is $00011011_\Delta$. If, on the other hand, we consider the 4-bit fractions $_\Delta 1001$ and $_\Delta 0011$, then the 8-bit product is $_\Delta 00011011$. In digital filtering applications, it is usually necessary to approximate the $2b$-bit product of two $b$-bit numbers by a $b$-bit result. In integer arithmetic this is difficult. With fractional arithmetic, on the other hand, this can be accomplished by truncating or rounding to the most significant $b$ bits. For multiplication with fractions, overflow can never occur since the product of two fractions is a fraction. Thus for the 4-bit example previously mentioned, the product $_\Delta 00011011$ can be approximated by $_\Delta 0001$ (truncation) or $_\Delta 0010$ (rounding).

An alternative to fixed-point arithmetic is a floating-point representation. In this case, a positive number $F$ is represented as $F=2^c M$, where $M$, the mantissa, is a fraction between 1/2 and 1, and $c$, the characteristic, can be either positive or negative. The product of two floating-point numbers is carried out by multiplying the mantissa as fixed-point fractions and adding the characteristics. Since the product of the mantissas will be between 1/4 and 1, a normalization of the mantissa and corresponding adjustment of the characteristic may be necessary. The sum of two floating-point numbers is carried out by scaling the mantissas of the smaller number to the right until the characteristics of the two numbers are equal and then adding the mantissas. For example, consider the sum of $F_1$ and $F_2$ with $F_1=4$ and $F_2=5/4$. Then in floating-point notation, $F_1=2^{c_1}M_1$, and $F_2=2^{c_2}M_2$ with

$$c_1 = 11_\Delta \quad (=3 \text{ decimal})$$
$$M_1 = {}_\Delta 1000 \quad (=0.5 \text{ decimal})$$
$$c_2 = 1_\Delta \quad (=1 \text{ decimal})$$
$$M_2 = {}_\Delta 1010 \quad (=5/8 \text{ decimal}).$$

In order to carry out the addition, $c_2$ must be changed to equal $c_1$ and $M_2$ must be adjusted accordingly. Thus first the representation of $F_2$ is changed to $F_2=2^{\hat{c}_2}\hat{M}_2$ with

$$\hat{c}_2 = 11_\Delta$$
$$\hat{M}_2 = {}_\Delta 00101$$

in which case the mantissas can now be added. The resulting sum is $F=2^c M$ with $c=11_\Delta$ and $M={}_\Delta 10101$. In this case the sum of $M_1$ and $\hat{M}_2$ is a fraction between 1/2 and 1 and therefore no further adjustment of $c$ has to be carried out. In a more general case, the sum may not be in that range, and consequently, $c$ would be adjusted to bring the mantissa into the proper range. From this example it should be clear that in general with floating-point arithmetic, the mantissa can exceed the register length and must therefore be truncated or rounded for both addition and multiplication whereas this is only necessary for multiplication in the fixed-point case. On the other hand, if the result of addition in the fixed-point case exceeds the register length, truncation or rounding will not help, i.e., the dynamic range has been exceeded. Thus while floating point introduces error due to arithmetic round-off, it provides much greater dynamic range than fixed point. As we will see later, both of these effects must be considered when comparing fixed-point and floating-point realizations of digital filters.

### B. *Representation of Negative Numbers*

There are three common means used for representing fixed-point negative numbers. The first, and most familiar, is sign and magnitude, i.e., the magnitude (which is of course positive) is represented as a binary number and the sign is represented by the leading binary digit which, if 0 corresponds to a + and if 1 corresponds to a − (or vice versa). Thus for example, in sign and magnitude $0_\Delta 0011$ represents 3/16 and $1_\Delta 0011$ represents $-3/16$. Two other related representations of negative numbers are often referred to as one's-complement and two's-complement representations. Considering all numbers to be fractions, a positive number is represented as before. For two's complement representation a negative number is represented by 2.0 minus its magnitude. For example $-(0_\Delta 0110)$ in sign and magnitude is represented as $1_\Delta 1010$ in two's-complement since $10_\Delta 000 - 0_\Delta 0100 = 1_\Delta 1010$. For one's-complement, the negative number is represented by subtracting the magnitude from the largest number representable in the register. Thus $-(0_\Delta 0110)$ is represented by

$(1_\Delta 1111)-(0_\Delta 0110)=1_\Delta 1001$. One's complement representation is equivalent to representing a negative number by the bit-by-bit complement of its magnitude. The choice of representation for negative numbers in a particular system is usually based almost entirely on hardware considerations.

For the representation of negative floating-point numbers there are a variety of conventions that have been used. In this paper we will consider the sign of the number to be associated with the mantissa so that the mantissa is a signed fraction. The representation of this signed fraction can of course be in sign and magnitude, one's-complement of two's-complement notation.

### C. A Model for Arithmetic Roundoff

In formulating a model for arithmetic roundoff, we shall consider both fixed-point numbers and mantissas of floating-point numbers to be represented as $b+1$-bit binary fractions, with the binary point just to the right of the highest order bit (or sign bit). This convention represents no loss of generality, and its convenience has been alluded to above. The numerical value (for positive numbers) of a one in the least significant bit is $2^{-b}$, and this quantity can be referred to as the width of quantization.

As indicated previously, the effect of finite register length on the result of arithmetic operations depends on whether fixed-point or floating-point arithmetic is used, and how negative numbers are represented. Let us consider first the effect of truncation and rounding in the fixed-point case. For sign and magnitude, one's-complement and two's-complement, the representation of positive numbers is identical and, consequently, so is the effect of truncation and rounding. If $E_T$ denotes the error due to truncation, i.e., the value after truncation minus the value before truncation, this error will always be negative for positive numbers. That is, the effect of truncation is to reduce the value of the numbers. More specifically, if $b_2$ denotes the number of bits (exclusive of sign) after truncation, and $b_1$ denotes the number of bits before truncation, then the result satisfies $0 \ge E_T \ge -(2^{-b_2}-2^{-b_1})$.

With sign and magnitude representation of negative numbers, truncation reduces the magnitude of the number and the error $E_T$ satisfies $0 \le E_T \le (2^{-b_2}-2^{-b_1})$. For a two's-complement negative number represented by the bit string $1_\Delta, a_1, a_2, \cdots, a_{b_1}$, the magnitude is given by

$$M_1 = 2.0 - x_1$$

where

$$x_1 = 1 + \sum_{i=1}^{b_1} a_j 2^{-j}.$$

Truncation to $b_2$ bits ($b_2 < b_1$) produces the bit string $1_\Delta, a_1, a_2, \cdots, a_{b_2}$, where now the magnitude is

$$M_2 = 2.0 - x_2$$

with

$$x_2 = 1 + \sum_{i=1}^{b_2} a_i 2^{-i}.$$

The change in magnitude is

$$\Delta M = M_2 - M_1 = \sum_{i=b_2+1}^{b_1} a_j 2^{-j}$$

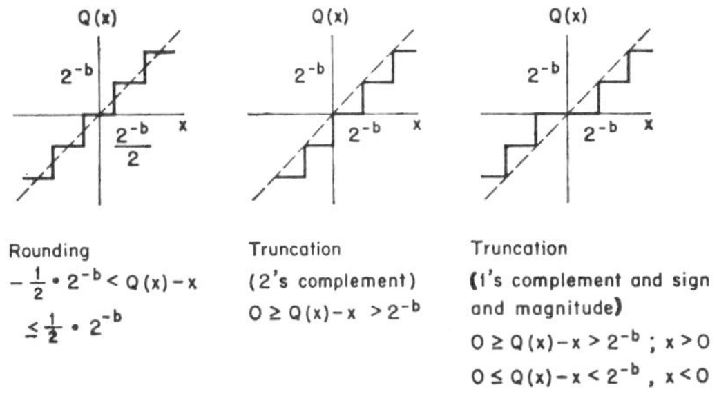


Fig. 1. Transfer characteristics for rounding and truncation.

and it is easily seen that

$$0 \le \Delta M \le 2^{-b_2} - 2^{-b_1}.$$

Hence the effect of truncation for two's-complement negative numbers is to *increase* the magnitude of the negative number; the truncation error is negative, and satisfies $0 \ge E_T \ge -(2^{-b_2} - 2^{-b_1})$.

For a one's-complement negative number represented by the bit string $1_\Delta, a_1, a_2, \cdots, a_{b_1}$, the magnitude is given by

$$M_1 = 2.0 - 2^{-b_1} - x_1$$

and truncation to $b_2$ bits yields a magnitude

$$M_2 = 2.0 - 2^{-b_2} - x_2$$

where $x_1$ and $x_2$ are as defined above. The change in magnitude is

$$\Delta M = M_2 - M_1 = \sum_{i=b_2+1}^{b_1} a_i 2^{-i} - (2^{-b_2} - 2^{-b_1})$$

and now

$$0 \ge \Delta M \ge -(2^{-b_2} - 2^{-b_1}).$$

Hence the effect of truncation for one's complement negative numbers is to *decrease* the magnitude of the negative number; the truncation error is positive, and satisfies $0 \le E_T \le 2^{-b_2} - 2^{-b_1}$.

The effect of rounding, of course, will be the same independent of how negative numbers are represented and the rounding error will always be greater than or equal to $(-1/2)2^{-b}$ and less than or equal to $(+1/2)2^{-b}$. The effect of truncation and rounding for the fixed-point case is summarized in Fig. 1, where $x$ represents the value before truncation or rounding and $Q(x)$ represents the value after. In the figure it is assumed that $x$ can take on a continuous range of values, corresponding to $b_1 = \infty$ in the discussion above, and that the quantized word length is $b$ bits plus sign.

For the case of floating-point arithmetic, the effect of truncation or rounding is reflected only in the mantissa. It is convenient in the floating-point case to describe the error in a multiplicative sense rather than in an additive sense as is done in fixed-point arithmetic. In other words, for a floating-point word, if $x$ represents the value before truncation or rounding and $Q(x)$ represents the value after, then we express $Q(x)$ as equal to $x(1+\epsilon)$. For the case of rounding, for example, the error in the mantissa is between $\pm 2^{-b}/2$, and consequently the error in the value of the floating-point word is

$$-2^c \cdot \frac{2^{-b}}{2} \le Q(x) - x \le 2^c \cdot \frac{2^{-b}}{2}$$

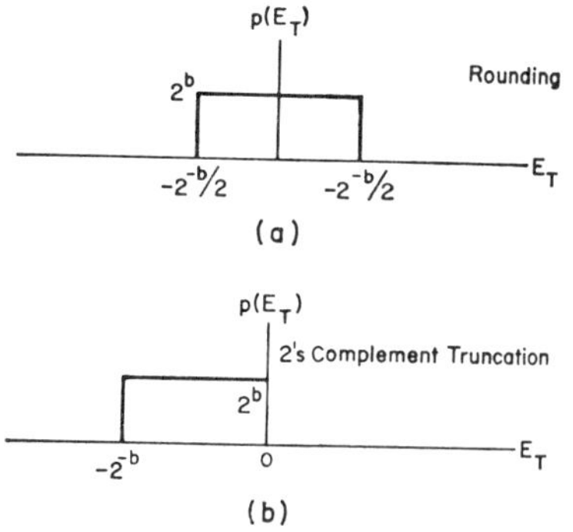


Fig. 2. (a) Probability density function for rounding noise. (b) Probability density function for noise due to two's-complement truncation.

or, since $Q(x)-x=\epsilon x$

$$-2^c \cdot \frac{2^{-b}}{2} \leq \epsilon x \leq 2^c \cdot \frac{2^{-b}}{2}$$

and since $2^{c-1} \leq x < 2^c$, we can write that for the case of rounding $-2^{-b} \leq \epsilon \leq 2^{-b}$. In a similar manner we can show that for one's-complement and for sign and magnitude truncation $0 \geq \epsilon \geq -2 \cdot 2^{-b}$. For two's-complement truncation

$$0 \geq \epsilon \geq -2 \cdot 2^{-b}, \qquad x > 0$$
$$0 \leq \epsilon \leq 2 \cdot 2^{-b}, \qquad x < 0.$$

*D. Statistical Model of Arithmetic Roundoff*

A convenient means for analyzing the effect of quantization is to represent the error statistically [1], [2]. In particular, for the case of fixed-point arithmetic and rounding $E_T$ is represented as a random variable with a probability density shown in Fig. 2(a). For the case of two's-complement truncation, the probability density is shown in Fig. 2(b).

In each of these cases, the assumption is that the random variable $E_T$ is independent of $x$. For one's complement and sign magnitude truncation, this assumption cannot be made since the mean value of the error is directly correlated with the sign of $x$. In the analysis that follows for fixed-point arithmetic, the discussion is phrased in terms of rounding. The results are easily modified for two's-complement truncation. In particular the variance of the noise is identical for both cases. However, for rounding the noise is zero mean and for two's-complement truncation it is not zero mean.

For the floating-point case, the parameter $\epsilon$ is considered to be a random variable which is independent of $x$. In that case the assumption of independence is reasonable for rounding, sign and magnitude truncation, and one's-complement truncation, but not for two's-complement truncation. The random variable $\epsilon$ is bounded by $-2^{-b} \leq \epsilon \leq 2^{-b}$. We will generally assume $\epsilon$ to be uniformly distributed in this range with a variance $\sigma_\epsilon^2=(1/3)2^{-2b}$. Empirical work has shown that the distribution is not quite uniform so that while $\sigma_\epsilon^2$ is proportional to $2^{-2b}$, the constant of proportionality is slightly less than 1/3. However, the interpretation of the results depends primarily on the proportionality to $2^{-2b}$.

## III. Finite Register Length Effects for Digital Filters [3]

*A. Introduction*

The basic arithmetic operations involved in implementation of a digital filter are multiplication by a constant and addition. For fixed-point arithmetic, roundoff is introduced only after the multiplication. Because of the possibility of overflow due to addition, there is a dynamic range limitation in fixed-point filters. In contrast, floating-point filter implementation has a much less severe dynamic range constraint, although arithmetic roundoff is introduced due to both multiplication and addition. In the next sections we will first develop the statistical analysis of arithmetic roundoff for fixed-point filters including dynamic range considerations. This is followed by a statistical analysis for floating-point arithmetic and a discussion of zero input limit cycle behavior for fixed-point arithmetic.

*B. Statistical Analysis of Fixed-Point Errors in a Digital Filter [4], [5]*

In many situations it is reasonable to model the effect of rounding in a digital filter by a simple statistical model. The approach is to model the effect of the rounding at each multiplier by a white-noise source uniformly distributed in amplitude between plus and minus $(1/2)2^{-b}$. Each of the noise sources is assumed to be linearly independent of each other and of the input. Experimentally these assumptions have been justified for a broad class of inputs including random signals, speech, etc. The model is clearly not valid for certain inputs, such as constant inputs. If the impulse response from the $k$th noise source to the output is $h_k(n)$ then the steady-state output noise variance due to the $k$th noise source is

$$\sigma_{ok}^2 = \sigma_\epsilon^2 \sum_{n=0}^{\infty} h_k^2(n) \tag{1}$$

where $\sigma_\epsilon^2=(1/12)2^{-2b}$. Since all the noise sources are assumed to be uncorrelated, the total output noise is

$$\sigma_o^2 = \sum_k \sigma_{ok}^2. \tag{2}$$

For example, if we consider the first-order filter in Fig. 3 one noise source is introduced. In this case, the impulse response from the noise source input to the output is $h(n)=a^n u_{-1}(n)$ where $u_{-1}$ denotes a unit step sequence, so that

$$\sigma_o^2 = \frac{1}{12} 2^{-2b} \frac{1}{1-a^2}. \tag{3}$$

For a second-order filter with one complex pole pair there are two noise sources as indicated in Fig. 4. The resulting output noise is

$$\sigma_o^2 = \frac{2}{12} 2^{-2b} \left( \frac{1+r^2}{1-r^2} \frac{1}{r^4+1-2r^2 \cos 2\theta} \right). \tag{4}$$

*C. Dynamic Range Considerations for Fixed-Point Filters*

As indicated previously, the possibility of overflow must be considered in the implementation of digital filters with fixed-point arithmetic. With the convention that each fixed-point register represents a signed fraction, each node in the filter must be constrained to maintain a magnitude less than unity in order to avoid overflow. Letting $x(n)$ denote the filter input and $y_k(n)$ and $h_k(n)$ denote the output and unit sample response for the $k$th node in the filter, then

$$y_k(n) = \sum_{r=0}^{\infty} h_k(r) x(n-r). \tag{5}$$

If $x_{max}$ denotes the maximum of the absolute value of the in-

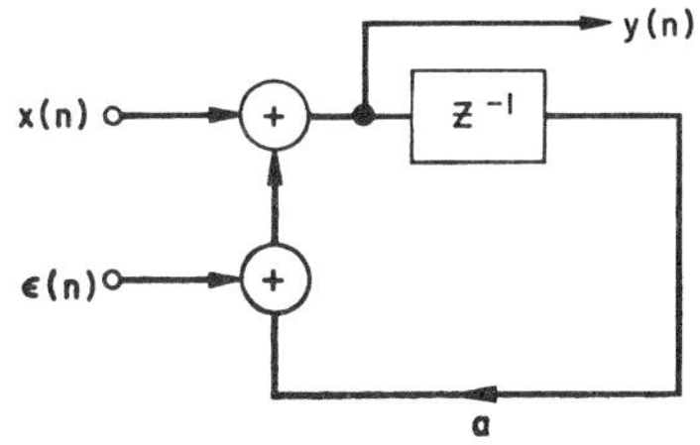


Fig. 3. Noisy first-order filter (fixed point).

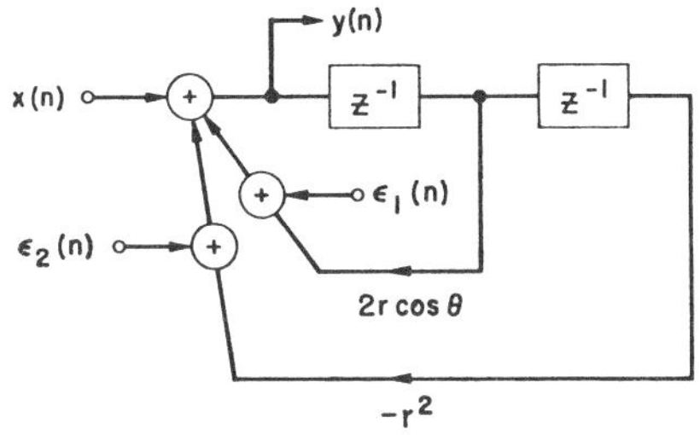


Fig. 4. Noisy second-order filter (fixed point).

put then

$$|y_k(n)| \leq x_{\max} \sum_{r=0}^{\infty} |h_k(r)|. \tag{6}$$

Thus, since we require that $|y_k(n)| < 1$, (6) requires that

$$x_{\max} < 1 \Big/ \sum_{r=0}^{\infty} |h_k(r)|, \qquad \text{for all } k. \tag{7}$$

Equation (7) thus provides an upper bound on the maximum value of the input to insure that no overflow occurs in the $k$th node. For a general input (7) in fact provides a least upper bound, i.e., if the maximum value of the input exceeds the bound, overflow can occur. This is a consequence of the fact that equality can be achieved in (6) with a sequence $x(n)$ for which at $n = n_o$, $x(n_o - r) = [\text{sgn } h_k(r)]$ for $r = 0$ to $\infty$. (Where sgn $(x) = 1$ for $x \geq 0$ and sgn $(x) = -1$ for $x < 0$.) Thus in the most general case, (7) is required to guarantee that no overflow occurs. The condition in (7) would generally be satisfied by applying attenuation to the signal at the filter input.

If we assume, for example, that the input $x(n)$ is a white-noise sequence with a uniform amplitude distribution, we would choose for the case of the first-order filter a maximal input amplitude of $(1-a)$. For this case, if $\sigma_x^2$ denotes the variance of the input signal, and $\sigma_y^2$ denotes the variance of the output signal, then

$$\sigma_x^2 = \left(\frac{1}{3}\right)(1-a)^2 \tag{8a}$$

$$\sigma_y^2 = \left(\frac{1}{3}\right)\left(\frac{(1-a)^2}{1-a^2}\right). \tag{8b}$$

For this example, we can then compute a *noise-to-signal* ratio as the ratio $\sigma_o^2/\sigma_y^2$ with the result

$$\frac{\sigma_o^2}{\sigma_y^2} = \frac{1}{4} 2^{-2b} \frac{1}{(1-a)^2}. \tag{9}$$

In a similar manner we can derive a noise-to-signal ratio for the second-order filter shown in Fig. 4. As in the first-order case, we restrict the maximum input in order to guarantee that the dynamic range of the registers is not exceeded. If we consider the input sequence to be uniformly distributed white noise, the resulting output noise-to-signal ratio will be

$$\frac{\sigma_o^2}{\sigma_y^2} = \frac{1}{2} 2^{-2b} \left(\sum_{n=0}^{\infty} |h_n|\right)^2 = \frac{1}{2} 2^{-2b} \left(\frac{1}{\sin\theta} \sum_{n=0}^{\infty} r^n |\sin[(n+1)\theta]|\right)^2. \tag{10}$$

While it is difficult to evaluate this expression exactly, it is possible to obtain an upper and lower bound. Since $\sum_{n=0}^{\infty}|h_n|$ is the largest possible output obtainable with an input that never exceeds unity, it must be larger than the response of the second-order filter to a sinusoid of unity amplitude at the resonant frequency. With this consideration, we can write that

$$\left(\sum_{n=0}^{\infty} |h_n|\right)^2 \geq 1/(1-r)^2(1+r^2-2r\cos 2\theta) \tag{11}$$

since the right-hand side of this inequality is the gain at resonance. Furthermore,

$$\left(\frac{1}{\sin\theta}\sum_{n=0}^{\infty} r^n |\sin[(n+1)\theta]|\right)^2 \leq \left(\frac{1}{\sin\theta}\sum_{n=0}^{\infty} r^n\right)^2. \tag{12}$$

Therefore, for the second-order case

$$\frac{1}{2} 2^{-2b} \frac{1}{(1-r)^2(1+r^2-2r\cos 2\theta)} \leq \frac{\sigma_o^2}{\sigma_y^2} \leq \frac{1}{2} 2^{-2b} \frac{1}{\sin^2\theta(1-r)^2}. \tag{13}$$

For both the first- and second-order filter an expression for the noise-to-signal ratio can be obtained which provides some insight into the behavior of the noise-to-signal ratio as the poles approach the unit circle. For the first-order filter let $\delta = 1 - a$ so that as $\delta \to 0$, the pole approaches the unit circle. Then in terms of $\delta$, the noise-to-signal ratio for the first-order filter is

$$\frac{\sigma_o^2}{\sigma_y^2} = \frac{1}{4} 2^{-2b} \frac{1}{\delta^2}. \tag{14}$$

For the second-order filter, let $\delta = 1 - r$ so that, again, as $\delta \to 0$ the poles approach the unit circle. Then if we assume that $\delta \ll 1$, we can approximate $(1 + r^2 - 2r\cos 2\theta)$ as

$$(1 + r^2 - 2r\cos 2\theta) \cong 4\sin^2\theta + \delta^2 \tag{15}$$

which for $4\sin^2\theta$ large compared with $\delta^2$ we will approximate as $4\sin^2\theta$. Consequently, incorporating this approximation,

$$\left(\frac{1}{2}\right) 2^{-2b} \frac{1}{4\delta^2\sin^2\theta} \leq \frac{\sigma_o^2}{\sigma_y^2} \leq \frac{1}{2} 2^{-2b} \frac{1}{\delta^2\sin^2\theta}. \tag{16}$$

Thus we observe that the noise-to-signal ratio as considered thus far can be considered to be proportional to $2^{-2b}/\delta^2$. We note from this dependence that if $\delta$ is halved, then to maintain the same noise-to-signal ratio $b$ must be increased by 1, i.e., one bit must be added to the register length. This dependence provides a convenient basis for comparison of different overflow strategies and different kinds of arithmetic.

In the above analysis, the filter input was assumed to be uniformly distributed white noise. As $\delta$ approaches zero the frequency response of both the first- and second-order filter becomes more selective so that more and more of the input energy is out of band. An alternative basis for determining the noise-to-signal ratio is for an input which is sinusoidal. For this choice of inputs, of course, we would not use the general condition of (7) to avoid overflow since we can determine exactly the maximum allowable input amplitude as a function of the filter parameters.

In particular, if the input is of the form $x(n) = x_{\max} \cos n\phi$ then the steady-state output is of the form $y(n) = y_{\max} \cos (n\phi + \psi)$. To prevent overflow, $y_{\max}$ must be less than unity and to maximize the output signal energy, $y_{\max}$ is chosen to be as large as possible. Thus the maximum noise-to-signal ratio is obtained when $x_{\max}$ is chosen so that $y(n) = \cos (n\phi + \psi)$. Note that in order to choose $x_{\max}$ in this way, the frequency of the input signal must be known. For an input sinusoid of unknown frequency $x_{\max}$ must be attenuated so that overflow will not occur even in the worst case, where the frequency of the input coincides with the peak gain in the filter's transfer function.

For fixed-point filters, within the validity of the statistical model for roundoff error, the output noise is independent of the form and amplitude of the input signal. Thus for this choice of inputs, the noise-to-signal ratio obtained for the first-order filter is

$$\frac{\sigma_o^2}{\sigma_y^2} = \frac{1}{24} 2^{-2b} \frac{1}{1 - a^2}. \tag{17}$$

If, as before, we let $a = 1 - \delta$, then for $\delta \ll 1$

$$\frac{\sigma_o^2}{\sigma_y^2} = \frac{1}{48} \frac{2^{-2b}}{\delta}. \tag{18}$$

Thus in this case, the noise-to-signal ratio is proportional to $1/\delta$ rather than $1/\delta^2$ so that if $\delta$ is multiplied by 1/4 and the register length is increased by one bit, the noise-to-signal ratio will remain constant. We can consider the second-order case in a similar manner. Again for a sinusoidal input, the output with maximum amplitude has the form $y(n) = \cos (n\phi + \psi)$ so that the noise-to-signal ratio in this case is

$$\frac{\sigma_o^2}{\sigma_y^2} = \frac{1}{12} 2^{-2b} \left( \frac{1 + r^2}{1 - r^2} \frac{1}{1 + r^4 - 2r^2 \cos 2\theta} \right). \tag{19}$$

Again, choosing $r = 1 - \delta$ with $\delta \ll 1$,

$$\frac{\sigma_o^2}{\sigma_y^2} \cong \frac{2^{-2b}}{4\delta \sin^2 \theta} \tag{20}$$

so that, as with the first-order filter, the noise-to-signal ratio is proportional to $1/\delta$ rather than $1/\delta^2$. The comparison in the noise-to-signal ratio for a white-noise input and a sinusoidal input serves to illustrate the dependence of the effect of dynamic range considerations on the particular form of the input. In some sense, the two cases considered represent extremes. As the input becomes more confined to a known narrow band of frequencies the above analysis with a sinusoidal input would be more representative, and as the input becomes more wide-band the above analysis with a white-noise input is more representative.

In the above discussion, the noise-to-signal ratio for the case of white-noise input was derived on the basis that overflow must be avoided. In a practical case, a scaling of the input on the basis of (7) can be considered to be somewhat pessimistic since the probability of equality being attained in (7) is extremely small. Furthermore, for many filters it is difficult to compute the sum in (7). Jackson [7] has formulated the dynamic range constraints on fixed-point digital filters in terms of $L_p$ norms. In particular, let $Y(\omega)$, $X(\omega)$, and $H(\omega)$ denote the Fourier transforms of the filter output, input, and system impulse response, respectively. Then it can be shown in general that

$$|y(n)| \leq \|H\|_p \|X\|_q \quad 1/p + 1/q = 1 \tag{21}$$

where $\|H\|_p$ and $\|X\|_q$ are the $L_p$ norm and $L_q$ norm of $H(\omega)$ and $X(\omega)$, respectively, where these norms are defined as

$$\|H\|_p = \left[ \frac{1}{2\pi} \int_{-\pi}^{\pi} |H(\omega)|^p \, d\omega \right]^{1/p}$$

and

$$\|X\|_q = \left[ \frac{1}{2\pi} \int_{-\pi}^{\pi} |X(\omega)|^q \, d\omega \right]^{1/q}.$$

For example, with $H(\omega)$ chosen as unity, a consequence of (21) is that

$$|x(n)| \leq \|X\|_q, \qquad \text{all } q \geq 1.$$

As another consequence, if we choose $p = 1$, $q = \infty$, and use the fact that the $L_\infty$ norm of $|X(\omega)|$ is the maximum value of $|X(\omega)|$ then we obtain the statement that

$$|y(n)| \leq \max [|X(\omega)|] \frac{1}{2\pi} \int_{-\pi}^{\pi} |H(\omega)| \, d\omega.$$

As an alternative, with $p = 2$, $q = 2$,

$$|y(n)| \leq \left[ \frac{1}{2\pi} \int_{-\pi}^{\pi} |H(\omega)|^2 \, d\omega \right]^{1/2} \cdot \left[ \frac{1}{2\pi} \int_{-\pi}^{\pi} |X(\omega)|^2 \, d\omega \right]^{1/2}.$$

To prevent overflow in the output we require that $|y(n)| < 1$ and to insure this from (21) we will require that $\|H\|_p \|X\|_q < 1$. Consequently, the input must be scaled in such a way that

$$\|X\|_q < 1/\|H\|_p. \tag{22}$$

This condition is somewhat less general than (7) but in many cases is easier to apply. According to (22) with $p = 2$, $q = 2$, the condition is in terms of the energy in the input signal and the energy in the system impulse response. For $q = 1$, $p = \infty$, (22) provides a bound in terms of the peak value of the magnitude of the transfer function, which is perhaps most appropriate for a sinusoidal input.

For the case of a random input (21) cannot be applied since the input and output do not have Fourier transforms. In this case the corresponding condition is phrased in terms of $\phi_{yy}(n)$ the autocorrelation function of the output, $\Phi_{xx}(\omega)$ the power density spectrum of the input, and $H(\omega)$ the magnitude of the system function. In particular, the inequality corresponding to (21) is

$$\phi_{yy}(n) \leq \|H^2\|_p \|\Phi_{xx}\|_q \tag{23a}$$

or equivalently

$$\phi_{yy}(n) \leq \|H\|_{2p}^2 \|\Phi_{xx}\|_q. \tag{23b}$$

Since, if the input is zero mean, $\phi_{yy}(0) = \sigma_y^2$ it follows that

$$\sigma_y^2 \leq \|H\|_{2p}^2 \|\Phi_{xx}\|_q. \tag{24}$$

Two particular cases of interest are $p=1$, $q=\infty$ and $p=\infty$, $q=1$ so that

$$\sigma_y^2 \leq \|H\|_2^2 \|\Phi_{xx}\|_\infty \tag{25}$$

and

$$\sigma_y^2 \leq \|H\|_\infty^2 \|\Phi_{xx}\|_1. \tag{26}$$

As Jackson points out, (25) implies the most stringent condition on the input spectrum $\Phi_{xx}(\omega)$ whereas (26) implies the most stringent condition on the transfer function. From (25), if the input spectrum is white so that $\Phi_{xx}(\omega)=\sigma_x^2$ for all $\omega$, then

$$\sigma_y^2 \leq \sigma_x^2 \|H\|_2^2 \tag{27}$$

with the input sequence Gaussian, then, the output will overflow no more often than the input overflows if

$$\|H\|_2 \leq 1. \tag{28}$$

More generally, (27) provides a basis for choosing the input variance to control the maximum percentage of time that the output can overflow.

*D. Statistical Analysis of Roundoff Errors with Floating-Point Arithmetic*

For the case of floating-point arithmetic, noise is introduced due both to the adds and the multiplies. In analyzing the effect of floating-point roundoff the effect of rounding will be represented multiplicatively so that if $[x]$ denotes rounding of the mantissa in a floating-point number, then

$$[x] = x(1+\epsilon). \tag{29}$$

To illustrate the analysis of roundoff errors with floating-point arithmetic let us consider a first-order filter. Let $w(n)$ denote the ideal response of the filter, that is, the response with no roundoff noise and let $y(n)$ denote the response of the filter in the presence of roundoff noise. Then following Liu and Kaneko [8] we can write that

$$w(n) = aw(n-1) + x(n) \tag{30}$$

$$y(n) = [ay(n-1)(1+\epsilon_n) + x(n)](1+\xi_n). \tag{31}$$

We assume that $\epsilon_n$ and $\xi_n$ are uniformly distributed between $-2^{-b}$ and $2^{-b}$, are uncorrelated from iteration to iteration, are independent of each other, and also are independent of the signal. Letting $E(n)$ represent the error in the output, so that $E(n)=y(n)-w(n)$, we can write from the above two equations that

$$E(n) - aE(n-1) = aw(n-1)(\epsilon_n+\xi_n) + x(n)\xi_n = u(n) \tag{32}$$

where we have neglected second-order terms in $\epsilon$, $\xi$, and $E$. Since $\epsilon$ and $\xi$ are statistically independent of $x$, and of $w(n-1)$, the term $u(n)$ is easily shown to be a white-noise sequence. Its variance, of course, depends on the excitation $x(n)$. The derivation of (32) with the second-order terms neglected corresponds to representing the roundoff noise as an additive noise source that is statistically independent of the signal but whose variance depends on the signal variance. Specifically, consider the first-order network drawn in Fig. 5 with the two noise sources $e_1(n)$ and $e_2(n)$. From the model for multiplier

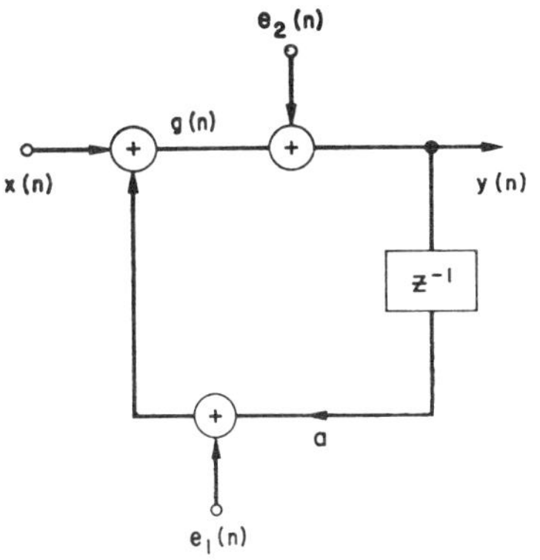


Fig. 5. Noisy first-order filter (floating point).

roundoff noise, the noise source $e_1(n)$ is given by

$$e_1(n) = ay(n-1)\epsilon_n \tag{33}$$

and the noise source $e_2(n)$ is given by

$$e_2(n) = g(n)\xi_n. \tag{34}$$

The analysis above in which we neglected second-order terms corresponds in this case to evaluating the variance of $e_1(n)$ and $e_2(n)$ by using the mean-square values for $y(n-1)$ and $g(n)$ that would result if no roundoff noise were present. Therefore, if we assume that $x(n)$ is a zero-mean white-noise input, with variance $\sigma_x^2$, then the variances of $e_1(n)$ and $e_2(n)$ are, respectively,

$$\sigma_{e_1}^2 = a^2\sigma_\epsilon^2\overline{y^2(n-1)} = a^2\sigma_\epsilon^2\sigma_x^2\frac{1}{1-a^2} \tag{35}$$

$$\sigma_{e_2}^2 = \sigma_\xi^2\overline{g^2(n)} = \sigma_\xi^2\sigma_x^2\frac{1}{1-a^2} \tag{36}$$

where the bar denotes expected value. Then, since $e_1(n)$ and $e_2(n)$ are independent, because $\epsilon_n$ and $\xi_n$ are independent, the output noise variance is

$$\sigma_o^2 = \sigma_\epsilon^2\sigma_x^2\frac{1+a^2}{(1-a^2)^2} = \sigma_\epsilon^2\sigma_y^2\frac{1+a^2}{1-a^2} \tag{37a}$$

where we have assumed again that $\sigma_\epsilon^2$ and $\sigma_\xi^2$ are equal. The output noise-to-signal ratio is

$$\frac{\sigma_o^2}{\sigma_y^2} = \sigma_\epsilon^2\frac{1+a^2}{1-a^2}. \tag{37b}$$

We can analyze the effect of roundoff noise in the second-order filter in a similar manner. In Fig. 6 is shown the network for a second-order filter with roundoff noise sources included. Note that since noise sources must be included due to addition, two summers are included to add the three variables in the feedback loop. The noise sources $e_3(n)$ and $e_4(n)$ represent the noise due to the multiplies and the noise sources $e_1(n)$ and $e_2(n)$ represent the noise due to the additions. With assumptions similar to those above in which we neglected second-order terms, we write that

$$\begin{aligned} e_1(n) &= y(n)\epsilon_1(n) \\ e_2(n) &= [y(n)-x(n)]\epsilon_2(n) \\ e_3(n) &= 2r\cos\theta y(n-1)\epsilon_3(n) \\ e_4(n) &= -r^2y(n-2)\epsilon_4(n) \end{aligned} \tag{38}$$

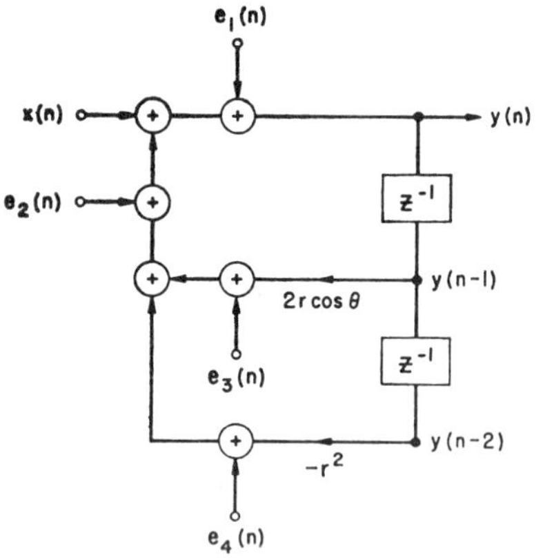


Fig. 6. Noisy second-order filter (floating point).

where $\epsilon_1$, $\epsilon_2$, $\epsilon_3$, and $\epsilon_4$ are independent random variables with equal variance $\sigma_\epsilon^2$. If as before, $x(n)$ is assumed to be a white random process with variance $\sigma_x^2$, then the output noise-to-signal ratio for the second-order case is

$$\frac{\sigma_o^2}{\sigma_y^2} = \sigma_\epsilon^2\left[1 + G\left(3r^4 + 12r^2\cos^2\theta - 16\frac{r^4\cos^2\theta}{1+r^2}\right)\right] \quad (39)$$

where

$$G = \frac{1+r^2}{1-r^2}\left(\frac{1}{r^4+1-4r^2\cos^2\theta+2r^2}\right). \quad (40)$$

For the high gain case, it is possible to compare fixed-point and floating-point arithmetic by approximating the expressions for the noise-to-signal ratio. For the first-order case, with $a=1-\delta$, and $\delta \ll 1$, the result (37b) for the first-order filter, can be approximated as

$$\frac{\sigma_o^2}{\sigma_y^2} \simeq \frac{1}{3}\,2^{-2b}\,\frac{1}{\delta}\,. \quad (41)$$

Similarly for the second-order filter

$$\frac{\sigma_o^2}{\sigma_y^2} \simeq \frac{1}{3}\,2^{-2b}\left(\frac{3+4\cos^2\theta}{4\delta\sin^2\theta}\right) \quad (42)$$

where in (41) and (42) we have taken $\sigma_\epsilon^2 = (1/3)2^{-2b}$.

For fixed-point arithmetic we recall that for a white-noise input the noise-to-signal ratio behaved as $1/\delta^2$ and for a sinusoidal input as $1/\delta$. Comparison of (41) and (42) with (14) and (16) and (18) and (20) indicates a slightly larger noise-to-signal ratio for floating-point arithmetic as compared with fixed-point arithmetic with a sinusoidal input of known frequency but a significantly smaller noise-to-signal ratio for floating-point arithmetic as compared with fixed-point arithmetic with a white-noise input. It is important to keep in mind, however, that the noise-to-signal ratios for the fixed-point filters were computed on the basis that the input signal was as large as possible. If the input signal level decreases, the noise-to-signal ratio will increase since the output noise variance is independent of the input signal level. For floating-point arithmetic, on the other hand, the output noise variance is proportional to the output signal variance and as the input level is scaled up or down so is the roundoff noise. It is also important to note that the comparison just discussed assumes that the floating-point mantissa is equal in length to the entire fixed point word, and does not account for the extra bits needed for the characteristic. The authors [6] have previously compared fixed- and floating-point filters on the basis of equal total word length. However, in completing such a comparison one must take account of the large difference in hardware complexity between implementing floating-point arithmetic, and adding a few bits to a fixed-point arithmetic element.

Oppenheim [9] has proposed a realization of recursive digital filters using block floating-point arithmetic. Here the input and filter states (i.e., the inputs to the delay registers) are jointly normalized before the multiplications and additions are performed in fixed-point arithmetic. The scale factor (or exponent) obtained during the normalization is then applied to the final output to obtain a fixed-point output. The roundoff noise properties of such a realization were studied, and the noise-to-signal ratio was found to lie between that for fixed and floating point.

### *E. Zero-Input Limit Cycle Behavior of Digital Filters for Fixed-Point Arithmetic*

In the preceding discussion the effect of arithmetic roundoff was modelled as an additive white-noise source, uncorrelated with the data. Justification of this model assumes that from iteration to iteration, the input can be expected to pass through several quantization levels. Consequently, this model is applied primarily when the input signal has a complicated behavior and cannot be expected to be valid in general. For example, consider a first-order filter for which the difference equation is

$$y_n = \alpha y_{n-1} + x_n \quad (43)$$

and for which the register length for the data is 4 bits and the coefficient $\alpha$ is 0.5. If the input $x_n$ is 7/8 and if rounding is applied after the arithmetic then on successive iterations of the filter, the output will be:

$$\begin{aligned} y_0 &= 7/8 \\ y_1 &= 1/2 \\ y_2 &= 1/4 \\ y_3 &= 1/8 \\ y_n &= 1/8, \qquad \text{for } n \geq 4. \end{aligned}$$

Thus due to rounding, the output reaches a steady-state nonzero value and since the ideal steady-state output is zero, this nonzero value represents roundoff error. Clearly this kind of roundoff error cannot be modelled as white noise, but in fact represents a limit cycle due to the nonlinearity corresponding to the quantizer which implements the rounding. Limit cycle behavior of this type was first noted by Blackman [10] who referred to the amplitude intervals within which these limit cycles are confined as "deadbands." Blackman considered only first-order limit cycles corresponding to a dc behavior in the deadband. More generally, Jackson [11] has considered limit cycle behavior in first- and second-order filter sections with an analysis based on the location of the "effective" poles in the filter due to roundoff. Following the approach presented by Jackson, consider a first-order filter with a difference equation of the form of (43). Due to the register length constraint, the product $\alpha y_{n-1}$ must be rounded. Let $(\cdot)'$ denote the operation of rounding. If the register length is $(b+1)$ bits and if data are represented as fractions then

$$\left|(\alpha y_{n-1})' - \alpha y_{n-1}\right| \leq (\tfrac{1}{2})2^{-b}. \quad (44)$$

If $y_{n-1}$ is such that $\left|(\alpha y_{n-1})'\right| = \left|y_{n-1}\right|$ then the magnitude of the effective value of the coefficient is unity corresponding to the pole of the filter being on the unit circle. The range of values for which this condition is met is

$$|y_{n-1}| - |\alpha y_{n-1}| \leq (\tfrac{1}{2})2^{-b} \tag{45}$$

or

$$|y_{n-1}| \leq \frac{(0.5)2^{-b}}{1 - |\alpha|}. \tag{46}$$

This range of values is referred to as the deadband. Due to rounding, of course, values within the deadband must be in steps of $2^{-b}$. For the first-order filter, when the filter state falls within this range and the input is zero, the effective pole is on the unit circle and the filter will support a limit cycle behavior. If the coefficient $\alpha$ is positive, as in the above example, the limit cycle response is dc, i.e., has constant magnitude and sign. For $\alpha$ negative the limit cycle behavior has constant magnitude but alternating sign.

For a second-order filter there is a larger variety of modes of limit cycle behavior. In particular, consider the second-order difference equation

$$y_n = x_n - \beta_1 y_{n-1} - \beta_2 y_{n-2}. \tag{47}$$

With $\beta_1^2 < +4\beta_2$ the filter poles occur as a complex conjugate pair and with $\beta_2 = 1$ the poles occur on the unit circle. The approach proposed by Jackson for examining the limit cycle behavior of the second-order filter corresponds to considering the filter behavior when the effect of rounding places the effective poles of the filter on the unit circle. With zero input the effective poles will be on the unit circle if

$$|y_{n-2}| - |(\beta_2 y_{n-2})| \leq \tfrac{1}{2}\, 2^{-b} \tag{48}$$

or

$$|y_{n-2}| \leq \frac{(0.5)2^{-b}}{1 - |\beta_2|}. \tag{49}$$

Thus if the output falls within this range the effective value of $\beta_2$ is unity so that the effective poles are on the unit circle. With the effective value of $\beta_2$ as unity, the effective value of $\beta_1$ controls the oscillation frequency.

A second mode of limit cycle behavior occurs in second-order filters when the effect of rounding is to place an effective pole at $z = \pm 1$. As shown by Jackson, the deadband corresponding to this mode is for values less than or equal to $1/(1-|\beta_1|+\beta_2)$ in steps of integer multiples of $2^{-b}$.

While this approach is somewhat heuristic, Jackson has found that these bounds are consistent with experimental results and hence he has hypothesized that they represent necessary and sufficient conditions. These bounds for second-order filters are summarized in Fig. 7, showing different deadband subregions in the $\beta_1$, $\beta_2$ plane. The number within an area in the $\beta_1$, $\beta_2$ plane represents the maximum magnitude of the limit cycle in multiples of $2^{-b}$ and the cross hatched region represents the region for which no limit cycles can occur.

Recently, Parker and Hess [12] have studied the limit cycle problem further, and found that these bounds are approximately correct and sufficient, but not necessary. In other words, there exist some limit cycles outside the regions specified by Fig. 7.

In addition to the above classes of limit cycles, a more severe type of limit cycle can occur due to overflow in filters implemented using one's-complement or two's-complement

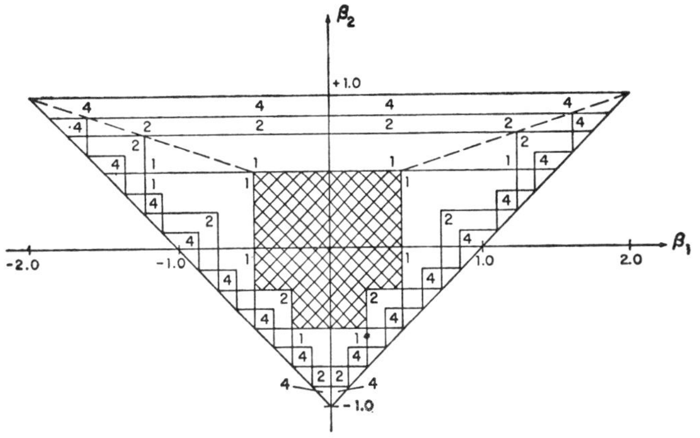

Fig. 7. Deadband subregions.

arithmetic. These limit cycles have been referred to as overflow oscillations [13] and can be avoided by using saturation arithmetic.

### *F. Effects of Parameter Quantization in Digital Filters*

In the preceding sections we focussed on the effects of arithmetic roundoff in digital filters. Another consequence of the requirement of finite register length is that the filter coefficients cannot be specified exactly. Classical design procedures generally lead to filter coefficients with arbitrary accuracy and the implementation of the filter then requires that the coefficients be modified to fit the available register length. For hardware realizations of digital filters it is, of course, desirable to keep the register length as small as possible.

One common approach to the problem of parameter quantization is the use of filter configurations or structures which in some sense are least sensitive to inaccuracies in the parameters. One of the difficulties in evaluating the sensitivity of filter structures is the choice of a meaningful measure of the sensitivity. Most commonly, the sensitivity of the filter is tied to the movement of the poles of the filter. For this choice Kaiser [14] has shown that for a filter with clustered poles a cascade or parallel combination of first- and second-order sections provides more accuracy in the pole positions than a direct form realization. This is basically a consequence of the fact that for a polynomial whose roots are clustered, the sensitivity of the roots to changes in the polynomial coefficients increases as the order of the polynomial increases. Thus the roots can be more accurately controlled if the polynomial is factored into first- and second-order factors.

Even within the choice of first- and second-order sections some flexibility remains. For a direct form implementation of a pole pair as shown in Fig. 8(a) the coefficients are $-r^2$ and $2r\cos\theta$. For a given quantization on the coefficients the poles must lie on a grid in the $z$ plane defined by the intersection of concentric circles, corresponding to quantization of $r^2$ and vertical lines, corresponding to quantization of $2r\cos\theta$. Such a grid is illustrated in Fig. 8(b). An alternative realization of a pole pair is the coupled form proposed by Rader and Gold [15], as shown in Fig. 9(a). In this case the coefficients are $r\cos\theta$ and $r\sin\theta$ and consequently the poles must lie on a rectangular grid as illustrated in Fig. 9(b). We note, for example, that for a given coefficient word length the direct form permits more accurate placement of poles with $r$ close to unity and $\theta$ large while the coupled form is more advantageous for $\theta$ small. There are, in theory, many other structures in addition to the direct and coupled forms for implementing pole

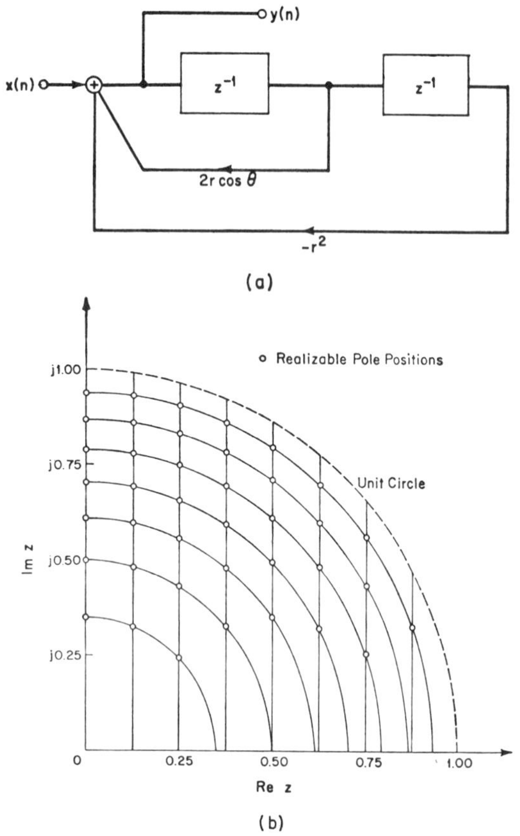


Fig. 8. (a) Direct form implementation of a pole pair. (b) Grid of allowable pole positions—direct form.

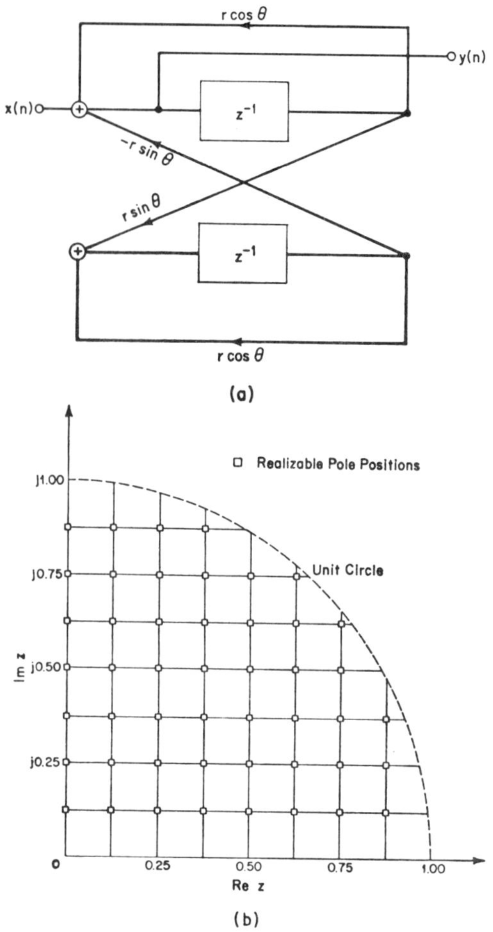


Fig. 9. (a) Coupled form implementation of a pole pair. (b) Grid of allowable pole positions—coupled form.

pairs although they are the most commonly considered [16]. Different structures, of course, imply different grids in the $z$ plane and generally it is advantageous to choose a structure for which the grid is dense in the region of the $z$ plane where the poles are to be located.

With a given choice of structure there remains the question as to how the pole locations on the grid should be chosen. A common procedure is to truncate or round the ideal coefficients. An alternative used by Avenhaus and Schussler [17] and also by Steiglitz [18] is to search over the grid in the vicinity of the ideal pole locations to select a grid point which locally minimizes the maximum error in the filter frequency response. As an alternative to the use of cascade or parallel connections of first- and second-order sections, more general filter structures can be considered. Digital wave filters, as proposed by Fettweis [19] and investigated by Bingham [20] and by Crochiere [21], appear to have much less sensitivity to parameter inaccuracies than the cascade form.

It would, of course, be desirable to incorporate the constraint of quantized coefficients into the design of digital filters. For nonrecursive filters, algorithmic design falls within the framework of integer linear programming. For recursive filters, however, the equations become nonlinear. In general, the development of design procedures with quantized coefficients remains an important area of research.

## IV. Effects of Arithmetic Roundoff in the FFT

### A. *Introduction*

The FFT algorithm [22] for computing the discrete Fourier transform (DFT) plays a central role in many signal processing applications [23]. As with the implementation of digital filters, it is important to understand the effect of finite register length arithmetic on the performance of the algorithm.

There are many forms of the FFT algorithm and the detailed effects of quantization will differ depending on the form used. The most commonly used forms of the algorithm are the radix-2 forms for which the size of the transform computed is an integer power of two. For the most part, the discussion below is phrased in terms of a particular form of the radix-2 FFT, commonly referred to as the decimation in time form of the algorithm; the results however are applicable with only minor modification to the decimation in frequency form. We feel that most of the ideas employed in the error analysis of the radix-2 forms of the algorithm can be utilized in other forms such as mixed radix, etc.

Our approach in analyzing noise in the FFT is basically statistical. In most cases, the predictions of the models are supported with experimental data (from Weinstein [24], unless otherwise stated). For floating and block floating point arithmetic, in order to simplify the analysis and obtain concrete results, it is convenient to assume a simple, white-noise model for the signal being transformed. Discussion of how the results might be expected to change for other types of signals is included, as are experimental noise measurements on FFT's of nonwhite signals.

### B. *The FFT Algorithm*

The FFT algorithm is directed toward computing the DFT of a finite duration sequence $f(n)$, defined as

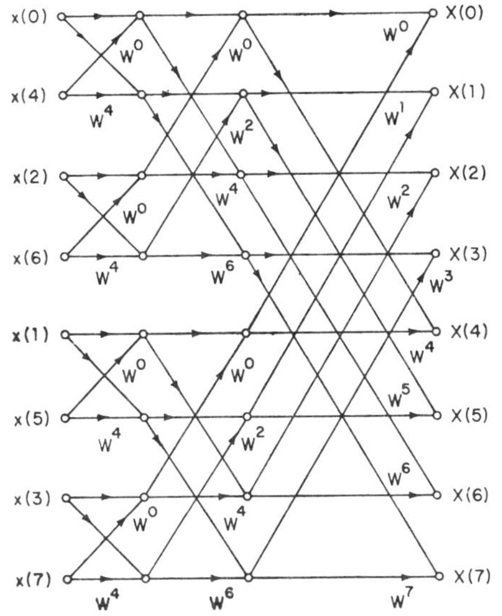


Fig. 10. FFT flow graph, $N=8$.

$$F(k) = \sum_{n=0}^{N-1} f(n)W^{nk}, \qquad W = e^{-j(2\pi/N)}. \tag{50}$$

A flow chart depicting the FFT algorithm for $N=8=2^3$ is shown in Fig. 10. A specific decimation in time algorithm is depicted. (An implementation of this particular form of the algorithm was used for the reported experimental work.) Some key aspects of this diagram, which are common to all standard radix-2 algorithms, are as follows. The DFT is computed in $\nu=\log_2 N$ stages. At each stage, the algorithm passes through the entire array of $N$ complex numbers, two at a time, generating a new $N$ number array. The $\nu$th array contains the desired DFT. The basic numerical computation operates on a pair of numbers in the $(m+1)$th array. This computation, referred to as a "butterfly" is

$$\begin{aligned} X_{m+1}(i) &= X_m(i) + \tilde{W}X_m(j) \\ X_{m+1}(j) &= X_m(i) - \tilde{W}X_m(j). \end{aligned} \tag{51}$$

Here, $X_m(i)$ and $X_m(j)$ represent a pair of numbers in the $m$th array, and $\tilde{W}$ is some appropriate integer power of $W$, that is

$$\tilde{W} = W^p = e^{-j2\pi p/N}. \tag{52}$$

The form of the butterfly computation is actually somewhat different for a decimation in frequency algorithm, where the computation is

$$\begin{aligned} X_{m+1}(i) &= X_m(i) + X_m(j) \\ X_{m+1}(j) &= [X_m(i) - X_m(j)]\tilde{W}. \end{aligned} \tag{53}$$

At each stage, $N/2$ separate butterfly computations are carried out to produce the next array. The integer $p$ varies with $i$, $j$, and $m$ in a manner which depends on the specific form of the FFT algorithm that is used. Fortunately, our analysis is not tied to the specific way in which $p$ varies. Also, the specific relationship between $i$, $j$, and $m$, which determines how we index through the $m$th array, is not important for the analysis. The details of the analysis for decimation in time and decimation in frequency differ somewhat due to the different butterfly forms, but the basic results for the dependence of noise-to-signal ratio on $N$ do not change significantly. In our analysis we will assume a butterfly of the form (51), corresponding to decimation in time.

### C. FFT Roundoff Noise with Fixed-Point Arithmetic

We will model the roundoff noise by associating an independent white-noise generator with each multiplier. This means that a noise source feeds into each node of the signal flow graph of Fig. 10 (excluding the initial array of nodes, since we are not considering A/D noise here). Since we are dealing with complex multiplications, these elemental noise sources are complex. Defining the complex variance $\sigma_B^2$ as the expected squared magnitude of such a noise source, we have

$$\sigma_B^2 = 4\,\frac{2^{-2b}}{12} \tag{54}$$

where it is assumed that each of the four real multiplications used to perform the complex multiplication is rounded separately. In Fig. 10, $3\times8=24$ such noise sources must be inserted. To add the effects of each of the noise sources in evaluating the total roundoff noise in the output, we note that the transmission function from any node in the flow graph to any other connected node is multiplication by a complex constant of unity magnitude. Since we assume that all noise sources are uncorrelated, the noise variance at any output node is equal to $\sigma_B^2$ times the number of noise sources that propagate to that node. The general result which is easily verified for the case $N=8$ by inspection of Fig. 10 is that $(N-1)$ noise sources propagate to each output node so that the output noise variance $\sigma_E^2$ is given by

$$\sigma_E^2 = (N-1)\sigma_B^2$$

which for large $N$ we take as

$$\sigma_E^2 \simeq N\sigma_B^2. \tag{55}$$

According to this result, the variance of the output noise is proportional to $N$, the number of points transformed. The effect of doubling $N$, or adding another stage in the FFT, is to double the output noise variance. Using the assumptions we have made thus far about the noise generators in the FFT (all uncorrelated, with equal variances), the output noise is white, i.e., the $N$ noise samples $E(k)$ are mutually uncorrelated, with independent real and imaginary parts. This follows from the fact that the output of any butterfly is white (two outputs uncorrelated with equal variance, real and imaginary parts uncorrelated) if the input is white. Since the noise sources in our system are white, and all connected to the output via some combination of butterfly computations, the output noise must also be white.

In order to simplify the analysis leading to (55), we have neglected some details. First, we have associated equal variance noise sources with all multipliers, including where $W=1$ and $j$. In many programmed FFT's these multiplications are performed noiselessly. If we assume in the analysis that these multiplications are noiseless, the output noise variance will no longer be uniform over the output array. For example, the zeroth output point would be noiseless. The average variance over the output array will be somewhat lower than the result in (55), but will retain a linear dependence on $N$. Second, the assumption that all noise sources are uncorrelated is contradicted by the fact that the two noise sources associated with a given butterfly are negatives of each other, and therefore completely correlated. This does not affect the result for output noise variance, since the two outputs of a butterfly connect to a disjoint set of output points. However, it implies that the output noise samples $E(k)$ are somewhat correlated. These details are worth mentioning, but not worth analyzing here at length, because they cloud the essential ideas of the analysis, are quite program-dependent, and do not change the essential character of the dependence of mean-squared output noise on $N$.

In implementing the FFT with fixed-point arithmetic we must insure against overflow. From (51) it follows that

$$\max\left[\,|X_m(i)|,\ |X_m(j)|\,\right] \le \max\left[\,|X_{m+1}(i)|,\ |X_{m+1}(j)|\,\right] \quad (56)$$

and also that

$$\max\left[\,|X_{m+1}(i)|,\ |X_{m+1}(j)|\,\right] \le 2\max\left[\,|X_m(i)|,\ |X_m(j)|\,\right]. \quad (57)$$

Equation (56) implies that the maximum modulus is nondecreasing from stage to stage so that, if the magnitude of the output of the FFT is less than unity then the magnitude of the points in each array must be less than unity,[1] i.e., there will be no overflow in any of the arrays.

In order to express this constraint as a bound on the input sequence, we note that the maximum possible output can be expressed in terms of the maximum input as

$$X(k)\big|_{\max} \le |x(n)|_{\max} \sum_{n=0}^{N-1} |W^{nk}| = N\,|x(n)|_{\max}. \quad (58)$$

Thus bounding the input sequence so that

$$|x(n)| < 1/N \quad (59)$$

will prevent overflow. To obtain an explicit expression for output signal variance, we assume $x(n)$ white, with real and imaginary parts each uniformly distributed in $(-1/\sqrt{2}N,\ 1/\sqrt{2}N)$. Then we have

$$\sigma_X^2 = \overline{|X(k)|^2} = N\sigma_x^2 = N\overline{|x(n)|^2} = \frac{1}{3N}. \quad (60)$$

Combining this with (55) yields

$$\frac{\sigma_E^2}{\sigma_X^2} = 3N^2\sigma_B^2. \quad (61)$$

The assumption of white input signal is not critical here. For example, if a complex sinusoid $x(n) = (1/N)\exp j(2\pi k_0 n/N + \phi)$ had been selected $\sigma_E^2/\sigma_X^2$ would still be proportional to $N^2$, which is the essential point of (61).

Equation (57) suggests an alternative procedure for preventing overflow. Since the maximum modulus increases by no more than a factor of two from stage to stage we can prevent overflow by requiring that $|x(n)| < 1$ and incorporating an attenuation factor of 1/2 at each stage. Using this step-by-step scaling, the attainable output signal level (for white input) is the same as in (60) since the output signal level does not depend on where the scaling is done, but only on how much overall scaling is done. However, the output noise level will be much less than in (55) since the noise introduced at early stages of the FFT will be attenuated by the scaling which takes place at the later array. Quantitatively for $N = 2^\nu$

$$\sigma_E^2 = \sigma_{B'}^2 \sum_{k=1}^{N/2} \frac{1}{k} \quad (62)$$

where $\sigma_{B'}^2$ represents the roundoff noise introduced due to multiplication by $W$ and scaling and will consequently be slightly higher than $\sigma_B^2$. In particular, if we assume that the scaling is accomplished with rounding, it can be shown

$$\sigma_{B'}^2 = \frac{5}{6}\,2^{-2b}. \quad (63)$$

For large $N$, (62) is approximately

$$\sigma_E^2 = 2\sigma_{B'}^2 \quad (64)$$

and thus is much less than the noise variance resulting when all of the scaling is carried out on the input data.

Now, we can combine (64) with (60) to obtain the output noise-to-signal ratio for the case of step-by-step scaling and white input. We obtain

$$\frac{\sigma_E^2}{\sigma_X^2} = 6N\sigma_{B'}^2 = (5N)2^{-2b} \quad (65)$$

a result proportional to $N$, rather than to $N^2$. An interpretation of (65) is that the rms output noise-to-signal ratio increases as $N$, or by half a bit per stage. This result was first obtained by Welch [25]. It is important to note that the assumption of white signal is not essential in the analysis. The basic result of half-a-bit-per-stage increase holds for a broad class of signals, with only the constant multiplier in (65) being signal-dependent. In particular, for a general input with scaling at each array, the output variance is related to the variance of the input array by

$$\sigma_X^2 = \frac{1}{N}\sigma_x^2 = \frac{1}{N}\overline{|x(n)|^2} \quad (66)$$

so that

$$\frac{\sigma_E^2}{\sigma_X^2} = \frac{\frac{5}{3}N2^{-2b}}{\sigma_x^2} \quad (67)$$

where, to reduce noise-to-signal ratio, we would like to make $\sigma_x^2$ as large as possible but are limited by the constraint $|x(n)| < 1$. The result (67) has been verified experimentally for both wide-band and narrow-band signals [24], [25].

We should also note that the dominant factor causing the increase of $\sigma_E^2/\sigma_X^2$ with $N$ is the decrease in signal level (required by the overflow constraint) as we pass from stage to stage. According to (63) and (64), very little noise (only a bit or two) is present in the final array. Most of the noise has been shifted off by the scalings. However, the mean-squared signal level has decreased by a factor of $1/N$ from its initial value, due to the scalings. Our output consists not of the DFT defined by (50) but of $1/N$ times this DFT.

We have assumed straight fixed point computation in this section, i.e., only preset attenuations were allowed, and we were not permitted to rescale on the basis of an overflow test. Clearly, if the hardware or programming facility are such that straight fixed point must be used, we should, if possible, incorporate attenuators of 1/2 at each array rather than using a large attenuation of the input array.

[1] Actually one should discuss overflow in terms of the real and imaginary parts of the data, rather than the magnitude. However, $|x| < 1$ implies that $|\mathrm{Re}\,(x)| < 1$ and $|\mathrm{Im}\,(x)| < 1$, and only a slight increase in allowable signal level is achieved by scaling on the basis of Re and Im parts.

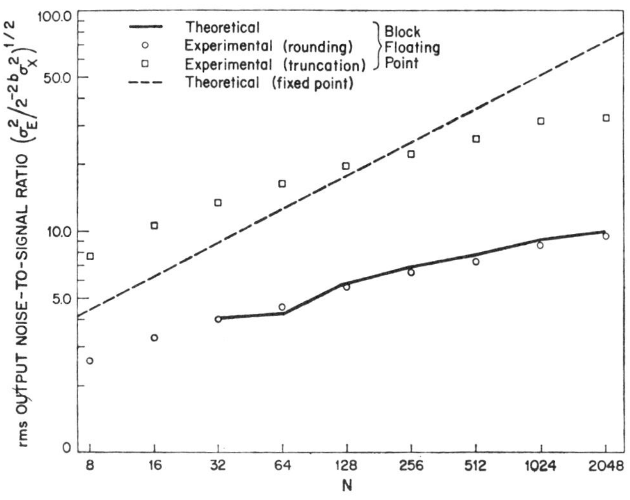


Fig. 11. Experimental and theoretical noise-to-signal ratios for block floating-point FFT.

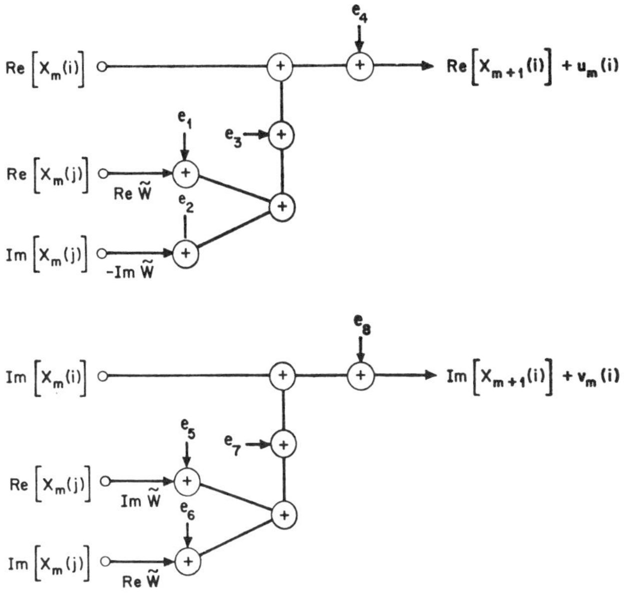


Fig. 12. Noisy butterfly computation (floating point).

A third approach to avoiding overflow is the use of block floating point. In this procedure the original array is normalized to the far left of the computer word, with the restriction that $|x(n)|<1$; the computation proceeds in a fixed point manner, except that after every addition there is an overflow test; if overflow is detected, the entire array is shifted right 1 bit and the computation continues. The number of necessary shifts are counted to determine a scale factor or exponent for the entire final array. The output noise-to-signal ratio depends strongly on how many overflows occur, and at what stages of the FFT they occur. The positions and timing of overflows are determined by the signal being transformed, and thus, in order to analyze noise-to-signal ratio in block floating FFT, one needs to know the signal statistics. This is in contrast to the fixed point analysis above, where it was not necessary to assume specific signal statistics.

The necessary number of right shifts of the array is related to the peakiness of the DFT of the signal being transformed. If the constant signal $x(n)=1$ or the single-frequency input $x(n)=\exp j(2\pi/N)k_o n$ is transformed, the output (with $k_o$ an integer) will consist of a single nonzero point and (for $N=2^\nu$) $\nu$ scalings of the array will be necessary, one for each stage.

A reasonable case to examine is the case of a white input signal; the DFT of a white signal is white, and one might expect (since the spectral energy is spread) that scalings at all stages would not be necessary, and a noise-to-signal ratio advantage over fixed point would be gained. This problem can be analyzed theoretically [24] but the analysis is quite involved and will be omitted. Instead, we will present some experimental results.

In Fig. 11 experimentally measured values of output noise-to-signal ratio are presented for block floating FFT's of white inputs, using rounded arithmetic. The quantity plotted is $(\sigma_E^2/2^{-2b}\sigma_x^2)^{1/2}$, the rms noise-to-signal ratio. For comparison, a theoretical curve representing fixed point noise-to-signal ratio (for rounded arithmetic) is also shown. We see that for white input block floating point provides some advantages over fixed point, especially for the larger transforms. For $N=2048$, the rms noise-to-signal ratio for block floating point is about 1/8 that of fixed point, representing a 3-bit improvement.

An experimental investigation was used to examine how the results for block floating point change, when truncation rather than rounding is used. The results of this experiment are also shown in Fig. 11. Noise-to-signal ratios are generally a bit or two worse than for rounding. The rate of increase of noise-to-signal ratio with $N$ seems to be about the same as for rounding.

### *D. FFT Roundoff Noise with Floating-Point Arithmetic*

The effect of arithmetic roundoff with floating-point arithmetic has been analyzed theoretically and experimentally by Gentleman and Sande [26], by Weinstein [27], and by Kaneko and Liu [28]. As with the statistical analysis of roundoff errors with fixed-point arithmetic, noise is introduced due to each butterfly computation. As with floating-point errors in digital filters, we neglect second-order error terms so that noise sources are introduced after each multiplication and addition that are assumed to be white but for which the variance is proportional to the variance of the signal at that node. Unless the input signal is assumed to be white, the analysis becomes quite complicated due to the variation of the variance of the signal and therefore of the noise sources within each array. Kaneko and Liu have obtained detailed formulas for a general stochastic model of the input signal. We will confine attention here to the case of white input signal, where the signal at any array in the FFT is also white, with constant variance across the array.

In Fig. 12 a typical butterfly computation (only top half) is indicated, including the noise sources due to multiplication and addition. The assumption of white input signal implies that

$$\overline{[\mathrm{Re}\,(X_m)]^2} = \overline{[\mathrm{Im}\,(X_m)]^2} = \tfrac{1}{2}\,\overline{|X_m|^2} \qquad (68)$$

and application of our floating-point noise model as in Section III-D yields the noise source variances

$$\sigma_{e_1}{}^2 + \sigma_{e_2}{}^2 = \sigma_{e_5}{}^2 + \sigma_{e_6}{}^2 = \sigma_{e_3}{}^2 = \sigma_{e_7}{}^2 = \tfrac{1}{2}\,\sigma_\epsilon{}^2\overline{|X_m|^2} \qquad (69)$$

$$\sigma_{e_4}{}^2 = \sigma_{e_8}{}^2 = \sigma_\epsilon{}^2\overline{|X_m|^2}. \qquad (70)$$

The variance of the complex noise source $U_m = u_m + jv_m$ is then

$$\overline{|U_m|^2} = 4\sigma_\epsilon^2 \overline{|X_m|^2} \tag{71}$$

so that the variance of the noise generated in computing the $(m+1)$th array is $4\sigma_\epsilon^2$ times the variance of the signal in the $m$th array. If the input (zeroth) array is white noise with variance $\sigma_x^2$ then the noise generated in the $(m+1)$th array is $2^m\sigma_x^2(4\sigma_\epsilon^2)$. If $\sigma_{om}^2$ is the output noise due to the noise generated in the $(m+1)$th array, then

$$\sigma_{om}^2 = 2^{\nu-(m+1)}2^m\sigma_x^2(4\sigma_\epsilon^2) = 2N\sigma_\epsilon^2\sigma_x^2. \tag{72}$$

Since the noise generated in each array is assumed to be independent, the total output noise variance $\sigma_E^2$ is

$$\sigma_E^2 = 2\nu N\sigma_\epsilon^2\sigma_x^2. \tag{73}$$

By noting that the output signal variance is related to the input signal variance by

$$\sigma_X^2 = N\sigma_x^2 \tag{74}$$

the result follows:

$$\frac{\sigma_E^2}{\sigma_X^2} = 2\sigma_\epsilon^2\nu. \tag{75}$$

A further result, which can be derived from our model, is an expression for the final expected output noise-to-signal ratio which results after performing an FFT and an inverse FFT on a white signal $x(n)$. The inverse FFT introduces just as much roundoff noise as the FFT itself, and thus the resulting output noise-to-signal ratio is

$$\frac{\sigma_E^2}{\sigma_x^2} = 4\sigma_\epsilon^2\nu \tag{76}$$

or just double the result in (75).

In order to see the implications of (75) or (76) in terms of register length requirements, it is useful to express these results in units of bits. We use

$$(\sigma_E^2/\sigma_X^2\sigma_\epsilon^2) \text{ bits} = \tfrac{1}{2}\log_2(2\nu) \tag{77}$$

to represent the number of bits by which the rms noise-to-signal ratio increases in passing through a floating point FFT. For example, for $\nu = 8$ this represents 2 bits and for $\nu = 11$ it represents 2.23 bits. The number of bits of rms noise-to-signal ratio increases as $\log_2(\log_2 N)$, so that doubling the number of points in the FFT produces a very mild increase in output noise, significantly less than the half-bit-per-stage increase for fixed-point computation. In fact, to obtain a half-bit increase in the result above, we would have to double $\nu$, or square $N$.

In the analysis leading to (75), we have not considered the fact that multiplications by 1 can be performed noiselessly. For a specified radix-2 algorithm, such as the decimation in time algorithm shown in Fig. 10, these reduced variances for $\tilde{W} = 1$ and $j$ can be included in the model to obtain a slightly reduced prediction for output noise-to-signal ratio. However, for reasonably large $N$, this modified noise analysis yields only slightly better predictions of output noise than does the simplified analysis above.

A consequence of our analysis leading to (75) is that the output noise is white. This follows from the fact that each

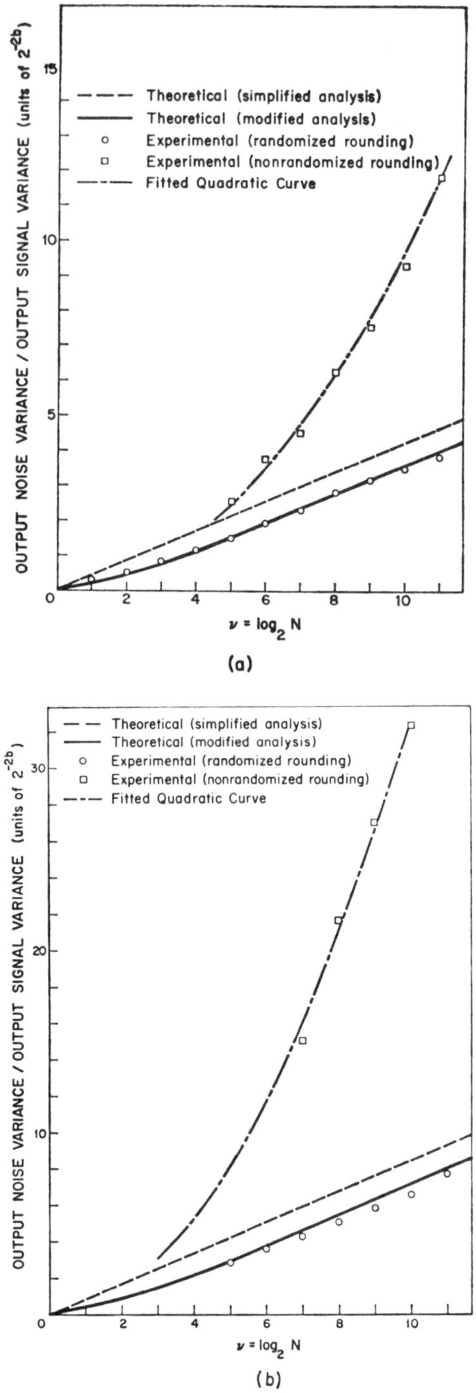


Fig. 13. (a) Experimental and theoretical noise-to-signal ratios for floating point FFT. (b) Experimental and theoretical noise-to-signal ratios for floating-point FFT and inverse.

array of noise sources is white. The reduced noise source variance for $\tilde{W} = 1$ and $j$ implies that for some arrays there will be a variation of noise source variance over the array. This implies a slight variation of output noise variance over the output array, and thus our modified noise analysis will only predict an average noise variance over the output array.

The results discussed above have been verified with excellent agreement as shown in Fig. 13(a) and (b). To obtain this agreement, however, it was necessary to use randomized rounding, i.e., randomly rounding up or down when the value of mantissa was exactly $(1/2)2^{-b}$. The modified theoretical

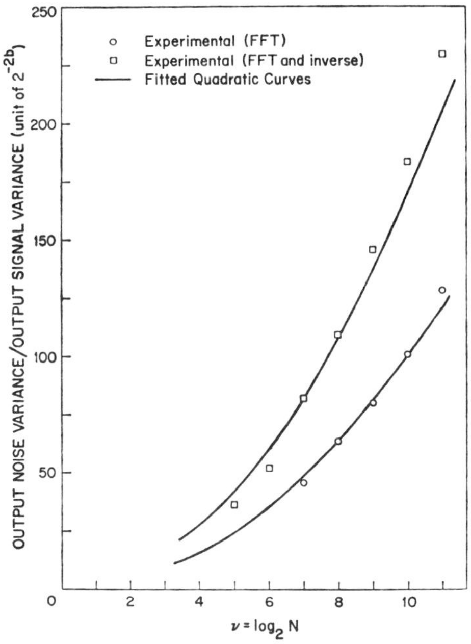


Fig. 14. Experimental noise-to-signal ratios for floating point FFT and FFT-inverse FFT; truncation used instead of rounding.

curve shown was obtained by taking into account reduced noise source variances for $\tilde{W}=1$ and $\tilde{W}=j$. Also shown are experimental results for nonrandomized rounding. These results were fitted empirically with a curve of the form $a\nu^2$, but this quadratic dependence was not established theoretically. Noise-to-signal ratios were also measured for the case where truncation rather than rounding was used in the arithmetic; the results, with empirically fitted quadratic curves, are shown in Fig. 14.

Our analysis, and all the above experiments, applied to the case of white signal. Some experimental investigation has been carried out as to whether the predictions are valid when the signal is nonwhite. Specifically, the noise introduced in computing an FFT was measured for sinusoidal signals of several frequencies, for $\nu=8$, 9, 10, and 11. The results, averaged over the input frequencies used, were within 15 percent of those predicted by (75). In these experiments, the "randomized" rounding procedure was used.

### E. Effects of Coefficient Quantization in the FFT

As with the implementation of digital filters, the implementation of the FFT algorithm requires the use of quantized coefficients. While a completely definitive study of the effects of coefficient quantization in the FFT remains to be done, two approaches have been pursued for which some results have been obtained.

Although the nature of coefficient quantization is inherently nonstatistical, Weinstein [24] has obtained some useful results by means of a rough statistical analysis. This statistical analysis corresponds to introducing random jitter in the coefficients and determining the output noise-to-signal ratio due to this noise. While the detailed effect due to coefficient error due to quantization is different than that due to jitter, it is reasonable to expect that in a gross sense the magnitude of the errors is comparable.

To develop this statistical analysis, we let $F(k)$ denote the DFT of a sequence $f(n)$ and $\hat{F}(k)$ the result of transforming $f(n)$ with a radix-2 FFT algorithm with jittered coefficients. Then

$$F(k) = \sum_{n=0}^{N-1} f(n)W^{nk} \tag{78}$$

and

$$\hat{F}(k) = \sum_{n=0}^{N-1} f(n)\Omega_{nk}. \tag{79}$$

Because of the form of the FFT algorithm each element $\Omega_{nk}$ will be a product of $\nu=\log_2 N$ quantized coefficients. Thus

$$\Omega_{nk} = \prod_{i=1}^{\nu} (W^{a_i} + \delta_i) \tag{80}$$

where

$$\prod_{i=1}^{\nu} W^{a_i} = W^{nk} \tag{81}$$

with $b$ bits for the real and imaginary parts of each of the coefficients, excluding sign, $|\delta_i|$ is less than or equal to $(\sqrt{2})2^{-b}$. If we assume that the real and imaginary parts of the jitter in the coefficients are uncorrelated and uniformly distributed between plus and minus $(1/2)2^{-b}$ then $\sigma_\delta^2$, the variance of $\delta_i$ is $\sigma_\delta^2=2^{-2b}/6$. The error in the computation of the DFT can be expressed as

$$E(k) = \hat{F}(k) - F(k) = \sum_{n=0}^{N-1} f(n)(\Omega_{nk} - W^{nk}). \tag{82}$$

From (80) and (81) we can express the factor $(\Omega_{nk}-W^{nk})$ as

$$(\Omega_{nk} - W^{nk}) = \sum_{i=1}^{\nu} \delta_i \prod_{\substack{j=1\\ j\neq i}}^{\nu} W^{a_i} + \text{higher order terms.} \tag{83}$$

If we neglect higher order error terms, and assume that $\delta_i$ are mutually uncorrelated then the variance of $(\Omega_{nk}-W^{nk})$ is equal to $\nu(2^{-2b}/6)$. Finally, assuming that all elements $\Omega_{nk}$ are uncorrelated with each other and with the input signal, the output error variance $\sigma_E^2$ is

$$\sigma_E^2 = \nu \frac{2^{-2b}}{6} \sum_{n=0}^{N-1} |f(n)|^2. \tag{84}$$

Since from Parseval's relation

$$\sum_{n=0}^{N-1} |f(n)|^2 = \frac{1}{N} \sum_{n=0}^{N-1} |F(k)|^2$$
$$= \text{mean-squared output signal} \tag{85}$$

the ratio of mean-squared output error to mean-squared output signal is thus

$$\sigma_E^2 \Big/ \left[\left(\frac{1}{N}\right)\left(\sum_{n=0}^{N-1} |F(k)|^2\right)\right] = \left(\frac{\nu}{6}\right) 2^{-2b}. \tag{86}$$

Although we would not expect (86) to predict with great accuracy the error in an FFT due to coefficient quantization, it is helpful as a rough estimate of the error. The key result of (86), which we would like to test experimentally, is that the error-to-signal ratio increases very mildly with $N$, being proportional to $\nu=\log_2 N$, so that doubling $N$ produces only a slight increase in the error-to-signal ratio.

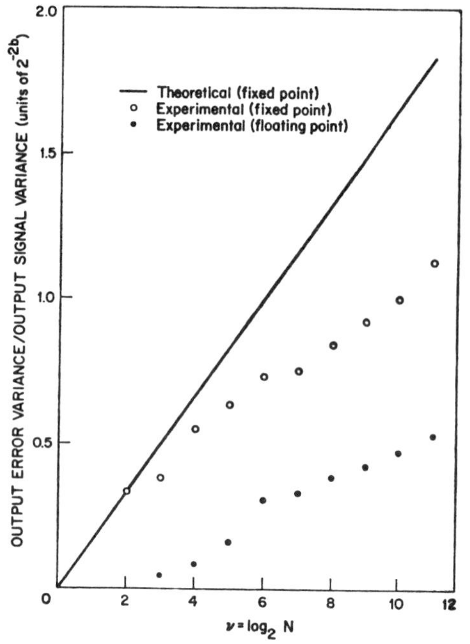


Fig. 15. Errors due to coefficient quantization in FFT.

To test this result, experimental measurements on errors due to coefficient quantization were made. In each run, a sequence $f(n)$—white in the sense that all $2N$ real numbers making up the $N$-point complex sequence were mutually uncorrelated, with zero means, and equal variances—was obtained using a random number generator. This sequence was transformed twice, once using a 36-bit coefficient table, and once using a coefficient table rounded to much shorter word length (e.g., 12 bits). For each transform, 36-bit accuracy was used in the arithmetic to make the effect of roundoff error negligible. The results were subtracted, squared, averaged over the output array, and divided by the output signal variance ($N$ times the input signal variance) to obtain an experimental output error-to-signal ratio. For each value of $N$, several random sequences were transformed and the results were averaged to obtain statistically convergent estimates.

These results are displayed in Fig. 15; the quantity plotted is $\sigma_E^2/2^{-2b}\sigma_F^2$ where $\sigma_F^2$ is the mean-squared output signal as defined in (85). The theoretical curve corresponding to (86) is shown, and the circles represent measured output error-to-signal ratio for the fixed-point case. We note that the experimental results generally lie below the theoretical curve. No experimental result differs by as much as a factor of two from the theoretical result, and since a factor of two in $\sigma_E^2/\sigma_F^2$ corresponds to only half-a-bit difference in the rms output error, it seems that (86) is a reasonably accurate estimate of the effect of coefficient errors. The experimental results do seem to increase essentially linearly with $\nu$, but with smaller slope than given in (86).

In the above, fixed-point arithmetic has been assumed. However, since a block floating-point FFT will generally use fixed-point coefficients, our results are valid for the block floating-point case also. With some slight modifications, it is possible to obtain similar results for the floating-point case. Except for a constant factor, the floating- and fixed-point results are the same. Experimental results for the floating-point case are represented by the solid dots in Fig. 15, and are observed to be slightly lower than the results for the fixed-point case.

A different approach to the characterization of FFT coefficient quantization has been taken by Tufts, Hersey, and Mosier [29]. In their analysis the effect of coefficient quantization is represented in terms of the level of spurious sidelobes introduced. In particular, a sequence $f(n) = u_o(n-n_o)$ where $u_o(n)$ denotes a unit sample, has a DFT with a purely sinusoidal real and imaginary part, i.e.,

$$F(k) = W^{n_o k} \tag{87}$$

and the inverse DFT of $F(k)$ should, of course, have only a single nonzero component. Due to coefficient quantization, however, the DFT obtained is

$$\hat{F}(k) = \Omega_{n_o k} \tag{88}$$

and since the real and imaginary parts are not exactly sinusoidal, the inverse DFT, with exact coefficients, of $F(k)$ will have spurious components. For each of the set of $N$ sequences

$$f_k(n) = u_o(n - n_k), \qquad k = 0, 1, \cdots, N-1. \tag{89}$$

Tufts *et al.* compute the DFT with quantized coefficients followed by the inverse DFT with accurate coefficients. At the output of the inverse DFT, the size and frequency locations of the spurious sidelobe components produced due to the quantized coefficients are observed. Since any function $f(n)$ can be constructed as a weighted sum

$$f(n) = \sum_{k=0}^{N-1} a_k f_k(n) \tag{90}$$

the spurious sidelobes produced for any $f(n)$ can in principle be determined by combining the responses due to a set of impulses. But carrying out such a combination is not practical for arbitrary $f(n)$. Tufts *et al.* have, however, tabulated the worse sidelobe levels encountered for any $f_k(n)$ as a function of the number of bits retained in the coefficients, for the case of a 64-point FFT and sign-magnitude representation of coefficients.

## V. Examples

### *A. Introduction*

In the preceding sections the effects of arithmetic roundoff have been analyzed for simple (first- and second-order) digital filters and the FFT. These algorithms are the basic building blocks in more complicated digital processing such as a higher order digital filter or a convolutional filter realized via the FFT. Examples will be presented in this section to indicate how some of the ideas developed above can be applied to analyze and to choose the most advantageous configuration for such systems. The first two examples concern the realization of higher order recursive filters and have borrowed from the work of other authors. The third example deals with an FFT filter.

### *B. Fixed-Point Digital Filter in Cascade and Parallel Form*

After a digital filter has been specified in terms of its poles and zeroes, and the type of arithmetic has been selected, a choice must still be made among the various possible configurations of the filter which will differ with respect to the effects of roundoff noise. An exhaustive study of the selection

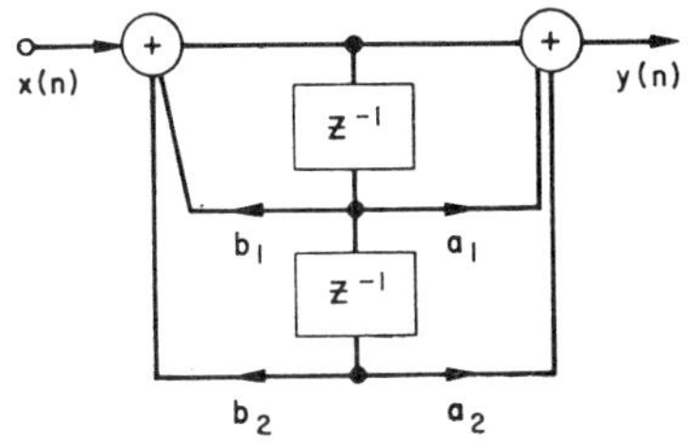


Fig. 16. Second-order section with poles preceding zeros. Used in Jackson's $1P$ parallel form with $a_2=0$. Also used in $1D$ direct form.

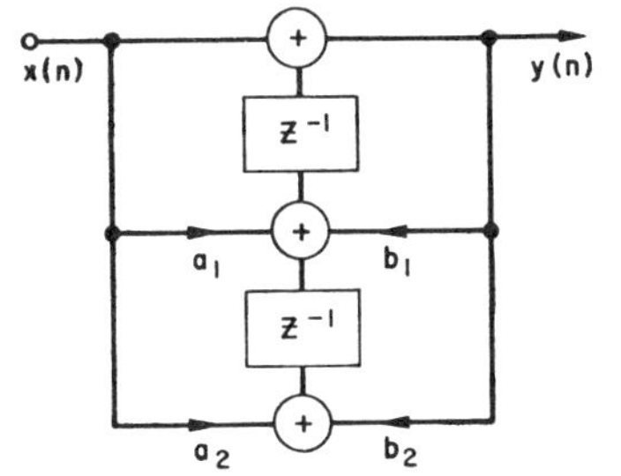


Fig. 17. Second-order section with zeros preceding poles. Used in $2P$ parallel form with $a_2=0$. Also used in $2D$ direct form.

of filter form is beyond the scope of this paper, but an excellent example of the necessary considerations is given by Jackson [30] in his analysis of roundoff noise for fixed-point digital filters realized in cascade and parallel form

Jackson considers two parallel form realizations: the $1P$ form where the individual second-order sections are realized, as shown in Fig. 16, with the poles preceding the zeros, and the $2P$ form where zeros precede poles in the individual sections, as shown in Fig. 17. (Figs. 16 and 17 do not show all the scaling coefficients needed to prevent overflow.) His analysis indicates that for a variety of scaling criteria (based on different $L_p$ norms[2] of the input signal) and for various measures of the output noise (such as its total power, or its peak spectral value), the output signal-to-noise ratios of the two forms are very close. Generally, a very slight advantage will be gained with the $1P$ form.

Comparison of the two parallel forms basically reduces to a comparison of the noise properties of the two forms of second-order sections, since the noises from the second-order sections are simply additive in the output of the parallel form. Hence the discussion above applies to a comparison of second-order section realizations. Another form of second-order section which could be considered is a coupled form as shown in Fig. 18. For the case $a_1=r\cos\theta$, $a_2=r\sin\theta$, this filter has poles at $z=re^{\pm j\theta}$ and a zero at $z=r\cos\theta$. The coupled form noise-to-signal ratio has been compared [24] to ratios for forms essentially the same as those in Figs. 16 and 17 for the

[2] If

$$F(\omega) = \sum_{n=-\infty}^{\infty} f(n) \exp\left(-j2\pi \frac{\omega}{\omega_s} n\right)$$

represents the Fourier transform of a signal or of a filter impulse response $f(n)$, then the corresponding $L_p$ norm is

$$\|F\|_p = \left(\frac{1}{\omega_s}\int_0^{\omega_s} |F(\omega)|^p\, d\omega\right)^{1/p}$$

where $\omega_s$ denotes sampling frequency.

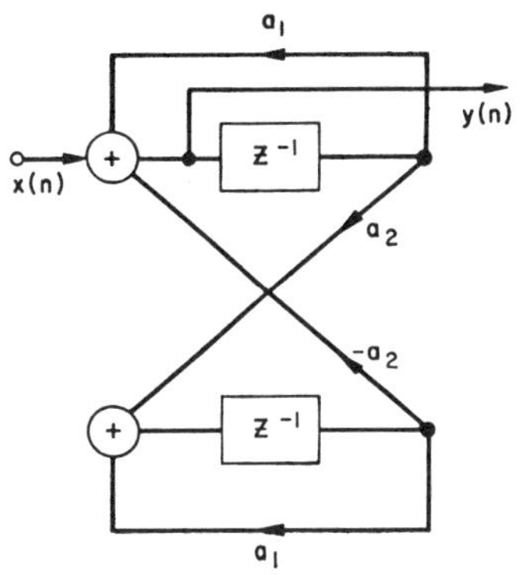


Fig. 18. Coupled form for second-order section.

case of white-noise input and an absolute overflow constraint (through the type of analysis given in Section III-C). The coupled form was found to have substantially lower noise-to-signal ratio for filters with high gain and low resonant frequencies. For $\delta=1-r$, and $\delta\ll 1$, the results vary as

$$\frac{\sigma_e^2}{\sigma_y^2} \sim 1/\delta^2 \sin^2\theta \qquad \text{(forms of Figs. 16 and 17)}$$

$$\frac{\sigma_e^2}{\sigma_y^2} \sim 1/\delta^2 \qquad \text{(coupled form).} \tag{91}$$

The implication of this result, together with the somewhat reduced coefficient sensitivity for the coupled form, is that this form may be a good choice in some situations, despite the fact that its implementation requires four multiplications instead of three for a pole pair and a single zero.

As stated above, Jackson found not much difference between the noise properties of the $1P$ and $2P$ parallel forms. However, the situation is more interesting in the case of the cascade form. Here he finds that large differences are possible between the roundoff noise outputs of the $1D$ (poles before zeros in individual sections) and $2D$ (zeros before poles in each section) forms. Also the ordering of the sections and the pairing of poles and zeros have important effect on the output signal-to-noise ratio. Jackson's analyses lead to several rules of thumb for selection of $1D$ or $2D$, for ordering of sections, and for pairing of poles and zeros.

In general, the choice of configuration depends on which $L_p$ norm of the scaled transfer functions is constrained to prevent overflow and on which $L_r$ norm of the output noise spectrum is used as a measure of performance. Two $L_p$ constraints on the filter are of particular interest: the $p=\infty$ case, where the peak value of the transfer function to each possible overflow node is constrained; and the $p=2$ case where the rms transfer function to each node is constrained. The choice $p=\infty$ is just slightly less stringent than the absolute overflow constraint (7), and prevents overflow even when the input is a narrow-band signal at resonance of the relevant transfer function. The $p=2$ constraint is more appropriate for preventing overflow when the input is wide-band in nature. Two $L_r$ norms on the output noise spectrum are of particular interest: the $r=1$ norm which measures the total output noise power, and the $r=\infty$ norm which measures the peak value of the spectrum of the output noise.

With regard to selection of $1D$ or $2D$ forms, Jackson's rule of thumb says to select $1D$ when $p=2$, $r=\infty$ and $2D$ when

$p=\infty$, $r=1$; for $p=2$, $r=1$ and for $p=\infty$, $r=\infty$, either form may be selected.

The choice of ordering of sections also depends on the norms which are selected. For $p=2$, $r=\infty$, the sections should be ordered in decreasing peakedness, where peakedness is defined as the peak gain of a section divided by its rms gain. For $p=\infty$, $r=1$, the sections should be ordered in increasing peakedness. For $p=2$, $r=1$ and for $p=\infty$, $r=\infty$, the choice of ordering depends on whether form $2D$ or $1D$ is chosen for the individual sections. Decreasing peakedness should be chosen with form $2D$, and increasing peakedness with form $1D$.

The rule for pairing of poles and zeros is as follows: Let $|H_n(\omega)|$ denote the magnitude of the frequency response of the $n$th section, and $M_n$ denote the maximum over $\omega$ of $|H_n(\omega)|$. Then the pairing should be chosen such that the maximum over $n$ of $M_n$ is minimized.

The above rules are illustrated by Jackson with a specific filter example—a sixth-order Chebyshev band rejection filter, and the results are in accord with his rules. He analyzes the output noise of this filter for parallel forms $1P$ and $2P$ and for all orderings of cascade forms $1D$ and $2D$ (with proper pole–zero pairing). Little difference is seen between the two parallel forms. For $p=2$, $r=\infty$ the peak output noise spectrum is 7–12 dB worse for the $2D$ cascade forms than for the $1D$ forms; while for $p=\infty$, $r=1$, the output noise power is 7–12 dB worse for $1D$ than for $2D$. The effects of pole–zero pairing and of ordering of sections also follow quite well the rules previously stated. The parallel forms turn out to be slightly superior to the best cascade forms with respect to roundoff noise.

As Jackson indicates, these rules of thumb have certain qualifications and are not always valid. However, they have been shown to be helpful in a variety of types of examples [31].

### C. Choice of Form for Floating-Point Digital Filter

By means of an example, Liu and Kaneko [8] have compared the direct, cascade, and parallel form realizations of a floating-point digital filter. The filter selected was an eighth-order low-pass elliptic filter. The noise-to-signal ratio for the parallel form was about 20 dB worse than for the direct form, while the cascade form was comparable to (about 1.5 dB worse than) the parallel form.

Various orderings of cascade form floating-point filters have not been studied in detail. Probably floating-point cascade filters are not too sensitive to ordering since large variations in signal level from stage to stage can be accommodated by the floating-point exponent.

A comparison of the noise-to-signal ratio properties of floating-point second-order sections where poles precede zeros (Fig. 16) and where zeros precede poles (Fig. 17) indicates that at least for white-noise inputs the behavior of the two forms is essentially identical. For a high-gain second-order section of low resonant frequency, a coupled form realization yields some noise-to-signal ratio advantage over both of these two forms.

### D. FFT Filter

The results of our roundoff noise analysis for fixed-point FFT will now be applied to obtain an expression for the output noise-to-signal ratio of a finite impulse response digital filter, implemented by means of the FFT. The overflow constraints of this type of filter will be accounted for in the analysis. Attention will be focussed on a prototype low-pass filter with 256-point impulse response and a cutoff frequency of 1/4 the half sampling frequency. Rounded arithmetic will be assumed.

Let us examine the basic steps in the filtering computation, tracing the buildup of noise variance as we proceed. First the FFT is used to compute the DFT of a section of input. In the implementation of a filter with 256-point impulse response, it is reasonable to compute a 512-point FFT, where the input consists of 256 data samples and 256 zeros. Actually 512 real input samples would be treated simultaneously, by placing sections of 256 real samples in both the real and imaginary parts of the input to the FFT. To guarantee against overflow, a scaling of 1/2 is needed at each stage of the FFT yielding an overall attenuation of 1/512. The samples of the input sequence must be less than unity in magnitude. The noise $E_1(k)$ at the output of this first DFT has variance

$$\sigma_{E_1}^2 = \overline{|E_1(k)|^2} = \frac{5}{3}2^{-2b}. \tag{92}$$

This noise variance is small, because most of the roundoff noise has been shifted off by the attenuations. However, the scalings have also caused the mean-squared signal to decrease by a factor of 1/512.

Next this computed transform is multiplied by a sequence $H(k)$ representing the DFT of a 512-point sequence $\bar{h}(n)$ consisting of the filter impulse response plus 256 zeros. This complex multiplication introduces roundoff noise of variance $2^{-2b}/3$. Assuming that we have chosen $|H(k)|<1$, the mean square of the noise $E_1$ becomes reduced by

$$\frac{1}{512}\sum_{k=0}^{511}|H(k)|^2 = B \tag{93}$$

a ratio of the filter bandwidth to the sampling frequency. Thus after the multiplication, the variance of the total noise $E_2(k)$ is

$$\sigma_{E_2}^2 = \frac{2^{-2b}}{3} + B\frac{5}{3}2^{-2b}. \tag{94}$$

This noise is not white, but has a component whose spectrum has been shaped by the filter.

For the example under consideration $B\approx 1/4$, so

$$\sigma_{E_2}^2 \approx \frac{3}{4}2^{-2b}. \tag{95}$$

Note that $\sigma_{E_2}^2$ is slightly less than $\sigma_{E_1}^2$ and represents only about a bit of noise. However, if the signal spectrum is flat, the mean-squared signal will also be reduced somewhat due to the multiplication by $H(k)$.

Now an inverse transform is computed to obtain a section of output. The noise variance at the output of this transform depends on how many scalings are necessary in the inverse FFT. In order to determine how many scalings are necessary, a bound on the output of the circular convolution is required [32]. For a particular filter, such a bound can be stated as

$$|y(n)| < \sum_{l=0}^{M-1}|h(l)| \tag{96}$$

where $y(n)$ is the output and $M$ is the length of the impulse response. The prototype filter has an impulse response

$$h(n) = \frac{2}{256}\left[\frac{1}{2} + \sum_{k=0}^{31} (-1)^k \cos \frac{2\pi kn}{256} + 0.7 \cos \frac{2\pi(32)n}{256} - 0.225 \cos \frac{2\pi(33)n}{256} + 0.01995 \cos \frac{2\pi(33)n}{256}\right] \quad (97)$$

and

$$\sum_{n=0}^{255} |h(n)| = 3.12. \quad (98)$$

Hence, only two scalings (at the first two arrays) are necessary in the inverse transform. Then, in propagating through the IFFT (inverse FFT), the variance of the noise $E_2(k)$ increases by a factor of 512/16. (The 512 represents the gain of the inverse DFT, and the 1/16 is due to the scalings.) The variance of the additional output noise $E_3(k)$ caused by roundoff in the IFFT can be estimated easily via the method of Section IV-C. The result is

$$\sigma_{E_3}^2 = \sigma_{B'}^2\left(\frac{512}{8} + \frac{512}{4}\right) + \sigma_B^2 \sum_{k=1}^{6} 2^k = (202)2^{-2b}. \quad (99)$$

The total mean-squared output noise is

$$\sigma_E^2 = \frac{512}{16}\sigma_{E_2}^2 + \sigma_{E_3}^2 = (226)2^{-2b} \quad (100)$$

or in units of bits of rms output noise

$$\tfrac{1}{2}\log_2 (2^{2b}\sigma_E^2) = 3.91 \text{ bits}. \quad (101)$$

The mean-squared output signal can be estimated if specific statistics are assumed for the input signal $x(n)$. As an example, assume that $x(n)$ is white with variance $\sigma_x^2 = 2/3$. This variance goes through an attenuation of 1/512 in the first FFT, an attenuation of $B = 1/4$ due to multiplication by $H(k)$, and a gain of 512/16 in the inverse transform. The mean-squared output signal is then

$$\sigma_y^2 = \left(\frac{1}{512}\right)\left(\frac{1}{4}\right)\left(\frac{512}{16}\right)\left(\frac{2}{3}\right) = \frac{1}{96} \quad (102)$$

and the output noise-to-signal ratio is

$$\frac{\sigma_E^2}{\sigma_y^2} \approx (22\,000)2^{-2b}. \quad (103)$$

Assuming an input noise-to-signal ratio (due to A/D noise) of $(1/4)2^{-2b}$, the noise-to-signal ratio has worsened by a factor of about 5500, or

$$\tfrac{1}{2}\log_2 5500 = 6.15 \text{ bits} \quad (104)$$

in passing through the FFT filter.

## References


[1] W. R. Bennett, "Spectra of quantized signals," *Bell Syst. Tech. J.*, vol. 27, pp. 446–472, 1948.

[2] B. Widrow, "Statistical analysis of amplitude-quantized sampled-data systems," *AIEE Trans. (Appl. Indust.)*, vol. 81, pp. 555–568, Jan. 1961.

[3] B. Liu, "Effect of finite word length on the accuracy of digital filters—A review," *IEEE Trans. Circuit Theory*, vol. CT-18, pp. 670–677, Nov. 1971.

[4] J. B. Knowles and R. Edwards, "Effect of a finite-word-length computer in a sampled-data feedback system," *Proc. Inst. Elec. Eng.*, vol. 112, pp. 1197–1207, 1965.

[5] B. Gold and C. M. Rader, "Effect of quantization noise in digital filters," in *Proc. Spring Joint Comput. Conf., AFIPS Conf. Proc.*, vol. 28, pp. 213–219, 1966.

[6] C. Weinstein and A. V. Oppenheim, "A comparison of roundoff noise in floating point and fixed point digital filter realizations," *Proc. IEEE* (Lett.), vol. 57, pp. 1181–1183, June 1969.

[7] L. B. Jackson, "On the interaction of roundoff noise and dynamic range in digital filters," *Bell Syst. Tech. J.*, vol. 49, pp. 159–184, 1970.

[8] B. Liu and T. Kaneko, "Error analysis of digital filters realized with floating-point arithmetic," *Proc. IEEE*, vol. 57, pp. 1735–1747, Oct. 1969.

[9] A. V. Oppenheim, "Realization of digital filters using block-floating-point arithmetic," *IEEE Trans. Audio Electroacoust.*, vol. AU-18, pp. 130–136, Jan. 1970.

[10] R. B. Blackman, *Linear Data-Smoothing and Prediction in Theory and Practice.* Reading, Mass.: Addison-Wesley, 1965.

[11] L. Jackson, "An analysis of limit cycles due to multiplication rounding in recursive digital filters," in *Proc. 7th Allerton Conf. Circuit and System Theory*, pp. 69–78, 1969.

[12] S. R. Parker and S. F. Hess, "Limit-cycle oscillations in digital filters," *IEEE Trans. Circuit Theory*, vol. CT-8, pp. 687–697, Nov. 1971.

[13] P. M. Ebert, J. E. Mazo, and M. C. Taylor, "Overflow oscillations in digital filters," *Bell Syst. Tech. J.*, vol. 48, pp. 2999–3020, 1969.

[14] J. F. Kaiser, "Some practical considerations in the realization of linear digital filters," in *Proc. 3rd Allerton Conf. Circuit and Systems Theory*, pp. 621–633, 1965.

[15] C. M. Rader and B. Gold, "Effects of parameter quantization on the poles of a digital filter," *Proc. IEEE* (Lett.), vol. 55, pp. 688–689, May 1967.

[16] E. Avenhaus, "An optimization procedure to minimize the word length of digital filter coefficients," presented at the London Conf. on Digital Filtering, Aug. 31, 1971.

[17] E. Avenhaus and W. Schuessler, "On the approximation problem in the design of digital filters with limited wordlength," *Arch. Elek. Übertragung*, vol. 24, pt. 12, pp. 571–572, 1970.

[18] K. Stieglitz, "Designing short-word recursive digital filters," in *Proc. 9th Allerton Conf. Circuit and System Theory* (Monticello, Ill.), pp. 778–788, Oct. 1971.

[19] A. Fettweis, "Some principles of designing digital filters imitating classical filter structures," *IEEE Trans. Circuit Theory* (Corresp.), vol. CT-18, pp. 314–316, Mar. 1971.

[20] J. Bingham, "A new type of digital filter with a reciprocal ladder configuration," in *Proc. 1970 IEEE Int. Symp. Circuit Theory, Dig. of Tech. Papers*, pp. 129–130.

[21] R. Crochiere, "Digital ladder filter structures and coefficient sensitivity," M.I.T. Res. Lab. of Electronics, Quart. Progr. Rep. 103, Oct. 15, 1971.

[22] J. W. Cooley and J. W. Tukey, "An algorithm for the machine calculation of complex Fourier series," *Math. Comput.*, vol. 19, pp. 297–301, 1965.

[23] B. Gold and C. M. Rader, *Digital Processing of Signals.* New York: McGraw-Hill, 1969.

[24] C. J. Weinstein, "Quantization effects in digital filters," M.I.T. Lincoln Lab. Tech. Rep. 468, ASTIA Doc. DDC AD-706862, Nov. 21, 1969.

[25] P. D. Welch, "A fixed-point fast Fourier transform error analysis," *IEEE Trans. Audio Electroacoust.*, vol. AU-17, pp. 153–157, June 1969.

[26] W. M. Gentleman and G. Sande, "Fast Fourier transforms—For fun and profit," in *Proc. Fall Joint Computer Conf., AFIPS Conf. Proc.*, pp. 563–578, 1966.

[27] C. J. Weinstein, "Roundoff noise in floating point fast Fourier transform computation," *IEEE Trans. Audio Electroacoust.*, vol. AU-17, pp. 209–215, Sept. 1969.

[28] T. Kaneko and B. Liu, "Accumulation of roundoff error in fast Fourier transforms," *J. Ass. Comput. Mach.*, vol. 17, pp. 637–654, Oct. 1970.

[29] D. W. Tufts, H. S. Hersey, and W. E. Mosier, "Effects of FFT coefficient quantization on bin frequency response," *Proc. IEEE* (Lett.), vol. 60, pp. 146–147, Jan. 1972.

[30] L. B. Jackson, "Roundoff-noise analysis for fixed-point digital filters realized in cascade or parallel form," *IEEE Trans. Audio Electroacoust.*, vol. AU-18, pp. 107–122, June 1970.

[31] L. B. Jackson, "An analysis of roundoff noise in digital filters," Sc.D. dissertation, Stevens Inst. Technol., Dep. Elec. Eng., Castle Point, Hoboken, N. J., 1969.

[32] A. V. Oppenheim and C. Weinstein, "A bound on the output of a circular convolution with application to digital filtering," *IEEE Trans. Audio Electroacoust.*, vol. AU-17, pp. 120–124, June 1969.

## BIBLIOGRAPHY


[33] J. E. Bertram, "The effect of quantization in sampled-feedback systems," *AIEE Trans. (Appl. Industry)*, vol. 77, pp. 177–182, July 1958.

[34] F. Bonzanigo, "Constant-input behavior of recursive digital filters," presented at IEEE Arden House Workshop in Digital Filtering, Harriman, N. Y., Jan. 1970.

[35] R. B. Blackman, *Linear Data-Smoothing and Prediction in Theory and Practice*. Reading, Mass.: Addison-Wesley, 1965, pp. 75–79.

[36] E. E. Curry, "The analysis of round-off and truncation errors in a hybrid control system," *IEEE Trans. Automat. Contr.* (Short Papers), vol. AC-12, pp. 601–604, Oct. 1967.

[37] L. D. Divieti, C. M. Rossi, R. M. Schmid, and A. E. Vereschkin, "A note on computing quantization errors in digital control systems," *IEEE Trans. Automat. Contr.* (Corresp.), vol. AC-12, pp. 622–623, Oct. 1967.

[38] R. Edwards, J. Bradley, and J. Knowles, "Comparison of noise performances of programming methods in the realization of digital filters," in *Proc. 1969 Polytech. Inst. Brooklyn Symp. on Computer Processing in Communications*.

[39] A. Fettweis, "A general theorem for signal-flow networks, with applications," *Arch. Elek. Übertragung*, vol. 25, pp. 557–561, Dec. 1969.

[40] W. A. Gardner, "Reduction of sensitivities in sampled-data filters," *IEEE Trans. Circuit Theory* (Corresp.), vol. CT-17, pp. 660–663, Nov. 1970.

[41] B. Gold and L. Rabiner, "Analysis of digital and analog formant synthesizers," *IEEE Trans. Audio Electroacoust.*, vol. AU-16, pp. 81–94, Mar. 1968.

[42] B. Gold and C. M. Rader, *Digital Processing of Signals*. New York: McGraw-Hill, 1969.

[43] R. M. Golden and S. A. White, "A holding technique to reduce number of bits in digital transfer functions," *IEEE Trans. Audio Electroacoust.*, vol. AU-16, pp. 433–437, Sept. 1968.

[44] D. Herrmann and W. Schuessler, "On the accuracy problem in the design of nonrecursive digital filters," *Arch. Elek. Übertragung*, vol. 24, pt. 11, pp. 525–526, 1970.

[45] J. F. Kaiser, "Digital filters," in F. F. Kuo and J. F. Kaiser, Eds., *System Analysis by Digital Computer*. New York: Wiley, 1966, ch. 7.

[46] T. Kaneko and B. Liu, "Roundoff error of floating-point digital filters," in *Proc. 6th Allerton Conf. on Circuit and System Theory*, pp. 219–227, Oct. 1968.

[47] J. Katzenelson, "On errors introduced by combined sampling and quantization," *IRE Trans. Automat. Contr.*, vol. AC-7, pp. 58–68, Apr. 1962.

[48] W. C. Kellogg, "Information rates in sampling and quantization," *IEEE Trans. Informat. Theory*, vol. IT-13, pp. 506–511, July 1967.

[49] J. B. Knowles and R. Edwards, "Simplified analysis of computational errors in a feedback system incorporating a digital computer," presented at S.I.T. Symp. on Direct Digital Control, London, England, Apr. 22, 1965.

[50] ——, "Complex cascade programming and associated computational errors," *Electron. Lett.*, vol. 1, pp. 160–161, Aug. 1965.

[51] ——, "Finite word-length effects in a multirate direct digital control system," *Proc. Inst. Elec. Eng.*, vol. 112, pp. 2376–2384, Dec. 1965.

[52] J. B. Knowles and E. M. Olcayto, "Coefficient accuracy and digital filter response," *IEEE Trans. Circuit Theory*, vol. CT-15, pp. 31–41, Mar. 1968.

[53] A. A. Kosyakin, "The statistical theory of amplitude quantization," *Avtomat. Telemekh.*, vol. 22, p. 722, 1969.

[54] I. M. Langenthal, "Coefficient sensitivity and generalized digital filter synthesis," in *EASCON 1968 Rec.*, pp. 386–392, 1968.

[55] C. E. Maley, "The effect of parameters on the roots of an equation system," *Comput. J.*, vol. 4, pp. 62–63, 1961–1962.

[56] P. E. Mantey, "Eigenvalue sensitivity and state-variable selection," *IEEE Trans. Automat. Contr.*, vol. AC-13, pp. 263–269, June 1968.

[57] R. K. Otnes and L. P. McNamee, "Instability thresholds in digital filters due to coefficient rounding," *IEEE Trans. Audio Electroacoust.*, vol. AU-18, pp. 456–463, Dec. 1970.

[58] C. Rader and B. Gold, "Digital filter design techniques in the frequency domain," *Proc. IEEE*, vol. 55, pp. 149–171, Feb. 1967.

[59] Q. I. Rahman, "The influence of coefficients on the zeros of polynomials," *J. London Math Soc.*, vol. 36, pt. 1, pp. 57–64, Jan. 1961.

[60] I. W. Sandberg, "Floating-point-roundoff accumulation in digital filter realization," *Bell Syst. Tech. J.*, vol. 46, pp. 1775–1791, Oct. 1967.

[61] ——, "A theorem concerning limit cycles in digital filters," in *Proc. 7th Ann. Allerton Conf. Circuit and System Theory*, pp. 63–68, 1968.

[62] J. B. Slaughter, "Quantization errors in digital control systems," *IEEE Trans. Automat. Contr.*, vol. AC-9, pp. 70–74, Jan. 1964.

[63] O. Sornmoonpin, "Investigation of quantization errors," M.Sc. thesis, Univ. of Manchester, England, 1966.

[64] T. G. Stockham, Jr., "A-D and D-A converters: Their effect on digital audio fidelity," presented at 41st Meeting of the Audio Engineering Society, New York, Oct. 5–8, 1971.

[65] D. W. Tufts, W. Knight, and D. W. Rorabacher, "Effects of quantization and sampling in digital correlators and in power spectral estimation," *Proc. IEEE* (Lett.), vol. 56, pp. 79–82, Jan. 1969.

[66] D. W. Tufts, D. W. Rorabacher, and W. E. Mosier, "Designing simple, effective digital filters," *IEEE Trans. Audio Electroacoust.*, vol. AU-18, pp. 142–158, June 1970.

[67] C. Weinstein, "Quantization effects in frequency sampling fiilters," in *NEREM Rec.*, p. 222, 1968.

[68] B. Widrow, "A study of rough amplitude quantization by means of Nyquist sampling theory," *IRE Trans. Circuit Theory*, vol. CT-3, pp. 266–276, Dec. 1956.

[69] J. H. Wilkinson, "Error analysis of floating-point comparison," *Numerisch. Math.*, vol. 2, pp. 319–340, 1960.

[70] ——, *Rounding Errors in Algebraic Processes*. Englewood Cliffs, N. J.: Prentice-Hall, 1963.

[71] E. P. F. Kan and J. K. Aggarwal, "Error analysis of digital filter employing floating-point arithmetic," *IEEE Trans. Circuit Theory*, vol. CT-18, pp. 678–686, Nov. 1971.

# Coefficient Accuracy and Digital Filter Response

J. B. KNOWLES, MEMBER, IEEE, AND E. M. OLCAYTO

*Abstract*—The frequency response of a digital filter realized by a finite word-length machine deviates from that which would have been obtained with an infinite word-length machine. An "ideal" or "errorless" filter is defined as a realization of the required pulse transfer function by an infinite word-length machine. This paper shows that quantization of a digital filter's coefficients in an actual realization can be represented by a "stray" transfer function in parallel with the corresponding ideal filter. Also, by making certain statistical assumptions, the statistically expected mean-square difference between the real frequency responses of the actual and ideal filters can be readily evaluated by one short computer program for all widths of quantization. Furthermore, the same computations may be used to evaluate the rms value of output noise due to data quantization and multiplicative rounding errors. Experimental measurements verify the analysis in a practical case. The application of the results to the design of the digital filters is also considered.

## I. Introduction

SYSTEMS which are used to spectrally shape information are known as filters. Until quite recently, linear analog elements have been used almost exclusively in this context. However, with the increasing flexibility of digital equipment, attention has been directed toward digital filtering techniques. In particular, methods[1],[2] have been developed which enable the conversion of certain well-established analog filter designs into digital filters having essentially the same frequency response over half the Nyquist interval. These techniques are significant in that the resulting digital filters largely maintain the frequency response of their analog "parents" even when relatively small sampling frequencies are employed. That is to say, frequency aliasing effects[1] are largely eliminated.

An electronic digital computer processes information in binary-number format. When the so-called "scientific programming" convention is employed in a fixed-point machine, the binary number $\gamma_0\gamma_1 \cdots \gamma_{M-1}$ has the significance

$$\gamma_0\gamma_1 \cdots \gamma_{M-1} = -\gamma_0 + \sum_{i=1}^{M-1} \gamma_i 2^{-i} \tag{1}$$

where

$$\gamma_i = 0 \quad \text{or} \quad 1 \quad \text{for all } j \tag{2}$$

and $M$ is the word-length of the computer. In connection with (1), the width of quantization ($q$) is defined as:

$$q = 2^{-(M-1)}. \tag{3}$$

By virtue of the finite computer word-length, the performance of a digital filter is inevitably degenerated. First, when continuous data are read into the computers, a quantization error[3] is incurred. Second, a roundoff error arises in the evaluation of each arithmetic product in the computation.[4],[5] Third, the coefficients of the difference equation which represents the digital filter are subject to amplitude quantization errors. In consequence, the frequency response of the actual filter realized deviates from that which would have been obtained with an infinite word-length machine. For the purposes of comparison, it is convenient to define "the ideal" or "errorless" filter as a realization of the required pulse transfer function by an infinite word-length computer.

The first two types of error described above have already been thoroughly investigated,[3]-[6] and this paper is concerned with the third. Although the problem of coefficient errors is not practically significant in the case of a digitally controlled feedback system, the loss of performance is important in digital filter applications due to the absence of overall negative feedback and greater digital system complexity. The effect of coefficient errors on digital filter performance was first investigated by Kaiser.[7] An absolute accuracy bound was derived for the difference equation coefficients within which the realization was still asymptotically stable. However, such an error bound is inevitably pessimistic, because coefficient quantization errors are intrinsically statistical. Furthermore, this analysis does not enable the computer word-length to be selected so that the degeneration in the real frequency response of the actual digital filter is maintained within acceptable practical limits.

This paper shows that the quantization of a digital filter's coefficients can be represented by a "stray" transfer function in parallel with the corresponding ideal filter. Also, by making certain justifiable statistical assumptions, the statistically expected mean-square difference between the real frequency responses of the actual and corresponding errorless filter can be readily evaluated by one short computer program for all widths of quantization. Further, the same computational results may be used to evaluate the rms value of the filter's output noise due to data quantization and multiplicative roundoff errors.[4],[5] Unfortunately, while the method is always applicable to direct and parallel programming, it is generally unsuitable for cascade programming.

Manuscript received July 3, 1967; revised October 3, 1967.
J. B. Knowles is with the Control and Instrument Group, United Kingdom Atomic Energy Authority, Winfrith, Dorset, England.
E. M. Olcayto was formerly with the University of Manchester Institute of Science and Technology, Manchester, England. He is now with the Turkish Post Office, Izmir, Turkey.

Reprinted from *IEEE Trans. Circuit Theory*, vol. CT-15, pp. 31–41, Mar. 1968.

*Justification for a Statistical Analysis*

In the transformation of an analog filter to a digital filter by Golden and Kaiser's method[1] a matrix of coefficients is derived. For representation in the computer these coefficients are quantized to a specific number of bits. That is to say, for representation within the computer these coefficients must be processed by the following gain characteristic.

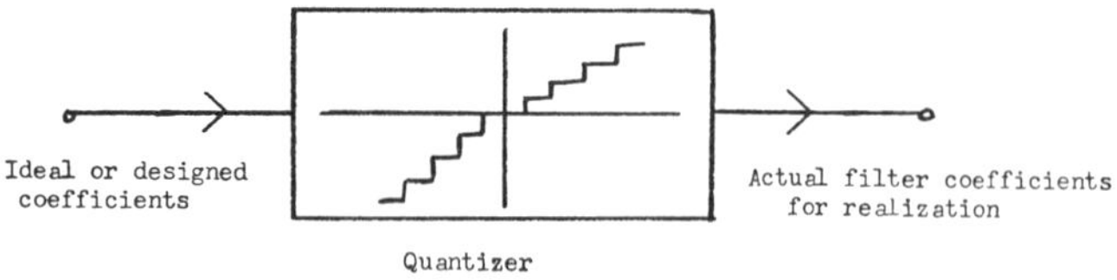


In other words, the justification for a statistical analysis of filter coefficient errors is the same as for a statistical analysis of quantization in A/D converters. From one viewpoint, the output of a quantizer to a specific input is deterministic and not really a statistical problem at all. Alternatively, the exact value of the actual input viewed in terms of the quantized data is uncertain. In fact, for a realized filter coefficient of $nq$, the true coefficient value lies in the interval $nq - \epsilon$ to $nq + \epsilon$, where $|\epsilon| < q/2$. It is this uncertainty regarding the exact value of the input to the quantizer that justifies, in the authors' opinion, the statistical analysis of all amplitude quantization effects. Such applications of statistical techniques seem to have a strong experimental confirmation which is a further justification for this form of analysis.

## II. Analysis for Direct Programming

In the following analysis, the errors on the difference equation coefficients will be assumed to be statistically independent and uniformly distributed with zero mean. These assumptions for the case of rounded quantization have been confirmed by careful experimental investigations.[5],[8] Thus, if the actual pulse transfer function realized is

$$H(z^{-1}) = \frac{A(z^{-1})}{B(z^{-1})} = \frac{\sum_{k=0}^{N} a_k z^{-k}}{1 + \sum_{k=1}^{N} b_k z^{-k}}, \tag{4}$$

then the numerator and denominator coefficients may be written as

$$a_k = \bar{a}_k + \alpha_k$$
$$b_k = \bar{b}_k + \beta_k \tag{5}$$

where $\bar{a}_k$ and $\bar{b}_k$ are the desired values of the coefficients, and the error quantities $\alpha_k$, $\beta_k$ are statistically independent. It should be observed that no error at all is involved in the representation of unity and zero magnitude coefficients within the computer. Neglecting multiplicative roundoff errors, the direct programming realization of (4) is specified by the recursion equation[1]

$$y'(n) = \sum_{k=0}^{N} a_k x(n-k) - \sum_{k=1}^{N} b_k y'(n-k) \tag{6}$$

where $x(i)$ and $y'(i)$ represent the input and output sequences of the actual filter. However, the ideal digital filter produces an output sequence $y(n)$ according to the recursion equation

$$y(n) = \sum_{k=0}^{N} \bar{a}_k x(n-k) - \sum_{k=1}^{N} \bar{b}_k y(n-k). \tag{7}$$

Defining the computational error quantity

$$e(n) = y'(n) - y(n), \tag{8}$$

then substituting (5), (6), and (7) into (8) yields

$$e(n) = \sum_{k=0}^{N} \alpha_k x(n-k) - \sum_{k=1}^{N} \bar{b}_k e(n-k) - \sum_{k=1}^{N} \beta_k y(n-k) - \sum_{k=1}^{N} \beta_k e(n-k), \tag{9}$$

and neglecting second-order quantities, one obtains

$$e(n) = \sum_{k=0}^{N} \alpha_k x(n-k) - \sum_{k=1}^{N} \bar{b}_k e(n-k) - \sum_{k=1}^{N} \beta_k y(n-k). \tag{10}$$

Taking the $z$ transformation of (10) gives

$$0 = \alpha(z^{-1})X(z^{-1}) - \beta(z^{-1})Y(z^{-1}) - E(z^{-1})B_\infty(z^{-1}) \tag{11}$$

where

$$\alpha(z^{-1}) = \sum_{k=0}^{N} \alpha_k z^{-k}, \qquad \beta(z^{-1}) = \sum_{k=1}^{N} \beta_k z^{-k}$$
$$A_\infty(z^{-1}) = \sum_{k=0}^{N} \bar{a}_k z^{-k}, \qquad B_\infty(z^{-1}) = 1 + \sum_{k=0}^{N} \bar{b}_k z^{-k} \tag{12}$$

and

$$E(z^{-1}) = \sum_{k=0}^{\infty} e(k) z^{-k}.$$

As

$$Y(z^{-1}) = H_\infty(z^{-1})X(z^{-1}), \tag{13}$$

then (11) may be reduced to

$$E(z^{-1}) = \frac{\alpha(z^{-1}) - \beta(z^{-1})H_\infty(z^{-1})}{B_\infty(z^{-1})} X(z^{-1}). \tag{14}$$

Hence, from (8) and (14) one obtains

$$Y'(z^{-1}) = \left[ H_\infty(z^{-1}) + \frac{\alpha(z^{-1}) - \beta(z^{-1})H_\infty(z^{-1})}{B_\infty(z^{-1})} \right] X(z^{-1}). \tag{15}$$

[1] If multiplicative roundoff errors are considered, then an error term $c(n)$ must be added to the right-hand side of (6) (see Knowles and Edwards[4],[6]). However, a more complete description of degenerative effects in a finite word-length computer is given in Section VI of this paper.

Thus, the physical significance of the analysis is that the actual filter may be represented as the ideal filter in parallel with the "stray" transfer function specified by (14). This is illustrated in Fig. 1.

A convenient measure of the degeneration in filter performance due to coefficient errors is given by the following statistical mean-square convergence criterion:

$$\overline{\sigma_\omega^2} = \overline{\frac{1}{2\pi/T}\int_0^{2\pi/T} |H^*(j\omega) - H_\infty^*(j\omega)|^2 \, d\omega} \tag{16}$$

where $H^*(j\omega)$ and $H_\infty^*(j\omega)$ represent the frequency responses of the actual and ideal filters and $T$ is the sampling period. Setting

$$z = e^{j\omega T} \tag{17}$$

and substituting (15) into (16) yields

$$\overline{\sigma_\omega^2} = \overline{\frac{1}{2\pi j}\oint_{|z|=1} \frac{E(z^{-1})}{X(z^{-1})}\frac{E(z)}{X(z)}\frac{dz}{z}}. \tag{18}$$

It is evident that the quantity $\overline{\sigma_\omega^2}$ only exists when the coefficient rounding errors are such that the actual filter $H(z^{-1})$ is still asymptotically stable. This limitation does not appear to represent a practical difficulty, because one is generally interested in determining the computer word-length for the actual filter frequency response to be within only a few percent of the ideal stable filter. Under these conditions, the stability of the filter will have been hardly affected by coefficient rounding. That is to say, it is contended that loss of stability in a realization will occur only after the deviation between the actual and ideal frequency responses has become intolerable. This is illustrated in Fig. 2 for a lowpass filter. Assuming filter stability, then the integrand of (18) can be reasonably expected to satisfy the conditions of the Lebesgue–Fubini theorem.[9] Hence, interchanging the order of integration and substituting (14) into (18), one obtains

$$\overline{\sigma_\omega^2} = \frac{1}{2\pi j} \cdot \oint_{|z|=1} \overline{\left[\frac{\alpha(z^{-1}) - \beta(z^{-1})H_\infty(z^{-1})}{B_\infty(z^{-1})}\right]\left[\frac{\alpha(z) - \beta(z)H_\infty(z)}{B_\infty(z)}\right]} \frac{dz}{z}. \tag{19}$$

Using the polynomial forms for $\alpha(z^{-1})$ and $\beta(z^{-1})$ and the stipulated statistical assumptions, (19) reduces to

$$\overline{\sigma_\omega^2} = \left[\sum_{k=0}^{N} \overline{\alpha_k^2}\right] \frac{1}{2\pi j}\oint_{|z|=1} \frac{1}{B_\infty(z^{-1})B_\infty(z)}\frac{dz}{z} + \left[\sum_{k=1}^{N} \overline{\beta_k^2}\right] \frac{1}{2\pi j}\oint_{|z|=1} \frac{A_\infty(z^{-1})A_\infty(z)}{[B_\infty(z^{-1})]^2[B_\infty(z)]^2}\frac{dz}{z}. \tag{20}$$

The quantities

$$\frac{1}{2\pi j}\oint_{|z|=1} \frac{1}{B_\infty(z^{-1})B_\infty(z)}\frac{dz}{z} \quad \text{and} \quad \frac{1}{2\pi j}\oint_{|z|=1} \frac{A_\infty(z^{-1})A_\infty(z)}{[B_\infty(z^{-1})]^2[B_\infty(z)]^2}\frac{dz}{z}$$

represent the sum of the squares of the impulse response terms for the digital filters $[1/B_\infty(z^{-1})]$ and $[A_\infty(z^{-1})]/[B_\infty(z^{-1})]^2$, respectively. Consequently, these integrals may be readily evaluated to any degree of accuracy using a short digital computer program based on Fig. 3.

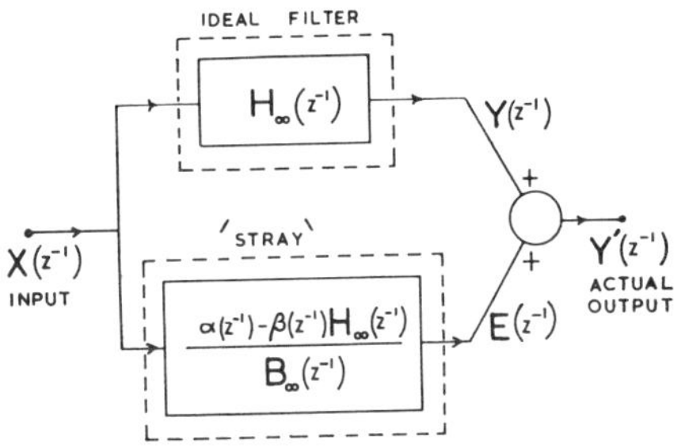


Fig. 1. "Stray" transfer function due to the coefficient rounding.

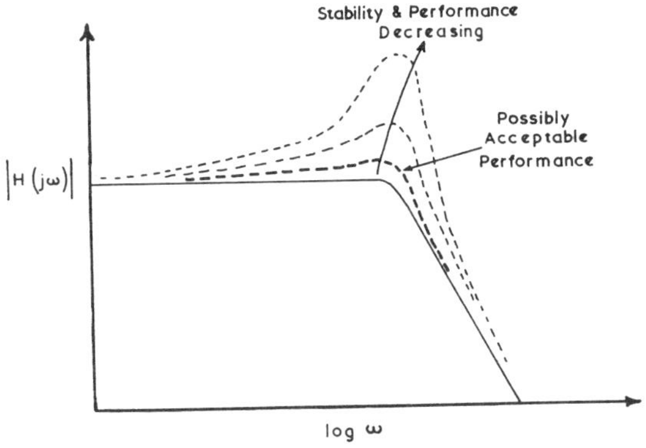


Fig. 2. Degeneration in filter frequency response due to the coefficient rounding.

On-line computers generally operate on a fixed-point basis, and in this case with rounded quantization, the error quantities in (5) lie in the interval

$$\left.\begin{aligned} |\alpha_k| &\le q/2 \\ |\beta_k| &\le q/2 \end{aligned}\right\} \quad \text{for all } k. \tag{21}$$

By virtue of the uniform statistical distribution assumed for these error quantities, it follows that the summations involved in (20) reduce to

$$\left.\begin{aligned} \sum_{k=0}^{N} \overline{\alpha_k^2} &= \mu \frac{q^2}{12} \\ \sum_{k=1}^{N} \overline{\beta_k^2} &= \nu \frac{q^2}{12} \end{aligned}\right\} \tag{22}$$

where $\mu$ and $\nu$ represent the number of nonzero and nonunity coefficients in the numerator and denominator, respectively. Thus, (20) to (22) specify $\overline{\sigma_\omega^2}$ for all widths of quantization.

Numbers in a floating-point machine are sometimes represented according to

$$x = U8^v \tag{23}$$

where $U$ and $v$ are sign conscious binary numbers. If the mantissa ($U$) has $M$ bits, then with rounded operations the error quantities in (5) satisfy

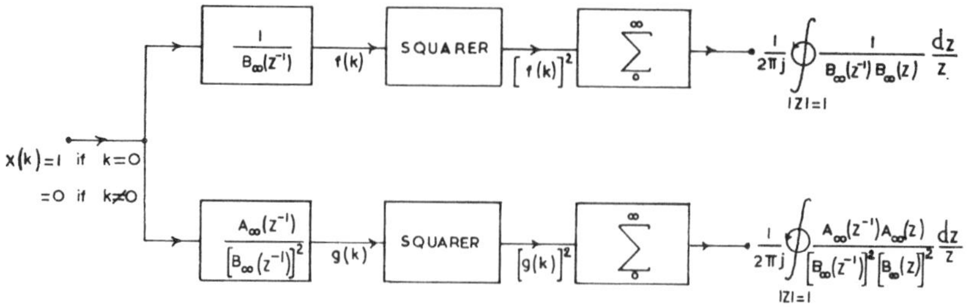


Fig. 3. A computational technique for evaluating

$$\frac{1}{2\pi j}\oint_{|z|=1}\frac{1}{B_\infty(z^{-1})B_\infty(z)}\frac{dz}{z}$$

and

$$\frac{1}{2\pi j}\oint_{|z|=1}\frac{A_\infty(z^{-1})A_\infty(z)}{[B_\infty(z^{-1})]^2[B_\infty(z)]^2}\frac{dz}{z}.$$

$$\left.\begin{aligned}|\alpha| &\le \tfrac{1}{2}2^{-(M-1)}8^{v'}\\ |\beta| &\le \tfrac{1}{2}2^{-(M-1)}8^{v'}\end{aligned}\right\} \qquad (24)$$

where $v'$ is the particular exponent value used in the representation of a given coefficient. Although the error on each coefficient depends now upon the magnitude of the coefficient, it is still relatively easy to evaluate the summations $\sum\overline{\alpha_k^2}$ and $\sum\overline{\beta_k^2}$ once the numerical values of the filter coefficients are specified.

## III. Analysis for Parallel Programming

If the pulse transfer function of the actual filter $H(z^{-1})$ does not contain multiple poles, then for parallel programming it is first expressed in the form

$$H(z^{-1}) = R(z^{-1}) + \sum_{i=1}^{N} H_i(z^{-1}) \qquad (25)$$

where

$$\left.\begin{aligned}R(z^{-1}) &= \sum_{i=0}^{P} r_i z^{-i}\\ H_i(z^{-1}) &= \frac{a_{0i}+a_i z^{-1}}{1+b_{1i}z^{-1}+b_{2i}z^{-2}}.\end{aligned}\right\} \qquad (26)$$

It should be observed, however, that $H_i(z^{-1})$ includes the case of a real pole by setting

$$a_{1i} = b_{2i} = 0. \qquad (27)$$

As shown in Ragazzini and Franklin[10] and Jury,[11] the actual parallel programming of $H(z^{-1})$ is implemented by direct programming each of the elementary transfer functions $R(z^{-1})$, $H_i(z^{-1})$ and then summing their responses. This programming technique is illustrated in Fig. 4.

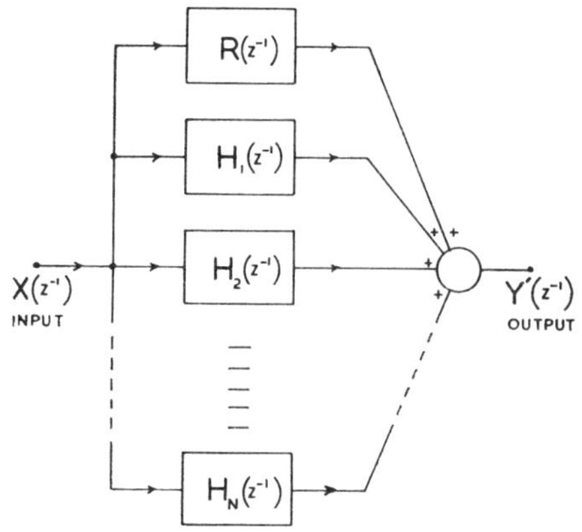


Fig. 4. Parallel programming of a digital filter.

By virtue of (15), which is illustrated in Fig. 1, it follows that the actual filter transfer function realized is

$$H(z^{-1}) = H_\infty(z^{-1}) + \sum_{i=0}^{P}\rho_i z^{-i} + \sum_{i=1}^{N}\frac{(\alpha_{0i}+\alpha_{1i}z^{-1}) - (\beta_{1i}z^{-1}+\beta_{2i}z^{-2})H_{i\infty}(z^{-1})}{B_{i\infty}(z^{-1})} \qquad (28)$$

where

$$\left.\begin{aligned}H_\infty(z^{-1}) &= R_\infty(z^{-1}) + \sum_{i=1}^{N} H_{i\infty}(z^{-1})\\ H_{i\infty}(z^{-1}) &= \frac{A_{i\infty}(z^{-1})}{B_{i\infty}(z^{-1})} = \frac{\bar{a}_{0i}+\bar{a}_{1i}z^{-1}}{1+\bar{b}_{1i}z^{-1}+\bar{b}_{2i}z^{-2}}\\ R_\infty &= \sum_{i=0}^{P}\bar{r}_i z^{-i}\\ a_{ki} &= \bar{a}_{ki}+\alpha_{ki}\\ b_{ki} &= \bar{b}_{ki}+\beta_{ki}\\ r_i &= \bar{r}_i+\rho_i.\end{aligned}\right\} \qquad (29)$$

Substituting (28) into (18), one obtains

$$\overline{\sigma_\omega^2} = \sum_{i=0}^{P}\overline{\rho_i^2} + \frac{1}{2\pi j}\oint_{|z|=1}\sum_{m=1}^{N}\sum_{n=1}^{N}\frac{\overline{[(\alpha_{0m}+\alpha_{1m}z^{-1}) - (\beta_{1m}z^{-1}+\beta_{2m}z^{-2})H_{m\infty}(z^{-1})][(\alpha_{0n}+\alpha_{1n}z) - (\beta_{1n}z+\beta_{2n}z^{+2})H_{n\infty}(z)]}}{B_{m\infty}(z^{-1})B_{n\infty}(z)}\frac{dz}{z}. \qquad (30)$$

Assuming that the systems are stable and that the coefficient errors are statistically independent with zero mean, (30) reduces to

$$\overline{\sigma_\omega^2} = \sum_{i=0}^{P} \overline{\rho_i^2} + \sum_{m=1}^{N} \left[ (\overline{\alpha_{0m}^2} + \overline{\alpha_{1m}^2}) \frac{1}{2\pi j} \oint_{|z|=1} \frac{1}{B_{m\infty}(z^{-1})B_{m\infty}(z)} \frac{dz}{z} \right]$$
$$+ \sum_{m=1}^{N} \left[ (\overline{\beta_{1m}^2} + \overline{\beta_{2m}^2}) \frac{1}{2\pi j} \oint_{|z|=1} \frac{A_{m\infty}(z^{-1})A_{m\infty}(z)}{[B_{m\infty}(z^{-1})]^2[B_{m\infty}(z)]^2} \frac{dz}{z} \right]. \quad (31)$$

The mean-square quantities $\overline{\rho_i^2}$, $\overline{\alpha_{ki}^2}$, and $\overline{\beta_{ki}^2}$ are calculated exactly as in Section II, assuming uniform probability density functions. It should be noted again that zero and unity filter coefficients may be represented in the computer without error. The contour integrals involved in (31) may be evaluated by the computing technique illustrated in Fig. 3. However, considerable computing time may be saved by using the following results, which include the case of a simple pole by virtue of (27).

$$\frac{1}{2\pi j} \oint_{|z|=1} \frac{1}{B_{i\infty}(z^{-1})B_{i\infty}(z)} \frac{dz}{z} = \frac{1 + \bar{b}_{2i}}{(1 - \bar{b}_{1i} + \bar{b}_{2i})(1 + \bar{b}_{1i} + \bar{b}_{2i})(1 - \bar{b}_{2i})}$$
$$\frac{1}{2\pi j} \oint_{|z|=1} \frac{A_{i\infty}(z^{-1})A_{i\infty}(z)}{[B_{i\infty}(z^{-1})]^2[B_{i\infty}(z)]^2} \frac{dz}{z} \quad (32)$$
$$= \frac{(B_0B_2 + B_1^2)(B_0^3A_3 + B_2^3A_0) + B_0^2B_2^2(B_0A_2 + B_2A_1)}{2(B_0B_1B_2)^3}$$

where

$$\left.\begin{aligned} B_0 &= 1 + \bar{b}_{1i} + \bar{b}_{2i} \\ B_1 &= 2(1 - \bar{b}_{2i}) \\ B_2 &= 1 - \bar{b}_{1i} + \bar{b}_{2i} \\ A_0 &= (\bar{a}_{0i} + \bar{a}_{1i})^2 \\ A_1 &= 3\bar{a}_{0i}^2 + 2\bar{a}_{0i}\bar{a}_{1i} + 3\bar{a}_{1i}^2 \\ A_2 &= 3\bar{a}_0^2 - 2\bar{a}_{0i}\bar{a}_{1i} + 3\bar{a}_{1i}^2 \\ A_3 &= (\bar{a}_{0i} - \bar{a}_{1i})^2. \end{aligned}\right\} \quad (33)$$

Further, it is apparent that for the case of a simple quadratic transfer function, coefficient rounding which results in $b_{2i} = 1$ causes loss of asymptotic stability.

## IV. Analysis for Cascade Programming

For cascade programming, the pulse transfer function of the actual filter $H(z^{-1})$ is expressed in the form

$$H(z^{-1}) = \prod_{i=1}^{N} H_i(z^{-1}) \quad (34)$$

where

$$H_i(z^{-1}) = \frac{a_{0i} + a_{1i}z^{-1} + a_{2i}z^{-2}}{1 + b_{1i}z^{-1} + b_{2i}z^{-2}}. \quad (35)$$

As previously shown,[10],[11] the cascade programming of $H(z^{-1})$ is implemented by the successive direct programming of each elementary transfer function, $H_i(z^{-1})$, with the output from $i$th transfer function forming the input to the $(i + 1)$th. This programming technique is shown in Fig. 5. By virtue of (15), it follows that, due

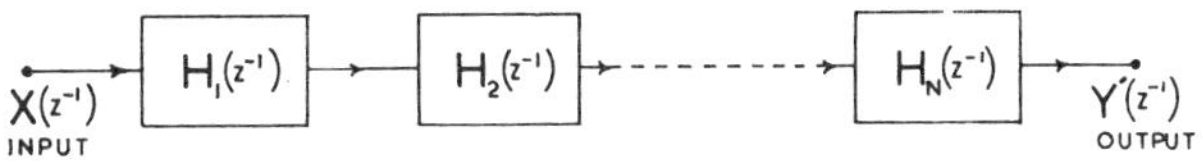


Fig. 5. Cascade programming of a digital filter.

to the filter coefficient errors, the actual transfer function realized is

$$H(z^{-1}) = \prod_{i=1}^{N} [1 + \epsilon_i(z^{-1})]H_{i\infty}(z^{-1}) \quad (36)$$

where

$$\left.\begin{aligned} \epsilon_i(z^{-1}) &= \frac{\alpha_{0i} + \alpha_{1i}z^{-1} + \alpha_{2i}z^{-2}}{A_{i\infty}(z^{-1})} - \frac{\beta_{1i}z^{-1} + \beta_{2i}z^{-2}}{B_{i\infty}(z^{-1})} \\ H_{i\infty}(z^{-1}) &= \frac{A_{i\infty}(z^{-1})}{B_{i\infty}(z^{-1})} = \frac{\bar{a}_{0i} + \bar{a}_{1i}z^{-1} + \bar{a}_{2i}z^{-2}}{1 + \bar{b}_{1i}z^{-1} + \bar{b}_{2i}z^{-2}} \\ a_{ki} &= \overline{a_{ki}} + \alpha_{ki} \\ b_{ki} &= \overline{b_{ki}} + \beta_{ki}. \end{aligned}\right\} \quad (37)$$

Since

$$H_\infty(z^{-1}) = \prod_{i=1}^{N} H_{i\infty}(z^{-1}), \quad (38)$$

then (36) can be written as

$$H(z^{-1}) = H_\infty(z^{-1}) \prod_{i=1}^{N} (1 + \epsilon_i(z^{-1})). \quad (39)$$

Substituting (39) into (18) yields

$$\overline{\sigma_\omega^2} = \frac{1}{2\pi j} \oint_{|z|=1} H_\infty(z^{-1})H_\infty(z) \cdot \overline{\left\{\prod_{m=1}^{N} (1 + \epsilon_m(z^{-1})) - 1\right\}\left\{\prod_{n=1}^{N} (1 + \epsilon_n(z)) - 1\right\}} \frac{dz}{z} \quad (40)$$

or

$$\overline{\sigma_\omega^2} = \frac{1}{2\pi j} \oint_{|z|=1} H_\infty(z^{-1})H_\infty(z) \cdot \left\{ \overline{\prod_{m=1}^{N} (1 + \epsilon_m(z^{-1}))(1 + \epsilon_m(z))} - \overline{\prod_{m=1}^{N} (1 + \epsilon_m(z^{-1})) - \prod_{n=1}^{N} (1 + \epsilon_n(z)) + 1} \right\} \frac{dz}{z}. \quad (41)$$

As the coefficient errors are assumed to be statistically independent with zero mean, it follows that (41) reduces to

$$\overline{\sigma_\omega^2} = \frac{1}{2\pi j} \oint_{|z|=1} H_\infty(z^{-1})H_\infty(z) \cdot \left\{ \left[ \prod_{m=1}^{N} (1 + \overline{\epsilon_m(z^{-1})\epsilon_m(z)}) \right] - 1 \right\} \frac{dz}{z}. \quad (42)$$

Substituting (37) into (42) one obtains

$$\overline{\sigma_\omega^2} = \frac{1}{2\pi j} \oint_{|z|=1} H_\infty(z^{-1})H_\infty(z) \cdot \left\{ \left[ \prod_{m=1}^{N} \left[ 1 + \frac{\sum_{k=0}^{2} \overline{\alpha_{km}^2}}{A_{m\infty}(z^{-1})A_{m\infty}(z)} + \frac{\sum_{k=1}^{2} \overline{\beta_{km}^2}}{B_{m\infty}(z^{-1})B_{m\infty}(z)} \right] \right] - 1 \right\} \frac{dz}{z} \tag{43}$$

or

$$\overline{\sigma_\omega^2} = \frac{1}{2\pi j} \oint_{|z|=1} \frac{1}{[B_\infty(z^{-1})]^2[B_\infty(z)]^2} \cdot \left[ \prod_{m=1}^{N} \left\langle A_{m\infty}(z^{-1})A_{m\infty}(z)B_{m\infty}(z^{-1})B_{m\infty}(z) + B_{m\infty}(z^{-1})B_{m\infty}(z) \cdot \left( \sum_{k=0}^{2} \overline{\alpha_{km}^2} \right) + A_{m\infty}(z^{-1})A_{m\infty}(z) \left( \sum_{k=1}^{2} \overline{\beta_{km}^2} \right) \right\rangle - A_\infty(z^{-1})A_\infty(z)B_\infty(z^{-1})B_\infty(z) \right] \frac{dz}{z}. \tag{44}$$

In order to automatically compute the contour integral in (43), it is first necessary to implement the factorization

$$\prod_{m=1}^{N} \left[ A_{m\infty}(z^{-1})A_{m\infty}(z)B_{m\infty}(z^{-1})B_{m\infty}(z) + B_{m\infty}(z^{-1})B_{m\infty}(z) \cdot \left( \sum_{k=0}^{2} \overline{\alpha_{km}^2} \right) + A_{m\infty}(z^{-1})A_{m\infty}(z) \left( \sum_{k=1}^{2} \overline{\beta_{km}^2} \right) \right] - A_\infty(z^{-1})A_\infty(z)B_\infty(z^{-1})B_\infty(z) = P(z^{-1})P(z). \tag{45}$$

The poles and zeros of $P(z^{-1})$ are evidently dependent upon the mean-square values of the coefficient errors. Therefore, the width of quantization is no longer a simple scaling factor as in the previous cases. As practical wide-band digital filters may involve polynomials of high degree (22 poles in Golden and Kaiser[11]), the factorization indicated by (45) is extremely laborious and generally not worthwhile. A more economic computational procedure is to measure the loss of performance directly by the technique adopted in Section V of this paper.

## V. Experimental Results

By a slight modification to the design procedure, it is possible to synthesize the bandstop filter example of Golden and Kaiser[1] in the form of (4). The relevant filter coefficients for this direct programming realization are shown in Table I. Assuming the coefficients to possess identical error bounds, as in a fixed-point machine, the degeneration in filter performance was calculated for all computer word-lengths from (20) and (22) as

$$\overline{\sigma_\omega^2} = (3.91 \times 10^{20})q^2. \tag{46}$$

The numerical value in (46) was checked by direct and cascade programming of the appropriate transfer functions of Fig. 3 using 80-bit accuracy. In both cases the same numerical value was obtained. Expressing (46) in terms of the word-length of a fixed-point machine, as defined in (3), one obtains

$$\sqrt{\overline{\sigma_\omega^2}} = 1.98 \times 10^{10} \cdot 2^{-(M-1)} \tag{47}$$

or

$$\log_{10} \sqrt{\overline{\sigma_\omega^2}} = 10.60 - 0.3M, \tag{48}$$

which is shown in Fig. 6. Using a previous analysis,[4],[5] the rms value of the filter's output noise due to multiplicative rounding errors was derived as

$$\log_{10} \sqrt{\overline{n_q^2}} = 10.62 - 0.3M, \tag{49}$$

which is also shown in Fig. 6.

As a means of verifying (48) the coefficients of the bandstop filter shown in Table I were appropriately rounded using the equations

$$a_k = q \times [\text{integral part } (\bar{a}_k/q + 0.5)] \tag{50}$$

$$b_k = q \times [\text{integral part } (\bar{b}_k/q + 0.5)],$$

and the quantity

$$\sigma_\omega^2 = \frac{1}{2\pi/T} \int_0^{2\pi/T} |H^*(j\omega) - H_\infty^*(j\omega)|^2 \, d\omega \tag{51}$$

was measured directly by means of the technique shown in Fig. 7.

Examination of Fig. 6 suggests that, for all practical purposes, 80-bit realization can be considered "ideal" or "errorless" relative to the realizations using less than 50 bits. In consequence, the "ideal" filter was realized using 80-bit coefficients. Further, all multiplications in both actual and ideal filters were performed with this accuracy. By this procedure, coefficient rounding errors were isolated from data and multiplicative rounding errors in the actual filter. The values of $\sigma_\omega$ thus computed are shown in Fig. 6 for coefficient word-lengths in the range of 39 to 47 bits. It is observed that $\sigma_\omega$ is never greater than twice its expected value $\sqrt{\overline{\sigma_\omega^2}}$ which confirms, to some extent, the analysis.

It is also possible to realize Golden and Kaiser's bandstop filter example in the parallel programmed form specified by (25). The relevant filter coefficients for this parallel form of programming are shown in Table II. Assuming the coefficients have identical error bounds, the degeneration in filter performance was calculated using double-length arithmetic for all computer word-lengths from (22) and (32) as

$$\sqrt{\overline{\sigma_\omega^2}} = 18.65q \tag{52}$$

or

$$\log_{10} \sqrt{\overline{\sigma_\omega^2}} = 1.57 - 0.3M, \tag{53}$$

which is shown in Fig. 8. Using a previous analysis[4],[5] the rms value of the output noise for parallel programming was derived as

$$\log_{10} \sqrt{\overline{n_q^2}} = 1.62 - 0.3M, \tag{54}$$

TABLE I

Direct Programming Coefficients

| $k$ | $\bar{a}_k$ | | | | | $\bar{b}_k$ | | | | |
|---|---|---|---|---|---|---|---|---|---|---|
| 0 | 6.14532 | 02400 | 45660 | 95935 | $10^{-1}$ | 1.00000 | 00000 | 00000 | 00000 | |
| 1 | 1.82862 | 83743 | 34944 | 30234 | | 2.84644 | 19689 | 74009 | 00222 | |
| 2 | 9.18598 | 76393 | 23345 | 37501 | | 1.36544 | 18024 | 06957 | 35998 | $10^{+1}$ |
| 3 | 2.01785 | 43858 | 39049 | 83632 | $10^{+1}$ | 2.86863 | 12827 | 03054 | 60931 | $10^{+1}$ |
| 4 | 5.65974 | 01304 | 61323 | 06076 | $10^{+1}$ | 7.68668 | 01556 | 51020 | 29537 | $10^{+1}$ |
| 5 | 9.77616 | 97277 | 77455 | 67929 | $10^{+1}$ | 1.26978 | 24238 | 74008 | 59583 | $10^{+2}$ |
| 6 | 1.95560 | 58902 | 77163 | 75480 | $10^{+2}$ | 2.42719 | 14138 | 13632 | 85357 | $10^{+2}$ |
| 7 | 2.74444 | 52924 | 17107 | 32360 | $10^{+2}$ | 3.25744 | 49025 | 55813 | 19723 | $10^{+2}$ |
| 8 | 4.27221 | 29208 | 22686 | 15701 | $10^{-2}$ | 4.84590 | 46018 | 12943 | 21722 | $10^{+2}$ |
| 9 | 4.95063 | 95407 | 14507 | 65312 | $10^{+2}$ | 5.36954 | 24793 | 46719 | 51875 | $10^{+2}$ |
| 10 | 6.24128 | 70904 | 97272 | 04930 | $10^{+2}$ | 6.46875 | 76125 | 18105 | 92119 | $10^{+2}$ |
| 11 | 6.00062 | 41335 | 12317 | 55061 | $10^{+2}$ | 5.94574 | 02388 | 97171 | 95950 | $10^{+2}$ |
| 12 | 6.24128 | 70905 | 09213 | 35302 | $10^{+2}$ | 5.90817 | 75064 | 19863 | 11885 | $10^{+2}$ |
| 13 | 4.95063 | 95407 | 34381 | 27910 | $10^{+2}$ | 4.47868 | 92604 | 45618 | 46370 | $10^{+2}$ |
| 14 | 4.27221 | 29208 | 46950 | 36973 | $10^{+2}$ | 3.69085 | 95309 | 78270 | 04033 | $10^{+2}$ |
| 15 | 2.74444 | 52924 | 38924 | 26919 | $10^{+2}$ | 2.26468 | 72853 | 40520 | 74943 | $10^{+2}$ |
| 16 | 1.95560 | 58902 | 95258 | 27851 | $10^{+2}$ | 1.54004 | 37815 | 20019 | 23587 | $10^{+2}$ |
| 17 | 9.77616 | 97278 | 91991 | 46783 | $10^{-1}$ | 7.34798 | 81860 | 20413 | 09382 | $10^{+1}$ |
| 18 | 5.65974 | 01305 | 31808 | 57200 | $10^{+1}$ | 4.05582 | 38655 | 76315 | 12082 | $10^{+1}$ |
| 19 | 2.01785 | 43858 | 69746 | 80847 | $10^{+1}$ | 1.37867 | 64204 | 74181 | 15588 | $10^{+1}$ |
| 20 | 9.18598 | 76394 | 60991 | 74744 | | 5.97585 | 13278 | 52671 | 08900 | |
| 21 | 1.82862 | 83743 | 68430 | 88781 | | 1.13257 | 95931 | 45398 | 63928 | |
| 22 | 6.14532 | 02401 | 45188 | 00124 | $10^{-1}$ | 3.61729 | 37140 | 82771 | 44243 | $10^{-1}$ |

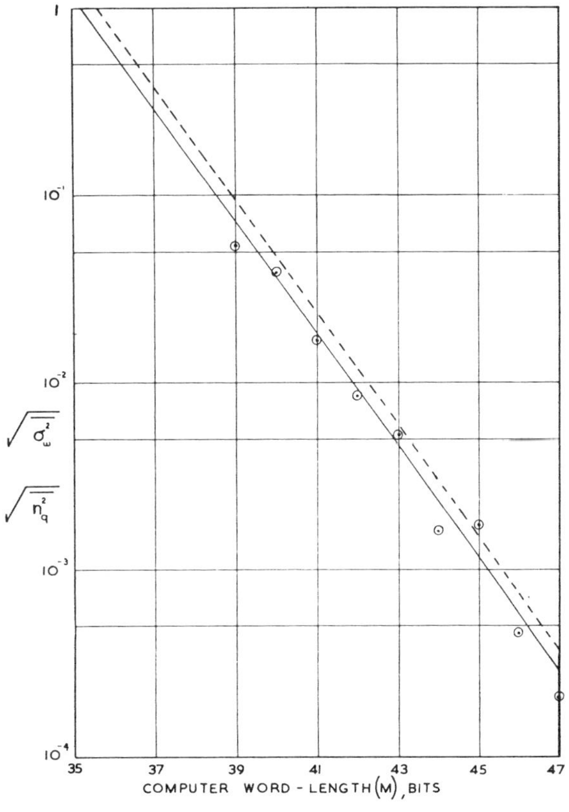


*Key*: —— $\sqrt{\overline{\sigma_\omega{}^2}}$; - - - $\sqrt{\overline{n_q{}^2}}$; ⊙ $\sigma_\omega$.

Fig. 6. Direct programming errors.

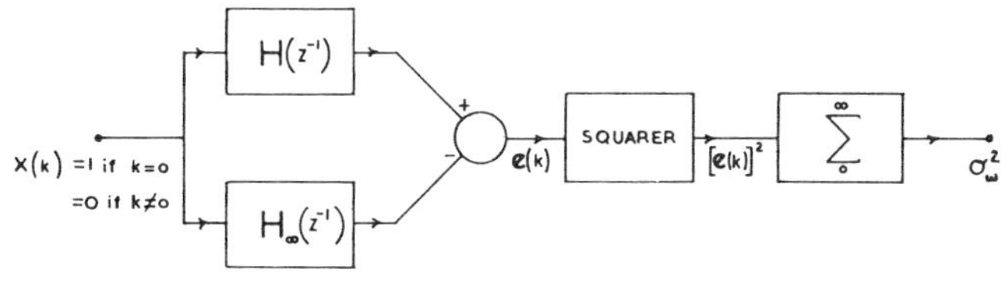


Fig. 7. A computational technique for evaluating $\sigma_\omega{}^2$.

TABLE II

PARALLEL PROGRAMMING COEFFICIENTS, $r_o$ = 1.6988724578

| $k$ | $\overline{a_{1k}}$ | $\overline{a_{0k}}$ | $\overline{b_{2k}}$ | $\overline{b_{1k}}$ |
|---|---|---|---|---|
| 1 | $-7.4670270246 \quad 10^{-4}$ | $7.3152391358 \quad 10^{-4}$ | $9.9890799167 \quad 10^{-1}$ | $4.2851754740 \quad 10^{-1}$ |
| 2 | $9.8551937100 \quad 10^{-4}$ | $7.4764471504 \quad 10^{-4}$ | $9.9888372720 \quad 10^{-1}$ | $1.1064725159 \quad 10^{-1}$ |
| 3 | $4.5563620191 \quad 10^{-3}$ | $-4.3569204420 \quad 10^{-4}$ | $9.9572786927 \quad 10^{-1}$ | $4.3175124685 \quad 10^{-1}$ |
| 4 | $-4.8884642202 \quad 10^{-3}$ | $-4.4273924710 \quad 10^{-4}$ | $9.9563067875 \quad 10^{-1}$ | $1.0637363408 \quad 10^{-1}$ |
| 5 | $-1.3323384702 \quad 10^{-2}$ | $-7.4675471331 \quad 10^{-3}$ | $9.8829189882 \quad 10^{-1}$ | $4.4149478013 \quad 10^{-1}$ |
| 6 | $1.1878029670 \quad 10^{-2}$ | $-7.6702309337 \quad 10^{-3}$ | $9.8800730189 \quad 10^{-1}$ | $9.4074003990 \quad 10^{-2}$ |
| 7 | $2.3334703327 \quad 10^{-2}$ | $4.0192757395 \quad 10^{-2}$ | $9.6614183349 \quad 10^{-1}$ | $4.6635062848 \quad 10^{-1}$ |
| 8 | $-1.3303541404 \quad 10^{-2}$ | $4.1424600464 \quad 10^{-2}$ | $9.651810179 \quad 10^{-1}$ | $6.1626308659 \quad 10^{-2}$ |
| 9 | $4.3661569186 \quad 10^{-2}$ | $-1.4357673940 \quad 10^{-1}$ | $8.7422935455 \quad 10^{-1}$ | $5.1860379640 \quad 10^{-1}$ |
| 10 | $-8.8447283670 \quad 10^{-2}$ | $-1.4766037472 \quad 10^{-1}$ | $8.6946006357 \quad 10^{-1}$ | $-2.0456474227 \quad 10^{-2}$ |
| 11 | $-1.1676074543 \quad 10^{-1}$ | $-8.6018371841 \quad 10^{-1}$ | $5.2836504970 \quad 10^{-1}$ | $2.0745921909 \quad 10^{-1}$ |

TABLE III

CASCADE PROGRAMMING COEFFICIENTS, Gain Constant = 0.001106

| $k$ | $\bar{a}_{2k} = \bar{a}_{0k}$ | $\bar{a}_{1k}$ | $\bar{b}_{2k}$ | $\bar{b}_{1k}$ |
|---|---|---|---|---|
| 1 | 0.9989079917 | 0.4183952392 | 0.9989079917 | 0.4285175474 |
| 2 | 1.1404846087 | 0.1380427242 | 0.9988837272 | 0.1106472516 |
| 3 | 0.9957278693 | 0.4129474324 | 0.9957278693 | 0.4317512468 |
| 4 | 1.2704693736 | 0.1592449261 | 0.9956306788 | 0.1063736341 |
| 5 | 0.9882918988 | 0.3989114203 | 0.9882918988 | 0.4414947801 |
| 6 | 1.7233973619 | 0.2358636857 | 0.9880073019 | 0.0940740040 |
| 7 | 0.9661418335 | 0.3666423085 | 0.9661418335 | 0.4663506285 |
| 8 | 3.6768896818 | 0.5950245813 | 0.9651814018 | 0.0616263087 |
| 9 | 0.8742293546 | 0.2928381810 | 0.8742293546 | 0.5186037964 |
| 10 | 24.9117164807 | 5.1673793969 | 0.8694600636 | −0.0204564742 |
| 11 | 2.9258251702 | 0.7942989865 | 0.5283650497 | 0.2074592191 |

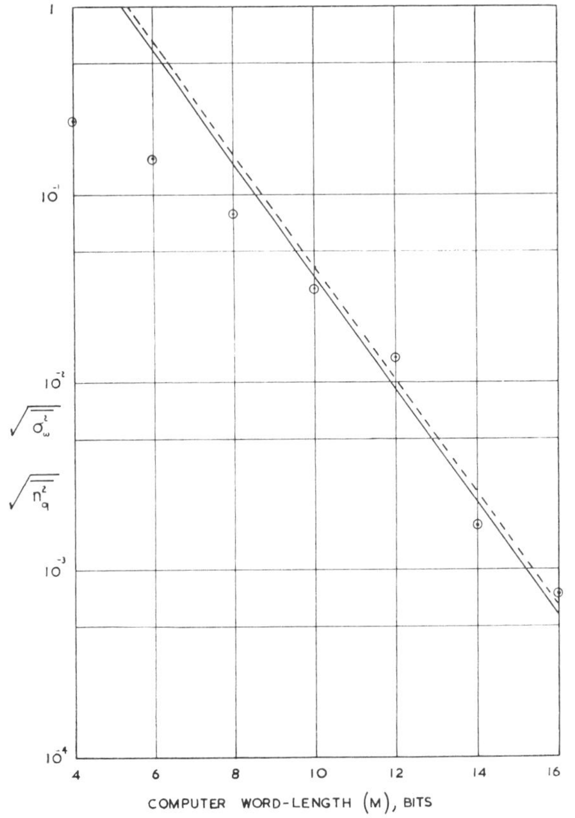


*Key:* —— $\sqrt{\overline{\sigma_\omega^2}}$; - - - $\sqrt{\overline{n_q^2}}$; ⊙ $\sigma_\omega$.

Fig. 8. Parallel programming errors.

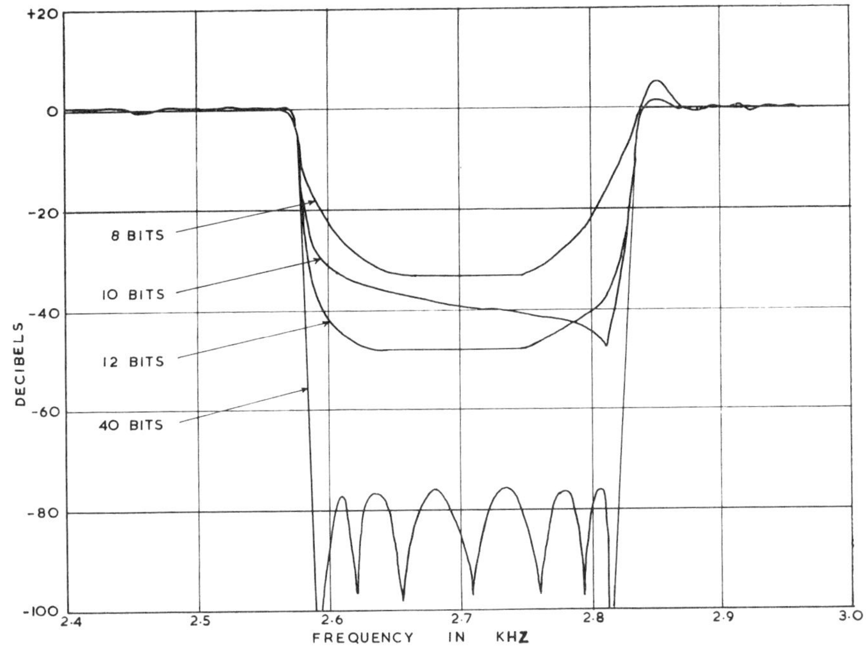


Fig. 9. Degeneration in the frequency response of a 22nd-order bandstop digital filter.

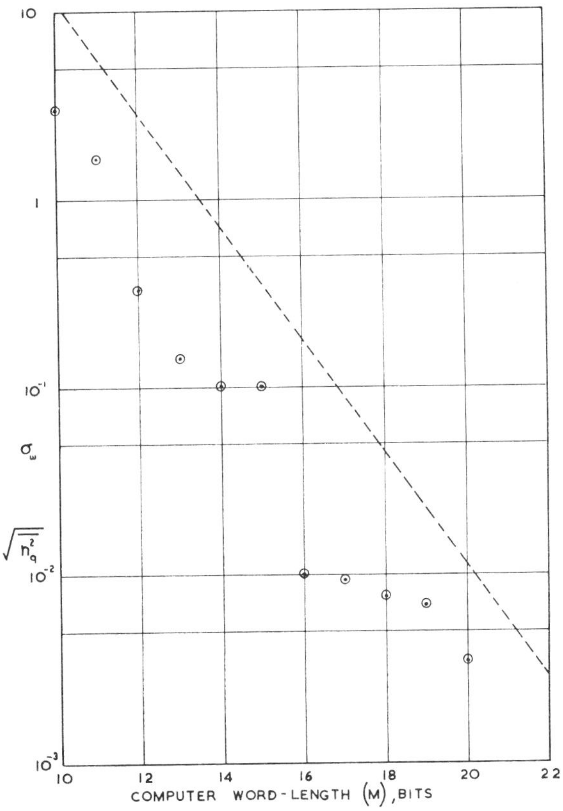


*Key*: - - - $\sqrt{\overline{n_q^2}}$; ⊙ $\sigma_\omega$.

Fig. 10. Cascade programming errors.

which is also shown in Fig. 8. Examination of Fig. 8 suggests that, for all practical purposes, a 40-bit realization may be considered "ideal" relative to less than 20-bit realizations. Further, coefficient rounding errors can be isolated from the data quantization and multiplicative rounding errors in the actual filter by performing all multiplications to a 40-bit accuracy. Using this experimental technique, the value of $\sigma_\omega$ was computed for coefficient word-lengths in the range of 4 to 16 bits and the results obtained are included in Fig. 8. It is observed that $\sigma_\omega$ is in close agreement with its expected value $\sqrt{\overline{\sigma_\omega^2}}$, except for word-lengths of less than 8 bits, where an apparently deterministic deviation of theoretical and experimental values occur. This phenomenon is attributed to the deletion of second-order quantities in (9). For illustrative purposes, the effect of coefficient accuracy on the real frequency response of the parallel programmed filter is shown in Fig. 9.

The cascade version of the digital filter under consideration is specified by the coefficients in Table III. For reasons of completeness, the degeneration in filter performance with this method of programming is shown in Fig. 10. In this case the ideal filter and all multiplications were performed to an accuracy of 40 bits, which may be justified in the same manner as before. It is seen that the actual degeneration for cascade programming is less than that for direct programming but greater than that for parallel programming.

## VI. Conclusions

Due to inherent quantization errors on the coefficients of the difference equation representing a digital filter, the frequency response of an actual realization deviates from that which would have been obtained using an infinite word-length computer. This paper specifies this loss of filter performance in terms of the statistically expected mean-square deviation in the frequency responses of the actual and ideal filters. Because the width of quantization for the filter coefficients appears as a simple scaling factor in the results for direct and parallel programming, one simple computer program may be used to evaluate the statistically expected degeneration in filter performance for all computer word-lengths. This computation appears to be both shorter and more significant than several evaluations of $\sigma_\omega$ for different widths of quantization using the technique illustrated in Fig. 7. Further, the computations involved may also be used to evaluate the rms value of the output noise due to data quantization and multiplicative rounding errors. However, for cascade programming, direct evaluation of $\sigma_\omega^2$ using the technique illustrated in Fig. 7 appears to be the simplest procedure in general, by virtue of the factorization problem involved. Experimental determinations of $\sigma_\omega$ as defined by (49) for the cases of direct and parallel programming are within twice its statistical expectation $\sqrt{\overline{\sigma_\omega^2}}$, as computed through (20) and (31). This agreement between theoretical and experimental results evidently provides confirmation of the analysis. An important aspect of the results obtained is to show that a significant improvement in the accuracy of a realization can be achieved by employing only one extra bit.

With the analysis just presented and that available elsewhere,[4],[5] it is possible to specify a specific linear model of the basically nonlinear effects occurring in the finite word-length representation of a pulse transfer function. In the case of a fixed-point operation with direct programming, this specific statistical model is as shown in Fig. 11. It will be observed that the parallel transfer function arrangement

$$\frac{\sqrt{\frac{\mu}{12}}\,q}{B_\infty(z^{-1})} + \frac{\sqrt{\frac{\nu}{12}}\,qA_\infty(z^{-1})}{[B_\infty(z^{-1})]^2}$$

gives the statistically expected mean-square deviation $(\overline{\sigma_\omega^2})$ between the actual and ideal frequency responses. As the width of quantization ($q$) would be chosen in practice so that the deviation in the real frequency responses of $H(z^{-1})$ and $H_\infty(z^{-1})$ is extremely small, then the noise shaping filter $1/B(z^{-1})$ may be evidently replaced by $1/B_\infty(z^{-1})$ with only second-order errors. Quantitatively, the power transmitted by $1/B_\infty(z^{-1})$ and

$$\frac{1}{B_\infty(z^{-1})} - \frac{\sum_{k=1}^{N} \beta_k z^{-k}}{[B_\infty(z^{-1})]^2}$$

can be considered identical for a satisfactory actual realization.

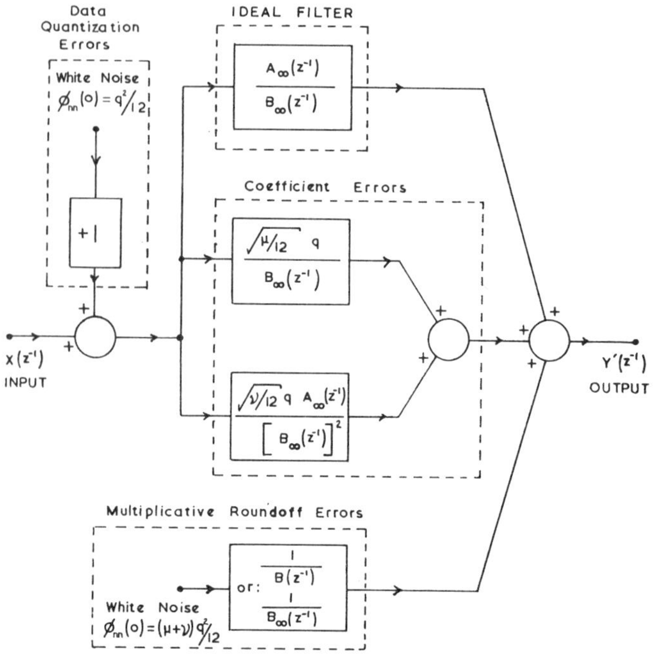

Fig. 11. Representation of finite word-length degenerate effects in a digital filter.

An apparently reasonable method of designing digital filters using $\overline{\sigma_\omega^2}$ as a criterion is as follows. A filter design by the Golden and Kaiser technique is generally best initially implemented in parallel programmed form on a general purpose computer. The development of a parallel realization involves only one factorization[2] in the $s$ plane as opposed to the two required with cascade programming. Further, parallel programmed filters are apparently less susceptible to multiplicative and coefficient quantization than directly programmed filters (see Knowles and Edwards,[4],[5] Kaiser,[7] and this paper).

After checking the real frequency responses of the analog and parallel programmed digital filters for reasonable agreement, a computation of $\overline{\sigma_\omega^2}$ can be effected using (31). These same calculations can also be utilized to obtain the rms value of the output noise due to data and multiplicative rounding errors. As $\overline{\sigma_\omega^2}$ is the mathematical expectation of

$$\frac{1}{2\pi/T}\int_0^{2\pi/T} |H^*(j\omega) - H_\infty^*(j\omega)|^2\,d\omega,$$

then it is evidently a measure of the deviation between the actual and the ideal filter responses at any frequency. For many bounded functions, plus or minus three times its rms value contains most, and in some cases all, probable observations. It seems reasonable, therefore, to select the computer word-length according to

$$3\sqrt{\overline{\sigma_\omega^2}} < \left|\begin{matrix}\text{Acceptable Gain}\\ \text{Fluctuation}\end{matrix}\right| \tag{55}$$

provided that the output noise due to data quantization and multiplicative roundoff errors is also acceptable with

[2] Private correspondence with Dr. J. F. Kaiser.

TABLE IV

| $M$ | Maximum Passband Ripple | Minimum Attenuation in Rejection Band |
|---|---|---|
| 18 | −0.4997 dB at 2140 Hz | −71.12 dB at 2610 Hz |
| 19 | −0.4996 dB at 2140 Hz | −75.48 dB at 2630 Hz |
| 20 | −0.4997 dB at 2140 Hz | −75.61 dB at 2730 Hz |
| 21 | −0.4997 dB at 2140 Hz | −76.43 dB at 2680 Hz |
| ∞ | −0.4997 dB at 2140 Hz | −76.50 dB at 2730 Hz |

this word-length. This procedure evidently enables the designer to assess whether a practical realization of the given digital filter can be made on a smaller word-length computer.

As an example of the use of this analysis in digital filter design, the number of bits required for the Golden-Kaiser filter to meet the following specification will be considered:

Ripple in passband = 0.5 dB.

Minimum attenuation in rejection band = −75 dB.

The ideal realization of this filter has a minimum attenuation of −76.5 dB in the rejection band. The allowance of 1.5-dB fluctuation about this particular point evidently constitutes a worst case design for the rejection band. Thus, the allowable gain fluctuations in the pass and the rejection bands are

$$\pm\left[\text{antilog}_{10}\left(\frac{0.5}{20}\right) - 1\right] = \pm 0.059$$

and

$$\pm\left[\text{antilog}_{10}\left(-\frac{75}{20}\right) - \text{antilog}_{10}\left(-\frac{76.5}{20}\right)\right] = \pm 0.266 \times 10^{-4},$$

respectively. Substituting these values into (55) yields

$$\sqrt{\overline{\sigma_\omega^2}} < 0.020$$

and

$$\sqrt{\overline{\sigma_\omega^2}} < 0.89 \times 10^{-5}.$$

For parallel programming, the minimum consistent number of bits is obtained from Fig. 8 as 19 bits. This result was confirmed by experimental frequency response measurements for computer word-lengths of 18, 19, 20, and 21 bits. The maximum passband ripple together with least rejection band attenuation is presented in Table IV.

## Acknowledgment

The authors wish to acknowledge the stimulus and encouragement received from their correspondence with Dr. J. F. Kaiser of Bell Telephone Laboratories, Murray Hill, N. J.

## References

[1] R. M. Golden and J. F. Kaiser, "Design of wideband sampled-data filters," *Bell Sys. Tech. J.*, vol. 43, pp. 1533–1546, July 1964.

[2] J. F. Kaiser, "Design methods for sampled-data filters," *Proc. 1st Allerton Conf. on Circuit and System Theory*, pp. 221–236, November 1963.

[3] B. Widrow, "Statistical analysis of amplitude-quantized sampled-data systems," *Trans. AIEE (Applications and Industry)*, pt. 2, vol. 79, pp. 555–568, January 1961.

[4] J. B. Knowles and R. Edwards, "Effects of finite-word-length computer in a sampled-data feedback system," *Proc. IEE* (London), vol. 112, pp. 1197–1207, June 1965.

[5] ——, "Finite word-length effects in a multirate direct digital control system," *Proc. IEE* (London), vol. 112, pp. 2376–2384, December 1965.

[6] C. M. Rader and B. Gold, "Digital filter design techniques in the frequency domain," *Proc. IEEE*, vol. 55, pp. 149–171, February 1967.

[7] J. F. Kaiser, "Some practical considerations in the realization of linear digital filters," *Proc. 3rd Allerton Conf. on Circuit and System Theory* (Monticello, Ill.), pp. 621–633, October 1965.

[8] O. Sornmoonpin, "Investigation of quantization errors," M.Sc. thesis, University of Manchester, England, 1966.

[9] L. M. Graves, *Theory of Functions of Real Variables*, 2nd ed. New York: McGraw-Hill, 1956.

[10] J. R. Ragazzini and G. F. Franklin, *Sampled-Data Control Systems*. New York: McGraw-Hill, 1958.

[11] E. I. Jury, *Sampled-Data Control Systems*. New York: Wiley, 1958.

# On the Design of Digital Filters with Coefficients of Limited Word Length

ERNST AVENHAUS

**Abstract**

**The cost of a digital filter, if implemented as a special-purpose computer, depends heavily on the word length of the coefficients. Therefore, it should be reduced as much as possible. On the other hand, a small word length causes large coefficient deviations that impair the wanted performance of the digital filter. The necessary word length may be reduced by choosing a suitable structure for the filter. Two methods for doing this will be presented, taking selective filters as a model. A further reduction of the word length may be won by optimizing the rounded filter coefficients in the discrete parameter space. A description of a modified univariate search will be given.**

## I. Introduction

Let the transfer function of a digital filter with constant coefficients be described by

$$H(z) = \sum_{\mu=0}^{m} b_\mu z^\mu \Big/ \sum_{\nu=0}^{n} c_\nu z^\nu, \qquad c_n = 1 \tag{1}$$

with

$$z = e^{sT} = e^{(\sigma + j\omega)T}.$$

We consider the magnitude of this function on the unit circle in the $z$ plane

$$|H(\Omega)| = |H(z)|\big|_{z=e^{j\Omega}}. \tag{2}$$

A wanted magnitude function $|H_w(\Omega)|$ and the tolerated deviation $\delta(\Omega)$ may be prescribed. As an example, Fig. 1 shows a tolerance scheme of a bandpass described by

$$|H_w(\Omega)| = \begin{cases} 1 & \text{in the passband (PB)} \\ 0 & \text{in the stopband (SB)} \end{cases} \tag{3}$$

and

$$\delta(\Omega) = \begin{cases} \delta_P & \text{in PB} \\ \delta_S & \text{in SB.} \end{cases} \tag{4}$$

A transfer function $|H(\Omega)|$ satisfies the tolerance conditions, if

$$\big|\,|H_w(\Omega)| - |H(\Omega)|\,\big| \le \delta(\Omega), \qquad \forall \Omega \tag{4a}$$

or

$$\epsilon(\Omega) = \frac{1}{\delta(\Omega)} \big|\,|H_w(\Omega)| - |H(\Omega)|\,\big| \le 1, \qquad \forall \Omega \tag{4b}$$

holds. In our example the normalized error function $\epsilon(\Omega)$ becomes

$$\epsilon(\Omega) = \begin{cases} \dfrac{1}{\delta_P} \big|1 - |H_N(\Omega)|\big|, & \text{in PB} \\ \dfrac{1}{\delta_S} \cdot |H_N(\Omega)|, & \text{in SB,} \end{cases} \tag{5}$$

where

$$H_N(z) = K \cdot H(z) \tag{6}$$

is the normalized transfer function. $K$ can be arbitrarily chosen such that

$$\max \epsilon = \max_{\Omega} \epsilon(\Omega) \tag{7}$$

becomes a minimum. Since there are no constraints in the transition ranges between passband and stopbands, $\epsilon(\Omega)$ is not defined in these intervals. In the following max $\epsilon$, defined as described above, will be considered as a figure of merit for judging the performance of an actual transfer function $H_N(z)$. At first it will be used to state the problem at hand in more detail.

According to (4b) max $\epsilon$ has to be smaller than, or at most equal to, one for satisfaction of the prescribed tolerance scheme. By minimizing max $\epsilon$ with an appropriate approximation procedure we get a set of coefficients of a filter with equal ripple behavior in the passband as well as in the stopband. In practical cases the parameters of this filter turn out to have about eight decimal digits accuracy, and the frequency response $|H_N(\Omega)|$ uses only a minimum fraction of the tolerance scheme, i.e., the performance will be described by a value max $\epsilon_0 < 1$, indicating that there is a certain margin which can be used for a reduction of the word length. In the case of the bandpass filter of Fig. 1, a filter of at least the eighth degree is necessary, leading to max $\epsilon_0 = 0.526054$. By reducing the word length of the coefficients step-by-step, a minimum increment $Q = 2^{-9}$ has been found to be necessary, in this particular example, for satisfaction of the tolerance scheme and we get max $\epsilon = 0.87$. A larger increment does not lead to an acceptable solution as has been indicated in Fig. 1.

In a more general approach we now investigate the word length sufficient to satisfy the specifications using the three well-known canonical forms shown in Figs. 2–4. In the direct form (Fig. 2), the coefficients $b_\mu$, $c_\nu$ of (1) will be used. For the cascade form (Fig. 3), the transfer function has to be written in product form as

$$H(z) = K \cdot \prod_{\lambda=1}^{l} \frac{b_{2\lambda} z^2 + b_{1\lambda} z + b_{0\lambda}}{z^2 + c_{1\lambda} z + c_{0\lambda}} \tag{8}$$

with

Manuscript received November 18, 1971.
The author is with the Institut für Nachrichtentechnik, Erlangen, Germany.

Reprinted from *IEEE Trans. Audio Electroacoust.*, vol. AU-20, pp. 206–212, Aug. 1972.

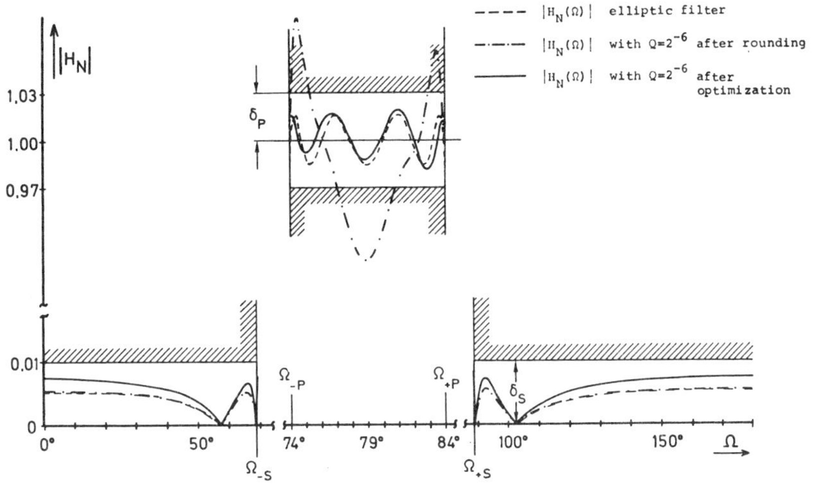


Fig. 1. The influence of rounding and optimization on the frequency response.

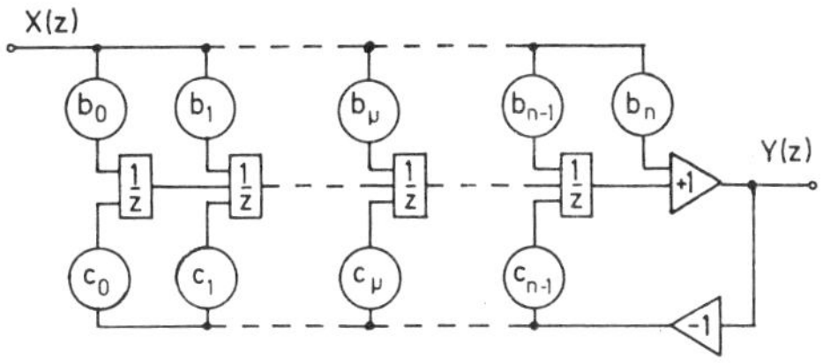


Fig. 2. Direct programming of a digital filter.

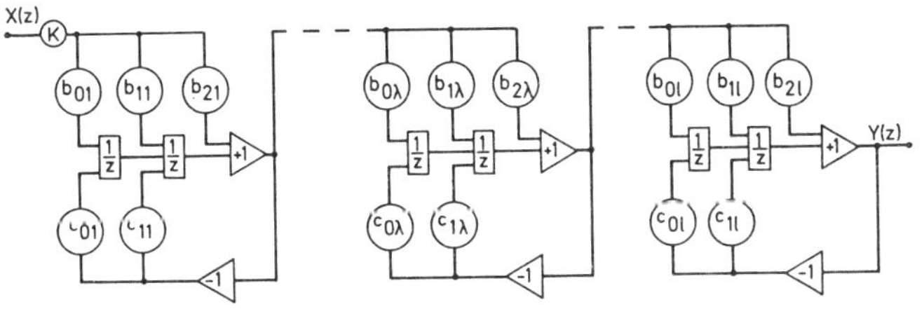


Fig. 3. Cascade programming of a digital filter.

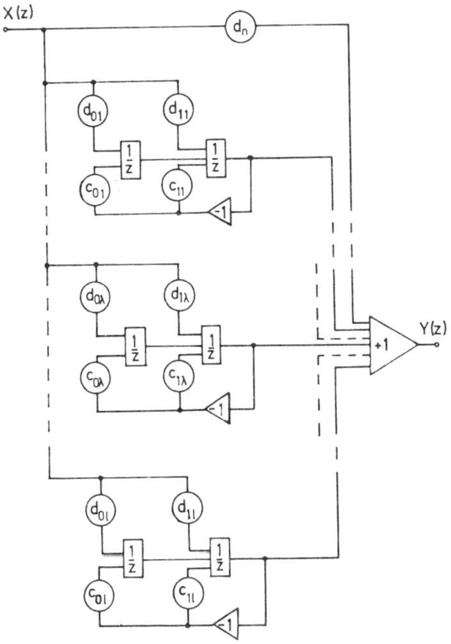


Fig. 4. Parallel programming of a digital filter.

$$l = \begin{cases} \dfrac{n}{2}, & \text{if } n \text{ is even} \\ \dfrac{n+1}{2}, & \text{if } n \text{ is odd.} \end{cases}$$

The coefficients of the parallel form (Fig. 4) will be won by the partial fraction expansion of (1) as

$$H(z) = d_0 + \sum_{\lambda=1}^{l} \frac{d_{1\lambda} z + d_{0\lambda}}{z^2 + c_{1\lambda} z + c_{0\lambda}}, \tag{9}$$

if we assume for simplicity all poles of $H(z)$ to be distinct.

The coefficients $a_\nu$ may be stored as fixed-point binary numbers as

$$a_\nu = \pm \sum_{i=i_M}^{i_L} \alpha_{\nu i} 2^i \tag{10}$$

where $\alpha_{\nu i}$ are 0 or 1, $\alpha_{\nu i_L}$ is the least significant bit, and $\alpha_{\nu i_M}$ is the most significant bit. Obviously, the range of coefficients is

$$-2 \cdot 2^{i_M} < a_\nu < 2 \cdot 2^{i_M}, \qquad \forall \nu \tag{11}$$

while the minimum increment, i.e., the quantization step size is defined as

$$Q = 2^{i_L}, \tag{12}$$

and the roundoff error is

$$|E| \leq \frac{Q}{2}. \tag{13}$$

Taking the one bit for the sign into account, the word length of the coefficients turns out to be

$$W = 2 + i_M - i_L. \tag{14}$$

## II. Valuation of the Sufficient Word Length for Different Filter Structures

If the deviation of the magnitude function holds the inequality relation

$$\Delta|H(\Omega)| = ||H(\Omega) + \Delta H(\Omega)| - |H(\Omega)|| \leq \delta(\Omega)(1 - \max \epsilon_0), \qquad \forall \Omega, \quad (15)$$

the tolerance conditions will be satisfied. According to (13) all coefficient errors are restricted by

$$|\Delta a_\nu| \leq \frac{Q}{2}, \qquad \forall \nu. \quad (16)$$

For a fixed frequency $\Omega$, let

$$\max_{\Delta a_\nu} \{ |H(\Omega) + \Delta H(\Omega, \Delta a_\nu)| \} \quad (17)$$

be the maximum magnitude of the transfer function if the coefficients are changed within the range limited by unequality (16) and let

$$\min_{\Delta a_\nu} \{ |H(\Omega) + \Delta H(\Omega, \Delta a_\nu)| \} \quad (18)$$

be the minimum magnitude at $\Omega$ in dependance of $\Delta a_\nu$. The deviation $\Delta|H(\Omega)|$ (15) is then bounded by

$$\Delta|H(\Omega)| \leq \max \{ (\max_{\Delta a_\nu} |H + \Delta H| - |H|), (|H| - \min_{\Delta a_\nu} |H + \Delta H|) \}. \quad (19)$$

Now let

$$H(\Omega) = \frac{N(\Omega)}{D(\Omega)}$$

be the transfer function, $|\Delta_N|$ be an upper bound of the deviation of the numerator, and $|\Delta_D|$ be an upper bound of the deviation of the denominator. Then the unequalities

$$\max_{\Delta a_\nu} |H(\Omega) + \Delta H(\Omega, \Delta a_\nu)| \leq \frac{|N(\Omega)| + |\Delta_N|}{|D(\Omega)| - |\Delta_D|} \quad (20a)$$

and

$$\min_{\Delta a_\nu} |H(\Omega) + \Delta H(\Omega, \Delta a_\nu)| \geq \frac{|N(\Omega)| - |\Delta_N|}{|D(\Omega)| + |\Delta_D|} \quad (20b)$$

are valid. With these results (19) becomes

$$\Delta|H(\Omega)| \leq \frac{|N(\Omega)| + |\Delta_N|}{|D(\Omega)| - |\Delta_D|} - |H(\Omega)|. \quad (21)$$

This can be verified easily, if $|D(\Omega)| > |\Delta_D|$, i.e., if the system stays stable after changing the coefficients. Since $|\Delta_D|$ and $|\Delta_N|$ depend on the quantization, with (15) and (21) we get an implicit expression for the sufficient value $Q_s$ for each particular frequency $\Omega$

$$\frac{|N(\Omega)| + |\Delta_N(Q_s(\Omega))|}{|D(\Omega)| - |\Delta_D(Q_s(\Omega))|} - |H(\Omega)| = \delta(\Omega)(1 - \max \epsilon_0). \quad (22)$$

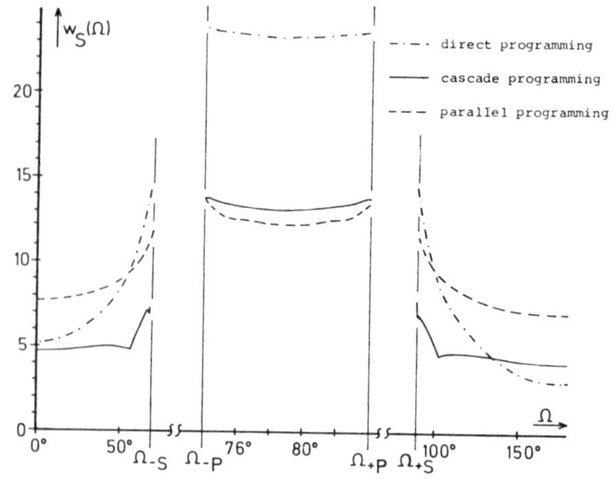


Fig. 5. Valuation of filter structures by sufficient word length functions.

The corresponding expressions for the cascade and parallel form are

$$\prod_{\lambda=1}^{l} \frac{|N_\lambda(\Omega)| + |\Delta_{N\lambda}(Q_s(\Omega))|}{|D_\lambda(\Omega)| - |\Delta_{D\lambda}(Q_s(\Omega))|} - \prod_{\lambda=1}^{l} \frac{|N_\lambda(\Omega)|}{|D_\lambda(\Omega)|} = \delta(\Omega)(1 - \max \epsilon_0) \quad (23)$$

and

$$\sum_{\lambda=1}^{l} \frac{|N_\lambda(\Omega)| + |\Delta_{N\lambda}(Q_s(\Omega))|}{|D_\lambda(\Omega)| - |\Delta_{D\lambda}(Q_s(\Omega))|} - \sum_{\lambda=1}^{l} \frac{|N_\lambda(\Omega)|}{|D_\lambda(\Omega)|} = \delta(\Omega)(1 - \max \epsilon_0). \quad (24)$$

Using $Q_s(\Omega)$, we define a word length function according to (14)

$$w_s(\Omega) = 2 + i_M - \mathrm{ld} Q_s(\Omega), \quad (25)$$

the maximum of which will be the sufficient word length for satisfying the tolerance scheme

$$W_s = \max_{\Omega} \{ w_s(\Omega) \}. \quad (26)$$

For example, in the direct form

$$|\Delta_N| = \tfrac{1}{2} M_N \cdot Q_s(\Omega)$$

and

$$|\Delta_D| = \tfrac{1}{2} M_D \cdot Q_s(\Omega) \quad (27)$$

are upper bounds of the magnitude deviation in numerator and denominator, where $M_N$ and $M_D$ are the numbers of the coefficients in the numerator and denominator, respectively. With (22) we get

$$Q_s(\Omega) = 2 \frac{\left| \sum_{\nu=0}^{n} c_\nu e^{j\nu\Omega} \right| \delta(\Omega)(1 - \max \epsilon_0)}{M_N + M_D |H(\Omega)| + M_D \delta(\Omega)(1 - \max \epsilon_0)} \cdot \quad (28)$$

Fig. 5 shows $w_s(\Omega)$ for the three canonical forms, using the bandpass of the eighth degree as an example. We see that for the direct form the sufficient word length is 24 bits, while 14 bits will suffice for the cascade or parallel form.

### III. A Statistical Method for the Estimation of the Word Length

A better estimation of the word length may be won by handling the coefficient errors as random variables. This approach leads to an ensemble of filters with coefficients being statistically distributed in a properly chosen range. The variation range of the filter performance of this ensemble will be calculated [2], [3]. The influence of coefficient errors on the transmission property has been investigated in the frequency domain by Knowles and Olcayto [2]. For convenience the approach will be recalled briefly. After rounding, the coefficients $a_\nu^\circ$ lie in the range $\pm\frac{1}{2}Q$ around the original values $a_\nu$

$$a_\nu - \tfrac{1}{2}Q \le a_\nu^\circ \le a_\nu + \tfrac{1}{2}Q, \tag{29}$$

leading to a transfer function $H^\circ(\Omega)$. An ensemble of transfer functions $\tilde{H}(\Omega)$ with coefficients $\tilde{a}_\nu$ being equally distributed in the range

$$a_\nu - \frac{Q}{2} \le \tilde{a}_\nu = a_\nu + \Delta\tilde{a}_\nu < a_\nu + \frac{Q}{2} \tag{30}$$

obviously contains the transfer function $H^\circ(\Omega)$. So by investigating the ensemble $\tilde{H}(\Omega)$, we get a statement on $H^\circ(\Omega)$.

Knowles and Olcayto used the mean value

$$\sigma_\Omega^2 = \overline{\frac{1}{2\pi}\int_0^{2\pi} |\tilde{H}(\Omega) - H(\Omega)|^2 d\Omega} \tag{31}$$

as a measure, but this expression seems to be less appropriate for filters with high attenuation required in the stopband, as will be demonstrated later by an example. Here we use the mean value and the variance of the power transfer functions

$$\tilde{P}(\Omega) = \tilde{H}(\Omega)\cdot\tilde{H}^*(\Omega) \tag{32}$$

instead, where $\tilde{H}^*(\Omega)$ is the complex conjugate of $\tilde{H}(\Omega)$. $\tilde{P}(\Omega)$ can be expressed approximately by the power transfer function $P(\Omega)$ of the original filter and the sensitivity functions. It is

$$\begin{aligned}\tilde{P}(\Omega) &= P(\Omega) + \Delta\tilde{P}(\Omega)\\ &\approx P(\Omega) + \sum_{\nu=1}^{\eta_e} \frac{\partial P(\Omega)}{\partial a_\nu}\Delta\tilde{a}_\nu,\end{aligned} \tag{33}$$

where $\eta_e$ is the total number of erroneous coefficients. An elliptic filter satisfies the unequality

$$\begin{aligned}(|H_w(\Omega)| &- \delta(\Omega)\cdot\max\epsilon_0)^2\\ &\le P(\Omega) \le (|H_w(\Omega)| + \delta(\Omega)\cdot\max\epsilon_0)^2, \qquad \forall\Omega.\end{aligned} \tag{34}$$

A member of the ensemble $\tilde{H}(\Omega)$ satisfies the tolerance constraints if the unequality

$$\begin{aligned}(|H_w(\Omega)| - \delta(\Omega))^2 &\le \tilde{P}(\Omega) = P(\Omega) + \Delta\tilde{P}(\Omega)\\ &\le (|H_w(\Omega)| + \delta(\Omega))^2, \qquad \forall\Omega\end{aligned} \tag{35}$$

holds. This condition is accomplished, if the unequality

$$\begin{aligned}|\Delta\tilde{P}(\Omega)| \le |&-2|H_w(\Omega)|\cdot\delta(\Omega)(1-\max\epsilon_0)\\ &+\delta^2(\Omega)(1-\max{}^2\epsilon_0)|, \qquad \forall\Omega\end{aligned} \tag{36}$$

is valid.

The mean value of the random variable $\tilde{P}(\Omega)$ is defined as

$$\mu_p(\Omega) = \int_{-\infty}^{+\infty} \tilde{P}(\Omega)\cdot p(\tilde{P}(\Omega))\, d\tilde{P}(\Omega) \tag{37}$$

with $p(\tilde{P}(\Omega))$ being the probability density function of $\tilde{P}(\Omega)$ at the fixed frequency $\Omega$. In correspondence, the variance of the power transfer functions is

$$\sigma_p^2(\Omega) = \int_{-\infty}^{+\infty} (\tilde{P}(\Omega) - \mu_p(\Omega))^2\cdot p(\tilde{P}(\Omega))\, d\tilde{P}(\Omega). \tag{38}$$

Equations (37) and (38) may be evaluated by means of the probability density functions

$$p_\nu(\Delta\tilde{a}_\nu) = \begin{cases}\dfrac{1}{Q}, & \text{if } -\dfrac{Q}{2} \le \Delta\tilde{a}_\nu \le \dfrac{Q}{2}\\ 0, & \text{otherwise}\end{cases} \tag{39}$$

of the random variables $\Delta\tilde{a}_\nu$, if the functions

$$\begin{aligned}\tilde{P}_\nu(\Omega) = \tilde{P}(\Omega, \Delta a_1, \cdots, \Delta a_\nu, \cdots,\\ \Delta a_\kappa, \cdots, \Delta a_{\eta_e})\Big|_{\substack{\Delta a_\kappa = \text{const}\\ \kappa\ne\nu}}\end{aligned} \tag{40}$$

are monotonous. This assumption holds, if $\tilde{P}(\Omega)$ is estimated by (33). $\mu_p(\Omega)$ and $\sigma_p(\Omega)$ can be shown to be [4]

$$\mu_p(\Omega) \approx P(\Omega)$$

and

$$\sigma_p^2(\Omega) \approx \frac{Q^2}{12}\sum_{\nu=1}^{\eta_e}\left(\frac{\partial P(\Omega)}{\partial a_\nu}\right)^2. \tag{41}$$

According to the finite value theorem [5], $p(\tilde{P}(\Omega))$ will approach a normal probability density function if $\eta_e$ increases, since $\Delta\tilde{P}(\Omega)$ is approximately a sum of equally distributed random variables. By this we can state that 95 percent of the power transfer functions $\tilde{P}(\Omega)$ vary less than $2\cdot\sigma_p(\Omega, Q)$ at $\Omega$.

$$|\Delta\tilde{P}(\Omega)| \le 2\sigma_p(\Omega, Q). \tag{42}$$

According to (36) and (42) we get a condition, which if satisfied, guarantees that 95 percent of the filters with the transfer functions $\tilde{H}(\Omega)$ meet the specifications at $\Omega$

$$\begin{aligned}2\sigma_p(\Omega, Q) \le |&-2|H_w(\Omega)|\cdot\delta(\Omega)(1-\max\epsilon_0)\\ &+\delta^2(\Omega)(1-\max{}^2\epsilon_0)|.\end{aligned} \tag{43}$$

For estimating the percentage of filters satisfying the tolerance scheme for all $\Omega$ it would be necessary to evaluate the conditional probability of hurting the tolerance limits as a function of $\Omega$. But the function $Q_p(\Omega)$, already won by (41) and (43), will allow us to judge a filter structure.

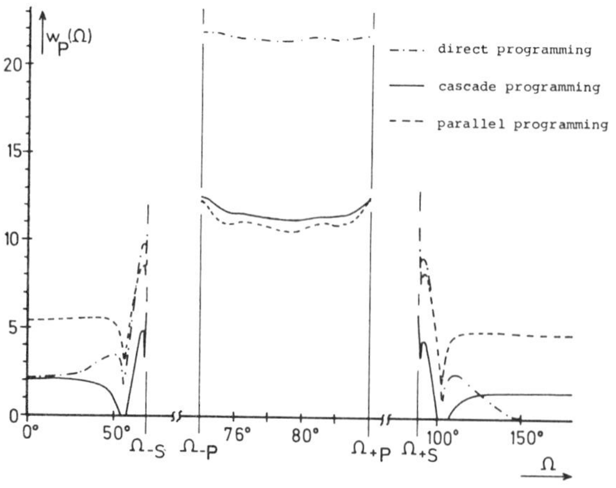


Fig. 6. Valuation of filter structures by statistical word length functions.

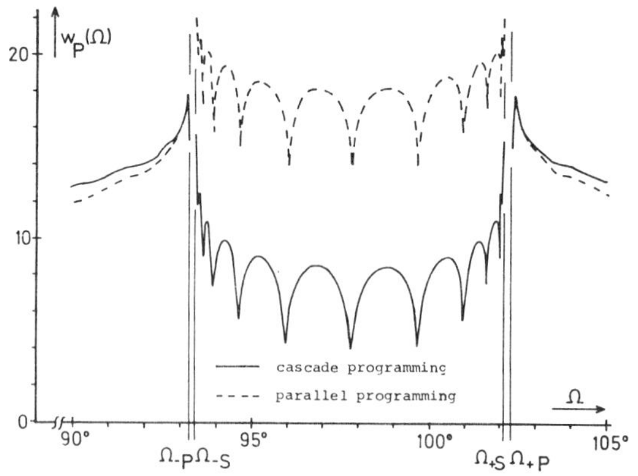


Fig. 7. Statistical word length function of a band rejection filter of the 22nd degree.

$$Q_p(\Omega) = \sqrt{3}\,|\,-2\,|\,H_w\,|\cdot\delta\cdot(1-\max\epsilon_0) + \delta^2(1-\max^2\epsilon_0)\,|\left(\sqrt{\sum_{\nu=1}^{\eta_e}\left(\frac{\partial P}{\partial a_\nu}\right)^2}\right)^{-1}. \quad (44)$$

With (14) we get the word length function

$$w_p(\Omega) = 2 + i_M - \mathrm{ld}\, Q_p(\Omega)$$

to estimate the word length

$$W_p = \max_{\Omega} w_p(\Omega). \quad (45)$$

Using the bandpass example, Fig. 6 shows the word length function $w_p(\Omega)$ of three canonical forms. According to this figure the word length estimations are reduced by about 1.5 bits in comparison to Fig. 5.

Fig. 7 shows the word length function $w_p(\Omega)$ of a band rejection filter of the 22nd degree first used by Kaiser [6] as an example. The tolerance scheme is given by

$$\Omega_{-P} = 93.20° \quad \Omega_{-s} = 93.43° \quad \Omega_{+s} = 102.10°$$
$$\Omega_{+P} = 102.33° \quad \delta_s = 2\cdot 10^{-4} \quad \delta_P = 0.037.$$

In the cascade structure the word length is determined by the passband behavior, while in the parallel structure the stopband requirements determine the word length. The difference is about four bits. In contrast to Knowles' and Olcayto's result, the cascade form should be favorable to parallel programming in this case.

## IV. Direct Calculation of a Sufficient Word Length

For particular examples $|H(\Omega)|$, and according to (7) and (5), max $\epsilon$ can be calculated using a set of coefficients of certain word length $W$. In order to determine max $\epsilon$, the extremal values of $|H(\Omega)|$ have to be calculated, and in addition the transfer function must be determined at the cutoff frequencies. Of course for each new set of coefficients the extremal values will be located at different frequencies. By checking whether max $\epsilon_w$ satisfies (4b) a sufficient word length can be calculated most easily. We introduce the function

$$d(W) = \mathrm{ld}\,\frac{\max\epsilon_W - \max\epsilon_0}{1-\max\epsilon_0}. \quad (46)$$

Using $d(W)$, the condition (4b) will be transformed into

$$d(W) \leq 0. \quad (47)$$

$d(W)$ has been calculated for the bandpass of the eighth degree, again using the three canonical structures (Fig. 8). It turns out that for the direct form a word length of 22 bits, for the two other forms 11 bits, is sufficient. Notice that $d(W)$ does not decrease monotonously. The results found by the two other methods are shown for comparison. The merits of the estimate are to be seen.

## V. Optimization in Discrete Parameter Space

### A. The Optimization Problem

If a magnitude function with rounded coefficients does not satisfy the tolerance scheme, we can try to vary the coefficients by an optimization procedure such that the figure of merit max $\epsilon$ becomes less than one. In doing so, only discrete coefficient values can be accepted.

$$a_\nu = l_\nu\cdot Q \quad (48)$$

where $l_\nu$ is an integer.

Generally speaking the normalized transfer function $H_N(\Omega)$ is determined by a set of

$$M_F = m + n \quad (49)$$

independent coefficients. Since in the case of selective filters the zeros are restricted to lie on the unit circle, the number of free parameters reduces to

$$M_F = \begin{cases} \dfrac{m}{2} + n, & m \text{ even} \\[2ex] \dfrac{m-1}{2} + n, & m \text{ odd.} \end{cases} \quad (50)$$

First, all optimization procedures based on evaluating the gradient can not be applied in this case. An approximate solution by integer linear programming is not possible

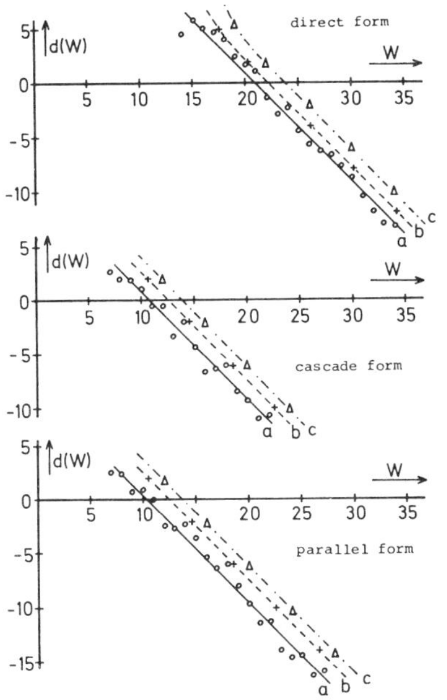


Fig. 8. Valuation of different structures. Line *a*: direct calculation *d*(*W*); line *b*: statistical method; line *c*: upper bound method.

either. The procedure of Gomory leads to a degenerate dual problem [7]. Three optimization procedures that have been investigated more thoroughly are described in the following section.

### B. Global Search

The global minimum of max $\epsilon$ for coefficients of prescribed word length will be found in any case, if all points of the grid will be controlled in a sufficiently large range around the best continuous parameter set **a**. But by choosing an $M_F$-dimensional cube with length $Q$ of one side, we have to investigate $2^{M_F}$-parameter sets. The bandpass of the eighth degree with $M_F=12$ free parameters, e.g., requires 4096 times the calculation of max $\epsilon$, which needs 45 min computing time on the CD 3300. Nevertheless, the optimum has not been found, since the chosen cube was too small. Therefore, this procedure is not appropriate due to the prohibitively long computing time.

### C. Univariate Search

We choose an initial parameter set $a_i'$ in the discrete parameter space, which may be found by rounding the coefficients of the elliptic filter, and compute max $\epsilon_i$ (Fig. 1). The procedure is characterized by always varying only one parameter $a_\nu'$ in both directions by step-size $\Delta a_\nu$, while the other parameters remain fixed. If we find a better parameter set by this way, it will become the new initial parameter set. The procedure is finished if all parameters have been varied without finding a better figure of merit. If we choose $\Delta a_\nu$ equal to the width of quantization $Q$, all parameter sets being investigated are points of the grid.

Fig. 9 shows max $\epsilon$ of the bandpass example after optimization at four different word lengths. In comparison with systems of rounded coefficients we observe an improvement. Nevertheless, we cannot find the global optimum by the aid of the univariate search.

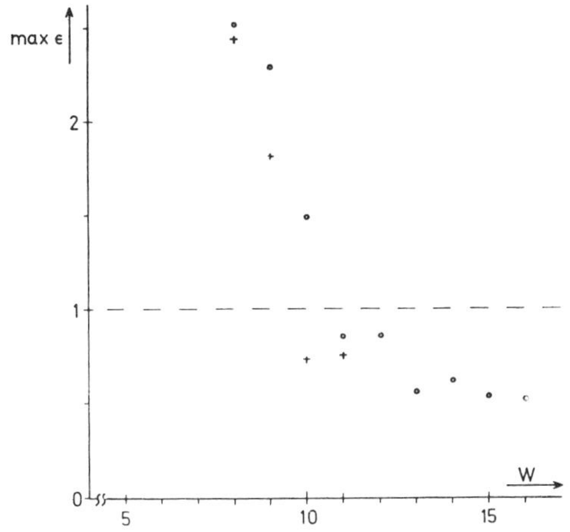


Fig. 9. The figure of merit as a function of the word length. 0: after rounding; +: after univariate search.

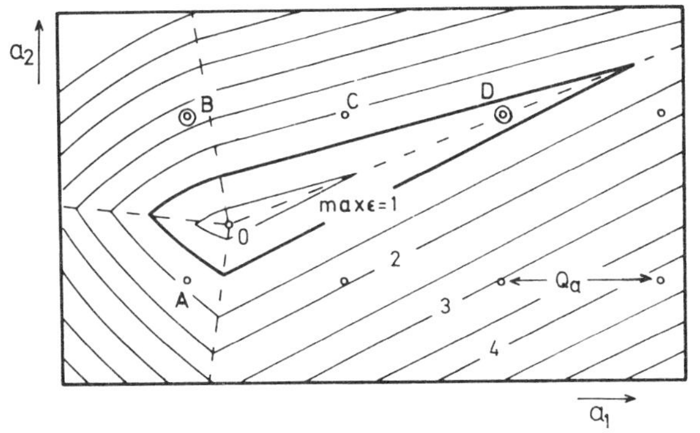


Fig. 10. The optimization problem in discrete parameter space.

### D. A Modified Univariate Search

A difficulty one has to expect in the univariate search will be explained in a two-dimensional parameter space by Fig. 10. The closed lines connect spots with max $\epsilon$ being constant. Moreover, the points of the two-dimensional grid have been drawn in the neighborhood of the optimal continuous parameter set **a** (point 0). The central range contains all solutions of the continuous parameter space with max $\epsilon \leq 1$.

Rounding the coefficients of **a**, point $A$ will become the initial solution of the univariate search. The global optimum (point $D$) will not be found in discrete parameter space, since variation of only one coefficient by $\Delta a_\nu = Q$ will not lead from point $A$ to a better figure of merit. Neither does the global search, if we choose a square with edge length $Q$.

On the other hand, if we reduce the word length by one bit, $B$ may be the initial point, and both procedures will lead to the global optimum. This explains the nonmonotonous dependence on word length and max $\epsilon$ mentioned in Section IV.

An optimization procedure suitable in discrete parameter space, which is based on the univariate search, has to overstep points in the grid with unfavorable max $\epsilon$, say from point $A$ via $B$ and $C$ to point $D$.

The univariate search has been modified such that in addition to the initial solution, some ten parameter sets with the

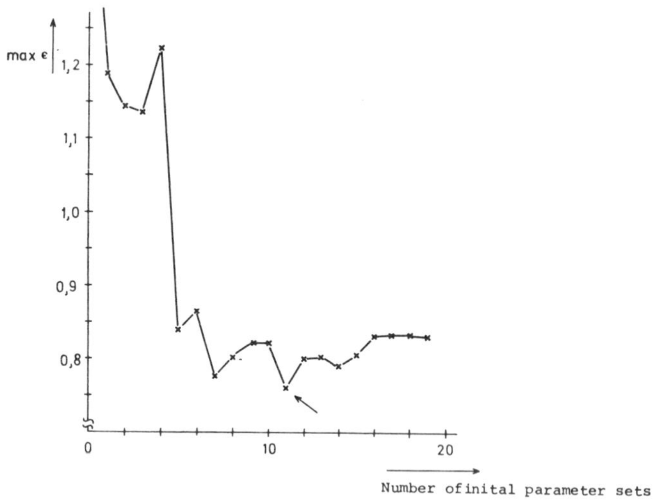


Fig. 11. Typical course of the figure of merit as a function of subsequent initial parameter sets.

next unfavorable max $\epsilon$ have been stored, not having been initial solutions so far. Now if no parameter variation of the initial solution will cause an improvement, the procedure will not stop; the parameter set with the next unfavorable figure of merit will become the new initial solution. By this way we can overstep points with higher levels. But since the stop criterion was lost, we have to fix a maximum number of initial points, say 20, which stops the procedure. Fig. 11 shows a typical course of max $\epsilon$ as a function of subsequent initial points.

It has been found that the procedure becomes much more efficient by combining it with the global search. We generate a matrix $S$ the elements of which are

$$S_{K\nu} = \left.\frac{\Delta\epsilon(\Omega_K)}{\Delta a_\nu}\right|_{\Delta a_\nu = Q}. \tag{51}$$

Here the $\Omega_K$ are the cutoff frequencies or the points, where $|H(\Omega)|$ has its extremal values. The $S_{K\nu}$ will be calculated during the univariate search anyway.

Using $S$ we arrange the parameters $a_\nu$ according to their influence on max $\epsilon$. Following this sequence we choose step-by-step four subsequent parameters and span a four-dimensional subspace for the global search with 2 $Q$ being the length of one side of the cube. To save computing time the matrix $S$ will be used to find the best max $\epsilon'$ in the subspace by

$$\boldsymbol{\epsilon}'(\Omega_\kappa) = \boldsymbol{\epsilon}(\Omega_\kappa) + \boldsymbol{S}\Delta a_\nu(\Omega_\kappa). \tag{52}$$

The result will be checked afterwards by computing the error function exactly.

Although it cannot be guaranteed that this method finds the global optimum, it turns out to be a great improvement if compared to the univariate search. It has been used to optimize the coefficients of ten different bandpass filters, having the same values $(\Omega_P - \Omega_{-P})$ and $(\Omega_S - \Omega_{-S})$ as the example in Fig. 1, but with different values $\Omega_P$. Fig. 12 indicates the results for rounding to a step $Q = 2^{-7}$ and for optimization by the univariate search with and without modification. By the new procedure, a satisfying transfer

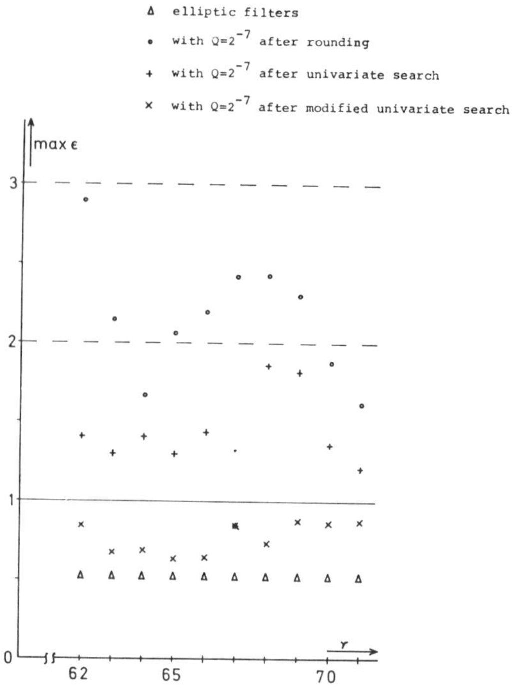


Fig. 12. Figures of merit of different bandpass filters.

function has been found in all examples, while the univariate search finds it only in one case.

Finally, Fig. 1 indicates how much can be won by this optimization technique. While $Q = 2^{-9}(W = 11)$ is needed, if the coefficients are rounded, only $Q = 2^{-6}(W = 8)$, i.e., three bits less are necessary, if the coefficients are optimized. The optimization of the bandpass filter takes about 10 min computing time.


## Acknowledgment

The author wishes to thank Prof. W. Schüßler for stimulating this investigation, for numerous valuable suggestions, and for his help with the preparation of the manuscript. Calculations were made by means of the computer CD 3300 of the computation center of the Erlangen-Nürnberg University.


## References


[1] W. Schüßler, "Zur allgemeinen Theorie der Verzweigungsnetzwerke," *Arch. Elek. Übertragung*, vol. 22, pp. 361–367, 1968.

[2] J. B. Knowles and E. M. Olcayto, "Coefficient accuracy and digital filter response," *IEEE Trans. Circuit Theory*, vol. CT-15, pp. 31–41, Mar. 1968.

[3] O. Mildenberger, "Die Genauigkeit der Lösung von Differenzengleichungen in Abhängigkeit von der Stellenzahl des verwendeten Rechners," *Arch. Elek. Übertragung*, vol. 25, pp. 90–100, Feb. 1971.

[4] E. Avenhaus, "Zum Entwurf digitaler Filter mit minimaler Speicherwortlänge für Koeffizienten und Zustandsgrößen," in *Ausgewählte Arbeiten über Nachrichtensysteme*, no. 13. Erlangen, Germany: Erlangen Univ., 1971.

[5] M. Fisz, *Wahrscheinlichkeitsrechnung und mathematische Statistik.* Berlin: VEB Deutscher Verlag der Wissenschaften, 1966, p. 181.

[6] R. M. Golden and J. F. Kaiser, "Design of wideband sampled-data filters," *Bell Syst. Tech. J.*, vol. 43, pp. 1533–1546, July 1964.

[7] S. J. Gass, *Linear Programming.* New York: McGraw-Hill, 1964 pp. 154–166.

# Analysis of Quantization Errors in the Direct Form for Finite Impulse Response Digital Filters

DAVID S. K. CHAN and LAWRENCE R. RABINER

*Abstract*—An analysis of the three possible types of quantization effects in the direct form realization of finite impulse response (FIR) digital filters is presented. These quantization effects include roundoff noise, A-D noise, and filter frequency response errors due to coefficient quantization. Since the analysis of roundoff noise and A-D noise for the direct form is straightforward, this paper concentrates on an analysis of the effects of quantized coefficients on the resulting filter frequency response. Based on this analysis, statistical bounds on the error incurred in the frequency response of a filter due to coefficient quantization are developed and verified by extensive experimental data. Using these bounds, a procedure for applying known techniques for FIR filter design to the design of filters with finite word length coefficients is presented. On the whole, the direct form is shown to be a very attractive structure for realizing FIR filters.

Manuscript received January 15, 1973; revised February 9, 1973.
The authors are with Bell Laboratories, Murray Hill, N.J. 07974.

## I. Introduction

Important developments in research in recent years have made it possible to readily design finite impulse response (FIR) digital filters with arbitrary frequency or time response characteristics [1]-[6]. However, little is as yet known concerning practical aspects in the implementation of these filters. In particular, although many structures [7], [8] have been proposed for realizing FIR filters, differences in the effects of quantization on these different structures are still not

Reprinted from *IEEE Trans. Audio Electroacoust.*, vol. AU-21, pp. 354–366, Aug. 1973.

well understood. In two recent papers [9], [10] the problem of roundoff noise in cascade realizations of FIR filters has been treated in depth. The subject of this paper is the implementation of FIR filters in the direct form.

An analysis of all possible types of quantization effects in the direct form is presented, together with a procedure for obtaining filters whose coefficients are represented by a given number of bits and which satisfy given specifications on their frequency response. Although the method is not optimum, it is simple to apply and yields useful results.

The direct form is usually avoided in the implementation of infinite impulse response (IIR) digital filters mainly because Kaiser [2] has shown that the sensitivity of filter response to the accuracy of representation of the denominator coefficients in the IIR direct form increases very rapidly with increases in filter order compared to either the cascade or the parallel form. However, in this paper it is shown that the same is not true for FIR digital filters. In fact, the direct form is shown to be a very attractive structure for the realization of FIR filters.

## II. Quantization Effects in the FIR Direct Form

There are three basic ways in which quantization affects the performance of an FIR digital filter.

1) Quantization of the results of arithmetic operations within the filter causes errors in the filter output, referred to as roundoff noise.

2) Quantization of input samples to the filter causes inaccuracies known as A-D noise.

3) Quantization of the filter coefficients alters the frequency response of the filter.

Overflows can also occur within the filter. However, if proper scaling procedures are used, such overflows can be eliminated. In addition, limit cycles can occur in recursive realizations of FIR filters (such as the frequency-sampling structure [7]). However, limit cycles cannot occur in the direct form, which is a nonrecursive structure. In the following, each of the three types of quantization effects listed above is analyzed for the direct form.

### A. Roundoff Noise

To analyze roundoff noise, the most important case of fixed-point arithmetic with rounding is assumed. Similar analyses follow for all other cases. Furthermore, the usual model used for the analysis of roundoff errors [9] is employed, viz., to each rounding point in the filter is associated a zero-mean white noise source of variance $Q^2/12$ ($Q$ = quantization step size) whose samples are uniformly distributed random variables on the interval $(-Q/2, Q/2)$, and all noise sources are assumed to be uncorrelated with each other and with the input signal. Using this model, the mean and variance of roundoff noise at the output of a filter can be easily computed.

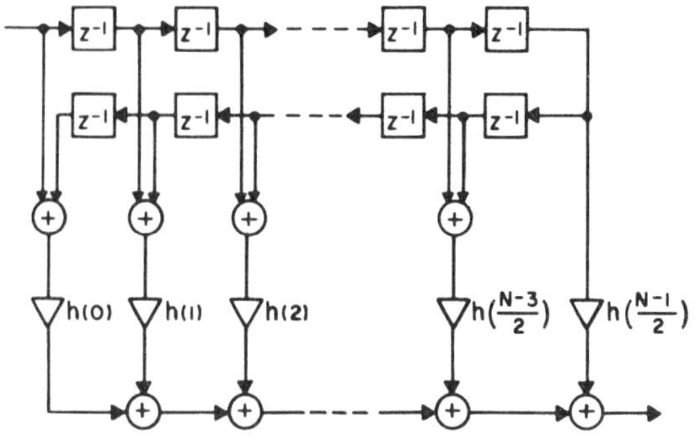

Fig. 1. Block diagram of linear phase direct form FIR filter.

The transfer function for an FIR filter can be written in the form

$$H(z) = \sum_{n=0}^{N-1} h(n) z^{-n} \qquad (1)$$

where $\{h(n)\}$ is the impulse response of the filter. The direct form is simply defined to be a straightforward implementation of (1). If the filter has linear phase,[1] then $\{h(n)\}$ satisfies [9]

$$h(n) = h(N-1-n), \qquad 0 \leq n \leq N-1. \qquad (2)$$

In this case, a slightly different version of the direct form, requiring fewer multipliers, can be derived. For simplicity only, the case of odd $N$ is considered. By (1) and (2),

$$\begin{aligned} H(z) &= \sum_{n=0}^{(N-3)/2} h(n) z^{-n} + \sum_{n=(N+1)/2}^{N-1} h(n) z^{-n} + h\left(\frac{N-1}{2}\right) z^{-(N-1)/2} \\ &= \sum_{n=0}^{(N-3)/2} h(n)\left[z^{-n} + z^{-(N-1-n)}\right] + h\left(\frac{N-1}{2}\right) z^{-(N-1)/2}. \end{aligned} \qquad (3)$$

Equation (3) is the defining equation for the linear phase direct form. It is seen that, given the filter order, the direct form for a linear phase filter requires approximately half as many multipliers as that required for an arbitrary phase filter. A block diagram of the linear phase direct form is shown in Fig. 1.

The statistics of the roundoff noise at the output of a direct form FIR filter depend, as expected, on the location of points in the filter where rounding is performed. Two possibilities are considered. First of all,

[1] The results presented in this paper can be extended in a straightforward manner to the case where the filter impulse response is antisymmetric, i.e., $h(n) = -h(N-1-n)$, $0 \leq n \leq N-1$. For convenience we shall neglect this case here.

if all multiplication products are represented exactly and rounding is performed only after they are summed, i.e., at the filter output, then only one noise source is present in the filter, and it superimposes noise directly onto the output signal.[2] Thus, independent of filter order and type of direct form (i.e., linear phase or arbitrary phase), the output roundoff noise is uniformly distributed between $-Q/2$ and $Q/2$, with zero mean and variance $Q^2/12$, where $Q$ is the quantization step size.

On the other hand, if all multiplication products are rounded before they are summed, then the arbitrary phase direct form and the linear phase direct form yield different results for a given $N$. Specifically, denoting by $\{e_i(n)\}$ the noise sequence produced by the $i$th noise source, and by $\{E(n)\}$ the noise sequence at the filter output, for the arbitrary phase direct form

$$E(n) = \sum_{n=0}^{N-1} e_i(n), \tag{4}$$

but for the linear phase direct form

$$E(n) = \sum_{n=0}^{(N-1)/2} e_i(n). \tag{5}$$

The mean of the output noise is zero in either case, but the variance for the arbitrary phase direct form is

$$\sigma_A^2 = N\frac{Q^2}{12}, \tag{6}$$

whereas for the linear phase direct form it is

$$\sigma_L^2 = \left(\frac{N+1}{2}\right)\frac{Q^2}{12}, \tag{7}$$

Thus for all the possible variations of the FIR direct form mentioned, the output roundoff noise variance always satisfies

$$\sigma_{RO}^2 \geqslant \frac{Q^2}{12}. \tag{8}$$

Note at this point that if the noise sources were statistically independent rather than merely uncorrelated, then by the central limit theorem of probability theory, in both (4) and (5) $E(n)$ would be essentially Gaussian distributed for all $n$ and sufficiently large $N$ since it would be a sum of independent, identically distributed random variables. In fact, the convergence as $N$ is made large would be very rapid since the $e_i(n)$'s are uniformly distributed. Thus, for all values of $N$ of practical interest, the output noise would be essentially Gaussian. Even though an independence assumption on the noise sources may not be valid, the output noise distribution can still be expected to resemble a Gaussian distribution.

[2] We are assuming throughout this discussion that all the filter coefficients are nontrivial, e.g., an impulse response consisting of $\pm 1$'s would have no roundoff noise. Although such cases do occur in practice, they are trivial to analyze, and hence will not be considered here.

### B. A-D Noise

Next, A-D noise is considered. Again, the quantization process is modeled as a white noise source whose samples are uniformly distributed on $(-Q/2, Q/2)$. Since this quantization noise is added to the input signal, its effect on the output signal of a filter is independent of the type of structure used to realize the filter. In particular, if $\{\epsilon(n)\}$ denotes the A-D noise sequence on input to an FIR filter as defined by (1), and $\{\mathcal{E}(n)\}$ denotes the output noise sequence, then

$$\mathcal{E}(n) = \sum_{k=0}^{N-1} h(k)\,\epsilon(n-k). \tag{9}$$

Thus the output A-D noise has zero mean and variance given by

$$\sigma_{AD}^2 = \frac{Q^2}{12}\sum_{n=0}^{N-1} h^2(n), \tag{10}$$

or using Parseval's theorem for discrete signals,

$$\sigma_{AD}^2 = \frac{Q^2}{12}\cdot\frac{1}{2\pi}\int_0^{2\pi} |H(e^{j\omega})|^2\, d\omega. \tag{11}$$

It is shown next that when the gain of a filter is determined by sum scaling, peak scaling, or $L_p$-norm scaling where $p \geqslant 2$, viz., the practically important scaling methods [9], then

$$\sigma_{AD}^2 \leqslant \frac{Q^2}{12}. \tag{12}$$

Consider first sum scaling, which requires

$$\sum_{n=0}^{N-1} |h(n)| = 1 \tag{13}$$

where $\{h(n)\}$ is the scaled filter impulse response. Clearly, (13) implies $|h(n)| \leqslant 1$ for all $n$; hence $h^2(n) \leqslant |h(n)|$ for all $n$; thus

$$\sum_{n=0}^{N-1} h^2(n) \leqslant \sum_{n=0}^{N-1} |h(n)| = 1. \tag{14}$$

Next (see [9]), for any $p \geqslant 2$,

$$\left[\frac{1}{2\pi}\int_0^{2\pi} |H(e^{j\omega})|^2\, d\omega\right]^{1/2}$$

$$\leqslant \left[\frac{1}{2\pi}\int_0^{2\pi} |H(e^{j\omega})|^p\, d\omega\right]^{1/p} \leqslant \max_{\omega} |H(e^{j\omega})|. \tag{15}$$

Hence if either $[1/2\pi \int_0^{2\pi} |H(e^{j\omega})|^p\, d\omega]^{1/p} = 1$, which defines $L_p$-norm scaling, for some $p \geqslant 2$, or

$\max |H(e^{j\omega})| = 1$, which defines peak scaling, then

$$\frac{1}{2\pi}\int_0^{2\pi} |H(e^{j\omega})|^2\, d\omega \leqslant 1. \tag{16}$$

Equations (10), (11), (14), and (16) show that $\sigma_{AD}^2 \leqslant Q^2/12$ for all these scaling methods; hence (12) is established.

Thus it is shown that for all practical filters realized in direct form, the A-D noise at the filter output has a variance that is smaller than or equal to that of the roundoff noise. In fact, for the often more economical scheme of rounding before summation, A-D noise is negligible compared to roundoff noise for filter orders greater than about 10.

It is instructive to point out that the output A-D noise is also essentially Gaussian for sufficiently large $N$ if different samples of the quantization error sequence were statistically independent of each other because, although the random variables that make up the sum in (9) are not identically distributed, they are all uniformly distributed with zero probability density outside a finite interval; hence they satisfy the Lindeberg condition [11] of the central limit theorem.

### C. Filter Response Errors

The effect of coefficient quantization on the frequency response of an FIR filter in direct form is now analyzed. The linear phase direct form is considered first. Again, $N$ is assumed to be odd, and rounding is used as the quantization process. From (3) the frequency response of a linear phase FIR filter is given by

$$\begin{aligned} H(e^{j\omega}) &= \sum_{n=0}^{(N-3)/2} h(n)\left[e^{-j\omega n} + e^{-j\omega(N-1-n)}\right] + h\left(\frac{N-1}{2}\right) e^{-j\omega(N-1)/2} \\ &= \left[\sum_{n=0}^{(N-3)/2} 2h(n)\cos\left[\left(\frac{N-1}{2} - n\right)\omega\right] + h\left(\frac{N-1}{2}\right)\right] e^{-j\omega(N-1)/2}. \end{aligned} \tag{17}$$

The factor $e^{-j\omega(N-1)/2}$ in (17) simply represents a pure delay of an integer number of samples that is unaffected by quantization of the coefficients $\{h(n)\}$. Hence this factor need not be considered, and only the change in the real function $\overline{H}(e^{j\omega}) = H(e^{j\omega}) \times e^{j\omega(N-1)/2}$ due to coefficient quantization is studied.

Following the approach of Knowles and Olcayto [12], let $\{h^*(n)\}$ be the sequence that results when $\{h(n)\}$ is rounded to a quantization step size of $Q$. Then $h^*(n) = h(n) + e(n)$ and $h^*(N-1-n) = h^*(n)$ for $0 \leqslant n \leqslant (N-1)/2$, where $e(n)$ for each $n$ is a number that satisfies $|e(n)| \leqslant Q/2$. Furthermore, let $H^*(z)$ be the $z$-transform of $\{h^*(n)\}$ and let $\overline{H}^*(e^{j\omega}) = H^*(e^{j\omega})\, e^{j\omega(N-1)/2}$. Finally, define an error function by

$$E_L(e^{j\omega}) = \overline{H}^*(e^{j\omega}) - \overline{H}(e^{j\omega}). \tag{18}$$

Thus

$$E_L(e^{j\omega}) = \sum_{n=0}^{(N-3)/2} 2e(n)\cos\left[\left(\frac{N-1}{2} - n\right)\omega\right] + e\left(\frac{N-1}{2}\right). \tag{19}$$

From (19) it is seen that $E_L(e^{j\omega})$ is simply the frequency response of a linear phase filter that has $\{e(n)\}$ as the first half of its impulse response (the other half by symmetry). The function $E_L(e^{j\omega})$ has the physical significance that a filter which has quantized coefficients can be represented as the parallel connection of the "infinite-precision" version of that filter with a filter whose frequency response is $E_L(e^{j\omega})\, e^{-j\omega(N-1)/2}$.

Since $|e(n)| \leqslant Q/2$, a bound on $E_L(e^{j\omega})$ that is independent of the $e(n)$'s can be derived as follows:

$$|E_L(e^{j\omega})| \leqslant \sum_{n=0}^{(N-3)/2} 2|e(n)| \left|\cos\left[\left(\frac{N-1}{2} - n\right)\omega\right]\right| + \left|e\left(\frac{N-1}{2}\right)\right| \leqslant \frac{Q}{2}\left[1 + 2\sum_{n=1}^{(N-1)/2} |\cos n\omega|\right]. \tag{20}$$

Or, independent of $\omega$,

$$|E_L(e^{j\omega})| \leqslant N\frac{Q}{2}. \tag{21}$$

Now let us consider the arbitrary phase direct form. From (1) the frequency response of an arbitrary phase FIR filter is given by

$$H(e^{j\omega}) = \sum_{n=0}^{N-1} h(n) e^{-j\omega n}. \tag{22}$$

If $H^*(e^{j\omega}) = \sum_{n=0}^{N-1} h^*(n)\, e^{-j\omega n}$ is the frequency response of the filter with quantized coefficients, where $h^*(n) = h(n) + e(n)$ and $|e(n)| \leqslant Q/2$ for $0 \leqslant n \leqslant N-1$, then define an error function by

$$\begin{aligned} E_A(e^{j\omega}) &= H^*(e^{j\omega}) - H(e^{j\omega}) \\ &= \sum_{n=0}^{N-1} e(n)\, e^{-j\omega n}. \end{aligned} \tag{23}$$

Again, the error function is the frequency response of an FIR filter that has impulse response $\{e(n)\}$. However, note that $E_A(e^{j\omega})$ is a complex function. The same bound as (21) also applies to $E_A(e^{j\omega})$ since

$$|E_A(e^{j\omega})| \leqslant \sum_{n=0}^{N-1} |e(n)||e^{-j\omega n}| \leqslant N\frac{Q}{2}. \tag{24}$$

Furthermore, the error in the magnitude of the frequency response due to coefficient quantization also satisfies the bound in (24) because of the well-known inequality for complex numbers:

$$\| H^*(e^{j\omega})| - |H(e^{j\omega})\| \leqslant |H^*(e^{j\omega}) - H(e^{j\omega})|. \quad (25)$$

Unfortunately, all of the bounds derived thus far are overly pessimistic, and thus are of little practical usefulness. In the next section, far more useful statistical bounds are derived.

### III. Statistical Analysis of Filter Response Errors

Because of the inherently hard-to-predict nature of quantization errors, a statistical analysis of the effect of coefficient quantization on filter response is appropriate, even though for a given filter the quantization process is performed only once, after which the filter response is exactly determined. The aim of such an analysis is to provide the filter designer with some means of predicting, without knowing the values of the coefficients of a filter, how much accuracy for the coefficients is required to obtain a desired filter response. In the following, such an analysis is presented for the case of coefficient rounding.

The statistical model [12] to be used is a very reasonable one that assumes that errors due to the quantization of different coefficients are statistically independent, and that each error is uniformly distributed between $-Q/2$ and $Q/2$, and thus has zero mean and variance $Q^2/12$, where $Q$ is the quantization step size.

The linear phase direct form is considered first. From (19) and the statistical model it is clear that for each $\omega$, the error in the frequency response due to coefficient rounding has zero mean and a variance given by

$$\overline{E_L(e^{j\omega})^2} = \sum_{n=0}^{(N-3)/2} 4\,\overline{e(n)^2} \cos^2\left[\left(\frac{N-1}{2} - n\right)\omega\right] + \overline{e\left(\frac{N-1}{2}\right)^2}$$
$$= \frac{Q^2}{12}\left[1 + 4\sum_{n=1}^{(N-1)/2} \cos^2 \omega n\right]. \quad (26)$$

Thus, defining a weighting function by

$$W_N(\omega) = \left[\frac{1}{2N-1}\left(1 + 4\sum_{n=1}^{(N-1)/2} \cos^2 \omega n\right)\right]^{1/2} \quad (27)$$

the standard deviation of the error is given by

$$\sigma_{EL}(\omega) = \overline{E_L(e^{j\omega})^2}^{1/2}$$
$$= \sqrt{\frac{2N-1}{3}} \cdot \frac{Q}{2} W_N(\omega). \quad (28)$$

$W_N(\omega)$ may be expressed in closed form by summing the series in (27), thus giving

$$W_N(\omega) = \left[\frac{1}{2N-1}\left(N + 2\cos\left(\frac{N+1}{2}\right)\omega\right)\frac{\sin\left(\frac{N-1}{2}\right)\omega}{\sin\omega}\right]^{1/2}, \quad (29)$$

or following some arithmetic manipulation,

$$W_N(\omega) = \left[\frac{1}{2} + \frac{1}{2N-1}\left(\frac{-1}{2} + \frac{\sin N\omega}{\sin\omega}\right)\right]^{1/2}. \quad (30)$$

Equation (30) shows $W_N(0) = W_N(\pi) = 1$, and $0 < W_N(\omega) \leqslant 1$ for all $N$. Thus

$$\sigma_{EL}(\omega) \leqslant \frac{Q}{2}\sqrt{\frac{2N-1}{3}}. \quad (31)$$

The behavior of $W_N(\omega)$ in the limit of large $N$ is readily seen from (30). In the range $0 < \omega < \pi$,

$$\lim_{N\to\infty} W_N(\omega) = \frac{1}{\sqrt{2}}, \quad (32)$$

whereas for $\omega = 0, \pm\pi, \pm 2\pi, \cdots$,

$$W_N(\omega) = 1, \quad \text{all } N. \quad (33)$$

The convergence of (32) is not uniform on $(0, \pi)$. However, on $[\epsilon, \pi - \epsilon]$ for any $\epsilon > 0$ the convergence is uniform. Fig. 2 shows $W_N(2\pi f)$ versus $f$ for $N = 7$ and $N = 67$. The significance of these plots and (32) is that they show $\sigma_{EL}(\omega)$ to be well described by its bound in (32), viz., $\sigma_{EL}(\omega)$ is given to better than a factor of 2 by its bound for all $\omega$.

From (19) it is seen that for any $\omega$, $E_L(e^{j\omega})$ is a sum of independent random variables whose probability density functions vanish outside some finite interval, and hence satisfy the Lindeberg condition [11] of the central limit theorem. Thus $E_L(e^{j\omega})$ is essentially Gaussian for sufficiently large $N$. Furthermore, the convergence to Gaussian distribution is expected to be quite rapid since the individual terms in the summation are all uniformly distributed random variables. Because of this tendency of the errors to be Gaussian, their mean and variance alone constitute an excellent description of their statistical behavior.

A bound similar to (31) is derived next for the arbitrary phase direct form. The complex error incurred in this case in the frequency response due to coefficient quantization is given in (23). Note that $E_A(e^{j\omega})$ is a two-dimensional random variable. Clearly, its mean is zero for all $\omega$. Furthermore, its probability density function is symmetric about the origin in the complex plane. Thus the second moment of the magnitude of $E_A(e^{j\omega})$ about the origin is a good measure of its deviation from the mean. Now

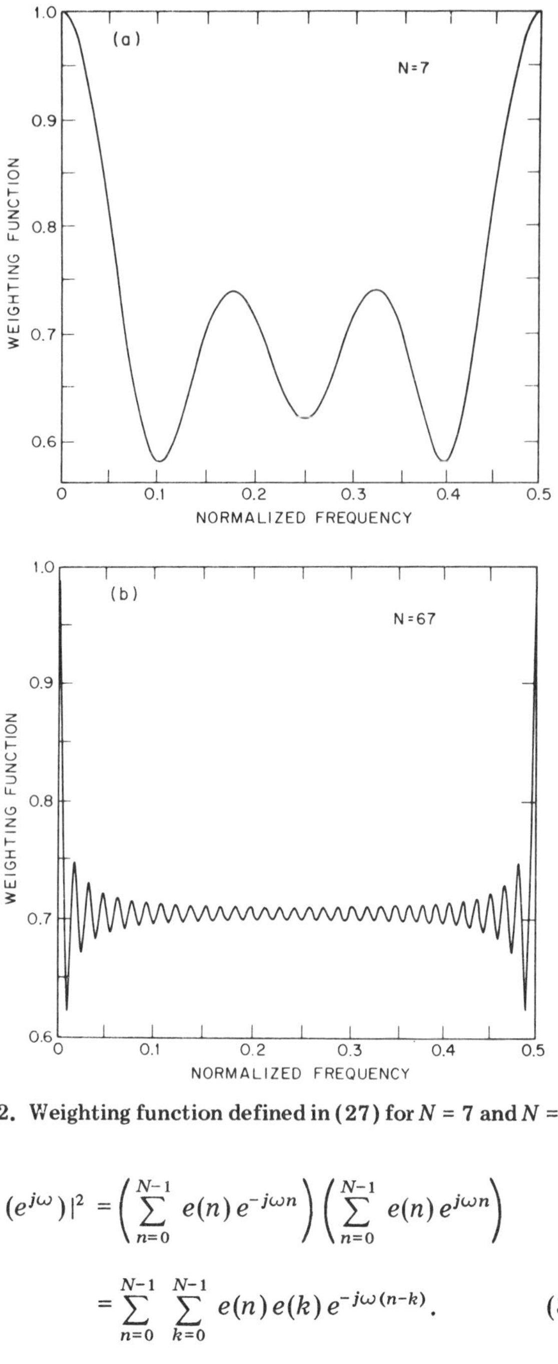


**Fig. 2.** Weighting function defined in (27) for $N = 7$ and $N = 67$.

$$|E_A(e^{j\omega})|^2 = \left(\sum_{n=0}^{N-1} e(n)\, e^{-j\omega n}\right)\left(\sum_{n=0}^{N-1} e(n)\, e^{j\omega n}\right)$$

$$= \sum_{n=0}^{N-1}\sum_{k=0}^{N-1} e(n)\, e(k)\, e^{-j\omega(n-k)}. \qquad (34)$$

Therefore

$$\overline{|E_A(e^{j\omega})|^2} = \sum_{n=0}^{N-1} \overline{e(n)^2} = N\frac{Q^2}{12} \qquad (35)$$

using the independence assumption. Thus a measure of the deviation of $|E_A(e^{j\omega}|)$ from zero can be defined as

$$\sigma_{EA} = \frac{Q}{2}\sqrt{\frac{N}{3}}. \qquad (36)$$

Since, as shown in (25), the error incurred in the magnitude of the frequency response is bounded by $|E_A(e^{j\omega})|$, $\sigma_{EA}$ is a useful bound on its expected amount of deviation from zero.

## IV. Implications of Analysis for Design

Considerable insight into the problem of implementing FIR filters in direct form can be gained from the analysis of the previous section if the results are phrased in terms of the usual filter design parameters. In this section, useful relationships between different parameters are derived for the linear phase direct form.

Most filter design problems of interest involve finding a filter with a frequency response that approximates some ideal response to within a specified amount of error. In general, the specified bound on the error is a function of frequency. However, for the important case of band-select filters, a single bound on the error is usually specified for each frequency band of interest, where bounds for different frequency bands may be different. For simplicity, the discussions of this paper are restricted to band-select filters. However, this is only a matter of convenience, and the ideas developed can be easily generalized.

Let $L(\omega)$ be some real, ideal band-select function that is desired to be approximated by the frequency response of a linear phase FIR filter. The usual design specifications consist of a set of disjoint frequency bands $\Omega_k \subset [0, \pi]$, $k = 1, \cdots, M$ and a set of error bounds $\delta_k > 0$, $k = 1, \cdots, M$ such that for each $k$, $L(\omega)$ is to be approximated to within an error of $\delta_k$ for all $\omega \in \Omega_k$. The frequency bands $\Omega_k$ are separated by transition bands where the filter frequency response is unconstrained. Quantizing the coefficients of a filter designed to meet these specifications can increase the approximation error for $\omega \in \cup_{k=1}^{M}, \Omega_k$ beyond their specified bounds. A statistical bound on the effect of such error increases is first developed.

Define, as in Section 11-C, $\overline{H}(e^{j\omega})$ and $\overline{H}^*(e^{j\omega})$ to be the frequency response, less the pure delay factor, of a linear phase FIR filter in direct form that has unquantized and quantized coefficients, respectively. Suppose $\overline{H}(e^{j\omega})$)has been designed to approximate $L(\omega)$ so that

$$\max_{\omega\in\Omega_k} |\overline{H}(e^{j\omega}) - L(\omega)| = \delta_k, \qquad k = 1, \cdots, M. \qquad (37)$$

Clearly, for all $\omega$

$$|\overline{H}^*(e^{j\omega}) - L(\omega)| \leqslant |\overline{H}^*(e^{j\omega}) - \overline{H}(e^{j\omega})| + |\overline{H}(e^{j\omega}) - L(\omega)|. \qquad (38)$$

Thus for all $\omega \in \Omega_k$,

$$|\overline{H}^*(e^{j\omega}) - L(\omega)| \leqslant \max_{\omega\in\Omega_k} |E_L(e^{j\omega})| + \delta_k \qquad (39)$$

where $E_L(e^{j\omega})$ is as defined in (18). With high probability $E_L(e^{j\omega})$ ought to be bounded by two to three times its standard deviation; thus for all $\omega$

$$|E_L(e^{j\omega})| \lessapprox 2\sigma_e \tag{40}$$

where $\lessapprox$ denotes "is with high probability less than or equal to" and $\sigma_e$ is defined as

$$\sigma_e = \max_{\omega} \sigma_{EL}(\omega) = \frac{Q}{2}\sqrt{\frac{2N-1}{3}} \tag{41}$$

with $\sigma_{EL}(\omega)$ defined in (28). Consequently,

$$|\bar{H}^*(e^{j\omega}) - L(\omega)| \lessapprox \underbrace{2^{-t}\sqrt{\frac{2N-1}{3}} + \delta_k}_{\epsilon_k}, \omega \in \Omega_k \tag{42}$$

where $t$ is the number of bits to which the coefficients of $\bar{H}^*(e^{j\omega})$ are rounded, so that $Q = 2^{-t}$ (sign bit not included).

Equation (42) says that, with high probability, $L(\omega)$ can be approximated on $\Omega_k$ by the frequency response of an $N$-point linear phase filter, with $t$-bit coefficients, to an error bounded by $\epsilon_k$ if the frequency response of the filter with infinite-precision coefficients $\bar{H}(e^{j\omega})$ can be designed so that for all $\omega \in \Omega_k$

$$|\bar{H}(e^{j\omega}) - L(\omega)| \leqslant \epsilon_k - 2^{-t}\sqrt{\frac{2N-1}{3}}. \tag{43}$$

Equations (42) and (43) are investigated further by rephrasing them in terms of in-band "rejection"[3] in decibels. Define the in-band rejection on $\Omega_k$ of the quantized and unquantized filter, respectively, as

$$DL_k^* = -20\log_{10}\left(\max_{\omega\in\Omega_k} |\bar{H}^*(e^{j\omega}) - L(\omega)|\right) \tag{44}$$

and

$$DL_k = -20\log_{10}\left(\max_{\omega\in\Omega_k} |\bar{H}(e^{j\omega}) - L(\omega)|\right). \tag{45}$$

Then from (37) and (42),

$$DL_k^* \gtrapprox -20\log_{10}\left(10^{-(DL_k/20)} + 2^{-t}\sqrt{\frac{2N-1}{3}}\right). \tag{46}$$

The lower bound in (46) is plotted in Fig. 3 as a function of $DL_k$ for $N = 129$ and several values of $t$. It is seen that for each $t$, there is a maximum value of in-band rejection that can be statistically guaranteed (i.e., with high probability) as a lower bound to the attainable in-band rejection of an $N$-point filter if its coefficients are rounded to $t$-bits, no matter how high

[3] The use of the term "rejection" is not meaningful in frequency regions where the filter is designed to pass the signal, i.e., the filter passbands. It is used here for consistency in describing the characteristics of the filter in *any* frequency band. It should also be noted that the rejection in any band is not referenced to a peak filter gain. Thus if all the filter coefficients are multiplied by a constant, the rejection in each band changes even though the filter is substantially the same.

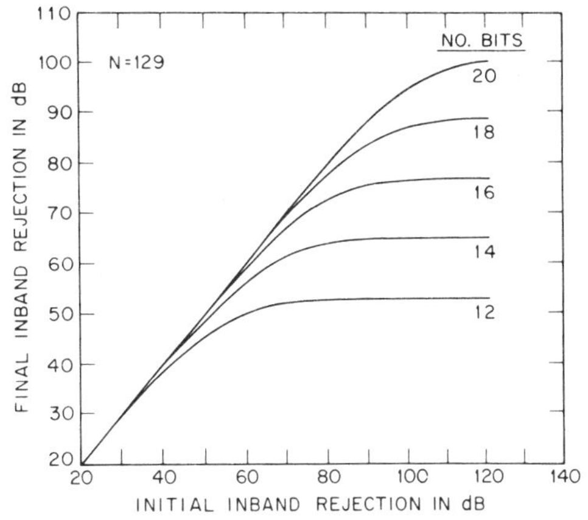


**Fig. 3. Lower bounds on in-band rejection obtained after coefficient quantization as a function of initial rejection.**

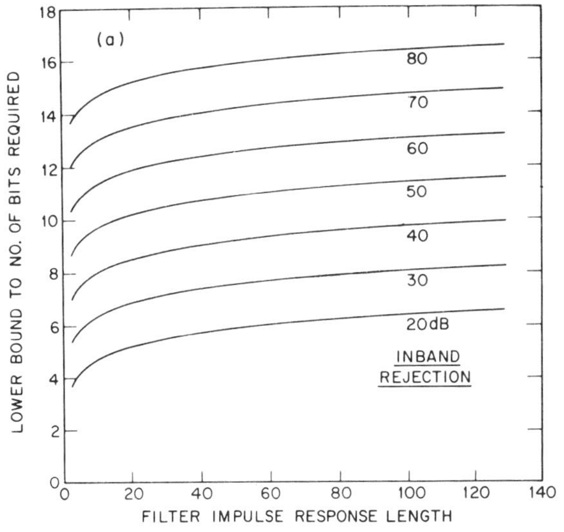


(a)

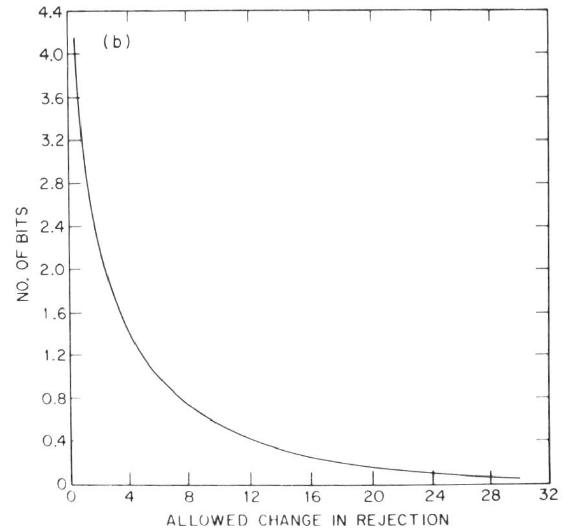


(b)

**Fig. 4. (a) Lower bound to number of bits required to realize a filter of given minimum in-band rejection. (b) Additional number of bits required in excess of lower bound in (a) if change in rejection due to quantization is constrained.**

a value of in-band rejection the unquantized filter has. Thus if the significance level chosen in (40) is used for design purposes, a minimum number of bits is required to implement any filter of a given order and a given minimum in-band rejection for the most tightly specified band if design specifications are to be met with as much probability as there exists for (40) to be satisfied. To find this lower bound for $t$, let $DL^*m_k$ equal the right-hand side of (46), so that $DL_k^* \gtrsim DL^*m_k$, and solve for $t$ to give

$$t \geqslant t_{\min} = -\log_2 \left[ \sqrt{\frac{3}{2N-1}} \left(10^{-(DL^*m_k/20)} - 10^{-(DL_k/20)}\right)\right]$$

$$= \frac{DL^*m_k + 10\log_{10}[(2N-1)/3]}{20\log_{10} 2} - \log_2\left(1 - 10^{(DL^*m_k - DL_k)/20}\right). \quad (47)$$

Defining $t_\infty$ to be the limit of $t_{\min}$ as $DL_k \to \infty$, viz.,

$$t_\infty = \frac{DL^*m_k + 10\log_{10}[(2N-1)/3]}{20\log_{10} 2} \cong \frac{DL^*m_k}{6} + \frac{5}{3}\log_{10}[(2N-1)/3], \quad (48)$$

(47) can be written as

$$t \geqslant t_{\min} = t_\infty - \log_2(1 - 10^{-(\Delta_k/20)}) \quad (49)$$

where $\Delta_k = DL_k - DL^*m_k$ is the maximum allowable change in in-band rejection on $\Omega_k$ due to coefficient quantization.

Thus $t_\infty$ is a lower bound (statistical) for $t$ given $N$ and $DL^*m_k$ (desired minimum rejection on $\Omega_k$ of quantized filter), but $t_{\min}$ is a tighter, more practical bound that is a function of $\Delta_k$. Equations (48) and (49) show that, for fixed $N$ and $\Delta_k$, $t$ is bounded to increase linearly with $DL^*m_k$ at a rate of 1 bit per 6 dB of increase in the in-band rejection desired for the quantized filter. Fig. 4(a) and (b) show plots of $t_\infty$ and $(t_{\min} - t_\infty)$, respectively. It is of interest to note that, from Fig. 4(b), approximately 1 bit above the number given by $t_\infty$ is required to realize any filter if $(DL_k - DL_k^*)$ for that filter is to be bounded by 6 dB, i.e., if the in-band approximation error on $\Omega_k$ is not to increase by more than double due to quantization.

A useful formula needed in the next section is derived from (46) as follows:

$$DL_k - DL_k^* \lesssim 20\log_{10}\left(1 + 10^{(DL_k/20)}\, 2^{-t}\sqrt{\frac{2N-1}{3}}\right) = 20\log_{10}\left(1 + 2^{((DL_k/6.02)-t)}\sqrt{\frac{2N-1}{3}}\right) \quad (50)$$

taking $20\log_{10} 2 = 6.02$.

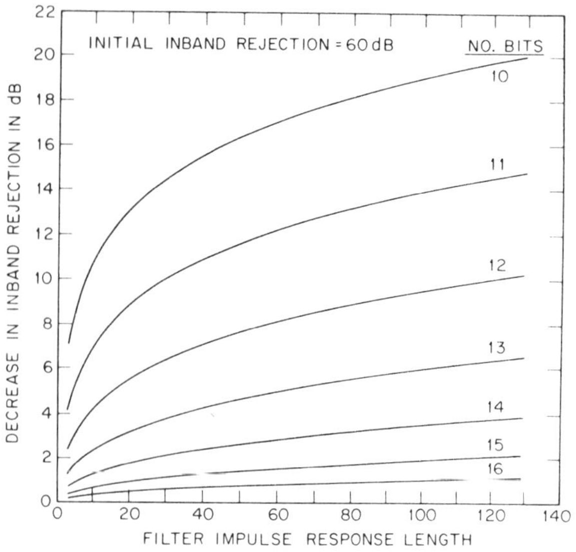


**Fig. 5. Bounds on the change of in-band rejection due to coefficient quantization.**

Equation (50) gives a statistical bound on the loss of in-band rejection due to coefficient quantization. This bound is plotted in Fig. 5 as a function of $N$ for $DL_k$ = 60 dB and $t$ = 10 to 16 bits.

### V. Experimental Verification and Discussion

To verify the validity of (40), the values of $DL_k - DL_k^*$ for $\Omega_k$ equal to both the passband and stopband were found for a large number of low-pass extraripple [5] filters with equal number of ripples in their passband and stopband and values of $N$ ranging from 15 to 127. Fig. 6(a) shows a plot of these values for filters with passband and stopband rejection of 40 and 60 dB, respectively, and coefficients quantized to 11 bits, together with the bound in (50) (solid curves) for these values of in-band rejection. Also, the bound of (50) derived using $\sigma_e$ rather than $2\sigma_e$ in (40) [equivalent to replacing $t$ by $(t + 1)$ in (50)] is plotted for both passband and stopband as dashed curves in Fig. 6(a). Similar results, except with 16-bit coefficients, are shown in Fig. 6(b). Clearly, (40) is well supported by these results.

It is important to point out that, contrary to claims by Herrmann and Schuessler [8], [13], the analysis presented in previous sections shows that the passband of a linear phase direct form filter is in no way favored over the stopband to have lower sensitivity to coefficient accuracy. It is only a difference in in-band rejection that makes one band more sensitive than another. For instance, in the example given by Herrmann and Schuessler [13], the absolute errors due to quantization added to the passband and stopband responses of the filter are actually comparable. However, since the ripples in the passband are 100 times larger than those in the stopband, percentagewise the errors in the passband are not seen.

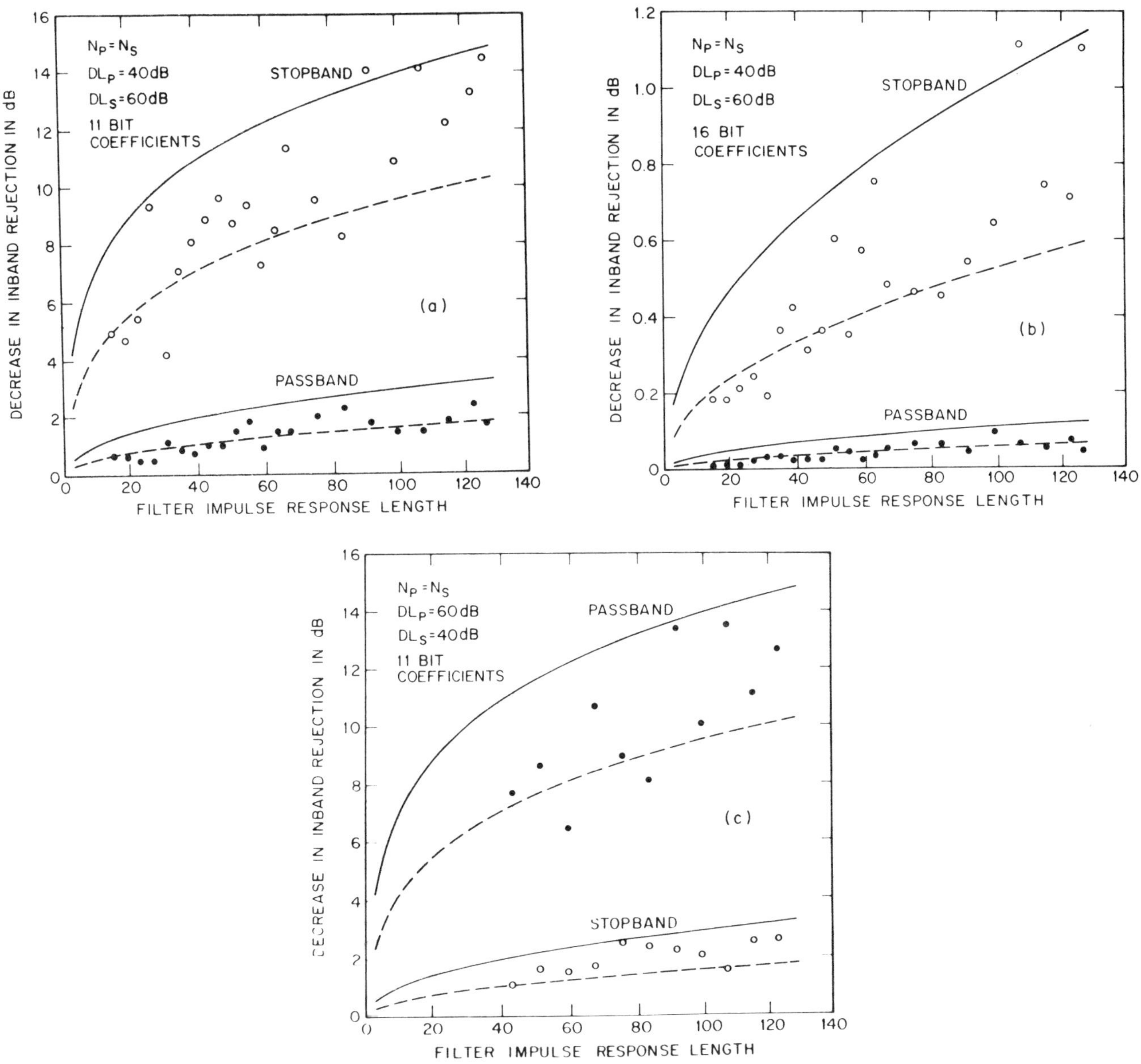


**Fig. 6. Comparisons of experimental data to predicted statistical bounds.**

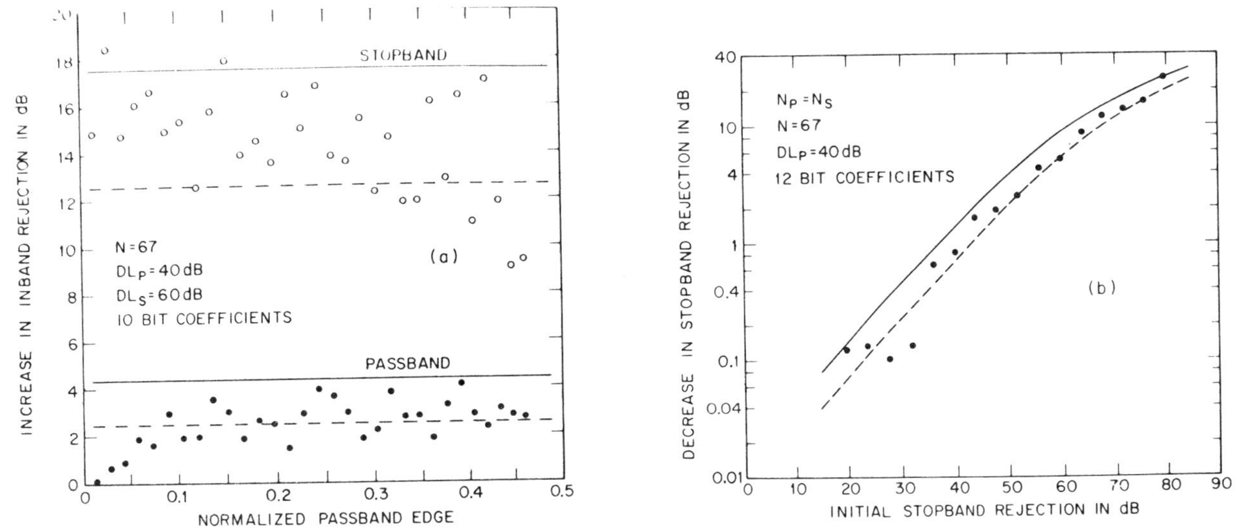


**Fig. 7. Comparisons of experimental data to predicted statistical bounds.**

To illustrate this point, the results of quantizing filters with values of passband and stopband rejection exactly reversed from those of the filters of Fig. 6(a), (b), viz., with 60 and 40 dB, respectively, are shown in Fig. 6(c). As can be seen, the roles of the passband and stopband are simply exactly interchanged. In many cases of interest, filters often do require greater rejection in the stopband than in the passband; thus in such cases it is true that the passband is less sensitive.

It is also a consequence of the presented analysis that the sensitivity of the response of a filter in any frequency band to coefficient quantization should not depend strongly on the width of the band. In Fig. 7(a), the results of coefficient quantization for low-pass extraripple 67-point filters of passband and stopband rejection of 40 and 60 dB, respectively, and all possible bandwidths (except the two extremes) are shown, together with the statistical bounds. It is seen that, indeed, the data points do not have any strong average dependence on bandwidth. Finally, Fig. 7(b) shows the statistical bound (50) and measured data points for the stopband of several 67-point filters, with 40-dB passband rejection, as a function of stopband rejection. All the results shown in Figs. 6 and 7 clearly demonstrate the validity and usefulness of (40).

### VI. Application to Design of Filters with Finite Word Length Coefficients

In Section V, the usefulness of the bound (40) for filter design has been discussed. Specifically, it enables a set of specifications for a filter with coefficients of a given word length to be converted to a set of specifications for the design of an "infinite word length" filter in such a way that when the coefficients of a filter designed to meet the new specifications are quantized, the filter meets the original specifications. In general, this process involves converting a given set of minimum in-band (i.e., referring to both passbands and stopbands) rejections $\{DL^*m_k\}$ into a new set $\{DL_k\}$. Clearly, in any given problem, a filter with the minimum $N$ that will satisfy the given specifications is desired, since both complexity and sensitivity increase with $N$. If $N$ can be determined *a priori*, the rule for conversion implied by the previous analysis is immediate from (46) as

$$DL_k = -20 \log_{10} \left(10^{-(DL^*m_k/20)} - 2^{-t} \sqrt{\frac{2N-1}{3}}\right)$$
$$= DL^*m_k - 20 \log_{10} \left[1 - 2^{((DL^*m_k/6.02)-t)} \cdot \sqrt{\frac{2N-1}{3}}\right] \quad (51)$$

taking $\log_{10} 2 = 0.301$. However, in general, for a given set of $\Omega_k$, $k = 1, \cdots, M$, the minimum $N$ is a function of $DL_k$, $k = 1, \cdots, M$. Thus, $\{DL_k\}$ is the solution of a set of $M$ simultaneous nonlinear equations when a solution exists. Unfortunately, the exact functional relationship among $N$, $\{DL_k\}$, and $\{\Omega_k\}$ is not known for any of the available design methods. However, a useful approximate relation has been found [14] for the case of low-pass optimal filters [4]-[6].

Let $DL_p$, $DL_s$, $F_p$, and $F_s$ denote the passband and stopband rejection, and passband and stopband cutoff frequency (with sampling frequency normalized to unity), respectively, of any low-pass filter. Using data from over 1500 extraripple filters with equal number of ripples in their passband and stopband and various values of $N$, $DL_p$, and $DL_s$, viz., $3 \leq N \leq 127$ and $6 \leq DL_p \leq DL_s \leq 100$ dB, it has been shown that for fixed $DL_p$ and $DL_s$, the product $(N-1)(F_s - F_p) \triangleq D_N(DL_p, DL_s)$ is essentially a constant independent of $N$ for sufficiently large $N$. Denoting this limiting value of the product by $D_\infty(DL_p, DL_s)$, it is found to be given to an excellent approximation by

$$D_\infty(DL_p, DL_s) = [-6.64 \times 10^{-7} (DL_p)^2 + 1.78 \times 10^{-4} DL_p + 0.0238] DL_s + [-6.65 \times 10^{-6} (DL_p)^2 + 0.0297 DL_p - 0.4278]. \quad (52)$$

A correction function is then added to improve the approximation for small values of $N$, with the result that $D_N(DL_p, DL_s)$ can be well approximated for all $N$ by

$$\hat{D}(DL_p, DL_s, F_s - F_p) = D_\infty(DL_p, DL_s) - f(DL_p, DL_s)(F_s - F_p)^2 \quad (53)$$

where

$$f(DL_p, DL_s) = 0.0256 (DL_s - DL_p) + 11.012. \quad (54)$$

Thus, since $D_N(DL_p, DL_s) = (N-1)(F_s - F_p)$, the minimum $N$ required for a filter with cutoff frequencies $F_p$ and $F_s$ and in-band rejections at least $DL_p$ and $DL_s$ can be found as

$$N_{\min} = \frac{D_{N_{\min}}(DL_p, DL_s)}{F_s - F_p} + 1 \approx \frac{\hat{D}(DL_p, DL_s, F_s - F_p)}{F_s - F_p} + 1. \quad (55)$$

The error in $N_{\min}$ incurred by substituting the approximate value of $N_{\min}$ given above in place of its actual value was found to be less than 1 for all $N$, $3 \leq N \leq 127$, and $20 \leq DL_p \leq DL_s \leq 100$ dB.

It has also been shown [14], [15] that given $DL_p$, $DL_s$, and $N$, for all optimal filters, $(F_s - F_p)$ is fairly independent of $F_p$ or $F_s$ if they are not too close to 0 or half the sampling frequency. Thus $D_N(DL_p, DL_s)$ should be fairly independent of the filter cutoff frequencies, so that the approximate equation for $N_{\min}$ given in (54) found for the case of extraripple filters

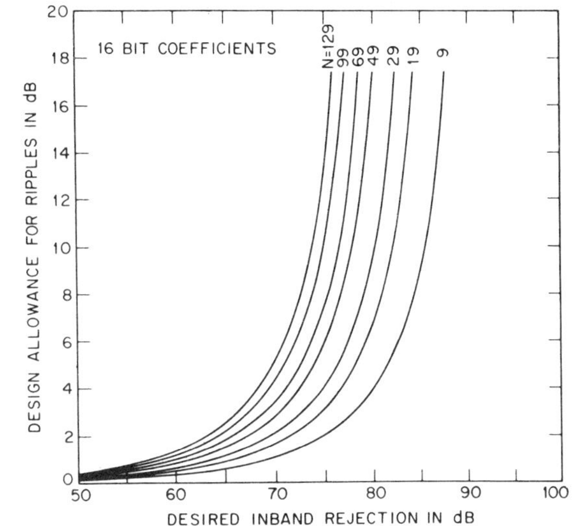


**Fig. 8. Design curves for conversion of specifications on in-band rejection.**

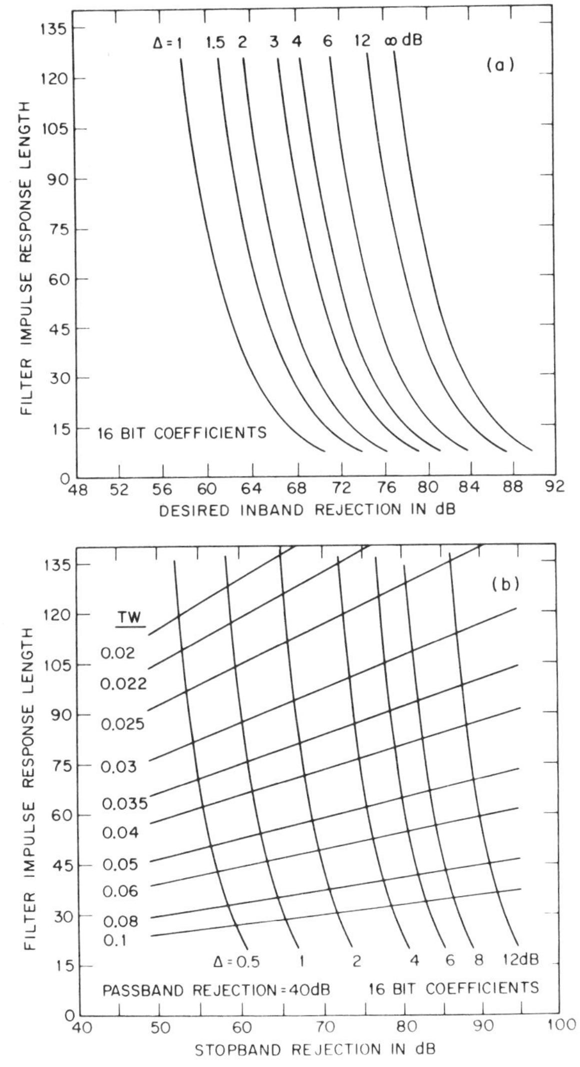


**Fig. 9. (a) Bounds on $N$ for design procedure to converge. (b) Design curves for conversion of stopband rejection of filter with 40-dB passband rejection.**

with an equal number of passband and stopband ripples should still be fairly good for other values of $F_p$ or $F_s$.

Although an expression explicit in $DL_p$ and $DL_s$ has been found for $N_{\min}$, substituting it into (51) still leaves that set of equations too complex to be solved in closed form. However, letting $DL_p^*$ and $DL_s^*$ be the desired minimum passband and stopband rejection for a $t$-bit filter, also denoting the right-hand side of (51) by $g_t(DL^* m_k, N)$, and writing $N_{\min}$ as $N_{\min}(DL_p, DL_s, F_s - F_p)$, the following simple iterative procedure can be used to find the specifications $DL_p$ and $DL_s$ for a $t$-bit filter.

*Step 1:*

Set $N_0 \leftarrow N_{\min}(DL_p^*, DL_s^*, F_s - F_p)$
$DL_p \leftarrow g_t(DL_p^*, N_0)$
$DL_s \leftarrow g_t(DL_s^*, N_0)$.

*Step 2:*

Set $N_1 \leftarrow N_{\min}(DL_p, DL_s, F_s - F_p)$
$\overline{DL}_p \leftarrow g_t(DL_p^*, N_1)$
$\overline{DL}_s \leftarrow g_t(DL_s^*, N_1)$.

*Step 3:*

If $|\overline{DL}_p - DL_p| > \epsilon$ or $|\overline{DL}_s - DL_s| > \epsilon$, then set $DL_p \leftarrow \overline{DL}_p$, $DL_s \leftarrow \overline{DL}_s$, and return to Step 2; otherwise stop.

The number $\epsilon$ is some tolerance level to be chosen.

The function $g_t(x, N) - x$ for $t = 16$ bits is plotted for several values of $N$ in Fig. 8. It is seen that the function is rather insensitive to $N$ for a given $x$ if the combination of $x$ and $N$ is not too large. Thus the algorithm described above will converge rapidly in this region. On the other hand, for a given $x$, there is a maximum value for $N$, however large, above which the algorithm cannot converge. This value of $N$ can be found from (51) as

$$N_{\max} = \frac{1}{2} + \frac{3}{2}\left(1 - 10^{-(\Delta/20)}\right)^2 \cdot 2^{2(t-(DL^* m_k/6.02))} \qquad (56)$$

where $\Delta$ is the maximum allowable change in in-band rejection due to quantization. Equation (56) is plotted in Fig. 9(a) for several values of $\Delta$ and $t = 16$ bits. The curve for $\Delta = \infty$ is the upper bound on $N$ required for convergence of the algorithm described. Clearly, the smaller $N$ is compared to this bound, the faster the convergence will be. The curves for other values of $\Delta$ are bounds on $N$ that must be satisfied if coefficient quantization is required to incur no more than a change of $\Delta$ dB in in-band rejection. Note that both the function $g_t(DL^* m_k, N) - DL^* m_k$ and $N_{\max}$ depend on $t$ and $DL^* m_k$ only through $(t - (DL^* m_k / 6.02))$. Thus the labels for the horizontal axes of Figs. 8 and 9(a) simply need to be shifted if plots of the functions for values of $t$ other than 16 bits are desired.

To further illustrate the procedure described, the important case of a low-pass filter whose stopband rejection is significantly higher than its passband rejection is considered. As an example, $DL_p^* = 40$ dB and $t = 16$ bits is chosen. As can be concluded from Fig. 8, 40 dB of rejection is hardly affected by quantization to 16 bits for at least all $N \leqslant 129$. Thus the passband need not be considered. In Fig. 9(b), (54) is plotted for $DL_p = 40$ dB and various values of $(F_s - F_p)$, the transition bandwidth [denoted by $TW$ in Fig. 9(b)]. Also plotted are curves that describe the functional relationship between $N$ and $DL_s$ for various values of the bound $\Delta$ on in-band rejection change due to quantization to 16 bits. Note that as opposed to Fig. 8 or 9(a), the horizontal coordinate of Fig. 9(b) is the rejection for the infinite-bit filter $(DL_s)$ rather than the minimum rejection obtainable after quantization $(DL_s^*)$. For a given $DL_s$ and $TW$ or $N$, the latter is equal to $DL_s - \Delta$, where $\Delta$ is found by observing which curve of constant $\Delta$ the given point $(DL_s, N)$ falls on.

Thus to design a filter with $t = 16$ bits, $DL_p^* = 40$ dB, and given $F_p$, $F_s$, and $DL_s^*$, Fig. 9(b) is used as follows. Choose a value of $\Delta$ that is desired not to be exceeded and find the bound on $N$, $N_\Delta$ from Fig. 9(a). Find the point with $DL_s = DL_s^*$ on the curve $TW = (F_s - F_p)$ in Fig. 9(b). Move to the right on that curve until either $N$ exceeds $N_\Delta$, in which case the method will not converge to a solution, or else a value of $\Delta$ is attained such that $DL_s - \Delta = DL_s^*$. The value of $DL_s$ at this point is the desired value. This method is reasonably accurate as long as $F_p$ and $F_s$ are not too close to 0 or half the sampling frequency. Since the $TW$ curves are only approximate, it may be necessary to try one or more adjacent values of the $N$ obtained if a satisfactory filter is not found.

To conclude this section, two examples are given to enlighten the procedures described.

*Example 1:* Find a low-pass transfer function with 14-bit coefficients in the direct form and $DL_p^* = 40 \pm 0.5$ dB, $DL_s^* \geqslant 56.5$ dB, $F_p = 0.246$, and $F_s = 0.272$.

First of all, by evaluating $N_{\min}(DL_p^*, DL_s^*, F_s - F_p)$, $N = 95$ is obtained as a first estimate of the minimum $N$ required. Clearly, even if the final $N$ becomes as high as 129, 40-dB rejection is not decreased by more than 0.5 dB by quantizing to 14 bits (see Fig. 8, noting that $x$-axis labels must be shifted 12 dB to the right to correspond to 14-bit coefficients). Thus one can set $DL_p = 40$ dB without further calculation. To find $DL_s$, the three-step algorithm is followed. Using $DL_s^* = 56.5$ dB, Step 1 gives $N_0 = 94.7$ and $DL_s = 59.9$ dB. Step 2 then gives $N_1 = 98.4$ and $\overline{DL}_s = 60$ dB. Using $\epsilon = 0.1$ dB in Step 3, the algorithm has already converged. The specifications $N = 99$, $DL_p = 40$ dB, $DL_s = 60$ dB, $F_p = 0.246$, and $F_s = 0.272$ are used to design an optimal infinite-precision transfer function. After quantizing the coefficients of this transfer function in direct form to 14 bits, the passband and stopband rejections are measured to be $DL_p^* = 39.8$ dB and $DL_s^* = 56.7$ dB.

*Example 2:* Find a low-pass transfer function with 16-bit coefficients in the direct form and $DL_p^* = 40$ dB $\pm$ 0.1 dB, $DL_s^* \geqslant 74$ dB, $F_p = 0.2480$, and $F_s = 0.2955$.

Clearly, $DL_p = 40$ dB can be specified *a priori* with very little possible error. Following the three-step algorithm yields in the end $DL_s = 80$ dB and $N = 67$. An infinite-precision optimal transfer function satisfying these specifications yields, when quantized to 16 bits (direct form), $DL_p^* = 40.0$ dB and $DL_s^* = 75.0$ dB.

In both of the above examples, the final transfer function meets specifications quite tightly. In general, however, much leeway may still be left for possible improvement. In that case, lower values of $N$ can be attempted. However, whether or not any improvement (or how much) is possible cannot, in general, be foreseen.

## VIII. Conclusions

A thorough statistical analysis of all three types of quantization effects in the direct form FIR filter has been presented. Resulting from the analysis of filter response errors due to coefficient quantization, a procedure for designing filters with finite word length coefficients has been proposed and tested using on the order of 500 filter examples. The direct form has been shown to be a very attractive structure for realizing FIR filters because it exhibits very low A-D and roundoff noise, and although its response is rather sensitive to the accuracy of its coefficients, this factor is more than adequately compensated for by the minimal number of multipliers it requires. For instance, for a given filter order, the linear phase direct form uses approximately only one-third as many multipliers as required for the cascade form [9], [10]. Thus a linear phase direct structure using 18-bit coefficients is comparable in complexity to a cascade structure using 6-bit coefficients. Clearly, an 18-bit direct form filter can realize a far greater range of transfer functions than can a 6-bit cascade form filter. It remains to be seen exactly how the direct structure compares with other types of structures for realizing FIR filters.

## Acknowledgment

The authors would like to acknowledge the helpful comments and criticisms of Dr. J. F. Kaiser on this paper. In particular, Dr. Kaiser noted a simple way to sum the series in (27).

## References

[1] L. R. Rabiner, "Techniques for designing finite-duration impulse-response digital filters," *IEEE Trans. Commun. Technol.*, vol. COM-19, pp. 188–195, Apr. 1971.

[2] J. F. Kaiser, "Digital filters," ch. 7 in *Systems Analysis by Digital Computer*, F. F. Kuo and J. F. Kaiser, Ed. New York: Wiley, 1966.

[3] L. R. Rabiner, B. Gold, and C. A. McGonegal, "An approach to the approximation problem for nonrecursive digital filters," *IEEE Trans. Audio Electroacoust.*, vol. AU-18, pp. 83–106, June 1970.

[4] L. R. Rabiner, "The design of finite impulse response digital filters using linear programming techniques," *Bell Syst. Tech. J.*, vol. 51, pp. 1177–1198, July–Aug. 1972.

[5] T. W. Parks and J. H. McClellan, "Chebyshev approximation for nonrecursive digital filters with linear phase," *IEEE Trans. Circuit Theory*, vol. CT-19, pp. 189–194, Mar. 1972.

[6] E. Hofstetter, A. V. Oppenheim, and J. Siegel, "A new technique for the design of nonrecursive digital filters," in *Proc. 5th Annu. Princeton Conf. Inform. Sci. Syst.*, 1971, pp. 63–72.

[7] L. R. Rabiner and R. W. Schafer, "Recursive and nonrecursive realizations of digital filters designed by frequency sampling techniques," *IEEE Trans. Audio Electroacoust.*, vol. AU-19, pp. 200–207, Sept. 1971.

[8] H. W. Schuessler, "On structures for nonrecursive digital filters," *Arch. Elek. Übertragung*, vol. 26, pp. 255–258, 1972.

[9] D. S. K. Chan and L. R. Rabiner, "Theory of roundoff noise in cascade realizations of finite impulse response digital filters," *Bell Syst. Tech. J.*, vol. 52, pp. 329–245, Mar. 1973.

[10] —, "An algorithm for minimizing roundoff noise in cascade realizations of finite impulse response digital filters," *Bell Syst. Tech. J.*, vol. 52, pp. 347–385, Mar. 1973.

[11] W. Feller, *An Introduction to Probability Theory and Its Applications*, vol. 2. New York: Wiley, 1968, pp. 491–492.

[12] J. B. Knowles and E. M. Olcayto, "Coefficient accuracy and digital filter response," *IEEE Trans. Circuit Theory*, vol. CT-15, pp. 31–41, Mar. 1968.

[13] O. Herrmann and H. W. Schuessler, "On the accuracy problem in the design of nonrecursive digital filters," *Arch. Elek. Übertragung*, vol. 24, pp. 525–526, 1970.

[14] O. Herrmann, L. R. Rabiner, and D. S. K. Chan, "Practical design rules for optimum finite impulse response lowpass digital filters," *Bell. Syst. Tech. J.*, vol. 52, pp. 769–799, July–Aug. 1973.

[15] T. W. Parks, L. R. Rabiner, and J. H. McClellan, "On the transition width of finite impulse response digital filters," *IEEE Trans. Audio Electroacoust.*, vol. AU-21, pp. 1–4, Feb. 1973.

# A New Statistical Approach to the Coefficient Word Length Problem for Digital Filters

R. E. CROCHIERE

*Abstract*—A statistical approach is proposed for estimating the necessary coefficient word length of a digital filter. With this statistical word length definition, an optimization procedure is then proposed for minimizing the statistical word length for a given filter structure and a given set of maximum error constraints. It is shown that by minimizing this statistical word length, the actual coefficient word length upon rounding of the coefficients can generally be reduced as well. Several examples are given and improvements of one to three bits in the actual coefficient word length are observed. The procedure does not necessarily lead to the actual global minimum coefficient word length.

## I. Introduction

An important factor in the cost of a digital filter is the coefficient word length necessary to implement the filter. Consequently, it is desired to keep this word length to a minimum. The problem is closely associated with the choice of the filter structure and its sensitivity properties. The use of statistical methods has played an important role in the study of these coefficient sensitivity properties and the effects of coefficient quantization.

One of the first attempts to use a statistical measure for comparison of digital filters was made by Knowles and Olcayto [1]. More recently, Avenhaus [2] proposed a statistical approach for estimating the coefficient word length necessary to keep the power transfer function of a digital filter within a prescribed error bound. He demonstrated that with this approach the necessary coefficient word length could be predicted to within several bits and that the concept of a statistical word length is a useful method for efficiently comparing filter structures in the initial design stages. A similar approach has been taken by Chan and Rabiner [3] for finite impulse response filters in direct form.

In this paper, a statistical estimate of the word length is defined such that the magnitude of the system function is maintained within a prescribed error. An optimization procedure is then presented for minimizing this statistical word length for a given filter realization. The procedure is proposed as an intermediate design phase toward choosing the filter design in such a way that the actual word length of the filter for coefficient rounding is reduced. The statistical word length is based on the magnitude of the system function instead of the power transfer function because this choice leads to a linear expression for the statistical quantization step size which greatly simplifies the development of the optimization procedure.

Manuscript received September 5, 1973; revised April 15, 1974. This work was supported by the National Science Foundation under Grant GK-31353.

The author is with the Research Laboratory of Electronics and the Department of Electrical Engineering, Massachusetts Institute of Technology, Cambridge, Mass. 02139.

## II. A Statistical Estimate of the Required Word Length

It will be assumed in this paper that the primary filter characteristic of interest is the magnitude of the system function. This function is designed to approximate an ideal magnitude response $H_I = H_I(\omega)$ with an allowed error deviation $\delta = \delta(\omega)$. When the filter coefficients are specified to infinite precision, the magnitude of the system function, $|T| = |T(e^{j\omega})|$, will be designated as $|T_0|$. Let $\Delta|T| = \Delta|T(e^{j\omega})|$ be the error incurred in $|T_0|$ when finite precision coefficients are used in the actual filter. It is desired that this magnitude response, $|T| = |T_0| + \Delta|T|$, also meet the error criterion, $H_I \pm \delta$.

A statistical approach can now be taken to obtain a conservative estimate of the magnitude of the error $\Delta|T|$. It will be assumed that the quantization step size for rounding of the coefficients in fixed-point arithmetic is $Q$, and that the quantization errors $\Delta c_i$ will have a uniform probability of lying between $-Q/2$ and $Q/2$ for an arbitrary set of coefficients $c_i$ ($i = 1,2,\cdots,m$). The statistical errors in the coefficients can then be characterized as having zero means, and variances of $Q^2/12$. The mean of $\Delta|T|$ is then zero and the variance of $\Delta|T|$, assuming that the errors $\Delta c_i$ are independent, can be determined approximately as

$$\sigma^2_{\Delta|T|} \cong \frac{Q^2 S^2}{12} \tag{1}$$

where $S^2$ is defined as

$$S^2 = S^2(\omega) \triangleq \sum_{i=1}^{m} \left(\frac{\partial|T_0|}{\partial c_i}\right)^2. \tag{2}$$

The partial derivatives in (2) can be computed efficiently for arbitrary filter structures by using the transpose or adjoint network approach [4], [5], [6]. For the case where some coefficients are quantized to greater precision than others, $Q$ can be chosen to be the smallest quantization step size, and the corresponding step sizes for the coefficients $c_i$ can be given in terms of weighted units of this step size in the form $k_i Q$.

Using an argument similar to that by Avenhaus [2], the distribution of $\Delta|T|$, by the central limit theorem, is approximately Gaussian. Consequently, it is possible to choose a value, $x\sigma_{\Delta|T|}$, such that $|\Delta|T||$ will be less than $x\sigma_{\Delta|T|}$ with probability $y$ for an arbitrary set of coefficients. That is,

$$P[|\Delta|T|| \leq x\sigma_{\Delta|T|}] \cong y \tag{3}$$

where $x$ and $y$ are related by the error function

$$y = \frac{2}{\sqrt{2\pi}} \int_0^x e^{-x^2/2}\, dx. \tag{4}$$

Reprinted from *IEEE Trans. Circuits Syst.*, vol. CAS-22, pp. 190–196, Mar. 1975.

This function is tabulated in numerous mathematical handbooks. Once an acceptable confidence factor, $y$, is selected, a corresponding value of $x$ can be determined from tables, and a statistical "bound," $x\sigma_{\Delta|T|}$, on $|\Delta|T||$ can be determined. This value, although referred to as a "bound," is not a true bound, but rather a conservative estimate in that it is possible with probability $1 - y$ for $|\Delta|T||$ to exceed this value.

In order to meet the specified error criterion, $H_I \pm \delta$, it is necessary to impose an upper limit on the error magnitude $|\Delta|T||$.

This can be written as

$$|\Delta|T|| \leq \delta - ||T_0| - H_I|. \tag{5}$$

In order to be assured that the filter requirements will be met under coefficient quantization with probability $y$, it is necessary to stipulate that the statistical "bound" on $|\Delta|T||$ satisfy the error criterion in (5). This gives the relationship

$$|\Delta|T|| \lesssim \frac{xQS}{\sqrt{12}} \leq \delta - ||T_0| - H_I|. \tag{6}$$

As we wish to satisfy this error criterion for all frequencies, $Q$ should be chosen as

$$Q = \min_{\omega} \frac{\sqrt{12}\,(\delta - ||T_0| - H_I|)}{xS}. \tag{7}$$

The coefficient word length of the digital filter can now be defined as

$$W = 1 + i_M - i_L \tag{8}$$

where $2^{i_M}$ is the power of 2 represented by the most significant bit and $2^{i_L}$ is the power of 2 represented by the least significant bit. This definition does not include the extra bit needed for representing the sign. For the definition of the statistical word length, $i_L$ is chosen to be

$$i_L = \log_2 Q. \tag{9}$$

The statistical word length as a function of frequency can also be defined as

$$w(\omega) = 1 + i_M - \log_2 \left[\frac{\sqrt{12}\,(\delta - ||T_0| - H_I|)}{xS}\right] \tag{10}$$

and $W$ can be written as

$$W = \max_{\omega} w(\omega). \tag{11}$$

This statistical word length is somewhat different than that proposed by Avenhaus [2], although both definitions yield approximately the same result. The difference is demonstrated by the bandpass filter example in Fig. 1. The magnitude $|T_0|$ in the passband ripples between $1 + \varepsilon_p$ and $1 - \varepsilon_p$ where $\varepsilon_p = 0.0157816$. The allowed passband error deviation, $\delta_p$, was selected to be $2\varepsilon_p$ and $y$ was chosen to be 0.95. In Fig. 1(b), the statistical word length $w(\omega)$ is plotted using both definitions for the case of a cascade realization. It can be observed from Fig. 1 that the statistical word length $w(\omega)$ defined in (10) has its maxima near the frequencies for which $|T_0|$ has its maximum error, $1 \pm \varepsilon_p$. This observation was made for other equiripple examples also and is treated in greater detail in [7]. This property is useful in computing $W$ for equiripple or weighted-equiripple designs as it implies that $w(\omega)$ has to be computed at only a relatively small set of maximum error frequencies of $|T_0|$ in order to determine the maximum, $W$. The same behavior can be observed in the definition by Avenhaus if $\max_\omega \varepsilon(\omega)$ is replaced by $\varepsilon(\omega)$ in his equation.

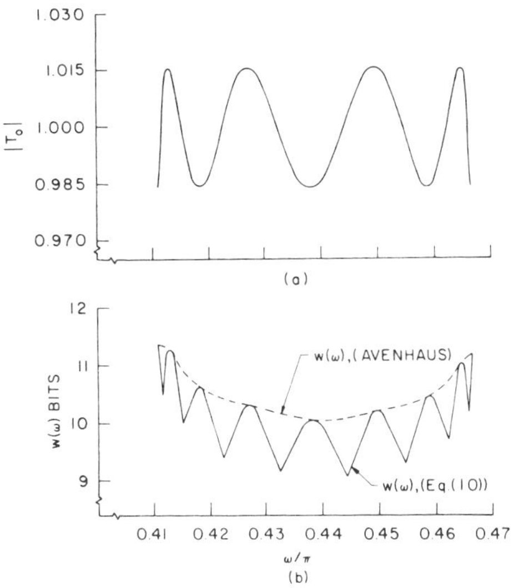


Fig. 1. (a) Passband frequency response for the bandpass filter example. (b) Statistical word length for a cascade filter realization with $\delta_p = 2\varepsilon_p$ and $y = 0.95$, $(x = 2)$.

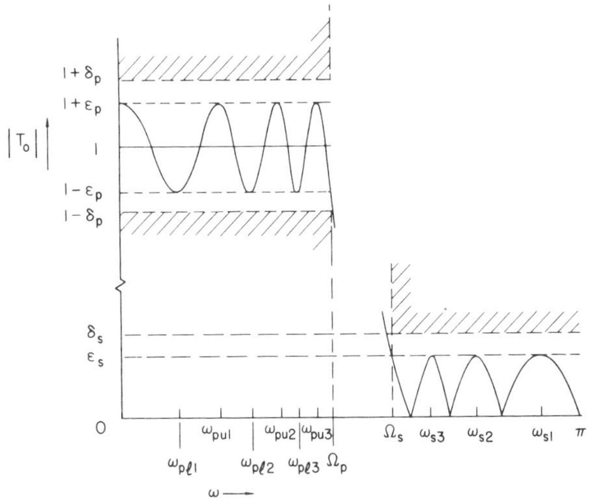


Fig. 2. Passband and stopband error constraints for a low-pass filter.

## III. An Optimization Technique for Minimizing the Statistical Word Length

Generally, in a filter design problem such as a low-pass filter, see (Fig. 2), $\delta_p$, $\delta_s$, $\Omega_p$, and $\Omega_s$ are prespecified, where $\delta_p$ and $\delta_s$ are the allowed passband and stopband errors, respectively, and $\Omega_p$ and $\Omega_s$ are the passband and stopband edge frequencies, respectively. From these specifications, the necessary filter order is calculated. Since the actual order of the filter must be an integer, the next largest integer, $n$, is then selected as the required order of the filter. Because of this selection, there exists a continuum of approximations, $\{|T|\}$, with passband and stopband errors $\varepsilon_p$ and $\varepsilon_s$, respectively, which satisfy the constraints $\delta_p \geq \varepsilon_p$ and $\delta_s \geq \varepsilon_s$. The question then arises as to which approximation in this set should be chosen such that the required coefficient word length for a given filter structure is minimized. In general, for the case of digital filters, this is a nonlinear integer programming problem and a solution is difficult to obtain.

In this paper a different approach is taken. Instead of minimizing the actual coefficient word length, we determine the optimal infinite precision approximation, $|T_0|$, which minimizes the statistical word length for a given structure. While this choice of $|T_0|$ does not guarantee that the actual coefficient word length will be minimized upon rounding of the coefficients, it has been observed in examples that the approach generally leads to a reduction of the actual coefficient word length. The result of the optimization procedure is to choose the approximation $|T_0|$ such that it is specifically matched to the coefficient sensitivity properties of the structure. That is, it chooses $|T_0|$ such that at frequencies where the coefficient sensitivity is high there is greater allowance for error and where it is low there is less allowance for error. Although we will specifically concentrate on piecewise-constant low-pass filter designs in the examples of this paper, the concept extends to other designs as well.

### *A. Two-Parameter Statistical Approach*

To demonstrate the statistical word length minimization procedure, consider first the two-parameter approach in which $\varepsilon_p$ and $\varepsilon_s$ are allowed to vary and the passband and stopband edge frequencies are fixed. We consider separate statistical word lengths for the stopband and passband and denote them as $W_s$ and $W_p$, respectively. For the equiripple example in Fig. 2 we can express $W_s$, using (10) and (11), as

$$W_s = 1 + i_M + \log_2 \left[ \frac{x \max\limits_{\omega_s} S(\omega_s)}{\sqrt{12}\,(\delta_s - \varepsilon_s)} \right] \tag{12}$$

where $H_I = 0$, $\omega_s$ is the set of maximum error frequencies in the stopband denoted as $\omega_s = \{\Omega_s, \omega_{s1}, \omega_{s2}, \cdots\}$, and $|T_0|$ equals its maximum value $\varepsilon_s$ at these frequencies (see Fig. 2). For the passband, $H_I = 1$, and a similar expression can be obtained. For this case, however, a slight improvement can be obtained by "statistically scaling" the approximation to equalize the statistical error at the upper and lower bounds. This leads to the expression [7]

$$W_p = 1 + i_M + \log_2 \left[ \frac{\frac{x}{2} \left[\max\limits_{\omega_{pu}} S(\omega_{pu}) + \max\limits_{\omega_{pl}} S(\omega_{pl})\right]}{\sqrt{12}\,(\delta_p - \varepsilon_p)} \right] \tag{13}$$

where $\omega_{pu} = \{0, \omega_{pu1}, \omega_{pu2}, \cdots\}$ and $\omega_{pl} = \{\omega_{pl1}, \omega_{pl2}, \cdots, \Omega_p\}$ are the upper and lower bound frequency sets, respectively, and $|T_0|$ has its maximum error of $1 \pm \varepsilon_p$, respectively, at these frequencies (see Fig. 2).

If $\varepsilon_p$ is decreased, the passband statistical word length $W_p$ will decrease. This is done at the expense of increasing $\varepsilon_s$ and the stopband statistical word length, $W_s$. The overall coefficient word length, $W$, of the digital filter must equal the maximum of $W_p$ and $W_s$ in order to satisfy both the passband and stopband constraints. As we desire to minimize $W$, the obvious choice is to specify $\varepsilon_p$ and $\varepsilon_s$ such that

$$W = W_p = W_s. \tag{14}$$

With the aid of (12) and (13), the necessary relationship between $\varepsilon_p$ and $\varepsilon_s$, in order to satisfy (14), can be derived as

$$\frac{\delta_s - \varepsilon_s}{\delta_p - \varepsilon_p} = B \tag{15}$$

where

$$B \triangleq \frac{\max\limits_{\omega_s} S(\omega_s)}{\frac{1}{2}\left[\max\limits_{\omega_{pu}} S(\omega_{pu}) + \max\limits_{\omega_{pl}} S(\omega_{pl})\right]}. \tag{16}$$

This expression is valid for monotonic response functions as well as equiripple responses. In the monotonic case, however, extra frequency points may have to be included in the sets $\omega_s$, $\omega_{pu}$, and $\omega_{pl}$ to assure that the proper maxima of $S$ are obtained. Another point to be noted in the preceding relationship is that the factor $x$ cancels, making the optimization procedure independent of the choice of the confidence factor. Consequently, for any choice of $y$ the same optimal infinite precision system function will be obtained as a result of minimizing the statistical word length.

A second relationship between $\varepsilon_p$ and $\varepsilon_s$ can be obtained from the approximation used in determining $|T_0|$. This relationship is dependent on the type of approximation chosen. By constraining the passband and stopband edge frequencies to be fixed, an intrinsic nonlinear relationship between $\varepsilon_p$ and $\varepsilon_s$ can be expressed in the form

$$F(\varepsilon_p, \varepsilon_s) = 0. \tag{17}$$

For the case of analytic approximations such as the Butterworth, Tchebyshev, or elliptic approximations, analytic expressions for (17) are generally available. For algorithmic designs the intrinsic relation in (17) is implied but an analytic expression may not be available. In this case, the intrinsic function may have to be determined numerically over the region of interest.

The objective of the optimization procedure is to solve (15) and (17) for the two unknowns $\varepsilon_p$ and $\varepsilon_s$. The procedure begins by computing an initial approximation $|T_0|_1$ which satisfies the passband and stopband constraints $\delta_p$ and $\delta_s$. This sets the order of the filter, $n$, and gives an initial trial solution $\varepsilon_{p,1}$ and $\varepsilon_{s,1}$. An appropriate filter structure is chosen for the realization, and the sensitivity ratio $B_1$ for this first trial is computed using (16). The solution to (15) and (17) is then obtained by an iterative procedure. On each trial, $j$, of the procedure, the sensitivity ratio $B_j$ is assumed to be invariant over the feasible solutions $\{|T|\}$. Using this assumption, we choose the value of $B$ in (15) to be $B_j$ and the new trial values $(\varepsilon_{p,j+1}, \varepsilon_{s,j+1})$ are determined by solving (15) and (17) for $(\varepsilon_p, \varepsilon_s)$. A new filter is synthesized with the parameters $(\varepsilon_{p,j+1}, \varepsilon_{s,j+1})$, using the same structure as that of the initial trial, and a new sensitivity ratio $B_{j+1}$ is computed from (16) for this realization. This new sensitivity ratio, $B_{j+1}$, should be an improved estimate of the actual value of $B$. The procedure is then repeated for trials $j+1, j+2, \cdots$, until it converges to the optimal solution which satisfies (15), (16), and (17) simultaneously, thus minimizing the statistical word length. The convergence of the procedure was found, on the basis of empirical observations, to be very rapid. For the examples analyzed, satisfactory solutions were generally obtained after three or four trials.

*Examples*

To demonstrate the method, an elliptic lowpass filter was chosen. For the elliptic approximation, the intrinsic function $F(\varepsilon_p, \varepsilon_s)$ is well known [8]–[10] and can be written in the form

$$F(\varepsilon_p, \varepsilon_s) = \frac{K(k_{1c})K(k)}{K(k_1)K(k_c)} - n = 0 \tag{18}$$

where

$n$ = the order of the filter,

$K(\cdot)$ = the complete elliptic integral of the first kind,

$k$ = the selectivity factor

$$= \frac{\tan(\Omega_{pc}/2)}{\tan(\Omega_{sc}/2)}$$

$$k_c = +\sqrt{1-k^2} \tag{19}$$

$$k_1 = \sqrt{\left[\left(\frac{1+\varepsilon_p}{1-\varepsilon_p}\right)^2 - 1\right] \Big/ \left[\left(\frac{1+\varepsilon_p}{\varepsilon_s}\right)^2 - 1\right]} \tag{20}$$

and

$$k_{1c} = +\sqrt{1-k_1^2}.$$

For the two-parameter statistical approach, the elliptic design edge frequencies $\Omega_{pc}$ and $\Omega_{sc}$ are assigned to be equal to the specified passband and stopband edge frequencies. That is, $\Omega_{pc} = \Omega_p$ and $\Omega_{sc} = \Omega_s$.

In order to compute the values $B$, $W_p$, and $W_s$ efficiently, we should know the maximum error frequencies $\omega_{pu}$, $\omega_{pl}$, and $\omega_s$. These frequencies can be determined, in the case of the elliptic approximation, from the elliptic functions; expressions for their evaluation are given in the Appendix.

To obtain the successive trial parameters it is necessary to solve (15) and (18) with $B = B_j$. This can be accomplished by expressing (15) in the parametric form

$$\varepsilon_{s,j+1} = \varepsilon_s = B_j t + \delta_s \tag{21}$$

$$\varepsilon_{p,j+1} = \varepsilon_p = t + \delta_p \tag{22}$$

where $t$ is a dummy variable. Upon substitution in (18), the intrinsic expression $F(\varepsilon_p, \varepsilon_s)$ can be written in terms of the single variable $t$ as $F(t) = 0$. Consequently, the problem is reduced to the simpler problem of determining a zero of a nonlinear one-dimensional function in $t$. Many elegant algorithms exist for solving this problem but as $F(t)$ is well behaved in the region of interest, the following simple approach, sometimes referred to as the bisection method [11], works well and does not require the evaluation of derivatives.

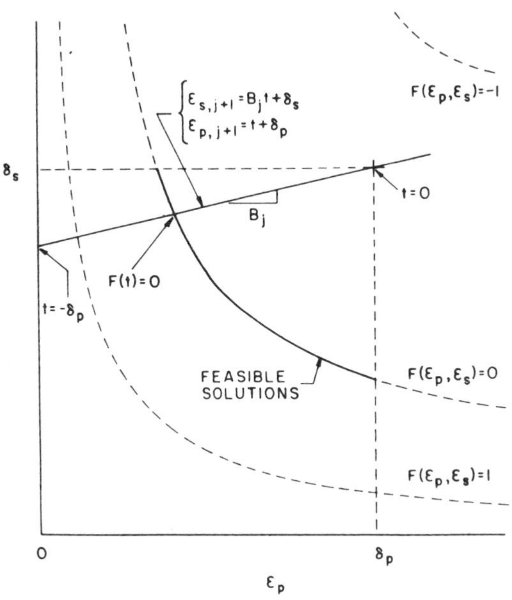


Fig. 3. Optimization problem illustrated in the $(\varepsilon_p, \varepsilon_s)$ parameter space.

The problem can be clearly illustrated in the $\varepsilon_p, \varepsilon_s$ parameter space (see Fig. 3). The parametric equations (21) and (22) trace a straight line in this space with positive slope $B_j$. The intrinsic relation traces a curve of feasible solutions over the region $\varepsilon_p \leq \delta_p$, $\varepsilon_s \leq \delta_s$. Above this curve $F(\varepsilon_p, \varepsilon_s) < 0$ and below it $F(\varepsilon_p, \varepsilon_s) > 0$. The desired solution is the intersection of the curve and the straight line. Choosing $t = t_a = 0$ corresponds to the point $(\varepsilon_s = \delta_s, \varepsilon_p = \delta_p)$ which lies above the curve $F(\varepsilon_p, \varepsilon_s) = 0$, and on the line. A second point on the line, below the curve, can be obtained by choosing

$$t = t_b = \max(-\delta_p, -\delta_s/B_j)$$

in which case either $\varepsilon_p = 0$ or $\varepsilon_s = 0$ and $F(t_b) = +\infty$. Between these two points $F(t)$ varies monotonically [8], [10]. The solution can now be obtained with a systematic trial-and-error procedure by choosing a new point $t_c = \frac{1}{2}(t_a + t_b)$ and evaluating $F(t_c)$. If $F(t_c) < 0$, assign $t_a = t_c$ and if $F(t_c) > 0$, assign $t_b = t_c$. The procedure is then repeated iteratively, halving the interval $[t_a, t_b]$ on each trial, until it converges. In practice, this procedure requires approximately 15 trials to reach an acceptable solution for which $-0.0001 \leq F(t) \leq 0$. The design parameters for the new filter trial can then be obtained from the solution $t$ with the aid of the parametric equations (21) and (22).

Two filter examples were optimized by these procedures. In the first example, the statistical word length was reduced by 1.54 bits (Table I). It was found that an actual improvement of 1 bit in the word length is obtained by using the

TABLE I

TWO-PARAMETER OPTIMIZATION EXAMPLE

FILTER PARAMETERS: $\delta_p = 0.006$, $\delta_s = 0.01$, $\Omega_p = 0.1\pi$, $\Omega_s = 0.13\pi$

FILTER STRUCTURE: CASCADE, $n = 6$

| TRIAL | $i$ = | 1 | 2 | 3 |
|---|---|---|---|---|
| OPTIMIZATION PARAMETERS | $\varepsilon_{p,i}$ | 0.0041751 | 0.0010926 | 0.0011100 |
| | $\varepsilon_{s,i}$ | 0.0050764 | 0.0099233 | 0.0098453 |
| STATISTICAL WORD LENGTH (BITS) $y = 0.95$ | $W_p$ | 14.90 | 13.29 | 13.30 |
| | $W_s$ | 7.41 | 14.39 | 13.36 |
| | $W$ | 14.90 | 14.39 | 13.36 |
| SENSITIVITY RATIO | $B_i$ | 0.015067 | 0.033380 | 0.033065 |
| ACTUAL WORD LENGTH | $W_a$ | 13 | 13 | 12 |

STATISTICAL WORD LENGTH IMPROVEMENT: 1.54 BITS
ACTUAL WORD LENGTH IMPROVEMENT: 1 BIT

statistically optimized design with coefficient rounding instead of the initial trial design with coefficient rounding. The second example resulted in a statistical word length reduction of 1.74 bits and an improvement of 2 bits in the actual word length. In both examples, the optimization procedure converged to the final solution in only three trials on $B_j$.

### *B. Four-Parameter Statistical Approach*

In the two-parameter approach, only the passband and stopband errors $\varepsilon_p$ and $\varepsilon_s$ are allowed to vary. For the case of equiripple functions, additional tradeoffs can often be made by shifting the passband and/or stopband edge frequencies, $\Omega_{pc}$ and $\Omega_{sc}$, of the approximation function, into or out of the transition band. This, in effect, creates positive or negative "sensitivity guard bands." The direction of the shift is dependent on the sensitivity at the edge frequency relative to the in-band sensitivity. Often the sensitivity at the edge frequencies is greater than the in-band sensitivity, as can be seen in Fig. 1(b), thus requiring positive "sensitivity guard bands" for word length minimization.

To demonstrate this approach, an equiripple passband and stopband low-pass filter magnitude function is assumed. For this case, there are four optimization parameters $\varepsilon_p$, $\varepsilon_s$, $\Omega_{pc}$, and $\Omega_{sc}$. An initial trial is chosen with parameters $(\varepsilon_{p,1},\varepsilon_{s,1},\Omega_{pc,1},\Omega_{sc,1})$ and, as in the two-parameter approach, new trials are successively generated until the statistical word length is minimized. For each interation new parameters $(\varepsilon_{p,j+1},\varepsilon_{s,j+1},\Omega_{pc,j+1},\Omega_{sc,j+1})$ must be determined for the next trial. The relations for determining these new trial parameters can be derived by defining four statistical word lengths for the passband, stopband, passband edge, and stopband edge, respectively, and requiring that they be equal. This leads to the set of parametric equations [7]

$$\varepsilon_{s,j+1} = B_j t + \delta_s \tag{23}$$

$$\varepsilon_{p,j+1} = t + \delta_p \tag{24}$$

$$\Omega_{sc,j+1} = \Omega_{sc,j} + \frac{1}{C_j}\left[|T_0(\Omega_s)|_j - \frac{\varepsilon_{s,j}}{B_j t + \delta_s} \cdot \left(\delta_s + \frac{S(\Omega_s)B_j t}{\max\limits_{\omega_s'} S(\omega_s')}\right)\right] \tag{25}$$

$$\Omega_{pc,j+1} = \Omega_{pc,j} + \frac{1}{D_j}\left[|T_0(\Omega_p)|_j - 1 + \frac{\varepsilon_{p,j}}{t+\delta_p}(\delta_p + G_j t)\right] \tag{26}$$

where

$$C_j \triangleq \left.\frac{\partial |T_0|_j}{\partial \omega}\right|_{\omega=\Omega_s} \tag{27}$$

$$D_j \triangleq \left.\frac{\partial |T_0|_j}{\partial \omega}\right|_{\omega=\Omega_p} \tag{28}$$

$$G_j = \frac{S(\Omega_p) + \frac{1}{2}[\max\limits_{\omega_{pu}'} S(\omega_{pu}') - \max\limits_{\omega_{pl}'} S(\omega_{pl}')]}{\frac{1}{2}[\max\limits_{\omega_{pu}'} S(\omega_{pu}') + \max\limits_{\omega_{pl}'} S(\omega_{pl}')]} \tag{29}$$

and

$$B_j = \frac{\max\limits_{\omega_s'} S(\omega_s')}{\frac{1}{2}[\max\limits_{\omega_{pu}'} S(\omega_{pu}') + \max\limits_{\omega_{pl}'} S(\omega_{pl}')]}. \tag{30}$$

The values $G_j$ and $B_j$ are determined from the filter realization for the $j$th trial. In addition to the preceding equations we need the intrinsic function

$$F(\varepsilon_p,\varepsilon_s,\Omega_{pc},\Omega_{sc}) = 0. \tag{31}$$

Upon substitution of (23)–(26) into (31) the problem again reduces to one of finding a zero of a nonlinear function $F(t)$ in the single variable $t$. The same algorithm that was described for the two-parameter approach applies here.

### *Examples*

To demonstrate the four-parameter approach the elliptic approximation was again chosen. The intrinsic function for (31) can be written in the same form as (18). Expressions for determining the maximum error frequencies $\omega_{pu}'$, $\omega_{pl}'$, and $\omega_s'$ are given in the Appendix.

Two examples were analyzed with the four-parameter approach. The filter specifications for the examples are the same as those used in the two-parameter optimization examples. For the first example, an improvement of 1.70 bits was observed in the statistical word length and an actual improvement of 3 bits was obtained using the statistically optimized design instead of the initial trial design (Table II). For the second example, a statistical word length improvement of 1.84 bits was observed and an actual improvement of 1 bit was achieved using the optimized design and coefficient rounding.

In comparison, slightly greater improvement in the statistical word length was obtained from the four-parameter approach. As can be seen from the examples, this does not necessarily guarantee that the actual word length will

TABLE II

FOUR-PARAMETER OPTIMIZATION EXAMPLE

| | | | | | |
|---|---|---|---|---|---|
| FILTER PARAMETERS: | | $\delta_p = 0.006$, | $\delta_s = 0.01$ | | |
| | | $\Omega_p = 0.1\pi$ | $\Omega_s = 0.13\pi$ | | |
| FILTER STRUCTURE: | | CASCADE, | $n = 6$ | | |
| TRIAL | $i$ = | 1 | 2 | 3 | 4 |
| OPTIMIZATION PARAMETERS | $\varepsilon_p$ | 0.0041750 | 0.0010925 | 0.0011533 | 0.0011601 |
| | $\varepsilon_s$ | 0.0050764 | 0.0099848 | 0.0099479 | 0.0099429 |
| | $\Omega_{pc}$ | $0.1\pi$ | $0.100055\pi$ | $0.100284\pi$ | $0.100309\pi$ |
| | $\Omega_{sc}$ | $0.13\pi$ | $0.129989\pi$ | $0.129974\pi$ | $0.129974\pi$ |
| STATISTICAL WORD LENGTH (BITS) $y = 0.95$ | $W_p$ | 14.88 | 13.19 | 13.22 | 13.22 |
| | $W_s$ | 5.81 | 15.13 | 13.35 | 13.22 |
| | $W_{\Omega p}$ | 14.92 | 13.37 | 13.23 | 13.22 |
| | $W_{\Omega s}$ | 7.41 | 14.62 | 13.25 | 13.22 |
| | $W$ | 14.92 | 15.13 | 13.35 | 13.22 |
| ACTUAL WORD LENGTH | $W_a$ | 13 | 13 | 13 | 10 |
| STATISTICAL WORD LENGTH IMPROVEMENT: 1.70 BITS | | | | | |
| ACTUAL WORD LENGTH IMPROVEMENT: 3 BITS | | | | | |

be improved by using the four-parameter approach instead of the two-parameter approach. In Table II, the growth of the "sensitivity guard bands" is evident as the new trials are generated.

*Bandpass Example*

To demonstrate how the statistical method might be applied to the case of bandpass filters, the bandpass example used by Avenhaus [2] was chosen (Fig. 1(a)). The filter approximation was synthesized from a low-pass, fourth-order, elliptic prototype, using the low-pass to bandpass transformation [12],

$$g(z^{-1}) = -\frac{z^{-1}(z^{-1} - \alpha)}{1 - \alpha z^{-1}}. \tag{32}$$

Optimization was performed on the low-pass prototype with the appropriate sensitivities taken from the bandpass implementation. The mapping parameter, $\alpha$, in (32) was fixed to the same value for all of the trials. The statistical word length was reduced by 1.23 bits for the two-parameter approach and by 1.78 bits for the four-parameter approach. In both cases, the actual coefficient word length was improved by one bit for coefficient rounding using the optimized design.

## IV. Conclusions and Additional Comments

From the examples it can be seen that minimizing the statistical word length usually leads to a reduction in the actual coefficient word length by approximately 1 to 3 bits for coefficient rounding. There is no assurance, however, that the actual global minimum word length will be obtained. This can be seen in the bandpass example, where an improvement of 1 bit in the actual coefficient word length was obtained with the statistical approach, whereas a 3-bit improvement for the same example was found by Avenhaus [2] who used a nonlinear integer programming approach.

As the statistical optimization procedure is very efficient computationally, we conjecture that this approach might be useful as an initial phase in a nonlinear integer programming approach if a further reduction of the coefficient word length is desired. The statistical approach could be used first to determine the global minimum statistical word length. This solution could then serve as an initial starting point for a nonlinear, integer, local search procedure to determine the actual minimum word length. This two-phase procedure would, it is hoped, result in a considerable reduction in the total computation time.

For the case of algorithmic synthesis techniques, it was proposed that the implied intrinsic function $F(\varepsilon_p,\varepsilon_s)$ be determined numerically over the range $\varepsilon_p \le \delta_p$, $\varepsilon_s \le \delta_s$. A more efficient technique might be to attempt to incorporate the sensitivity relations directly into the algorithmic synthesis procedure. This concept might lead to some interesting possibilities.

The statistical optimization procedure proposed in this paper should also be applicable, with slight modifications, to the problem of sensitivity minimization or tolerance maximization in analog filters. As the sensitivity problem for analog filters is truly a statistical one, this method might be even more suited to this problem than to the digital case.

It is conjectured that a savings of 1–2 bits in the digital case would correspond to factors of approximately 2–4 in tolerance improvement in a similar analog case.

## Appendix

The frequencies at which $|T_0|$ has its maximum error for the low-pass elliptic approximation can be determined from the elliptic functions [10]. Expressions for evaluating the frequencies for the two- and four-parameter optimization approaches are given. It is assumed that the $z$-plane elliptic approximations are obtained from $s$-plane designs by the bilinear transformation [9].

For the two-parameter approach, the sets of frequencies $\omega_{pu}$, $\omega_{pl}$, and $\omega_s$ are given as

$$\omega_{pu} = \begin{cases} \{0,\omega_{pui}\} & n \text{ odd} \\ \{\omega_{pui}\} & n \text{ even} \end{cases} \tag{A-1}$$

$$\omega_{pl} = \begin{cases} \{\omega_{pli},\Omega_p\} & n \text{ odd} \\ \{0,\omega_{pli},\Omega_p\} & n \text{ even} \end{cases} \tag{A-2}$$

and

$$\omega_s = \begin{cases} \{\Omega_s,\omega_{si}\} & n \text{ odd} \\ \{\Omega_s,\omega_{si},\pi\} & n \text{ even} \end{cases} \tag{A-3}$$

where the sets of frequencies $\omega_{pui}$, $\omega_{pli}$, and $\omega_{si}$ are given in (A-7), (A-8), and (A-9), respectively.

For the four-parameter approach, $\omega_{pu}'$, $\omega_{pl}'$, and $\omega_s'$ are given as

$$\omega_{pu}' = \begin{cases} \{0,\omega_{pui}\} & n \text{ odd} \\ \{\omega_{pui}\} & n \text{ even} \end{cases} \tag{A-4}$$

$$\omega_{pl}' = \begin{cases} \{\omega_{pli}\} & n \text{ odd} \\ \{0,\omega_{pli}\} & n \text{ even} \end{cases} \tag{A-5}$$

and

$$\omega_s' = \begin{cases} \{\omega_{si}\} & n \text{ odd} \\ \{\omega_{si},\pi\} & n \text{ even.} \end{cases} \quad \text{(A-6)}$$

The sets of frequencies $\omega_{pui}$, $\omega_{pli}$, and $\omega_{si}$ are given as

$$\omega_{pui} = 2\tan^{-1}\left[\tan\left(\frac{\Omega_{pc}}{2}\right)\text{sn}\left(\frac{(2i-1+v)K(k)}{n}, k\right)\right],$$

$$\text{for } i = 1,2,3,\cdots,\frac{n-v}{2} \quad \text{(A-7)}$$

$$\omega_{pli} = 2\tan^{-1}\left[\tan\left(\frac{\Omega_{pc}}{2}\right)\text{sn}\left(\frac{(2i-v)K(k)}{n}, k\right)\right],$$

$$\text{for } i = 1,2,3,\cdots,\frac{n-2+v}{2} \quad \text{(A-8)}$$

and

$$\omega_{si} = 2\tan^{-1}\left[\frac{\tan\left(\frac{\Omega_{sc}}{2}\right)}{\text{sn}\left(\frac{(2i-v)K(k)}{n}, k\right)}\right],$$

$$\text{for } i = 1,2,\cdots,\frac{n-2+v}{2} \quad \text{(A-9)}$$

where

sn $(u,k)$ = the elliptic sine function,

$K(k)$ = the complete elliptic integral of the first kind,

$n$ = the order of the filter,

$k$ = the selectivity factor (see (19)),

and

$$v = \begin{cases} 1 & n \text{ odd} \\ 0 & n \text{ even.} \end{cases}$$

An algorithm for computing $K(k)$ is given by Hastings [13] and algorithms for computing sn $(u,k)$ are given by Gold and Rader [9] and by Byrd and Friedman [14]. These computations are not as formidable as they first appear. The entire optimization procedure could easily be performed on a programmable desk calculator.

## REFERENCES

[1] J. B. Knowles and E. M. Olcayto, "Coefficient accuracy and digital filter response," *IEEE Trans. Circuit and Systems*, vol. CAS-15, pp. 31–41, Mar. 1968.
[2] E. Avenhaus, "On the design of digital filters with coefficients of limited word length," *IEEE Trans. Audio Electroacoust.*, vol. AU-20, pp. 206–212, Aug. 1972.
[3] D. S. K. Chan and L. R. Rabiner, "Analysis of quantization errors in the direct form for finite impulse response digital filters," *IEEE Trans. Audio Electroacoust.*, vol. AU-21, pp. 354–366, Aug. 1973.
[4] R. E. Crochiere, "Computational methods for sensitivity analysis of digital filters," *Quarterly Progress Rep. No.* 109, Research Laboratory of Electronics, M.I.T., pp. 113–123, Apr. 15, 1973.
[5] R. E. Seviora and M. Sablatash, "A Telegen's theorem for digital filters," *IEEE Trans. Circuit and Systems*, vol. CAS-18, pp. 201–203, Jan. 1971.
[6] A. Fettweis, "A general theorem for signal-flow networks, with applications," *Arch. Elekt. Übertrag.*, vol. 25, pp. 557–561, 1971, see also *Digest of Technical Papers*, London, 1971, and *IEEE International Symposium on Electrical Network Theory*, pp. 3–4.
[7] R. E. Crochiere, "Digital network theory and its application to the analysis and design of digital filters," Ph.D. dissertation, Dep. Elec. Eng., M.I.T., Cambridge, Mass., 1974.
[8] L. R. Rabiner, J. F. Kaiser, O. Herrmann, and M. Dolan, "Some comparisons between FIR and IIR filters," *Bell Syst. Tech. J.*, Feb. 1974.
[9] B. Gold and C. M. Rader, *Digital Processing and Signals.* New York: McGraw-Hill, 1969.
[10] D. C. Youla, "A tutorial exposition of some key network-theoretic ideas underlying classical insertion-loss filter design," *Proc. IEEE*, vol. 59, pp. 760–799, May 1971.
[11] R. W. Hamming, *Numerical Methods for Scientists and Engineers.* New York: McGraw-Hill, 1962.
[12] A. G. Constantinides, "Spectral transformations for digital filters," *Proc. IEEE*, vol. 117, pp. 1585–1590, Aug. 1970.
[13] C. Hastings, Jr., *Approximations for Digital Computers.* Princeton: Princeton University Press, 1955.
[14] P. F. Byrd and M. D. Friedman, *Handbook of Elliptic Integrals for Engineers and Physicists.* Berlin: Lange, Maxwell, and Springer, 1954.

# DESIGNING SHORT-WORD RECURSIVE DIGITAL FILTERS*

K. STEIGLITZ
Department of Electrical Engineering
Princeton University
Princeton, New Jersey 08540

ABSTRACT

This paper deals with the problem of designing recursive digital filters with limited word length coefficients to meet magnitude specifications in the frequency domain. Schuessler and Avenhaus have reported results on this problem obtained by rounding the coefficients of an optimal filter and applying a modified univariate search. We describe here another search method, based on the Pattern Search of Hooke and Jeeves, and relate some computational experience with the problem.

## I. INTRODUCTION

Schuessler and Avenhaus [1-3] have reported results on the problem of designing recursive digital filters with limited word length coefficients to meet magnitude specifications in the frequency domain. They begin with a filter optimal in the minimax sense, round its coefficients, and apply a discrete optimization method to obtain a short-word filter which meets specifications. Their optimization algorithm consists of univariate search with two important modifications: (1) a backtracking feature which returns, after a local optimum has been found, to previous next-best points in parameter space; and (2) a more powerful local search which entails generating a four-dimensional subspace determined by the most sensitive parameters.

We shall describe here another algorithm, based on the well-known Pattern Search of Hooke and Jeeves [5], which yields similar results, and has the advantages of being easy to program and flexible in trading off computing time with quality of solutions. The Pattern Search is a univariate search with an acceleration feature, that has been modified in two ways: (1) the search is randomized so that a variety of locally optimal solutions can be generated; and (2) the neighborhood ex-

* This work was supported by NSF Grant GJ-965, and U.S. Army Research Office - Durham under contract DAHCO4-69-C-0012.

Reprinted with permission from *Proc. 9th Annu. Allerton Conf. on Circuit and Syst. Theory*, Oct. 6-8, 1971, pp. 778-788.

plored at each point can be expanded, as in the Schuessler-Avenhaus method.

One purpose of this paper is to show that the problem is amenable to the fairly standard optimization algorithm of Hooke and Jeeves. Another purpose is to demonstrate that for a given problem a wide variety of locally optimum short-word filters exist, and that a randomized procedure for producing an assortment of solutions seems useful. In this respect the problem is similar to other combinatorial problems, such as the traveling salesman problem [8] and the problem of designing low cost redundant communication networks [9].

## II. STATEMENT OF THE PROBLEM

We consider the cascade form transfer function

$$Y(\omega) = \prod_{k=1}^{K} \frac{1 + a_k z^{-1} + z^{-2}}{1 + b_k z^{-1} + c_k z^{-2}} \tag{1}$$

where (1) $z = e^{j\pi\omega}$, $\omega$ is the frequency in fractions of the Nyquist frequency;

(2) $|a_k| < 2$, so that the zeros are complex and on the unit circle;

(3) $c_k < 1$, $1 + c_k + b_k > 0$, $1 + c_k - b_k > 0$, so that the poles are inside the unit circle;

(4) $2^Q a_k$, $2^Q b_k$, and $2^Q c_k$ are integers.

From (2), (3) and (4) we can conclude that $a_k$ and $b_k$ require $Q+2$ bits (a sign bit, a unit bit, and Q fraction bits); and that $c_k$ requires Q+1 bits (a sign bit, and Q fraction bits). The coefficient quantization step-size is $2^{-Q}$.

We are given an ideal desired magnitude characteristic $Y^d$. Define the error function

$$e(\omega) = \frac{1}{\delta(\omega)} \left| A|Y(\omega)| - Y^d \right|, \quad \omega \in W \tag{2}$$

where A is a positive gain constant, $W \subset [0,\pi]$ is a closed set of frequencies of interest, and $\delta(\omega)$ is a tolerance function. The problem is to choose the coefficients $a_k$, $b_k$, and $c_k$ so that

$$F(a_k, b_k, c_k) = \min_{A} \max_{\omega \in W} e(\omega) \tag{3}$$

is as small as possible. In particular, whenever $F \leq 1$, we say we have a design which meets the tolerance specifications $\delta(\omega)$.

Two remarks: First, the gain A is not constrained in precision, since a wide range of overall gain is usually tolerable. Second, the zeros are restricted to the unit circle since this reduces the complexity of the filter and is a good assumption for filters with bands of desired rejection, such as lowpass, highpass, bandpass, etc. This assumption can be relaxed for other cases with only trivial changes in what follows.

## III. HOOKE AND JEEVES' PATTERN SEARCH [5]

Figures 1 and 2 show flowcharts of the basic pattern search, and Table 1 summarizes the variables used. The flowcharts are of a specialized type called D-charts [6], and have no explicit _go-to_ control statements. Two control statements are used: an _if_ (condition) _then_ A _else_ B, indicated by a simple branching; and a _do_ A _while_ (condition), indicated by an iteration node "⊗" with the condition on a branch to the right. The first time the condition is violated at the node, execution continues downward. All variables are global.

The basic strategy of the main program (MAIN in Fig. 1) is the following: the current value of the parameter vector is $\Psi$, and $\varphi$ is the exploratory value. Subroutine EXPLOR (Fig. 2) performs a univariate search by incrementing and decrementing each component of $\varphi$, accepting any improvements in the function F. If an improvement is found, $\theta$ is set to $\Psi$, saving the old value; $\Psi$ is set to $\varphi$, updating the current value; and $\varphi$ is set to $2\varphi-\theta$, thereby attempting to extrapolate the pattern of the successful move. EXPLOR is then invoked for a local exploration about $\varphi$, and the process continued until no improvements are found. If at least one improvement was found during this process $(S>0)$, the strategy is applied again in an attempt to build up a new pattern. When finally no improvements are found in this attempt, $(S=0)$, the stepsize $\Delta$ is reduced by a factor of 2 and the process repeated. The algorithm terminates when $\Delta$ falls below some prescribed value $\delta$.

In order that condition (4) above hold throughout, it is sufficient that it hold for the initial value of the parameter vector, that $\Delta_0$ be an integer power of 2 no less than $2^{-Q}$, and that $\delta$ be at least $2^{-Q}$. Generally, it was found effective to take $\Delta_o = 2^{-Q+1}$ or $2^{-Q}$, and $\delta = 2^{-Q}$. Too large a value of $\Delta_o$ tends to encourage prematurely large steps early in the algorithm, which leaves the parameter vector trapped at poor local optima.

## IV. RANDOMIZATION

The original Pattern Search of Hooke and Jeeves enumerates the coordinates in the natural order. The course of the algorithm is sensitive to the order in which the coordinates are explored, however, since improvements are retained whenever encountered. It was found that using a random ordering of coordinates produced a variety of final local optima, the best of which is selected as the final solution. In this way, more computer time can be invested in the hope of improving the solution. The set $\{1,\ldots,N\}$ is randomly shuffled by using a pseudorandom number generator with different initial seeds, and an algorithm described by Knuth [10].

## V. TWO-OPT

As reported by Avenhaus [3], the univariate search is rather weak, and often leads to local optima which are relatively poor. For this reason, a stronger version of EXPLOR was developed, using the routine TWO-OPT shown in a dotted box in EXPLOR. Fig. 3 shows a D-chart for TWO-OPT, which works as follows: if no improvement is found by changing coordinate i = LIST (k), two coordinates at a time are changed: i and j = LIST ($\ell$), $\ell > k$. In this way, stronger local optima are found, at the expense of increasing the size of the neighborhood search. This approach is similar to the use of 3-opt instead of 2-opt for the traveling salesman problem [8].

## VI. COMPUTATIONAL RESULTS

(The values of F given are taken from a finite grid of frequency points, and may be slightly optimistic.)

Example 1 This is the bandpass example of Schuessler and Avenhaus [1-3]:

$$K = 4,\ N = 12,\ W = W_p \cup W_s, \text{ where}$$

$W_p$ = [.411111, .466666] is the passband, and $W_s$ = [0.,.383333]$\cup$[.494444,1.] is the stopband.

$$Y^d = 1 \quad \omega \in W_p$$
$$0 \quad \omega \in W_s$$

$$\delta(\omega) = .03 \quad \omega \in W_p$$
$$.01 \quad \omega \in W_s$$

The starting point was obtained by rounding the coefficients of the elliptic filter used by Avenhaus [4]. The complete algorithm with TWO-OPT was run 3 times with 3 different initializations of the random number generator, $\Delta_o = 2^{-5}$ and $\delta = 2^{-6}$. An unacceptable solution with F = 1.80 was produced once, and twice the following acceptable local optimum with F = .802 was found (in hexadecimal):

| a | b | c |
|---|---|---|
| .70 | -.48 | .E8 |
| -1.18 | -.74 | .E8 |
| - .04 | -.30 | .F8 |
| - .BC | -.90 | .F8 |

This differs from Avenhaus' solution [4] with F = .732 by $2^{-6}$ in $a_2$, and this discrepancy is due to the fact that a finite and fixed grid of frequency points was used, rather than a search for minima and maxima of F. Each run required between 1788 and 2650 evaluations of F.

Other runs, without TWO-OPT, and with various $\Delta_o$, all failed to produce acceptable solutions, thus showing the necessity of using the more powerful version of EXPLOR.

Example 2 This is another example due to Schuessler and Avenhaus [2,3], and data provided by Avenhaus [4] provided the lowpass elliptic filter rounded to obtain a starting point.

K = 5, N = 15, $W = W_p \cup W_s$, where

$W_p$ = [0.,.333333] and $W_s$ = [.376111,1.]

$$Y^d = 1 \quad \omega \in W_p$$
$$0 \quad \omega \in W_s$$
$$\delta(\omega) = .01 \quad \omega \in W_p$$
$$.001 \quad \omega \in W_s$$

Five different random starts with $\Delta_o = \delta = 2^{-7}$, produced 5 distinct solutions with F from .629 to 1.040, using between 1158 and 2302 evaluations of F. The best solution is not quite as good as Avenhaus' solution [4] with F = .581, but is acceptable. In hexadecimal it is

| a | b | c |
|---|---|---|
| -.BE | -1.12C | .60 |
| -.AA | -1.11A | .8C |
| -.60 | -1.104 | .BC |
| .52 | - .F6 | .E0 |
| 1.A | - .F4 | .F6 |

Three other runs with $\Delta_o = 2^{-6}$, $\delta = 2^{-7}$ produced similar results, with F from .752 to .917. EXPLOR without TWO-OPT also produced some acceptable local optima, with the best of 4 local optima having F = .845, and using 356 evaluations of F.

Example 3 This the elliptic lowpass filter given by Gold and Rader [7]:

$$K = 4,\ N = 12,\ W = W_p \cup W_s, \text{ where}$$

$W_p = [0.,.4]$ and $W_s = [.44,1.]$

$$Y^d = 1 \quad \omega \in W_p$$
$$\quad\ \ 0 \quad \omega \in W_s$$

$$\delta(\omega) = .03 \quad \omega \in W_p$$
$$\qquad\ \ .001 \quad \omega \in W_s$$

The margin for improvement here is not great, since the infinite precision filter has $F \cong .877$. One run with $\Delta_o = 2^{-8}$, $\delta = 2^{-9}$ produced a local optimum with F = .935. Three runs with $\Delta_o = 2^{-7}$, $\delta = 2^{-8}$ produced a best solution with F = 1.189. Two runs with $\Delta_o = \delta = 2^{-8}$ produced a best solution with F = 1.080. Two runs with $\Delta_o = \delta = 2^{-7}$ produced a best solution with F = 1.018. Notice that this best 7-bit filter is slightly better than the best 8-bit filter, an anomoly due to the suboptimal nature of the solutions. The best 7-bit filter is

| a | b | c |
|---|---|---|
| 1.1A | -1.12C | .6C |
| .5A | - .E2 | .AA |

| a | b | c |
|---|---|---|
| -.32 | -.AA | .DA |
| -.5C | -.98 | .F6 |

## CONCLUSION

It is an open question whether a practical and exact algorithm can be found to find optimal solutions to this problem. On the other hand, heuristically motivated search techniques, such as the Schuessler-Avenhaus algorithm and the one described here, provide practical solutions with reasonable computing costs.


## ACKNOWLEDGEMENTS

The author is indebted to Prof. Dr. W. Schuessler of the University Erlangen-Nuernberg and Dr. J.F. Kaiser of Bell Telephone Laboratories for helpful discussions about this problem, and to Dipl.-Ing. E. Avenhaus for kindly supplying the author with data.


## REFERENCES


1. E. Avenhaus and W. Schuessler, "On the Approximation Problem in the Design of Digital Filters with Limited Wordlength," Archiv Der Elektrischen Vebertragung, Vol. 24, No. 12, pp. 571-572; December 1970.

2. W. Schuessler,"On the Approximation Problem in the Design of Digital Filters," Proc. Fifth Annual Princeton Conference on Information Sciences and Systems, Princeton, N.J., pp. 54-63; March 25-26, 1971.

3. E. Avenhaus, "An Optimization Procedure to Minimize the Wordlength of Digital Filter Coefficients," 1971 Imperial College Symposium on Digital Filtering, Imperial College, London; August 31, September 1,2, 1971.

4. E. Avenhaus, private communication.

5. R. Hooke and T.A. Jeeves,"'Direct Search' Solution of Numerical and Statistical Problems," JACM, Vol. 8, No. 2, pp. 212-229; April 1961.

6. J. Bruno and K. Steiglitz, "The Expression of Algorithms by Charts," to appear in CACM.

7. B. Gold and C.M. Rader, Digital Processing of Signals, McGraw-Hill, 1969. See Chapter 3, Design Example 1, pp. 86-90.

8. S. Lin, "Computer Solutions of the Traveling Salesman Problem," BSTJ, pp. 2245-2269; December 1965.

9. K. Steiglitz, P. Weiner, and D.J. Kleitman, "The Design of Minimum-Cost Survavable Networks," IEEE Trans. on Circuit Theory, Vol. CT-16, No. 4, pp. 455-460; Nov. 1969.

10. D.E. Knuth, The Art of Computer Programming, Vol. 2, "Seminumerical Algorithms," Addison-Wesley, 1969. See Section 3.4.2, Algorithm P, p. 125.

Table 1 Description of Variables

N dimension of parameter space

Δ step-size

$\Delta_0$ initial step-size

δ minimum step-size (for terminating search)

φ current value of parameter vector

ψ exploratory value of parameter vector

θ previous value of parameter vector

F( ) function to be minimized, see Eq. (3)

FPSI F(Ψ)

FPHI F(φ)

S current number of consecutive successful steps

FNEW F( ) at exploratory point

LIST( ) pseudorandom permutation of {1,...,N}

k,ℓ counting indexes

i,j dimension indexes

SAVE, SAVEi, SAVEj temporary variables to store old values of parameters

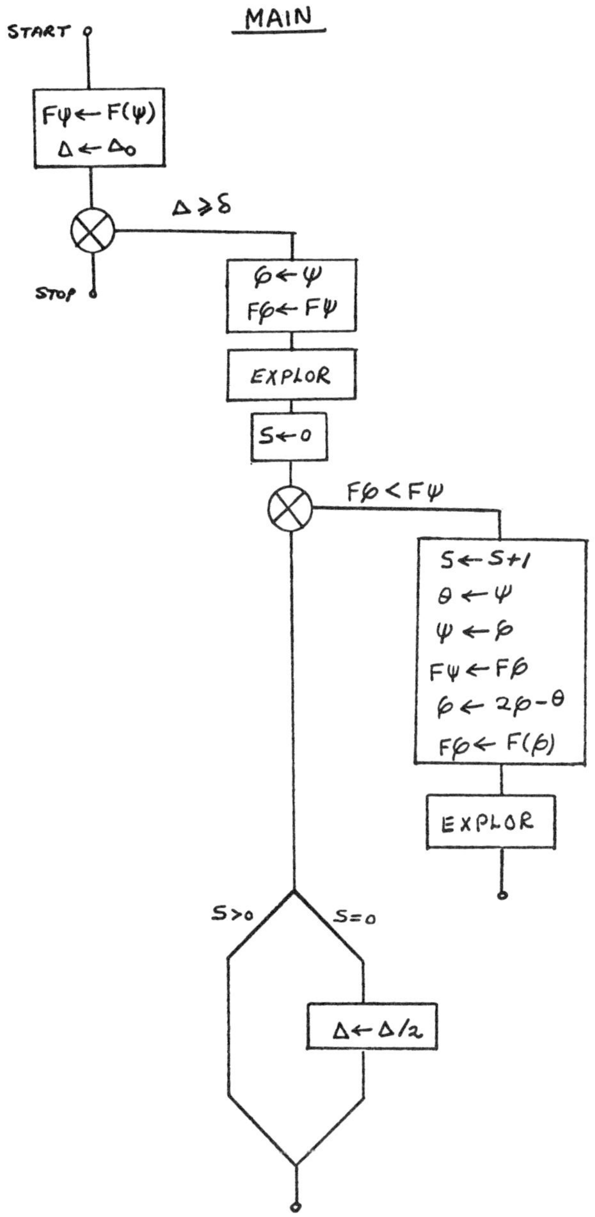
MAIN
START
Fψ ← F(ψ)
Δ ← Δ₀
Δ ≥ δ
STOP
φ ← ψ
Fφ ← Fψ
EXPLOR
S ← 0
Fφ < Fψ
S ← S+1
θ ← ψ
ψ ← φ
Fψ ← Fφ
φ ← 2φ−θ
Fφ ← F(φ)
EXPLOR
S>0
S=0
Δ ← Δ/2

Figure 1

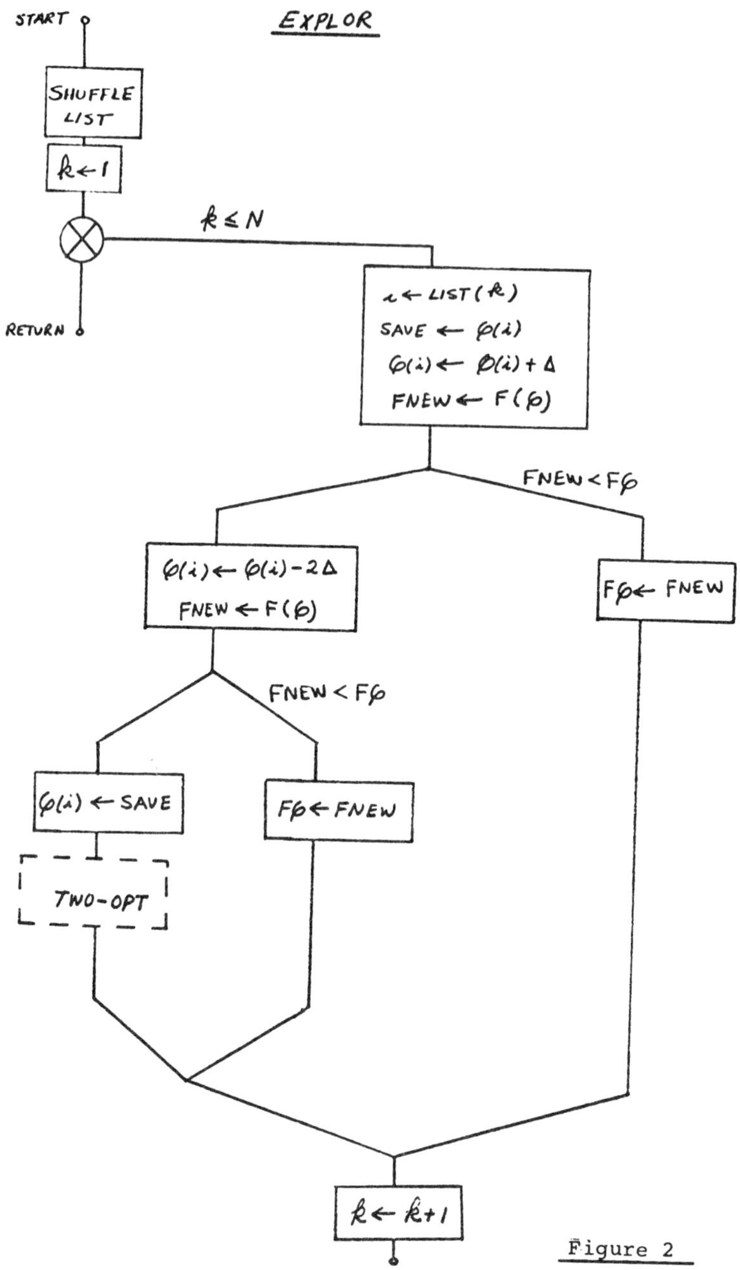
EXPLOR
START
SHUFFLE LIST
k ← 1
k ≤ N
RETURN
i ← LIST(k)
SAVE ← φ(i)
φ(i) ← φ(i) + Δ
FNEW ← F(φ)
FNEW < Fφ
φ(i) ← φ(i) − 2Δ
FNEW ← F(φ)
Fφ ← FNEW
FNEW < Fφ
φ(i) ← SAVE
Fφ ← FNEW
TWO-OPT
k ← k+1

Figure 2

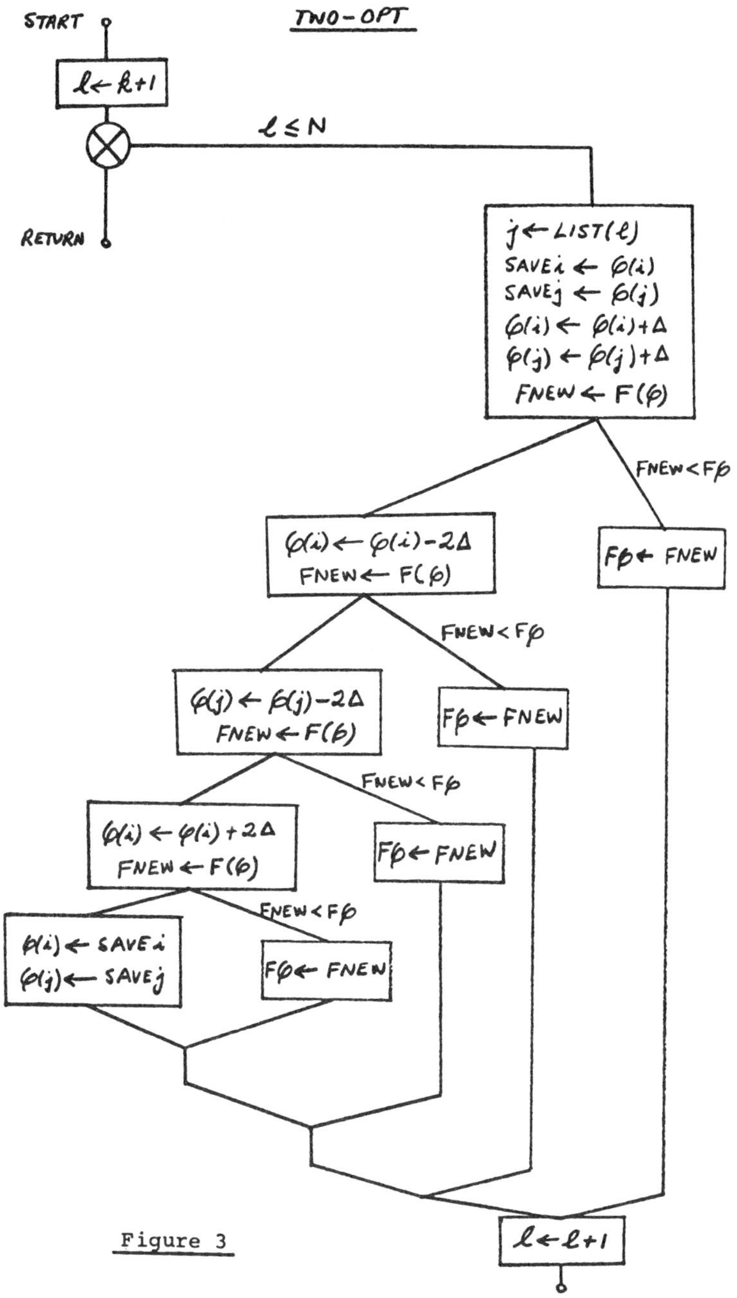
START
TWO-OPT
ℓ ← k+1
ℓ ≤ N
RETURN
j ← LIST(ℓ)
SAVEi ← φ(i)
SAVEj ← φ(j)
φ(i) ← φ(i)+Δ
φ(j) ← φ(j)+Δ
FNEW ← F(φ)
FNEW < Fφ
Fφ ← FNEW
φ(i) ← φ(i)−2Δ
FNEW ← F(φ)
FNEW < Fφ
Fφ ← FNEW
φ(j) ← φ(j)−2Δ
FNEW ← F(φ)
FNEW < Fφ
Fφ ← FNEW
φ(i) ← φ(i)+2Δ
FNEW ← F(φ)
FNEW < Fφ
Fφ ← FNEW
φ(i) ← SAVEi
φ(j) ← SAVEj
ℓ ← ℓ+1

Figure 3

# Optimization of Recursive Digital Filters with Finite Word Lengths

CHRISTAKIS CHARALAMBOUS AND MICHAEL J. BEST

*Abstract*—**The application of the "branch and bound" technique for nonlinear discrete optimization, due to Dakin, to the problem of finding the coefficients of a recursive digital filter with prescribed number of bits, to meet arbitrary response specifications of the magnitude characteristic, is investigated. Due to the fact that the objective function is nonlinear and the stability constraints are linear with respect to the parameters, the recent algorithm for nonlinear programming due to Best and Ritter is used. Based on the ideas presented, a general computer program has been developed. Numerical experience with the present approach is also presented.**

Manuscript received December 20, 1973; revised April 6, 1974. This work was supported by the National Research Council of Canada under Grant A8189 and through a Postdoctorate Fellowship to C. Charalambous.

The authors are with the Department of Combinatorics and Optimization, University of Waterloo, Waterloo, Ont., Canada.

## INTRODUCTION

OVER the past few years a number of papers have been published for the design of recursive digital filters with the coefficients having infinite word length (this is called continuous problem) [1]–[4].

Recently, some "heuristic" techniques have been developed for the design of recursive digital filters with limited word length coefficients to meet magnitude specifications in the frequency domain [5]–[7].

This paper describes the application of the "branch and bound" technique for nonlinear discrete optimization due to Dakin [8], for choosing the coefficients of a digital filter with a given finite word length to meet arbitrary

Reprinted from *IEEE Trans. Acoust., Speech, Signal Processing*, vol. ASSP-22, pp. 424–431, Dec. 1974.

response specifications of the magnitude characteristic. For recursive digital filters the objective function is nonlinear, and the stability constraints are linear with respect to the parameters and therefore the efficient algorithm due to Best and Ritter [9], [10] was used to solve the continuous subproblems.

The results obtained by our method were significantly better in all cases, than those obtained by Suk and Mitra [7]. The digital computer used was an IBM 360/75.

## THEORY

### *The Form of Transfer Function* [1]

The transfer function of a recursive filter under consideration, is given by

$$H(z) = A \prod_{k=1}^{K} \frac{1 + a_k z^{-1} + b_k z^{-2}}{1 + c_k z^{-1} + d_k z^{-2}} \tag{1}$$

and can be realized in a cascade form which is the most commonly used form of implementation.

For our problem

$$z = \exp\,(j\pi\psi) \tag{2}$$

where

$$\psi = f/f_s. \tag{3}$$

$f_s$ is the sampling frequency, and $f$ denotes the frequency.

For stability of the filter, the poles of the transfer function must lie within the unit circle in the $z$-domain and therefore [5],

$$\left.\begin{aligned} d_k &< 1 \\ 1 + c_k + d_k &> 0 \\ 1 - c_k + d_k &> 0 \end{aligned}\right\} \quad k = 1,\cdots,K. \tag{4}$$

Also, since the coefficients must have finite word length

$$2^Q a_k, 2^Q b_k, 2^Q c_k, 2^Q d_k, 2^Q A = \text{Integers for } k = 1,\cdots,K \tag{5}$$

where $2^{-Q}$ is the coefficient quantization step size, and $Q + 1$ is the word length.

The amplitude of the filter is given by [4]

$$|H(x,\theta)| = A \prod_{k=1}^{K} \frac{N_k}{D_k}$$

$$= A \prod_{k=1}^{K} \left( \frac{1 + a_k^2 + b_k^2 + 2b_k(2\cos^2\theta - 1) + 2a_k(1 + b_k)\cos\theta}{1 + c_k^2 + d_k^2 + 2d_k(2\cos^2\theta - 1) + 2c_k(1 + d_k)\cos\theta} \right)^{1/2} \tag{6}$$

where

$$\theta = \pi\psi. \tag{7}$$

$$x = [a_1 b_1 c_1 d_1 \cdots a_K b_K c_K d_K A]^T. \tag{8}$$

Thus $x$ is a $4K + 1$ dimensional vector.

### *The Error Function* [11]

Define real error functions related to the "upper" and "lower" specifications, respectively as follows:

$$e_u(x,\psi) \triangleq w_u(\psi)(F(x,\psi) - S_u(\psi))$$

$$e_l(x,\psi) \triangleq w_l(\psi)(F(x,\psi) - S_l(\psi)) \tag{9}$$

where $F(x,\psi)$ is the approximating function, $\psi$ is an independent variable for example frequency, $S_u(\psi)$ and $S_l(\psi)$ are upper and lower response specifications, $w_u(\psi)$ and $w_l(\psi)$ are upper and lower positive weighting functions and $x$ is a vector containing the $n$ independent parameters (in our case $n = 4K + 1$). If $S_u = S_l = S$, $w_u = w_l = w$, then we have a single specified function and $e_u = e_l = e$.

As usual we will sample all the functions to obtain for example $e_{ui}(x)$ for $i \in I_u$ and $e_{li}(x)$ for $i \in I_l$ where $I_u \triangleq \{1,2,\cdots,n_u\}$ and $I_l \triangleq \{n_{u+1},n_{u+2},\cdots,n_u + n_l\}$. $n_u$ and $n_l$ are the number of sample points over the upper and lower specifications respectively.

### *The Objective Function* [11], [12]

Let us define for convenience

$$f_i(x) \triangleq \begin{cases} e_{ui}(x) & i \in I_u \\ -e_{li}(x) & i \in I_l \end{cases} \tag{10}$$

$$I \triangleq I_u \cup I_l. \tag{11}$$

The following objective function proposed by Bandler and Charalambous [11], [12] will be used.

$$U(x) = M(x)\Big(\sum_{i \in J(x)} [f_i(x)/M(x)]^q\Big)^{1/q} \tag{12}$$

where

$$M(x) \triangleq \max_{i \in I} f_i(x) \tag{13}$$

$$J(x) \triangleq \begin{cases} \{i \mid f_i(x) \geq 0\} & \text{for } M(x) > 0 \\ I & \text{for } M(x) < 0 \end{cases} \tag{14}$$

$$q = p \operatorname{sgn} M(x) \begin{cases} 1 < p < \infty & \text{for } M(x) > 0 \\ 1 \leq p < \infty & \text{for } M(x) < 0. \end{cases} \tag{15}$$

Differentiating (12) with respect to $\boldsymbol{x}$,

$$\nabla U(\boldsymbol{x}) = \{ \sum_{i \epsilon J(\boldsymbol{x})} [f_i(\boldsymbol{x})/M(\boldsymbol{x})]^q \}^{(1/q)-1} \cdot \sum_{i \epsilon J(\boldsymbol{x})} [f_i(\boldsymbol{x})/M(\boldsymbol{x})]^{q-1} \nabla f_i(\boldsymbol{x}) \quad (16)$$

where

$$\nabla \triangleq [(\partial/\partial x_1)(\partial/\partial x_2) \cdots (\partial/\partial x_n)]^T.$$

$\nabla f_i(\boldsymbol{x}) = w_{ui} \nabla F_i(\boldsymbol{x})$ for $i \in I_u$ and $\nabla f_i(\boldsymbol{x}) = -w_{li} \nabla F_i(\boldsymbol{x})$ for $i \in I_l$. Differentiating (6) with respect to $a_j$, $b_j$, $c_j$ and $d_j$,

$$\frac{\partial \mid H(\boldsymbol{x},\theta) \mid}{\partial a_j} = \frac{\mid H(\boldsymbol{x},\theta) \mid}{N_j^2} [a_j + (1 + b_j) \cos \theta] \quad (17a)$$

$$\frac{\partial \mid H(\boldsymbol{x},\theta) \mid}{\partial b_j} = \frac{\mid H(\boldsymbol{x},\theta) \mid}{N_j^2} [-1 + 2 \cos^2 \theta + a_j \cos \theta + b_j] \quad (17b)$$

$$\frac{\partial \mid H(\boldsymbol{x},\theta) \mid}{\partial c_j} = \frac{-\mid H(\boldsymbol{x},\theta) \mid}{D_j^2} [c_j + (1 + d_j) \cos \theta] \quad (17c)$$

$$\frac{\partial \mid H(\boldsymbol{x},\theta) \mid}{\partial d_j} = \frac{-\mid H(\boldsymbol{x},\theta) \mid}{D_j^2} [-1 + 2 \cos^2 \theta + c_j \cos \theta + d_j]. \quad (17d)$$

Therefore, our objective is to minimize (12) subject to the constraints (4) and (5). Note that very little extra effort is required to calculate the gradient vector once $\mid H(\boldsymbol{x},\theta) \mid$ is calculated.

*Dakin's Tree-Search Algorithm for discrete Optimization Problems* [8]

The *nonlinear discrete optimization problem* can be stated as follows.

Minimize $U(\boldsymbol{x})$ subject to

$$g_i(\boldsymbol{x}) \geq 0, \qquad i = 1,2,\cdots,m \quad (18)$$

plus the additional constraints

$$x_i/q = \text{Integer}, \qquad i = 1,2,\cdots,n \quad (19)$$

where $q$ is the discrete step size. If $q = 1$, then we have the *integer programming problem*. If (19) is not satisfied, it is a *nondiscrete solution*. If (19) is relaxed, then we have the *continuous problem*.

In 1966 Dakin [8] proposed an algorithm to solve the above problem, which consists of a number of applications of a continuous algorithm.

Let us suppose that we have found the optimum point to the continuous problem and the parameter $x_j$ is not discrete. Then, the discrete part of the parameter $x_j$ is given by

$$d_j = [x_j/q]q \quad (20)$$

where $[x_j/q]$ denotes the integer part of $x_j/q$. Therefore

$$x_j = d_j + f_j \quad (21)$$

where $0 < f_j < q$. Let

$$x_{uj} \triangleq \begin{cases} d_j & \text{if } x_j > 0 \\ d_j - q & \text{if } x_j < 0 \end{cases} \quad (22a)$$

$$x_{lj} \triangleq \begin{cases} d_j + q & \text{if } x_j > 0 \\ d_j & \text{if } x_j < 0 \end{cases} \quad (22b)$$

then, the range $x_{uj} < x_j < x_{lj}$ is inadmissible, and we may divide all solutions to the constraints (18) and (19) into *two nonoverlapping groups*,

$$x_j \leq x_{uj} \quad (23)$$

$$x_j \geq x_{lj}. \quad (24)$$

We now add each of these constraints in turn to the continuous problem and solve these augmented problems. We then repeat the procedure for each of the two solutions so obtained. The logical structure is that of a "tree." A typical tree is shown in Fig. 1 which is obtained by minimizing

$$U(\boldsymbol{x}) = x_1^2 + 6x_2^2$$

subject to

$$g_1(\boldsymbol{x}) = x_1 + 2x_2 - 1.2 \geq 0$$

and $q = 1$. Each node represents a solution to an augmented continuous problem, and is shown with the additional constraint which was applied to reach the solution. The tree will always terminate in one of two ways: 1) if a discrete solution is obtained (denoted by "$I$") or 2) if the current set of constraints has no solution (denoted by NS). The best solution is then selected from among all the feasible discrete solutions. The example considered the best solution occurs at the point $\check{\boldsymbol{x}} = [2 \quad 0]^T$ and $U(\check{\boldsymbol{x}}) = 4$. Once a feasible discrete solution is reached, its corresponding objective function value is used as an upper bound so that any new branch yielding a larger objective function value would be terminated, because further addition of parameter constraints would increase the objective function value. If we have a feasible discrete point to start with we can insert the value of the objective function at this point at the beginning of the algorithm.

In practice, the number of nodes generated by the algorithm is greatly affected by the value of the upper bound on the objective function. In formulating the algorithm, two approaches were considered. For approach 1, the discrete point closest to the continuous optimum at the first node was obtained. If this point was feasible, the corresponding objective value was then used as the initial upper bound. This upper bound was then updated whenever a discrete solution was obtained from a continuous subproblem at a subsequent node. For approach 2, the discrete point closest to the continuous optimum at *every* node was obtained. If this point was feasible and gave a smaller objective value, then it was used to update

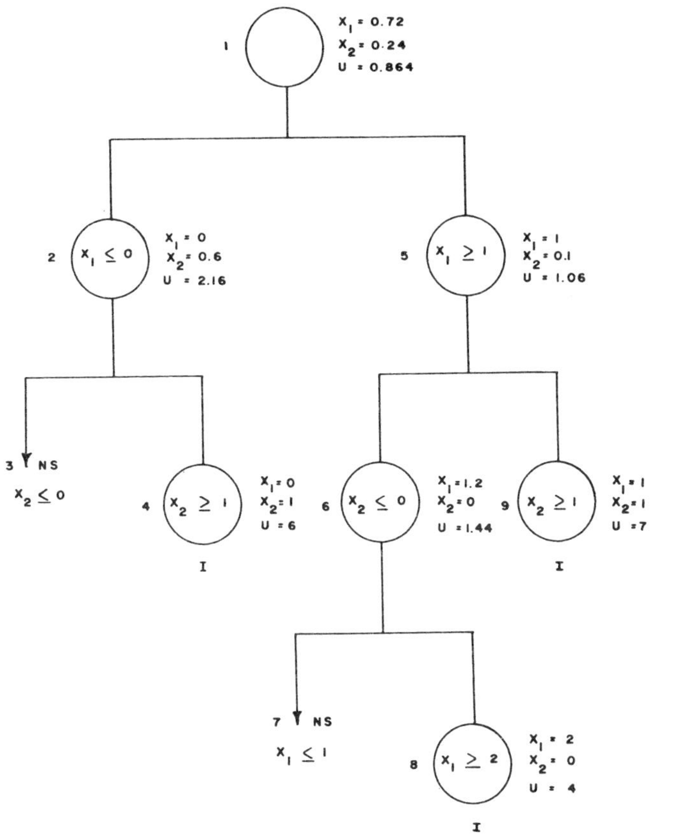


Fig. 1. Tree representation of search procedure.

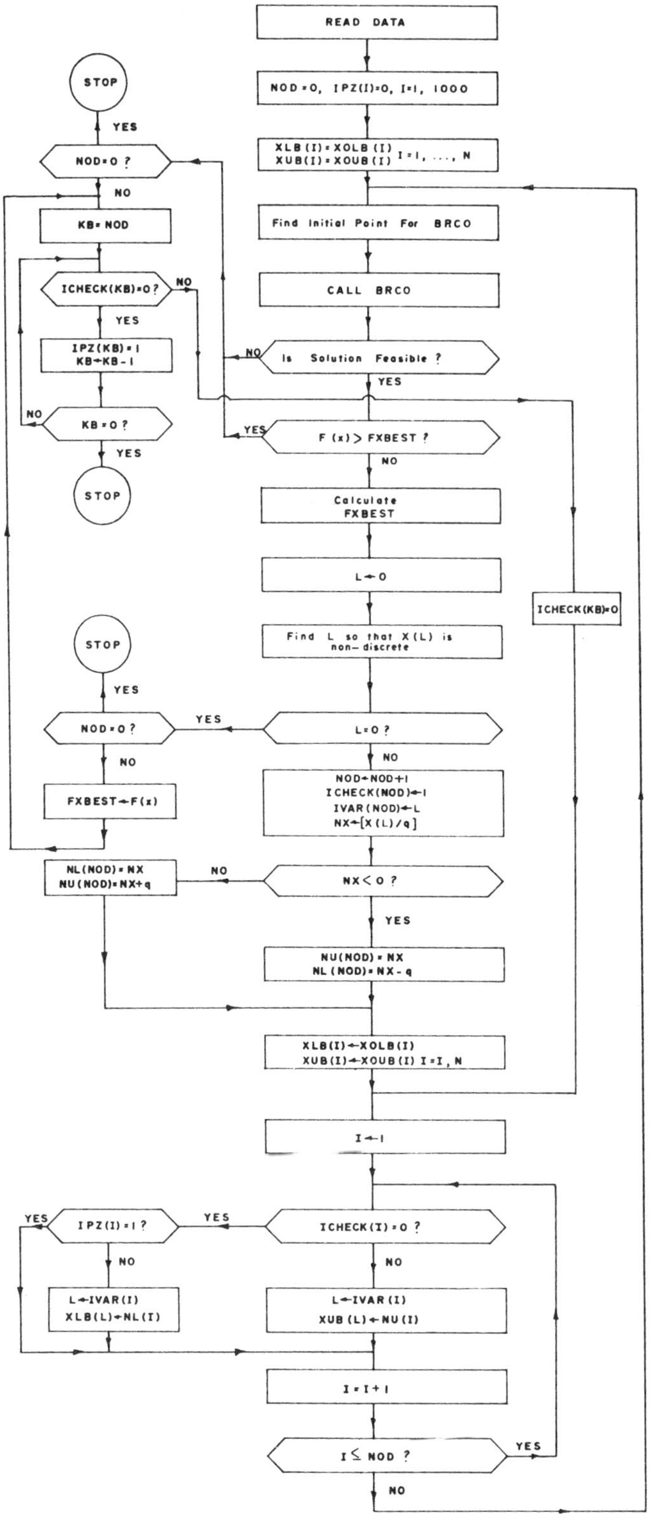


Fig. 2. Flow diagram for Dakin's discrete optimization algorithm.

the upper bound. Obviously the second approach is at least as effective as the first.

Our specific problem has two important properties.

*Property 1:* The stability constraints are linear with respect to the parameters and therefore we need an efficient algorithm which can handle linear constraints. The recent algorithm due to Best and Ritter [9], [10], can handle efficiently linear constraints, has a fast rate of convergence and also requires only one optimization to reach the continuous solution, unlike the well-known barrier methods (e.g., Fiacco and McCormick [13]) which require many continuous optimizations to closely approximate the continuous constrained optimum.

*Property 2:* The discrete step size $q$ is very small (of the order $2^{-6}$), and generally the objective function behaves like a quadratic function in the feasible region. Therefore, we need an algorithm which can handle linear constraints and also, can terminate in a finite number of iterations when the objective function is quadratic.

The algorithm due to Best and Ritter has this additional property and therefore it was used as a subalgorithm.

Fig. 2 shows the flow diagram for the computational procedure.

*The Algorithm of Best and Ritter* [9], [10]

The algorithm of Best and Ritter finds the minimum of a nonlinear function subject to linear constraints. It is a feasible direction method where at each iteration a direction finding linear program is solved. This linear program contains constraints which require the current search direction to be orthogonal to certain differences of gradiants which results in approximately conjugate directions. A method is given in Best [9] to efficiently solve consecutive linear programs. A cycle is terminated when no further decrease in the objective function is possible or when a previously inactive constraint becomes active. If the former case applies, then a special step is used to accelerate the rate of convergence to the optimal point. The orthogonality constraints are then dropped and a new cycle is started with the final point of the previous cycle.

If $n$ denotes the number of variables for the problem and $q_c$ denotes the number of linear constraints which are active (tight or binding) at the optimum, then a cycle will consist of at most $(n - q_c + 1)$ iterations and the rate of convergence of the algorithm will be $(n - q_c + 1)$ step cubic. This means that the number of correct figures in the objective function will approximately triple throughout each cycle. For example, if $n = 5$ and $q_c = 2$ and the starting point given to the algorithm contained 3 correct figures, then after $4(= 5 - 2 + 1)$ iterations approximately 9 correct figures would be obtained.

The procedure outlined in the Appendix for obtaining a starting point for the new optimization close to the optimal solution of the previous optimization is crucial in utilizing the rapid convergence rate of the algorithm.

The algorithm will solve a quadratic program in a finite number of iterations.

If the starting point used for the algorithm is in fact the optimal point, then the algorithm will recognize this and terminate after just one iteration. This is in contrast to penalty function algorithms which would initially move away from the optimum and eventually return to it.

## EXAMPLES

For all the examples considered, we used the value of $p = 2$.

*Example 1: A Wide-Band Differentiator* [1], [7]

Let us suppose that we have the following design specifications for the magnitude characteristic:

$$S(\psi) = \psi, \qquad 0 \leq \psi \leq 1$$

$w(\psi) = 1$ and we want to find the discrete optimum for $Q = 6$ and $Q = 9$ for one-section design.

*Solution:* It was decided to use 21 uniformly spaced sample points of $\psi$.

Starting point:

$$x^0 = [-0.33 \quad -0.67 \quad 0.86 \quad 0.102 \quad 0.366]^T.$$

Steiglitz [1] obtained the following set of coefficients with infinite word length:

$$\begin{aligned} a_1 &= -0.32917379 \\ b_1 &= -0.67082621 \\ c_1 &= 0.85938970 \\ d_1 &= 0.10210106 \\ A &= 0.36637364. \end{aligned}$$

The original rounded coefficients, the optimal coefficients obtained by our approach and the so called optimal obtained by Suk and Mitra [7] are compared in Table I. Tables II and III show the intermediate discrete solutions obtained by the two approaches adopted. Rows one and two show results obtained by using the first and second approach, respectively. The letter $D$ denotes that the objective function obtained by using the closest discrete point has been used as an upper bound. It can be seen from these results that the second approach is considerably more efficient than the first approach. From Tables II and III it can be seen that using our second approach the algorithm took only 96 function evaluations and 49 gradient evaluations for $Q = 6$ and only 75 function evaluations and 41 gradient evaluations for $Q = 9$ to find a better discrete point than the one obtained by Suk and Mitra.

The first algorithm took 82 s and 74 s to search the entire tree structure for $Q = 6$ and $Q = 9$, respectively, while the second algorithm took only 43 s and 42 s to search the entire tree structure with $Q = 6$ and $Q = 9$, respectively.

Figs. 3 and 4 show the optimum error curve for $Q = 6$ and $Q = 9$, respectively.

*Example 2: Low-Pass Filter* [7]

Design a low-pass recursive digital filter of the cascade form, for 10 kHz sampling rate, whose ideal magnitude response is specified by

$$\begin{array}{ll} f = 0, \ 900 \ (100) & S = 1 \\ f = 1000 & S = 1/\sqrt{2} \\ f = 1200 & S = 0 \\ f = 1500, \ 5000, \ (500) & S = 0. \end{array}$$

$w(\psi) = 1$ and we want to find the discrete optimum for $Q = 6$ and $Q = 9$ for one-section design.

*Solution:* The specifications can be prescribed as:

$$\begin{array}{lll} \psi = 0, \ 0.18 \ (0.02); & S(\psi_i) = 1, & i = 1,\cdots,10 \\ \psi = 0.2; & S(\psi_i) = 0.1, & i = 11 \\ \psi = 0.24; & S(\psi_i) = 0, & i = 12 \\ \psi = 0.3, \ 1 \ (0.1); & S(\psi_i) = 0, & i = 13,\cdots,20. \end{array}$$

Starting point:

$$x^0 = [0 \quad 1 \quad -1 \quad 0.5 \quad 0.1]^T.$$

The results obtained by Suk and Mitra [7] and by our approach are shown in Table IV. Tables V and VI show the intermediate discrete results obtained by the two approaches adopted. Again, the second approach was more efficient than the first approach. From Table V it can be seen that although the second approach found the optimum in 189 function evaluations, it took 2502 function evaluations to search the entire tree structure. The results obtained by $Q = 6$ are better than the results obtained by Suk and Mitra with $Q = 9$.

The first algorithm took 42 s and 35 s to search the entire tree structure with $Q = 6$ and $Q = 9$, respectively, while algorithm 2 took 35 s and 31 s to search the entire tree structure with $Q = 6$ and $Q = 9$, respectively.

## CONCLUSIONS

The algorithm due to Dakin for discrete optimization in conjunction with the continuous algorithm due to Best and Ritter was found quite efficient in optimizing recursive digital filters with the coefficients having finite

TABLE I
COEFFICIENTS FOR DIGITAL WIDE-BAND DIFFERENTIATOR

| | Q = 6 Rounded | Q = 6 Suk and Mitra | Q = 6 New Approach | Q = 9 Rounded | Q = 9 Suk and Mitra | Q = 9 New Approach |
|---|---|---|---|---|---|---|
| $2^Q a_1$ | −21 | −22 | −19 | −169 | −169 | −167 |
| $2^Q b_1$ | −43 | −45 | −47 | −343 | −342 | −341 |
| $2^Q c_1$ | 55 | 56 | 59 | 440 | 438 | 439 |
| $2^Q d_1$ | 7 | 7 | 8 | 52 | 51 | 52 |
| $2^Q A$ | 23 | 23 | 23 | 188 | 188 | 188 |
| $U^2(x) \times 10^3$ | 5.8425 | 0.64305 | 0.52233 | 0.38747 | 0.286925 | 0.278135 |

TABLE II
PROGRESS OF THE ALGORITHM, SHOWING INTERMEDIATE DISCRETE RESULTS WITH $Q = 6$, FOR EXAMPLE 1

| | Number of Nodes | Function Evaluations | Gradient Evaluations | $U^2(x) \times 10^3$ |
|---|---|---|---|---|
| D | 1 | 25 | 14 | 5.84269 |
| D | 1 | 25 | 14 | 5.84269 |
| | 20 | 417 | 225 | 5.40648 |
| D | 2 | 71 | 36 | 3.04873 |
| | 21 | 433 | 235 | 5.26012 |
| D | 3 | 96 | 49 | 0.52798 |
| | 24 | 501 | 274 | 4.99535 |
| | 16 | 546 | 296 | 0.52798 |
| | 25 | 506 | 278 | 3.93481 |
| | 50 | 1880 | 1007 | 0.52233 (optimal) |
| | 31 | 618 | 345 | 3.74253 |
| (Approach 2 completed) | 62 | 2319 | 1239 | 0.52233 (optimal) |
| | ⋮ | ⋮ | ⋮ | ⋮ |
| | 151 | 4115 | 2278 | 0.52233 (optimal) |

TABLE III
PROGRESS OF THE ALGORITHM, SHOWING INTERMEDIATE DISCRETE RESULTS WITH $Q = 9$, FOR EXAMPLE 1

| | Number of Nodes | Function Evaluations | Gradient Evaluations | $U^2(x) \times 10^3$ |
|---|---|---|---|---|
| D | 1 | 25 | 14 | 0.387425 |
| D | 1 | 25 | 14 | 0.387425 |
| | 27 | 487 | 268 | 0.382260 |
| D | 3 | 55 | 31 | 0.306297 |
| | 28 | 514 | 284 | 0.377711 |
| D | 4 | 75 | 41 | 0.280035 |
| | 36 | 651 | 362 | 0.369385 |
| | 11 | 301 | 165 | 0.280035 |
| | 37 | 665 | 371 | 0.359649 |
| D | 25 | 697 | 380 | 0.279822 |
| | ⋮ | ⋮ | ⋮ | ⋮ |
| | 172 | 3957 | 2183 | 0.278135 (optimal) |
| | 79 | 2285 | 1206 | 0.278135 |

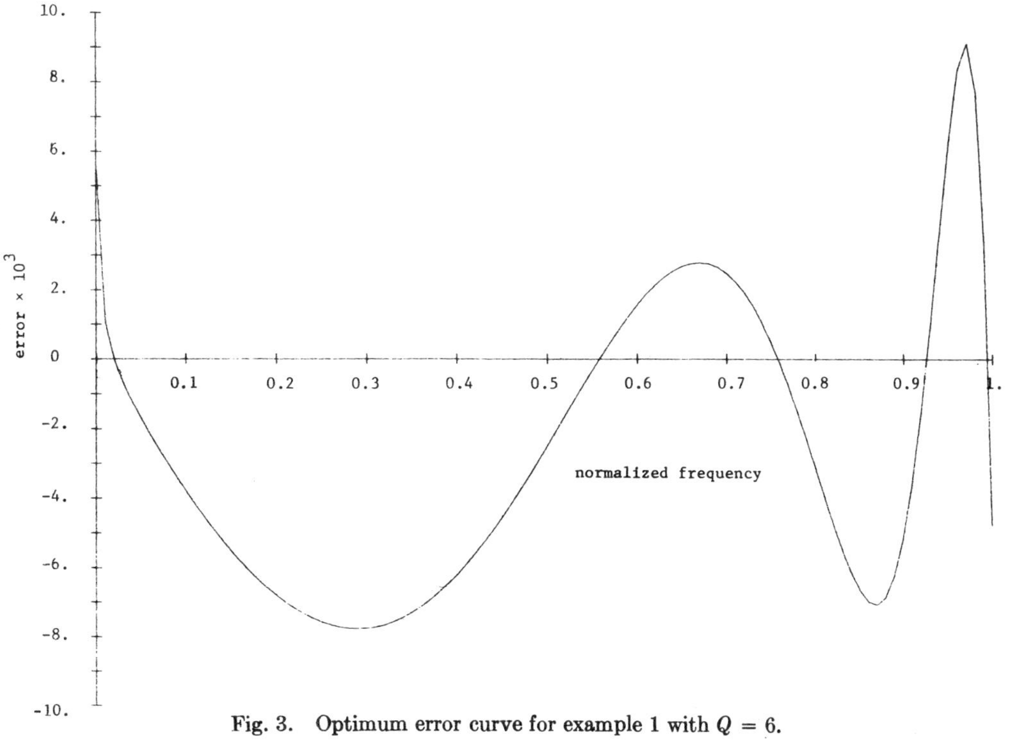


Fig. 3. Optimum error curve for example 1 with $Q = 6$.

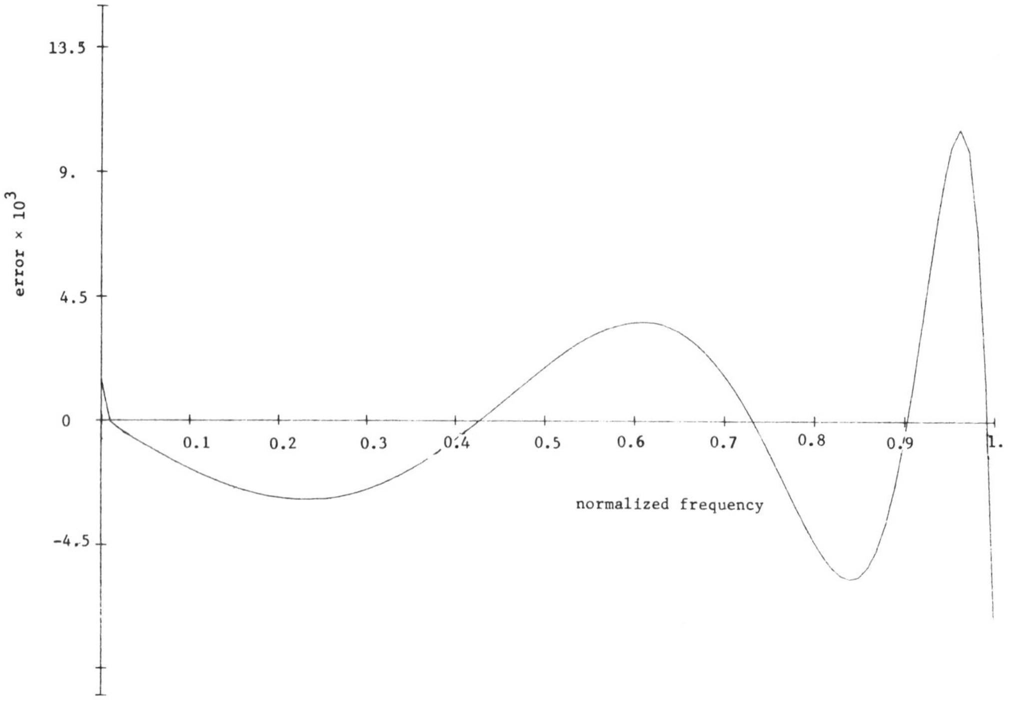


Fig. 4. Optimum error curve for example 1 with $Q = 9$.

TABLE IV
COEFFICIENTS FOR DIGITAL LOW-PASS FILTER

| | $Q = 6$ | | $Q = 9$ | |
|---|---|---|---|---|
| | Suk and Mitra | New Approach | Suk and Mitra | New Approach |
| $2^Q a_1$ | −16 | −21 | −104 | −185 |
| $2^Q b_1$ | 84 | 65 | 566 | 510 |
| $2^Q c_1$ | −92 | −93 | −728 | −740 |
| $2^Q d_1$ | 42 | 42 | 322 | 333 |
| $2^Q A$ | 6 | 7 | 52 | 59 |
| $U^2(\mathbf{x})$ | 0.31535 | 0.29059 | 0.29487 | 0.28999 |

TABLE V
PROGRESS OF THE ALGORITHM, SHOWING INTERMEDIATE DISCRETE RESULTS WITH $Q = 6$, FOR EXAMPLE 2

| | Number of Nodes | Function Evaluations | Gradient Evaluations | $U^2(x)$ |
|---|---|---|---|---|
| $D$ | 1 | 128 | 56 | 0.301527 |
| $D$ | 1 | 128 | 56 | 0.301527 |
| | 11 | 343 | 170 | 0.295938 |
| $D$ | 3 | 189 | 85 | 0.290592 (optimal) |
| | 15 | 433 | 221 | 0.295155 |
| | 6 | 259 | 124 | 0.290592 |
| | 16 | 470 | 241 | 0.294909 |
| (Approach 2 completed) | 27 | 2116 | 1006 | 0.290592 |
| | 22 | 639 | 330 | 0.294740 |
| | 23 | 645 | 334 | 0.290592 |
| | 43 | 2502 | 1216 | 0.290592 |

TABLE VI
PROGRESS OF THE ALGORITHM, SHOWING INTERMEDIATE DISCRETE RESULTS WITH $Q = 9$, FOR EXAMPLE 2

| | Number of Nodes | Function Evaluations | Gradient Evaluations | $U^2(x)$ |
|---|---|---|---|---|
| $D$ | 1 | 128 | 56 | 0.290453 |
| $D$ | 1 | 128 | 56 | 0.290453 |
| | 6 | 208 | 102 | 0.290108 |
| $D$ | 2 | 148 | 67 | 0.290031 |
| | 12 | 344 | 177 | 0.290107 |
| $D$ | 5 | 283 | 142 | 0.290009 |
| | 14 | 359 | 188 | 0.290106 |
| $D$ | 6 | 289 | 146 | 0.290008 |
| | 16 | 401 | 211 | 0.290055 |
| $D$ | 9 | 339 | 172 | 0.29000798 |
| | 23 | 523 | 281 | 0.290031 |
| | 12 | 409 | 210 | 0.29000798 |
| | 28 | 626 | 335 | 0.290008 |
| $D$ | 32 | 988 | 512 | 0.290000 |
| | ⋮ | ⋮ | ⋮ | ⋮ |
| | 70 | 2055 | 1036 | 0.289994 (optimal) |
| | 53 | 1817 | 898 | 0.289994 |

word length. Unlike most of the other methods published to solve the problem stated previously, our approach is based on mathematical basis and also on the use of an efficient gradient algorithm which can handle linear constraints. As the results indicate, algorithm 2 gives, in general, good discrete results in a short time, although it might require much more time to search the entire tree structure. Although the value of $p = 2$ was used in the examples, any value of $p$ in the range $1 < p < \infty$ can be used, but in general it will require much more time to reach the discrete optimum. The next stage of the research is to investigate a number of modifications to the algorithm with the goal of reducing the overall computation time.

## APPENDIX

The following is a list of the important variables and subroutines used in the algorithm flow chart shown in Fig. 2.

| | |
|---|---|
| DATA | Starting point, initial upper and lower bounds on the parameters, initial value of FXBEST, stability constraints value of $q$ and number of variables. |
| XOLB(I),XOUB(I) | Original lower and upper bounds on component I. |
| FXBEST | Current best upper bound for the objective function. |
| XLB(I),XUB(I) | Lower and upper bound on component I given to BRCO, for the current continuous subproblem. |
| BRCO | Best–Ritter continuous optimization algorithm. |
| Initial Point for BRCO | The optimal solution for the last continuous subproblem is used as tentative initial point. If some component violates one of the bounds for the current continuous subproblem then that component is replaced by that bound. The point so obtained is then checked for feasibility with respect to the stability constraints. If it is feasible then it is used as the starting point for BRCO. If not, a starting point is obtained using a phase 1 simplex procedure. |
| Calculate FXBEST | Update FXBEST if the discrete point closest to the continuous optimum is feasible and the objective function value at this point is smaller than the previous FXBEST. |

## ACKNOWLEDGMENT

The authors wish to thank Dr. J. W. Bandler, B. Bardakjian, and J. Chen of McMaster University, Hamilton, Ont., Canada, for useful discussions.

## REFERENCES

[1] K. Steiglitz, "Computer-aided design of recursive digital filters," *IEEE Trans. Audio Electroacoust.* (*Special Issue on Digital Filtering*), vol. AU-18, pp. 123–129, June 1970.

[2] A. G. Deczky, "Synthesis of recursive digital filters using the minimum $p$-error criterion," *IEEE Trans. Audio Electroacoust.* (*Special Issue on Digital Filtering*), vol. AU-20, pp. 257–263, Oct. 1972.

[3] J. W. Bandler and B. L. Bardakjian, "Least $p$th optimization of recursive digital filters," *IEEE Trans. Audio Electroacoust.* vol. AU-21, pp. 460–470, Oct. 1973.

[4] C. Charalambous, "Minimax design of recursive digital filters," *Comput. Aided Des.*, vol. 6, pp. 73–81, Apr. 1974.

[5] K. Steiglitz, "Designing short-word recursive digital filters," in *Proc. 9th Annu. Allerton Conf. Circuit and Systems Theory* (Urbana, Ill.), Oct. 1971, pp. 778–788.

[6] E. Evanhaus, "On the design of digital filters with coefficients of limited word length," *IEEE Trans. Audio Electroacoust.*, vol. AU-20, pp. 206–212, Aug. 1972.

[7] M. Suk and S. K. Mitra, "Computer-aided design of digital filters with finite word length," *IEEE Trans. Audio Electroacoust.*, vol. AU-20, pp. 356–363, Dec. 1972.

[8] R. J. Dakin, "A tree-search algorithm for mixed integer programming problems," *Comput. J.*, vol. 8, pp. 250–255, 1966.

[9] M. J. Best, "A feasible conjugate direction method to solve linearly constrained optimization problems," *J. Optimiz. Theory Appl.*, vol. 16, pp. 25–38, July 1975.

[10] M. J. Best and K. Ritter, "An accelerated conjugate direction method to solve linearly constrained minimization problems," Univ. of Waterloo, Waterloo, Ont., Canada, Res. Rep. CORR 73-16, 1973; also, *J. Comput. Syst. Sci.*, to be published.

[11] J. W. Bandler and C. Charalambous, "Practical least $p$th optimization of networks," *IEEE Trans. Microwave Theory Tech.*, vol. MTT-20, pp. 834–840, Dec. 1972.

[12] C. Charalambous, "Nonlinear least $p$th approximation and nonlinear programming with applications in the design of networks and systems," Ph.D. dissertion, McMaster Univ., Hamilton, Ont., Canada, Feb. 1973.

[13] A. V. Fiacco and G. P. McCormick, "The sequential unconstrained optimization technique for nonlinear programming, a primal-dual method," *Management Sci.*, vol. 10, pp. 360–366, Jan. 1964.

# A Bound on Limit Cycles in Fixed-Point Implementations of Digital Filters

IRWIN W. SANDBERG, Member, IEEE
JAMES F. KAISER, Senior Member, IEEE
Bell Telephone Lab., Inc.
Murray Hill, N. J. 07974

Abstract

**We present an upper bound on the rms value of self-sustained limit cycles in fixed-point implementations of digital filters. The bound can be easily evaluated for the important special case of sections of order two, and simulation results for this case show that the bound is a useful tool.**

## I. Introduction

The primary purpose of this paper is to present an upper bound on the rms value of self-sustained limit cycles due to roundoff in fixed-point implementations of digital filters. We consider in greatest detail the important practical case of second-order sections. However, our main result, the theorem proved in the Appendix, is applicable to sections of arbitrary order.

## II. Second-Order Sections

If $\{y_n\}_{-2}^{\infty}$ is the sequence of machine numbers that corresponds to the zero-input response of a second-order digital filter, then

$$y_n = -ay_{n-1} - by_{n-2} + \delta_n, \qquad n \geq 0 \tag{1}$$

in which $a$ and $b$ are fixed real constants and $\{\delta_n\}_0^{\infty}$ takes into account the effects of roundoff. If, for example, $q(\cdot)$ denotes the standard quantization function defined by the condition that for all real $x$, $q(x)$ is equal to the integer part of $[x+0.5 \text{ sign } (x)]$, and if

$$y_n = q(-ay_{n-1}) + q(-by_{n-2}), \qquad n \geq 0 \tag{2}$$

(i.e., if each product is rounded and the rounded products are added) then $|\delta_n| \leq 1$ for all $n \geq 0$. Alternatively, if

$$y_n = q(-ay_{n-1} - by_{n-2}), \qquad n \geq 0, \tag{3}$$

then it is clear that $|\delta_n| \leq 0.5$ for all $n \geq 0$. It is assumed throughout that the infinite-precision counterpart of the digital filter is asymptotically stable (i.e., that $1+az^{-1}+bz^{-2} \neq 0$ for all $|z| \geq 1$).

Let us suppose that the numbers $y_{-2}, y_{-1}, y_0, y_1, \cdots$ of (1) approach[1] a limit cycle of period $(K+1)$ as $n \to \infty$, in which $K$ is an arbitrary nonnegative integer. More explicitly, let $(y_n - p_n) \to 0$ as $n \to \infty$ and $(\delta_n - s_n) \to 0$ as $n \to \infty$ in which $\{p_n\}_{-\infty}^{\infty}$ and $\{s_n\}_{-\infty}^{\infty}$ satisfy $p_n = p_{n+K+1}$ and $s_n = s_{n+K+1}$ for all $n$. Let $\|P\|$ and $\|S\|$ denote, respectively,

$$\left(\frac{1}{K+1}\sum_{n=0}^{K} p_n^2\right)^{1/2} \quad \text{and} \quad \left(\frac{1}{K+1}\sum_{n=0}^{K} s_n^2\right)^{1/2}$$

the root-mean-square values of the response sequence $\{p_n\}_{-\infty}^{\infty}$ and the roundoff sequence $\{s_n\}_{-\infty}^{\infty}$.

According to a special case of the theorem proved in the Appendix, we have

$$\min_l \phi(l)\|S\| \leq \|P\| \leq \max_l \phi(l)\|S\| \tag{4}$$

in which $l$ varies over $0, 1, \cdots, K$ and

$$\phi(l) = |H\{\exp[i2\pi l/(K+1)]\}|,$$

and $H(z) = (1+az^{-1}+bz^{-2})^{-1}$ for all $|z| = 1$. The left inequality of (4) is included for completeness. We shall not discuss it further here.

For the simple special cases in which $K=0$ and $K=1$ (i.e., for the dc and $\frac{1}{2}f_s$ situations), respectively,

$$\max_l \phi(l) = (1+b+a)^{-1} \tag{5}$$

and

$$\max_l \phi(l) = (1+b-|a|)^{-1}. \tag{6}$$

The expressions that appear on the right sides of (5) and (6) appear also in related upper bounds [1], [2] on the maximum absolute value of $p_n$ for the cases $K=0$ and $K=1$. While bounds for such special situations are of interest, it is ordinarily not the case that the period $(K+1)$ of the limit cycle is known at the outset. Therefore, we proceed as follows.

Since it is certainly true that

$$\max_l \phi(l) \leq \max_{\theta \in [0, 2\pi]} |H(e^{i\theta})|$$

for all $K$, and since it can be shown that

$$\max_{\theta \in [0, 2\pi]} |H(e^{i\theta})| = \rho(a, b)$$

in which

$$\rho(a,b) = \begin{cases} \dfrac{1}{(1-b)\left(1-\dfrac{a^2}{4b}\right)^{1/2}}, & \text{for } b>0 \text{ and } |a| \leq \dfrac{4b}{1+b} \\[2ex] \dfrac{1}{1+b-|a|}, & \text{for either } b \leq 0 \text{ or } b>0 \text{ and } |a| \geq \dfrac{4b}{1+b} \end{cases}$$

we see that

$$\|P\| \leq \rho(a,b)\|S\|, \tag{7}$$

which is our main result concerning second-order sections. Referring to (1), if $\delta$ is the smallest positive number such that

Manuscript received August 27, 1971.

[1] Of course, since the digital filter is a finite-state machine, the limit cycle will in fact be reached in a finite number of sampling periods.

Reprinted from *IEEE Trans. Audio Electroacoust.*, vol. AU-20, pp. 110–112, June 1972.

$|\delta_n| \le \delta$ for all $n \ge 0$, then obviously $\|S\| \le \delta$, and hence[2]

$$\|P\| \le \delta\rho(a, b). \tag{8}$$

By considering certain simulation results reported by Jackson [1] and by Parker and Hess [3], we can obtain some understanding of the extent to which our bound (7) is not grossly pessimistic. For the case of complex-conjugate poles of $H(z)$, $\delta = 1$, and rounding after each multiplication as in (2), Jackson, using an "effective-value model," was led to the "bound"

$$\max_n |p_n| \le \text{integer part of } [0.5(1-b)^{-1}] \tag{9}$$

which he reported was never exceeded in his simulations by more than unity (i.e., by more than one level). Recently Parker and Hess [3] have presented an example in which the right side of (9) is 11 and $\max_n |p_n|$ is 17. They have suggested, on the basis of their simulation studies, that (9) should be replaced by

$$\max_n |p_n| \le 1.5(1-b)^{-1}. \tag{10}$$

The ratio of the right side of (8) for $b > 0$ and $|a| \le 4b(1+b)^{-1}$ and with $\delta = 1$ to the right side of (10) is

$$(1.5)^{-1}\left(1 - \frac{a^2}{4b}\right)^{-1/2}. \tag{11}$$

This ratio of two different types of bounds does not differ from unity by more than a factor of, say, three or four for a large range of useful design values of $a^2/4b$. Thus, on the strength of this basically qualitative argument, we see that our bound is a useful bound for at least a large class of design parameters.

## Appendix

### The Theorem

#### Notation and Definitions

Let $m$ be a positive integer, and let $c_1, c_2, \cdots, c_m$ be (real or complex) constants such that

$$1 + \sum_{k=1}^{m} c_k z^{-k} \ne 0, \qquad \text{for } |z| \ge 1. \tag{12}$$

Let $\{y_n\}_{-m}^{\infty}$ and $\{\delta_n\}_0^{\infty}$ be sequences of constants such that

$$y_n + \sum_{k=1}^{m} c_k y_{n-k} = \delta_n, \qquad n \ge 0 \tag{13}$$

and $(\delta_n - s_n) \to 0$ as $n \to \infty$ for some sequence $\{s_n\}_{-\infty}^{\infty}$ with the property that $s_n = s_{n+K+1}$ for $n = 0, \pm 1, \pm 2, \cdots$, in which $K$ is a nonnegative constant.

Let $\{p_n\}_{-\infty}^{\infty}$ denote the sequence with the properties that $p_n = p_{n+K+1}$ for $n = 0, \pm 1, \pm 2, \cdots$ and $(y_n - p_n) \to 0$ as $n \to \infty$, and let $\|S\|$ and $\|P\|$ denote

$$\left(\frac{1}{K+1}\sum_{n=0}^{K} |s_n|^2\right)^{1/2} \quad \text{and} \quad \left(\frac{1}{K+1}\sum_{n=0}^{K} |p_n|^2\right)^{1/2},$$

respectively. Finally, let $\mathcal{R} = \{0, 1, 2, \cdots, K\}$, and

$$H(e^{i\theta}) = \sum_{n=0}^{\infty} h_n e^{-in\theta}$$

for $\theta \in [0, 2\pi]$ in which $\{h_n\}_0^{\infty}$ is the inverse $z$ transform of

$$\left(1 + \sum_{k=1}^{m} c_k z^{-k}\right)^{-1}.$$

*Theorem:*

$$\min_{l \in \mathcal{R}} \phi(l)\|S\| \le \|P\| \le \max_{l \in \mathcal{R}} \phi(l)\|S\|$$

in which $\phi(l) = |H\{\exp[i2\pi l/(K+1)]\}|$.

*Proof:*[3]

From (12) and (13) it is clear that

$$y_n = r_n + \sum_{k=0}^{n} h_{n-k}\delta_k, \qquad n \ge 0 \tag{14}$$

in which $r_n \to 0$ as $n \to \infty$. Since $s_n = s_{n+K+1}$ for $n = 0, \pm 1, \pm 2, \cdots$, we have

$$s_k = \sum_{l=0}^{K} \hat{s}_l \exp\left(\frac{i2\pi lk}{K+1}\right)$$

with

$$\hat{s}_l = (K+1)^{-1} \sum_{n=0}^{K} s_n \exp\left(-\frac{i2\pi ln}{K+1}\right)$$

and

$$\sum_{n=0}^{K} |s_n|^2 = (K+1) \sum_{n=0}^{K} |\hat{s}_n|^2.$$

Moreover, we can show, using (12) and (14), that

$$p_n = \sum_{k=-\infty}^{n} h_{n-k} s_k$$

for all $n$.[4] Therefore

$$\begin{aligned} p_n &= \sum_{k=-\infty}^{n} h_{n-k} \sum_{l=0}^{K} \hat{s}_l \exp\left(\frac{i2\pi lk}{K+1}\right) \\ &= \sum_{l=0}^{K} \hat{s}_l \sum_{k=-\infty}^{n} h_{n-k} \exp\left(\frac{i2\pi lk}{K+1}\right) \\ &= \sum_{l=0}^{K} \hat{s}_l \exp\left(\frac{i2\pi ln}{K+1}\right) \sum_{m=0}^{\infty} h_m \exp\left(-\frac{i2\pi lm}{K+1}\right) \\ &= \sum_{l=0}^{K} H\left[\exp\left(\frac{i2\pi l}{K+1}\right)\right] \hat{s}_l \exp\left(\frac{i2\pi ln}{K+1}\right). \end{aligned}$$

Thus

[2] If the sequence $\{s_n\}_{-\infty}^{\infty}$ is approximately the closely spaced samples of a sinusoid (as is often observed in simulations), then $\|S\|$ is approximately equal to $(0.707)\delta$. This estimate of $\|S\|$ can often be used to improve the engineering utility of (7).

[3] The proof given here is similar to a portion of a proof given in [4, proof of Theorem 2]. Our theorem is similar to results proved earlier [5].

[4] We note in passing that

$$\max_n |p_n| \le \max_n |s_n| \sum_{n=0}^{\infty} |h_n|,$$

which might be of use in other work.

$$\begin{aligned}\|P\|^2 &= \sum_{l=0}^{K} \phi(l)^2 \,|\, \hat{s}_l \,|^2 \\ &= \max_{l \in \mathcal{R}} \phi(l)^2 \sum_{l=0}^{K} |\, \hat{s}_l \,|^2 \\ &\leq \max_{l \in \mathcal{R}} \phi(l)^2 \|S\|^2\end{aligned}$$

and, similarly,

$$\|P\|^2 \geq \min_{l \in \mathcal{R}} \phi(l)^2 \|S\|^2$$

which proves the theorem.

## References


[1] L. B. Jackson, "An analysis of limit cycles due to multiplication rounding in recursive digital (sub) filters," in *Proc. 7th Annu. Allerton Conf. Circuit and Syst. Theory*, Oct. 1969, pp. 69–78.

[2] F. Bonzanigo, "Constant-input behavior of recursive digital filters," presented at the IEEE Arden House Workshop on Digital Filtering, Harriman, N. Y., Jan. 1970.

[3] S. R. Parker and S. F. Hess, "Limit-cycle oscillations in digital filters," *IEEE Trans. Circuit Theory*, vol. CT-18, pp. 687–697, Nov. 1971.

[4] I. W. Sandberg, "Some properties of a classic numerical integration formula," *Bell Syst. Tech. J.*, vol. 46, pp. 2061–2080, Nov. 1967.

[5] ——, "Two theorems on the accuracy of numerical solutions of systems of ordinary differential equations," *Bell Syst. Tech. J.*, vol. 46, pp. 1243–1266, July–Aug. 1967.

# An Absolute Bound on Limit Cycles Due to Roundoff Errors in Digital Filters

JAMES L. LONG and TIMOTHY N. TRICK

*Abstract*—An absolute bound on limit cycle oscillations in fixed-point digital filter implementations due to roundoff errors is presented. Periodicity of the limit cycles is assumed in the derivation. Useful design results are explicitly given for the case of second-order filter sections. In addition it is shown that this bound is equal to the rms bound of Sandberg and Kaiser for real roots, and is never more than a factor of two greater than the rms bound for complex roots for second-order filters.

## I. Introduction

In the design of fixed-point digital filters one is particularly interested in determining the minimum bit-per-word requirement for the internal arithmetic. Bounds on the maximum value of the error due to roundoff induced limit cycles are of prime importance in making that determination. In this paper, a new bound on the maximum amplitude of limit cycles in fixed-point digital filters is presented.

## II. Derivation of an Absolute Bound for Periodic Limit Cycles

Consider the $N$th-order nonlinear difference equation (1) representing the direct form digital-filter realization

$$y(n) = -[\alpha_1 y(n-1)]_r - [\alpha_2 y(n-2)]_r - \cdots - [\alpha_N y(n-N)]_r + x(n) \quad (1)$$

where $[\cdot]_r$ denotes the standard quantization function. The coefficients $\alpha_1, \alpha_2, \cdots, \alpha_N$ and the numbers $y(-N), y(-N+1), \cdots, y(-1), y(0), \cdots$ are assumed to be fixed-point machine numbers. Roundoff after multiplication of these numbers accounts for the need of quantization functions in the otherwise linear equation (1). Each product is rounded before addition takes place, as is often the case in practical filters, to alleviate the need for double precision registers. Throughout this paper the independent forcing term $x(n)$ will be taken to be zero and the coefficients in (1) are such that the equation without quantizers is asymptotically stable.

Equation (1) may be replaced by a linear difference equation driven by roundoff induced error sequences $e_1(n), e_2(n), \cdots, e_N(n)$ as follows:

$$y(n) = -\alpha_1 y(n-1) + e_1(n) - \alpha_2 y(n-2) + e_2(n) - \cdots - \alpha_N y(n-N) + e_N(n). \quad (2)$$

Assuming a normalized quantization width of 1, the error terms in (2) are bounded in magnitude of $\frac{1}{2}$. A single error term may be defined as follows:

$$e(n) \triangleq e_1(n) + e_2(n) + \cdots + e_N(n) \quad (3)$$

where $|e(n)| \leq N/2$.

Now suppose that the limit cycle is periodic of period $M$; then the error input sequence has the property that

$$e(k) = e(k+M) \quad (4)$$

for all $k$. The steady-state output due to the roundoff induced error $e(n)$ is given by the discrete convolution

$$y(n) = \sum_{k=-\infty}^{n} h(n-k)e(k) \quad (5)$$

which can be decomposed in a manner directly analogous to the method of Trick and Anderson [3] for the case of a continuous convolution. First, (5) is expressed as a double summation

$$y(n) = \sum_{i=0}^{\infty} \left[ \sum_{k=n-(i+1)M+1}^{k=n-iM} h(n-k)e(k) \right]. \quad (6)$$

Letting $p \triangleq n - iM - k$ one obtains

$$y(n) = \sum_{i=0}^{\infty} \left[ \sum_{p=0}^{p=M-1} h(p+iM)e(n-iM-p) \right]. \quad (7)$$

Interchanging the summation in (7) and noting that $e(n-p-iM) = e(n-p)$ we have

$$y(n) = \sum_{p=0}^{M-1} e(n-p) \left[ \sum_{i=0}^{\infty} h(p+iM) \right]. \quad (8)$$

The desired bound follows directly from (8) as

$$|y(n)|_{\max} = \frac{N}{2} \sum_{p=0}^{M-1} \left| \sum_{i=0}^{\infty} h(p+iM) \right| \quad (9)$$

for any value of $n$.

A bound on the right-hand side of (9) for any $M$ and the asymptotic behavior for large $M$ are of particular interest. Considering (9) it is obvious that

$$\sum_{p=0}^{M-1} \left| \sum_{i=0}^{\infty} h(p+iM) \right| \leq \sum_{p=0}^{M-1} \sum_{i=0}^{\infty} |h(p+iM)|. \quad (10)$$

Letting $k \triangleq p + iM$ and interchanging order of summation in (10) gives

Manuscript received July 28, 1972; revised September 28, 1972. This work was supported by the Joint Services Electronics Program (U. S. Army, U. S. Navy, and U. S. Air Force) under Contract DAAB-07-67-C-0199.

The authors are with the Coordinated Science Laboratory, University of Illinois, Urbana, Ill. 61801.

Reprinted from *IEEE Trans. Audio Electroacoust.*, vol. AU-21, pp. 27–30, Feb. 1973.

$$\sum_{p=0}^{M-1} \sum_{i=0}^{\infty} | h(p + iM) | = \sum_{i=0}^{\infty} \sum_{k=iM}^{k=(i+1)M-1} | h(k) | = \sum_{k=0}^{\infty} | h(k) | . \quad (11)$$

Combining (9) and (11) one gets

$$| y(n) |_{\max} \leq \frac{N}{2} \sum_{k=0}^{\infty} | h(k) | \quad (12)$$

which is independent of the value of $n$ and $M$. Utilizing the asymptotic stability property assumed on $h(p)$ one can show that

$$\lim_{M\to\infty} \sum_{p=0}^{M-1} \left| \sum_{i=0}^{\infty} h(p + iM) \right| = \sum_{p=0}^{\infty} | h(p) | . \quad (13)$$

### III. Absolute Bound for Second-Order Filters

Equation (9) can be reduced to an explicit function of the filter coefficients for the commonly used second-order filter. Consider the second-order transfer function

$$H(z) = \frac{z^2}{z^2 + az + b}, \quad (14)$$

where letters $a$ and $b$ are used in place of $\alpha_1$ and $\alpha_2$ to agree with standard notation. The inversion formula for (14) is

$$h(p) = \frac{1}{2\pi j} \int_c \frac{z^2}{z^2 + az + b} \cdot z^{p-1} dz \quad (15)$$

where $c$ is taken to be the unit circle; therefore,

$$\sum_{i=0}^{\infty} h(p + iM) = \sum_{i=0}^{\infty} \left\{ \frac{1}{2\pi j} \int_c \frac{z^{p+iM+1}}{z^2 + az + b} dz \right\} . \quad (16)$$

Combining the results obtained by evaluating the right-hand side of (16) by the residue theorem with (9), one gets for the case of distinct roots of the equation $z^2+az+b$

$$| y(n) |_{\max} = \sum_{p=0}^{M-1} \left| \frac{1}{2\sqrt{\left(\frac{a}{2}\right)^2 - b}} \left[ \frac{\left(-\frac{a}{2} + \sqrt{\left(\frac{a}{2}\right)^2 - b}\right)^{p+1}}{1 - \left(-\frac{a}{2} + \sqrt{\left(\frac{a}{2}\right)^2 - b}\right)^M} - \frac{\left(-\frac{a}{2} - \sqrt{\left(\frac{a}{2}\right)^2 - b}\right)^{p+1}}{1 - \left(-\frac{a}{2} - \sqrt{\left(\frac{a}{2}\right)^2 - b}\right)^M} \right] \right| \quad (17)$$

and for the repeated root case

$$| y(n) |_{\max} = \sum_{p=0}^{M-1} \left| \left( -\frac{a}{2} \right)^p \left[ \frac{p}{\left[ 1 - \left( -\frac{a}{2} \right)^M \right]} + \frac{1 + \left( -\frac{a}{2} \right)^M (M - 1)}{\left[ 1 - \left( -\frac{a}{2} \right)^M \right]^2} \right] \right| . \quad (18)$$

Note that (17) and (18) are computationally more attractive than the "matrix" method of Parker and Hess [2].

Two important special cases of the "periodic" bound (17) are when $M=1$ and $M=2$, corresponding to frequencies of dc and one-half the sampling frequency:

$$| y(n) |_{\max} = \frac{1}{1 + a + b}, \qquad M = 1 \quad (19)$$

$$| y(n) |_{\max} = \frac{1}{1 + b - | a |}, \qquad M = 2. \quad (20)$$

These bounds are the same as those derived by Jackson [5] for each of the above special cases, and they are also identical to the results obtained by Parker and Hess [2] with their matrix bound and by Kaiser and Sandberg [1] with their rms bound.

In general, the period of the limit cycle is not known, which makes the summation of the absolute value of the impulse response an attractive bound since it is the least upper bound on (17) independent of $M$. The evaluation of $\sum_{n=0}^{\infty} |h(n)|$ can be done exactly for the cases of real roots by straightforward summation of terms yielding the following:

$$\sum_{n=0}^{\infty} | h(n) | = \frac{1}{1 - | a | + b} \quad (21)$$

for distinct real roots, and

$$\sum_{n=0}^{\infty} | h(n) | = \frac{1}{(1 - \sqrt{b})^2} \quad (22)$$

for a double real root where $a^2=4b$.

Obtaining an exact expression for the case of complex roots is difficult, since the impulse response of a second-order filter with complex roots has the form

$$h(n) = \frac{r^n}{\sin \theta} \sin (n + 1)\theta \quad (23)$$

where $r=\sqrt{b}$, $\theta=\cos^{-1}(-a/2\sqrt{b})$, and $4b>a^2$. In this case the least upper bound on the absolute value of the roundoff error is

$$\sum_{n=0}^{\infty} | h(n) | = \sum_{n=0}^{\infty} \left| \frac{r^n}{\sin \theta} \sin (n + 1)\theta \right| . \quad (24)$$

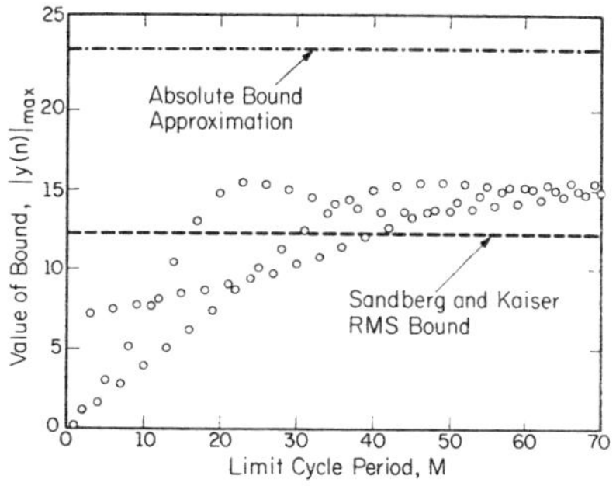


**Fig. 1. Amplitude bound for $a=1.09$ and $b=0.9$.**

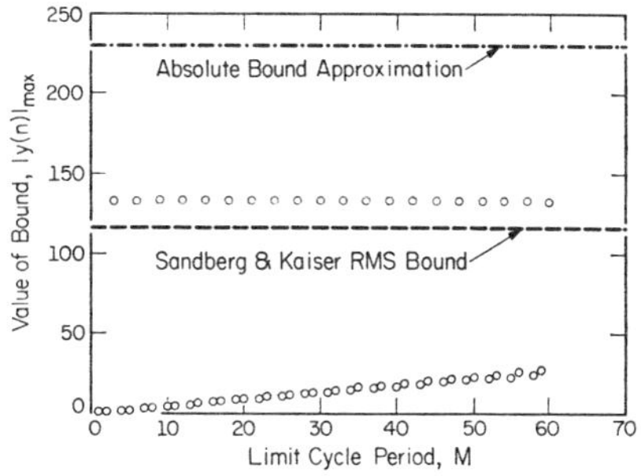


**Fig. 2. Amplitude bound for $a=1.0$ and $b=0.99$.**

The following two estimates of (24) can be useful for design purposes:

$$\sum_{n=0}^{\infty} | h(n) | \leq \sum_{n=0}^{\infty} \left| \frac{r^n}{\sin \theta} \right| = \frac{1+\sqrt{b}}{(1-b)\sqrt{1-\frac{a^2}{4b}}} \qquad (25)$$

or

$$\sum_{n=0}^{\infty} | h(n) | \leq \sum_{n=0}^{\infty} (n+1) r^n = \frac{1}{(1-\sqrt{b})^2}. \qquad (26)$$

Thus the absolute value of the roundoff error is bounded by

$$| y(n) |_{\max} \leq \begin{cases} \dfrac{1}{1-|a|+b}, & \text{for either } b \leq 0 \text{ or } b > 0 \text{ and } 2\sqrt{b} \leq |a| \qquad (27) \\ \dfrac{1}{(1-\sqrt{b})^2}, & \text{for } b > 0 \text{ and } 2b\sqrt{\dfrac{2}{\sqrt{b}}-1} \leq |a| \leq 2\sqrt{b} \qquad (28) \\ \dfrac{(1+\sqrt{b})}{(1-b)\sqrt{1-\dfrac{a^2}{4b}}}, & \text{for } b > 0 \text{ and } |a| \leq 2b\sqrt{\dfrac{2}{\sqrt{b}}-1}. \qquad (29) \end{cases}$$

## IV. Comparison of Results

A computer program was used to evaluate the bound (17) for the case $M=7$ and $b=0.9$ while the parameter $a$ was stepped from $-1.7$ to 1.7. The results of this computation were identical to those displayed graphically for the same parameter values by Hess [4] using his "matrix" method. Figs. 1 and 2 were also constructed using (17). Fixed values of $a$ and $b$ were used while $M$ was varied. The rms bound of Kaiser and Sandberg [1] as well as the absolute bound derived in this paper are displayed on the graphs. One can see by the asymptotic behavior for large $M$ that the absolute bound for complex roots is not the least upper bound due to the approximations made in (25) and (26). Values of the bounds in Figs. 1 and 2 can also be compared to the heuristic bound

$$| y(n) |_{\max} \leq \text{integer part of} \left[ \frac{0.5}{1-b} \right] \qquad (30)$$

which was derived by Jackson [5] using an "effective-value model." This bound yields numerical values of 5 for the example illustrated in Fig. 1 and 50 for the example illustrated in Fig. 2. However, Parker and Hess [2] have shown by example that (30) does not always bound the limit cycle, whereas (27)–(29) do.

Interesting comparisons can also be made between the absolute bounds (27)–(29) and the rms bounds obtained by Sandberg and Kaiser. Their results are as follows (using notation consistent with this paper):

$$\left(\frac{1}{M}\sum_{n=0}^{M-1} y^2(n)\right)^{1/2} \leq \rho(a, b) \tag{31}$$

for any $M>0$ and where

$$\rho(a,b) = \begin{cases} \dfrac{1}{1+b-|a|}, & \text{for either } b \leq 0 \text{ or } b > 0 \text{ and } |a| \geq \dfrac{4b}{1+b} \quad (32) \\ \dfrac{1}{(1-b)\left(1-\dfrac{a^2}{4b}\right)^{1/2}}, & \text{for } b > 0 \text{ and } |a| < \dfrac{4b}{1+b}. \quad (33) \end{cases}$$

One should make the important observation that (32) is identical to (27) over the region where $a^2-4b \geq 0$. Therefore, the rms bound (32) is also an absolute bound over the region $a^2-4b \geq 0$. Comparison of (29) and (33) over the region $b>0$ and $|a| < 4b/(1+b)$ shows that they differ by a factor of $(1+\sqrt{b})$. Note that (29) cannot exceed (33) by more than a factor of two over the specified region and for small values of $b$ they are approximately the same. The following important result can be shown by careful comparison of the bounds:

$$\frac{|y(n)|_{\max}}{\rho(a,b)} \leq 2 \tag{34}$$

for any $a$ and $b$ for which the linear filter is asymptotically stable. Realizing this fact, the designer using the rms bound can be certain that the maximum value of the limit cycle will not exceed this bound by a factor of more than two.

## References

[1] I. W. Sandberg and J. F. Kaiser, "A bound on limit cycles in fixed-point implementations of digital filters," *IEEE Trans. Audio Electroacoust.*, vol. AU-20, pp. 110–112, June 1972.

[2] S. R. Parker and S. F. Hess, "Limit-cycle oscillations in digital filters," *IEEE Trans. Circuit Theory*, vol. CT-18, pp. 687–697, Nov. 1971.

[3] T. N. Trick and D. R. Anderson, "Stability of nonlinear networks," *IEEE Trans. Circuit Theory*, vol. CT-16, pp. 302–311, Aug. 1969.

[4] S. F. Hess, "A deterministic analysis of limit cycle oscillations in recursive digital filters due to quantization," Ph.D. dissertation, Naval Postgraduate School, Monterey, Calif., Dec. 1970.

[5] L. B. Jackson, "An analysis of limit cycles due to multiplication rounding in recursive digital (sub) filters," in *Proc. 7th Annual Allerton Conf. Circuit and System Theory*, pp. 69–78, Oct. 1969.

# SOME REMARKS ON THE CLASSIFICATION OF LIMIT CYCLES IN DIGITAL FILTERS

by T. A. C. M. CLAASEN, W. F. G. MECKLENBRÄUKER
and J. B. H. PEEK

**Abstract**

Classification schemes for limit cycles in digital filters are considered. In particular the scheme based on symmetry considerations is completed. A classification scheme based on the concept of the accessibility of limit cycles is introduced. With this scheme several problems concerning limit cycles are solved, among which a lower bound on the maximum amplitude of limit cycles in second-order filters with two rounding nonlinearities is given.

## 1. Introduction

The various signals in a digital filter are stored with a finite accuracy and a fixed- or floating-point binary arithmetic is used representing these signals. The wordlength after a multiplier or an adder is in general larger than the original wordlength. A recursive digital filter can only be realized by using a feedback of the signals. Consequently the wordlengths of the signals in the feedback loop have to be limited somewhere to their original wordlength.

This limitation of wordlength can be obtained by either rounding a digital word of say $N+K$ bits to $N$ bits or suppressing several of the least significant bits (magnitude truncation). These operations are equivalent to a quantization of the signals and hence the resulting filter will be nonlinear. For a more detailed explanation of these two arithmetical operations, see ref. 1. Quantization of the signals can lead to so-called limit-cycle oscillations, in case no input signal is present, and can seriously limit the application of the recursive filters in practice [3]).

On the other hand the resulting number after an arithmetic operation can exceed the maximum number a register can contain. In order to use feedback in this case the most significant bits have to be modified (e.g. saturation arithmetic). This can have even more-serious consequences and can result in

Reprinted with permission from *Philips Res. Rep.*, vol. 28, pp. 297-305, Aug. 1973.

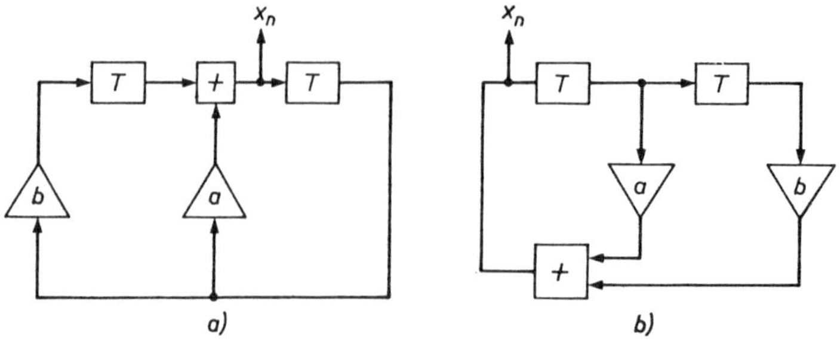


Fig. 1. The canonical forms of the recursive part of a second-order digital filter.

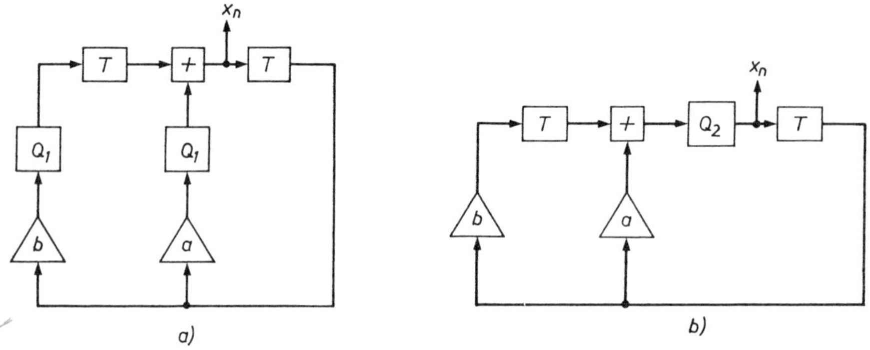


Fig. 2. The two possible implementations of the quantizers in the configuration of fig. 1*a*.

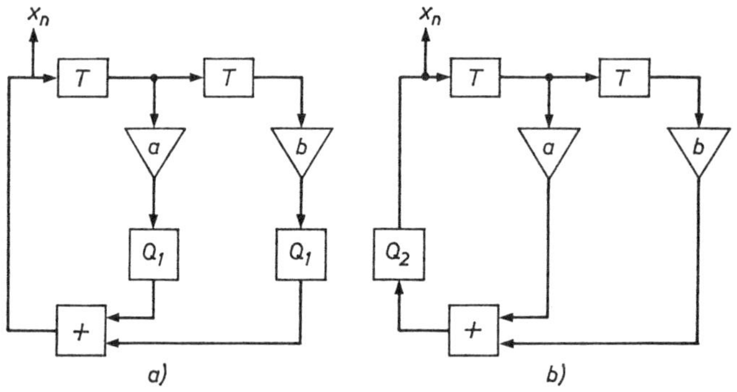


Fig. 3. The two possible implementations of the quantizers in the configuration of fig. 1*b*.

overflow oscillations which are specific types of limit cycles with large amplitudes.

This paper intends to give the following two classification schemes for these limit cycles:

(1) classification of limit cycles based on symmetry considerations as proposed in ref. 1;

(2) classification of limit cycles based on accessibility considerations.

Using these classification schemes it will be possible to give some general properties of limit cycles such as sufficient conditions for the existence of limit cycles in second-order filters using rounding.

The possible canonical configurations of the recursive part of a second-order digital-filter section that will be dealt with are depicted in figs 1*a* and *b*. Since only limit cycles will be studied here, no input terminal is indicated in the figures. There are two different ways of implementing the unavoidable quantizations in fig. 1*a* as given in figs 2*a* and *b*. Note that the two registers in fig. 2*b* have different lengths. Similarly, figs 3*a* and *b* show the two ways to implement the quantizers in the canonical configuration of fig. 1*b*.

It is easy to see that the behaviour of the filters given in figs 2*a* and 3*a* is the same, and likewise the behaviour of the filters given in figs 2*b* and 3*b*. Thus only the configurations of fig. 3 will be discussed. Both systems given in fig. 3 are described by the difference equation

$$x_n = Q_2(Q_1(a\,x_{n-1}) + Q_1(b\,x_{n-2}))$$

or (1)

$$x_n = [[a\,x_{n-1}]_{Q_1} + [b\,x_{n-2}]_{Q_1}]_{Q_2},$$

where $x_n$ denotes the filter output signal after $n$ time intervals each with a duration of $T$ s, and $Q(x) = [x]_Q$. For the system of fig. 3*a*, $Q_2(x) = x$, and for the system of fig. 3*b*, $Q_1(x) = x$.

The limit cycles are described by the successive values of the state of the digital filter, which can be the numbers stored in the two shift registers in fig. 3. In this case the limit cycles are described by the successive values of the output $x_n$. For a limit cycle of length $N$ it follows that $x_n = x_{n+N}$ for all $n$.

## 2. Classification of limit cycles based on symmetry considerations

As regards symmetry, one can in general distinguish two types of limit cycles.

*Type A: Symmetric limit cycles*

The length $N$ of the limit cycle is even and the limit cycle has half-wave symmetry:

$$x_{n+N/2} = -x_n \quad \text{for all } n.$$

A further distinction is type A1: $N/2$ is odd, and type A2: $N/2$ is even.

*Type B: Asymmetric limit cycles*

All limit cycles which are not of type A are called asymmetric limit cycles. These can be subdivided into type B1: $N$ is odd, and type B2: $N$ is even.

Both the rounding- and magnitude-truncation characteristics are described by odd functions. Consequently, if there is a limit cycle of type B in the filter: $x_1, \ldots, x_N$, then a limit cycle of type B can also be maintained in the same filter described by $-x_1, \ldots, -x_N$, which is different from the former. This classification scheme follows the line developed in ref. 1, but type B2 seems not to have been noticed. For example an asymmetric limit cycle with length $N = 20$ has been observed in a second-order filter with one rounding operation and coefficient values $a = 1{\cdot}90$ and $b = -0{\cdot}98$

$$(-1-4-7-9-10-10-9-7-4-1+2+5+8+10+11+11+10+8+5+2).$$

With this classification method a relation can be given between the limit cycles of a second-order digital filter described by eq. (1) with the coefficients $(a, b)$ and a second-order digital filter that has the same structure but coefficients $(-a, b)$. This relation is given in the following table (which has partly been derived by using lemmas 1 and 2 of ref. 1). From this table it can be seen that a limit cycle of type A1 with a certain amplitude transforms into two limit cycles of type B1 with the same amplitudes as that of type A1, but with lengths that have been halved.

| filter coeff. | type | length | no. of l.c. | type | length | no. of l.c. | type | length | no. of l.c. | type | length | no. of l.c. |
|---|---|---|---|---|---|---|---|---|---|---|---|---|
| a, b | A1 | $N$ | 1 | A2 | $N$ | 1 | B1 | $N$ | 2 | B2 | $N$ | 2 |
| ⇕ | | | | | | | | | | | | |
| $-a, b$ | B1 | $N/2$ | 2 | A2 | $N$ | 1 | A1 | $2N$ | 1 | B2 | $N$ | 2 |

**3. Classification of limit cycles based on accessibility considerations**

Limit cycles can also be classified as follows:

(1) those which only appear if the filter is started with initial conditions pertaining to that limit cycle;

(2) those which can be reached from initial conditions that do not pertain to that limit cycle.

In the latter case there has to be at least one state $S_n$ of the filter corresponding to a point of the limit cycle, which state can be reached from at least two different states $S_{n-1}$ and $S'_{n-1}$, the predecessors of $S_n$. Thus both states have as their successor the state $S_n$. The point corresponding to such a state will be called a *branch point*, see ref. 2. Limit cycles that have at least one branch point will be called *accessible* limit cycles. Thus the following two types of limit cycles can be distinguished:

(1) inaccessible limit cycles (also called pure cycles in ref. 8); these have no branch points;

(2) accessible limit cycles; these have at least one branch point.

Accessible limit cycles will more frequently be observed in the digital filter. If the filter is started with randomly chosen initial conditions it is unlikely that these will correspond to a point on a limit cycle. Thus it is quite certain that if in the end a limit cycle occurs it will be an accessible one. Some phenomena observed by Kieburtz [3]) can be explained in this way.

In the specific case of a second-order digital-filter section described by eq. (1), the state $S_n$ is given by $(x_n, x_{n-1})$ and the predecessor $S_{n-1}$ is given by $(x_{n-1}, x_{n-2})$. If $S_n$ is a branch point then there must be another predecessor $S'_{n-1}$ which is given by $(x_{n-1}, x'_{n-2})$ with $x'_{n-2} \neq x_{n-2}$. This gives the following two equations:

$$\begin{aligned} x_n &= [[a\, x_{n-1}]_{Q_1} + [b\, x_{n-2}]_{Q_1}]_{Q_2} \\ &= [[a\, x_{n-1}]_{Q_1} + [b\, x'_{n-2}]_{Q_1}]_{Q_2}. \end{aligned} \tag{2}$$

The following problems concerning limit cycles in a second-order digital filter will now be solved with the concepts just introduced.

*Problem* 1: Under what conditions do limit cycles necessarily occur?

In other words, if the filter is started with initial conditions which differ from (0, 0), under what conditions will the point (0, 0) never be reached? The state (0, 0) is always a stationary point and hence one predecessor is the state (0, 0) itself. Consequently if (0, 0) is not a branch point it cannot be reached from other states and limit cycles must necessarily exist. Therefore the question can be reformulated again: under what conditions is the point corresponding to the state (0, 0) not a branch point? This question can be answered for filters with different quantization characteristics.

In appendix 1 it is shown that in case one or two rounding operations are used, (0, 0) is a branch point only if $|b| \leqslant 0{\cdot}5$. Therefore such filters with multiplication coefficient $b$ (see fig. 3) for which $|b| > 0{\cdot}5$ will always exhibit limit cycles. This result is derived in ref. 1 for two rounding operations by using the idea that the state (0, 0) cannot be reached.

It is also shown in appendix 1 that where one or two magnitude-truncation operations are used, (0, 0) is always a branch point for $|b| < 1$, and limit cycles do not necessarily exist.

In the foregoing we investigated whether the state (0, 0) was a branch point or not. In the next two problems the concept of a branch point will be applied to arbitrary states of the filter.

*Problem* 2: What can be said about the amplitude of limit cycles?

Various authors have derived upper bounds for the amplitudes of limit cycles that occur in a second-order digital-filter section [1,4,5,6]), exact ones, which may be rather pessimistic, and approximate ones. Of this latter type is the upper bound $0{\cdot}5/(1-|b|)$ derived by Jackson [4]). In ref. 1 it was already found that this bound is too low in certain cases.

In appendix 2 it is shown that where there are two roundoff quantizers (fig. 3*a*) the value derived by Jackson is in fact a lower bound for the maximum amplitude of the possible limit cycles. This means that there must exist at least one limit cycle with an amplitude larger than or equal to $[0{\cdot}5/(1-|b|)]_{\text{entier}}$, and this limit cycle is the smallest accessible one.

Before stating the last problem the following brief introduction is necessary.

In the case of two nonlinearities and magnitude truncation in the second-order digital-filter section, Kao [7]) has shown that stationary points and limit cycles of length 2 are only possible for $|a| > 1$, $b < 0$. For both cases the maximum amplitude is [7])

$$|x_n| \leqslant \frac{1}{1-|a|-b}. \tag{3}$$

Along similar lines as used in appendix 2 it can be shown that the predecessors of all branch points are given by the states $S_{n-1} = (x_{n-1}, x_{n-2})$ with

$$|x_{n-2}| = \begin{cases} 0 \\ 1 \\ \left[\dfrac{M}{1-|b|}\right]_{\text{entier}} \qquad \text{for} \quad M = 1, 2, \ldots . \end{cases} \tag{4}$$

*Problem* 3: For which values of $(a, b)$ are only limit cycles with $|x_n| \leqslant 1$ accessible?

From (3) and (4) it is seen that this region is given by

$$\frac{1}{1-|a|-b} < \frac{1}{1-|b|} = \frac{1}{1+b} \qquad (b < 0),$$

which is equivalent to $|a| < -2b$. The corresponding regions are indicated in fig. 4 where the inside of the triangle is the region of stability of the linear filter. It can be seen that for high-$Q$ poles ($b \gtrsim -1$) in particular, limit cycles giving values greater than unity are inaccessible. Since it is unlikely that the filter is started on such an inaccessible limit cycle it is quite certain that in these regions only accessible limit cycles with $|x_n| \leqslant 1$ will occur. Although a limit cycle of amplitude $|x_n| \leqslant 1$ has no serious consequences in practical applications if only one second-order filter is used, it can have more-serious consequences if a cascade of second-order filters is used.

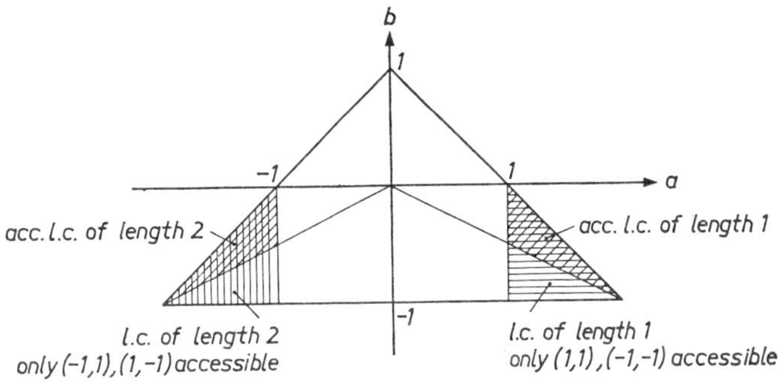


Fig. 4. The regions where only limit cycles of lengths 1 and 2 exist. The parallel-hatched areas indicate where only the limit cycles with amplitude $\leqslant 1$ are accessible. In the cross-hatched areas also limit cycles with larger amplitudes are accessible.

**Acknowledgement**

We would like to thank Professor Dr F. L. H. M. Stumpers for carefully reading the manuscript and making helpful suggestions.

*Eindhoven, July 1973*

**Appendix 1**

Because the state (0, 0) is a stationary point of eq. (1) one of its predecessors is (0, 0) itself. For (0, 0) to be a branch point there must be at least another predecessor (0, $K$), where the integer $K \neq 0$, assuming that the quantization step is normalized to 1. With (2) this gives the following equation for $K$:

$$[[b\,K]_{Q_1}]_{Q_2} = 0. \tag{A.1}$$

This equation can be simplified, owing to the fact that we have in fig. 3 either the two nonlinearities $Q_1$ or only one nonlinearity $Q_2$. In both cases eq. (A.1) is given by

$$[b\,K]_Q = 0, \tag{A.2}$$

where $Q$ can be either roundoff or magnitude truncation. For roundoff, eq. (A.2) has solutions $K \neq 0$ if

$$|b\,K| \leqslant 0{\cdot}5, \qquad |b| \leqslant \frac{0{\cdot}5}{K}.$$

Thus, since $K$ is an integer, we have $|b| \leqslant 0{\cdot}5$ and this is also the condition for (0, 0) to be a branch point. For $|b| > 0{\cdot}5$ there is only the solution $K = 0$. For magnitude truncation eq. (A.2) has solutions $K \neq 0$ if $|b\,K| < 1$, or $|b| < 1$. In this case (0, 0) is always a branch point.

**Appendix 2**

From appendix 1 it follows that for roundoff and $|b| > 0{\cdot}5$ the state (0, 0) is not a branch point. If, however, the filter is started with initial conditions that do not pertain to a limit cycle, the filter must reach an accessible limit cycle because the state (0, 0) is not accessible. Therefore at least one of the points corresponding to this limit cycle must be a branch point.

Now let $S_n$ be a branch point, thus one of its predecessors $S_{n-1} = (x_{n-1}, x_{n-2})$ or $S'_{n-1} = (x_{n-1}, x'_{n-2})$ must also lie on the limit cycle. If $|x_{n-2}| < |x'_{n-2}|$ holds good we can conclude that the maximum amplitude $A_{\max}$ of this limit cycle must be greater than or equal to $|x_{n-2}|$:

$$A_{\max} \geqslant |x_{n-2}| = X.$$

In the case of one nonlinearity (fig. 3*b*) $X$ cannot be calculated for arbitrary values of $a$ and $b$. However, for two nonlinearities (fig. 3*a*) this can easily be done.

With eq. (2) we have $[b\, x_{n-2}]_R = [b\, x'_{n-2}]_R$, where the operator $R$ denotes rounding. Assume $0{\cdot}5 < |b| < 1$ and solve this equation for $|b|\, X$. Defining $|x'_{n-2}| = X + K$, $K > 0$, gives

$$[|b|\, X]_R = [|b|\, X + |b|\, K]_R. \tag{A.3}$$

From $|b| < 1$ it follows that $[|b|\, X]_R = X - M$ with $M \geqslant 0$. Equation (A.3) implies that both numbers $|b|\, X$, $|b|\, X + |b|\, K$ must lie in the same roundoff interval and this situation is depicted in fig. 5.

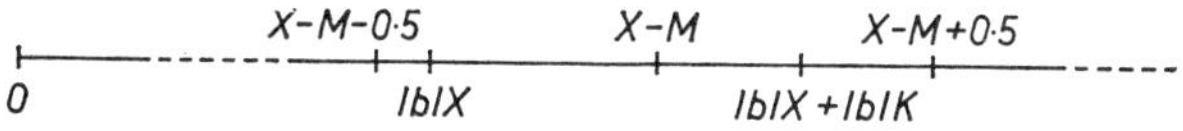


Fig. 5. The two numbers $|b|\, X$ and $|b|\, (X + K)$ in the same roundoff interval.

Equation (A.3) has a solution if the following two inequalities hold simultaneously:

$$X - M - 0{\cdot}5 \leqslant |b|\, X,$$
$$|b|\, X + |b|\, K < X - M + 0{\cdot}5,$$

which gives the interval

$$\frac{|b|\,(K-1)}{1-|b|} + \frac{M+0{\cdot}5}{1-|b|} - 1 < X \leqslant \frac{M+0{\cdot}5}{1-|b|}. \tag{A.4}$$

The lower bound of (A.4) is a minimum for $K = 1$ and $M = 0$ and the minimum for $|x_{n-2}| = X$ is given by

$$\text{entier}\left(\frac{0{\cdot}5}{1-|b|}\right).$$

The smallest possible accessible limit cycle must attain this value and hence

$$A_{\max} \geqslant \left[\frac{0{\cdot}5}{1-|b|}\right]_{\text{entier}}.$$

## REFERENCES


[1]) S. R. Parker and S. F. Hess, IEEE Trans. Circuit Theory **CT-18**, 687-697, 1971.
[2]) S. W. Golomb, Shift register sequences, Holden Day, San Francisco, 1967.
[3]) R. B. Kieburtz, An experimental study of roundoff effects in a tenth order recursive digital filter, Proc. of the IEEE Intern. Conf. on Communications, 19-21 June 1972, Philadelphia.
[4]) L. B. Jackson, An analysis of roundoff noise in digital filters, Sc.D. dissertation, Stevens Inst. Techn., Hoboken, N.Y., 1969.
[5]) I. W. Sandberg and J. F. Kaiser, IEEE Trans. Audio Electroacoustics **AU-20**, 110-112, 1972.
[6]) L. J. Long and T. N. Trick, IEEE Trans. Audio Electroacoustics **AU-21**, 27-30, 1973.
[7]) C. Kao, An analysis of limit cycles due to sign-magnitude-truncation in multiplication in recursive digital filters, 5th Asimolar Conference on Circuit and System Theory, Pacific Grove, 1971.
[8]) A. Gill, Linear sequential circuits, McGraw-Hill, New York, 1966, p. 109.

# SECOND-ORDER DIGITAL FILTER WITH ONLY ONE MAGNITUDE-TRUNCATION QUANTISER AND HAVING PRACTICALLY NO LIMIT CYCLES

*Indexing terms: Digital filters, Quantisation*

A 2nd-order digital filter with only one magnitude-truncation quantiser is discussed, and results are given that indicate that limit cycles are almost absent in this filter.

*Introduction:* Any digital filter can be realised by cascading a number of 2nd-order digital-filter sections. To realise the recursive part of a digital filter, signal feedback has to be used. After a multiplier, the word length is generally greater than before, and this implies that the word length has to be limited to the original word length to close the feedback loop. The word length can be limited either by rounding off or by disregarding a number of the least significant bits. This is equivalent to a quantisation nonlinearity that has either the roundoff or the magnitude-truncation characteristic (for a more detailed explanation of these two arithmetical operations, see Reference 1). Consequently, the realised filter will be nonlinear and, for a zero input, so-called limit-cycle oscillations will be possible.[1]

In principle, there are two different realisations for these quantisers (Figs. 1*a* and *b*). The quantisers Q have either the roundoff or the magnitude-truncation characteristic. Between the two realisations there is the following difference: in Fig. 1*a*, the word length is limited immediately after the multipliers, whereas, in Fig. 1*b*, full precision is maintained in the feedback loop as long as possible, and the necessary limitation of the word length occurs only once.

In the literature, the two quantisers shown in Fig. 1*a* have been extensively discussed.[1,2] In particular, it has been shown that two quantisers, having the magnitude-truncation characteristic, have certain advantages over roundoff with respect to the occurrence of limit cycles.[2]

The existence of limit cycles can be a nuisance in designing fixed- or floating-point-arithmetic digital filters.[3,4] It is important therefore to find methods that avoid, as much as possible, the occurrence of limit cycles.

*Results:* Where only one quantiser is involved, one can choose either roundoff or magnitude truncation. The results of our analysis show that, regardless of whether one or two roundoff operations with fixed-point arithmetic are used, limit cycles must occur for $|b| > 0{\cdot}5$ (Reference 5), where $b$ is one of the two multiplier coefficients (see Fig. 1).

For only one magnitude-truncation quantiser (Fig. 1*b*), we found that limit cycles occur for only a very few values of the multiplier coefficients $a$ and $b$. As is well known,[6] the area of absolute stability of the linear filter is given by $1-b > |a|$ and $|b| < 1$. This is the triangle in the $(a, b)$ plane, shown in Fig. 2.

The following results, in particular, have been obtained for one magnitude-truncation quantiser:

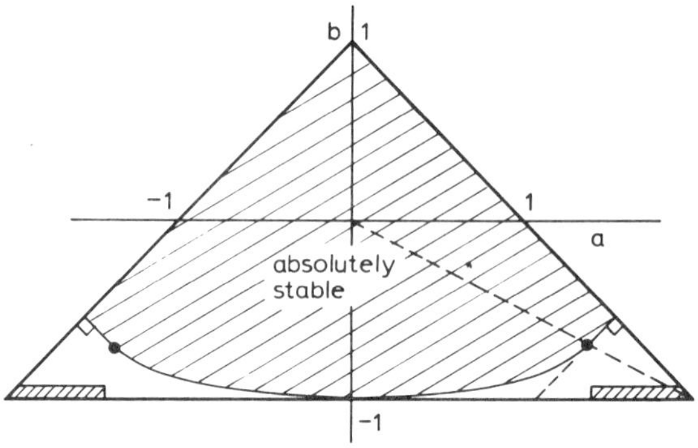


**Fig. 2** *Region of absolute stability for one magnitude-truncation quantiser and regions where limit cycles can occur*

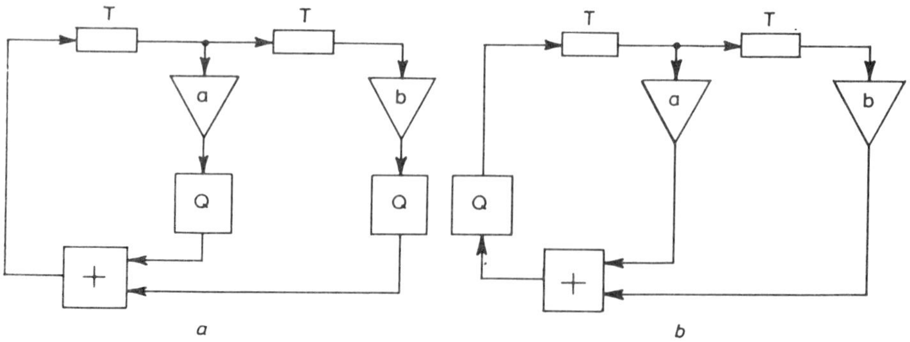


**Fig. 1** *Two ways to implement the quantisers in a 2nd-order filter*



Reprinted with permission from *Electron Lett.*, vol. 9, pp. 531–532, Nov. 1, 1973.

(*a*) Limit cycles of lengths 1, 2 and 4 cannot occur for *a* and *b* inside the triangle.

(*b*) It has been proved that the area of absolute stability of the nonlinear filter is the shaded area in Fig. 2. This area, where no limit cycles can occur, is bounded by the left- and right-hand sides of the triangle, by a part of the ellipse $a^2+8b(b+1)=0$ and by the two straight lines $|a|=b+2$. The result holds good for fixed- and floating-point arithmetic.

(*c*) In the area remaining, where it was not possible to prove absolute stability, limit cycles have been found with simulations using fixed-point arithmetic, only in the two trapezoid areas indicated in Fig. 2. These trapezoid areas are defined by the intervals $-1<b\leqslant-0{\cdot}94$ and $1{\cdot}42\leqslant|a|<2$. A detailed analysis has shown that these areas also have a fine structure such that, in these areas, the filter will not exhibit limit cycles for all parameters *a* and *b*.

(*d*) By using the describing-function method, the frequencies of the limit cycles that occur can be given approximately by

$$\omega_l T = \cos^{-1}\left\{\frac{1-(a^2+b^2)}{2ab}\right\}$$

where $T$ is the sampling period used in the filter.

*1st October 1973*

T.A.C.M. CLAASEN
W. F. G. MECKLENBRÄUKER
J. B. H. PEEK

*Philips Research Laboratories*
*Eindhoven, Netherlands*

## References

1 PARKER, S. R., and HESS, S. F.: 'Limit cycle oscillations in digital filters', *IEEE Trans.*, 1971, **CT-18,** pp. 687–697
2 KAO, C.: 'An analysis of limit cycles due to sign-magnitude truncation in multiplication in recursive digital filters'. 5th asimolar conference on circuit and system theory, Pacific Grove, 1971
3 KIEBURTZ, R. B.: 'An experimental study of roundoff effects in a tenth order recursive digital filter'. Proceedings of the IEEE international conference on communication, Philadelphia, USA, 1972
4 KANEKO, T.: 'Limit cycle oscillations in floating-point digital filters', *IEEE Trans.*, 1973, **AU-21,** pp. 100–106
5 CLAASEN, T. A. C. M., MECKLENBRÄUKER, W. F. G., and PEEK, J. B. H.: 'Some remarks on the classification of limit cycles in digital filters'. *Philips Res. Rep.* (to be published)
6 JURY, E. I.: 'Theory and application of the *z*-transform method' (Wiley, 1964)

# Frequency Domain Criteria for the Absence of Zero-Input Limit Cycles in Nonlinear Discrete-Time Systems, with Applications to Digital Filters

T. CLAASEN, W. F. G. MECKLENBRÄUKER, AND J. B. H. PEEK

*Abstract*—**On the basis of a theorem of Barkin, a number of criteria are derived which give sufficient conditions for the absence of zero-input limit cycles in discrete-time and especially digital systems. These criteria are formulated in the frequency domain and provide a possibility of investigating the absence of a zero-input limit cycle of a specific length $N$. Depending on the particular characteristics of the nonlinearities and the number of nonlinearities in the system, different criteria are obtained. Application of the criteria results in most cases in a linear programming problem. The solutions to this problem for the case of a second-order digital filter with one and with two quantization nonlinearities are discussed.**

## I. Introduction and Summary of Results

SIGNALS in digital systems can only be stored with a finite precision. After an arithmetical operation, the wordlength of the new signal is in general larger than before. If this signal arises in a feedback loop, as will be the case in recursive filters, it is necessary to quantize this signal somewhere before it can be stored in a register again, and precautions must be taken in the case of overflow. Because of these nonlinear operations, so-called limit cycles[1] can occur which may seriously limit the application of recursive filters in practice [1]. It is therefore of interest to study conditions under which these limit cycles cannot occur.

A particular type of limit cycle is a zero-input limit cycle, which is the periodic output of the filter in case the input is zero.

Absence of limit cycles can be studied by means of criteria that guarantee the asymptotic stability of a system. Up till now this has been done by means of the Lyapunov method [2], [3]. The method is not very appropriate for dealing with the specific quantization nonlinearities encountered in digital filters which have highly discontinuous characteristics.

In control theory, frequency domain stability criteria for sampled-data systems with nonlinear feedback are well known. They are discrete analogons of the Popov or the circle criteria familiar from continuous-time systems.

Although an extensive literature exists on this subject, the following abridgment is given for orientation.

To a system with only one nonlinearity, a criterion derived by Tsypkin is applicable [4]. Tsypkin has also derived some generalizations for the case of a nonlinearity with a monotonic characteristic [5], [6]. For systems with several nonlinearities there also exist criteria, such as those derived by Jury and Lee [7] and by Sandberg [8].

On the basis of a criterion proposed by Garber [9], Barkin [10] has given sufficient conditions for the absence

Manuscript received March 11, 1974; revised August 7, 1974.

The authors are with Philips Research Laboratories, Eindhoven, The Netherlands.

[1] The terminology used in this paper is that recommended by the IEEE Committee on Digital Signal Processing [19].

Reprinted from *IEEE Trans. Circuits Syst.*, vol. CAS-22, pp. 232–239, Mar. 1975.

of zero-input limit cycles of a specified length in discrete-time systems. Generalizations of this criterion will be derived in the present paper.[2] Application of these criteria gives two main advantages.

1) A digital filter has only a finite number of states, since the signals are quantized and bounded in amplitude. Moreover, since the sequence of the states is uniquely determined by the filter structure, it follows that the autonomous behavior of a digital filter ends up in either a limit cycle or the zero-state. Therefore, the absence of zero-input limit cycles necessarily implies that the output signal becomes zero after a finite time. Consequently, these criteria provide sufficient conditions for stability of digital filters.

2) Where stability cannot be proved, these criteria may still give information about the absence of zero-input limit cycles of a specified length (e.g., stationary points, or limit cycles of length 2).

After a short discussion of Barkin's criterion, theorems will be derived which apply to systems with one nonlinearity. It will be explained in Section III that, depending on some specific properties of the characteristic of the nonlinearity (such as time-invariance, monotony, and symmetry), different criteria can be applied.

Each of these criteria is illustrated by an example in which the absence of limit cycles in a second-order recursive digital filter is examined and values of the filter coefficients *a,b* are found for which the filter with one nonlinear element is stable.

The application of Theorems 2 and 3 leads to a linear programming problem. The resulting areas where no limit cycles can occur are given for a digital filter with one magnitude-truncation quantizer. As was to be expected, the area becomes larger as more constraints are imposed on the nonlinear characteristic.

In Section IV the generalizations of these theorems for systems with several nonlinearities are given and the absence of limit cycles in a second-order digital filter with two quantization nonlinearities is investigated. The resultant area of stability is the same as that which can be derived by Sandberg's theorem 2′ [8].

## II. The Barkin Criterion

Barkin [10] has derived two criteria by means of which the absence of zero-input limit cycles of a specific length $N$ can be investigated in discrete-time systems with one nonlinearity. In this paper we are interested mainly in Barkin's second criterion, because this is the only one that can be used for discontinuous nonlinearities, which occur frequently in digital systems.

Any system with one nonlinearity can be modeled as shown in Fig. 1, where $W$ is a time-invariant, linear system and $Q$ the nonlinearity for which $Q(x_n) = y_n$. Since the Barkin criterion investigates the absence of zero-input limit cycles, $u_n = 0$. For the existence of zero-input limit cycles,

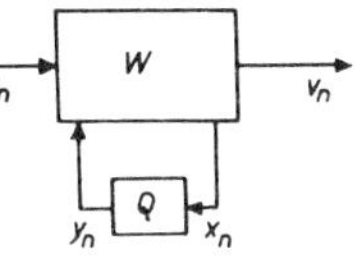


Fig. 1. Nonlinear system divided into a linear part $W$ and a nonlinearity $Q$.

only the transmission from $y_n$ to $x_n$ is of interest. Therefore, in considering zero-input limit cycles, the linear system $W$ can be described by the transmission function $W(z)$:

$$X(z) = W(z) \cdot Y(z) \tag{1}$$

where $X(z)$ and $Y(z)$ are the $z$ transforms of $x_n$ and $y_n$, respectively. The nonlinearity $Q$ must satisfy the following conditions to enable the Barkin criterion to be applied:

$$Q(0) = 0 \tag{2}$$

$$0 \leq \frac{Q(x)}{x} \leq k, \qquad \forall x \neq 0 \tag{3}$$

$$(Q(x+h) - Q(x)) \cdot h \geq 0, \qquad \forall x, h. \tag{4}$$

This means that the characteristic of the nonlinearity $Q$ lies in the sectors shown in Fig. 2 and is a nondecreasing function of the argument.

The explicit formulation of Barkin's criterion will not be given here, because it is a special case of the more general criteria which will be derived in this paper. The interested reader is referred to Barkin's original paper [10].

## III. Sufficient Conditions for the Absence of Zero-Input Limit Cycles in Discrete-Time Systems with One Nonlinearity

In this section, sufficient conditions for the absence of limit cycles in discrete-time systems with one nonlinearity will be derived. Throughout this section it is assumed that the nonlinearity satisfies conditions (2) and (3). With these restrictions a criterion will be formulated in Section III-A. It will be shown that more specialized criteria can be given if the nonlinearity satisfies additional restrictions. This will be done in Section III-B for a time-invariant, nondecreasing characteristic and in Section III-C it is assumed that the characteristic is also symmetrical.

### *A. Time-Invariant, Nonmonotonic, or Time-Varying Characteristic*

For a system in which the nonlinear characteristic fulfils the sector condition ((2) and (3)) the following theorem can be given.

*Theorem 1:* Limit cycles of length $N$ are absent from the discrete-time system in Fig. 1, consisting of a linear part, described by the transmission function $W(z)$ which must be finite for $|z| = 1$, and one nonlinearity satisfying (2) and (3), if for $l = 0(1)\lfloor N/2 \rfloor$,[3]

$$\operatorname{Re} W(z_l) - \frac{1}{k} < 0 \tag{5}$$

[2] Since we only investigate zero-input limit cycles, we will sometimes use the shorter term, limit cycles, instead.

[3] $\lfloor r \rfloor$ indicates the integer part of $r$ and $l = 0(1)L$ denotes $l = (0, 1, \ldots, L)$.

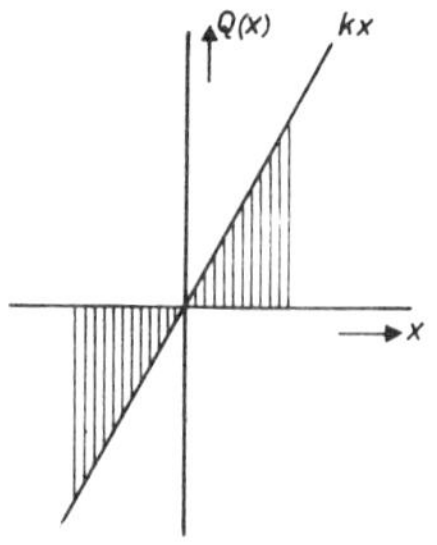


Fig. 2. Sector of the $(x, Q)$-plane in which the nonlinear characteristics must lie.

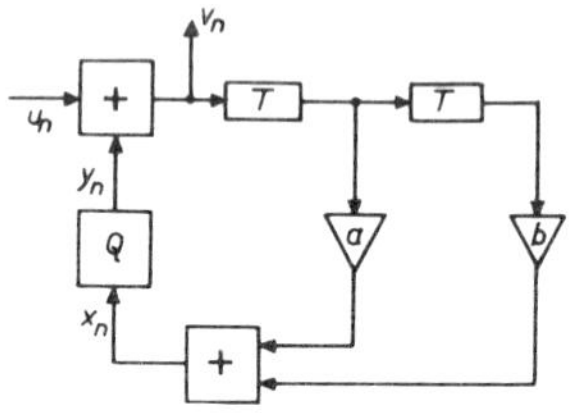


Fig. 3. Second-order recursive digital filter with one nonlinearity.

where

$$z_l = e^{j(2\pi/N)l}.$$

The proof of Theorem 1 will be given in Appendix I.

*Remarks:* 1) For time-invariant nonlinearities, conditions (2) and (3) imply that the characteristic must lie in the indicated sectors in Fig. 2. For time-varying nonlinearities, the condition that

$$Q(0,n) = 0 \qquad (2')$$

$$0 \leq \frac{Q(x,n)}{x} \leq k, \qquad \forall n,x \qquad (3')$$

must be fulfilled at every instant $n$.

2) The special boundaries of these sectors are not essential restrictions on the nonlinear characteristic. Any nonlinearity whose characteristic lies in the sectors $k_{\min} \leq Q(x)/x \leq k_{\max}$ can be separated into a linear part $k_{\min} \cdot x$ and a nonlinear part which satisfies condition (3) with $k = k_{\max} - k_{\min}$. If $k_{\min}$ is negative this separation is necessary to satisfy condition (3), if $k_{\min}$ is positive this may yield better results.

*Example 1:* To illustrate the usefulness of Theorem 1, the second-order digital filter depicted in Fig. 3 is considered, where the nonlinearity fulfils the sector condition for a certain value of $k$. If $Q$ is a quantizer, $k = 2$ for roundoff and $k = 1$ for magnitude-truncation [12].

The linear part of this structure is described by

$$W(z) = az^{-1} + bz^{-2}. \qquad (6)$$

With this function inequality (5) is written as

$$a \cos \frac{2\pi}{N} l + b \cos \frac{4\pi}{N} l - \frac{1}{k} < 0. \qquad (7)$$

From this it follows that limit cycles of length $N = 1$

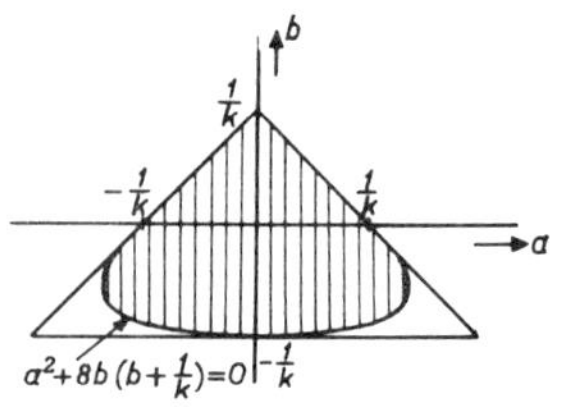


Fig. 4. Area of asymptotic stability of the digital filter of Fig. 3 derived by Theorem 1.

(stationary points) cannot exist if

$$a + b - \frac{1}{k} < 0. \qquad (8)$$

Limit cycles of length $N = 2$ are absent if, in addition to (8), the following inequality also holds

$$-a + b - \frac{1}{k} < 0. \qquad (9)$$

This process can be continued for increasing values of $N$ yielding the areas from which the corresponding limit cycles are absent.

In the area where

$$a \cos \phi + b \cos 2\phi - \frac{1}{k} < 0 \qquad (10)$$

holds for $0 \leq \phi \leq \pi$, inequality (7) is satisfied for all values of $N$ and $l$. This region is easily found to be the region shown in Fig. 4 and limit cycles cannot exist in a filter with coefficients $a$ and $b$ situated in this area.

### *B. Time-Invariant Nondecreasing Characteristic*

Next we assume the nonlinearity to be time-invariant and we put the additional condition of inequality (4) on the characteristic of the nonlinearity. Now Theorem 1 can be extended to yield Theorem 2.

*Theorem 2:* Limit cycles of length $N$ are absent from the discrete-time system in Fig. 1, consisting of a linear part described by the transmission function $W(z)$ which must be finite for $|z| = 1$, and one nonlinearity satisfying (2), (3), and (4), if $\exists \alpha_p \geq 0$ such that for $l = 0(1)\lfloor N/2 \rfloor$:

$$\mathrm{Re}\left\{W(z_l)\left[1 + \sum_{p=1}^{N-1} \alpha_p(1 - z_l^{\,p})\right]\right\} - \frac{1}{k} < 0 \qquad (11)$$

where $z_l = e^{j(2\pi/N)l}$

This theorem will be proved in Appendix II.

*Remark:* Taking $\alpha_p = 0$ for $p = 2(1)N - 1$ or for $p = 1(1)N - 2$, hence taking only $\alpha_1$ or $\alpha_{N-1}$ into account, the Barkin criterion mentioned in Section II is obtained.

In order to be able to apply this criterion, the following steps must be taken.

1) Determine the transmission function $W(z)$ of the linear part of the system.

2) Find the value of $k$ corresponding to the nonlinear characteristic.

3) With $W(z_l) = R_l + jI_l$, find solutions $\alpha_p \geq 0$ such

that the following set of inequalities is true:

$$\sum_{p=1}^{N-1} \alpha_p[R_l(1 - \cos pl\omega_N) + I_l \sin pl\omega_N] - \frac{1}{k} < -R_l,$$

$$l = 0(1)\left\lfloor \frac{N}{2} \right\rfloor \quad (12)$$

where $\omega_N = 2\pi/N$.

One of the methods of finding these solutions is to transform this set of inequalities into a linear programming problem. For this an objective function $f$ is required that can be obtained by introducing additional variables.

Very often one is interested in finding regions for the system parameters in which limit cycles cannot occur. In this case, one of the parameters can be extracted from the function $W(z)$ and introduced as a variable into the problem. The objective function for the linear programming procedure is then given by requiring the value of this parameter to be extreme.

*Example 2:* Once more the second-order filter of Fig. 3 is considered, where the nonlinearity $Q$ performs the necessary quantization of the signals. For the quantization we will take either rounding or magnitude-truncation, giving a value of $k = 2$ and $k = 1$, respectively. Both quantization characteristics are nondecreasing functions. The transmission function of the linear part is described by (6). The objective function $f$ of the linear programming problem is obtained by extracting $|a|$ from $W(z)$ of (6) taking sign $(a)$ and $b/|a|$ as fixed. Defining $R_l'$ and $I_l'$ by

$$W(z_l) = |a|(R_l' + jI_l') \quad (13)$$

and $\alpha_N$ by

$$\alpha_N = \frac{1}{|a|k} \quad (14)$$

the problem can be restated as

$$\text{Maximize } f = -\alpha_N \quad (15)$$

subject to

$$\sum_{p=1}^{N-1} \alpha_p[R_l'(1 - \cos pl\omega_N) + I_l' \sin pl\omega_N] - \alpha_N \leq -R_l' \quad (16)$$

for $l = 0(1)\lfloor N/2 \rfloor$ and

$$\alpha_p \geq 0 \quad (17)$$

for $p = 1(1)N$.

This is precisely a linear programming problem. The solutions can be found by applying existing algorithms [13]. By introducing the equality sign in (16) we find the boundary of the wanted region. On the boundary the absence of limit cycles is, however, not assured.

From the continuity of the transmission function $W(e^{j\phi})$, it follows that the region of asymptotic stability can be approximated by taking a large value of $N$ (e.g., $N = 70$). In this specific example, we used the MINIT-algorithm [14] and found the region depicted in Fig. 5 for $k = 1$. This applies to the filter with one magnitude-truncation quantizer. A comparison with Fig. 4 shows that the area where the absence of limit cycles has been proved is enlarged. It can be seen, however, that the region is asymmetric with respect to the line $a = 0$. On the other hand, it can be proved [12], [15] that, if for certain values of $a$ and $b$ a limit cycle exists in the filter with a symmetric quantization characteristic, a limit cycle must also exist in a filter with coefficients $-a$ and $b$. From this it follows that the region for which no limit cycles can occur must also be symmetrical with respect to $a = 0$. A symmetrical region is obtained by the application of Theorem 3 to this example. This criterion refers especially to this type of characteristic.

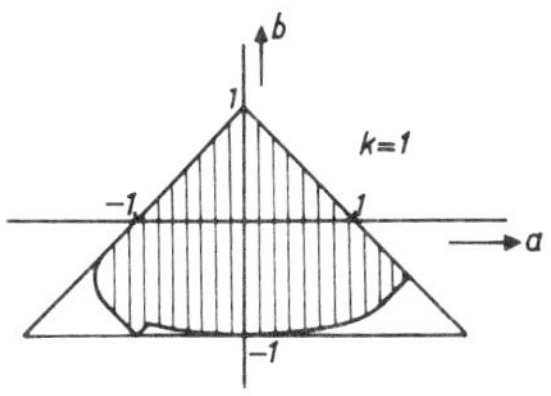


Fig. 5. Area of asymptotic stability of the digital filter of Fig. 3 derived by Theorem 2.

### C. Time-Invariant Nondecreasing Symmetrical Characteristic

Usually, nonlinearities occurring in digital systems have a symmetrical characteristic, i.e.,

$$Q(-x) = -Q(x). \quad (18)$$

This is the case with a quantizer or a saturation characteristic for overflow. With this additional restriction on the characteristic the following sufficient conditions for the absence of limit cycles can be obtained.

*Theorem 3:* Limit cycles of length $N$ are absent from the discrete-time system in Fig. 1 consisting of a linear part, described by the transmission function $W(z)$ which must be finite for $|z| = 1$, and one nonlinearity satisfying (2), (3), (4), and (18), if $\exists \alpha_p, \beta_p \geq 0$ such that for $l = 0(1)\lfloor N/2 \rfloor$

$$\text{Re}\left\{W(z_l)\left[1 + \sum_{p=1}^{N-1} \{\alpha_p(1 - z_l^p) + \beta_p(1 + z_l^p)\}\right]\right\} - \frac{1}{k} < 0 \quad (19)$$

where $z_l = e^{j(2\pi/N)l}$.

The proof of this theorem is given in Appendix III.

The implementation of this criterion is the same as for Theorem 2. The solutions can be found by a linear programming procedure having a larger number of parameters than in the previous case. While such a procedure gives the maximum area that can be found by this criterion, it is sometimes advantageous to reduce the number of parameters drastically by taking all parameters $\alpha_p$ and $\beta_p$, except one, equal to zero. Doing so, it is sometimes possible to find an analytical expression for the corresponding boundary. A comparison between these two solutions is made below.

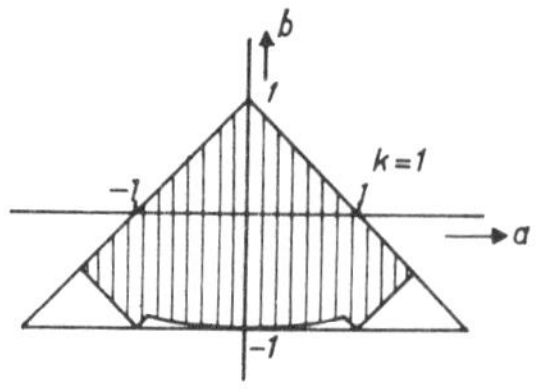


Fig. 6. Area of asymptotic stability of the digital filter of Fig. 3 derived by Theorem 3.

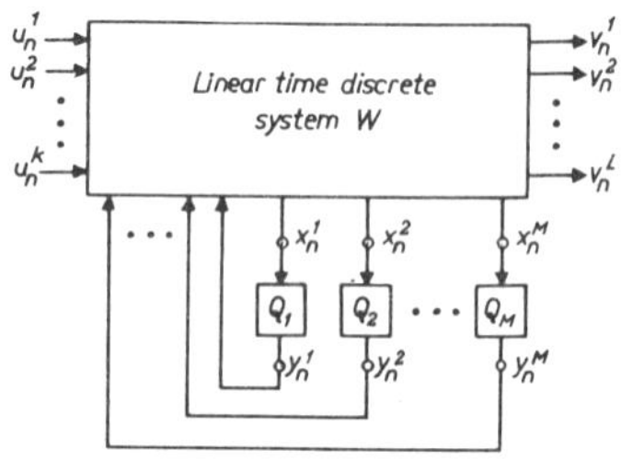


Fig. 7. Nonlinear system divided into a linear part $W$ and several nonlinearities $Q_m$.

*Example 3:* Since the quantization nonlinearities considered in Example 2 also satisfy (18), it is possible to apply Theorem 3 in this case, too. If only $\alpha_1$ or $\beta_1$ is taken unequal to zero, an analytical solution is found which is given in [16]. Applying the MINIT-algorithm we find the region indicated in Fig. 6. It can be seen that the region of absolute stability has been enlarged by using all the degrees of freedom of the criterion, but the gain obtained in this example is small.

In [18] besides the regions of asymptotic stability also the regions are given in which limit cycles have been found in second-order digital filters with either one magnitude-truncation quantizer or one roundoff quantizer. From this it follows that the criteria provide fairly good results.

## IV. Generalization of the Criteria for Systems with Several Nonlinearities

Discrete-time systems can generally have several nonlinearities. It is always possible to separate the nonlinearities from a linear part so as to obtain a structure of the kind as shown in Fig. 7.

The criteria derived in Section III for the case of one nonlinearity will now be extended to the case of several nonlinearities. Each of the nonlinear characteristics is now assumed to fulfil the same conditions as in Section III. The sector in which every characteristic must lie may, however, be different for every nonlinearity. Thus we can write

$$Q_m(0) = 0 \tag{20}$$

$$0 \le \frac{Q_m(x)}{x} \le k_m, \qquad \forall x \ne 0 \tag{21}$$

for $m = 1(1)M$, where $M$ is the total number of nonlinearities. With regard to limit cycles, the linear part $W$ can be described by a matrix transmission function $\boldsymbol{W}(z)$, which is a square matrix of dimension $M$.

It is shown in Appendix IV that with these definitions the generalized form of Theorem 1 is given by Theorem 4.

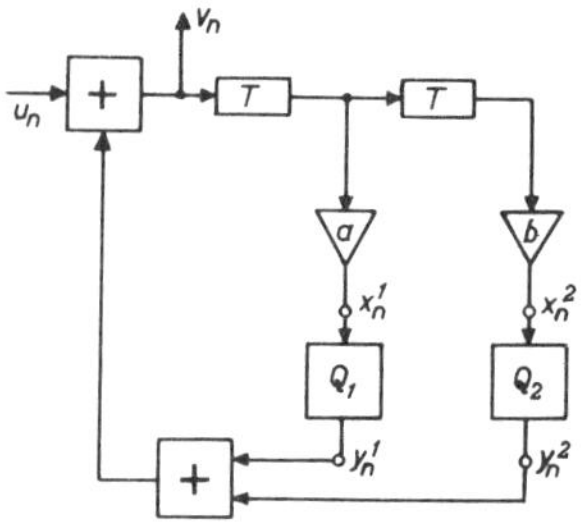


Fig. 8. Second-order recursive digital filter with two nonlinearities.

*Theorem 4:* Limit cycles of length $N$ are absent from the discrete-time system in Fig. 7 consisting of a linear part, described by the matrix transmission function $\boldsymbol{W}(z)$ which must be finite for $|z| = 1$, and $M$ nonlinearities $Q_m$, each of which satisfies (20) and (21), if for $l = 0(1)\lfloor N/2 \rfloor$, the Hermitian part of

$$\boldsymbol{W}(z_l) - \textbf{diag}\left(\frac{1}{k_m}\right) \tag{22}$$

is negative definite, where $z_l = e^{j(2\pi/N)l}$.

*Remark:* The restrictions put on the characteristics of the nonlinearities are the same as in Theorem 1. Thus it is possible to apply this criterion to systems with time-varying or time-invariant nonmonotonic characteristics of the nonlinearities.

*Example 4:* By means of Theorem 4 the absence of limit cycles in a second-order digital filter with two quantizers can be investigated. This system is depicted in Fig. 8. The linear part is described by the transmission matrix

$$\boldsymbol{W}(z) = \begin{pmatrix} az^{-1} & az^{-1} \\ bz^{-2} & bz^{-2} \end{pmatrix}. \tag{23}$$

For two roundoff quantizers we have $k_1 = k_2 = k = 2$, and for two magnitude-truncation quantizers we have $k_1 = k_2 = k = 1$.

The Hermitian part of $\boldsymbol{W}(z_l) - (1/k)\mathbf{1}_2$ is given by

$$\frac{1}{2}\begin{pmatrix} az_l^{-1} + az_l - \dfrac{2}{k} & az_l^{-1} + bz_l^2 \\ az_l + bz_l^{-2} & bz_l^{-2} + bz_l^2 - \dfrac{2}{k} \end{pmatrix}. \tag{24}$$

From Theorem 4 it follows that limit cycles of length $N = 1$ (stationary points) cannot exist if the matrix

$$\begin{pmatrix} a - \dfrac{1}{k} & \dfrac{a+b}{2} \\ \dfrac{a+b}{2} & b - \dfrac{1}{k} \end{pmatrix} \tag{25}$$

is negative definite. The corresponding region in the $(a,b)$ plane is shown in Fig. 9, for $k = 1$, where the region inside the triangle gives the area of asymptotic stability of the linear filter. Note that limit cycles of length $N = 1$ have been found for every value of $a$ and $b$ in the crosshatched area in Fig. 9 for a filter with two magnitude-truncation quantizers [17].

In the same manner the area can be computed for which limit cycles of length $N = 2$ are nonexistent. The absence

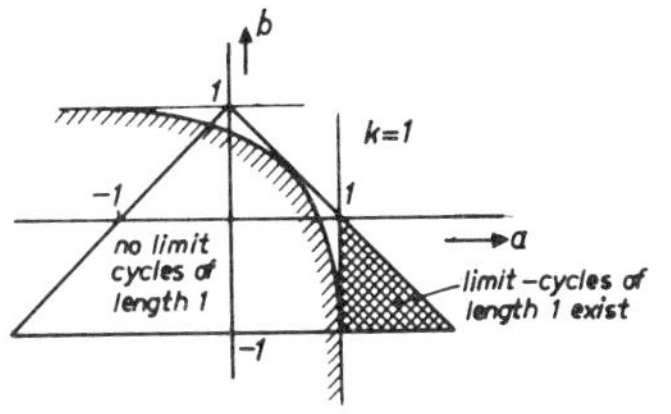


Fig. 9. Area where limit cycles of length 1 cannot occur in the filter of Fig. 8 and area where these limit cycles have been found.

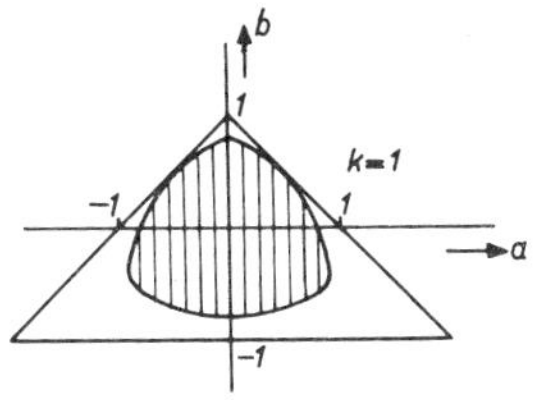


Fig. 10. Area of asymptotic stability of the digital filter of Fig. 8 derived by Theorem 4.

of limit cycles of any length is guaranteed in the area of the $(a,b)$ plane where the Hermitian matrix

$$\frac{1}{2}\begin{pmatrix} 2a\cos\phi - \frac{2}{k} & ae^{-j\phi} + be^{j2\phi} \\ ae^{j\phi} + be^{-j2\phi} & 2b\cos 2\phi - \frac{2}{k} \end{pmatrix} \tag{26}$$

is negative definite for $0 \leq \phi \leq \pi$. This area is shown in Fig. 10.

For completeness we will now give the generalizations of Theorems 2 and 3, which apply to time-invariant nonlinearities and which characteristics must be nondecreasing:

$$(Q_m(x + h) - Q_m(x)) \cdot h \geq 0, \qquad \forall x, h \tag{27}$$

for $m = 1(1)M$.

*Theorem 5:* Limit cycles of length $N$ are absent from the discrete-time system of Fig. 7 consisting of a linear part, described by the matrix transmission function $W(z)$, which must be finite for $|z| = 1$, and $M$ nonlinearities $Q_m$, each of which satisfying (20), (21), and (27) if for $l = 0(1)\lfloor N/2 \rfloor$ the Hermitian part of

$$\left[\mathbf{1}_M + \mathbf{diag}\left(\sum_{p=1}^{N-1} \{\alpha_{p,m}(1 - z_l^p) + \beta_{p,m}(1 + z_l^p)\}\right)\right] \cdot W(z_l) - \mathbf{diag}\left(\frac{1}{k_m}\right)$$

is negative definite, where $z_l = e^{j(2\pi/N)l}$ and $\alpha_{p,m} \geq 0$ and $\beta_{p,m} = 0$.

If, in addition, the characteristics are symmetrical

$$Q_m(x) = -Q_m(-x)$$

for $m = 1(1)M$, then $\alpha_{p,m} \geq 0$ and $\beta_{p,m} \geq 0$.

The proof of this theorem follows the same lines as that of Theorems 3 and 4.

The parameters $\alpha_{p,m}$ and $\beta_{p,m}$ appear in a nonlinear way in the inequalities resulting from Theorem 5. Up till now no attempts have been made to solve the corresponding nonlinear programming problem and thus applications of this criterion will not be discussed.

## Appendix I
## Proof of Theorem I

This proof follows the same lines as the proof of the Barkin criterion [10]. The following function is introduced:

$$\rho = \frac{2}{N}\sum_{n=0}^{N-1} Q(x_n)\left[x_n - \frac{1}{k}Q(x_n)\right]. \tag{28}$$

From (2) and (3) it follows that each term of the sum in (28) is nonnegative and it should be remarked that this is also the case, if the nonlinearity is time-varying. Thus

$$\rho \geq 0. \tag{29}$$

If now it is assumed that $x_n$ is periodic with period length $N$ then from Parseval's theorem of the discrete Fourier transform $\rho$ can be written as

$$\rho = \sum_{l=0}^{N-1} Y_l^* \left(X_l - \frac{1}{k} Y_l\right) \tag{30}$$

where $X_l = (2/N)\sum_{n=0}^{N-1} x_n z_l^{-n}$ $(l = 0(1)N - 1)$ is the discrete Fourier transform of $x_n$ and similarly $Y_l$ of $y_n = Q(x_n)$. Moreover, we have from (1)

$$X_l = W(z_l) \cdot Y_l. \tag{31}$$

With this relation we must make the restriction that $W(z)$ may not have poles for $z = z_l$. To avoid a longer discussion, we therefore require $W(z)$ to be finite on the unit circle $|z| = 1$. Combining (30) and (31), we find

$$\rho = \sum_{l=0}^{N-1} |Y_l|^2 \left(W(z_l) - \frac{1}{k}\right). \tag{32}$$

$\rho$ is clearly real from (28) and so must be the sum in (32). If the real part of every term is negative, then $\rho$ must be negative, which contradicts (29). Thus limit cycles of length $N$ cannot exist.

## Appendix II
## Proof of Theorem 2

For the proof of Theorem 2 the following function is considered:

$$\rho(\alpha_1, \cdots, \alpha_{N-1}) = \frac{2}{N}\left[\sum_{n=0}^{N-1} Q(x_n)\left(x_n - \frac{1}{k}Q(x_n)\right) + \sum_{p=1}^{N-1} \alpha_p \sum_{n=0}^{N-1} Q(x_n) \cdot (x_n - x_{n+p})\right]. \tag{33}$$

Here again, it will be proved that $\rho$ is nonnegative if $x_n$ is periodic with period length $N$.

The first sum is nonnegative (see Appendix I). To prove this fact for the second sum with $\alpha_p \geq 0$, we will make use of a result of Barkin [10]:

$$\sum_{n=0}^{N-1} Q(\xi_n) \cdot (\xi_n - \xi_{n+1}) \geq 0 \tag{34}$$

which holds for every sequence $\xi_n$ with $\xi_0 = \xi_N$ and a nondecreasing function $Q(x)$.

Furthermore, the term

$$s_p = \sum_{n=0}^{N-1} Q(x_n) \cdot (x_n - x_{n+p}), \qquad p > 0 \tag{35}$$

is considered. From the periodicity of $x_n$ it follows that

$$ps_p = \sum_{n=0}^{pN-1} Q(x_n)\cdot(x_n - x_{n+p}) = \sum_{i=0}^{p-1}\sum_{k=0}^{N-1} Q(x_{kp+i})\cdot(x_{kp+i} - x_{(k+1)p+i}). \quad (36)$$

With the definition $\xi_k(p,i) = x_{kp+i}$, the interior sum can be written as

$$\sum_{k=0}^{N-1} Q(\xi_k(p,i))(\xi_k(p,i) - \xi_{k+1}(p,i)).$$

Since $\xi_k$ is periodic with period length $N$,

$$\xi_k(p,i) = x_{kp+i} = x_{kp+i+pN} = \xi_{k+N}(p,i)$$

it follows from (34) and (36) that $s_p \geq 0$ and, therefore,

$$\rho(\alpha_1,\cdots,\alpha_{N-1}) \geq 0. \quad (37)$$

Applying Parseval's theorem to (33) gives

$$\rho(\alpha_1,\cdots,\alpha_{N-1}) = \sum_{l=0}^{N-1} |Y_l|^2 \left[\operatorname{Re} W(z_l)\left\{1 + \sum_{p=1}^{N-1}\alpha_p(1 - z_l^p)\right\} - \frac{1}{k}\right]. \quad (38)$$

If the conditions of Theorem 2 are fulfilled then $\rho(\alpha_1,\cdots,\alpha_{N-1})$ is negative and this contradicts (37).

## Appendix III
## Proof of Theorem 3

In order to prove Theorem 3 we add to the function $\rho(\alpha_1,\cdots,\alpha_{N-1})$, defined by (33), the sum

$$\sum_{p=1}^{N-1}\beta_p\cdot\sum_{n=1}^{N-1} Q(x_n)\cdot(x_n + x_{n+p}). \quad (39)$$

By proving that this sum is nonnegative for $\beta_p \geq 0$ and $Q(x) = -Q(-x)$, inequality (19) is obtained by using Parseval's theorem as in Appendix II.

Define

$$s_p = \sum_{n=0}^{N-1} Q(x_n)\cdot(x_n + x_{n+p}).$$

Then,

$$2p\cdot s_p = \sum_{i=0}^{p-1}\sum_{k=0}^{2N-1} Q(x_{kp+i})\cdot(x_{kp+i} + x_{(k+1)p+i})$$

and with $\xi_k(p,i) = (-1)^k x_{kp+i}$, this becomes

$$\sum_{i=0}^{p-1}\sum_{k=0}^{2N-1}(-1)^k Q(\xi_k(p,i))((-1)^k\xi_k(p,i) + (-1)^{k+1}\xi_{k+1}(p,i)) = \sum_{i=0}^{p-1}\sum_{k=0}^{2N-1} Q(\xi_k(p,i))(\xi_k(p,i) - \xi_{k+1}(p,i)).$$

Since $\xi_k(p,i) = \xi_{k+2N}(p,i)$ this sum is nonnegative, as follows from (34).

## Appendix IV
## Proof of Theorem 4

For the proof of Theorem 4 the following $m$ functions are introduced:

$$\rho_m = \frac{2}{N}\sum_{n=0}^{N-1} Q_m(x_n^m)\left(x_n^m - \frac{1}{k_m}Q_m(x_n^m)\right). \quad (40)$$

From Appendix I it follows that

$$\rho_m \geq 0 \qquad m = 1(1)M. \quad (41)$$

Now consider the function

$$\rho = \sum_{m=1}^{M}\rho_m \quad (42)$$

and from (41) it follows that

$$\rho \geq 0. \quad (43)$$

In order to simplify the notation, the signals will be represented by vectors. Thus the input signals $x_n^m$ of the nonlinearities are combined in the vector

$$\boldsymbol{x}_n = (x_n^1,\cdots,x_n^M)^T$$

and the output signals $y_n^m$ in the vector

$$\boldsymbol{y}_n = (y_n^1,\cdots,y_n^M)^T.$$

Assuming a limit cycle of length $N$ we can apply the discrete Fourier transform to each of these signals and get the vectors $\boldsymbol{X}_l$ and $\boldsymbol{Y}_l$, respectively.

The linear part of the system is described by the equation

$$\boldsymbol{X}_l = \boldsymbol{W}(z_l)\cdot\boldsymbol{Y}_l. \quad (44)$$

From Parseval's theorem applied to (42) and with (44) we obtain similarly to Appendix I

$$\rho = \sum_{l=0}^{N-1}\tilde{\boldsymbol{Y}}_l\left[-\textbf{diag}\left(\frac{1}{k_m}\right) + \boldsymbol{W}(z_l)\right]\boldsymbol{Y}_l \quad (45)$$

where $\tilde{\boldsymbol{Y}}_l$ is the conjugate complex transpose of $\boldsymbol{Y}_l$. With the definition of

$$\boldsymbol{A}(z_l) = -\textbf{diag}\left(\frac{1}{k_m}\right) + \boldsymbol{W}(z_l)$$

this can be written as

$$\rho = \sum_{l=0}^{N-1}\tilde{\boldsymbol{Y}}_l\cdot\boldsymbol{A}(z_l)\cdot\boldsymbol{Y}_l = \sum_{l=0}^{N-1}\tilde{\boldsymbol{Y}}_l\cdot\boldsymbol{A}_H(z_l)\cdot\boldsymbol{Y}_l \quad (46)$$

where $\boldsymbol{A}_H(z_l)$ is the Hermitian part of $\boldsymbol{A}(z_l)$:

$$\boldsymbol{A}_H(z_l) = \tfrac{1}{2}(\boldsymbol{A}(z_l) + \tilde{\boldsymbol{A}}(z_l)).$$

Equation (46) follows from the fact that $\rho$ is real.

If, for every $l$, $\boldsymbol{A}_H(z_l)$ is negative definite, then it follows from (46) that

$$\rho < 0.$$

This contradicts (43) and hence no limit cycles of length $N$ can occur.

## Acknowledgment

The authors would like to thank N. A. M. Verhoeckx for carefully reading the manuscript and for valuable comments.

## References

[1] R. B. Kieburtz, "An experimental study of round-off effects in a 10th order, recursive digital filter," *Proc. IEEE Int. Conf. Communications*, June 1972, pp. 17–21.

[2] A. N. Willson, Jr., Some effects of quantization and adder overflow on the forced response of digital filters. *Bell Syst. Tech. J.*, vol. 51, pp. 863–887, Apr. 1972.

[3] ——, "Limit cycles due to adder overflow in digital filters," *IEEE Trans. Circuit Theory*, vol. CT-19, pp. 342–346, July 1972.

[4] Ya.Z. Tsypkin, "Frequency criteria for the absolute stability of nonlinear sampled data systems," *Automat. Remote Contr.*, vol. 25, pp. 261–267, Mar. 1964.

[5] ——, "Absolute stability of a class of non-linear automatic sampled data systems," *Automat. Remote Contr.*, vol. 25, pp. 918–923, July 1964.

[6] ——, "A criterion for absolute stability of automatic pulse-systems with monotonic characteristics of the non-linear element," *Sov. Phys. Dokl.*, vol. 9, pp. 263–266, Oct. 1964.

[7] E. I. Jury and B. W. Lee, "The absolute stability of systems with many non-linearities," *Automat. Remote Contr.* vol. 26, pp. 943-961, June 1965.

[8] I. W. Sandberg, "On the boundedness of solutions of non-linear integral equations," *Bell Syst. Tech. J.*, vol. 44, pp. 439–453, Mar. 1965.

[9] E. D. Garber, "Frequency criteria for the absence of periodic responses," *Automat. Remote Contr.*, vol. 28, Nov. 1967.

[10] A. I. Barkin, "Sufficient conditions for the absence of auto-oscillations in pulse-systems," *Automat. Remote Contr.*, vol. 31, pp. 942–946, June 1970.

[11] P. Vidal, *Non-Linear Sampled Data Systems.* New York: Gordon and Breach, 1969.

[12] S. R. Parker and S. F. Hess, "Limit cycle oscillations in digital filters," *IEEE Trans. Circuit Theory*, vol. CT-18, pp. 287–297, Nov. 1971.

[13] R. W. Llewellyn, *Linear Programming.* New York: Holt, Rinehart and Winston, 1964.

[14] R. C. Salazar and S. K. Sen, "MINIT-algorithm for linear programming," *Commun. Ass. Comput. Mach.*, vol. 11, pp. 433–440, June 1968.

[15] T. A. C. M. Claasen, W. F. G. Mecklenbräuker, and J. B. H. Peek, "Some remarks on the classification of limit cycles in digital filters," *Philips Res. Rep.*, vol. 28, pp. 297–305, Aug. 1973.

[16] ——, "Second order digital filter with only one magnitude-truncation quantizer and having practically no limit-cycles," *Electron. Lett.*, vol. 9, pp. 531–532, Nov. 1973.

[17] C. Kao, "An analysis of limit cycles due to sign-magnitude truncation in multiplication in recursive digital filters," 5th Asimolar Conf. Circuit and System Theory. Pacific Grove, 1971.

[18] T. A. C. M. Claasen, W. F. G. Mecklenbräuker, and J. B. H. Peek, "A comparison between the stability of second order digital filters with various arithmetics," in *Proc. 1974 European Conf. Circuit Theory and Design*, 1974, pp. 354–358.

[19] L. R. Rabiner, *et al.*, "Terminology in digital signal processing," *IEEE Trans. Audio Electroacoust.*, vol. AU-20, pp. 322–337, Dec. 1972.

# An Experimental Study of Roundoff Effects in a Tenth-Order Recursive Digital Filter

R. B. KIEBURTZ

*Abstract*—This paper reports experimental measurements of limit-cycle amplitude and period in digital recursive filters. The results that were obtained on two cascaded forms of a tenth-order filter show good agreement with existing theoretical bounds. (The bounds only apply to single second-order sections because analysis of cascaded sections is difficult.) Roundoff noise of the filter was measured and it agreed both in spectral shape and total power with that predicted by computer simulation. Some preliminary observations on relative magnitude of limit cycle and roundoff noise were made. In addition, the gradual buildup of a signal through a limit cycle was observed and 1 bit of noise in the least significant bit (LSB) in the filter was seen to break up limit cycles.

## INTRODUCTION

A topic of current theoretical and experimental interest is the limit-cycle behavior of cascaded recursive digital filters under zero input and small input conditions. In particular, it has been possible to estimate bounds on the amplitude and length of limit cycle of a single second-order section with zero input in terms of the filter coefficient values [1]–[3]. For filters of higher order realized in parallel form the output is just the sum of the limit cycles of the individual second-order sections; but for those realized in cascade form there is no adequate theoretical model, since all but the first section can be and ordinarily are driven by a nonzero limit cycle of the previous section. This appears to be a difficult theoretical problem. It has been speculated that the filter behaves essentially as though each section were producing its own maximum amplitude limit cycle or else is filtering the limit cycle from the previous stage, whichever is the larger output [1].

A cascade form recursive digital filter has been constructed as a part of an experimental study of a digital *A*-type channel bank [4]–[6]. The purpose of this paper is to discuss the limit-cycle behavior of this filter under a variety of conditions that include two different orderings of the second-order sections and two different values of scaling between sections. Limit-cycle behavior in response to impulse, sinusoidal, and stochastic inputs was studied. The filter under study is a tenth-order elliptic low-pass filter whose performance requirements are shown in Fig. 1.

At the point in the design cycle where coefficients exist that satisfy the filter specifications, the next step is to consider scaling and ordering to minimize some type of roundoff effects. In a tenth-order filter there are at most $(5!)^2$ different possible orderings, since pole and zero parts of each second-order section can be arranged independently. The filter design can be optimized in the direction of minimizing either peak or total roundoff noise energy. It has been found that peak noise can be minimized by arranging sections approximately in order of decreasing $Q$ [7]. In this case the high $Q$ section, which has the largest peak response and hence will tend to amplify the roundoff noise most, is placed early in the filter and encounters the maximum filtering actions. Also this arrangement requires the largest amount of prescaling before entering the filter to prevent overflow. In the somewhat more common case of desiring to minimize total noise energy, the sections are arranged roughly in order of increasing $Q$ [7]. This is the "optimum" condition in our case.

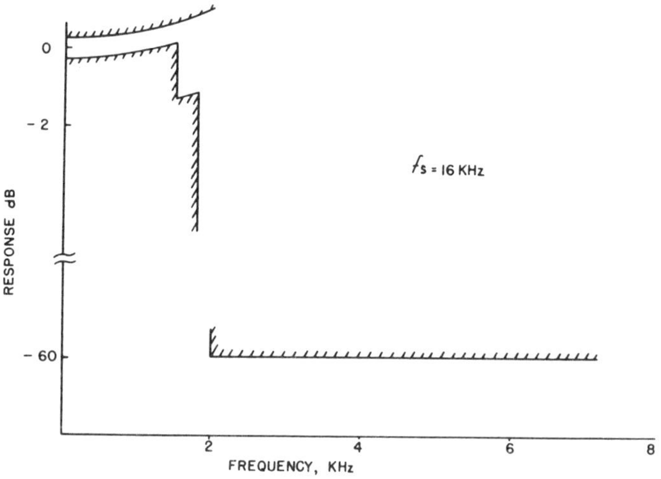


Fig. 1. Recursive filter characteristics.

The design procedure, after one selects an ordering on a somewhat imprecise basis, is to calculate the resulting roundoff noise relative to the least significant bit (LSB) by replacing each of the three real multipliers per second-order section by an ideal multiplier plus an additive white noise source to account for rounding, and sweep a full load sine wave from 0 to $\frac{1}{2}$ the sampling frequency $(f_s/2)$ at the input, scaling so as to prevent adder overflow at any critical point [7]. Most practical filters accomplish "power of 2" scaling by shifting the data word toward the binary point to prevent overflow. The size of the roundoff noise relative to the limit-cycle noise (which also derives from roundoff effects) cannot be stated precisely, but from qualitative arguments, they can be expected to be roughly the same size.

Two filter orderings were obtained: an optimum ordering for minimizing total noise energy (high-$Q$ pole last) and the other

Paper approved by the Data Communications Committee of the IEEE Communications Society for publication without oral presentation. Manuscript received November 14, 1972; revised Febraury 5, 1973.
The author is with Bell Laboratories, Holmdel, N.J. 07733.

Reprinted from *IEEE Trans. Commun.*, vol. COM-21, pp. 757–763, June 1973.

ordering primarily for comparison (high-$Q$ pole first). The filter was realized by constructing a single second-order section and time multiplexing it to make possible a cascade of five second-order sections. Read only memories were used to provide coefficients to the multipliers.

## Theoretical Bounds

In 1969, Jackson [1] made a rough estimate of the maximum limit-cycle amplitude based on an "effective value model" in which the pole locations in a second-order section were considered to move to the unit circle. The basic expression was

$$A_{\max} = \left[\frac{0.5}{1 - \beta_2}\right] \tag{1}$$

where the brackets refer to the integer part of the quantity, $A_{\max}$ is in units of quantization step size, and $|\beta_2|$ is the feedback multiplier coefficient ($\leqslant 1$) as shown in Fig. 2.

This bound, along with several others, was discussed by Parker and Hess [2]. Recently, a more precise estimate was made by Sandberg and Kaiser [3] for a second-order section without a feed-forward path. It was

$$A_{\max} = \frac{1}{(1-\beta_2)(1-(\beta_1^2/4\beta_2))^{1/2}}, \quad \text{for } \beta_2 > 0 \text{ and } |\beta_1| \leqslant \frac{4\beta_2}{1+\beta_2}$$

$$= \frac{1}{1+\beta_2 - |\beta_1|}, \quad \text{for } \beta_2 \leqslant 0 \text{ or } \beta_2 > 0 \text{ and } |\beta_1| \geqslant \frac{4\beta_2}{1+\beta_2} \tag{2}$$

where $\beta_1$ is the second feedback multiplier coefficient ($\leqslant 2$) as shown in Fig. 2. In addition, Kaiser [8] has estimated the maximum length of a limit cycle for a single second-order section to be

$$L_{\max} \leqslant 16\, A_{\max}. \tag{3}$$

## Experimental Results

A common way of depicting limit cycles has been to use a quasi-phase plane plot in which each output sample $y_n$ is plotted against its immediate predecessor $y_{n-1}$. This rapidly gets difficult to read for the common case of limit cycles that have many sign reversals per period. Instead, the author has used time domain plots of the filter outputs and wave analyzer plots of their spectra. The time plots make it easy to exploit the eye's natural abilities to spot symmetries and periodicity. Limit-cycle waveforms often repeat around a particular time with even symmetry. Sometimes portions of waves will be approximately identical under polarity reversal and translation in time. Sometimes two limit cycles will be negatives of each other. The spectral plots make it easy to see the harmonic content of each limit cycle, and sometimes one can see the filtering action of the filter upon the limit cycles.

The basic experimental arrangement was to start the filter with initial conditions zero, then impulse excite it by applying a single PCM input word, ranging in magnitude from one LSB to full scale. It was possible to obtain point-by-point agreement between the observed limit cycles and those generated by a computer program that simulated the filter. One useful observation was that under impulse excitation, the filter produces a number of different limit cycles, of the order of ten. Most of the different limit cycles came from excitations in the least significant half of the input word, while impulse excitations from the most significant half of the word often lead to the same limit cycle. In general, this most common limit cycle was worst case either in length or amplitude, and most often in both senses. It is possible but inefficient to search for the worst case limit cycles by the computer program; they are easily found by trial and error using the laboratory filter. The limit cycles to be discussed later were all the highest amplitude worst cases.

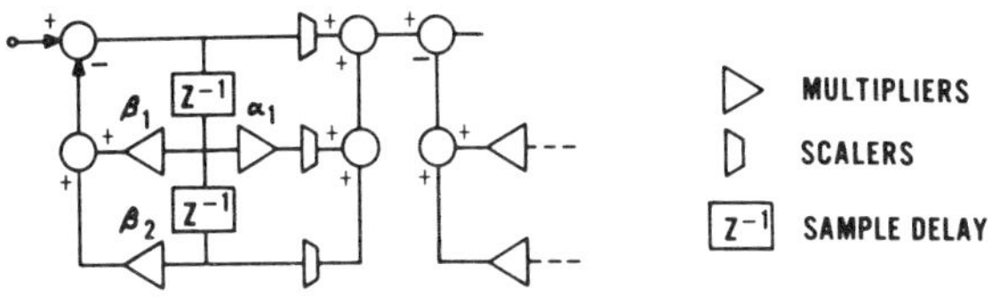


| | SECTION 1 | 2 | 3 | 4 | 5 |
|---|---|---|---|---|---|
| $\beta_2$ | .63 | .46 | .82 | .92 | .96 |
| $\beta_1$ | −1.4 | −1.3 | −1.5 | −1.5 | −1.5 |
| $\alpha_1$ | −.65 | .78 | −1.2 | −1.4 | −1.4 |
| SCALE | 1/8 | 1/8 | 1/2 | 1/4 | 1 |
| LIMIT CYCLE PERIOD | dc | dc | 10 FRAMES | 50f | 150f |
| $\Delta f$ | | | 1600Hz | 320Hz | 107Hz |
| LEVELS INSIDE | −2 | −14 | | | |
| OUTSIDE | −2 | −6 | −6TO−8 | −2TO−5 | −2TO−7 |
| COMMENTS | LEVELS CHECK | | NEW LINES AT nx 1600Hz | NEW LINES AT mx 320 Hz OLD LINES CHECK AT nx 1600 | NEW LINES AT px 107Hz OLD LINES CHECK AT nx 1600 mx 320 Hz |

Fig. 2. Filter with high-$Q$ pole last.

One striking feature of the limit cycles as they progressed through the filter was the integral relation of their periods. This is summarized in Table I. The existence of this integral relation is given in a theorem by Storzbach [9], the proof of which employs feedback shift-register theory. Other observations made from Table I are that even very long period limit cycles can have small amplitude; that scaling always reduces the amplitude of the largest limit cycle; that with the high-$Q$ pole first, the system starts out with a long period limit cycle; that with the high-$Q$ pole last, the system has a tendency to start with short period limit cycles and to progress to the final long period limit cycle in several steps.

A linear analysis of the system of course will not predict the appearance of limit cycles, but it is useful in predicting the response of the remaining sections of the filter to existing limit cycles. Fig. 2 is a diagram of the optimized tenth-order filter, including values of the multiplier coefficients and scaling coefficients. The basic sampling rate of the filter is 16 kHz and the limit cycles at the third to fifth sections of 10, 50, and 150 sample intervals or frames, lead to fundamental frequencies of 1600, 320, and 107 Hz, respectively. The limit cycle at the output of the third section produces, in addition to the fundamental of 1600-Hz, second to fifth harmonics of approximately equal amplitude, with very little energy in between. At the output of the fourth section, the levels of all these tones check well with what linear analysis would predict, but there are in addition many new lines due to the 50-frame limit-cycle fundamental and its harmonics ($\Delta f = 320$ Hz). In turn, the fifth section attenuates the 10 and 50 frame lines according to linear theory but also produces new components at frequency interval $\approx 107$ Hz that corresponds to the 150-frame limit cycle. Of approximately two dozen representative frequency points checked, only two produced levels

TABLE I
CHARACTERISTICS OF LIMIT CYCLES

| | HIGH Q POLE LAST (OPTIMIZED) | | | | | HIGH Q POLE FIRST (NON-OPT) | | | | |
|---|---|---|---|---|---|---|---|---|---|---|
| | SCALED (G) | | | UNSCALED (H) | | SCALED (J) | | | UNSCALED (K) | |
| SECTION | PERIOD | LEVELS | SCALING | PERIOD | LEVELS | PERIOD | LEVELS | SCALING | PERIOD | LEVELS |
| 1 | dc | -2 | 1/8 | dc | 0 | 132 | 0 TO -3 | 1/4 | 44 | +3 TO -3 |
| 2 | dc | -6 | 1/8 | dc | +6 | 132 | +2 TO -5 | 1 | 132 | +8 TO -8 |
| 3 | 10 | -6 TO -8 | 1/2 | 10 | +11 TO +15 | 132 | +5 TO -11 | 1 | 132 | ≈ +24TO-20 |
| 4 | 50 | -2 TO -5 | 1/4 | 70 | ≈ +12 TO +21 | 132 | ≈ +10TO-29 | 1/2 | 132 | ≈ +80TO-80 |
| 5 | 150 | -2 TO -7 | 1 | 70 | ≈ +13 TO +30 | 132 | ≈-48TO-224 | 1/4 | 132 | ≈ ± 1280 |

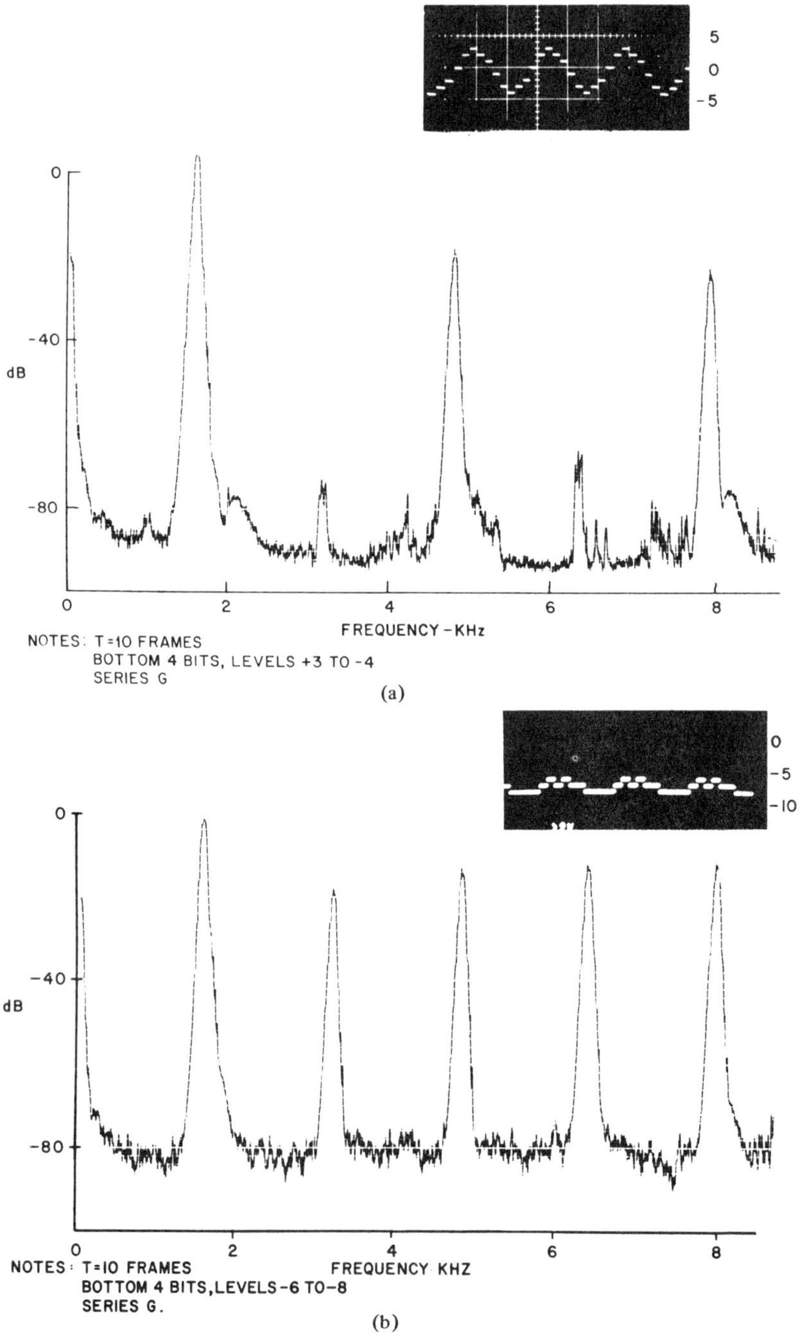


Fig. 3. (a) Limit cycle inside third section. (b) Limit cycle outside third section.

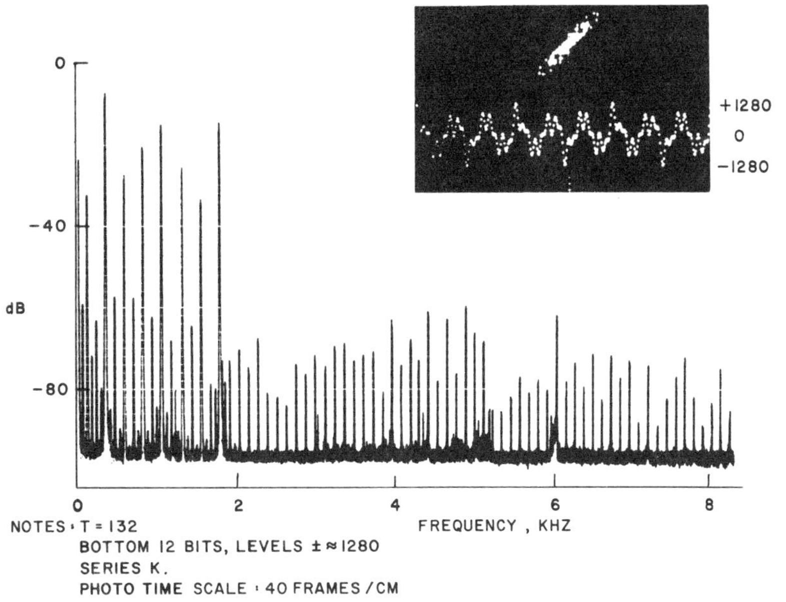


Fig. 4. Limit cycle waveform and spectrum at filter output.

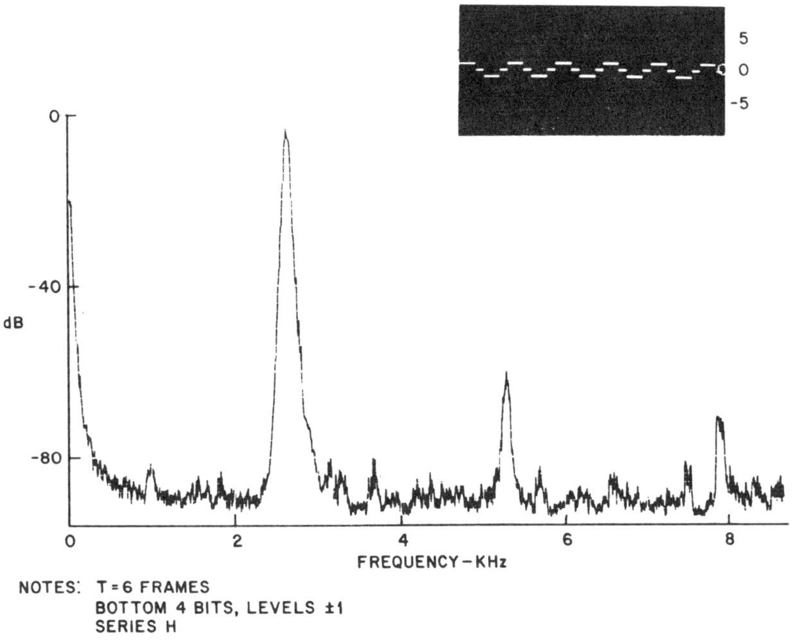


(a)

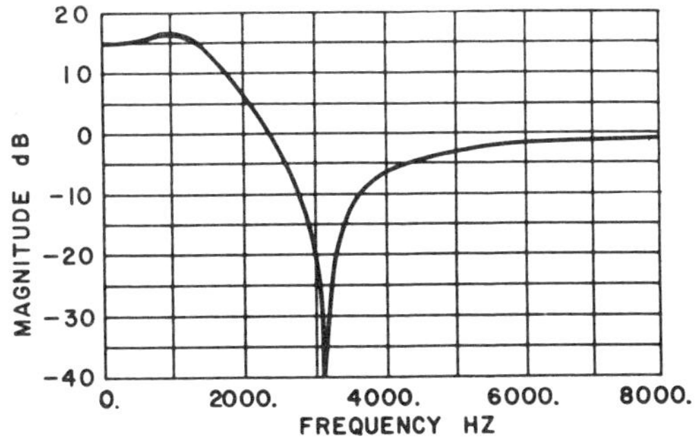


(b)

Fig. 5. (a) Limit cycle inside first section. (b) Frequency response of first section (high $Q$ pole last).

lower than predicted, and those only by 5 and 8 dB. Another simple check was successfully made by noting the progression of the dc limit cycles through the first two sections. Inside the first section a signal of −2 LSB levels was noted. After multiplying by the feed-forward multiplier coefficient and scaling by $\frac{1}{8}$ and rounding appropriately, a dc level of −2 was predicted and confirmed between the first and second sections. This was extended to level −14 inside the second section and −6 at its output. After that, the ac limit cycle began and no such easy check is possible.

An interesting feature is that sometimes at the section where the limit cycle is generated, the fundamental and almost all the harmonics are present at the section output, and all at nearly the same level. The spectral peaks are very clean and sharp. Sometimes inside a section the even harmonics are depressed by as much as 40 or 50 dB, but are present at the output. An example of this is shown in Fig. 3(a) and (b). Fig. 3(a) is the spectrum inside the third section, with the inset showing the waveform. Fig. 3(b) shows the spectrum and waveform at the output of the section.

Evidence that the spectral lines are sharp is further given by Fig. 4, which shows the time series, the quasi-phase-plane plot, and the spectrum of the limit cycle at the output of the last section of the filter with the high-$Q$ section first and no scaling (right-most columns in Table I). The period of the limit cycle is 132 frames and the frequency interval between harmonics is about 120 Hz. Also in this plot the filtering action can be seen, as the peaks in the stop band are 40–50 dB lower than in the 0–2000-Hz passband. This example of a large amplitude limit cycle consumes 12 bits of dynamic range.

Another interesting anomaly is sometimes seen. In the unscaled version of the filter with high-$Q$ pole last, an ac limit cycle of period 6 frames ($f \cong 2600$ Hz) and covering levels 0, ±1 is present inside the first section [see Fig. 5(a)], but at the output a dc level of 0 exists. The explanation is that there is a transmission zero in that section just close enough to 2600 Hz to quench the limit cycle as shown in Fig. 5(b).

In general, but not always, the amplitude of the limit cycle is smaller after passing through the feed-forward part of the second-order section. This relates to the estimate that length of limit cycle should be less than 16 times the maximum amplitude. In two cases (fifth stage of series $G$ and first stage of series $J$ from Table I) the inequality (3) fails using the amplitude at the section output, but checks in all cases using the amplitude inside the section.

A question of practical interest is what happens to a relatively complex cascaded filter that has an existing limit cycle when an input signal is slowly increased from zero. This was tried using the series $K$ filter (Table I), which gave a rather large limit cycle to work with. The limit cycle does not immediately break up, but while the input signal is still below the first quantization level, some signal structure can be seen in the output. By the time the input signal is about 2 quantization levels peak to peak, the signal structure is clearly visible, but is still slightly distorted by the limit cycle. The appearance of a signal building up through a limit cycle is shown in Fig. 6(a)–(c). The input quantization step size is 5 mV, and input signals of 2, 4, and 8 mV are shown. The effective gain of the filter as studied was quite large, to enable detailed examination of the limit cycle.

The same configuration filter was driven from a noise generator whose amplitude was adjusted to 1 LSB at the input A/D converter. In this case the limit-cycle structure appeared to be broken up. A spectral plot of the output (under the same conditions as Fig. 4) is shown in Fig. 7. The random input signal appears to eliminate the limit cycle.

The effect of scaling on the limit cycles is to break up the fine structure of the limit cycles because low-order bits are deleted, thus restricting the number of levels available. How-

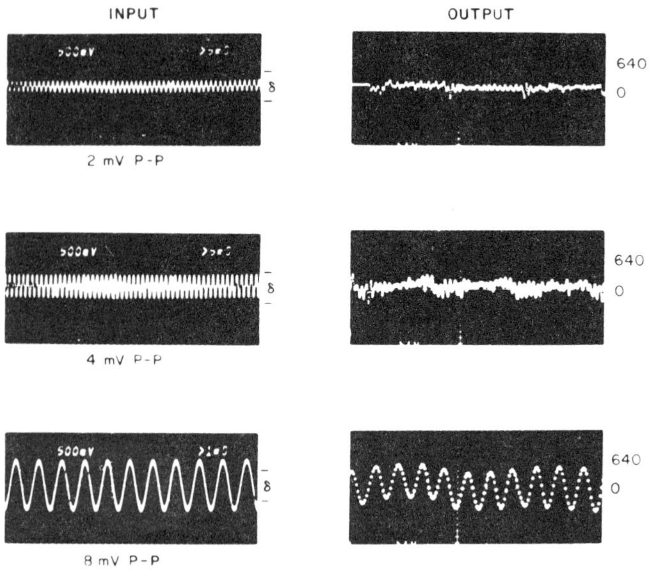


Fig. 6. Buildup of signal through limit cycle.

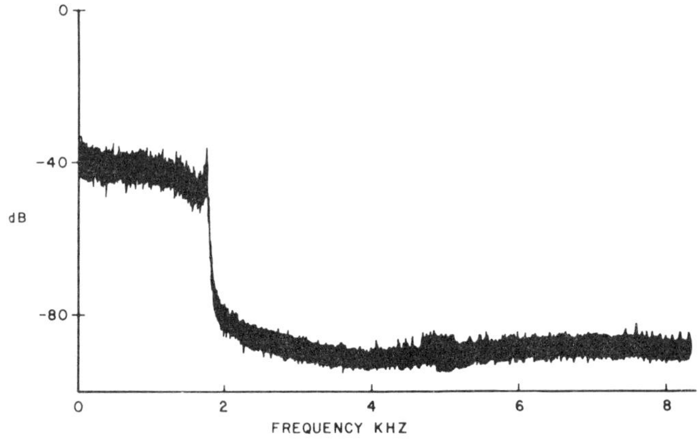


Fig. 7. Spectrum with noise input (1 LSB).

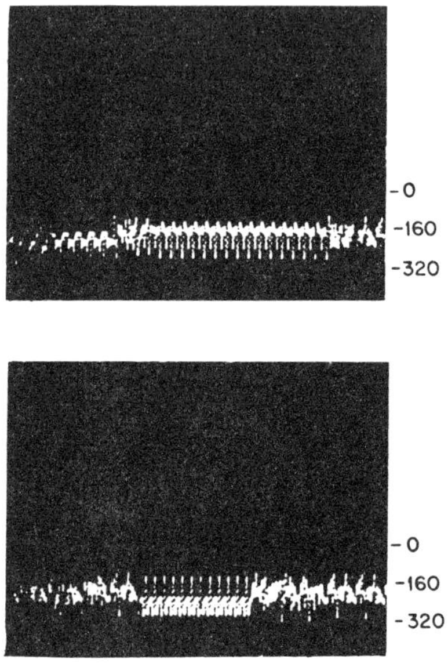


PHOTO TIME SCALE: 200 FRAMES/cm

Fig. 8. Limit cycle hopping.

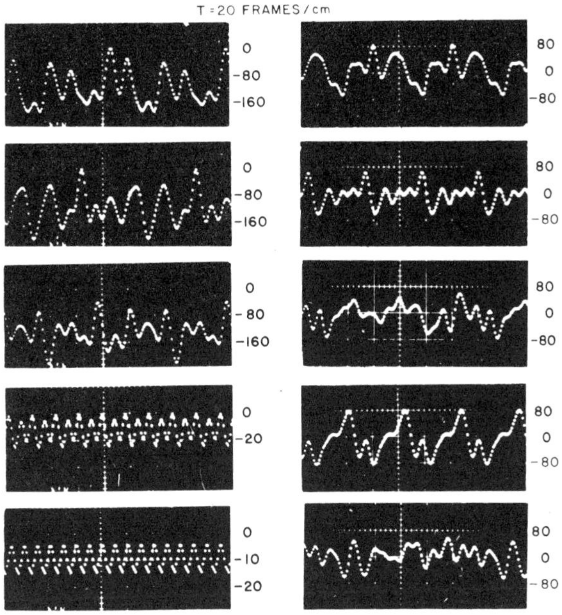


Fig. 9. Gallery of limit cycles.

TABLE II
COMPARISON OF LIMIT CYCLE AND ROUNDOFF NOISE

| | High-$Q$ Pole Last (Optimum) | High-$Q$ Pole First (Nonoptimum) |
|---|---|---|
| Scaled | | |
| Estimated roundoff noise | 32 dB ≈ 6 b | 47 dB ≈ 8 b |
| Limit cycle noise | 32 dB ≈ 6 b | 44 dB ≈ 8 b |
| No scaling between sections | | |
| Estimated roundoff noise | 71 dB ≈ 12 b | 67 dB ≈ 12 b |
| Limit cycle noise | 45 dB ≈ 8 b | 59 dB ≈ 10 b |

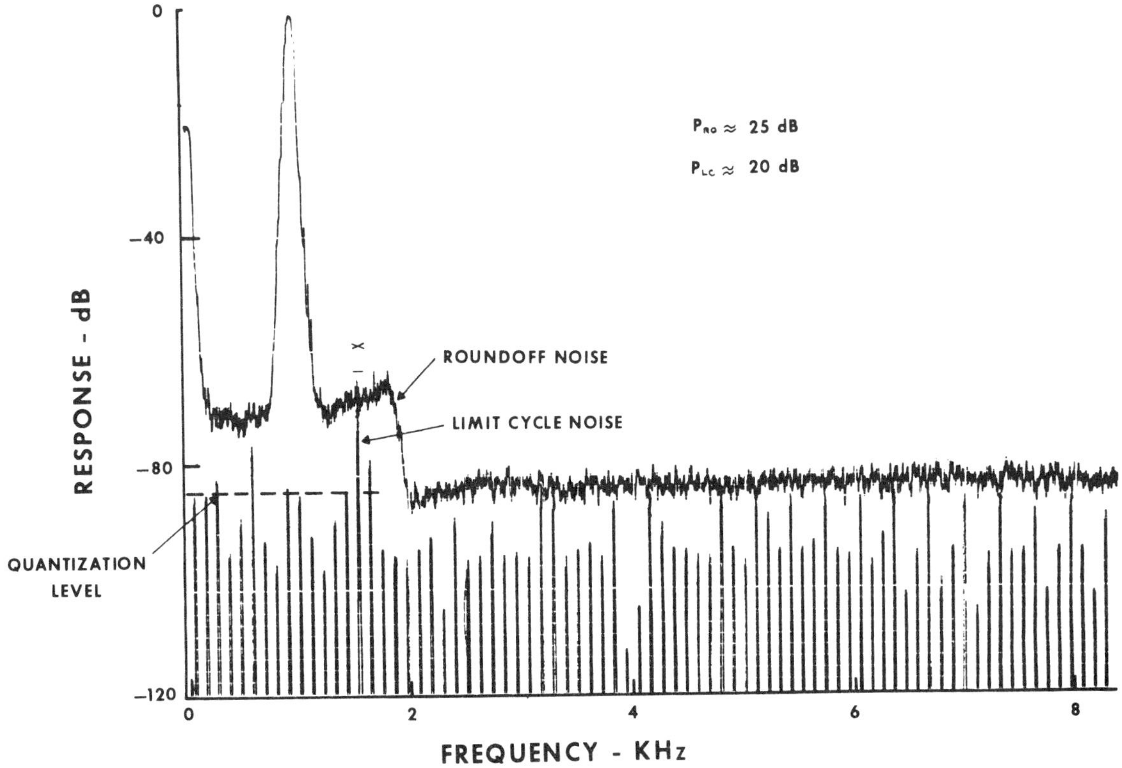


Fig. 10. Comparison of limit cycles and roundoff noise.

ever, this does not seem to lead to shorter periods because some limit cycles have been observed with periods over 100 frames, which use only levels −1, −2, and −3.

An interesting behavior exhibited by the digital filter we tested was a tendency to "hop" from one limit cycle to another if a small signal of the order of a step size was applied. Fig. 8 shows two examples of this behavior.

To give an idea of the rich variety of waveform, amplitude, and dc offset possible in limit cycles in the same digital filter, started with different initial conditions, Fig. 9 is included. All of these limit cycles occurred in the same filter. The procedure was to zero the filter by opening the feedback loop and then impulse excite it with a single PCM word whose amplitude could be set to any desired value. Generally, out of the whole population of possible limit cycles, one will tend to recur much more often than any other.

## Summary and Conclusions

Table II summarizes one of the results of the study. The limit-cycle noise measured under zero input conditions never exceeded the roundoff noise predicted by computer simulation under nonzero input conditions. The comparisons were made on an rms power basis relative to the LSB. Further evidence on the relative magnitude of roundoff and limit-cycle noise is shown for the specific "optimum" filter of interest in Fig. 10. This shows, on a comparable basis, the spectrum of the roundoff noise measured by driving the filter with a 1-kHz signal and the spectrum of the most common limit cycle for this configuration. Neglecting the 1-kHz peak caused by the driving signal, the roundoff noise spectrum looks continuous and agrees with computer simulations. The limit-cycle spectrum is linelike, with one line dominating (about 1600 Hz). The influence of all the other lines increases the total power by less than 2 dB (horizontal dash above strongest line). The strength of this line could be increased another 5 dB (crossmark) before its power equaled that of the roundoff noise. In other words, the total roundoff noise power in this case is 25 dB and the total limit-cycle power is 20 dB, referred to the LSB of internal arithmetic. This may be a typical result.

The lower part of the table, representing no scaling between sections, might be of interest if scaling were done at the end of the filter instead of between sections, or if limit cycles posed a problem during zero input conditions. The figures show that limit cycles can be reduced relative to the roundoff noise; however there is a penalty in dynamic range if this is done. One way of looking at this is that scaling between sections actually increases limit-cycle noise relative to roundoff noise.

The integral relation between limit cycles of driven sections has been confirmed experimentally. The smooth growth of a signal up through a limit cycle has been observed. We checked Jackson's suggestion that either a second-order section filters its input or else it produces its own maximum amplitude limit cycle, whichever leads to the larger output. Actually, a filter section can be doing both at the same time. However, a 1-b random noise input was observed to break up the limit-cycle behavior of a filter that had been previously triggered into a limit cycle.

## References


[1] L. B. Jackson, "An analysis of limit cycles due to multiplication rounding in recursive digital (sub) filters," in *Proc. 7th Annu. Allerton Conf. System Theory*, Oct. 8–10, 1969, pp. 69–78.

[2] S. R. Parker and S. F. Hess, "Limit-cycle oscillations in digital filters," *IEEE Trans. Circuit Theory*, vol. CT-18, pp. 687–697, Nov. 1971.

[3] I. S. Sandberg and J. F. Kaiser, "A bound on limit cycles in fixed-point implementations of digital filters," *IEEE Trans. Audio Electroacoust.*, vol. AU-20, pp. 110–112, June 1972.

[4] S. L. Freeny, R. B. Kieburtz, K. V. Mina, and S. K. Tewksbury, "An exploratory terminal for translating between analog frequency division and digital time division signals," in *1971 Proc. Int. Conf. Communications*, Montreal, P.Q., Canada, pp. 22-31–22-36.

[5] —, "Design of digital filters for an all digital frequency division multiplex-time division multiplex translator," *IEEE Trans. Circuit Theory*, vol. CT-18, pp. 702–711, Nov. 1971.

[6] —, "Systems analysis of a TDM-FDM translator/digital *A*-type channel bank," *IEEE Trans. Commun. Technol.*, vol COM-19, pp. 1050–1059, Dec. 1971.

[7] L. B. Jackson, "On the interaction of roundoff noise and dynamic range in digital filters," *Bell Syst. Tech. J.*, vol. 49, pp. 159–184, Feb. 1970.

[8] J. F. Kaiser, private communication.

[9] W. W. Storzbach, Ph.D. dissertation, Columbia Univ., New York, N.Y., 1972; also, in *Proc. Int. Symp. Circuit Theory*, Los Angeles, Calif., Apr. 18–21, 1972, pp. 233–236.

# Limit-Cycle Oscillations in Floating-Point Digital Filters

TOYOHISA KANEKO

*Abstract*—In a digital filter realized with fixed-point arithmetic, there is a peculiar phenomenon known as limit-cycle oscillation, which is due to roundoff errors. For floating-point arithmetic, it has been conjectured that its amplitude is negligibly small, if it does exist. This paper shows that limit-cycle oscillations can exist in floating-point digital filters and that their amplitude can be large. Also, conditions for the existence of limit-cycle oscillations are derived.

## I. Introduction

Since all digital processors are implemented with a finite number of elements, there are inherent error problems due to the finite word length. The finite arithmetic involves rounding or truncating operations that are essentially nonlinear. Such a nonlinear operation will sometimes generate sustained oscillations known as limit-cycle oscillations if a feedback loop exists in an algorithm.

Manuscript received April 19, 1972.

The author is with the IBM T. J. Watson Research Center, Yorktown Heights, N. Y. 10598.

In digital filters, three sources of errors due to the finite word length have been identified: 1) input quantization error due to quantizing an input sequence, 2) coefficient truncation error due to truncating coefficients in the algorithm, and 3) roundoff error due to rounding intermediate results or final results. They have been studied extensively in the past. A recent paper by Liu [1] is an excellent review. Very recently, the phenomenon of limit-cycle oscillations has attracted much attention. A limit-cycle oscillation in a digital filter was first reported by Blackman [2], who called it a "deadband effect." It is a peculiar phenomenon caused by roundoff errors, and takes place only in a recursive filter. With a zero or constant value input, the output of a digital filter reaches and sustains a constant value different from the predicted value. Sometimes these are constant amplitude sinusoidal oscillations.

In the past, limit-cycle oscillations in digital filters have been studied only for fixed-point arithmetic. Jackson [3] derived bounds on the amplitude of zero-input limit-cycle oscillations by linearizing the nonlinear effect, and gave sufficient conditions on coefficients of first- and second-order digital filters to generate limit-cycle oscillations. A paper by Parker and Hess [4] derived three more exact bounds on the amplitude of limit-cycles using Lyapunov's direct method and a general matrix formulation.

There has been no work on limit-cycle oscillations in floating-point digital filters. It has been conjectured [1], [3] that their amplitude is negligibly small if they do exist. This conjecture seems to be founded on the fact that the floating-point quantizer has a very small

Reprinted from *IEEE Trans. Audio Electroacoust.*, vol. AU-21, pp. 100–106, Apr. 1973.

step size unlike a constant (not small) step size in the fixed-point quantizer.

This paper studies limit-cycle oscillations in floating-point digital filters. It will be shown that there can exist limit-cycle oscillations and that their amplitude can be large. Section II describes floating-point arithmetic and its properties involving roundoff errors. Limit-cycle oscillations in first- and second-order digital filters are analyzed in Sections III and IV, respectively. Section V describes some basic theorems on limit-cycle oscillations. Conditions for limit-cycle oscillations not to exist are described in Section VI. Section VII summarizes the results, and discusses similarities and differences between limit-cycle oscillations in fixed- and floating-point digital filters.

## II. Floating-Point Arithmetic and Roundoff Errors

Before going into the analysis of limit-cycle oscillations in floating-point digital filters, we will review floating-point arithmetic and its roundoff error properties. A floating-point number is represented in the form

$$(\text{sgn})\cdot(\alpha)\cdot\beta^m$$

where (sgn) is the sign, $\alpha$ is the mantissa or fractional part, $\beta$ is the base, and $m$ is the integer exponent. It is assumed throughout this paper that $t$ digits are alloted to the mantissa, and that $m$ is bounded as $-N<m\leq M$, and $M$ and $N$ are large enough so that there are neither overflows nor underflows.[1]

When two floating-point numbers with $t$-digit mantissas are multiplied, the mantissa of the exact product has a length of either $2t$ or $(2t-1)$ digits. The part beyond $t$ significant digits is rounded, generating a roundoff error. There is, in general, a roundoff error in the addition of two floating-point numbers, or in shortening the length of the mantissa. It is shown by Wilkinson [5] that

$$fl(x\cdot y) = (x\cdot y)(1+\epsilon), \tag{1}$$

$$fl(x+y) = (x+y)(1+\delta), \tag{2}$$

$$fl(x) = x(1+\mu), \tag{3}$$

and

$$|\epsilon|, |\delta|, |\mu| \leq \tfrac{1}{2}\beta\cdot\beta^{-t} \tag{4}$$

where the notation $fl(\cdot)$ [5] denotes the machine number resulting from performing the arithmetic operation described in the parenthesis in some designated order, and $fl(x)$ is the result after shortening $x$. The statistical properties of the errors $\epsilon$, $\delta$, and $\mu$ were studied by Kaneko and Liu [6] recently. It is to be noted that a roundoff error due to the addition $\delta$ has a large probability at $\delta=0$, especially when the base $\beta$ is large, say, $\beta=16$.

[1] Overflows and underflows occur in floating-point arithmetic when the exponents go beyond the upper bound, namely, $M$ in the above example, and the lower bound, namely, $-N$, respectively.

## III. Limit-Cycle Oscillations in First-Order Digital Filters

Limit-cycle oscillations in first- and second-order digital filters will be studied in this and subsequent sections, respectively. These filters are especially important because any higher order filter can be realized with a combination (cascade or parallel or both) of first- and second-order filters, and also because the arrangement is preferable in terms of roundoff error accumulation and coefficient truncation errors.

A first-order digital filter is specified by

$$w_n = x_n + a w_{n-1}, \qquad n > 0 \tag{5}$$

with $w_0 = W$ where $\{x_n\}$ and $\{w_n\}$ are the input and output sequences, respectively, $a$ is the coefficient of the filter, and $W$ is an initial value. For the filter operation specified by (5) to be stable, we need

$$|a| < 1. \tag{6}$$

The so-called $Q$ of the filter is defined by

$$Q = 1/(1-|a|). \tag{7}$$

When the filtering algorithm specified by (5) is carried out in a machine and the input is zero, i.e., $x_n=0$, we have

$$y_n = fl(a y_{n-1}) \tag{8}$$

where $y_n$ is the actual output sequence in the machine, and the notation $fl(\ )$ is already defined in Section II. It is well known [7] that the filter specified by (8) has a limit-cycle oscillation of zero frequency if we have

$$u = fl(au) \tag{9}$$

for some $u(\neq 0)$. Equation (9) implies the existence of unity positive feedback in the operation. If it is negative, we have an alternating sign oscillation that will be discussed in Section V. Now we investigate solutions for $u$ in (9) given $a$. When the machine has floating-point arithmetic with base $\beta$ and length of the mantissa $t$, the filter[2] with the highest $Q$ is given by

$$a = 1 - \beta^{-t}. \tag{10}$$

Although $a$ is specified by the difference of two numbers 1 and $\beta^{-t}$, $a$ is actually a number with the exponent zero and with the $t$-digit mantissa of the form

$$\underbrace{(\beta-1)(\beta-1)\cdots(\beta-1)}_{t\text{-digits}}. \tag{11}$$

Using (1), (9) becomes

$$u = au(1+\epsilon)$$

or

[2] If one is allowed to use two numbers in specifying a value, a higher $Q$ filter can be specified by, for example, the difference of 1 and $\beta^{-m}$:

$$a = 1 - \beta^{-m}$$

where $m$ is the exponent and $m>t$. In this filter, the subsequent analysis favors the existence of a limit-cycle oscillation.

$$a = \frac{1}{1+\epsilon}, \qquad \text{if } u \neq 0$$

$$\epsilon = \frac{1-a}{a}. \tag{12}$$

If $a$ is given by (10), then

$$\epsilon = \frac{\beta^{-t}}{1-\beta^{-t}} \approx \beta^{-t} + \beta^{-2t} + \beta^{-3t} + \cdots. \tag{13}$$

The right-hand side of (13) is larger than $\beta^{-t}$. However, $|\epsilon| \leq 2^{-t}$ for $\beta = 2$ from (4). Therefore, (13) does not hold with $\beta = 2$. That is, there are no limit cycles in a first-order digital filter for binary arithmetic.

Now we determine the amplitude of a limit-cycle oscillation for $\beta > 2$. Consider $u = (\beta^{-1}/2)\beta^m$, where $m$ is an integer. Then for $u$ to be a machine number implies that $\beta$ must be an even integer:

$$fl(au) = fl[(1-\beta^{-t})(\beta^{-1}/2)\beta^m].$$

The mantissa of the exact product is

$$\underbrace{(\beta/2 - 1)(\beta - 1) \cdots (\beta - 1)}_{t\text{-digits}}\underset{t+1.}{(\beta/2)}$$

After rounding the $(t+1)$th digit, the above becomes $(\beta^{-1}/2)$ again. So

$$fl(au) = u.$$

Now take $u = (\alpha)\beta^m$ where $\alpha$ is the $t$-digit mantissa specified by

$$\alpha = .\alpha_1\alpha_2 \cdots \alpha_t$$

and

$$1 < \alpha_1 < \beta/2 \text{ or } \alpha_2\alpha_3 \cdots \alpha_t \neq 0, \qquad \text{if } \alpha_1 = 1.$$

The mantissa of the exact product $au$ is

$$\begin{aligned}(1-\beta^{-t})\alpha &= .\alpha_1\alpha_2 \cdots \alpha_t - \beta^{-t}(\alpha_1\alpha_2 \cdots \alpha_t)\\ &= .\underbrace{\alpha_1\alpha_2 \cdots (\alpha_t - 1)}_{t\text{-digits}}\\ &\quad \cdot \underbrace{(\beta-1-\alpha_1)(\beta-1-\alpha_2) \cdots (\beta-\alpha_t)}_{t\text{-digits}}\end{aligned} \tag{14}$$

where $\alpha_t \geq 1$ is assumed. If $\alpha_t = 0$, borrow from the first nonzero $\alpha_i$, $i = t-1, t-2, \cdots, 2$. Since the $(t+1)$th digit $(\beta - 1 - \alpha_1) \geq \beta/2$, the result after rounding is again

$$.\alpha_1\alpha_2 \cdots \alpha_t.$$

Therefore we have

$$fl(au) = u.$$

The behavior of the filter specified by (5) with $a = 1 - \beta^{-t}$ is as follows. If $(\beta^{-1}/2)\beta^m < W \leq \beta^m$, $\{y_n\}$ decreases until it is in $(\beta^{m-2}, (\beta^{-1}/2)\beta^m]$, and stays at the first value in the domain. If $\beta^{m-2} < W \leq (\beta^{-1}/2)\beta^m$, $\{y_n\}$ stays at the initial value $W$. This is illustrated in Fig. 1.

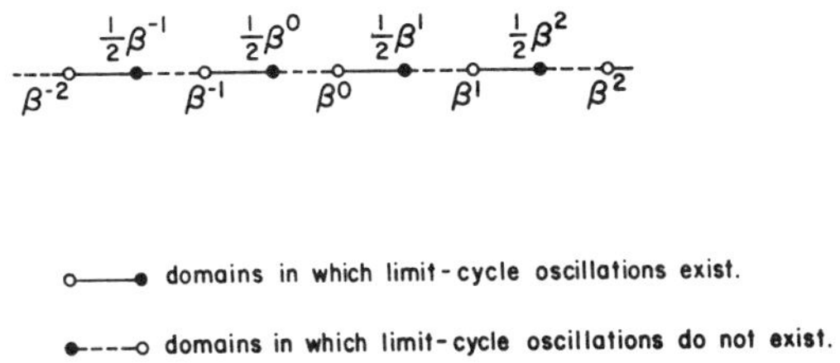


**Fig. 1.** Domains of limit-cycle oscillations in the first-order digital filter $w_n = (1-\beta^{-t})w_{n-1}$.

The number of domains where limit cycles exist is $2(M+N)$ if the exponent ranges as $-N < m \leq M$.

So far, only the filter specified by $a = 1 - \beta^{-t}$ has been studied. We now investigate a first-order filter specified by $a = 1 - k\beta^{-t}$ where $k$ is a positive integer. With this filter, the $(t+1)$th digit of (14) becomes

$$\beta - 1 - k\alpha_1. \tag{15}$$

If $\beta - 1 - k\alpha_1 \geq \beta/2$, 1 is added to the $t$th digit after rounding, and we have $fl(au) = u$ again.

Therefore

$$\begin{aligned} k &\leq \beta/2 - 1, && \text{if } t \geq 2\\ k &\leq \tfrac{1}{2}(\beta/2 - 1), && \text{if } t = 1 \end{aligned} \tag{16}$$

because $\alpha_1$ can be 1 if $t \geq 2$, but $\alpha_1 \geq 2$ if $t = 1$. Then limit cycles exist in the domain

$$\left(\beta^{m-1}, \frac{\beta}{2k}\beta^{m-1}\right]. \tag{17}$$

The expression given in (17) implies that the range of the limit cycle domain decreases as $k$ increases or as $Q$ decreases.

*Example:* Take $\beta = 10$, $t = 2$. Then $k \leq 4$ from (16). Therefore $a = 0.96$ is the lowest $Q$ filter with limit cycles. From (17), the domain of limit cycles for $a = 0.96$ is

$$(0.10 \times 10^m, 0.12 \times 10^m]$$

or

$$0.11 \times 10^m \quad \text{and} \quad 0.12 \times 10^m.$$

Also, a transient behavior is given in Table I using the above example and the initial value $W = 0.20$.

## IV. Limit-Cycle Oscillations in Second-Order Digital Filters

A second-order digital filter is specified by

$$w_n = x_n - aw_{n-1} - bw_{n-2}, \qquad n \geq 2 \tag{18}$$

with

$$w_0 = W_0 \quad \text{and} \quad w_1 = W_1$$

where $\{x_n\}$ and $\{w_n\}$ are the input and output sequences, respectively, and $a$ and $b$ are the coefficients.

TABLE I

The Transient Behavior of a Filter $y_n = +ay_{n-1}$ where $a=0.96$, $\beta=10$, $t=2$, $W=y_0=0.20$

| $n$ | Exact ($a \times y_{n-1}$) | $y_n$ |
|---|---|---|
| 0 | | 0.20 |
| 1 | 0.96×0.20=0.192 | 0.19 |
| 2 | 0.96×0.19=0.1824 | 0.18 |
| 3 | 0.96×0.18=0.1728 | 0.17 |
| 4 | 0.96×0.17=0.1632 | 0.16 |
| 5 | 0.96×0.16=0.1536 | 0.15 |
| 6 | 0.96×0.15=0.1440 | 0.14 |
| 7 | 0.96×0.14=0.1344 | 0.13 |
| 8 | 0.96×0.13=0.1248 | 0.12 |
| 9 | 0.96×0.12=0.1152 | 0.12 |
| 10 | 0.96×0.12=0.1152 | 0.12 |

The pulse transfer function $H(z)$ is given in the $Z$ domain as

$$H(z) = 1/D(z)$$

and

$$D(z) = 1 + az^{-1} + bz^{-2}. \tag{19}$$

We will consider only a filter possessing two conjugate roots in $D(z)$. So

$$a^2 - 4b < 0. \tag{20}$$

For the filter to be stable, the two roots should be inside the unit circle:

$$0 < b < 1. \tag{21}$$

Also, we impose

$$a \neq 0 \tag{22}$$

because (18) with $a=0$ specifies only every other term and reduces to two decoupled first-order system already treated. The following two types of computational methods are commonly used for the filter specified by (18).

*Type 1:* The multiplications and addition in (18) are all carried out exactly with double precision accuracy, and the result is rounded into a $t$-digit floating-point number at the final step.

*Type 2:* The result at each arithmetic step in (18) is rounded into a $t$-digit floating-point number.

Using the notation $fl[\ ]$, the expression of (18) can be written as

$$y_n = fl[(-ay_{n-1}) + (-by_{n-2})] \text{ for type 1} \tag{23a}$$

$$y_n = fl[fl[(-a)y_{n-1}] + fl[(-b)y_{n-2}]] \text{ for type 2} \tag{23b}$$

where $\{y_n\}$ is the actual machine result.

Now let us first examine limit-cycle oscillations at zero frequency. The set $y_n = y_{n-1} = y_{n-2} = u$. For type 1, the same result can be obtained by multiplying $(-a-b)$ and $u$. Therefore, (23a) becomes

$$u = fl[\{-(a+b)\}u] \tag{24}$$

where is the same equation encountered in Section III. Therefore, $-(a+b)$ should satisfy (12).

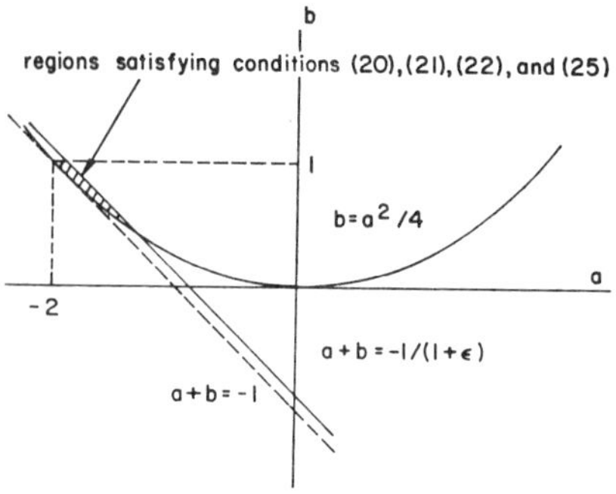


Fig. 2. Regions satisfying the conditions (20), (21), (22), and (25).

$$-a - b = \frac{1}{1+\epsilon}\cdot \tag{25}$$

Conditions (20), (21), (22), and (25) are shown in Fig. 2.

As is seen in Fig. 2, the coefficients $a$ and $b$ satisfying the above conditions are situated close to $-2$ and 1, respectively. Hence, $a$ and $b$ may be written in the following form:

$$a = -(2 - k\beta^{-t+1})$$

and

$$b = 1 - l\beta^{-t} \tag{26}$$

where $k$ and $l$ are positive integers. Substituting $a$ and $b$ in (25), it can be shown that there is no $\epsilon$ satisfying (4). This is due to the fact that the relative accuracy of $a$ is not as good as that of $b$ because the first digit of $a$ is assigned to be 1. To improve the accuracy of $a$, a second-order filter can be realized with[3]

$$w_n = x_n - 2(aw_{n-1}) - bw_{n-2} \tag{27}$$

which requires one more multiplication than (18). Then it can be easily shown that limit-cycle oscilations exist with $b = -a = 1 - k\beta^{-t}$ [$k$ satisfies (16)].

Now let us examine zero frequency limit cycles for type 2. It is to be noted that a difference operation is carried out between the first and second terms on the right-hand side in (23b) because $a$ is negative and $b$ is positive. As was mentioned in Section II, roundoff error in the difference is often zero, especially for large $\beta$, say, 16. So roundoff error at this step is not considered. Let

$$fl(-au) = c_1 u$$

and

$$fl(bu) = c_2 u. \tag{28}$$

Then the condition that limit cycles exist is

$$c_1 - c_2 = 1. \tag{29}$$

An obvious solution is given by $c_2 = 1$ and $c_1 = 2$. But

[3] Also, the following form improves the accuracy of $a$.

$$w_n = x_n - (2 - a)w_{n-1} - bw_{n-2}.$$

Existence of limit cycles can be shown for this form as well.

$c_1 = 2$ is not realizable in an $a$ given in (26). In general, it is difficult to solve (28) for $u$, $a$, and $b$ given $c_1$ and $c_2$. A somewhat heuristic method is illustrated by the following example.

*Example:* Take $\beta = 10$ and $t = 2$. First fix $u$ and $a$ or $b$. Say $a = -1.9$ and $u = 0.33$. Then the $b$ with which limit cycles exist should satisfy [from (23b)]

$$fl(b \times 0.33) = fl(1.9 \times 0.33) - 0.33 = 0.30.$$

Then $b = 0.90$, 0.91, 0.92. $b = 0.90$ should be excluded since it does not satisfy (20).

*Example:* Take $\beta = 2$, $t = 6$, $a = 1.11011$, and $u = 1.11000$. Then the $b$ that generates a limit-cycle oscillation is given by

$$b = 0.110111$$

and it can be shown that $a^2 - 4b < 0$.

As shown in this example, such limit cycles exist for a relatively low $Q$ filter.

From (1) and (2), (23b) with $y_n = y_{n-1} = y_{n-2} = u$ can be written as

$$u = [-au(1+\xi) - bu(1+\eta)](1+\delta) \tag{30}$$

where

$$|\xi|, |\eta|, |\delta| \le (\beta/2)\beta^{-t}.$$

Or

$$-a(1+\xi) - b(1+\eta) = \frac{1}{1+\delta}. \tag{31}$$

If $a$ and $b$ are given, it is possible to examine the existence of limit cycles using (30) and (31) [and (20) and (21)].

More interesting oscillation modes in a second-order digital filter occur for other than zero frequency. If the pulse transfer function is given in (19), a sufficient condition for the existence of a limit cycle is

$$b = 1.$$

In terms of the time domain, the coefficient of $y_{n-2}$ is $-1$. For type 1, this condition is

$$b(1 + \epsilon_n) = 1 \tag{32}$$

where $\epsilon_n$ is the relative rounding error of the result $(-ay_{n-1} - by_{n-2})$. Since $b$ is fixed, $\epsilon_n$ should be constant as well. Therefore $(-ay_{n-1} - by_{n-2})$ can take values that result in the same rounding error. One class of values that satisfy the above condition is a set of three numbers $\{p, 0, -p\}$. Note that zero is always a trivial solution. Then the following limit cycle exists with $a = -b$:

$$p, p, 0, -p, -p, 0, p, p, \cdots.$$

*Example:* Take $\beta = 10$, $t = 1$, $W_0 = W_1 = 0.2$, 0.3, 0.4, or 0.5. $w_n = 0.9\, w_{n-1} - 0.9\, w_{n-2}$.

Then

$$y_n = W_0, W_0, 0, -W_0, -W_0, 0, W_0, W_0, \cdots.$$

For type 2, condition (32) becomes

$$fl(by_{n-2}) = y_{n-2}, \tag{33}$$

which was studied extensively in Section III. That is, solutions exist for $b = 1 - k\beta^{-t}$ where $k$ satisfies (16), and $\{y_n\}$ is in the domain specified by (17).

A significant difference for the solution of (33) between fixed- and floating-point arithmetics is as follows. The values of $y_{n-2}$ that satisfy (33) are in the range $[-d, d]$ for fixed-point arithmetic, where $d$ is the positive integer. However, for floating-point arithmetic, the solutions are in the multiple ranges specified by

$$(\beta^{m-1}, (\beta/2k)\beta^m] \tag{34}$$

or

$$[-(\beta/2k)\beta^m, -\beta^{m-1})$$

where $b = 1 - k\beta^{-t}$ and $m$ is an integer. Also, zero is a trivial solution. Fig. 3(a) and (b) illustrate the above situations. For fixed-point arithmetic, the frequency of a limit-cycle oscillation is determined only by the coefficient $a$, and can go as low as zero frequency without any further restrictions. However, for floating-point arithmetic, $\{y_n\}$ cannot pass through domains other than those specified by (34). If this should be attempted, the amplitude is attenuated and the limit cycle diminishes. This peculiar situation implies that the frequency of a limit cycle has a lower bound. To determine a lower bound, first consider a situation shown in Fig. 4(a), where all points of a limit cycle belong to either $(\beta^{m-1}, (\beta/2k)\beta^{m-1}]$ or $[-(\beta/2k)\beta^{m-1}, -\beta^{m-1})$ for some $m$. As a matter of fact, $m$ is determined by initial values. Consider a sinusoidal wave specified by

$$A \sin 2\pi f_0 s$$

where $A$ is the amplitude, $f_0$ is the frequency, and $s$ is the (continuous) time. For the limit cycle to remain in the domains specified by (34), a necessary condition is [see Fig. 4(a)]

$$A \sin 2\pi f_0(\tfrac{1}{2}) \ge \beta^{m-1}.$$

Here, the frequency of the sample sequence is normalized to 1. Therefore, we have

$$f_0 \ge \frac{1}{\pi} \sin^{-1}(\beta^{m-1}/A). \tag{35}$$

If $A = (\beta/2k)\beta^{m-1}$, then

$$f_0 \ge \frac{1}{\pi} \sin^{-1}(2k/\beta).$$

Another oscillation mode is shown in Fig. 4(b), where most points belong to the above domains, but the other points pass through some inside domains. Then the frequency can go lower than the bound given by (35). If

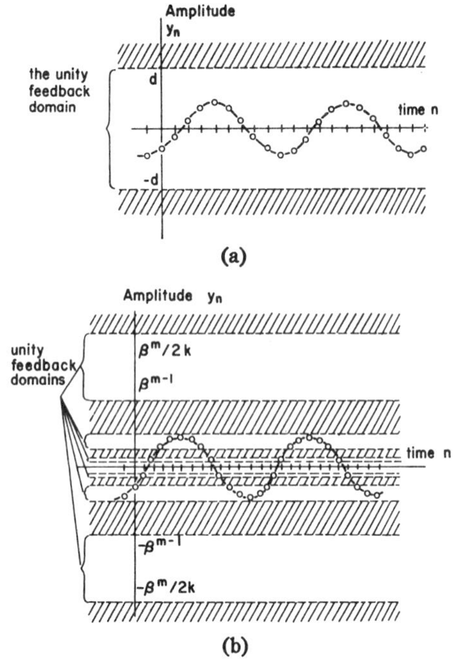


Fig. 3. (a) Limit-cycle oscillation in a second-order digital filter realized with fixed-point arithmetic. (b) Limit-cycle oscillation in a second-order digital filter realized with floating-point arithmetic.

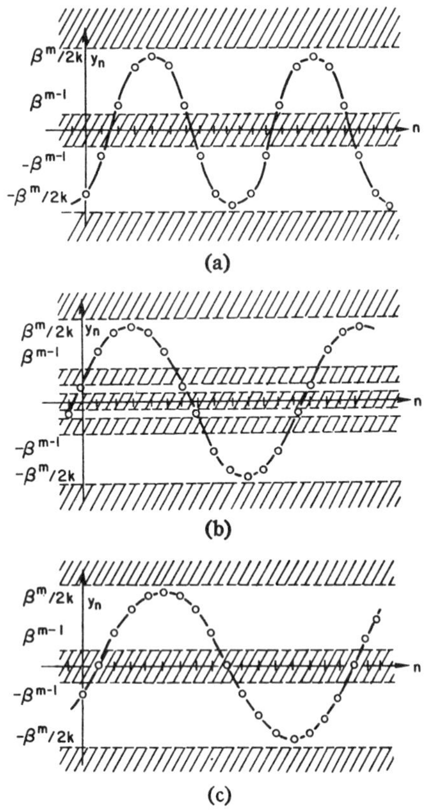


Fig. 4. (a) Limit-cycle oscillation existing only in two (positive and negative) domains. (b) Limit-cycle oscillation existing in more than two domains. (c) Limit-cycle oscillation that includes zero.

zero is included in the limit cycle, as is shown in Fig. 4(c), then the lower bound can be a half of (35):

$$f_0 \geq \frac{1}{2\pi} \sin^{-1} (\beta^{m-1}/A).$$

## V. Some Basic Properties of Limit-Cycle Oscillations in Floating-Point Digital Filters

Consider an $N$th order digital filter specified by

$$w_n = x_n - \sum_{i=1}^{N} a_i w_{n-i} \tag{36}$$

where $\{x_n\}$ and $\{w_n\}$ are the input and output sequences, respectively.

The pulse transfer function is given by

$$H(z) = 1/D(z) \tag{37}$$

and

$$D(z) = 1 + \sum_{i=1}^{N} a_i z^{-1}.$$

*Theorem 1:* When the digital filter specified by (36) utilizes floating-point arithmetic, a limit-cycle oscillation is described with a finite number of values $\{u_0, u_1, \cdots, u_{k-1}\}$ (therefore, the frequency is $1/k$). Then another limit-cycle oscillation exists that can be described by $\{u_0\beta^m, u_1\beta^m, \cdots, u_{k-1}\beta^m\}$ where $m$ is an integer.

*Proof:* Since $\{u_0, \cdots, u_{k-1}\}$ are solutions, $\{y_n, y_{n+1}, \cdots, y_{n+k-1}\}$ can be computed by substituting those numbers where $n$ is an arbitrary positive integer. Then multiply $\beta^m$ in the equations. Because these multiplications do not change the relative amount of the roundoff errors, $\{u_0\beta^m, u_1\beta^m, \cdots, u_{k-1}\beta^m\}$ are also solutions.

*Theorem 2:* If $u_0$ is a limit-cycle solution of (36), then the filter specified by

$$w_n = x_n - \sum_{i=1}^{N} (-1)^i a_i w_{n-i}, \qquad x_n = 0 \tag{38}$$

generates the alternating sign oscillation

$$u_0, \; -u_0, \; u_0, \; -u_0, \; u_0, \; \cdots.$$

*Proof:* Again write the exact computational procedure including roundoff errors for (36) by substituting $u_0$. Since the equation $u_0 = -\sum a_i u_0$ holds, we have $(-1)^n u_0 = -\sum (-1)^i a_i (-1)^{n-i} u_0$, which implies the existence of the alternation sign oscillation $y_n = (-1)^n u_0$ in (38).

## VI. Conditions for Limit-Cycle Oscillations not to Exist

From a designer's view, it is important to obtain conditions on coefficients that do not generate any limit-cycle oscillations. In general, limit-cycle oscillations do not exist in a low $Q$ filter. More specifically, in a floating-point filter realized on a $\beta$-ary and $t$-digit machine, there are no limit-cycle oscillations for a first-order digital filter if (12) does not hold. So we have

$$\left|\frac{1-a}{a}\right| > \frac{1}{2}\beta^{1-t}. \tag{39}$$

Considering only a positive $a$, we have

$$a < 1/(1 + \tfrac{1}{2}\beta^{1-t}) \tag{40}$$

$$Q < (1 + \tfrac{1}{2}\beta^{1-t})2\beta^{t-1} \approx 2\beta^{t-1}. \tag{41}$$

Similarly, limit-cycle oscillations at zero frequency can be avoided in second-order digital filters of types 1 and 2 if (25) and (31), respectively, do not hold. Thus from (25)

$$-(a + b) < 1/(1 + \tfrac{1}{2}\beta^{1-t}). \tag{42}$$

Equation (31) can be rewritten as

$$-a\xi - b\eta + \frac{\delta}{1+\delta} = a + b + 1$$

where $|\xi|$, $|\eta|$, and $|\delta|$ are bounded by $\frac{1}{2}\beta^{1-t}$.

The above equality does not hold if

$$|a + b + 1| > (|a| + |b| + 1/(1 - \tfrac{1}{2}\beta^{1-t}))\tfrac{1}{2}\beta^{1-t}. \tag{43}$$

For limit-cycle oscillations at other than zero frequency, we obtain the following condition from (32):

$$|b| < 1/(1 + \tfrac{1}{2}\beta^{1-t}). \tag{44}$$

## VII. Summary and Discussion

The existence of limit-cycle oscillations has been demonstrated for first- and second-order digital filters realized with floating-point arithmetic, whereas previously it had been conjectured that either no limit-cycle oscillations exist or, if they do, they are negligibly small. For first- and second-order digital filters realized with both fixed- and floating-point arithmetics, there exists a constant value limit-cycle oscillation(s). For the former arithmetic, there can be only one domain that includes zero. However, for the latter, there are many limit-cycle oscillations in separated domains. The zero value is an isolated point. In control theory terminology, limit cycles in fixed-point digital filters are stable, while those in floating-point filters are semistable.

For a second-order floating-point digital filter, there exist limit-cycle oscillations other than those of zero frequency or the alternative sign. For a limit cycle to be stable, its values should remain in domains where negative unity feedback can be obtained from $y_{n-2}$. This special condition imposes a lower bound for the frequency of the limit-cycle oscillation.

## References


[1] B. Liu, "Effect of finite word length on the accuracy of digital filters—A review," *IEEE Trans. Circuit Theory*, vol. CT-18, pp. 670–677, Nov. 1971.

[2] R. B. Blackman, *Linear Data-Smoothing and Prediction in Theory and Practice.* Reading, Mass.: Addison-Wesley, 1965, pp. 75–79.

[3] L. B. Jackson, "An analysis of limit cycles due to multiplication rounding in recursive digital filters," in *Proc. 7th Annu. Allerton Conf. Circuit and Syst. Theory*, 1969, pp. 69–79.

[4] S. R. Parker and S. F. Hess, "Limit-cycle oscillations in digital filters," *IEEE Trans. Circuit Theory*, vol. CT-18, pp. 687–697, Nov. 1971.

[5] J. H. Wilkinson, *Rounding Errors in Algebraic Processors.* Englewood Cliffs, N. J.: Prentice-Hall, 1964.

[6] T. Kaneko and B. Liu, "On local round-off errors in floating-point arithmetic," *J. Ass. Comput. Mach.*, to be published.

[7] J. R. Ragazzini and G. F. Franklin, *Sampled-Data Control Systems.* New York: McGraw-Hill, 1958, pp. 93–105.

# Part 5
# Network Structures

As the field of digital signal processing has matured, there has been greater emphasis on the construction of special-purpose hardware specifically for digital filtering applications. The cost of such hardware is directly affected by considerations such as the required signal and coefficient word lengths,scaling requirements, and the total number of arithmetic operations required. These considerations are, in turn, strongly affected by the choice of a network structure that is used for the implementation. As a result, there has been great interest recently in the design and analysis of structures of many types in search of those that might be most suitable for practical hardware considerations. From the many papers that have been recently published in this area, we have attempted to select those that appear to have the most promising characteristics or that exposed new and potentially useful network concepts around which further research may evolve. The papers by Avenhaus and by Szczupak and Mitra pertain to recent extensions to the popular classes of cascade structures with second-order sections. The paper by Gray and Markel deals with a new class of structures known as lattice or ladder structures, which are especially useful in speech research. The next five papers (by Gray and Markel, by Fettweis, by Sedlmeyer and Fettweis, a second paper by Fettweis, and by Fettweis and Meerkötter) cover the topic of wave digital filter structures, which have received much attention lately. These papers draw strongly on the well-developed theory of classical analog networks and show how this rich body of concepts can be applied to the design of digital structures as well. The paper by Schuessler concerns the design of nonrecursive structures as opposed to recursive structures and shows how some classes of structures can be inferred from classical interpolation formulas. The final paper, by Crochiere and Oppenheim, deals with a range of topics related to the analysis and comparison of some of these filter structures and summarizes some general network properties of these structures.

# A Proposal to Find Suitable Canonical Structures for the Implementation of Digital Filters with Small Coefficient Wordlength

By **Ernst Avenhaus**

*A Report from the Institut für Nachrichtentechnik, Universität Erlangen-Nürnberg*

## 1. Introduction

The coefficient accuracy of recursive digital filters has been investigated by Knowles and Olcayto [1] some years ago. It has been shown that filters being composed of basic building blocks of second order are well appropriate for an implementation in the cascade form as well as in the parallel form. Jackson has reported in his thesis [2] on several methods to implement a second order subsystem, thus obtaining structures with different properties. Additional canonical forms have been found by Hess [3] which do not differ in their coefficient accuracy. The multitude of possible structures seems to be so large, however, that they cannot be investigated systematically for their property in a special case. It will be shown by three examples in this paper, how a suitable structure may be developed for implementation of a special transfer function with small coefficient wordlength.

## 2. The Density of Allowable Root Positions as a Measure for the Valuation of a Structure

Let

$$H(z) = \prod_{\lambda=1}^{l} k_\lambda \frac{b_{2\lambda} z^2 + b_{1\lambda} z + b_{0\lambda}}{z^2 + c_{1\lambda} z + c_{0\lambda}} \tag{1}$$

with $l = \begin{cases} n/2 & \text{if } n \text{ is even,} \\ \dfrac{n+1}{2} & \text{if } n \text{ is odd,} \end{cases}$

be the transfer function of a digital filter of degree $n$. The coefficients $b_{\nu\lambda}$ and $c_{\nu\lambda}$ as well as the scaling factors $k_\lambda$ are implemented directly, if subsystems of the form shown in *Fig. 1* are cascaded. The roots of the polynomial

$$z^2 + a_{1\lambda} z + a_{0\lambda} = 0 \tag{2a}$$

will be the zeros $z_{0\lambda}$ of $H(z)$, if

$$a_{1\lambda} = \frac{b_{1\lambda}}{b_{2\lambda}}; \qquad a_{0\lambda} = \frac{b_{0\lambda}}{b_{2\lambda}}; \qquad b_{2\lambda} \neq 0. \tag{2b}$$

They will be the poles $z_{\infty\lambda}$ of $H(z)$, if

$$a_{1\lambda} = c_{1\lambda}, \quad a_{0\lambda} = c_{0\lambda}. \tag{2c}$$

$$z_{\lambda 1/2} = -\frac{1}{2} a_{1\lambda} \pm \mathrm{j} \sqrt{a_{0\lambda} - \frac{1}{4} a_{1\lambda}^2}. \tag{3}$$

The coefficients $b_{2\lambda}$ of complex zero pairs may be chosen to be one. In the following Eqn. (2a) will be used only. The coefficients $a_\nu$ may be stored as fixed-point binary numbers by

$$a_\nu = \pm \sum_{i=i_M}^{i_L} \alpha_{\nu i} 2^i \tag{4}$$

where $\alpha_{\nu i}$ are zero or one and $\alpha_{\nu i_L}$ is the least significant bit; $\alpha_{\nu i_M}$ is the most significant bit. Obviously, the range of coefficients is

$$-2 \cdot 2^{i_M} < a_\nu < 2 \cdot 2^{i_M} \qquad \forall \nu \tag{5}$$

while the minimum increment, i. e. the quantization step size is defined as

$$Q = 2^{i_L}. \tag{6}$$

Taking into account the one bit for the sign, the wordlength of the coefficients turns out to be

$$W = 2 + i_M - i_L. \tag{7}$$

The finite wordlength restricts the $a_\nu$ to discrete parameter values, i. e.

$$a_\nu = l_\nu \cdot Q \tag{8}$$

with the $l_\nu$ being integers. And therefore with respect to unequality (5) the root locations belong to a finite set in the $z$-plane.

In [4] structures have been valuated by the sensitivity of the root positions. For the present a similar measure may serve for assessing filter structures: the density of allowable root positions in the $z$-plane. This criterion, however, only allows to estimate the suitability of a certain structure indirectly. The maximum deviation $\Delta z_\lambda$ of a root locus $z_\lambda$ according to rounding errors can be derived from the density of allowable root positions. The deviation $\Delta H(z, z_\lambda + \Delta z_\lambda)$ of the transfer function $H(z)$ is approximately proportional to $\Delta z_\lambda$. From

$$H(z) = \prod_{\nu=1}^{n} \frac{z - z_{0\nu}}{z - z_{\infty\nu}} \tag{9}$$

we get

$$\frac{\Delta H(z, z_\lambda + \Delta z_\lambda)}{H(z, z_\lambda)} = \frac{H(z, z_\lambda) - H(z, z_\lambda + \Delta z_\lambda)}{H(z, z_\lambda)} = = s \cdot \frac{\Delta z_\lambda}{z - z_\lambda} \tag{10}$$

with $s = 1$ if $z_\lambda = z_{0\lambda}$,
$s \approx -1$ if $z_\lambda = z_{\infty\lambda}$
and $|\Delta z_\lambda| \ll |z - z_{\infty\lambda}|$.

In other words, the influence of rounding errors on the frequency response

$$H(z)\,|_{z = \mathrm{e}^{\mathrm{j}\Omega}} = H(\mathrm{e}^{\mathrm{j}\Omega}) \tag{11}$$

increases, if the distance between $z_\lambda$ and the point $\mathrm{e}^{\mathrm{j}\Omega}$ in the $z$-plane decreases.

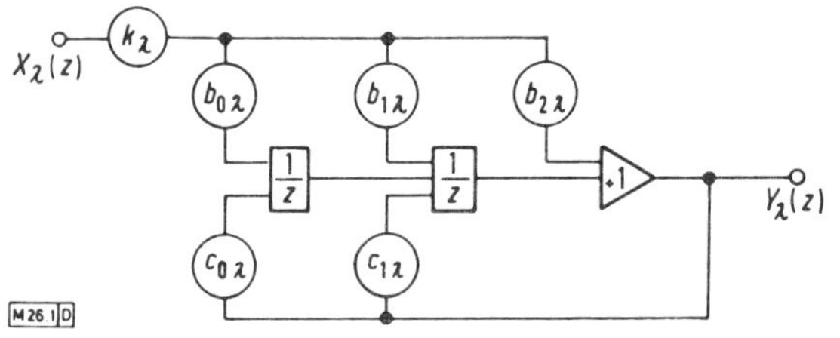


Fig. 1.
Building Block of second order.

Reprinted with permission from *Nachrichtentech. Z.*, vol. 25, pp. 377–382, Aug. 1972.

Considering stable systems with minimum phase only, we have to require that the allowable root positions in the neighbourhood of the unit circle are denser than around the origin of the $z$-plane. The circuit A, shown in Fig. 3 as well as the structures being derived in [3], meet this requirement, as can be seen in *Fig. 2.* But there is a disadvantage in this structure, because the density of allowable root locations is strongly reduced around the points $z = 1$ and $z = -1$. These are the regions, where the poles and zeros of low pass and high pass filters with very small passbands have to be positioned. Circuit A of *Fig. 3* seems to be less appropriate for implementing such a system because only the reduction of the quantization step size of the coefficients yields an increasing density of the allowable root locations. There are many structures with different distributions of allowable pole positions. Circuit B in Fig. 3 exibits an example. *Fig. 4* shows all possible pole positions in a section of the $z$-plane for the circuits A and B.

In the following some structures will be developed being better qualified for the implementation of special transfer functions.

### 3. Development of Circuits with Different Densities of Possible Root Locations

As this paper deals with the reduction of the cost of digital filters we consider canonical structures only. In the literature, however, there exists no general definition for a "canonical form" till now. Therefore we state some properties of a "canonical structure".

a) A structure is canonical with respect to the state storage-locations if the number of the storage-locations is equal to the degree of the filter.

b) A structure is canonical with respect to the coefficients if the number of coefficient storage-locations is equal to the degree of the numerator plus the degree of the denominator of any transfer function with real coefficients. Scaling factors are not counted.

c) A structure is canonical with respect to the multipliers, if their number is equal to the number of coefficients and if the structure is canonical with respect to the coefficients.

d) The number of accumulating registers will not be regarded, because they are less expensive if compared with multipliers.

#### 3.1. *Development of Circuits with Constant Density of Possible Root Locations*

In the first example we consider the problem to develop structures with constant density of the possible root locations in the whole $z$-plane. In this case the roots have to be cross-points of a grid with quadratic meshes. Additionally, we want the pole and zero positions to be independent of each other. Such a structure will be canonical only, as far as state storage-locations and coefficients are concerned, i. e., we have to have more multipliers than coefficients. It may be of interest if the multipliers are multiplexed.

In order to realize the grid with quadratic meshes the coefficients of polynomial Eqn. (2a) will be implemented by other coefficients, called $e_0$ and $e_1$. Let $z_R$ be the real and $z_I$ be the imaginary part of the root, then the quadratic meshes will be obtained by

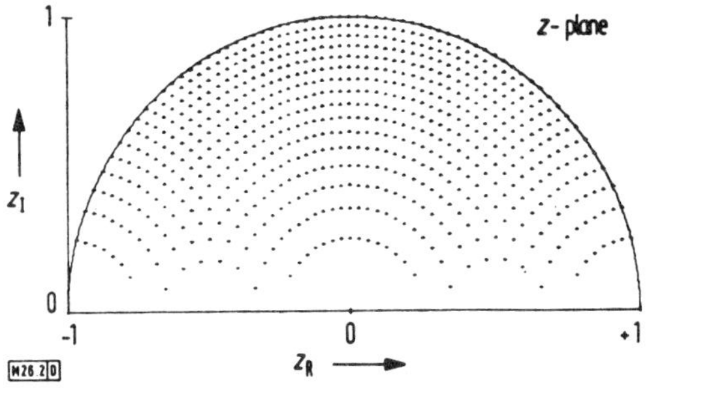


Fig. 2.
All possible root positions with positive imaginary part inside the unit circle of the polynomial equation $z^2 + a_1 z + a_0 = 0$.

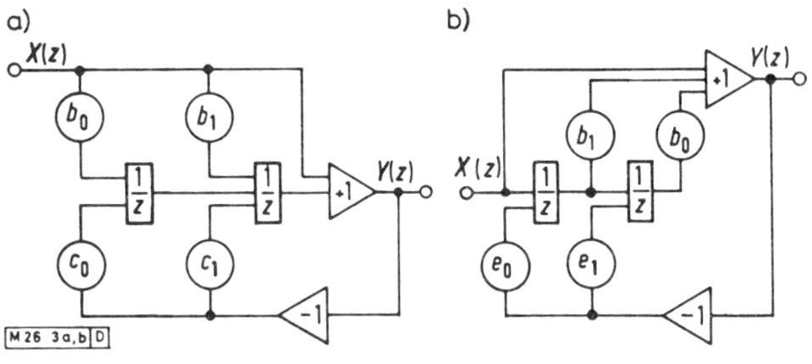


Circuit A

$$H(z) = \frac{z^2 + b_1 z + b_0}{z^2 + c_1 z + c_0}$$

Circuit B

$$H(z) = \frac{z^2 + b_1 z + b_0}{z^2 + z(b_1 e_0 + b_0 e_1) + b_0 e_0}$$

Fig. 3.
Two building blocks with different density of possible pole positions.

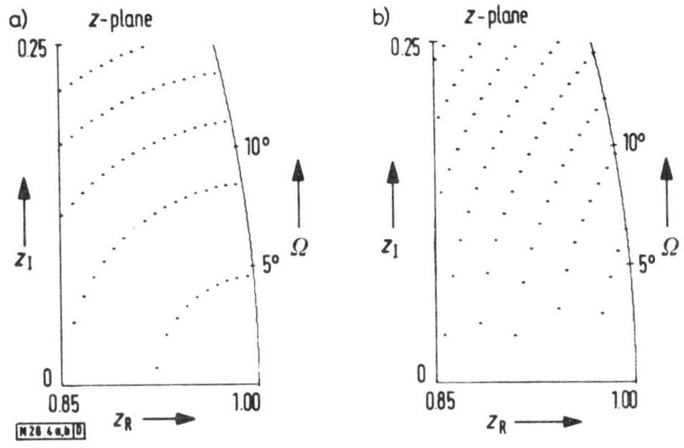


Circuit A
$z^2 + c_1 z + c_0 = 0$

Circuit B
$z^2 + z(b_1 e_0 - b_0 e_1) + b_0 e_0 = 0$
$b_1 = 1{,}8; \quad b_0 = 1$

Fig. 4.
Possible pole positions of circuit A and circuit B in the neighbourhood of the point $z = 1$.
Quantization step size $Q = 2^{-6}$.

$$z_R = \alpha e_1 + \beta e_0 , \qquad z_I = \beta e_1 - \alpha e_0 \tag{12}$$

with parameters $\alpha$ and $\beta$ to be properly chosen. According to Eqn. (8) only discrete values

$$e_0 = l_0 \cdot Q \quad \text{and} \quad e_1 = l_1 \cdot Q \tag{13}$$

can be accepted with $l_0$ and $l_1$ being integers. Considering a fixed value of $e_1$, all roots are located on a straight

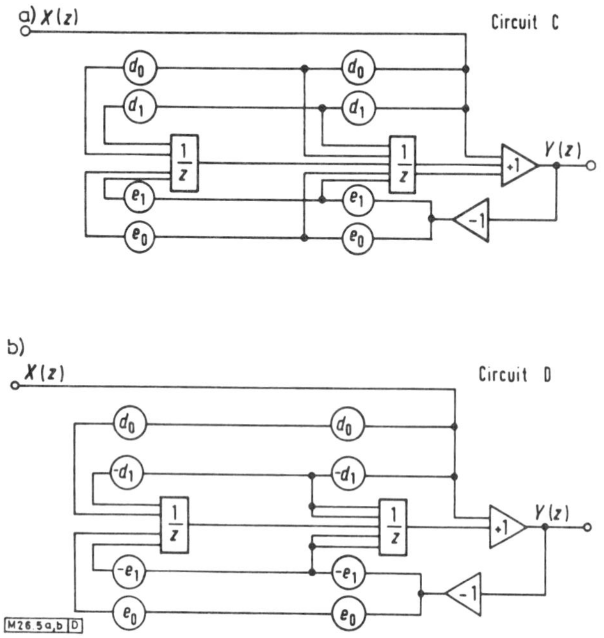


Fig. 5.
Circuits with constant density of possible root locations.

a) $H(z) = \dfrac{z^2 + 2\,(d_1 + d_0)\,z + 2\,(d_1^2 + d_0^2)}{z^2 + 2\,(e_1 + e_0)\,z + 2\,(e_1^2 + e_0^2)}$

b) $H(z) = \dfrac{z^2 - 2\,d_1\,z + d_1^2 + d_0^2}{z^2 - 2\,e_1\,z + e_1^2 + e_0^2}$

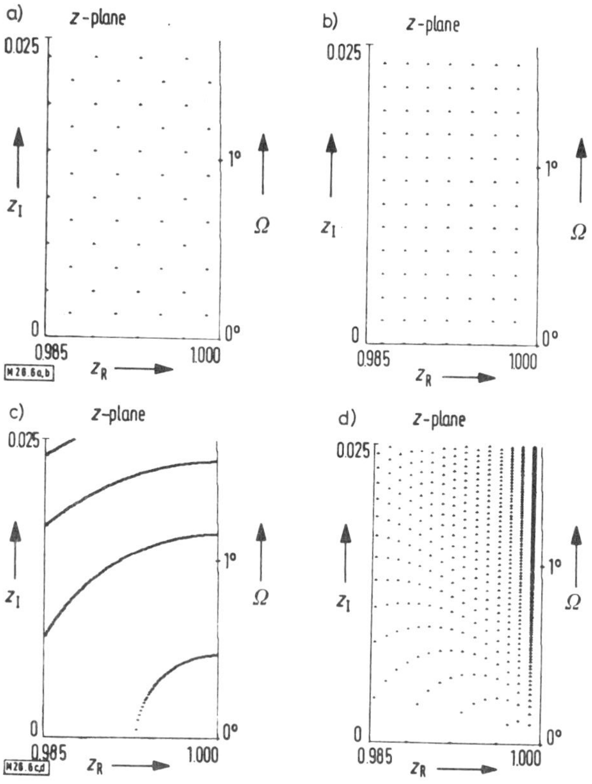


Fig. 6.
Allowable root positions of different circuits in the neighbourhood of the point $z = 1$.

a) $z^2 + 2\,(e_1 + e_0)\,z + 2\,(e_1^2 + e_0^2) = 0$
Circuit C: $Q = 2^{-9}$

b) $z^2 + 2\,e_1\,z + e_1^2 + e_0^2 = 0$
Circuit D: $Q = 2^{-9}$

c) $z^2 + c_1\,z + c_0 = 0$
Circuit A: $Q = 2^{-12}$

d) $z^2 + (e_1 - 2)\,z + 1 - e_1\,e_0 = 0$
Circuit E: $Q = 2^{-9}$

line with gradient $m_0 = -\alpha/\beta$ in constant spacing $|\Delta z| = \sqrt{\alpha^2 + \beta^2} \cdot |\Delta e_0|$. Correspondingly for a fixed value of $e_0$, all roots are positioned on a straigth line with $m_1 = \beta/\alpha = -1/m_0$ in constant spacing $|\Delta z| = \sqrt{\alpha^2 + \beta^2} \cdot |\Delta e_1|$, with $|\Delta e_0| = |\Delta e_1| = Q$. That means, we have a grid with quadratic meshes.

From Eqn. (12) we get the polynomial equation of second order

$$\begin{aligned}(z - z_R - \mathrm{j}\,z_I)\,(z - z_R + \mathrm{j}\,z_I) &= \\ = z^2 - 2\,z(\alpha\,e_1 + \beta\,e_0) &+ (\alpha\,e_1 + \beta\,e_0)^2 + \\ + (\beta\,e_1 - \alpha\,e_0)^2 &= 0\,.\end{aligned} \tag{14}$$

$\alpha = \beta = -1$ yields the denominator polynomial

$$z^2 + 2\,(e_1 + e_0)\,z + 2\,(e_1^2 + e_0^2) = 0 \tag{15}$$

and the numerator polynomial

$$z^2 + 2\,(d_1 + d_0)\,z + 2\,(d_1^2 + d_0^2) = 0\,. \tag{16}$$

Thus, the transfer function of the second order building block is given by

$$H(z) = \frac{z^2 + 2\,(d_1 + d_0)\,z + 2\,(d_1^2 + d_0^2)}{z^2 + 2\,(e_1 + e_0)\,z + 2\,(e_1^2 + e_0^2)}\,. \tag{17}$$

The transfer function Eqn. (17) may be implemented by circuit C of *Fig. 5*.

All possible root locations of Eqn. (15) are shown in *Fig. 6* for $Q = 2^{-9}$ in the neighbourhood of $z = 1$ inside the unit circle. Using circuit A of Fig. 3 with $Q = 2^{-9}$ no complex pole can be positioned in the region shown in Fig. 6. Therefore, all possible root locations of Eqn. (2a) have been drawn for comparision with $Q = 2^{-12}$ in Fig. 6c. Considering a fixed quantization step size $Q$, the allowable root positions of Eqn. (15) are much denser than those of Eqn. (2a) in the region shown in Fig. 6.

We get a similar structure, if we choose $\alpha = 1$ and $\beta = 0$. In this case, $e_1$ is the real part and $e_0$ is the imaginary part of the root of Eqn. (14). The density of allowable root locations is shown in Fig. 6b. The corresponding transfer function

$$H(z) = \frac{z^2 - 2\,d_1\,z + d_1^2 + d_0^2}{z^2 - 2\,e_1\,z + e_1^2 + e_0^2} \tag{18}$$

may be realized by circuit D of Fig. 5. Rader and Gold [5] have published a structure with these properties for the denominator already. As has been mentioned above these structures are not canonical with respect to the multipliers, but they are with respect to the coefficients.

### 3.2. *Development of Circuits for Low Pass Filters and High Pass Filters with Narrow Passbands*

Now we look for a structure with high density of possible root locations around the point $z = 1$ and $z = -1$ respectively. We demand the circuits to be canonical as far as the coefficients and the multipliers are concerned, and we want the coefficients of the numerator and the denominator to be independent of each other.

The distance of a root location from point $z = 1$ with respect to Eqn. (2a) is

$$R = \sqrt{(z_R - 1)^2 + z_I^2} = \sqrt{1 + a_1 + a_0}\,. \tag{19}$$

Looking at quantized parameters we may write

$$\begin{aligned} a_1 &= -2 + l_1 \cdot Q\,, \\ a_0 &= \phantom{-}1 - l_0 \cdot Q \end{aligned} \tag{20}$$

and get

$$R = \sqrt{(l_1 - l_0)\,Q}\,. \tag{21}$$

That means the distance from the complex root next to point $z = 1$ is

$$R_1 = \sqrt{Q} = 2^{\frac{1}{2} i_L}\,. \tag{22}$$

On the other hand, the distance between point $z = \mathrm{j}$ and a possible root next to it is

$$R_\mathrm{j} \approx \frac{Q}{2} = \frac{1}{2} \cdot 2^{i_L}\,. \tag{23}$$

Thus, we nearly have to double the wordlength of the coefficients, if we want the same density of root positions around $z = 1$ as we get it around $z = \mathrm{j}$ by the circuit A of Fig. 3.

To get the same density of pole positions at $z = 1$ without doubling the wordlength, we have to replace the coefficients $a_0$ and $a_1$ in Eqn. (19) by coefficients $e_0$ and $e_1$ so that $R_1$ becomes proportional to $Q$. To that end it is necessary that the radicand in Eqn. (19) becomes a product or a sum of products of $e_0$ and $e_1$. Let us set

$$R = \sqrt{e_1 \cdot e_0} = \sqrt{l_1 \cdot Q \cdot l_0 \cdot Q}\,. \tag{24}$$

In addition the absolute values of $l_1$ and $l_0$ must be small; i. e.

$$|e_1| \ll 1 \tag{25}$$

and

$$|e_0| \ll 1\,. \tag{26}$$

Now the coefficients $e_0$ and $e_1$ have to be chosen adequately. If we take

$$e_1 = a_1 + 2 \tag{27}$$

Eqn. (25) is satisfied, because $a_1$ is about $-2$ if the roots of Eqn. (2a) are positioned in the neighbourhood of the point $z = 1$. Replacing this expression in Eqn. (19) we get

$$a_0 - 1 - e_1 + e_1 e_0\,. \tag{28}$$

Eqn. (26) is satisfied because $a_0$ is about 1.

The density of the possible root locations of the polynomial equation

$$z^2 + (e_1 - 2)\,z + 1 - e_1 + e_1 e_0 = 0 \tag{29}$$

with quantization step size $Q = 2^{-9}$ is shown in Fig. 6d.

The numerator polynomial may be realized by Eqn. (29) as well as the denominator. But considering low pass and high pass filters only, the roots of the numerator are positioned on the unit circle. Because the absolute values of the coefficients $b_{0\lambda}$ are zero or one in these cases, only one numerator coefficient is required to be stored in a building block. Thus we choose the numerator polynomial to be

$$z^2 - (d_1 + 2)\,z + 1\,. \tag{30}$$

The circuit E implementing the corresponding transfer function

$$H(z) = \frac{z^2 - (d_1 + 2)\,z + 1}{z^2 + (e_1 - 2)\,z + 1 - e_1 + e_1 e_0}$$

is shown in *Fig. 7*, which is well suited for low pass filters. For implementing high pass filters, one may

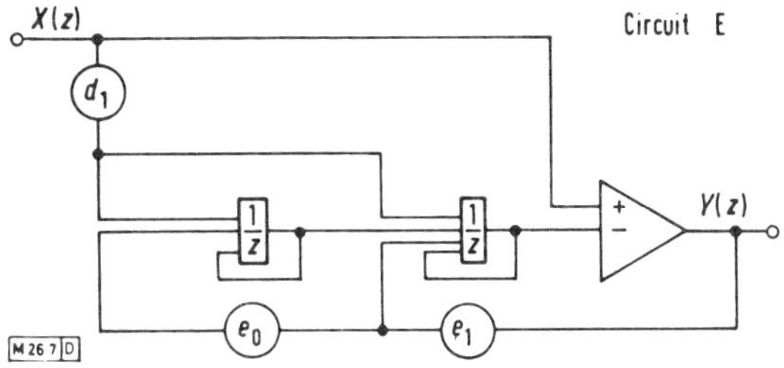


Fig. 7.
Building block of second order suitable for low pass filters.

$$H(z) = \frac{z^2 + z\,(d_1 - 2) + 1}{z^2 + (e_1 - 2)\,z + 1 - e_1 + e_1 e_0}$$

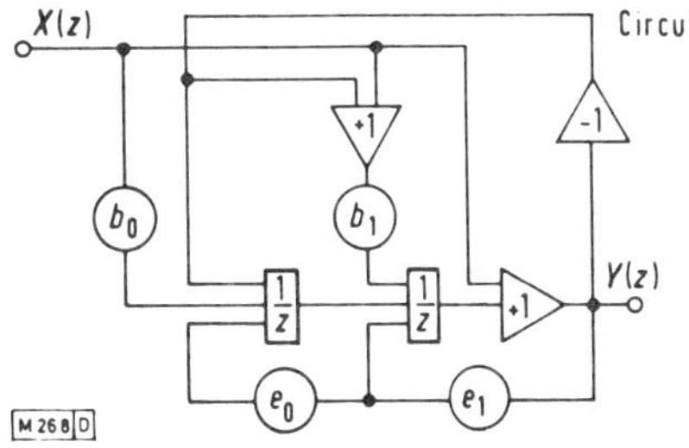


Fig. 8.
Building block of second order suitable for band pass and band stop filters.

$$H(z) = \frac{z^2 + b_1 z + b_0}{z^2 + z\,(b_1 - e_1) + 1 - e_1 e_0}$$

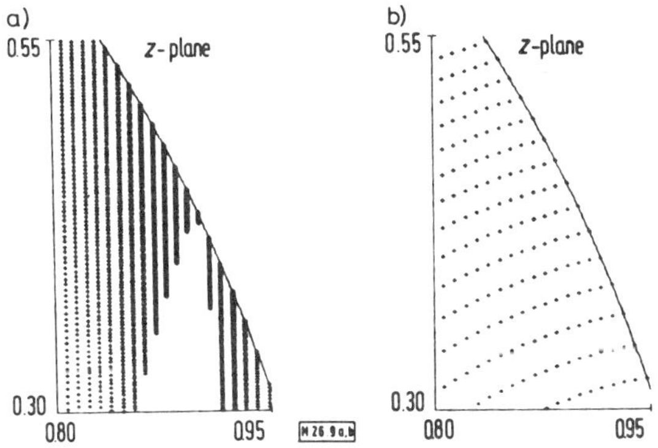


Fig. 9.
Possible pole positions of circuit E and circuit A.

a) $z^2 + z\,(b_1 - e_1) + 1 - e_1 e_0 = 0$ Circuit E: $Q = 2^{-6}$ $b_1 = -1{,}8$; $b_0 = 1$

b) $z^2 + c_1 z + c_0 = 0$ Circuit A: $Q = 2^{-6}$

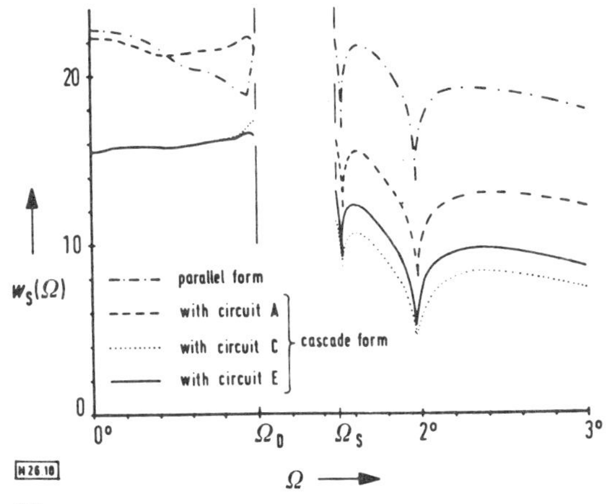


Fig. 10.
"Statistical wordlength functions" of a low pass filter with different structures.

realize the transfer function

$$H(z) = \frac{z^2 + (2 - d_1)\,z + 1}{z_2 + (2 - e_1)\,z + 1 - e_1 + e_1 e_0}\,. \tag{31}$$

### 3.3. *A Suitable Structure for Bandpass Filters with Narrow Passbands and Band Stop Filters with Narrow Stop Bands*

Considering band pass filters with narrow passbands and band stop filters with narrow stop bands, we generally find the properties that

1. the poles are positioned near the unit circle and
2. the distances between poles and zeros are smaller than one.

In the following we only handle second order transfer functions with

a) $$b_0 = 1\,, \tag{32a}$$

b) $$|b_1 - c_1| \ll 1\,, \tag{32b}$$

c) $$1 - c_0 \ll 1\,. \tag{32c}$$

The poles and zeros which can be implemented by the circuit A of Fig. 3 are positioned

1. on straight lines parallel to the imaginary axis,
2. on circles around the point $z = 0$.

Using

$$c_0 = 1 - g_0 \tag{33}$$

we find in the neighbourhood of the unit circle ($c_0 \approx 1$) possible pole positions on circles with the radius

$$r = \sqrt{1 - g_0} = 1 + \Delta r \approx 1 - \frac{1}{2}\,g_0 = 1 - \frac{1}{2}\,l \cdot Q\,. \tag{34}$$

The allowable pole locations can be expected to be denser if we choose coefficients $e_1$ and $e_0$ to get $\Delta r$ proportional to $Q^2$. With the expression

$$c_0 = 1 - e_1 e_0 \tag{35}$$

the radius turns out to be

$$r = \sqrt{1 - e_1 e_0} \approx 1 - \frac{1}{2}\,e_1 e_0 = 1 - \frac{1}{2}\,l_1 \cdot Q \cdot l_0 \cdot Q\,. \tag{36}$$

$$\Delta r \approx -\frac{1}{2}\,l_1\,l_0\,Q^2\,.$$

To get high density of pole positions both coefficients have to satisfy the inequalities

$$|e_1| \ll 1\,, \qquad |e_0| \ll 1\,, \tag{37}$$

which is possible because of inequality (32c). Since we want a canonical structure $e_1$ or $e_0$ have to define the linear coefficient of the denominator polynomial. It becomes small if we use inequality (32b) to get

$$e_1 = b_1 - c_1\,. \tag{38}$$

The corresponding transfer function

$$H(z) = \frac{z^2 + b_1 z + b_0}{z^2 + z(b_1 - e_1) + 1 - e_1 e_0} \tag{39}$$

may be implemented by the building block shown in *Fig. 8*.

The coefficients of numerator and denominator are not independent of each other in this case. *Fig. 9a* shows the possible pole positions of Eqn. (39) for $b_1 = -1{,}8$ and $b_0 = 1$ in the neighbourhood of the zero of the numerator, in comparison with the allowable roots of Eqn. (2a) with $Q = 2^{-6}$; *Fig. 9b*. The absolute values of the coefficients have been restricted to

$$|e_1| \leqq 2\,; \quad |e_0| \leqq 2\,. \tag{40}$$

By means of three examples it has been shown, that for special applications a suitable structure with high density of allowable root locations may be developed. In the last chapter a numerical example of a low pass demonstrates the advantage of these new circuits.

## 4. Assessment of the New Circuits by a Lowpass Example

There exist some direct methods for assessing the suitability of structures of special purpose digital filters, proposed in [6, 7]. Here we want to use one of them, the statistical method for estimating the coefficient wordlength being sufficient to satisfy a given tolerance scheme. Let us choose a low pass example with $\Omega_P$, $\Omega_S$ being the bandlimits and $\delta_P$, $\delta_S$ being the tolerated deviations in the passband and in the stopband respectively.

The tolerance scheme

$$\Omega_P = 1°\,; \quad \Omega_S = 1.5°\,; \quad \delta_P = 0.03\,; \quad \delta_S = 0.001 \tag{41}$$

can be satisfied by a system of sixth order. The "statistical wordlength functions" $w_P(\Omega)$ have been drawn in *Fig. 10* for the circuits A, C and E.

$$W_P = \operatorname*{Max}_{\Omega}\,\{w_P(\Omega)\} \tag{42}$$

gives a measure of the wordlength being sufficient for satisfying the tolerance conditions by a probability of about 95%. Looking at Fig. 10 it can be expected that the coefficient wordlength of the cascade form, using circuit E for a building block, may be equal or less than 16 bits. On the other hand, using circuit A in the cascade or in the parallel form, we only can expect that the wordlength being sufficient for satisfying the tolerance conditions may be equal or less than 23 bits.

After optimization by a modified univariate search [6, 7], the sufficient wordlength turned out to be

$$\left.\begin{aligned} W &= 20 \text{ bits with circuit A} \\ W &= 13 \text{ bits with circuit C} \\ W &= 10 \text{ bits with circuit E} \end{aligned}\right\}\text{ in cascade form}\,.$$

That means in this example we can save up to 10 bits by using a suitable structure for the building blocks of second order.

## 5. Conclusion

If the coefficients of a digital filter are quantized (4) and if they are restricted to finite values (5), there exists only a finite set of possible root locations in the $z$-plane. The distribution of these roots may be arranged adequately by the choice of a suitable structure in order to reduce coefficient sensitivity due to rounding errors, as has been developed by three examples. The low pass filter example shows, that the 10 bit saving justifies a close look at an adequate structure for implementing digital filters.

## 6. Acknowledgement

The author would like to record his gratitude to Professor Schüßler, who stimulated and sponsored investigations on wordlength reduction in digital filtering.

**References**

[1] *Knowles, J. B.; Olcayto, E. M.:* Coefficient accuracy and digital filter response. IEEE Trans. CT—15, No. 1 (March 1968) pp. 31 to 41.

[2] *Jackson, L. B.:* An analysis of roundoff noise in digital filters. Ph. D. Thesis. Stevens Institute of Technology, Castle Point, Hoboken, New Jersey 1969.

[3] *Hess, S.:* A deterministic analysis of limit cycle oscillations in recursive digital filters due to quantization. Ph. D. Thesis. United States Naval Post graduate School, Monterey, California 1970.

[4] *Mantey, P. E.:* Eigenvalue sensitivity and state-variable selection. IEEE Trans. AC—13, No. 10 (June 1968) pp. 263 to 269.

[5] *Rader, C. M.; Gold, B.:* Effects of parameter quantization on the poles of a digital filter. Proc. IEEE 55 (May 1967) No. 5, pp. 688 to 689.

[6] *Avenhaus, E.:* Zum Entwurf digitaler Filter mit minimaler Speicherwortlänge für Koeffizienten und Zustandsgrößen. Ph. D. Thesis. Institut für Nachrichtentechnik der Universität Erlangen—Nürnberg 1971, Ausgewählte Arbeiten über Nachrichtensysteme, Nr. 13.

[7] *Avenhaus, E.:* On the design of digital filters with coefficients of limited word-length. Will be published by the IEEE Trans. Audio & Electroacoustics.

(Manuscript received: April 14, 1972)

# Digital Filter Realization Using Successive Multiplier-Extraction Approach

JACQUES SZCZUPAK, STUDENT MEMBER, IEEE, AND SANJIT K. MITRA, FELLOW, IEEE

*Abstract*—**A new approach to the digital filter realization problem is proposed in this paper. In this approach multipliers are extracted one at a time successively in the development of all realizations. The approach is illustrated by finding all possible realizations of a second-order transfer function containing two delays, four multipliers, and four two-input adders with the added restriction that no products of multipliers appear in the expression for the transfer function.**

## I. INTRODUCTION

THE subnetwork extraction approach to realize a specified network function has been widely used in analog network synthesis. There are two basic ways this approach can be adapted for digital filter realization: 1) the delay-extraction approach and 2) the multiplier-extraction approach. In the delay-extraction approach, assuming the total number of delays in the realization to be $n$, the final realization is considered as a $(n+1)$-pair $N$ whose $n$ pairs of input and output terminals are constrained by the $n$ delays. The $(n+1)$-pair $N$ contains only multipliers and adders. If $n$ is equal to the order of the digital transfer function, this approach is equivalent to realization based on state-variable representation of the transfer function. The delay extraction approach was used by Jackson [1] and later by Parker and Hess [2] to develop a number of canonic realizations of biquadratic transfer functions such that all of them have the same transfer function under a fixed point realization. This condition is met in networks if no products of multipliers appear in the coefficients of the transfer function. Note that if no constraints are imposed, then this approach leads to an infinite number of equivalent realizations.

In the multiplier-extraction approach [3], the realization is considered as a $(m+1)$-pair $M$ of which $m$ pairs of input and output terminals are constrained by $m$ multipliers. The $(m+1)$-pair $M$ is assumed to contain only delays and adders. This second approach, however, leads to a finite number of equivalent realizations and has been used successfully to generate all possible canonic and noncanonic multiplier-invariant equivalent realizations of digital all-pass networks [4], notch filters [5], and other cases [6]. In all of these realizations, the total number of multipliers were restricted to at most three.

For a larger number of multipliers, it is more convenient to use a successive extraction procedure which is illustrated in this paper. This alternate approach reduces progressively the complexity of the remaining network to the point where one is able to identify the network's topology by simple inspection of its mathematical description. The proposed variation is also quite general in the sense that the extraction procedure may generate any realization of a given transfer function.

## II. PROBLEM STATEMENT

The successive multiplier-extraction approach is illustrated in this paper for a particular case. A general procedure is easily inferred from this example allowing one to develop from specific goals a more convenient strategy to adapt the method.

The problem solved here is the minimal realizations of second-order transfer functions with no product of multipliers. It should be noted that a minimal realization of a biquadratic transfer function uses four multipliers, two delays, and four two-input adders. It has been shown that the successive multiplier-extraction approach leads to a number of new such structures in addition to those already developed by Jackson [1]. The requirements of not allowing products of multipliers enables different realizations to have identical transfer functions eliminating the small variations in coefficients caused by the use of a fixed number of digits.

## III. STRUCTURES WITH NO PRODUCT OF MULTIPLIERS

Let $N$ be the realization of the transfer function $T_1(z) = Y_1(z)/X_1(z)$. A partition of $N$ with multiplier $\alpha_1$ removed is shown in Fig. 1, where the two-pair $N_1$ is characterized by a transfer matrix $\boldsymbol{T}_2$:

$$\begin{bmatrix} Y_1 \\ Y_2 \end{bmatrix} = \begin{bmatrix} t_{11}^0 & t_{12}^0 \\ t_{21}^0 & t_{22}^0 \end{bmatrix} \begin{bmatrix} X_1 \\ X_2 \end{bmatrix} = \boldsymbol{T}_2 \cdot \begin{bmatrix} X_1 \\ X_2 \end{bmatrix}. \qquad (1)$$

In (1), the $z$ dependence of the terminal variables has not been shown but is implied. From (1) and the constraining equation

$$X_2 = \alpha_1 Y_2 \qquad (2)$$

Manuscript received June 14, 1974; revised November 18, 1974. This work was supported in part by the National Science Foundation under Grant ENG74-08879 and in part by Conselho Nacional de Pesquisas and Copertide-UFRJ of Brazil.

J. Szczupak is with the Department of Electrical Engineering, University of California, Davis, Calif. 95616, on leave from COPPE, Universidade Federal do Rio de Janeiro, Brazil.

S. K. Mitra is with the Department of Electrical Engineering, University of California, Davis, Calif. 95616.

Reprinted from *IEEE Trans. Acoust., Speech, Signal Processing*, vol. ASSP-23, pp. 235–239, Apr. 1975.

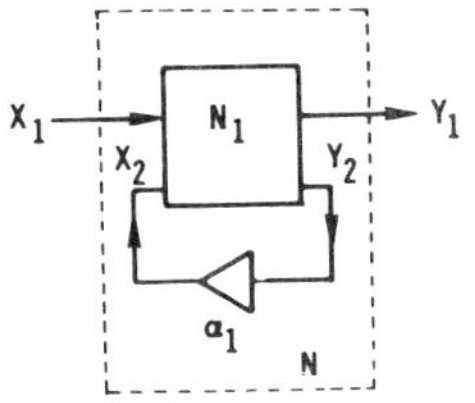


Fig. 1. Digital two-pair constrained by one multiplier.

we readily arrive at

$$T_1(z) = \frac{Y_1}{X_1} = \frac{\alpha_1(t_{12}^0 t_{21}^0 - t_{11}^0 t_{22}^0) + t_{11}^0}{1 - \alpha_1 t_{22}^0}. \tag{3}$$

Some properties of the transfer parameters $t_{ij}^0$ follow from (3) and the multiplier-product restrictions imposed on the realization of $T_1(z)$. Denoting $t_{ij}^0 = N_{ij}^0/D_{ij}^0$, we conclude the following.

*Condition 1:* $N_{22}^0$ has no multiplier.

*Condition 2:* det $(\boldsymbol{T}_2)$ has no multiplier in the numerator.

*Condition 3:* If $N_{11}^0$ has a multiplier, then either $N_{12}^0$ or $N_{21}^0$ must also have a multiplier. (Without any loss of generality, we assume that all multipliers be in $N_{12}^0$ and none in $N_{21}^0$. Note that the converse case can be handled readily using the transpose realization [1].)

Next, consider the extraction of the multiplier $\alpha_2$ from $N_1$ as shown in Fig. 2. Here now the subnetwork $N_2$ is mathematically described by the transfer matrix

$$\boldsymbol{T}_3 = \begin{bmatrix} t_{11}' & t_{12}' & t_{13}' \\ t_{21}' & t_{22}' & t_{23}' \\ t_{31}' & t_{32}' & t_{33}' \end{bmatrix}. \tag{4}$$

As was done for the first multiplier extraction, using the relation $Y_3 = \alpha_2^{-1}X_3$, it is possible to obtain $\boldsymbol{T}_2$ from $\boldsymbol{T}_3$ after a reduction in the resulting third-order matrix

$$T_1(z) = \frac{-\alpha_1\alpha_2 \det(\boldsymbol{T}_3) + \alpha_1(t_{12}'t_{21}' - t_{11}'t_{22}') + \alpha_2(t_{13}'t_{31}' - t_{11}'t_{33}') + t_{11}'}{1 - \alpha_1 t_{22}' - \alpha_2 t_{33}'}. \tag{14}$$

$$\boldsymbol{T}_2 = \begin{bmatrix} t_{11}' - \dfrac{t_{13}'t_{31}'\alpha_2}{t_{33}'\alpha_2 - 1} & t_{12}' - \dfrac{t_{13}'t_{32}'\alpha_2}{t_{33}'\alpha_2 - 1} \\ t_{21}' - \dfrac{t_{23}'t_{31}'\alpha_2}{t_{33}'\alpha_2 - 1} & t_{22}' - \dfrac{t_{23}'t_{32}'\alpha_2}{t_{33}'\alpha_2 - 1} \end{bmatrix}. \tag{5}$$

From (5) and restrictions similar to conditions 1 and 2, one obtains the following relations:

$$t_{22}'t_{33}' - t_{23}'t_{32}' = 0 \tag{6}$$

$$t_{22}^0 = -\frac{t_{22}'}{t_{33}'\alpha_2 - 1} \tag{7}$$

$$t_{21}'t_{33}' - t_{23}'t_{31}' = 0 \tag{8}$$

$$t_{21}^0 = -\frac{t_{21}'}{t_{33}'\alpha_2 - 1}. \tag{9}$$

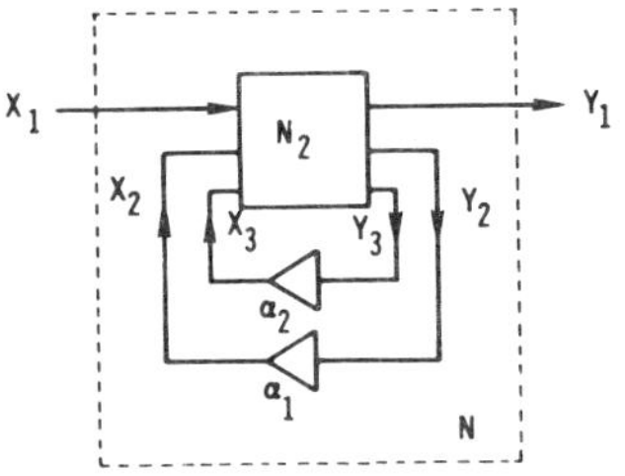


Fig. 2. Digital three-pair constrained by two multipliers.

If the roles of $\alpha_1$ and $\alpha_2$ are reversed in such a way that the reduction to a second-order matrix is done using the row and column corresponding to $\alpha_1$, it yields

$$\hat{\boldsymbol{T}}_2 = \begin{bmatrix} t_{11}' - \dfrac{t_{12}'t_{21}'\alpha_1}{t_{22}'\alpha_1 - 1} & t_{13}' - \dfrac{t_{12}'t_{23}'\alpha_1}{t_{22}'\alpha_1 - 1} \\ t_{31}' - \dfrac{t_{32}'t_{21}'\alpha_1}{t_{22}'\alpha_1 - 1} & t_{33}' - \dfrac{t_{12}'t_{23}'\alpha_1}{t_{22}'\alpha_1 - 1} \end{bmatrix} = \begin{bmatrix} \hat{t}_{11}^0 & \hat{t}_{12}^0 \\ \hat{t}_{21}^0 & \hat{t}_{22}^0 \end{bmatrix}. \tag{10}$$

Applying the same type of reasoning used for the discussion of $\boldsymbol{T}_2$, we arrive at the following additional relations:

$$\hat{t}_{22}^0 = -\frac{t_{33}'}{t_{22}'\alpha_1 - 1} \tag{11}$$

and either

$$t_{13}'t_{22}' - t_{12}'t_{23}' = 0 \tag{12}$$

or

$$t_{31}'t_{22}' - t_{21}'t_{32}' = 0. \tag{13}$$

The complete reduction of the multipliers rows in the third-order matrix yields (14).

Therefore, using the multiplier product condition, it follows that

$$\det(\boldsymbol{T}_3) = 0. \tag{15}$$

From (6), (8), and (15), we obtain

$$t_{13}'(t_{21}'t_{32}' - t_{22}'t_{31}') = 0. \tag{16}$$

Up to this point we have developed a set of necessary conditions to be fulfilled by the third-order matrix elements. Two of them, given by (12) and (13), were found to be exclusive. Our problem now reduces to determining which one of these two relations should hold. For this purpose we assume initially that (13) is not valid. From the necessary conditions it follows that in this case (12) must be satisfied together with (16). However, comparing (16) and (13) and using our initial assumption about the last equation, we obtain the condition $t_{13}' = 0$. Taking this value into (12) results in $t_{12}'t_{23}' = 0$. Clearly, $t_{13}'$ and $t_{12}'$ cannot be zero simultaneously as the resulting network would be degenerate. The same reasoning shows $t_{13}'$ and $t_{23}'$ cannot both be zero as here the multiplier $\alpha_2$ will not contribute to $T_1(z)$. Therefore $t_{12}'t_{23}' \neq 0$ and (12) is not true. As a result, (13) must be satisfied. One may notice that when the first multiplier extraction was performed, $t_{12}^0$ could have multipliers in the numerator but not $t_{21}^0$; this is the basis for the choice between (12) and (13). It is also clear that when we take the transpose of these realizations the situation is reversed and (12) is satisfied.

Our basic interest is in networks having transfer functions with no product of multipliers for any value they can assume. Networks in which a specific choice of multiplier values satisfy the no multiplier product condition will be disregarded.

Conditions 1, 2, and 3, together with (3), (7), (9), (11) and the substitution of (13) into (10), lead to the absence of multiplier in the numerator of $t_{22}'$, $t_{33}'$, $t_{21}'$, and $t_{31}'$. Putting together the proportionality between the second and third rows of $\boldsymbol{T}_3$, given by (6), (8), and (13), with the absence of multipliers just mentioned, the class of networks being sought will have $\boldsymbol{T}_3$ matrices in which all elements not in the first row do not have multiplier in the numerator.

Notice that Conditions 1, 2, and 3 derived for $\boldsymbol{T}_2$ were extended to $\boldsymbol{T}_3$. They now will be generalized for higher order matrices.

Basically the general rule for a matrix $\boldsymbol{T}_{k+1}$, obtained by the extraction of one multiplier from a realization of a matrix $\boldsymbol{T}_k$, is that the reduction of the row and column corresponding to the multiplier must satisfy the necessary conditions for the $\boldsymbol{T}_k$ matrix. This gives us a recursive approach to generate necessary conditions for higher order matrices from, for instance, the matrix $\boldsymbol{T}_3$. Exceptions to this rule are the transposed realizations from these matrices or equivalently, the transpose of the matrices.

For instance, if we extract 3 multipliers from $T_1$ generating a matrix $\boldsymbol{T}_4 = [\hat{t}_{ij}]_{4\times 4}$, its reduction to $\boldsymbol{T}_3$ yields

$$\boldsymbol{T}_3 = \begin{bmatrix} \hat{t}_{11} + \dfrac{\hat{N}_{11}}{\hat{d}} & \hat{t}_{12} + \dfrac{\hat{N}_{12}}{\hat{d}} & \hat{t}_{13} + \dfrac{\hat{N}_{13}}{\hat{d}} \\ \hat{t}_{21} + \dfrac{\hat{N}_{21}}{\hat{d}} & \hat{t}_{22} + \dfrac{\hat{N}_{22}}{\hat{d}} & \hat{t}_{23} + \dfrac{\hat{N}_{23}}{\hat{d}} \\ \hat{t}_{31} + \dfrac{\hat{N}_{31}}{\hat{d}} & \hat{t}_{32} + \dfrac{\hat{N}_{32}}{\hat{d}} & \hat{t}_{33} + \dfrac{\hat{N}_{33}}{\hat{d}} \end{bmatrix} \tag{17}$$

where

$$\hat{d} = 1 - \alpha_3 \hat{t}_{44} \tag{18}$$

$$\hat{N}_{ij} = \alpha_3 \hat{t}_{i4} \hat{t}_{4j}, \qquad i,j = 1,2,3. \tag{19}$$

Using the requirement of no multiplier in the numerator of the elements in the second and third rows of $\boldsymbol{T}_3$, the result is

$$\boldsymbol{T}_3 = \begin{bmatrix} \hat{t}_{11} + \dfrac{\hat{N}_{11}}{\hat{d}} & \hat{t}_{12} + \dfrac{\hat{N}_{12}}{\hat{d}} & \hat{t}_{13} + \dfrac{\hat{N}_{13}}{\hat{d}} \\ \dfrac{\hat{t}_{21}}{\hat{d}} & \dfrac{\hat{t}_{22}}{\hat{d}} & \dfrac{\hat{t}_{23}}{\hat{d}} \\ \dfrac{\hat{t}_{31}}{\hat{d}} & \dfrac{\hat{t}_{32}}{\hat{d}} & \dfrac{\hat{t}_{33}}{\hat{d}} \end{bmatrix} \tag{20}$$

with

$$\hat{t}_{ij}\hat{t}_{44} - \hat{t}_{i4}\hat{t}_{4j} = 0, \qquad i = 2,3; \; j = 1,2,3, \tag{21}$$

the $\hat{t}_{ij} (i = 2,3;\; j = 1,2,3)$ with no multipliers in the numerator and the second and third row of $\boldsymbol{T}_3$ proportionality, originated from (6), (8), and (13). Using the reduction of other multiplier's rows and columns results in having $\hat{t}_{ij} (i = 2,\cdots,4,\; j = 1,\cdots,4)$ with no multiplier in the numerator and proportionality between rows 2, 3, and 4 of $\boldsymbol{T}_4$.

The extraction of 4 multipliers results in a fifth-order matrix $\boldsymbol{T}_5 = [t_{ij}]_{5\times 5}$. An analogous approach may be applied to the $t_{ij}$ reducing $\boldsymbol{T}_5$ to $\boldsymbol{T}_4$. It yields similar requirements as before: no multiplier in the numerator of any element in rows other than the first row and proportionality between rows excluding the first.

A further reduction of $\boldsymbol{T}_5$ to $\boldsymbol{T}_3$, taking advantage of the relations between the $t_{ij}$, yields

$$\boldsymbol{T}_3 = \begin{bmatrix} t_{11} + \dfrac{N_{11}}{d} & t_{12} + \dfrac{N_{12}}{d} & t_{13} + \dfrac{N_{13}}{d} \\ \dfrac{t_{21}}{d} & \dfrac{t_{22}}{d} & \dfrac{t_{23}}{d} \\ \dfrac{t_{31}}{d} & \dfrac{t_{32}}{d} & \dfrac{t_{33}}{d} \end{bmatrix} \tag{22}$$

where

$$d = 1 - \alpha_3 t_{44} - \alpha_4 t_{55} \tag{23}$$

and

$$N_{1j} = \alpha_3 t_{14} t_{4j} + \alpha_4 t_{15} t_{5j}, \qquad j = 1,2,3. \tag{24}$$

The necessary conditions developed for the matrices $\boldsymbol{T}_k$ may be transformed into sufficient conditions by careful choice of the matrix elements. This is shown in the following application.

## IV. APPLICATION

Consider the realization of $T_1(z)$

$$T_1(z) = \frac{\alpha z^{-2} + \beta z^{-1} + \gamma}{az^{-2} + bz^{-1} + 1}, \qquad \gamma = 0,1. \tag{25}$$

Minimal realization of such a transfer function uses 4 2-input adders and 4 multipliers. To these constraints will be added the no product of the multipliers property.

A procedure for realizing those functions follows directly from the recursive definition for the $\boldsymbol{T}_k$ matrices. Knowing

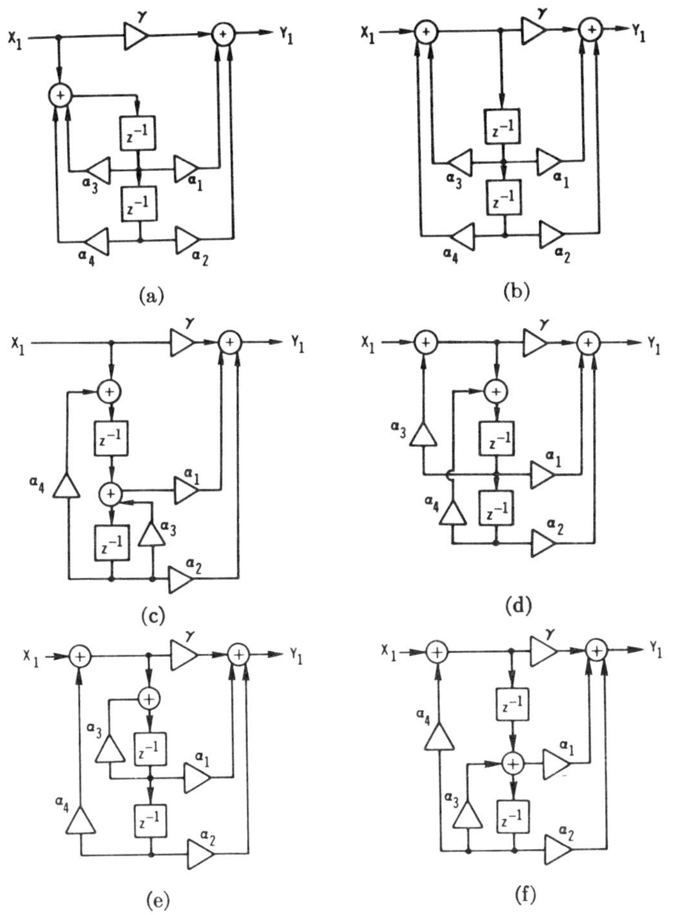


Fig. 3. Minimal second-order realizations with $t_{22} = 0$, $t_{33} = 0$, $t_{44} \neq 0$, and $t_{55} \neq 0$.

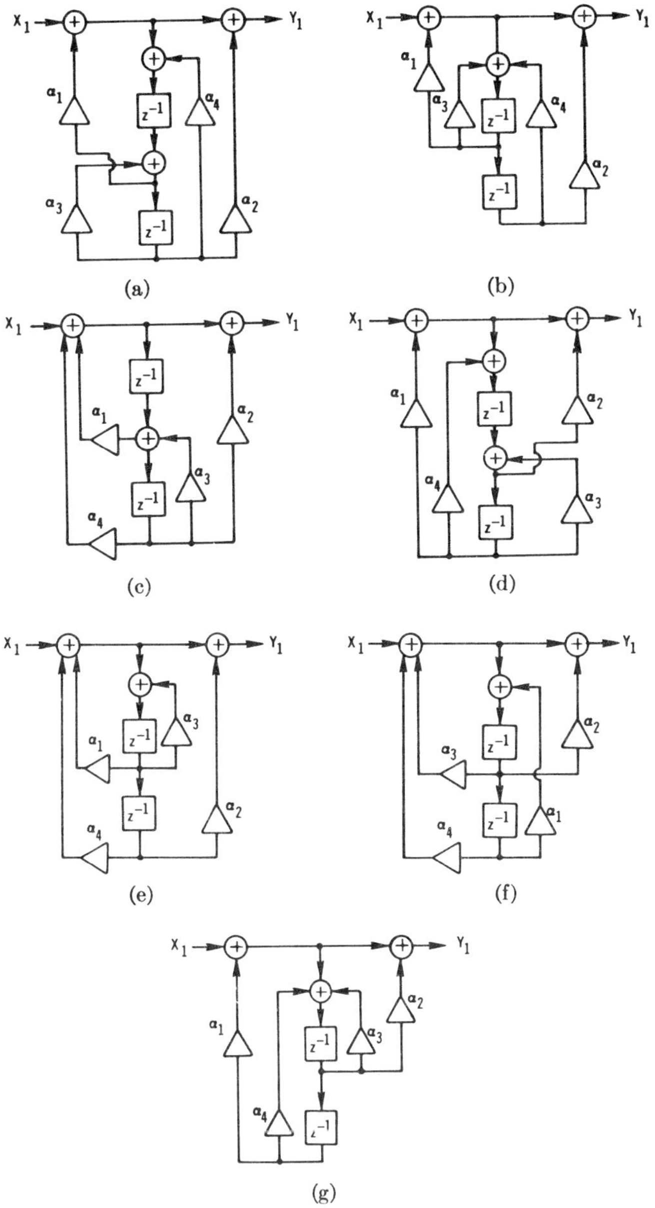


Fig. 4. Minimal second-order realizations with $t_{22} \neq 0$, $t_{33} = 0$, $t_{44} \neq 0$, and $t_{55} \neq 0$.

$T_1(z)$, it is possible to find all possible matrices $\boldsymbol{T}_2$, satisfying the necessary and sufficient conditions. From there the $\boldsymbol{T}_3$ matrices are developed and so on. Another approach to be followed here is to generate directly all possible matrices of some order higher than two. For this case the $\boldsymbol{T}_3$ matrices were chosen. The adder limitation will help in restricting the search for all possible $\boldsymbol{T}_3$ matrices. Furthermore, a matrix of that order will give enough information about its realizations, not requiring further search for higher order matrices.

Comparing (4) and (22) and using (14) yields

$$T_1(z) = \frac{\alpha_1[t_{12}'(t_{21}/d) - t_{11}'(t_{22}/d)] + \alpha_2[t_{13}'(t_{31}/d) - t_{11}'(t_{33}/d)] + t_{11}'}{1 - \alpha_1(t_{22}/d) - \alpha_2(t_{33}/d)} \tag{26}$$

where $t_{1j}'$, $j = 1,2,3$, is the element in the $1,j$ entry of (22). Noticing that at least two of the four $t_{ii}(i = 2,3,4,5)$ must be different from zero in order to get the general $T_1(z)$ of (25), the search may be divided into three major branches:

$$t_{22} = t_{33} = 0, \quad t_{44} \neq 0, \quad t_{55} \neq 0 \tag{27a}$$

$$t_{22} \neq 0, \quad t_{33} = 0, \quad t_{44} \neq 0, \quad t_{55} \neq 0 \tag{27b}$$

$$t_{22} \neq 0, \quad t_{33} \neq 0, \quad t_{44} \neq 0, \quad t_{55} \neq 0. \tag{27c}$$

Proper choice of the other $t_{ij}$'s follows in order to match (26) with (25). The results are given in Figs. 3–5, for the corresponding cases (27a), (27b), and (27c). The transpose realizations of these structures may be obtained by straightforward procedure [1] and are thus omitted. It should be noted that the configurations of Figs. 3(a), and (b), 4(e), and (f), and 5(b) were previously advanced by Jackson [1]. The remaining ten are new realizations. It should be noted that the circuits of Figs. 4 and 5 require that $\gamma$ be equal to one. The explicit expressions for the transfer function of each structure in Figs. 3–5 appear in Table I.

To illustrate the development of the configurations in

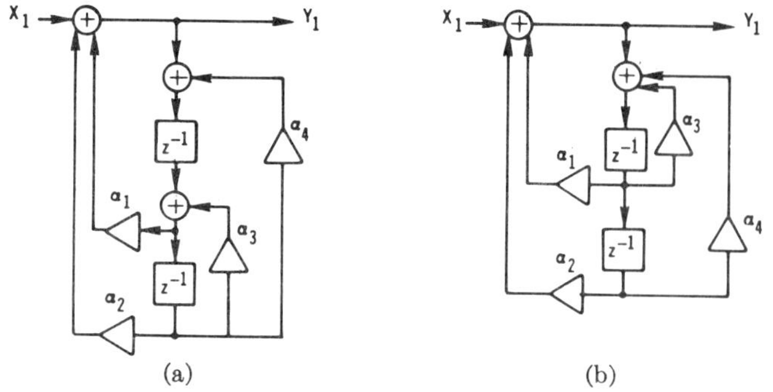


Fig. 5. Minimal second-order realizations with $t_{22} \neq 0$, $t_{33} \neq 0$, $t_{44} \neq 0$, and $t_{55} \neq 0$.

TABLE I

| Figures | Transfer Functions |
|---|---|
| 3(a) and (c) | $\dfrac{\gamma + (\alpha_1 - \gamma\alpha_3)z^{-1} + (\alpha_2 - \gamma\alpha_4)z^{-2}}{1 - \alpha_3 z^{-2} - \alpha_4 z^{-2}}$ |
| 3(b) | $\dfrac{\gamma + \alpha_1 z^{-1} + \alpha_2 z^{-2}}{1 - \alpha_3 z^{-1} - \alpha_4 z^{-2}}$ |
| 3(d) | $\dfrac{\gamma + \alpha_1 z^{-1} + (\alpha_2 - \gamma\alpha_4)z^{-2}}{1 - \alpha_3 z^{-1} - \alpha_4 z^{-2}}$ |
| 3(e) and (f) | $\dfrac{\gamma + (\alpha_1 - \gamma\alpha_3)z^{-1} + \alpha_2 z^{-2}}{1 - \alpha_3 z^{-1} - \alpha_4 z^{-2}}$ |
| 4(a) and (b) | $\dfrac{1 - \alpha_3 z^{-1} + (\alpha_2 - \alpha_4)z^{-2}}{1 - (\alpha_1 + \alpha_3)z^{-1} - \alpha_4 z^{-2}}$ |
| 4(c) and (e) | $\dfrac{1 - \alpha_3 z^{-1} + \alpha_2 z^{-2}}{1 - (\alpha_1 + \alpha_3)z^{-1} - \alpha_4 z^{-2}}$ |
| 4(d) and (g) | $\dfrac{1 + (\alpha_2 - \alpha_3)z^{-1} - \alpha_4 z^{-2}}{1 - \alpha_3 z^{-1} - (\alpha_1 + \alpha_4)z^{-2}}$ |
| 4(f) | $\dfrac{1 + \alpha_2 z^{-1} - \alpha_1 z^{-2}}{1 - \alpha_3 z^{-1} - (\alpha_1 + \alpha_4)z^{-2}}$ |
| 5(a) and (b) | $\dfrac{1 - \alpha_3 z^{-1} - \alpha_4 z^{-2}}{1 - (\alpha_1 + \alpha_3)z^{-1} - (\alpha_2 + \alpha_4)z^{-2}}$. |

Figs. 3–5, a sample realization is developed next to outline the steps involved in details. For the case given by (27a), if

$$\delta_2 = (t_{12}'t_{21} - t_{11}'t_{22}) \quad (28)$$

$$\delta_3 = (t_{13}'t_{31} - t_{11}'t_{33}) \quad (29)$$

and

$$t_{11}' = t_{11} + N_{11}/d, \quad (30)$$

it follows from (26) that

$$d = 1 - \alpha_3 z^{-1} - \alpha_4 z^{-2} = 1 + bz^{-1} + az^{-2}. \quad (31)$$

This last equation gives the values for $\alpha_3$ and $\alpha_4$:

$$\alpha_3 = -b \quad (32\text{i})$$

$$\alpha_4 = -a. \quad (32\text{ii})$$

The numerator of (26) yields

$$\delta_2\alpha_1 + \delta_3\alpha_2 + t_{11}(1 - \alpha_3 z^{-1} - \alpha_4 z^{-2}) + N_{11} = \alpha z^{-2} + \beta z^{-1} + \gamma. \quad (33)$$

A proper choice of parameters is done in order to satisfy the adders limitation, and the results are given by

$$t_{11} = \gamma, \quad \delta_3 = z^{-2}, \quad \delta_2 = z^{-1} \quad (34\text{a})$$

$$t_{11} = 0, \quad N_{11} = \gamma, \quad \delta_3 = z^{-2}, \quad \delta_2 = z^{-1} \quad (34\text{b})$$

$$t_{11} = \gamma, \quad N_{11} = \alpha_3 z^{-1} + \alpha_4 z^{-2}, \quad \delta_3 = z^{-1}, \quad \delta_2 = z^{-1} \quad (34\text{c})$$

$$t_{11} = \gamma, \quad N_{11} = \alpha_3 z^{-1}, \quad \delta_2 = z^{-1}, \quad \alpha_3 = z^{-2} \quad (34\text{d})$$

$$t_{11} = \gamma, \quad N_{11} = \alpha_4 z^{-2} \quad (34\text{e})$$

where

$$\gamma = 0,1.$$

The corresponding realizations are found in Fig. 3.

## V. CONCLUSION

The multiplier extraction method is a procedure for systematic development of all possible multiplier-invariant realizations for a given transfer function. Furthermore, due to the resultant simple algebraic relations, additional conditions, and properties may easily be incorporated in the development of the matrix expansion process. The minimal realizations of second-order transfer functions with no product of multipliers has been used to illustrate the procedure. The method's generality is shown in the additional realizations (Figs. 3(c)–(f), 4(a)–(d), 4(g), and 5(a)) that resulted from its application to the above second-order realization problem, complementing previous results of Jackson [1]. Finally, if the restrictions on the number of adders and delays are relaxed, additional realizations of second-order digital filters due to precision arithmetic can be readily obtained using the approach suggested in this paper.


## ACKNOWLEDGMENT

The authors would like to thank Professor L. B. Jackson for his comments.


## REFERENCES


[1] L. B. Jackson, "An analysis of roundoff noise in digital filters," Sc.D. thesis, Stevens Institute of Technology, Hoboken, N. J., 1967.

[2] S. R. Parker and S. Hess, "Canonic realizations of second-order digital filters due to finite precision arithmetic," *IEEE Trans. Circuit Theory* (Corresp.), vol. CT-19, pp. 410–413, July 1972.

[3] S. K. Mitra and K. Hirano, "Multiplier extraction approach—A new method of digital filter realization," presented at the IEEE Arden House Workshop on Digital Signal Processing, Harriman, N. Y., Jan. 1974.

[4] ——, "Digital all-pass networks," *IEEE Trans. Circuits Syst.*, vol. CAS-21, pp. 688–700, Sept. 1974.

[5] K. Hirano, S. Nishimura, and S. K. Mitra, "Design of digital notch filters," *IEEE Trans. Circuits Syst.*, vol. CAS-21, pp. 540–546, July 1974.

[6] ——, "Approximation technique for the time-sharing realization of digital band-elimination filters," *Mem. Fac. Eng. Kobe Univ.*, no. 20, 1974.

# Digital Lattice and Ladder Filter Synthesis

A. H. GRAY, JR., Member, IEEE, and
JOHN D. MARKEL, Member, IEEE

*Abstract*—There is evidence that in addition to standard digital filter forms such as the direct, parallel, and cascade forms, digital lattice and ladder filters may play an important role in finite word length implementation problems.

In this paper, techniques are developed in detail for efficiently synthesizing digital lattice and ladder filters from any stable direct form. In one form, a lattice filter canonic in terms of multiplies and delays is obtained. An internal scaling procedure is also introduced that will be of importance for optimizing one of the lattice forms for finite word length implementation.

Manuscript received May 15, 1973; revised August 9, 1973. This work was supported under ONR Contract N00014-67-C-0118.

A. H. Gray, Jr. is with the Department of Electrical Engineering and Computer Science, University of California, Santa Barbara, Calif. 93106.

J. D. Markel is with Speech Communications Research Laboratory, Santa Barbara, Calif. 93109.

## Introduction

Techniques for synthesizing digital ladder structures [1], [2] have recently been studied because of their apparent frequency response insensitivity with respect to the more standard direct, parallel, and cascade forms [3]. Recent studies by Fettweis [4], [5] show that digital ladder filters can be patterned after classical analog structures. Crochiere has demonstrated that, at least for one filter, the coefficient sensitivity of the digital ladder filter is at least several bits less than a cascade implementation of the same filter [6]. It is certainly desirable to know how to directly transform a direct form into a corresponding ladder or lattice structure. Notable steps in this direction have been made by Mitra and Sherwood [1], [2]. Unfortunately, their results are restrictive in that only certain classes of stable filters can be synthesized. In addition, their procedure requires the solving of a set of linear simultaneous equations

Reprinted from *IEEE Trans. Audio Electroacoust.*, vol. AU-21, pp. 491–500, Dec. 1973.

in order to obtain the tap gains (gain terms for implementing the numerator without additional delays).

In this paper, a general solution to the synthesis of digital lattice and ladder structures from direct forms is presented. In one form, the resultant filter is canonical in the number of multipliers and delays. The procedure follows from an application of an inner product formulation first developed by Itakura and Saito [7] and later expanded upon by Markel and Gray [8].

The solution is easily developed and programmed as a recursive procedure for the filter design. Furthermore, the procedure is extremely efficient. No simultaneous equation solution for tap gains is necessary. What "falls out" of the procedure is a "two-multiplier" lattice form. A trivial transformation results in a "three-multiplier" ladder form. From the two-multiplier form a new generalized "one-multiplier" form, canonic in both multipliers and delays, is developed. With the generalized one-multiplier model, a new technique is presented for optimally scaling the filter structure with respect to finite word length implementation.

### Lattice Filter Formulation

We wish to design a digital filter that will implement the direct form transfer function

$$G = G(z) = P_M(z)/A_M(z) \tag{1}$$

where $P_M(z)$ and $A_M(z)$ are $M$th order polynomials in $z^{-1}$, of the form

$$P_m = P_m(z) = \sum_{n=0}^{m} p_{m,n} z^{-n} \tag{2}$$

and

$$A_m = A_m(z) = \sum_{n=0}^{m} a_{m,n} z^{-n} \tag{3}$$

where $m = M$. Without loss of generality, it is assumed that the leading coefficient of $A_M(z)$ is one, $a_{M,0} = 1$.[1] If the filter is implemented in the direct form, then the parameters representing the filter are $2M + 1$ in general, $p_{M,0}, p_{M,1}, \cdots, p_{M,M}, a_{M,1}, a_{M,2}, \cdots, a_{M,M}$.

The $2M + 1$ parameters of the lattice filter that will be designed are the $M$ $k$-parameters $k_0, k, \cdots, k_{M-1}$ and the $M + 1$ tap parameters $\nu_0, \nu_1, \cdots, \nu_M$. These parameters are recursively obtained by starting from $A_M(z)$ and $P_M(z)$ as

$$zB_m(z) = A_m(1/z)z^{-m} \tag{4}$$

$$k_{m-1} = a_{m,m} \tag{5}$$

$$A_{m-1}(z) = [A_m(z) - k_{m-1} zB_m(z)]/(1 - k_{m-1}^2) \tag{6}$$

$$\nu_m = p_{m,m} \tag{7}$$

$$P_{m-1}(z) = P_m(z) - zB_m(z)\nu_m \tag{8}$$

for $m = M, M - 1, \cdots, 1$ with $\nu_0 = p_{0,0}$.

The only division that arises is in (6). This will never be a division by zero, for it is known that the $k$-parameters will always have a magnitude less than one for a stable filter [7], [8]. Should any $k_m$ equal plus or minus one, or have a magnitude larger than one, then $A_M(z)$ will not have all its roots within the unit circle and the filter will be unstable. In terms of the new variables, $P_M(z)$ is equivalent to

$$P_M(z) = \sum_{m=0}^{M} \nu_m zB_m(z). \tag{9}$$

Therefore, an equivalent representation of $G(z)$ is given as

$$G(z) = \sum_{m=0}^{M} \nu_m \frac{zB_m(z)}{A_M(z)}. \tag{10}$$

It remains to be shown that this formulation results in a digital lattice structure. First, the general proof of these equations will be presented.

### Proof of Formulation

The results stated here are essentially applications of the inner product formulation published elsewhere [7], [8].

First, we define

$$R(z) = [A_M(z)A_M(z^{-1})]^{-1}, \tag{11}$$

which will be positive on the unit circle. An inner, or scalar, product of functions of $z$ can be then defined by

$$\langle F(z), G(z)\rangle = \int_{-\pi}^{\pi} R(e^{j\theta})F^*(e^{j\theta})G(e^{j\theta})\frac{d\theta}{2\pi} \tag{12}$$

where the asterisk denotes a complex conjugate. Using this notation, it can be shown [8] that by defining $zB_m(z) = A_m(1/z)z^{-m}$, both the polynomials $A_m(z)$ and $B_m(z)$ are orthogonal to the powers of $z^{-1}$, from the 1st through the $m$th; that the polynomials $zB_m(z)$ form an orthogonal set; and that the polynomials are related by the recursion matrix

$$\begin{bmatrix} A_{m+1}(z) \\ B_{m+1}(z) \end{bmatrix} = \begin{bmatrix} 1 & k_m \\ k_m z^{-1} & z^{-1} \end{bmatrix} \begin{bmatrix} A_m \\ B_m \end{bmatrix} \tag{13}$$

for $m = 0, 1, \cdots, M - 1$ with $A_0(z) = 1$ and $B_0(z) = z^{-1}$.

In addition, the coefficients $\alpha_m$ can be found from

$$\alpha_m = \langle A_m(z), A_m(z)\rangle = \langle B_m(z), B_m(z)\rangle = \langle 1, A_m(z)\rangle, \tag{14}$$

or through a recursion relation

[1] Throughout the paper, for notational simplicity, capital letters will be used to refer to the corresponding $z$ transforms.

$$\alpha_{m+1} = \alpha_m(1 - k_m^2) \qquad \text{for } m = 0, 1, \cdots, M-1 \quad (15)$$

with $\alpha_M = 1$. The $k$-parameters can also be expressed in terms of scalar products as $k_m = -\beta_m/\alpha_m$ where

$$\beta_m = \langle A_m(z), B_m(z)\rangle = \langle 1, B_m(z)\rangle. \quad (16)$$

Equation (6) is obtained by simply computing the matrix inverse from (13) and replacing $m$ by $m-1$. Equations (7) and (8) are obtained by noting that if $\nu_m$ is defined as $P_{m,m}$, then $P_m - zB_m\nu_m$ results in a polynomial reduced in order by 1 since $zB_m$ has a final coefficient of unity for all $m$ (from (4) and the fact that $a_{m,0} = 1$).

From the recurrence relations, it can be seen that each $P_m(z)$ will be of the form of (2) and that each $A_m(z)$ will be of the form of (3) with $a_{m,0} = 1$. Equation (9) is proven from (8) by simply summing $P_m - P_{m-1}$ from $m = 0$ to $M$ with $P_{-1} = 0$.

### A "Two-Multiplier" Lattice Filter

In this section, the general form of the filter as a block diagram is first developed. If $X = X(z)$ is the input to $G = G(z)$, and $Y = Y(z)$ is the output, from (10),

$$Y = \sum_{m=0}^{M} \nu_m \frac{zB_mX}{A_M}. \quad (17)$$

The recursions specified by (13) and (14) can be rewritten in the form

$$\frac{A_mX}{A_M} = \frac{A_{m+1}X}{A_M} - k_m \frac{B_mX}{A_M} \quad (18)$$

$$\frac{zB_{m+1}X}{A_M} = k_m \frac{A_mX}{A_M} + \frac{B_mX}{A_M}. \quad (19)$$

Noting that $A_MX/A_M = X$ and $A_0X/A_M = zB_0X/A_M$, a block diagram for the filter is shown in Fig. 1. The details of a prototype section $G_m$ are obtained from (18) and (19) as shown in Fig. 2. The overall implementation of $G(z)$ with these particular sections is referred to as a "two-multiplier model" since each section contains two multiplies. By making proper analogies to classical network theory [9], this filter would be defined as a lattice type of filter. It is trivial, however, to obtain a ladder representation from the lattice. Substituting (18) into (19) results in a "three-multiplier" form as shown in Fig. 3. By inspection of the figure it is possible to easily reduce it to a two-multiplier ladder form.

As can be noted from Fig. 1, each of the terms multiplying the tap parameters in the summation of (10) appear explicitly; direct multiplication by these tap parameters and a single summation will fully implement the filter in the form of (10).

The forms presented are canonical with respect to delay only. By further manipulation it is possible to obtain a form that is also canonical with respect to multiplications. It was evidently first shown by Itakura and Saito that (18) and (19) can be transformed into a one-multiplier model. In the next section the one-multiplier model is derived with a generalization that includes a sign parameter definition. The potential value of this generalization will be discussed in a later section.

### One-Multiplier Implementation

Here we define a new set of sign parameters, $\epsilon_0, \epsilon_1, \cdots, \epsilon_{M-1}$. Each of these will be equal to plus one or minus one. How the sign parameters are chosen will be the topic of a later section. These are then used to obtain a set of modified tap parameters, $\hat{\nu}_0, \hat{\nu}_1, \cdots, \hat{\nu}_M$, and a set of modified polynomials by the relations

$$\hat{\nu}_m = \nu_m/\pi_m \quad (20)$$

$$\hat{A}_m(z) = \pi_m A_m(z), \quad \hat{B}_m(z) = \pi_m B_m(z) \quad (21)$$

where

$$\pi_m = \begin{cases} 1 & \text{for } m = M \\ \prod_{n=m}^{M-1} (1 + \epsilon_n k_n) & \text{for } m = 0, 1, \cdots, M-1. \end{cases} \quad (22)$$

The recursion relations of (13) then become

$$(1 + \epsilon_m k_m)\hat{A}_{m+1}(z) = \hat{A}_m(z) + k_m\hat{B}_m(z) \quad (23)$$

$$z(1 + \epsilon_m k_m)\hat{B}_{m+1}(z) = k_m\hat{A}_m(z) + \hat{B}_m(z). \quad (24)$$

From (23), $\hat{A}_m(z)$ is obtained by inspection as

$$\hat{A}_m(z) = \hat{A}_{m+1}(z) - k_m[\hat{B}_m(z) - \epsilon_m\hat{A}_{m+1}(z)]. \quad (25)$$

Substituting (25) into (24) and noting that $(1 - k_m^2) = (1 + \epsilon_m k_m)(1 - \epsilon_m k_m)$ and $\epsilon_m^2 = 1$ results in

$$z\hat{B}_{m+1}(z) = \hat{B}_m(z) - \epsilon_m k_m[\hat{B}_m(z) - \epsilon_m\hat{A}_{m+1}(z)]. \quad (26)$$

Equations (25) and (26) can be implemented as indicated in Fig. 4, where Fig. 4(a) shows the blocks in detail for $\epsilon_m = +1$ and Fig. 4(b) shows the blocks in detail for $\epsilon_m = -1$. Note that the general form of Fig. 1 is still valid with the $A_m$, $B_m$, and $\nu_m$ terms hatted since by substituting (20) through (22) into (10), $G(z)$ is equivalent to

$$G(z) = \sum_{m=0}^{M} \hat{\nu}_m \frac{z\hat{B}_m(z)}{\hat{A}_M(z)}. \quad (27)$$

From Fig. 1 (with the replacement of hatted terms), it is seen that each term needed to multiply the modified tap parameters, as indicated in (27), appears explicitly in the model; all that needs to be added are the multiplications by the tap parameters and the summation indicated.

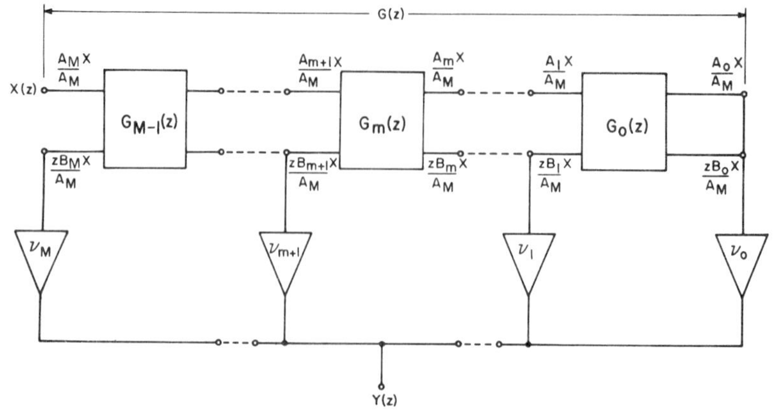


Fig. 1. Ladder and lattice form implementation.

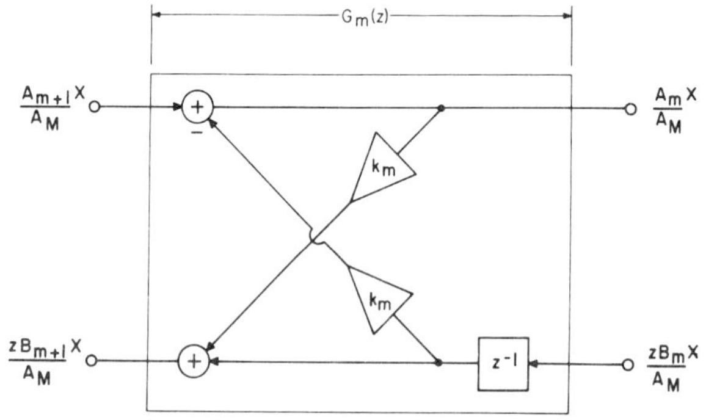


Fig. 2. Prototype filter for two-multiplier lattice model.

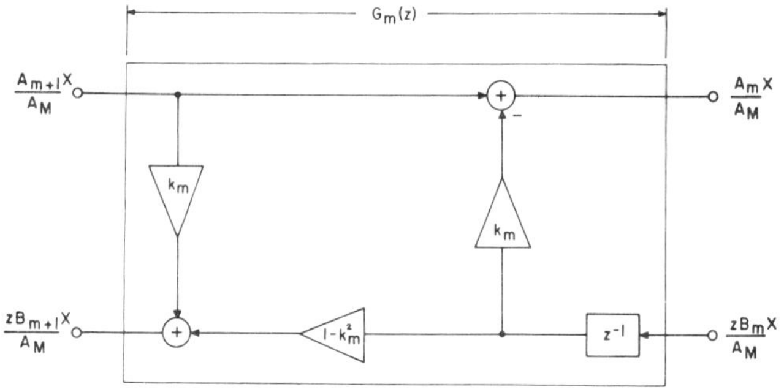


Fig. 3. Prototype filter for three multiplier ladder model.

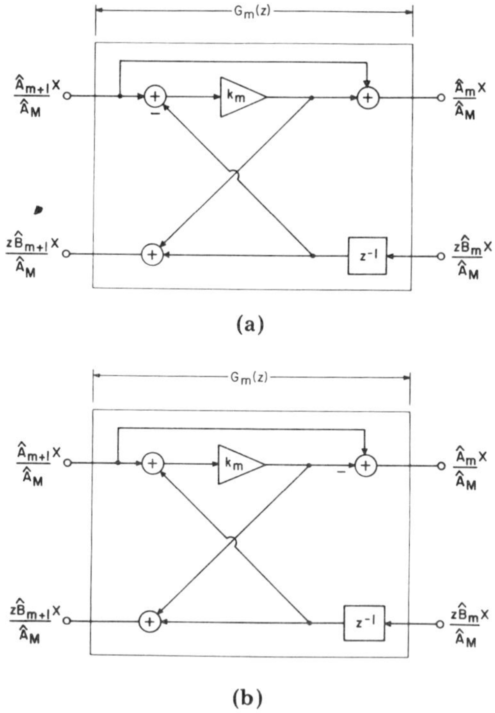


Fig. 4. Prototype filters for generalized one-multiplier lattice model. (a) $\epsilon_m = +1$. (b) $\epsilon_m = -1$.

## Sensitivity

It has been noted that when poles of a filter lie at low frequencies and are near the unit circle, direct implementation of a digital filter by using a recursive approach is much worse than factoring the filter into a complex pole pair and using a cascade approach [3]. The meaning of the term "much worse" is related to how sensitive the angular location of the complex pole is to small percentage changes in the filter parameters.

Due to the complexity of the problem, no attempt is made here to derive relationships for sensitivity of the pole or zero locations with respect to small percentage variations in the filter parameters. Rather, we shall determine which calculations are the most sensitive in changing the gross behavior of the filter and, then, show how to choose the sign parameters so as to make those the calculations that are performed with the numbers having the largest magnitudes.

In linear prediction synthesis of speech, it has been observed that it is those $k$-parameters having the largest magnitude that are the most important. In dealing with voiced speech signals it usually is true that $k_0$ and $k_1$ have the largest effect on the end results.

Filter sensitivity to $k$-parameters near unit magnitude can be noted by considering the case of two separate filters being implemented, whose only difference lies in the $l$th $k$-parameter. One of the filters uses $k_l$ and the other filter uses $k_l' = k_l + \Delta k_l$. The polynomials for the filter might be denoted by $A_n(z)$ and $A'_n(z)$ so that the polynomials are equal for $n = 0, 1, \cdots, l$. For $n$ larger than $l$ they will differ, and the effect of this difference will first show up in the $n = l + 1$ polynomial. It is shown in Appendix I that the difference of the log magnitudes of $A'_{l+1}(e^{j\theta})$ and $A_{l+1}(e^{j\theta})$ will oscillate between the values $\log[1 + \Delta k_l/(1 + k_l)]$ and $\log[1 - \Delta k_l/(1 - k_l)]$. Thus, the closer that the magnitude of $k_l$ is to one, the more significant the effects of nonzero $\Delta k_l$.

As a result of this, it can be noted that the $k$-parameters are inherently scaled in a desirable fashion, that is, the ones that affect the filter most are the ones with the largest magnitudes. In the following sections the problem of scaling is studied in more detail.

### Relative Sizes of Numerical Values

For purposes of normalization we shall define a set of $z$ transforms $U_m(z)$, $V_m(z)$, and $W_m(z)$ by the relations

$$U_m(z) = A_m(z)X(z)/[A_M(z)\sqrt{\alpha_m}], \quad (28)$$

$$V_m(z) = B_m(z)X(z)/[A_M(z)\sqrt{\alpha_m}], \quad (29)$$

and

$$W_m(z) = [V_m(z) - \epsilon_m U_m(z)]/\sqrt{2(1 + \epsilon_m k_m)}, \quad (30)$$

and $X(z)$ represents the $z$ transform of the input to the filter. The particular definitions correspond to normalized node values and multiplier input values for the various filter forms. The motivation for these particular normalization constant choices will be presented shortly. While it is impossible to judge the sizes of the time sequences associated with $U_m(z)$, $V_m(z)$, and $W_m(z)$ for all possible inputs, some statements can be made about their sizes based upon their spectra, their log spectra, and their rms value when the input is zero mean uncorrelated noise.

First, in terms of their spectra, it is shown in Appendix II that each of the three has the unit energy or total squared integral when $X(z) = 1$. Second, when the input time sequence is a random, stationary, uncorrelated, and zero mean process having a variance or mean square equal to $\sigma_x^2$, then it is shown in Appendix II that the time sequences associated with $U_m(z)$, $V_m(z)$, and $W_m(z)$ will all be random, stationary, zero mean process each having a variance or mean square equal to $\sigma_x^2$. Thus, in a qualitative sense, one can say that these time sequences, on the average, will have identical rms values. Third, the rather surprising property also shown in Appendix III is that the $U_m(z)$ and $V_m(z)$ form two separate sequences of mini-max approximations in terms of their log spectra.

In the two-multiplier model of Fig. 2, the node values used in the $m$th block are

$$\frac{A_m(z)}{A_M(z)} X(z) = U_m(z)\sqrt{\alpha_m} \quad (31)$$

and

$$\frac{B_m(z)}{A_M(z)} X(z) = V_m(z)\sqrt{\alpha_m}. \quad (32)$$

These also represent the inputs to the multipliers as seen from Fig. 2. From (15), one should note that the $\alpha_m$ form a sequence that decreases with increasing $m$. Thus the largest multiplier inputs will be in the blocks with the smallest $m$, those furthest from the input. This is useful in speech synthesis where $k_0$ and $k_1$ are the $k$-parameters closest to unit magnitude. However if the $\alpha_m$ cover a wide range $[\alpha_0 >> \alpha_M = 1]$, then the multiplier inputs in the blocks with large $m$ will be a great deal smaller in size than those with small $m$.

In the one-multiplier models of Fig. 4, the node values are specified by

$$\frac{\hat{A}_m(z)}{\hat{A}_M(z)} X(z) = \pi_m\sqrt{\alpha_m}\, U_m(z) \quad (33)$$

and

$$\frac{\hat{B}_m(z)}{\hat{A}_m(z)} X(z) = \pi_m\sqrt{\alpha_m}\, V_m(z). \quad (34)$$

From (26) and Fig. 4, the multiplier input in the $m$th block is easily seen to be

$$\frac{\hat{B}_m(z) - \epsilon_m \hat{A}_{m+1}(z)}{\hat{A}_M(z)} X(z) = \frac{\pi_{m+1}[B_m(z) - \epsilon_m A_m(z)]}{A_M(z)} X(z) = \pi_m\sqrt{\alpha_m}\sqrt{2/(1 + \epsilon_m k_m)}\, W_m(z). \quad (35)$$

The product $\pi_m\sqrt{\alpha_m}$ appears in each of these expressions and is obtained by recursion from (15) and (22) as

$$\pi_m\sqrt{\alpha_m} = \prod_{n=m}^{M-1} \sqrt{(1 + \epsilon_n k_n)/(1 - \epsilon_n k_n)}. \quad (36)$$

As will be seen in the next section, appropriate choices of the sign parameters can be utilized to make this product largest when $k$-parameters are near one in magnitude, without requiring a monotonically decreasing sequence with increasing $m$, as was the case with the $\sqrt{\alpha_m}$ terms in the two-multiplier model.

As in the case of showing that $U_m(z)$, $V_m(z)$, and $W_m(z)$ represent $z$ transforms of unit energy signals (when $X(z) = 1$) and that their time sequences will have variances equal to $\sigma_x^2$ for random inputs with variance equal to $\sigma_x^2$, it is also shown in Appendix II that the impulse response of the filter contains a total amount of energy given by

$$\int_{-\pi}^{\pi} |G(e^{j\theta})|^2 \frac{d\theta}{2\pi} = \sum_{n=0}^{M} \nu_n^2 \alpha_n, \tag{37}$$

and that when the input is the random, zero mean, uncorrelated stationary sequence with variance $\sigma_x^2$, the filter output $y_n$ will be random and zero mean, with a variance given by

$$\sigma_y^2 = \sigma_x^2 \sum_{n=0}^{M} \nu_n^2 \alpha_n. \tag{38}$$

### Sign Parameter Choices

Numerous criteria could be utilized to pick optimal values of the sign parameters, based upon the user's definition of optimal. In order to obtain an effective filter over differing types of input, we shall choose our criteria in terms of rms values for the case where the input signal is random, uncorrelated, stationary, and zero mean. If the variance of the input signal is $\sigma_x^2$, so that its rms value is $\sigma_x$; then the results of Appendix II show that the rms values of the time sequences associated with $\hat{A}_m(z)X(z)/\hat{A}_M(z)$ and $\hat{B}_m(z)X(z)/\hat{A}_M(z)$ will both be given by $\pi_m\sqrt{\alpha_m}$, the rms value of the multiplier input in the $m$th section will be given by $\pi_m\sqrt{\alpha_m}\,\sigma_x\sqrt{2/(1+\epsilon_m k_m)}$, and that the rms value of the output of the overall filter from (38) is

$$\sigma_y = \sigma_x \left[ \sum_{n=0}^{M} \nu_n^2 \alpha_n \right]^{1/2}. \tag{39}$$

One optimization criterion based upon rms values is to require that $\pi_m\sqrt{\alpha_m}$ be largest for the $k$-parameter nearest unity in magnitude. Then the sign parameters are recursively found by requiring that $\pi_m\sqrt{\alpha_m}$ be as large as possible without exceeding the maximum value. If that maximum occurs for $m = l$, then the recursion relation proceeds for $m = l-1, l-2, \cdots, 0$ and again for $m = l, l+2, \cdots, M-1$. This relation is simple to effect because of the fact that

$$\frac{\pi_m\sqrt{\alpha_m}}{\pi_{m+1}\sqrt{\alpha_{m+1}}} = \sqrt{(1+\epsilon_m k_m)/(1-\epsilon_m k_m)}. \tag{40}$$

By simply changing the sign parameter, this ratio can always be made smaller than or larger than one.

Once the sign parameters are chosen, from the aforementioned or some other rule, the rms values of the multiplier inputs are simply evaluated, in addition to the rms value of the output from (39). Scaling of the input signal can be carried out by observing the maximum rms value appearing in the filter and applying some sort of rule to relate rms value to probability of this peak value exceeding one. For example, if the input is Gaussian and scaled so that the peak rms value appearing in the filter is $\frac{1}{3}$, then normal distribution tables show that the probability of that signal's magnitude exceeding unity will be given by 0.0026.

To implement this approach, the following algorithm is applied. Assume that $k_l$ has the largest magnitude among the $k_m$, $m = 0, 1, \cdots, M-1$. Define the quantities

$$Q_m = \frac{\pi_m\sqrt{\alpha_m}}{\pi_l\sqrt{\alpha_l}} \tag{41}$$

and

$$q_m = \sqrt{(1+|k_m|)/(1-|k_m|)}. \tag{42}$$

Then $Q_l = 1$. Each $Q_m$ should be as large as possible without exceeding $Q_l$.

By combining (39) through (42),

$$Q_m/Q_{m+1} = \begin{cases} q_m & \text{if } \epsilon_m = \operatorname{sgn}(k_m) \\ 1/q_m & \text{if } \epsilon_m = -\operatorname{sgn}(k_m). \end{cases} \tag{43}$$

For $m = l-1, l-2, \cdots, 0$, choose $\epsilon_m = \operatorname{sgn}(k_m)$ if $Q_{m+1} < 1/q_m$. Otherwise choose $\epsilon_m = -\operatorname{sgn}(k_m)$. For $m = l, l+1, \cdots, M-1$, choose $\epsilon_m = -\operatorname{sgn}(k_m)$ if $Q_m < 1/q_m$, otherwise choose $\epsilon_m = \operatorname{sgn}(k_m)$. The effect of these choices is to maximize $Q_m$ at each node without exceeding unity.

### An Example

As an illustrative example, the third-order Chebyshev low-pass filter used by Mitra and Sherwood [2] is synthesized in a lattice form. In this case, $M = 3$ and

$$P_3(z) = 0.0154 + 0.0462\,z^{-1} + 0.0462\,z^{-2} + 0.0154\,z^{-3}, \tag{44}$$

and

$$A_3(z) = 1 - 1.990\,z^{-1} + 1.572\,z^{-2} - 0.4583\,z^{-3}. \tag{45}$$

In order to illustrate the simplicity of the approach, and also to provide enough details so that a user could check a computer program, all calculations were carried out using only paper, pencil and a calculator (HP-35).

First, from (5) and (7),

$$k_2 = -0.4583 \text{ and } \nu_3 = 0.0154.$$

From (4),

$$zB_3(z) = -0.4583 + 1.572\,z^{-1} - 1.990\,z^{-2} + z^{-3}.$$

From (6) and (8) with $m = 3$,

$$P_2(z) = 0.02245782 + 0.0219912\,z^{-1} + 0.076846\,z^{-2}$$

and

$$A_2(z) = 1 - 1.607107469\,z^{-1} + 0.835462647\,z^{-2}.$$

Repeating the procedure from (4) through (8) with $m = 2$,

$$k_1 = 0.835462647$$

and

$$\nu_2 = 0.076846$$

with the reduced order polynomials

$$P_1(z) = -0.0417441466 + 0.1454909806\, z^{-1}$$

and

$$A_1(z) = 1 - 0.8755871285\, z^{-1}.$$

Using $m = 1$ in the recursion relations gives

$$k_0 = -0.8755871285$$

and

$$\nu_1 = 0.1454909806$$

with the final polynomials

$$P_0(z) = 0.0856458833$$

$$A_0(z) = 1$$

so that $\nu_0 = 0.0856458833$. This completes the evaluation of the $k$-parameters and the tap parameters. Since all the $k$-parameters have magnitudes less than one, the filter is stable. These results define a two-multiplier lattice form. Next, the one-multiplier model with optimal sign choice is presented.

To obtain the sign parameters, the algorithm of the preceding section is applied. Since $k_0$ has the largest magnitude, (41) and (42) result in $l = 0$,

$$Q_0 = 1$$

and

$$q_0 = 3.882719036.$$

As $Q_0$ is more than $1/q_0$, choose $\epsilon_0 = \text{sgn}(k_0) = -1$. Therefore

$$Q_1 = Q_0/q_0 = 0.2575514712.$$

From (42),

$$q_1 = 3.339954283.$$

As $Q_1$ is less than $1/q_1$, choose $\epsilon_1 = -\text{sgn}(k_1) = -1$. Therefore

$$Q_2 = Q_1 q_1 = 0.8602101393.$$

From (42),

$$q_2 = 1.640756072.$$

As $Q_2$ is more than $1/q_2$, choose $\epsilon_2 = \text{sgn}(k_2) = -1$.

Having found the sign parameters, the multipliers $\pi_m$ can be obtained from (22):

$$\pi_3 = 1$$
$$\pi_2 = 1.4583$$
$$\pi_1 = 0.2399448219$$
$$\pi_0 = 0.4500374194.$$

Using these with (20), the modified tap parameters are obtained as

$$\hat{\nu}_3 = 0.0154$$
$$\hat{\nu}_2 = 0.0526956044$$
$$\hat{\nu}_1 = 0.6063518248$$
$$\hat{\nu}_0 = 0.1903083602.$$

This completes the design procedure for the one-multiplier model. The implementation is shown in Fig. 5.

In order to obtain an idea about relative sizes of the numbers being utilized, one can utilize results of the previous section and Appendix II. In particular, at stage $m$ the nodes $X\hat{A}_m/\hat{A}_M$, $X\hat{B}_m/\hat{A}_M$, and $Xz\hat{B}_m/\hat{A}_M$ will all have rms values given by $\pi_m\sqrt{\alpha_m}\,\sigma_x$ for a random input with variance equal to $\sigma_x^2$.

The values of $\alpha_m$ are most easily evaluated from (15). Starting with $\alpha_M = 1$,

$$\alpha_3 = 1$$
$$\alpha_2 = 1.265885102$$
$$\alpha_1 = 4.191642467$$
$$\alpha_0 = 17.96311599.$$

This then gives the results for the node values

$$\pi_3\sqrt{\alpha_3}\,\sigma_x = \sigma_x$$
$$\pi_2\sqrt{\alpha_2}\,\sigma_x = 1.640563849\,\sigma_x$$
$$\pi_1\sqrt{\alpha_1}\,\sigma_x = 0.4912510573\,\sigma_x$$
$$\pi_0\sqrt{\alpha_0}\,\sigma_x = 1.907389831\,\sigma_x.$$

The multiplier inputs were shown to have rms values given by $\pi_m\sqrt{\alpha_m}\,\sigma_x\sqrt{2/(1+\epsilon_m k_m)}$. Therefore,

$$\pi_0\sqrt{\alpha_0}\,\sigma_x\sqrt{2/(1+\epsilon_0 k_0)} = 1.969635389\,\sigma_x$$
$$\pi_1\sqrt{\alpha_1}\,\sigma_x\sqrt{2/(1+\epsilon_1 k_1)} = 1.712719502\,\sigma_x$$
$$\pi_2\sqrt{\alpha_2}\,\sigma_x\sqrt{2/(1+\epsilon_2 k_2)} = 1.921253614\,\sigma_x.$$

The rms value of the output in this case is found simply from (39) to be given by

$$\sigma_y = 0.4777060586\,\sigma_x.$$

## Conclusions

Several different digital lattice and ladder forms have been developed starting from the standard direct form. The procedure was shown to have several desirable features as follows.

1) All stable recursive filters can be transformed into the forms presented. Other procedures developed thus far have undesirable restrictions such as all poles being required to be in the left half of the $z$-plane.

2) A built-in stability test exists within the synthesis process. If any $k$-parameter magnitude exceeds or equals unity, the filter is unstable. Otherwise, it is stable.

3) The synthesis procedure is very efficient. Only on the order of $M^2$ operations are necessary for the synthesis. No polynomial root solvers or explicit simultaneous equation solvers are necessary.

A new procedure was also introduced for scaling nodes of a particular one-multiplier lattice form.

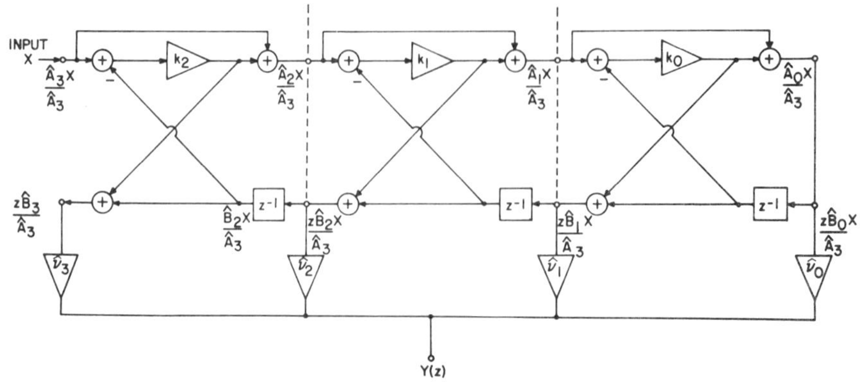

Fig. 5. One-multiplier implementation of the example.

Presently, the application of this scaling procedure is being studied for finite word length digital filter implementation.

### Appendix I

Let $A_m(z)$ satisfy the recursion relation (13) where $B_m(z)$ is from (4) given by

$$B_m(z) = z^{-(m+1)} A_m(1/z), \qquad \text{(I-1)}$$

$A_0(z) = 1$, and $m = 0, 1, \cdots, M - 1$. Let $A'_m(z)$ satisfy a similar set of recursion relations whose only difference lies in the fact that $k_l$ is replaced by $k'_l = k_l + \Delta k_l$. Thus $A'_m(z) = A_m(z)$ for $m = 0, 1, \cdots, l$. Combining (13) and (I-1) with $z = e^{j\theta}$, one finds that on the unit circle

$$\frac{A'_{l+1}(e^{j\theta})}{A_{l+1}(e^{j\theta})} = \frac{A_l(e^{j\theta}) + k'_l e^{-(l+1)j\theta} A_l(e^{-j\theta})}{A_l(e^{j\theta}) + k_l e^{-(l+1)j\theta} A_l(e^{-j\theta})}.$$

This can be rewritten as

$$\frac{A'_{l+1}(e^{j\theta})}{A_{l+1}(e^{j\theta})} = \frac{1 + k'_l e^{-j\Psi}}{1 + k_l e^{-j\Psi}} \qquad \text{(I-2)}$$

where $\Psi$ is given by

$$\Psi = \Psi(\theta) = (l + 1)\theta + 2 \arg [A_l(e^{j\theta})]. \qquad \text{(I-3)}$$

By using the form of $A_l(z)$ from (3) and the fact that it has all of its zeros within the unit circle, one can show that

$$\Psi(0) = 0 \qquad \text{(I-4)}$$

$$\Psi(\pi) = (l + 1)\pi \qquad \text{(I-5)}$$

$$\frac{d\Psi}{d\theta} > 1. \qquad \text{(I-6)}$$

The ratio of (I-2) has a magnitude that oscillates between the values $(1 + k'_l)/(1 + k_l)$ when $\Psi$ is an even multiple of $\pi$, and $(1 - k'_l)/(1 - k_l)$ when $\Psi$ is an odd multiple of $\pi$. As $\Psi$ is a monotonically increasing function of $\theta$, these extreme values will all be met.

In terms of the log spectra, this yields the result that

$$\log |A'_{l+1}(e^{j\theta})| - \log |A_{l+1}(e^{j\theta})|$$

will oscillate between the values

$$\log [(1 + k'_l)/(1 + k_l)] = \log [1 + \Delta k_l/(1 + k_l)]$$

and

$$\log [(1 - k'_l)/(1 - k_l)] = \log [1 - \Delta k_l/(1 - k_l)].$$

### Appendix II

Let $U_m(z)$, $V_m(z)$, and $W_m(z)$ be as given in (28)–(30). When $X(z) = 1$, one can apply (11) and (12) to obtain their energies in terms of inner products. In this way we find that

$$\int_{-\pi}^{\pi} |U_m(e^{j\theta})|^2 \frac{d\theta}{2\pi} = \langle A_m(z)/\sqrt{\alpha_m}, A_m(z)/\sqrt{\alpha_m} \rangle = \frac{\langle A_m(z), A_m(z) \rangle}{\alpha_m}.$$

From (14) one should note that this is simply one. In exactly the same manner, one can show that

$$\int_{-\pi}^{\pi} |V_m(e^{j\theta})|^2 \frac{d\theta}{2\pi} = 1.$$

One more step is needed in the handling of $W_m(z)$, for by the same applications of scalar products one finds that

$$\int_{-\pi}^{\pi} |W_m(e^{j\theta})|^2 \frac{d\theta}{2\pi} = \frac{\langle B_m(z) - \epsilon_m A_m(z), B_m(z) - \epsilon_m A_m(z) \rangle}{2\alpha_m (1 + \epsilon_m k_m)}.$$

If the numerator terms are multiplied out and use is made of (14) and (16), one finds that this quantity also equals unity. As a result,

$$\int_{-\pi}^{\pi} |U_m(e^{j\theta})|^2 \frac{d\theta}{2\pi} = \int_{-\pi}^{\pi} |V_m(e^{j\theta})|^2 \frac{d\theta}{2\pi} = \int_{-\pi}^{\pi} |W_m(e^{j\theta})|^2 \frac{d\theta}{2\pi} = 1. \qquad \text{(II-1)}.$$

In the same manner, one can obtain the energy of the overall filter output when $X(z) = 1$, by expressing the energy of $G(z)$ as a scalar product. From (1), (11), and (12)

$$I = \int_{-\pi}^{\pi} |G(e^{j\theta})|^2 \frac{d\theta}{2\pi} = \langle P_M(z), P_M(z) \rangle.$$

If $P_M(z)$ is expressed as in (9), the orthogonality of the polynomials $zB_m(z)$ can combine with (14) to give

$$I = \sum_{m=0}^{M} \nu_m^2 \alpha_m. \tag{II-2}$$

The symbolic notation $H(z)X(z)$ represents the $z$ transform of the output of a filter whose input time sequence has a $z$ transform and whose transfer function is $H(z)$. When we speak of the time sequence associated with $H(z)X(z)$ we imply the convolution

$$s_k = \sum_{n=-\infty}^{\infty} h_n x_{k-n} \tag{II-3}$$

where $h_k$ represents the unit sample response to the filter and its $z$ transform is $H(z)$. Equation (II-3) is considered to apply, even in cases where the input sequence, $x_k$, does not have a convergent $z$ transform. In such cases, referring to a time sequence associated with $H(z)X(z)$ will be understood to imply (II-3), even when $X(z)$ is not convergent.

If the input sequence is random and uncorrelated, then the output sequence will also be random. Using the symbol $E$ to denote the expected value (expectation or mean) of a random variable, one finds that the mean square of the sequence $s_k$ will be given by

$$\sigma_s^2 = E[s_k^2] = \sum_{n=-\infty}^{\infty} \sum_{m=-\infty}^{\infty} h_n h_m E[x_{k-n}x_{k-m}]. \tag{II-4}$$

If the input is stationary and uncorrelated,

$$E[x_{k-n}x_{k-m}] = \sigma_x^2 \delta_{nm}$$

where $\sigma_x^2$ is the input mean square and $\delta_{nm}$ is the Kronecker delta. In this case, (II-4) can be reduced to a single summation, and by applying $z$-transform properties one finds

$$\sigma_s^2 = \sigma_x^2 \sum_{n=-\infty}^{\infty} h_n^2 = \sigma_x^2 \int_{-\pi}^{\pi} |H(e^{j\theta})|^2 \frac{d\theta}{2\pi}. \tag{II-5}$$

By taking the definitions of $U_m(z)$, $V_m(z)$, and $W_m(z)$, along with the unit energy results of (II-1), one can then use (II-5) to note that when the input time sequence is random, stationary, and uncorrelated, the time sequences associated with $U_m(z)$, $V_m(z)$, and $W_m(z)$ will all have a mean square equal to the input mean square and, thus, rms values equal to the input rms value, $\sigma_x$. In addition, the rms output value of the overall filter is equal to $\sigma_x\sqrt{I}$ where $I$ is found from (II-2).

These results can be utilized with the various expressions for the $z$ transforms of the node values and multiplier inputs by applying (31) through (35). These are summarized in Table I.

**TABLE I**
**The rms Values at Nodes and Multiplier Inputs for One- and Two-Multiplier Models**

| Model | Notation | rms Value |
|---|---|---|
| Two multiplier Node values and multiplier inputs | $A_m X/A_M$<br>$B_m X/A_M$<br>$zB_m X/A_M$ | $(\alpha_m \sigma_x^2)^{1/2}$ |
| One multiplier Node values | $\hat{A}_m X/\hat{A}_M$<br>$\hat{B}_m X/\hat{A}_M$<br>$z\hat{B}_m X/\hat{A}_M$ | $\Pi_m (\alpha_m \sigma_x^2)^{1/2}$ |
| One multiplier multiplier input | $(\hat{B}_m - \epsilon_m \hat{A}_{m+1})X/\hat{A}_m$ | $\Pi_m \left(\frac{2\alpha_m \sigma_x^2}{1+\epsilon_m k_m}\right)^{1/2}$ |

## Appendix III

We shall consider, here, only the $U_m(z)$, given by (28), for in the frequency domain, the only difference between $U_m(e^{j\theta})$ and $V_m(e^{j\theta})$ lies in the phase since

$$B_m(e^{j\theta}) = e^{-j(m+1)\theta} A_m^*(e^{j\theta}). \tag{III-1}$$

The ratio of any two successive $U_m(z)$ terms can be simply expressed as

$$\frac{U_{m+1}(z)}{U_m(z)} = \sqrt{\alpha_m/\alpha_{m+1}} \, \frac{A_{m+1}(z)}{A_m(z)}.$$

This can be expressed in the frequency domain by applying (I-2) with $l = m$, $k_l' = 0$, so that $A_{l+1}'(e^{j\theta}) = A_m(e^{j\theta})$. Combining this result with the recursion relation of (18), one obtains

$$\frac{U_{m+1}(e^{j\theta})}{U_m(e^{j\theta})} = \frac{1 + k_m e^{-j\Psi}}{[1 - k_m^2]^{1/2}} \tag{III-2}$$

where $\Psi$ is given by (I-3) with $l = m$.

The magnitude of the ratio of (III-2) oscillates between $\sqrt{(1+k_m)/(1-k_m)}$ when $\Psi$ is an even multiple of $\pi$, and $\sqrt{(1-k_m)/(1+k_m)}$ when $\Psi$ is an odd multiple of $\pi$. From Appendix I, it should be noted that $\Psi$ is a monotonically increasing function of $\theta$, going from $\Psi(0) = 0$ to $\Psi(\pi) = (m+1)\pi$. Thus, as $\theta$ goes from zero to $\pi$, the ratio of (III-2) will hit its extreme values at exactly $m + 2$ points, two of them being the end points $\theta = 0$ and $\theta = \pi$.

In terms of the log spectrum, this yields the result that the difference

$$\log |U_{m+1}(e^{j\theta})| - \log |U_m(e^{j\theta})|$$

hits the extreme values

$$\pm\tfrac{1}{2} \log [(1 + k_m)/(1 - k_m)]$$

exactly $m + 2$ times. Thus the surprising result is

shown that $U_m(z)$, and similarly $V_m(z)$, form min-max sequences. In fact, this result can be used to show that $A_m(z)/\sqrt{\alpha_m}$ is the $m$th order polynomial having all its roots inside the unit circle which is the mini-max approximation in terms of the log spectrum to $A_{m+1}(z)/\sqrt{\alpha_{m+1}}$.

The phase of the ratio of (III-2) can also be studied, and it can be shown that the phase of the ratio will hit its extreme values, $\pm \sin^{-1}(k_m)$, exactly $m + 1$ times as $\theta$ goes from 0 to $\pi$.

### References

[1] S. K. Mitra and R. J. Sherwood, "Canonic realizations of digital filters using the continued fraction expansion," *IEEE Trans. Audio Electroacoust.*, vol. AU-20, pp. 185–194, Aug. 1972.

[2] ——, "Digital ladder networks," *IEEE Trans. Audio Electroacoust.*, vol. AU-21, pp. 30–36, Feb. 1973.

[3] B. Gold and C. M. Radar, *Digital Processing of Signals*. New York: McGraw-Hill, 1969.

[4] A. Fettweis, "Digital filter structures related to classical filter networks," *Arch. Elek. Übertragung.*, vol. 25, pp. 79–89, Feb. 1971.

[5] ——, "Some principles of designing digital filters imitating classical filter structures," *IEEE Trans. Circuit Theory*, vol. CT-18, pp. 314–316, Mar. 1971.

[6] R. Crochiere, "Digital ladder filter structures and coefficient sensitivity," Res. Lab. Electronics Rep. 103, Mass. Inst. Tech., Cambridge, Oct. 15, 1971.

[7] F. Itakura and S. Saito, "Digital filtering techniques for speech analysis and synthesis," presented at the 7th Int. Congr. Acoust., Paper 25C-1, Budapest, 1971.

[8] J. D. Markel and A. H. Gray, Jr., "On autocorrelation equations with applications to speech analysis," *IEEE Trans. Audio Electroacoust.*, vol. AU-21, pp. 69–79, Apr. 1973.

[9] E. A. Guillemin, *Synthesis of Passive Networks*. New York: Wiley, 1957.

# Digital Filter Structures Related to Classical Filter Networks

by Alfred Fettweis *

DK 621.372.54

It is well-known that the transfer properties of filters are very sensitive to variations of the coefficients of their transfer function. While for $LC$-filters this only represents a computational problem (the sensitivity to element variations for the usual ladder filters is rather small), it amounts to a realization problem in the design of digital filters. In order to overcome this difficulty, digital filters should thus be modeled after certain structures used for classical filters. It is shown in this paper that suitable structures for this purpose are certain resistively terminated unit element ladder filters. In order to arrive at realizable signal flow diagrams, wave quantities must then be used as signals. The resulting wave digital filters require at worst a moderately higher number of multipliers than for a conventional design. This disadvantage should be more than made up for by the reduction of the number of digits which have to be used.

**Digitalfilterstrukturen zur Nachbildung klassischer Filternetzwerke**

Bekanntlich sind die Übertragungseigenschaften von Filtern sehr empfindlich gegen Abweichungen der Koeffizienten ihrer Übertragungsfunktion. Während dies bei $LC$-Filtern jedoch lediglich Schwierigkeiten bei der Berechnung verursacht — die Empfindlichkeit gegen Bauelementeschwankungen ist bei den üblichen Abzweigfiltern ziemlich klein —, läuft dies bei Digitalfiltern auf ein Realisierungsproblem hinaus. Um diese Schwierigkeiten zu überwinden, sollten auch Digitalfilter nach den für klassische Filter üblichen Strukturen gebaut werden. In der vorliegenden Arbeit wird gezeigt, daß gewisse aus Einheitselementen aufgebaute, resistiv abgeschlossene Abzweigschaltungen hierzu geeignet sind. Um auf geeignete Signalflußdiagramme zu kommen, müssen dann Wellengrößen als Signale verwendet werden. Die so erhaltenen Wellendigitalfilter erfordern im ungünstigsten Fall eine leicht höhere Anzahl Multiplizierer als bei bisherigen Entwürfen. Dieser Nachteil dürfte durch die wesentliche Verringerung der erforderlichen Stellenzahl mehr als ausgeglichen werden.

## 1. Introduction

Digital filters have so far been built by making use of structures whose element (multiplier) values can be determined immediately from the required transfer function [1]—[5]. An obvious advantage of such a procedure is the fact that it does not require involved realizability theories or lengthy numerical computations, as is the case for the insertion loss design of conventional $LC$-filters. This is paid for by the disadvantage that these digital filter structures show a very high sensitivity of their transfer properties to element variations. For a digital realization, this does not only imply a deviation of the actual frequency characteristic from the wanted one, due to coefficient truncation, but also a certain noise due to roundoff accumulation [6]—[10].

The usual way to overcome this difficulty is to first factorize the given transfer function and to realize the filter accordingly as a simple cascade of first and second degree sections. One may wonder, however, whether this is the only solution to the problem, i.e. whether there exist other types of structures which also reduce the tolerance sensitivity, possibly by offering advantages not obtainable from the simple cascading principle. In particular, one might be interested in obtaining also for digital filters the good stopband insensitivity of ladder structures combined with the good passband insensitivity which resistance terminated $LC$-filters are known to have due to the fact that their effective loss (transducer loss) can never become negative [11], [12].

The purpose of this paper is to show that structures of the wanted type for digital filters actually do exist. In particular, we shall show that it is possible to design digital filters according to structures which are closely related to those used for the realization of so-called unit-element filters. As filters of this latter type can be built in such a way that they fulfill all the requirements mentioned above, the same is then also true for the corresponding digital filters. The theory of unit element filters is well-known [13], [14] and similar to the theory of classical $LC$-filters. Our present problem can thus be reduced to a problem whose solution is already known. In fact, we now have even some additional flexibility: As we shall see, gyrators and circulators can be built in the circuits to be described with the same ease as reciprocal elements, and we are not restricted by any desideratum for structures using only grounded unit elements [27].

## 2. General Principles

Consider first an ordinary $LC$-ladder filter. Its transfer function $H(p)$, which is a rational function of the complex frequency $p$, corresponds to a single differential equation between the input variable $x$ and the output variable $y$. Instead of using an $LC$-structure, it would be possible in theory to realize this differential equation directly on an analog computer, in which case the coefficients of $H(p)$ would each correspond to a multiplication by a constant. In reality, such a procedure would be completely

* Prof. Dr. A. Fettweis, Lehrstuhl für Nachrichtentechnik der Universität, D-463 Bochum, Postfach 2148.

Reprinted with permission from *Arch. Elek. Übertragung.*, vol. 25, pp. 79–89, Feb. 1971.

impractical since for good filters, the coefficients of $H(p)$ must be known with extreme accuracy in order to determine with only mild accuracy the zeros and poles of $H(p)$, and thus the actual transfer behavior. Despite this difficulty, $LC$-filters, and even quite complicated ones, are known to exist. This implies that in the usual $LC$-ladder structures the sensitivities to element variations are very much smaller. Mathematically, this is related to the fact that a ladder structure cannot fully be described by a single differential equation but only by a system of differential equations whose detailed structure not only corresponds to the overall input/output performance but also to the precise topology of the actual network. The low tolerance sensitivity of (terminated) ladder structures thus implies that the sensitivity to variations in the coefficients of the above mentioned system of differential equations is very much lower than the sensitivity to variations of the coefficients of the overall differential equation obtained after elimination of all internal variables. These peculiarities are, of course, also at the origin of the fact that filter computations often have to be carried out with a very high number of digits in order to end up with a number of digits for the elements which is just within the reach of technological possibilities.

For digital filters, the situation is very similar except that the numerical problem just described now corresponds to a realization problem due to the earlier mentioned phenomena of coefficient truncation and roundoff accumulation. The transfer function $H$ can now be written as a rational function of the variable $z = e^{pT}$, where $F = 1/T$ is the rate of operation of the filter. This transfer function corresponds to a single difference equation between the input and output variables. Here too, however, a high degree of improvement can be obtained by breaking up the single difference equation into an appropriate system of difference equations. One way of doing this corresponds to the usual factorization of $H$ and the subsequent realization by means of a simple cascade of low order sections. The way to be described here corresponds to trying to find a system of difference equations which in some sense imitates a classical terminated ladder filter structure. The analogy with the above discussion of the classical filter design problem shows that a solution obtained this way must certainly represent a considerable improvement. It would be interesting to know whether even better solutions do exist. In any case, due to their high relative cost, the number of multipliers has to be kept as low as possible, while adders may be used more generously.

In order to guarantee that a structure can be realized as digital filter, the following requirement must always be fulfilled: Every feedback loop contained in the signal flow diagram describing the structure must contain at least one delay element, i.e. may *not* be *delay-free*. One checks easily that this *rule* is fulfilled for any of the digital filter structures known so far. Violation of the rule would make it impossible to find a sequence in which the various computations can be carried out.

As frequency variable for establishing the analogy with the theory of unit-element filters we shall use the "complex frequency" $\psi$ defined by

$$\psi = \tanh(pT/2) = \frac{1 - e^{-pT}}{1 + e^{-pT}}, \tag{1}$$

which is also used e.g. in the theory of resonant-transfer circuits [15]—[17]. For $p = j\omega$, we thus can write

$$\psi = j\varphi, \quad \varphi = \tan(\omega T/2). \tag{2}$$

The choice of $\psi$ as frequency variable seems to be the only one appropriate to our problem. It has an immediate consequence which we briefly want to discuss.

If we use as signals in the flow diagrams the usual voltages and currents, it becomes impossible to observe the earlier given rule concerning feedback loops. In order to show this, let us consider an arbitrary element, e.g. an inductance; the equation to be realized is $V = \psi R I$ where $R$ is a positive constant. Due to eq. (1), this would require for the corresponding difference equation (3)

$$v(nT) + v[(n-1)T] = \{i(nT) - i[(n-1)T]\} R,$$

which shows that the computation of $v$ at the instant $t = nT$ requires not only e.g. the knowledge of $v$ at the previous sampling instant but also at the present one. Hence, any flow diagram representation of eq. (3) such as the one shown in Fig. 1 contains a delay-free path leading from the input to the output. The same is true for the current-voltage or voltage-current flow diagrams of all other elements. Hence, the interconnection of elements inevitably leads to delay-free feedback loops, thus making any physical implementation impossible.

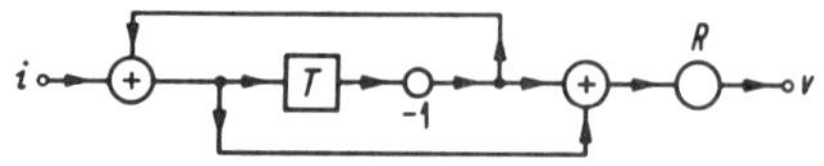


Fig. 1. Current-voltage flow diagram for an inductance in the $\psi$-domain.

It turns out that this dilemma can be avoided by basing all signal flow diagrams not on voltages and currents but on wave quantities. In fact, one may use indifferently voltage waves or current waves; we have adopted here a description by voltage waves. As will briefly be discussed in subsection 4.1, use of power waves would be less appropriate as this would lead to a larger number of multipliers required. If waves are used, we shall usually speak of a *wave flow diagram* rather than a signal flow diagram. Also, the filters to be described here may appropriately be called *wave digital filters*. A wave flow diagram which obeys the above mentioned rule will also be called *realizable*; otherwise it is called *unrealizable*.

As we shall see later, wave digital filters usually require two separate sequences of time instants at which the filter is operative. These two sequences are shifted by a time interval $T/2$ with respect to one another, as is also the case for the digital filters described by FJÄLLBRANT [18]. In any case, it will often be necessary to write difference relations without assuming that one of the pulses of the time sequence occurs at $t = 0$. In the discussion of elements and sources given in the next section, a general time variable $t$ is therefore used, i.e. no reference is made to any particular instants at which pulses occur.

## 3. Elements and Sources

In the present section, we shall establish the wave flow diagrams of the various elements and sources to be used. Wherever possible, the required wave quantities will be defined by means of a port resistance equal to the resistance constant $R$ occuring in the definition of the element or source under consideration[1]. This way, the wave equations do not contain $R$ at all anymore. We shall see in sections 4 and 5 what consequences this has when elements and source have to be interconnected in order to obtain the required filter diagram.

### *3.1. Reactive elements*

*Inductance*

The steady-state voltage-current relation to be realized is (Fig. 2a)

$$V = \psi R I \tag{4}$$

where $R$ is a positive constant. Using eq. (1), eq. (4) can be written

$$B = -\frac{1}{z} A \tag{5}$$

where $z$ is defined as usual by

$$z = e^{pT}, \tag{6}$$

and where

$$A = V + RI \quad \text{and} \quad B = V - RI \tag{7a, b}$$

are the incident and the reflected waves, respectively, the instantaneous values $a = a(t)$ and $b = b(t)$ of these waves being related to the instantaneous values of the voltage $v = v(t)$ and the current $i = i(t)$ by

$$a = v + Ri, \quad b = v - Ri. \tag{8a, b}$$

Hence, the difference equation for $a$ and $b$ is given by

$$b(t) = -a(t - T); \tag{9}$$

with

$$a = A e^{pt}, \quad b = B e^{pt}, \tag{10}$$

where $A$ and $B$ are assumed to be constants, the steady-state solution of eq. (9) is indeed given by eq. (5). The wave flow diagrams corresponding to eqs. (5) and (9) are the *wave one-ports* shown in Fig. 2b and c, respectively. The first one consists of a *transmittance* (transfer function) equal to $-1/z$, and the second one of a *delay* $T$ combined with a *sign inverter*.

*Capacitance*

The steady-state voltage-current relation to be realized is (Fig. 3a)

[1] Whenever required, the port resistance will be indicated either under the lower port terminal or at least in its neighbourhood.

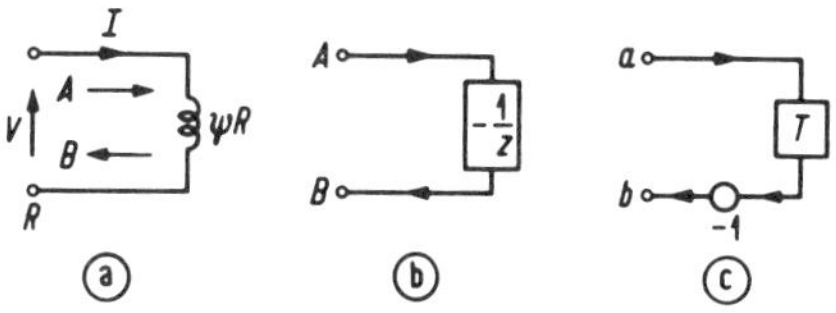


Fig. 2. (a) Inductance of impedance $\psi R$,
(b) its steady-state wave flow diagram consisting of a transmittance $-1/z$,
(c) its instantaneous wave flow diagram consisting of a delay $T$ and a sign inverter.

$$V = \frac{R}{\psi} I \tag{11}$$

where $R$ is a positive constant. Using eqs. (1), (6) and the definition (7), eq. (11) can be written

$$B = \frac{1}{z} A. \tag{12}$$

Hence, the difference equation for the instantaneous wave quantities defined by eq. (8) is given by

$$b(t) = a(t - T); \tag{13}$$

it reduces to eq. (12) if we make use of eq. (10). The wave flow diagrams are the wave one-ports shown in Fig. 3b and c; they consist of a transmittance $1/z$ and a delay $T$, respectively.

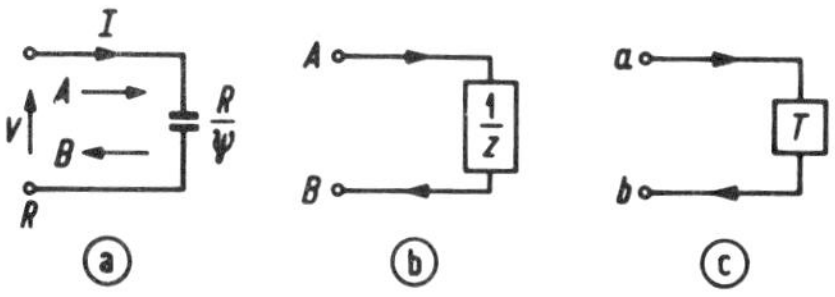


Fig. 3. (a) Capacitance of impedance $R/\psi$,
(b) its steady-state wave flow diagram consisting of a transmittance $1/z$,
(c) its instantaneous wave flow diagram consisting of a delay $T$.

### *3.2. Elements with frequency-independent behavior*

For reasons of simplicity we shall write all equations relating to elements with frequency-independent behavior for instantaneous quantities only. It is obvious that the corresponding equations for steady-state quantities have the same structure.

Fig. 4.
(a) Resistance $R$,
(b) its wave flow diagram consisting of a wave sink.

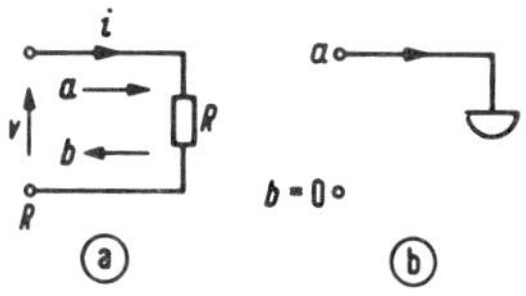


*Resistance*

The equation to be realized is (Fig. 4a)

$$v = Ri,$$

i.e., using eq. (8),

$$b = 0. \tag{14}$$

Hence, the wave flow diagram consists of a *wave sink*; we represent it as shown in Fig. 4b.

*Open-circuit*

We have $i = 0$ (Fig. 5a). Hence, choosing an arbitrary positive constant $R$ and using eq. (8), we obtain

$$b = a. \tag{15}$$

The wave flow diagram is thus a simple through connection (Fig. 5b).

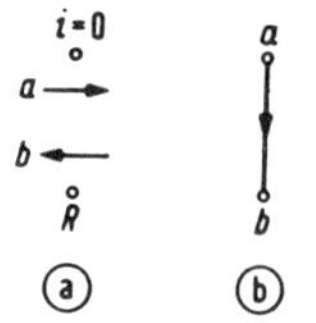


Fig. 5. (a) Open-circuit, (b) its wave flow diagram.

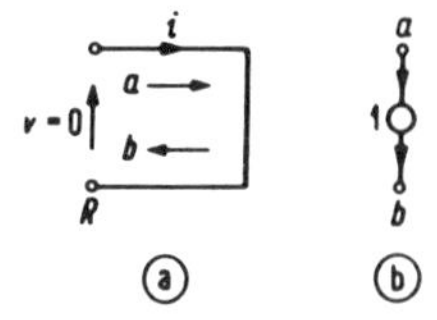


Fig. 6. (a) Short-circuit, (b) its wave flow diagram.

*Short-circuit*

We have $v = 0$ (Fig. 6a). Hence, choosing an arbitrary positive constant $R$ and using eq. (8), we obtain

$$b = -a. \tag{16}$$

The wave flow diagram consists thus of a sign inversion (Fig. 6b).

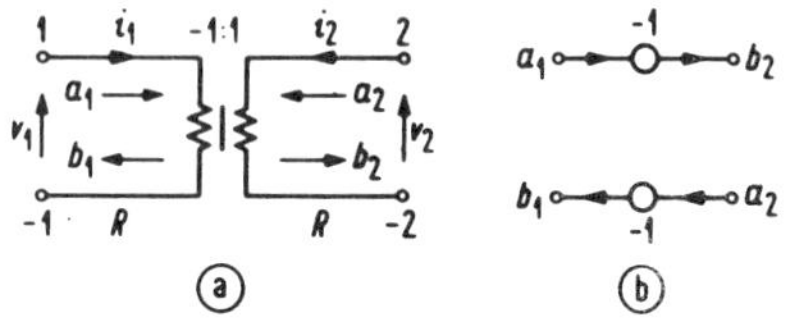

Fig. 7. (a) Ideal transformer of ratio $-1:1$, (b) its wave flow diagram.

*Ideal transformer of ratio* $-1/1$

The equations to be realized are (Fig. 7a)

$$i_1 = i_2, \quad v_1 = -v_2.$$

Hence, choosing an arbitrary positive constant $R$ and using the definitions

$$a_k = v_k + Ri_k, \quad b_k = v_k - Ri_k, \tag{17a, b}$$

$$k = 1, 2, \tag{17c}$$

we obtain $\quad b_1 = -a_2, \quad b_2 = -a_1. \qquad$ (18a, b)

The wave flow diagram is now the *wave two-port* shown in Fig. 7b.

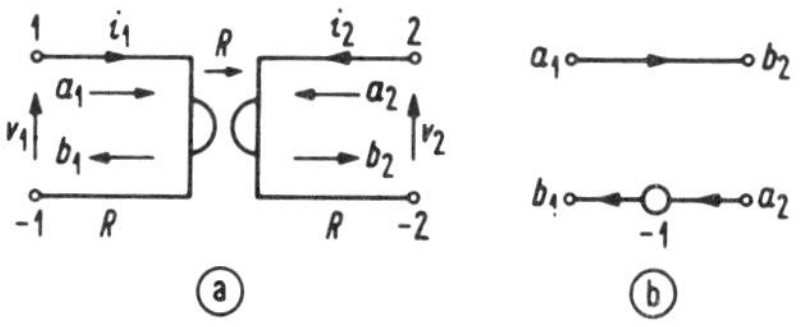

Fig. 8. (a) Gyrator of gyration resistance $R$, (b) its wave flow diagram.

*Gyrator*

The equations to be realized are (Fig. 8a)

$$v_1 = -Ri_2, \quad v_2 = Ri_1, \tag{19a, b}$$

where the gyration resistance $R$ is a positive constant. Using the definitions (17), we obtain

$$b_1 = -a_2, \quad b_2 = a_1. \tag{20a, b}$$

The corresponding wave two-port is shown in Fig. 8b.

*Circulator*

We consider first the three-port circulator of Fig. 9a whose circulation resistance $R$ is a positive constant. Using the definitions (17a, b) with $k = 1, 2,$ and 3, we have

$$b_1 = a_3, \quad b_2 = a_1, \quad b_3 = a_2. \tag{21a–c}$$

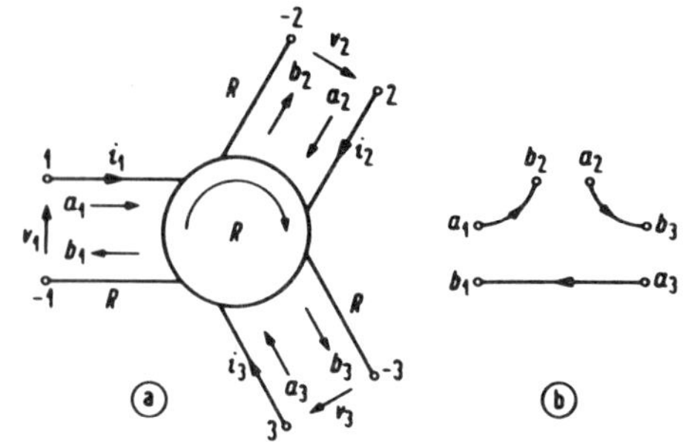

Fig. 9. (a) Three-port circulator of circulation resistance $R$, (b) its wave flow diagram.

The wave flow diagram is now the *wave three-port* shown in Fig. 9b. It is obvious that this result can easily be extended to $n$-port circulators, in which case eq. (21) has to be generalized to

$$b_1 = a_n, \quad b_2 = a_1, \ldots, b_n = a_{n-1}.$$

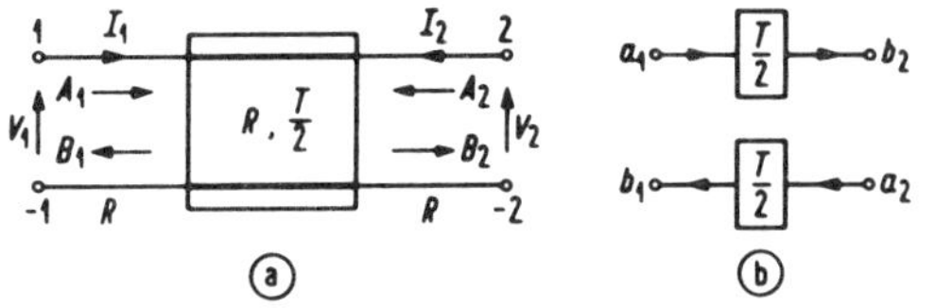

Fig. 10. (a) Unit element of delay $T/2$ and characteristic resistance $R$, (b) its wave flow diagram.

*3.3. Unit element*

The steady-state voltage-current relations to be realized are (Fig. 10a)

$$B_1 = A_2 e^{-pT/2}, \quad B_2 = A_1 e^{-pT/2} \tag{22a, b}$$

where

$$A_k = V_k + RI_k, \quad B_k = V_k - RI_k, \tag{23a, b}$$

$$k = 1, 2. \tag{23c}$$

Hence, the difference equations for the instantaneous waves defined by eq. (17) are given by

$$b_1(t) = a_2(t - T/2), \quad b_2(t) = a_1(t - T/2); \tag{24a, b}$$

with $\qquad a_k = A_k e^{pt}, \quad b_k = B_k e^{pt},$

where $A_k$ and $B_k$ are constants, the steady-state solution of eq. (24) is indeed given by eq. (22). The (instantaneous) wave flow diagram is now the wave two-port shown in Fig. 10b. The steady-state wave flow diagram is similar but with the delays $T/2$ replaced by transmittances $z^{-1/2}$.

In the symbol of Fig. 10a, the delay $T/2$ of the unit element has been indicated explicitly. We shall henceforth omit doing this since confusion can certainly not arise.

*3.4. Sources*

We shall again limit ourselves to considering instantaneous quantities only, as was also done in subsection 3.2.

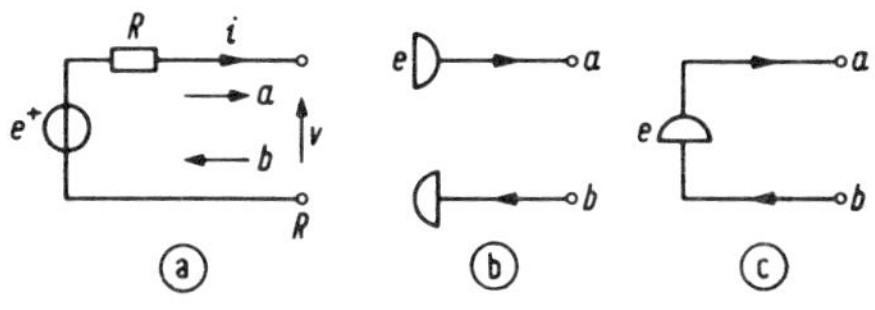

Fig. 11. (a) Source of voltage $e$ in series with a resistance $R$, (b), (c) two equivalent representations of its wave flow diagram.

*Resistive source*

It is sufficient to consider a voltage source of voltage $e = e(t)$ with a series resistance $R$ (Fig. 11a). The equation to be realized is

$$e = v + Ri,$$

i.e., using eq. (8),

$$a = e.$$

Hence, the wave flow diagram is a *wave source*; we represent it as shown in Fig. 11b or c.

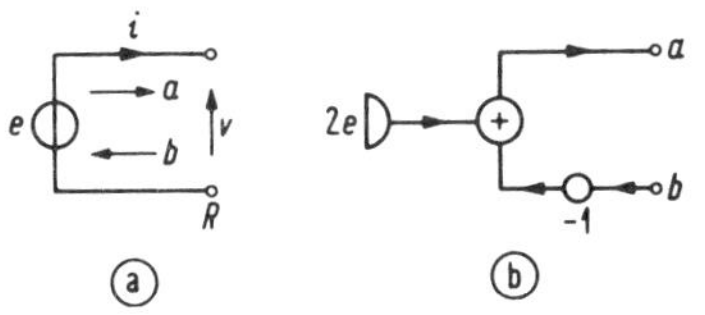

Fig. 12.
(a) A voltage source of voltage $e$,
(b) its wave flow diagram.

The situation is slightly less simple for a pure voltage source. The equation to be realized is $v = e = e(t)$; hence, choosing an arbitrary positive constant $R$ and using the definition (8), we must have

$$a = 2e - b.$$

The corresponding wave flow diagram consists of a wave source $2e$, an adder and a sign inverter (Fig. 12b).

For a pure current source (Fig. 13), one obtains in a similar way

$$a = 2e + b,$$

where the intensity of the current source has been chosen equal to $e/R$. The corresponding wave flow diagram is shown in Fig. 13b.

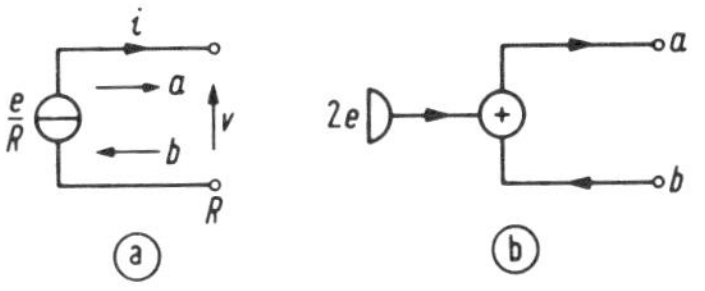

Fig. 13.
(a) A current source of current $e/R$,
(b) its wave flow diagram.

## 4. Interconnection of Ports

Circuits are obtained in principle by interconnecting the ports of elements and sources. A difficulty then arises in the wave flow diagram since in most cases discussed in section 3 the port resistance $R$ is not arbitrary but imposed by the element or source to which the port belongs. Hence, when interconnecting ports, we must have available a possibility of changing the port resistance. This can be achieved by means of the various *adaptors* to be discussed hereafter.

### *4.1. Simple change of reference resistance*

Consider two ports 1 and 2, with port resistances $R_1$ and $R_2$, respectively. The waves are related to the voltages and currents by

$$a_k = v_k + R_k i_k, \quad b_k = v_k - R_k i_k, \qquad (25\text{a, b})$$
$$k = 1, 2. \qquad (25\text{c})$$

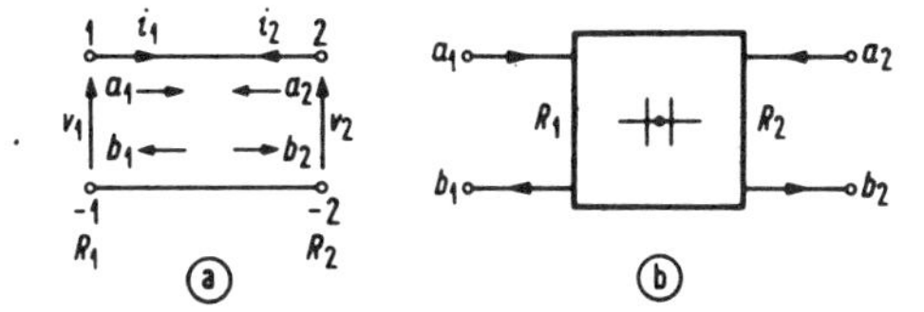

Fig. 14. (a) Interconnection of two ports with port resistances $R_1$ and $R_2$, respectively,
(b) Schematic representation of the corresponding wave two-port (two-port adaptor).

If these ports are simply interconnected (Fig. 14a), we have

$$v_1 = v_2, \quad i_1 = -i_2. \qquad (26\text{a, b})$$

Eliminating the $v_k$ and $i_k$ between eqs. (25) and (26), we thus obtain (27a, b)

$$b_1 = a_2 + \alpha(a_2 - a_1), \quad b_2 = a_1 + \alpha(a_2 - a_1)$$

where $\quad \alpha = (R_1 - R_2)/(R_1 + R_2). \qquad (28)$

These equations define a wave two-port which we call a *two-port adaptor*. We represent it schematically as shown in Fig. 14b, the symbols inside of the box referring to the fact that Fig. 14a corresponds in a sense as well to a parallel as to a series connection (cfr. Figs. 17 and 19). According to eq. (27), a detailed wave flow diagram corresponding to this two-port adaptor can be drawn as shown in Fig. 15. This

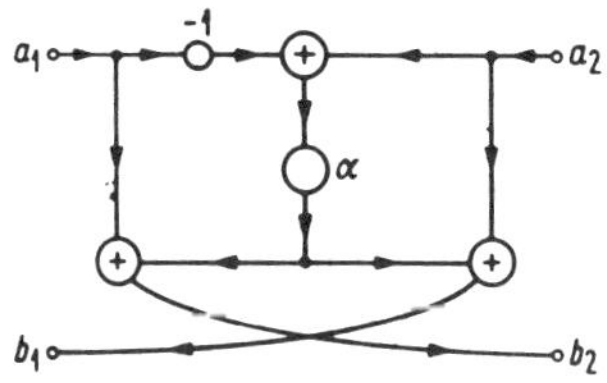

Fig. 15.
A possible realization of the two-port adaptor of Fig. 14b requiring one multiplier, $\alpha$.

wave flow diagram requires only one multiplier, $\alpha$. It contains a path leading from $a_1$ to $b_1$ and also a path leading from $a_2$ to $b_2$, but no feedback loop. By writing eq. (27) in some equivalent form, other equivalent wave flow diagrams can be obtained; not all of these, however, require only one multiplier.

Equation (28) shows that for $R_1 > 0$ and $R_2 > 0$, we have $|\alpha| < 1$. Vice versa, if this latter inequality is fulfilled, we have $R_1/R_2 > 0$, as can be seen from

$$R_1/R_2 = (1 + \alpha)/(1 - \alpha).$$

This shows that we will then always have $R_2 > 0$ as soon as $R_1 > 0$ and vice versa.

The connection under discussion allows us already to explain the advantage of using voltage (or current) waves rather than power waves. Indeed, a general representation of scattering equations such as eq. (27) would be (29a, b)

$$b_1 = \alpha_{11} a_1 + \alpha_{12} a_2, \quad b_2 = \alpha_{21} a_1 + \alpha_{22} a_2$$

where the scattering coefficients are related to one another by the linear relations

$$-\alpha_{11} = \alpha_{22} = 1 - \alpha_{12} = \alpha_{21} - 1 = \alpha .$$

If we had used power waves, we could still write equations like eq. (29), but the scattering coefficients would then be related by (see e.g. [19])

$$\alpha_{11}^2 + \alpha_{12}^2 = 1 , \quad \alpha_{12} = \alpha_{21} , \quad \alpha_{11} = -\alpha_{22} .$$

As the first of these expressions is nonlinear, it would thus not be possible to give a wave flow diagram of eq. (29) which would require only one multiplier.

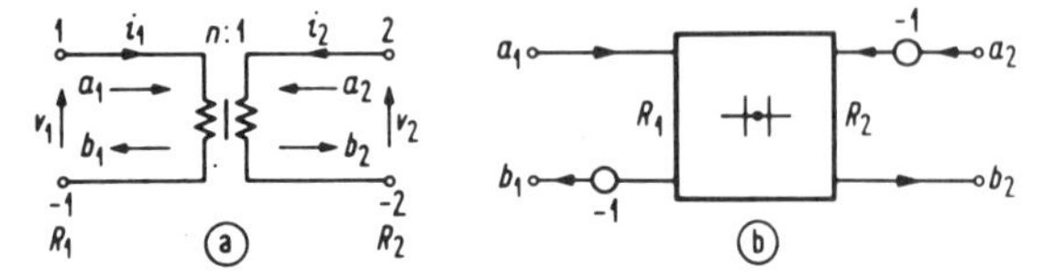


Fig. 16. (a) Transformer of ratio $n:1$, with $n > 0$, (b) its realization based on $n = R_1/R_2$.

As a simple application, we consider the realization of a transformer of ratio $n:1$ (Fig. 16a). If we had attempted to do this in a direct way, as was done for other elements in section 3, we would have needed two multipliers. In fact, however, such a transformer can be realized as a cascade of two gyrators having gyration resistances $R_1$ and $R_2$ such that $n = R_1/R_2$. Thus, using the result of Fig. 8, we obtain the wave flow diagram of Fig. 16b where one of the resistances $R_1$ and $R_2$ may be chosen arbitrarily. In order to conform with our assumption $R > 0$ made in connection with Fig. 8, we must assume at present that $n > 0$; due to Fig. 7, however, the corresponding result for $n < 0$ is easily obtained.

### 4.2. *Parallel connection*

Consider $n$ ports $1, 2, \ldots, n$, with port resistances $R_1, R_2, \ldots, R_n$, respectively. The waves are related to the voltages and currents by eq. (25a, b), but with

$$k = 1, 2, \ldots, n .$$

If these ports are connected in parallel (Fig. 17a), we have (30a, b)

$$v_1 = v_2 = \cdots = v_n , \quad i_1 + i_2 + \cdots + i_n = 0 .$$

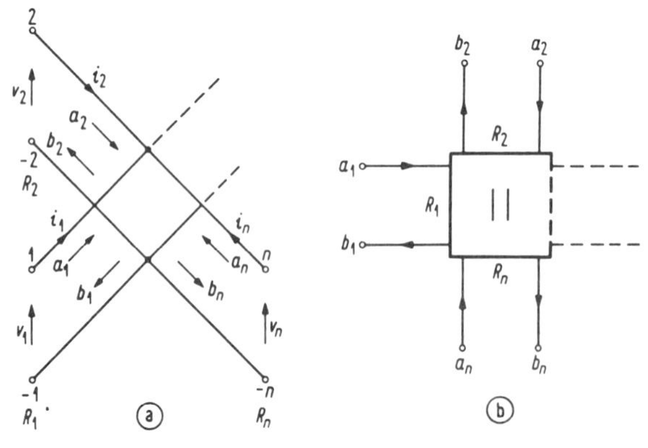


Fig. 17. (a) Parallel connection of $n$ ports with port resistances $R_1, R_2, \ldots, R_n$, respectively,
(b) schematic representation of the corresponding wave $n$-port ($n$-port parallel adaptor).

Eliminating the $v_k$ and $i_k$ between eqs. (25a, b) and (30), we obtain

$$b_k = a_0 - a_k , \quad a_0 = \sum_{k=1}^{n} \alpha_k a_k \qquad (31\text{a, b})$$

where (32a, b)

$$\alpha_k = 2G_k/(G_1 + G_2 + \cdots + G_n) , \quad G_k = 1/R_k .$$

These equations define a wave $n$-port which we call a *parallel adaptor*, or more precisely, a *n-port parallel adaptor*. We represent it schematically as shown in Fig. 17b, the two parallel lines inside of the box referring to "parallel connection."

If we would use eq. (31) as such, we would need altogether $n$ multipliers. In fact, we have

$$\alpha_1 + \alpha_2 + \cdots + \alpha_n = 2 \qquad (33)$$

so that one of the $\alpha_k$, say $\alpha_n$, can be eliminated. This way, eq. (31b) can be written

$$a_0 = 2a_n + \sum_{k=1}^{n-1} \alpha_k (a_k - a_n) , \qquad (34)$$

thus requiring only $n-1$ multipliers; port $n$ is then called the *dependent port*. The number of multipliers can further be reduced if some of the $\alpha_k$ are equal, i.e. if some of the $R_k$ are equal. The factor 2 appearing in eq. (34) does not require any special multiplier since a multiplication by 2 is a very elementary process. As an example, the detailed wave flow diagram for $n = 3$ is given, according to eqs. (31a) and (34), in Fig. 18. For every $k = 1, 2, 3$ it contains a path leading from $a_k$ to $b_k$ but it does not contain any feedback loop; these conclusion remains obviously true for arbitrary values of $n$.

If the $G_k$ are given positive constants, all the $\alpha_k$ also turn out positive; they satisfy eq. (33). Vice versa, if all the $\alpha_k$ and one of the $G_k$, say $G_k'$, are given, the remaining $G_k$ can be computed by means of

$$G_k/G_k' = \alpha_k/\alpha_k' ,$$

which is easily derived from eq. (32a). Thus, if $G_k'$ as well as $\alpha_1, \alpha_2, \ldots, \alpha_{n-1}$ are given positive constants and if $\alpha_n$ derived by eq. (33) is also positive, all $G_k$ turn out positive.

If port $n$ is the dependent port, as we have assumed, the coefficient $\alpha_n$ is implicitly determined by the remaining $\alpha_k$ through eq. (33). Thus, if the sum of these remaining $\alpha_k$ is close to 2, i.e. if $\alpha_n$ is small, it may be determined this way only within a large error. In order to ensure the highest possible precision one should therefore choose as dependent port the one whose corresponding $\alpha_k$ is the largest.

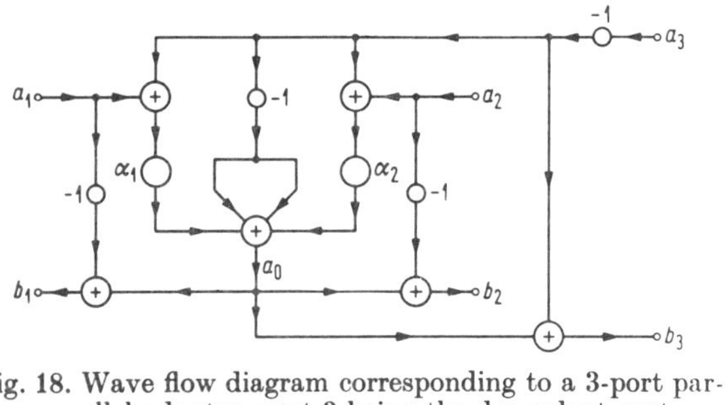

Fig. 18. Wave flow diagram corresponding to a 3-port parallel adaptor, port 3 being the dependent port.

### 4.3. *Series connection*

We consider $n$ ports as in subsection 4.2, but we now connect them in series (Fig. 19a). We then have (35a, b)

$$i_1 = i_2 = \cdots = i_n, \quad v_1 + v_2 + \cdots + v_n = 0 .$$

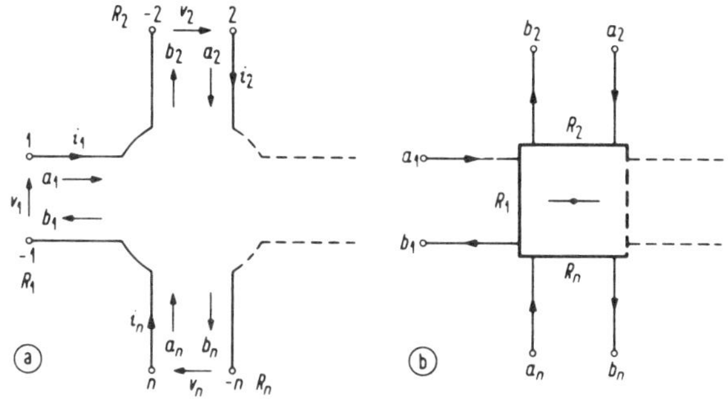


Fig. 19. (a) Series connection of $n$ ports with port resistances $R_1, R_2, \ldots, R_n$, respectively,
(b) schematic representation of the corresponding wave $n$-port ($n$-port series adaptor).

Eliminating the $v_k$ and $i_k$ between eqs. (25a, b) and (35), we obtain

$$b_k = a_k - \beta_k a_0 , \quad a_0 = \sum_{k=1}^{n} a_k \qquad (36\text{a, b})$$

where

$$\beta_k = 2\,R_k/(R_1 + R_2 + \cdots + R_n) . \qquad (37)$$

These equations define a wave $n$-port which we call a *series adaptor*, or more precisely, a *$n$-port series adaptor*. We represent it schematically as shown in Fig. 19b, the symbol inside of the box referring to "*series connection*".

If we would use eq. (36) as such, we would again need altogether $n$ multipliers. In fact, we have

$$\beta_1 + \beta_2 + \cdots + \beta_n = 2 \qquad (38)$$

so that one of the $\beta_k$, say $\beta_n$, can be eliminated. This can be done e.g. by using eq. (36) for $k = 1, 2, \ldots, n-1$, and then computing $b_n$ by means of the expression derived from eqs. (36) and (38),

$$b_n = -\,a_0 - \sum_{k=1}^{n-1} b_k ; \qquad (39)$$

the port $n$ is then again called the *dependent port*. The number of multipliers can further be reduced if some of the $\beta_k$ are equal, i.e. if some of the $R_k$ are equal. For the $\beta_k$, similar remarks can be made as for the $\alpha_k$ at the end of subsection 4.2.

As an example, the detailed wave flow diagram for $n = 3$ is given, according to eqs. (36a) and (39), in Fig. 20. For every $k = 1, 2, 3$ it contains a path leading from $a_k$ to $b_k$ but it does not contain any feedback loops; these conclusions remain obviously true for arbitrary values of $n$.

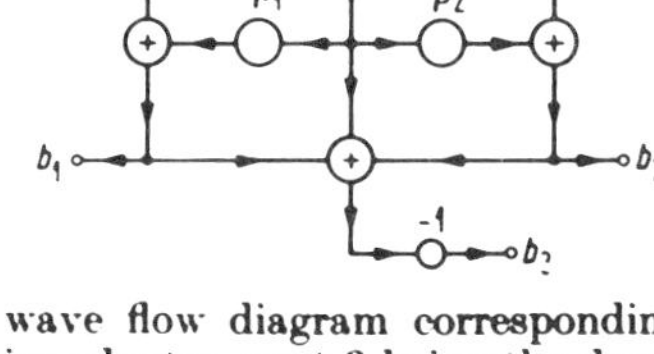


Fig. 20. A wave flow diagram corresponding to a 3-port series adaptor, port 3 being the dependent port.

Note: It is easily checked that for $n = 2$ the parallel and the series adaptor both become equivalent to the adaptor discussed in subsection 4.1.

## 5. Realization of Circuits

### 5.1. *General principles*

In sections 3 and 4 we have seen how various types of elements, sources, and adaptors can be built which serve as building blocks for the wave flow diagrams to be realized. When interconnecting these building blocks, the following principles must be observed:

1. The building blocks must be interconnected port by port, i.e. the two wave terminals of one wave port must be connected with the two wave terminals of precisely one other wave port.
2. For every two wave terminals which are joined together the two corresponding waves must be *compatible*, i.e. they must flow in the same direction.
3. The rule mentioned in section 2 must be respected, i.e. the resulting wave flow diagram must be realizable.

The first two of these principles, obviously, are relatively easy to comply with. The third one, however, has far reaching consequences. We have seen, indeed, that for every port of an adaptor, there always exists an inner path leading from the incident wave to the corresponding reflected wave. Hence, to a wave-port of an adaptor we may certainly always connect any of the wave one-ports shown in Figs. 2c, 3c, 4b, or 11c, but never those shown in Figs. 5b, 6b, 12b, or 13b. Similarly, we may always connect to it one of the wave two-ports shown in Fig. 10b, but never another adaptor. Finally, wave two-ports and multiports such as those of Figs. 7b, 8b, and 9b are conditionally acceptable, i.e. they will not by themselves cause the appearance of delay-free feedback loops but may do so if they are not properly terminated at the remaining port(s).

From this discussion we conclude that within the theory given here two consecutive adaptors must always be separated in some way or another by a wave two-port corresponding to a unit element. In order to examine the consequences of this conclusion, we study in subsection 5.2 the realization of one-port reactances, and in subsections 5.3 and 5.4 the realization of reactance two-ports.

Wherever possible, and in particular in all wave flow diagrams, we shall exclusively use instantaneous wave quantities such as $a_k$ and $b_k$. It is clear that we could equally well use the complex wave amplitudes $A_k$ and $B_k$, but then a delay $T$ should be replaced by a transfer function $z^{-1} = e^{-pT}$, and a delay $T/2$ by a transfer function $z^{-1/2}$.

### 5.2. *Realization of reactances*

*Series resonant circuit*

Assume that we want to realize the series resonant circuit of Fig. 21 a. If it forms part of a series arrangement for which a series adaptor can be used,

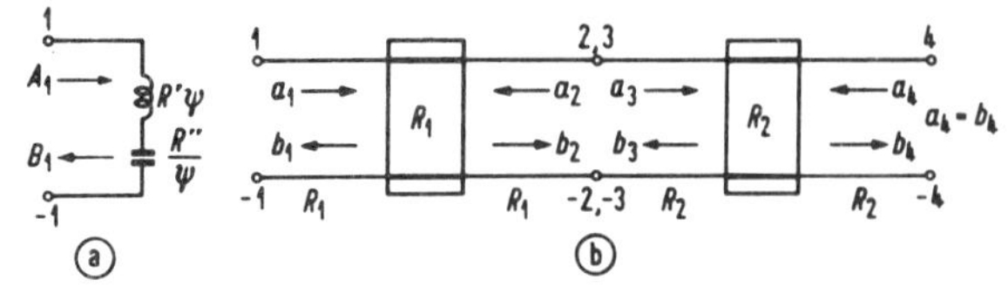

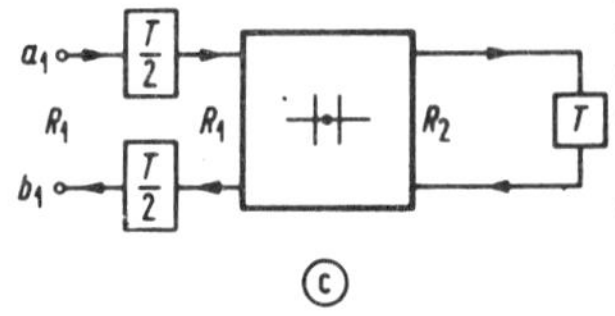

Fig. 21.
(a) A series resonant circuit,
(b) an equivalent circuit using two unit elements,
(c) the corresponding wave flow diagram.

there does not exist any problem: We simply have to terminate two of the wave ports of this adaptor by wave one-ports corresponding to an inductance of impedance $R'\psi$ and a capacitance of impedance $R''/\psi$, respectively. If, however, the series resonant circuit has to be placed in a shunt branch, as is frequently the case, such a method of realization is not acceptable since it would require connecting directly a series adaptor to a shunt adaptor.

This difficulty can be avoided by making use of the equivalence of a series resonant circuit with a cascade of two unit elements which are open-circuited at the output port 4 (Fig. 21 b). The characteristic resistances $R_1$ and $R_2$ of these unit elements are related to $R'$ and $R''$ by

$$R_1 = R' + R'', \quad R_2 = (R' + R'')R''/R'. \qquad (40\text{a, b})$$

The waves $a_1$, $b_1$, $a_2$, and $b_2$ are defined for a port resistance $R_1$ while the waves $a_3$, $b_3$, $a_4$, and $b_4$ are defined for a port resistance $R_2$. Consequently, the wave flow diagram corresponding to Fig. 21 b must make use of a two-port adaptor for changing from $R_1$ to $R_2$. Hence, using the results of Figs. 5, 10, and 14, and assuming that the waves $a_1$ and $b_1$ remain defined for a port resistance $R_1$, this wave flow diagram becomes as shown in Fig. 21 c. It thus requires only one multiplier, the corresponding parameter $\alpha$ being given by eq. (28), i.e., due to eq. (40), by

$$\alpha = (R' - R'')/(R' + R''). \qquad (41)$$

The wave flow diagram of Fig. 21 c, obviously, can be used not only for the realization of a shunt branch but also in any series arrangement. It is true that it contains a feedback loop but this does not make it unrealizable since this loop comprises the delay $T$. Furthermore, any additional feedback loop created by connecting Fig. 21 c to any other wave port is also acceptable since it necessarily contains the two delays $T/2$.

*Parallel resonant circuit*

The problem of realizing a parallel resonant circuit is in a sense the dual of the one just discussed. We may thus be very brief and simply refer to

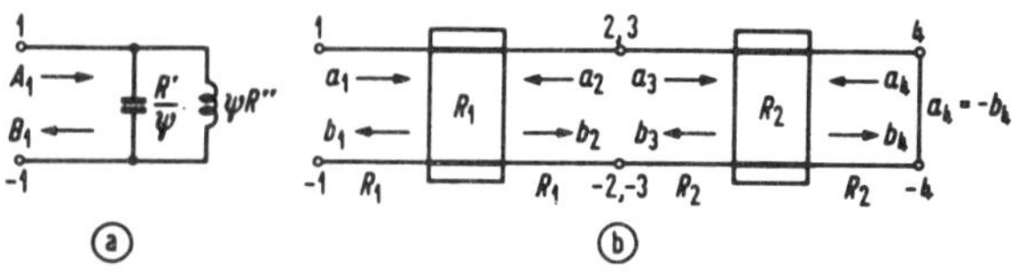

Fig. 22.
(a) A parallel resonant circuit,
(b) an equivalent circuit thereof using two unit elements,
(c) the corresponding wave flow diagram.

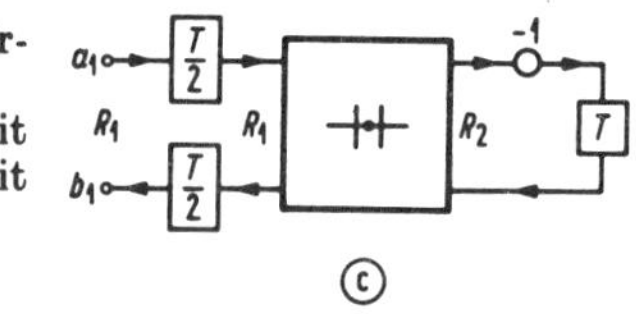

Fig. 22 which should be self-explanatory. The final conclusions are the same. Note, however, that the eqs. (40 a, b) have to be replaced by

$$R_1 = R'R''/(R' + R''), \quad R_2 = R''^2/(R' + R''),$$

although eq. (41) remains valid.

*Arbitrary reactances*

As has been shown by RICHARDS [20], any reactance of degree $n$ can be realized by means of either an open-circuited or a short-circuited cascade of $n$ unit elements. By generalizing the results of Figs. 21 and 22 it is easily seen that a wave flow diagram corresponding to such a cascade can always be realized by means of $n-1$ two-port adaptors, thus requiring $n-1$ multipliers. Such a wave flow diagram always contains $n-1$ feedback loops, all of which obey the rule of section 2. Furthermore, it remains true that the input wave port may freely be connected to any further wave port without violating this rule.

### 5.3. *Realization of all-pass circuits*

A unilateral first degree all-pass section can always be realized by means of a three-port circulator together with a capacitance or an inductance [19]. It is sufficient to consider the first of these cases (Fig. 23 a). Using the results of Figs. 3 and 9, we immediately obtain the corresponding wave two-port shown in Fig. 23 b; it contains one two-port adaptor, thus requiring one multiplier.

A unilateral second degree all-pass section can be realized in a similar way by making use e.g. of the result of Fig. 21. This leads to the result of Fig. 24 where $R_1$ and $R_2$ are given by eq. (40). It is clear that the same procedure can be used for the realiza-

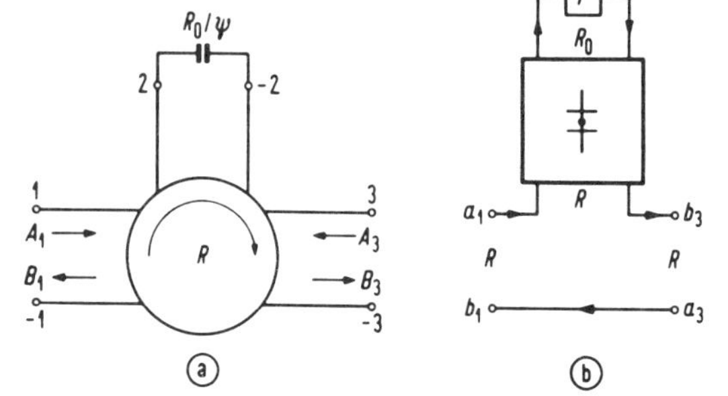

Fig. 23. (a) A unilaterial first degree all-pass section, (b) the corresponding wave two-port.

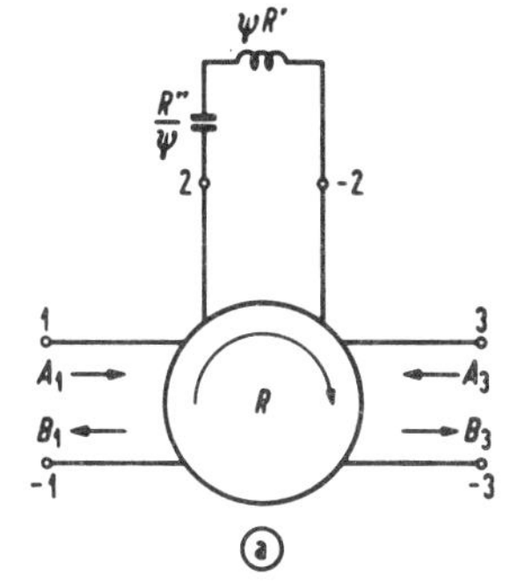


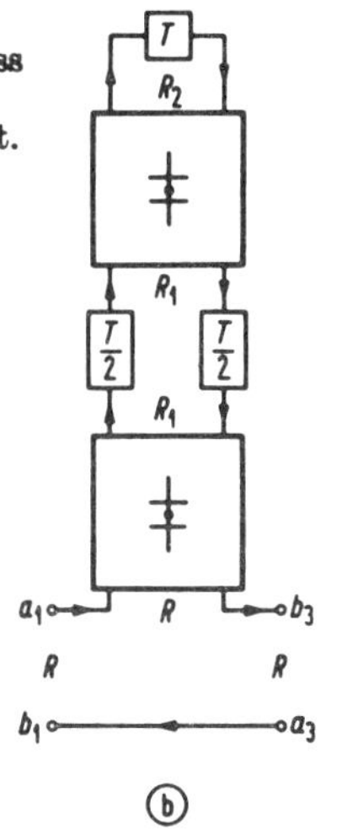


Fig. 24.
(a) A unilateral second degree all-pass section,
(b) the corresponding wave two-port.

tion of unilateral all-pass sections of any degree by using correspondingly higher degree reactances. Finally, bilateral all-pass sections can always be obtained by using four-port rather than three-port circulators.

An all-pass of arbitrary degree can also be obtained by cascading first and second degree sections. As all individual sections must then have the same circulation resistance $R$, no adaptors will be required between these sections. Finally, if the all-pass to be realized is unilateral and if we connect to it a source with a resistance $R$ as well as a load resistance also equal to $R$, the results of Figs. 11 and 4 show that no signal flows in the return path so that it may simply be dropped. An example is shown in Fig. 25, where $R_1$ and $R_2$ are still given by eq. (40).

### 5.4. *Realization of filter circuits*

From the theory of microwave transmission line filters we know that it is always possible to build

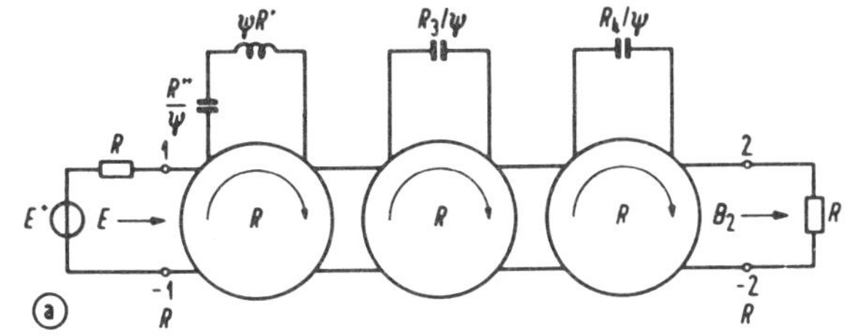


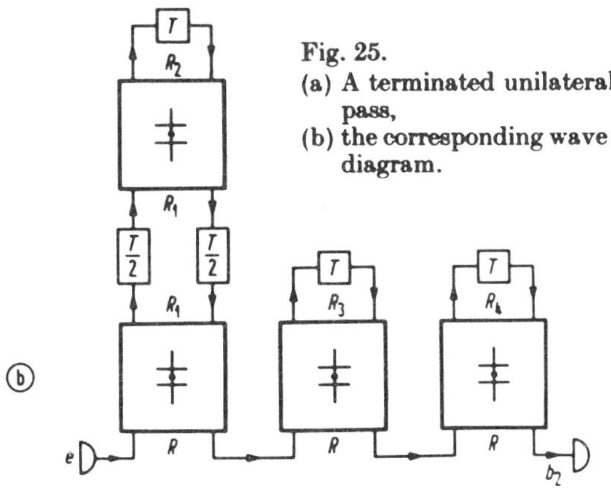


Fig. 25.
(a) A terminated unilateral all-pass,
(b) the corresponding wave flow diagram.

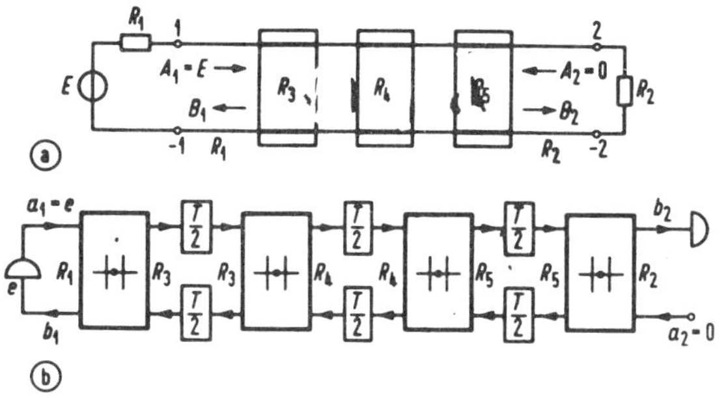


Fig. 26. (a) Filter built as a cascade of three unit elements, (b) the corresponding wave flow diagram.

filters by cascading a certain number of unit elements with different characteristic impedances [21]. Such filters always lead to realizable wave flow diagrams. If $n$ is the number of unit elements, the number of multipliers required is $n+1$. An example is shown in Fig. 26.

Of particular interest in microwave theory are more general filters which can be derived from a reactance two-port in cascade with a sufficient number of unit elements. By making use of the equivalence transformations due to Kuroda and Levy [13], [21]–[25], such structures can be transformed into new cascade structures, henceforth called *unit element ladder structures*, in which unit elements alternate with either series or shunt reactance arms. In fact, in microwave circuits, one is mainly interested for technological reasons in that subclass of such structures in which only shunt reactance arms appear. For our purpose, however, no such restriction exists since, as we have seen in section 4, it is just as easy to imitate series connections as shunt connections. With other words, any unit element ladder structure leads to a realizable wave flow diagram. An example corresponding to a band-pass filter is shown in Fig. 27, $R_7$ and $R_8$ being given by equations similar to eq. (40).

Let us designate by $m$ the number of reactive elements and by $n$ the number of unit elements in the unit element ladder structure; the number of multipliers required is then equal to $m+n+1$. The fact that we thus have to spend one multiplier per unit element does not simply represent an expense for which no filtering effect is obtained. Indeed, when designing the original filter we may always fully make use of the filtering capabilities inherent to unit elements themselves [13], [14] (cfr. Fig. 26). The number $m+n+1$ is usually also equal to the number of independent coefficients in the transfer function (transmittance) of the network. In this sense we may say that the number of multipliers is canonic.

Attention should also be drawn to the fact that it is nowhere necessary to design first explicitly the first mentioned structure from which the unit element ladder structure has been derived. In fact the synthesis process can be conducted in such a way that this latter structure is immediately obtained [13], [14]. It is not possible within the limitations of this paper to enter more deeply into the design aspects of unit element ladder structures.

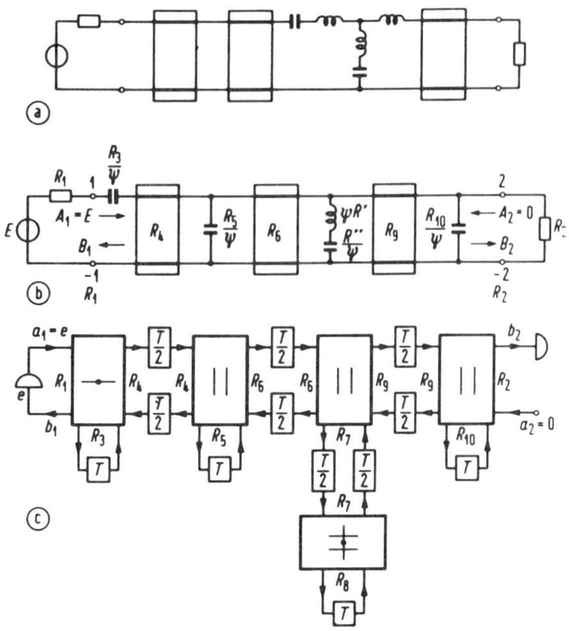

Fig. 27. (a) Band-pass filter composed of a reactance two-port in cascade with three unit elements,
(b) the corresponding unit element ladder structure,
(c) wave flow diagram derived from (b).

*5.5. Additional comments*

1. As a unit element requires two delays $T/2$ while an inductance or a capacitance requires only one delay $T$, it may seem that a unit element is more expensive to realize. In fact, however, a delay $T$ must always be realized by a tandem connection of two delays $T/2$ in order to allow an intermediate storage; this is necessary since the writing and reading processes cannot simultaneously be done at the same storage position. For a unit element, however, the writing and reading processes never coincide since they are always at a time $T/2$ apart; hence a subdivision of the delay $T/2$ is not required. We conclude from this that for a digital filter having $m$ reactive elements and $n$ unit elements, the number of delays $T/2$ (i.e. the number of storage positions) which are required is equal to $2(m+n)$.
2. In section 4 we have seen that the resistances remain positive as long as the $\alpha$, $\alpha_k$, and $\beta_k$ satisfy some quite uncritical conditions. Hence, except for extreme deviations of the multipliers, the circuit remains passive; in particular, it cannot normally become unstable, and its very low sensitivity to element variations discussed in the introduction normally remains valid.
3. So far, we have always assumed that the resistance parameters corresponding to the various elements are always positive. In fact, the various equations which we have obtained remain valid even if any number of these resistances are negative. If is true that in this case our argument for guaranteeing the low passband sensitivity is no longer necessarily valid. It is also known, however, that in $LC$-structures, in particular in $LC$-ladder structures, negative elements may appear which, when remaining within certain ranges, leave the overall structure passive. Note that inequalities such as
$$-1<\alpha<1, \quad 0<\alpha_k<2, \quad 0<\beta_k<2,$$
which were always valid before, do not have to hold anymore if some of the $R_k$ are negative, and the same is then obviously also true for conclusions derived from such inequalities.
4. As a unit element requires one multiplier and two delays $T/2$, just like an inductance or a capacitance, its cost is essentially the same too. It is well-known, however, that the filtering capability of a cascaded unit element is smaller than that of a shunt or series reactive element. Hence, one might be interested in economizing the multiplier of a unit element without having to give up its delays since these are necessary to make the wave flow diagram realizable. This goal can be achieved by forcing the characteristic resistance of the unit element to be equal to one of the other resistances belonging to an adjacent adaptor. We have seen, indeed, in subsections 4.2 and 4.3 that such a procedure enables us to reduce the number of multipliers which are required. It is true, of course, that this amounts to renouncing completely the filtering capabilities of the unit elements but such a procedure may nevertheless well be economical. It is also true that the classical filter design methods may then not be used any more in a straightforward way since such methods will in practice never by themselves furnish the required constraint. Hence, some optimization technique will have to be used in the design process.
5. In ladder structures, resonant frequencies corresponding to attenuation poles are determined by the product of two element values, and the same will usually be true for the wave digital filters described above. Due to the limited number of digits available for the coefficients, only a discrete pattern of resonant frequencies can, however, be achieved. Nevertheless, a precise tuning of the attenuation poles can be obtained by choosing for the filter computation only such values of the resonant frequencies which appear in this discrete pattern of realizable values.
6. In addition to the elements studied in section 3 it is also possible to realize quasi-reciprocal lines (QUARL's) and the related gyrating and circulating devices known from resonant-transfer theory [15]—[17]. It does not seem, however, that this would give us any advantage since the gyrator and the circulators discussed in subsection 3.2 offer even greater flexibility.
7. In order to realize a digital filter imitating a $LC$-filter one might also think of using the following procedure which apparently is much simpler than the one described above. We start from a given terminated lossless two-port comprising $m$

inductances and capacitances plus a certain number of frequency independent elements. If we remove these inductances and capacitances as well as the terminations, we are left with a frequency independent $(m+2)$-port. To each port, we attribute as port resistance the resistance of the corresponding element or source. We then express the $m+2$ reflected waves in terms of the $m+2$ incident waves. This determines an $(m+2)$-port adaptor, which in general is neither of the parallel nor the series type. Nevertheless, it does not contain any inner feedback loops so that we obtain a realizable wave flow diagram by connecting to its wave ports the wave one-ports of the individual elements and the source. An obvious disadvantage, however, is the fact that the $(m+2)$-port adaptor will in general require the large number of $(m+2)^2$ multipliers. Nevertheless, such a procedure may be used to synthesize certain smaller parts of a larger structure. In principle, the procedure could also be used if the two-port also contains resistances, although this should be of little practical importance.

8. The two-ports which we have discussed all concern filters which are resistively terminated at both ports. It should be clear, however, that the same principles can be used to design digital filters imitating classical open-circuit or short-circuit filters.
9. Similar to what is true for ordinary digital filters [26], wave digital filters can be transformed into corresponding digital $N$-path filters by simply adding a delay $(N-1)T$ to every delay $T$ and a delay $(N-1)T/2$ to every delay $T/2$, while at the same time keeping the rate of operation of the filter at $F = 1/T$.

## 6. Conclusions

We have seen that digital filters can be built which in many respects are similar to reactance ladder filters, or more precisely to unit element ladder filters. It is to be expected that digital filters of this type have similar low passband and stopband sensitivity to element (multiplier) variations as the aforementioned filters. From this point of view, they should thus be considerably better than conventional digital filters. Their signal flow diagram is based on the use of wave quantities.

Depending on the design procedure these wave digital filters require a moderately higher number (if a simple design procedure as for classical filters is used) or about the same number (if a certain more complicated design procedure is used) of multipliers as conventional digital filters. The number of adders will, however, be larger, and the same may be true for the number of delay elements (storage places); this is no serious disadvantage because the realization of adders and delays is comparatively cheap. In any case, due to the reduced sensitivity, the multipliers can be built with fewer digits and this should in many cases lead to a reduced overall cost.

These advantages of wave digital filters have been inferred so far only from the analogy of their structures with those of conventional resistively terminated ladder filters. Work should now be done to furnish theoretical and experimental evidence, in particular with the aid of computer simulation, of the extent to which our conclusions are valid. In particular, it should be examined which among the various alternate solutions are more advantageous than others.

(Received July 23rd, 1970.)

### References

[1] Kaiser, J. F., Digital filters; published in F. F. Kuo and J. F. Kaiser, System analysis by digital computer. J. Wiley & Sons, New York 1966, pp. 218–285.

[2] Special Issue on Digital Filters. Transact. Inst. Elect. Electron. Engrs. AU-**16** [1968], No. 3.

[3] Gold, G. and Rader, C. M., Digital processing of signals. McGraw-Hill Book Co., New York 1969.

[4] Kuntz, W., Schüssler, W. and Winkelnkemper, W., Untersuchungen über Eigenschaften, Entwurf und Realisierung digitaler Filter. Ausgewählte Arbeiten über Nachrichtensysteme, Nr. 10, Erlangen 1969.

[5] Vich, R., Synthesis of digital filters. Proceedings of the Summer School on Circuit Theory, Czechoslovak Academy of Sciences, Prague 1968.

[6] Sandberg, I. W., Floating-point round-off accumulation in digital filter realization. Bell Syst. tech. J. **44** [1967], 1775–1791.

[7] Knowles, J. B. and Olcayto, E. M., Coefficient accuracy and digital filter response. Transact. Inst. Elect. Electron. Engrs. CT-**15** [1968], 31–41.

[8] Weinstein, C. and Oppenheim, A. V., A comparison of roundoff noise in floating point and fixed point digital filter realizations. Proc. Inst. Elect. Electron. Engrs. **57** [1969], 1181–1183.

[9] Liu, B. and Kaneko, T., Error analysis of digital filters realized with floating-point arithmetic. Proc. Inst. Elect. Electron. Engrs, **57** [1969], 1735–1747.

[10] Mildenberger, O., Die Genauigkeit der Lösung von Differenzengleichungen in Abhängigkeit von der Stellenzahl des verwendeten Rechners (to be published).

[11] Orchard, H. J., Inductorless filters. Electron. Letters **2** [1966], 224–225.

[12] Holmes, W. H., Heinlein, W. E. and Grützmann, S., Sharp-cutoff low-pass filters using floating gyrators. J. Solid-State Circuits, Inst. Elect. Electron. Engrs. SC-**4** [1969], 38–50.

[13] Fraiture, L. and Neirynck, J., Theory of unit element filters. Revue HF **7** [1969], 325–340.

[14] Fettweis, A., Cascade synthesis of lossless two-ports by transfer matrix factorization. NATO Advanced Study Institute on Network Theory, Knokke, Belgium, 1969.

[15] Fettweis, A., Theory of resonant-transfer circuits; published in G. Biorci, Network and switching theory. Academic Press, New York 1968.

[16] Fettweis, A., Theory of resonant-transfer circuits and their application to the realization of frequency-selective networks. Proceedings of the Summer School on Circuit Theory, Czechoslovak Academy of Sciences, Prague 1968.

[17] Fettweis, A., Some properties of pulse impedances and pulse impedance matrices. AEÜ **24** [1970], 506–512.

[18] Fjällbrant, T., Digital filters with a number of shift sequences in each pulse repetition interval. IEEE International Symposium on Circuit Theory, San Francisco, December 8–10, 1969.

[19] Belevitch, V., Classical network theory. Holden-Day, San Francisco 1968.

[20] Richards, P. I., Resistor transmission-line circuits. Proc. Inst. Radio Engrs. **36** [1948], 217–220.

[21] Ozaki, H. and Ishii, J., Synthesis of transmission-line networks and the design of UHF filters. Transact. Inst. Radio Engrs. CT-**2** [1955], 325–336.

[22] Ikeno, N., Principles of designing distributed networks. Polytechnic Institute of Brooklyn, Report No. PIBMRI-1003-62, October 1962.

[23] Wenzel, R. J., Exact design of TEM microwave networks using quarter-wave lines. Transact. Inst. Elect. Electron. Engrs. MTT-**12** [1964], 94–111.

[24] Levy, R., A general equivalent circuit transformation for distributed networks. Transact. Inst. Elect. Electron. Engrs. CT-**12** [1965], 457–458.

[25] Horton, M. C. and Wenzel, R. J., Realization of microwave filters with equal ripple response in both pass and stop bands. Proceedings of the Symposium on Generalized Networks. Polytechnique Press of the Polytechnic Institute of Brooklyn, Brooklyn, N.Y., 1966, pp. 257–288.

[26] Fettweis, A. and Wupper, H., A solution to the balancing problem in $N$-path filters. Transact. Inst. Elect. Electron. Engrs. CT-**18** [1971] (in print).

[27] Fettweis, A., Entwurf von Digitalfiltern in Anlehnung an Verfahren der klassischen Netzwerktheorie. NTZ-Report der NTG-Fachtagung „Analyse und Synthese von Netzwerken", Stuttgart, 15./16. 9. 1970, S. 17–20.

# DIGITAL FILTERS WITH TRUE LADDER CONFIGURATION

AXEL SEDLMEYER AND ALFRED FETTWEIS

*Ruhr-Universität Bochum, Lehrstuhl für Nachrichtentechnik, Federal Republic of Germany*

## SUMMARY

A drawback of all ladder digital filters discussed hitherto is the requirement for including unit elements to separate consecutive ladder arms. It is shown how this requirement can be circumvented. By means of the new method, any LC-ladder filter can very simply be translated into a corresponding digital structure requiring only as many multipliers as corresponds to the number of degrees of freedom in the original LC-filter.

## INTRODUCTION

In previous papers,[1-11] the realization of wave digital filters and their advantages have been discussed. Among these filters, a certain subclass, which has also been called ladder digital filters, is of particular importance.

The ladder digital filters which have been analysed so far have one disadvantage: in order to satisfy the realizability condition for the resulting signal-flow diagram,[1,3-5] direct interconnection of adaptors should be avoided. For the equivalent unit element filter, from which the wave digital filter is derived, this implies that each series or shunt branch has always to be followed by a unit element before the next series or shunt branch may appear. Obviously, filters of that type do not really correspond to what are commonly considered to be true ladder filters.

It is true that the unit elements introduced above can be used perfectly well for improving the filtering capabilities,[12,13] but in this case they do not generally contribute as much as a regular L or C element. On the other hand, it has been shown how these unit elements can be introduced into the structure without requiring additional multipliers,[3] but this excludes the use of conventional filter design methods and requires very lengthy direct optimization procedures to be employed.

It is the purpose of this paper to show how these restrictions can quite easily be overcome. As we shall see, special, although very simple, choices of certain available parameters actually allow arbitrary adaptors to be directly interconnected. This way, any LC-ladder filter can very simply be translated into a corresponding digital structure. The resulting digital filters thus correspond to true ladder filters in the strict conventional sense. All known filter design methods, including those by means of tables, immediately become available for the design of these digital filters. The number of multipliers required for such a digital filter is always equal to the number of degrees of freedom in the original LC-filter.

## DIRECT INTERCONNECTION OF ADAPTORS

An *adaptor*[1,3-5] has $n$ ports, with $n \geqslant 2$, which we may number $\nu = 1, 2, \ldots, n$. To each port $\nu$, there correspond an arbitrary port resistance $R_\nu$, which is a positive constant, an incident signal (wave) $a_\nu$, and a reflected signal (wave) $b_\nu$. Instead of the instantaneous signal quantities $a_\nu$ and $b_\nu$, the corresponding steady-state signals $A_\nu$ and $B_\nu$ may also be used; for reasons of simplicity, we restrict ourselves hereafter to considering instantaneous signals only.

A *parallel adaptor*[1,3-5] serves to interconnect $n$ ports in parallel [Figure 1(a)]. It is described by the equations

$$b_\nu = a_0 - a_\nu \tag{1a}$$

$$a_0 = \sum_{\nu=1}^{n} \alpha_\nu a_\nu \tag{1b}$$

*Received 21 July 1972*

Reprinted with permission from *Int. J. Circuit Theory and Appl.*, vol. 1, no. 1, pp. 5–10, Mar. 1973. Published by John Wiley & Sons Ltd., Baffins Lane, Chichester, Sussex, England P019 IVD.

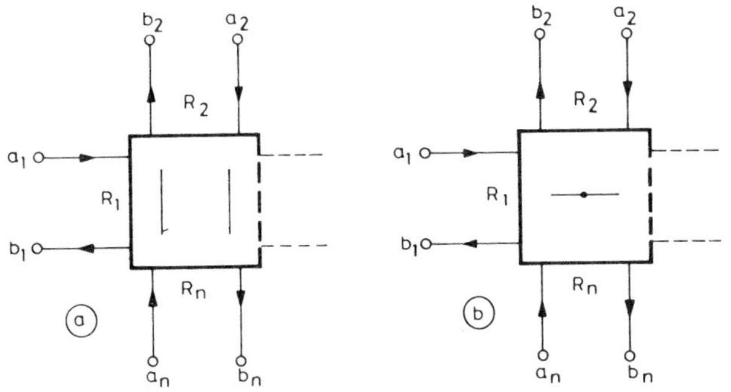


Figure 1. Schematic representation of an $n$-port parallel adaptor (a) and an $n$-port series adaptor (b)

where

$$\alpha_\nu = 2G_\nu/G \quad (2a)$$

$$G_\nu = 1/R_\nu \quad (2b)$$

$$G = G_1 + G_2 + \ldots + G_n \quad (3)$$

A *series adaptor*[1,3–5] serves to interconnect $n$ ports in series [Figure 1(b)]. It is described by the equations

$$b_\nu = a_\nu - \alpha_\nu a_0 \quad (4a)$$

$$a_0 = a_1 + a_2 + \ldots + a_n \quad (4b)$$

where

$$\alpha_\nu = 2R_\nu/R \quad (5a)$$

$$R = R_1 + R_2 + \ldots + R_n \quad (5b)$$

In both cases, the $\alpha_\nu$ satisfy the relations

$$\alpha_\nu > 0 \quad (6a)$$

$$\alpha_1 + \alpha_2 + \ldots + \alpha_n = 2 \quad (6b)$$

Obviously, each reflected signal $b_\nu$ depends, in general, on all the signals incident to the corresponding adaptor, and thus in particular on that $a_\nu$ whose subscript $\nu$ is the same as for the $b_\nu$ under consideration. Hence, in the signal-flow diagrams corresponding to equations (1) and (4), a direct path leads, in general, from any incident signal $a_\nu$ to the corresponding reflected signal $b_\nu$.

Assume then that two arbitrary adaptors are directly interconnected, as shown in Figure 2 (the symbols indicating the particular nature of the adaptors have been omitted because they are irrelevant for our present purpose). According to the known rules[1,3–5] which apply for any interconnection of wave $n$-ports, the port resistances of the two interconnected ports must be the same, and the reflected wave of the one port must become the incident wave of the other. In particular, a closed loop not containing any delay would thus normally appear (Figure 2), whence the resulting signal-flow diagram would be unrealizable, i.e. no

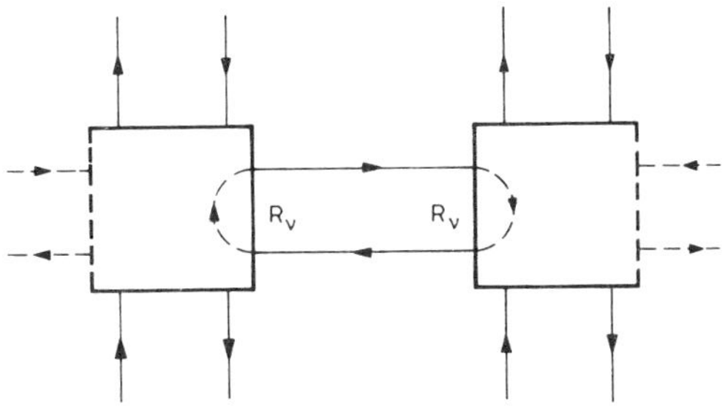


Figure 2. Direct interconnection of two ports of two arbitrary adaptors

properly ordered sequence for the actually required computations could be given.[1,3–5,14] This drawback has been overcome in the earlier publications[1–5] by requiring a delay to be inserted, possibly corresponding to a cascaded unit element as discussed in the introduction.

A much simpler solution, however, arises as follows. Assume that for an adaptor, one of the $\alpha_\nu$ may be chosen equal to unity, say

$$\alpha_n = 1$$

which, according to equations (2), (3) and (5), implies for the parallel adaptor

$$G_n = G_1 + G_2 + \ldots + G_{n-1} \tag{7}$$

and for the series adaptor

$$R_n = R_1 + R_2 + \ldots + R_{n-1} \tag{8}$$

In this case, for the parallel adaptor, equation (1a) can, for $\nu = n$, be written in the form

$$b_n = \alpha_1 a_1 + \alpha_2 a_2 + \ldots + \alpha_{n-1} a_{n-1}$$

and, for the series adaptor, equation (4a), for $\nu = n$, in the form

$$b_n = -a_1 - a_2 - \ldots - a_{n-1}$$

Consequently, in both cases, the expression for $b_n$ becomes independent of $a_n$, i.e. in the signal-flow diagram of the adaptor, the direct path normally leading from $a_n$ to $b_n$ is interrupted. Hence, the port $n$ may now be connected to a port of another adaptor without danger for a forbidden closed loop to arise. In particular, for this last port, no requirement of the type just discussed does exist.

The conditions (7) and (8) clearly satisfy the requirements $R_n > 0$ and $G_n > 0$ as soon as the remaining $R_\nu$ and $G_\nu$ are positive. Furthermore, equation (6) may now be replaced by

$$\alpha_\nu > 0, \qquad \alpha_1 + \alpha_2 + \ldots + \alpha_{n-1} = 1$$

The last relation still allows one of the $\alpha_1$ to $\alpha_{n-1}$ to be expressed in terms of the other, i.e. the corresponding multiplier to be eliminated, as is also the case for the adaptors in the general theory.[1–5] The port corresponding to the eliminated multiplier may again be called the *dependent port*. An example of a three-port series adaptor with $\alpha_3 = 1$, for which port 2 is the dependent port, is shown in Figure 3.

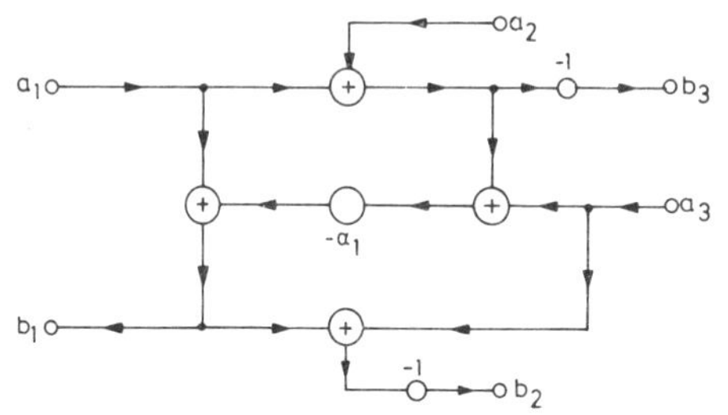


Figure 3. Signal-flow diagram corresponding to a three-port series adaptor for which port 2 is the dependent port and $\alpha_3 = 1$

It may seem at first that a requirement such as (7) or (8) freezes one degree of freedom which must be available for realizing the element values of the original filter. In reality, however, if two ports of two adaptors are interconnected, no further circuit or element, clearly, will be connected to these ports (cf. the rules for the realization of circuits.[1,3–5] Hence, the common value of their port resistances is not imposed by the original filter from which the digital filter is to be derived and can thus be chosen freely. In particular, it may always be chosen according to equation (7) for a parallel adaptor and according to equation (8) for a series adaptor.

## REALIZATION OF TRUE LADDER STRUCTURES

From the discussion given in the previous section, it is clear that adaptors may be interconnected in a large variety of ways. For filter applications, the most important of these ways of interconnection are those which correspond to ladder structures. For illustration, we may thus restrict ourselves to this case.

Circuits corresponding to ladder structures are obtained if we connect alternately, in cascade, any number of parallel and series adaptors, as shown schematically in Figure 4. For the interconnections to be permissible, it is then sufficient that for any two interconnected ports the $\alpha_\nu$ corresponding to one of them be equal to unity; in the cascade arrangement, obviously, this can always be achieved.

For all adaptors in Figure 4, the number of ports not entering the cascading operation is arbitrary, although it will usually be 1 or 2, or at most 3 (in Figure 4, two such ports are always shown). To each of

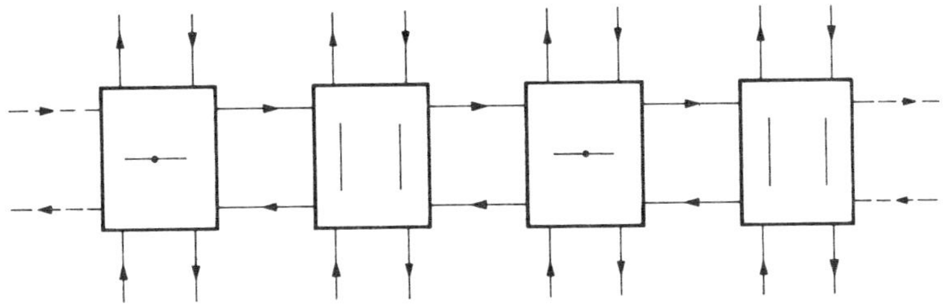

Figure 4. Alternating cascade connection of series and parallel adaptors to form the full equivalent of a ladder structure

these ports, we may connect a simple delay $T$ (corresponding to a capacitance,[1–5] a delay $T$ with a sign inversion (corresponding to an inductance[1–5]), or a variety of other arrangements, including arrangements comprising further adaptors, e.g. such ones which correspond to arbitrary reactances;[1,3–5] even arrangements beginning with an additional adaptor are not excluded.

As an example, we consider the band-pass LC-filter shown in Figure 5(a). For each element in Figure 5(a), the corresponding.impedance value is given, the parameter $\psi$ corresponding to the equivalent complex

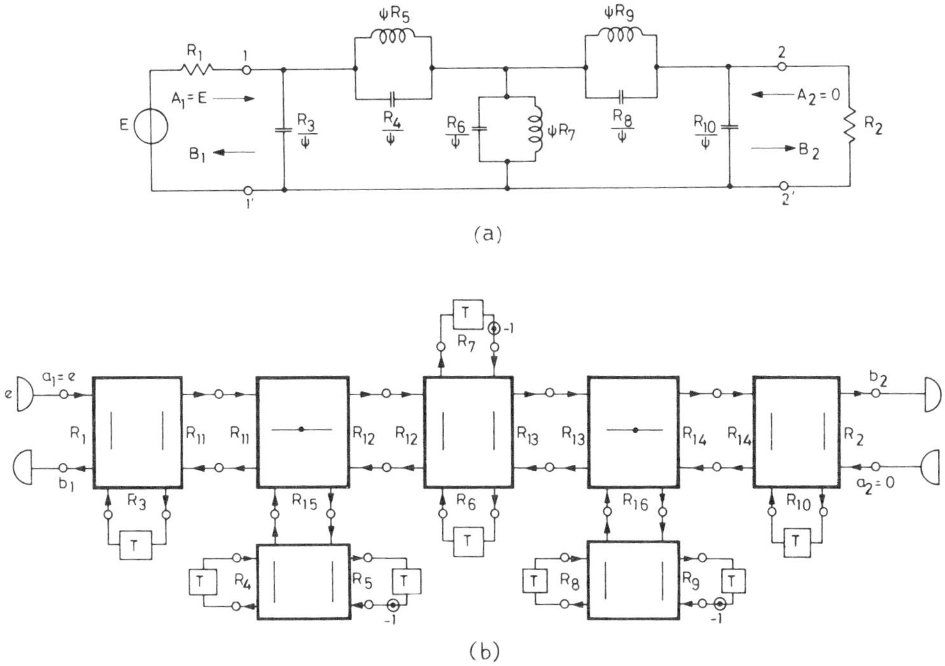


Figure 5. (a) A band-pass ladder LC-filter. (b) The corresponding digital filter

frequency defined by

$$\psi = \tanh(pT/2)$$

where $p$ is the actual complex frequency and $T$ the operating period of the circuit. A ladder digital filter (realizable signal-flow network) corresponding to Figure 5(a) is shown in Figure 5(b). For $R_7 = \infty$, this filter reduces to a low-pass.

The resistances $R_1$ to $R_{10}$ are imposed by the design of the LC-filter. There are various possibilities of choosing the resistances $R_{11}$ to $R_{16}$ in accordance with the requirements corresponding to equations (7) and (8). If we choose, at each interconnection of adaptor ports, the left-hand port to correspond to $\alpha_v = 1$, we obtain

$$G_{11} = G_1 + G_3, \quad R_{12} = R_{11} + R_{15}, \quad G_{15} = G_4 + G_5$$

$$G_{13} = G_6 + G_7 + G_{12}, \quad R_{14} = R_{13} + R_{16}, \quad G_{16} = G_8 + G_9$$

and if we choose the right-hand port,

$$G_{14} = G_2 + G_{10}, \quad R_{13} = R_{14} + R_{16}, \quad G_{16} = G_8 + G_9$$

$$G_{12} = G_6 + G_7 + G_{13}, \quad R_{11} = R_{12} + R_{15}, \quad G_{15} = G_4 + G_5$$

where in all cases, $G_v = 1/R_v$.

The total number of adaptor ports in Figure 5(b) is 22. In order to arrive at the total number of multipliers we have to deduct from this the number of dependent ports, i.e. 7 (equal to the number of adaptors), and the number of ports whose $\alpha_v$ is equal to unity, i.e. 6 (equal to the number of direct interconnections between adaptors). Hence, the structure of Figure 5(b) requires a total of $22-7-6 = 9$ multipliers. This corresponds precisely to the number of degrees of freedom in the original LC-structure of Figure 5(a) (number of resistive parameters minus one). These results are easily checked to see that they hold generally.

The validity of the theory has been verified by a simple numerical example, in the frequency and time domains, and detailed numerical results are planned to be published.

## COMPLEMENTARY REMARKS

In this paper, only true ladder filters have been discussed. It should be clear, however, that the present procedures can be used in quite arbitrary combinations with those described in earlier publications.[1–5] In other words, digital filters can be designed in a way that certain substructures inside them are realized as true ladder structures while others include interconnecting unit element elements. Hence, a large variety of mixed structures are available. For all these, the basic properties of pseudo-passivity, pseudo-losslessness and stability, and thus also the excellent properties with respect to attenuation sensitivity and roundoff noise,[1–11] obviously remain valid. Further studies should show whether among all possible structures thus available, certain ones lead to even better results than others. Note that requirements such as $\alpha_v > 0$ are not strictly necessary, as has also been discussed earlier.[3]

## REFERENCES

1. A. Fettweis, 'Entwurf von Digitalfiltern in Anlehnung an Verfahren der klassischen Netzwerktheorie', *NTZ-Report of the NTG-Fachtagung, 'Analyse und Synthese von Netzwerken,'* Stuttgart, 1970.
2. J. A. Bingham, 'A new type of digital filter with a reciprocal ladder configuration', *IEEE Int. Symp. Circuit Theory*, Digest of Technical Papers, 129–130 (1970).
3. A. Fettweis, 'Digital filter structures related to classical filter networks', *Arch. Elektronik Übertragungstechnik*, **25**, 79–89 (1971).
4. A. Fettweis, 'Some principles of designing digital filters imitating classical filter structures', *IEEE Trans. Circuit Theory*, CT-**18**, 314–316 (1971).
5. A. Fettweis, 'Wave digital filters', *Proc. Summer School Circuit Theory*, Institute of Radio Engineering and Electronics, Czechoslovak Academy of Sciences, 11-1 to 11-10 (1971).
6. A. Fettweis, 'On the connection between multiplier word length limitation and roundoff noise in digital filters', *IEEE Trans. Circuit Theory*, CT-**19**, 486–491 (1972).
7. A. Fettweis, 'Pseudo-passivity, sensitivity, and stability of wave digital filters', *IEEE Trans. Circuit Theory*, CT-**19** (1972).
8. A. Fettweis, 'Roundoff noise and attenuation sensitivity in digital filters with fixed-point arithmetic', *IEEE Trabs. Circuit Theory*, CT-**20** (1973).

9. R. E. Crochiere, 'Digital ladder structures and coefficient sensitivity', *IEEE Trans. Audio Electroacoustics*, AU-**20**, 240–246 (1972).
10. A. Chu and R. E. Crochiere, 'Comments and experimental results on optimal digital ladder structures', *IEEE Trans. Audio Electroacoustics*, AU-**20**, 317–318 (1972).
11. A. Fettweis, 'Scattering properties of wave digital filters', *Proc. Seminar Digital Filters*, 1–8, Florence, 1972.
12. L. Fraiture and J. Neirynck, 'Theory of unit-element filters', *Revue HF*, **7**, 325–340 (1969).
13. A. Fettweis, 'Cascade synthesis of lossless two-ports by transfer matrix factorization', *Network Theory* (Ed. R. Boite), 43–103, Gordon and Breach, New York, 1972.
14. B. Gold and C. M. Rader, *Digital Processing of Signals*, 151–152, McGraw-Hill, New York, 1969.

# Pseudopassivity, Sensitivity, and Stability of Wave Digital Filters

ALFRED FETTWEIS

*Abstract*—Two basic properties of wave digital filters, their property of requiring fewer digits for the multiplier coefficients (to which the property of roundoff noise reduction is related) and their stability, had so far only been inferred from certain analogies with unit element filters. These properties are now proved in a direct way by introducing the concepts of pseudopower, pseudopassivity, and pseudolosslessness, corresponding to the familiar concepts of power, passivity, and losslessness in classical circuits.

## I. Introduction

The structures of wave digital filters have been obtained by analogy with unit element filters and resonant-transfer filters [1]–[6]. Wave digital filters can be designed in such a way that they imitate the behavior of lossless filters between resistive terminations. In this way, it is possible for them to achieve the extremely low sensitivity of the attenuation characteristic with respect to parameter variations, which is known to be of great practical importance for the realization of classical filters.

Although tolerance problems as such do not exist for digital filters, there is even there a definite interest for obtaining structures with low sensitivities to parameter variations [7]–[9]. The reason for this is twofold. First, structures with low sensitivity are less affected by coefficient truncation, i.e., shorter word lengths suffice for the realization of their multiplier coefficients. Second, as there exists a relation between the sensitivity of the attenuation with respect to multiplier variations and the roundoff noise at the output of a digital filter [5], [10], any improvement in this sensitivity results in a corresponding roundoff noise improvement.

In previous papers [1]–[4], the basic sensitivity property for passband frequencies had simply been inferred by making use of appropriate analogies with the theory of the unit element filters from which the wave digital filters are derived, and the same is true concerning the stability considerations. The purpose of the present paper is to establish direct proofs that do not require any such analogies. Such direct proofs may also be of considerable importance for further research; indeed, from the insight thus gained it may be possible to find some answer to the question whether or not other types of structures exist which have similar or even better properties than wave digital filters.

Manuscript received November 19, 1971; revised May 9, 1972.
The author is with the Department of Electrical Engineering, University of Bochum, Bochum, Germany.

Digital filters are a special case of a more general type of circuits called signal-flow networks [11]. In particular, wave digital filters are a special case of wave-flow networks. Although we shall usually speak of wave digital filters, it should be clear that most results remain valid for the corresponding more general wave-flow networks. Finally, it should be clear that the results of Section II are directly analogous to corresponding results known in scattering parameter theory of conventional $n$-ports [12], [13].

## II. Conservation of Pseudopower in Wave Digital Filters

### *A. Definition of Pseudopower*

The building blocks (elements, sources, and adaptors) of wave digital filters are wave multiports or, more specifically, wave $n$-ports, with $n=1, 2, \cdots$. At each wave port $\nu=1, 2, \cdots, n$ of a wave $n$-port, an incident wave $a_\nu$ is flowing into it and a reflected wave $b_\nu$ out of it (Fig. 1); to the instantaneous wave quantities $a_\nu$ and $b_\nu$ correspond the steady-state wave quantities $A_\nu$ and $B_\nu$. Furthermore, to each wave port $\nu$ is attributed a port resistance $R_\nu$ that is a *positive* constant (Fig. 1). For a detailed description of the most important wave multiports, we refer to the literature [1], [3]–[5].

Let us now define the *instantaneous pseudopower* transmitted through a wave port $\nu$ of port resistance $R_\nu$ by means of

$$p_\nu = (a_\nu^2 - b_\nu^2)G_\nu \tag{1}$$

and the corresponding *steady-state pseudopower* for steady-state conditions at angular frequency $\omega$ by means of

$$P_\nu = (|A\nu|^2 - |B_\nu|^2)G\nu \tag{2}$$

where

$$G_\nu = 1/R_\nu. \tag{3}$$

In fact, the concept of steady-state pseudopower may even be extended to steady-state conditions at arbitrary complex frequencies

$$p = \sigma + j\omega \tag{4}$$

in which case we still use definition (2) together with (3). The direction in which the pseudopower is assumed to flow is defined to be the one determined by $a_\nu$ and $A_\nu$.

Reprinted from *IEEE Trans. Circuit Theory*, vol. CT-19, pp. 668–673, Nov. 1972.

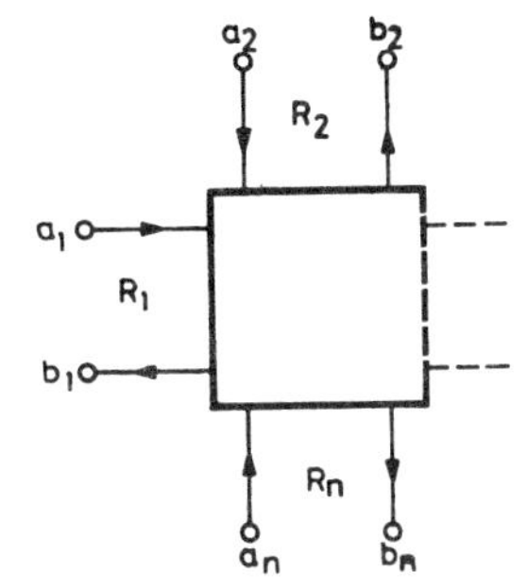


Fig. 1. General representation of a wave multiport (wave $n$-port).

In accordance with these definitions, we also define the instantaneous pseudopower absorbed by a wave $n$-port by means of[1]

$$p = \sum_{\nu=1}^{n} (a_\nu^2 - b_\nu^2)G_\nu \tag{5}$$

and the corresponding steady-state pseudopower at an arbitary complex frequency (4) by means of

$$P = \sum_{\nu=1}^{n} (|A_\nu|^2 - |B_\nu|^2)G_\nu. \tag{6}$$

### *B. Conservation of Pseudopower*

Wave digital filters are obtained by interconnecting wave multiports in a suitable way. The following *rules* are then always respected.

1) The building blocks are interconnected port by port, i.e., the two wave terminals of one wave port are connected with the two wave terminals of precisely one other wave port.

2) The waves corresponding to any two wave terminals that are joined together are compatible, i.e., they flow in the same direction.

3) The port resistances of two wave ports that are interconnected are the same.

We first consider a terminated wave digital filter, i.e., a wave digital filter terminated in such a way that no wave port remains unconnected to some other wave port. We assume it to consist of $K$ building blocks, numbered $k=1, 2, \cdots, K$. Let $p_k$ be the instantaneous pseudopower absorbed by building block $k$, and $P_k$ the corresponding steady-state pseudopower at an arbitrary complex frequency $p$. We then have Theorem 1.

*Theorem 1*

The total instantaneous pseudopower absorbed by all building blocks of a terminated wave digital filter is zero, i.e.,

$$\sum_{k=1}^{K} p_k = 0. \tag{7}$$

[1] The fact that the same symbol is used for the complex frequency as well as for the instantaneous power should not cause confusion.

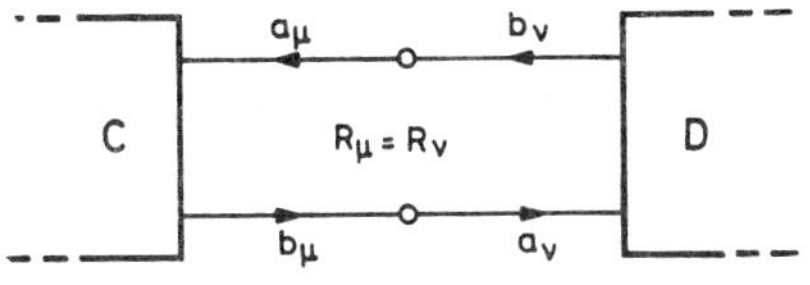


Fig. 2. Interconnection of two wave ports.

Similarly, the total steady-state pseudopower absorbed by all building blocks of a terminated wave digital filter at an arbitrary complex frequency is zero, i.e.,

$$\sum_{k=1}^{K} P_k = 0. \tag{8}$$

This theorem expresses the conservation of instantaneous and steady-state pseudopower, respectively.

*Proof:* The proof of Theorem 1 follows immediately from the above three rules. Indeed, if a port $\nu$ of one building block $C$ is connected to a port $\mu$ of another building block $D$, we have (Fig. 2)

$$R_\nu = R_\mu \qquad a_\nu = b_\mu \qquad b_\nu = a_\mu$$

and thus, according to (1) and (3),

$$p_\nu = -p_\mu.$$

Hence complete cancellation of terms will occur two by two in (7). The proof is similar for (8).

Next we consider wave digital filters that are again obtained from a certain number of building blocks according to the three rules, but for which a certain number $n$ of wave ports are not connected to other wave ports. These $n$ wave ports may be considered to be accessible from the outside so that the total circuit may again be represented as a wave $n$-port as shown in Fig. 1. The pseudopower absorbed by this wave $n$-port may then again be defined by means of (5) and (6), respectively. Under these conditions, we can state Theorem 2.

*Theorem 2*

The pseudopower (instantaneous or steady state) absorbed by a wave digital $n$-port composed of individual building blocks is equal to the sum of the corresponding powers absorbed by these building blocks. The *proof* of this statement is very similar to the proof of Theorem 1 and needs no further explanation.

## III. Pseudopower Absorbed by Various Types of Building Blocks

From now on, we shall restrict ourselves to considering the steady-state pseudopower; for reasons of brevity, we shall simply call it the pseudopower $P$. All defining equations of the building blocks will simply be written in terms of steady-state wave quantities. Unless specified otherwise, the complex frequency $p$ may be arbitrary. For all frequency-independent building blocks, the same relations as those derived hereafter remain valid for the instantaneous pseudopower.

### A. Elements and Sources

For any element, the port resistances at all ports are equal to a same constant $R$. This amounts to writing, in (6),

$$G_1 = G_2 = \cdots = G_n = 1/R. \quad (9)$$

Whenever the behavior of an element or a source is characterized by a resistance parameter $R$, the port resistances are chosen equal to this particular parameter. Hereafter, subscripts for wave quantities will be omitted in case of one-ports.

A *resistance* $R$ is described by the equation of a wave *sink*:

$$B = 0.$$

Hence the pseudopower absorbed by it is given by

$$P = |A|^2/R$$

and we have

$$P > 0, \qquad \text{for } R > 0.$$

An *open circuit* and a *short circuit* are described by

$$B = A \quad \text{and} \quad B = -A$$

respectively, and an *ideal transformer* of ratio $-1/1$ by[2]

$$B_1 = -A_2 \qquad B_2 = -A_1.$$

There is no resistance parameter characterizing these elements. Hence we have in these cases, for arbitrary values of $R$,

$$P = 0. \quad (10)$$

A *gyrator* of gyration resistance $R$ is described by

$$B_1 = A_2 \qquad B_2 = A_1$$

and an $n$-port *circulator* of circulation resistance $R$ by

$$B_1 = A_n, \quad B_2 = A_1, \cdots, \quad B_n = A_{n-1}.$$

Hence (10) also holds in these cases.

A *capacitance* of impedance $R/\psi$ and an *inductance* of impedance $R\psi$, where

$$\psi = \tanh (pT/2)$$

with $T$ as the sampling period, are described by

$$B = Ae^{-pT} \quad \text{and} \quad B = -Ae^{-pT}$$

respectively. Hence we obtain in both cases

$$P = |A|^2(1 - e^{-2\sigma T})/R \quad (11)$$

with $\sigma$ as the real part of $p$ as defined by (4). From (11) we conclude

$$P = 0, \qquad \text{for } \sigma = 0$$
$$P > 0, \qquad \text{for } \sigma > 0 \quad \text{and} \quad R > 0. \quad (12)$$

[2] Ideal transformers of arbitrary ratio are best realized by means of gyrators and thus need not be considered separately.

A *unit element* of characteristic impedance $R$ is described by

$$B_1 = A_2e^{-pT/2} \qquad B_2 = A_1e^{-pT/2}$$

from which

$$P = (|A_1|^2 + |A_2|^2)(1 - e^{-\sigma T})/R. \quad (13)$$

More generally, a QUARL (quasi-reciprocal line) of characteristic impedance $R$ is described by

$$B_1 = A_2e^{-pT_{21}} \qquad B_2 = A_1e^{-pT_{12}}$$

with

$$T_{12} + T_{21} = T \quad (14a)$$
$$T_{12} \geq 0 \quad (14b)$$
$$T_{21} \geq 0 \quad (14c)$$

from which

$$P = |A_1|^2(1 - e^{-2\sigma T_{12}})/R + |A_2|^2(1 - e^{2\sigma T_{21}})/R. \quad (15)$$

We thus conclude from (13), (14b), (14c), and (15) that (12) also applies for unit elements and QUARL's.

A resistive *source* with resistance $R$ and source intensity $E$ ($E$ is a given constant) is described by[3] the equation of a wave source

$$B = E$$

from which we obtain for the pseudopower absorbed by the source

$$P = (|A|^2 - |E|^2)/R.$$

Hence we may have in this case

$$P \gtreqless 0.$$

### B. Adapters

As two-port adapters are special cases of parallel and series adapters, we limit ourselves to considering these latter types.

A *parallel adapter* serves to interconnect, in parallel, $n$-ports with arbitrary port resistances $R_1, R_2, \cdots, R_n$. It is described by

$$B_\nu = A_0 - A_\nu \quad (16a)$$

$$A_0 = \sum_{\nu=1}^{n} \alpha_\nu A_\nu \quad (16b)$$

where

$$\alpha_\nu = 2G_\nu/G \qquad G_\nu = 1/R_\nu \quad (17a)$$
$$G = G_1 + G_2 + \cdots + G_n. \quad (17b)$$

The $\alpha_\nu$ are related by

$$\alpha_1 + \alpha_2 + \cdots + \alpha_n = 2. \quad (18)$$

[3] For the present purpose, it is better to designate, also for a source, the incident wave by $A$ and the reflected wave by $B$, rather than the other way around, as was done in previous papers [1], [3]–[5].

For positive values of the port resistances, the $\alpha_\nu$ are restricted to the range

$$0 < \alpha_\nu < 2.$$

If any of the $\alpha_\nu$ become negative, there is at least one of the port resistances which is negative.

If we express $B_\nu$ and $G_\nu$ in (6) by means of (16a) and (17a), we obtain for the pseudopower absorbed by the adapter

$$P = \frac{G}{2} \sum_{\nu=1}^{n} \alpha_\nu (A_0 A_\nu^* + A_0^* A_\nu - |A_0|^2).$$

Hence making use of (16b) and (18), we obtain

$$P = 0. \tag{19}$$

This result shows that the pseudopower absorbed by a parallel adapter is zero for any values of the real parameters $\alpha_\nu$ satisfying (18).

A *series adapter* serves to interconnect, in series, $n$-ports with arbitrary port resistances $R_1, R_2, \cdots, R_n$. It is described by

$$B_\nu = A_\nu - \beta_\nu A_0 \qquad A_0 = \sum_{\nu=1}^{n} A_\nu$$

where

$$\beta_\nu = 2R_\nu / R \qquad R = R_1 + R_2 + \cdots + R_n.$$

The $\beta_\nu$ are related by

$$\beta_1 + \beta_2 + \cdots + \beta_n = 2. \tag{20}$$

For positive values of the port resistances, the $\beta_\nu$ are restricted to the range

$$0 < \beta_\nu < 2.$$

If any of the $\beta_\nu$ become negative, there is at least one of the port resistances which is negative.

Here, too, it is easy to show by means of the above equations that for the pseudopower absorbed by a series adapter, (19) again applies for all values of the real parameters $\beta_\nu$ satisfying (20).

## IV. Sensitivity to Multiplier Variations and Stability of Wave Digital Filters

### A. Pseudopassivity and Pseudolosslessness

With the availability of the concept of pseudopower, in complete analogy with the concept of power in classical circuit theory, we may also define *pseudopassivity* and *pseudolosslessness* for wave digital filters, again in complete analogy with the concepts of passivity and losslessness in classical networks. In this sense, all building blocks discussed in Section III, except for the source, are *pseudopassive* since for all of them we have $P \geq 0$ for Re $p \geq 0$. In particular, gyrators, circulators, short circuits and open circuits, capacitances, inductances, unit elements, QUARL's, and all adapters are even *pseudolossless* since for all of these we have $P = 0$ at least for $p = j\omega$. A wave digital circuit composed exclusively of pseudopassive or pseudolossless elements will itself be called pseudopassive or pseudolossless, respectively.

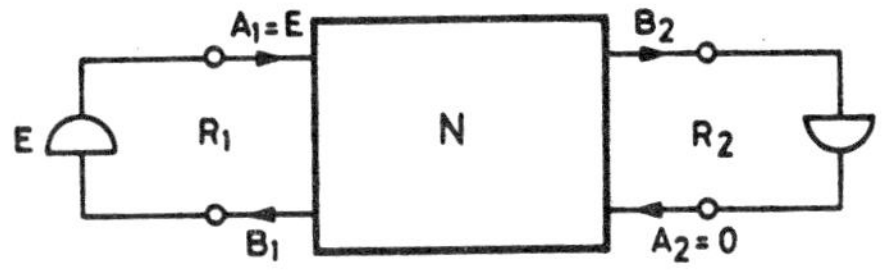


Fig. 3. Wave digital filter consisting of a source-free wave two-port $N$ terminated by a source and a sink.

We now consider a wave digital filter that consists of a source-free wave two-port $N$ with port resistances $R_1$ and $R_2$, respectively, terminated at port 1 by a source of intensity $E$ and resistance $R_1$, and at port 2 by a resistance $R_2$. This is shown in Fig. 3 where, for reasons of simplicity, $R_1$ and $R_2$ have each been indicated only once. The pseudopower $P$ absorbed by $N$ through ports 1 and 2 is given by

$$P = (|A_1|^2 - |B_1|^2)/R_1 + (|A_2|^2 - |B_2|^2)/R_2$$

i.e., since

$$A_1 = E \quad \text{and} \quad A_2 = 0$$

by

$$P = (|E|^2 - |B_1|^2)/R_1 - |B_2|^2/R_2. \tag{21}$$

Let the transfer function from port 1 to port 2 be defined by

$$S_{21} = \frac{B_2}{E} \cdot \sqrt{\frac{R_1}{R_2}}. \tag{22}$$

We assume, of course, that

$$R_1 > 0 \quad \text{and} \quad R_2 > 0. \tag{23}$$

According to (21) and (22), we thus have

$$|S_{21}|^2 = 1 - \frac{|B_1|^2}{|E|^2} - P\frac{R_1}{|E|^2}. \tag{24}$$

Hence if Re $p \geq 0$ and if $N$ is pseudopassive, in which case $P \geq 0$, we conclude that

$$|S_{21}| \leq 1. \tag{25}$$

Let us assume next that the circuit $N$ is pseudolossless and that $p = j\omega$. In this case, we have $P = 0$ so that (24) reduces to

$$|S_{21}|^2 = 1 - |B_1|^2/|E|^2.$$

Hence if the circuit is designed in such a way that we have $B_1 = 0$ for certain values of $\omega$, the limit

$$|S_{21}| = 1 \tag{26}$$

which is imposed by (25), is actually reached for these particular $\omega$.

Properties such as (25) and (26) are familiar for transfer functions in classical circuit theory. Note in particular that the requirement for pseudolosslessness is just as mildly restrictive as the requirement for losslessness in classical circuits. It leaves a wide range of choices which includes in particular all circuit arrangements previously discussed [1]–[5].

*B. Senstivity to Multiplier Variations in Wave Digital Filters*

We are now in a position to give the direct proof for the low-multiplier sensitivity announced in the Introduction, using the same argument that is familiar from the theory of gyrator–capacitor filters [14]–[16]. We assume throughout in this subsection that $p=j\omega$. Let $A$ be the attenuation[4] corresponding to $S_{21}$, i.e.,

$$A = -\ln |S_{21}|.$$

We obtain from (25),

$$A \geq 0 \tag{27}$$

and from (26),

$$A = 0. \tag{28}$$

Assume then that a wave digital filter $N$ (Fig. 3) is pseudolossless and that (28) is valid for a certain value of $\omega$, say $\omega_0$. Consider furthermore one of the multiplier coefficients, say $c$, of the circuit; this multiplier coefficient corresponds to one of the coefficients $\alpha_\nu$ or $\beta_\nu$ in one of the adapters. As we have seen in Subsection III-B, the pseudolosslessness of an adapter is independent of the particular values of its $\alpha_\nu$ or $\beta_\nu$. Let us thus modify the value of $c$ while maintaining $\omega$ still equal to $\omega_0$. In this case, (28) will, in general, no longer be true, but (27) certainly remains valid, whether we actually increase $c$ or decrease it, and this in turn implies that we have

$$\frac{\partial A}{\partial c} = 0, \qquad \text{for } \omega = \omega_0.$$

This property, of course, is no longer valid if $A \neq 0$ for $\omega=\omega_0$. For a well-designed filter, however, we have $A \approx 0$ for any value of $\omega$ inside the passband. Hence we then also have, inside the passband,

$$\frac{\partial A}{\partial c} \approx 0.$$

It is clear that these arguments are valid insofar as it is sufficient to consider first-order sensitivities only, as is usually the case. Their validity, in particular concerning the corresponding reduction in the number of bits for the multiplier coefficients in a digital filter realization, has been confirmed by the numerical results of various workers in the field; of these results, those by Bingham and Crochiere are available in the published literature [2], [17].

[4] Here, too, the fact that the same symbol $A$ is used as a wave quantity and as attenuation should not cause confusion.

*C. Stability of Wave Digital Filters*

We consider again the circuit of Fig. 3 where $N$ is still assumed to be pseudopassive. We have found that for such a circuit the transfer function $S_{21}$ is bounded in the entire right half of the $p$-plane [cf. (25)] and thus has no pole for Re $p \geq 0$. This proves that the wave digital filters under consideration are always *stable.*

In fact, except if $S_{21}$ is independent of $p$, i.e., of $\omega$, the result (25) can even be expressed in the stronger form

$$|S_{21}| < 1, \qquad \text{for Re } p > 0.$$

The proof of this follows again easily, e.g., by means of the maximum modulus theorem, just as in the case of classical networks.

*D. Range of Values of the Multiplier Coefficients*

It may seem at first that the conclusions reached in Subsections IV-B and IV-C are independent of the values of the multiplier coefficients $\alpha_\nu$ and $\beta_\nu$, since the property of pseudolosslessness of the adapters holds for any real $\alpha_\nu$ and $\beta_\nu$ and for any $p$, as long as conditions (18) and (20) remain satisfied. In reality, however, the requirement that the port resistances be positive is not independent of these coefficients.

For the conclusions of Subsection IV-B to be valid, it is sufficient that (23) holds, at least if the circuit $N$ in Fig. 3 is pseudolossless. As the coefficients $\alpha_\nu$ and $\beta_\nu$ are related to the port resistances, the conditions (23) impose certain restrictions, although relatively mild ones, upon these coefficients. These restrictions may become slightly more stringent if there are also resistances (sinks) present inside of $N$, since the corresponding resistance values must also be positive. In any case, it is sufficient, although not necessary, to insure that all $\alpha_\nu$ and $\beta_\nu$ are nonnegative since then *all* port resistances are positive.

The restrictions become more severe if the stability requirement is taken into account. Indeed, the stability proof of Subsection IV-C required essentially that the pseudopower absorbed by any of the building blocks inside of $N$ is nonnegative in the entire right half-plane. Owing to the results of Subsection III-B [cf. (11), (13), and (15)], this requires that the port resistances of all capacitances, inductances, unit elements, and QUARL's, i.e., of all elements implying delays, are nonnegative. This in turn requires that all port resistances in the complete digital filter be positive since due to the basic rule discussed in the earlier references [1], [3]–[5], [18], adapters may never be interconnected without being separated by some delay. Hence the requirement that all $\alpha_\nu$ and $\beta_\nu$ in a wave digital filter be nonnegative is necessary for the ability to apply the stability proof.

The requirements that equations such as (18) and (20) must hold strictly are easily satisfied in a digital realization. In fact, these equations are even automatically fulfilled if one of the ports in each adapter is chosen as the dependent port, as explained in the earlier papers [1], [3]–[5].

It may be noted that since the $\alpha_\nu$ and $\beta_\nu$ depend only on ratios of port resistances we could equally well use only negative resistances rather than positive ones, in which case $P$ in (24) would also change its sign. There is, however, no advantage to be gained this way.

## V. Conclusions

Although the concept of power and the related concepts of passivity and losslessness are meaningless for digital filters, we have seen how closely similar concepts, called pseudopower, pseudopassivity, and pseudolosslessness, can be defined for wave digital filters. It has been shown that the wave digital filters described in earlier papers [1]–[5] are pseudopassive and usually even pseudolossless. As a conclusion, it has been confirmed that wave digital filters can indeed be designed in such a way that they are stable and that their first-order sensitivities to changes in multiplier coefficients are zero; these properties had so far only been inferred by analogy with classical unit element filters.

## References


[1] A. Fettweis, "Entwurf von Digitalfiltern in Anlehnung an Verfahren der klassischen Netzwerktheorie," in *NTG-Fachtagung "Analyse und Synthese von Netzwerken"* (*NTZ-Report*). Stuttgart, Germany: Oct. 1970, pp. 17–20.

[2] J. A. Bingham, "A new type of digital filter with a reciprocal ladder configuration," in *1970 IEEE Int. Symp. Circuit Theory, Dig. Tech. Papers*, pp. 129–130.

[3] A. Fettweis, "Digital filter structures related to classical filter networks," *Arch. Elek. Übertragung*, vol. 25, pp. 79–89, Feb. 1971.

[4] ——, "Some principles of designing digital filters imitating classical filter structures," *IEEE Trans. Circuit Theory* (Corresp.), vol. CT-18, pp. 314–316, Mar. 1971.

[5] ——, "Wave digital filters," in *Proc. Summer School on Circuit Theory 1971* (Institute of Radio Engineering and Electronics, Czechoslovak Academy of Sciences), pp. 11-1–11-10.

[6] R. Boite, "Discrete-time systems with interwoven transfers," *Proc. Inst. Elec. Eng.*, vol. 118, pp. 312–316, Feb. 1971.

[7] H. W. Bode, *Network Analysis and Feedback Amplifier Design.* New York: Van Nostrand, 1945.

[8] R. Tomovic, *Sensitivity Analysis of Dynamic Systems.* New York: McGraw-Hill, 1964.

[9] K. Géher, *Theory of Network Tolerances.* Budapest, Hungary: Hungarian Academy of Sciences, 1971.

[10] A. Fettweis, "On the connection between multiplier word length limitation and roundoff noise in digital filters," *IEEE Trans. Circuit Theory*, vol. CT-19, pp. 486–491, Sept. 1972.

[11] ——, "A general theorem for signal-flow networks, with applications," *Arch. Elek. Übertragung*, vol. 25, pp. 557–561, Dec. 1971; also in *1971 IEEE Int. Symp. Electrical Network Theory, Dig. Tech. Papers*, (London, England), pp. 3–4. To be published in *Digital Signal Processing* (IEEE Press Selected Reprint Series), L. L. Rabiner and C. M. Rader, Eds.

[12] H. C. Carlin and A. B. Giordano, *Network Theory.* Englewood Cliffs, N. J.: Prentice-Hall, 1964.

[13] V. Belevitch, *Classical Network Theory.* San Francisco, Calif.: Holden-Day, 1968.

[14] H. J. Orchard, "Inductorless filters," *Electron. Lett.*, vol. 2, pp. 224–225, June 1966.

[15] W. H. Holmes, W. E. Heinlein, and S. Grützmann, "Sharp-cutoff low-pass filters using floating gyrators," *IEEE J. Solid-State Circuits*, vol. SC-4, pp. 38–50, Feb. 1969.

[16] H. J. Orchard and D. F. Sheahan, "Inductorless bandpass filters," *IEEE J. Solid-State Circuits*, vol. SC-5, pp. 108–118, June 1970.

[17] R. Crochiere, "Digital ladder filter structures and coefficient sensitivity," M.I.T. Res. Lab. Electron., Quart. Prog. Rep. 103, Oct. 15, 1971, pp. 129–137.

[18] B. Gold and C. M. Rader, *Digital Processing of Signals.* New York: McGraw-Hill, 1969.

# Suppression of Parasitic Oscillations in Wave Digital Filters

A. FETTWEIS, SENIOR MEMBER, IEEE, AND K. MEERKÖTTER

*Abstract*—**Parasitic oscillations can occur in digital filters due to the granularity of the signal and to overflow. It is shown that for wave digital filters, the earlier concept of pseudopower leads to a function called stored pseudopower, from which a simple criterion guaranteeing absence of parasitic oscillations can be derived. For fixed-point two's-complement arithmetic, this criterion can be taken into account by particularly simple requirements for the adaptor output signals. A special type of calculus is developed which makes it possible to derive the corresponding simple requirements for the chopping operations at the various stages of the signal processing in the adaptors. The method is extended to cover scaling.**

## I. Introduction

DUE TO the finiteness of the word length available for representing the signal (data) parameter, essentially two types of parasitic nonlinear oscillations (limit cycle oscillations) can occur in a digital filter under zero input conditions: *overflow oscillations* and *granularity oscillations*. Overflow oscillations can occur due to the possibility that results of arithmetic operations can exceed the available range. Granularity oscillations can originate due to the granular nature (fine structure) of the signal.

In conventional digital filters, overflow oscillations can be eliminated by using, e.g., a saturation arithmetic [1], [2] while various bounds for the amplitudes of granularity oscillations have been computed (for recent references, see [3], [4]). Due to the complexity of the mathematics involved, the most important results have so far been limited to sections of order not exceeding two. This is sufficient, in principle, since conventional digital filters are built by cascading first- and second-order sections.

The excellent sensitivity behavior of wave digital (WD) filters [5]–[14], however, can be obtained only if the entire filter is built as one complete block, i.e., as an intricate high-order feedback structure. Nevertheless, as will be shown in this paper, stability conditions guaranteeing the absence of both types of parasitic oscillations can be derived for WD filters. These conditions lead to requirements for the arithmetic operations in the filter which can easily be taken into account in hardware implementations.

The possibility of deriving such general stability conditions is obtained through judicious applications and generalization of the concepts of pseudopower and pseudopassivity, which had been introduced for proving the desired properties for the linear behavior of WD filters [8], [9]. In particular, we shall introduce a quadratic positive function called stored pseudopower which is necessarily decreasing if the above-mentioned stability conditions are fulfilled. In mathematical terminology, this pseudopower plays the role of a Lyapunov function.

At present, the most important way for implementing digital filters is by fixed-point two's-complement technology [15], [16]. For such an implementation of a WD filter,

Manuscript received November 28, 1973; revised May 24, 1974.
The authors are with the Department of Electrical Engineering, University of Bochum, Bochum, Germany.

Reprinted from *IEEE Trans. Circuits Syst.*, vol. CAS-22, pp. 239–246, Mar. 1975.

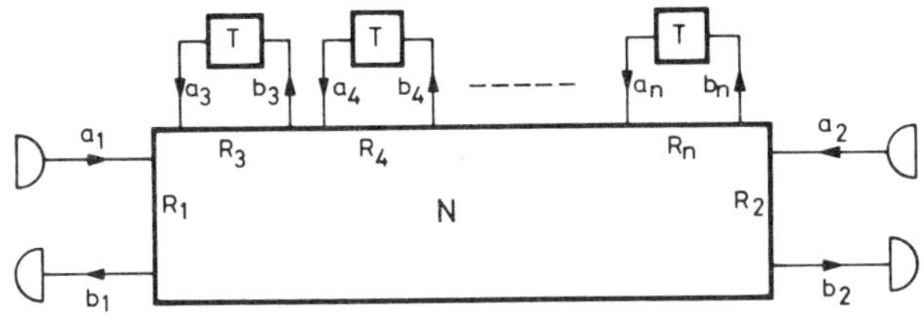


Fig. 1. A full-synchronic WD filter.

it is possible to derive, from the stability criteria, particularly simple design rules. The application of these rules becomes particularly simple by introducing a special type of calculus for describing the various chopping operations.

## II. Stability of WD Filters

### A. Stored Pseudopower

WD filters are either of *full-synchronic* or *half-synchronic* type. By full-synchronic, we mean that all arithmetic operations are being carried out, at least in principle, simultaneously at periodically recurring instants. By half-synchronic, we mean that the various arithmetic operations, although still all being carried out at a same rate, do not even in principle take place simultaneously. WD filters with true ladder configuration, just as most conventional digital filters, are full-synchronic, while WD filters derived from reference filters incorporating unit elements or quasi-reciprocal lines (QUARL's) can be half-synchronic. In the present paper, we consider only full-synchronic WD filters while the extension of our results to the half-synchronic case will be given elsewhere [17].

A full-synchronic WD filter can be represented as shown in Fig. 1 where we have assumed, as usual, that it is accessible only through two ports (this simplifying assumption is irrelevant for the validity of the present theory; cf. Section VI, comment 6), i.e., those marked 1 and 2. The boxes marked $T$ are delays having all the same delay $T > 0$. The $n$-port $N$ does not contain delays, and all arithmetic operations inside of $N$ are assumed to be carried out instantaneously at the operating instants

$$t_m = t_0 + mT, \qquad m = \cdots -2,-1,0,1,2,\cdots, \tag{1}$$

the rate of operation being thus

$$F = 1/T. \tag{2}$$

For each port $\nu = 1,2,\cdots,n$, there is indicated the instantaneous incident signal $a_\nu = a_\nu(t)$, the instantaneous reflected signal $b_\nu = b_\nu(t)$, and the port resistance[1] $R_\nu$ (under normal circumstances, the value of $b_1$ is irrelevant and $a_2 = 0$). To each $R_\nu$, there corresponds a port conductance $G_\nu = 1/R_\nu$. The signals $a_1$ and $a_2$ may be called the input signals, and $b_1$ and $b_2$ the output signals.

Consider first an individual delay connected to a port of port conductance $G = 1/R$ (Fig. 2). We define the pseudopower $p(t_m)$ stored in the delay during the time interval $(t_m,t_{m+1})$ by means of[2]

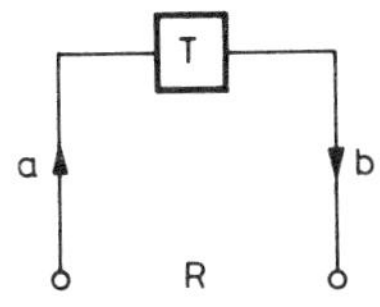


Fig. 2. A delay $T$ connected to a port of port resistance $R$.

$$p(t_m) = Ga^2(t_m). \tag{3}$$

Correspondingly, we define the pseudopower stored during $(t_m,t_{m+1})$ in the delays of the filter of Fig. 1 (note that the symbol $a_\nu$ now refers to a signal incident to $N$) by means of

$$p(t_m) = \sum_{\nu=3}^{n} G_\nu b_\nu^2(t_m). \tag{4}$$

In WD filters, the port resistances are usually positive, i.e.,

$$R_\nu > 0, \qquad G_\nu > 0, \qquad \nu = 1,2,\cdots,n. \tag{5}$$

Henceforth, we assume this condition to hold (cf. Section VI, remark 5). Thus $p(t)$ is a nonnegative function of time, defined for time values $t = t_m$ given by (1).

The increase of $p(t)$ at $t_m$ is given by

$$\Delta p(t_m) = p(t_m) - p(t_{m-1}) \tag{6}$$

i.e., since

$$a_\nu(t_m) = b_\nu(t_{m-1}), \qquad \nu = 3,4,\cdots,n \tag{7}$$

by

$$\Delta p(t_m) = \sum_{\nu=3}^{n} G_\nu[b_\nu^2(t_m) - a_\nu^2(t_m)]. \tag{8}$$

On the other hand, the pseudopower absorbed by the $n$-port $N$ at $t_m$ is defined by [8]

$$p_N(t_m) = \sum_{\nu=1}^{n} G_\nu[a_\nu^2(t_m) - b_\nu^2(t_m)]. \tag{9}$$

Hence, we obtain from (8) and (9),

$$\Delta p(t_m) = -p_N(t_m) + \sum_{\nu=1}^{2} G_\nu[a_\nu^2(t_m) - b_\nu^2(t_m)]. \tag{10}$$

In particular, in the absence of outside signals feeding into the circuit, i.e., for $a_1 = a_2 = 0$, we have

$$\Delta p(t_m) = -p_N(t_m) - G_1b_1^2(t_m) - G_2b_2^2(t_m). \tag{11}$$

Thus, if $N$ is pseudopassive, i.e., if

$$p_N(t_m) \geq 0, \qquad \text{for all } m \tag{12}$$

we obtain, for all $m$,

$$\Delta p(t_m) \leq -G_1b_1^2(t_m) - G_2b_2^2(t_m) \leq 0. \tag{13}$$

[1] It should be stressed that all port resistances, as always in the theory of WD filters, are those corresponding to the actual (i.e., finite word length) multiplier coefficients.

[2] According to the theory for deriving WD filters from classical filters, this expression has the dimension of a power. We therefore prefer using the term "pseudopower" rather than "pseudoenergy."

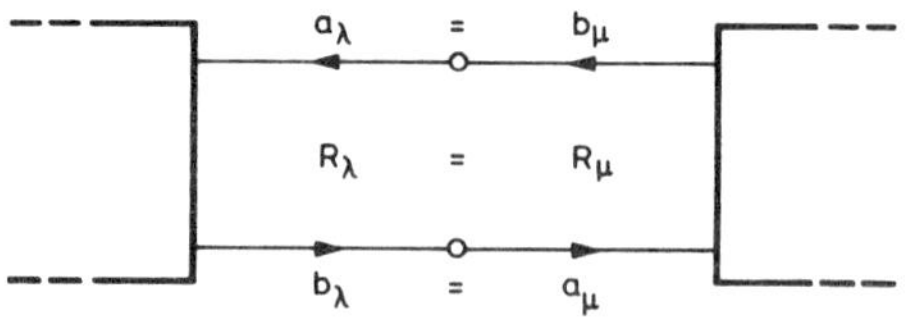


Fig. 3. Interconnection of two ports $\lambda$ and $\mu$.

In practice, the $n$-port $N$ is composed of a certain number of individual multiports $N_k$, with $k = 1,2,\cdots,K$. Let us designate by $p_k(t_m)$ the psuedopower absorbed at $t_m$ by $N_k$; it is given by an expression corresponding to (9) but referring to $N_k$ rather than $N$. It follows from [8, theorem 2] that for any $t_m$

$$p_N(t_m) = \sum_{k=1}^{K} p_k(t_m). \tag{14}$$

This is known to be a consequence of the fact that for any two ports $\lambda$ and $\mu$ which are interconnected we have (Fig. 3)

$$R_\lambda = R_\mu \qquad a_\lambda = b_\mu \qquad b_\lambda = a_\mu. \tag{15}$$

It follows from (14) that (12) is always fulfilled if the $N_k$ are pseudopassive, i.e., if

$$p_k(t_m) \geq 0, \qquad \text{for all } k = 1 \text{ to } K \text{ and all } m. \tag{16}$$

The building blocks $N_k$ occurring in practice are adaptors, ideal transformers, gyrators, and circulators. All these are *nonenergic*, i.e., not only delay-free (as we have assumed throughout) but also *pseudolossless* [8], thus such that $p_k(t_m)$ is always zero. Hence, the equality sign in (12) then holds.

### B. Stability Criteria

At the end of the previous subsection, we implicitly assumed ideal linear operating conditions. In reality, due to the finite number of digits available for representing the signal values, certain nonlinear modifications of the various arithmetic operations are required. These modifications do not affect the operations "delay," i.e., (7), and thus (8) as well as (10) still hold. They have nothing to do either with the port resistances, for which we may thus retain the original values, and they do not affect the signal equalities expressed by Fig. 3; hence (15) and thus (14) remain valid.

We now consider again a WD filter such as the one of Fig. 1. We assume that the nonlinear modifications to which we have just referred are applied to it. We assume furthermore that, at a certain initial time instant, say $t_0$, some arbitrary initial conditions are established in the filter, and that, from $t_0$ on, no signals are fed into it from the outside, i.e., according to (1), that $a_1(t_m) = a_2(t_m) = 0$ for $m \geq 0$ (autonomy condition). We then say that the filter is *output stable* if, after some finite time, $b_1(t_m)$ and $b_2(t_m)$ become permanently zero. On the other hand, we say that the filter is *completely stable* if, again after some finite time, *all* $b_\nu(t_m)$, i.e., for all $\nu = 1$ to $n$, become permanently zero.

Clearly, output stability implies freedom from observable oscillations. On the other hand, complete stability implies

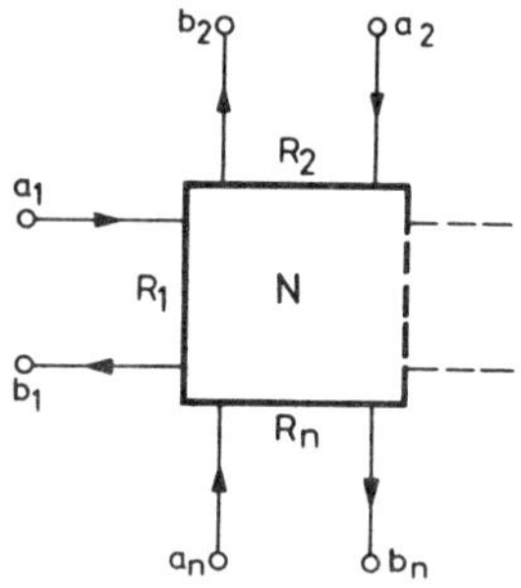


Fig. 4. The $n$-port $N$ in Fig. 1 (Section II) or one of its building blocks (Section III).

freedom from any observable or unobservable oscillation. Indeed, a realizable digital filter structure contains no delay-free loops [5]–[7]. Hence the $n$-port $N$, which is known not to contain any delay, does not contain any closed loop. It can thus not sustain any oscillation which would not be detectable through at least one of the $b_\nu$.

For the case of output stability, we can now state the following.

*Theorem 1:* A WD filter such as the one of Fig. 1 is output stable if its $n$-port $N$ is pseudopassive.

*Proof:* We first observe that a digital filter can only take a finite number of states. Thus, under autonomy conditions, it must, after a finite number of periods, return to a state already reached before. Hence any of its signal parameters must, after some finite time, either be carrying out periodic oscillations or be permanently zero (which corresponds also to a periodically repetitive condition). On the other hand, we conclude from (13) that if one of the output signals would be carrying out periodic oscillations with nonzero amplitude, $p(t_m)$ would be decreasing during each oscillation cycle by at least some fixed positive amount. This would imply that $p(t_m)$ would in the long run be decreasing below zero, which is excluded by (4) and (5).

Q.E.D.

We thus have obtained a first important stability criterion. Consider next the $n$-port $N$ by itself, as shown in Fig. 4. There is no need for specifying the time instant $t_m$, i.e., instead of (9), we may write for the pseudopower absorbed by $N$,

$$p_N = \sum_{\nu=1}^{n} (a_\nu^2 - b_\nu^2)G_\nu. \tag{17}$$

We assume that the $b_\nu$ are the output signals of $N$ under actual finite arithmetic conditions. On the other hand, let us designate by $b_{\nu o}$ the corresponding output signals under ideal linear conditions, i.e., if the additions and multiplications inside of $N$ would, for the same values of the $a_\nu$, be carried out exactly. We also assume that $N$ is pseudopassive under these ideal conditions, i.e., that $p_N \geq 0$ if the $b_\nu$ in (17) are replaced by the corresponding $b_{\nu o}$. According to (17), a *sufficient* condition for insuring the continued pseudopassivity of $N$ is thus to carry out the nonlinear modifications in such a way that, for the same $a_\nu$, we have

$$|b_\nu| \leq |b_{\nu o}|, \qquad \nu = 1 \text{ to } n. \tag{18}$$

Indeed, this condition is still valid under the extreme (but usual) situation that $N$ is pseudolossless ($p_N = 0$) under ideal linear conditions. We summarize this result as follows.

*Theorem 2:* A WD filter (Fig. 1) is output stable if its $n$-port $N$ (Fig. 4) is pseudopassive under ideal linear conditions and if the nonlinear modifications required by the finite arithmetic are carried out in such a way that, with reference to Fig. 4, (18) holds for each arbitrary set of signals $a_1$ to $a_n$ incident to $N$.

We can now also formulate a criterion for complete stability, as follows.

*Theorem 3:* A WD filter (Fig. 1) is completely stable if, in addition to the requirements for Theorem 2, it is, under ideal linear conditions, free from any oscillations (thus including unobservable ones).

*Proof:* Assume first that there would exist an oscillation for which $b_\nu(t_m) = b_{\nu o}(t_m)$ for $\nu = 1$ to $n$ and all $m$. In this case, all signal values coincide with those under ideal linear conditions, i.e., nonlinear modifications do not occur. For such a situation, however, oscillations have been excluded by hypothesis. Assume thus that we have $|b_\nu(t_m)| < |b_{\nu o}(t_m)|$ for at least one of the $\nu$ and at least one $t_m$ per oscillation cycle. In this case, according to (9), $p_N(t_m)$ would take a certain, but fixed positive value at least once per oscillation cycle. Similar as for the proof of Theorem 1, we then conclude from (11) that $p(t_m)$ would, even for $b_1(t_m)$ and $b_2(t_m)$ continuously zero, be decreasing below zero in the long run, which is impossible. Q.E.D.

Finally, as mentioned at the beginning of the present subsection, (14) still holds if $N$ is actually composed of individual building blocks $N_k$. Hence we can also state the following.

*Corollary:* Assume that $N$ (Figs. 1 and 4) is composed of individual multiports $N_k$ such that (15) holds. Each of the Theorems 1–3 then still holds if the conditions which it assumes to hold for $N$ are required to hold for each $N_k$ individually (note in particular that for Theorems 2 and 3 we thus require, e.g., (18) to hold not only for the ports of $N$ but for all the ports of each $N_k$).

## III. Pseudopassivity of Ideally Nonenergic Building Blocks Under Finite Arithmetic Conditions

### *A. General Considerations*

We consider the multiport $N$ of Fig. 4. We assume it to be not necessarily the $n$-port $N$ in Fig. 1 but any of the individual building blocks of the latter. We also assume it to be pseudopassive under ideal linear conditions. According to Section II-B, the condition we want to impose on $N$ is expressed by (18).

This condition is quite easy to satisfy in practice. Let indeed $m$ be the number of bits available for representing the signals between building blocks, i.e., in particular for the $a_\nu$ and $b_\nu$. By properly increasing the number of bits for the signals inside of $N$ (possibly as well towards more significant bits as towards less significant ones), we can make sure that the arithmetic operations required inside of $N$ are carried out exactly, i.e., that the output signals actually take the values $b_{\nu o}$. From these $b_{\nu o}$, one can always, by means of simple operations, determine numbers $b_\nu$ which satisfy (18) and which can fully be represented within the available word length $m$, preferably with the smallest possible error.

The increase in the number of signal bits which is required inside of an adaptor does not have to be very large. This is due to the fact that in practice a signal path in the building block $N$ traverses at most one multiplier and a few adders. Usually, the increase in complexity can even be restricted to a limited part of $N$ (cf. Section IV).

### *B. Two's Complement Arithmetic*

The results so far discussed hold for any type of finite arithmetic. We now assume that we use fixed point arithmetic with two's-complement representation [15], [16] for the signals between building blocks. If there are $m$ bits $\delta_0, \delta_1, \cdots, \delta_{m-1}$ in a number, its value is given by

$$-\delta_0 + \sum_{\mu=1}^{m-1} \delta_\mu 2^{-\mu}. \tag{19}$$

The highest value which this number can take is $1 - 2^{-m+1}$, and the lowest is $-1$. If we increase the number of bits, we may replace (19) by

$$x = -\delta_{-k} 2^k + \sum_{\mu=-k+1}^{m+l-1} \delta_\mu 2^{-\mu} \tag{20}$$

with $k \geq 0$ and $l \geq 0$. In this case the highest value which $x$ can take becomes $2^k(1 - 2^{-k-l-m+1})$, and the lowest $-2^k$.

In order to be able to present the subsequent results in a simple way, we now introduce a special notation and terminology. Let $x$ be a number given by (20). We designate by $x_*$ the number

$$x_* = -\delta_{-k} 2^k + \sum_{\mu=-k+1}^{m-1} \delta_\mu 2^{-\mu} \tag{21}$$

obtained by *rear chopping* (equivalent to two's-complement truncation in case of $k = 0$), i.e., by dropping the excess bits from $\mu = m$ to $m + l - 1$. Similarly, we designate by $x^*$ the number

$$x^* = -\delta_0 + \sum_{\mu=1}^{m+l-1} \delta_\mu 2^{-\mu} \tag{22}$$

obtained by *front chopping*, i.e., by dropping the excess bits from $\mu = -k$ to $-1$. Finally, we designate by $x_*^*$ the number obtained by front and rear chopping combined, i.e., the one given by (19).

Consider first a signal $b_{\nu o}$ given by an expression of the form (20), with $k = 0$ and $l > 0$. A signal $b_\nu$ of the form (19) and satisfying (18) with as small an error as possible is then clearly obtained by sign-magnitude truncation [18], i.e., by

$$b_\nu = (b_{\nu o})_* + \gamma_\nu \tag{23}$$

where

$$\gamma_\nu = 0, \qquad \text{for } \delta_0 = 0 \tag{24}$$

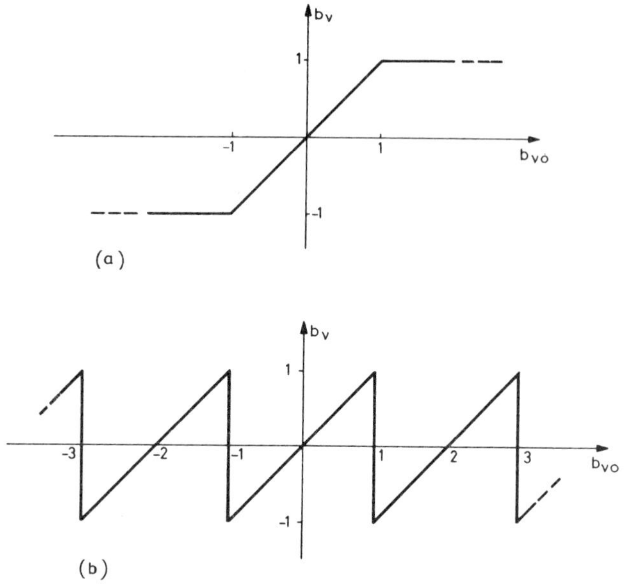


Fig. 5. Overflow reduction of a number $b_{vo}$ to a number $b_v$ in the acceptable range. (a) By a saturation characteristic. (b) By a "modulo 2" characteristic.

and, for $\delta_0 = 1$,

$$\gamma_v = \begin{cases} 0, & \text{if } \delta_\mu = 0 \text{ for } \mu = m \text{ to } m + l - 1 \quad (25a) \\ 2^{-m+1}, & \text{otherwise.} \quad (25b) \end{cases}$$

In fact, the choice $\gamma_v = 2^{-m+1}$ is compatible with (18) also for the case that $\delta_0 = 1$ and $\delta_\mu = 0$ for $\mu = m$ to $m + l - 1$, although a slightly increased error is then introduced. We may thus replace (25a) by the slightly more general condition, still valid for $\delta_0 = 1$,

$$\gamma_v = 0 \text{ or } 2^{-m+1} \text{ if } \delta_\mu = 0, \qquad \text{for } \mu = m \text{ to } m + l - 1. \tag{26}$$

The choice between the two possibilities offered by (26) is thus arbitrary, but should be made in practice according to which one leads to the simplest hardware.

Assume next that $b_{vo}$ is given by (20) but with either $b_{vo} \geq 1$ or $b_{vo} < -1$. In either case, any number $b_v$ of the form (19) satisfies (18) and is thus acceptable. The error is smallest if we replace $b_{vo}$ in the former case by $1 - 2^{-m+1}$ and in the latter by $-1$, i.e., neglecting the granularity, if we make use of the saturation characteristic represented in Fig. 5(a).

The operations required are often simplest, however, if we use instead the saw-tooth characteristic of Fig. 5(b), which is, e.g., automatically achieved in a two's-complement adder [15]. The calculations are then carried out *modulo 2*, i.e., a multiple of 2 is added or subtracted such that the resulting number is located in the acceptable interval $[-1,1)$. In the notation defined by (22), this amounts to replacing $b_{vo}$ by $(b_{vo})^*$. If not only $k > 0$ but also $l > 0$, we have, in addition, to apply rear chopping.

Altogether, we may thus conclude that, for $k \geq 0$ and $l \geq 0$, a suitable and simple way of deriving signals $b_v$ which satisfy (18) is by means of

$$b_v = c_v + \gamma_v \qquad c_v = (b_{vo})_*^* \tag{27}$$

where $\gamma_v$ is given by (24), (25a) (or alternately (26)), and (25b).

## C. Mathematics of Chopping Operations

In the previous subsection, we have seen how, in principle, signals $b_v$ satisfying (18) can be derived by suitable operations at the output of a building block $N$ (Fig. 4). In practice, however, the required operations can already be performed at earlier stages inside of $N$. As will be shown, this implies considerable hardware simplifications. These results can be derived most easily by means of the rules listed hereafter. Most of these rules are self-explanatory and need no explicit proof. The numbers $x_v$ for $v = 1,2,\cdots,n$ are assumed to be numbers of the form (20).

*Rule 1:* The order in which front and rear chopping are carried out is irrelevant, i.e.,

$$x_*{}^* = (x_*)^* = (x^*)_*.$$

*Rule 2:* If $x_{v*} = x_v$ for all $v = 1,2,\cdots,n$, then

$$(x_1 + x_2 + \cdots + x_n)_* = x_1 + x_2 + \cdots + x_n.$$

*Rule 3:* If $x_{v*} = x_v$ for $v = 2$ to $n$, then

$$(x_1 + x_2 + \cdots + x_n)_* = x_{1*} + x_2 + x_3 + \cdots + x_n.$$

*Rule 4:* The value of

$$y = (x_1 + x_2 + \cdots + x_n)^*$$

remains unchanged if we replace any of the $x_v$ by $x_v{}^*$.

Let us now designate by $\bar{x}$ the number obtained by replacing each $\delta_\mu$ in (20) by its complement $\bar{\delta}_\mu = 1 - \delta_\mu$, thus

$$\bar{x} = -\bar{\delta}_{-k}2^k + \sum_{\mu=-k+1}^{m+l-1} \bar{\delta}_\mu 2^{-\mu}. \tag{28}$$

Care is needed for properly carrying out this latter operation. Indeed, if $x = x_*$, i.e., if $\delta_\mu = 0$ for $\mu = m$ to $m + l - 1$ ($l > 0$), we have $\bar{x} \neq \overline{x_*}$. This can be seen by comparing (28) with the following expression derived from (21),

$$\overline{x_*} = -\bar{\delta}_{-k}2^k + \sum_{\mu=-k+1}^{m-1} \bar{\delta}_\mu 2^{-\mu}. \tag{29}$$

*Rule 5:* We have

$$(-x)_* = \begin{cases} -x_* - 2^{-m+1} = \overline{x_*}, & \text{if } x_* \neq x \quad (30a) \\ -x_* = \overline{x_*} + 2^{-m+1}, & \text{if } x_* = x. \quad (30b) \end{cases}$$

The proof of this result follows by noticing that

$$-x = -\bar{\delta}_{-k}2^k + \sum_{\mu=-k+1}^{m+l-1} \bar{\delta}_\mu 2^{-\mu} + 2^{-m-l+1} \tag{31}$$

and

$$-x_* = -\bar{\delta}_{-k}2^k + \sum_{\mu=-k+1}^{m-1} \bar{\delta}_\mu 2^{-\mu} + 2^{-m+1}.$$

*Rule 6:* We have

$$(-x)^* = -x^*.$$

This is easily checked by means of (22) and (31).

*Rule 7:* The operation

$$y = (x_1 + x_2 + \cdots + x_n)^*$$

designates the modulo 2 addition of the $x_\nu$ such that $|y| \leq 1$. Thus, if the $x_\nu$ are numbers for which $x_\nu = x_\nu^*$, this is precisely the result which is obtained by means of two's-complement arithmetic.

## IV. Pseudopassivity of Adaptors with Two's-Complement Arithmetic

### *A. Nonenergic Building Blocks*

We want to examine the application of the results of Section III to the ideally nonenergic building blocks used in WD filters. *Circulators* need no consideration since they do not involve arithmetic operations [5]–[7]. *Gyrators* and *ideal transformers* of ratio $-1/1$ only involve multiplications by $-1$ and are thus covered by Rules 5 and 6 of Section III-C.

For an $n$-port *parallel adaptor*, the reflected signals obtained under ideal conditions, $b_{\nu o}$ (cf. Section II-B), are given by [5]–[7]

$$b_{\nu o} = a_0 - a_\nu \qquad a_0 = \alpha_1 a_1 + \alpha_2 a_2 + \cdots + \alpha_n a_n \tag{32a, b}$$

$$\alpha_\nu = 2G_\nu/G \qquad G = G_1 + G_2 + \cdots + G_n \tag{33a, b}$$

and for an $n$-port *series adaptor* by [5]–[7]

$$b_{\nu o} = a_\nu - \alpha_\nu a_0 \qquad a_0 = a_1 + a_2 + \cdots + a_n \tag{34a, b}$$

$$\alpha_\nu = 2R_\nu/R \qquad R = R_1 + R_2 + \cdots + R_n \tag{35a, b}$$

where the multiplier coefficients $\alpha_\nu$ satisfy, in both cases, the relation

$$\alpha_1 + \alpha_2 + \cdots + \alpha_n = 2. \tag{36}$$

Ideal transformers of ratio $\neq -1$ can be obtained by means either of a two-port adaptor [5]–[7] or of two simple multipliers (Section V).

### *B. Adaptors Containing Only One Multiplier*

The conclusions obtained in Section IV are particularly easy to apply in case of adaptors containing only one multiplier. As will be seen in the present subsection, the increase in complexity is usually very slight. These results should also be seen in the light of the fact that any adaptor can be decomposed into adaptors containing only one multiplier each [19].

In a *three-port parallel adaptor* with a reflection-free port [12], [13], one of the multiplier coefficients $\alpha_\nu$ is equal to unity, say $\alpha_3 = 1$ (Fig. 6(a)). According to (36), we then have

$$\alpha_1 + \alpha_2 = 1. \tag{37}$$

Thus, due to (5) and (33), one might even assume without

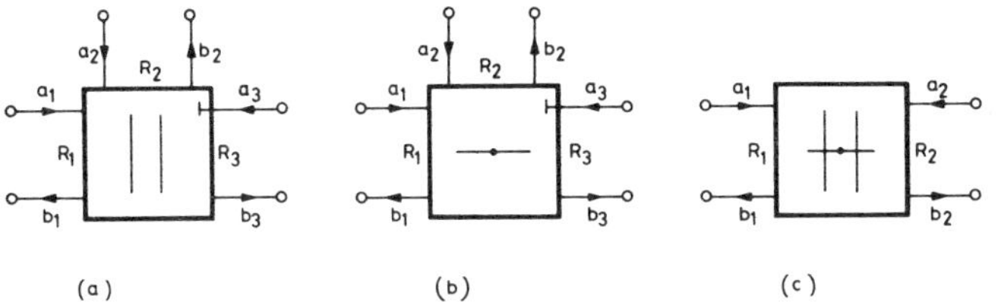


Fig. 6. (a) Three-port parallel adaptor whose port 3 is reflection-free. (b) Three-port series adaptor whose port 3 is reflection-free. (c) Two-port adaptor.

restriction (i.e., by proper numbering) that

$$0 < \alpha_1 \leq \tfrac{1}{2}. \tag{38}$$

The equations derived from (32)

$$b_{10} = c - a_1 + a_2 + a_3 \qquad b_{20} = c + a_3 \qquad b_{30} = c + a_2 \tag{39}$$

$$c = \alpha_1(a_1 - a_2) \tag{40}$$

require only one multiplication.

We thus conclude from (27) and (39), that pseudopassivity is assured if the $c_\nu$ are determined according to

$$c_1 = (c - a_1 + a_2 + a_3)^*_* \qquad c_2 = (c + a_3)^*_*$$
$$c_3 = (c + a_2)^*_*.$$

Since $a_{\nu *} = a_\nu$ for all $\nu$, we obtain by means of Rules 1, 3, 4, and 5 (30b) of Section III-C,

$$c_1 = (c_*^* - a_1 + a_2 + a_3)^* \qquad c_2 = (c_*^* + a_3)^*$$
$$c_3 = (c_*^* + a_2)^*. \tag{41}$$

According to Rule 7 of Section III-C, once $c_*^*$ is known, all operations required by these equations amount to simple two's-complement additions.

The determination of $c_*^*$ must be done with care. According to (40), we must first determine, exactly, the difference $a_1 - a_2$. We must thus provide one additional front bit, i.e., choose $k = 1$ for the representation (20) of this difference, while $l = 0$ is still sufficient. Since only $c_*^*$ is required in (41), both front and rear chopping may be performed immediately after the product $\alpha_1(a_1 - a_2)$ is formed. In fact, considerable simplifications are thus possible already during the computation of the product. Rear chopping does not require all excess bits to be stored [15], while front chopping implies that the bits to be dropped do not even have to be calculated. Note that if (38) holds, front chopping even never occurs.

For a *three-port series adaptor* whose port 3 is reflection-free [12], [13] (Fig. 6(b)), we have again $\alpha_3 = 1$ and thus (37), and (38) might then again be assumed to hold. From (34), we obtain

$$b_{10} = a_1 - c \qquad b_{20} = -a_1 - a_3 + c \qquad b_{30} = -a_1 - a_2$$
$$c = \alpha_1 a_o \qquad a_o = a_1 + a_2 + a_3.$$

For the quantities $c_\nu$ to be used in (27), we thus find

$$c_1 = -[-a_1 + c_*^*]^* \qquad c_2 = [-a_1 - a_3 + c_*^*]^*$$
$$c_3 = -(a_1 + a_2)^*.$$

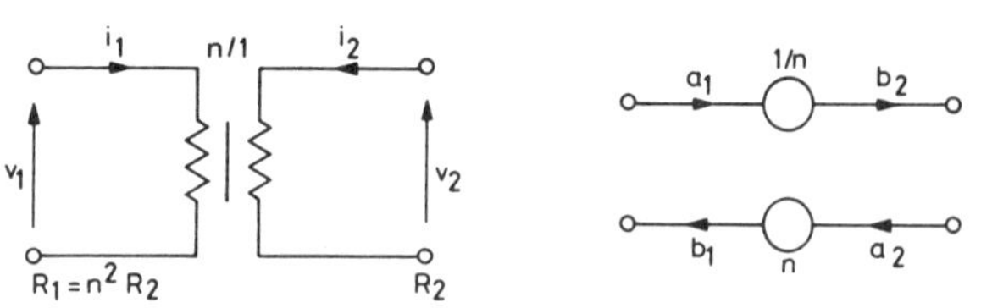


Fig. 7. A transformer of ratio $n/1$ and a corresponding wave-flow diagram obtained for $R_1/R_2 = n^2$.

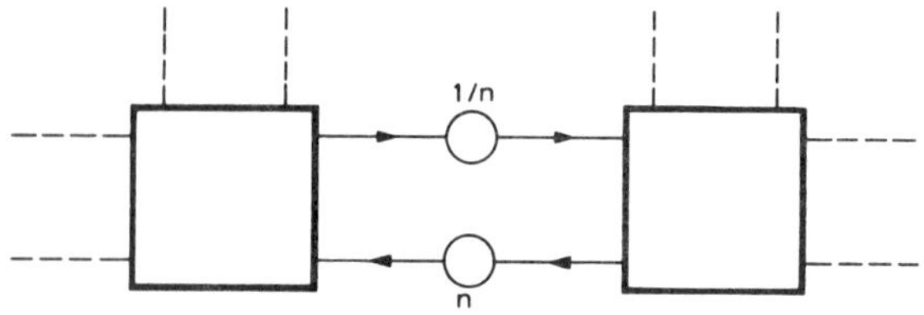


Fig. 8. Insertion of a transformer of ratio $n/1$ between two building blocks.

The representation of $a_0$ now requires two extra front bits, and even if (38) holds, front chopping of $c$ will, in general, be necessary. It remains true, however, that the extra front bits do not have to be calculated and that, once $c_*{}^*$ is known, only two's-complement additions are required.

A *two-port adaptor* (Fig. 6(c)) [5]–[7] is best described by the equations

$$b_{10} = a_2 + c \qquad b_{20} = a_1 + c \qquad c = \alpha(a_2 - a_1) \tag{42}$$

where

$$\alpha = (R_1 - R_2)/(R_1 + R_2).$$

Due to (5), we have

$$-1 < \alpha < 1.$$

For the quantities $c_1$ and $c_2$ to be used in (27), we obtain from (42),

$$c_1 = (a_2 + c_*{}^*)^* \qquad c_2 = (a_1 + c_*{}^*)^*.$$

The conclusions to be drawn are similar to those for (41).

### C. Adaptors Containing Several Multipliers

In view of the discussions already given in Section IV-B, it will not be necessary to go again into details. All relevant equations can be derived from (32), (34), and (36) by applying the same principles as before.

## V. Scaling

Scaling in wave digital filters can, e.g., be achieved by inserting ideal transformers of ratio $n/1$ at various places in the reference filter from which the wave digital filter is derived. Such an ideal transformer can be realized by a wave-flow diagram incorporating one two-port adaptor, i.e., one multiplier. For our present purpose, however, this would lead to unrealizable signal-flow diagrams, but a realization according to Fig. 7, i.e., requiring two multipliers, $n$ and $1/n$, is suitable [20]. This realization follows from

$$v_1 = nv_2 \qquad i_2 = -ni_1 \qquad a_\nu = v_\nu + R_\nu i_\nu$$

$$b_\nu = v_\nu - R_\nu i_\nu, \qquad \nu = 1 \text{ and } 2$$

if we choose

$$R_1/R_2 = n^2 \tag{43}$$

(rather than $R_1/R_2 = n$, as for the realization with one multiplier). Elimination of $v_1$, $v_2$, $i_1$, and $i_2$ indeed leads to

$$b_1 = na_2 \qquad b_2 = a_1/n. \tag{44}$$

In the wave digital filter structure, two-ports defined by equations of the form (44) may appear, e.g., between consecutive adaptors, as shown schematically in Fig. 8. The presence of two multipliers rather than one is no serious disadvantage since for scaling purposes one may usually choose $n$ equal to a (positive or negative) power of 2. The required hardware operations are then very simple.

Even if $n$ is equal to a power of 2, the operations required by (44) may still imply the need for front and/or rear chopping. On the other hand, the two-port defined by (44) is again pseudolossless since we have, using (43),

$$(a_1{}^2 - b_1{}^2)/R_1 + (a_2{}^2 - b_2{}^2)/R_2 = 0.$$

In order to be able to apply (27), we should thus again, in (44), write $b_{10}$ and $b_{20}$ rather than $b_1$ and $b_2$, respectively. Consequently, the quantities $c_1$ and $c_2$ appearing in (27) should be chosen according to

$$c_1 = (na_2)_*^* \qquad c_2 = (a_1/n)_*^*.$$

## VI. Complementary Remarks

1) We have seen that for all adaptors any multiplier $\alpha$ which has to be implemented may be assumed to be such that $|\alpha| < 1$. If $|\alpha| > \frac{1}{2}$, we thus have $|\beta| \le \frac{1}{2}$ either for $\beta = 1 - \alpha$ or for $\beta = 1 + \alpha$. Hence, only multiplier coefficients not exceeding $\frac{1}{2}$ in modulus are required.

2) It should be pointed out again that (18), from which all subsequent conditions have been derived, is only sufficient, not necessary. In fact, one can expect that it is possible to limit the introduction of additional pseudopassivity to certain ports only, i.e., to omit adding the correcting term $\gamma_\nu$ (cf. (27)) to the remaining ports. This has been confirmed in practice.

3) For the same reason as mentioned under 2), it may not be necessary at all, due to the inherent pseudopassivity introduced by the terminating resistances, to take measures against granularity oscillations if the order of the filter and/or the signal word length are sufficiently high.

4) Again for the reason mentioned under 2), it may often be possible to ignore the exception expressed by (30b) in Rule 5 of Section III-B.

5) The theory in this paper is essentially based on the assumption that all port resistances are positive. This assumption is always fulfilled as soon as the resistive parameters corresponding to the various elements in the reference filter (resistances, inductances, capacitances, unit elements, etc.) are all positive [5]–[7], [12], [13], [19], [21].

6) Throughout this paper we have considered only WD two-ports. Clearly, all considerations and conclusions

remain completely valid for WD multiports having an arbitrary number of access ports [9].

## References


[1] P. M. Ebert, J. E. Mazo, and M. G. Taylor, "Overflow oscillations in digital filters," *Bell Syst. Tech. J.*, vol. 48, pp. 2999–3020, Nov. 1969.

[2] A. N. Willson, Jr., "Limit cycles due to adder overflow in digital filters," *IEEE Trans. Circuit Theory*, vol. CT-19, pp. 342–346, July 1972.

[3] J. F. Kaiser, "Quantization effects in digital filters," in *Proc. 1973 IEEE Symp. Circuit Theory*, Toronto, Ont., Canada, pp. 415–417.

[4] S. Yakowitz and S. R. Parker, "Computation of bounds for digital filter quantization errors," *IEEE Trans. Circuit Theory*, vol. CT-20, pp. 391–396, July 1973.

[5] A Fettweis, "Digital filter structures related to classical filter networks," *Arch. Elek. Übertragung*, vol. 25, pp. 79–89, Feb. 1971.

[6] ——, "Some principles of designing digital filters imitating classical filter structures," *IEEE Trans. Circuit Theory*, vol. CT-18, pp. 314–316, Mar. 1971.

[7] ——, "Wave digital filters," *Proc. Tále Summer School on Circuit Theory*, Inst. Radio Eng. Electron., Czechoslovak Acad. Sciences, Prague, Czechoslovakia, pp. 11-1–11-10, 1971.

[8] ——, "Pseudopassivity, sensitivity and stability of wave digital filters," *IEEE Trans. Circuit Theory*, vol. CT-19, pp. 668–673, Nov. 1972.

[9] ——, "Scattering properties of wave digital filters" in *Proc. Florence Seminar on Digital Filtering*, Florence, Italy, Sept. 1972, pp. 1–8.

[10] R. E. Crochiere, "Digital ladder structures and coefficient sensitivity," *IEEE Trans. Audio Electroacoust.*, vol. AU-20, pp. 240–246, Oct. 1972.

[11] A. Chu and R. E. Crochiere, "Comments and experimental results on optimal digital ladder structures," *IEEE Trans. Audio Electroacoust.*, vol. AU-20, pp. 317–318, Oct. 1972.

[12] A. Sedlmeyer and A. Fettweis, "Digital filters with true ladder configuration," *Int. J. Circuit Theory Appl.*, vol. 1, pp. 5–10, 1973.

[13] ——, "Realization of digital filters with true ladder configuration," in *Proc. 1973 Int. Symp. Circuit Theory*, Toronto, Ont., Canada, pp. 149–152.

[14] K. Renner and S. C. Gupta, "On the design of wave digital filters with low sensitivity properties," *IEEE Trans. Circuit Theory*, vol. CT-20, pp. 555–567, Sept. 1973.

[15] J. B. Jackson, J. F. Kaiser, and H. S. McDonald, "An approach to the implementation of digital filters," *IEEE Trans. Audio Electroacoust.*, vol. AU-16, pp. 413–421, Sept. 1968.

[16] S. K. Tewksbury, "Special purpose hardware implementation of digital filters," in *Proc. 1973 IEEE Int. Symp. Circuit Theory*, Toronto, Ont., Canada, pp. 418–421.

[17] A. Fettweis and K. Meerkötter, "Suppression of parasitic oscillations in half-synchronic wave digital filters," to be published.

[18] B. Liu and M. E. Van Valkenburg, "On round-off error of fixed-point digital filters using sign-magnitude truncation," in *Digest Tech. Papers, London 1971 IEEE Int. Symp. Electrical Network Theory*, pp. 68–69.

[19] A. Fettweis, "On adaptors for wave digital filters," to be published.

[20] ——, "Reciprocity, interreciprocity, and transposition in wave digital filters," *Int. J. Circuit Theory Appl.*, vol. 1, pp. 323–337, Dec. 1973.

[21] A. Fettweis, H. Levin, and A. Sedlmeyer, "Wave digital lattice filters," *Int. J. Circuit Theory Appl.*, vol. 2, pp. 203–211, June 1974.

# On Structures for Nonrecursive Digital Filters

by WILHELM SCHÜSSLER *

Communication from the Institut für Nachrichtentechnik der Universität Erlangen-Nürnberg

*Dedicated to Professor Dr.-Ing. V. Aschoff on the occasion of his 65th birthday.*

Structures for nonrecursive digital filters can be found by specializing the general canonical structures of digital filters. Some new noncanonic structures will be presented as a result of expressing the transfer function in different forms.

**Strukturen für nichtrekursive digitale Filter**

Strukturen nichtrekursiver Filter lassen sich durch eine Spezialisierung der bekannten allgemeinen kanonischen Formen gewinnen. Es werden einige neue nichtkanonische Strukturen behandelt, die sich aus anderen Darstellungen der Übertragungsfunktion als Interpolationsausdruck ergeben.

## 1. Introduction

As is well known there are some basic structures for digital filters, the direct form, the cascade form and the parallel form (e.g. [1], [2]). They all are canonic, as far as the number of delay elements is concerned, but they differ, as far as the number of multipliers, the coefficient sensitivity and the rounding noise are concerned ([3] to [6]). The first two, the direct form and the cascade form can be specialized to become nonrecursive filters. Beside that there is a third structure for nonrecursive filters, the frequency sampling form, which does not have the minimum number of delay units [7], [8]. The question arises, whether there is a common viewpoint for these different structures and whether it is possible to find other structures, which might have other useful properties.

## 2. The Direct and Cascade Structures

A transfer function of a nonrecursive filter is a polynomial in $z^{-1}$.

$$H(z) = \frac{Y(z)}{U(z)} = \sum_{\mu=0}^{n} h_0(\mu)\, z^{-\mu}. \tag{1}$$

In order to get the minimum number of multiplications this polynomial can be written in form of the Horner scheme,

$$H(z) = h_0(0) + z^{-1}\{h_0(1) + + z^{-1}\langle h_0(2) + z^{-1}\,[h_0(3) + z^{-1}\ldots]\rangle\}. \tag{2}$$

The structure going with this representation is the direct form, the signal flow graph of which will be shown in Fig. 1 for the most interesting case of a filter with linear phase and $n = 2N$:

$$H(z) = \frac{1}{z^N}\left\{h_0(N) + \sum_{\mu=0}^{N-1} h_0(\mu)\,[z^{(N-\mu)} \pm z^{-(N-\mu)}]\right\} = \frac{1}{z^N}\, H_0(z), \tag{3}$$

$$H_0(\Omega) = h_0(N) + 2\sum_{\mu=0}^{N-1} h_0(\mu)\cos(N-\mu)\,\Omega \tag{4a}$$

or

$$H_0(\Omega) = 2\,\mathrm{j}\sum_{\mu=0}^{N-1} h_0(\mu)\sin(N-\mu)\,\Omega. \tag{4b}$$

There is no possibility to modify the structure. As has been demonstrated elsewhere [9] the coefficient sensitivity is rather high. But since the structure can be implemented such that there is only one rounding point, the noise is a minimum.

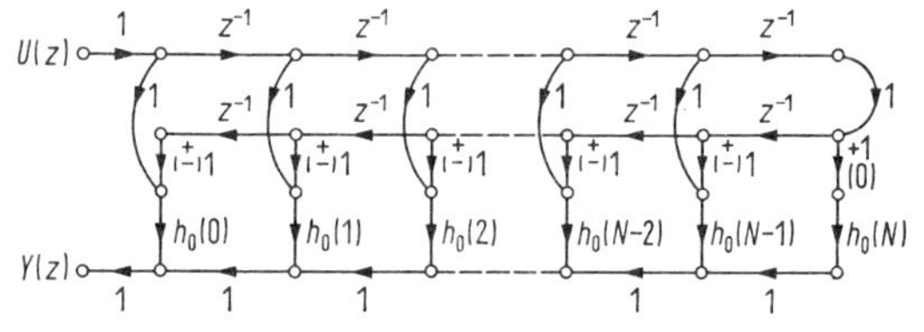


Fig. 1. Direct structure.

By expressing $H(z)$ as a product of transfer functions of lower order

$$H(z) = \prod_{\lambda=1}^{l} H_\lambda(z) \tag{5}$$

a representation will be found the structure of which is the cascade form (Fig. 2). In order to guarantee the linear phase two cases are of interest

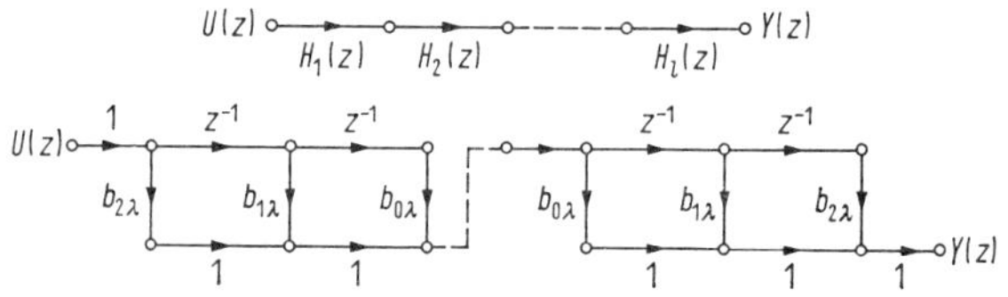


Fig. 2. Cascade structure.

i) Each subsystem has linear phase.

In this case $H_\lambda(z)$ can be of first or second order, if its zeros are located on the unit circle or on the real axis,

* Prof. Dr. W. SCHÜSSLER, Institut für Nachrichtentechnik der Universität, D-852 Erlangen, Egerlandstrasse 5.

Reprinted with permission from *Arch. Elek. Übertragung.*, vol. 26, pp. 255–258, June 1972.

$$H_\lambda(z) = \frac{1}{z}(z \pm 1),$$
$$H_\lambda(z) = \frac{1}{z^2}(z^2 + b_{1\lambda} z + 1). \tag{6a}$$

For zeros lying inside and — in mirror images — outside the unit circle, a system of at least fourth order is required:

$$H_\lambda(z) = \frac{1}{z^4}(z^4 + b_{1\lambda} z^3 + b_{0\lambda} z^2 + b_{1\lambda} z + 1). \tag{6b}$$

ii) Each subsystem is at most of second order.

Instead of systems with the transfer function (6b) pairs of systems of second order are required. For each

$$H_{\lambda 1}(z) = \frac{1}{z^2}(b_{2\lambda} z^2 + b_{1\lambda} z + b_{0\lambda}) \quad \text{with} \quad b_{2\lambda} \neq b_{0\lambda}$$

there is one

$$H_{\lambda 2}(z) = \frac{1}{z^2}(b_{0\lambda} z^2 + b_{1\lambda} z + b_{2\lambda}). \tag{7}$$

Especially in the stopband of selective filters, the cascade structure is less sensitive then the direct form as has been demonstrated in [9]. On the other hand there arise serious noise problems, since there are as many rounding points as there are subsystems. In order to minimize the influence of the rounding, the sequence of the subsystems can be varied and the coefficients of each block can be multiplied by a constant factor as has been done in the recursive case [10]. But in all cases the noise will be higher then in the direct form.

### 3. Lagrange Structure

The two structures considered so far are the result of two different representations of a polynomial. It can be expected, that all different expressions of a polynomial lead to different structures with — in general — different merits.

The superiority of the cascade structure as far as sensitivity is concerned, is due to the fact that each pair of zeros is caused by one subsystem of second order only. Beside special cases it can be guaranteed by satisfying the condition $b_{2\lambda} = b_{0\lambda}$, that wanted zeros on the unit circle stay on it even after rounding. Therefore the poles of attenuation might change their position but they are still poles. The question arises whether it is possible to prescribe any other values of the transfer function — generally speaking — at any other point of the $z$-plane. That leads to an interpolation problem and the different solutions known for it. First the Lagrange interpolation formula will be considered.

Let

$$H(z) = \frac{1}{z^n} \sum_{\mu=0}^{n} h_0(\mu) z^{n-\mu}$$

be prescribed in the points $z_\nu$ to be $H(z_\nu)$, where $\nu = 0(1)n$. With a slight variation of the Lagrange formula [11] $H(z)$ can be expressed as

$$H(z) = \frac{1}{z^n} \prod_{\lambda=0}^{n} (z - z_\lambda) \sum_{\nu=0}^{n} \frac{B_\nu}{z - z_\nu} \tag{8a}$$

with

$$B_\nu = \frac{z_\nu^n H(z_\nu)}{\prod\limits_{\substack{\lambda=0 \\ \lambda \neq \nu}}^{n} (z_\nu - z_\lambda)}. \tag{8b}$$

The structure for this expression is a combination of a cascade form consisting of $n+1$ nonrecursive blocks of first order and a parallel form of recursive filters of first order, in this general case all with complex coefficients. In some respect the Lagrange structure is a generalization of the cascade form, which is the result if $H(z_\nu) = 0$, $\nu = 1(1)n$ and $H(z_0) \neq 0$.

If the points $z_\nu$ and the prescribed values $H(z_\nu)$ are such that for any complex $z_\nu$ and $H(z_\nu)$ there is one $z_\nu^*$ with the prescribed value $H(z_\nu^*) = H^*(z_\nu)$ eq. (8a) can be modified such that there are real coefficients only. In the following the case of a system with linear phase will be considered again. In the special case

$$z_0 = 1, \quad z_\nu = e^{j\Omega_\nu} \quad \text{and} \quad H(z_\nu) = e^{-j\frac{n}{2}\Omega_\nu} H_0(\Omega_\nu),$$
$$z_\nu^* = e^{-j\Omega_\nu} \quad \text{and} \quad H(z_\nu^*) = e^{+j\frac{n}{2}\Omega_\nu} H_0(\Omega_\nu) \tag{9}$$

eq. (8a) becomes

$$H(z) = \frac{1}{z}(1 - z) \prod_{\lambda=1}^{N} \frac{1}{z^2}(z^2 + e_\lambda z + 1) \times \left(\frac{d_1 z}{z-1} + \sum_{\nu=1}^{N} \frac{d_{2\nu} z^2 + d_{1\nu} z}{z^2 + e_\nu z + 1}\right) \tag{10a}$$

with

$$d_1 = \frac{-H(1)}{\prod\limits_{\lambda=1}^{N} 2(1 - \cos \Omega_\lambda)},$$
$$d_{1\nu} = -2\,\mathrm{Re}(B_\nu e^{-j\Omega_\nu}), \quad d_{2\nu} = 2\,\mathrm{Re}(B_\nu),$$
$$e_\lambda = -2 \cos \Omega_\lambda. \tag{10b}$$

Fig. 3 shows the structure for this case.

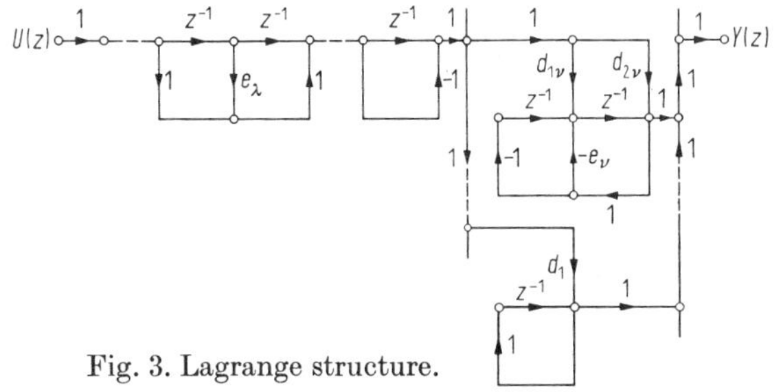

Fig. 3. Lagrange structure.

By a further restriction the points $\Omega_\nu$ will be picked equally spaced on the unit circle

$$\Omega_\nu = \nu \frac{2\pi}{n+1}.$$

In this case eq. (10a) becomes

$$H(z) = (1 - z^{-(n+1)})\left(\frac{d_1 z}{z-1} + \sum_{\nu=1}^{N} \frac{d_{2\nu} z^2 + d_{1\nu} z}{z^2 + e_\nu z + 1}\right) \tag{11}$$

and Fig. 3 reduces to the well known frequency sampling structure the special feature of which is the reduction of the cascade part of the structure to a very simple circuit consisting of a delay line with $n+1$ elements and a bypass [7]. This structure has a practical difficulty [7]. Due to the finite word-length of the coefficients the poles of the recursive part are not exactly the same as the zeros of the nonrecursive part. This does not occur in the general case described by eq. (10a), which on the other hand leeds to a more complicated structure. The possibility of extending the frequency sampling structure to the general form has been considered in [8].

## 4. The Newton Structure and Related Forms

Another structure will be found by expressing the polynomial using Newtons interpolation formula. Again the case of a linear phase system with prescribed values of the frequency response will be considered, especially for the case described by eq. (4a). With the transformation

$$\cos\Omega = 1 - 2y \tag{12}$$

the intervall $0 \leqq \Omega \leqq \pi$ will be mapped on the intervall $0 \leqq y \leqq 1$. $H_0(\Omega)$ will be transformed into a polynomial $G(y)$ of degree $N$ with the prescribed values $G(y_\nu) = H_0(\Omega_\nu)$, where

$$y_\nu = (1 - \cos\Omega_\nu)/2 .$$

$G(y)$ can be expressed as

$$G(y) = d_0 + \sum_{\nu=1}^{N} d_\nu \prod_{\lambda=0}^{\nu-1} (y - y_\lambda) . \tag{13}$$

The coefficients $d_\nu$ are calculated by the scheme (e.g. [11])

$$\begin{array}{lllll}
G(y_0)\ (= d_0) & & & & \\
 & [y_1 y_0]\ (= d_1) & & & \\
G(y_1) & & [y_2 y_1 y_0]\ (= d_2) & & \\
 & [y_2 y_1] & & [y_3 y_2 y_1 y_0]\ (= d_3) & \\
G(y_2) & & [y_3 y_2 y_1] & \cdot & (14) \\
 & [y_3 y_2] & \cdot & \cdot & \\
G(y_3) & \cdot & \cdot & \cdot & \\
\cdot & \cdot & \cdot & & \\
\cdot & \cdot & & & \\
\cdot & & & &
\end{array}$$

using the notations

$$[y_1 y_0] = \frac{G(y_1) - G(y_0)}{y_1 - y_0} ,$$

$$[y_2 y_1 y_0] = \frac{[y_2 y_1] - [y_1 y_0]}{y_2 - y_0}$$

etc.

The inverse transformation leads to

$$H_0(\Omega) = d_0 + \sum_{\nu=1}^{N} d_\nu \prod_{\lambda=0}^{\nu-1} \frac{-1}{2} (\cos\Omega - \cos\Omega_\lambda)$$

and

$$H(z) = d_0 z^{-N} + \sum_{\nu=1}^{N} d_\nu z^{-(N-\nu)} \prod_{\lambda=0}^{\nu-1} \frac{-1}{4z^2} (z^2 + e_\lambda z + 1) . \tag{15}$$

Fig. 4 shows the signal flow graph of the structure, described by eq. (15). Essentially it is a cascade form with zeros at the points $\Omega_\lambda$. The prescribed values $H(\Omega_\nu)$ are got by an appropriately chosen bypass.

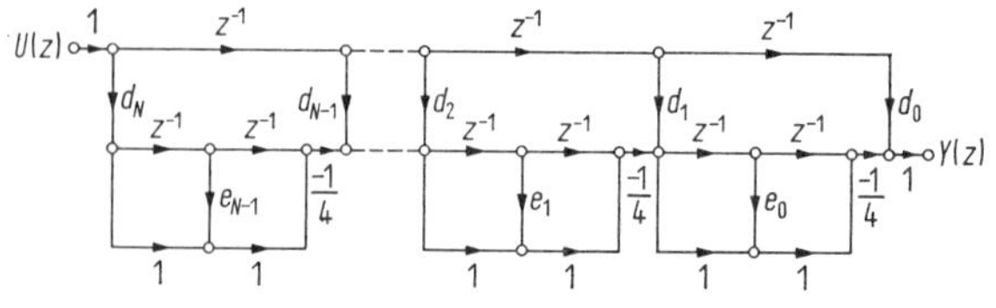


Fig. 4. Newton structure.

Special properties of the structure will be shown using a nonrecursive filter with an equal ripple frequency response as an example (Fig. 5). The points $\Omega_\nu$ will be chosen such that

$$H_0(\Omega_\nu) = 1 , \quad \nu = 0(1)3$$

and

$$H_0(\Omega_\nu) = 0 , \quad \nu = 4(1)7 .$$

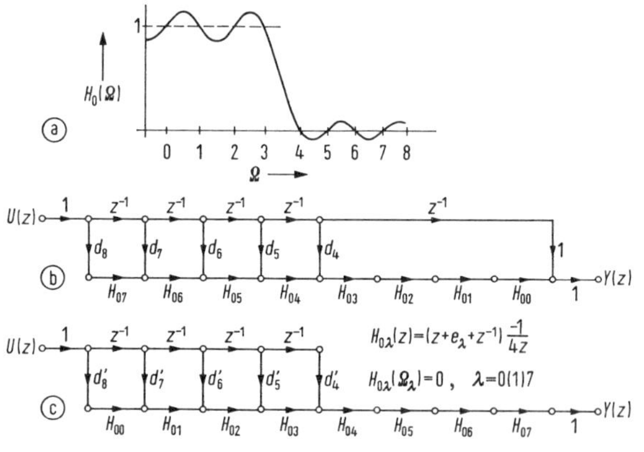


Fig. 5. Two Newton structures for the representation of a given transfer function.

First the corresponding values $G(y_\nu)$ will be introduced into eq. (14) such that the first four prescribed values are one. That leads to $d_0 = 1$, $d_1 = d_2 = d_3 = 0$ and for this special sequence to the signal flow graph shown in Fig. 5b. Obviously the value $H_0(\Omega_\nu) = 1$ will be guaranteed at certain points as were the values zero in the cascade form. It can be expected the sensitivity in the passband to be small. If on the other hand the values $G(y_\nu)$ are ordered such that the first four prescribed values are zero, the first coefficients turn out to be

$$d_0' = d_1' = d_2' = d_3' = 0 .$$

As far as the prescribed zeros are concerned the structure in Fig. 5c has the same property as the cascade form. Furthermore the rounding of a coefficient $d_\nu$ impairs only $H(\Omega_\nu)$. Its influence on other points $\Omega_\lambda$ can be eliminated by calculating the next column of the scheme (14) after rounding the coefficient $d_\nu$. If necessary the influence of rounding

$e_\nu = -2\cos\Omega_\nu$ can be eliminated as well by prescribing the frequency response for a frequency corresponding to the rounded value of $e_\nu$.

An extension of the Newton interpolation formula is the Hermite interpolation in the sense that not only certain values of the frequency response but in addition its derivatives at certain points will be prescribed [11]. The corresponding extension of the structure will be described using a lowpass with maximally flat frequency response as an example. In this case the following values are prescribed

$$H_0(0) = 1\,, \quad \left(\frac{\mathrm{d}^\mu H_0}{\mathrm{d}\Omega^\mu}\right)_{\Omega=0} = 0\,, \quad \mu = 1\,(1)\,N_0\,,$$
$$H_0(\pi) = 0\,, \quad \left(\frac{\mathrm{d}^\mu H_0}{\mathrm{d}\Omega^\mu}\right)_{\Omega=\pi} = 0\,, \quad \mu = 1\,(1)\,N_1\,, \tag{16}$$

where $N_0 + N_1 = N - 1$.

The solution for this problem expressed in form of the corresponding polynomial $G(y)$ is [12]

$$G(y) = (1-y)^{N_1+1} \sum_{\nu=0}^{N_0} \binom{N_1+\nu}{\nu} y^\nu\,. \tag{17}$$

For the special case $N_0 = 2$, $N_1 = 1$ eq. (17) leads to

$$G(y) = (1-y)^2 (1 + 2y + 3y^2)$$

and (18)

$$H(z) = \left(\frac{z+1}{2z}\right)^4 \left[\frac{1}{z^2} - \frac{2}{z}\left(\frac{z-1}{2z}\right)^2 + 3\left(\frac{z-1}{2z}\right)^4\right].$$

Fig. 6 shows the corresponding signal flow graph having two different subsystems with zeros at $z = 1$ or $z = -1$ respectively.

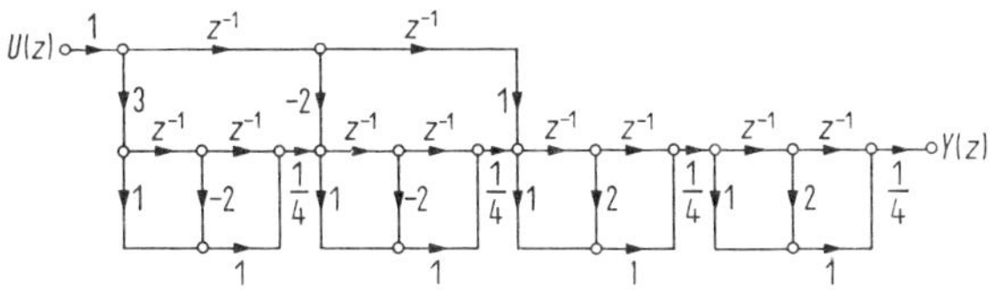


Fig. 6. Variation of the Newton structure (Hermite structure) applied for a lowpass with maximally flat frequency response.

Still another structure will be found if only point $\Omega_0$ has been picked and if $H_0(\Omega_0)$ as well as the first $N$ derivatives at this point are prescribed. Using again the same transformation (12) the polynomial $G(y)$ turns out to be

$$G(y) = \sum_{\mu=0}^{N} d_\mu (y - y_0)^\mu$$

with

$$y_0 = \frac{1}{2}(1 - \cos\Omega_0)\,, \quad d_\mu = \frac{1}{\mu!} G^{(\mu)}(y_0)\,.$$

The $G^{(\mu)}(y_0)$ have to be calculated using the prescribed derivatives

$$\left(\frac{\mathrm{d}^\lambda H_0(\Omega)}{\mathrm{d}\Omega^\lambda}\right)_{\Omega=\Omega_0} \text{ and the values } \left(\frac{\mathrm{d}^\lambda \Omega}{\mathrm{d}y^\lambda}\right)_{\Omega=\Omega_0} \quad \text{for } \lambda = 1\,(1)\,\mu\,.$$

Here the frequency response and the transfer function become

$$H_0(\Omega) = \sum_{\mu=0}^{N} (-1)^\mu \frac{d_\mu}{2^\mu} (\cos\Omega - \cos\Omega_0)^\mu\,,$$

$$H(z) = \sum_{\mu=0}^{N} (-1)^\mu \frac{d_\mu}{z^{N-\mu}} \left(\frac{z^2 + e_0 z + 1}{4z^2}\right)^\mu$$

with $e_0 = -2\cos\Omega_0$.

The corresponding structure will be shown in Fig. 7. It consists of $N$ identical subsystems of second order and $N+1$ coefficients determining the derivatives at the point $\Omega_0$ separately.

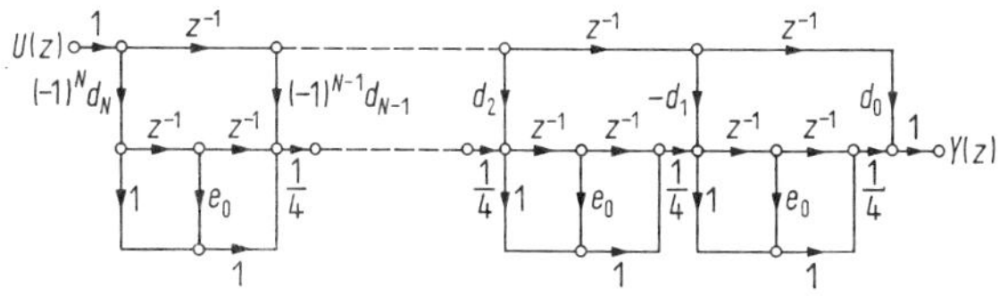


Fig. 7. Taylor structure.

Further investigations will be necessary to find the merits of the new structures as far as coefficient sensitivity and noise are concerned.

(Received March 2nd, 1972.)

**References**

[1] KAISER, J. F., Digital filters; Chap. 7 in "System analysis by digital computers". J. Wiley & Sons, New York 1966.

[2] SCHÜSSLER, W., Zur allgemeinen Theorie der Verzweigungsnetzwerke. AEÜ **22** [1968], 361–367.

[3] KNOWLES, J. B. and EDWARDS, R., Effects of a finite-word-length computer in a sampled data feedback system. Proc. Instn. Elect. Engrs. **112** [1965], 1197–1207.

[4] KNOWLES, J. B. and OLCAYTO, E. M., Coefficient accuracy and digital filter response. Transact. Inst. Elect. Electron. Engrs. CT-**15** [1968], 31–41.

[5] EDWARDS, R., BRADLEY, J., and KNOWLES, J. B., Comparison of noise performance of programming methods in the realization of digital filters. Proc. Symposium on Computer Processing in Communications. Vol. 19 of the Polytechn. Institute of Brooklyn Symposia Series, 1969, pp. 295–311.

[6] AVENHAUS, E., Zum Entwurf digitaler Filter mit minimaler Speicherwortlänge für Koeffizienten und Zustandsgrößen. Ausgewählte Arbeiten über Nachrichtensysteme Nr. 13, herausgegeben von W. SCHÜSSLER, Erlangen 1971.

[7] RADER, C. M. and GOLD, B., Digital filter design techniques in the frequency domain. Proc. Inst. Elect. Electron. Engrs. **55** [1967], 149–171.

[8] RABINER, L. R., GOLD, B., and MCGONEGAL, C. A., An approach to the approximation problem for nonrecursive digital filters. Transact. Inst. Elect. Electron. Engrs. AU-**18** [1970], 83–106.

[9] HERRMANN, O. und SCHÜSSLER, W., On the accuracy problem in the design of nonrecursive digital filters. AEÜ **24** [1970], 525–526.

[10] AVENHAUS, E., Zur Realisierung digitaler Filter mit günstigem Nutz-Störsignalverhältnis. Nachrichtentech. Z. **23** [1970], 217–219.

[11] ZURMÜHL, R., Praktische Mathematik für Ingenieure und Physiker. Springer-Verlag, Berlin 1965.

[12] HERRMANN, O., On the approximation problem in nonrecursive digital filter design. Transact. Inst. Elect. Electron. Engrs. CT-**18** [1971], 411–413.

# Analysis of Linear Digital Networks

RONALD E. CROCHIERE AND ALAN V. OPPENHEIM, SENIOR MEMBER, IEEE

*Invited Paper*

*Abstract*—A framework is presented for the analysis, representation, and evaluation of digital filter structures. Based on the notation of linear signal-flow graphs and their equivalent matrix representation, a set of general linear digital network properties are reviewed, including precedence relations computability, Tellegen's theorem, interreciprocity, and network transposition. These properties are then utilized in developing time and frequency domain analysis techniques and sensitivity analysis techniques. These techniques, in turn, are applied to the comparison of several basic digital filter structures.

## I. Introduction

When implementing a digital filter on a general-purpose computer, it is often unnecessary to pay close attention to the structure used for the implementation. In contrast, when implementing a filter in special-purpose hardware, such factors as speed, cost, wordlength, etc., are closely related to the network structure used. Because of the increasing practicality of hardware implementations of digital filters, there has been considerable effort directed over the last several years toward the development and investigation of digital filter structures. As part of this effort, there has been the development of techniques for analyzing filter structures efficiently and also the theoretical investigation of general properties of structures. Thus, in addition to the development of specific classes of structures, a set of properties and theorems for filter structures has evolved over the last several years. Much of this theory has a counterpart in classical analog network theory and, in fact, in many instances the application of specific results to digital filter structures was motivated by the success of similar techniques in analog network theory.

In this paper we suggest a framework for the analysis and representation of digital filter structures, and their evaluation with respect to several considerations. We have found it convenient to use the notation of linear signal-flow graphs or the equivalent matrix representation. This notation serves as the basis for a general network representation for digital networks, and through this representation many of the theorems and methods of signal-flow graphs can be specifically adapted for digital networks. In particular, in Section II we summarize the signal-flow graph and matrix notation, and in Section III we present a number of general network properties. Sections IV and V are directed toward a discussion of computer-aided analysis of structures. This includes a discussion of the use of the properties developed in Section III for time and frequency domain analysis and the analysis of sensitivity to coefficient quantization. A comparison of a number of structures with regard to coefficient sensitivity for one specific filter is also given.

There are a variety of considerations that must be taken into account in choosing or comparing structures. In Section VI we suggest, as one such consideration, that of the amount of parallelism inherent in a structure. The ideas discussed in that section represent only an initial, but we feel promising, attempt to quantify the consideration of parallelism in structures.

This paper is intended in part as a tutorial review and as such we have attempted to cover a broad range of topics. In many cases, details that are available through the references have been omitted. In particular, many of these details can be found in [1].

Manuscript received August 15, 1974; revised November 11, 1974. This work was supported in part by the National Science Foundation under Grant GK-31353 and in part by the Advanced Research Projects Agency of the Department of Defense, monitored by ONR under Contract N00014-67-A-0204-0064.

R. E. Crochiere is with Bell Laboratories, Murray Hill, N.J. 07974.
A. V. Oppenheim is with the Research Laboratory of Electronics and the Department of Electrical Engineering, Massachusetts Institute of Technology, Cambridge, Mass. 02139.

## II. Signal-Flow Graph and Matrix Representation of Digital Networks

In this paper we will find it convenient to represent digital networks in terms of linear signal-flow graphs or the equivalent matrix representation. This is motivated in part by the properties of signal-flow graphs which are conveniently exploited in the discussion of digital networks. Thus, in this section, we first summarize the notation of signal-flow graphs and the equivalent matrix equations to be used in this paper.

A signal-flow graph is a collection of nodes and directed branches [2], [3]. Associated with each node is a node signal and associated with each branch is a branch signal. For the case of digital networks, the node and branch signals are discrete-time signals or their $z$ transforms.

In representing digital networks with signal-flow graphs, we will find it convenient to distinguish between source branches and network branches. A *network branch* originates from a network node and terminates at a network node. The input to the branch is taken as the signal value at the node where the branch originates.

For the representation of digital filters, we restrict the network branches to correspond either to a constant gain (a coefficient branch) or to a constant gain in cascade with a unit delay (a coefficient-delay branch), and assume also that only one network branch of each kind is permitted between any two nodes. The output of the coefficient branch from node $j$ to node $k$ will be denoted as $w_{cjk}(n)$ and the output of the delay branch from node $j$ to node $k$ will be denoted as $w_{djk}(n)$. The node signal value at node $j$ will be denoted as $y_j(n)$. The branch outputs are then related to the branch inputs by

$$w_{cjk}(n) = f_{cjk}\, y_j(n) \qquad (1a)$$

and

$$w_{djk}(n) = f_{djk}\, y_j(n-1) \qquad (1b)$$

where $f_{cjk}$ and $f_{djk}$ are constant real coefficients. Alternatively, we can express the node values and branch outputs in terms of

Reprinted from *Proc. IEEE*, vol. 63, pp. 581–595, Apr. 1975.

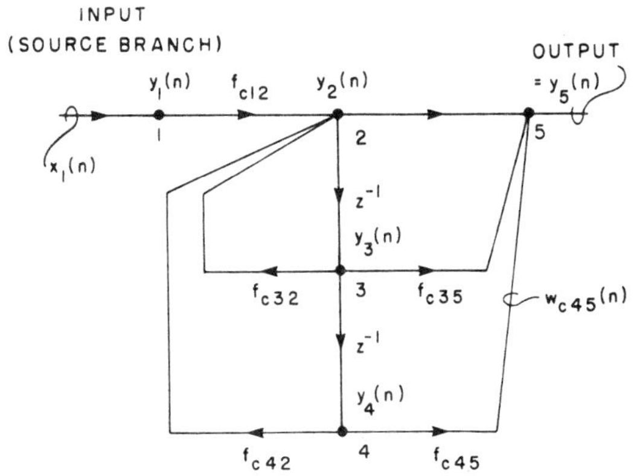


Fig. 1. Example of digital network expressed in signal-flow graph notation.

their $z$ transforms—in which case, (1a) and (1b) are written as

$$W_{cjk}(z) = F_{cjk}(z)Y_j(z) \tag{2a}$$

$$W_{djk}(z) = F_{djk}(z)Y_j(z) \tag{2b}$$

where

$$F_{cjk}(z) = f_{cjk} \tag{3a}$$

and

$$F_{djk}(z) = z^{-1}f_{djk}\,. \tag{3b}$$

The terms $F_{cjk}(z)$ and $F_{djk}(z)$ correspond to the *branch transmittances* of the coefficient and the coefficient-delay branches, respectively. It will generally be convenient to assume that there is both a coefficient and a coefficient-delay branch between each pair of network nodes, although some of the branch coefficients may be zero.

*Source branches* represent the injection of external signal sources into the network. They will be assumed to have unity transmittance so that the source branch output corresponds to the value of the external signals entering the network at that point. It will be assumed that there is at most one source branch associated with each node in the network. The source branch signal value associated with the $k$th node is designated as $x_k(n)$ or equivalently $X_k(z)$ for its $z$ transform. For convenience, we will assume that a source branch connects to each network node, although at some nodes the source branch output may be zero.

Using the foregoing signal-flow graph notation, linear digital networks of any arbitrary configuration can be represented. An example of such a digital network is given in Fig. 1. It corresponds to a conventional second-order direct form (canonical) structure. The input of the network is via a source branch entering node 1 and the output is taken as the node signal $y_5(n)$ of node 5. Coefficients $f_{c25}$, $f_{d23}$, and $f_{d34}$ are assumed to be unity and all other branches not shown are assumed to have coefficient values of zero. Nodes 2 and 5 act as both summing points and branch points in the network. The branch signal $w_{c45}(n)$, for example, is obtained from the multiplication of the node signal $y_4(n)$ by the branch coefficient $f_{c45}$.

The node signal value at each node in a network is the sum of the branch signals entering the node. Thus

$$Y_k(z) = X_k(z) + \sum_{j=1}^{N} \left[W_{cjk}(z) + W_{djk}(z)\right], \qquad k = 1, 2, \cdots, N \tag{4}$$

where it is assumed that there are $N$ nodes in the network. Alternatively, with the aid of (2) and (3), this can be written as

$$Y_k(z) = X_k(z) + \sum_{j=1}^{N} \left[f_{cjk} + f_{djk}\, z^{-1}\right] Y_j(z), \qquad k = 1, 2, \cdots, N. \tag{5}$$

As there are $N$ nodes in the network, there are $N$ corresponding equations and they constitute a total description of the network. Inputs to the network are described by means of the source branches and outputs of the network are obtained as the signal values of one or more of the network nodes.

The $N$ equations in (5) can be expressed more compactly in matrix form as

$$\boldsymbol{Y}(z) = \boldsymbol{X}(z) + \boldsymbol{f}_c^t\boldsymbol{Y}(z) + \boldsymbol{f}_d^t\boldsymbol{Y}(z)z^{-1} \tag{6}$$

where

- $\boldsymbol{Y}(z)$ a column vector of the $N$ node signal values $Y_k(z)$ ($k = 1, 2, \cdots, N$),
- $\boldsymbol{X}(z)$ a column vector of the $N$ branch signal values $X_k(z)$ ($k = 1, 2, \cdots, N$) of the source branches,
- $\boldsymbol{f}_c$ an $N \times N$ matrix of coefficients for network branches of the first kind (coefficient branches),
- $\boldsymbol{f}_d$ an $N \times N$ matrix of coefficients for network branches of the second kind (coefficient-delay branches).

The superscript $t$ in (6) denotes the matrix transpose, which is used for consistency between the subscript conventions of signal-flow graphs and matrices. An element $f_{djk}$ of matrix $\boldsymbol{f}_d$, for example, corresponds to the coefficient value of the coefficient-delay branch from node $j$ to node $k$. If no such branch exists in the network, then $f_{djk} = 0$. As an example, the matrix representation for the digital network in Fig. 1 is

$$\begin{bmatrix} Y_1(z) \\ Y_2(z) \\ Y_3(z) \\ Y_4(z) \\ Y_5(z) \end{bmatrix} = \begin{bmatrix} X_1(z) \\ 0 \\ 0 \\ 0 \\ 0 \end{bmatrix} + \begin{bmatrix} 0 & 0 & 0 & 0 & 0 \\ f_{c12} & 0 & f_{c32} & f_{c42} & 0 \\ 0 & 0 & 0 & 0 & 0 \\ 0 & 0 & 0 & 0 & 0 \\ 0 & 1 & f_{c35} & f_{c45} & 0 \end{bmatrix} \begin{bmatrix} Y_1(z) \\ Y_2(z) \\ Y_3(z) \\ Y_4(z) \\ Y_5(z) \end{bmatrix} + z^{-1} \begin{bmatrix} 0 & 0 & 0 & 0 & 0 \\ 0 & 0 & 0 & 0 & 0 \\ 0 & 1 & 0 & 0 & 0 \\ 0 & 0 & 1 & 0 & 0 \\ 0 & 0 & 0 & 0 & 0 \end{bmatrix} \begin{bmatrix} Y_1(z) \\ Y_2(z) \\ Y_3(z) \\ Y_4(z) \\ Y_5(z) \end{bmatrix}$$

The response of the node signal values $\boldsymbol{Y}(z)$ of a network due to the signal inputs $\boldsymbol{X}(z)$ can be expressed in terms of a matrix inversion as

$$\boldsymbol{Y}(z) = \boldsymbol{T}^t(z)\boldsymbol{X}(z) \tag{7}$$

where

$$\boldsymbol{T}^t(z) = [\boldsymbol{I} - \boldsymbol{f}_c^t - \boldsymbol{f}_d^t z^{-1}]^{-1} \tag{8}$$

and $\boldsymbol{I}$ is defined as the $N \times N$ identity matrix. The $N \times N$ matrix $\boldsymbol{T}(z)$ will be referred to as the transfer function matrix of the network. An element $T_{jk}(z)$ of this matrix can be identified as the transfer function from node $j$ to node $k$ and it characterizes the response of the network at node $k$ due to a single source branch input at node $j$. That is, with all source branch inputs but $X_j(z)$ set to zero,

$$T_{jk}(z) = \frac{Y_k(z)}{X_j(z)}. \tag{9}$$

In general, if more than one signal is injected into the network by means of source branches, the response at a particular node $k$ can be expressed, using superposition, as

$$Y_k(z) = \sum_{j=1}^{N} T_{jk}(z) X_j(z). \tag{10}$$

If the source branch at a specific node $a$ is defined as the network input and a specific node signal at node $b$ is defined to be taken as the output of the network, then the transfer function $T_{ab}(z)$ is the system function of the network. For example, in the network of Fig. 1, the system function would correspond to $T_{15}(z)$.

The matrix equation (6) can alternatively be expressed in the time domain. Specifically, the inverse $z$ transform of (6) corresponds to the time domain matrix difference equation

$$y(n) = x(n) + f_c^t y(n) + f_d^t y(n-1) \tag{11}$$

which is similar in some respects to a state-space description of a digital network [4]. It differs, however, in a number of important respects. The number of variables in (11) corresponds to the number of nodes in the network. This is in contrast to a state-space description where the number of variables is equal to the number of essential states in the network, i.e., the number of network nodes with one or more delay branches entering them. The matrix difference equation (11) has the property that the coefficients in the matrices $f_c^t$ and $f_d^t$ correspond in a one-to-one fashion with the branch transmittances in the flow graph. In a state-space representation, this is not generally true.

## III. General Network Properties

In the preceding section, we presented the representation of digital networks in terms of flow graph and matrix notation. Based on this representation, there are a number of network properties which we will want to make use of in the remaining section. The first of these is the notion of node precedence relations. As we will see, there is inherent in any given network structure an order of precedence in which node values are computed. The node precedence relations become particularly important in Section VI where we will discuss parallelism in structures.

A second general property that we discuss is Tellegen's theorem for digital networks. This simple but elegant theorem provides a useful framework for developing a number of other properties of digital networks. We conclude this section with a discussion of the notions of reciprocity, interreciprocity, and network transposition. These ideas are useful in computer-aided analysis of network structures and play a particularly important role in the sensitivity analysis of structures as presented in Section V.

### *Precedence Relations for Digital Networks*

The implementation of a digital network generally corresponds to a direct evaluation of the difference equation (11). We observe that (11) expresses each node value as a linear combination of the inputs and present and past values of the node values. If we do not order the computation of the nodes, it is possible that the computation of one node value requires other node values that have not yet been computed. Thus there is an implied set of precedence relations associated with this set of $N$ scalar equations and they are solely a consequence of the coefficient branch topology in the structure.

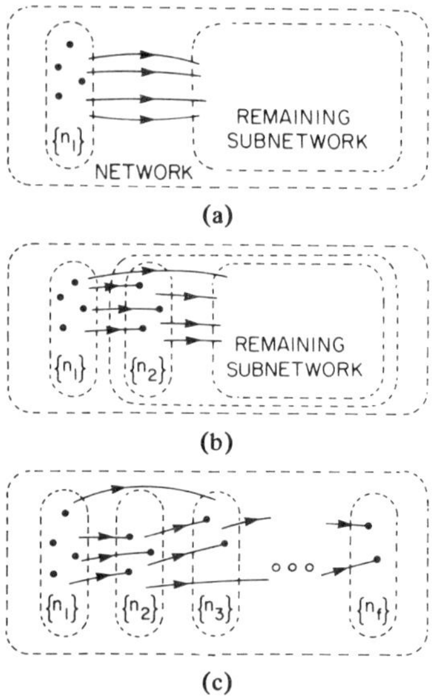


Fig. 2. Algorithmic procedure for generating precedence form of a network. (a) Extracting first node set. (b) Extracting second node set. (c) Final representation of network in precedence form (only coefficient branches are shown).

It is of interest to investigate the implications of these precedence relations. We first note that the precedence relations are solely a consequence of the coefficient branch topology in the structure and do not depend at all on the coefficient-delay branches. An algorithmic procedure for laying out a structure in a form which reveals these precedence relations will be given. The procedure begins by searching the network for all nodes which do not have coefficient branches entering them. These nodes are then separated from the network and assigned to node set $\{n_1\}$, as depicted in Fig. 2(a). It is apparent that these node values can be computed without knowledge of any of the node values in the network for time index $n$. Their evaluation may require only past node values for time $n-1$ due to coefficient-delay branches, or present source input values at time $n$.

Next, the subnetwork is considered which contains all nodes and branches in the network *except* those nodes which belong to the node set $\{n_1\}$ and those branches which connect these nodes. A search is made over this subnetwork for all nodes which do not have coefficient branches from other nodes in the subnetwork entering them. These nodes are assigned to node set $\{n_2\}$, as indicated in Fig. 2(b). It is apparent that each of the nodes in set $\{n_2\}$ must have at least one coefficient branch entering it which comes from node set $\{n_1\}$ when considering the overall network, since otherwise the node would have been contained in set $\{n_1\}$. This procedure is repeated, as depicted in Fig. 2(c), until at the last stage none of the remaining nodes of the final subnetwork have coefficient branches entering them. These nodes are assigned to node set $\{n_f\}$. The resulting configuration of the network, as given in Fig. 2(c), is referred to as the *precedence form* of the network as it depicts all of the precedence requirements of the network in terms of node evaluations. An example of the precedence form of the

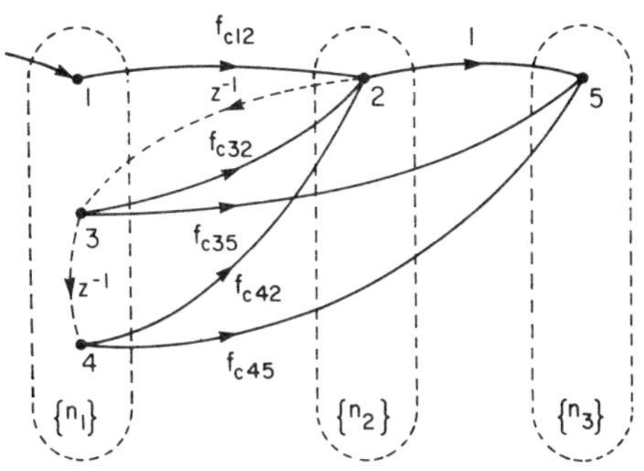


Fig. 3. Precedence form of network in Fig. 1.

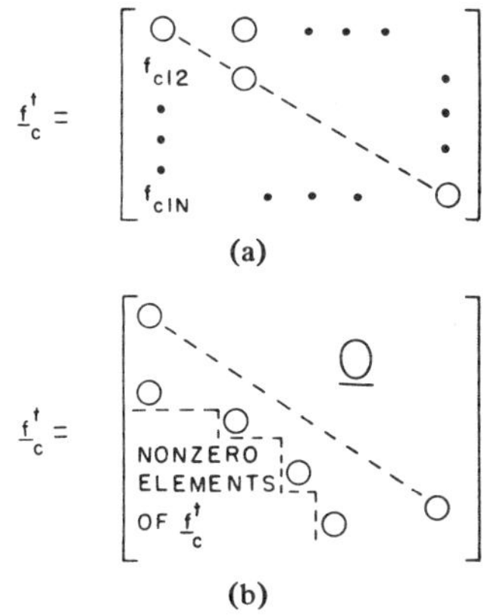


Fig. 4. (a) Form of matrix term $f_c^t$ for matrix representation in a computable form. (b) General form of $f_c^t$ when node sets contain more than one node each and nodes are numbered according to algorithmic procedure.

network in Fig. 1 is given in Fig. 3. In this figure, we have indicated the coefficient branches as solid lines and the coefficient-delay branches as dotted lines to emphasize the fact that the precedence relations are a function only of the coefficient branch topology.

If the foregoing algorithmic procedure does not proceed to completion, then the precedence relations are contradictory and the network is defined as being *noncomputable*. That is, if at some stage all nodes in the remaining subnetwork have coefficient branches entering them, then this subnetwork contains one or more closed loops without delay in it and none of the nodes can be evaluated without requiring the value of at least one other node in the subnetwork. If the algorithm does proceed to completion and produces a network in the form depicted in Fig. 2(c), then there are no closed loops without delay in the network and it is defined as being *computable*. Generally, computable networks are the only type of digital networks that are of interest in digital signal processing. It should be noted that if a network is noncomputable, the network equation (11) may still be solvable. However, they are not solvable by computing successive node values in the network.

The network property of computability can be represented explicitly in terms of the matrix representation. Specifically, the matrix representation in (6) or (11) for a computable network can be written in a form such that the matrix $f_c^t$ is zero on and above the main diagonal [5], as depicted in Fig. 4(a). Furthermore, it can be shown that if a network is noncomputable, then the matrix representation cannot be expressed in this form [1]. A procedure for numbering the nodes of a computable network in order to obtain a matrix representation

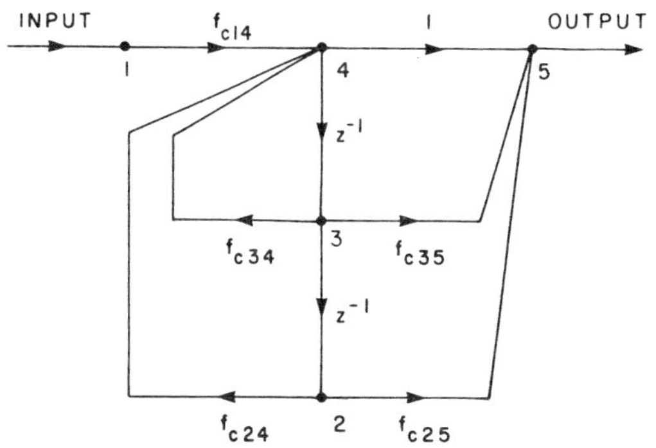


Fig. 5. Network of Fig. 1 with nodes numbered according to precedence form numbering scheme.

of this form can be determined with the aid of the precedence form of the network. To obtain this form, the network nodes can be numbered from 1 to $N$ consecutively, first numbering all nodes in set $\{n_1\}$, then all nodes in sets $\{n_2\}$, $\{n_3\}, \cdots$, $\{n_f\}$, until all of the nodes have been numbered. With the node numbers corresponding to the subscripts of the matrix equation, this leads to a form of the matrix $f_c^t$ given in Fig. 4(b). If the network is computable, then there is at least one node in each node set and consequently $f_c^t$ is at least of the form given in Fig. 4(a) and can generally be expressed in a form depicted in Fig. 4(b).

If this numbering scheme is applied to the network example of Fig. 1, it is apparent from the precedence form in Fig. 3 that the original node numbers 2 and 4 must be interchanged. This results in the new node numbering scheme given in Fig. 5. The matrix representation for this network then becomes

$$\begin{bmatrix} Y_1(z) \\ Y_2(z) \\ Y_3(z) \\ Y_4(z) \\ Y_5(z) \end{bmatrix} = \begin{bmatrix} X_1(z) \\ 0 \\ 0 \\ 0 \\ 0 \end{bmatrix} + \begin{bmatrix} 0 & 0 & 0 & 0 & 0 \\ 0 & 0 & 0 & 0 & 0 \\ 0 & 0 & 0 & 0 & 0 \\ f_{c14} & f_{c24} & f_{c34} & 0 & 0 \\ 0 & f_{c25} & f_{c35} & 1 & 0 \end{bmatrix} \begin{bmatrix} Y_1(z) \\ Y_2(z) \\ Y_3(z) \\ Y_4(z) \\ Y_5(z) \end{bmatrix} + z^{-1} \begin{bmatrix} 0 & 0 & 0 & 0 & 0 \\ 0 & 0 & 1 & 0 & 0 \\ 0 & 0 & 0 & 1 & 0 \\ 0 & 0 & 0 & 0 & 0 \\ 0 & 0 & 0 & 0 & 0 \end{bmatrix} \begin{bmatrix} Y_1(z) \\ Y_2(z) \\ Y_3(z) \\ Y_4(z) \\ Y_5(z) \end{bmatrix}$$

where $f_c^t$ is now of the form such that it is zero on and above the main diagonal.

If a network is nonrecursive and computable, i.e., if it has no loops (with or without delay) in its topology, then the matrix representation can be expressed in a form such that both $f_c^t$ and $f_d^t$ are simultaneously zero on and above their main diagonals [1], as expressed in Fig. 4(a). The procedure for generating this form of the matrix representation is similar to that for generating the computable form of the matrix, with the exception that both coefficient and coefficient-delay branches are used in the construction of a network form analogous to that of Fig. 2(c). An interesting consequence of this form of the matrix representation is that it leads to a transposed transfer function matrix $T^t(z)$ which is lower triangular. This follows from (8) since the inverse of a lower triangular matrix is lower triangular.

*Tellegen's Theorem for Digital Networks*

Tellegen's theorem for analog networks has played an important role in developing and understanding network properties [6], [7]. Because digital networks, and more generally signal-flow graphs, are not subject to Kirchhoff's laws, Tellegen's theorem in its most general form does not apply. It does apply, however, in a somewhat restricted form. Tellegen's theorem, as it applies to digital networks, provides a useful framework for developing and interpreting a number of important and useful network properties. In particular, we will utilize Tellegen's theorem to show that the matrix of transfer functions for a network and its transpose are interreciprocal. As we will see, this property is useful for computer-aided analysis of filter structures. Tellegen's theorem also leads to a number of useful relations which we utilize in Section V for sensitivity analysis of digital filter structures. While in many cases these properties can be developed in alternate ways without utilizing Tellegen's theorem directly, Tellegen's theorem is appealing in part because it provides a unifying basis.

Several variations of Tellegen's theorem for signal-flow graphs have been proposed, including those by Seviora and Sablatash [8], Fettweis [9], Blostein [10], and Lee [11], [12]. The form of the theorem and its related concepts which are presented here is perhaps most similar to that proposed by Fettweis, and is stated as follows.

*Tellegen's Theorem:* Consider two networks, each of which contains $N$ nodes. For the first network, the source branch signals, coefficient branch signals, coefficient-delay branch signals, and node signals are denoted as $X_k(z)$, $W_{cjk}(z)$, $W_{djk}(z)$, and $Y_k(z)$, respectively; for the second network, they are denoted as $X'_k(z)$, $W'_{cjk}(z)$, $W'_{djk}(z)$, and $Y'_k(z)$, respectively. Then Tellegen's theorem states that

$$\sum_{j=1}^{N}\sum_{k=1}^{N} [Y_k(z)\,(W'_{cjk}(z) + W'_{djk}(z)) - Y'_k(z)\,(W_{cjk}(z) + W_{djk}(z))] + \sum_{k=1}^{N} [Y_k(z)X'_k(z) - Y'_k(z)X_k(z)] = 0 \quad (12)$$

which follows in a straightforward manner by substituting (4) into the identity

$$\sum_{k=1}^{N} [Y_k(z)Y'_k(z) - Y'_k(z)Y_k(z)] = 0. \quad (13)$$

Alternatively, Tellegen's theorem can be expressed in the time domain as

$$\sum_{j=1}^{N}\sum_{k=1}^{N} [y_k(n)\,(w'_{cjk}(n) + w'_{djk}(n)) - y'_k(n)\,(w_{cjk}(n) + w_{djk}(n))] + \sum_{k=1}^{N} [y_k(n)x'_k(n) - y'_k(n)x_k(n)] = 0. \quad (14)$$

Three closely associated concepts which follow directly from Tellegen's theorem are those of reciprocity, interreciprocity, and network transposition. These concepts have long histories in classical network theory. They have been developed specifically for linear digital networks through the efforts of Jackson [13] and Fettweis [9]. They are also closely related to Mason's theorem on flow graph reversal [2], [3] and to the adjoint signal-flow graph concepts proposed by Seviora and Sablatash [8], Blostein [10], and Lee [11], [12]. The discussion in the next two sections follows most closely the approach taken by Fettweis.

*Reciprocity and Interreciprocity in Digital Networks*

A network is defined to be reciprocal if, for any two signal distributions in the network, denoted by primed and unprimed variables, it is possible to write [9]

$$\sum_{k=1}^{N} [Y_k(z)X'_k(z) - Y'_k(z)X_k(z)] = 0. \quad (15)$$

For linear shift-invariant networks, it follows by substituting (10) into (15) that this is equivalent to the condition

$$T_{jk}(z) = T_{kj}(z) \quad (16)$$

for all pairs of nodes $j$ and $k$ in the network. That is, the matrix of transfer functions $T(z)$ for a reciprocal network is symmetric, so that for any pair of nodes $j$ and $k$, the transfer function from $j$ to $k$ is identical to the transfer function from $k$ to $j$. In practice, digital networks generally do not have this property.

A more important, but closely related, concept is that of interreciprocity between two networks [9]. If two networks are considered, each with $N$ nodes, the first with signal variables $X_k(z)$ and $Y_k(z)$ and the second with signal variables $X'_k(z)$ and $Y'_k(z)$, then the two networks are interreciprocal if

$$\sum_{k=1}^{N} [Y_k(z)X'_k(z) - Y'_k(z)X_k(z)] = 0. \quad (17)$$

This relation is identical to that of (15), except that the primed and unprimed variables in this case correspond to sets of variables in two different networks. By substituting (10) into (17), it follows that for linear shift-invariant networks, this condition is equivalent to the condition

$$T_{jk}(z) = T'_{kj}(z) \quad (18)$$

for all pairs of nodes $j$ and $k$ in the two networks. That is to say, the transfer function matrix of the unprimed network is equal to the transpose of the transfer function matrix of the primed network. The applications of this concept will become apparent shortly.

*Network Transposition*

Another important concept in network theory is that of network transposition [9], [13]. The transposition of a network is defined as the operation of reversing the direction of all of the branches in the network. The resulting network is referred to as the transpose network. In the matrix representation, this corresponds to the operations

$$f'_c = f_c^t \quad (19a)$$

and

$$f'_d = f_d^t \quad (19b)$$

where the primed matrices correspond to those of the transpose network.

An important relationship between networks and their transposes is that they are interreciprocal. This can be shown by utilizing (19), together with Tellegen's theorem (12). Specifically, for the original network,

$$W_{cjk}(z) = F_{cjk}(z)\, Y_j(z) \tag{20a}$$

$$W_{djk}(z) = F_{djk}(z)\, Y_j(z) \tag{20b}$$

while, for the transposed (primed) network,

$$W'_{cjk}(z) = F'_{cjk}(z)\, Y'_j(z) = F_{ckj}(z)\, Y'_j(z) \tag{20c}$$

$$W'_{djk}(z) = F'_{djk}(z)\, Y'_j(z) = F_{dkj}(z)\, Y'_j(z). \tag{20d}$$

Equation (17) then follows from substitution of (20) into (12) and appropriate manipulations of the resulting summations.

## IV. Time and Frequency Domain Analysis of Digital Networks

A particularly important area of application of digital network theory is in the development of general techniques for the analysis of arbitrary digital networks. The need for such techniques can be recognized when one considers the large variety of possible structures recently proposed as candidates for filter implementation. Some of these types of structures are illustrated in Section VI. With a computer-aided network analysis package, the process of analyzing and comparing such structures can be reduced to a much simpler and more manageable task.

The general matrix representation for digital networks presented in Section II provides for a convenient way of uniquely expressing networks of any configuration on a computer. The process of describing these networks and analyzing them can then be interpreted in terms of performing appropriate manipulations on the computer.

### *Frequency Response Analysis*

One approach to obtaining the frequency response of an arbitrary network is to perform the matrix inversion in (8) at each frequency. In general, this approach is not very desirable as it requires an excessive amount of computation and can be prone to numerical errors [14]. A more desirable approach for computing the frequency response can be realized by expressing the matrix representation (6) in the form

$$[I - f_c^t - f_d^t z^{-1}]\, Y(z) = X(z) \tag{21}$$

where $I$ is the $N \times N$ identity matrix. This matrix equation corresponds to $N$ simultaneous linear scalar equations in the $N$ unknowns $Y_i(z), i = 1, 2, \cdots, N$. By assuming that the source branch inputs are all zero, except for that connected to node $a$, the solution to this set of equations provides the transfer functions $T_{aj}(z)$, $j = 1, 2, \cdots, N$. The solution of the $N$ simultaneous linear equations can be obtained by a procedure such as Gaussian elimination using complex arithmetic. Generally, the matrices $f_c^t$ and $f_d^t$ will be relatively sparse. Various techniques were recently proposed for taking advantage of this sparsity to minimize the amount of computation necessary to solve equations of this type [15], [16]. Although many of these techniques have been developed for computer-aided analysis of conventional analog circuits, many of them apply equally well for the case of digital networks.

### *Time Domain Analysis*

One approach for computation of the unit-sample response is to first express the time domain matrix representation (11) in a computable form and then solve the $N$ scalar equations corresponding to this matrix equation consecutively at each time increment $n$. As discussed in Section III, the computable form of the matrix representation is that form in which $f_c^t$ is zero on and above the main diagonal. To perform the network analysis, the initial conditions $y(-1) = \mathbf{0}$ are chosen and the input signals

$$x_j(n) = \begin{cases} 1, & j = a \text{ and } n = 0 \\ 0, & \text{otherwise} \end{cases}$$

are used. The unit-sample responses from a given input node $a$ to all other nodes in the network can then be computed by solving the $N$ scalar equations in the computable form of (11) for $y_j(n)$ for $j = 1, 2, \cdots, N$ consecutively and for $n = 0, 1, 2, 3, \cdots$ consecutively. As this set of calculations corresponds directly to that which must be performed in the actual implementation of the structure, simulation of effects such as limit cycles and overflow can also be incorporated into this procedure by duplicating on the computer the actual arithmetic operations which would be performed in the hardware implementation of the network.

An alternate approach to time domain analysis is to transform (11) into the form

$$y(n) = Ay(n - 1) + Bx(n) \tag{22}$$

where

$$B = [I - f_c^t]^{-1} \tag{23}$$

and

$$A = Bf_d^t. \tag{24}$$

This can be accomplished if the initial matrix inversion in (23) can be performed as it can for computable networks since in that case $\det [I - f_c^t] = 1$. The solution of the unit sample response can then be obtained by solving (22) successively for $y(n)$ with the same initial conditions and input signals already used. As this is no longer the same set of equations as that represented by the structure, the simulation of limit cycles or overflow effects cannot be made with this method.

### *Analysis of Transpose Networks*

The preceding frequency response and time response analyses yield the transfer functions or unit-sample responses, respectively, from the network input to all points internal to the network with a single analysis. In some cases, it is desired to analyze such responses from various points internal to the network to the output. For example, in the analysis of the effects of computational roundoff noise, the roundoff errors occur at various points within the network and can be modeled as an injection of white noise sources at these points in the network [13], [17], [18]. To determine the effect of these errors on the network output, it is necessary to determine the transfer functions from various internal points of the network to the output. With the analysis procedures described, this would require several analyses at each frequency. Alternatively, this can be performed with a single analysis at each frequency of interest by exploiting the interreciprocal property of transpose networks discussed in Section III. For example, the frequency response analysis of the transpose network corresponds to the analysis of the $N$ simultaneous linear equations given by

$$[I - f_c - f_d z^{-1}]\, Y'(z) = X'(z) \tag{25}$$

where the unprimed variables correspond to those of the

original network and the primed variables correspond to those of the transpose network. With $X'_b(z)$ chosen to be unity and $X'_j(z)$ chosen to be zero for $j \neq b$, the solution gives all of the transfer functions

$$T_{jb}(z) = T'_{bj}(z) = Y_j(z), \quad j = 1, 2, \cdots, N.$$

That is, it gives all of the transfer functions from all internal points of the network to the output for a single analysis of the transpose network. A similar computational savings can be obtained in a time domain analysis where the unit-sample responses from more than one internal point of the network to the output are desired.

## V. Sensitivity and Derivative Analysis of Networks

The implementation of digital filters is subject to the constraint of finite register length, the effects of which manifest themselves in two ways. One is the introduction of error due to arithmetic roundoff and the second is a change in the overall filter characteristics due to coefficient quantization. As indicated in the previous section, the analysis of the effects of arithmetic roundoff can utilize the properties of transpose networks to simultaneously compute the transfer functions from all of the roundoff noise sources to the output.

The effects of coefficient quantization can also be analyzed utilizing many of the ideas introduced in the previous section. The sensitivity of filter characteristics to quantization of the coefficients is highly dependent on the particular structure used to represent the filter and consequently, in evaluating filter structures, it is important to carry out an analysis of the sensitivities of the system function with respect to the branch coefficients or branch transmittances in the network. In this section, techniques for such a sensitivity analysis will be discussed. In addition, two other types of analysis, which are sometimes desired in general computer-aided analysis systems, are considered. The first of these is the evaluation of the group delay associated with the system function of a network and the second is the slope of the magnitude of the frequency response.

### *A First-Order Network Sensitivity Relation*

A convenient expression for evaluating the sensitivities of the system function, with respect to the branch coefficients, has been proposed by various authors, including Seviora and Sablatash [8], Fettweis [9], and Lee [11], [12]. The proof of this expression is very elegant in that it involves the use of Tellegen's theorem and the concepts of interreciprocity and network transposition [8], [9], [19]. It is a rather lengthy proof, however, and will not be repeated here. The sensitivity expression can be defined by considering the single-input-single-output network indicated in Fig. 6. The system function of interest in this network, $H(z)$, is equal to the transfer function $T_{ab}(z)$ from node $a$ to node $b$:

$$H(z) = T_{ab}(z). \tag{26}$$

The sensitivity of $H(z)$ with respect to a branch transmittance $F_{nm}(z)$ of a branch (either a coefficient or a coefficient-delay branch) from some node $n$ to some node $m$ can be given as

$$\left\{\begin{matrix}\text{sensitivity of } H(z) \\ \text{with respect to } F_{nm}(z)\end{matrix}\right\} \triangleq \frac{\partial T_{ab}(z)}{\partial F_{nm}(z)} = T_{an}(z)\, T_{mb}(z). \tag{27}$$

Thus the sensitivity of $H(z)$ with respect to $F_{nm}(z)$ can be expressed in terms of the product of two transfer functions in

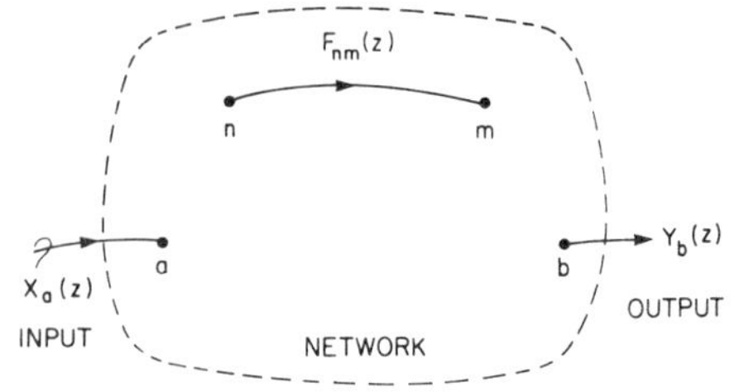


Fig. 6. Single-input-single-output network.

the network—the transfer function from node $a$ to node $n$ and the transfer function from node $m$ to node $b$.

In some cases, the sensitivities of the magnitude of the system function with respect to the branch coefficients $f_{cnm}$ and $f_{dnm}$ may be desired. These sensitivities can be derived with the aid of (27) and are given as [1]

$$\frac{\partial |H(z)|}{\partial f_{cnm}} = \text{Re}\left[\frac{|H(z)|}{H(z)} T_{an}(z)\, T_{mb}(z)\right] \tag{28}$$

and

$$\frac{\partial |H(z)|}{\partial f_{dnm}} = \text{Re}\left[\frac{|H(z)|}{H(z)} T_{an}(z)\, T_{mb}(z)\, z^{-1}\right] \tag{29}$$

where it has been assumed that $f_{cnm}$ and $f_{dnm}$ are real. The sensitivity relations (27)–(29) are expressed in terms of the system function and two transfer functions $T_{an}(z)$ and $T_{mb}(z)$ in the network. The first transfer function is always one which is directed from the input of the network to an internal point of the network and the second transfer function is always one which is directed from an internal point in the network to the output. Thus the evaluation of network sensitivities is performed via the computation of network transfer functions and they can be determined with only two analyses of the network at each frequency. The first set of transfer functions $T_{an}(z)$ is obtained from the analysis of the original network and the second set of transfer functions $T_{mb}(z)$ is obtained from the analysis of the transpose network as discussed in Section IV. The system function $T_{ab}(z)$ can be obtained from either of the two analyses. If the sensitivity of the system function with respect to more than one coefficient in the network is desired, as is often the case, these sensitivities can all be obtained from the results of the same two network analyses with essentially no extra cost in computation. This then constitutes a highly efficient technique for the analysis of multiparameter network sensitivities. The technique is similar to the adjoint network approach by Director and Rohrer [7] for evaluating sensitivities of analog networks.

### *Higher Order and Large-Change Network Sensitivities*

In studying the effects of variations of the system function $H(z)$ due to variations of coefficients or branch transmittances, we may, in some cases, desire to know the effect of changes in $H(z)$, denoted as $\Delta H(z)$, due to changes $\Delta F_{nm}(z)$ in a branch transmittance $F_{nm}(z)$, where $\Delta F_{nm}(z)$ may be more than an incremental change. These changes can be expressed with the aid of the Taylor's series expansion in the form

$$\Delta H(z) = \frac{\partial H(z)}{\partial F_{nm}(z)} \Delta F_{nm}(z) + \frac{1}{2} \frac{\partial^2 H(z)}{\partial F_{nm}^2(z)} (\Delta F_{nm}(z))^2 + \cdots \tag{30}$$

where $H(z) = T_{ab}(z)$, as denoted in Fig. 6. The higher order sensitivities in this expression can be obtained with the aid of the first-order network sensitivity relation in (27) and the chain rule for differentiation. This approach leads to a general expression for the higher order network sensitivities, which for the $r$th-order sensitivity can be given in the form [5]

$$\frac{\partial^r H(z)}{\partial F_{nm}^r(z)} = r!\, T_{mn}^{r-1}(z)\, T_{an}(z)\, T_{mb}(z). \tag{31}$$

By applying this expression to the Taylor's series expansion of (30), a closed-form solution can be obtained:

$$\Delta H(z) = \frac{T_{an}(z)\, T_{mb}(z)\, \Delta F_{nm}(z)}{1 - T_{mn}(z)\, \Delta F_{nm}(z)} \tag{32}$$

where the transfer functions $T_{an}(z)$, $T_{mn}(z)$, and $T_{mb}(z)$ are evaluated for $\Delta F_{nm}(z) = 0$. Various forms of equations similar to (32) have been widely used in feedback control theory involving the use of signal-flow graphs [20], [21]. It is sometimes referred to as the bilinear theorem.

An interesting consequence of the large-change sensitivity relation (32) can be observed for the class of nonrecursive computable networks. For these structures it has been observed in Section III that if $F_{nm}(z) \neq 0$, then $T_{mn}(z) = 0$. Consequently, for this class of structures, (32) takes the form

$$\Delta H(z) = T_{an}(z)\, T_{mb}(z)\, \Delta F_{nm}(z) = \frac{\partial H(z)}{\partial F_{nm}(z)}\, \Delta F_{nm}(z). \tag{33}$$

Thus the change in the system function varies linearly with respect to a change in a branch transmittance in nonrecursive computable networks [5].

### *Derivative Analysis of Networks*

Two other types of analysis, which are sometimes desired in general computer-aided analysis systems, are the evaluation of the group delay of the system function of a network and the slope of the magnitude of the frequency response. For these analyses, the frequency response can be defined as

$$H(z)\big|_{z=e^{j\omega}} = T_{ab}(e^{j\omega}) = |H(e^{j\omega})|\, e^{j\theta(\omega)} \tag{34}$$

where $\omega$ is the angular frequency and $\theta(\omega)$ is the phase. The group delay $\tau$ is then given by

$$\tau = -\frac{\partial \theta(\omega)}{\partial \omega} \tag{35}$$

and the slope of the magnitude of the frequency response is given by $\partial |H(e^{j\omega})| / \partial \omega$. General expressions for these quantities can be derived in terms of appropriate system functions in the network and can be expressed in the forms [1], [22]

$$\tau = \sum_{n=1}^{N} \sum_{m=1}^{N} \mathrm{Re} \left[ \frac{f_{dnm}\, T_{an}(z)\, T_{mb}(z)\, z^{-1}}{H(z)} \right] \Bigg|_{z=e^{j\omega}} \tag{36}$$

$$\frac{\partial |H(e^{j\omega})|}{\partial \omega} = \frac{1}{|H(e^{j\omega})|} \sum_{n=1}^{N} \sum_{m=1}^{N} \cdot \mathrm{Im} \left[ \frac{f_{dnm}\, T_{an}(z)\, T_{mb}(z)\, z^{-1}}{H(z)} \right] \Bigg|_{z=e^{j\omega}} \tag{37}$$

where Re $[\cdot]$ and Im $[\cdot]$ correspond to the operations of taking the real and the imaginary parts, respectively. The double sums in (36) and (37) correspond to the operation of summing over all possible coefficient-delay branches in the network. Alternatively, this sum may be carried out only over those coefficient-delay branches which have nonzero coefficient values $f_{dnm}$ as all other terms in the summation for which $f_{dnm} = 0$ will be zero. Similar analysis techniques for the analysis of group delay and the slope of the magnitude of the frequency response for analog networks have been proposed by Pinel and Blostein [23], [24].

Some of these general analysis techniques have been utilized in a digital network analysis package called CADNAP [1]. It is capable of analyzing networks of arbitrary configuration for system functions, unit-sample responses, and coefficient sensitivities. This package was used for the analysis of the networks given in Section VI.

### *A Statistical Measure of Word Length*

In evaluating and comparing various structures, an important consideration is the required word length. In principle, for any given design, the required word length can be determined exactly by evaluating the frequency response for successively smaller values of coefficient word length until the filter specifications are exceeded. An alternative approach is to base the analysis of coefficient errors on a statistical representation. One of the first attempts to use a statistical measure of coefficient errors for comparison of digital filters was made by Knowles and Olcayto [25]. More recently, a statistical approach for estimating the coefficient word length necessary to keep the magnitude of the system function of a digital filter within a prescribed error bound was proposed by Avenhaus [26] and modified by Crochiere [27]. A similar measure for finite impulse response filters in direct form has been proposed by Chan and Rabiner [28]. The resulting measure of word length by these methods, referred to as statistical word length, is a conservative estimate and can generally predict the necessary coefficient word length to within a couple of bits. This measure is generally more efficient to compute than the actual word length. Consequently, it is useful in studies where general comparisons of coefficient sensitivity properties of structures are desired but where exact values are unnecessary. Such a comparison is presented in Section VI.

A second application of statistical word length is as an aid in optimizing the design of a digital filter for minimum coefficient word length. The statistical word length is useful in this application because it is a more convenient function to minimize than the actual word length which is a nonlinear integer function of the filter design parameters. Improvements of one to three bits have been observed using this approach [27], [29].

To define the statistical word length, assume that it is of interest to design a filter for which the magnitude of the ideal system function is $H_I(\omega)$ with a given tolerance function $\delta(\omega)$. When the filter coefficients are specified to infinite precision, the magnitude of the system function will be denoted as $|H_0(\omega)|$. The magnitude error incurred when finite precision coefficients are actually used will be denoted by $\Delta|H(\omega)|$ so that the true magnitude response $|H(\omega)|$ will be

$$|H(\omega)| = \Delta|H(\omega)| + |H_0(\omega)|.$$

The filter must be designed in such a way that $|H(\omega)|$ approximates the ideal transfer function to within the allowed error deviation.

For a given filter structure, the error $\Delta|H(\omega)|$ can be expressed in terms of a linear first-order approximation:

$$\Delta|H(\omega)| \cong \sum_{i=1}^{m} \frac{\partial|H_0(\omega)|}{\partial c_i} \Delta c_i. \tag{38}$$

where the values $c_i (i = 1, 2, \cdots, m)$ represent the ideal infinite precision values for the filter. Now let $Q$ define the quantization step size for rounding of the coefficients and assume that the quantization errors $\Delta c_i$ have a uniform probability of lying between $-Q/2$ and $Q/2$. The statistical errors in the coefficients then have zero mean and variance $\sigma^2_{\Delta c_i} = Q^2/12$. Assuming that the errors $\Delta c_i$ are independent, the variance of $\Delta|H(\omega)|$ is

$$\sigma^2_{\Delta|H(\omega)|} \cong \frac{Q^2 S^2(\omega)}{12} \tag{39}$$

where

$$S^2(\omega) = \sum_{i=1}^{m} \left[\frac{\partial|H_0(\omega)|}{\partial c_i}\right]^2. \tag{40}$$

Assuming that $\Delta|H(\omega)|$ has approximately a Gaussian distribution with variance given by (39), we can choose a value $x\sigma_{\Delta|H(\omega)|}$ such that

$$P[|\Delta|H(\omega)|| \leqslant x\sigma_{\Delta|H(\omega)|}] = y$$

or

$$P\left[|\Delta|H(\omega)|| \leqslant \frac{xQS(\omega)}{\sqrt{12}}\right] = y. \tag{41}$$

We will be assured with probability $y$ that

$$|\Delta|H(\omega)|| \leqslant \delta(\omega) - ||H_0(\omega)| - H_I(\omega)|$$

if we choose $Q$ so that at all frequencies of interest

$$\frac{xQS(\omega)}{\sqrt{12}} \leqslant \delta(\omega) - ||H_0(\omega)| - H_I(\omega)|. \tag{42}$$

Then with $q(\omega)$ denoting the maximum allowable quantization step size as a function of frequency,

$$q(\omega) = \frac{\sqrt{12}\,(\delta(\omega) - ||H_0(\omega)| - H_I(\omega)|)}{xS(\omega)}. \tag{43}$$

The coefficient word length of the filter can be defined as

$$W = 1 + i_M - i_L \tag{44}$$

where $2^{i_M}$ is the value of the most significant bit and $2^{i_L}$ the least significant bit (the sign bit is not counted in (44)); $i_M$ is dictated by the magnitude of the coefficients and $i_L$ by the coefficient step size. In particular, using (43), $i_L$ is approximated as

$$i_L \cong \log_2 \left[\min_{\omega} \frac{\sqrt{12}\,(\delta(\omega) - ||H_0(\omega)| - H_1(\omega)|)}{xS(\omega)}\right] \tag{45}$$

and the statistical word length $W$ is given by

$$W = 1 + i_M - \log_2 \left[\min_{\omega} \frac{\sqrt{12}\,(\delta(\omega) - ||H_0(\omega)| - H_I(\omega)|)}{xS(\omega)}\right]. \tag{46}$$

To compute the statistical word length, it is necessary to compute the function $S(\omega)$ as given by (40). This can be done using the techniques discussed previously in this section.

For the case of equiripple filters, it is generally possible to make use of the fact that $q(\omega)$ will usually be a minimum at those frequencies for which $(\delta(\omega) - ||H_0(\omega)| - H_I(\omega)|)$ is a minimum, i.e., the frequencies at which the ripples occur. In this case, it is generally sufficient to evaluate (43) only at those frequencies in order to evaluate (46).

## VI. Digital Filter Structures and Sensitivity Comparisons

The choice of a filter structure is an important consideration in the design of a digital filter. This choice directly affects the word-length requirements for the signals and coefficients in the filter and determines its characteristics such as roundoff noise, limit cycle behavior, and dynamic range. In response to this problem, many types of structures were recently proposed by various authors as candidates for filter implementation. Some of these structures are examined in this section with regard to coefficient sensitivity for the case of a specific bandpass filter example used by Avenhaus [26]. The comparison is made on the basis of statistical word length and the number of required multiplies and adds. In Section VII, a further comparison will be made on the basis of parallelism and serialism. The results should not be taken as final or conclusive for all examples or applications but, rather, they should be used to gain insight into the types of structures that appear to be desirable and look promising for further investigation.

### *The Structures*

The filter specifications used for the comparison are given in Fig. 7. Six different structures were synthesized for this example. No attempt will be made to explain the details for synthesis of each of these different types of structures as these details are readily available in the references. Further examples also can be found in [1] and [30].

The first three types of structures which were examined are the well-known direct-form II structure, the cascade-form structure (with second-order direct-form II sections), and the parallel-form structure, given in Figs. 8, 9, and 10, respectively. These structures correspond to representations of the system function as a ratio of polynomials, a product of ratios of second-order polynomials, and a partial-fraction expansion, respectively. They were perhaps the first classes of structures to be considered in digital signal processing. Their properties have been studied extensively and are well known [19], [30]–[32]. Another way of expressing system functions is by means of continued-fraction expansions. Mitra and Sherwood [33], [34] have proposed several classes of structures which can be realized by this approach. Fig. 11 gives one such structure.

A major category of structures which have received attention recently are the ladder or lattice structures. Unlike the previous categories, these structures are not synthesized by representing the system function in terms of some factorization or expansion. Instead, the mechanisms behind the synthesis procedures for these structures are closely related to the conventional two-port network procedures used in analog circuit theory. As a consequence, these structures are composed of building blocks which take the form of two-input, two-output networks. An example of this class of structures is given in Fig. 12. It is synthesized according to procedures suggested by Gray and Markel [35].

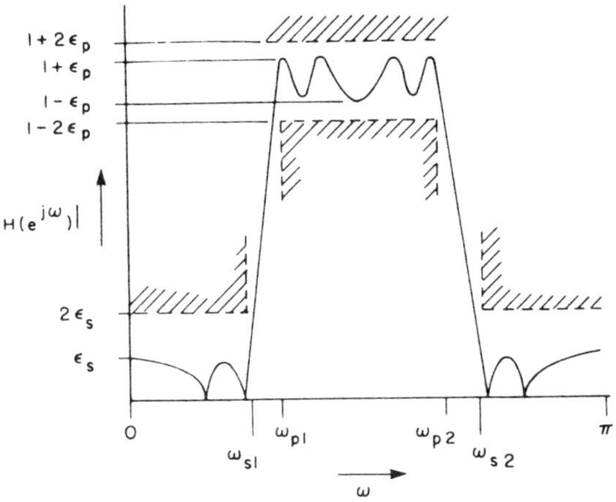


$\epsilon_p = 0.0157816$ $\epsilon_s = 0.00526063$

$\omega_{p1} = 0.411111\ \pi$ $\omega_{s1} = 0.3847015\ \pi$

$\omega_{p2} = 0.4666667\ \pi$ $\omega_{s2} = 0.4944444\ \pi$

Fig. 7. Filter design specifications for eighth-order bandpass example.

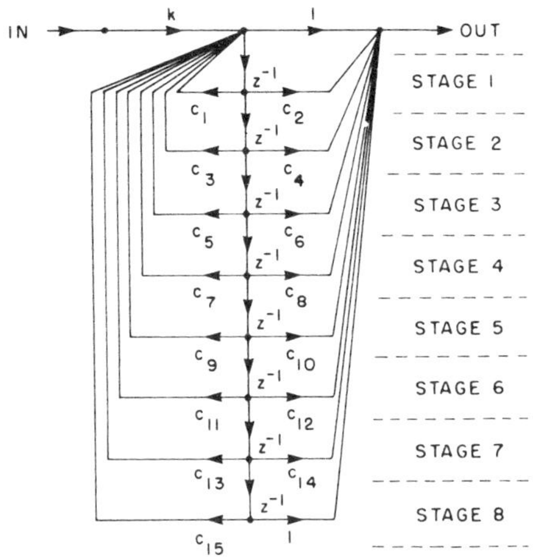


$k = 0.005656462366$

| | |
|---|---|
| $c_1 = 1.479757145$ | $c_2 = -1.387818861$ |
| $c_3 = -4.548129731$ | $c_4 = 3.990834175$ |
| $c_5 = 4.352241828$ | $c_6 = -3.801090558$ |
| $c_7 = -6.802822262$ | $c_8 = 5.978059098$ |
| $c_9 = 4.094876357$ | $c_{10} = -3.801090558$ |
| $c_{11} = -4.026604318$ | $c_{12} = 3.990834175$ |
| $c_{13} = -0.7833447265$ | $c_{14} = -1.387818861$ |
| $c_{15} = 1.232007442$ | |

Fig. 8. Direct-form II structure (Example 1).

Another class of structures synthesized in a manner similar to that of classical analog structures are the wave digital filters of Fettweis [36], [37]. This class of structures recently received considerable attention [38]–[41]. The synthesis procedure is similar to that of classical analog circuits; in fact, at present, the easiest way to design them is via the design of appropriate analog prototype structures. The digital structures are obtained by converting these analog designs to corresponding digital designs. The resulting digital structures are discrete simulations of the wave-flow diagrams of the analog structures. As such, digital structures designed in this way exhibit properties similar to their analog counterparts. Just as there are many ways to synthesize analog structures, there are also many

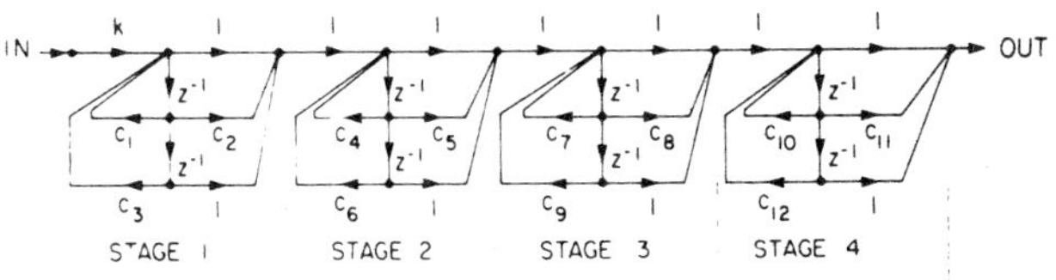


$k = 0.005656462366$

| | | | |
|---|---|---|---|
| $c_1 = 0.2855823838$ | $c_4 = 0.4457342317$ | $c_7 = 0.196901665$ | $c_{10} = 0.5515388641$ |
| $c_2 = 0.4512591755$ | $c_5 = -1.09869455$ | $c_8 = -0.0099665157$ | $c_{11} = -0.7304169706$ |
| $c_3 = -0.9116860616$ | $c_6 = -0.913078673$ | $c_9 = -0.9695353354$ | $c_{12} = -0.9705899009$ |

Fig. 9. Cascade-form (direct-form II sections) structure (Example 2).

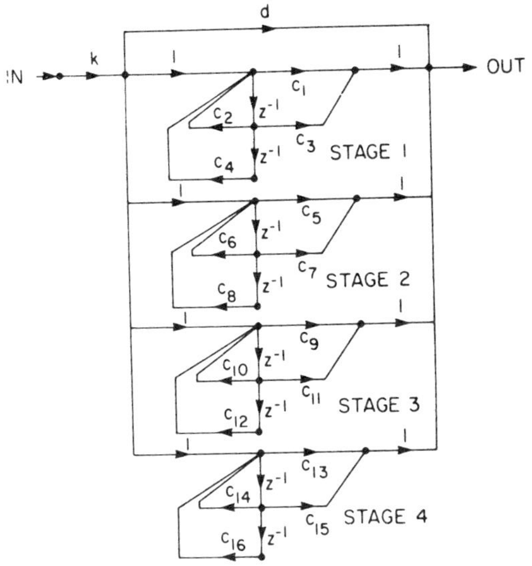


| | |
|---|---|
| $k = 0.10000$ | $d = 0.07220910762$ |
| $c_1 = 0.2034508401$ | $c_9 = -0.2052671178$ |
| $c_2 = 0.2855823838$ | $c_{10} = 0.196901665$ |
| $c_3 = 0.9026252904$ | $c_{11} = -0.2413054652$ |
| $c_4 = -0.9116860616$ | $c_{12} = -0.9695353354$ |
| $c_5 = 0.182701196$ | $c_{13} = -0.1965294023$ |
| $c_6 = 0.4457342317$ | $c_{14} = 0.5515388641$ |
| $c_7 = -0.9457352763$ | $c_{15} = 0.298888793$ |
| $c_8 = -0.913078673$ | $c_{16} = -0.9705899009$ |

Fig. 10. Parallel-form structure (Example 3).

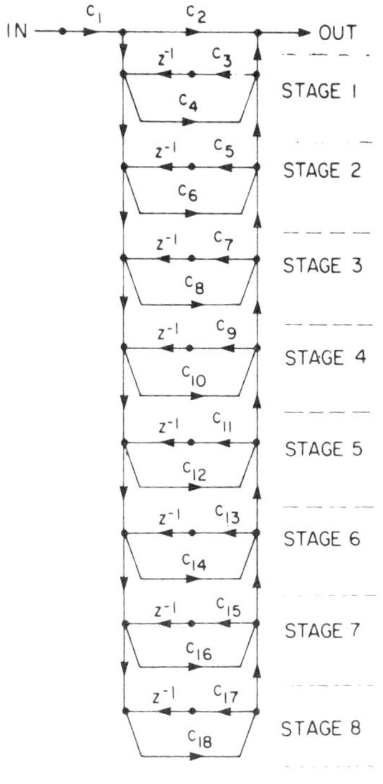


| | |
|---|---|
| $c_1 = 0.0002828231183$ | $c_2 = 25.53154355$ |
| $c_3 = -0.2117906668$ | $c_4 = 1.017015349$ |
| $c_5 = 0.1736820909$ | $c_6 = -12.7837888$ |
| $c_7 = 0.4649449432$ | $c_8 = -1.606209516$ |
| $c_9 = -0.2767502193$ | $c_{10} = 7.610142205$ |
| $c_{11} = 3.939247317$ | $c_{12} = -0.4301416747$ |
| $c_{13} = -0.9905202523$ | $c_{14} = 0.6779657016$ |
| $c_{15} = -21.82668197$ | $c_{16} = -0.4193825017$ |
| $c_{17} = 0.2481853405$ | $c_{18} = 0.4028556869$ |

Fig. 11. Continued-fraction expansion structure type 1B (Example 4).

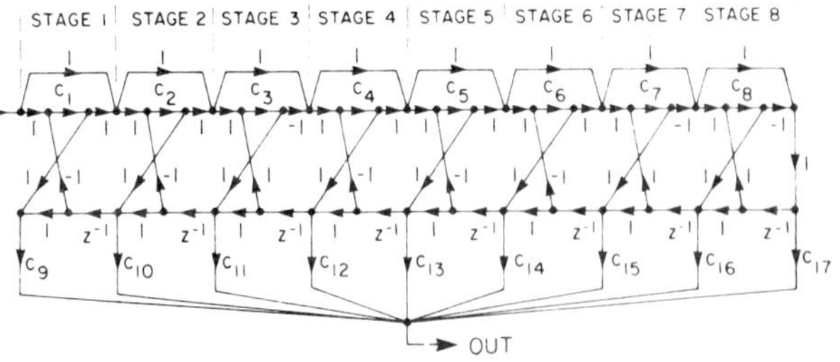


$c_1$ = 0.7833447265
$c_2$ = -0.1885428229
$c_3$ = 0.9843741951
$c_4$ = -0.1920962931
$c_5$ = 0.9936376827
$c_6$ = -0.1890252997
$c_7$ = 0.9914182038
$c_8$ = -0.1964790194
$c_9$ = 0.005656462366
$c_{10}$ = 0.0002916124024
$c_{11}$ = -0.001699657657
$c_{12}$ = -0.06980454614
$c_{13}$ = 0.2031623313
$c_{14}$ = 0.01061747099
$c_{15}$ = -0.01641846145
$c_{16}$ = -1.031187912
$c_{17}$ = 0.8446328267

Fig. 12. Ladder structure by Gray and Markel [35] (Example 5).

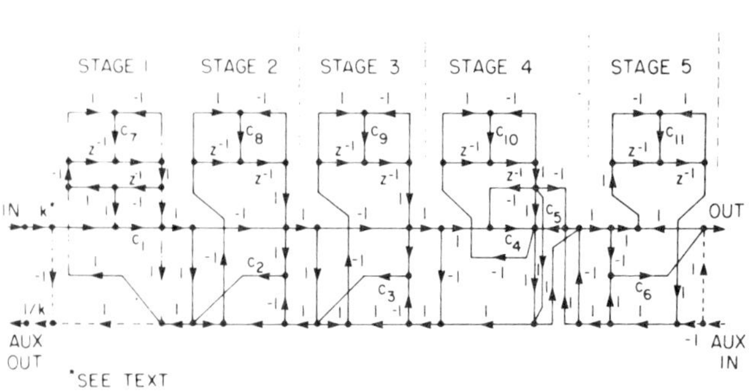


$k$ = 1.31194961
$c_1$ = 0.06502370945
$c_2$ = 0.2235692269
$c_3$ = 0.5809284606
$c_4$ = 0.1626985222
$c_5$ = 0.007805575218
$c_6$ = 0.0574957168
$c_7$ = -0.1915378541
$c_8$ = 0.0003565321277
$c_9$ = -0.3698270573
$c_{10}$ = -0.1915378541
$c_{11}$ = 0.1915378541

Fig. 13. One type of wave digital filter (Example 6).

TABLE I

| Example | Network Description | Statistical Word Length (bits) Passband $W_p$ | Stopband $W_s$ | $i_M$ | No. of Multiplies | No. of Adds | Bit × Multiply Product |
|---|---|---|---|---|---|---|---|
| 1 | direct-form II | 20.86 | 10.85 | 2 | 16 | 16 | 334 |
| 2 | cascade (direct-form II) | 11.33 | 5.76 | 0 | 13 | 16 | 147 |
| 3 | parallel form | 10.12 | 7.95 | −1 | 18 | 16 | 182 |
| 4 | continued-fraction 1B | 22.61 | 14.46 | 4 | 18 | 16 | 408 |
| 5 | ladder (Gray and Markel [35]) | 13.97 | 10.81 | 0 | 17 | 32 | 238 |
| 6 | wave digital filter (Fettweis [36], [37]) | 11.35[a] | 5.67 | −1[a] | 12/11[a] | 31 | 136/125 |

[a]See text.

ways to synthesize the wave digital filters. An example is given in Fig. 13.

*Comparisons of the Structures*

Comparisons of these structures have been made on the basis of statistical word length and the number of additions and multiplications required. The results are tabulated in Table I. The statistical word lengths were evaluated according to (46), with $x = 2$ (e.g., $y = 0.95$). This measure represents a conservative estimate of the coefficient word length (for fixed-point arithmetic) necessary to meet the filter specifications in Fig. 7.

Separate evaluations of the statistical word lengths were made for the passband and stopband specifications in order to obtain a more general picture of the sensitivity characteristics of each type of structure. Obviously, the largest of the two word-length estimates must be chosen to meet the specifications at all frequencies.

The statistical word lengths are generally conservative estimates of the necessary word length (for coefficient rounding) by approximately 2 bits. For example, the cascade (direct-form II) structure of Example 2 requires 10-bit coefficients to meet the specifications in Fig. 7, whereas the statistical word length predicts 11.33 bits. These word lengths do not include the extra bit required for representing the sign. One exception to this occurs for the case of the wave digital filters. For this class of structures, the statistical word length for meeting the passband constraints appears, from observations, to be conservative by approximately 4 bits. For example, the actual word length required for the wave digital filter to meet the passband requirements was 7 bits, whereas the statistical word length was 11.35 bits. This can be attributed to the special pseudo-passive and pseudolossless properties of these structures which are inherited from their analog counterparts [40], [41].

The value of $i_M$ for each structure (see Table I) corresponds to the power of 2 represented by the most significant bit in the binary representation of the coefficients. It is chosen such that the largest coefficient magntiude (except the scale factor) in the structure is within the range $2^{i_M}$ and $2^{i_M+1}$.

The number of multiplications and additions required for each structure are tabulated in Table I. It should be emphasized that these numbers are for the specific bandpass elliptic filter discussed and are not necessarily representative for arbitrary system functions. Also indicated in Table I is the bit–multiplier product for each structure. It represents a rough measure of the total amount of computation which must be performed in each structure.

The comparisons of the structures in this section have been made primarily on the basis of coefficient sensitivity and the number of multiplies and adds. These attributes are clearly important in the choice of a structure, particularly since they vary so widely from one structure to another, as can be seen in Table I. Of equal importance in the choice of a structure are the issues of roundoff noise, limit cycle effects, and dynamic range. Although these characteristics have not been considered in this comparison, the authors feel that it is important to mention them in order that these results may be viewed in a proper perspective to the overall issues of filter structure design.

## VII. Parallelism and Serialism in Structures

In the previous section we carried out a comparison of some structures based on coefficient sensitivity. There are, however, many other considerations to be taken into account in choosing a filter structure. In this section we suggest one such additional consideration, that of the potential parallelism inherent in the structure. In some applications it is of interest to implement a digital filter at extremely high sampling rates, requiring parallel computation; in such cases, a structure must be chosen which permits the required degree of parallelism. These considerations are similar in some respects to those of computation structures for multiprocessor design [42]-[44].

Our approach to discussing parallelism in structures is through the notion of precedence, as presented in Section III. The precedence form of a structure reveals the order in which nodes must be evaluated and, consequently, the smaller the number of precedence levels, the higher the degree of parallelism in the structure. By beginning from the node precedence relations, it is possible to develop a coefficient precedence graph which specifies the number of multiplier precedence levels. One of the major fundamental operations generally involved in the implementation of a digital filter structure is that of multiplication. Since the time needed to perform a multiply is often much greater than that needed for an addition or for other operations in the structure, the time necessary to compute one filter cycle can be roughly approximated in terms of multiplier cycle times (e.g., the time necessary to compute a multiply). For this situation, the parallel/serial tradeoffs of a structure (e.g., speed/cost tradeoffs) can be roughly approximated in terms of its multiplier precedence relations. To develop the multiplier precedence relations, it is convenient to draw the network such that all delay branches have unity coefficients and, consequently, do not require multiplies. All multiplications necessary for the implementation of the structure are then represented by the coefficient branches in the structure and appear in the precedence form of the network, as illustrated in Fig. 14. This form of network is identical to that in Fig. 2(c) and the precedence relations for coefficients are interpreted in a manner similar to that for nodes.

A precedence graph for those multiplies which must be performed in the implementation of the structure can now be extracted from the foregoing coefficient precedence relations. It must be recognized that some of the coefficients associated with the coefficient sets in Fig. 14 may not be required to be implemented as multiplies, namely, those which are unity. Consequently, the coefficient precedence relations must be modified to account for these situations. For example, if a coefficient which belongs to coefficient set $\{c_2\}$ is preceded only by unity gain coefficients in set $\{c_1\}$, then the multiply associated with this coefficient can be performed in the same time slot as the multiplies for coefficients in $\{c_1\}$. With these situations taken into account, the procedure leads to a set of multiplier precedence relations, as depicted by the multiplier precedence graph of Fig. 15. The circles in this graph represent the multiplies which must be performed and the arrows merely indicate the precedence relations. The multiplies associated with set $\{m_1\}$ can all be performed simultaneously without conflict in approximately the same time slot. Once these multiplies have been performed, all of the multiplies in set $\{m_2\}$ can be performed simultaneously in the next time slot, etc. Coefficients in set $\{m_1\}$ correspond to all of the coefficients in $\{c_1\}$ which must be implemented as multiplies plus all of the coefficients in sets $\{c_2\}, \{c_3\}, \cdots, \{c_{f-1}\}$ which are preceded only by coefficients that do not have to be implemented as multiplies. Coefficients in set $\{m_2\}$ correspond to those remaining coefficients in set $\{c_2\}$ plus any

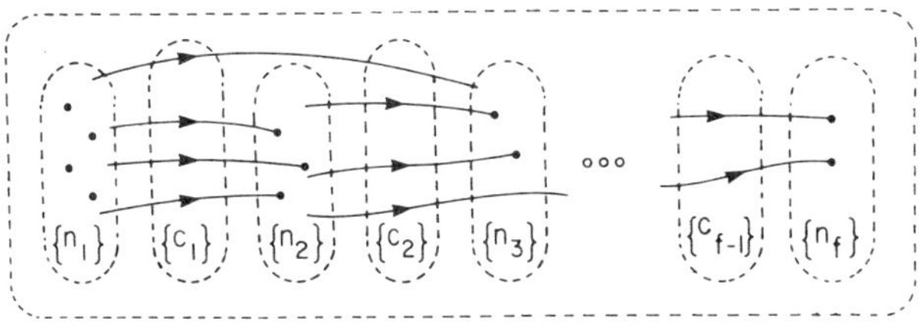


Fig. 14. Precedence form of a network illustrating precedence of coefficients.

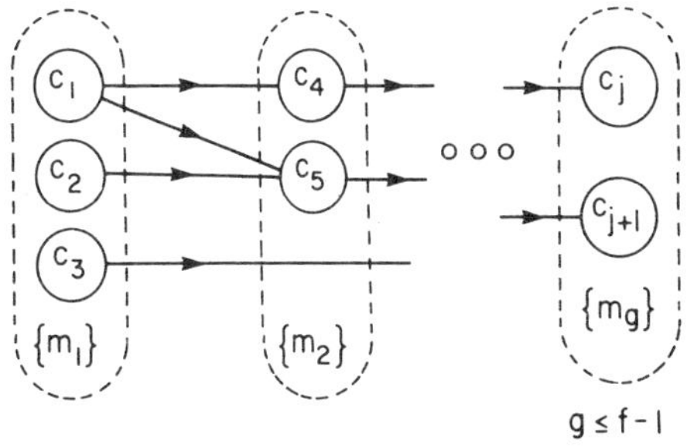


Fig. 15. Multiplier precedence graph.

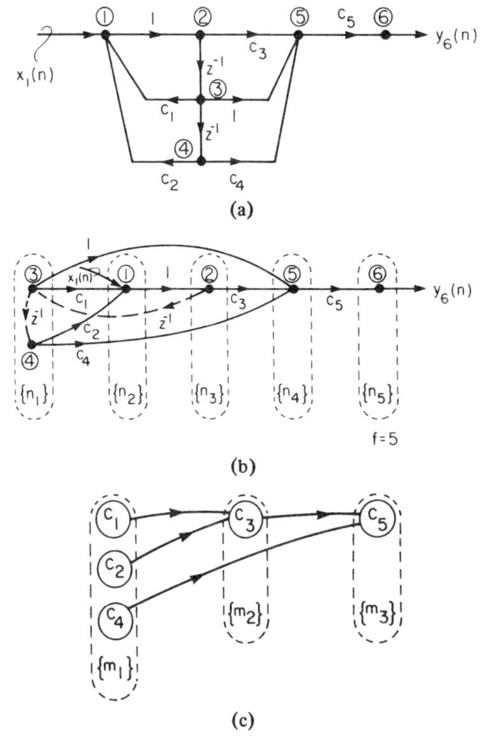


Fig. 16. (a) Example of digital network. (b) Its precedence form. (c) Its multiplier precedence graph.

coefficients in sets $\{c_3\}, \{c_4\}, \cdots, \{c_{f-1}\}$ which can be moved up without conflict in the multiplier precedence relations. By continuing this procedure, the multiplier precedence graph can be generated.

An example of a network, its precedence form, and its multiplier precedence graph are illustrated in Fig. 16. There are five levels of serialism with respect to nodes in this network and three levels of serialism with respect to the multiplies.

The multiplier precedence graph is a useful device for analyzing the tradeoffs between the number of multipliers $M$ used in the implementation and the approximate minimum computation time $t_m$ (in multiplier cycles) that it takes to

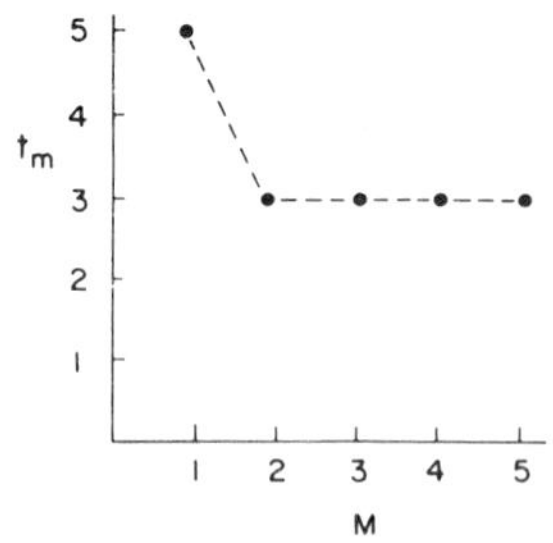


Fig. 17. Minimum multiplier cycle time $t_m$ versus number of multiplies $M$ for structure in Fig. 16.

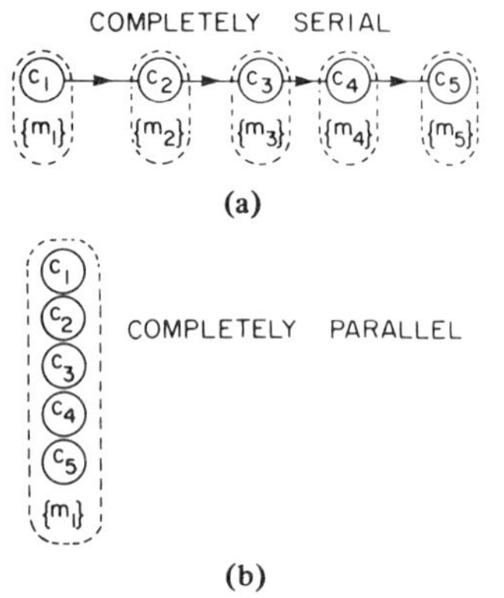


Fig. 18. Examples of multiplier precedence graphs. (a) Completely serial. (b) Completely parallel.

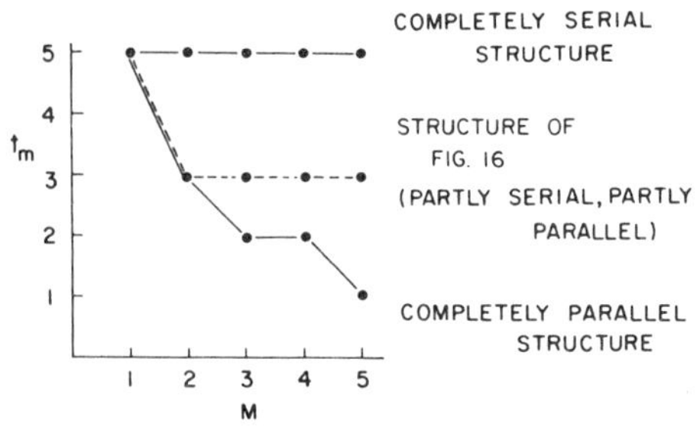


Fig. 19. Comparison between $t_m$ versus $M$ graphs for completely serial and completely parallel structures.

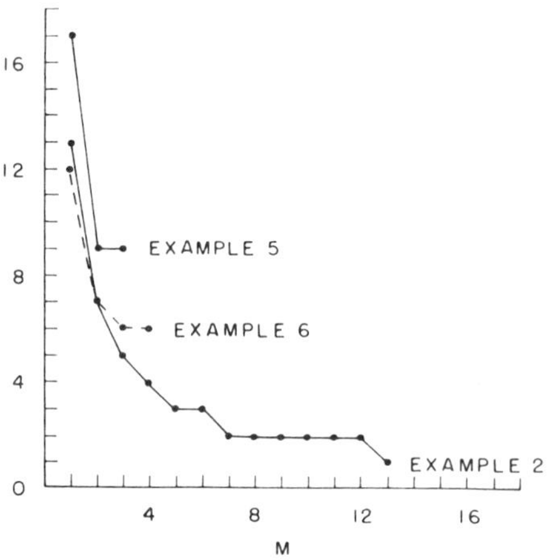


Fig. 20. Comparisons of $t_m$ versus $M$ graphs for some circuit examples given in Section VI.

compute one filter cycle. A useful way to illustrate this tradeoff is to plot $t_m$ against $M$. As an example (in the network of Fig. 16), if $M = 1$, then all of the multiplies must be computed serially and consequently $t_m \simeq 5$. If $M = 2$, coefficients $c_1$ and $c_2$ can be implemented in the first multiplier time slot, $c_3$ and $c_4$ in the second time slot, and $c_5$ in the third time slot. Therefore, $t_m \simeq 3$. As this is the inherent level of serialism in this particular structure, it is impossible to do better than this by using more multipliers in the implementation. The corresponding plot of $t_m$ versus $M$ is given in Fig. 17.

It is interesting to compare the $t_m$ versus $M$ plot for a structure against other possible structures that can realize the same system function. In particular, two extremes can be considered. A structure will be defined to be *completely serial* with respect to its multiplies if every level $\{m_i\}$ in its multiplier precedence graph contains only one possible multiply, as illustrated in Fig. 18(a). Alternatively, a structure will be defined to be *completely parallel* with respect to its multiplies if its multiplier precedence graph contains only one level $\{m_1\}$ as indicated by Fig. 18(b). Plots of the $t_m$ versus $M$ for these two extremes and for the network of Fig. 16 are given in Fig. 19.

For a completely serial structure, the $t_m$ versus $M$ plots correspond to the relation

$$t_m \simeq C \tag{47}$$

where $C$ is the total number of multiplies that must be performed in the structure. For a completely parallel structure, this relation is

$$t_m \simeq \text{Int}\left[\frac{C}{M}\right] \tag{48}$$

where Int [·] corresponds to the operation of rounding up the number in brackets to the next largest integer. As suggested by Fig. 19, many practical structures may have $t_m$ versus $M$ plots that lie somewhere between these two limits. Fig. 20 presents a comparison of $t_m$ versus $M$ plots for a number of the structures discussed in Section VI.

In some cases it may be possible to increase the amount of parallelism in the computation of a network by appropriately computing different node values (or multiplies) in the network simultaneously for different time indices $n$, i.e., by using *pipeline processing*. This can permit a speedup in the effective cycle time in the computation of a structure at the expense of introducing delay into some of the transfer functions of the network. This can be achieved, for example, in the network of Fig. 16 by performing the multiplies for $c_1$, $c_2$, and $c_4$ for time index $n + 2$ in the same time slot that multiply $c_3$ is performed for time index $n + 1$ and multiply $c_5$ is performed for time index $n$. This technique appears, at first, to violate the preceding concepts of inherent serialism and parallelism of a network but, in fact, it does not.

An alternate way of viewing pipelining is that it is a technique by which delays are inserted into appropriate paths in the structure in order to increase its parallelism or reduce its serialism. In effect, by using the pipeline scheme in the example given, the actual structure being computed is not that of Fig. 16 but instead the structure depicted in Fig. 21(a), where the computations for all of the multiplies are being performed in the same time slot for time index $n$. It is easy to see from Fig. 21(b) and (c) that this form of the structure has a greater amount of parallelism than the structure of Fig. 16 and that, in fact, it is completely parallel with respect to its multiplies. Its system function, however, has an extra delay of two unit delays associated with it over that for the network in Fig. 16.

As already illustrated, pipelining can be interpreted as a process by which extra unit delays are added in appropriate parts

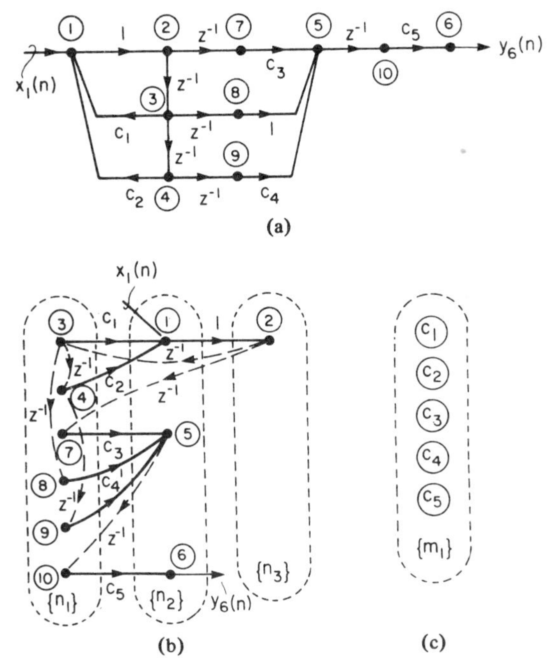


Fig. 21. (a) Pipelined form of network in Fig. 16. (b) Its precedence form. (c) Its multiplier precedence graph.

of a structure in order to reduce its inherent serialism at the expense of introducing extra delay in certain transfer functions of the structure. These delays can be inserted only into those parts of a structure which are not associated with loops or feedback paths. Also, if a structure contains several feedforward paths in parallel and a unit delay is added in one of these paths, then a unit delay must also be added in each of the paths which are in parallel with it in order to preserve the feedforward cancellations in the structure. As a result, nonrecursive structures can always be completely pipelined as they contain no loops of any kind. Recursive structures may be only partially pipelined because they contain loops. Only those parts of a recursive structure which are not associated with loops may be pipelined.

## VIII. Conclusions

We have reviewed the analysis, representation, and evaluation of digital filter structures. We have centered the discussion around the notation of linear signal-flow graphs and the associated matrix representation. Using this notation, we have identified a number of basic network properties and have applied them to an analysis of filter structures.

The basic network properties presented can in some sense be interpreted as contributing to a network theory applicable to digital filters. In some instances, the parallel with analog networks is very strong; in other instances, it is not. However, it is our feeling that just as network theory for analog filters has been an important framework for the analysis and synthesis of filter structures, a rich network theory exists which can serve the same purpose for digital filters. Hopefully, the ideas presented in this paper represent a start in this direction.

In Section VI we have presented a comparison of a number of filter structures based on coefficient sensitivity. The development of this comparison represents a useful application of many of the theoretical ideas presented in the paper. It is important to stress, however, that this comparison is based on only one example. Furthermore, it is important to keep in mind that coefficient sensitivity is only one of many factors to be taken into account in comparing structures. In Section VI we have tried to address ourselves to another factor, that of the degree of parallelism inherent in a structure. Our discussion represents only a preliminary attempt to include this consideration in the evaluation of structures. Over the next several years, we anticipate considerable activity directed toward a refinement and quantification of the parameters involved in comparing digital filter structures.

## Acknowledgment

The authors wish to thank Prof. H. W. Schuessler, Dr. L. R. Rabiner, and Dr. J. F. Kaiser for their careful reading of this manuscript during its preparation and their helpful comments.

## References

[1] R. E. Crochiere, "Digital network theory and its application to the analysis and design of digital filters," Ph.D. dissertation, Dep. Elec. Eng., M.I.T., Cambridge, Mass., May 1974.
[2] S. J. Mason, "Feedback theory—Some properties of signal flow graphs," *Proc. IRE*, vol. 41, pp. 1144-1156, Sept. 1953.
[3] —, "Feedback theory—Further properties of signal flow graphs," *Proc. IRE*, vol. 44, pp. 920-926, July 1956.
[4] P. M. DeRusso, R. J. Roy, and C. M. Close, *State Variables for Engineers*. New York: Wiley, 1965.
[5] R. E. Crochiere, "Some network properties of digital filters," Res. Lab. Electron., M.I.T., Cambridge, Mass., Quart. Progr. Rep. 107, pp. 103-112, Oct. 15, 1972; also in *Proc. 1973 IEEE Int. Symp. Circuit Theory*, pp. 146-148.
[6] P. Penfield, Jr., R. Spence, and S. Duinker, *Tellegen's Theorem and Electric Networks*. Cambridge, Mass.: M.I.T. Press, 1970.
[7] S. W. Director and R. A. Rohrer, "The generalized adjoint network and network sensitivities," *IEEE Trans. Circuit Theory*, vol. CT-16, pp. 318-323, Aug. 1969.
—, "Automated network design—The frequency-domain case," *ibid.*, vol. CT-16, pp. 330-337, Aug. 1969.
[8] R. E. Seviora and M. Sablatash, "A Tellegen's theorem for digital filters," *IEEE Trans. Circuit Theory*, vol. CT-18, pp. 201-203, Jan. 1971.
[9] A. Fettweis, "A general theorem for signal-flow networks, with applications," *Arch. Elek. Übertragung*, vol. 25, pp. 557-561, Dec. 1971; also in *Dig. Tech. Papers, 1971 IEEE Int. Symp. Electrical Network Theory*, pp. 3-4.
[10] M. L. Blostein, "On the application of certain network concepts to arbitrary systems," in *Proc. 1972 IEEE Int. Symp. Circuit Theory*, pp. 213-217.
[11] A. Y. Lee, "Topological properties of the summing and branching matrices of signal-flow graphs and the application to sensitivity analysis," in *Proc. 6th Asilomar Conf. Circuits and Systems*, Nov. 1972.
[12] —, "Computer-aided equation formulation and sensitivity analysis for discrete systems," in *Proc. 16th Midwest Symp. Circuit Theory*, Apr. 1973.
[13] L. B. Jackson, "On the interaction of roundoff noise and dynamic range in digital filters," *Bell Syst. Tech. J.*, pp. 159-184, Feb. 1970.
[14] G. E. Forsythe and C. B. Moler, *Computer Solution of Linear Algebraic Systems*. Englewood Cliffs, N.J.: Prentice-Hall, 1967.
[15] *Proc. IEEE (Special Issue on Computer-Aided Design)*, vol. 55, Nov. 1967.
[16] *IEEE Trans. Circuit Theory (Special Issue on Computer-Aided Circuit Design)*, vol. CT-18, Jan. 1971.
[17] A. V. Oppenheim and C. J. Weinstein, "Effects of finite register length in digital filtering and the fast Fourier transform," *Proc. IEEE*, vol. 60, pp. 957-976, Aug. 1972.
[18] L. B. Jackson, "Roundoff-noise analysis for fixed-point digital filters realized in cascade or parallel form," *IEEE Trans. Audio Electroacoust.*, vol. AU-18, pp. 107-122, June 1970.
[19] A. V. Oppenheim and R. W. Schafer, *Digital Signal Processing*, in press.
[20] J. Truxal, *Automatic Feedback Control Synthesis*. New York: McGraw-Hill, 1955.
[21] I. M. Horowitz, *Synthesis of Feedback Systems*. New York: Academic Press, 1963.
[22] R. E. Crochiere, "Computational methods for sensitivity analysis of digital filters," Res. Lab. Electron., M.I.T., Cambridge, Mass., Quart. Progr. Rep. 109, pp. 113-123, Apr. 15, 1973.
[23] J. F. Pinel and M. L. Blostein, "Computer techniques for the frequency analysis of linear electrical networks," *Proc. IEEE*, vol. 55, pp. 1810-1819, Nov. 1967.
[24] M. L. Blostein, "Some bounds on sensitivity in RLC networks,"

in *Proc. 1st Allerton Conf. Circuits and Systems*, Oct. 1963.

[25] J. B. Knowles and E. M. Olcayto, "Coefficient accuracy and digital filter response," *IEEE Trans. Circuit Theory*, vol. CT-15, pp. 31-41, Mar. 1968.

[26] E. Avenhaus, "On the design of digital filters with coefficients of limited word length," *IEEE Trans. Audio Electroacoust.*, vol. AU-20, pp. 206-212, Aug. 1972.

[27] R. E. Crochiere, "A new statistical approach to the coefficient word length problem for digital filters," *IEEE Trans. Circuits Syst.* (*Special Issue on Digital Filtering and Image Processing*); also in *Proc. 1974 IEEE Int. Symp. Circuits and Systems*, pp. 1-3.

[28] D. S. K. Chan and L. R. Rabiner, "Analysis of quantization errors in the direct form for finite impulse response digital filters," *IEEE Trans. Audio Electroacoust.*, vol. AU-21, pp. 354-366, Aug. 1973.

[29] G. Dehner, "On the design of digital Cauer filters with coefficients of limited word length," submitted to *Arch. Elek. Ubertragung.*

[30] R. E. Crochiere, "A comparison of digital filter structures on the basis of coefficient word length," Res. Lab. Electron., M.I.T., Cambridge, Mass., Progr. Rep. 115, Jan. 1975.

[31] J. F. Kaiser, "Digital filters," in *Systems Analysis by Digital Computer*, F. F. Kuo and J. F. Kaiser, Eds. New York: Wiley, 1966, ch. 7.

[32] B. Gold and C. M. Rader, *Digital Processing of Signals.* New York: McGraw-Hill, 1969.

[33] S. K. Mitra and R. J. Sherwood, "Canonic realizations of digital filters using the continued fraction expansion," *IEEE Trans. Audio Electroacoust.*, vol. AU-20, pp. 185-194, Aug. 1972.

[34] —, "Digital ladder networks," *IEEE Trans. Audio Electroacoust.*, vol. AU-21, pp. 30-36, Feb. 1973.

[35] A. H. Gray and J. D. Markel, "Digital lattice and ladder filter synthesis," *IEEE Trans. Audio Electroacoust.*, vol. AU-21, pp. 491-500, Dec. 1973.

[36] A. Fettweis, "Some principles of designing filters imitating classical filter structures," *IEEE Trans. Circuit Theory*, vol. CT-18, pp. 314-316, Mar. 1971.

[37] —, "Digital filter structures related to classical filter networks," *Arch. Elek. Übertragung*, vol. 25, pp. 79 ff, Feb. 1971.

[38] A. Sedlmeyer and A. Fettweis, "Realization of digital filters with true ladder configuration," in *Proc. 1973 IEEE Int. Symp. Circuit Theory*, pp. 149-152; also *Int. J. Circuit Theory Appl.*, vol. 1, no. 1.

[39] A. Fettweis and K. Meerkötter, "Suppression of parasitic oscillations in wave digital filters," in *Proc. 1974 IEEE Int. Symp. Circuits and Systems*, pp. 682-686.

[40] R. E. Crochiere, "Digital ladder structures and coefficient sensitivity," *IEEE Trans. Audio Electroacoust.*, vol. AU-20, pp. 240-246, Oct. 1972.

[41] A. Fettweis, "Pseudopassivity, sensitivity and stability of wave digital filters," *IEEE Trans. Circuit Theory*, vol. CT-19, pp. 668-673, Nov. 1972.

[42] J. L. Baer, "A survey of some theoretical aspects of multiprocessing," *Computing Surveys (ACM)*, vol. 5, pp. 31-80, Mar. 1973.

[43] J. Allen and R. Gallager, *Computation Structures*, Notes for a course at M.I.T. on Computation Structures, Course 6.032.

[44] R. T. Prosser, "Applications of Boolean matrices to the analysis of flow diagrams," Lincoln Lab., M.I.T., Lexington, Mass., Tech. Rep. 217, Jan. 22, 1960.

# Part 6
# Hardware

The preceding part contains papers which investigate new network structures with potentially desirable properties or implementation. Actual hardware techniques for realizing digital filters with reduced complexity or increased performance in certain cases are described in this part.

The motivation for much of the significant investment in signal processing research (although perhaps not the motivation of many of the researchers) is related to the ultimate realization of signal processing techniques in hardware. The past few years have seen the emergence of numerous signal processors as commercial devices, and new techniques are constantly being devised. Unfortunately, much more work is being conducted in this area than is reported in the literature and thus only a sampling of the current activity in hardware implementation is included here.

The first two papers, by Peled and Liu, and Eckhardt and Winkelnkemper, describe realization for the second-order sections of the cascade form for recursive infinite impulse response (IIR) filters; while the next two papers, by Heute, and Peled and Liu, consider nonrecursive finite impulse response (FIR) filter implementation.

Fast Fourier transform (FFT) processor structures were treated in the predecessor volume, *Digital Signal Processing*, where Bergland [1] presents a good overview of possible FFT processor structures, and Groginsky and Works [2] describe a particular realization of one of these structures.

In the present volume, Gold and Bially continue the structural investigation, focusing on the interaction between top-level FFT structure and the practical considerations of memory or processor speed and efficiency. Other papers from the literature of special architectures include Veenkant's paper [3], which describes an FFT processor based on Pease's perfect shuffle algorithm [4], as well as [5] and others.

### References

[1] G. D. Bergland, "Fast Fourier transform hardware implementations—An overview," *IEEE Trans. Audio Electroacoust.*, vol. AU-17, pp. 104–108, June 1969.

[2] H. L. Groginsky and G. A. Works, "A pipeline fast Fourier transform," *IEEE Trans. Comput.*, vol. C-19, pp. 1015–1019, Nov. 1970.

[3] R. L. Veenkant, "A serial minded FFT," *IEEE Trans. Audio Electroacoust.*, vol. AU-20, pp. 180–185, Aug. 1972.

[4] M. C. Pease, "An adaptation of the FFT for parallel processing," *J. Ass. Comput. Mach.*, vol. 15, Apr. 1968.

[5] A. M. Despain, "Fourier transform computers using CORDIC iterations," *IEEE Trans. Comput.*, vol. C-23, pp. 993–1001, Oct. 1974.

# A New Hardware Realization of Digital Filters

ABRAHAM PELED, STUDENT MEMBER, IEEE, AND BEDE LIU, FELLOW, IEEE

*Abstract*—**A new approach to the implementation problem of digital filters is presented. This approach capitalizes on recent advances in semiconductor memory technology and is shown to offer significant reductions in cost and power consumption for the same speed of operation as that of existing realizations. Furthermore, this approach makes possible speeds of operation which cannot be achieved by existing realizations. The proposed approach yields a very flexible hardware configuration and a discussion of the various options is presented together with a comparison to existing realizations. The mean-squared error resulting from the use of finite word length is analyzed.**

## I. INTRODUCTION

DIGITAL FILTERS have become an increasingly attractive replacement for analog filters due to recent advances in semiconductor technology. As the speed of operation increases, either to permit real-time processing of wide-band signals or to time share the arithmetic unit, there is a rapid increase in hardware complexity, as measured by the number of IC's used, and in power consumption. The major factor causing this increase lies with the high-speed multipliers.

Jackson *et al.* have proposed an approach to the implementation of digital filters that is well suited to large-scale integrated (LSI) construction [1]. They also propose a very efficient serial multiplier that produces a rounded binary number, and lends itself particularly well to multiplexed operation. Using current TTL technology this multiplier can accommodate a bit rate of approximately 25 MHz.

In this paper we propose a new approach for the hardware implementation of fixed point arithmetic digital filters. The new realization calls for the storing of the finite number of possible outcomes of an intermediate arithmetic operation, and using them to obtain the next output sample through repeated addition and shifting operations. No multiplications are needed. This hardware implementation is highly modular and uses only standard available IC's. We show that the proposed realization offers significant savings, for the same speed of operation as existing realizations, in terms of hardware complexity and power consumption. Furthermore, with this new realization, filters can be constructed to operate at speeds that are difficult or impossible to achieve with existing realizations. As in all realizations, the use of finite word lengths in the processor results in roundoff errors. These are analyzed and expressions are derived that allow the designer to choose the design parameters to achieve a specified performance.

## II. A NEW APPROACH TO IMPLEMENTING DIGITAL FILTERS

A digital filter is characterized by an input–output relationship of the form

$$y_n = \sum_{k=0}^{N} a_k x_{n-k} - \sum_{k=1}^{N} b_k y_{n-k} \tag{1}$$

where $\{x_n\}$ is the input sequence, $\{y_n\}$ the output sequence, and $\{a_k\}_{k=0}{}^N$, $\{b_k\}_{k=1}{}^N$ are the filter coefficients. For hardware realizations, it is convenient to consider the $z$-domain transfer function of the digital filter $H(z)$, given by

$$H(z) = \sum_{j=0}^{N} a_j z^{-j} / \left(1 + \sum_{j=1}^{N} b_j z^{-j}\right) \tag{2}$$

Manuscript received February 8, 1974; revised July 1, 1974. This work was supported by the Air Force Office of Scientific Research, USAF, under Grant AF-AFOSR 71-2101, and the National Science Foundation under Grant GK-24187.

A. Peled was with the Department of Electrical Engineering, Princeton University, Princeton, N.J. 08540. He is now with the IBM Thomas J. Watson Research Laboratories, Yorktown Heights, N.Y. 10598.

B. Liu is with the Department of Electrical Engineering, Princeton University, Princeton, N. J. 08540.

Reprinted from *IEEE Trans. Acoust., Speech, Signal Processing,* vol. ASSP-22, pp. 456–462, Dec. 1974.

where $z^{-1}$ is the unit delay operator. It has been long recognized [2]–[5] that out of the numerous equivalent configurations that realize (2), the most advantageous are the cascade and parallel forms shown in Fig. 1. The cascade form corresponds to a factorization of the numerator and denominator polynomials of (2) to bring forth an $H(z)$ of the form

$$H(z) = a_0 \prod_{j=1}^{M} \left[ \frac{(\alpha_{2j}z^{-2} + \alpha_{1j}z^{-1} + 1)}{(\beta_{2j}z^{-2} + \beta_{1j}z^{-1} + 1)} \right] \tag{3}$$

where $M$ is the least integer greater than or equal to $N/2$. The parallel form results from a partial fraction expansion of (2),

$$H(z) = \gamma_0 + \sum_{j=1}^{M} \left[ \frac{(\gamma_{1j}z^{-1} + \gamma_{0j})}{(\beta_{2j}z^{-2} + \beta_{1j}z^{-1} + 1)} \right]. \tag{4}$$

The various aspects of realizing higher order filters using basic second-order sections, including appropriate scaling to avoid overflow and ordering for minimum roundoff noise have been discusses in detail in [4]–[6]. Let us focus on the implementation of a second-order section specified by the input–output relationship

$$y_n = a_0x_n + a_1x_{n-1} + a_2x_{n-2} - b_1y_{n-1} - b_2y_{n-2} \tag{5}$$

with $a_2 = 0$ in the case of parallel realization. Let us assume that all signals are bounded by $\pm 1$ and that $B$ binary bits including sign bit in 2's complement code are used to represent the data, that is

$$x_k = -x_k^0 + \sum_{j=1}^{B-1} x_k^j 2^{-j} \qquad x_k^j = 0 \text{ or } 1. \tag{6}$$

Equation (5) can be rewritten as

$$y_n = a_0\Big(\sum_{j=1}^{B-1} x_n^j 2^{-j} - x_n^0\Big) + a_1\Big(\sum_{j=1}^{B-1} x_{n-1}^j 2^{-j} - x_{n-1}^0\Big)$$
$$+ a_2\Big(\sum_{j=1}^{B-1} x_{n-2}^j 2^{-j} - x_{n-2}^0\Big) - b_1\Big(\sum_{j=1}^{B-1} y_{n-1}^j 2^{-j} - y_{n-1}^0\Big)$$
$$- b_2\Big(\sum_{j=1}^{B-1} y_{n-2}^j 2^{-j} - y_{n-2}^0\Big). \tag{7}$$

Define a function $\varphi$ with five binary arguments as follows:

$$\varphi(x^1,x^2,x^3,x^4,x^5) = a_0x^1 + a_1x^2 + a_2x^3 - b_1x^4 - b_2x^5. \tag{8}$$

Then we can rewrite (7) as

$$y_n = \sum_{j=1}^{B-1} 2^{-j}\varphi(x_n^j,x_{n-1}^j,x_{n-2}^j,y_{n-1}^j,y_{n-2}^j)$$
$$- \varphi(x_n^0,x_{n-1}^0,x_{n-2}^0,y_{n-1}^0,y_{n-2}^0). \tag{9}$$

The function $\varphi$ can take on only $2^5 = 32$ distinct values depending on the binary vector that forms its arguments. It can be realized by a combinatorial network or by a read only memory (ROM) addressed by the arguments. Thus (9) suggests a possible mechanization of (5) requiring only addition and shifting operations. This approach is illustrated in Fig. 2. Data enter serially into shift registers

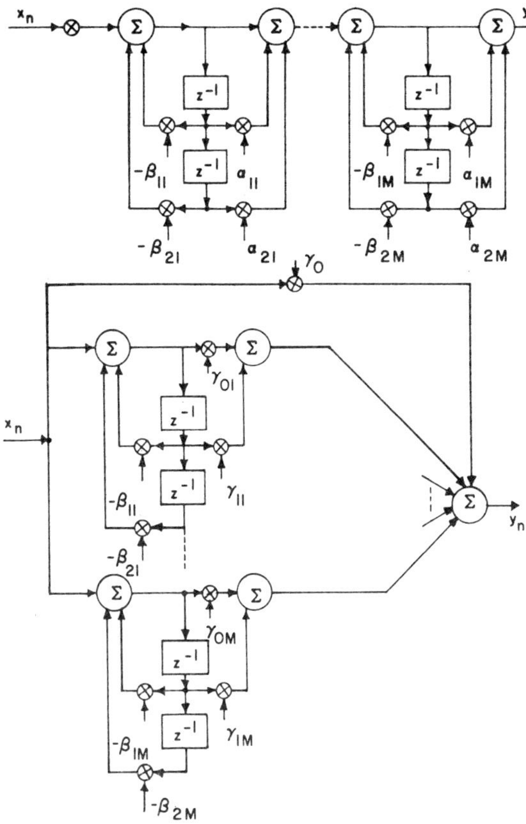


Fig. 1. (a) Cascade realization. (b) Parallel realization.

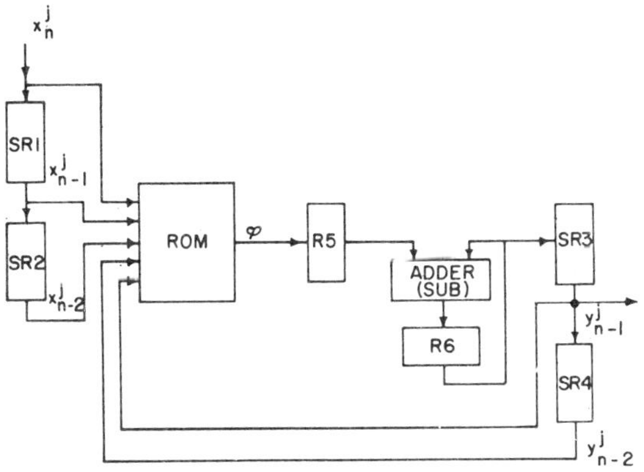


Fig. 2. New hardware realization of a second-order section of a digital filter.

$SR1$–$SR4$, with the least significant bit leading. At each shift, a new vector $\{x_n^j,x_{n-1}^j,x_{n-2}^j,y_{n-1},y_{n-2}^j\}$ appears at the input of the circuit realizing $\varphi$. The output $\varphi$ is loaded into register $R5$ which is connected to one of the two inputs of the accumulator with a possible sign change for $j = 0$. The other input of the accumulator is hardwired to the output register ($R6$) with a 1 bit right shift. After $B$ such shifts, the value in register $R6$ is rounded and the accumulator cleared. This rounded value is $y_n$, which is shifted serially into $SR3$, and the processor is ready to compute the next sample $y_{n+1}$.

Fig. 3 gives an example of a typical second-order section and its corresponding function $\varphi$ defined by its truth table with $B = 8$. This table can be used directly to program a ROM of 256 bits organized as $32 \times 8$ bit

MEMORY MAP FOR SECOND-ORDER SECTION

| MEMORY ADDRESS | CONTENTS |
|---|---|
| 0 0 0 0 0 | 0 0 0 0 0 0 0 0 |
| 0 0 0 0 1 | 1 1 0 0 0 1 1 0 |
| 0 0 0 1 0 | 0 1 1 1 0 1 0 0 |
| 0 0 0 1 1 | 0 0 1 1 1 0 0 1 |
| 0 0 1 0 0 | 0 0 0 0 0 1 1 0 |
| 0 0 1 0 1 | 1 1 0 0 1 1 0 0 |
| 0 0 1 1 0 | 0 1 1 1 1 0 1 0 |
| 0 0 1 1 1 | 0 0 1 1 1 1 1 1 |
| 0 1 0 0 0 | 1 1 1 1 0 1 0 1 |
| 0 1 0 0 1 | 1 0 1 1 1 0 1 1 |
| 0 1 0 1 0 | 0 1 1 0 1 0 0 1 |
| 0 1 0 1 1 | 0 0 1 0 1 1 1 1 |
| 0 1 1 0 0 | 1 1 1 1 1 0 1 1 |
| 0 1 1 0 1 | 1 1 0 0 0 0 0 1 |
| 0 1 1 1 0 | 0 1 1 0 1 1 1 1 |
| 0 1 1 1 1 | 0 0 1 1 0 1 0 1 |
| 1 0 0 0 0 | 0 0 0 0 0 1 1 0 |
| 1 0 0 0 1 | 1 1 0 0 1 1 0 0 |
| 1 0 0 1 0 | 0 1 1 1 1 0 1 0 |
| 1 0 0 1 1 | 0 0 1 1 1 1 1 1 |
| 1 0 1 0 0 | 0 0 0 0 1 1 0 0 |
| 1 0 1 0 1 | 1 1 0 1 0 0 1 0 |
| 1 0 1 1 0 | 0 1 1 1 1 1 1 1 |
| 1 0 1 1 1 | 0 1 0 0 0 1 0 1 |
| 1 1 0 0 0 | 1 1 1 1 1 0 1 1 |
| 1 1 0 0 1 | 1 1 0 0 0 0 0 1 |
| 1 1 0 1 0 | 0 1 1 0 1 1 1 1 |
| 1 1 0 1 1 | 0 0 1 1 0 1 0 1 |
| 1 1 1 0 0 | 0 0 0 0 0 0 1 0 |
| 1 1 1 0 1 | 1 1 0 0 0 1 1 1 |
| 1 1 1 1 0 | 0 1 1 1 0 1 0 1 |
| 1 1 1 1 1 | 0 0 1 1 1 0 1 1 |

sign bit ↑ ↑ binary point

Fig. 3. Typical second-order section and its $\varphi$.

words. In this example, $a_1 = 0.095$, $a_2 = -0.1665478$, $a_3 = 0.095$, $b_1 = -1.8080353$, and $b_2 = 0.9129197$. The five columns of the "memory address" correspond to the five binary arguments of the $\varphi$ function, i.e., $(x_n{}^j, x_{n-1}{}^j, x_{n-2}{}^j, y_{n-1}{}^j, y_{n-2}{}^j)$. The first bit in the contents is the sign bit and the binary point is to the right of the sign bit. Here $\varphi$ has been scaled down by 2.

Let us compare briefly this approach with existing realizations which use multipliers in the arithmetic unit as [1]. The realization illustrated in Fig. 2 substitutes a multiplier and storage for 5 filter coefficients with a combinatorial network or ROM and a shift register. Another difference is that our method requires no code conversions between 2's complement and sign magnitude [1], [7]. However, we hasten to mention that by fixing the function $\varphi$ we lost the flexibility inherent in using multipliers. Each additional second-order section to be realized requires a distinct function $\varphi$ and therefore an additional network. Nevertheless, we will show in Section IV that as a result of the recent advances in semiconductor memory technology our approach is significantly more economical in terms of power consumption and hardware complexity then existing realizations.

## III. OPTIONS AND PERFORMANCE OF THE NEW APPROACH

In this section we discuss the various options of hardware implementation of second-order sections via (9), and analyze their performances. Higher order filters will be discussed in Section IV.

We have already mentioned that the function $\varphi$ can be realized by a combinatorial network or by a ROM programmed appropriately. Although it may prove more economical in many instances to design a combinatorial network to realize $\varphi$, especially if high-density logic devices such as programmable logic arrays are used, the exact cost and performance would be heavily dependent on the specific coefficients and the intended volume of production. A most convenient way to realize $\varphi$, however, is to use ROM or programmable read only memory (PROM). In this paper we assume that ROM's are used to realize $\varphi$ which makes our analysis independent of the filter coefficients and therefore applicable to every digital filter.

Now let us proceed to evaluate the performance of the second-order section implemented as in Fig. 2. Essentially two steps have to be completed before we can shift a new data bit into the data registers $SR1$ to $SR4$.

*Step* 1: Evaluation of the function $\varphi$. The time needed is the access time of the ROM—$t_\varphi$.

*Step* 2: Adding two $B$ bit numbers. The time needed is the addition time—$t_\alpha$.

The two steps just mentioned can be performed concurrently, provided that the accumulator has an additional shift register for storing the addend. Let

$$t_R = \max \{t_\alpha, t_\varphi\} \tag{10}$$

then $1/t_R$ will be the maximum bit rate at which the filter can operate.

As an example let us consider $B = 12$. Using standard available TTL integrated circuits and bipolar memory we can get $t_\alpha \approx 40$ ns, $t_\varphi \approx 50$ ns. Then the second-order section illustrated in Fig. 2 will operate at a bit rate of approximate 20 MHz, have a package count of 20 IC's, and consume 9.6 W. Assuming a 12-bit word, this bit rate implies that this section can operate in real time on a signal having a bandwidth in excess of 800 kHz or it can be multiplexed between some 200 signals having a bandwidth of 4 kHz each. We will also mention that should Fig. 2 be implemented with emitter-coupled logic (ECL) IC's the bit rate would be at least 50 MHz, the package count 33 IC's and power consumption 26.4 W.

Let us return for a moment to (9) which incorporates the basic approach we propose. Its mechanization as suggested in Fig. 2 is but one of many possible. Fig. 4 depicts another possible mechanization of (9) for the case of 8-bit data. Here data are loaded in parallel into $R1$ to $R5$, and there are 8 separate but identical ROM's storing the values of the function $\varphi$. The outputs of the ROM's, $\varphi^0$ to $\varphi^7$, are added in a tree like structure with a proper number of shifts hardwired, using 7 adders in this case. Again, by providing each adder with two storage registers, concurrent operation of all levels is possible. It is clear from the above that the number of bits used for the data will only determine the number of levels needed but will not affect the throughput rate.

As an example let us consider $B = 8$. Using standard TTL·IC's and bipolar memory, a word rate of 20 MHz for the second-order section in Fig. 4 can be achieved. The package count is 60 IC's and the power consumption

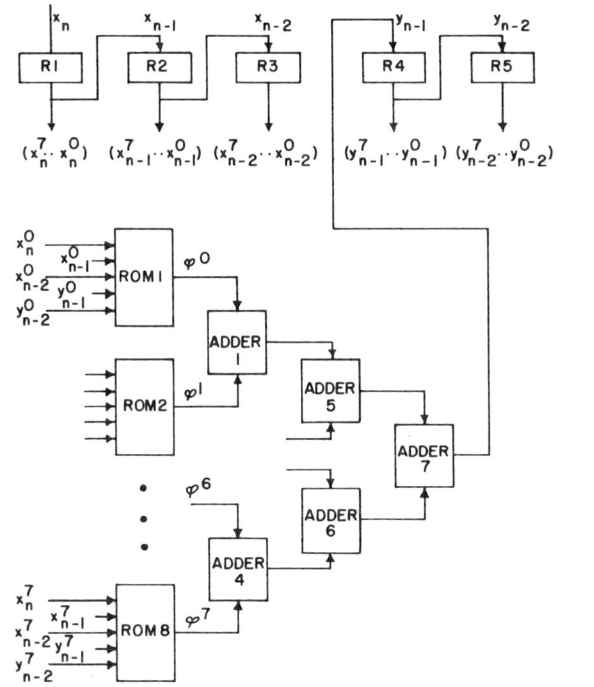

Fig. 4. High-speed realization of the second-order section.

24 W. This word rate implies that the section can operate in real time on a signal with a 10 MHz bandwidth. It should be noted that to achieve such a speed using multipliers would be very difficult or impossible unless several ECL multipliers are used. Such multipliers dissipate considerably more power and have a high-package count (e.g., a 9 × 9-bit multiplier described in [8] performs the multiplication in 35 ns and has 36 IC's dissipating 12.6 W). If ECL IC's are used to implement the section of Fig. 4, it is possible to realize a 50 MHz word rate, an operating speed unachievable using present multipliers.

Clearly, the two mechanizations of (9) illustrated in Figs. 2 and 4 represent two extreme cases. In the first one the data bits are processed serially, while in the second one all data bits are processed in parallel. Configurations that fall between these two extremes are possible and offer operating speeds between 2 MHz to 20 MHz word rate with a package count between 20 IC's to 60 IC's.

## IV. HIGH-ORDER DIGITAL FILTERS—A COMPARISON WITH EXISTING REALIZATIONS

In this section we discuss the realization of high-order digital filters and multiplexed digital filters using second-order sections implemented as described in Section III, and attempt to compare their performance with the implementations proposed in [1], [7].

It is rather difficult to compare between the implementation we propose and the existing realization without referring to a specific speed to operation and filter order. Therefore we chose to present four illustrative examples: 1) an eight-order digital filter realized in parallel operating at a high speed (1 MHz word rate), 2) an eight-order digital filter realized in cascade operating at a medium speed (250 kHz word rate), 3) a multiplexed digital filter, where a second-order section is shared between 128 low-speed channels of 8 kHz word rate each and 4) a 96 channel, tenth-order filter each with a word rate of 8 kHz.

Fig. 5 illustrates a possible mechanization of an eight-order digital filter realized in parallel using the configuration of Fig. 2 for the second-order section. Since $a_2 = 0$ in parallel form, the ROM's needed are of an organization of $2^4 \times B$ or $16 \times B$. The contents of ROM $J$ at address $\{x_n{}^j, x_{n-1}{}^j, v_{n-1}{}^j, v_{n-2}{}^j\}$ be

$$c[\text{ROM } J] = \gamma_{0J}x_n{}^j + \gamma_{1J}x_{n-1}{}^j - \beta_{1J}v_{n-1}{}^j - \beta_{2J}v_{n-2}{}^j \tag{11}$$

where $\{\gamma_{0J},\gamma_{1J},\beta_{1J},\beta_{2J}\}$ are the coefficients or the $J$th section of (4). The additional memory ROM 5 is used to combine the output of the sections and to allow for any necessary scaling to accommodate the dynamic range [4], [5].

Let us assume that the filter in Fig. 5 operates with $B = 12$. As mentioned in Section III, the second-order section in this case can operate at a word rate considerably faster than 1 MHz, if standard TTL IC's are used. The package count for the entire filter is approximately 72 IC's and the power consumption 28 W. The realization suggested in [1] uses a serial multiplier which can be constructed with approximately 20 standard TTL IC's operating at a word rate of 1 MHz (for 12 bit data) and consuming 8 W. Three such multipliers are needed for each second-order section if the filter is to operate at a 1 MHz word rate. At least 12 multipliers are needed for this eight-order filter. Thus the package count of the multipliers alone will be 240 IC's and their power dissipation in 96 W. To this, one has to add registers for the data, adders, and code converters.

The parallel TTL fast multiplier which uses 4 bit by 4 bit multipliers such as Texas Instruments SN74284, SN74285 can multiply 8 bit by 12 bit numbers in 90 ns. It has 24 IC's consuming 10 W. Although the speed is higher, additional hardware is required to parallel operation. Consequently, the final figures of package count and power consumption are similar to those obtained by the use of the serial multiplier.

Fig. 6 shows an eight-order digital filter realized in cascade form. Since each section operates successively, the speed of operation for the filter is reduced by a factor of 4 to a 250 kHz word rate. Again let us assume $B = 12$. Although the memory requirement is $32 \times 12$ bits for each section, a higher density memory such as $128 \times 8 = 1024$ bits can be used since each section is addressed at a different time. Therefore the package count of the arithmetic unit will be 33 IC's consuming approximately 14 W. The realization suggested in [1] operating at the same speed would need only 3 multipliers which will be shared between the sections. Thus the package count for the multipliers alone, in that case will be 60 IC's consumming 24 W.

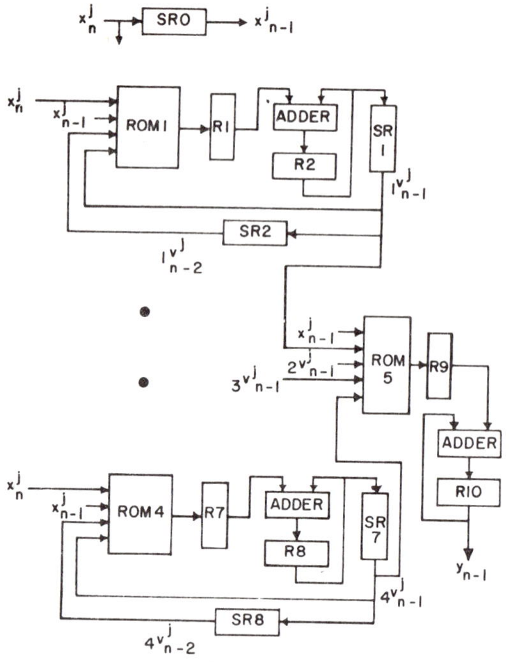


Fig. 5. Eighth-order filter realized in parallel form.

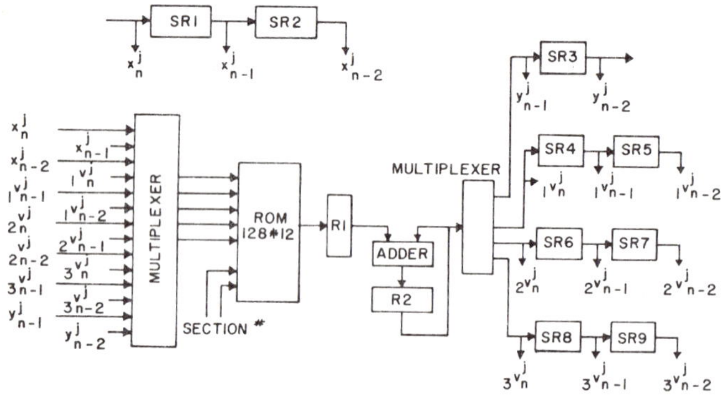


Fig. 6. Eighth-order digital filter realized in cascade form.

Consider a multiplexed digital filter where the arithmetic unit of the second-order section is shared between 128 second-order filters operating at a 8 kHz word rate each with $B = 12$. Intuitively, it would seem that the new approach proposed in this paper would be inferior to existing realizations due to the large amount of memory needed. However a closer examination shows that this is not the case. Fig. 7 shows the arithmetic unit of such a multiplexed filter. The data registers are not shown since the same number would be required in the existing realization. Assuming a parallel configuration for the second-order section a total of $128 \times (16 \times 12)$ bits of storage will be needed if $B = 12$. If we use a high density storage of 4096 bits organized as $512 \times 8$ bits (such as Signetics 8205) then only 6 IC's will be needed for the memory. This brings the total package count for the arithmetic unit to 18 IC's consuming about 10 W. Again 3 multipliers are needed by a realization such as [1] to achieve the word rate. Thus a total of 60 IC's consuming 24 W are needed for the multipliers alone. To this, one has to add also storage for 512 filter coefficients adders, etc.

Finally, let us consider a tenth-order digital filter to

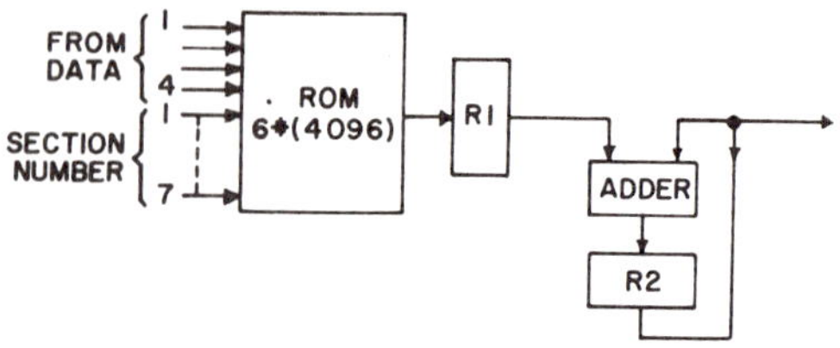


Fig. 7. Arithmetic unit of a multiplexed filter.

be multiplexed between 96 low-speed channels as the one mentioned in [7]. There, the same serial multiplier as in [1] is used and it is suggested that by using off-the-shelf medium-scale integrated (MSI) $T^2L$ circuits, the tenth-order filter would require about 190 IC's and consume approximately 100 W. If we assume a cascade realization for the tenth-order filter as the one in Fig. 6 which can operate at 400 kHz word rate based on a 2 MHz second-order section of Fig. 2, two such sections are needed to accommodate the 96 channels. By using 4096-bit ROM's, the total package count using standard IC's will be 54 IC's consuming 22 W.

All the above comparisons hold only for the case that the filters to be implemented are a part of a fixed system and not programmable. In the case of programmable filters, i.e., filters where characteristics are changed by inserting different coefficients, our approach would call for the use of random access memory (RAM) for implementing the function $\varphi$. Consequently, the programming would be considerably more time consuming, since up to $2^5/5 = 6.4$ times more data must be inserted then in the existing realization.

The examples we have presented clearly show that the proposed approach will offer significant savings in package count and power consumption over existing realizations. These savings further translate in reduced cost of parts, reduced assembly costs and increased reliability.

## V. ERROR ANALYSIS

In the previous sections we have not discussed the errors due to the finite word length used to represent the data and the values of the function $\varphi$. The effect of finite word length on digital filters using a multiplier has been investigated, and methods are well known to compute the mean-squared error [9], [10]. These results can be applied to the hardware configurations we proposed in this paper. In this section we derive expressions that will enable the system designer to evaluate the expected performance of the various hardware configurations proposed, in terms of the mean-squared error committed.

To begin with, a scaling of $\varphi$ is often necessary to prevent overflow. For the second order described by (5), the scaling of the coefficients $a_0$, $a_1$, and $a_2$ is straightforward. For stability reason, we must have $|b_1| < 2$ and $|b_2| < 1$. According to (8), it is then possible for $\varphi$ to be larger than 1. Therefore, it may be necessary to store in the ROM the value of $\varphi$ scaled by $2^{-K}$. In the majority of cases, $K = 1$ would be adequate. Effectively, this scaling corresponds to modifying (8) and (9) to

$$\varphi(x^1,x^2,x^3,x^4,x^5) = 2^{-K}(a_0x^1 + a_1x^2 + a_2x^3 - b_1x^4 - b_2x^5) \quad (12)$$

and

$$y_n = \sum_{j=1}^{B-1} 2^{-j+K}\varphi(x_n{}^j,x_{n-1}{}^j,x_{n-2}{}^j,y_{n-1}{}^j,y_{n-2}{}^j) - 2^K\varphi(x_n{}^0,x_{n-1}{}^0,x_{n-2}{}^0,y_{n-1}{}^0,y_{n-2}{}^0). \quad (13)$$

Although the desired output is given by (5), due to finite word length used, the actual computed output is different. Let $\bar{x}_n$ be the value of $x_n$ rounded to $B$ bits, and let $\bar{x}_n{}^j$ be its $j$th bit. The roundoff error $e_n' = \bar{x}_n - x_n$ is an uncorrelated sequence with zero mean and the variance of $e_n'$ is $2^{-2B}/3$ [9]. Let $\bar{\varphi}(\cdot,\cdot,\cdot,\cdot,\cdot)$ be $\varphi(\cdot,\cdot,\cdot,\cdot,\cdot)$ defined by (14) but rounded to $B$ bits. Let $v_n$ be the actual computed output and let $v_n{}^j$ be its $j$th bit, $j = 0,1,\cdots, B-1$. $v_n$ is the round value of

$$\sum_{j=1}^{B-1} 2^{-j+K}\bar{\varphi}(\bar{x}_n{}^j,\bar{x}_{n-1}{}^j,\bar{x}_{n-2}{}^j,v_{n-1}{}^j,v_{n-2}{}^j) - 2^K\bar{\varphi}(\bar{x}_n{}^0,\bar{x}_{n-1}{}^0,\bar{x}_{n-2}{}^0,v_{n-1}{}^0,v_{n-2}{}^0). \quad (14)$$

As described in Section II, $v_n$ is calculated by adding the shifted $\bar{\varphi}$'s. Only $B$ bits needed to be kept in the accumulation, and rounding occurs at the last addition. Let the roundoff error introduced be denoted by $e_n'''$. Then $e_n'''$ are uncorrelated each with zero mean and variance $2^{-2B}/3$. Therefore,

$$v_n = \sum_{j=1}^{B-1} 2^{-1+K}\bar{\varphi}(\bar{x}_n{}^j,\bar{x}_{n-1}{}^j,\bar{x}_{n-2}{}^j,v_{n-1}{}^j,v_{n-2}{}^j) - 2^K\bar{\varphi}(\bar{x}_n{}^0,\bar{x}_{n-1}{}^0,\bar{x}_{n-2}{}^0,v_{n-1}{}^0,v_{n-2}{}^0) + e_n'''. \quad (15)$$

Each $\bar{\varphi}$ is related to the unrounded value $\varphi$ by

$$2^K\bar{\varphi}(\bar{x}_n{}^j,\bar{x}_{n-1}{}^j,\bar{x}_{n-2}{}^j,v_{n-1}{}^j,v_{n-2}{}^j) = \varphi(\bar{x}_n{}^j,\bar{x}_{n-1}{}^j,\bar{x}_{n-2}{}^j,v_{n-1}{}^j,v_{n-2}{}^j) + e_{n,j}'' \quad (16)$$

where $e_{n,j}''$ is due to roundoff. The $e_{n,j}''$ are zero mean, uncorrelated, and each have variance $2^{-2B}/3$.

On using (13), (15), and (16), we have

$$v_n = e_n''' + \sum_{j=1}^{B-1} 2^{-j}e_{n,j}'' - e_{n,0}'' + a_0e_n' + a_1e_{n-1}' + a_2e_{n-2}'. \quad (17)$$

Define the roundoff error at the output of the second-order section by

$$e_n = v_n - y_n. \quad (18)$$

By using (5), (17), and (18), we have

$$e_n + b_1e_{n-1} + b_2e_{n-2} = a_1e_n''' + a_2e_{n-1}' + a_3e_{n-2}' + \delta_n \quad (19)$$

where

$$\delta_n = e_n''' + \sum_{j=1}^{B-1} 2^{-j}e_{n,j}'' - e_{n,0}'' \quad (20)$$

is an uncorrelated zero mean sequence and the variance of $\delta_n$ can be easily shown to be $(2^{-2B}/3)[1 + (1 - 2^{-2B})/(1 - 2^{-2})]$ or approximately $(7/9)2^{-2B}$.

The error $\{e_n\}$ can be regarded as the output of a linear system with two inputs $\{e_n'\}$ and $\{\delta_n\}$ as illustrated in Fig. 8 where $B(z) = 1/(1 + b_1z^{-1} + b_2z^{-2})$. It is straightforward to show that the $e_n$ has zero mean and mean-square value given by

$$E\{\epsilon_n{}^2\} = \frac{2^{-2B}}{3}\frac{1}{2\pi i}\oint_{|z|=1} [H(z)H(1/z) + (7/3)B(z)B(1/z)]\,(dz/z). \quad (21)$$

Given the filter coefficients, such integrals can be evaluated, either numerically or algebraically, by a computer program or table [11]. To evaluate the mean-squared error of higher order filters, realized in either parallel or cascade form, we assume that each second-order section contributes an uncorrelated error component as the one described above and the total output error is obtained by summing these various errors weighted by the transfer function from their point of injection to the output. For the parallel form, the output error is simply the sum of the various errors from the second-order sections, Fig. 1(b). That is

$$E\{(e_{P,n})^2\} = \sum_{j=1}^{M} E\{(e_n{}^j)^2\} \quad (22)$$

where $e_{P,n}$ is the error in parallel form realization and $M$ is the number of second-order sections. Equation (22) does not take into account the additional error committed in the final scaling of the outputs of the second-order sections before summing them up. However, if the scaling coefficients are known this can be easily accounted for.

For the cascade form realization, the output mean-squared error can be easily computed using the error model given in Fig. 8.

$$E\{(e_{C,n})^2\} = \sum_{j=1}^{M-1} \frac{E\{(e_n{}^j)^2\}}{2\pi i}\oint H^j(z)H^j(1/z)z^{-1}\,dz \quad (23)$$

where $e_{C,n}$ is the error in the cascade form realization, $M$ is the number of second-order sections, and $H^j(z)$ is the transfer function between the output of the $j$th second-order section and the final output. We see that the common factor in the error expression (21) is the exponential decrease as a function of the word length $B$, as is the case in existing realizations. Equations (21)–(23) allow the system designer to choose the word length $B$ according to the errors that can be tolerated.

## VI. CONCLUSION

We have proposed a new approach to the implementation problem of digital filters, which offers significant savings in cost and power consumption over existing realizations and permits operation at higher speeds than those achievable by existing realizations. The savings made possible by the proposed approach are due to the fact that the flexibility inherent in using a multiplier and thus permitting an infinite number of transfer functions

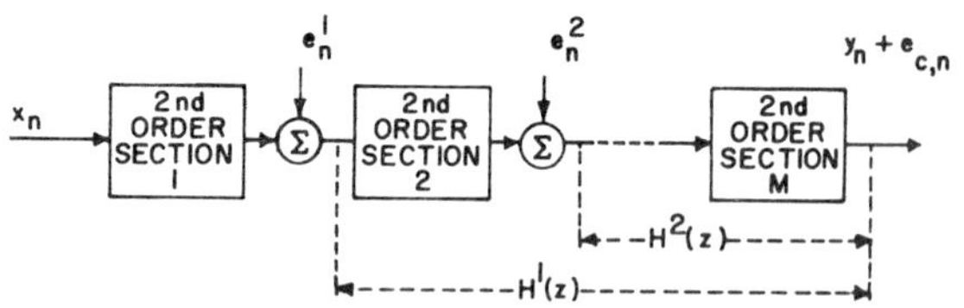


Fig. 8. Model for the mean squared error of a digital filter realized in cascade form.

to be realized is not utilized in practice when implementing filters for fixed systems, and our approach eliminates that over capacity. An error analysis was carried out and expressions derived which will allow the designer to choose the word length which will insure a specified performance.

The continuing emergence of technologies which further reduce the power requirements and increase the devsity of semiconductor memory such as silicone on saphire (SOS) appears to suggest that in the future the exchange between multiplier logic and memory bits for digital filters will only become more economic and offer even more savings than what has been pointed out by us in this paper.

## ACKNOWLEDGMENT

The authors wish to thank the reviewers for their valuable comments and suggestions. One of the reviewers also pointed out that the basic configuration of Fig. 2 has been disclosed recently in [12].

## REFERENCES

[1] L. B. Jackson, J. F. Kaiser, and H. S. McDonald, "An approach to the implementation of digital filters," *IEEE Trans. Audio Electroacoust.* (*Special Issue on Digital Filters: The Promise of LSI Applied to Signal Processing*), vol. AU-16, pp. 413–421, Sept. 1968.

[2] J. F. Kaiser, "Digital filters," in *Systems Analysis by Digital Computer*, F. F. Kuo and J. F. Kaiser, eds. New York: Wiley, 1966, ch. 7.

[3] J. B. Knowles and R. Edwards, "Effects of a finite-word-length computer in a sampled-data feedback system," *Proc. Inst. Elec. Eng.* (*London*), vol. 112, pp. 1197–1207, June 1965.

[4] L. B. Jackson, "On the interaction of roundoff noise and dynamic range in digital filters," *Bell Syst. Tech. J.*, vol. 49, pp. 159–184, Feb. 1970.

[5] ——, "Roundoff-noise analysis for fixed-point digital filters realized in cascade or parallel form," *IEEE Trans. Audio Electroacoust.* (*Special Issue on Digital Filtering*), vol. AU-18, pp. 107–122, June 1970.

[6] W. S. Lee, "Optimazation of digital filters for low roundoff noise," in *Proc. 1973 IEEE Int. Symp. Circuit Theory*, Toronto, Ont., Canada, Apr. 1973, pp. 418–421.

[7] S. K. Tewksbury, "Special purpose hardware implementation of digital filters," in *Proc. 1973 IEEE Int. Symp. Circuit Theory*, Toronto, Ont., Canada, Apr. 1973.

[8] S. D. Peraris, "A 40-ns 17-bit by 17-bit array multiplier," *IEEE Trans. Comput.* (Short Notes), vol. C-20, pp. 442–447, Apr. 1971.

[9] B. Liu, "Effect of finite word length on the accuracy of digital filters—a review," *IEEE Trans. Ciruit Theory* (*Special Issue on Active and Digital Networks*), vol. CT-18, pp. 670–677, Nov. 1971.

[10] A. V. Oppenheim and C. J. Weinstein, "Effects of finite register length in digital filtering and the fast Fourier transform," *Proc. IEEE*, vol. 60, pp. 957–976, Aug. 1972.

[11] K. J. Åström, E. I. Jury, and R. G. Agniel, "A numerical method for the evaluation of complex integrals," *IEEE Trans. Automat. Contr.* (Short Papers), vol. AC-13, pp. 468–471, Aug. 1970.

[12] A. Croisier, D. J. Esteban, M. E. Levilion, and V. Rizo, "Digital filter for PCM encoded signals," U. S. Patent 3 777 130, Dec. 3, 1973.

# Implementation of a Second Order Digital Filter Section with Stable Overflow Behaviour

By **Bernd Eckhardt** and **Wolfgang Winkelnkemper**

*Communication from the Institut für Nachrichtentechnik, Universität Erlangen-Nürnberg*

The usual structures for recursive digital filters are the cascade or parallel arrangement of second order sections. These sections may be implemented in the first or second canonical form; *Fig. 1* [2]. To the authors' best knowledge up to now any recursive digital filter is implemented in the cascade or parallel form with second order sections in the second canonical form, using serial fixed point arithmetic and 2's complement coding of the state variables.

For this structure Ebert et al. [3] discussed the inherent threat of oscillations due to overflow, assuming that with the given wordlength numbers of magnitude less than or equal to unity are representable. They gave the necessary and sufficient condition for stability:

$$|y^{\mathrm{L}} + y^{\mathrm{C}}| \geqq 2 \qquad (1)$$

where $y^{\mathrm{L}}$ denotes the output value in the linear case and $y^{\mathrm{C}}$ the corrected value at the output of the nonlinearity (the nonlinearity due to rounding is neglected in this paper). *Fig. 2* shows the allowed region for nonlinear graphs according to Eqn. (1).

The main point in this approach was the assumption that there is only one nonlinearity in the feedback loop which is true for the second canonical form. It is located at point d in Fig. 1b.

The probably simplest and usual way to secure stability is to detect an overflow at point d and to replace the state variable by +1 (in case of positive overflow) or −1 (in case of negative overflow), thus implementing the so called "saturation arithmetic".

As Jackson [4] discussed in detail, there are cases of required filter performance where the first canonical

Reprinted with permission from *Nachrichtentech. Z.*, vol. 26, pp. 282–284, June 1973.

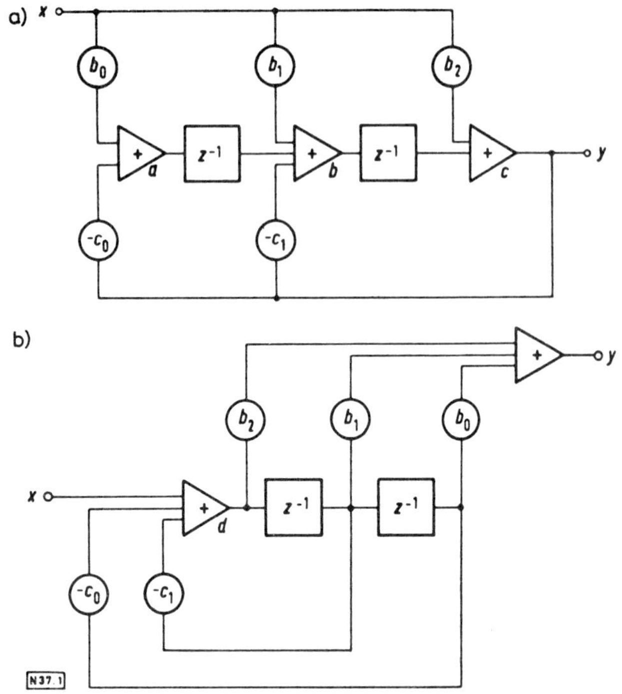


Fig. 1a.
First canonical form for a second order section.

Fig. 1b.
Second canonical form for a second order section.

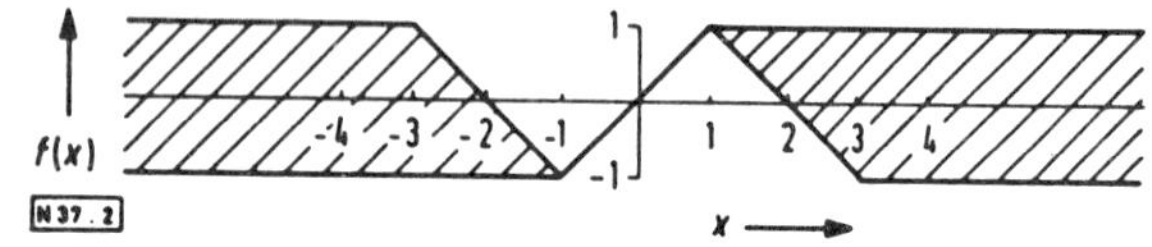


Fig. 2.
Allowable area (shaded region) for the graph of the nonlinearity in a second order filter.

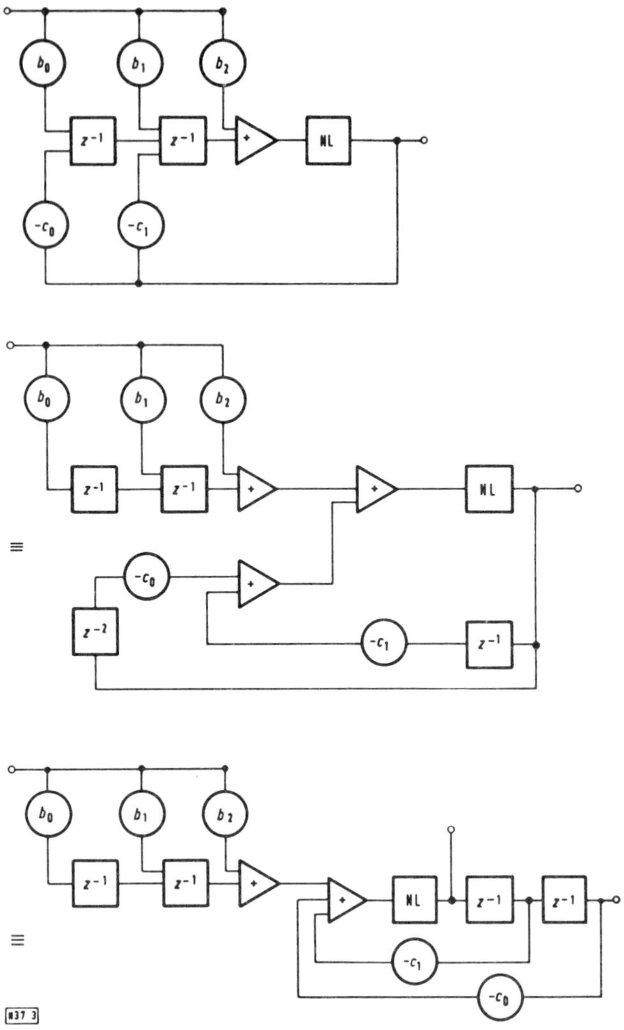


Fig. 3.
Rearrangement of a second order filter in the first canonical form with only one nonlinearity.

form is superior. For this structure the stability problem turns out to be a little more complicated. As derived in *Fig. 3,* Eqn. (1) holds for stability too, if there is only one nonlinearity, located at point c, since in this case the feedback loops are identical for the first and second canonical form. It has to be taken into account, however, that nonlinearities are located at the outputs of all the adders if the system is excited, since the adders may overflow.

It is an advantage of 2's complement coding that internal overflows at points a or b are allowable and do not affect linearity if the output value at point c is of magnitude less than unity. But this advantage causes a further difficulty: an actual overflow at the output adder may be not only allowable but even necessary to compensate a preceeding overflow at the adders a or b. An output value at the adder c without overflow may be not correct if there has been an overflow already at a preceeding adder.

So, two problems arise which have to be solved with minimum expense:

1. An overflow at the output has to be detected correctly, avoiding the mentioned errors.

2. Based on the information from 1. corrections have to be made in case of overflow in such a way that there is

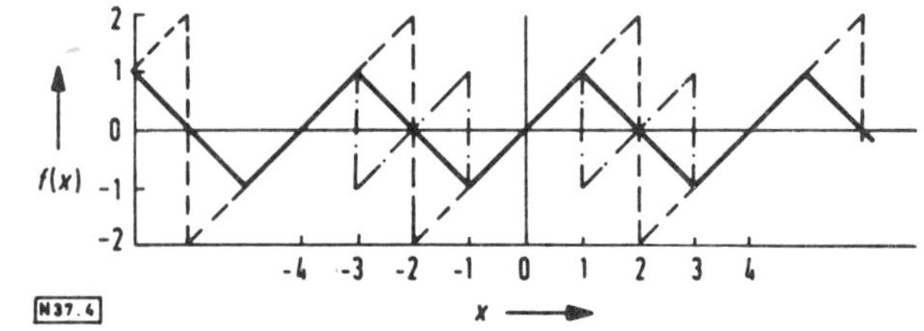


Fig. 4.
— — — graph of the lengthened state variable
— · — · graph of the state variable after cutting off the CB
——— resulting graph after complementing.

only one nonlinearity in the feedback loop which is located at point c and the graph of which is in accordance with Fig. 2.

Obviously, for both problems further information synchronous with the output value has to be kept ready at point c about what happened one sample time intervall earlier at point b and two intervalls earlier at point a.

Referring to Fig. 2 we can see that for the graph of the nonlinearity restrictions are given for values of magnitude up to 3. So for both problems pointed out above two control bits seem to be necessary. But of all the graphs satisfying Fig. 2 (an infinite number) there is just one which is periodic in the intervall 4, namely the triangular graph, described by the function:

$$f(x) = \begin{cases} x - 4\lambda & \text{for } 4\lambda - 1 \leqq x \leqq 4\lambda + 1 \\ -x + 4\lambda + 2 & \text{for } 4\lambda + 1 \leqq x \leqq 4\lambda + 3 \end{cases} \quad (2)$$

with $\lambda$ being an arbitrary integer.

The graph is shown by the solid line in *Fig. 4*. This nonlinearity can be implemented with only one control bit, for it is of no interest which $\lambda$ in Eqn. (2) would be true to describe the correct linear value; only two cases have to be distinguished.

We can simply derive the correction rule for this type of nonlinearity: Since the terms $-4\lambda$ and $+4\lambda+2$ don't appear in the given wordlength for $x$, $f(x)$ is given by

$$f(x) = x$$

or (3)

$$f(x) = -x$$

depending on one additional control bit which still has to be defined.

At this point, before we derive how to obtain the control bit, it is of some interest to consider where to put it. The first idea is to store the necessary information in flipflop elements that bypass the shift register delay elements for the state variables. The other possibility is to lengthen the state variables one bit. For parallel arithmetic these two versions are almost identical. For serial arithmetic, and this is of main interest today, they differ. If the organization of the filter is such that there is no empty time slot between two following samples, the first solution seems to be better. In the other case the organization has to be changed so that there is one empty position in the input and output sequence which gives places to insert the control bit internally. If the section is multiplexed, however, and the multiplexing degree is variable by external control, the bypassing storage would require the same adjustable organization as necessary for the state variables' delay elements. In this case it will be easier to lengthen the state variables, although the cycle time is lengthened too. For theoretical explanation we can assume that the state variables are lengthened, independent of where the additional bit is stored.

The necessary control bit can be obtained in a simple way: we have to carry out the addition at points a, b, c with one bit extra (broken line in Fig. 4).

This bit is actually the most significant bit of the lengthened state variable. We call it the control bit (CB). The next one in significance we call the MSB, since it is the most significant bit in the original state variable. The control bit has to be generated at the outputs of the coefficient multipliers. If the coefficient is of magnitude less than or equal to unity, the CB is obtained by repeating the MSB which is the last one in time in serial arithmetic. If the coefficient is of magnitude greater than unity, the CB already exists.

Summarizing what has been derived for the implementation of the triangular type nonlinear graph we get the following rule: Internally all arithmetic operations from all the coefficient multiplier outputs to the filter section output have to be performed with one additional bit in front of the original state variables, that means with double dynamic range. At the output this extra bit is cut off again and a decision is made whether or not this control bit is equivalent to the original MSB. If not, the original state variable has to be changed in sign, that means complemented. Fig. 4 shows the corresponding graph.

Looking into implementation problems in more detail, two more difficulties arise. At first, the 2's complement code is not symmetrical, since the "$-1$" has no positive counterpart. If the correct value is "$+1$", after cutting off the CB and complementing we obtain "$-1$". But in this case the rule fails, since according to the graph "$+1$" must not be complemented. The second problem is that the state variable has to be multiplied with the coefficients $-c_0$ and $-c_1$ immediately, that is, when the LSB appears. The decision, whether or not the state variable has to be complemented, can be done not before the CB appears.

Both problems are overcome with the following arrangement: Since the coefficient multiplication is a linear operation, we can complement the product instead of the input variable. If the state variable is "$+1$", after cutting off the CB the actual multiplier input is "$-1$". But the CB and the MSB are not equivalent, so the products $(-c_0)\cdot(-1)$ and $(-c_1)\cdot(-1)$ will be complemented and are correct again. The serial-parallel multiplier is affected with a delay of at least the wordlength times the clock period, so the time problem is avoided. In most cases the complementation can be performed easily in combination with the following addition, by inverting the product and putting the carry-flipflop to 1 instead of 0 at the beginning.

The described overflow nonlinearity was implemented in a very flexible programmable digital filter [5] with best success.

The authors wish to thank the Deutsche Forschungsgemeinschaft which supported the project.

* * *

**References**

[1] *Avenhaus, E.:* Zum Entwurf digitaler Filter mit minimaler Speicherwortlänge für Koeffizienten und Zustandsgrößen. Ausgewählte Arbeiten über Nachrichtensysteme Nr. 13, herausgegeben von W. Schüßler, Erlangen 1971.

[2] *Schüßler, W.:* Zur allgemeinen Theorie der Verzweigungsnetzwerke. Arch. elektr. Übertr. 22 (1968) pp. 361 to 367.

[3] *Ebert, P. M.; Mazo, J. E.; Taylor, M. G.:* Overflow oscillations in digital filters. Bell System techn. J. 48 (1969) pp. 2999 to 3020.

[4] *Jackson, L. B.:* Roundoff-noise analysis for fixed-point digital filters realized in cascade or parallel form. IEEE Trans. AU–18 (1970) pp. 107 to 122.

[5] *Eckhardt, B.; Winkelnkemper, W.:* Ein Beitrag zum Schaltungsentwurf digitaler Filter für den Tonfrequenzbereich. To be published.

(Manuscript received on 29th January 1973)

Anschriften der Autoren s. S. 274.

# Hardware Considerations for Digital FIR Filters Especially with Regard to Linear Phase

by Ulrich Heute *

Report from the Institut für Nachrichtentechnik der Universität Erlangen-Nürnberg

The hardware realization of digital FIR filters in the direct form is considered. Starting from a review of the well known multiplier-multiplex structure, two main points are aimed at:

- Speed problems, occurring within the hardware blocks themselves as well as emerging from the sampling rate requirements, are solved by parallel use of more than one multiplex units, in two different ways.
- A multiplex realization structure for linear phase filters is developed. The solution of the speed problems in this special case is shown. Detailed operation descriptions are given.

**Überlegungen zum praktischen Aufbau nichtrekursiver Digitalfilter insbesondere im Hinblick auf lineare Phase**

Die praktische Realisierung digitaler nichtrekursiver Filter in direkter Form wird betrachtet. Nach einem Rückblick auf die bekannte Multiplizierer-Multiplexstruktur zielt die Arbeit im wesentlichen auf zwei Punkte:

- Geschwindigkeitsprobleme treten sowohl innerhalb der Hardware-Blöcke selbst als auch nach außen hin auf Grund der Anforderungen an die Abtastfrequenz auf. Sie werden auf zwei verschiedenen Wegen gelöst, indem mehr als eine Multiplexeinheit benützt wird.
- Eine Multiplexrealisierung für linearphasige Filter wird entwickelt. Die Lösung der Geschwindigkeitsprobleme für diesen Sonderfall wird gezeigt. Die Arbeitsschritte der Strukturen werden im Detail angegeben.

## 1. The Direct Form Multiplex Structure

### *1.1. Introduction*

Among the two canonical structures of digital FIR filters in direct form (Fig. 1), the "second canonical" version [1] is advantageous, in two aspects: Without larger efforts, the output noise is kept low by rounding after accumulating, and for the case of flexible filters, it is easy to implement the filter with one multiplexed multiplier. It is well

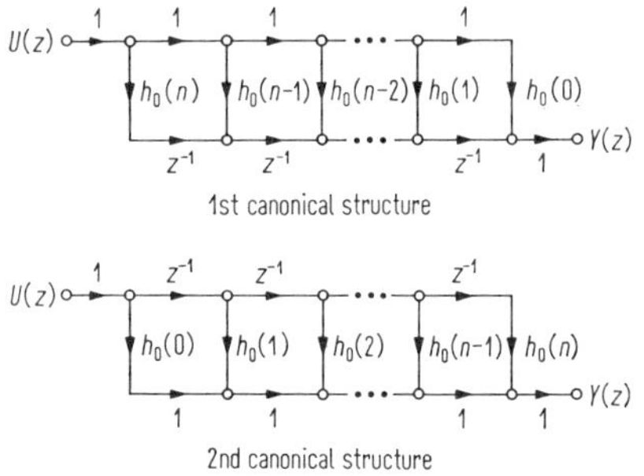


Fig. 1. First and second canonical versions of digital FIR filters.

known that this multiplex structure renders it possible to use LSI- (MOS-) registers. Details, concerning clock schemes and state variable processing, are given by Winkelnkemper (chapter 6 in [1]) and Gold (chapter 9 in [2]). In Fig. 2, the structure is shown, and the processing steps are pointed out, briefly. They form the well known convolution sum

$$y(k+d) = \sum_{\nu=0}^{n} h_0(\nu)\, u(k-\nu)\,,$$

where $k$ is the free variable, and $d$ a delay, imposed by the multiplex technique. For simplicity, the coefficients and the state variables are indicated in the figures by $h_\nu$ and $u_\nu$, only.

In Fig. 2, the cycle of total duration $T$ starts with $u(k) \triangleq u_0$ being entered and then the multiplication $h_0(n) \cdot u(k-n)$ being started. This looks quite natural (and gives us $d = 1$), but it is not necessary: We may begin with any of the products $h_0(\nu) \cdot u(k-\nu)$, and we may enter $u(k)$ at any moment which allows us to get the desired sample ordering in the register (e.g., one might start with $h_0(n) \cdot u(k-n)$ and enter $u(k)$ *after* this multiplication, throwing out $u(k-n)$; this would save one word of the memory). This fact will prove to be useful in a later part of the paper.

### *1.2. Speed problems*

Now, we shall have a closer look at the main hardware blocks in Fig. 2, on one hand, and at the conditions imposed by a possible user of the filter, on the other.

The (MOS-) memory will have a limited shift frequency $f_s = 1/\tau_s$. The multiplication can be carried out with a maximum speed $f_m = 1/\tau_m$; normally, we have $f_m \leqq f_s$.

A user will ask for a certain degree $n$ and a sampling rate $f = 1/T$. Then, everything is all right, if

* Dipl.-Ing. U. Heute, Institut für Nachrichtentechnik, Universität, D-852 Erlangen, Cauerstrasse 7.

Reprinted with permission from *Arch. Elek. Übertragung.*, vol. 29, pp. 116–120, Mar. 1975.

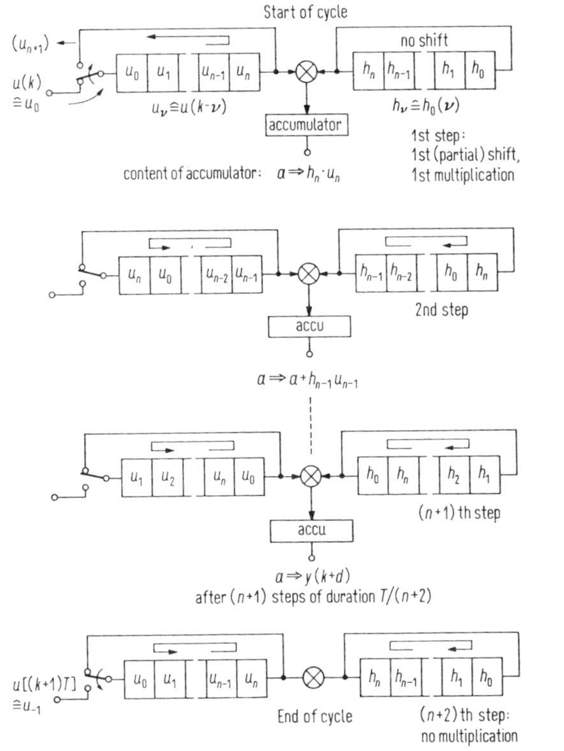


Fig. 2. General multiplex structure; processing steps. (In all figures describing the processing order, a "step" starts with a shift, which may be "partial", if only some of the memories move their contents; the shift is followed by a multiplication, in general. In all figures, the situation after the shift is drawn.)

we have

$$T \overset{(>)}{=} (n+2)\,\tau_s = (n+2)\,\tau_m\,.$$

But it may happen that this equation does not hold, and the following questions arise:

(i) What is to be done, if

$$T = \frac{n+2}{p_s}\,\tau_s = \frac{n+2}{p_s}\,\tau_m$$

with $p_s > 1$, i.e., if our hardware, as a whole, is too slow for the required sampling rate?

(ii) How can we cope with

$$T = (n+2)\,\tau_s = (n+2)\,\frac{\tau_m}{q_m}$$

with $q_m > 1$, i.e. a multiplier which is too slow?

(iii) What is needed, if

$$T = \frac{n+2}{p_s}\,\tau_s = \frac{n+2}{p_s}\,\frac{\tau_m}{q_m}$$

with $p_s > 1$ and $q_m > 1$, i.e. if the register is too slow, and the multiplier is even slower?

### *1.3. Different kinds of parallelism*

As far as point (i) is concerned, the answer is a "parallelism" [2] of $p$ "filter units", as they were shown in Fig. 2; of course, $p = \langle p_s \rangle$ [1] has to be the appropriate integer. The resulting structure (for $p=2$) and the first processing steps are given in Fig. 3. The following steps are found easily, as both parts of the filter perform the same procedure, as outlined in Fig. 2. In the upper half, $n$ is replaced by $n_1 = (n-1)/2$; so, $(n_1+2)$ steps have to be carried out.

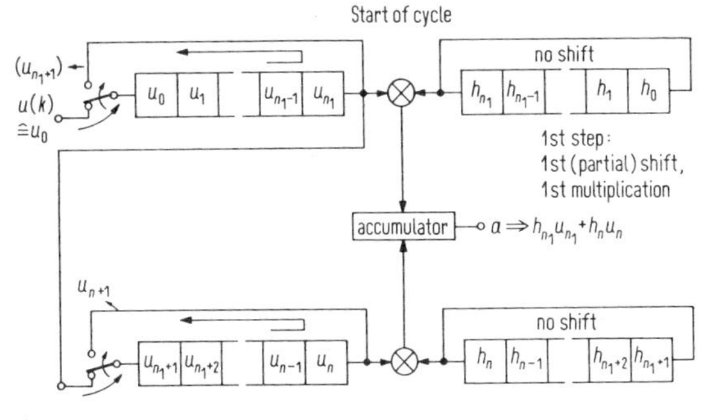


Fig. 3. "Parallelism" of $p(=2)$ filter units (of equal size, if $(n+1)/p$ is integer).

Regarding problem (ii), the solution is similar: $(q-1)$ single (TTL-) registers follow the "long" (MOS-) memory chain. They contain one state variable, each, and give us $(q-1)$ additional taps of the delay line. So, $q$ (with $q = \langle q_m \rangle$) multipliers can be used in parallel (Fig. 4 for $q=2$). The differences of the multiplication and the shift frequencies are equalized, in this way, because $q$ shifts are carried out during one multiplication (requiring multipliers with a memory, of course).

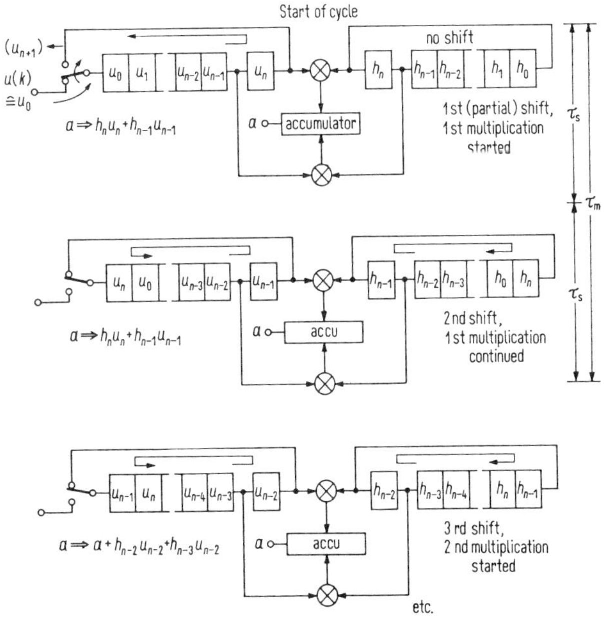


Fig. 4. "Speed-adjusting parallelism" of $q(=2)$ multipliers.

It is easy to answer question (iii), now. We can adjust the multiplier and register speeds by the method shown above, and then apply Gold's parallelism [2], in addition. This solution, outlined in Fig. 5 for $p=q=2$, may be rather interesting, if a group of filters are to be built with different degrees,

[1] $\langle r \rangle$ is the "smallest integer $\geqq r$".

but for the same sampling frequency: The main part is a "building block" with adjusted speeds, as in Fig. 4; the parallelism of $p$ of these blocks allows an increase of the filter degree by, roughly, a factor of $p$.

To some extent, these considerations are part of a widespread knowledge on digital filter hardware. The question of the speed adjustment *inside* a filter block, however, was not examined before, and, what is more important, it seems to be not so well known that analogous structures do exist for linear phase FIR filters, as well.

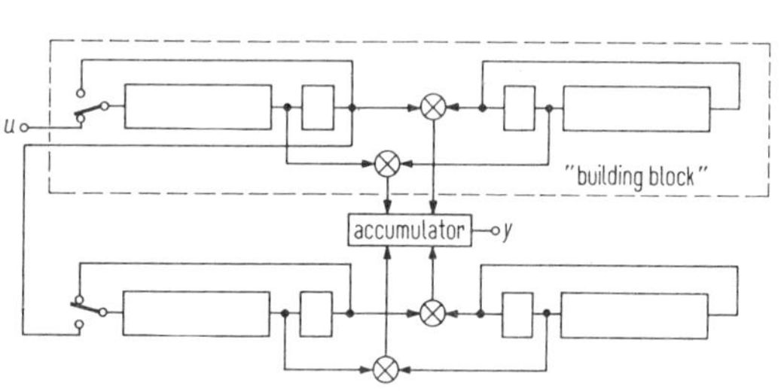


Fig. 5. Combined parallelism and "building block" structure.

They will be derived in the following sections of the paper. An even filter degree $n = 2N$ (i.e.: an impulse response of odd length) is assumed.

## 2. Multiplier Multiplex Realizations of Linear Phase Filters

### 2.1. *Principal considerations*

Linear phase filters have a symmetric impulse response:

$$h_0(\nu) = \pm h_0(2N - \nu), \qquad \nu = 0(1)(N-1).$$

Thus, the size of the coefficient memory can be reduced by (roughly) a factor of 1/2. The same applies to the number of multiplications, if an addition is carried out before each multiplication; this can be seen from the convolution equation

$$y(k+d) = h_0(N) \cdot u(k-N) + \sum_{\nu=0}^{N-1} h_0(\nu)\,[u(k-\nu) \pm u(k-2N+\nu)].$$

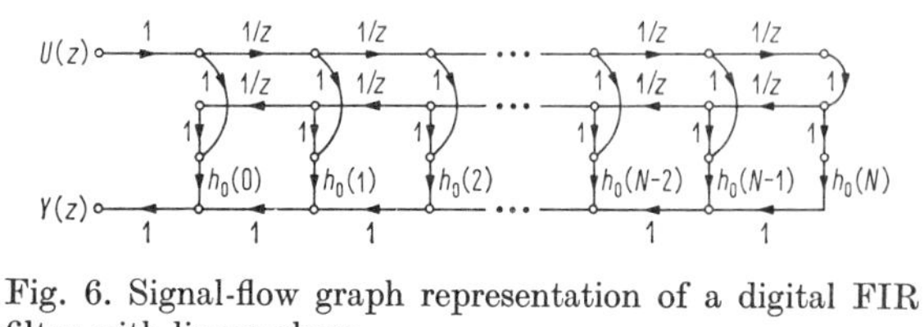


Fig. 6. Signal-flow graph representation of a digital FIR filter with linear phase.

In the following, only the plus sign will be regarded, without restricting the generality. The resulting structure is well known ([1], Fig. 6).

Now, the multipliers and the state variable adders can be realized by one unit, to be multiplexed $(N+1)$ times. The steps of operation within one sampling period are to be described as follows:

$$\begin{array}{lll}
0: & \text{add } \{u(k) + u(k-2N)\}, & \text{multiply by } h_0(0), \\
1: & \text{add } \{u(k-1) + u(k-2N+1)\}, & \text{multiply by } h_0(1), \\
\vdots & & \\
N-1: & \text{add } \{u(k-N+1) + u(k-N-1)\}, & \text{multiply by } h_0(N-1), \\
N: & \text{add } \{u(k-N) + 0\}, & \text{multiply by } h_0(N).
\end{array}$$

All results are accumulated and, finally, rounded.

The signal-flow graph (Fig. 6) and this scheme indicate, what has to be changed in the "general" multiplex structure of Fig. 2: Two state variables are needed at the same moment. Therefore, the register is divided into two parts. The corresponding block diagram (for one cycle) is shown in Fig. 7, and the location of the $u(k-\nu)$ and $h_0(\nu)$ at the beginning of the calculation are indicated as well as their movement during the cycle.

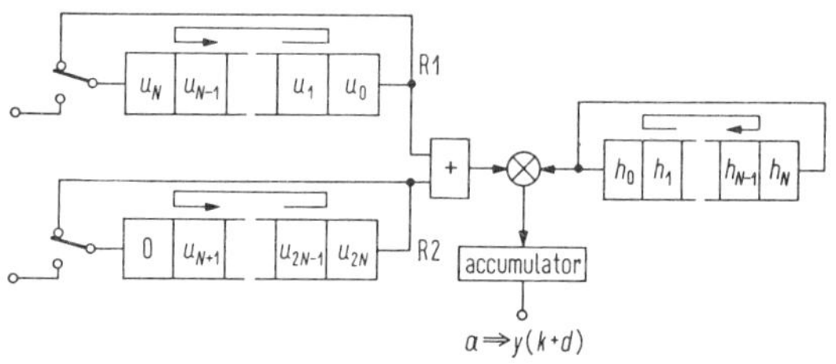


Fig. 7. Block diagram of a linear phase filter (for one cycle).

Obviously, the data chains are different in an important aspect: In the first part of the register (R1), the indices $\nu$ have to *decrease*, whereas they must *increase* in the second half (R2), from left to right.

At a first glance, this means a difficulty: LSI-MOS-registers, usually, are not able to act as universal shift left — shift right devices. But it looks like we have to feed $u_0$ into the right end of R1, before the next cycle, and use the left end as an output to deliver $u_N$ to R2. However, this is not true.

### 2.2. *Generation of different sample orders in two circular registers*

What might be called the "usual way" of feeding a circular register, was shown in the block diagrams of Fig. 2. This results in the "natural" order of the $u_\nu$, as needed in R2 of Fig. 7.

However, we can choose a different moment within a cycle, to enter a new value $u_0 = u(k)$. In Fig. 8, it can be seen what happens, if the new sample is fed in just one shift period *before* the end of a cycle: After $m$ shift times we end up with the same reversed ordering of the samples, which has been presumed in the beginning. Now $\mu$ has been replaced by $\mu - 1$.

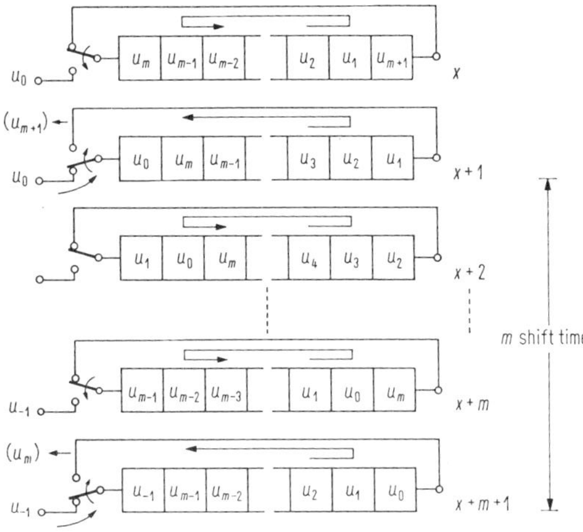


Fig. 8. Generation of a reverse sample ordering in a register of length $(m + 1)$.

So, returning to Fig. 7, we can feed $u_N$ into R 2 in the usual manner, after $(N + 1)$ shifts, and feed $u_0$ into R 1 in the way outlined above, after $N$ shifts, in order to get the desired constellation of the data.

Of course, there is no simple rule such as "End of cycle — enter $u_0$ — forward $u_{N+1}$", any more. But, as mentioned in the beginning of the paper, this is not important (and could be avoided, for the price of time); it means nothing but a change of the clock generator logic, as the two parts of the memory need different clock pulse trains.

Different clocks, however, are necessary anyway: When a result is complete, a *relative* shift of the data by two places is needed, since in the next cycle, $h_0 \cdot (u_{-1} + u_{2N-1})$, $h_1 \cdot (u_0 + u_{2N-2})$, etc., will be calculated.

### 2.3. *Detailed operation description*

The order of the steps involved in the calculation of one output sample $y(k+d)$ is to be seen in a survey of block diagrams (Fig. 9). It is worthwhile to stress some points: a) It turns out to be possible indeed to use all registers in a simple "shift-one-way-manner" — the somewhat awkward sample ordering is generated by appropriate clocking, only. b) In section 1.1, it was said that it is not necessary to combine the end of a cycle with the entering of new data; here, $u(k+1)$ is entered with the $(N+1)$th shift. c) It was said in section 1.1, too, that there is no prescription of the first multiplication to be carried out; here, we start with

$$h_0(1)\,[u(k-1) + u(k-2N+1)]\,,$$

and the last operation is that one using $u(k) \triangleq u_0$, i.e. step "0" in the scheme in section 2.1. The third ("0"-) input for the second half register, of course, is implemented by a change of the switching clocks.

## 3. Parallelism in Linear Phase Filter Realizations

Naturally, the speed problems discussed in section 1.2 for the general case, will arise here again.

The solutions, again, consist of different types of parallelism of more than one arithmetic units. They are depicted in Fig. 10 for the case that a) the hard-

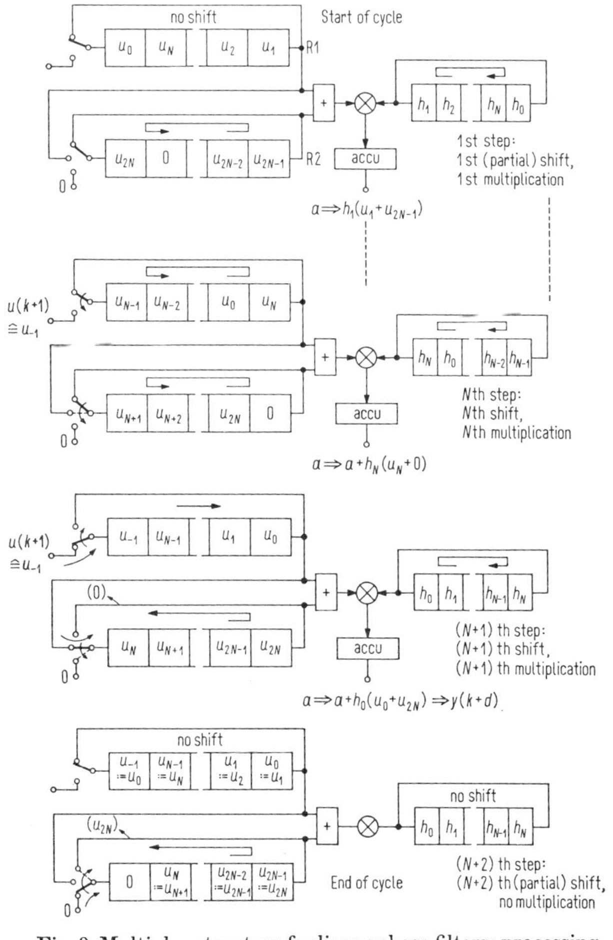


Fig. 9. Multiplex structure for linear phase filters; processing steps.

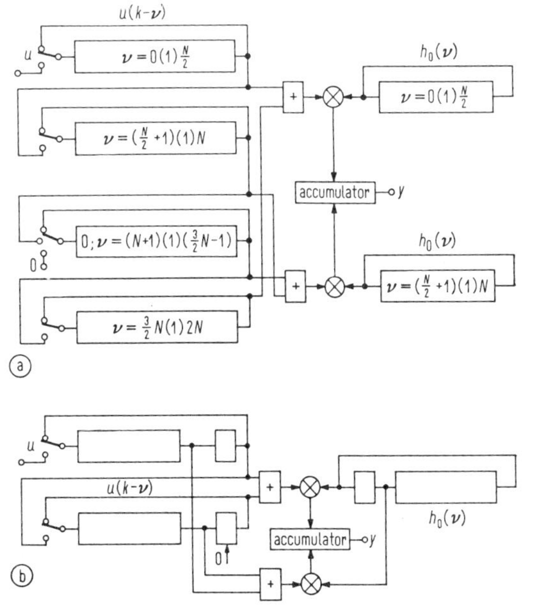


Fig. 10. Parallelisms for linear phase filters: (a) Increasement of the degree by a factor $p$ $(= 2)$; $N$ even; (b) speed adjustment (for $q = 2$).

ware is too slow for the required degree and frequency range (Gold's parallelism for the linear phase filter), and b) the multiplication time is too large for the memory; $p=q=2$ and $N$ even are assumed.

In Fig. 10a the $p$ filter sections show the same properties as the devices explained in sections 2.3 and 3.2, as far as the data chains in the registers are concerned. So, a close examination of the different shift clocks is necessary to achieve the appropriate "natural" and "inverse" sample orders; but no details do need to be dealt with here, as the results are similar to those obtained before. In Fig. 10b the hardware looks even more awkward than in the "general" case (Fig. 5). However, a detailed look at the processing steps of such a device also reveals some advantages, provided by the additional hardware: The "0" input for R2, in Fig. 9, may be replaced by the clear facility of the TTL registers; the "shift $q$ times — multiply once" scheme allows us to do the additional (relative) shift of the data *without* loosing any time.

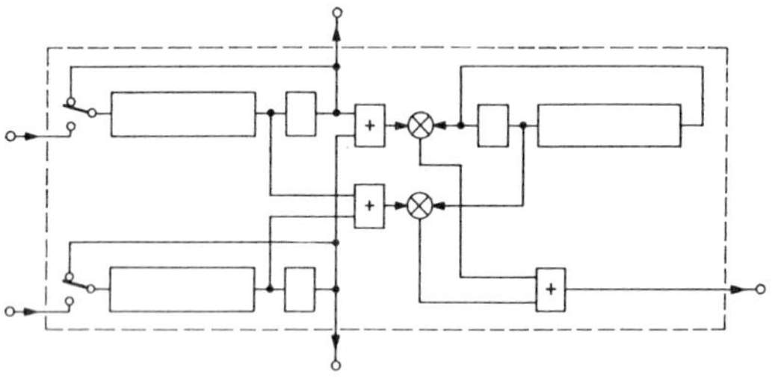

Fig. 11. "Building block" for linear phase filters with an internal speed adjustment (for $q = 2$).

Both solutions may be combined, if necessary. This results in a building block concept, again, as suggested for the case of a general phase filter, in section 1.3. A block consists of a coefficient memory, two state variable registers and $q$ multipliers, thus being "speed-adjusted" (Fig. 11).

The necessary interconnections of the block inputs and outputs for different values of $p$ may be seen easily, if the cases explained above are regarded (Fig. 9, Fig. 10b).

## 4. Conclusion

The well known one-multiplier-multiplex structure for the realization of digital FIR filters in the direct form [1] was recapitulated. The possibility to increase the filter degree (or the sampling rate) by the parallelism, proposed in literature [2], was explained briefly, and a new type of parallelism was suggested, adjusting memory and multiplier speeds.

In part two of the paper, a realization structure for linear phase filters was proposed, which makes use of multiplier multiplexing. It was shown that the same kind of LSI memories can be used here, which makes the arbitrary phase filter multiplex realization attractive, as the inverse motion of two data chains can be accomplished by applying appropriate clocks. Instead of $(n+1) = (2N+1)$ multiplications, $N+1$ arithmetic operations are necessary, only, and $N+2$ shifting clock periods have to pass. It has to be mentioned at this point, that some ideas of this solution were outlined, for a special case, in an unpublished note already in 1971 [3].

Finally, block parallelism and speed-adjusting parallelism were shown to be possible for this case, as well.

*Acknowledgement*

The final construction of the paper emerged from many helpful discussions with Prof. Dr.-Ing. H. W. Schüßler. The author wishes to express his thanks to him as well as to Dipl.-Ing. B. Eckhardt for his useful hints and corrections.

(Received May 31st, 1974.)

### References

[1] Schüßler, H. W., Digitale Systeme zur Signalverarbeitung. Springer-Verlag, Berlin 1973.

[2] Rabiner, L. R. and Gold, B., Theory and applications of digital signal processing. To be published.

[3] Deigner, A., Preliminary design considerations for a digital filter. Private communication and unpublished drafts, 1971.

# A New Approach to the Realization of Nonrecursive Digital Filters

ABRAHAM PELED and BEDE LIU, Fellow, IEEE

*Abstract*—A new realization of nonrecursive digital filters that are used to operate on analog signals is proposed. This realization requires no multiplications, and exploits the relative simplicity of delta modulation as a means for analog to digital conversion. This realization also permits a mechanization as a "very fast" digital filter, using read-only memory (ROM). An evaluation of this realization in terms of computation time storage requirements and mean-squared error is presented. These characteristics are compared with their counterparts for existing realization methods of nonrecursive digital filters. Computer simulation results that tend to confirm the theoretical results of the error analysis are included.

## I. Introduction

Nonrecursive digital filters possess many distinct features that make them attractive for a variety of applications. Their excellent phase characteristic, inherent stability, and conceptual simplicity are some of the causes for the extensive research efforts that have been invested in their design and analysis. Efficient methods are known for the design of nonrecursive digital filters using various approximation methods [1], and some common designs have been tabulated [2].

Nonrecursive filters require relatively longer computation time per output sample. Although this can be improved by using the fast convolution [3], [4], such an approach is only suitable when the filter is implemented on a computer or when a fast Fourier transform (FFT) box is available.

In this paper we present a new approach for nonrecursive digital filtering of analog signals that requires very simple hardware and does not require multiplication. We also suggest a particular hardware realization that permits very fast digital filtering, using read-only memory (ROM) so that each output sample requires only one addition.

Manuscript received March 26, 1973; revised June 20, July 17, and August 1, 1973. This work was supported by the Air Force Office of Scientific Research, USAF, under Grant AF-AFOSR 71-2101, and by the National Science Foundation under Grant GK-24187.

The authors are with the Department of Electrical Engineering, Princeton University, Princeton, N. J. 08540.

The proposed scheme uses a delta modulator [5] for analog-to-digital conversion at the input. The one-bit stream at the output of the delta modulator is then fed into a digital processor that involves no multiplication, but only additions and sign changes. We proceed to compare the efficiency and to study the various tradeoffs involved in the proposed realization versus existing realization schemes. An analysis is performed to show that the error for the proposed realization is due to two sources; one introduced by the delta modulation of the input and the other resulting from the use of finite word length to represent the filter coefficients. A simple expression is derived to predict the performance of the proposed realization. The influence of various design parameters on the resulting error is discussed and some guidelines are given to select the parameters. Finally, to confirm the validity of the error analysis computer simulation results are presented for various input signals.

Delta modulation has been used by Goodman [6] to convert an analog signal into a pulse-code modulated (PCM) signal. In his scheme one of the structures discussed in the present paper was used for minimum mean squared error estimation of the pulse-amplitude modulation (PAM) sample. Lockhart [7] also used delta modulation for filtering an analog signal. However, his realization is hybrid, employing analog multiplications and analog additions.

## II. A New Approach for Implementing Nonrecursive Digital Filters

A nonrecursive digital filter is characterized by an input-output relationship of the form

$$y_n = \sum_{k=0}^{N-1} a_k \, x_{n-k} \tag{1}$$

where $\{x_n\}$ is the input sequence, $\{y_n\}$ the output sequence, and

$$\{a_k\}_{k=0}^{N-1}$$

are the filter coefficients. Consider the processing of an analog signal by such a filter realized with $b$-bit word hardware. The conventional implementation of this filter is illustrated in Fig. 1. The analog signal is sampled and each sample is quantized and converted to a $b$-bit word by a PCM. The output sequence of the PCM is fed to a digital processor that computes the output samples $\{y_n\}$ according to (1). The hardware and computational requirements can be easily determined from the block diagram. We need a $b$-bit PCM, storage for $N$ filter coefficients, and a digital processor capable of performing addition and multiplication. The processor performs $N$ multiplications and $N - 1$ additions per output sample.

The approach proposed in this paper is illustrated

Reprinted from *IEEE Trans. Audio Electroacoust.*, vol. AU-21, pp. 477–484, Dec. 1973.

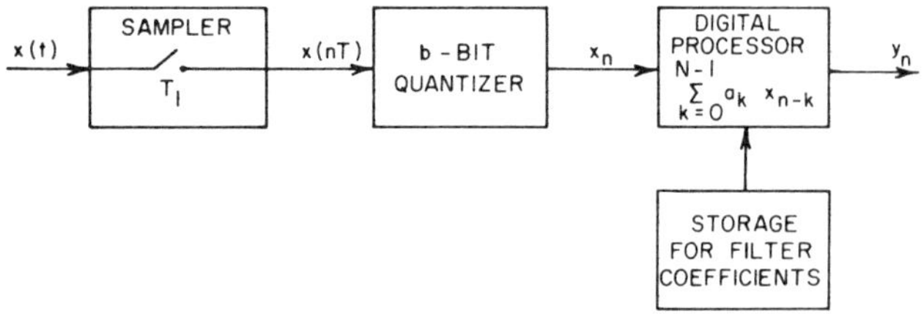


Fig. 1. Realization using PCM.

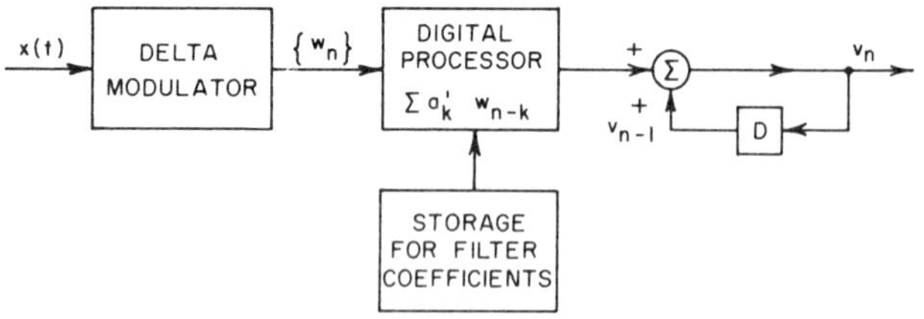


Fig. 2. Realization using delta modulation.

in Fig. 2. The input signal enters a delta modulator whose output is the sequence $\{w_n\}$ with each $w_n$ taking on the values +1 or -1. This sequence enters a digital processor that computes

$$\sum_{k=0}^{N-1} a_k' w_{n-k}$$

and adds this sum to the content of an accumulator to obtain the output sample $v_n$. This operation is described by

$$v_n = v_{n-1} + \sum_{k=0}^{N-1} a_k' w_{n-k}. \tag{2}$$

We first show that if the error introduced by the delta modulator is negligible, the coefficients $\{a_k'\}$ can be chosen so that the output $\{v_n\}$ agrees with $\{y_n\}$. An error free delta modulator is characterized by the following relation between $w_n$ and samples of the input signal $x(t)$ [5],

$$x(nT) = x[(n-1)T] + \Delta_x w_n \tag{3}$$

where $T$ is the sampling interval, and $\Delta_x$ the step size of the delta modulation. Substituting (3) into (2), we obtain

$$v_n = \sum_{k=0}^{N-1} \frac{a_k'}{\Delta_x} x[(n-k)T] + C \tag{4}$$

where $C$ is a constant, which can easily shown to be zero if we require that the output be zero when no signal is present at the input. With that assumption we immediately see that $v_n$ equals the $y_n$ given by (1) provided that

$$a_k' = a_k \Delta_x. \tag{5}$$

In the preceding discussion, we have assumed that no roundoff or quantization error occurs in the arithmetic operations. Thus we see that at least in an idealized situation, the new implementation scheme has the same capability as the conventional approach of Fig. 1.

Let us pursue this idea for a moment, since we know that the error introduced by the delta modulator and by the arithmetic operation can be made very small. A comparison of the realizations proposed in Figs. 1 and 2 reveals immediately two advantages of the delta modulation realization over the PCM realization. The first is the relative simplicity of a delta modulator as compared to a PCM. The second is a simpler processor due to the absence of multiplications. However, it is well known that in order to obtain a good fidelity with delta modulation, we have to sample much faster than the Nyquist rate, which is sufficient for the PCM realization. In the remainder of the paper, we will discuss the performance of the proposed realization and compare it to an equivalent PCM realization.

## III. Possible Hardware Realizations

Two possible mechanizations of (2) are presented in Figs. 3 and 4. Both configurations will yield exactly the same output, but differ in storage requirements and speed of operation.

In Fig. 3, the input analog signal $x(t)$ is passed through a single integration delta modulator [8] whose output is $\{w_n\}$ with each $w_n$ taking on +1 or -1 only. The sequence $\{w_n\}$ enters the arithmetic unit that computes in a straightforward manner the increment of the output according to

$$\Delta v_n = \sum_{j=0}^{N-1} a_j' w_{n-j}. \tag{6}$$

This increment is added to the previous value of the output to form the current output,

$$v_n = v_{n-1} + \Delta v_n. \tag{7}$$

The configuration of Fig. 4 has no arithmetic unit. Since the increment $\Delta v_n$ can take on only a finite number $(2^N)$ of values, depending on the array $\{w_{n-j}\}_{j=0}^{N-1}$, these $2^N$ possible values of $\Delta v_n$ can be stored in a ROM at addresses that correspond to the binary word obtained from $\{w_{n-j}\}_{j=0}^{N-1}$ by substituting 0 for -1. The value of the increment $\Delta v_n$ can be obtained once $w_n$ is in the shift register. So to obtain the output $v_n$, only one addition is needed. The increase in the speed of operation is significant. However, there is a serious shortcoming in this configuration, since the exponential dependence of the storage capacity needed on the number of filter coefficients makes the storage requirement considerable even for a modest number of filter coefficients. It should be pointed out, that for linear phase filters, the coefficients $\{a_n\}_{n=0}^{N-1}$ are symmetric [9]. Therefore, only $2^{(N+1)/2}$ distinct values of the increment

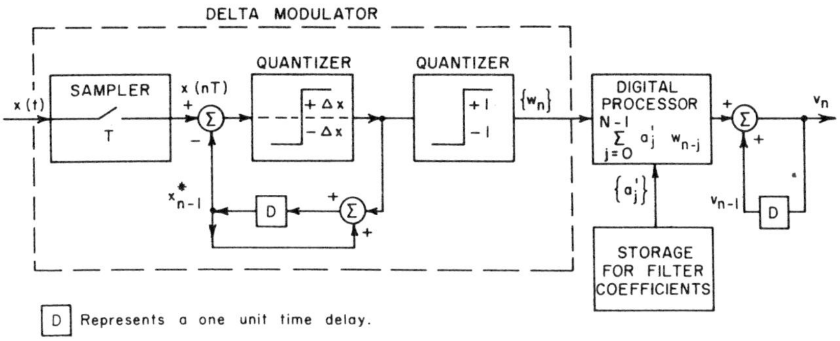


Fig. 3. Hardware realization of nonrecursive digital filter using delta modulation.

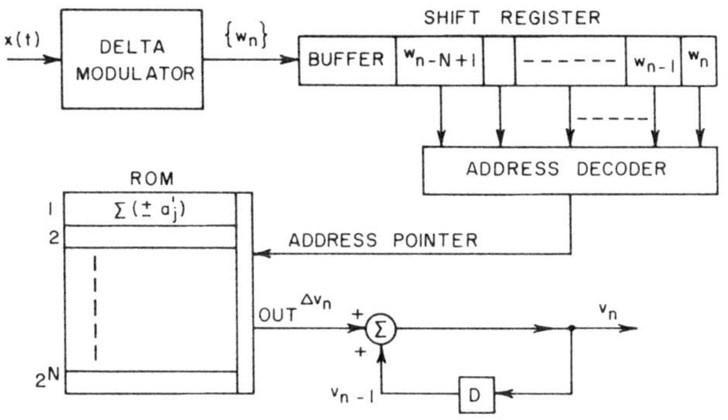


Fig. 4. A very fast realization of a nonrecursive digital filter.

$\Delta v_n$ need to be stored. The storage requirements will be discussed further in the next section.

### IV. Computation Time and Storage Considerations

In this section we attempt to compare the two equivalent realizations of a nonrecursive digital filter, the one using PCM as shown in Fig. 1 and the one using delta modulation as shown in Fig. 3. As mentioned previously, we have to sample much faster than the Nyquist rate in the case of delta modulation. Let $K$ be the ratio of the sampling frequency in the delta modulation to the Nyquist frequency used in PCM.

In the case of the PCM realization the computation time per output sample is given by

$$C_{\mathrm{PCM}} = MP\alpha + (M - 1)\alpha \tag{8}$$

where $M$ is the number of filter coefficients of the PCM realization, $\alpha$ is the time required for one full word addition, and $P\alpha$ is the time required for multiplication.

In the case of the delta modulation (DM) realization the computation time per output sample is

$$C_{\mathrm{DM}} = N\alpha \tag{9}$$

where $N$ is the number of filter coefficients, and is in general different from $M$, because of the different sampling rates. In fact, in order for the two realizations to achieve the same or approximately the same frequency characteristic, it is reasonable to assume that $N = KM$. If we use the Fourier series method to design the filter, then it can be shown [10] that the number of coefficients needed to achieve a given transition region $\Delta\omega$ in the frequency characteristic is $N = 2\pi/T\Delta\omega$ in the case of Hamming weights or $N = 2\sqrt{R^2 + \pi^2}/T\Delta\omega$ in the case of Kaiser weights, where $R$ is related to the overshoot. In both cases decreasing the sampling interval $K$ times will require a $K$ fold increase in the number of coefficients needed to achieve the same frequency characteristic. For other design methods $N = KM$ is a very good approximation.

We need to compare $KC_{\mathrm{DM}}$ with $C_{\mathrm{PCM}}$ because of the different sampling rates. From the previous discussion, the comparison to be made is, assuming $M >> 1$,

$$MK^2\alpha \overset{?}{\lessgtr} M(P + 1)\alpha \tag{10}$$

or equivalently

$$K^2 \overset{?}{\gtrless} P + 1. \tag{11}$$

It is therefore obvious that in order to save computation time through the delta modulation realization, one must have $K^2 < P + 1$.

If special purpose hardware is used to perform multiplication then the value of $P$ would be approximately[1] 3, which requires $K < 2$. This results in an oversampling factor insufficient for good delta modulation for most practical signals. If the multiplication is realized by software the value of $P$ is approximately[1] 50, which implies $K < 7$, a value that will allow good quality delta modulation for most practical

[1] According to a survey article by Butler [11] on 45 minicomputers by 27 manufacturers, the ratio $P$ for fixed point hardware ranges between 2.0 and 12.5 with an average of 4.78. If the multiplication is done by software only, the ratio $P$ was between 7.8 and 281 with an average of 91.7. Of the minicomputers surveyed, 22.2 percent had standard hardware multipliers, 35.5 percent had such multipliers as optional equipment, and 33.3 percent had no such option.

signals [8]. Thus, the implementation of Fig. 3 offers a saving in computation time only if the equivalent PCM realization uses software to perform multiplication. Thus the main advantage of this new form of realization is the simplicity in hardware, and the price paid is a slower speed of operation.

For the configuration of Fig. 4, the computation of each output sample requires one shift, one readout, and one actual addition. It is safe to assume that the total time needed is less than $2\alpha$, and the comparison with the implementation of Fig. 1 is

$$2K\alpha \overset{?}{\lessgtr} M(P+1)\alpha. \tag{12}$$

Thus this form of delta modulation realization would always offer a significant saving in computation time.

The realization proposed in Figs. 3 and 4 are really two extreme solutions to the mechanization of (2), one requires the performing of all computations each time, and the other requires the storing of all possible outcomes of the computation and using them when needed. We have seen that the later requires a large amount of storage even for a modest number of coefficients. However a very good compromise between the two is possible. To illustrate the idea, let us consider a 15 tap filter. The realization using ROM as described by Fig. 4 requires a storage of $2^{15}$ words. Now, suppose we rewrite

$$\sum_{j=0}^{14} a_j' w_{n-j}$$

as

$$\sum_{j=0}^{4} a_j' w_{n-j} + \sum_{j=5}^{9} a_j' w_{n-j} + \sum_{j=10}^{14} a_j' w_{n-j}$$

and compute separately each partial sum by using a ROM to store all its possible outcomes. For each partial sum, we need $2^5$ words, so the total storage needed will be $3 \times 2^5 = 96$ words as compared to $2^{15} = 32\,768$ words in the straightforward ROM realization. However, two more additions are needed for each output sample. The tradeoff of an exponential decrease in storage for a linear increase in the number of additions needed is obviously attractive.

In general, let $N$ be the number of filter coefficients and $L$ be the number of coefficients in a partial sum; the last partial sum may contain less than $L$ terms if $L$ does not divide $N$. The computation time needed is $3\alpha + \left[\frac{N}{L}\right]_I \alpha$ and the storage requirement is $\left(\left[\frac{N}{L}\right]_I + 1\right)2^L$. Further reduction in storage is possible by taking advantage of the symmetry of the filter coefficients for linear phase filters. It is possible in most cases to use modest storage and achieve a saving in computation time over the PCM realization.

## V. Error Analysis

The desired output of a nonrecursive digital filter applied to an analog signal is

$$y_n = \sum_{j=0}^{N-1} a_j x[(n-j)T] \tag{13}$$

where $a_j$ are the filter coefficients, and $x[kT]$ are samples of the input signal. The proposed realization through delta modulation introduces two errors. One is due to the representation of each filter coefficient by a finite number of bits, and the second is due to delta modulation. The actual computed output is

$$v_n = v_{n-1} + \sum_{j=0}^{N-1} (a_j')^Q w_{n-j} \tag{14}$$

where $(a_j')^Q$ is $a_j'$ quantized to $b$-bits and $w_k$ is given by

$$w_k = (x_n^* - x_{n-1}^*)/\Delta_x \tag{15}$$

and $x_n^*$ is the $n$th sample value of the reconstructed staircase signal approximating $x(t)$. Let $e_{D_n}$ be the error

$$e_{D_n} = x(nT) - x_n^*. \tag{16}$$

Let $\epsilon_n$ be the error between the desired output and the actual output at time $nT$,

$$\epsilon_n = y_n - v_n. \tag{17}$$

One possible fidelity criterion of the approximation of $y_n$ by $v_n$ is the mean-squared error $E\{\epsilon_n^2\}$ where $E$ denotes expectation.

In Appendix A an expression is derived for the mean-squared error

$$E\{\epsilon_n^2\} = \sum_{k=0}^{N-1}\sum_{j=0}^{N-1} a_k a_j E\{e_{D_{n-k}} e_{D_{n-j}}\} + \frac{q}{\Delta_x^2}\sum_{j=0}^{N-1} \cdot [E\{x_{n-j}^{*2}\} + E\{x_{n-j}^* x_{n-N+j+1}\}] \tag{18}$$

where $q$ is the variance of the quantization error given by

$$q = \frac{2^{-2(b-1)}}{3}. \tag{19}$$

$b$ is the number of bits including the sign bit used in fixed point representation of the filter coefficient. The conversion is done using rounding. The first term of (18) is due to the delta modulation and the second term is due to roundoff.

The delta modulation involves a nonlinear feedback system that makes the analysis impossible for arbitrary signal statistics. For Gaussian input signals with rational spectra, the statistics of the error introduced by delta modulation are well understood and theoretical and experimental results are available [12]. The statistics of the signal $\{x_n^*\}$ are known only for Gaussian input signal and when the step size $\Delta_x$ is chosen to prevent overload, say by the so called $4\sigma$ rule [13]. Under these assumptions an expression has been found by Goodman [13] for the autocorrelation function of $\{x_n^*\}$. However this results in rather complicated equations and no insight into the behavior of the error can be gained.

In order to obtain a simple expression for the mean-squared error, some simplifying assumptions are made. It is assumed that the delta modulation error $\{e_{D_n}\}$ is approximately white and has a power $N_D$. Thus the first term on the right side of (18) can easily be shown to be $N_D \sum_{j=0}^{N-1} a_j^2$. As for the second term, we have from (16)

$$E\{x_k^* x_j^*\} = E\{x(kT)x(jT)\} + E\{e_{D_k} e_{D_j}\} \quad (20)$$

when it is substituted into (18), the term $E\{e_{D_k} e_{D_j}\}$ can be dropped since it produces a second-order effect.

Thus assuming a wide-sense stationary input we can rewrite (18) as

$$E\{\epsilon_n^2\} = N_D A + \frac{q}{\Delta_x^2}\left[NR_{xx}(0) + \sum_{j=0}^{N-1} \cdot R_{xx}(N - 2j + 1)\right] \quad (21)$$

where

$$A = \sum_{j=0}^{N-1} a_j^2 .$$

Furthermore since $R_{xx}(0) \geqslant |R_{xx}(k)|$ for all $k$ we can write the output error as

$$E\{\epsilon_n^2\} = N_D A + \frac{\beta q}{\Delta_x^2} NR_{xx}(0) \quad (22)$$

where $\beta$ is some number, $1 \leqslant \beta \leqslant 2$. Or equivalently the signal to noise ratio in decibels at the output is

$$(S/N)_{0\text{dB}} = 10 \log\left(\frac{Y}{E\{\epsilon_n^2\}}\right) = 10 \log \cdot \left(\frac{Y}{N_D A + \frac{\beta q}{\Delta_x^2} NR_{xx}(0)}\right) \quad (23)$$

where $Y = E\{y_i^2\}$ is the output signal power. Fig. 5 illustrates the qualitative behavior of the output signal to noise ratio as a function of the word length used to represent the filter coefficients.

We can see from the figure that there are two regions. One where the error is dominated by the first term in (22), that is the error due to delta modulation, and the second region corresponding to short word length $b$, where the second term of (22) dominates and the performance deteriorates rapidly.

From (22) we can derive a good estimate of the number of bits $b$ needed to prevent deterioration of the performance, i.e., operation to the left of the knee in the signal to noise ratio curve given in Fig. 5.

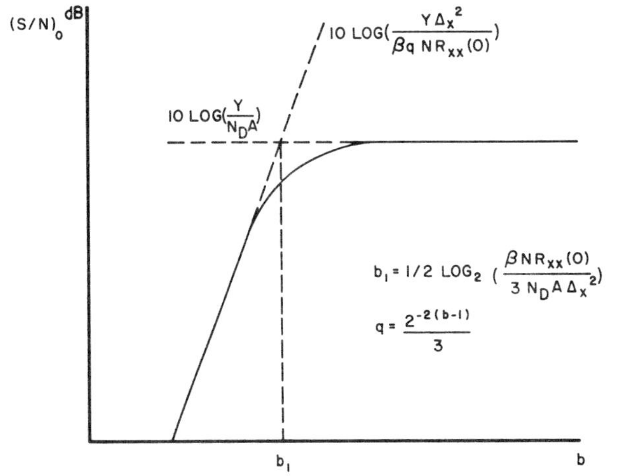


**Fig. 5. Behavior of the output signal to noise ratio (23).**

Let $b$ be the number of bits for which the two terms on the right side of (22) equal (taking into account that $b$ can be only an integer)

$$b_1 = \tfrac{1}{2}\left[\log_2\left(\frac{\beta NR_{xx}(0)}{3N_D A\Delta_x^2}\right)\right]_I + 1 \quad (24)$$

where $[\ ]_I$ denotes the integer part of the quantity inside the brackets. (Thus for $b > b_1$, the error is determined solely by the delta modulation, and for $b < b_1$ the performance deteriorates rapidly.) The $b_1$ given in (24) can be regarded as the minimum number of bits needed for satisfactory operation of the filter. It is worth mentioning that for most common filters the value of $A$ is less than 1, since it represents the area under the filter transfer function relative to an ideal low pass up to the sampling frequency. This means that, for a sufficient word length, the output error will be smaller than the error caused by the delta modulation. The justification for the simplifying assumptions made earlier and the validity of the error expression (21) lies in the experimental verification, and that is the subject of the next section.

## VI. Computer Simulation Results

The realization for nonrecursive digital filters through delta modulation of the input can be conveniently simulated on a digital computer. The delta modulator used was exactly the one in Fig. 3, that is a single integration, ideal integrator delta modulator. In practice it would be preferable to use a leaky integrator [8]. We used three types of input signals, for which theoretical and experimental results are known regarding their performance under delta modulation. These are sine wave with random phase, band-limited white Gaussian noise, and band-limited white Gaussian noise passed through an $RC$ filter. The delta modulation step used was the optimum one as given in [8], [14], and [15]. We used the method described by

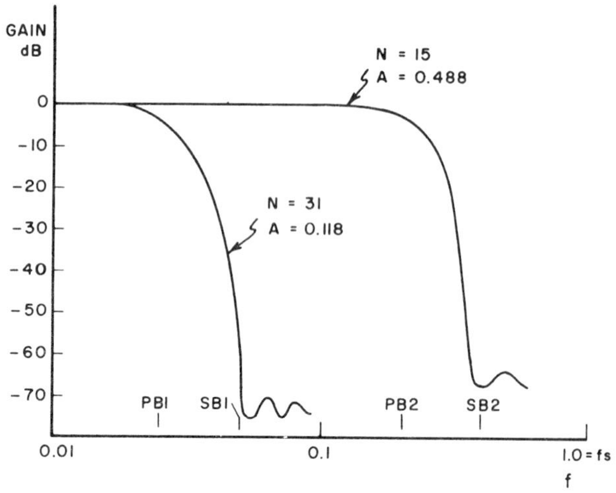


Fig. 6. Frequency response of the two filters used.

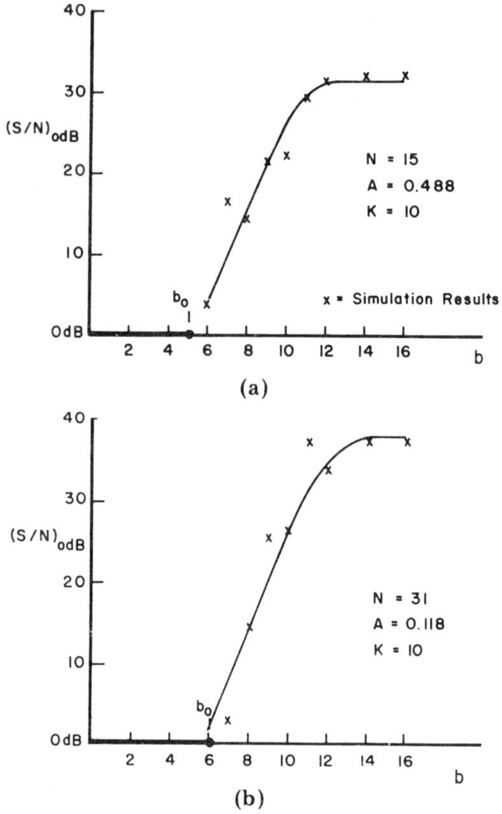


Fig. 7. Sine wave input. (a) $N$ = 15. (b) $N$ = 31.

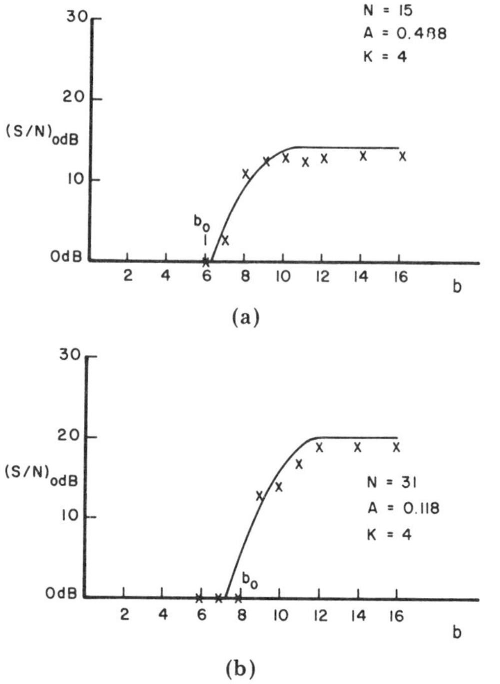


Fig. 8. Band-limited white Gaussian noise input, unity bandwith, zero mean, 0.25 standard deviation. (a) $N$ = 15. (b) $N$ = 31. *Note:* $X$ = simulation results.

O'Neal [8] to obtain white band-limited Gaussian noise sampled $K$ times the Nyquist rate, and we used the digital equivalent of an $RC$ filter to obtain the $RC$-Gaussian input signal. Two digital filters were tested, one with 15 and the other with 31 coefficients, both have unity gain in the passband and at least 60-dB attenuation in the stopband. Their frequency response is plotted in Fig. 6. In order to examine the validity of the analysis presented in Section V, we computed the mean-squared error between the error-free output as given by (13), and the output with error as given by (14), using various word lengths for the filter coefficients. The input was scaled between -1 and 1. No overflow occurred because of the unity gain of the filter. We used fixed point with rounding to represent the filter coefficients. The signal to noise ratio was computed rather than the mean-squared error itself:

$$(S/N) = \frac{\sum_{n=1}^{L} y_n^2}{\sum_{n=1}^{L} (y_n - v_n)^2}. \tag{25}$$

$L$ is the number of points used. In Figs. 7-9 the results of the simulations are plotted for the different input signals and filters used. The ordinate represents the signal to noise ratio in decibels and the abscissa the number of bits used for the representation of the filter coefficients. To get the theoretical curve corresponding to (23) we computed the actual mean-squared error introduced by the delta modulation in our simulation, that is we constructed the staircase signal $x_n^*$ and set

$$N_D = \frac{1}{L} \sum_{n=1}^{L} [x(nT) - x_n^*]^2. \tag{26}$$

The second term in (22) was computed using (19), the autocorrelation values of the signals used, and the delta modulation step size.

It is worth mentioning a practical consideration that is the reason for the straight line corresponding

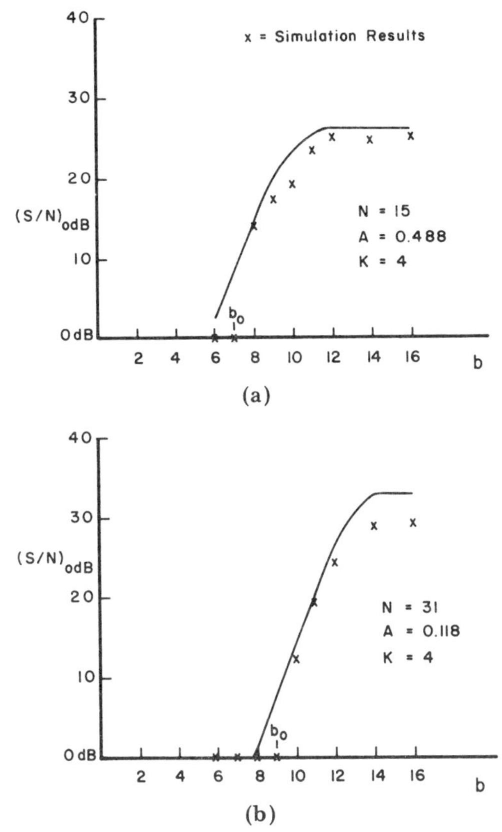


Fig. 9. Band-limited white Gaussian noise passed through *RC* filter, zero mean, $1/RC = 0.25$. (a) $N = 15$. (b) $N = 31$.

to a signal to noise ratio of 0 dB for very short word length. The reason for this is the fact that $\{a'_j\}_{j=0}^{N-1}$ are numbers between $-1$ and $+1$, and usually much smaller than 1 in absolute value. If the word length used $b_0$ is such that

$$\max |a'_j| < 2^{-b_0} \tag{27}$$

then all filter coefficients will be quantized to zero and the output $v_n$ will be identically zero that corresponds to a signal to noise ratio of 0 dB.

It is evident from the experimental results that (23) is indeed a useful expression that will allow the designer to predict the approximate output signal to noise ratio for the realization through delta modulation. Furthermore, (24) gives a good estimate of the minimum word length needed to achieve the best possible signal to noise ratio. As we mentioned earlier theoretical and experimental results for the delta modulation error are known for a variety of input signals and can be directly applied to evaluate the expected performance of this realization of nonrecursive digital filters.

## VII. Conclusion

We have proposed a new realization of nonrecursive digital filters using delta modulation. The main advantage of the scheme is the simplicity of the hardware, as a result of the elimination of multiplications in the processor, and of a simpler analog to digital conversion method. We have also suggested a realization permitting very fast filtering operation using ROM. The efficiency of the proposed realization was analyzed and compared to existing realization methods and some of the tradeoffs involved were discussed. An error analysis was carried out and a useful expression, which allows the system designer to choose the design parameters and predict the expected performance, was derived. Experimental results that confirm the error analysis were presented. In summary, the proposed realization requires very simple hardware and with a sufficient word length for the filter coefficients the output signal to noise ratio is determined by that of the delta modulation.

## Appendix A

We derive here an expression for the mean-squared error between the desired output $y_n$ and the computed output $v_n$ of the new implementation. Equations (13)-(17) are repeated here:

$$y_n = \sum_{j=0}^{N-1} a_j x[(n-j)T] \tag{A1}$$

$$v_n = v_{n-1} + \sum_{j=0}^{N-1} (a'_j)^Q w_{n-j} \tag{A2}$$

$$x_n^* = x_{n-1}^* + \Delta_x w_n \tag{A3}$$

$$e_{D_n} = x(nT) - x_n^* \tag{A4}$$

$$\epsilon_n = y_n - v_n. \tag{A5}$$

If the actual filter coefficients $(a'_j)^Q$ are obtained from $a_j \Delta x$ by rounding to $b$ bits we have $a'_j$.

$$(a'_j)^Q = a_j \Delta_x - e_{Q_j} \tag{A6}$$

where rounding noise $e_{Q_j}$ is known to have the following properties [16]:

$$E\{e_{Q_j}\} = 0 \tag{A7}$$

$$E\{e_{Q_j} e_{Q_k}\} = q\delta_{jk} + q\delta_{j,N-k-1} \tag{A8}$$

where $q = 2^{-2(b-1)}/3$. The second term in (A8) is due to the fact that for linear phase filters the coefficients are in pairs $a_j = a_{N-j-1}$. Using the previous we can rewrite (A2) as

$$v_n = v_{n-1} + \sum_{j=0}^{N-1} a_j x_{n-j}^* - \sum_{j=0}^{N-1} a_j x_{n-j-1}^* - \sum_{j=0}^{N-1} e_{Q_j} w_{n-j}. \tag{A9}$$

Combining (A1), (A4), (A5), and (A9) gives

$$\epsilon_n = \epsilon_{n-1} + \sum_{j=0}^{N-1} a_j e_{D_{n-j}} - \sum_{j=0}^{N-1} a_j e_{D_{n-1-j}} + \sum_{j=0}^{N-1} e_{Q_j} w_{n-j}. \quad \text{(A10)}$$

Equation (A10) provides a difference equation for the error in terms of the delta modulation error and quantization error. We assume the following initial conditions.

$$w_o = 0; \quad \epsilon_o = 0; \quad e_{D_o} = 0; \quad \text{(A11)}$$

with these assumptions the solution is

$$\epsilon_n = \sum_{j=0}^{N-1} a_j e_{D_{n-j}} + \sum_{j=0}^{N-1} e_{Q_j} \left( \sum_{i=1}^{n-j} w_i \right). \quad \text{(A12)}$$

Using the fact that summation over the past of the sequence $\{w_i\}$ gives us the value of $x_n^*$, the staircase function approximating the input, that is

$$x_n^* = \Delta_x \left( \sum_{i=1}^{n} w_i \right). \quad \text{(A13)}$$

On substituting (A13) into (A12) we finally get

$$\epsilon_n = \sum_{j=0}^{N-1} a_j e_{D_{n-j}} + \frac{1}{\Delta_x} \sum_{j=0}^{N-1} e_{Q_j} x_{n-j}^*. \quad \text{(A14)}$$

The mean-squared error $E\{\epsilon_n^2\}$ is given by

$$E\{\epsilon_n^2\} = E\left\{ \left[ \sum_{j=0}^{N-1} a_j e_{D_{n-j}} + \frac{1}{\Delta_x} \sum_{j=0}^{N-1} \cdot e_{Q_j} x_{n-j}^* \right]^2 \right\}. \quad \text{(A15)}$$

Using (A7) and (A8) we can see that the cross term is zero if we assume independence of $e_{Q_k}$ and $e_{D_j}$, which is obviously true. Thus,

$$E\{\epsilon_n^2\} = \sum_{k=0}^{N-1} \sum_{j=0}^{N-1} a_k a_j E\{e_{D_{n-k}} e_{D_{n-j}}\} + \frac{q}{\Delta_x^2} \sum_{j=0}^{N-1} \cdot [E\{x_{n-j}^{*2}\} + E\{x_{n-j}^* x_{n-N+j+1}^*\}] \quad \text{(A16)}$$

which is (18) in Section V.

## References


[1] L. R. Rabiner, "Techniques for designing finite-duration impulse response digital filters," *IEEE Trans. Commun. Technol.*, vol. COM-19, pp. 188–195, Apr. 1971.

[2] R. W. Hankins, "Design procedures for equiripple nonrecursive digital filters," Tech. Rep. 485, Mass. Inst. Tech., Cambridge, May 1972.

[3] T. G. Stockham, "High speed convolution and correlation," in *1966 Spring Joint Computer Conf., AFIPS Conf. Proc.*, vol. 28, pp. 229–233.

[4] H. D. Helms, "Fast Fourier transform method of computing difference equations and simulating filters," *IEEE Trans. Audio Electroacoust.*, vol. AU-15, pp. 85–90, June 1967.

[5] K. W. Cattermole, *Principles of Pulse Code Modulation*. New York: American Elsevier, 1969, pp. 198–218.

[6] D. J. Goodman, "The application of delta modulation to analog-to-PCM encoding," *Bell Syst. Tech. J.*, vol. 40, pp. 321–343, Feb. 1969.

[7] G. B. Lockhart, "Digital encoding and filtering using delta modulation," in *Proc. Conf. Digital Processing of Signals in Communications*, University of Technology, Loughborough, England, Apr. 1972, pp. 43–52.

[8] J. B. O'Neal, "Delta modulation quantizing noise, analytical and computer simulation results for Gaussian and television input signals," *Bell Syst. Tech. J.*, vol. 45, pp. 117–141, Jan. 1966.

[9] J. F. Kaiser, "Digital filters," in *System Analysis by Digital Computer*, F. F. Kuo and J. F. Kaiser, Eds. New York: Wiley, 1966, ch. 7.

[10] J. V. Wait, "Digital filters," in *Active Filters: Lumped, Distributed, Integrated, Digital, and Parametric*, L. P. Huelsman, Ed. New York: McGraw-Hill, 1970, p. 243.

[11] J. L. Butler, "Comparative criteria for minicomputers," *Instrum. Technol.*, vol. 17, pp. 67–82, Oct. 1970.

[12] D. Slepian, "On delta modulation," *Bell Syst. Tech. J.*, vol. 51, pp. 2101–2137, Dec. 1972.

[13] D. J. Goodman, "Delta modulation granular quantizing noise," *Bell Syst. Tech. J.*, vol. 48, pp. 1197–1217, June 1969.

[14] A. A. Reqvicha and A. Voelcker, "Design of nonrecursive digital filters," *IEEE Trans. Audio Electroacoust.*, vol. AU-18, pp. 464–470, Dec. 1970.

[15] F. B. Johnson, "Calculating delta modulator performance," *IEEE Trans. Audio Electroacoust.*, vol. AU-16, pp. 121–129, Mar. 1968.

[16] B. Liu, "Effect of finite word length on accuracy of digital filters—A review," *IEEE Trans. Circuit Theory*, vol. CT-18, pp. 670–677, Nov. 1971.

# Parallelism in Fast Fourier Transform Hardware

BEN GOLD and THEODORE BIALLY

*Abstract*—The fast Fourier transform algorithm is derived by means of successive fracturing of one-dimensional data strings into two-dimensional arrays. Using this formulation, a diagrammatic representation of mixed radix and highest radix FFT algorithms is derived. Using this representation, two broad classes of FFT hardware are explored, from the point of view of speed, parallelism, radix number, and type of memory.

## I. Introduction

Over the past few years, a large body of knowledge has been generated on the subject of fast Fourier transform algorithms [6]–[9], [12], [14]. While most of this work has been concerned with the radix-2 algorithm, substantial information has also been made available on higher radix and mixed radix algorithms [2], [16], [19]. All of these FFT algorithms have been used in both programming and hardware applications [1], [3]–[5], [10], [11], [17], [18].

FFT programs naturally have to be conditioned by the architecture of the particular computer. On the other hand, the designer of an FFT hardware device has the freedom to match the hardware design and the FFT algorithm. Such freedom is of particular advantage, for example, in designing high-speed systems, such as matched filters for wide bandwidth radar systems. The material in this paper is directed especially at such problems, where the designer wants to configure his digital structure so as to squeeze all the speed possible from his system.

The following example illustrates the difference between software and hardware methodology in FFT design. Consider first a general-purpose computer with a register length such that two registers are required to hold a single complex datum. Furthermore, the computer has the capability of performing a single arithmetic function at a time, such as an addition or multiplication. A radix-2 algorithm would require eight memory cycles, four multiply cycles, and six add cycles for each of $N/2 \log_2 N$ elementary computations. Since all of these are performed sequentially, a very restrictive upper limit is established on the speed capability of such a computer performing the FFT.

Manuscript received May 4, 1972. This work was sponsored by the Department of the Air Force.
The authors are with M.I.T. Lincoln Laboratory, Lexington, Mass. 02173.

Now, consider a piece of special-purpose FFT hardware. It is possible to design such hardware so that memory and arithmetic cycles overlap. Furthermore, both memory and arithmetic can be paralleled in such a way as to cause each elementary computation to be performed in one cycle. Thus, it can be seen that hardware design can be specifically tailored to the FFT algorithm to give substantial speed increases. Obviously, a general-purpose computer can also be designed where special emphasis is given to FFT performance, and such computers have been built [20], but in these structures it is difficult to attain the optimum speed while maintaining the desired generality. In this paper, we concentrate on the representation of FFT structures with both memory and arithmetic parallelism plus control, which leads to optimum speed.

## II. Representation of FFT Structures

The FFT computes the discrete Fourier transform (DFT) of $N$ input variables $x(0), x(1), \cdots, x(N-1)$ where, by definition,

$$X(k) = \text{DFT}\,(x(n)) = \sum_{n=0}^{N-1} x(n)W^{nk},$$

$$W = \exp\left(-j\,\frac{2\pi}{N}\right), \qquad n = 0, 1, \cdots, N-1. \quad (1)$$

The FFT gains speed by factoring the composite number $N$ into the product of integers:

$$N = r_1 r_2 \cdots r_m. \quad (2)$$

Equation (2) corresponds to fracturing the one-dimensional signal $x(n)$ into an $m$-dimensional signal. For example, if $N=64$ and $M=3$, we might have $r_1=r_2=r_3=4$, and the 64-point transform is translated into a three-dimensional $4\times4\times4$ transform. If $N$ is a prime number so that factorization of $N$ is not possible, the original signal can usually be augmented with zeros, and a new composite value of $N$ is attained.

A convenient way of performing the FFT algorithm is as follows. Begin by representing $N$ as the product of two numbers, say, $N=N_1N_2$. If $N_1$ and $N_2$ are both composite, they can further be broken down: $N_1=N_{11}N_{12}$, $N_2=N_{21}N_{22}$. In this way, each iteration consists of going through the mechanics of a two-dimensional transform to achieve a one-dimensional transform. The procedure for performing this latter operation is the following.

1) Arrange the one-dimensional data into a two-dimensional array, as shown for example, on the left of Fig. 1.

2) Do DFT's on each row individually.

3) Multiply each term of the resultant matrix by $W_1^{ij}$ where $i$ is the row index and $j$ is the column index and $W_1=\exp\,(-j(2\pi/N))$.

4) Transform the resultant matrix column by column to obtain the result. Thus, if we demonstrate that a

Reprinted from *IEEE Trans. Audio Electroacoust.*, vol. AU-21, pp. 5–16, Feb. 1973.

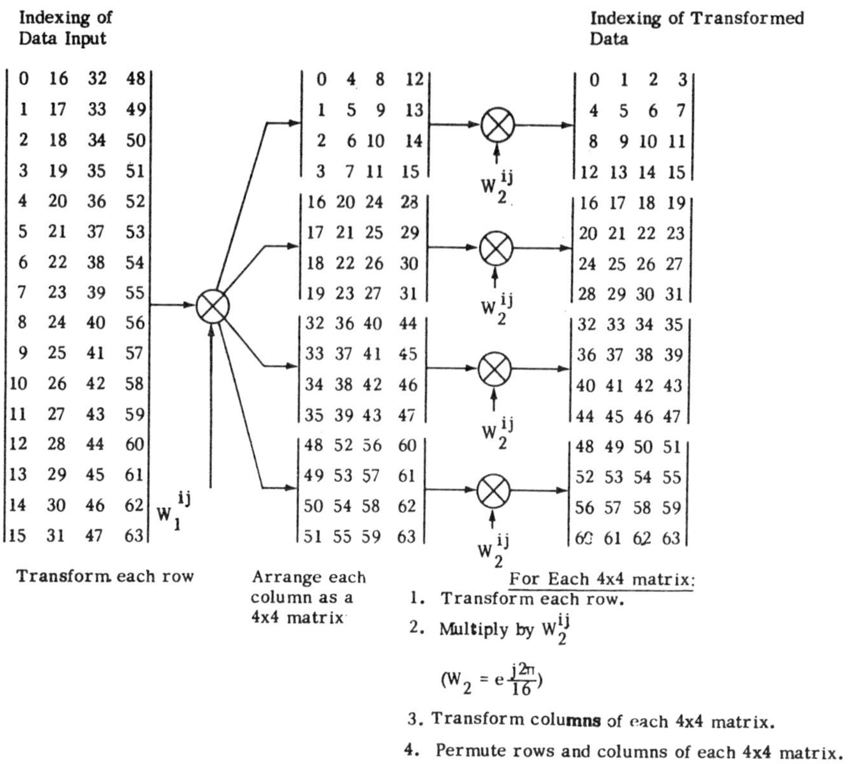


Fig. 1. Illustration of a 64-point DFT by successive fracturing of one-dimensional data into two dimensions.

given one-dimensional FFT can be reformulated in two dimensions, we will have effectively shown how to do the FFT in general. Beginning with (1), if $N$ is composite, it can be written as $N=LM$. As an example, let $L=3$, $M=5$ so that, given $x(0)$, $x(1) \cdots x(14)$, we need to find $X(0)$, $X(1) \cdots X(14)$. Our plan is to first arrange the $x(n)$'s and $X(k)$'s into rectangular arrays as follows:

$$\begin{array}{cccccc} m \rightarrow & & m = 0, 1 \cdots M-1 & & & \\ l & x(0) & x(1) & x(2) & x(3) & x(4) \\ \downarrow & x(5) & x(6) & x(7) & x(8) & x(9) \\ l = 0, 1 \cdots & x(10) & x(11) & x(12) & x(13) & x(14) \\ L-1 & & & & & \end{array}$$

$$\begin{array}{cccccc} r \rightarrow & & r = 0, 1 \cdots M-1 & & & \\ s & X(0) & X(3) & X(6) & X(9) & X(12) \\ \downarrow & X(1) & X(4) & X(7) & X(10) & X(13) \\ s = 0, 1 \cdots & X(2) & X(5) & X(8) & X(11) & X(14). \\ L-1 & & & & & \end{array}$$

Notice that the orderings of the sequence and its transform are orthogonal. Given this, the variables $n$ and $k$ can be represented as

$$n = Ml + m, \quad k = Lr + s. \tag{3}$$

Substituting (3) into (1), we obtain

$$X(Lr+s) = \sum_{m=0}^{M-1} \sum_{l=0}^{L-1} x(Ml+m) W^{(Ml+m)(Lr+s)}. \tag{4}$$

Expanding $W^{(Ml+m)(Lr+s)}$, observing that $W^{MLrl} = W^{Nlr} = 1$, and properly associating the indices with the summation terms, we can rearrange (4):

$$X(Lr+s) = \sum_{m=0}^{M-1} W^{Lmr} W^{ms} \sum_{l=0}^{L-1} x(Ml+m) W^{Msl}. \tag{5}$$

Now, let us define

$$q(s, m) = \sum_{l=0}^{L-1} x(Ml+m)(W^M)^{sl}. \tag{6}$$

For any fixed $m$, $q(s, m)$ can be interpreted as a column transform on the $x(n)$ array. For example, if $m=1$, $M=5$, $L=3$, then

$$q(s, 1) = x(1)(W^5)^0 + x(6)(W^5)^1 + x(11)(W^5)^2, \tag{7}$$

which we observe to be the DFT of column 1 of the $x(n)$ array; $W^5$ is the correct kernel for this three-point transform because $W = \exp(j2\pi/15)$, so that $W^5 = \exp(j2\pi/3)$. Substituting (6) into (5) yields

$$X(Lr+s) = \sum_{m=0}^{M-1} q(m, s) W^{ms} (W^L)^{mr}. \tag{8}$$

Equation (8) is seen to be a row transform of the array of products $q(m, s)W^{ms}$, with $W^L$ representing the kernel of this transform. Using the indexing arrangement shown next to the arrays, we can thus compute (4) by successive applications of (1) and (8), leading precisely to the procedure we initially outlined. The $W^{ms}$ of (8) can be identified as the twiddle factors of the algorithm.

Accordingly, the complete FFT algorithm will consist of successively breaking down one-dimensional sig-

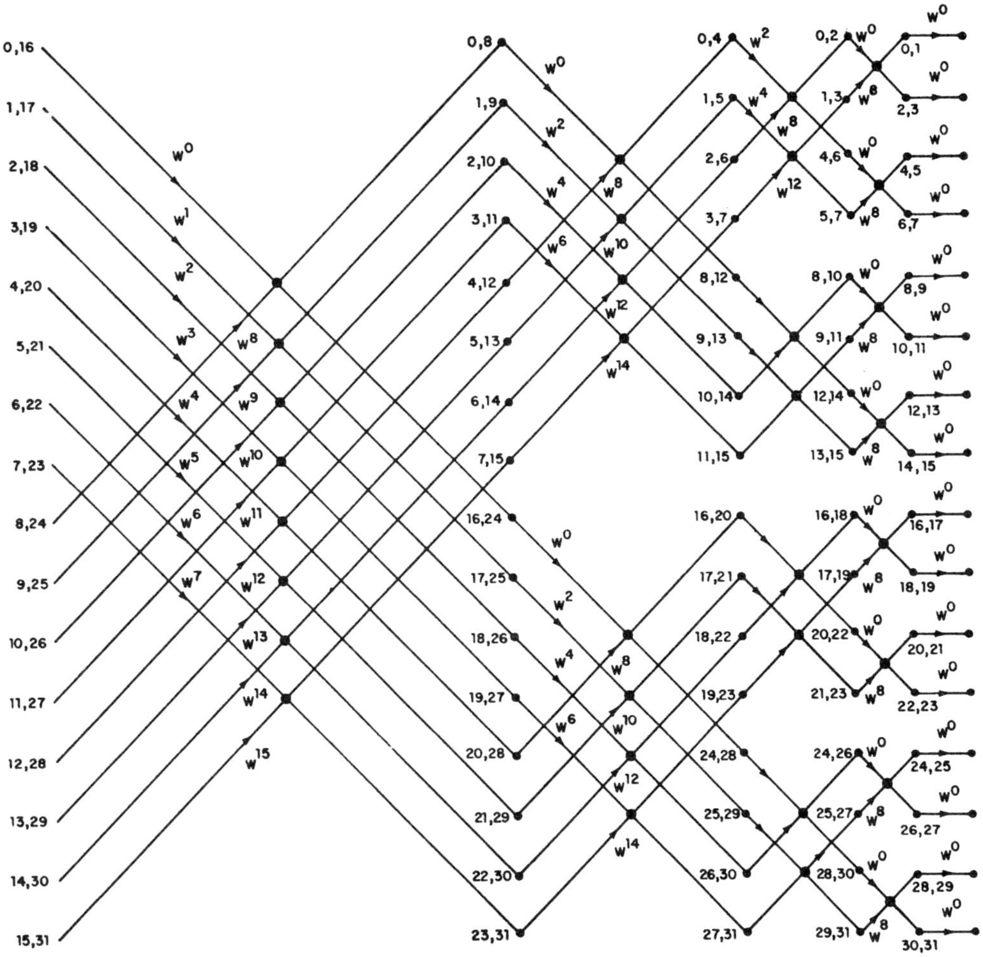


**Fig. 2. Flow diagram for radix-2 32-point FFT with two complex words per register.**

nals into their two-dimensional counterparts. An illustration of this technique is given in Fig. 1. We begin with $N=64$ and first represent it as a $4\times16$ array, as shown in Fig. 1. After performing steps 1), 2), and 3), we are ready for step 4). However, rather than explicitly performing four 16-point column transforms, we choose to represent each column by a $4\times4$ array. Now steps 1)–4) are performed on each individual $4\times4$ array. Notice that the second set of twiddle factors are based on 16-point transforms, so that $W_2=\exp(j2\pi)/16$. This particular illustration can be interpreted as a fracturing of $N=4\times4\times4$, since there are three levels of computation, each level corresponding to the computation of a set of 16 4-point DFT's plus twiddle factors. Such a computation is called a radix-4 computation. More generally, if $N$ can be expressed as $N=r^m$, and if the algorithm is carried out via a succession of $r$-point transforms, the resultant FFT is called a radix-$r$ FFT.

In a radix-$r$ FFT, we denote an "elementary computation" (EC) by an $r$-point DFT followed by multiplication of each of the $r$ results by the appropriate twiddle factor $W^{ij}$. It is easy to show that the number of EC's required is

$$C_r = \frac{N}{r}\log_r N. \tag{9}$$

Equation (9) shows that $C_r$ decreases as $r$ increases; in fact, when $r=N$, $C_r=1$. But, of course, the complexity of an EC increases with increasing $r$. For example, for $r=2$, EC consists of a single complex multiplication and a pair of complex additions, while for $r=4$, EC consists of three complex multiplications and some extra additions. Thus, the "best" value of $r$ is not easily arrived at, and depends greatly on the particular problem.

### III. Nomenclature for FFT's Using Parallelism

Most FFT algorithms exist as programs on general-purpose computers where the memory word length usually contains one data sample. In discussing high-speed FFT hardware, it is important to include the possibility of parallelism both in the EC's and in the memory cycles. We therefore introduce a flow diagram nomenclature slightly different from the Rader diagrams of the radix-2 FFT usually found in the literature [12]. In this nomenclature, the memory parallelism is contained in the data assignments as seen in Fig. 2. The arrows and associated $W$'s represent the parameters of the EC's, and the nodes (heavy dots) show how the outputs must be distributed in order to be read for the next computation level. Notice that there are two pairs of data points entering and leaving each node in the 32-point transform of Fig. 2. This derives from the fact

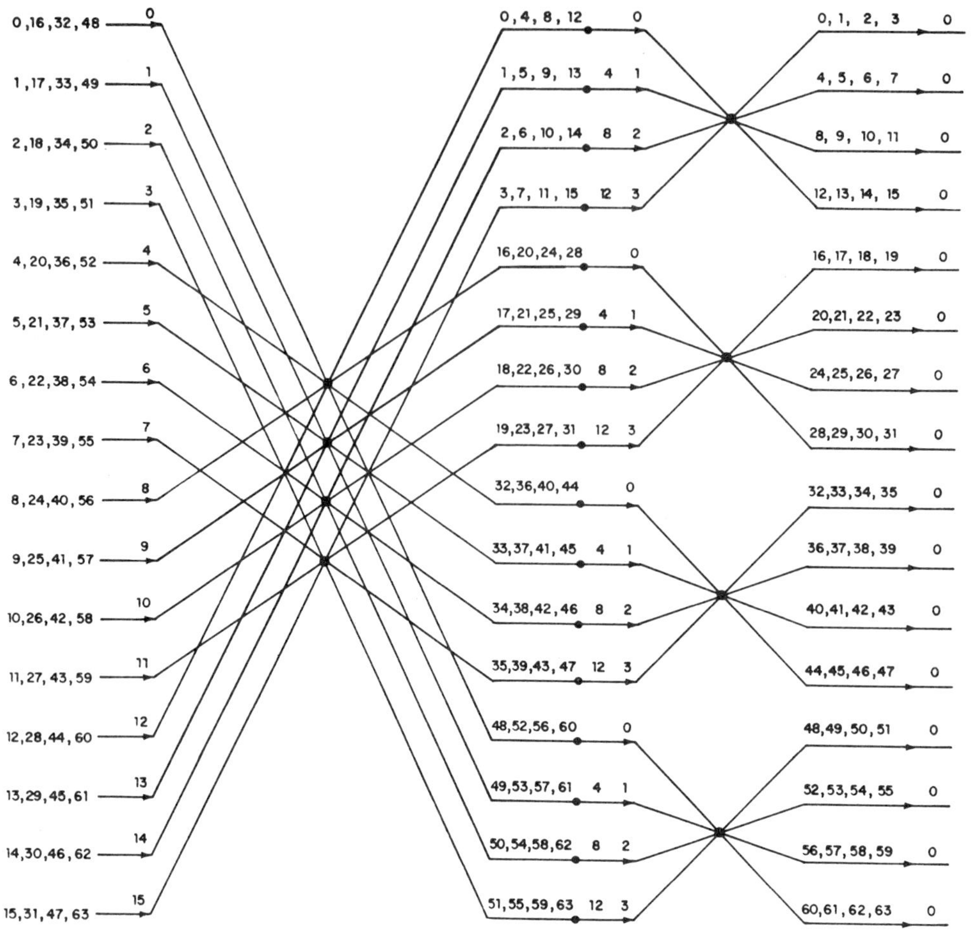


Fig. 3. 64-point FFT, radix-4 with quadrapole word memory.

that each line in the diagram represents a complete EC of the transform, and it implies that the computational model must effectively reorder (permute the rows and columns) of a $2\times 2$ matrix after each EC. Here we have the in-place idea extended to four data points. This is the natural extension of the in-place algorithm in which two data points are computed, and the two output results are entered back into the original memory location.

The nomenclature of Fig. 2 can be extended to the radix-4 case of Fig. 1. The result is seen in Fig. 3. Now we have a quadruplet of inputs to an EC (represented by the arrow). The number near the arrow represents the four twiddle factors of each EC as follows. The number $k$ means that the twiddle factors are $W^0$, $W^k$, $W^{2k}$, $W^{3k}$. Each node accepts a $4\times 4$ matrix of EC outputs and distributes them via a row–column permutation.

## IV. EC's for $r=2$ and $r=4$

For the "decimation in time" algorithm [12], the EC is

$$A' = A + CW^i$$
$$C' = A - CW^i \qquad (10)$$

where $A$ and $C$ are the complex data points, $W=\exp(j2\pi/N)$, and $i$ as an index which is a function of how many steps have been completed in the algorithm.

For the "decimation in frequency" algorithm,

$$A' = A + C$$
$$C' = (A - C)W^i. \qquad (11)$$

Thus, both EC's always involve two input data points $A$ and $C$, a coefficient $W^i$, and two output data points $A'$ and $C'$.

A radix-4 EC corresponds to a four-point DFT followed by multiplication of each output by a complex "twiddle" factor. Thus there are four inputs, $A$, $B$, $C$, and $D$, and four outputs, $A'$, $B'$, $C'$, and $D'$, given by

$$A' = (A + B + C + D)W^0$$
$$B' = (A + jB + C - jD)W^k$$
$$C' = (A - B + C - D)W^{2k}$$
$$D' = (A - jB - C + jD)W^{3k}. \qquad (12)$$

In (12), $j=\sqrt{-1}$ and $k$ is an algorithm-position-dependent index; note, however, that the first "twiddle" is always $W^0=1$.

## V. Parallel Memory—Pipeline Arithmetic Configuration of Radix-$r$ FFT

In Section III we showed that a radix-$r$ FFT can be implemented in a parallel way by performing a total of $r$

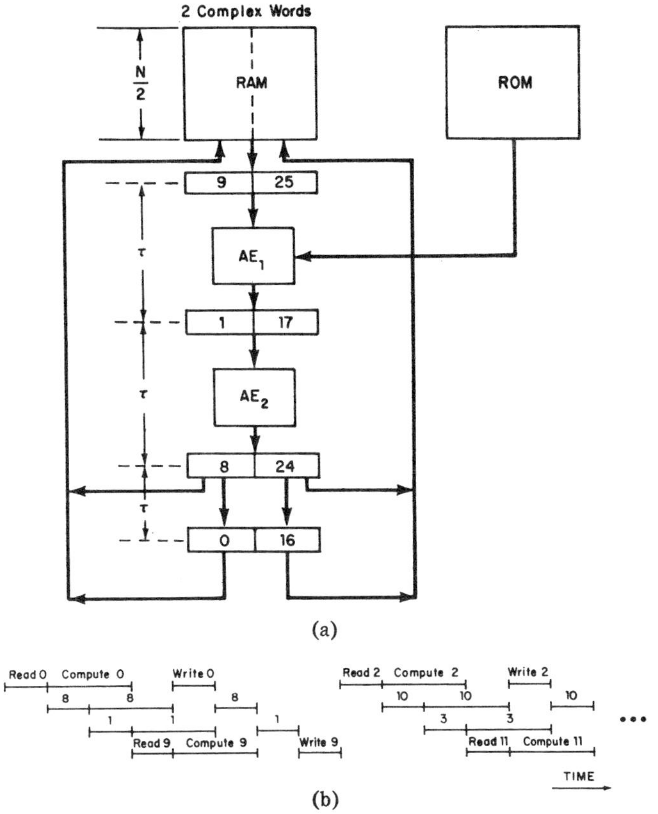


Fig. 4. Structure and timing diagram for radix-2 FFT with parallel memory. (a) Structure. (b) Timing.

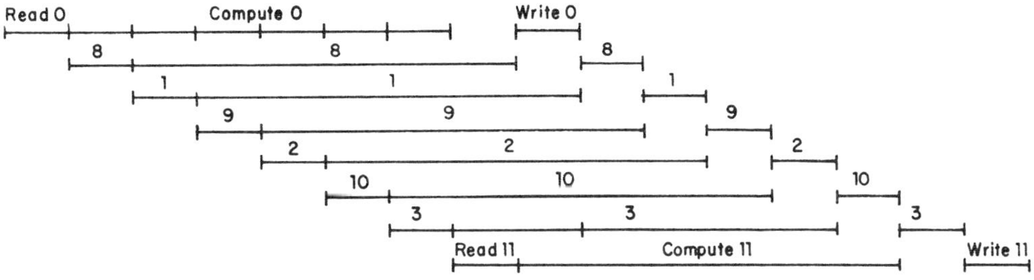


Fig. 5. Alternate timing diagram for structure of Fig. 4.

EC's, and then permuting the rows and columns of the result before doing the next set of $r$ EC's. In this section, a specific computing structure that can perform such FFT's will be examined; this structure has the following features.

1) Memory cycles and computation cycles are completely overlapped; we can think of such a structure as "matched" in the sense that the system is designed to be both memory limited and arithmetically limited simultaneously.

2) The data are contained in a random access memory with a register length equal to $r$ complex data words.

A radix-2 example of such a structure and its associated (partial) timing diagram is shown in Fig. 4. The RAM consists of $N/2$ registers, each of the length of two complex data words. The numbers in the timing diagram correspond to the 32-point transform depicted in Fig. 2, which also shows how the addressing is to be sequenced. In order to realize a matched system, the second read cycle must, without pause, follow the first read cycle, etc Thus, subsequent read cycles must be overlapped with the EC. Furthermore, since writing must follow the completion of at least two EC's, we can see from Fig. 4's timing diagram that four successive read cycles are required before writing can occur. This means that the AE *must* contain pipelining, that is, a second EC begins before the first is finished. Fig. 4 shows the pipelining needed when it is assumed that an EC takes two memory cycles; the state of the machine shown corresponds in time to the arrow in the timing diagram.

Should the designer be unable to construct an AE that performs its function in two memory cycles, a longer pipeline, of six memory cycles, will also result in a matched system; this timing is depicted in Fig. 5. The resulting structure will be a lengthened version of Fig. 4, with a total of six pipelining buffers entered between six partial arithmetic elements; if the computation can pro-

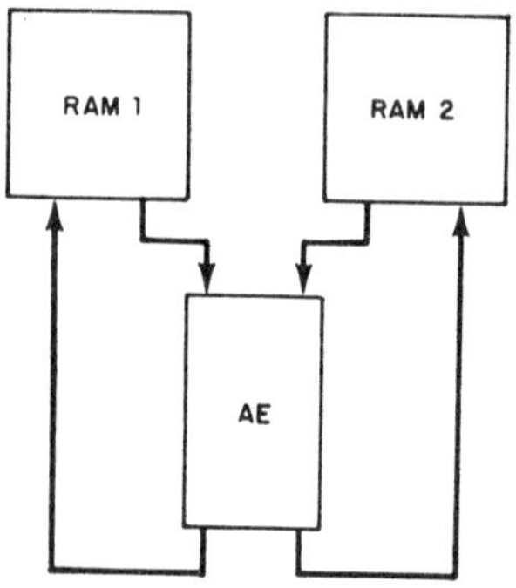


(a)

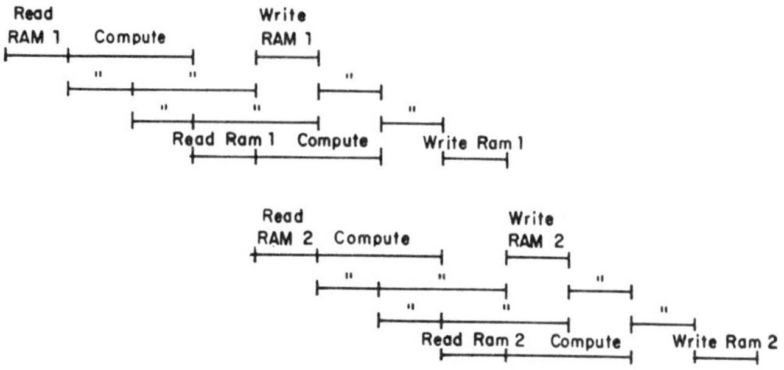


(b)

Fig. 6. Two RAM's doing two FFT's on one AE. (a) Structure. (b) Timing.

ceed more rapidly, some of these partial AE's will be "nulls," simply wires between adjacent buffers.

The timing diagram of Fig. 4 shows that whereas the memory is being continuously exercised, the AE is only working half the time; during memory read cycles, the AE pipeline buffer registers are being filled, and during the write cycles, these registers are being emptied. McHugh [2] has suggested that another memory can be serviced in an interleaved way by the one AE; the situation and timing is depicted in Fig. 6. While the results of the first four computations are being written into RAM 1, RAM 2 is keeping the pipeline filled by reading into the AE.

These ideas can be extended to higher radices; Fig. 7 shows a radix-4 and its associated timing. Here the computation is fractured into four units (1 unit = 1 memory cycle time) and eight reads are followed by eight writes. The pipeline culminates in a $4 \times 4$ permutation matrix; this matrix is represented in Fig. 7 by four registers, each containing the result of a radix-4 EC. Since these registers keep getting new information clocked in every unit, permutation clearly involves a form of buffer storage that we have not shown in the figure. In the same way as in the radix-2 case, two RAM's can be serviced by a single AE.

The structure just described satisfies (9) in the sense that the complete FFT takes exactly $C_r$ memory cycle times. Table I shows the value of $C_r$ for various $N$'s and $r$'s. We see that large time savings are possible by recourse to higher radix FFT's, but beyond $r=4$, the price one pays in extra EC and permutation hardware appears to be prohibitive. As an example of the bandwidth attainable by these structures, assume that a memory cycle time is 50 ns. A radix-4 computation of a 256-point transform yields $C_r = 1024$; since both a read and write cycle are needed for each EC, we get $1024 \times 100$ ns $= 102.4$ $\mu$s, which corresponds to a 2.5-MHz data rate. It is interesting to note that the speed requirements on the hardware are specified almost exclusively by the input bandwidth and not the FFT size to be processed, whereas the memory size requirements are only a function of the FFT size. These comments are pertinent to the design of digital pulse compression systems with differing time and bandwidth requirements.

A variation on the above configurations can be made by increasing the parallelism beyond the radix size. Thus, in Fig. 8 we show a 32-point radix-2 transform wherein the memory register length is 4 rather than 2 (as in Fig. 2). The coefficients may be found by comparing with Fig. 2. In this case, a matched system is attained by employing two EC's in parallel and performing the permutations indicated in the flow diagram. For any degree of parallelism, radix-2 implies that no greater than two rows of results must be permuted. In this system, $C_r$ is $C_2$, but since two EC's are done in parallel, the FFT time is halved; since $C_4 = 1/4\ C_2$, a radix-4 computation using the same memory format would be twice as fast.

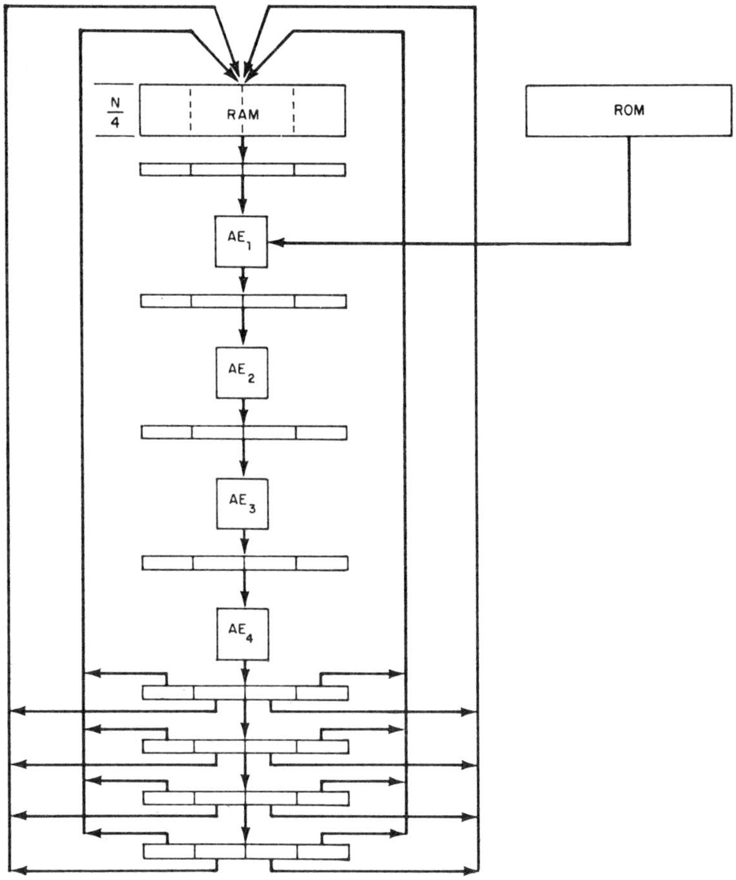


(a)

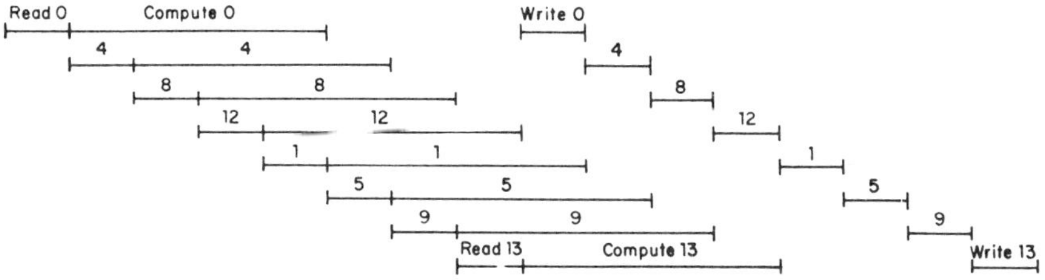


(b)

Fig. 7. Structure and timing for radix-4 FFT. (a) Structure. (b) Timing.

TABLE I
Tabulation of $C_r$ for Various $N$ and $r$

| $N$ | $r$ 2 | 4 | 8 | 16 |
|---|---|---|---|---|
| 64 | 192 | 48 | 16 | — |
| 128 | 448 | — | — | — |
| 256 | 1024 | 1024 | — | 32 |
| 512 | 2304 | — | 192 | — |
| 1024 | 5120 | 1280 | — | — |
| 2048 | 11 264 | — | — | — |
| 4096 | 25 576 | 6144 | 2048 | 768 |

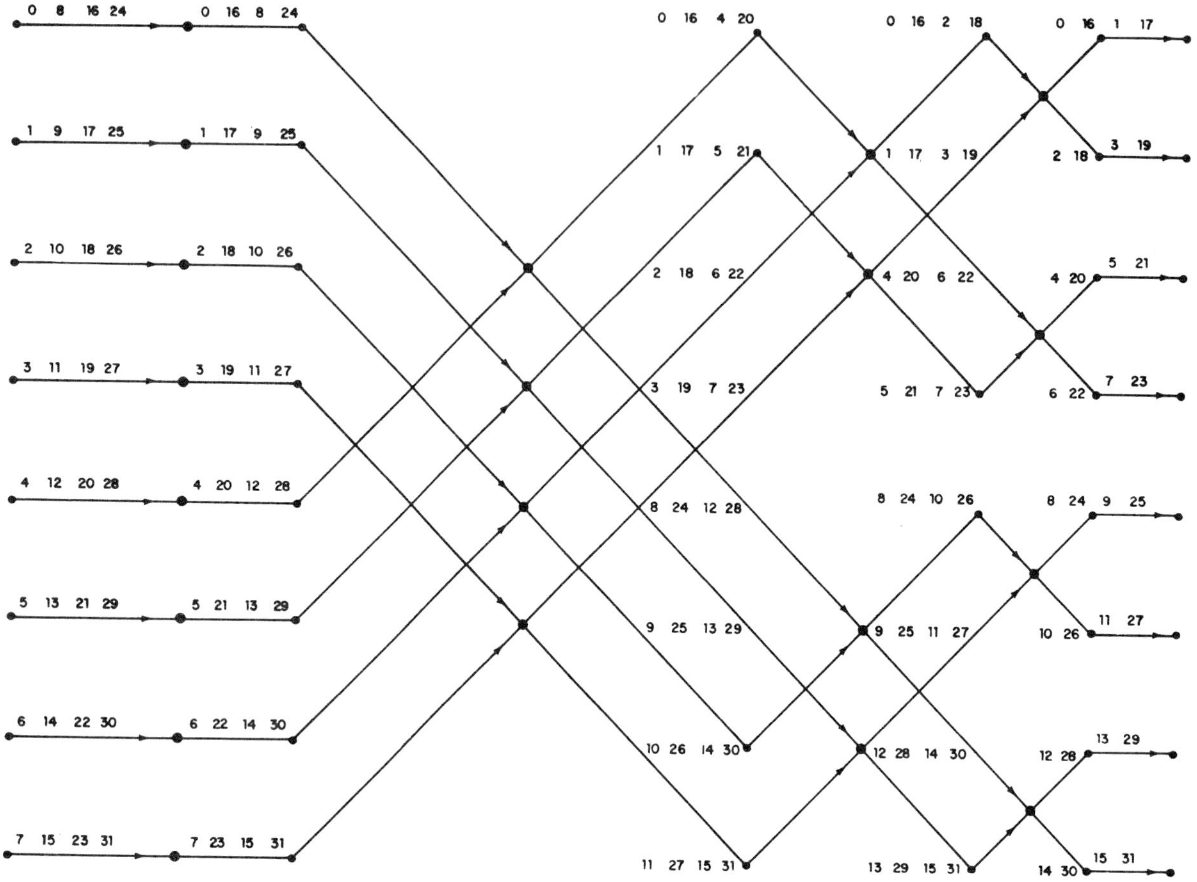


Fig. 8. Radix-2 32-point FFT with parallelism of 4.

## VI. Pipeline FFT Structure of Radix-$r$

An FFT hardware architecture that is of particular interest is the pipeline processor [13]. The term "pipeline FFT" processor refers to the particular structure described in this section. It has also been referred to as a "cascade" processor [15]. It is unfortunate that we have also used the term "pipeline" in Section V to denote a particular AE structure. Although there is good reason to use the term "pipeline" in both cases, it is not intended that the structures discussed are thereby similar. In this device, log $N$ identical arithmetic elements are operated in parallel in a way that requires surprisingly simple and straightforward control functions. Although a number of variations of the basic design exist, all of these implement the FFT via a radix-2 algorithm. Some reference has been made to higher radix pipeline structures in the literature [15], but the specific architecture of these machines has not been presented in any detail. In this section, we extend the FFT pipeline concept to the implementations of arbitrary radix, and derive some basic parameters of these processors. It will be shown that when several waveforms are to be processed simultaneously, one can achieve significant savings in both arithmetic hardware and total memory size through the use of higher radix pipeline structures.

The basic computational element of a radix-$r$ pipeline processor is a device that computes an $r$-point DFT and effects an arbitrary complex weighting on $r-1$ of the resultant $r$ output data (Fig. 9). The input sequence is applied to the computational element in a parallel format via $r$ input ports, while the outputs are produced simultaneously, after a finite delay, at $r$ output ports. Note that this computational element, a "CE," effects the basic EC of the radix-$r$ transform. An alternate element could be constructed in which weighting (twiddle factors) is applied to the inputs instead of to the outputs. With the exception of the sequence in which coefficients will be required, the resultant machines will be completely equivalent.

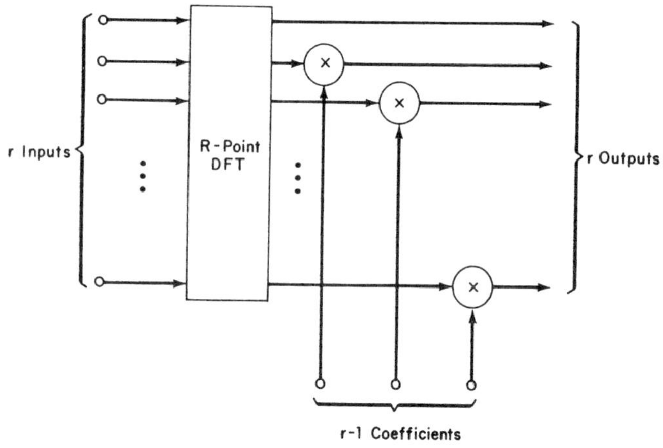


Fig. 9. Computational element for radix-5 pipeline processor.

We assume that the size $N$ of the transform to be performed is an integral power of $r$, i.e.,

$$N = r^m, \qquad m = 1, 2, \cdots.$$

With this restriction, the transform is computed in $m$ stages in which, at each stage, the computational element of Fig. 9 is exercised $N/r$ times. In the pipeline processor, $m$ separate computational elements are used, and each one is dedicated to a particular level of the algorithm.

Referring to the flow diagram of either Figs. 2 or 3, it is observed that in the first level of a radix-$r$ computation, each EC is performed on points that in the original data set are separated by $N/r$ locations. In general, the $k$th level consists of EC's that are performed on data that are separated by $N/r^k$ positions. In the pipeline processor, these separations of input data points are obtained through the use of shift register memories, as indicated for the first level of the transform in Fig. 10. Assuming that the signal to be transformed is applied to the commutator switch at the left of the diagram, it is easy to verify that the correct data sequences will appear at the inputs to the CE if the commutator is advanced by one position after each block of $N/r$ points is received. We assume for simplicity that the CE performs an EC instantaneously, so that the outputs occur simultaneously with the inputs. In practice, of course, some integral number of clock periods will elapse between inputs and outputs, but the throughput rate will remain the same.

Although the $r$ inputs of the CE are in parallel, the element operates at the same rate at which incoming data are clocked. The timing is such that, as it is shown in Fig. 10, the processor remains idle during the time that the first $r-1/r$ $N$ data occur, and then it computes the first level of the transform during the receipt of the last $N/r$ points. Clearly, a more efficient mode of operation would use the CE continuously. This efficiency is possible if one were to simultaneously process $r$ waveforms, where each is delayed from the next by $N/r$ samples. In this case, an $r$-pole commutator, as indicated in Fig. 11, would replace the switch of Fig. 9. Fig. 12 illustrates the timing of this stage. Note that while each of the $r$ inputs is devoted to a separate and independent data block, the $r$ output terminals simultaneously deliver data from the same block. In the illustration, the $ij$th output segment is the $j$th portion of the output due to the $i$th input sequence.

Although the above discussion refers specifically to the first stage of the transform, it applies equally to all successive stages. Referring again to Figs. 2 or 3, it is easily observed that after having completed the first level, the remainder of the transform can be thought of as being composed of $r$ completely separate $N/r$ point FFT's, at least in terms of the flow paths in the diagram. Twiddle factors, of course, are chosen based on the $N$-point case. Further, it turns out that each $N/r$-point transform that originates in the second level is to be effected on a block of data that at the first stage output occurs at a single output terminal. That is to say, each of the blocks $ij$ in Fig. 12 is to be treated as an independent block of data and transformed accordingly.

In effect, one can think of the second stage of the algorithm as being the first stage of a new set of FFT's of length $N/r$. Note that $r$ of these $N/r$-point transforms are to be effected simultaneously. Our experience with the first level indicates that $r$ transforms can be handled simultaneously if the inputs are skewed in time as in Fig. 12. Extending this concept to the second level of the algorithm, we follow the outputs of the first CE

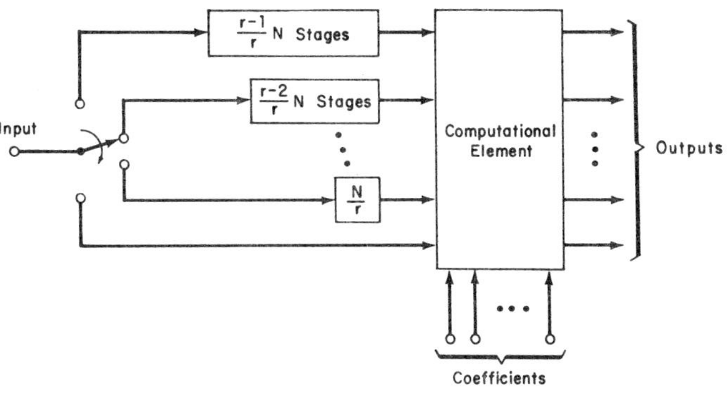


Fig. 10. First stage of radix-5 pipeline.

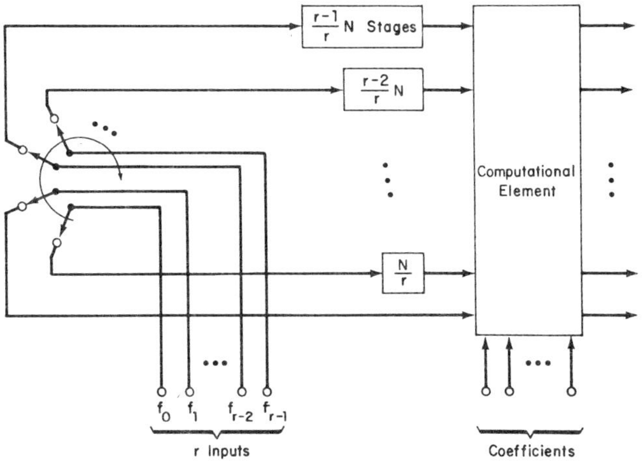


Fig. 11. Multiple input switching scheme.

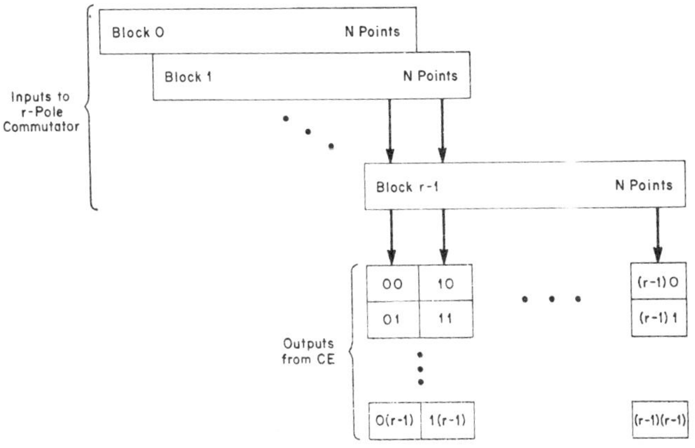


Fig. 12. Timing of first pipeline.

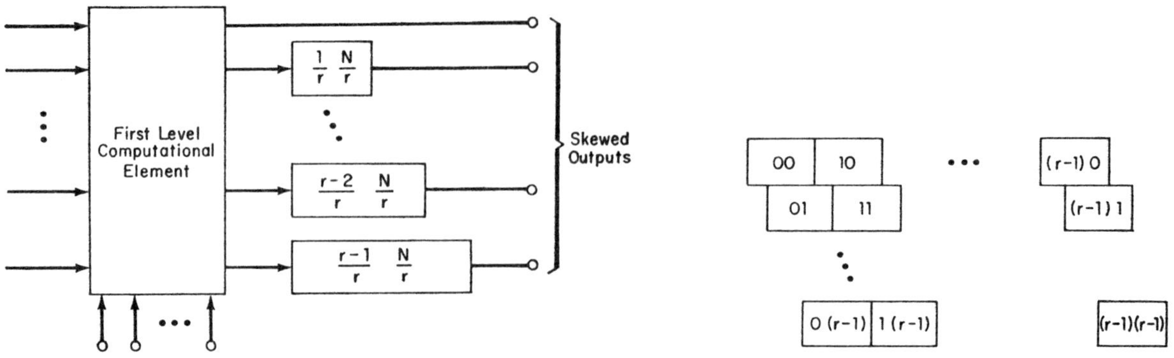


**Fig. 13.** Output memories. (a) Memory arrangement. (b) Time relationships.

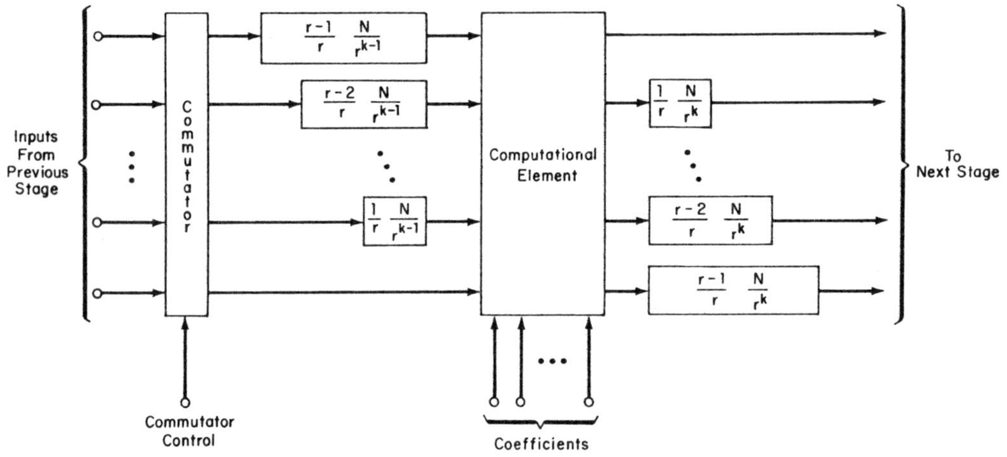


**Fig. 14.** *k*th coefficients stage of radix-*r* pipeline.

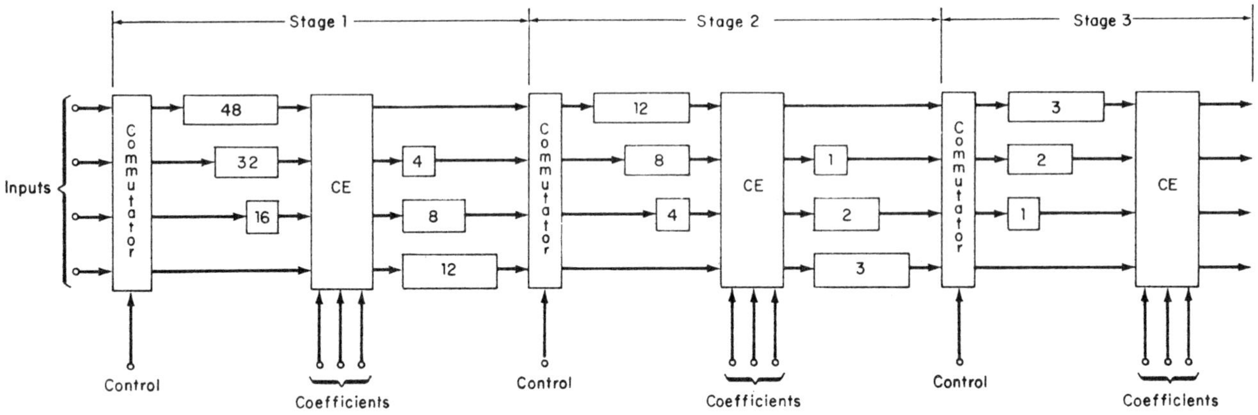


**Fig. 15.** Radix-4 pipeline $N=64$.

by a set of shift register delay lines that misalign the output groups (Fig. 13). Note that these shift registers are $1/r$ as long as those that are used at the input of the computational element. Merging Figs. 12 and 13, we obtain a structure that accepts an $r$ block of data length $N$, skewed by $N/r$ points, effects one level of an FFT on them, and delivers a succession of $r$ blocks of data of length $N/r$, skewed by $N/r^2$. It is a simple matter to extend this concept to the $k$th stage of a radix-$r$ FFT, in which one accepts $r$ blocks of data of length $N/r^{k-1}$, skewed by $N/r^k$, and delivers a succession of $r$ blocks of data of length $N/r^k$, skewed by $N/r^{k+1}$. Thus, the $k$th stage of a radix-$r$ pipeline FFT processor has the structure shown in Fig. 14. A complete radix-$r$ pipeline for a 64-point transform ($m=3$) is shown in Fig. 15. This device handles four simultaneous transforms.

It might be pointed out that the commutator switches are not operated at the same rate. Each successive switch is operated $r$ times faster than the one before it, with the last switch changing its state at the system sampling rate. The last stage delivers $r$ output points at once, with all $r$ referring to the same input data block.

The $N$ outputs for any one input sequence are delivered in $N/r$ successive groups of $r$. Note that no provision for serializing these values has been included in Fig. 15. This could be accomplished with additional memory if necessary.

Several observations and generalizations can be made about pipeline FFT processors of the type described. First, it has already been noted that a radix-$r$ machine will handle $r$ transforms simultaneously. It is of interest to compare the amount of arithmetic hardware and total memory that the various implementations require per computed transform.

### A. Arithmetic Hardware

A computational element for a pipeline processor consists of a number of adders and multipliers. Assuming that CE's are designed in as efficient a way as possible, the following is true.

| | | |
|---|---|---|
| Radix-2: | 2 | Adders |
| | 1 | Multiplier |
| Radix-4: | 8 | Adders |
| | 3 | Multipliers |
| Radix-8: | 24 | Adders |
| | 9 | Multipliers |

In the above, all elements operate on complex quantities, i.e., an adder consists of two separate (real) adders, and a multiplier contains four real multipliers and two real adders. In the radix-8 case, two of the multipliers operate with fixed coefficients. This might result in some hardware economy. Since a radix-$r$ pipeline requires $\log_r N$ CE's, we obtain the results given in Table II.

Of course, the fact that more transforms are performed in a higher radix pipeline means that the total amount of hardware that one uses increases with increasing radix. The economies of higher radix implementations become real only when the transforms have to be performed at sufficiently high rates that hardware parallelism would be required in implementations of lower radix. From Table II it is obvious that radix-4 pipeline structures offer a net savings of 25 percent in multiplier hardware. Radix-8 structures seem to offer little, if any, advantage over the radix-4 case.

### B. Memory Requirements

From Fig. 13 it may be verified that the $k$th stage of a radix-$r$ pipeline uses

$$\sum_{m=1}^{r-1} \frac{m}{r}\left(\frac{N}{r^{k-1}} + \frac{N}{r^k}\right) \tag{13}$$

stages of shift-register memory. Further, there are $\log_r N$ stages so that the total memory requirement $M$ is given by

$$M = \sum_{k=1}^{\log_r N-1} \sum_{m=1}^{r-1} \frac{n}{r}\left(\frac{N}{r^{k-1}} + \frac{N}{r^k}\right)$$

or

TABLE II

| $N$ | $r$ | Arithmetic requirements per transform |
|---|---|---|
| 256 | 2 | 8 Adders, 4 Multipliers |
| 1024 | 2 | 10 Adders, 5 Multipliers |
| 4096 | 2 | 12 Adders, 6 Multipliers |
| 256 | 4 | 8 Adders, 3 Multipliers |
| 1024 | 4 | 10 Adders, $3\frac{3}{4}$ Multipliers |
| 4096 | 4 | 12 Adders, $4\frac{1}{2}$ Multipliers |
| 4096 | 8 | 12 Adders, $4\frac{1}{4}$ Multipliers |

$$M = \frac{N(r+1)}{r}\left(\sum_{k=1}^{\log_r N-1} \frac{1}{r^k}\right)\left(\sum_{m=1}^{r-1} m\right). \tag{14}$$

Now

$$\sum_{m=1}^{r-1} m = \frac{r(r-1)}{2}, \tag{15}$$

and if the pipeline is reasonably long, then

$$\sum_{k=1}^{\log_r N-1} \frac{1}{r^k} \simeq \frac{r}{r-1} \tag{16}$$

so that

$$M = \frac{N(r+1)}{2}. \tag{17}$$

On a per-transform basis, the pipeline requires $N(r+1)/2r$ complex registers of memory. This is a monotonically decreasing function of $r$, equal to 3/4 $N$ for $r=2$ and approaching $N/2$ for large $r$. As in the arithmetic case, if several very-high-speed transforms are to be done simultaneously, there will be a net savings of memory if higher radix implementations are employed.

## Appendix

### Representation of a Mixed Radix FFT

Although the FFT representations shown in Figs. 2 and 3 were done for radix-2 and radix-4, similar representations can be drawn for mixed radices. An example is given in Fig. 16 of a 60-point FFT expressed as a $5\times3\times4$ matrix. The first stage of the process shows a $5\times12$ two-dimensional array. Each five-number row undergoes a five-point DFT, and the five results are multiplied by the corresponding row of twiddle factors. For example, row 3 is multiplied by $W^0$, $W^3$, $W^6$, $W^9$, $W^{12}$ which is represented in the figure by an arrow with $W^3$. Since twiddles in the $r$th column have an exponent that is always $r$ times the exponent in column 1, one exponent serves to identify all twiddles for that row. The inputs and outputs from the heavy node points express the fact that the data points from the input registers are redistributed among the output registers via row–column permutations. Thus, the 15 numbers in registers 0, 4, and 8 of the first stage become the 15 numbers of registers 0, 4, 8, 12, 16 of the second stage. To perform the 12-point column DFT's, each column is expressed as a $4\times3$ array. We then move on to the next

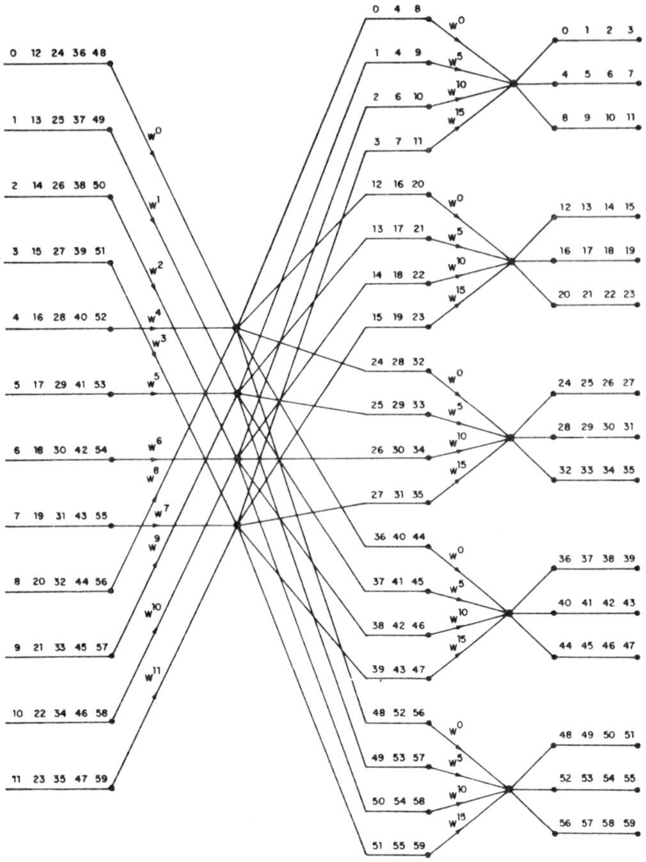


Fig. 16

stage wherein three-point DFT's are performed on the rows, then twiddled, and finally, four-point DFT's are performed on the columns (which are expressed as rows in the last stage). This structure results in successive row DFT's interspersed with twiddling.

Because of the mixed radix, the size of the rows and columns keep varying from stage to stage. This makes it awkward to perform the algorithm in-place, as we were able to do when a fixed radix was used.

## References


[1] H. Andrews, "A high-speed algorithm for the computer generation of Fourier transforms," *IEEE Trans. Comput.* (Short Notes), vol. C-17, pp. 373–375, Apr. 1968.

[2] G. D. Bergland, "A fast Fourier transform algorithm using base 8 iterations," *Math. Comput.*, vol. 22, pp. 275–279, Apr. 1968.

[3] ——, "Fast Fourier transform hardware implementations—A survey," *IEEE Trans. Audio Electroacoust.*, vol. AU-17, pp. 109–119, June 1969.

[4] ——, "Fast Fourier transform hardware implementations—An overview," *IEEE Trans. Audio Electroacoust.*, vol. AU-17, pp. 104–108, June 1969.

[5] G. D. Bergland and D. E. Wilson, "An FFT algorithm for a global, highly parallel processor," *IEEE Trans. Audio Electroacoust.*, vol. AU-17, pp. 125–127, June 1969.

[6] W. T. Cochran *et al.*, "What is the fast Fourier transform?," *IEEE Trans. Audio Electroacoust.*, vol. AU-15, pp. 45–55, June 1967.

[7] J. W. Cooley and J. W. Tukey, "An algorithm for the machine computation of complex Fourier series," *Math. Comput.*, vol. 19, pp. 297–301, Apr. 1965.

[8] J. W. Cooley, P. Lewis, and P. Welch, "Application of the fast Fourier transform to computation of Fourier integrals, Fourier series, and convolution integrals," *IEEE Trans. Audio Electroacoust.*, vol. AU-15, pp. 79–84, June 1967.

[9] ——, "Historical notes on the fast Fourier transform," *IEEE Trans. Audio Electroacoust.*, vol. AU-15, pp. 76–79, June 1967.

[10] M J. Corinthios "The design of a class of fast Fourier transform computers," *IEEE Trans. Comput.*, vol. C-20, pp. 617–623, June 1971.

[11] ——, "A fast Fourier transform for high-speed signal processing," *IEEE Trans. Comput.*, vol. C-20, pp. 843–846, Aug. 1971.

[12] B. Gold and C. M. Rader, *Digital Processing of Signals.* New York: McGraw-Hill, 1969.

[13] H. L. Groginsky and G. A. Works, "A pipeline fast Fourier transform," *IEEE Trans. Comput.*, vol. C-19, pp. 1015–1019, Nov. 1970.

[14] R. Klahn, R. R. Shively, E. Gomez, and M. J. Gilmartin, "The time-saver: FFT hardware," *Electronics*, pp. 92–97, June 24, 1968.

[15] G. C. O'Leary, "Nonrecursive digital filtering using cascade fast Fourier transformers," *IEEE Trans. Audio Electroacoust.*, vol. AU-18, pp. 177–183, June 1970.

[16] M. C. Pease, "An adaptation of the fast Fourier transform for parallel processing," *J. Ass. Comput. Mach.*, vol. 15, pp. 252–264, Apr. 1968.

[17] R. R. Shively, "A digital processor to perform a fast Fourier transform," in *Proc. 1st IEEE Comput. Conf.*, 1967, pp. 21–24.

[18] ——, "A digital processor to generate spectra in real time," *IEEE Trans. Comput.*, vol. C-17, pp. 485–481, May 1968.

[19] R. C. Singleton, "An algorithm for computing the mixed radix fast Fourier transform," *IEEE Trans. Audio Electroacoust.*, vol. AU-17, pp. 93–103, June 1969.

[20] B. Gold, I. L. Lebow, P. G. McHugh, and C. M. Rader, "The FDP, A fast programmable signal processor," *IEEE Trans. Comput.*, vol. C-20, pp. 33–38, Jan. 1971.

# Part 7
# Terminology

# Terminology in Digital Signal Processing

LAWRENCE R. RABINER, JAMES W. COOLEY,
HOWARD D. HELMS, LELAND B. JACKSON,
JAMES F. KAISER, CHARLES M. RADER,
RONALD W. SCHAFER, KENNETH STEIGLITZ,
and CLIFFORD J. WEINSTEIN

*Abstract*—The committee on Digital Signal Processing of the IEEE Group on Audio and Electroacoustics has undertaken the project of recommending terminology for use in papers and texts on digital signal processing. The reasons for this project are twofold. First, the meanings of many terms that are commonly used differ from one author to another. Second, there are many terms that one would like to have defined for which no standard term currently exists. It is the purpose of this paper to propose terminology which we feel is self-consistent, and which is in reasonably good agreement with current practices. An alphabetic index of terms is included at the end of the paper.

## Introduction

As an aid to classifying the different types of terms to be defined, we have placed each term in one of the following groups:

1) Introductory Terms—General Definitions
2) Discrete Systems—Block Diagram Terminology
3) Relations Between Discrete and Continuous Signals
4) Theory and Design of Digital Filters
5) Finite Word Length Effects—A/D, D/A Conversion
6) Discrete Fourier Transforms and the FFT
7) Discrete Convolution and Spectrum Analysis.

In the above mentioned sections of this paper we will be discussing terminology related to the processing of one-dimensional signals. For convenience, we will assume that this dimension is time—although the definitions apply equally well to any single dimension.

## I. Introductory Terms—General Definitions

1) In discussing waveform processing problems, the distinctions analog versus digital and continuous time versus discrete time are often made. Although they are often used interchangeably, different meanings should be attributed to the two sets of terms.

2) The term *analog* generally describes a waveform that is continuous in time (or any other appropriate independent variable) and that belongs to a class that can take on a continuous range of amplitude values. Examples of *analog waveforms* or *analog signals* are those derived from acoustic sources. Such signals are represented mathematically as functions of a continuous variable. The functions sin $(\omega t)$ and the step function $au_{-1}(t)$ are examples of common mathematical functions that could describe "analog signals." The use of the term "analog" in this context appears to stem from the field of analog computation, where a current or voltage waveform serves as a physical analog of some variable in a differential equation.

3) The term *continuous time* implies that only the independent variable necessarily takes on a continuous range of values. In theory the amplitude may, but need not, be restricted to a finite or countable infinite set of values (i.e., the amplitude may be quantized). Therefore, analog waveforms are continuous-time waveforms with continuous amplitude. In practice, however, "continuous-time waveforms" and "analog waveforms" are equivalent. Since most signal processing problems have nothing to do with analogs as such, the use of the term analog waveform is often ambiguous at the least and may in fact be misleading. Thus, the term continuous-time waveform is preferable.

4) *Discrete time* implies that time (the independent variable) is quantized. That is, discrete-time signals are defined only for discrete values of the independent variable. Such signals are represented mathematically as *sequences* of numbers. Those discrete-time signals that take on a continuum of values are referred to as *sampled-data* signals.

5) The term *digital* implies that both time and amplitude are quantized. Thus a *digital system* is one in which signals are represented as sequences of numbers which take on only a finite set of values. Thus one uses *digital* when discussing actual physical realizations (as hardware or programs) of discrete-time signal processing systems, whereas the term *discrete time* is a better modifier when considering mathematical abstractions of such systems in which the effects of amplitude quantization are ignored. A *digital signal* or *digital waveform* is a sequence produced, for example, by digital circuitry or by an analog-to-digital converter which is sampling a continuous-time waveform. In digital signal processing these terms are commonly shortened to *signal* or *waveform*. Sometimes the term signal is restricted to being a desirable component of a sequence instead of being used interchangeably with waveform. *Noise* is either defined as a) an undesirable component of a sequence, or b) a sequence of random variables.

6) (*Digital*) *simulation* is the exact or approximate representation of a given system (discrete or contin-

Manuscript received August 1, 1972.
L. R. Rabiner, J. F. Kaiser, and R. W. Schafer are with Bell Telephone Laboratories, Inc., Murray Hill, N. J. 07974.
J. W. Cooley is with the IBM T. J. Watson Research Center, Yorktown Heights, N. Y. 10598.
H. D. Helms is with Bell Telephone Laboratories, Inc., Whippany, N. J. 07981.
L. B. Jackson is with Rockland Systems Corporation, West Nyack, N. Y. 10994.
C. M. Rader and C. J. Weinstein are with M.I.T. Lincoln Laboratory, Lexington, Mass. 02173. (Operated with support from the U. S. Air Force.)
K. Steiglitz is with Princeton University, Princeton, N. J. 08540.

Reprinted from *IEEE Trans. Audio Electroacoust.*, vol. AU-20, pp. 322–337, Dec. 1972 uous) called the source system, by a (digital) system called the *object system.*

7) *Next-state simulation* is a method of digital simulation whereby the values of the digital system signals are represented by nodes in a block diagram representation. Usually, there is a close correlation between blocks in the object system and elements of the source system. The method entails ordering the calculations in the digital system so that all the inputs to each block at a given sample time are computed before the output is computed.

8) A *real-time process* is one for which, on the average, the computing associated with each sampling interval can be completed in a time less than or equal to the sampling interval. A program running in 100 times real time requires 100 times as long to process the same number of samples; i.e., it is 100 times too slow for real time operation. A program ten times as fast as it needs to be could be said to run in 1/10 real time. Obviously, the extra speed can only be used if other computing can be done in the interstices, or if the complete sequences to be processed have been stored beforehand.

9) *Throughput rate* is the total rate at which digital information is processed by a discrete-time system, measured in bits per second or samples per second. In a multiplexed system, where several signals are processed, we may refer to the *throughput rate per signal,* measured in bits per second per signal or samples per second per signal. Thus, a multiplexed system which processes 10 signals, each at 1000 bits/s, has a throughput rate of 10 000 bits/s, and a throughput rate/signal of 1000 bits/s/signal.

10) A *multirate system* is a discrete-time system in which there are signals sampled at different intervals which are usually integer multiples of some basic or fundamental interval.

## II. Discrete Systems

1) The *z transform* plays a role in discrete-time system theory analogous to that of the Laplace transform in continuous-time system theory. Two view points regarding the $z$ transform are common. One is based on what may be termed the one-sided $z$ transform, which is defined as

$$X(z) = \sum_{n=0}^{\infty} x(n) z^{-n} \tag{1}$$

regardless of the value of $x(n)$ for $n<0$.[1] One application of the *one-sided z transform* is in the solution of linear difference equations with constant coefficients. Solutions are obtained for the interval $0 \leq n < \infty$ subject to prescribed initial conditions. These solutions are obtained with the aid of the equations

$$y(n) \equiv x(n-m), \qquad m>0 \tag{2}$$

$$Y(z) = z^{-m}\left[X(z) + \sum_{i=1}^{m} x(-i)z^{i}\right]. \tag{3}$$

In (3), $X(z)$ appears multiplied by $z^{-m}$. This result for $m=1$ accounts for the fact that $z^{-1}$ is often termed the *unit delay operator,* since a delay of the sequence by one sample is equivalent to multiplication of the $z$ transform by $z^{-1}$. ( Similarly, $z$ is often called the *unit advance* operator.)

2) In many cases, sequences are defined over both positive and negative values of $n$. In such cases, a somewhat more general point of view is called for. In general, the $z$ transform is written as

$$X(z) = \sum_{n=-\infty}^{\infty} x(n) z^{-n}. \tag{4}$$

It should be noted that a common usage is to call (1) simply the $z$ transform, and (4) the *two-sided z transform.* Since (4) is most general, it would seem preferable to refer to (4) as the $z$ transform, and the special case, (1), as the *one-sided z transform.*

3) It is possible to think of the $z$ transform as simply a formal series whose properties can be tabulated, and which never need be summed. However, it is generally preferable to realize that if certain convergence conditions are met, both (1) and (4) are Laurent series in the complex variable $z$. As such, all the properties of the Laurent series apply. For example, if the series in (1) converges, it must converge in a region $|z|>R_+$. If the series of (4) converges, it must converge in an annular region $R_+<|z|<R_-$, where $R_+$ may be zero and $R_-$ may be infinity. The coefficients of a Laurent series are determined by an integral relationship. In the context of the $z$ transform, this relation is

$$x(n) = \frac{1}{2\pi j}\oint_C X(z) z^{n-1}\, dz, \tag{5}$$

where $C$ is a closed contour inside the region of convergence of the power series and enclosing the origin. Equation (5) is referred to as the *inverse z transform.*

4) In the region of convergence of the series, both (1) and (4) represent analytic functions of the complex variable $z$. These functions can often be extended by analytic continuation everywhere except at certain singular points (poles). Since these singularities of the $z$ transform are characteristic of the particular sequence, it is common to plot their locations in the *z plane,* (i.e., the complex plane determined by the real and imaginary parts of $z$). It should be noted that it is often convenient, because of the special functional form which characterizes exponential sequences, to plot singularities in the *$z^{-1}$ plane.* Furthermore, some authors define the $z$ transform as

$$\tilde{X}(z) = \sum_{n=-\infty}^{\infty} x(n) z^{n}. \tag{6}$$

[1] The notation $x(n)$ is used rather than $x_n$ or $x(nT)$ to denote a sequence because of the ease of handling complicated indices, e.g., $x(N-1/2)$.

Clearly, (6) is related to our definition by

$$\tilde{X}(z) = X(z^{-1}). \tag{7}$$

If either (6) or the $z^{-1}$ plane is encountered, it is a simple matter to replace $z$ with $z^{-1}$ in order to relate the $z$-transform definitions and to note that the region inside the unit circle of the $z$ plane corresponds to the region outside the unit circle of the $z^{-1}$ plane.

5) A *discrete-time impulse at* $k = k_0$ is a discrete-time signal $x(k)$ such that $x(k) = 0$ unless $k = k_0$, in which case $x(k) = 1$. This is an analogy with an *impulse* at time $t_0$ in the continuous-time case, where $x(t) = \delta(t - t_0)$, the Dirac delta function. The response of a digital filter to a *discrete-time impulse* at $k = 0$ is called its *impulse response*, or sometimes, the *unit sample response*. Other terms generally used for *digital impulse* are *unit sample*, *unit pulse*, or simply *impulse*.

6) A *sample value* is the value or number associated with one member of a sequence that represents a discrete-time signal. This term is generally used regardless of whether or not the value represents a sample of a continuous-time signal.

7) A *discrete-time linear time-invariant system* or a *discrete-time linear filter* is characterized by its impulse-response sequence $h(n)$.

8) *Discrete-time convolution* is the operation on a signal or sequence $g(n)$ by the impulse response sequence $h(n)$ to yield a digital signal (or sequence) $f(n)$; the operation is defined by the expression

$$f(n) = \sum_{k=-\infty}^{\infty} h(k)g(n-k). \tag{8}$$

This expression is the discrete-time counterpart of the convolution integral for continuous-time systems.

9) An alternate characterization of a discrete-time linear system is the $z$ transform of $h(n)$:

$$H(z) = \sum_{n=-\infty}^{\infty} h(n)z^{-n}. \tag{9}$$

The complex function $H(z)$ is called the *system function* or transfer function. The values taken by $H(z)$ when evaluated on the unit circle in the $z$ plane give the *frequency response*. Each point on the unit circle, characterized by its angle only, corresponds to a particular frequency.[2] Several different units of frequency are in common use. Some authors express frequency in the conventional units of Hz, kHz, etc. Others use a system of rad/s. Still other authors use a normalized frequency with each frequency expressed as a fraction of the sampling frequency $(f/f_s)$ or half the sampling frequency $(f/(f_s/2))$. Finally, some authors express the normalized frequency in rad/sample. The following table relates the units of frequency to the corresponding angle in the $z$-plane ($T = 1/f_s$ is the sampling period).

| Unit of Frequency | $z$-Plane Substitution | Range of Frequency Around Unit Circle |
|---|---|---|
| $f$ in Hz | $z = e^{j2\pi fT}$ | $0 \leq f \leq f_s$ |
| $\omega$ in rad/s | $z = e^{j\omega T}$ | $0 \leq \omega \leq 2\pi f_s$ |
| $\hat{f} = (f/f_s)$, a fraction of the sampling frequency | $z = e^{j2\pi\hat{f}}$ | $0 \leq \hat{f} \leq 1$ |
| $\tilde{f} = (f/0.5f_s)$, a fraction of half the sampling frequency | $z = e^{j\pi\tilde{f}}$ | $0 \leq \tilde{f} \leq 2$ |
| $\theta$ in rad/sample | $z = e^{j\theta}$ | $0 \leq \theta \leq 2\pi$ |

When the frequency response is expressed in polar form, its magnitude as a function of frequency is called the *amplitude response*, and its angle as a function of frequency is called the *phase response*.

10) One class of linear time-invariant discrete-time filters is characterized by system functions of the form

$$H(z) = \frac{\sum_{k=0}^{M} b_k z^{-k}}{1 + \sum_{k=1}^{N} a_k z^{-k}}. \tag{10}$$

Such filters have a *recursive realization* in the form of the difference equation

$$y(n) = -\sum_{k=1}^{N} a_k y(n-k) + \sum_{k=0}^{M} b_k x(n-k), \tag{11}$$

where $y$ is the output sequence and $x$ is the input sequence. The system function is generally written in terms of powers of $z^{-1}$ as in (10) because it places in evidence the form of the difference equation given in (11), i.e., in terms of delays. $H(z)$ can also be written as

$$H(z) = \frac{z^{N-M} \sum_{k=0}^{M} b_k z^{M-k}}{z^N + \sum_{k=1}^{N} a_k z^{N-k}} \tag{12}$$

where it is assumed that the zeros of the numerator [zeros of $H(z)$] are distinct from the zeros of the denominator [poles of $H(z)$]. This form places in evidence the fact that regardless of the relative values of $M$ and $N$, $H(z)$ has the same number of poles as zeros. In most cases—especially digital filters derived from analog designs—$M$ will be less than or equal to $N$, and there will be at least $N - M$ zeros at $z = 0$. Systems of this type are called $N$th-order systems. When $M > N$, the order of the system is no longer unambiguous. Here, $N$ gives the order as the term is used to characterize the mathematical properties of the difference equations. $M$ gives the order used to characterize the complexity of a realization of the system. There is no general agreement as to which of $M$ or $N$ best characterizes the system when $M > N$.

[2] As will be clear from the discussion in Section III, each point on the unit circle, in fact, corresponds to an infinite set of uniformly spaced frequencies that are indistinguishable in a digital system.

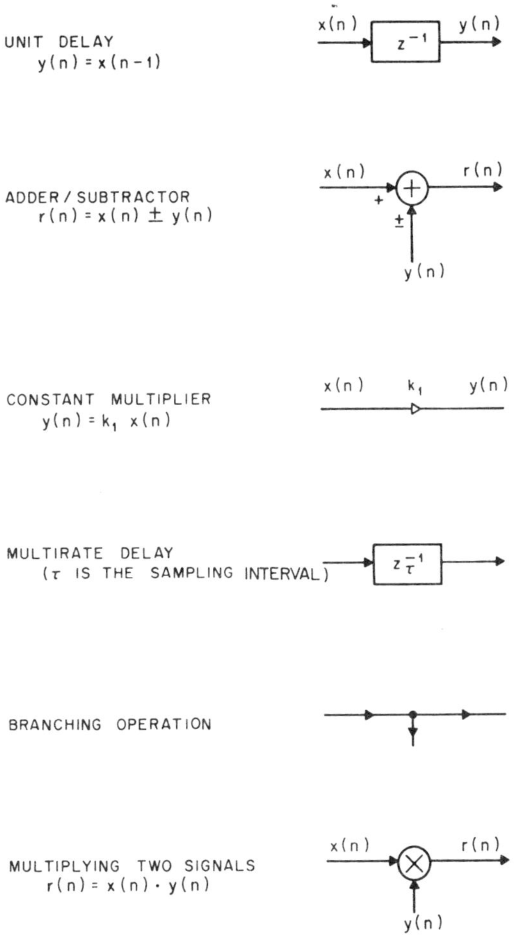


**Fig. 1.** Recommended terminology for use in block diagrams of digital systems.

11) Since the purpose of a *block diagram* is to graphically depict the way in which a particular system is realized, the terminology shown in Fig. 1 is recommended.

## III. Relations Between Discrete and Continuous Signals

1) If a sequence arises as the result of periodic sampling of a continuous-time signal $x_c(t)$, i.e., $x(n)=x_c(nT)$ where $T$ is the sampling period, then $X(z)$, the $z$ transform of $x(n)$, is related to $X_c(s)$, the Laplace transform of $x_c(t)$ by the relationship

$$X(z)\big|_{z=e^{sT}} = \sum_{n=-\infty}^{\infty} x_c(nT)e^{-snT}$$

$$= \frac{1}{T}\sum_{k=-\infty}^{\infty} X_c\left(s+j\frac{2\pi}{T}k\right). \tag{13}$$

This implies that the $s$ plane and the $z$ plane are related. It should be emphasized that (13) makes clear the fact that the $s$ plane is *not* mapped into the $z$ plane in a one-to-one manner. The actual relationship of $X(z)$ to $X_c(s)$ is depicted in Fig. 2, which shows the $s$ plane divided into an infinite number of horizontal strips of

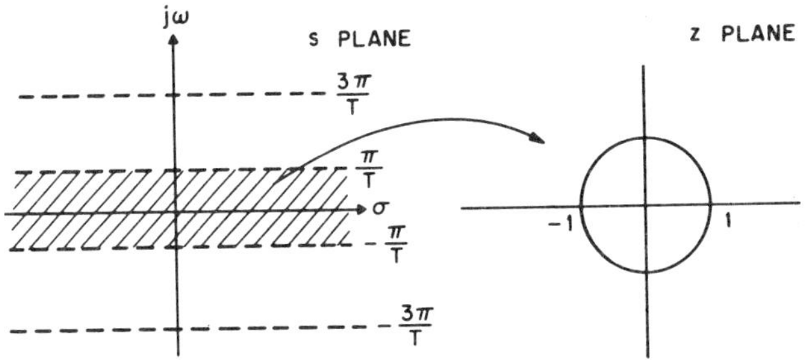


**Fig. 2.** The mapping of the $s$ plane to the $z$ plane implied by sampling a continuous-time signal.

width $2\pi/T$, each of which maps into the entire $z$ plane. The contributions from each strip are added to produce $X(z)$.

2) The $j$ axis of the $s$ plane corresponds to the unit circle of the $z$ plane. For this reason, the point $z=1$ is often casually referred to as the *DC point* since it corresponds to the point $s=0$ of the $s$ plane. The $z$ transform evaluated on the *unit circle* ($|z|=1$) is of particular interest in digital filtering of sampled signals. For example, there are some sequences for which the $z$ transform does not converge (does not exist) except on the unit circle, e.g., the ideal lowpass digital filter and the ideal digital differentiator for band-limited waveforms. The $z$ transform evaluated on the unit circle is called the *Fourier transform* of the sequence. This definition is consistent with the classical terminology for continuous-time systems. Evaluating (4) and (5) of Section II on the unit circle, yields

$$X(e^{j\theta}) = \sum_{n=-\infty}^{\infty} x(n)e^{-jn\theta} \tag{14}$$

$$x(n) = \frac{1}{2\pi}\int_{-\pi}^{\pi} X(e^{j\theta})e^{jn\theta}\,d\theta. \tag{15}$$

This pair of equations is a Fourier transform pair for the sequence $x(n)$. Alternatively, any of the frequency units of the table in Section II can be used in place of $\theta$.

3) If a continuous-time waveform, $x_c(t)$, is band-limited to a frequency $f_0$, i.e., $X_c(j2\pi f)$, the Fourier transform of $x_c(t)$, is zero for $|f|>f_0$, then $x_c(t)$ can be recovered exactly from its samples, i.e., $x_c(nT)$, $-\infty<n<\infty$ if $T<1/2f_0$. This is clear from (13) where we can see that

$$X(e^{j2\pi fT}) = \frac{1}{T}X_c(j2\pi f), \qquad -\frac{1}{2T}<f<\frac{1}{2T} \text{if } T<1/2f_0.$$

$f_N=2f_0$ is called the *Nyquist rate*, and is the lowest rate at which $x_c(t)$ can be sampled, and still be recovered.[3] $T_N=1/f_N=1/2f_0$ is often called the *Nyquist interval*. Additional terms which are often used are *sampling frequency* or *sampling rate* for $f_s$, the rate at which $x_c(t)$ is actually sampled in a particular system; and $T=1/f_s$

[3] This discussion ignores the possible reduction in sampling rate which can be obtained for bandpass signals.

is called the *sampling interval*. The highest frequency present in $x_c(t)$, (defined above) $f_0$, is called the *Nyquist frequency*. The Nyquist frequency is sometimes called the *folding frequency*. It is recommended that the term Nyquist frequency be avoided because of the general confusion with the term Nyquist rate. Furthermore, we recommend that the term folding frequency refer to half of the actual sampling frequency (see Fig. 3).

4) The relationship (13) between the Fourier transform of a sequence of samples $x_c(nT)$ and the Fourier transform of the continuous time signal $x_c(t)$ is depicted in Fig. 4. Part (a) of this figure depicts a band-limited Fourier transform $X_c(j2\pi f)$. In Fig. 4(b) and (c) the sampling rate is greater than or equal to the Nyquist rate and we note that the form of $X_c(j2\pi f)$ is preserved to within a constant multiplier $1/T$ in the frequency range $-1/2T<f\leq 1/2T$. However, in Fig. 4(d) the signal $x_c(t)$ is *undersampled*, i.e., sampled at a rate below the Nyquist rate. In this case the Fourier transform of the sequence obtained by sampling is not equal to $X_c(j2\pi f)/T$ due to the fact that some of the other terms in (13) such as $X_c(j2\pi f-j(2\pi/T))$ are nonzero in the frequency range $-1/2T\leq f\leq 1/2T$. One way of viewing this is to say that a set of frequencies in $X_c(j2\pi f)$ is indistinguishable from a different set of frequencies in $X_c(j2\pi f-j2\pi/T)$. These frequencies are called *aliases* of one another and the process of confounding frequencies as in Fig. 4(d) is called *aliasing*.

5) Suppose we have a sampled waveform $x(n)$ with $z$ transform $X(z)$. We define a new sampled waveform $y(n)$ using one of every $M$ samples as the samples of the new waveform, i.e., $y(n)=x(Mn)$, with $M$ any positive integer. Clearly, this process is equivalent to sampling at a lower rate, and it is to be expected that aliasing may occur. When the aliasing occurs due to "sampling" a discrete-time signal it is called *digital aliasing*. It is readily shown that $Y(z)$ can be written in terms of $X(z)$ as

$$Y(z) = \frac{1}{M}\sum_{l=0}^{M-1} X(z^{1/M}e^{-j(2\pi/M)l}). \qquad (16)$$

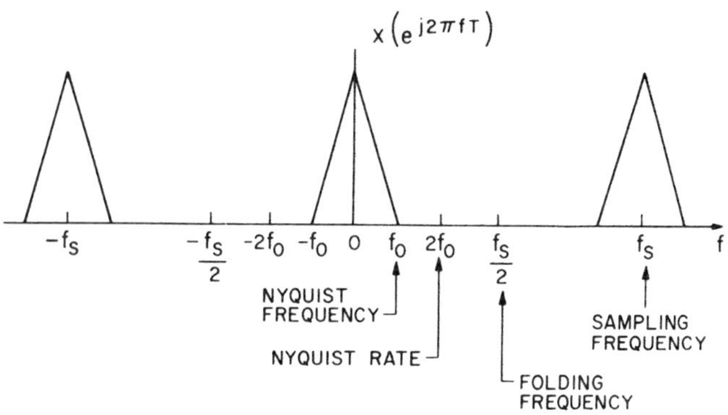


Fig. 3. Labeling of terminology concerned with frequencies related to the sampling process.

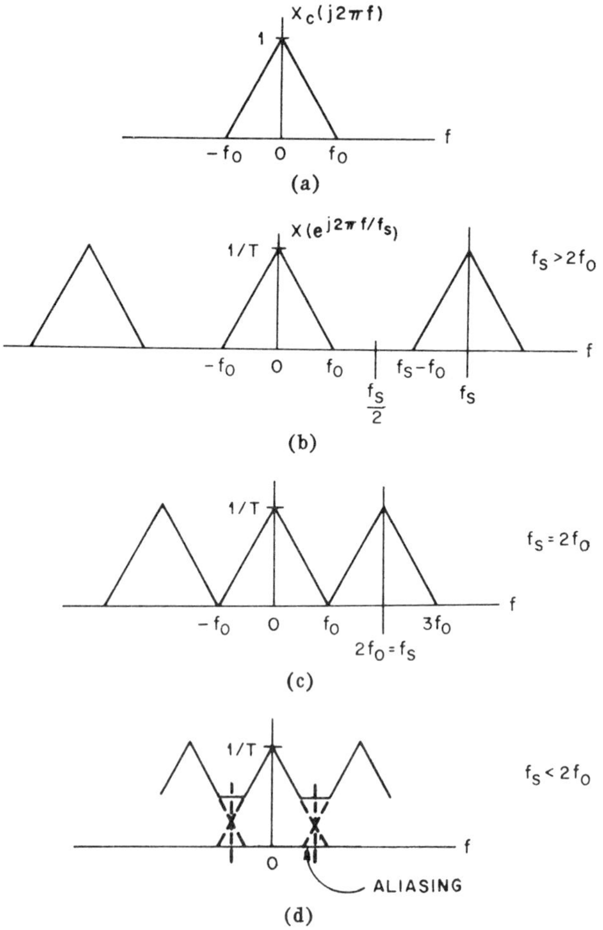


Fig. 4. An example of the effects of various sampling frequencies on the frequency response of the digital signal.

## IV. Theory and Design of Digital Filters

1) *Discrete filters* may be divided into two classes on the basis of whether the signal values can take on a continuum of values (sampled-data filters) or a finite set of values (digital filter). Thus we have the following.

a) A *sampled-data filter* is a computational process or algorithm by which a sampled-data signal acting as an input is transformed into a second sampled-data signal termed the output. The sampled-data signal is considered only at a set of points (usually equally spaced in time or space as the independent variable); at these points the signal can take on a continuum of values.

b) A *digital filter* is a computational process or algorithm by which a digital signal or sequence of numbers (acting as input) is transformed into a second sequence of numbers termed the output digital signal. The numbers are limited to a finite precision. The algorithm may be implemented in software as a computer subroutine for a general-purpose machine or in hardware as a special-purpose computer. The term digital filter is then applied to the specific routine in execution or to the hardware.

2) Further complexity of filtering action may be obtained by switching. Thus, a *switched filter* is one in

which the input and output are simultaneously switched in a definite pattern among a group of input and output ports. The filter being switched may be either of the continuous or discrete types. Examples of switched filters are commutating filters or $n$-path filters.

3) A *multiplexed filter* is a restricted form of a switched filter; commonly a single discrete filter which by means of a switching action is made to perform the function of several discrete filters virtually simultaneously. The multiplexing is most commonly done in a time-division manner whereby the input to the discrete filter is sequentially switched from a number of input signals and the filter output sequentially switched in synchronism to a corresponding set of output signal lines. Thus a single filter may be made to do the work of many filters by this time division multiplexing.

4) A *recursive filter* is a discrete-time filter which is realized via a recursion relation, i.e., the output samples of the filter are explicitly determined as a weighted sum of past output samples as well as past and/or present input samples. For example, $y(n)=b_0x(n)+b_1x(n-1)+b_2x(n-2)-a_1y(n-1)-a_2y(n-2)$.

5) A *nonrecursive filter* is a discrete-time filter for which the output samples of the filter are explicitly determined as a weighted sum of past and present input samples only. For example, $y(n)=b_0x(n)+b_1x(n-1)+b_2x(n-2)$.

6) A *finite impulse response* (FIR) filter is a filter whose impulse response $h(n)$ is zero outside some finite limits, i.e., $h(n)=0$, for $n>N_1$ and $n<N_2$ with $N_1 \geq N_2$.

7) An *infinite impulse response* (IIR) filter is a filter for which either $N_1=\infty$ or $N_2=-\infty$ or both, in 6). Thus the duration of the filter's impulse response is infinite.

8) It should be noted that the poles of FIR filters are restricted to $z=0$ or $z=\infty$, whereas there are no such restrictions on the positions of either the poles or zeros of IIR filters.

9) The terms recursive and nonrecursive are recommended as descriptions for how a filter is *realized* and not whether or not the filter impulse response is of finite duration. (Although IIR filters are generally realized recursively, and FIR filters are generally realized nonrecursively, IIR filters can be realized nonrecursively and FIR filters can be realized recursively.)

10) A *transversal filter* is a filter (either continuous or discrete) in which the output signal is generated by summing a series of delayed versions of the input signal weighted by a set of weights termed the *tap gains*. If the signal delays are accomplished by a tapped delay line then the filter is termed a *tapped delay line* filter.

11) A *comb filter* is a filter comprised of the sum or difference of input and output of a delay of $M$ units and unit gain yielding a transfer characteristic $H(z)=1\pm z^{-M}$ (see Fig. 5); this filter has $M$ zeros of transmission equally spaced on the unit circle in the $z$ plane thus giving rise to a frequency characteristic having $M$ equal peaks and $M$ real frequency zeros.

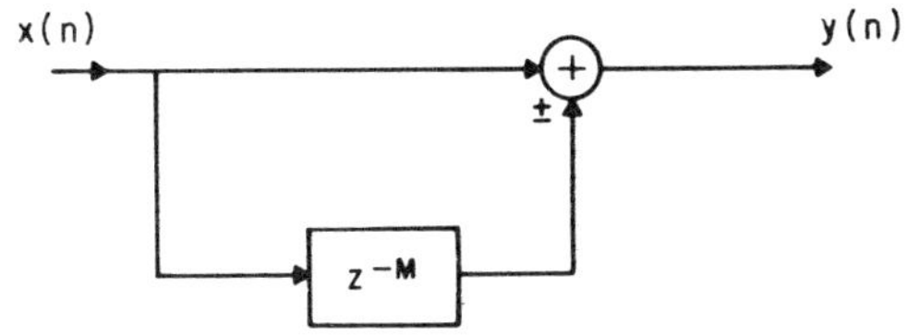


**Fig. 5. Block diagram representation of comb filter.**

12) A *frequency-sampling filter* is an FIR filter which is designed by varying one or more of its DFT[4] coefficients (called *frequency samples*) to minimize some aspects of the filter's frequency response. For example, the DFT coefficients of a frequency-sampling lowpass filter are 1.0 in the passband, 0.0 in the stopband, and variable in the transition band. One design criterion would be to choose the variable coefficients to minimize peak stopband ripple.

13) An *extraripple filter* (also called *maximal ripple filter*) is an FIR filter whose frequency response is equiripple in both the passband and stopband, and whose frequency response contains the maximum possible number of ripples.[5] There is no general agreement as to the appropriateness of this term, and as such, no recommendation as to its usage is made.

14) An *equiripple* (*optimal*) *filter* is an FIR filter which is the unique best approximation in the minimax sense to some specified frequency response characteristic over any closed subset of the frequency interval. For the lowpass filter case the optimal filter may be an extraripple filter, an equiripple filter with one less than the maximum possible number of ripples, or a filter with the maximum possible number of ripples all except one of which are of equal amplitude.

15) A *frequency sampling realization* is a means of realization of an FIR filter of duration $N$ samples as a cascade of a comb filter and a parallel bank of $N$ complex pole resonators. The filter output is obtained as a weighted sum of the outputs of each of the parallel branches; the multiplier on the $k$th branch being the $k$th DFT coefficient of the filter impulse response.

16) A *Kalman filter* (*discrete time*) is a linear, but possibly time-varying discrete-time filter with the property that it provides a least mean-square error estimate of a (possibly vector-valued) discrete-time signal based on noisy observations. The statistical description of the problem is such that the Kalman filter has a recursive implementation, using a linear combination of new observations and old estimates. The filter design may be based on a more general error criterion, using a nonquadratic loss function. Its essential features are that its design is based on a statistical criterion in the time domain, and that it is, in general, time varying. If the filter is further restricted to be time invariant it becomes the *Wiener filter*.

[4] See Section VI-1 for a definition of DFT.
[5] See Section IV-29 for a definition of ripple.

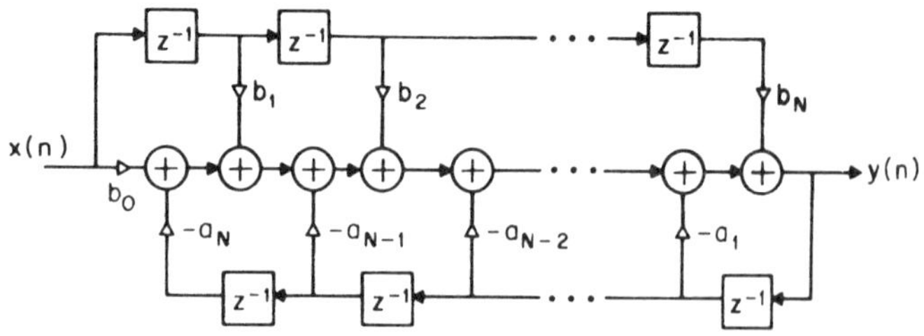


Fig. 6. Block diagram representation of direct form 1 for an $N$th-order system.

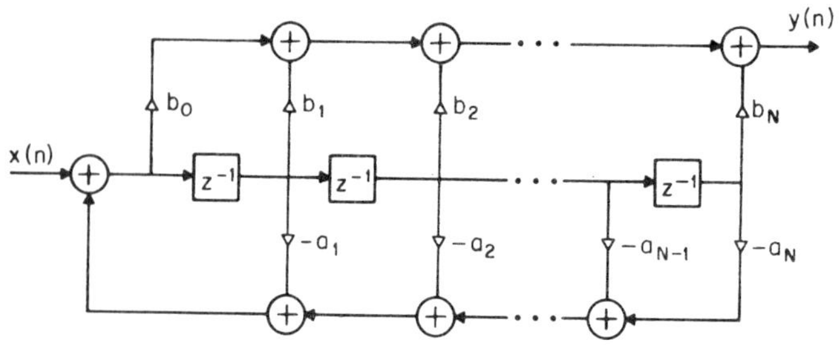


Fig. 7. Block diagram representation of direct form 2 for an $N$th-order system.

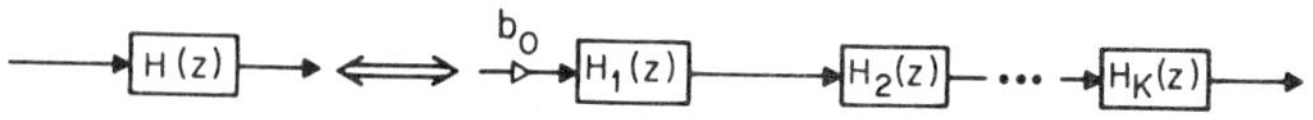


Fig. 8. Block diagram representation of the cascade form.

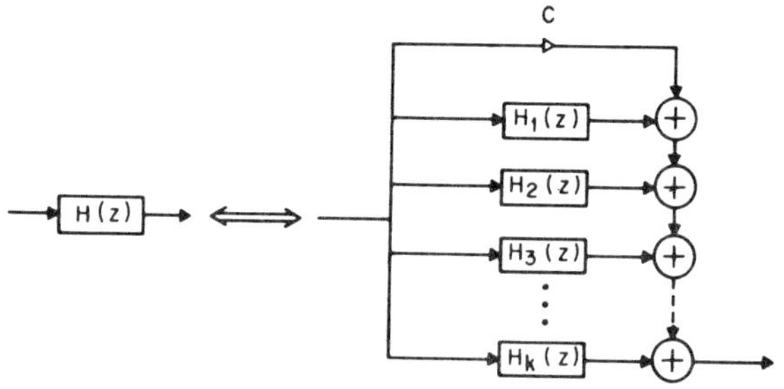


Fig. 9. Block diagram representation of the parallel form.

17) The forms for realizing digital filters include the following:

a) *Direct form 1* (shown in Fig. 6) where

$$H(z) = \frac{\sum_{i=0}^{N} b_i z^{-1}}{\sum_{i=0}^{N} a_i z^{-1}}, \qquad a_0 = 1. \tag{17}$$

For convenience in showing the realization, the order of the numerator and denominator are set to be the same. Direct form 1 uses separate delays for both the numerator polynomial and the denominator polynomial. In certain cases, e.g., floating-point additions, the results may depend on the exact ordering in which the additions are performed.

b) *Direct form 2* is shown in Fig. 7. Direct form 2 has been called the *canonic form* because it has the minimum number of multiplier, adder, and delay elements, but since other configurations also have this property, this terminology is not recommended.

c) *Cascade canonic form* (or *series form*), which is shown in Fig. 8, where

$$H(z) = b_0 \prod_{i=1}^{K} H_i(z) \tag{18}$$

and $H_i(z)$ is either a *second-order section*, i.e.,

$$H_i(z) = \frac{1 + b_{1i}z^{-1} + b_{2i}z^{-2}}{1 + a_{1i}z^{-1} + a_{2i}z^{-2}}, \tag{19}$$

or a *first-order section*, i.e.,

$$H_i(z) = \frac{1 + b_{1i}z^{-1}}{1 + a_{1i}z^{-1}}, \tag{20}$$

and $b_0$ is implicitly defined in (17), where $K$ is the integer part of $(N+1)/2$.

d) *Parallel canonic form*, which is shown in Fig. 9, where

$$H(z) = C + \sum_{i=1}^{K} H_i(z) \tag{21}$$

where $H_i(z)$ is either a *second-order section*, i.e.,

$$H_i(z) = \frac{b_{0i} + b_{1i}z^{-1}}{1 + a_{1i}z^{-1} + a_{2i}z^{-2}}, \tag{22}$$

or a *first-order section*, i.e.,

$$H_i(z) = \frac{b_{0i}}{1 + a_{1i}z^{-1}}, \tag{23}$$

and $K$ = integer part of $(N+1)/2$ and $C$ is proportional to $b_N$ as defined in (17).

18) The individual second- and first-order sections of the cascade and parallel forms are generally realized in one of the direct forms.

19) *Transpose configurations* for all of the above forms can be obtained by reversing the directions of all signal flow (i.e., by reversing the directions of all arrows) and by interchanging all branch nodes and summing junctions. The resulting circuits have the same transfer functions but different roundoff noise and overflow properties.

20) When the transfer function of a high-order filter is decomposed into a cascade connection of lower order filter sections by distributing the pole and zero factors among the lower order sections, then *pairing* refers to the associating of a specific zero factor with a specific pole factor to form an elemental or individual section. *Ordering* refers to the sequence or order in which the individual sections are connected in cascade to form the composite higher order filter. Varying the pairing and ordering can dramatically change the noise properties and dynamic range of both discrete and continuous filters. As an example, if

$$H(z) = \frac{\prod_{j=1}^{l} N_j(z)}{\prod_{j=1}^{m} D_j(z)} \tag{24}$$

and $l = m = 5$ then a possible detailed realization would be

$$H(z) = \frac{N_1(z)}{D_2(z)} \times \frac{N_3(z)}{D_1(z)} \times \frac{N_2(z)}{D_5(z)} \times \frac{N_4(z)}{D_3(z)} \times \frac{N_5(z)}{D_4(z)} \quad (25)$$

where the pairing is $N_1$ with $D_2$, $N_3$ with $D_1$, $N_2$ with $D_5$, $N_4$ with $D_3$, and $N_5$ with $D_4$. The implied ordering is $N_1/D_2$ first, followed by $N_3/D_1$, $N_2/D_5$, $N_4/D_3$, and finally $N_5/D_4$.

21) Two important properties of digital filters are *stability* and *causality*. The definition of *stability* most often used in digital filtering is as follows: a system is stable if every bounded (finite) input produces a bounded (i.e., finite) output. For linear time-invariant digital filters, a necessary and sufficient condition for stability is

$$\sum_{n=-\infty}^{\infty} |h(n)| < \infty. \quad (26)$$

22) A system is said to be *causal* if the output for $n = n_0$ is dependent only on values of the input for $n \leq n_0$. For linear time-invariant digital filters, this implies that the unit sample response sequence (i.e., the impulse response) is zero for $n < 0$. For the case of most interest, i.e., causal linear time-invariant filters with rational transfer functions, stability implies that all the poles of $H(z)$ must be inside the unit circle in the $z$ plane.

23) The *gain of a discrete filter* is the steady-state ratio of the peak magnitude (or any other consistent measure like root-mean-square, for example) of the output to the peak magnitude (or other consistent measure) of the input signal to the discrete filter. The usual input signals are either periodic sequences, e.g., sine waves, or pseudorandom sequences.

24) The *frequency-scale factor* is the factor by which all the poles and zeros of a normalized filter (cutoff frequency of 1 rad/s) must be multiplied to yield the actual filter pole and zero values, i.e., the ratio of the unnormalized to the normalized frequency scale of a filter.

25) The *filter bandwidth* is the width, in units of frequency, between the two points that define the edges of the passband of the frequency characteristics of a filter. The frequency points are usually defined as those values of frequency at which the attenuation or loss is a specified amount and beyond which the essential filter characteristic changes from pass (small attenuation) to stop (larger attenuation).

26) A commonly specified frequency point is the 3-dB or half-power point. For the elliptic and Chebyshev filters the frequency points are the highest and lowest frequencies at which the filter attenuation satisfies the equiripple passband attenuation limits. For other filters the frequency points may be defined in terms of the asymptotic intersections of the passband and stopband logarithmic asymptotes. Some examples of typical filter characteristics are shown.

27) Typical magnitude-square characteristics for several of the standard forms of continuous-time filters are given below using the following terminology.

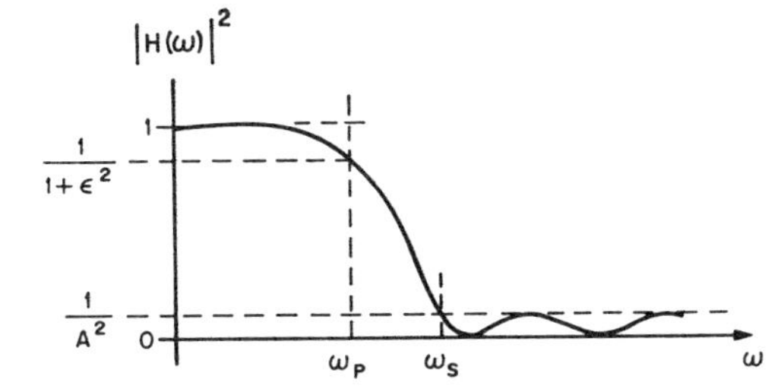


**Fig. 10. An example of the magnitude-squared characteristics of a typical filter.**

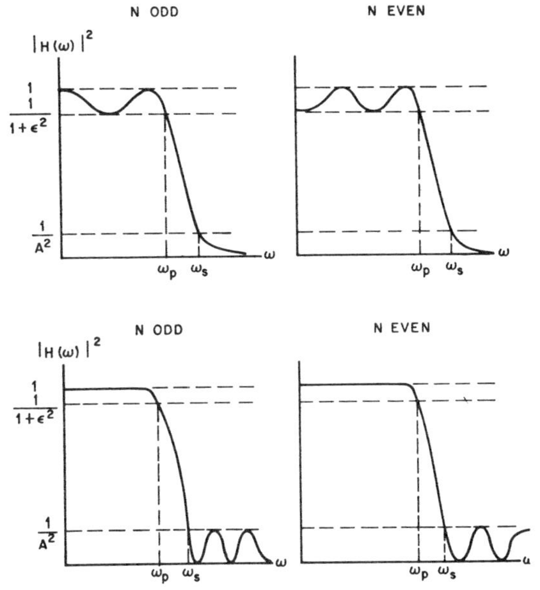


**Fig. 11. The magnitude-squared characteristics for even and odd order Chebyshev types I (top) and II (bottom) filters.**

$|H(\omega)|^2 \triangleq$ Magnitude-squared characteristic (frequency in rad/s).
$\omega_p$ Passband edge frequency.
$\omega_s$ Stopband edge frequency.

A typical response is shown in Fig. 10.

a) *Butterworth filter*: Maximally flat magnitude at $\omega = 0$

$$|H(\omega)|^2 = \frac{1}{1 + (\omega/\omega_p)^{2N}}. \quad (27)$$

b) *Chebyshev filters*:

Type I—Equiripple passband, monotone stopband:

$$|H(\omega)|^2 = \frac{1}{1 + \epsilon^2 C_N^2(\omega)}. \quad (28)$$

Type II—Equiripple stopband, passband maximally flat at $\omega = 0$:

$$|H(\omega)|^2 = \frac{1}{1 + \epsilon^2 \left[\dfrac{C_N(\omega_s)}{C_N(\omega_s/\omega)}\right]^2} \quad (29)$$

where the $C_N(\omega)$ are the Chebyshev polynomials. Fig. 11

shows the response for the two types of Chebyshev filters for $N$ odd and even.

c) *Elliptic (Cauer) filters*—equiripple passband and stopband:

$$|H(\omega)|^2 = \frac{1}{1+\epsilon^2\psi_N^2(\omega)} \tag{30}$$

where the $\psi_N(\omega)$ is a rational Chebyshev function involving elliptic functions. Fig. 12 shows the response of elliptic filters for $N$ both odd and even.

28) The term *transition band* is used to describe an interval of frequencies where a filter characteristic changes from one kind of behavior to another, one example being the transition band from a pass to a stop characteristic. The *transition ratio* is a relative measure of the passband width to the sum of the widths of the passband and the adjacent transition band(s). It can also be defined for a single-transition band-passband pair provided the width of the passband is defined. For the filter shown in Fig. 13 the transition ratios are defined as

$$\text{transition ratio} = \frac{\omega_c - \omega_{l2}}{\omega_c - \omega_l} \text{ (lower region)} \tag{31}$$

$$\text{transition ratio} = \frac{\omega_{u1} - \omega_c}{\omega_u - \omega_c} \text{ (upper region)} \tag{32}$$

where $\omega_c$ may be defined as either the arithmetic mean of the band-edge frequencies, i.e.,

$$\omega_c = \frac{\omega_{u1} + \omega_{l2}}{2} \tag{33}$$

or as the geometric mean of these same two frequencies, i.e.,

$$\omega_c = \sqrt{\omega_{u1}\omega_{l2}}. \tag{34}$$

Thus the transition ratio is bounded on the upper side by unity. Transition ratios near unity imply sharp cutoff filters.

29) The nature of a filter's response characteristic that approximates a desired characteristic by being alternatively greater than and less than the desired response as the independent variable is increased is called the *ripple*. The ripple may be expressed as the ratio of the maximum to the minimum of the response in a specified range, e.g., the passband of a filter. In this case, the ripple is usually expressed in percent or in decibels by taking 20 $\log_{10}$ of the ratio. Alternatively, the ripple may be expressed relative to some specified level of response such as plus or minus a fixed number of units. For example, consider the magnitude response shown in Fig. 14, where *passband ripple* $= 2.268/2.160 = 1.05$ which implies a $(2.268)/\sqrt{2.268 \times 2.160} = 1.0247$ or $\pm 2.47$ percent variation about the geometric mean; thus passband ripple expressed in dB (=) 20 $\log_{10}$ (1.05) $= 0.424$ dB overall or $\pm 0.212$ dB ripple about the geometric mean.

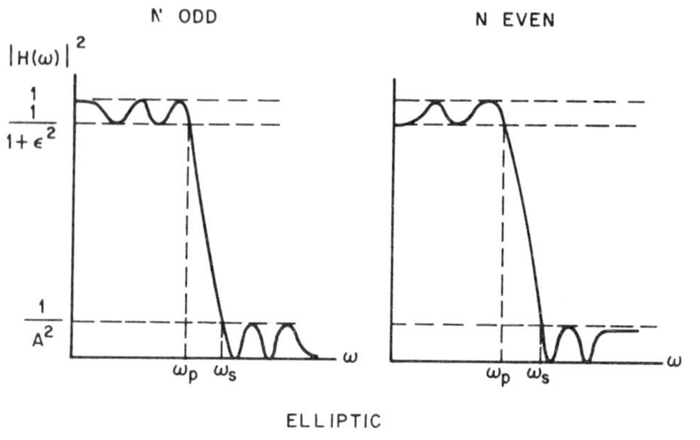


Fig. 12. The magnitude-squared characteristics for even and odd order elliptic filters.

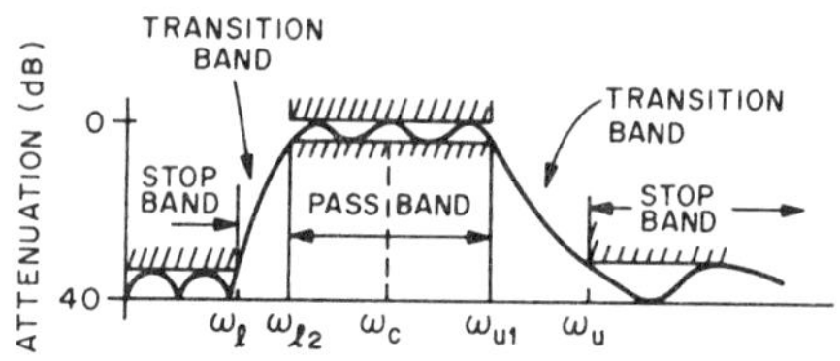


Fig. 13. The attenuation characteristics of a typical bandpass filter showing passband, stopbands, and transition bands.

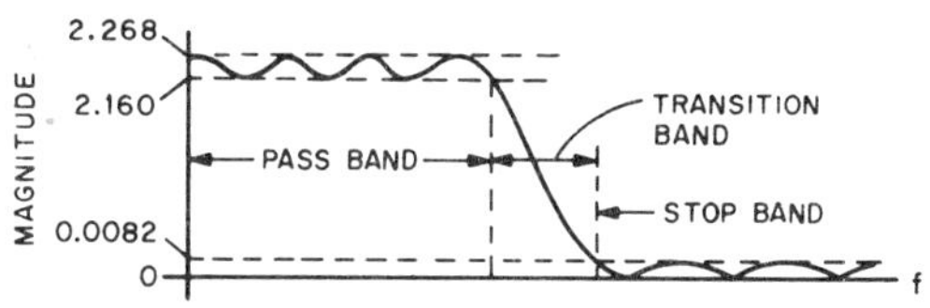


Fig. 14. The magnitude characteristic of a typical filter showing its ripple characteristics.

30) The passband ripple is also termed the *in-band ripple*. The terms *stopband ripple* and *out-of-band ripple* have also been used when the out-of-band frequency response has the characteristic of a ripple; numerically this has been used to express the ratio of the minimum out-of-band attenuation to the mean in-band attenuation. We recommend that this ratio be termed *minimum stopband attenuation* and that the terms stopband ripple and out-of-band ripple not be used except qualitatively. In the example, minimum stopband attenuation $= 0.0082 (=) -41.7$ dB and the relative minimum stopband attenuation $= (0.0082)/\sqrt{2.268 \times 2.160} = 0.003705$ $(=) -48.6$ dB.

### Methods for Designing Digital Filters

31) An important class of techniques for designing infinite impulse response filters to be realized recursively is based on a transformation of a continuous-time filter. This class consists of at least three techniques.

a) *Impulse invariance* (also called the standard $z$

transformation or standard $z$) is a technique in which the impulse response of the derived digital filter is identical to the sampled impulse response of a continuous-time filter. If the continuous-time filter has a transfer function[6]

$$H_c(s) = \frac{\sum_{k=0}^{M} d_k s^k}{\sum_{k=0}^{N} c_k s^k} = \sum_{k=1}^{N} \frac{A_k}{s + \alpha_k}, \qquad M < N \quad (35)$$

then the requirement that

$$h(n) = h_c(t)\big|_{t=nT}, \qquad 0 \le n \le \infty \quad (36)$$

implies that $H(z)$ is obtained from the partial fraction expansion of $H_c(s)$ by the substitution

$$\frac{1}{s+\alpha_k} \rightarrow \frac{1}{1 - e^{-\alpha_k T} z^{-1}}. \quad (37)$$

It can be shown that

$$H(z)\big|_{z=e^{sT}} = \frac{1}{T} \sum_{k=-\infty}^{\infty} H_c\left(s + j\frac{2\pi}{T}k\right). \quad (38)$$

Thus, impulse invariance is only satisfactory when $H_c(s)$ is band limited. If as in most instances, $H_c(s)$ is not sufficiently band limited, $H(z)$ is an aliased version of $H_c(s)$. Therefore, this technique is primarily used for narrowband filter designs or else the transformation is applied to the cascade combination of a *guard filter* and $H_c(s)$.

Another important point is clear from (38). Due to the $1/T$ multiplier, digital filters derived by impulse invariance have a gain approximately $1/T$ that of the continuous-time filter. This is generally compensated by multiplying each factor in the partial fraction expansion by $T$, so that the digital filter will have approximately the same gain as the continuous-time filter from which it was derived.

b) *Bilinear transformation* (also called the bilinear $z$ transform, the bilinear $z$ transformation or $z$ form) is a technique used to circumvent the aliasing problem of the impulse invariant technique. This approach uses the algebraic transformation

$$s = \frac{2}{T}\,\frac{1 - z^{-1}}{1 + z^{-1}}, \quad (39)$$

to derive the system function of the digital filter as

$$H(z) = H_c(s)\big|_{s=(2/T)(1-z^{-1})/(1+z^{-1})}. \quad (40)$$

This transformation has the effect of mapping the entire $s$ plane into the $z$ plane in such a way that the left-half $s$ plane maps into the inside of the unit circle and the right-half $s$ plane maps to the outside of the unit circle. This results in a nonlinear warping of the frequency scale according to the relation

$$\frac{\omega_c T}{2} = \tan\frac{\omega_d T}{2} \quad (41)$$

where $\omega_c$ is the continuous-time frequency variable and $\omega_d$ is the discrete-time frequency variable. Because of this warping of the frequency scale, this design technique is most useful in obtaining digital designs of filters whose frequency response can be divided into a number of pass and stop bands in which the response is essentially constant. Generally it is necessary to take appropriate account of the warping of the frequency scale.

c) *Matched z transform* (also called the matched $z$ transformation, mapping poles and zeros, or matched $z$) is a technique based on mapping the poles and zeros of the continuous-time filter by the substitution

$$s - s_i \rightarrow 1 - e^{s_i T} z^{-1}. \quad (42)$$

This means that the poles of $H(z)$ will be identical to those obtained by impulse invariant method, however the zeros will not correspond.

32) In the context of designing a discrete-time system and especially a digital filter, an *optimization technique* is a procedure for minimizing a prescribed performance function based on design requirements. An example is the design of a discrete-time filter to have the minimum mean-square deviation from a desired frequency-domain characteristic. An *iterative optimization technique* is a procedure for generating successive approximations converging (hopefully) to an optimum. This is opposed to an *analytical design technique*, which yields a closed form solution, such as the Chebyshev design for a lowpass filter.

## V. Finite Word Length Effects—A/D, D/A Conversion

1) A *digital-to-analog (D/A) converter* is a device which operates on a digital input signal $s(nT)$ to produce a continuous-time output signal $s(t)$ ideally defined by

$$s(t) = \sum_n s(nT)h(t - nT) \quad (43)$$

where $h(t)$ characterizes the particular converter. For example, $h(t)$ is a square pulse of duration $T$ for a zero order hold $D/A$ converter. The $D/A$ converter is usually followed by a linear time-invariant low-pass continuous-time filter called a *postfilter*. The combination of D/A converter and postfilter is called a *reconstruction device* or *reconstruction filter*.

2) An *analog-to-digital (A/D) converter* is a device which operates on a continuous-time waveform to pro-

[6] This formulation assumes that all poles are distinct. Appropriate modifications can be made to deal with multiple order poles.

duce a digital output consisting of a sequence of numbers each of which approximates a corresponding sample of the input waveform. Expressing the numerical equivalent of each sample by a finite number of bits (instead of the infinite number required to completely specify each sample) is the *quantizing* inherent in the conversion process. The error produced by quantizing is called *quantizing noise* or *A/D conversion noise*.

### Representation of Numbers

3) Various systems are used to represent the numbers in a digital filter. In *fixed-point number* representation, the position of the binary (or decimal) point is assumed fixed. The bits to the right of the (fixed) binary (or decimal) point represent the fraction part of the number and the bits to the left represent the integer part. For example, the binary number 011.001 has the value $0\times 2^2+1\times 2^1+1\times 2^0+0\times 2^{-1}+0\times 2^{-2}+1\times 2^{-3}$.

4) A *floating-point number* is formed by two fixed point numbers, the *mantissa*[7] and the *exponent*. The floating-point number is equal to the product of the mantissa with the quantity resulting when a given *base* is raised to the power denoted by the exponent. The base is the same for every floating-point number in the digital filter. Consequently, the numerical value of an entry in a specified position in the mantissa is determined by the exponent. The mantissa is generally normalized to be as large as possible but less than some number (e.g., 1.0). For example, $0.1\times 10^2$ is legitimate, whereas $0.01\times 10^3$ and $10.0\times 10^0$ are usually considered to be illegitimate floating-point decimal representations of the number 10. The most commonly used base is two (*binary representation*). The base 16 (*hexadecimal representation*) is used in some general purpose computers. The base 8 is called *octal representation*.

5) The representation of *block floating-point numbers* is determined by examining all numbers in a block (i.e., array). The largest number is represented as an ordinary floating-point number with a normalized mantissa. The remaining numbers in the block use the exponent associated with this largest number. This use of a single exponent for the whole block saves memory. This type of arithmetic is popular in realizations of the fast Fourier transform.

### Representation of Negative Numbers

6) The discussion so far has dealt with the representation of nonnegative numbers. There are three common systems used for representing signed numbers. The representation of positive numbers is the same in these three systems. The first, and most familiar, is *sign and magnitude*, i.e., the magnitude (which is, of course, positive) is represented as a binary number and the sign is represented by an additional binary digit in the leading position which, if 0 corresponds to a + and if 1 corresponds to a − (or vice versa). Thus, for example, in sign and magnitude 0.0011 represents 3/16 and 1.0011 represents −3/16. Two related representations of signed numbers are ones complement and twos complement. In each of these systems a positive number is represented as in sign and magnitude. For *twos-complement* representation the negative of a particular positive number is obtained by complementing all the bits and adding one unit in the position of the least significant bit. For example, −(0.0110) would be represented in twos complement as (1.1001)+(0.0001)=1.1010. A carry out of the sign bit is neglected in the addition, so that −(0.0000)=(1.1111)+(0.0001)=0.0000. For *ones-complement* representation the negative of a given positive number is obtained simply by complementing all the bits.

7) The choice of representation for negative numbers in a particular system is based almost entirely on hardware considerations. With ones-complement and twos-complement numbers, subtraction can be performed conveniently with an adder. For example, in twos complement, the difference $A-B$ is formed by simply adding to $A$ the twos complement of $B$.

### Finite Word Length Effects

8) Even though the input to a digital filter is represented with finite word length (e.g. through A/D conversion), the result of processing will naturally lead to values requiring additional bits fo, their representation. For example, a $b$-bit data sample multiplied by a $b$-bit coefficient results in a product which is $2b$ bits long. If in a recursive realization of a filter we do not quantize the result of arithmetic operations, the number of bits required will increase indefinitely, since after the first iteration $2b$ bits are required, after the second iteration $3b$ bits are required, etc. Two common methods are used to eliminate the lower order bits resulting from arithmetic operations in a digital filter.

a) *Truncation* is accomplished by discarding all bits (or digits) less significant than the least significant bit (or digit) which is retained.

b) *Rounding* of a number to $b$ bits, when the number is initially specified to more than $b$ bits, is accomplished by choosing the rounded result as the $b$-bit number closest to the original unrounded quantity. When the unrounded quantity lies equidistant between two adjacent $b$-bit numbers, a random choice ought to be made as to which of these numbers to round to. For example, 0.01011 rounded to three bits would be 0.011; but 0.01010 rounded to three bits can be chosen as either 0.011 or 0.010, and the choice should be random. In many situations, however, one can choose to always round up in this midway situation with negligible effect on the accuracy of the computation.

9) *Roundoff error* (or *roundoff noise*) or *truncation error* (or *truncation noise*) is caused by rounding off or truncating the products formed within the digital filter.

[7] The term mantissa as defined here is unfortunately not the same as the term mantissa commonly used in logarithm tables. The definition presented here is dictated by its extensive occurrence in the literature.

The roundoff or truncation error is sometimes well modeled as a random process. On the other hand, if the data sequence to a recursive realization of a digital filter consists of constants (e.g., zero) or some other periodically repeating samples, the roundoff or truncation error is periodic and causes a *deadband effect* or *limit cycle* in the filter output. A common type of limit cycle is a *zero-input limit cycle* where the output of a digital filter remains periodic and nonzero, after the input has been set to zero. *Dither* is a sequence of numbers that is added to the input to a recursive digital filter to ameliorate the deadband effect. Even though dither increases the mean-square error in the output, it can disrupt the pattern of roundoff errors causing the deadband effect, thereby permitting the output to return to zero.

10) *Overflow* occurs when a digital filter computes a number that is too large to be represented in the arithmetic used in that filter. If no compensation is made for the overflow then large errors in the filter output can result either in the form of transients or of *overflow oscillations*. A technique used to compensate in part for overflow is *saturation arithmetic* where a sum that is too large to be represented is set equal to the largest representable number in the filter.

11) *Dynamic range* is the ratio between the largest and smallest signals which can be represented in the filter with a given fidelity criterion. Unfortunately the fidelity criterion is often vague or unspecified.

12) The roundoff or truncation noise introduced within a digital filter produces a resultant noise at the output of the digital filter. A *signal-to-noise ratio* can be defined in this context, for example, as the ratio of the ideal mean-squared output signal (filter output in the absence of any rounding) to the mean-squared output noise due to rounding or truncation. Expressing this ratio in bits, as $(1/2)\log_2$ of the ratio, gives an approximate indication of the number of accurate bits in the filter output. Different definitions of signal-to-noise ratio may be appropriate in different contexts.

13) Another effect of finite word length is *coefficient quantization error* (or *parameter quantization error*), which occurs when the coefficients of a digital filter, initially specified with unlimited accuracy, are quantized by rounding or truncation. Coefficient quantization error appears as error in the digital filter's response (e.g., impulse response, transfer function, frequency response, etc.).

## VI. Discrete Fourier Transforms and the FFT

1) For a sequence of $N$ numbers, possibly complex, the *discrete Fourier transform* (*DFT*) is another sequence of exactly $N$ numbers which are the values of the $z$ transform of the original finite sequence for $N$ values of $z$, specifically

$$z = e^{j(2\pi/N)k}, \qquad k = 0, 1, \cdots, N-1. \tag{44}$$

*Discrete Fourier transformation* is the operation of computing, or otherwise forming, the discrete Fourier transform of the sequence.

2) From the definition, the sequence $\{f(0), f(1), \cdots, f(N-1)\}$ has the DFT[8] $\{F(0), F(1), \cdots, F(N-1)\}$

$$F(k) = \sum_{n=0}^{N-1} f(n)e^{-j(2\pi/N)nk}. \tag{45}$$

3) It is possible to recover the original sequence from its DFT by the operation

$$f(n) = \frac{1}{N}\sum_{k=0}^{N-1} F(k)e^{j(2\pi/N)nk}, \tag{46}$$

giving a sequence of $N$ samples $\{f(0), f(1), \cdots, f(N-1)\}$ as the inverse discrete Fourier transform of $\{F(0), F(1), \cdots, F(N-1)\}$. The operation is called *inverse discrete Fourier transformation* or IDFT, and is remarkably similar in form to the discrete Fourier transformation.

4) Some authors have defined the DFT in related but different ways, involving $e^{j(2\pi/N)nk}$, or a multiplicative factor of $1/N$ or $1/\sqrt{N}$. By considering the expressions for the DFT and IDFT it is evident that the constants $1/N$ or $1/\sqrt{N}$ and the possible use of $e^{j(2\pi/N)nk}$ in other definitions of the DFT can easily be compensated for in the other definitions of the IDFT.

5) Suppose for an $N$-point sequence we are interested in computing its DFT, and suppose $N$ is a composite integer

$$N = r_1 \times r_2 \times \cdots \times r_\mu \tag{47}$$

where the $r_i$ are a set of factors of $N$, not necessarily prime factors. Of the various algorithms for computing such a DFT, some require a number of operations *proportional* to $N\sum_{i=1}^{\mu} r_i$ (since the word proportional allows considerable latitude, it is not necessary to be too specific about the meaning of "operation"); such algorithms are called *fast Fourier transforms* (FFT).[9] An important special case is when

$$r_1 = r_2 = \cdots = r_\mu = 2$$

so that

$$\sum_{k=1}^{\mu} r_k \rightarrow 2\log_2 N.$$

For fast Fourier transforms in this case, the proportionality is to $N\log_2 N$.

6) A subclass of FFT algorithms is known which use high speed convolution techniques to compute the DFT of a sequence through a formula in which it is expressed as a convolution. Examples of such algorithms are the chirp $z$ transform and the prime algorithm.

7) In order to classify different FFT algorithms and

[8] For convenience, the notation of $F(k)$ is used to denote DFT coefficients rather than $F(e^{j(2\pi/N)k})$.

[9] The word transform instead of algorithm is embedded ineradicably in the literature.

to relate them to one another it is useful to consider the following procedure applied to a sequence $f(n)$ of length $N$ where $N$ has a factorization $P \times Q \times R$. The reader should generalize to more complicated factorizations. Let us replace the index $n$ by the triplet $(n_0, n_1, n_2)$ where

$$(n_0, n_1, n_2) = n = n_0 + Rn_1 + QR\,n_2 \tag{48}$$

and

$$0 \le n_0 < R$$
$$0 \le n_1 < Q$$
$$0 \le n_2 < P.$$

Similarly, we replace the index $k$ by the triplet $(k_0, k_1, k_2)$ where

$$(k_0, k_1, k_2) = k = k_0 + Pk_1 + PQ\,k_2 \tag{49}$$

and

$$0 \le k_0 < P$$
$$0 \le k_1 < Q$$
$$0 \le k_2 < R.$$

Then (45) can be manipulated into the form

$$F(k_0, k_1, k_2) = \sum_{n_0=0}^{R-1}\left(\left\{\sum_{n_1=0}^{Q-1}\left(\left\{\sum_{n_2=0}^{P-1} f(n_0, n_1, n_2) \cdot e^{-j(2\pi/P)n_2k_0}\right\}\phi_A\right)e^{-j(2\pi/Q)n_1k_1}\right\}\phi_B\right)e^{-j(2\pi/R)n_0k_2}. \tag{50}$$

Here $\phi_A$ and $\phi_B$ are of unit magnitude and have arguments dependent on the indices. Equation (50), if followed as a recipe, suggests a way of computing an $N$-point DFT as a collection of smaller DFT's. There are: $QR$ DFT's of $P$-point sequences; $PR$ DFT's of $Q$-point sequences; and $PQ$ DFT's of $R$-point sequences.

The only other operations called for in (50) are the multiplications by $\phi_A$, $\phi_B$. These have been called *twiddle factors*, *phase factors*, and *rotation factors* by various authors.

8) To save multiplications (50) is commonly modified in one of two ways. The first way is to combine the factors $\phi_A$ and $e^{-j(2\pi/P)n_2k_0}$ inside the sum over $n_2$ and the factors $\phi_B$ and $e^{-j(2\pi/Q)n_1k_1}$ inside the sum over $n_1$. The algorithm so produced has been called a *decimation-in-frequency* algorithm or a *Sande–Tukey* algorithm.

9) A second way to save multiplications is to combine the factors $\phi_A$ and $e^{-j(2\pi/Q)n_1k_1}$ inside the sum over $n_1$ and the factors $\phi_B$ and $e^{-j(2\pi/R)n_0k_2}$ inside the sum over $n_0$. The resulting form of the algorithm is called a *Cooley–Tukey* or *decimation-in-time* algorithm.

10) An algorithm with the twiddle factors explicitly present, as separate multiplications, is neither Cooley–Tukey nor Sande–Tukey.

11) There is a third way to save multiplications, which works only when the factors $P$, $Q$, $R$ are relatively prime. By permuting the data sequence $f(n)$ and accepting a permuted transform sequence, a formula like (50) can be derived in which $\phi_A = \phi_B = 1$. This algorithm is called the *prime factor algorithm*.

12) Any of the basic forms can be programmed so that each summation is computed and the result stored in the memory formerly occupied by its input data as soon as that input data is no longer needed. In the case of (50), the amount of extra storage over the $N$ cells needed for the sequence itself is only the greater of $P$, $Q$, or $R$ cells.

A programmed version of an algorithm which takes advantage of this possibility is called *in place*. Unfortunately, when storage use is minimized in this way, either the input sequence or the output sequence must be unusually ordered. An example of this unusual order might be that

$$F(k_0 + Pk_1 + PQk_2) \tag{51}$$

ends up in storage location $k_2 + Rk_1 + QRk_0$. The unusual ordering may also take place in the use of weights in the algorithm. All such effects are called "digit reversal." If $N$ is a power of two, and $P = Q = R = 2$, the effect is called *bit reversal*.

13) If the factors of $N$ are equal, say to $r$, the algorithm is called a *base-r* or *radix-r* algorithm and, if the factors are different, it is called a *mixed-radix* algorithm.

14) A common notation is to let $W$ or $W_N$ represent the reciprocal of the $N$th principle root of unity,

$$W = W_N = \exp\left(-j\frac{2\pi}{N}\right). \tag{52}$$

Some authors have used

$$W = \exp\left(j\frac{2\pi}{N}\right). \tag{53}$$

15) For the base-2 Cooley–Tukey form of the FFT algorithm, the most fundamental operation is of the form

$$X = A + W^kB$$
$$Y = A - W^kB. \tag{54}$$

The Sande–Tukey form can be obtained by solving for $A$ and $B$ in terms of $X$ and $Y$,

$$A = 0.5(X + Y)$$
$$B = 0.5(X - Y)W^{-k}. \tag{55}$$

This gives an elementary operation of an inverse transform. The Sande–Tukey form of the forward transform (DFT) is obtained by eliminating the 1/2 and replacing $W^{-k}$ by $W^k$. From the appearance of the system flow graph for these operations, each operation is called a *butterfly*.

16) For diagramming the flow of processing data for the FFT, a flow graph notation is used. The flow graph for a trivial eight-point FFT appears as follows:

$$F(k) = \sum_{n=0}^{7} f(n)W^{nk}, \qquad k = 0, 1, \cdots, 7 \tag{56}$$

where

$$W = \exp\left(-j\frac{\pi}{4}\right). \tag{57}$$

The flow graph is shown in Fig. 15.

17) Each node represents a variable and the arrows terminating at that node originate at the nodes whose variables contribute to the value of the variable at that node. The contributions are additive, and the weight of each contribution, if other than unity, is indicated by the constant written close to the arrowhead of the contribution. Each node is assigned a pair of indices $n$, $L$. Variables at nodes in row $n$ replace each other as they are computed and are stored in the cell with index $n$. All nodes in a column $L$ are computed on iteration $L$. In this form of the algorithm, the exponent of $W$ on the line entering node $(n; L)$ is $n \cdot 2^{3-L}$ (mod 8). It can also be seen that for each pair of operands, the second $W$ is $W^4 = -1$ times the first and that the "butterfly" operation is indeed given by (54).

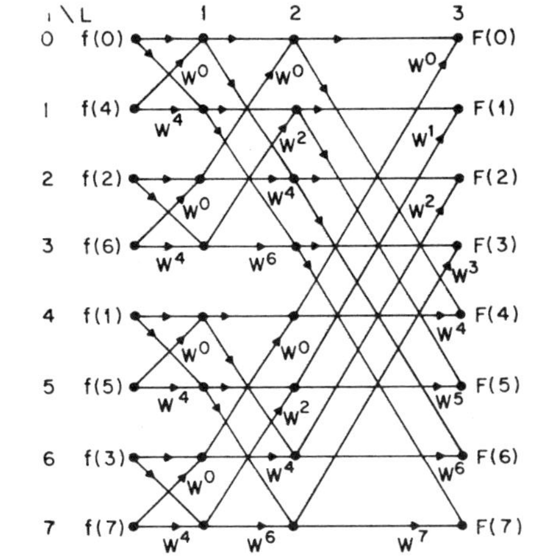


**Fig. 15. The flow graph for an eight-point FFT.**

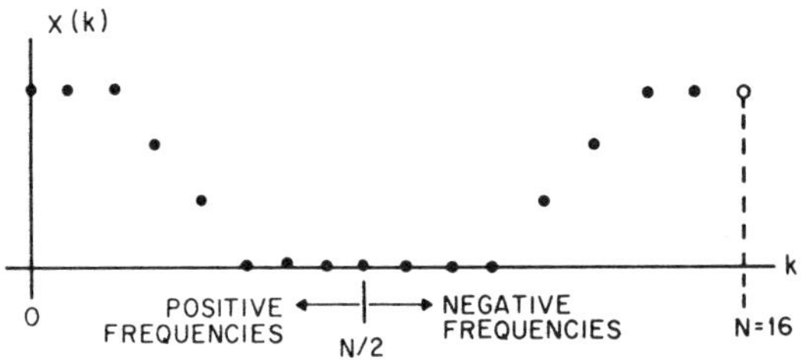


**Fig. 16. The locations (in frequency) of the DFT points for a 16-point transform.**

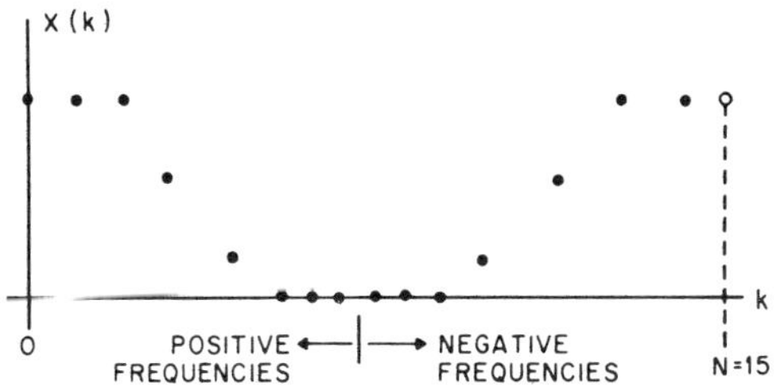


**Fig. 17. The locations (in frequency) of the DFT points for a 15-point transform.**

## VII. Discrete Convolution and Spectrum Analysis

1) The Fourier transform (i.e., the $z$ transform evaluated for $z = e^{j2\pi fT}$) of a sampled waveform is periodic in frequency, i.e., $X(e^{j2\pi T(f+f_s)}) = X(e^{j2\pi Tf})$ where $f_s = 1/T$ = sampling frequency. For this reason it is convenient to represent a *negative frequency* as the equivalent *positive frequency* below the sampling frequency, i.e., $f = -f_s/20$ is equivalent to $f = -f_s/20 + f_s = 19/20\, f_s$. In this manner one need only describe the spectrum in the positive frequency range of $0 \le f < f_s$. For the DFT this convention also holds. Thus a typical spectrum of a 16-point DFT is shown in Fig. 16. For the above example the number of DFT points was even (16). The first DFT point, $X(0)$, corresponds to the Fourier transform evaluated at 0 frequency. The $(N/2+1)$st DFT point, $X(N/2)$, corresponds to the Fourier transform evaluated at half the sampling frequency. The index $k$ corresponds to a frequency $f = k/TN$ in $X(e^{j2\pi Tf})$.

2) If the number of points in the DFT were odd, say $N = 15$, a typical spectrum would be as shown in Fig. 17. In this case there is no DFT point which corresponds to evaluating the Fourier transform at half the sampling frequency.

3) The *discrete convolution* of two sequences can be computed from the inverse discrete Fourier transform of the product of the discrete Fourier transforms of the two sequences. Thus, if $X(k)$ and $Y(k)$ are the discrete Fourier transforms of $x(n)$ and $y(n)$, the inverse discrete Fourier transform of the product of these discrete Fourier transforms produces a *periodic discrete convolution*, also called a *cyclic or circular discrete convolution* or simply a cyclic convolution.

4) The cyclic convolution can be written algebraically as

$$\sum_{m=0}^{N-1} x((n-m))y((m)) \tag{58}$$

where $((m))$ means the index is taken modulo $N$. Equation (58) can be written as

$$\sum_{m=0}^{n} x((n-m))y((m)) + \sum_{m=n+1}^{N-1} x((n-m))y((m)), \tag{59}$$

or, alternately

$$\sum_{m=0}^{n} x((m))y((n-m)) + \sum_{m=n+1}^{N-1} x((m))y((n-m+N)). \tag{60}$$

The simultaneous presence of both summations is generally undesirable for computing ordinary convolutions. For example, if the first summation in (60) is chosen to represent the desired convolution, then the second summation represents an error term. By augmenting both sequences with zeros so that they have the same length $N$, which is at least as great as one less than the sum of the lengths of the two sequences, cyclic convolution can be made to yield the same result as ordinary convolution.

5) This use of the fast Fourier transform to compute discrete convolutions is sometimes called *fast convolution* or *FFT convolution*. This technique can be easily adapted for computing acyclic (i.e., nonperiodic) correlation functions. In this form, it is called *fast correlation or FFT correlation*.

6) If one of the two sequences is much shorter than the other, the longer sequence can be *sectioned* into pieces whose discrete convolutions can be computed separately. These discrete convolutions can be combined to produce the discrete convolution of the whole sequence. (*Sectioning* is used because it reduces the required amounts of computation and memory.) One of these sectioning techniques (*overlap-save* or *select-save*) involves computing the inverse DFT of the product of the DFT's of a) $N$ samples of the input sequence, and b) the shorter sequence augmented with a sufficient number of zeros so that its sequence contains $N$ samples. (Usually $N$ is at least twice as large as the length of the shorter sequence.) Some of the members of the sequence resulting from the inverse DFT are members of the sequence formed by the desired acyclic (i.e., nonperiodic) convolution. (This number of members equals one more than the number of zeros originally augmenting the shorter sequence.) The longer original sequence is advanced by this number of members. Iterating this process gives the whole convolution. The *overlap-add* technique for sectioning uses a similar technique but additionally requires adding shifted sequences of partial convolutions.

7) A *window* is a finite sequence, each element of which multiplies a corresponding element of the main sequence. (This is called *windowing*.) The sequence of products formed by this element-by-element multiplication is often more useful than the main sequence. The Fourier transform of a typical window (sometimes called the *spectral window*) consists of a *mainlobe*, which usually contains a large percentage of the energy in the window, and *sidelobes* which contain the remaining energy in the window.

8) Windows can be used in estimating power spectra. In the *direct method*, the power spectrum is estimated by computing the square of the absolute value of the DFT of the windowed sequence. The DFT of the windowed sequence is the convolution of the DFT's of the window and the original sequence. This convolution smoothes the input power spectrum, consequently values of the power spectrum at frequencies separated by less than the width of the mainlobe of the spectral window cannot be *resolved*. In addition to this limit on *resolution*, the estimate of the power spectrum may contain significant *leakage*, i.e., erroneous contributions from components of the power spectrum at frequencies possibly distant from the frequency of interest because of the nonzero energy in the spectral window sidelobes.

9) Windows are useful in determining the coefficients of a finite impulse response digital filter. In this case, the original sequence consists of samples of the impulse response corresponding to a transfer function which is approximated by the Fourier transform of the sequence of pairwise products; the product sequence is used as the coefficients of the finite impulse response digital filter.

10) Windows are used also in the *indirect method* of computing a power spectrum. In this method, the sequence consisting of samples of the autocorrelation function is multiplied by the window. The DFT of the resulting sequence is an estimate of the power spectrum.

11) Windows can be used also in estimating cross spectra where the estimates are obtained by multiplying the products of the DFT's of two or more distinct sequences.

12) The determination of a finite impulse response described by an ordinary convolution is called *deconvolution* or *FIR identification*.

**Acknowledgment**

The authors would like to acknowledge the comments and criticisms of this paper provided by G. D. Bergland, C. H. Coker, D. L. Favin, B. Gold, O. Hermann, S. Lerman, A. V. Oppenheim, H. O. Pollak, and H. F. Silverman.

**Alphabetic Index of Terms**

# Author Index

# Subject Index